Molecular Applications of Quantum Defect Theory

Molecular Applications of Quantum Defect Theory

Edited by Ch Jungen

Laboratoire Aimé Cotton du CNRS,
Université de Paris-Sud, Orsay, France

With an introductory essay by
M Seaton FRS

Institute of Physics Publishing
Bristol and Philadelphia

British Library Cataloguing-in-Publication Data

A catalogue record for this book is available from the British Library.

ISBN 0 7503 0162 7

Library of Congress Cataloging-in-Publication Data are available

Published by Institute of Physics Publishing, wholly owned by The Institute of Physics, London
Institute of Physics Publishing, Techno House, Redcliffe Way, Bristol BS1 6NX, UK
US Editorial Office: Institute of Physics Publishing, The Public Ledger Building, Suite 1035, 150 South Independence Mall West, Philadelphia, PA 19106, USA

Printed in the UK by J W Arrowsmith Ltd, Bristol

Contents

3. Relativistic theory

4. Eigenchannel approach

5. Empirical analysis

Preface

This volume collects together 39 papers, whose publication dates span a period of 39 years, dealing with quantum defect theory and its applications. The selection includes many of the fundamental papers on the subject for the reader who is interested in seeing how the concepts developed over the years. In addition to the more historical aspects, the purpose of the collection is to provide in one volume all the elements necessary for someone who would like to do some work in the area of quantum defect theory. Devised originally for the description of atomic Rydberg states, the quantum defect method has subsequently been put to use in ever more diverse circumstances to the point that sometimes there is disagreement as to what should be called 'QDT' and what should be given a different name. Another purpose of the present book is that the selection of papers should reflect the versatility and vast possibilities of application of the quantum defect method, showing that this is a vigorous and expanding field. The volume is introduced by an essay of Mike Seaton, one of the pioneers in the field, giving an account of the origins of QDT and a discussion of some applications in atomic physics. An introduction by myself follows; this puts the main emphasis on recent work involving applications in molecular physics, but also includes some further discussion of the atomic work. The reprinted papers are followed by a bibliography listing a little over 400 papers that appeared between 1982 and 1996 in the field of atomic and molecular physics and that are linked in one way or another to quantum defect theory. This bibliography updates and complements the one provided by M J Seaton in his 1983 review which is also reprinted in this volume. However, while I tried to make the bibliography as reasonably complete as possible, it is inevitable that some contributions may have been overlooked. I apologize in advance for such omissions. Finally I should like to thank M Aymar, A Suzor-Weiner, C H Greene, S C Ross and M J Seaton for their critical comments regarding the selection of papers, the introductory remarks and/or the bibliography. My especial thanks go to Mike Seaton for contributing his essay.

Ch Jungen

An introductory essay—the development of QDT

M J Seaton

Department of Physics and Astronomy, University College London, UK

Note

** is used in referring to papers reproduced in the present collection.
* is used in referring to papers cited by Seaton (1983)**, or included in the bibliography of the present book.

In the present essay I consider some problems relating to the structure and origins of quantum defect theory (QDT). Many later developments and applications of the theory are discussed by Christian Jungen in the following article.

My 1983** review, which I will refer to as S83, includes an account of the early history of QDT: in 1885* Balmer published a formula for the wavelengths of hydrogen lines involving an integer n known later as a *quantum number*; Rydberg's formula of 1889* for series in alkali spectra differed from that of Balmer in that n was replaced by $(n - \mu)$ where μ is now known as the *quantum defect*; in 1913* Bohr obtained Balmer's formula using classical mechanics and imposing quantization conditions; in 1916* Sommerfeld showed that for alkali atoms the closed elliptical orbits of the Bohr theory are replaced by orbits which precess and that μ is proportional to the rate of precession; in 1928* Hartree gave a theory of quantum defects using the 'new' quantum mechanics. Hartree's paper involved an analysis of the mathematical properties of Coulomb functions, that is to say the wave functions for the motion of a particle in a Coulomb potential, and laid the foundations for all subsequent work on quantum defect theory.

The conceptual structure of the theory can be described as follows. Consider an atomic or molecular system containing $(N + 1)$ electrons and assume that one can define a radius r_0 such that:

- interactions between all particles in the system are important in the inner region of $r < r_0$;
- only one electron can be in the outer region of $r > r_0$;
- an electron in the outer region moves in a simple Coulomb potential.

Assume, further, that one can construct inner-region functions $\Psi_i(E)$ which are bounded for all $r \leqslant r_0$ and which, for small values of r, are slowly varying functions of the total energy E. Outer-region functions, $\Psi_O(E)$, are constructed as products of functions $\psi_i(E_i)$ for N electrons in the inner region multiplied by functions $\phi_i(e_i)$ for the outer electron. In the language of collision theory, the ψ_i are *target functions* and i specifies a *channel.* The energies are: E_i for the target; e_i for the outer electron; $E = E_i + e_i$ for the whole system. Channel i is said to be *open* if $e_i > 0$ and *closed* if $e_i < 0$.

For each channel we may define two linearly independent Coulomb functions, f_i and g_i, which are analytic functions of the energy e_i. At $r = r_0$ we match the inner-region functions Ψ_I to the outer-region functions Ψ_O, and then express the channel functions ϕ_i as linear combinations of the f_i and g_i: we obtain $\phi_i = F_i(E)f_i + G(E)g_i$ where the coefficients F_i and G_i will be slowly varying functions of E. The number of such linearly independent solutions is equal to the total number of channels. Let there be J channels and let the solutions be Ψ_j, $j = 1$ to J. However, when some channels are closed such solutions will not be bounded in the limit of $r \to \infty$. For the closed channels one must introduce Coulomb functions θ_i which decay exponentially in the limit of $r \to \infty$. The θ_i can be expressed as linear combinations of the f_i and g_i in which the coefficients are known functions of the energies e_i. Finally, one can form linear combinations of the functions Ψ_j so as to obtain functions Ψ in which the ϕ_i for all closed channels are proportional to θ_i. These may be referred to as the physical solutions. When some channels are open, the number of linearly independent physical solutions is equal to the number of open channels. When all channels are closed, physical solutions exist only at the discrete eigenvalues of E.

The final result is that required physical quantities, such as positions of energy levels, radiative transition probabilities and collision cross-sections, can be expressed in terms of the $F_i(E)$ and $G_i(E)$ which are slowly varying functions of E (and which are determined by the potentials in the inner region), and known functions of E which may vary much more rapidly.

Let us illustrate the above ideas for the case of the one-channel theory. We use atomic units for r and e and consider the Coulomb potential $v(r) = -z/r$ with $z > 0$. We introduce scaled variables $\rho = zr$ and $\epsilon = 2e/z^2$ and we put $\epsilon = \kappa^2$ for $\epsilon > 0$ and $\epsilon = -1/\nu^2$ for $\epsilon < 0$: ν is referred to as the *effective quantum number.* It is also necessary to specify the angular momentum quantum number l for the electron. Let $s(\epsilon, l; \rho)$ be the regular Coulomb function, bounded at the origin $\rho = 0$. For $\epsilon > 0$ that function has the asymptotic form $s \propto \sin(\zeta)$ where ζ is the usual Coulomb phase function. We take the normalization of s to be such that

$$s(\epsilon, l; \rho) \sim (\pi\kappa)^{-1/2} \sin(\zeta) \qquad (\epsilon > 0, \quad \rho \to \infty) \tag{1}$$

and we take the irregular solution $c(\epsilon, l; \rho)$ to be such that

$$c(\epsilon, l; \rho) \sim (\pi\kappa)^{-1/2} \cos(\zeta) \qquad (\epsilon > 0, \quad \rho \to \infty). \tag{2}$$

The definitions of s and c can be continued to the region of $\epsilon < 0$ but in that region they are, in general, both exponentially increasing as $r \to \infty$. The functions s and c are not, strictly speaking, analytic functions of ϵ but for many purposes they may serve in place of the analytic functions.

For $\epsilon < 0$ we introduce the Whittaker function $\theta(\epsilon, l; \rho)$ which has asymptotic form

$$\theta(\epsilon, l; \rho) = (2\rho/\nu)^{\nu} \exp(-\rho/\nu)\Theta(\epsilon, l; \rho) \tag{3}$$

where Θ has an asymptotic expansion in powers of $1/\rho$:

$$\Theta(\epsilon, l; \rho) = 1 + (1/\rho)b_1(\epsilon, l) + \cdots. \tag{4}$$

The relation between θ and s and c, which was given by Hartree, is

$$(-1)^l \left(\frac{\nu^3}{2}\right)^{1/2} \mathcal{K}\theta = c\sin(\pi\nu) - s\cos(\pi\nu), \tag{5}$$

where

$$\mathcal{K} = [\nu^2\Gamma(\nu + l + 1)\Gamma(\nu - l)]^{-1/2}. \tag{6}$$

Let us now consider an electron moving in a potential $V(r)$ which is equal to the Coulomb potential $v(r)$ for $r \geqslant r_0$. Using $V(r)$ we obtain a function $F(\epsilon, l; \rho)$ which is regular at the origin and which, for $r \geqslant r_0$, can be equated to a linear combination of s and c. It is often convenient to take F to be normalized in such a way that

$$F(\epsilon, l; \rho) = s(\epsilon, l; \rho) + K(\epsilon, l)c(\epsilon, l; \rho) \qquad \text{for } r \geqslant r_0. \tag{7}$$

The quantity $K(\epsilon, l)$ is the *reactance matrix* (a one-by-one matrix for the one-channel case!). We put

$$K(\epsilon, l) = \tan(\pi\mu(\epsilon, l)) \tag{8}$$

which defines the quantum defect $\mu(\epsilon, l)$ as a continuous function of ϵ. It is also a slowly varying function. We then have

$$F = (\cos(\pi\mu))^{-1}\{\cos(\pi\mu)s + \sin(\pi\mu)c\}. \tag{9}$$

For the bound states, with energies $\epsilon_n = -1/\nu_n^2$ where n is an integer, we put

$$\nu_n = n - \mu(\epsilon, l). \tag{10}$$

From (5) and (9) we then obtain $F \propto \theta$. That completes Hartree's derivation of the Rydberg formula.

Text books of the 1930s made little mention of quantum defects, except as empirical quantities used by experimental spectroscopists. Renewed interest was kindled by the work of Bates and Damgaard (1949)* who gave a simple procedure for estimating atomic oscillator strengths, and of Kuhn and Van Vleck (1950)* who considered applications in solid-state physics.

The work of Bates and Damgaard was concerned with an electron of known energy e moving in a potential with asymptotic Coulomb form. They argued that the radiative dipole integrals depend mainly on the wave functions for large r and can be estimated using the expansions (4) for Θ, truncated in such a way as to obtain convergent integrals. Some 47 years later, calculations based on their work are still used extensively.

Agnette Damgaard was the first research student of David Bates. I came second. In the early 1950s I became interested in some spectrum lines in nebulae produced by mechanisms of radiative capture. In order to estimate the capture rates I needed information on near-threshold photo-ionization cross-sections, and I asked myself how the method of Bates and Damgaard, for bound–bound transitions, could be generalized to the bound–free case. For $\epsilon > 0$ one may put $F \sim (\pi\kappa)^{-1/2}\sin(\zeta + \delta)$ where δ is the phase shift produced by short-range non-Coulomb potentials. In 1955* I was able to show that the phase shift δ is equal to π times the extrapolated quantum defect μ. That result follows immediately from equations (1), (2) and (9).

Meanwhile the work of the solid-state physicists had led to important developments in the mathematical theory culminating in the publication of Ham's 1955** paper which I am pleased to see included in the present volume. The Coulomb functions are, of course, special cases of confluent hypergeometric functions and one might have thought that they had been studied in exhaustive detail by applied mathematicians. However, a number of properties important for QDT had not been stated clearly. I think it may be worthwhile to summarize some properties elucidated by Ham, omitting technical details and using the notation of S83. Firstly, there are two functions $f(\epsilon, l; \rho)$ and $g(\epsilon, l; \rho)$ analytic in ϵ (power series absolutely and uniformly convergent for all finite ϵ). The expansion for the regular function f is quite straightforward but that for g is a good deal more complicated. Ham introduces a third function $h(\epsilon, l; \rho)$ which has an asymptotic expansion in powers of ϵ and is useful for $| \epsilon |$ not too large. The function h has properties simpler than those of g and is used extensively. In solid-state work one is concerned with atoms having electrons which are tightly bound but, since the atoms are in a crystal lattice, they have energies different from those of free atoms. For a given l, the hydrogen atom has a lowest level with $n = l + 1$, and for non-hydrogenic atoms there is a similar lower termination of each series of levels. Using the functions f and h, Ham was able to study properties of non-hydrogenic atoms in the vicinities of their most tightly bound states. For higher energies, such that $\nu \gg l$, one may use the simpler functions, s and c, in place of f and h.

Building on the work of Ham, in 1958** I published a fairly complete account of one-channel QDT (the work on photo-ionization was published with Burgess in 1960*). I considered both positive and negative values of ϵ (Ham had considered only negative values). The paper included results for the normalization of the bound-state functions and for the calculation of elastic-scattering cross-sections using extrapolated quantum defects.

For what range of energies can QDT be used? Ham considered a model problem, of one electron in a central potential $V(r)$ which is equal to $-2z/r$ for $r \geqslant r_0$ with r_0 finite. For that problem he gave a theory which is mathematically rigorous, involves only analytic functions, and is valid for all finite ϵ. His proof fails, for large negative ϵ, as soon as one includes exchange terms (the Hartree–Fock problem in place of the Hartree problem). The reason, from a physical standpoint, is that if all the electrons are indistinguishable and all tightly bound, it makes little sense to consider an outer region containing only one electron. The proof also fails for a real physical system if one goes to large values of E, since one must then include many excited N-electron states ψ_i and the radius r_0 becomes increasingly large. Eventually one reaches the point where the system can have two or more electrons in the continuum, and failure is complete. In practice there is a range of E for which the theory can be very useful: but I think one must abandon rigorous formulations and be content with vaguer concepts, such as 'functions which vary slowly…'.

What are the origins of the many-channel theory? In my description of the conceptual structure of the theory I have anticipated the many-channel case, that for which one has more than one target state ψ_i. Many aspects of the conceptual structure are common to other theories: all complicated interactions within a finite volume; inner-region functions having a simple energy dependence on the boundary; only one particle in the outer region, moving in a simple potential. Many of these theories originated in nuclear physics and they have various names, the most common being *effective range theory* (Bethe 1949, Jackson and Blatt 1950, Ross and Shaw 1961). I consider effective range theory to be the more general term and QDT to be a special case for which the outer-region potential has an attractive Coulomb form. Some authors, however, have taken a different view, and refer to theories with non-Coulomb potentials as generalizations of QDT.

While having much in common with effective range theory, R-matrix theory (Wigner 1946, Wigner and Eisenbud 1946, Lane and Thomas 1958) has a separate distinguishing feature, expressing the behaviour of the functions at the boundary in terms of a matrix $\mathbf{R}(E)$ and giving a prescription for the calculation of that matrix. Although it originated and has been used extensively in nuclear physics, the R-matrix method as a computational tool has been most successful in atomic and molecular physics, where the Hamiltonians are known (Burke and Robb 1975).

A special feature of the attractive Coulomb potential is that it gives an infinite Rydberg series of bound states, and infinite series of resonances converging to each new threshold. The behaviour of elastic cross-sections just below and just above new thresholds was studied by Baz' (1959), more general studies of near-threshold cross-sections were made by Fonda and Newton (1959), and important formulae for averaging cross-sections over resonances were given by Gaĭlitis (1963)**. My own interest in many-channel QDT was stimulated by the work of Damburg and Gaĭlitis (1962)** who applied the effective range theory of Ross and Shaw to the problem of electron scattering by atomic hydrogen. Our first paper on the subject (Bely, Moores and Seaton 1964*), concerned with both bound-state and collision problems, was presented at ICPEAC III.

There are two types of users of QDT: those who employ the formulae for empirical fitting of data from experiments or computations, and those who obtain all quantities in the formulae from *ab initio* calculations. Some of the most impressive developments have been in applications to molecular systems, initiated by Fano (1970)**. Further discussion of the molecular work is given in the following article, written by Christian Jungen whose knowledge of that work is far greater than my own. I find it gratifying to see how neatly the theory can be carried over to the relativistic case, in the work of Johnson and Cheng (1979)** and Lee and Johnson (1980)**. Some remarks about di-electronic recombination given in S83 were not entirely correct but they were, hopefully, corrected in my 1985** paper with Bell.

I conclude this essay by mentioning an awkward problem that refuses to go away. For real physical systems it is never completely correct to take the outer-region potential to be of pure Coulomb form. One has off-diagonal dipole potentials ($1/r^2$) which arise from dipole couplings between target states; quadrupole potentials ($1/r^3$) which may be on or off the diagonal; and polarization potentials ($1/r^4$) arising from second-order dipole interactions. The boundary radius, r_0, is always taken to be large compared with the mean radii of all target states included and it may be shown to follow that these additional long-range potentials are always small, in the outer region, compared with the Coulomb potential. They are not, however, always negligible. The most difficult cases can arise for molecular systems, for which there is not any obvious unique choice for the origin of the radial co-ordinate r.

The worst case is one which can be handled. For a hydrogenic target the nL target levels are degenerate with respect to L (I use L for the target angular momenta, l for those of the added electron) and the $(1/r^2)$ potentials connecting nL, $n(L \pm 1)$ are particularly important (cancellation effects reduce the importance of multipole potentials connecting target states of different energy). What saves the day is that one can diagonalize all $(1/r^2)$ potentials for a given level, including the centrifugal terms. One then obtains $-\lambda(\lambda+1)/r^2$ in place of the usual $-l(l+1)/r^2$, where λ is not an integer and may not even be real. This technique was first applied to electron collisions with neutral hydrogen by Gaĭlitis and Damburg (1963) who showed that the $(1/r^2)$ potentials produce some remarkable effects (cross-sections infinitely oscillatory as thresholds are approached). It was applied to collisions with He^+ by Bely (1963)* who used Coulomb functions with arguments $(\epsilon, \lambda; \rho)$, and by Dubau (1978)* who used functions with better linear independence. The procedure used was to calculate complex quantum defects in the region just above the He^+ 2s, 2p threshold and, on making extrapolations, to obtain the positions and widths of the resonances below that threshold. It turns out that the theory for the $(e + He^+)$ problem, with inclusion of excited sates of He^+, is satisfactory when the additional long-range potentials are included, but not when they are omitted. It might have been thought that the $(e + He^+)$ system would provide the simplest problem for applications of QDT. But that is certainly not the case.

For non-hydrogenic targets the multipole potentials can still be important. With a choice of r_0 typically used in a calculation, neglect of $1/r^2$ potentials for $r > r_0$ can lead to errors in cross-sections for optically allowed transitions which may be of the order of 30 per cent or so, increasing with increasing projectile energy. Use of QDT formulae, implying neglect of the dipole potentials in the outer region, can therefore lead to significant errors.

The problems which arise from the presence of multipole potentials are not so serious for empirical procedures: thus for bound states one can use the usual formulae, even though they may be difficult to derive. It is in *ab initio* work that the problems arise: if, for example, one wishes to use the formulae of Gaĭlitis for averaging over resonances, then the outer-region multipole potentials must be 'switched off'. I have tried to overcome these problems using a perturbation treatment which is consistent to first order in the long-range multipole potentials. I obtained some formulae but never published them; they were too complicated and all the charm of QDT was lost.

References

Baz' A I 1959 *Sov. Phys.–JETP* **9** 1256
Bethe H A 1949 *Phys. Rev.* **76** 38
Burke P G and Robb W D 1975 *Adv. Atom. Mol. Phys.* **11** 143
Fonda L and Newton R G 1959 *Ann. Phys. (USA)* **7** 133
Gaĭlitis M and Damburg R 1963 *Proc. Phys. Soc.* **82** 192
Jackson J D and Blatt J M 1950 *Rev. Mod. Phys.* **22** 77
Lane A M and Thomas R G 1958 *Ann. Phys. (USA)* **30** 257
Ross M H and Shaw G L 1961 *Ann. Phys. (USA)* **13** 147
Wigner E P 1946 *Phys. Rev.* **70** 15
Wigner E P and Eisenbud L 1947 *Phys. Rev.* **72** 29

Molecular applications of quantum defect theory

Ch Jungen
Laboratoire Aimé Cotton du CNRS, Université de Paris-Sud, 91405 Orsay, France

Note

** is used in referring to papers reproduced in the present collection.
* is used in referring to papers cited by Seaton (1983)**, or included in the bibliography of the present book.

Rydberg states have not only been observed in atoms but also in molecules, although not yet in clusters. While a great number of such states have been studied in diatomic molecules, the available information on triatomic systems is much more limited, while for larger polyatomic molecules detailed observations of Rydberg states are still scarce. For example, the observation of fine-structure resolved high Rydberg states ($n \approx 100$) in the benzene molecule was only achieved in 1995 by Neuhauser and Neusser (1996). Balmer established his famous empirical formula describing the energy levels of the hydrogen atom in 1885. Only four years were to pass until 1889 when Rydberg was able to establish the modified formula which bears his name, thereby implicitly introducing the concept of the 'quantum defect'. Rau and Inokuti (1996) recently pointed out that the term 'quantum defect' appeared for the first time in a paper published by Schrödinger in 1921. The first mention of MQDT, 'many channel quantum defect theory', can be found in a paper by Bely, Moores and Seaton published in 1964*. As every physics student knows, Rydberg's discovery was based on the sodium atom in which he observed the s, p, d and f series of states. It took no less than 77 years before a correspondingly 'complete' system of Rydberg states was established in a molecule: in 1966 Miescher published an energy level diagram of the NO molecule and a list of corresponding quantum defects, which turned out to be quite similar to those known for the Na atom.

The reason for this long delay is that, unlike in atoms, molecular Rydberg series are rarely directly observed as such. Rather they emerge from careful and extended fine-structure analyses which take account of the nuclear degrees of freedom of the system studied and the molecular vibrations and rotations. In polyatomic molecules the rotation–vibration structure is often so dense that it completely masks the Rydberg structure. An example is the water molecule where Rydberg series near the ionization threshold only emerged once the water vapour could be produced in a supersonically cooled jet so that the rotational population distribution was restricted to a very few asymmetric top rotational levels (Page *et al* 1988). Similarly it was a cooling technique which led to the observation in 1970 of Rydberg levels with $n \approx 40$ in H_2—the first such observation for any molecule (Herzberg 1969, Takezawa 1970, Herzberg and Jungen 1972*).

As Miescher already recognized in his pioneering work on NO, the complications due to the presence of the nuclear degrees of freedom in molecular systems are not limited to the sheer effect of crowding of numerous levels. More seriously, there is always a coupling of electronic and rotational–vibrational degrees of freedom which disrupts the regularity of atom-like Rydberg structure as the ionization limit of a molecule is approached. This breakdown of the separability of electronic and nuclear motion is not accidental but is systematic. This was stated clearly by Mulliken, who between 1964 and 1969 published a series of fundamental papers on Rydberg states of molecules. In the first of these papers he writes:

"In most discussions on molecular wave functions, the validity of the Born–Oppenheimer approximation is assumed. This approximation is most nearly accurate when the frequencies of motion, which can be gauged by energy level spacings, are much larger for the electronic than for the nuclear motions. In a Rydberg state series, as n increases, the frequencies for the Rydberg electron become smaller and smaller relative to those of nuclear vibration and rotation. This leads to more or less radical changes in coupling relations. ...In the B.O. approximation, the wave function for the lowest-n Rydberg states of a diatomic molecule with a closed-shell core takes the form

$$\Psi \approx \Psi_{el}\Psi_{vib}\Psi_{rot} = (\mathcal{A}\Psi_{core}\psi_{Ryd})\Psi_{vib}\Psi_{rot}, \tag{1}$$

where the operator $\mathcal{A}$ in equation (1) makes Ψ antisymmetric in all the electrons. As n increases and the inner loops of the Rydberg molecular orbital become less and less important relative to the outer loops, the Rydberg electron withdraws more and more from influencing the rotational and vibrational motions of the nuclei and for large n the wavefunction takes the form

$$\Psi \approx \mathcal{A}[(\Psi_{core}\Psi_{rot}\Psi_{vib})\psi_{Ryd}]. \tag{2}$$

Here ψ_{Ryd} is no longer included within the electronic factor of the B.O. approximation."

Mulliken referred to equations (1) and (2) as Rydberg-coupled and Rydberg-uncoupled wavefunctions, respectively.

Ever since Mulliken's work, rovibronic coupling, or as it is sometimes called, rotation/vibration-electron coupling, has remained a central aspect of the study of molecular Rydberg states. Indeed, the successful quantitative treatment of the coupling beween nuclear and electronic motion may be considered as the main achievement of the molecular applications of multichannel quantum defect theory.

These developments were initiated by a paper due to Fano (1970)** which is reprinted in this volume. It deals with rotation-electron coupling in the H_2 molecule and was spurred by the experimental work of Herzberg mentioned earlier. Fano showed that the frame transformation method developed for electron molecule scattering is ideally suited for use in combination with MQDT. In classical terms, the Rydberg electron has a large amount of kinetic energy as it approaches the boundary of a positively charged Rydberg core. The collision with the latter is therefore 'sudden', leaving no time for the heavy nuclei to move during the process. This means that, if anything, it is only electronic excitation that can occur, whereas the nuclear motion is unaffected—as anticipated in the Born–Oppenheimer approximation. Changes of the rotational and vibrational quantum numbers still occur of course, but they are induced by the uncoupling of the Rydberg electron from the molecular frame as it moves away from the core. As is suggested by equations (1) and (2), this transition from the Rydberg-coupled to the Rydberg-uncoupled regime is basically geometrical in nature, and in many cases can be treated *a priori* and analytically. The elements of the electron–ion core reactance matrix **K** are correspondingly written as

$$K_{ij} = \sum_{\alpha=1}^{N} U_{i\alpha} \tan \pi\mu_\alpha U_{\alpha j}^{\dagger}. \tag{3}$$

N is the total number of channels, i.e. $N = N_c + N_o$ where N_c channels are closed and N_o channels are open. **U** is the unitary eigenvector matrix of **K** and the $\tan \pi\mu_\alpha$ are its N eigenvalues. If the level structure of the ion core (expressed here in Rydberg energy units) is very closely spaced,

$$\Delta E_{core} \ll 1 \tag{4}$$

then the eigenvector matrix **U** can be taken to be equal to the frame transformation matrix. This is what is done in the molecular applications of QDT since in that case the ΔE_{core} correspond to the core rotational–vibrational fine-structure spacings which satisfy condition (4). The N so-called eigenquantum defects μ_α or eigenphaseshifts $\pi\mu_\alpha$ are defined in the molecule-fixed frame which is the natural working frame for quantum chemical calculations.

One of the advantages of this concept in practice is that the number of 'dynamical' parameters in a specific molecular Rydberg problem is greatly reduced and equals the number N of eigenquantum defects μ_α, as compared to the number $N(N-1)/2$ of independent elements of the 'geometric' frame transformation **U**. Thus for $N = 2$ two of the three independent elements of the scattering matrix are 'dynamical' and depend on the problem studied, whereas one element is 'geometric' and is known *a priori*. For $N = 100$ there are 100 dynamical parameters in the molecular QDT problem while at the same time the number of analytic geometrical quantities has increased to 4950. The power of the frame transformation method thus becomes evident. The presence of a large number of Rydberg and/or dissociation channels is the rule rather than the exception in molecular systems and this is again related to the existence of the vibrational and rotational degrees of freedom. Thus each vibrational mode of a molecular ion gives rise to as many ionization thresholds of the neutral molecule as there are vibrational levels corresponding to that mode. Further, if we consider an asymmetric top polyatomic molecule with a given total angular momentum J and an associated Rydberg electron with a single well defined ℓ value, we see that there will be $(2J+1)(2\ell+1)$ different rotational–electronic Rydberg channels. For instance in SO_2 the most probable rotational level at room temperature corresponds to $J \approx 18$, so that an electronic p channel for this J value will give rise to no less than 111 different asymmetric top rotational–electronic Rydberg channels. The close spacing of the rotational Rydberg limits in a relatively heavy molecule is also at the origin of the striking 'fringe' effect observed by Labastie *et al* (1984)** in the sodium dimer.

Several of the articles collected in this volume deal with various aspects of eigenchannel theory and frame transformations in molecules. The paper by Lu and Fano (1970)**, which appeared just before Fano's paper, discussed the graphical analysis of perturbed Rydberg series. It introduced the well known Lu–Fano plot which provides a striking visual demonstration of the eigenchannel concept. Lee and Lu (1973)** presented a summary and refinement of the atomic eigenchannel theory developed previously by Lu (1971)* and described the application to the Ar atom. Dill (1973)** extended these ideas to predict resonance behaviour of the photoelectron angular distributions in Xe. A few years later Jungen and Atabek (1977)** gave a comprehensive account of quantum defect theory for bound states in rotating/vibrating diatomic molecules, followed by a detailed application to *ungerade* Rydberg states of low n in H_2 and its isotopes. The paper by Jungen and Dill (1980)** extended this approach to treat rotational/vibrational autoionization of the same molecule. A further and much more recent extension by Ross and Jungen (1994)** addressed the electronic multichannel case of a molecule. Both ℓ-mixing induced by the non-spherical field of the core, and mixing of electron configurations induced by electron repulsion were considered. Specifically the effects of doubly excited electron configurations in H_2 were described, which are important in the *gerade* states and cause the appearance of potential energy curves with double minima and associated strong vibronic coupling.

Clearly the spectacular increase of the number of channels with which one has to deal in molecules as compared to atoms stems in part from the lower symmetry of molecular systems. A lowering of the symmetry (and hence an increase of the effective number of channels) is also brought about by external perturbations such as external fields or perturbing atoms. Several papers reprinted here deal with extending QDT to include these situations. These are the papers by Fano (1981)** and Sakimoto (1986)**, (1989)** which treat the Stark effect, and the papers by O'Mahony and Taylor (1986)** and Du and Greene (1987)** which treat, respectively, the quadratic Zeeman effect in non-hydrogenic atoms and the effect of neutral perturbers on the spectrum of Rydberg atoms. Zoller and his co-workers (Alber and Zoller 1984*, 1988*, Giusti-Suzor and Zoller (1987)*) have shown how the interaction of Rydberg electrons with strong laser fields can also be cast in the form of a quantum defect theory.

The Lu–Fano plot mentioned earlier has been the basis for much empirical work especially in atoms where one frequently has multichannel situations involving effectively just two series limits. The paper by Wynne and Armstrong (1979)** is a representative example which summarizes numerous multiphoton excitation experiments carried out by the IBM group on Ca, Sr and Ba. The systematic behaviour of the channel interactions in these atoms thereby becomes visually evident. Aymar (1984)**, in a paper on Ba, extended the empirical MQDT analysis to finer details of the spectra such as Landé g factors, hyperfine structures and lifetimes, which yield information on the wavefunction which is difficult to extract from the energy levels alone. In 1984 Giusti and Fano** devised a generalized 'universal' version of the Lu–Fano plot which is obtained from the original plot through an adequate superposition of the base pair of Coulomb functions f and g (corresponding to s and c in equation (5) of M Seaton's introductory essay).

In general the reactance matrix $\mathbf{K}$ of equation (3) includes open as well as closed channels—this is why it typically exhibits a smooth variation with energy on a gross scale. This 'smooth' $\mathbf{K}$ matrix is to be distinguished from the 'physical' matrix $\mathbf{K}(E)$

of scattering theory whose dimension corresponds to the number N_o of open channels at the given energy E and which has poles near the positions of bound levels. The passage from **K** to $\mathbf{K}(E)$ is achieved through the procedure of 'elimination' of the closed channels by means of the application of the required boundary condition at infinity to the closed channel components of the wavefunction. Their effect is thereby incorporated analytically into the energy-dependence of $\mathbf{K}(E)$. $\mathbf{K}(E)$ is given by the following well known expression (see Seaton 1983**):

$$\mathbf{K}(E) = \mathbf{K}^{oo} - \mathbf{K}^{oc}[\mathbf{K}^{cc} + \tan\beta(E)]^{-1}\mathbf{K}^{co}. \tag{5}$$

Here the smooth matrix **K** has been partitioned into four submatrices representing, respectively, interactions between open, open and closed and closed channels. $\beta(E)$ is the accumulated phase parameter which equals $\pi\nu(E)$ where ν is the effective principal quantum number for each channel. Eissner and Seaton (1974)** developed a procedure for parametrizing $\mathbf{K}(E)$ near resonances in the case where **K** is not known. The procedure preserves the accuracy of the full multichannel treatment and takes advantage of the smoothness of the full **K**. Lecomte (1987)** obtained a parametrization which in a sense is the inverse of that of Eissner and Seaton: here the open channels are eliminated while the focus is on an effective closed channel reactance matrix.

A remark concerning notation appears appropriate at this point. Different authors often tend to use different notation. Seaton's (1983)** review includes a table (his table 1) giving the relationships between the symbols used by various authors to denote the base pair of linearly independent regular and irregular radial Coulomb functions used to describe the motion of the Rydberg electron. The following Table indicates how the reactance and related scattering matrices as well as their extensions to include closed channels are denoted in a few of the key papers.

Table 1. Relations between notations used in various papers.

Author	Reactance	matrix	Scattering	matrix
	$N \times N$	$N_o \times N_o$	$N \times N$	$N_o \times N_o$
Seaton (1983)**	$\mathcal{R}$ or Y	$\mathbf{R}$	χ	$\mathbf{S}$
Fano (1978)**	$K^{(s)}$	$K(E)$		
Mies (1979)**	$\mathbf{Y}(E)$	$\mathbf{Y}^T_{oo}$	$\mathbf{S}(E)$	$\mathbf{S}_{oo}$
Greene and Jungen (1985)*	$\mathbf{K}$	$\mathbf{K}(E)$	$\mathbf{S}$	$\mathbf{S}(E)$

The exact meanings of the symbols may in addition depend on the base pair of radial functions used. Thus Seaton's (1983)** $\mathcal{R}$ corresponds to use of the μ quantum defect whereas Y corresponds to use of the 'smooth' η defect originally introduced by Ham (1955)**. However the **K** matrix of Ross and Jungen (1994)** refers to the latter whereas **K** in the review by Greene and Jungen (1985)* refers to the former.

Quantum defect theory lends itself to both *ab initio* calculation and empirical analysis. The present collection includes a few papers whose aim was the direct calculation from first principles of quantum defect parameters. Eissner and Seaton (1972)** described the closed coupling method for atomic systems. Burke and Seaton (1971)* give more detailed descriptions of the close coupling and R-matrix methods which have been used extensively by the groups at Queen's University Belfast and at University College

London. Raseev and Le Rouzo (1983)* and Greene (1983)** formulated a variational eigenchannel R-matrix procedure. This latter was applied by Greene and Aymar (1991)** to Sr, Ba and Ra in an approach where in a first step the eigenchannel R-matrix calculation yielded the 'smooth' $\mathbf{K}$ of equation (3), which subsequently was used for extensive MQDT calculations of bound and autoionizing spectra. It is probably fair to say that the empirical QDT work carried out under Aymar's guidance, and the R-matrix work done by her group as well as by Greene and co-workers, had a profound effect on the atomic spectroscopy of the last decade.

A typically molecular phenomenon, which has no counterpart in the physics of isolated atoms, is the possibility for a molecule to fragment into a pair of heavy particles—atoms or molecules—in addition to fragmentation into an ion and an electron. In other words, in addition to ionization, molecules also have the possibility of dissociation. Similarly, a collision between an electron and a molecular ion which populates the elastic and inelastic electron + ion channels may also lead to the formation of two neutral heavy particles—this is the process of dissociative recombination. Since the beginning of the 1980s a considerable amount of work has been spent on incorporating molecular dissociation channels into the framework of multichannel quantum defect theory.

Giusti (1980)** was the first to obtain a consistent multichannel theory achieving this goal. Giusti's paper is reprinted in this volume. The theory is adapted to a molecular variant of a situation which also occurs in atoms in which a Rydberg series is perturbed by a single valence state. Of course in the atomic case the interaction with the single perturber does not induce dissociation, but merely modifies the Rydberg level structure and the ionization cross-section. The atomic problem was treated by Mies (1979)** in a paper which is also reprinted here. Mies evaluated the perturbed, strongly energy-dependent, quantum defect in terms of the position of the perturber state and the Rydberg-valence state interaction energy. In the related molecular problem the valence state is unstable with respect to dissociation. The coupling between the electronic and nuclear degrees of freedom leads to dissociation of bound Rydberg levels. For fixed internuclear distance Giusti's treatment reduces to the expression given by Mies. The vibration-electron coupling is accounted for by the frame transformation. Remember that the molecular problem involves a large (in principle infinite) number of ionization channels since the presence of the vibrational degree of freedom turns a single atomic Rydberg channel into a large number of molecular Rydberg channels each corresponding to a particular vibrational quantum number v^+ of the ion core. Giusti's method has been instrumental in the understanding of the dissociative recombination of interstellar molecules. As an example, the article by Schneider, Dulieu and Giusti-Suzor (1991)** on dissociative recombination of H_2^+ is reprinted here.

A rather different treatment of interfering ionization and dissociation channels is due to Jungen (1984)**. Here dissociation is not induced by a channel external to the Rydberg series but instead comes from within the channel itself: vibration-electronic coupling between a dissociated Rydberg state with low principal quantum number and higher members of the same series can induce dissociation in the latter. The theoretical treatment in this case invokes R-matrix theory for the vibrational (dissociation) coordinate whereas the quantum defect treatment for the radial electron coordinate is preserved. An attractive aspect of this theory is that a single function, the nuclear coordinate dependent quantum defect $\mu(R)$, not only accounts for Rydberg level positions and their perturbations, but also autoionization and predissociation.

The idea of generalizing quantum defect theory to potential fields other than the Coulomb attraction between two point charges arises naturally in molecular physics where the motion of the nuclei is governed by electronic potentials which exhibit a wide variety of different shapes. In the opinion of M Seaton, as expressed in his introductory essay, this generalization should not be referred to as 'generalized quantum defect theory'. Rather he views quantum defect theory as a special case of 'effective range' theory. One might nevertheless take the alternative view that 'effective range' theories, such as used in nuclear physics, do not treat series of bound states as 'scattering' states at negative energy as is characteristic of QDT. The 'general' forms of quantum defect theory developed in the papers by Greene, Rau and Fano (1982)** and Mies (1984)** do concentrate on this aspect. They define quantum defects that extrapolate smoothly across fragmentation thresholds corresponding to ionization or to dissociation and where the fragments may or may not carry electric charge.

A particularly elegant development is the phase-amplitude method of Greene, Rau and Fano (1982)**. The main feature of this theory is that the effective principal quantum number ν of QDT is replaced by the accumulated phase parameter β which, for a Coulomb field, corresponds to $\pi(\nu - \ell)$. β/π measures the (generally non-integral) energy-dependent number of half-oscillations (half-wavelengths) which accumulate in the potential well before the wavefunction reaches its (generally divergent) asymptotic form. For a Coulomb field the relationship between $\beta/\pi = (\nu - \ell)$ and the energy ϵ (in Rydbergs) is simply that given by the Rydberg equation, $\beta/\pi = (-\epsilon)^{-\frac{1}{2}} - \ell$. The elegance of QDT rests largely on the fact that given this expression the radial electron wave functions never actually need to be evaluated explicitly. In the generalized phase-amplitude formulation this simplicity of QDT is lost, but the accumulated phase $\beta(\epsilon)$ can still be evaluated by straightforward numerical evaluation of Milne's equation (1930), which is an inhomogeneous version of the one-particle radial Schrödinger equation at the given energy ϵ. Raoult and Balint-Kurti (1988)** showed how generalized QDT leads to a much more efficient description, in numerical terms, of vibrational predissociation in the Van der Waals complex ArH_2 than the customary coupled channel approach. This is because by defining a suitable 'core' or 'reaction' zone they were able to separate short-range coupling from long-range potential effects, thereby obtaining quantum defect matrices that vary slowly with energy on the scale of the phenomenon studied. In general, however, there are still far less applications of these concepts than of the QDT for Coulomb fields.

One question which often comes up in discussions concerning the use of QDT is how this method relates to the more familiar configuration treatments. Speaking a different language often renders communication difficult. QDT as a variant of scattering theory basically operates 'on the energy shell', at a given fixed total energy. Configuration mixing theory involves the superposition of zero-order states extending over the full energy spectrum, whereby the zero-order states correspond to an approximate Hamiltonian and their mixing is brought about by taking account of the full Hamiltonian. Fano (1978)** gave a general derivation of the connection between configuration and quantum defect treatments both in the discrete range and the continuum based on use of the Green's function. Mies (1979)** in the work mentioned already, demonstrated how a single discrete isolated state perturbing a Rydberg channel can be incorporated into the latter whose quantum defect now acquires a strong energy dependence. At a much more phenomenological level Herzberg and Jungen (1972)* and later Raoult and Jungen (1981)* showed the equivalence of Hamiltonian matrix diagonalization and quantum defect techniques. Their simple analytical expressions for rotational and vibrational autoionization in molecules have been rather widely used, such as in the discussions of the conspicuous lifetime lengthening that occurs in very high Rydberg states for $n \geqslant 100$ due to the presence of weak stray fields and collisions (Chupka 1993).

References

Balmer J J 1885 *Ann. d. Phys. Chem.* **25** 80
Chupka W A 1993 *J. Phys. Chem. Phys.* **98** 4520
Herzberg G 1969 *Phys. Rev. Lett.* **23** 1081
Miescher E 1966 *J. Mol. Spectrosc.* **20** 130
Milne W E 1930 *Phys. Rev.* **35** 863
Mulliken R S 1964 *J. Am. Chem. Soc.* **86** 3183
Mulliken R S 1966 *J. Am. Chem. Soc.* **88** 1849
Mulliken R S 1969 *J. Am. Chem. Soc.* **91** 4615
Neuhauser R and Neusser H J 1996 *Chem. Phys. Lett.* **253** 151
Page R H, Larkin R J, Shen Y R and Lee Y T 1988 *J. Chem. Phys.* **88** 2249
Rau A R P and Inokuti M 1996 *Am. J. Phys.* at press
Rydberg J R 1890 *K. Svenska Vetenskaps Akad. Handlinger* **23** 1
Schrödinger E 1921 *Z. Phys.* **4** 347
Takezawa S 1970 *J. Chem. Phys.* **52** 2575, 5793

Part A

Atomic theory

1955 *Solid State Phys.* **1** 127–92

The Quantum Defect Method[1]

Frank S. Ham[2]

Department of Physics, University of Illinois, Urbana, Illinois

I. Introduction

Recent attempts to determine theoretically the wave functions and energy levels of electrons in solids have demonstrated clearly that the amount of labor involved in such a study is enormous, even if one endeavors to simplify the problem by making quite restrictive assumptions concerning the form of the interaction potential and that of the many-particle wave function. It is, therefore, of interest when a procedure is discovered which makes short-cuts in the calculations possible without destroying the reasonableness of the model employed. The present article is concerned with the presentation of such a procedure.

The Quantum Defect Method proceeds from the assumption that when calculating wave functions of electrons in the valence and conduc-

[1] This article is an expansion of a paper by H. Brooks and F. S. Ham which has been submitted for publication to *The Physical Review* (1955). It has been drawn for the most part from the present author's doctoral thesis (Harvard University, Cambridge, Mass., 1954, unpublished).

[2] National Research Council Postdoctoral Fellow, 1954–1955.

tion bands it is a sufficient approximation in many solids to take as a one-electron Hamiltonian an operator which, near any one of the ions in the lattice, is the same as that appropriate to a valence electron in the corresponding free atom. It then takes advantage of the fact, to be discussed in detail, that the regular solution of the resulting Hartree-Fock equation near one of the ions of the lattice can be obtained explicitly, for arbitrary energies and for radial distances greater than the outer radius of the ion core, from the observed spectroscopic term values of the free atom. Since the electron energy levels in the solid can be determined from these solutions of this equation outside the core, it follows that the energy band structure of the solid can be determined without explicit knowledge of the one-electron Hamiltonian within the core and without the necessity of integrating the Hartree-Fock equation through a complicated "potential," even if this "potential" is known. We may, then, dispense with the numerical work involved in solving the equation when the potential is known, and in addition we may calculate the band structure of certain solids for which an atomic potential either is not known or for some reason cannot be constructed successfully, as is the case with the heavy elements.

The present form of the Quantum Defect Method[3] is a development of a procedure originated by Kuhn and Van Vleck.[4] The principal modifications from the original proposal are the work of H. Brooks,[5,6] who has made a systematic application of the method in studying the lattice constant, cohesive energy, compressibility, and g factor of the alkali metals. The present author, in collaboration with Brooks, has used the method to calculate energy eigenvalues at symmetry points in the Brillouin zone for sodium and has assisted in putting the arguments for the method in the form given below.[7,8] Also, Kambe has investigated the noble metals from this point of view.[9] It is expected that the method will be used by Brooks and the author in the near future in a detailed investigation of the band structure of the alkali metals, the solids which satisfy best the conditions for its simple application, and in an examination of the main features of the bands of the divalent and trivalent metals.

In the following pages, we shall attempt to give a full discussion of the way in which the necessary information is extracted from the

[3] We shall refer to the Quantum Defect Method by the abbreviation QDM throughout most of this paper.

[4] T. S. Kuhn and J. H. Van Vleck, *Phys. Rev.* **79,** 382 (1950).

[5] H. Brooks, *Phys. Rev.* **91,** 1027 (1953), and unpublished work.

[6] J. H. Van Vleck, *Proc. 1953 Intern. Conf. Theoret. Phys., Tokyo,* p. 640 (1954).

[7] F. S. Ham, Ph.D. thesis, Harvard University, Cambridge, Mass., 1954 (unpublished).

[8] H. Brooks and F. S. Ham, *Phys. Rev.* (to be published) (1955).

[9] K. Kambe, *Phys. Rev.* **99,** 419 (1955).

spectroscopic data and of the theoretical justification of this procedure. We shall not attempt to give a complete account of the application of the Quantum Defect Method. We shall, however, indicate briefly that it provides sufficient information for use in connection with several of the techniques available for the calculation of energy bands and shall report some of the results that have so far been obtained. Further details concerning application and the results of more complete calculations will be reported in subsequent publications when they are completed.

We shall first discuss the use of a one-electron Hamiltonian in solid-state calculations, with particular reference to the Hartree-Fock[10] equation. We shall make use of the arguments given by Wigner and Seitz[11] in support of the assumption that a one-electron operator which, near an ion of the lattice, equals the Hamiltonian appropriate to the valence electron in the corresponding free atom is sufficiently accurate for many solids. Assuming at the start that we can approximate the interaction operators in the one-electron Hamiltonian by a simple potential function, we shall introduce the fundamental features of QDM, with which the solutions of the resulting one-electron Schrödinger-like equation involving this potential are obtained, by reviewing the original proposal of Kuhn and Van Vleck. We shall then discuss briefly several of the methods for calculating energy bands, notably those of Wigner and Seitz,[11] Bardeen,[12] Howarth and Jones,[13] and Kohn and Rostoker,[14] for which QDM can be used.

In developing QDM, we shall first prove that, for a certain class of ion potentials, the solution of the Schrödinger equation which is regular at the origin is determined for arbitrary energy and radii outside the ion core by the eigenvalue spectrum. We shall show that this solution may be written as a linear combination of two standard confluent hypergeometric functions outside the core. The coupling constant, which determines the relative amplitude of the two standard functions, involves a single parameter which depends on the nature of the core potential. We shall show by the WKB method that this parameter may be identified with the "extrapolated quantum defect," which is obtained by putting a smooth curve through the experimental values of the quantum defect of the spectral terms when the values are plotted against energy. This identification is valid for energies above approximately the lowest eigen-

[10] F. Seitz, "Modern Theory of Solids," p. 234 and Appendix, p. 677. McGraw-Hill, New York, 1940.

[11] E. Wigner and F. Seitz, *Phys. Rev.* **43,** 804 (1933); **46,** 509 (1934).

[12] J. Bardeen, *J. Chem. Phys.* **6,** 367, 372 (1938).

[13] D. J. Howarth and H. Jones, *Proc. Phys. Soc.* **A65,** 355 (1952).

[14] W. Kohn and N. Rostoker, *Phys. Rev.* **94,** 1111 (1954).

value of the hydrogenic potential for the angular momentum under consideration and also above the energy at which the outer WKB turning point equals the radius of the ion "core." At lower energies, we shall show that a second parameter, the "η-defect," is extrapolated more safely. In particular, this extrapolation is more reliable if data obtained from energy levels of electrons in the core are utilized. In support of these arguments, we shall present the results of exact calculations of the parameters using a simplified model of a core potential.

We shall show that QDM takes reasonably accurate account of exchange and correlation effects in the interaction between the valence and core electrons in the heavy elements. The method is, therefore, particularly convenient when these effects are large enough to complicate seriously the construction of an effective one-electron potential for the valence electron. We shall show that it is necessary to make appropriate changes in the experimental quantum defect data before applying QDM in the solid if we wish to take account of polarization of the ion core by the valence electron or wish to make corrections for small overlap of neighboring ions. Tables of data necessary for the application of QDM to the alkali metals are presented in the last sections of this article. The results of some of the calculations on these metals are also listed.

II. The One-Electron Potential and the Hartree-Fock Equation

In seeking the wave functions and energies of possible states of a solid, we should like to solve the many-body Schrödinger equation for the interacting electrons and nuclei. Since an exact solution is out of the question, we must be content to make simplifying assumptions which bring the problem into manageable form. We therefore assume, following Fock,[15] that it is a good approximation to take the wave function for the N electrons of the system in the form of a single determinant of one-electron functions having the form

$$\Psi = \frac{1}{\sqrt{N!}} \begin{vmatrix} \varphi_1(\mathbf{r}_1) & \varphi_2(\mathbf{r}_1) & \cdots & \varphi_N(\mathbf{r}_1) \\ \varphi_1(\mathbf{r}_2) & \varphi_2(\mathbf{r}_2) & \cdots & \varphi_N(\mathbf{r}_2) \\ \cdot & \cdot & \cdots & \cdot \\ \cdot & \cdot & \cdots & \cdot \\ \cdot & \cdot & \cdots & \cdot \\ \varphi_1(\mathbf{r}_N) & \varphi_2(\mathbf{r}_N) & \cdots & \varphi_N(\mathbf{r}_N) \end{vmatrix}, \tag{2.1}$$

in which the $\varphi_j(\mathbf{r}_i)$ are functions both of positional coordinates x_i, y_i, z_i, and of spin, and are normalized to unity. This function satisfies the requirement of Fermi statistics, for it is antisymmetrical under permu-

[15] V. Fock, *Z. Physik* **61**, 126 (1930).

tations of the electron coordinates and obviously vanishes if two of the φ_j are the same, as required by the Pauli principle. If we neglect spin-dependent interactions, the Hamiltonian of the system is

$$\mathcal{H} = \sum_i^N \{-(\hbar^2/2m)\nabla_i^2 + V_i(\mathbf{r}_i)\} + (\tfrac{1}{2}) \sum_{i\neq j} (e^2/r_{ij}), \tag{2.2}$$

in which m is the electron mass, $\hbar$ is Planck's constant divided by 2π, r_{ij} is the distance between the positions of electrons i and j, and $V_i(\mathbf{r}_i)$ is the potential function representing the interaction of electron i with the nuclei. Consistent with our spin-independent Hamiltonian, we assume that the one-electron functions may be factored into a function of position $\psi_j(x_i,y_i,z_i)$ and an eigenfunction of spin $\eta_j(\zeta_i)$; moreover, we assume that each orbital function ψ_j appears twice among the $\varphi_j(\mathbf{r}_i)$ of (2.1), once associated with an "up" spin, once with "down." This latter assumption would not be valid for a ferromagnetic material, but it is a reasonable approximation to the situation in the ground state of other solids. If we now require that the expectation value of the Hamiltonian (2.2) for the state (2.1) be stationary with respect to variations in the $\psi_j(x_i,y_i,z_i)$, we find that the "best" one-electron functions must satisfy Fock's equation,

$$-(\hbar^2/2m)\nabla_1^2\varphi_i(\mathbf{r}_1) + \left[V(\mathbf{r}_1) + \sum_j e^2 \int \frac{|\varphi_j(\mathbf{r}_2)|^2}{r_{12}}\, d\tau_2\right]\varphi_i(\mathbf{r}_1) - \sum_j e^2 \left[\int \frac{\varphi_j^*(\mathbf{r}_2)\varphi_i(\mathbf{r}_2)}{r_{12}}\, d\tau_2\right]\varphi_j(\mathbf{r}_1) = \sum_j \lambda_{ij}\varphi_j(\mathbf{r}_1). \tag{2.3}$$

The integrals in this equation imply summation over spin coordinates as well as integration over those of position. If the spin summation is carried out, the first sum involving j continues to extend over all the N occupied orbitals, that is each $\psi_j(\mathbf{r}_2)$ appears twice. The second sum involving j includes only those orbitals having spin functions which are the same as that appearing in $\varphi_i(\mathbf{r}_2)$. Thus the second sum extends only once over the $\psi_j(\mathbf{r}_2)$. We may remark that the term for $j = i$ in the first sum is canceled by that in the second. These terms are left explicitly in the summation for the sake of symmetry. We may put $\lambda_{ii} = \epsilon_i$ in (2.3). By Koopman's theorem[16] this parameter represents the energy required to remove the ith electron from the solid. The λ_{ij} are the Lagrange multipliers introduced in the variation procedure in order to insure normalization and orthogonality of the $\psi_j(\mathbf{r}_i)$.

[16] F. Seitz, p. 313 in ref. 10.

We define the Dirac density matrix[17] by the relation

$$\rho(\mathbf{r}_1,\mathbf{r}_2) = \sum_j \varphi_j^*(\mathbf{r}_1)\varphi_j(\mathbf{r}_2). \tag{2.4}$$

In terms of this, the first summation in (2.3) becomes

$$U\varphi_i(\mathbf{r}_1) = \left[e^2 \int \frac{\rho(\mathbf{r}_2,\mathbf{r}_2)}{r_{12}}\,d\tau_2\right]\varphi_i(\mathbf{r}_1). \tag{2.5}$$

The quantity in brackets is the Coulomb interaction of the ith electron with the *entire* electronic charge distribution, including its own. The second summation may be written in the form

$$A\varphi_i(\mathbf{r}_1) = -e^2 \int \frac{\rho(\mathbf{r}_2,\mathbf{r}_1)\varphi_i(\mathbf{r}_2)}{r_{12}}\,d\tau_2. \tag{2.6}$$

This equation defines the Dirac exchange operator A, which, like other operators in the Fock equation, is Hermitean. It follows that we can write (2.3) in terms of a Hermitean operator $\mathcal{H}^F$, the Fock one-electron Hamiltonian, which is the same for all the N wave functions φ_i:

$$\begin{aligned} \mathcal{H}^F\varphi_i(\mathbf{r}) &= \epsilon_i\varphi_i(\mathbf{r}), \\ \mathcal{H}^F &\equiv -(\hbar^2/2m)\nabla^2 + V + U + A. \end{aligned} \tag{2.7}$$

We can show, moreover, that it is consistent to assume that $\mathcal{H}^F$ is a periodic operator, being unchanged by a displacement through any one of the vectors $\mathbf{R}_j$ defining the periodic structure of the lattice and that the $\varphi_i(\mathbf{r}_j)$ have the Bloch form[18]

$$\varphi_i(\mathbf{r}_j) = e^{(i\mathbf{k}\cdot\mathbf{r}_j)}u_i(\mathbf{r}_j). \tag{2.8}$$

Here $u_i(\mathbf{r}_j)$ has the periodicity of the lattice and $\mathbf{k}$ is a real vector.

In writing (2.7), we have set all the λ_{ij} with $i \neq j$ equal to zero. This is not possible in general, for the λ_{ij} must be chosen in such a way that the set of Eqs. (2.3) have solutions for all i which are normalized and mutually orthogonal. In general, this condition requires that some of the λ_{ij} be different from zero. However, solutions of (2.7) corresponding to different ϵ_i are automatically orthogonal because $\mathcal{H}^F$ is Hermitean and the same for all the φ_i. In addition to this, the functions $\varphi_i(\mathbf{r})$ in the Bloch form (2.8) are orthogonal if they correspond to different $\mathbf{k}$. Moreover, functions which possess the same $\mathbf{k}$ but which transform according to different irreducible representations, or belong to different rows of the

[17] P. A. M. Dirac, *Proc. Cambridge Phil. Soc.* **26**, 376 (1930).
[18] F. Bloch, *Z. Physik* **52**, 555 (1928).

same representation of the group of the vector **k**,[19] are orthogonal. Thus, we can take the λ_{ij} to be zero for $i \neq j$ except perhaps in the case of accidental degeneracy, that is, when two solutions φ_i and φ_j of (2.7) which have the same transformation properties also have the same ϵ_i. We shall assume that such accidental degeneracies are sufficiently infrequent that we may omit the complications to which they give rise.

Since we do not yet know the $\varphi_i(\mathbf{r})$, and, indeed, wish to determine them from (2.7), we must introduce reasonable approximations for the operators U and A. We could eventually recalculate these quantities once the $\varphi_i(\mathbf{r})$ are known if we should wish to check or improve the self-consistency of the procedure. We have already noted that U is a simple potential operator. The exchange operator A is more complicated, being an integral operator. However, $A\varphi_i(\mathbf{r}_1)$ may be regarded as representing the potential interaction of the ith electron with a *positive* distribution of charge containing one electronic unit and possessing a spin parallel to that of φ_i. Equivalently, this quantity may be viewed as the correction to (2.5) arising from a hole scooped out of the entire distribution of electrons in the neighborhood of $\mathbf{r}_1$ possessing parallel spin.[11,20] To see this, we write (2.6) in the form

$$A\varphi_i(\mathbf{r}_1) = -e^2\left[\int \frac{\rho(\mathbf{r}_2,\mathbf{r}_1)\varphi_i(\mathbf{r}_2)}{\varphi_i(\mathbf{r}_1)r_{12}}\,d\tau_2\right]\varphi_i(\mathbf{r}_1). \tag{2.9}$$

The "charge density" evidently is $e[\rho(\mathbf{r}_2,\mathbf{r}_1)\varphi_i(\mathbf{r}_2)/\varphi_i(\mathbf{r}_1)]$. When we integrate this over the entire crystal, we find the total charge

$$Q = +e\sum_j \int \frac{\varphi_j^*(\mathbf{r}_2)\varphi_j(\mathbf{r}_1)\varphi_i(\mathbf{r}_2)}{\varphi_i(\mathbf{r}_1)}\,d\tau_2 = e\sum_{\substack{j\\(\text{spins }\parallel)}} \frac{\delta_{ij}\varphi_j(\mathbf{r}_1)}{\varphi_i(\mathbf{r}_1)} = +e. \tag{2.10}$$

Restricting our attention for the moment to a monovalent metal, we see that the foregoing exchange charge density is most appropriately viewed as a hole in the distribution of conduction electrons if $\varphi_i(\mathbf{r}_1)$ in (2.9) represents a state of the conduction band, for the contribution of the core states to this charge density vanishes if $\mathbf{r}_1$ lies outside the ion cores. The core states contribute nothing to the total charge (2.10), so that in this case the conduction electrons contribute the whole amount. Moreover, their contribution to the charge density is concentrated within a sphere not much larger than the atomic cell. This conclusion follows from the integration condition (2.10) and the fact that the exchange charge density for $\mathbf{r}_2 = \mathbf{r}_1$ arising from the conduction electrons is equal to the density of conduction electrons of a given spin at that point. The

[19] L. P. Bouckaert, R. Smoluchowski, and E. Wigner, *Phys. Rev.* **50,** 58 (1936).
[20] J. C. Slater, *Phys. Rev.* **81,** 385 (1951).

exchange density presumably falls off monotonically for increasing $\mathbf{r}_1 - \mathbf{r}_2$, as Wigner and Seitz have proved to be the case for free electrons.[11] The part of the exchange term (2.6) arising from the summation in (2.4) involving conduction states therefore cancels that part of the Coulomb interaction (2.5) arising from conduction electrons of parallel spin in a sphere centered on r_1 and roughly the size of the atomic cell. The contribution to the exchange operator arising from the core states may be regarded as resulting from a rearrangement of the core electrons which occurs whenever the conduction electron penetrates a core and which originates in the repulsion of electrons with parallel spin. In the following discussion, we must always include this part of the exchange operator in our representation of the interaction between a conduction electron and an ion core.

The repulsive correlation of electrons having parallel spin is the result of the Pauli exclusion principle, that is, of the use of the determinental wave function (2.1). No analogous statistical correlation in the distribution of electrons having spin antiparallel to $\varphi_i(\mathbf{r})$ is included in the Fock equation. Thus, we still should take account of the presence of this charge within the cell when approximating the operator U. However, we know that we have not taken complete account of the general tendency of electrons to avoid each other as a result of their electrostatic repulsion. It is reasonable to believe that we should find a hole in the distribution of electrons of antiparallel spin as well, if we could take into account such correlations.[21] Therefore, because of exchange and electrostatic repulsion, it would seem to be reasonable for a monovalent metal to assume that we may disregard terms in (2.7) that correspond to penetration of conduction electrons into the cell in which $\mathbf{r}_1$ is located. We shall assume, moreover, that ion cores in other cells are screened by the conduction electron distribution. This should be true to good approximation if the atomic cell contains a single ion and is not too far from spherical in shape. Furthermore, we shall assume that the atomic cell is sufficiently large, relative to the ion core, that the latter is distorted negligibly from its form in the free ion, an assumption valid at normal pressures for the alkali metals as well as for divalent and trivalent metals having rare gas cores. This assumption is not correct for the noble or transition metals. To summarize, we may assume, when treating a monovalent metal with a small ion core, that the operator $U + A$ in the Fock equation may be approximated by the corresponding operator for the free ion located in a given cell of the lattice whenever $\mathbf{r}_1$ is within that atomic cell.

The foregoing discussion, which is essentially that given by Wigner and Seitz,[11] is most appropriate to the alkali metals. As has been re-

[21] E. Wigner, *Phys. Rev.* **46**, 1002 (1934).

marked, there is appreciable overlap between the cores of neighboring ions in the noble metals. In such cases, the Fock equation for the conduction electron wave function within one cell should contain not only $U + A$ for the free ion in the given cell but an additional term which accounts for the penetration of the core electrons of the nearby ions into the cell. For multivalent metals, on the other hand, it clearly would be wrong to suppose that correlation and exchange effects prevent more than one conduction or valence electron from being simultaneously in the neighborhood of the same ion, for, on the average, there must be as many electrons in a single cell as there are valence electrons in the free atom. Consequently, we must add to the ion core contribution to $U + A$ other terms that take into account the additional valence electrons in the cell. We may be able to neglect such terms, however, when $\mathbf{r}_1$ is *within* the ion core, a possibility that we shall mention again later. The situation is similar in more complicated nonmetallic solids, such as diamond, germanium, and silicon. In ionic crystals the cell surrounding each ion is not electrically neutral and, therefore, contributes significantly to the potential in nearby cells.

We shall consider primarily the simple situation that applies approximately to the alkali metals in presenting the Quantum Defect Method below. That is, we shall usually assume that the operator $U + A$ in the Fock equation for a conduction electron may be replaced by the corresponding operator for the free ion in the same cell. We shall note briefly, however, that QDM may be applied to a somewhat more complicated model, appropriate to other solids, in which $U + A$ for the free atom is used only within the ion core and a more elaborate operator, which takes into account the interaction with other valence electrons, is used for the region between the core and the surface of the appropriate cell surrounding the ion.

The Fock equation, as simplified above, evidently takes account of the interactions between the valence electrons rather crudely. It is customary to make corrections for this simplification once the energy bands and wave functions in this approximation have been obtained. We shall not be concerned with such corrections in the present work. They have been discussed by Wigner and Seitz,[11,21,22] Herring,[23] Pines,[24] Macke,[25] and Raimes.[26]

[22] F. Seitz, *Phys. Rev.* **47,** 400 (1935).
[23] C. Herring, *Phys. Rev.* **82,** 282 (1951).
[24] D. Pines, *Phys. Rev.* **92,** 626 (1953); also previous papers in the series, and Solvay Conference Report, 1954 (to be published).
[25] W. Mäcke, *Z. Naturforsch.* **5a,** 192 (1950).
[26] S. Raimes, *Proc. Symposium Radar Research Establishment Malvern*, p. 16 (1954).

III. Fundamental Features of the Quantum Defect Method—the Work of Kuhn and Van Vleck

In calculating the energy bands of a solid, using the simplifications of the last section, one encounters the problem of approximating $U + A$ suitably in the free atom. We shall assume, for the present, that the effect of the integral operator A on the wave functions of interest can be approximated by a spherically symmetric potential function. We could represent $U + A$ approximately by making a self-consistent calculation of the wave functions of the core and valence electrons of the atom in its ground state, using the Fock equation, and using the result to compute the various terms in (2.7), which might then be approximated by an effective potential. A simpler procedure for a monovalent atom is to construct a spherically symmetric potential function which is Coulombic outside the ion core and which reproduces to reasonable accuracy that part of the free atom spectrum which arises from transitions of the valence electron alone. The latter procedure has been used by Prokofjew[27] for sodium and Seitz for lithium.[28] The potentials reproduce the s, p, and d spectra to within 1% for sodium and considerably better for lithium. Gorin[29] was less successful with potassium, for he was unable to find a single potential that would reproduce both the s and p spectra, although he could reproduce each separately to within several per cent.

Once such an effective potential has been found, one uses it in place of $U + A$ in (2.7) and obtains the approximate Fock equation in one of the atomic cells:

$$-(\hbar^2/2m)\nabla'^2\varphi + V'(r')\varphi = E\varphi. \tag{3.1}$$

If we expand the wave function within the same cell in spherical harmonics with the nucleus of the ion as the local origin, so that

$$\varphi(\mathbf{r}') = \frac{1}{r'}\sum_{L,M} A_{LM}Y^{LM}(\theta,\varphi)U^L(r'), \tag{3.2}$$

we find, on separating (3.1), that the radial differential equation for angular momentum L is

$$\frac{d^2U^L}{dr^2} + \left[\epsilon - V(r) - \frac{L(L+1)}{r^2}\right]U^L = 0. \tag{3.3}$$

[27] W. Prokofjew, *Z. Physik* **58,** 255 (1929).

[28] F. Seitz, *Phys. Rev.* **47,** 400 (1935). This potential was published incorrectly in Seitz's paper, although Seitz used the correct potential in his calculations. The correct potential has been recently published by Kohn and Rostoker.[14]

[29] E. Gorin, *Physik. Z. Sowjetunion* **9,** 328 (1936).

This equation must be integrated for various values of ϵ subject to the boundary condition $U^L(0) = 0$. To determine the energy eigenvalues corresponding to points in the Brillouin zone, we must then determine ϵ in such a way that (3.2) satisfies appropriate boundary conditions on the surface of the cell. We have introduced atomic units, according to the equations $E = me^4\epsilon/2\hbar^2$, $\mathbf{r}' = (\hbar^2/me^2)\mathbf{r}$, in writing (3.3). Thus, the unit of energy is the ionization potential of the hydrogen atom (1 Rydberg unit) and the unit of length the radius of the smallest Bohr orbit in hydrogen.

We notice that the potential is Coulombic outside of the ion core, so that $V(r) = -2Z/r$. Here Z is the difference between the absolute value of the nuclear charge and that of the electronic core, in units of the charge of the electron ($Z = 1$ for a monovalent atom). Hence, outside of the core, the solution of (3.3) which vanishes at the origin may be expressed as a linear combination of any two linearly independent solutions of the Coulomb differential equation. The usual procedure for determining this combination requires that one first use the free atom spectrum and the Fock equation for the atom to construct the effective core potential. This potential is then used in Fock's equation for the crystal, with the boundary condition at the origin, to obtain the radial wave function for a given L and for an arbitrary energy, and in the Coulomb region the appropriate combination of Coulomb functions may then be determined from the value of the logarithmic derivative of this radial function. Kuhn and Van Vleck[4] pointed out that one would be able to avoid the difficulties of constructing the potential and of integrating the radial equation if it were possible to go *directly* from knowledge of the spectrum to the knowledge of this combination of Coulomb functions outside the core. Such a procedure would be particularly useful if the surface of the atomic cell, on which one must satisfy periodicity boundary conditions, were outside the core and if convenient solutions of the Coulomb differential equation could be found for the ranges of r and ϵ of interest. Kuhn[30] has given formulas for such solutions by extending the work of Wannier[31] and Jastrow[32] (Appendix). An oversight in Kuhn's analysis has recently been corrected by the author,[33,34] who has prepared a more accurate table of the necessary functions.

Kuhn and Van Vleck found that the logarithmic derivative $(1/U^L)\cdot$

[30] T. S. Kuhn, *Quart. Appl. Math.* **9,** 1 (1951).
[31] G. H. Wannier, *Phys. Rev.* **64,** 358 (1943).
[32] R. Jastrow, *Phys. Rev.* **73,** 70 (1948).
[33] F. S. Ham, *Phys. Rev.* (to be published) (1955).
[34] F. S. Ham, Tables for the Calculation of Coulomb Wave Functions, Tech. Rept. No. 204, Cruft Laboratory, Harvard University, Cambridge, Mass., 1955.

(dU^L/dr) of the radial function which vanishes at the origin is a reasonably smooth function of the energy for sufficiently small values of the radius r. Moreover, this function vanishes at $r = \infty$ when ϵ corresponds to an eigenvalue of the free atom, so that, in the Coulomb region for such an energy, $U^L(r)$ must equal that single solution of the Coulomb differential equation which has the same property. From this Coulomb function, we can then calculate the logarithmic derivative of $U^L(r)$ at the eigenvalue for any value of r outside the core. If there exists a radius R lying *in* the Coulomb region and sufficiently small that the logarithmic derivative at R is a smooth function of the energy, we can obtain the logarithmic derivative for arbitrary energies within a reasonable range by interpolating it between the eigenvalues. This procedure enables us to determine the correct combination of Coulomb functions at energies intermediate to the eigenvalues.

In applying this procedure to the alkali metals, one finds that the logarithmic derivative is not as smooth a function of energy as might be desired for a radius R clearly outside the core because of singularities occurring when $U^L(R) = 0$. This makes interpolation and extrapolation rather uncertain. Moreover the analytical argument given by Kuhn and Van Vleck for the smoothness of $(1/U^L)(dU^L/dr)$ at small R is applicable only if R is smaller than the radius associated with the first radial node of the wave function. Therefore, the underlying theorems can give no information about the behavior outside the core of the logarithmic derivative of a wave function having a form such as that of a $2s$ or $3p$ eigenfunction. Thus, in principle at least, one can say that the observation that the logarithmic derivative is sufficiently smooth as a function of energy at a value of R near the edge of the core for this procedure to work is essentially an empirical one.

One clearly would like to have another quantity which reflects accurately the extent of the deviation of the core potential from the Coulomb form $-2/r$ but which can be shown to be a smoother function of energy than the logarithmic derivative. Kuhn[35] has pointed out that the quantum defect is such a parameter. It is well known from the Rydberg-Ritz equation[36] that the term values of the optical spectrum of an alkali atom (in Rydberg units) may be written in the form

$$\epsilon = -(m - \delta_m)^{-2}. \tag{3.4}$$

Here m is an integer which increases by unity between adjacent terms of the same series, and δ_m is the quantum defect which varies slowly and

[35] T. S. Kuhn, *Phys. Rev.* **79**, 515 (1950).

[36] H. E. White, "Introduction to Atomic Spectra," p. 16. McGraw-Hill, New York, 1934.

in a regular manner in a single series. Kuhn has shown, moreover, that the solution of (3.3) which is regular at the origin may be approximated in the Coulomb region outside the core by a linear combination of two functions of the WKB type given by Imai,[37] which are approximate solutions of (3.3). The coupling constant between the solutions depends on a parameter which may be identified with the "extrapolated quantum defect" obtained by drawing a smooth curve through the experimental values of the quantum defect when these are plotted on an energy scale. This procedure works quite well for functions of zero angular momentum. Kuhn has obtained values of the ground-state energy for the alkali metals, corresponding to $\mathbf{k} = 0$ in the conduction band, which are in good agreement with the values obtained from the procedure of Kuhn and Van Vleck. However, the method is not satisfactory for p functions.[6] To determine the effective mass in the metal, one needs the solutions of (3.3) for $L = 1$ at the ground-state energy. In the alkali metals, these energies are below or very near the minimum at $\epsilon = -(1/(1 + \frac{1}{2})^2) = -0.444$ of the effective radial potential $-2/r + (L + \frac{1}{2})^2/r^2$ appropriate for use in the WKB approximation (Section VI). Consequently the turning points of the hydrogenic potential on the real axis are either very close together or absent entirely. The WKB approximation to the solutions of (3.3) obtained by Imai for the Coulomb region outside the core is defined with respect to the outer turning point in the hydrogenic potential. This approximation is accurate only if the turning points are well separated. Hence, the approximation cannot be applied in the usual manner for p functions having low energy. It follows that Kuhn's procedure is not applicable. Kambe has investigated this situation but has not found a satisfactory solution.

In subsequent sections, we shall present the Quantum Defect Method in its present form. This shares with Kuhn's method the advantage of using a slowly varying quantity which can be extrapolated safely from the experimental values. However, it does not have the disadvantage of relying upon functions like those developed by Imai which become inaccurate at low energies. Moreover, the method is simpler to use than Kuhn's, once tables of the Coulomb functions are available. The foregoing discussion of Kuhn's procedure reveals, however, the basic idea behind QDM. From experimental data, involving the spectrum of the free atom, it is possible to obtain an explicit expression for that solution of the radial differential equation (3.3) which satisfies the condition $U^L(0) = 0$, for values of r in the Coulomb region outside the core, provided one limits interest to values of the energy not too far removed from the eigenvalues of the free atom.

[37] I. Imai, *Phys. Rev.* **74**, 113 (1948).

IV. Calculation of Energy Bands with QDM

We have pointed out in the last section that QDM gives us that solution of (3.3) which satisfies $U^L(0) = 0$ for radii outside the ion core and for arbitrary energies near the atomic eigenvalues. Consequently, in combining QDM with the various methods of calculating energy bands once the solutions of the radial differential equation are known, we can make use only of those methods which require no information about the solutions within the core or about the core potential. This restriction evidently excludes the method of orthogonalized plane waves,[38] which requires the evaluation of matrix elements of the lattice potential with plane waves that have been made orthogonal to the ion core wave functions. Such orthogonalized plane waves usually do not satisfy the Hartree-Fock equation. Fortunately, however, a number of other procedures may be used. We will give a brief description of the most important of these below.

The method of Wigner and Seitz[11] replaces the actual polyhedral unit cell in the lattice by a sphere of equal volume on the surface of which appropriate boundary conditions must be satisfied by the wave function. In the polyhedron, these boundary conditions, obtained from the Bloch form of the wave function (2.8), are

$$\begin{aligned} \varphi_j(\mathbf{r}_s) &= e^{[i\mathbf{k}\cdot(\mathbf{r}_s - \mathbf{r}_s')]}\varphi_j(\mathbf{r}_s'), \\ \frac{\partial\varphi_j(\mathbf{r}_s)}{\partial n} &= -e^{[i\mathbf{k}\cdot(\mathbf{r}_s - \mathbf{r}_s')]}\frac{\partial\varphi_j(\mathbf{r}_s')}{\partial n'}. \end{aligned} \tag{4.1}$$

Here $\mathbf{r}_s$ is a point on the surface of the polyhedron and $\mathbf{r}_s'$ is a second point on the surface related to $\mathbf{r}_s$ by $\mathbf{r}_s' = \mathbf{r}_s + \mathrm{R}$, where R is one of the vectors describing the translational symmetry of the lattice. The directional derivative is to be taken along the outward normal to the surface. In the equivalent sphere and for $\mathbf{k} = 0$, (4.1) is replaced by

$$\begin{aligned} \varphi_j(\mathbf{r}_s) &= \varphi_j(\mathbf{r}_s'), \\ \frac{\partial\varphi_j(\mathbf{r}_s)}{\partial r} &= -\frac{\partial\varphi_j(\mathbf{r}_s')}{\partial r}. \end{aligned} \tag{4.2}$$

Here $\mathbf{r}_s$ and $\mathbf{r}_s'$ now lie at opposite ends of a diameter. If the wave function at $\mathbf{k} = 0$ is unchanged by inversion, (4.2) yields the boundary condition for the solution of (3.3)

$$\left[\frac{dU^L}{dr} - \frac{1}{r}U^L\right]_{r_s} = 0. \tag{4.3}$$

This condition clearly requires a knowledge of $U^L(r)$ only at the radius

[38] C. Herring, *Phys. Rev.* **57**, 1169 (1940).

of the equivalent sphere. Thus, QDM clearly provides all the necessary information to determine the energy eigenvalue for $\mathbf{k} = 0$ if the sphere is outside the core.

Bardeen[12] has shown, in the same spherical approximation, that the energy for nonzero $\mathbf{k}$ may be written to order k^2 as

$$\epsilon(k) = \epsilon_0 + \alpha k^2, \tag{4.4}$$

where

$$\alpha = \gamma \left[r \frac{1}{P} \frac{dP}{dr} - 1 \right]_{r_s}, \tag{4.5a}$$

$$\gamma = \frac{(\frac{1}{3}) r_s [R(r_s)]^2}{\int_0^{r_s} [R(r)]^2 dr}. \tag{4.5b}$$

R is the appropriate solution of (3.3) for $L = 0$ which satisfies (4.3) and the condition $U^L(0) = 0$. P is the solution at the same energy for $L = 1$. It would appear from (4.5b) that the evaluation of γ would necessitate a knowledge of R in the core region. It has been proved by Silverman,[39] however, that

$$\int_0^{r_s} R^2 dr = - \left[rR(r) \frac{\partial}{\partial r} \left(\frac{1}{r} \frac{\partial R}{\partial \epsilon} \right) \right]_{r_s, \epsilon_0}. \tag{4.6}$$

Thus, since QDM enables us to find $U^0(r_s)$ as a continuous function of ϵ in the neighborhood of ϵ_0 for the solution which satisfies $U^0(0) = 0$, we can evaluate the expression in (4.6) with the method.

The approximation of replacing the polyhedral cell by the equivalent sphere is most accurate when the Bloch function appropriate to $\mathbf{k} = 0$ has the full symmetry of the lattice, so that it is most nearly s-like in the neighborhood of an ion. In the spherical approximation, the wave function then actually is an s function in each cell. The approximation also may be expected to be good when the $\mathbf{k}$ vector is not too close to the surface of the Brillouin zone under these circumstances.[40] These condi-

[39] R. A. Silverman, *Phys. Rev.* **85**, 227 (1952). A similar formula was also used by W. Bowers, Doctoral thesis, Cornell University, Ithaca, N.Y., 1943.

[40] Brooks has given a proof (unpublished) based on Eq. (4.7) which shows that the sphere of equal volume yields the best result for the energy when the boundary conditions (4.2) are used, provided the wave function is an s function in the spherical approximation and provided both the correct wave function and the potential are nearly constant in the volume between the surface of the equivalent sphere and the surface of the polyhedral cell. The present author has used the spherical approximation and (4.2) to calculate the eigenvalues for the empty lattice[41] when the vector $\mathbf{k}$ has the value $(2\pi/a, 2\pi/a, 0)$, which is equivalent to $\mathbf{k} = (0,0,0)$ in the reduced zone scheme of the body-centered cubic lattice. He obtained the following approximate energy eigenvalues for the twelve functions which have the same energy as the possible combinations of twelve plane waves associated with this $\mathbf{k}$ vector and transform in the same way under the cubic symmetry operators. The

tions are satisfied by the electrons in the occupied region of the conduction band in the alkali metals. Brooks[5] has used this approximation and QDM in his work on these metals. For **k** vectors near the surface of the zone or for excited bands, however, one must use the correct polyhedral cell and (4.1) to obtain accurate eigenvalues.[41,43] One procedure of this type has been discussed by Howarth and Jones.[13] They have used a finite expansion of the wave function in spherical harmonics, as in (3.2), taking into account the symmetry of the **k** vector whenever possible in order to restrict the possible independent combinations of spherical harmonics that may enter. They have then determined the coefficients in this expansion and the corresponding energy by satisfying (4.1) at an appropriately chosen set of points. This method only requires values of the wave function and its derivatives on the surface of the polyhedron, so that it may be used with QDM. However, the scatter in eigenvalues obtained by satisfying boundary conditions at different sets of points often is larger than one would wish, for reasons which are not entirely understood. It is usually necessary to correct the approximate eigenvalues by a technique developed by Howarth and Jones,[7] which makes use of the following exact formula, derived by methods indicated in their article,[13]

$$(\epsilon - \epsilon_0) \int_v \psi^* \psi_0 d\tau = (\tfrac{1}{2}) \int_s \left\{ \psi^*(\mathbf{r}) \left[\frac{\partial \psi_0(\mathbf{r})}{\partial n} + e^{(i\mathbf{k}\cdot(\mathbf{r}-\mathbf{r}'))} \frac{\partial \psi_0(\mathbf{r}')}{\partial n'} \right] - \frac{\partial \psi^*(\mathbf{r})}{\partial n} [\psi_0(\mathbf{r}) - e^{(i\mathbf{k}\cdot(\mathbf{r}-\mathbf{r}'))} \psi_0(\mathbf{r}')] \right\} ds. \quad (4.7)$$

Here the first integral extends over the volume of the polyhedron, the second over its surface. The approximate eigenvalue and wave function

---

notation is that of ref. 19. The number in the parentheses indicates the number of functions of each symmetry type occurring among the 12 plane waves.

Wave function symmetry	Approximate eigenvalue	Error
Γ_1 (*s*) (1)	$83.3/a^2$	5%
Γ_{15} (*p*) (3)	$83.3/a^2$	5%
Γ_{12}, Γ_{25}' (*d*) (2,3)	$46.1/a^2$	40%
Γ_{25} (*f*) (3)	$201.5/a^2$	150%
Correct value	$78.96/a^2$	

The unexpectedly inaccurate result for the five *d*-like functions Γ_{12}, Γ_{25}' throws considerable doubt upon the accuracy of Sternheimer's calculation of the position of the bottom of the *d* band in cesium.[42] This calculation could be repeated profitably using QDM and better boundary conditions.

[41] W. Shockley, *Phys. Rev.* **51,** 129 (1937).

[42] R. Sternheimer, *Phys. Rev.* **78,** 238 (1950).

[43] F. Von der Lage and H. Bethe, *Phys. Rev.* **71,** 612 (1947).

are, respectively, ϵ_0 and ψ_0, determined from an appropriate set of boundary conditions as indicated above. The quantities ϵ and ψ are the exact eigenvalue and wave function, and are, of course, unknown. However, if ψ_0 is approximately equal to ψ, the error in ϵ is of the second order in $(\psi - \psi_0)$, provided $(\epsilon - \epsilon_0)$ is approximated by substituting ψ_0 for ψ in the various integrals. This procedure, which is closely related to a variational method proposed by Kohn,[44] and which we shall call the "correction integral method," can be used with QDM, for the integrals on the right of (4.7) involve values of ψ_0 and its derivatives on the surface of the cell, whereas the volume integral on the left can be expressed in terms of a surface integral, once ψ^* has been replaced by ψ_0^*. The transformation from volume to surface integral involves a procedure similar to that used by Silverman in proving (4.6).[45] This is the procedure used by the author with QDM in calculating eigenvalues for sodium at points on the surface of the Brillouin zone (Section XI). It appears to be accurate, if the correction integrals are used; however, evaluation of the surface integrals indicated in (4.7) is tedious.

Kohn and Rostoker[14] have recently proposed a method which is based on a variational procedure derived from an integral equation for the Bloch function. This method assumes that the lattice potential is spherically symmetric about each ion within a sphere inscribed in the atomic cell and is constant in the volume outside the spheres. The calculation of the band structure for a particular ion potential requires only the knowledge of the logarithmic derivative of the radial wave functions of different L on the surface of this sphere. This information can, of course, be obtained from QDM. It appears that the combination of QDM and the Kohn-Rostoker procedure may be very satisfactory because of the rapid convergence of the latter. Moreover, once certain "structure constants" which appear in the method have been calculated for a given type of lattice, they may be used with different potentials without much additional effort.

The method of "augmented plane waves" developed by Slater[46] and

[44] W. Kohn, *Phys. Rev.* **87**, 472 (1952).

[45] This transformation actually is not necessary. Suppose one of the coefficients in the expansion (3.2) is arbitrarily set equal to unity. If a fixed set of $N - 1$ boundary conditions is used to determine $N - 1$ other coefficients for several values of the energy in the neighborhood of the eigenvalue, the resulting series of N terms may be substituted for ψ_0 and ψ on the right side of (4.7). The energy ϵ_1, at which the right side of this expression, thus evaluated, is equal to zero, may be taken to be the "corrected" eigenvalue. This procedure is accurate because the expression for ψ_0 obtained from the $N - 1$ boundary conditions plus an Nth one that yields $\epsilon_0 = \epsilon_1$ would have given $\epsilon = \epsilon_1$ when substituted into the correction integral.

[46] J. C. Slater, *Phys. Rev.* **92**, 603 (1953).

modified by Slater and Saffren[47] probably can also be used with QDM. This method, however, requires values of the logarithmic derivative of solutions of the radial differential equations for different L for energies up to two or three Rydberg units above the ground state.[48] Extrapolation of the quantum defect to energies so far from the eigenvalue range may introduce significant errors in the resulting functions. This point could be investigated profitably in the course of future work. We should also mention that QDM may be used in a formalism recently suggested by Leigh[49] which is a modification of an earlier proposal by Slater,[50] and is qualitatively similar to the augmented plane wave method. It has, however, the advantage from the point of view of QDM that it does not require the use of energies far from the eigenvalue range.

It is clear from this discussion that there is a variety of methods available for the calculation of the band structure of solids with which QDM can profitably be used. It is also evident, however, that the knowledge of the crystal wave functions, which arises out of combination of the methods, is less complete than that of the band structure, for QDM can yield explicit expressions for the wave functions only outside the ion core. It may be noted that the functions can be normalized with the use of formulas similar to (4.6).[51] It follows that the information obtained concerning the wave functions can be used to calculate further properties of the crystal, such as electrical conductivity or optical absorption, only if the appropriate quantities can be expressed in terms of surface integrals over the atomic cell, or if suitable approximations can be found to continue the wave function into the core region, or if further relevant information can be extracted from the spectral data of the free atom.

V. Proof of the Sufficiency of the Assumptions of QDM

We shall now prove two mathematical theorems which will be used to show that the solution $U^L(r)$ which satisfies the condition $U^L(0) = 0$ is determined outside the core for an arbitrary energy by the eigenvalue spectrum of $V(r)$ for the same L, provided the ion potential $V(r)$ in (3.3) satisfies certain conditions. This is, of course, the fundamental principle upon which QDM is based. We shall then discuss the relevance of this

[47] J. C. Slater and M. M. Saffren, *Phys. Rev.* **92,** 1126 (1953).

[48] D. J. Howarth, *Proc. Symposium Radar Research Establishment Malvern*, p. 32 (1954).

[49] R. S. Leigh, *Proc. Symposium Radar Research Establishment Malvern*, p. 40 (1954).

[50] J. C. Slater, *Phys. Rev.* **51,** 846 (1937).

[51] Brooks has shown in unpublished work that it is possible to extend QDM to a calculation of the amplitude of the wave function at the nucleus if data on atomic hyperfine splitting are available or if an approximate Hartree-Fock potential is known. No calculations have as yet been carried out using this proposal. Details of the procedure will be reported at a future date.

mathematical framework to the problem of determining $U^L(r)$ from experimental data involving the spectrum of an actual atom.

We consider a potential $V(r)$ of the following form. In the region $r > r_0$, $V(r) = -2/r$, that is, the potential is an attractive Coulomb potential. In the range $0 < r < r_1$, $V(r) = -2Z/r + g(r)$, where r_1 may be arbitrarily small but not zero. Here Z is a constant ≥ 1, and $g(r)$ is an analytic function of the complex variable r for $|r| < r_1$. Finally, in the range $r_1 < r < r_0$, $V(r)$ is piecewise continuous but otherwise arbitrary. Except in the region $|r| < r_1$, we consider only real positive values of r. We shall now prove the following theorem:

Theorem I: If $V(r)$ has the foregoing form, there exists a solution of (3.3) for real $r \geq 0$, which, together with its derivative, is a continuous function of the real variable r, which satisfies the boundary condition that $U^L(0) = 0$, and which is an entire analytic function of the energy parameter ϵ for $|\epsilon| < \infty$.

To prove this theorem, we first consider the neighborhood of $r = 0$. Since $g(r)$ is analytic for $|r| < r_1$, we can apply the Fuchsian analysis to (3.3) in the neighborhood of the equation's singularity at $r = 0$.[52] We find that, for $|r| < r_1$, the solution of (3.3) which is zero at the origin can be expressed as

$$U^{L,\epsilon}(r) = r^{L+1} \sum_{s=0}^{\infty} a_s r^s. \tag{5.1}$$

For convenience, we choose $a_0 = 2^{L+1}/(\Gamma(2L + 2))$. Thus, the series is completely determined if $g(r)$ is known.

In the region $r_1 \leq r \leq r_0$, $V(r)$ is piecewise continuous. Equation (3.3), therefore, has a continuous solution in any subinterval of this range in which $V(r)$ is continuous. The solution is determined uniquely if arbitrary values are assigned to it and its derivative at any point of the interval.[53] We may, therefore, match this solution, $U^{L,\epsilon}(r)$, and its derivative, at the inner end point of each sub-interval to the corresponding values at the outer point of the preceding interval. Since (5.1) determines the function out to r_1, we can extend the function and its derivative continuously in this way to the Coulomb region, $r > r_0$. We may then use Coulomb functions to continue it as much farther as we wish.

To prove that $U^{L,\epsilon}(r)$ is an analytic function of the complex variable ϵ, we must show that $U^{L,\epsilon}(r)$ is defined for every value of ϵ and that it has

[52] E. T. Whittaker and G. N. Watson, "A Course of Modern Analysis," 4th ed., Chapter 10. Cambridge U. P., New York, 1952.

[53] E. L. Ince, "Ordinary Differential Equations," p. 73. Longmans, Green, London, 1927.

a first derivative with respect to ϵ everywhere.[54] Since the foregoing discussion depends in no way upon the value of the complex number ϵ ($< \infty$), the first condition is proved. To establish the second condition, we must prove the existence of the limit

$$\lim_{h \to 0} \frac{U^{L,\epsilon+h}(r) - U^{L,\epsilon}(r)}{h} \tag{5.2}$$

for all ϵ and r. Writing (3.3) as

$$F_L(\epsilon) U^{L,\epsilon}(r) = 0, \tag{5.3}$$

and defining

$$W^L(\epsilon,h,r) = U^{L,\epsilon+h}(r) - U^{L,\epsilon}(r), \tag{5.4}$$

we have

$$F_L(\epsilon) W^L(\epsilon,h,r) = hU^{L,\epsilon+h}(r). \tag{5.5}$$

Now there exists a second independent solution of (3.3),[55] which, for $|r| < r_1$ and integer L, as we now assume, has the general form

$$U_2^{L,\epsilon}(r) = A\left[\sigma U^{L,\epsilon}(r) \ln(r) + r^{-L} \sum_{s=0}^{\infty} b_s r^s\right]. \tag{5.6}$$

This function may be continued to $r > r_1$ by the process used above for $U^{L,\epsilon}(r)$. If we put $b_0 = 2^{-L-1}\Gamma(2L+1)$ and $A = 1$ in (5.6), the constants σ and b_s are determined by the differential equation. Moreover, with this choice,

$$U_2^{L,\epsilon}(r) \frac{d}{dr} U^{L,\epsilon}(r) - U^{L,\epsilon}(r) \frac{d}{dr} U_2^{L,\epsilon}(r) = 1. \tag{5.7}$$

Thus, a general solution of (5.5) may be written as[56]

$$hU^{L,\epsilon}(r) \int_a^r U_2^{L,\epsilon}(r') U^{L,\epsilon+h}(r') dr' - hU_2^{L,\epsilon}(r) \int_b^r U^{L,\epsilon}(r') U^{L,\epsilon+h}(r') dr', \tag{5.8}$$

where a and b are arbitrary constants. We require from (5.4) and the boundary condition $U^{L,\epsilon}(0) = 0$, however, that $W^L(\epsilon,h,0) = 0$. Moreover, since we have selected a_0 in (5.1) to be the same for all ϵ, we see that

$$\lim_{r \to 0} \frac{U^{L,\epsilon+h}(r)}{U^{L,\epsilon}(r)} = 1, \tag{5.9}$$

[54] K. Knopp, "Theory of Functions," Part I, p. 65. Dover, New York, 1945.

[55] Whittaker and Watson, p. 201 in ref. 52.

[56] N. F. Mott and H. S. W. Massey, "The Theory of Atomic Collisions," 2nd ed., p. 108. Oxford U. P., New York, 1949.

which imposes the further restriction

$$\lim_{r\to 0} \frac{W^L(\epsilon,h,r)}{U^{L,\epsilon}(r)} = 0. \tag{5.10}$$

These conditions suffice to determine a and b in (5.8) in such a way that

$$W^L(\epsilon,h,r) = h\left\{U^{L,\epsilon}(r)\int_0^r U_2{}^{L,\epsilon}(r')U^{L,\epsilon+h}(r')dr' - U_2{}^{L,\epsilon}(r)\int_0^r U^{L,\epsilon}(r')U^{L,\epsilon+h}(r')dr'\right\}. \tag{5.11}$$

We may show, either from the series (5.1) and (5.6) and our continuing process for $r > r_1$ *together with our conventions concerning the choice of constants not fixed by the differential equation*, or from (5.11) and a similar formula for the difference $U_2{}^{L,\epsilon+h}(r) - U_2{}^{L,\epsilon}(r)$, that both $U_2{}^{L,\epsilon}(r)$ and $U^{L,\epsilon}(r)$ are continuous functions of ϵ. Therefore, dividing both sides of (5.11) by h and taking the limit $h \to 0$, we see that the right-hand side of the resulting equation exists for all ϵ and r. Since the left-hand side equals (5.2), the proof of the existence of $(dU^{L,\epsilon}(r)/d\epsilon)$ is complete. We conclude that $U^{L,\epsilon}(r)$ is an analytic function of ϵ in the entire ϵ plane, excluding the point at infinity. The proof of Theorem I is thereby complete.

We now make use of the discussion of the Coulomb functions given in the Appendix and define

$${}^0U_c{}^{L,n}(r) = (z/2)J_{2L+1}{}^n(z), \tag{5.12}$$

$${}^1\tilde{U}_c{}^{L,n}(r) = (z/2)N_{2L+1}{}^n(z) - G_L(L,n)(z/2)J_{2L+1}{}^n(z), \tag{5.13}$$

where $z = (8r)^{\frac{1}{2}}$. These functions satisfy the Wronskian relation

$${}^0U_c{}^{L,n}(r)\frac{d}{dr}{}^1\tilde{U}_c{}^{L,n}(r) - {}^1\tilde{U}_c{}^{L,n}(r)\frac{d}{dr}{}^0U_c{}^{L,n}(r) = \frac{2}{\pi}. \tag{5.14}$$

Both ${}^0U_c{}^{L,n}(r)$ and ${}^1\tilde{U}_c{}^{L,n}(r)$ can be expanded in convergent power series in $1/n^2 = -\epsilon$ for all finite values of ϵ, so that both are entire functions of ϵ in the finite ϵ plane.

The two functions are a pair of linearly independent solutions of (3.3) when $V(r) = -2/r$. Hence, for $r > r_0$, we can express $U^{L,\epsilon}(r)$ as

$$U^{L,\epsilon}(r) = \beta(\epsilon){}^0U_c{}^{L,n}(r) + \lambda(\epsilon){}^1\tilde{U}_c{}^{L,n}(r). \tag{5.15}$$

We shall now prove a second theorem.

Theorem II: Under the conditions required for the validity of Theorem I, the ratio $\lambda(\epsilon)/\beta(\epsilon)$, associated with a particular value of L, is determined completely for all values of ϵ by the eigenvalue spectrum of $V(r)$ for the given L.

To prove this theorem, we multiply (3.3) by ${}^0U_c^{L,n}(r)$ and subtract from the result the product of $U^{L,\epsilon}(r)$ and the Coulomb radial equation satisfied by ${}^0U_c^{L,n}(r)$ (Appendix (A.1)). We obtain upon integrating,

$$\left[{}^0U_c^{L,n}(r)\frac{d}{dr}U^{L,\epsilon}(r) - U^{L,\epsilon}(r)\frac{d}{dr}{}^0U_c^{L,n}(r)\right]_0^r - \int_0^r {}^0U_c^{L,n}(r')U^{L,\epsilon}(r')[V(r') + (2/r')]dr' = 0. \quad (5.16)$$

Making use of (5.15), (5.14), and the form assumed for $V(r)$ for $r > r_0$, we derive the relation

$$\lambda(\epsilon) = (\pi/2)\int_0^{r_0} {}^0U_c^{L,n}(r')U^{L,\epsilon}(r')[V(r') + (2/r')]dr'. \quad (5.17)$$

In a similar manner, we may derive an expression resembling (5.16), but differing in the sense that ${}^1\tilde{U}_c^{L,n}(r)$ replaces ${}^0U_c^{L,n}(r)$ everywhere. When evaluating the second expression, we must determine the limit of the bracketed quantity as $r \to 0$, for this is now different from zero. We recall from (5.1) and our choice for a_0 that, in this limit, $U^{L,\epsilon}(r) \sim (2r)^{L+1}/\Gamma(2L+2)$. Moreover, we ascertain from (5.13), (A.4), and the dominant term in the series for the Bessel function, $J_{-2L-1}(z)$ as $z \to 0$, that ${}^1\tilde{U}_c^{L,n}(r) \sim -(1/\pi)\Gamma(2L+1)(z/2)^{-2L}$. Thus, we obtain finally

$$\beta(\epsilon) = 1 - (\pi/2)\int_0^{r_0} {}^1\tilde{U}_c^{L,n}(r')U^{L,\epsilon}(r')[V(r') + (2/r')]dr'. \quad (5.18)$$

The integrals in (5.18) and (5.17) are finite, and the integrands are analytic functions of ϵ. Therefore, $\beta(\epsilon)$ and $\lambda(\epsilon)$ are analytic functions of ϵ. At least one of the functions will not vanish in a region near $\epsilon = 0$, so that the ratio of the other function to this one should be an analytic function of ϵ in the given region near $\epsilon = 0$. The ratio may be continued analytically throughout the whole ϵ plane, except at isolated points where it possesses a pole because the function in the denominator vanishes. To determine the ratio uniquely, it suffices to specify its value at an infinite sequence of points in the ϵ plane having $\epsilon = 0$ as a limit point. Such a sequence is provided by the eigenvalues of (3.3) because $V(r) = -2/r$ for $r > r_0$. We shall now show that the value of the ratio $\beta(\epsilon)/\lambda(\epsilon)$ is determined at these eigenvalues.

At an eigenvalue, $U^{L,\epsilon}(r)$ falls exponentially to zero as $r \to \infty$. Hence, in the range $r > r_0$, $U^{L,\epsilon}(r)\alpha W_{n,L+\frac{1}{2}}(2r/n)$,[57] for the latter is the solution of the Coulomb radial equation which has this property. However, $W_{n,L+\frac{1}{2}}(2r/n)$ is given in terms of $(z/2)J_{2L+1}{}^n(z)$ and $(z/2)N_{2L+1}{}^n(z)$ by

[57] Whittaker and \ son, Chapter 16 in ref. 52.

(A.12). Comparing with (5.15) and (5.13), we see that *at eigenvalues*

$$\frac{\beta(\epsilon)}{\lambda(\epsilon)} = \frac{\Gamma(n+L+1)}{n^{2L+1}\Gamma(n-L)\tan\pi(n-L-1)} + G_L(L,n). \qquad (5.19)$$

We recall that $\epsilon = -1/n^2$, in which n is chosen positive for negative real ϵ. The right-hand side of (5.19) is, therefore, known when the eigenvalue is known. Hence, the eigenvalues of $V(r)$ for given L determine $\beta(\epsilon)/\lambda(\epsilon)$ for that L, and Theorem II is proved. It follows from (5.15) that $U^{L,\epsilon}(r)$ is determined explicitly for $r > r_0$ and all ϵ, apart from an arbitrary constant factor.

In attempting to apply Theorem II to the determination of $U^{L,\epsilon}(r)$ for a real atom from the experimental spectrum, we must recognize that the actual atomic potential does not satisfy the conditions of Theorem I exactly. Thus, there is no radius r_0 beyond which $V(r) = -2/r$, for there always is a term $-2\alpha'/r^4$ in the potential arising from the polarizability of the core. In addition, there are terms which fall off exponentially with increasing r as a result of the small but finite probability that a core electron will be found more distant from the nucleus than r. However, it is not necessary to know $U^{L,\epsilon}(r)$ at values of r greater than the distance from the nucleus to the most distant corner of the polyhedral cell for calculations of wave functions in a solid. Considering the other uncertainties in the lattice potential, one sees that it is adequate to approximate the free ion potential within the cell by a potential identically equal to $-2/r$ beyond, say, the radius r_i of the sphere inscribed in the cell. This approximation to the correct potential is consistent with the requirement that the approximate potential satisfy the conditions of Theorem I. In principle, the eigenvalues of this new potential can be determined from those of the correct potential by perturbation theory. The new eigenvalues determine uniquely the solutions of the approximate radial differential equation for $r > r_i$ at all energies. We see, therefore, that the radial solutions of interest in our solid-state calculations are determined uniquely by the eigenvalue spectrum, even if Theorems I and II cannot be applied directly to the correct ion potential.

We must observe that, in practice, we know from experiment the approximate values of only a finite number of the eigenvalues. This information is not sufficient to determine the solutions uniquely for all energies. However, we shall show in the following sections that, in the energy range of practical importance, the most important quantities are slowly varying functions of energy which can be determined once a few eigenvalues are known. In practice, this determination is ambiguous only if one attempts to extrapolate the quantities far from the range of energy in which the atomic eigenvalues lie.

Finally, we must remark that the conclusion of this section does not contradict an interesting theorem proved by Jost and Kohn.[58] Given a potential with m bound states for a specified L, the theorem states that there is an m-dimensional manifold of equivalent potentials which yields the same eigenvalues and positive energy phase shifts for the specified L. Two such equivalent potentials would certainly lead to different radial solutions; however, at most, one of the many equivalent potentials satisfies the conditions of Theorem I. Hence, the uniqueness of determination of $\beta(\epsilon)/\lambda(\epsilon)$ from the spectrum is preserved. In proving this, let us designate by $f_m(r)$ one of the eigenfunctions of a potential $V(r)$ satisfying the conditions of Theorem I. An equivalent potential $V'(r)$ differs from $V(r)$ by one or more terms of the form

$$\frac{\lambda}{N} f_m(r) f_m'(r) + \frac{\lambda^2}{8N^2} [f_m(r)]^4, \tag{5.20}$$

according to the theorem of Jost and Kohn. Here

$$N = 1 + \frac{\lambda}{4} \int_r^\infty [f_m(r')]^2 dr',$$

and λ is an arbitrary real positive number. Such terms approach zero exponentially as $r \to \infty$ but are never identically zero. Hence, the equivalent potentials are excluded in Theorem I by the requirement that $V(r) = -2/r$ for $r > r_0$. Of course, if $f_m(r)$ corresponds to a core state of low energy, $V'(r)$ can differ only very slightly from $V(r)$ in the range $r > r_0$ and yet be very different within the core. This could make ambiguous the choice of an approximate potential appropriate to Theorem I which reproduces fairly accurately an experimental spectrum. However, the smoothness of the quantities used for extrapolation in QDM, demonstrated in the following sections, indicates that the values of $\beta(\epsilon)/\lambda(\epsilon)$ obtained for two such nearly equivalent potentials would differ significantly only in a range in which the eigenvalues are separated by an interval of several Rydberg units. There would be no significant difference in the range of practical importance within half a Rydberg unit of the closely spaced eigenvalues of the valence electron.

VI. The Extrapolated Quantum Defect and the WKB Justification of QDM

We have already remarked that the quantities $\lambda(\epsilon)$ and $\beta(\epsilon)$ appearing in (5.15), and given by the integrals (5.17) and (5.18), are analytic functions of ϵ for all finite ϵ. Hence, they can be expanded in power series in ϵ. We expect both to be slowly varying functions of ϵ. In other words,

[58] R. Jost and W. Kohn, *Phys. Rev.* **87,** 977 (1952); **88,** 382 (1952).

we expect that a few terms in the series expansion will suffice to give both quite accurately in the range of ϵ of interest to us. This expectation is strengthened when we examine tables for the Coulomb functions ${}^0U_c^{L,n}(r)$ and ${}^1\tilde{U}_c^{L,n}(r)$, for a few terms in the power series expansions represent both quite accurately for values of r corresponding to the ion cores $(r < r_0)$. The same is probably true for the atomic function $U^{L,\epsilon}(r)$. We conclude, from the integral representations of $\beta(\epsilon)$ and $\lambda(\epsilon)$, that the same is true for these quantities and, therefore, is true for the ratio $\beta(\epsilon)/\lambda(\epsilon)$ in a region where it is analytic. We can infer that this ratio should not oscillate wildly and should be reasonably well-behaved; however, we cannot conclude much more. The ratio does vary by more than is convenient in certain regions, especially in the vicinity of a zero of $\lambda(\epsilon)$. The situation resembles closely that found by Kuhn and Van Vleck in their attempts to extrapolate the logarithmic derivative of the radial function.

In order to obtain a quantity which may be extrapolated more reliably, we write

$$\begin{aligned} \alpha(n) &= \beta(\epsilon) - \lambda(\epsilon)G_L(L,n), \\ \gamma(n) &= \lambda(\epsilon). \end{aligned} \tag{6.1}$$

Then, from (5.13) and (5.15), we have

$$U^{L,\epsilon}(r) = \alpha(n)(z/2)J_{2L+1}{}^n(z) + \gamma(n)(z/2)N_{2L+1}{}^n(z). \tag{6.2}$$

We now set

$$\frac{\alpha(n)}{\gamma(n)} = -\frac{\Gamma(n+L+1)}{n^{2L+1}\Gamma(n-L)\tan\pi\nu(n)}, \tag{6.3}$$

in which $\nu(n)$ is understood to be a function of n defined by (6.3) for all values of n $(\epsilon = -1/n^2)$. From (5.19) we obtain the relation valid only at an eigenvalue, namely,

$$\frac{\alpha(n)}{\gamma(n)} = \frac{\Gamma(n+L+1)}{n^{2L+1}\Gamma(n-L)\tan\pi(n-L-1)}. \tag{6.4}$$

Comparing this with the previous formula, we see that, at an eigenvalue,

$$\nu(n) = L + 1 - n + m'. \tag{6.5}$$

Here m' is an undetermined integer. This equation may be transposed to

$$\epsilon = -(1/n^2) = -(m' + L + 1 - \nu(n))^{-2}. \tag{6.6}$$

Therefore, $\nu(n)$ differs from the experimental quantum defect δ_m of (3.4) by at most an integer at an eigenvalue. The experimental quantum defect is defined in terms of (6.6) by assigning the value $m' = 0$ to the lowest eigenvalue of angular momentum L, which is often associated with a core state, and increasing m' by unity in passing to each higher eigenvalue.

It is tempting to set the extrapolated quantum defect $\nu(n)$ equal to the experimental defect at the eigenvalues and interpolate it smoothly for intermediate energies. This is essentially the procedure that we shall adopt, except when extrapolating to energies below the eigenvalues of the valence electron. We must justify this procedure, however, for it is by no means obvious that it is legitimate.

We shall provide such justification with the help of the WKB method, extending our definition of $V(r)$ to the complex r plane. Since $V(r)$ is not an analytic function of r, being defined differently in different regions of the real axis, we divide the r plane into three annular regions as follows: (I) in the range $|r| < r_1$, we take $V(r) = -2Z/r + g(r)$; (II) for $r_1 < |r| < r_0$, we choose $V(r)$ to be the extension of the segmented function defined under Theorem I of Section 5 along the positive real axis; (III) finally, in the range $|r| > r_0$, we select $V(r) = -2/r$. The quantities r_1, Z, and $g(r)$ have been defined in connection with Theorem I above. The solution $U^L(r)$ of (3.3) is an analytic function of r within regions I and III of the r plane. It is to be joined continuously along the real axis, with a continuous first derivative also along this axis, through region II.

We now consider a negative value of energy sufficiently high that the two turning points of the Schrödinger equation for the real radial variable are sufficiently well separated that a WKB approximation is legitimate. Moreover, we assume that the outer turning point r_2 lies in the Coulomb region, $r_2 > r_0$. Following Kuhn,[35] we introduce the change of variable $x = \ln(-\epsilon r) = \ln(r/n^2)$, $u(x) = \exp(-x/2)U^L$ in the radial differential equation (3.3). We obtain

$$\frac{d^2u}{dx^2} + P(x)u = 0, \tag{6.7}$$

where

$$P(x) = [-n^2e^{2x} - n^4e^{2x}V(n^2e^x) - (L + \tfrac{1}{2})^2]. \tag{6.8}$$

If r lies in region III and is not too close to the turning point r_2, the WKB approximation provides that any solution of (6.7) is well approximated by

$$u \sim [P(x)]^{-\frac{1}{4}}\{Ae^{iz} + Be^{-iz}\} \tag{6.9}$$

for the corresponding value of x. The constants A and B are to be appropriately chosen. Moreover,

$$z = \int_{x_2}^{x} [P(x)]^{\frac{1}{2}}dx, \tag{6.10}$$

in which x_2 is the value of x corresponding to r_2. The path of integration lies in the region of the x plane corresponding to region III. It is necessary to choose different values of A and B in various parts of the complex plane in order to represent accurately a given single-valued solution of (6.7).[59,60]

We assume that the turning points are sufficiently far apart in the x plane that they may be treated separately. We may divide the complex plane near x_2 in the manner indicated in Fig. 1. On line T_1, which represents the real axis between the turning points, z is real, as it is on T_2 and T_3. We may choose the phase of z so that it is positive on T_1. However, z is imaginary on S_1, S_2, and S_3, the so-called Stokes's lines. Thus, one of the terms in the WKB expression for u is exponentially damped and

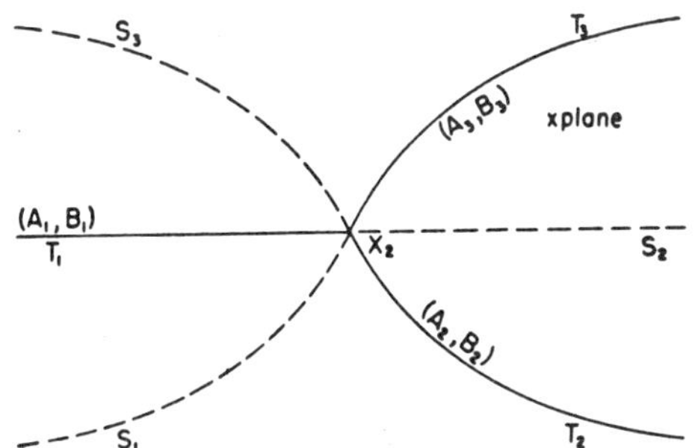

FIG. 1. The complex x plane near the outer WKB turning point x_2. The lines S_1, S_2, and S_3 are the Stokes's lines on which z is imaginary. On the lines T_1, T_2, and T_3, z is real, and z is positive on T_1, which represents the real x axis between the turning points.

mathematically meaningless on S_1, S_2, and S_3. Furry has investigated this situation[61] and has found that, whenever a single-valued function is represented on T_1 with use of the constants A_1 and B_1, the corresponding constants to be used on T_2 are

$$\begin{aligned} A_2 &= A_1 + iB_1, \\ B_2 &= B_1. \end{aligned} \tag{6.11}$$

Hence, if on T_1, far from x_2, we write

$$u \sim C[P(x)]^{-\frac{1}{4}} \cos (z - \pi/4 + \mu), \tag{6.12}$$

or

$$\begin{aligned} A_1 &= (\tfrac{1}{2})Ce^{i\mu - i(\pi/4)}, \\ B_1 &= (\tfrac{1}{2})Ce^{-i\mu + i(\pi/4)}, \end{aligned} \tag{6.13}$$

[59] E. C. Kemble, "The Fundamental Principles of Quantum Mechanics," Section 21. McGraw-Hill, New York, 1937.

[60] G. N. Watson, "A Treatise on the Theory of Bessel Functions," 2nd ed., p. 201. Cambridge U. P., New York, 1948.

[61] W. H. Furry, *Phys. Rev.* **71**, 360 (1947).

we have the following relations on T_2:

$$A_2 = Ce^{i(\pi/4)} \sin \mu,$$
$$B_2 = (\tfrac{1}{2})Ce^{i(\pi/4)-i\mu}. \tag{6.14}$$

Thus, near T_2 we have

$$u \sim [P(x)]^{-\frac{1}{4}}(C/2)e^{i(\pi/4)}\{e^{iz}2 \sin \mu + e^{-iz-i\mu}\}. \tag{6.15}$$

We can make a similar WKB development about the inner turning point x_1; however, here we have the boundary condition that the wave function must be zero at $r = 0$, that is, it must approach zero as $x \to -\infty$. Thus, the quantity corresponding to μ in the development about x_1 must be zero or an integral multiple of π. Continuing the WKB solution from x_1 to x_2 along the real x axis, and making use of the requirement that the expansions about the two turning points agree in the region between them, we obtain for the quantity μ above

$$\mu = (p + \tfrac{1}{2})\pi - \int_{x_1}^{x_2} [P(x)]^{\frac{1}{2}}dx. \tag{6.16}$$

Here p is an integer which we choose once and for all in defining μ. We now define[35]

$$\pi\delta_w = \int_{x_1}^{x_2} [P(x)]^{\frac{1}{2}}dx - \int_{x_1'}^{x_2'} [P_h(x)]^{\frac{1}{2}}dx, \tag{6.17}$$

where $P_h(x)$ is the function corresponding to a pure Coulomb potential. The second integral can be carried out exactly. We obtain

$$\mu = \pi(p + L + 1 - n - \delta_w). \tag{6.18}$$

The wave function must approach zero at an eigenvalue as r or $x \to +\infty$. This requires that $\sin \mu = 0$, or that μ be an integral multiple of π. Therefore, at an eigenvalue,

$$\epsilon = -n^{-2} = -(m - \delta_w)^{-2} \tag{6.19}$$

We see from (6.16) that m is an integer which increases by unity from one eigenvalue to the next. Thus, as Kuhn has shown,[61a] δ_w can be regarded as the WKB approximation to the experimental quantum defect δ_m at the eigenvalues. Moreover, δ_w is a slowly varying function of the energy, for x_2 and x_2' in (6.17) coincide, since x_2 is assumed to lie in the Coulomb region $r > r_0$. It follows that the integrands in (6.17) are equal for $r > r_0$. The inner turning points are quite insensitive to the energy. In the core region, where $P(x)$ and $P_h(x)$ differ, both integrands are slowly varying functions of ϵ. Therefore, if x_2 lies in the Coulomb region,

[61a] This conclusion was first demonstrated, in essence, by N. Bohr [*Proc. London Phil. Soc.* **35**, 296 (1923)]. It has also been discussed by Kemble [in ref. 59, p. 478].

δ_w is a slowly varying function of ϵ. We may note, however, that this conclusion is no longer correct at an energy sufficiently low that x_2 is within the core. Thus, δ_w evidently will vary more rapidly with ϵ the lower the energy, down to the point where δ_w, defined in (6.17), vanishes. We shall now show that we may identify δ_w with the quantity $\nu(n)$ used in (6.3) to represent the exact solution (6.2) in the form of a linear combination of the two Coulomb functions.

To do this, we shall compare the asymptotic form of the WKB solution in the vicinity of the line T_2 of Fig. 1 with that of the exact solution expressed in terms of Coulomb functions. On line T_2 exp (iz) is oscillatory. We shall find it advantageous to use a trick. We have assumed that the outer turning point x_2 lies in the Coulomb region. Thus we could evaluate exp (iz) by carrying out the integral from x_2 to a value of x far out along T_2. If we were to do this for a pure Coulomb potential, for which $U^L(r) = (z/2)J_{2L+1}{}^n(z)$ is the correct solution, we would find that the asymptotic form of the WKB solution is correct in the limit $|n| \to \infty$. Otherwise, the error is $0(1/|n|)$. This is the accuracy we expect of the WKB method. Moreover, we know that the method usually gives a poorer approximation to the wave function than to the eigenvalues.[61] However, we notice that, whenever x_2 is in the Coulomb region, the integral defining z is exactly the same regardless of the core potential. Thus, we may "evaluate" it by using a pure Coulomb potential and replacing exp (iz) by the "correct" value derived from the asymptotic form of the exact solution. We obtain in this manner

$$e^{2iz} = e^{\frac{2r}{n}} \left(\frac{2r}{n}\right)^{-2n} \frac{\Gamma(n + L + 1)\Gamma(n - L)}{2\pi}. \tag{6.20}$$

Employing this value in the asymptotic expansion for an atom with a nonhydrogenic core, so that $\delta_w \neq 0$, and comparing the result with the asymptotic expansion[62] of (6.2) for large $|x|$ in the neighborhood of line T_2, we obtain[63]

$$\frac{\alpha(n)}{\gamma(n)} = -\frac{\Gamma(n + L + 1)}{n^{2L+1}\Gamma(n - L)\tan \pi\delta_w}. \tag{6.21}$$

Comparing this with (6.3), and recalling that we *defined* $\nu(n)$ in terms of the ratio $\alpha(n)/\gamma(n)$ for *all* values of ϵ in terms of this expression, we see that the WKB analysis allows us to identify $\nu(n)$ with δ_w. This supports our assertion that $\nu(n)$ may be obtained by putting a smooth curve through the quantum defect values obtained experimentally.

[62] Whittaker and Watson, p. 343 in ref. 52.
[63] See Eqs. (A.10) and (A.11).

At positive energies, where $n = i|n|$ and $\epsilon = 1/|n|^2$, a similar WKB analysis yields the following asymptotic formula for large positive real values of r:

$$U^L \sim \cos\left\{\int_{x_1}^{x} [P_h(x)]^{\frac{1}{2}}dx + \pi\delta_w - (\pi/4)\right\}. \tag{6.22}$$

Here x_1 is the inner turning point for the pure Coulomb potential, and, at positive energies, it is the only one on the real x axis. Again δ_w is defined by (6.17), where now the upper limit of integration is x. Comparing this with the asymptotic expansion of (6.2) along the positive real r axis, we obtain

$$\frac{\alpha(n)}{\gamma(n)} = \frac{\Gamma(i|n| + L + 1)}{|n|^{2L+1}\Gamma(i|n| - L)}\left\{\frac{-\cos\pi\delta_w e^{\pi|n|}}{2\sin\pi(i|n| - L)\sin\pi\delta_w} + \frac{e^{-\pi|n|-i\pi(2L+\frac{1}{2})}}{2\sin\pi(i|n| - L)}\right\}. \tag{6.23}$$

If we neglect $\exp(-2\pi|n|)$ in this expression, because it is small compared with unity, we obtain (6.21) exactly. Hence, $\nu(n)$ in (6.3) may be extrapolated from the eigenvalue range to positive energies. For higher positive energies, for which $\exp(-2\pi|n|)$ is not negligible, we should replace (6.3) by (6.23) before using the extrapolated $\nu(n)$. However, this correction can become significant only when $\epsilon > 1$ Rydberg unit. We shall rarely be interested in such values in solid-state calculations. We should also note that (6.23) has a small imaginary part, which cancels a corresponding part of $N_{2L+1}{}^n(z)$ when $n = i|n|$. It follows that the resulting $U^L(r)$ is real, as it should be.

For negative energies sufficiently low that the outer turning point penetrates the non-Coulomb core region, the WKB analysis is less satisfactory. This corresponds to energies below about -0.9 Rydberg units for sodium when $L = 0$ and $r_0 \sim 2$. It corresponds to energies below -0.7 Rydberg units for rubidium, when $L = 0$ and $r_0 \sim 3$. For $L = 1$, the limit is -0.4 Rydberg units for all the alkalies. Inserting a factor $(1 + \Delta)$ on the right side of (6.20), to take into account the fact that $V(r)$ differs from $-2/r$ over a small part of the range of integration of z, we obtain, in place of (6.21),

$$\frac{\alpha(n)}{\gamma(n)} = -\frac{\Gamma(n + L + 1)}{n^{2L+1}\Gamma(n - L)}\left\{\frac{\cos\pi\delta_w - e^{i\pi(n-L+\frac{1}{2})}\Delta\sin\pi(n - L + \delta_w)}{\sin\pi\delta_w - e^{i\pi(n-L-1)}\Delta\sin\pi(n - L + \delta_w)}\right\}. \tag{6.24}$$

Comparing this with (6.3), we observe that $\nu(n)$ evidently differs from δ_w, as defined in (6.17), by a quantity which is proportional to Δ and vanishes at the eigenvalues. However, this quantity has an imaginary part which, when substituted in (6.2), would yield a value of $U^L(r)$ which can not be made real by multiplication with a complex constant, except at the eigenvalues, where the term in Δ vanishes. The exact $U^L(r)$ is certainly real.[64] This difficulty suggests that we should not take the WKB result seriously at such low energies. This opinion is confirmed by our discussion of the "forced zeroes" of tan $\pi\nu(n)$ in Section VII, of which the WKB approximation gives no hint. We conclude that the WKB approximation validates the smooth extrapolation of $\nu(n)$ only at high negative energies and at positive energies. It provides no information at energies so low that the outer turning point penetrates the core. It is also unreliable at energies somewhat below the lowest eigenvalue of the hydrogenic potential for the value of L under consideration, that is, for energies below the value $\epsilon = \dfrac{-1}{(L+1)^2}$. This is true even if the outer turning point is outside the core at this energy, as happens for large values of L. It is just in this range that account must be taken of the forced zeroes when extrapolating $\nu(n)$, as discussed in Section VII. As mentioned above, the zeroes are not predicted by the WKB analysis.

VII. Extrapolation at Low Energies—the "η-Defect"

For $L > 0$ we find from (6.2) and (6.3) that $\nu(n)$ is not a smooth function of energy at low negative energies and is not accurately approximated by the δ_w given by the WKB approximation. This follows from the circumstance that the factor

$$\frac{\Gamma(n+L+1)}{n^{2L+1}\Gamma(n-L)} = \frac{(n^2-L^2)(n^2-(L-1)^2)\cdots(n^2-1)}{n^{2L}} \tag{7.1}$$

in (6.3) vanishes for $n = 1, 2, 3, \ldots L$. The function $\alpha(n)$ in (6.2) is not zero at the corresponding energies for all potentials, so that (6.3) requires than $\tan(\pi\nu(n))$ vanish and $\nu(n)$ equal an integer or zero at such values of n. Consequently, $\nu(n)$ must either oscillate as a function of energy in this range or change its value at least by unity between two of the energies. This situation makes extrapolation into the region quite unreliable. The WKB approximation gave no hint that $\nu(n)$ should be inde-

[64] Had we not used the device of evaluating exp $(2iz)$ in (6.20) by comparison with the exact solution, we should have obtained a factor $(1 + \Delta)$ on the right of (6.20) by direct integration of (6.10). Here Δ is a number of magnitude $0(1/n)$. This leads again to (6.24) and introduces into $U^L(r)$ an imaginary part which vanishes as $|n| \to \infty$. Such an effect evidently is the result of inaccuracy introduced in evaluating the wave function by the WKB method and should not be taken seriously. However, it serves as a warning that the justification of QDM based on the WKB, while convincing at the higher energies, is not rigorous.

pendent of the form of $V(r)$ at these energies. Since the energies lie below the minimum at $\epsilon = -1/(L+\frac{1}{2})^2$ in the effective radial potential in (6.8) when $V(r) = -2/r$, either there are no turning points on the real r axis, or the outer turning point is within the core region. This situation supports the point stressed in the last section to the effect that the WKB approximation can not be trusted in such circumstances.

The behavior described above causes no particular difficulty in the extrapolation of $\nu(n)$ for $L = 1$ in the case of the alkali metals. When a quadratic function of the energy is fitted to the experimental quantum defect values of the three lowest eigenvalues of the valence electron, the extrapolated function is very nearly equal to an integer at $\epsilon = -1$. For sodium, for example, the value so obtained is 1.033. Evidently, one can improve the extrapolation by fitting a cubic function to the appropriate integer at -1 and to the experimental points used above. In the case of the alkali metals, this improvement makes an almost insignificant difference in the value of $\nu(n)$ at the energies which are important in the band structure calculations. It is essential that the improved extrapolation be used in calculations for the noble metals, however, according to work by Kambe.[9]

Extrapolation of $\nu(n)$ is not satisfactory for $L = 2$ even for the alkali metals. The forced zeroes are at -0.25 and -1 Rydberg units, whereas the energies of interest in band calculations are as low as -0.6 Rydberg units. The eigenvalues lie between 0 and -0.15 Rydberg units, so that the oscillation forced upon $\nu(n)$ makes the rather large extrapolation quite unreliable. The situation clearly is worse for $L = 3$.

We are led, consequently, to define a quantity $\eta(n)$, which, for convenience, we shall call the "η-defect," in terms of the equation

$$\frac{\alpha(n)}{\gamma(n)} = -\frac{1}{\tan \pi\eta(n)}. \tag{7.2}$$

Comparing this relation with (6.3), we see that $\tan(\pi\eta(n))$ has no forced zeroes. Moreover, for energies somewhat above the highest forced zero, (7.1) increases monotonically to unity as $\epsilon \to 0$, so that $\eta(n)$ should be as smooth a function of energy as $\nu(n)$ in this region. Since it has no forced zeroes, we may expect $\eta(n)$ to be a more slowly varying function of energy over the range $-1 < \epsilon < 0$ than $\nu(n)$ is. Hence, it should be more suitable for extrapolation from experimentally determined values.

We find, indeed, that the η defect *is* remarkably smooth for the alkali metals and varies less rapidly than $\nu(n)$ in all cases. To verify this supposition further, we have calculated $\eta(n)$, $\nu(n)$, and $\alpha(n)/\gamma(n)$ for $L = 1$, using the potential

$$V(r) = \begin{cases} -2Z/r & r < r_0, \\ -2/r & r > r_0, \end{cases} \tag{7.3}$$

in which $Z = 1.44$, $r_0 = 3.125$ atomic units. The radial potential $V(r) + L(L+1)/r^2$ has a minimum value of -1.037 Rydberg units at $r = 1.389$, in contrast to the minimum of -0.5 Rydberg units, at $r = 2$, for the hydrogenic potential $V(r) = -2/r$. Of course, (7.3) is scarcely a reasonable approximation to any actual ionic potential because of the discontinuity in $V(r)$ at r_0. However, this discontinuity should have no important effect on the relative variation of $\eta(n)$ and $\nu(n)$. The calculations were carried out with the use of a revised version of Kuhn's tables[30]

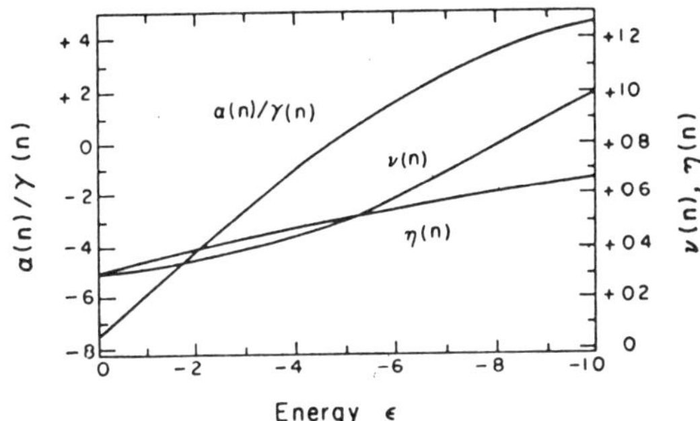

FIG. 2. Calculated values of $\nu(n)$, $\eta(n)$, and $\alpha(n)/\gamma(n)$, plotted as functions of energy, for $L = 1$. The model potential given by Eq. (7.3), with $Z = 1.44$ and $r_0 = 3.125$, has been used. Note that the energies are negative.

of the coefficients $U_k^{(1)}(z)$ and $D_k^{(1)}(z)$ of the Appendix. The solution of the differential equation for $r < r_0$ was matched at r_0 to the solution (6.2) outside r_0. Figure 2 shows the results for negative energies. Similar calculations for $L = 2$ and 3 were made with $Z = 1.291$, $r_0 = 3.4848$. The corresponding curves of $\nu(n)$ and $\eta(n)$ are plotted in Fig. 3. In all cases, $\eta(n)$ is a less rapidly varying function of ϵ than either $\nu(n)$ or $\alpha(n)/\gamma(n)$, as anticipated.

As a check on the error made in extrapolating $\eta(n)$, we have fitted a quadratic function of ϵ to the values of $\eta(n)$ for $L = 1$ which are calculated from the potential (7.3) at $\epsilon = -0.1$, -0.2, -0.3. We list in Table I the values obtained at neighboring energies by extrapolation, as well as the exact values and the percentage error. We also list the results obtained by extrapolation using a cubic function of ϵ fitted to the calculated value at $\epsilon = -1.0$, as well as at the foregoing three points. Evidently, the accuracy of the extrapolation to low energies can be improved if additional values of $\eta(n)$ can be obtained at energies below the range in which the eigenvalues of the valence electrons lie. We shall show later

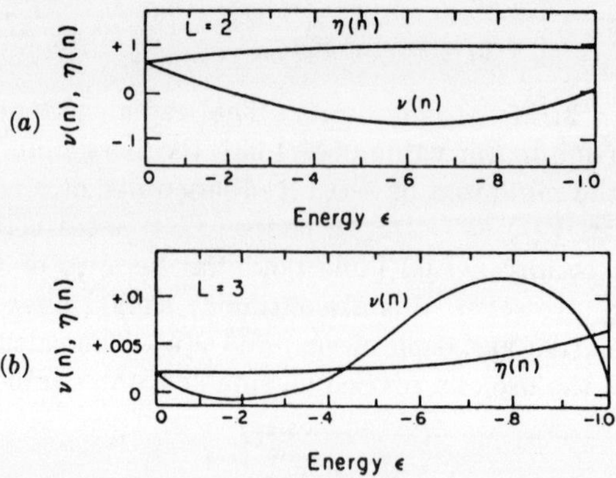

FIG. 3. Calculated values of $\nu(n)$ and $\eta(n)$, plotted as functions of energy. The model potential of Eq. (7.3), with $Z = 1.291$ and $r_0 = 3.4848$, has been used. The energies are negative, as in Fig. 2. (a) $L = 2$ (b) $L = 3$.

in this section that such information can be obtained from a knowledge of the eigenvalues corresponding to the core electrons.

We can obtain further information about the behavior of $\nu(n)$ and $\eta(n)$ at low energies which may be used to improve the accuracy with which $\eta(n)$ can be extrapolated. The improvements have made only

TABLE I. COMPARISON OF EXTRAPOLATED $\eta(n)$ WITH THE EXACT VALUE FOR THE POTENTIAL OF EQ. (7.3), $L = 1$, $Z = 1.48$, $r_0 = 3.125$

ϵ	Exact $\eta(n)$	Quadratic extrapolation		Cubic extrapolation	
		$\eta(n)$	Error	$\eta(n)$	Error
0	0.29323	0.28629	2.36%	0.28729	2.03%
−0.1	0.33254	exact		exact	
−0.2	0.37825	exact		exact	
−0.3	0.42342	exact		exact	
−0.4	0.46728	0.46805	0.16%		
−0.5	0.50786	0.51214	0.84%	0.50813	0.05%
−0.6	0.54383	0.55569	2.13%	0.54567	0.34%
−1.0	0.64034	0.72449	13.1%	exact	

slight changes in values of $\eta(n)$ for the energies of interest in calculations of the band structure of the alkali metals. Such improvements are important, however, in calculations like those for the noble metals[9] involving considerable extrapolation. In Figs. 4 and 5 we show, therefore, typical graphs of $\eta(n)$ and $\nu(n)$ for $L = 1$, 2, and 3. The curves are the same as those of Figs. 2 and 3; however, they are plotted against n instead of ϵ.

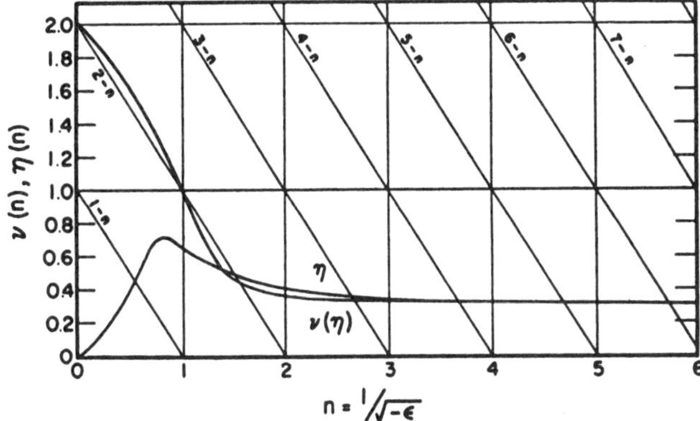

FIG. 4. Calculated values of $\nu(n)$ and $\eta(n)$ for $L = 1$, plotted as functions of n, for negative energies. The model potential of Eq. (7.3) with $Z = 1.44$ and $r_0 = 3.125$ has been used. The curves are plotted accurately for $n > 1$ and are sketched in for $n < 1$ in accordance with the rules of Section VII.

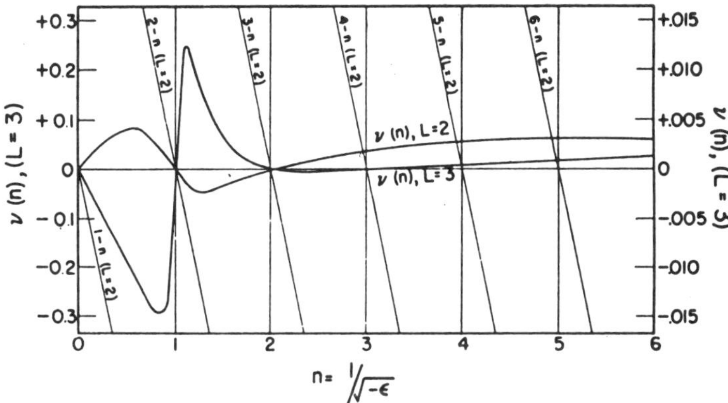

FIG. 5. Calculated values of $\nu(n)$ for $L = 2$ and 3, plotted as a function of n, for negative energies. The model potential of Eq. (7.3) with $Z = 1.291$ and $r_0 = 3.4848$ has been used. The lines $y = (m - n)$ are appropriate to $L = 2$; those for $L = 3$ would be almost vertical and are not shown. The curves are plotted accurately for $n > 1$ and are sketched in for $n < 1$ in accordance with the rules of Section VII.

The advantage of plotting the curves in this fashion is made clear by (6.5), for the eigenvalues of the potential are given by the points at which the curve $y = \nu(n)$ intersects one of the straight lines $y = (m - n)$, in which m is an arbitrary integer. The only exceptions to this rule are the crossings at $n = 1, 2, 3 \ldots L$ which are "forced" by the requirement that $\nu(n)$ be an integer at these points. Eigenvalues occur at the latter crossings only if the curve $y = \nu(n)$ is tangent to the line at the point of intersection. This may be shown from the asymptotic expansion of U^L in (6.2) if it is required that the term which becomes infinite as $r \to \infty$ be identically zero. We shall now give a discussion, based upon the graphs, which will set forth certain general properties of $\eta(n)$ and $\nu(n)$.

We consider first a sequence of potentials, starting with the pure Coulomb potential $V(r) = -2/r$ and passing by small changes to an ionic potential of the form discussed in Section IV. We shall follow the change of $\eta(n)$ and $\nu(n)$ through the sequence. It clearly is required that the quantities $\alpha(n)$ and $\gamma(n)$ and the location of the eigenvalues change continuously throughout this process. Both $\eta(n)$ and $\nu(n)$ are zero for the Coulomb potential. Since the potential is depressed at small r as we follow the sequence towards the ionic potential, we see, from perturbation theory, that the eigenvalues are lowered. Consequently, if $\nu(n)$ for $L = 1$ remains smooth, it has the form shown in Fig. 6a. The forced zero at $n = 1$ requires that the function change sign at this point. From (6.3) and (7.2), we see that $\eta(n)$ has the same sign as $\nu(n)$ for $n > 1$ and the opposite sign for $n < 1$. Hence, $\eta(n)$ has a form like that shown in the same figure. At $n = 0$, where $\epsilon = -\infty$, the perturbation has no effect upon the solution of the radial equation. Thus, $\eta(n)$ and $\nu(n)$ are zero here, as they are at infinite *positive* energy.[65] Since $\eta(n)$ must vary continuously through all perturbations, we see that $\eta(n)$ should always equal zero both at $n = 0$ and at the other "end" of the graph, that is, at infinite positive energy. Its maximum deviation from zero should take place at energies within a few Rydberg units of $\epsilon = 0$. The curves obtained both from calculations with the model potentials and from the experimental spectra of the alkalies suggest that $\eta(n)$ has a maximum somewhere in this range and is a slowly varying function of the energy. We shall postulate that this is a universal form for physically reasonable ionic potentials.

Although $\eta(n)$ changes continuously and evidently preserves the same general form throughout the potential sequence, the behavior of $\nu(n)$ is more complicated. Suppose the maximum of $\eta(n)$ increases beyond $\frac{1}{2}$, at which point $|\tan \pi\eta(n)| = |\tan (\pi\nu(n))| = \infty$ and $\gamma(n) = 0$. Then, as one of the points $\eta(n) = \frac{1}{2}$ approaches $n = 1$, the curve $\nu(n)$ is "stretched," because of the forced zero, in the manner illustrated in Fig. 6b. This stretching breaks the curve $y = \nu(n)$ when the point $\eta(n) = \frac{1}{2}$ coincides with $n = 1$ (parts A and B of Fig. 6c). Since $\tan(\pi\nu(n))$ is unchanged, however, if $\nu(n)$ is increased by unity, the break in the curve may be

[65] These conclusions can be demonstrated algebraically by expressing $\alpha(n)$ as an integral of the form (5.18) with $(z/2)N_{2L+1}{}^n(z)$ replacing ${}^1\tilde{U}_c{}^{(L,n)}(r)$, and considering the limiting form of the ratio $\alpha(n)/\gamma(n)$ from these integrals as $|n| \to 0$.

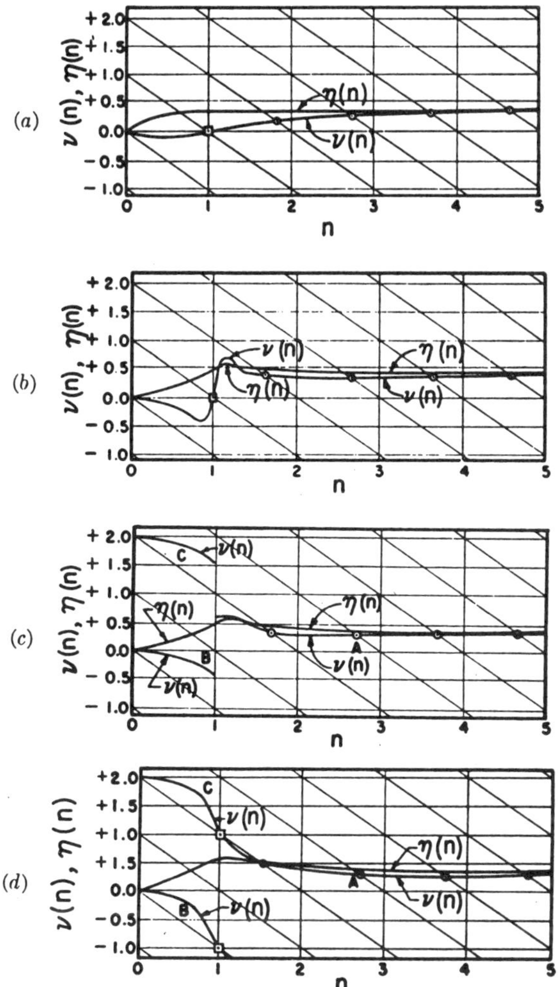

FIG. 6. The variation of $\nu(n)$ and $\eta(n)$ for $L = 1$ through a sequence of potentials leading from the pure Coulomb potential to that of an ion with a non-Coulombic core. Eigenvalues are marked with circles, forced zeroes of $\tan(\pi\nu(n))$ with squares.

(a) Sketch of $\nu(n)$ and $\eta(n)$ for a potential differing slightly from the pure Coulomb potential.

(b) Sketch of $\nu(n)$ and $\eta(n)$ showing the "stretching" of $\nu(n)$ caused by the forced zero when the maximum of $\eta(n)$ is slightly greater than $\frac{1}{2}$.

(c) Sketch of $\nu(n)$ and $\eta(n)$ when the point at which $\eta(n) = \frac{1}{2}$ coincides with $n = 1$. The curve of $\nu(n)$ is broken at $n = 1$ by the coincidence of the forced zero and the pole of $\tan(\pi\eta(n))$.

(d) Sketch of $\nu(n)$ and $\eta(n)$ when the points at which $\eta(n) = \frac{1}{2}$ straddle $n = 1$. The continuity of the $\nu(n)$ curve is restored by replacing part B by part C.

restored after $\eta(n) = \frac{1}{2}$ has passed over $n = 1$ if part B of the curve in Fig. 6c is replaced by part C. This process evidently may occur again if the change in the potential is continued. Thus, although $\eta(n)$ has a relatively simple form, $\nu(n)$ may have a complicated behavior. In general, $\nu(n)$ will equal an integer different from zero at $\epsilon = -\infty$ if it is regarded as a continuous function of n.

We may write down the following rules concerning the form of $\eta(n)$ and $\nu(n)$ and the relations between the two functions. These rules, together with the foregoing discussion, serve to determine $\eta(n)$ and $\nu(n)$ with little uncertainty once the eigenvalue spectrum of a potential is known.

1. As $n \to 0$ and $\to i0$ ($\epsilon \to \pm\infty$), $\eta(n) \to 0$ and, therefore, $\nu(n)$ approaches an integer.

2. The curve $y = \nu(n)$ can cross the lines $y = (m - n)$ only at eigenvalues or at "forced" crossings. It must cross one of the lines at such a point.

3. Consider the case $\eta(n) = m + \frac{1}{2}$, where m is an integer. Then, $\nu(n) = m' + \frac{1}{2}$ for the same value of n, where m' is an integer not necessarily equal to m. The converse relation also is valid. This rule follows from (6.3) and (7.2) if examined at points where $\tan(\pi\nu(n))$ and $\tan(\pi\eta(n))$ have poles. We also see that $\nu(n) = m'$ when $\eta(n) = m$. The converse also is true, except at the "forced" integers of $\nu(n)$.

4. At energies below the lowest eigenvalue corresponding to a particular L, $\gamma(n)$ is negative and never zero. Therefore, $\eta(n)$ and $\nu(n)$ cannot assume integer values for such energies, an exception occurring in the case of $\nu(n)$, at a "forced" value. The first conclusion follows from (5.17) and (6.1) and from the fact that $U^L(r)$ has nodes only at the origin at such energies, if we assume that $V(r) \leq -2/r$, as is expected for an actual ionic potential. We may recall from Section 5 that $U^L(r)$ was positive for very small r.

5. Whenever the potential is changed infinitesimally by an amount ΔV which is everywhere negative or zero, $\nu(n)$ is changed by an amount $\Delta\nu$ which is zero only at the points where $\nu(n)$ is forced to be an integer or zero. Moreover, the eigenvalues are lowered. Thus, given a function $\nu(n)$, associated with a potential, it must be possible to draw a nearby curve $\nu'(n)$ which crosses $\nu(n)$ only at the forced integers and which has all its eigenvalues below those of $\nu(n)$.

6. Consider the range of energies below the lowest eigenvalue, or the range for which $n < (L(L+1))^{\frac{1}{2}}$, depending upon which condition is the more stringent. We find from (6.2), from the asymptotic form of the Coulomb functions, and from the fact $U^L(r)$ has no nodes at such energies, that $\alpha(n)$ is positive and $|\tan(\pi\nu(n))| < |\tan(\pi(n - L - 1))|$ in the range

$2m + L + \frac{1}{2} < n < 2m + L + \frac{3}{2}$. Combined with rule 4, this condition fixes the sign of $\tan(\pi\nu(n))$ in this range and puts a bound on its magnitude.

In our previous discussion of the practical problem of obtaining $\eta(n)$ from the experimental spectrum of an atom, we have used data only for those term values which correspond to the states of the atom in which there is a single valence electron outside the core of closed shells. However, the foregoing rules, particularly rule 2, suggest that, when extrapolating $\eta(n)$ to low energies, we could use eigenvalues of core electrons to improve the accuracy. We shall show in Section IX that the eigenvalues most appropriate for this extrapolation are the energy parameters appearing in the Hartree-Fock equation for the ionized core. This procedure is most satisfactory if, as is true for the alkali metal atoms,[66] the core wave functions of the ion are not very different from those of the neutral atom. Fortunately, the extrapolated values of $\eta(n)$ of interest in band calculations are not particularly sensitive to the exact location of the core eigenvalues, since these eigenvalues ordinarily lie considerably below the valence band energies of the solid. The highest *d* shell of the noble metals represents an exception. Thus, a precise knowledge of the appropriate core eigenvalues usually is not required. It appears to be most suitable to determine the eigenvalues from Hartree-Fock calculations of the ionic wave functions if they are available. Otherwise, one may use, instead, the experimental values of the energy necessary to remove a given core electron from the ion. The latter would equal the Hartree-Fock energy parameter if Koopman's theorem[16] were applicable to an atomic system; however, the difference has made no significant uncertainty in $\eta(n)$ in the band calculations we have examined in cases in which both sets of data were available (Section X).

The rules listed above have been used in checking the reasonableness of the curves of $\eta(n)$ and $\nu(n)$ obtained from the data on the alkali metal atoms and presented in Section X. In particular, they are useful in establishing the appropriate corrections in extrapolations to either side of the range of eigenvalues associated with the valence electron. As remarked earlier, the corrections, particularly the use of the core eigenvalues, are important when using QDM with the noble metals.[9]

VIII. Corrections for Core Polarization, Overlap, and Valence Electron Interaction in Multivalent Atoms

We have remarked in Section V that a potential function which represents the interaction between a valence electron and the core as accu-

[66] D. R. Hartree and W. Hartree, *Proc. Roy. Soc.* **A193**, 299 (1948); *Proc. Cambridge Phil. Soc.* **34**, 550 (1938).

rately as possible does not satisfy the conditions for the use of QDM, for it differs from the pure Coulomb potential $-2/r$ at large r, both by a polarization term and by terms arising from the tails of the core wave functions. In band calculations, however, we are interested in $U^L(r)$ for values of r within the atomic cell. Since the boundary condition $U^L(0) = 0$ must always be satisfied in integrating (3.3), the value of the function in this range does not depend upon the form of the ionic potential $V(r)$ for values of r outside the cell. Moreover, QDM gives us an explicit expression for $U^L(r)$ only in the Coulomb region where $V(r) = -2/r$. Thus, in using QDM, we clearly wish to replace the actual ionic potential by one that equals $-2/r$ for values of r greater than a convenient radius r_i, which may be taken near the radius of the inscribed sphere or of the equivalent sphere.[67] We shall now show how the quantum defect $\nu(n)$ for this modified potential may be obtained from the experimental term values of the free atom.

If the difference between the two potentials is small for $r > r_i$, as is usually the case, perturbation theory can be used, the correct wave functions being approximated for $r > r_i$ by Coulomb functions that are zero at $r = \infty$. Such functions are the correct solutions for $r > r_i$ for the modified potential at the shifted eigenvalues, which are, of course, unknown. However, a simpler and more direct procedure makes use of the WKB method to approximate the change in the quantum defect. This change is given directly by (6.17):

$$\pi\Delta\nu(n) = \int_{r_i}^{r_2}\left[-\frac{1}{n^2} - V(r) - \frac{(L+\frac{1}{2})^2}{r^2}\right]^{\frac{1}{2}} dr, \\ - \int_{r_i}^{r_2'}\left[-\frac{1}{n^2} + \frac{2}{r} - \frac{(L+\frac{1}{2})^2}{r^2}\right]^{\frac{1}{2}} dr, \quad (8.1)$$

where $V(r)$ is the original ionic potential, and r_2 and r_2' are the outer turning points in the ionic and Coulomb potentials, respectively. The

[67] Of course, the actual crystal potential will be slightly different from $-2/r$ near the surface of the cell because of exchange and correlation effects and the influence of nearby ions. This difference from the ion potential is difficult both to estimate and to include in calculating the wave functions, and it is usually omitted. The additional uncertainty concerning the proper choice of r_i causes an error in the wave functions and eigenvalues that should be no larger than that arising from this omission. If such errors are significant, the corresponding energy band calculations cannot be considered reliable. There is good reason to believe that the ground-state energy and wave function are not very sensitive to small changes in the potential near the surface of the cell, but calculations for states of higher energy are probably less reliable. The author intends to examine this sensitivity in future work.

energy $\epsilon = -1/n^2$ must be sufficiently high that the WKB approximation to $\nu(n)$ can be expected to be accurate. We assume, of course, that we can obtain an explicit approximation to $V(r)$ in this external region, so that (8.1) can be evaluated. We then subtract from the experimental δ_m's the values of $\Delta\nu(n)$ so obtained at the energy eigenvalues of the ionic potential. We thus obtain the values of the modified extrapolated quantum defect *at the same energies*, which do not coincide with the eigenvalues of the modified potential. We may obtain $\eta(n)$ at these energies from Eqs. (6.3) and (7.2) and may extrapolate $\eta(n)$ in the usual manner, still using the core eigenvalues of the unmodified potential, since they are affected negligibly by a small change in the potential for $r > r_i$ if r_i is greater than the core radius.[67a] The results of such calculations of $\Delta\nu(n)$ for the alkali metal atoms are given in Section X.

Account may be taken in a similar manner of overlap between the core wave functions of neighboring ions in the solid and of the penetration of other conduction electrons into the cell. Thus, we can approximate the actual crystalline potential within the cell by assuming that this potential equals that of the ion within a sphere $r < R$. For $R < r < r_i$, in which r_i is, for example, the radius of the inscribed sphere or that of the equivalent sphere, we assume the potential differs somewhat from that of the ion but is still spherically symmetric. We can first calculate from (8.1) the quantity $\Delta_1\nu(n)$ arising from the difference between the free ion potential and a potential which deviates from it by being equal to $-2/r$ for $r > r_i$. We can then calculate an additional correction, $\Delta_2\nu(n)$, from a similar formula in order to obtain $\nu(n)$ for the potential which differs

[67a] Although the core eigenvalues and, consequently, the value of $\eta(n)$ at the eigenvalues are not altered significantly by the change in the potential outside the core, the change does affect $\eta(n)$ at energies intermediate to the core eigenvalues. At such energies, the function $U^L(r)$, which satisfies the boundary condition at the origin, increases exponentially for large values of r, instead of decreasing exponentially, as at the eigenvalues. If we consider the process of integrating (3.3) numerically outward from the origin, it is clear that the combination of Coulomb functions, which is obtained in a Coulomb region $r > r_0$, will depend on the nature of $V(r)$ for $r_i < r < r_0$, even though r_i lies in the region in which the core eigenfunctions decrease exponentially. This conclusion follows also from Eqs. (5.17), (5.18), and (7.2). It was suggested by the WKB formula (6.24). Evidently $\eta(n)$ may oscillate about a smooth curve through the eigenvalues at these low energies, for a change in the potential outside the core may change $\eta(n)$ at energies between the eigenvalues. This oscillation makes the extrapolation of $\eta(n)$ far below the valence eigenvalue range uncertain. However, from the data on the alkali metal atoms presented in Section X, it does not appear that this uncertainty is appreciable for s and p functions at interesting energies less than 0.5 Rydberg units below the lowest valence eigenvalue. The uncertainty for the d functions, most notably those in potassium, is considerably greater.

from the latter by being equal to the approximate crystalline potential for $R < r < r_i$. This value of $\nu(n)$ may then be used to calculate $\eta(n)$, which can be extrapolated and used in the band calculations. This procedure requires that we be able to approximate both the ionic and crystalline potentials for $r > R$. However, it does not require that we know the former explicitly for the range $r < R$ if the spectrum of the free atom obeys the Rydberg-Ritz law sufficiently accurately that a quantum defect which varies regularly can be obtained from the data.

It is more often customary, in solid-state calculations, to correct the potential for core overlap by renormalizing the core functions in such a way that the square of the amplitude of each function integrates to unity over the cell. The form of the functions obtained from Hartree-Fock calculations for the free atom is preserved. The crystalline potential is then calculated from the modified charge distribution within each cell. We could, of course, use this procedure in correcting the experimental quantum defect by setting $R = 0$ above, and we could perhaps estimate the difference between the crystalline and ionic potentials from Hartree Fock wave functions more accurately than we can approximate either of the potentials.

The use of QDM with a multivalent element is not quite as simple as with monovalent substances if we wish to take into account the presence of more than one valence electron in the atomic cell, for the term values of the neutral atom do not follow the Ritz law with the accuracy found in the alkali metal atoms. This leads to an erratic experimental quantum defect. We can obtain an alkali-like spectrum, of course, by considering the ionized atom possessing a single valence electron outside of closed shells; however, it is then necessary to correct the crystalline potential for the presence of the other valence electrons in the cell. We can still use QDM to express the solutions of the radial wave equation outside the core in terms of Coulomb wave functions appropriate to the potential $-2Z/r$, in which Z is the ionic charge, for the transformations $r' = rZ$ and $n' = nZ$ convert the radial equation containing this potential to the standard form involving $-2/r'$. Moreover, we can probably approximate the potential arising from the other valence electrons in the cell by that of a uniform charge distribution having the appropriate density. Since this part of the potential is proportional to r^2 it probably can be neglected with safety within the core if the core is reasonably small compared with the cell. Consequently, we can obtain the wave function at the surface of the core from QDM, using the alkali-like spectrum of the multiply ionized ion. We can then continue the function to larger r by integrating the radial differential equation with the known potential formed by combining $-2Z/r$ with the term in r^2. This process clearly is more complicated than that necessary for a monovalent atom. No attempt has so far been made to apply it in an actual calculation.[67b]

IX. Correlation and Exchange Interaction between Valence and Core Electrons

In our previous discussion of the justification of QDM, we made the assumption that the interaction between the valence and core electrons in both the atom and the solid could be represented accurately by a one-electron potential function which is spherically symmetric. It is the purpose of this section to show that, in fact, QDM takes more accurate account of exchange and correlation effects than is possible in a simple potential representation of the interaction.[68]

We have seen that, in the approximate Hartree-Fock procedure, the exchange operator A of (2.6) is an integral operator which is the same for all wave functions of the different one-electron states in a single configuration. Writing (2.6) in the form (2.9), however, we see that $A\varphi_i(\mathbf{r})$ can be represented as the product of $\varphi_i(\mathbf{r})$ and a potential function

$$V_i(\mathbf{r}) = -e^2\left[\int \frac{\rho(\mathbf{r},\mathbf{r}')\varphi_i(\mathbf{r}')}{\varphi_i(\mathbf{r})|\mathbf{r} - \mathbf{r}'|}\,d\tau'\right]. \qquad (9.1)$$

The potential function depends on the wave function $\varphi_i(\mathbf{r})$.

Let us now examine a sequence of electronic configurations of a single atom possessing a single valence electron outside a number of tightly bound closed shells, as in the alkali metal atoms. The configurations differ in having the valence electron in different eigenstates of the same angular momentum quantum numbers L and M. We shall compare the $V_i(\mathbf{r})$ of the different configurations. In each case i corresponds to the valence wave function. We first remove that term in $\rho(\mathbf{r},\mathbf{r}')$ of Eq. (2.4) for which $j = i$. This term in $A\varphi_i(\mathbf{r})$ cancels an identical one in $U\varphi_i(\mathbf{r})$ of (2.5).

[67b] Seitz (p. 357 in ref. 10) has calculated the wave functions and cohesive energies of lithium, sodium, and potassium when an r^2 term, arising from a uniform distribution of the charge of the conduction electrons, is included in the potential. The wave functions of occupied states differ by at most a few per cent from those obtained with the omission of this term, and the cohesive energies are changed by less than 1 kg cal/mole. However, we may expect the effect of this term in the potential to be greater in a multivalent atom, for the corresponding charge density is greater. Moreover, wave functions at points in **k**-space near the surface of the Brillouin zone should be more strongly perturbed by such a term than those near the energy minimum. Such wave functions correspond to occupied states in multivalent metals.

[68] This view is contrary to that expressed by Kuhn and Van Vleck,[4] who asserted that the method assumes a "best central field" but avoids the difficulties of determining this field and of using it in the band calculations.

We shall designate the potential we obtain from (9.1) in this way by $V_i^*(\mathbf{r})$. Then, $V_i^*(\mathbf{r})$ vanishes outside the core, since each term in its expansion contains $\varphi_j(\mathbf{r})$, the wave function of a core state. Moreover, the core wave functions $\varphi_j(\mathbf{r})$ should not change markedly in the sequence of configurations, for although they do depend on the valence state through the Coulomb and exchange terms in the Hartree-Fock equation, the interaction involving the valence electron should be much weaker than that involving the nucleus and the other core electrons. Consequently, we may regard the core $\varphi_j(\mathbf{r})$ as approximately the same throughout the configuration sequence.[66] Furthermore, if this sequence corresponds to an energy range of roughly 0.4 Rydberg units, or less, as is the case for the alkali metal atoms, the valence function $\varphi_i(\mathbf{r})$ will have approximately the same form within the core for different states of the same L and M, even though the $\varphi_i(\mathbf{r})$ are completely different outside the core and differ in amplitude within the core. This can be shown most easily by approximating the radial part of the wave function with the use of the WKB method. Assuming that we can neglect the angular variation of $V_i^*(\mathbf{r})$, we calculate the change in the wave function at a point within the core resulting from a change in energy which is small compared with the average of the difference between the energy and the radial potential, $V(r) + (L + \frac{1}{2})^2/r^2$, inside the core. It then follows that $V_i^*(\mathbf{r})$ is approximately independent of the energy for a given L, or of the valence electron state, for $V_i^*(\mathbf{r})$ depends on the form of $\varphi_i(\mathbf{r})$ within the core and not on its amplitude. Aside from this, $V_i(\mathbf{r})$ depends only on the core functions. For different L, $V_i^*(\mathbf{r})$ may be quite different, however, for the corresponding valence wave functions are not at all alike within the core.

We have presented this discussion, in which the exchange operator is approximated by a potential function, in order to show that it is reasonable to use a potential to demonstrate that the extrapolated quantum defect $\nu(n)$ is a slowly varying function of energy in the range of energy near the eigenvalues of the valence electron. We have not attempted to take into account the angular variation of $V_i^*(\mathbf{r})$. Actually, QDM takes a more accurate account of exchange than is possible with a potential function, since $\nu(n)$ is obtained from the *experimental* eigenvalues. Thus, we may regard the functions obtained outside the core from QDM at arbitrary energies as the solutions of the Hartree-Fock equation *including the integral exchange operator*. Indeed, we can assert that QDM gives us the solutions of the more general problem in which the correlation between core and valence electrons is included. It is gratifying that the experimental spectra of the alkali metal atoms yield quantum defects and η-defects which vary slowly for the more complicated situation, which includes the dependence of the core wave functions on the state of the valence electron. In fact, the functions vary about as slowly with energy, in this case, as those we would expect to find if the interaction could be represented exactly by a simple potential.

We must now consider the choice of the core eigenvalues to be used in extrapolating the η-defect to low energies. One is tempted at first to use the core eigenvalues obtained in a Hartree-Fock calculation for the neutral atom, for the Hartree-Fock Hamiltonian in (2.7) is the same for all the $\varphi_i(\mathbf{r})$ of a given configuration. However, (2.7) has quite a different form for a core wave function $\varphi_m(\mathbf{r})$ than for a valence function. In particular, it does not reduce to the differential equation for an electron in a simple Coulomb potential when $\mathbf{r}$ lies outside the core because of the Coulomb and exchange terms involving the valence electron. We shall show below that QDM is applicable to the calculation of Bloch functions in a crystal if we can neglect the dependence of the core wave functions on the state of the valence electron. Consequently, in solving the Hartree-Fock equation (2.7) for the valence electron function $\varphi_i(\mathbf{r})$ of the neutral atom, we wish to use the same core $\varphi_j(\mathbf{r})$ for all possible valence wave functions. We regard the functions obtained from QDM for arbitrary energies as the solutions of the integro-differential equation resulting from this approximation. Canceling the term with $j = i$ in $A\varphi_i(\mathbf{r})$ against the identical term in $U\varphi_i(\mathbf{r})$, and making the substitution $f(\mathbf{r}) = \varphi_i(\mathbf{r})$, we find that the resulting equation for f is of exactly the same form as that obtained by substituting $f = \varphi_m(\mathbf{r})$ in the Hartree-Fock equation for the core function $\varphi_m(\mathbf{r})$ in the *ionized* atom. This is true provided that the core functions appearing in $\rho(\mathbf{r}_2,\mathbf{r}_1)$ and $\rho(\mathbf{r}_2,\mathbf{r}_2)$ in the operators A and U are taken to be identical with those appropriate to the neutral atom and provided that the functions φ_m and $\varphi_m{}^*$ appearing in A and U are set equal to the *ordinary* core functions and *not* identified with f and f^*. Neglecting thus the dependence of the core functions on the valence state, we conclude that the core eigenvalues to be used in extrapolating $\eta(n)$ for a given L in QDM evidently are those associated with the Hartree-Fock equation of the ion rather than of the neutral atom, for the same L. The accuracy of this procedure depends upon the extent to which the extrapolated $\eta(n)$ is sensitive to small changes in the core eigenvalues and upon the extent to which the valence electron influences the shape of the core wave functions. As remarked in Section 8, the procedure appears to be quite accurate for the alkali metals. The Hartree-Fock core eigenvalues may be obtained approximately from experimental measurements on spark spectra; however, the accuracy of this step is probably less than that of neglecting the change in the core functions induced by a change of state of the valence electron, because of the relatively great coupling between the different core functions. Consequently, it appears preferable to use calculated Hartree-Fock eigenvalues when these are available.

We have remarked above that the valence exchange potential $V_i^*(\mathbf{r})$ may be quite different for different L. This is a well-known fact which has recently been discussed by Herman *et al.*[69] They have given tables of approximate exchange potentials for different L for germanium. The difference is greatest for the heavy elements and is responsible for Gorin's[29] failure to find a single potential function which would reproduce both the s and p term values of the spectrum of neutral potassium in spite of the fact that he found potentials which reproduced either the s *or* the p term values to within several per cent. Automatically QDM takes account of this difference, of course. We wish to show now that it is essentially correct to combine the functions obtained from QDM for different L into a single Bloch function for the crystal, despite the difference in the potentials which are used to approximate the exchange operator for different L.

It is not obvious immediately that this procedure is correct, for if we first approximate the exchange operator by a potential and then expand the Bloch function in spherical harmonics, as in (3.2), it might appear that we should use an average potential for all L rather than a different potential for the various terms in the series. Let us keep the exchange operator in its original integral form in (3.1) and, as usual, cancel the term with $j = i$ in $A\varphi_i(\mathbf{r})$ against the identical term in $U\varphi_i(\mathbf{r})$. We then see that within the atomic cell of the crystal the operator which acts upon $\varphi_i(\mathbf{r})$ in the approximate Fock equation is linear. Moreover, it is identical with the operator which acts upon $\varphi_i(\mathbf{r})$ for any L in the free atom, provided that we can neglect the difference in the core functions $\varphi_j(\mathbf{r})$ for the different situations. We have observed above that such neglect should be justified if the core electrons are tightly bound, as in the alkali metal atoms. With this approximation, we see that QDM provides us with functions for different L which can be combined into the Bloch functions and which may be regarded as satisfying the Fock equation in which the exchange operator is in integral form. Hence, QDM is a more general procedure than one that starts by approximating the valence-core interaction by a potential even if one uses a different potential for different L. Accordingly, it should be more accurate in calculations of the band structure of the heavy elements.

We can make the additional observation that QDM should take approximate account of the correlation between the valence and core electrons. Thus, we can improve the determinental approximation (2.1) for the wave function of a configuration of a monovalent atom by assum-

[69] F. Herman, J. Callaway, and F. S. Acton, *Phys. Rev.* **95**, 371 (1954).

ing that, whenever $r_1 > r_0$, where r_0 is the core radius beyond which there is negligible probability of finding two electrons in a monovalent atom, the wave function can be written as

$$\Psi = \varphi(\mathbf{r}_1)\Psi_c(\mathbf{r}_2,\mathbf{r}_3 \ . \ . \ . \ \mathbf{r}_N). \tag{9.2}$$

Here $\Psi_c(\mathbf{r}_2, \ . \ . \ . \ \mathbf{r}_N)$ denotes the wave function of the core electrons when $r_1 > r_0$.[70] We need make no assumption concerning the form of Ψ_c beyond requiring that it be normalized, corresponding to the condition that the probability of finding $N - 1$ electrons in the core be unity, and that it be spherically symmetric under simultaneous rotation of all the coordinates $\mathbf{r}_2 \ . \ . \ . \ \mathbf{r}_N$. In writing (9.2), we are, for simplicity, neglecting the polarization of the core by the valence electron. This is a long-range correlative effect.[70a] We have no need for information about the form of the wave function Ψ when $r_1 < r_0$, beyond the knowledge that it must connect smoothly to (9.2), and, therefore, must have the same normalization and the same transformation properties under the rotation of the coordinates $\mathbf{r}_1 \ . \ . \ . \ \mathbf{r}_N$. We can then show from the variational principle, essentially in the manner used in the derivation of Fock's equation, that $\varphi(\mathbf{r}_1)$ in (9.2) satisfies the Schrödinger equation for a Coulomb potential for $r_1 > r_0$. Hence, from this point of view, one obtains $\varphi(\mathbf{r}_1)$ for $r_1 > r_0$ from QDM.

In the solid, we assume that the N-electron Hamiltonian within the cell is the same as that of the atom and that the Bloch function can be written in the form (9.2) for $r_1 > r_0$. Then, if $\Psi_c(\mathbf{r}_2 \ . \ . \ . \ \mathbf{r}_N)$ is the same for all L in the atom and is also the same for the Bloch function, we clearly can combine atomic functions of different L and the same ϵ, obtained from QDM, to form the Bloch function for $r_1 > r_0$. The continuation of these functions into the core provides the continuation of the Bloch function. Hence, in this situation, QDM clearly furnishes us with the means of constructing the correct $\varphi(\mathbf{r}_1)$ for the Bloch function outside the core. This φ takes account of complicated correlations between the valence and core electrons when the valence electron penetrates the core. On the other hand, if $\Psi_c(\mathbf{r}_2 \ . \ . \ . \ \mathbf{r}_N)$ is significantly different for different L and for the Bloch function, we cannot form the Bloch function as a simple linear combination of atomic functions obtained from

[70] Ψ should be antisymmetrical under permutations of all the electron coordinates. Hence, if Ψ_c in (9.2) is antisymmetrical, Ψ should contain $N - 1$ additional terms in which $\mathbf{r}_1$ is interchanged with $\mathbf{r}_2 \ . \ . \ . \ \mathbf{r}_N$. When $\mathbf{r}_1$ lies outside the core, however, as is assumed in writing (9.2), the additional terms have negligible amplitude.

[70a] We may include the effect of core polarization by assuming that Ψ_c depends parametrically on $\mathbf{r}_1$. With the appropriate parametric dependence, we find from the variational principle that $\varphi(\mathbf{r}_1)$ satisfies the Schrödinger equation for a Coulomb potential to which a polarization term has been added.

QDM. This situation is the same as that which would occur in connection with the Hartree-Fock equation if the wave functions of the core states were significantly different for different valence wave functions, so that the interaction operators U and A were different for different valence states. Evidently, QDM cannot give an entirely correct result in such situations.[70b] Indeed, it is to be expected that the core wave functions will depend slightly on the valence state. As remarked earlier, however, it should be a sufficiently accurate approximation to neglect such differences if the core electrons are tightly bound. Therefore, QDM should provide an excellent means of taking account of effects arising from valence-core exchange and correlation in band calculations.

X. Quantum Defect Data for the Alkali Metals[70c]

In Table II we present data for the use of QDM with the alkali metals.[71] We tabulate first the lowest five or six energy levels available to the valence electron for $L = 0, 1, 2, 3$. In the case of doublets, a weighted mean of the two levels is used. We also tabulate the corresponding η-defect, $\eta(n)$, calculated from the experimental quantum defect δ_m. We tabulate δ_m instead of $\eta(n)$ for all f terms and for the d terms of lithium and sodium. The quantum defects for these terms are small and evidently arise, for the most part, from the core polarization term in the Schrödinger equation for $\mathbf{r} > \mathbf{r}_i$ (see Table III). We shall see below that we may set $\eta(n) = 0$ for these terms in calculating the band structure of the metals. In several of the tables, we have also listed the highest core eigenvalues of the ion, obtained from Hartree-Fock calculations for Li^+, Na^+, and K^+,[66,72,73] and from estimates based on the spark spectra, and from Hartree and Hartree's calculations for Rb^+ and Cs^+.[74]

The important data required for the practical use of QDM are the coefficients a, b, c, d in the interpolation and extrapolation formulas for $\eta(n)$, which equals $\nu(n)$ for $L = 0$ but otherwise is different:

$$\eta(n) = a + b(1/n)^2 + c(1/n^2)^2 + d(1/n^2)^3. \tag{10.1}$$

[70b] A similar approximation of using core wave functions appropriate to the ground state of the free atom or ion is usually employed in calculating energy bands by methods which make use of an explicit crystalline potential. These methods take less accurate account than does QDM of exchange and correlation interaction between valence and core electrons, however.

[70c] This section has been taken almost verbatim from ref. 8.

[71] Data for Li, Na, K, and Rb were obtained from C. E. Moore, *Natl. Bur. Standards (U.S.) Circ.* **467**, Vols. I and II (1949, 1952). For Cs the data are from R. F. Bacher and S. Goudsmit, "Atomic Energy States." McGraw-Hill, New York, 1932.

[72] D. R. Hartree and W. Hartree, *Proc. Roy. Soc.* **A166**, 450 (1938).

[73] V. Fock and M. Petrashen, *Physik. Z. Sowjetunion* **6**, 398 (1934); **8**, 547 (1936).

[74] D. R. Hartree and W. Hartree, *Proc. Roy. Soc.* **A151**, 96 (1935); **A143**, 506 (1934).

Here $(1/n^2)$ is the negative of the energy. Usually both a quadratic and a cubic formula are given. The former is obtained from the lowest three valence eigenvalues, unless otherwise noted, and should be used for energies higher than the lowest valence eigenvalue. The cubic relation is fitted to the same three eigenvalues and to the highest core eigenvalue of the same L. It should be used for extrapolation to energies below the lowest valence state. It is important to note that, whenever the coefficient c in the quadratic expression is large, or of order unity, this formula cannot be employed reliably for extrapolation too far (e.g., more than 0.5 Rydberg units) above or below the eigenvalues used to determine the coefficients. Whenever such extrapolation is necessary, it preferably should be done graphically, keeping in mind the rules of Section VIII, rather than with the use of the formula. This is particularly true of the η-defects for $L = 2$.

Finally, we list in Table II the values of $\eta(n)$ obtained from the quadratic interpolation formula for several valence eigenvalues above those used to determine the coefficients in (10.1). These should be compared with the experimental values in order to estimate the accuracy of the procedure. It should be noted, however, that the experimental quantum defects of the eigenvalues become progressively less accurate the closer ϵ gets to the series limit $\epsilon = 0$, since the quantum defect is the difference between the square root of the reciprocal of the term value and the principal quantum number. Moreover, a small error in the series limit affects the values obtained for the quantum defect of the higher eigenvalues much more than it does those at lower energy. It is for these reasons that we have usually used the lowest three valence eigenvalues to determine the coefficients in (10.1).

The Hartree-Fock equations have not been solved for Rb^+ or Cs^+, so that we have taken the ionization potential of the once-ionized atom, obtained from the spark spectra,[71] for the highest p eigenvalues of the core for these ions, since the corresponding values for Na^+ and K^+ are in sufficiently good agreement with the Hartree-Fock values. We have estimated the s and d core eigenvalues of Rb^+ and Cs^+ from the calculations of Hartree and Hartree[74] without exchange, the spark data for the p eigenvalue, and the relative values of the s and p eigenvalues of Na^+ and K^+. The error in these eigenvalues should be less than 10%. The corresponding uncertainty in the value of $\eta(n)$ in the range $-0.6 < \epsilon < 0$ should be less than 0.5% of the difference between $\eta(n)$ and the nearest integer.

It will be noticed that the c term in the extrapolation formulas for $\eta(n)$ for Na, K, Rb, and Cs, when $L = 2$, is about ten times larger than the corresponding term in the expressions for $L = 0$ or 1. This relatively rapid variation, which makes algebraic extrapolation unreliable, as remarked above, evidently is due to the effect that inverts many of the d doublets in the alkali spectra, namely, configuration interaction with an excited core state. This interaction has been reviewed by White[75] in connection with the doublet inversion. There is little point in attempting to subtract it since we wish to retain such correlation effects in QDM. It is worth noting, however, that this interaction, rather than experimental inaccuracy or an unexplained breakdown in QDM, is responsible for the variation in $\eta(n)$.

The data of Table II have not been modified to take into account any correction of the type discussed in Section VII if core polarization is omitted from the potential for $r > r_i$. This correction is somewhat uncertain, for we have some freedom in choosing r_i. Moreover, we do not know the value of the ionic polarizability accurately. Thus, it has seemed preferable to present the uncorrected data, together with a second table giving a reasonable estimate of the correction. The latter is presented in Table III. The polarization term in the potential is[76] $-2\alpha'/r^4$, in atomic units. Here $\alpha' = (3m^3e^6\kappa/8\pi n^6L)$, in which L is Avogadro's number, m and e are the electron mass and charge, and κ is the ionic molar polarizability. We have used Pauling's values for κ, as reproduced by Van Vleck,[76] since a more recent table[77] of κ for the alkalies gives values which lead to quantum defects for $L = 3$ larger than those observed experimentally. Pauling's values give quantum defects which are more nearly right. We have chosen r_i to agree approximately with the radius of the sphere inscribed in the atomic cell at the observed lattice spacing; however, for convenience in the calculations, we used the same r_i for Li and Na, and for K and Rb. For the most part we have calculated $\Delta\nu(n)$ from (8.1) by expanding the right-hand side of the equation in powers of α', a procedure which is inaccurate near the turning point but which has proved sufficiently accurate in several test cases in which it was compared with a direct numerical integration of (8.1). From the two values of $\Delta\nu(n)$ given in Table III and the data of Table II, we have calculated the corresponding values of $\Delta\eta(n)$. We have used these, along with the additional requirement that $\Delta\eta(n)$ vanish at the highest core eigenvalue of the appropriate L (Section VIII), to determine a quadratic expression for $\Delta\eta(n)$, namely,

$$\Delta\eta(n) = \alpha + \beta(1/n^2) + \gamma(1/n^2)^2. \tag{10.2}$$

This should be subtracted from the data of Table II to obtain the extrapo-

[75] White, Chapter 19 in ref. 36.

[76] J. H. Van Vleck, "The Theory of Electric and Magnetic Susceptibilities," Chapter 8. Oxford U. P., New York, 1932.

[77] J. R. Tessman, A. H. Kahn, and W. Shockley, *Phys. Rev.* **92,** 890 (1953).

lation formula for $\eta(n)$ which is appropriate for the potential modified by the omission of the long-range polarization beyond r_1.

The experimental quantum defect is very small for the f terms of all the alkali metal atoms and for the d terms of sodium and lithium; as is clear from Table II. Therefore, the wave functions are equal, for all practical purposes, to the corresponding hydrogenic eigenfunctions. In accordance, we have determined $\Delta\nu(n)$ at the eigenvalues by evaluating the expectation value of $-2\alpha'/r^4$, with use of the formula

$$\Delta\nu(n) = (1.34)\kappa\left[\frac{3}{2(L-\frac{1}{2})L(L+\frac{1}{2})(L+1)(L+\frac{3}{2})} - \frac{1}{2n^2(L-\frac{1}{2})(L+\frac{1}{2})(L+\frac{3}{2})}\right] \quad (10.3)$$

given by Van Vleck.[78] This is legitimate for these terms because the corresponding wave functions have only a small fraction of their amplitude within the sphere $r = r_i$. Values calculated in this way are labeled in the table. By comparing them with the experimental δ_m of Table II, we obtain justification for the opinion that it suffices to take the quantum defect identically zero for all f and higher terms of the alkali metals in band calculations. This may also be done for the d terms of lithium and sodium.

Tables II and III follow.

[78] J. H. Van Vleck, p. 216, Eq. (12), in ref. 76.

TABLE II. UNMODIFIED QUANTUM DEFECT DATA FOR THE ALKALI METALS[a]

(a) Lithium

Energy levels, $\eta(n)$, and interpolation coefficients

Energy levels

m	s	p	d	f
1	−5.56			
2	−0.3963153	−0.2604905		
3	−0.1483756	−0.1144805	−0.1112205	
4	−0.0772371	−0.0639544	−0.0625524	−0.062490
5	−0.0472774	−0.0407517	−0.0400282	−0.039941
6	−0.0318927	−0.0282182	−0.0277946	
7			−0.020420	

m	$\eta(n)$	$\eta(n)$	δ_m	δ_m
1	0.576			
2	0.411528	0.054773		
3	0.403916	0.050139	0.00148	
4	0.401786	0.048822	0.00168	−0.0003
5	0.400896	0.048268	0.00176	−0.0037
6	0.40043	0.048354	0.00182	
7			0.0020	

Interpolation coefficients for $\eta(n)$

s − Quadratic	p − Quadratic	s − Cubic
a = 0.399501	a = 0.047366	a = 0.399503
b = 0.029405	b = 0.02092	b = 0.029352
c = 0.00238	c = 0.02886	c = 0.00270
		d = −0.00052

Values of $\eta(n)$ from interpolation formulas

m	s − Quadratic	p − Quadratic
5	0.400896	0.048267
6	0.40044	0.04798

(b) Sodium

Energy levels, $\eta(n)$, and interpolation coefficients

Energy levels

m	s	p	d	f
2	−6.15	−3.60		
3	−0.3777259	−0.2231017	−0.1118769	
4	−0.1431615	−0.1018731	−0.0628869	−0.062524
5	−0.0751717	−0.0583926	−0.0402139	−0.040024
6	−0.0462661	−0.0378395	−0.0279069	−0.027790
7	−0.0313263	−0.0265072	−0.0204921	−0.020417

[a] No polarization correction is included in the values of $\eta(n)$ in this table.

TABLE II. UNMODIFIED QUANTUM DEFECT DATA FOR THE ALKALI METALS[a]

(b) Sodium

(Continued)

	$\eta(n)$	$\eta(n)$	δ_m	δ_m'
2	1.597	0.570		
3	1.372910	0.853367	0.01028	
4	1.357063	0.85379	0.01232	0.0008
5	1.35269	0.85426	0.01332	0.0015
6	1.35090	0.85443	0.01390	0.0013
7	1.35004	0.85452	0.01435	0.0015

Interpolation coefficients for $\eta(n)$

s − Quadratic	p − Quadratic	s − Cubic	p − Cubic
a = 1.347970	a = 0.855151	a = 1.347980	a = 0.855176
b = 0.06197	b = −0.01791	b = 0.06173	b = −0.01870
c = 0.01071	c = 0.04443	c = 0.01223	c = 0.05173
		d = −0.0026	d = −0.01904

Values of η from interpolation formulas

m	s − Quadratic	p − Quadratic
6	1.35086	0.85454
7	1.34992	0.85471

(c) Potassium

Energy levels, $\eta(n)$, and interpolation coefficients

Energy levels

m	s	p	d	f
3	−3.93	−2.34	−0.1227898	
4	−0.3190371	−0.2003552	−0.0693696	−0.062715
5	−0.1274241	−0.0938238	−0.0439638	−0.040135
6	−0.0688848	−0.0547213	−0.0302014	−0.027856
7	−0.0431538	−0.0358754	−0.0219739	−0.020461
8	−0.0295695	−0.0253387	−0.0166894	−0.015662

m	$\eta(n)$	$\eta(n)$	$\eta(n)$	δ_m
3	2.496	1.695	0.266336	
4	2.229567	1.730406	0.26568	0.00686
5	2.198606	1.71978	0.26857	0.0084
6	2.189883	1.71633	0.27109	0.0084
7	2.18617	1.71469	0.27211	0.0090
8	2.18462	1.71386	0.27291	0.0096
9	2.18337			

Interpolation coefficients for $\eta(n)$

s − Quadratic	p − Quadratic	d − Quadratic[b]	s − Cubic	p − Cubic
a = 2.180059	a = 1.711928	a = 0.27691	a = 2.180112	a = 1.711987
b = 0.13915	b = 0.07617	b = −0.2475	b = 0.13779	b = 0.07416
c = 0.0502	c = 0.0801	c = 1.3145	c = 0.0600	c = 0.1002
			d = −0.0190	d = −0.0577

[b] These coefficients have been obtained from the experimental values of $\eta(n)$ at $m = 3,5,7$. The formula obtained with the usual choice of $m = 3,4,5$ is in less satisfactory over-all agreement with the experimental values.

TABLE II. UNMODIFIED QUANTUM DEFECT DATA FOR THE ALKALI METALS[a]

(c) Potassium

(*Continued*)

Values of $\eta(n)$ from interpolation formulas

m	s – Quadratic	p – Quadratic	d – Quadratic[b]
3			exact
4			0.26607
5			exact
6			0.27063
7	2.18616	1.71476	exact
8	2.18422	1.71391	0.27315
9	2.18308		

(d) Rubidium

Energy levels, $\eta(n)$, and interpolation coefficients

Energy levels

m	s	p	d	f
3			−10.0	
4	−3.39	−2.04	−0.1306385	−0.0628691
5	−0.307017	−0.1909450	−0.0727987	−0.0402185
6	−0.123545	−0.0904362	−0.0455865	−0.0279146
7	−0.067248	−0.0531507	−0.0310738	−0.0204991
8	−0.042321	−0.0350247	−0.0224988	−0.0156884
9	−0.029087	−0.0248282	−0.0170291	
10	−0.021220	−0.0185195	−0.0133324	
	$\eta(n)$	$\eta(n)$	$\eta(n)$	δ_m
3			0.999	
4	3.457	2.706	1.362468	0.01176
5	3.19524	2.679816	1.35303	0.01360
6	3.15497	2.661585	1.35044	0.01473
7	3.1438	2.65514	1.34933	0.01554
8	3.1390	2.65198	1.34882	0.01618
9	3.1365	2.65033	1.34855	
10	3.1352	2.64932	1.34842	

Interpolation coefficients for $\eta(n)$

s – Quadratic	p – Quadratic	d – Quadratic
a = 3.13119	a = 2.646250	a = 1.348754
b = 0.18164	b = 0.163965	b = 0.000532
c = 0.08790	c = 0.06192	c = 0.79947
s – Cubic	**p – Cubic**	**d – Cubic**
a = 3.13130	a = 2.646318	a = 1.348790
b = 0.17873	b = 0.161570	b = −0.001015
c = 0.1096	c = 0.08678	c = 0.81997
d = −0.0435	d = −0.07434	d = −0.08234

TABLE II. UNMODIFIED QUANTUM DEFECT DATA FOR THE ALKALI METALS[a]

(d) Rubidium

(Continued)

Values of $\eta(n)$ from interpolation formulas

m	s – Quadratic	p – Quadratic	d – Quadratic
7			1.34954
8	3.1390	2.65207	1.34917
9	3.1365	2.65036	1.34899
10	3.1351	2.64931	1.34890

(e) Cesium

Energy levels, $\eta(n)$, and interpolation coefficients

Energy levels

m	s	p	d	f
4			−7.3	−0.063177
5	−2.83	−1.73	−0.153516	−0.040397
6	−0.286181	−0.181072	−0.080100	−0.028020
7	−0.117271	−0.086739	−0.048698	−0.020561
8	−0.064589	−0.051410	−0.032675	−0.015724
9	−0.040961	−0.034066	−0.023427	
10	−0.028300	−0.024243		

m	$\eta(n)$	$\eta(n)$	$\eta(n)$	δ_m
4			1.996	0.02148
5	4.406	3.710	2.4828	0.02461
6	4.13070	3.6258	2.4791	0.02595
7	4.07985	3.5961	2.4758	0.02605
8	4.06522	3.5852	2.4730	0.02527
9	4.0590	3.5793	2.4704	
10	4.0556	3.5757		

Interpolation coefficients for $\eta(n)$

s – Quadratic	p – Quadratic	d – Quadratic
a = 4.04808	a = 3.5696	a = 2.4686
b = 0.2586	b = 0.3019	b = 0.1742
c = 0.1054	c = 0.0479	c = −0.532

s – Cubic	p – Cubic	d – Cubic
a = 4.04822	a = 3.5697	a = 2.4686
b = 0.2548	b = 0.2983	b = 0.1756
c = 0.1353	c = 0.0871	c = −0.552
d = −0.0638	d = −0.123	d = 0.071

Values of $\eta(n)$ from interpolation formulas

m	s – Quadratic	p – Quadratic	d – Quadratic
8			2.4737
9	4.0588	3.5799	2.4724
10	4.0555	3.5769	

TABLE III. POLARIZATION CORRECTIONS TO QUANTUM DEFECT DATA FOR THE ALKALI METALS*

(a) Lithium

	$\Delta\nu(n)$	
ϵ	s^a	p^a
0.0		0.00061
−0.1	0.00063	
−0.2		0.00098
−0.3	0.00087	
m	d^b	f^b
3	0.0015	
4	0.0016	0.00024
5	0.0017	0.00026
6	0.0018	0.00028
7		0.00029

Interpolation coefficients for $\Delta\eta(n)$

s	p
$\alpha =$ 0.00050	$\alpha =$ 0.00061
$\beta =$ 0.0013	$\beta =$ 0.0030
$\gamma =$ −0.00025	

r_i = 3.10 a.u., κ = 0.074, α' = 0.099

(b) Sodium

	$\Delta\nu(n)$	
ϵ	s^a	p^a
0.0		0.00379
−0.1	0.00387	
−0.2		0.00605
−0.3	0.00539	
m	d^b	f^b
3	0.0090	
4	0.0102	0.0015
5	0.0107	0.0016
6	0.0110	0.0017
7	0.0112	0.0018

Interpolation coefficients for $\Delta\eta(n)$

s	p
$\alpha =$ 0.00307	$\alpha =$ 0.00379
$\beta =$ 0.0082	$\beta =$ 0.0171
$\gamma =$ −0.0014	$\gamma =$ −0.0050

r_i = 3.10 a.u., κ = 0.457, α' = 0.611

* The corrections $\Delta\nu(n)$ have been obtained by:
[a] Expanding (8.1) in powers of α' and evaluating the term linear in α'.
[b] Evaluating formula (10.3).
The entries in these tables are labeled according to the procedure used in calculating them.

TABLE III. POLARIZATION CORRECTIONS TO QUANTUM DEFECT DATA FOR THE ALKALI METALS* (*Continued*)

(c) Potassium

$\Delta\nu(n)$

ϵ	s^a	p^a
0.0		0.0076
−0.1	0.0086	
−0.2		0.0136
−0.3	0.0124	

ϵ	d^a	m	f^b
0.0	0.0103	4	0.0067
−0.08	0.0156	5	0.0075
		6	0.0080
		7	0.0082

Interpolation coefficients for $\Delta\eta(n)$

s	p	d
$\alpha =$ 0.0065	$\alpha =$ 0.0076	$\alpha =$ 0.0103
$\beta =$ 0.0213	$\beta =$ 0.0322	$\beta =$ 0.0838
$\gamma =$ −0.0059	$\gamma =$ −0.0152	

$r_i = 4.25$ a.u., $\kappa = 2.12$, $\alpha' = 2.83$

(d) Rubidium

$\Delta\nu(n)$

ϵ	s^a	p^a
0.0		0.0128
−0.1	0.0146	
−0.2		0.0230
−0.3	0.0209	

ϵ	d^a	m	f^b
0.0	0.0174	4	0.0113
−0.08	0.0263	5	0.0127
		6	0.0135
		7	0.0139

Interpolation coefficients for $\Delta\eta(n)$

s	p	d
$\alpha =$ 0.0111	$\alpha =$ 0.0128	$\alpha =$ 0.0174
$\beta =$ 0.0362	$\beta =$ 0.0468	$\beta =$ 0.0605
$\gamma =$ −0.0116	$\gamma =$ −0.0260	$\gamma =$ −0.0062

$r_i = 4.25$ a.u., $\kappa = 3.57$, $\alpha' = 4.77$

TABLE III. POLARIZATION CORRECTIONS TO QUANTUM DEFECT DATA FOR THE ALKALI METALS* (*Continued*)

(e) Cesium

s^a		p^a		d^a	
ϵ	$\Delta\nu(n)$	ϵ	$\Delta\nu(n)$	ϵ	$\Delta\nu(n)$
−0.28618	0.0244	−0.18107	0.0266	−0.02343	0.0206
−0.11727	0.0187	−0.02424	0.0158	−0.08010	0.0279

f

ϵ	$\Delta\nu(n)^b$
−0.06318	0.0195
−0.04040	0.0219
−0.02802	0.0231
−0.02056	0.0239
−0.01572	0.0244

Interpolation coefficients for $\Delta\eta(n)$

	s		*p*		*d*
$\alpha =$	0.0130	$\alpha =$	0.0142	$\alpha =$	0.0185
$\beta =$	0.0450	$\beta =$	0.0566	$\beta =$	−0.0107
$\gamma =$	−0.0175	$\gamma =$	−0.0375	$\gamma =$	0.0011

$r_i = 4.95$ a.u., $\kappa = 6.15$, $\alpha' = 8.22$

XI. Results of Band Calculations with QDM for the Alkali Metals

In Table IV, we list the values of the ground-state energy ϵ_0 and the parameters α and γ which appear in Bardeen's theory (Section IV). They have been calculated by Brooks for several values of the equivalent sphere radius r_s using QDM and the Wigner-Seitz spherical approximation. Unfortunately, at the time at which this article is being prepared the revised values of these quantities obtained from the up-to-date form of QDM presented here are not available.[78a] Consequently we list the results of Brooks's earlier work,[5] in which the quantum defect $\nu(n)$ was extrapolated for $L = 1$ instead of the η-defect $\eta(n)$ and in which core eigenvalues and polarization corrections were not used. We expect that the revised calculations using these modifications will yield values differing slightly from these but that the differences will not be great.

We also list in Table IV the values of ϵ_0, α, and γ obtained by Kuhn and Van Vleck using the extrapolated logarithmic derivatives.[4] Brooks found that his values of ϵ_0 for sodium did not differ significantly from

[78a] The revision of the calculations is being carried out by H. Brooks, and his results will be published in *The Physical Review*.

TABLE IV. VALUES OF THE GROUND-STATE ENERGY ϵ_0 AND THE PARAMETERS α AND γ FOR THE ALKALI METALS

Element	Author	r_s	$z_s = (8r_s)^{\frac{1}{2}}$	$-\epsilon_0$	γ	α
Li	Brooks[a]		4.5	0.6809	1.197	0.471
			5.0	0.6927	1.052	0.674
			5.5	0.6411	0.993	0.819
Na	Brooks[a]		5.0	0.6644	1.206	1.060
			5.5	0.6235	1.046	1.017
			6.0	0.5666	0.935	1.015
K	Brooks[a]		5.5	0.5281	1.272	1.114
			6.0	0.5070	1.108	1.074
			6.5	0.4679	0.986	1.057
Rb	Brooks[a]		6.0	0.4933	1.182	1.129
			6.5	0.4584	1.036	1.099
			7.0	0.4202	0.901	1.073
Cs	Brooks[a]		6.0	0.4703	1.372	1.383
			6.5	0.4418	1.156	1.229
			7.0	0.4067	1.027	1.188
Na	Kuhn and Van Vleck[b]		5.0	0.6644	1.120	0.962
			5.5	0.6235	1.020	0.980
			6.0	0.5666	0.930	1.013
K	Kuhn and Van Vleck[b]		5.5	0.5281	0.835	0.745
			6.0	0.5070	0.930	0.909
			6.5	0.4679	1.043	1.122
Rb	Kuhn and Van Vleck[b]		6.0	0.4933	1.011	0.132
			6.5	0.4584	0.964	0.427
			7.0	0.4202	1.153	0.781
Li	Silverman and Kohn[c]	3.2094	5.0671	0.6832		0.7270
Li	Kohn and Rostoker[d]	3.2094	5.0671	0.6832		0.723
Na	Bardeen[e]	3.16	5.028	0.6619	1.1294	1.1392
		3.80	5.514	0.6225	1.0287	1.079
		4.44	5.960	0.5722	0.9305	1.041

[a] H. Brooks, *Phys. Rev.* **91,** 1027 (1953), and unpublished work (see text and footnote 80[a] of Section XI).

[b] T. S. Kuhn and J. H. Van Vleck, *Phys. Rev.* **79,** 382 (1950); J. H. Van Vleck, *Proc. 1953 Intern. Conf. Theoret. Phys., Tokyo* p. 640 (1954).

[c] R. A. Silverman and W. Kohn, *Phys. Rev.* **80,** 912 (1950).

[d] W. Kohn and N. Rostoker, *Phys. Rev.* **94,** 1111 (1954).

[e] J. Bardeen, *J. Chem. Phys.* **6,** 367, 372 (1938).

those of Kuhn and Van Vleck, and he consequently used these authors' values of ϵ_0 for sodium, potassium, and rubidium in his calculations. Brooks's values of α are different from those of Kuhn and Van Vleck, partly because his extrapolation of $\nu(n)$ for $L = 1$ is more reliable than their use of the extrapolated logarithmic derivative, and partly because Kuhn and Van Vleck used an approximate evaluation of the normalization integral in (4.5), instead of formula (4.6) suggested subsequently by Silverman. Also, we give values of the parameters obtained by Bardeen,[12] Silverman and Kohn,[79] and Kohn and Rostoker[14] on the basis of the explicit potential functions derived by Prokofjew for sodium and Seitz for lithium. All authors but Kohn and Rostoker employed the spherical approximation. These values should agree reasonably well with the QDM results, as they do, for the valence-core interaction in the lighter elements can be represented accurately by a single potential function. This is not the case for the heavy elements.

In Table V we compare theoretical values of the equilibrium lattice constant and cohesive energy, derived for zero pressure and the absolute zero of temperature by various authors, with the values obtained from experiment. We also give the theoretical pressure at 0°K. corresponding to the compression which is observed experimentally at 10,000 atmos.[80] The values attributed to Brooks have been calculated from the data of Table IV, and appropriate plane wave values of the Coulomb, exchange, and correlation energy for the mutual interaction of the valence electrons.[80a] The total energy has been fitted to the expression used by Bardeen[12]

$$E_{\text{coh}} = Ar_s^{-1} + Br_s^{-2} + Cr_s^{-3}, \tag{11.1}$$

the constants being determined from the theoretical data at three values of r_s near the equilibrium radius r_{s0}. This expression has then been minimized to obtain r_{s0}, the equilibrium value of r_s, as well as the corresponding lattice constant and cohesive energy. From the formula $P = -\left(\frac{\partial E}{\partial V}\right)_p$ and (11.1), one obtains an expression for the pressure as a function of r_s/r_{s0}. The values given in Table V have been calculated for the values of r_s/r_{s0} observed by Swenson[80] at 4.2°K for a pressure of 10,000 atmos-

[79] R. A. Silverman and W. Kohn, *Phys. Rev.* **80,** 912 (1950).

[80] C. A. Swenson, *Phys. Rev.* **99,** 423 (1955).

[80a] The values for the theoretical pressure were calculated by the author using data for the constants in Eq. (11.1) kindly supplied by Brooks. The values of the lattice constants and cohesive energies of Li, K, and Rb were given by Brooks,[6] and those for Na and Cs were calculated by Brooks in an unpublished revision of that work. In these revised calculations, account was taken of errors discovered[7,34] in Kuhn's tables.[30]

TABLE V. COMPARISON OF THEORETICAL AND EXPERIMENTAL VALUES OF EQUILIBRIUM LATTICE CONSTANT, COHESIVE ENERGY, AND COMPRESSIBILITY FOR THE ALKALI METALS (0°K)

Element	Author	Lattice Constant (in Å)	Cohesive Energy (in kcal/mole)	$(r_s/r_{s0})_{10,000}$[a]	Pressure[b] (atmos.)
Li	Brooks[c]	3.40	38.7		10,500
	Silverman and Kohn[d]		35.7		
	Seitz[e]	3.52	33.8		
	Experiment[f]	3.43	36.5	.9792	10,000
Na	Brooks[c]	4.26	26.3		9,000
	Kuhn and Van Vleck[g]	4.14	25.9		
	Bardeen[h]	4.53	23.0		
	Wigner and Seitz[i]	4.74	23.2		
	Experiment[f]	4.22	26.0	.9637	10,000
K	Brooks[c]	5.16	22.2		10,500
	Kuhn and Van Vleck[g]	4.51	27.9		
	Gorin[j]	5.8	14.5		
	Experiment[f]	5.20	22.6	.9371	10,000
Rb	Brooks[c]	5.45	20.8		7,200
	Kuhn and Van Vleck[g]	5.20	24.2		
	Experiment[f]	5.56	18.9	.9299	10,000
Cs	Brooks[c]	5.93	19.8		13,100
	Experiment[f]	5.92	18.8	.9033	10,000

[a] Ratio of observed r_s at 10,000 atmospheres to the value (r_{s0}) observed at zero pressure, 4.2°K. [C. A. Swenson, *Phys. Rev.* **99**, 423 (1955).]

[b] Pressure required to compress lattice to observed value of $(r_s/r_{s0})_{10,000}$.

[c] H. Brooks, *Phys. Rev.* **91**, 1027 (1953), and unpublished work (see text and footnote 80[a] of Section XI).

[d] R. A. Silverman and W. Kohn, *Phys. Rev.* **80**, 912 (1950).

[e] F. Seitz, *Phys. Rev.* **47**, 400 (1935).

[f] Experimental values of lattice constant and $(r_s/r_{s0})_{10,000}$ are from Swenson (see footnote *a*); those for the cohesive energy have been taken from Brooks (see footnote *c*).

[g] T. S. Kuhn and J. H. Van Vleck, *Phys. Rev.* **79**, 382 (1950); J. H. Van Vleck, *Proc. 1953 Intern. Conf. Theoret. Phys., Tokyo* p. 640 (1954).

[h] J. Bardeen, *J. Chem. Phys.* **6**, 367, 372 (1938).

[i] E. Wigner and F. Seitz, *Phys. Rev.* **43**, 804 (1933); **46**, 509 (1934).

[j] E. Gorin, *Physik. Z. Sowjetunion* **9**, 328 (1936).

pheres. The experimental values of the lattice constant are due to Swenson; the values of the cohesive energy have been collated previously by Brooks.[5] The agreement between Brooks's theoretical results and experiment is fairly good. The results for the heavier alkalies are in much better agreement than those found in the earlier theories. This is to be expected, of course, since QDM takes more general account of the valence core interaction than is possible with a simple potential.[80b]

Table VI contains the results of preliminary work by the author using QDM to determine eigenvalues at points of symmetry on the

TABLE VI. EIGENVALUES OF SODIUM

Wave function	Ham[a] (a = 8.048 a.u.)		Howarth-Jones[b] (a = 8.138 a.u.)	
	Sodium	Empty lattice	Sodium	Empty lattice
H_s	+0.136 ± 0.005	−0.001	+0.0997	−0.012
H_d	+0.006 ± 0.02	−0.001	+0.0172	−0.012
H_p	−0.018 ± 0.01	−0.001	−0.0135	−0.012
N_s	−0.287 ± 0.01	−0.305	−0.315	−0.310
N_p	−0.292 ± 0.01	−0.305	−0.268	−0.310
Γ_s	−0.6097[c]	−0.6097	−0.608	−0.608

[a] F. S. Ham, Ph.D. thesis, Harvard University, Cambridge, Mass., 1954 (unpublished).
[b] D. J. Howarth and H. Jones, *Proc. Phys. Soc.* **A65**, 355 (1952).
[c] From Brooks's values of ϵ_0 (Table IV).

surface of the Brillouin zone for sodium. This work has been carried out using the Howarth-Jones procedure (Section IV) of satisfying various combinations of point boundary conditions on the surface of the polyhedral atomic cell and correcting the approximate eigenvalues so obtained by means of the correction integral method. The corresponding eigenvalues obtained by Howarth and Jones,[13] using the explicit Prokofjew potential, are also included in Table VI, as are the values for the empty lattice. The zero in the energy scale of the latter has been adjusted so

[80b] The revised calculations using the form of QDM presented in this paper and the recent values of correlation and exchange energy calculated by Pines[24] undoubtedly will change slightly the theoretical values of Table V. The values of the theoretical pressure will be altered the most, for these values are particularly sensitive to small changes in the constants in Eq. (11.1). Hence in Table V the theoretical pressure values are probably significant only to the extent that they are correct in order of magnitude. A more exacting comparison of theory and experiment must await the completion of the revised calculations. These will include the contribution of the k^4 term in the energy and will make use of a more accurate expression for the pressure than that derived from (11.1), which is not very satisfactory from either a theoretical or experimental[80] point of view.

that the ground-state energy [illegible] (Γ_s) coincides with the value for sodium obtained from Brooks's work (Table IV). The notation for the different states is the same as that used by Howarth and Jones, H, N, and Γ denoting the points $k = (2\pi/a)(1,0,0)$, $(2\pi/a)(\frac{1}{2},\frac{1}{2},0)$, and $(0,0,0)$, respectively. The subscript indicates the transformation character of the associated wave function. Due to an oversight, the author's calculations were made for a lattice parameter a slightly different from that used by Howarth and Jones; however, the eigenvalues which would be obtained from QDM for the Howarth-Jones value probably differ from the values given by not more than 0.01 Rydberg units. This is the difference in the empty lattice eigenvalues in the two cases. Significant differences between the author's values and those of Howarth and Jones occur for H_s, N_s, and N_p. The differences in the values at N result from our use of the correction integral in obtaining the eigenvalue, and not from differences in the wave functions obtained from QDM and from numerical integration using the Prokofjew potential. Howarth and Jones took as their final eigenvalue the mean of a number of uncorrected eigenvalues obtained from the point boundary conditions. An identical averaging procedure, using QDM and the same sets of boundary conditions, gave the values -0.274 for N_p and -0.319 for N_s, in excellent agreement with Howarth and Jones. The difference at H_s may be of similar origin in part; however, it appears to be largely the result of differences in the radial wave functions. The indicated possible error in the QDM eigenvalues is an estimate based upon the scatter of the several determinations of each eigenvalue. Evidently, the agreement between the present work and that of Howarth and Jones is good. This was to be expected since the Prokofjew potential reproduces the sodium eigenvalues to an accuracy of 1%. The principal differences arise from the method used to obtain the eigenvalues from the boundary conditions. The values of Table VI cannot be compared at this time with experimental data. It remains to be seen how sensitive the values are to slight changes in the potential near the surface of the cell. The author hopes to make such a study in the near future. The results indicate, however, that QDM is a useful tool in the study of the band structure of the heavier alkali metals and is by no means limited to calculations using the spherical approximation.

Other applications of QDM are currently in progress and will be reported when they are completed. In addition to work on the divalent and trivalent metals initiated by Brooks, we should mention his calculation of the g factor and spin-lattice relaxation time for the alkali metals[81] taking spin-orbit coupling into account. The results of this work are being revised to include the modifications of QDM which have occurred

[81] H. Brooks, *Phys. Rev.* **94**, 1411 (1954).

since the calculation was started. We should also mention a proposal by Brooks (unpublished) whereby QDM can be extended by use of the WKB approximation, "corrected," as in Section VI, by comparison with exact Coulomb wave functions. This procedure can yield the ratio of the amplitude of the wave function very near the nucleus to that outside the core if data on atomic hyperfine splitting is available or if the constant term in the one-electron potential is known approximately, at very small radius, from a Hartree-Fock calculation. It is anticipated that this proposal will be applied to a calculation of the Knight shift in the alkali metals in the near future.

XII. Acknowledgment

The author would like to express his gratitude to Professor Harvey Brooks for the interest he has taken in this work and the many suggestions and criticisms he has offered concerning it. The author is indebted to Professor Brooks for suggesting the preparation of this article as an expanded version of the joint paper[8] to be published in *The Physical Review*. The author would like to thank both Professor Brooks and Professor Frederick Seitz for their criticisms of the manuscript, and also Dr. Swenson and Dr. Kambe for making available the results of their work prior to publication.

The author would also like to thank the National Research Council for the award of a postdoctoral fellowship, during the tenure of which this article has been prepared.

Appendix: Solutions of the Hydrogenic Wave Equation[82]

The hydrogenic radial wave equation

$$\frac{d^2U^L}{dr^2} + \left[-\frac{1}{n^2} + \frac{2}{r} - \frac{L(L+1)}{r^2} \right] U^L = 0 \tag{A.1}$$

has solutions in the form

[82] This section is based upon Kuhn's analysis of the Coulomb wave functions[30] and upon a recent paper by the author[33] which corrects an oversight in Kuhn's work. The second paper develops both convergent and asymptotic series in $(1/n^2)$ for the irregular functions. The reader is referred to the papers for proofs of statements made above.

Tables of the coefficients $U_k^{(L)}(z)$ in (A.2) have been prepared by the author for the series expansions of $(z/2)J_{2L+1}{}^n(z)$ and $(z/2)N_{2L+1}{}^n(z)$. These are valid for $L = 0$ and for values of z of interest in band calculations. The tables, together with recurrence relations connecting functions of different L, make possible accurate and relatively simple computations of the Coulomb functions needed in such work.[34] The tables are more detailed than those published by Kuhn. They may be obtained from the author or from Cruft Laboratory, Harvard University.

$$U^{L,n}(r) = \sum_{k=0}^{\infty} n^{-2k} U_k^{(L)}(z), \tag{A.2}$$

where $z = (8r)^{\frac{1}{2}}$. In particular, Kuhn has shown that one such function which vanishes at $z = 0$, if $Re(L) > -1$, is

$$^0U^{L,n}(r) = (z/2)J_{2L+1}{}^n(z) = (z/2) \sum_{k=0}^{\infty} n^{-2k} \sum_{q=2k}^{3k} a_{kq}(L)(z/2)^q J_{2L+1+q}(z). \tag{A.3}$$

Here the coefficients $a_{kq}(L)$ are functions of the (complex) variable L which can be determined, and $J_{2L+1+q}(z)$ is the Bessel function of the first kind of order $(2L + 1 + q)$. A second solution of (A.1), which is independent of $(z/2)J_{2L+1}{}^n(z)$ for all values of L, has been defined by Wannier:

$$(z/2)N_{2L+1}{}^n(z) = \frac{\dfrac{\Gamma(n + L + 1)}{n^{2L+1}\Gamma(n - L)} J_{2L+1}{}^n(z) \cos \pi(2L + 1) - J_{-2L-1}{}^n(z)}{\sin \pi(2L + 1)}, \tag{A.4}$$

where, for an integral value of $(2L + 1)$, $(z/2)N_{2L+1}{}^n(z)$ is defined by the limit of the expression on the right as $2L + 1 \rightarrow$ integer. The functions satisfy the Wronskian relation

$$[(z/2)J_{2L+1}{}^n(z)]\left[(z/2)\frac{d}{dz}(z/2)N_{2L+1}{}^n(z)\right] - [(z/2)N_{2L+1}{}^n(z)]\left[(z/2)\frac{d}{dz}(z/2)J_{2L+1}{}^n(z)\right] = z^2/4\pi. \tag{A.5}$$

The quantity $(z/2)N_{2L+1}{}^n(z)$, which is irregular at $z = 0$, cannot be expanded in a convergent series of the form (A.2); however, it has the following asymptotic expansion, valid for $|\arg(n)| \leq \pi - \Delta < \pi$ when $|n| \rightarrow \infty$:

$$\frac{\Gamma(n - L)n^{2L+1}}{\Gamma(n + L + 1)} N_{2L+1}{}^n(z) \sim \sum_{k=0}^{\infty} n^{-2k} \sum_{q=2k}^{3k} a_{kq}(L)(z/2)^q Y_{2L+1+q}(z) \tag{A.6}$$

Here $Y_{2L+1+q}(z)$ is the Bessel function of the second kind of order $(2L + 1 + q)$. If L equals a positive integer or zero, however, we can conveniently define a second irregular function by the relation

$$(z/2)Q_{2L+1}{}^n(z,L) = (z/2)N_{2L+1}{}^n(z) - G_L(L,n)(z/2)J_{2L+1}{}^n(z), \tag{A.7}$$

where

$$G_L(L,n) = \frac{1}{2\pi}\left\{\frac{\Gamma(n+L+1)}{n^{2L+1}\Gamma(n-L)}[\Psi(n+L+1)+\Psi(n-L)-2\ln n] - \sum_{p=0}^{L} n^{-2p}\left[\frac{d}{dx}b_p(x)\right]_{x=L}\right\}. \quad \text{(A.8)}$$

The function $\Psi(x) = [(d/dx)\ln\Gamma(x)]$ is the logarithmic derivative of the gamma function. The quantities $b_p(L)$ are polynomials in L appearing in the asymptotic expansion, in terms of n, valid for $|\arg(n)| \leq \pi - \Delta < \pi$ and arbitrary L:

$$\frac{\Gamma(n+L+1)}{n^{2L+1}\Gamma(n-L)} \underset{|n|\to\infty}{\sim} \sum_{p=0}^{\infty}\frac{b_p(L)}{n^{2p}}. \quad \text{(A.9)}$$

The function $Q_{2L+1}{}^n(z,L)$ can be expanded in a convergent power series of the form (A.2). This is more complicated than (A.3), however, and we shall not give it here. Moreover, $Q_{2L+1}{}^n(z,L)$ clearly satisfies the Wronskian relation (A.5) with $J_{2L+1}{}^n(z)$.

The various functions of this section may be expressed in terms of standard confluent hypergeometric[57] functions by means of the relations

$$(z/2)J_{2L+1}{}^n(z) = \frac{n^{L+1}}{\Gamma(2L+2)}M_{n,L+\frac{1}{2}}(z^2/4n), \quad \text{(A.10)}$$

$$(z/2)N_{2L+1}{}^n(z) = \frac{n^{-L}e^{i\pi n}}{\pi}\{e^{-i\pi(L-\frac{1}{2})}\sin\pi(n-L)\Gamma(L+1-n)W_{n,L+\frac{1}{2}}(z^2/4n) - \Gamma(L+1+n)\cos\pi(n-L)W_{-n,L+\frac{1}{2}}(e^{i\pi}z^2/4n)\}. \quad \text{(A.11)}$$

Also,

$$W_{n,L+\frac{1}{2}}(z^2/4n) = \{\Gamma(n+L+1)n^{-L-1}(z/2)J_{2L+1}{}^n(z)\cos\pi(n-L-1) + \Gamma(n-L)n^L(z/2)N_{2L+1}{}^n(z)\sin\pi(n-L-1)\}. \quad \text{(A.12)}$$

Asymptotic expansions of the functions, valid for large absolute value of $(z^2/4n)$, may be obtained from these relations and the asymptotic formulas for the Whittaker functions.[62]

1958 *Mon. Not. R. Astron. Soc.* **118** 504–18

THE QUANTUM DEFECT METHOD

M. J. Seaton

(Received 1958 May 27)

Summary

The Quantum Defect Method uses interpolated or extrapolated quantum defects to determine the asymptotic forms of atomic wave functions. The method may be used in the calculation of atomic transition probabilities and photo-ionization cross sections, in electron-ion collision calculations and also in connection with solid state problems.

The paper gives a summary of previous work on the fundamental ideas of the method and presents some new results for positive energy states and for the normalization of bound-state wave functions. Some applications of the method are discussed.

Introduction.—When calculating atomic properties one may consider semi-empirical methods which make use of known energy-level data. One approach is to construct potentials which will reproduce a number of observed levels (**1, 2, 3**) but this has the disadvantage that each individual case has to be treated separately and that the calculations may be very laborious. Hylleraas (**4**) and Bates and Damgaard (**5**) have developed more general methods for transition probability calculations. The transition integrals are tabulated directly as functions of the effective quantum numbers* ν_{nl} defined by

$$T_{nl}=\frac{Rz^2}{\nu_{nl}^2}, \tag{1}$$

where T_{nl} is the energy, measured in cm^{-1}, required to remove the nl electron, $R=109\,737\ \mathrm{cm}^{-1}$ is the Rydberg wave-number and z is the residual charge ($z=1$ for neutral atoms).

The quantum defect μ_{nl}, defined by $\mu_{nl}=n-\nu_{nl}$, usually varies slowly in a spectral series and tends to a definite limit $\mu_{\infty l}$ at the spectral head. For collisions between electrons and positive ions one requires the phase differences between the wave functions for the electron in the ion field and the corresponding wave function in a pure Coulomb field. It has been shown previously (**6**) that the zero energy phase is equal to $\pi\mu_{\infty l}$. This result has been used in various electron–ion collision problems (**7, 8, 9, 10**).

Similar methods have been developed for the study of solid-state problems (Ham (**11**)); for these problems one requires the asymptotic forms of the wave functions for negative energies other than those corresponding to eigenvalues of the free atom. The use of interpolated or extrapolated quantum defects will be referred to as the Quantum Defect Method. In the present paper the general theory of the method is developed by considering a simplified mathematical problem. Some of our analysis is a summary of the work of Ham (**11**) and of some much

* For effective quantum numbers we use ν in place of the more familiar n^*.

earlier work by Hartree (12). These authors were concerned with the negative energy states. In the present paper new results are given for the positive energy states and also for the normalization of bound state wave functions. We discuss the use of the method in collision problems and also its practical limitations. In a later paper (Burgess and Seaton (13)) the Quantum Defect Method will be used for the calculation of photo-ionization cross-sections.

1. *Valence-electron states.*—We consider an N-electron atom with nuclear charge Z and put $z=(Z-N+1)$. An approximate radial function for a valence-electron may be obtained on solving an equation of the type

$$\left[\frac{d^2}{dr^2}-\frac{l(l+1)}{r^2}-V(r)+E\right]P(E,l;r)=0 \tag{2}$$

where r is in atomic units and E and V are in Rydberg units (13·60 eV or 109 737 cm^{-1}). The potential $V(r)$ is such that $rV(r)=-2Z$ for r small and $rV(r)=-2z$ for r large.

For $E<0$ solutions of (2) which are everywhere bounded and continuous exist only for the discrete eigenvalues E_{nl} of E. The quantum number n may be defined by stating that: n may take on the values $(l+1)$, $(l+2)$, ...; $n=(l+1)$ corresponds to the lowest eigenvalue; $E_{n'l}$ is greater than E_{nl} for $n'>n$. An equivalent definition is that the number of nodes in $P(E_{nl},l;r)$, excluding the origin and infinity, is $(n-l-1)$. We put $E_{nl}=-z^2/\nu_{nl}^2$ and $\mu_{nl}=n-\nu_{nl}$. Then $\mu_{nl}=0$ for the pure Coulomb field, $V(r)=-2z/r$.

For $E>0$ we put $E=k^2$. The solution $P(k^2,l;r)$ which is bounded at the origin will have asymptotic form

$$P(k^2,l;r)\underset{r\to\infty}{\sim}\text{const.}\times\sin\left[kr-\tfrac{1}{2}l\pi+\frac{z}{k}\ln(2kr)+\arg\Gamma\left(l+1-\frac{iz}{k}\right)+\delta_l\left(\frac{k^2}{z^2}\right)\right]. \tag{3}$$

The phase δ_l is zero for the pure Coulomb field.

2. *Coulomb functions.*—We consider the equation

$$\left[\frac{d^2}{dr^2}-\frac{\lambda(\lambda+1)}{r^2}+\frac{2z}{r}+E\right]y=0 \tag{4}$$

where λ is not necessarily an integer. Introducing

$$\rho=zr,\qquad \epsilon=\frac{E}{z^2}=-\frac{1}{\kappa^2} \tag{5}$$

we obtain

$$\left[\frac{d^2}{d\rho^2}-\frac{\lambda(\lambda+1)}{\rho^2}+\frac{2}{\rho}-\frac{1}{\kappa^2}\right]y=0. \tag{6}$$

For $\epsilon<0$ we put $\kappa=\nu$ with ν real and positive and for $\epsilon>0$ we put $\kappa=i\gamma$ with γ real and positive. Then $\gamma=z/k$.

Entire analytic functions.—Let $y(\kappa,\lambda;\rho)$ be a solution of (6). Following Ham (11) we shall say that $y(\kappa,\lambda;\rho)$ is an entire analytic function of $\epsilon=-1/\kappa^2$ if, for ρ infinite, $y(\kappa,\lambda;\rho)$ is a single-valued analytic function of ϵ in the entire ϵ-plane save for the point at infinity. Such a function may be represented by an expansion

$$y(\kappa,\lambda;\rho)=\sum_{p=0}^{\infty}\epsilon^p f_p(\lambda;\rho) \tag{7}$$

absolutely and uniformly convergent for $\epsilon \leqslant |\mathscr{E}| < \infty$, $\rho \leqslant |\mathscr{R}| < \infty$. Conversely, y will be an entire analytic function if such an expansion exists.

For the theory of the Quantum Defect Method we require, for the case of $\lambda = l = 0, 1, 2, \ldots$, two linearly independent solutions of (6) both of which are entire analytic functions of ϵ. We also require the asymptotic forms of these solutions for ρ large.

Solutions of the Coulomb equation.—We consider the following solutions of (6):

$$y_1(\kappa, \lambda; \rho) = \frac{\kappa^{\lambda+1}}{\Gamma(2\lambda+2)} M_{\kappa, \lambda+1/2}(2\rho/\kappa), \tag{8}$$

$$\begin{aligned} y_2(\kappa, \lambda; \rho) &= y_1(\kappa, -\lambda-1; \rho) \\ &= \frac{\kappa^{-\lambda}}{\Gamma(-2\lambda)} M_{\kappa, -\lambda-1/2}(2\rho/\kappa), \end{aligned} \tag{9}$$

$$y_3(\kappa, \lambda; \rho) = \frac{A(\kappa, \lambda)\cos\pi(2\lambda+1) y_1(\kappa, \lambda; \rho) - y_2(\kappa, \lambda; \rho)}{\sin\pi(2\lambda+1)}, \tag{10}$$

$$y_4(\kappa, \lambda; \rho) = y_3(\kappa, \lambda; \rho) - G(\kappa, \lambda) y_1(\kappa, \lambda; \rho), \tag{11}$$

$$y_5(\kappa, \lambda; \rho) = W_{\kappa, \lambda+1/2}(2\rho/\kappa). \tag{12}$$

The functions $M_{\kappa, \lambda+\lambda/2}$ and $W_{\kappa, \lambda+1/2}$ are defined by Whittaker and Watson (**14**) and

$$A(\kappa, \lambda) = \frac{\Gamma(\kappa+\lambda+1)}{\kappa^{2\lambda+1}\Gamma(\kappa-\lambda)}, \tag{13}$$

$$\begin{aligned} G(\kappa, \lambda) &= \frac{1}{2\pi}\frac{\partial}{\partial\lambda} A(\kappa, \lambda) \\ &= \frac{A(\kappa, \lambda)}{2\pi}\{\psi(\kappa+\lambda+1) + \psi(\kappa-\lambda) - 2\ln(\kappa)\} \end{aligned} \tag{14}$$

where $\psi(x) = d\,[\ln\Gamma(x)]/dx$.

The definition of y_1 may be written

$$y_1(\kappa, \lambda; \rho) = (2\rho)^{\lambda+1} e^{-\rho/\kappa} \Phi(\lambda+1-\kappa, 2\lambda+2; 2\rho/\kappa) \tag{15}$$

where

$$\Phi(\lambda+1-\kappa, 2\lambda+2; 2\rho/\kappa) = \sum_{\sigma=0}^{\infty} \frac{\Gamma(\lambda+1-\kappa+\sigma)\Gamma(2\lambda+2)}{\Gamma(\lambda+1-\kappa)\Gamma(2\lambda+2+\sigma)\sigma!}\left(\frac{2\rho}{\kappa}\right)^{\sigma} \tag{16}$$

The series for Φ and for $\exp(-\rho/\kappa)$ converge absolutely and uniformly for $|\rho/\kappa| \leqslant |B| < \infty$. Since $\Gamma(\lambda+1-\kappa+\sigma)/\Gamma(\lambda+1-\kappa)$ is a polynomial in κ of order σ one may express y_1 as a series of powers of $(1/\kappa)$. By Kummer's transformation (**14**), $y_1(-\kappa, \lambda; \rho) = y_1(\kappa, \lambda; \rho)$ and therefore only even powers of $(1/\kappa)$ will occur. Therefore y_1 is an entire analytic function of ϵ. In the same way one may show that y_2 is an entire analytic function of ϵ.

For $\lambda \neq l$ $(l = 0, 1, 2, \ldots)$, y_1 and y_2 provide two linearly independent solutions of (6) but on letting $\lambda \to l$ one obtains

$$y_2(\kappa, l; \rho) = -A(\kappa, l) y_1(\kappa, l; \rho). \tag{17}$$

Since

$$\lim_{|\kappa|\to\infty} y_1(\kappa, \lambda; \rho) = (2\rho)^{1/2} J_{2\lambda+1}(\sqrt{(8\rho)}), \tag{18}$$

$$\lim_{|\kappa|\to\infty} y_2(\kappa, \lambda; \rho) = (2\rho)^{1/2} J_{-2\lambda-1}(\sqrt{(8\rho)}), \tag{19}$$

$$\lim_{|\kappa|\to\infty} A(\kappa, \lambda) = 1, \tag{20}$$

in the limit of large κ (17) corresponds to the Bessel function relation

$J_{2l+1} = -J_{-2l-1}$. The Weber function is defined by

$$Y_{2\lambda+1} = \frac{\cos \pi(2\lambda+1) J_{2\lambda+1} - J_{-2\lambda-1}}{\sin \pi(2\lambda+1)} \tag{21}$$

and by the limit of (21) for $\lambda \to l$. Then $J_{2\lambda+1}$ and $Y_{2\lambda+1}$ are linearly independent for all λ. Wannier (**15**) introduced the function y_3 by analogy with (21). From equations (18) to (21) one obtains

$$\lim_{|\kappa| \to \infty} y_3(\kappa, \lambda; \rho) = (2\rho)^{1/2} Y_{2\lambda+1}(\sqrt{(8\rho)}). \tag{22}$$

Defining $y_3(\kappa, l; \rho)$ as the limit of equation (10) for $\lambda \to l$ one obtains

$$y_3(\kappa, l; \rho) = \frac{1}{2\pi} \left\{ A(\kappa, l) \frac{\partial}{\partial \lambda} y_1(\kappa, \lambda; \rho) + \frac{\partial}{\partial \lambda} y_2(\kappa, \lambda; \rho) \right.$$
$$\left. + y_1(\kappa, l; \rho) \frac{\partial}{\partial \lambda} A(\kappa, \lambda) \right\}_{\lambda = l}. \tag{23}$$

Since $y_1(\kappa, \lambda; \rho)$ and $y_2(\kappa, \lambda; \rho)$ are entire analytic functions for all λ, $\partial y_1/\partial \lambda$ and $\partial y_2/\partial \lambda$ are also entire analytic functions. Also, since

$$A(\kappa, l) = \frac{(\kappa+l)(\kappa+l-1)\ldots(\kappa-l)}{\kappa^{2l+1}} \tag{24a}$$

$$= \frac{[\kappa^2 - l^2][\kappa^2 - (l-1)^2]\ldots[\kappa^2 - 1]}{\kappa^{2l}} \tag{24b}$$

and $A(\kappa, 0) = 1$, $A(\kappa, l)$ is a polynomial in $1/\kappa^2$ of order l. But $\partial A(\kappa, \lambda)/\partial \lambda$ cannot be expressed as a convergent power series in $1/\kappa^2$ and therefore $y_3(\kappa, l; \rho)$ is not an entire analytic function of ϵ. The entire analytic function $y_4(\kappa, l; \rho)$ is obtained (**11**) on subtracting from $y_3(\kappa, l; \rho)$ the term in $\partial A/\partial \lambda$.

The required entire analytic functions are therefore $y_1(\kappa, l; \rho)$ and $y_4(\kappa, l; \rho)$. The expansions of these functions are discussed in the Appendix.

Properties of the function $G(\kappa, l)$.—It is convenient to put

$$G(\kappa, l) = \mathscr{G}(\kappa, l) + i\mathscr{H}(\kappa, l) \tag{25}$$

where $\mathscr{G}$ and $\mathscr{H}$ are real. From (14) $G(\nu, l)$ is real for ν real and positive: therefore

$$\mathscr{H}(\nu, l) = 0 \qquad (\nu \text{ real and positive}). \tag{26}$$

For γ real and positive $G(i\gamma, l)$ is complex: using

$$\psi(x) - \psi(1-x) = -\pi \cot(\pi x)$$

one obtains

$$\mathscr{H}(i\gamma, l) = \frac{A(i\gamma, l)}{[e^{2\pi\gamma} - 1]} \qquad (\gamma \text{ real and positive}). \tag{27}$$

Relations involving the Whittaker function.—It is convenient to express y_3 and y_4 in terms of y_1 and the Whittaker function y_5. Using the relations

$$W_{\kappa, \lambda+1/2} = \frac{\Gamma(-2\lambda-1)}{\Gamma(-\lambda-\kappa)} M_{\kappa, \lambda+1/2} + \frac{\Gamma(2\lambda+1)}{\Gamma(\lambda+1-\kappa)} M_{\kappa, -\lambda-1/2} \tag{28}$$

(**14**) valid for $\lambda \neq l$ and

$$\Gamma(x)\Gamma(1-x) = \frac{\pi}{\sin(\pi x)} \tag{29}$$

one obtains

$$y_5 = \frac{\pi}{\sin(2\pi\lambda)} \left\{ \frac{\kappa^{-\lambda-1} y_1}{\Gamma(-\lambda-\kappa)} - \frac{\kappa^{\lambda} y_2}{\Gamma(\lambda+1-\kappa)} \right\}. \tag{30}$$

We use (30) to eliminate y_2 in (10) and then let $\lambda \to l$. This gives

$$y_3(\kappa, l; \rho) = -A(\kappa, l) \cot(\pi\kappa) y_1(\kappa, l; \rho) - \frac{\Gamma(l+1-\kappa)}{\pi\kappa^l} y_5(\kappa, l; \rho) \tag{31}$$

and from (11)

$$y_4(\kappa, l; \rho) = -[A(\kappa, l) \cot(\pi\kappa) + G(\kappa, l)] y_1(\kappa, l; \rho) - \frac{\Gamma(l+1-\kappa)}{\pi\kappa^l} y_5(\kappa, l; \rho). \tag{32}$$

Asymptotic expansions.—In the limit of $\rho \to \infty$, $y_1(\kappa, l; \rho)$ ceases to be a single-valued function of κ. We therefore consider $\kappa = \nu + i\gamma$ with $\nu \geqslant 0$, $\gamma \geqslant 0$. The asymptotic expansion for y_1 is then (16)

$$y_1(\kappa, l; \rho) \underset{\rho \to \infty}{\sim} \kappa^{l+1} e^{-i\frac{1}{2}\pi(l-\kappa)} \left\{ \frac{(2\rho/\kappa)^{-\kappa} \exp[\rho/\kappa + i\frac{1}{2}\pi(l-\kappa)]}{\Gamma(l+1-\kappa)} - \frac{(2\rho/\kappa)^{\kappa} \exp[-\rho/\kappa - i\frac{1}{2}\pi(l-\kappa)]}{\Gamma(l+1+\kappa)} \right\}. \tag{33}$$

It is seen that $y_1(\nu, l; \rho) \to \infty$ for $\rho \to \infty$, except when $\nu = n = (l+1), (l+2), \ldots$; when $\nu = n$, $\Gamma(l+1-\nu)$ is infinite and $y_1(n, l; \rho) \to 0$ for $\rho \to \infty$.

Substitution of $\kappa = i\gamma$ in (33) gives

$$y_1(i\gamma, l; \rho) \underset{\rho \to \infty}{\sim} \frac{2\gamma^{l+1} e^{-\pi\gamma/2}}{|\Gamma(l+1-i\gamma)|} \sin(x) \tag{34}$$

with

$$x = \frac{\rho}{\gamma} - \tfrac{1}{2}l\pi + \gamma \ln(2\rho/\gamma) + \arg\Gamma(l+1-i\gamma). \tag{35}$$

Using (29) one may show that

$$|\Gamma(l+1-i\gamma)|^2 = \frac{\pi\gamma^{2l+1} A(i\gamma, l)}{\sinh(\pi\gamma)} \tag{36}$$

and therefore

$$y_1(i\gamma, l; \rho) \underset{\rho \to \infty}{\sim} \left(\frac{2\gamma}{\pi}\right)^{1/2} \left[\frac{1 - e^{-2\pi\gamma}}{A(i\gamma, l)}\right]^{1/2} \sin(x). \tag{37}$$

The asymptotic form of the Whittaker function is

$$y_5(\kappa, l; \rho) \underset{\rho \to \infty}{\sim} (2\rho/\kappa)^{\kappa} e^{-\rho/\kappa}. \tag{38}$$

Therefore, for all ν, $y_5(\nu, l; \rho) \to 0$ for $\rho \to \infty$. From (38) one obtains

$$\frac{\Gamma(l+1-i\gamma)}{\pi(i\gamma)^l} y_5(i\gamma, l; \rho) \underset{\rho \to \infty}{\sim} \left(\frac{2\gamma}{\pi}\right)^{1/2} \left[\frac{A(i\gamma, l)}{1 - e^{-2\pi\gamma}}\right]^{1/2} e^{ix} \tag{39}$$

and, using (25), (27), (32) and (36),

$$y_4(i\gamma, l; \rho) \underset{\rho \to \infty}{\sim} -\left(\frac{2\gamma}{\pi}\right)^{1/2} \left\{ \left[\frac{1 - e^{-2\pi\gamma}}{A(i\gamma, l)}\right]^{1/2} \mathscr{G}(i\gamma, l) \sin(x) + \left[\frac{A(i\gamma, l)}{1 - e^{-2\pi\gamma}}\right]^{1/2} \cos(x) \right\} \tag{40}$$

and therefore

$$[y_4(i\gamma, l; \rho) + \mathscr{G}(i\gamma, l) y_1(i\gamma, l; \rho)] \underset{\rho \to \infty}{\sim} -\left(\frac{2\gamma}{\pi}\right)^{1/2} \left[\frac{A(i\gamma, l)}{1 - e^{-2\pi\gamma}}\right]^{1/2} \cos(x). \tag{41}$$

3. *The modified Coulomb field.*—We consider

$$\left[\frac{d^2}{d\rho^2} - \frac{l(l+1)}{\rho^2} + u(\rho) + \frac{2}{\rho} - \frac{1}{\kappa^2}\right] \mathscr{Y}(\kappa, l; \rho) = 0. \tag{42}$$

Let ρ_0, ρ_1 be such that $0 < \rho_0$, $\rho_0 \leqslant \rho_1 < \infty$ and let $u(\rho)$ be such that: $\rho u(\rho)$ analytic for $\rho<\rho_0$; $u(\rho)$ piecewise continuous for $\rho_0<\rho<\rho_1$; $u(\rho)=0$ for $\rho>\rho_1$. Let $\mathscr{F}(\kappa,l;\rho)$ be a continuous solution of (42) satisfying

$$\mathscr{F}(\kappa,l;0)=0, \tag{43}$$

$$\mathscr{F}(\kappa,l;\rho)=y_1(\kappa,l;\rho)-\beta(\epsilon)y_4(\kappa,l;\rho) \qquad (\rho>\rho_1). \tag{44}$$

It is shown by Ham (11) that $\mathscr{F}(\kappa,l;\rho)$ will be an entire analytic function of $\epsilon=-1/\kappa^2$. Since y_1 and y_4 are entire analytic functions it follows that $\beta(\epsilon)$ will be an entire analytic function and from the fact that y_1 and y_4 are real for ϵ real it may be shown to follow that $\beta(\epsilon)$ will be real for ϵ real. Therefore $\beta(\epsilon)$ may be represented by an expansion

$$\beta(\epsilon)=\sum_{p=0}^{\infty} B_p\epsilon^p \tag{45}$$

convergent for $|\epsilon|\leqslant|\mathscr{E}|<\infty$ and the coefficients B_p will be real.

The negative energy states.—From (32) and (44)

$$\mathscr{F}(\kappa,l;\rho)=\{1+\beta(\epsilon)[A(\kappa,l)\cot(\pi\kappa)+G(\kappa,l)]\}y_1(\kappa,l;\rho) + \beta(\epsilon)\frac{\Gamma(l+1-\kappa)}{\pi\kappa^l}y_5(\kappa,l;\rho) \qquad (\rho>\rho_1) \tag{46}$$

and since $G(\nu,l)=\mathscr{G}(\nu,l)$,

$$\mathscr{F}(\nu,l;\rho)=\{1+\beta(\epsilon)[A(\nu,l)\cot(\pi\nu)+\mathscr{G}(\nu,l)]\}y_1(\nu,l;\rho) + \beta(\epsilon)\frac{\Gamma(l+1-\nu)}{\pi\nu^l}y_5(\nu,l;\rho) \qquad (\rho>\rho_1). \tag{47}$$

Let $\epsilon_n=-1/\nu_n{}^2$ be an eigenvalue of ϵ; then $\mathscr{F}(\nu_n,l;\rho)\to 0$ for $\rho\to\infty$. If $u(\rho)$ is not identically zero we may assume $\nu_n\neq n$ and therefore $y_1(\nu_n,l;\rho)\to\infty$ for $\rho\to\infty$. The condition $\mathscr{F}(\nu_n,l;\rho)\to 0$ for $\rho\to\infty$ then requires that the coefficient of y_1 should vanish in (47); this requirement is (11)

$$\beta(\epsilon_n)=-[A(\nu_n,l)\cot(\pi\nu_n)+\mathscr{G}(\nu_n,l)]^{-1} \tag{48}$$

or, in terms of the quantum defect defined by

$$\mu(\epsilon_n)=n-\nu_n, \tag{49}$$

it is

$$\beta(\epsilon_n)=[A(\nu_n,l)\cot(\pi\mu(\epsilon_n))-\mathscr{G}(\nu_n,l)]^{-1}. \tag{50}$$

Suppose all the eigenvalues ϵ_n to be known. Then $\beta(\epsilon_n)$ is known for an infinite sequence of values of ϵ_n and, in consequence of the possibility of the expansion (45), $\beta(\epsilon)$ is uniquely determined for all finite ϵ. A quantum defect $\mu(\epsilon)$ may therefore be defined for all finite ϵ by

$$\beta(\epsilon)=[A(\kappa,l)\cot(\pi\mu(\epsilon))-\mathscr{G}(\kappa,l)]^{-1} \tag{51}$$

together with the condition that $\mu(\epsilon)$ should be a continuous function and should satisfy (49) at the eigenvalues. Since $\mathscr{G}$ is real and $\beta(\epsilon)$ and $A(\kappa,l)$ are real for real ϵ it follows that* $\mu(\epsilon)$ will be real for real ϵ.

The η-defect.—Letting $\nu\to m$ where $m=0, 1, 2, \ldots l$ we find that $\mathscr{G}(\nu,l)$ remains finite but that $A(\nu,l)$ tends to zero. However $\beta(-1/\nu^2)$, as determined from the eigenvalue spectrum, does not normally tend† to $-1/\mathscr{G}(m,l)$. Therefore as $\nu\to m$, $\mu(-1/\nu^2)$ must tend to an integer in such a way that $A(\nu,l)\cot(\pi\mu)$

* The definition of $\mu(\epsilon)$ used by Ham (11) is obtained on replacing $\mathscr{G}$ by G in (51). For ϵ real and negative this is identical with our definition but for ϵ real and positive $\mu(\epsilon)$ as defined by Ham will be complex.

† This may be shown (11) by solving the radial equation for various functions $u(\rho)$.

remains finite (11). This is confirmed by examination of observed energy levels. Fig. 1 shows the values of μ for the *nd* series in K I and in Ca II; in the former case $\mu \rightarrow 0$ and in the latter $\mu \rightarrow 1$ as $\nu \rightarrow 2$.

Ham introduces the η-defect defined by

$$A(\kappa, l) \cot \pi\mu = \cot \pi\eta. \tag{52}$$

We consider that $\eta(\epsilon)$ is a continuous function and that $\eta(0)=\mu(0)$. Since η does not necessarily tend to an integer as $\nu \rightarrow m$ we may expect that η will usually vary more slowly than μ. Values of η are shown in Fig. 1.

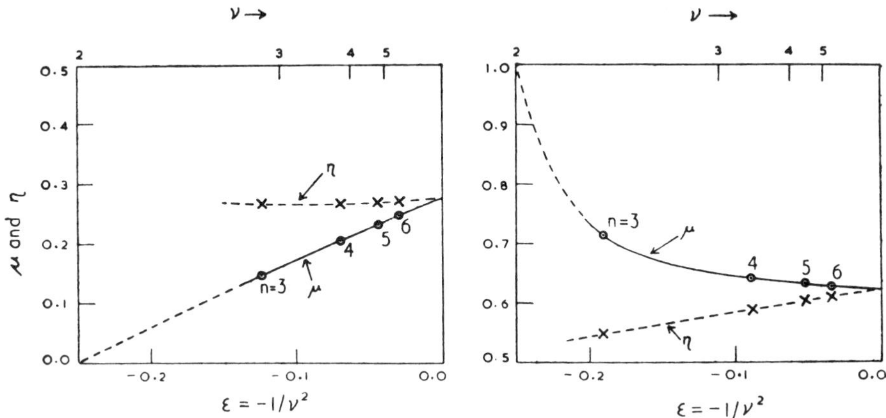

The nd series of K I. *The nd series of* Ca II

FIG. 1.—*Quantum defects and* η*-defects.*

The positive energy states.—Equation (44) may be written

$$\mathscr{F} = [1 + \beta\mathscr{G}]y_1 - \beta[y_4 + \mathscr{G}y_1] \qquad (\rho > \rho_1). \tag{53}$$

Using the asymptotic forms (37) and (41) we obtain

$$\mathscr{F}(i\gamma, l; \rho) \underset{\rho\to\infty}{\sim} \left(\frac{2\gamma}{\pi}\right)^{1/2} \left[\frac{1 - e^{-2\pi\gamma}}{A(i\gamma, l)}\right]^{1/2} \beta(\epsilon) \left\{\left[\frac{1}{\beta(\epsilon)} + \mathscr{G}(i\gamma, l)\right] \sin(x) + \left[\frac{A(i\gamma, l)}{1 - e^{-2\pi\gamma}}\right] \cos(x)\right\} \tag{54}$$

or

$$\mathscr{F}(i\gamma, l; \rho) \sim \left(\frac{\gamma}{\pi}\right)^{1/2} C(\epsilon) \sin[x + \delta(\epsilon)] \tag{55}$$

where

$$\tan\delta = \frac{[A/(1 - e^{-2\pi\gamma})]}{[(1/\beta) + \mathscr{G}]} \tag{56}$$

and

$$C = (2)^{1/2} \left[\frac{1 - e^{-2\pi\gamma}}{A}\right]^{1/2} \beta \left\{\left[\frac{1}{\beta} + \mathscr{G}\right]^2 + [A/(1 - e^{-2\pi\gamma})]^2\right\}^{1/2}. \tag{57}$$

Substitution of $(1/\beta) + \mathscr{G} = A\cot(\pi\mu)$ from (51) gives

$$\tan\delta(\epsilon) = \frac{\tan\pi\mu(\epsilon)}{(1 - e^{-2\pi\gamma})} \tag{58}$$

and

$$C = (2)^{1/2}[A(1 - e^{-2\pi\gamma})]^{1/2}\beta\{[\cot\pi\mu]^2 + [1 - e^{-2\pi\gamma}]^{-2}\}^{1/2}. \tag{59}$$

To the phase $\delta(\epsilon)$ defined by (55) we may add any integral multiple of π. Consistent with (58) we may define $\delta(\epsilon)$ uniquely by requiring that $\delta(\epsilon)$ be continuous and that

$$\delta(0) = \pi\mu(0). \tag{60}$$

In most applications we may assume $2\pi\gamma$ to be large. Neglecting* $\exp(-2\pi\gamma)$ in (58) and (59) we obtain

$$\delta(\epsilon) = \pi\mu(\epsilon) \qquad (2\pi\gamma \gg 1), \tag{61}$$

and

$$C = (2A)^{1/2}\beta/\sin\delta \qquad (2\pi\gamma \gg 1). \tag{62}$$

The zero-energy phase and the number of bound states.—For an attractive potential $V(r)$ which, for large r, tends to zero more rapidly than the Coulomb potential a general theorem (**17**) states that $\delta(0) = \pi N$ where N is the number of bound states obtainable with the potential V. For a potential tending asymptotically to the attractive Coulomb form the number of bound states is infinite. In such a case one may define $N(\epsilon')$ as the number of bound states with ϵ_{ll} less than some small negative value ϵ'. Then $N(\epsilon')$ will increase by unity when $\mu(\epsilon')$ increases by unity. One may thus regard $\mu(0)$ as the increase in the number of bound states which results from the non-Coulomb part of the potential $u(r)$; this despite the fact that $\mu(0)$ is generally not an integer. The relation $\delta(0) = \pi\mu(0)$ may therefore be regarded as a generalization of the relation $\delta(0) = \pi N$.

4. *Normalization of the radial functions.*—We put

$$\left.\begin{aligned} \mathscr{F} &= \mathscr{F}(\kappa, l; \rho) \\ \mathscr{F}' &= \mathscr{F}(\kappa', l; \rho) \\ D &= \frac{d}{d\rho} \end{aligned}\right\}. \tag{63}$$

Then from (42)

$$\int_0^\rho \mathscr{F}\left[D^2 - \frac{l(l+1)}{\rho^2} + u(\rho) + \frac{2}{\rho} - \frac{1}{\kappa'^2}\right]\mathscr{F}'\,d\rho = 0. \tag{64}$$

Integrating by parts and using the equation satisfied by $\mathscr{F}$ one obtains

$$(\mathscr{F}D\mathscr{F}' - \mathscr{F}'D\mathscr{F})\Big|_0^\rho + \left(\frac{1}{\kappa^2} - \frac{1}{\kappa'^2}\right)\int_0^\rho \mathscr{F}\mathscr{F}'\,d\rho = 0. \tag{65}$$

Assuming $D\mathscr{F}$ and $D\mathscr{F}'$ to be finite for $\rho = 0$ it follows from (43) that $(\mathscr{F}D\mathscr{F}' - \mathscr{F}'D\mathscr{F})$ is zero at the lower limit and therefore

$$\int_0^\rho \mathscr{F}\mathscr{F}'\,d\rho = \frac{(\mathscr{F}'D\mathscr{F} - \mathscr{F}D\mathscr{F}')}{(1/\kappa^2 - 1/\kappa'^2)}. \tag{66}$$

We consider this expression in the limit of $\rho \to \infty$.

Equation (66) is the standard expression used for the normalization of positive energy radial functions. One obtains (**18, 19**)

$$\int_0^\infty \mathscr{F}(i\gamma, l; \rho)\mathscr{F}(i\gamma', l; \rho)\,d\rho = [C(\epsilon)]^2\boldsymbol{\delta}(\epsilon - \epsilon') \tag{67}$$

where $\boldsymbol{\delta}(\epsilon - \epsilon')$ is the Dirac δ-function.

* The condition $(2\pi\gamma) \gg 1$ is equivalent to $k \ll (2\pi z)$ where k^2 is the electron kinetic energy in units of 13·60 eV.

To obtain the normalization for bound states we put

$$\mathscr{F} = \mathscr{F}(\nu, l; \rho) \tag{68}$$

and

$$\mathscr{F}' = \mathscr{F}_n = \mathscr{F}(\nu_n, l; \rho) \tag{69}$$

where $\epsilon_n = -1/\nu_n^2$ is an eigenvalue. The integral

$$\Delta_n = \int_0^\infty \mathscr{F}_n^2 \, d\rho \tag{70}$$

is evaluated as *

$$\Delta_n = \lim_{\rho\to\infty} \lim_{\nu\to\nu_n} \left\{ \frac{\mathscr{F}_n D\mathscr{F} - \mathscr{F} D\mathscr{F}_n}{(1/\nu^2 - 1/\nu_n^2)} \right\}. \tag{71}$$

Using (33), (38) and (47) we have

$$\mathscr{F} \sim a\rho^\nu \, e^{-\rho/\nu} + b\rho^{-\nu} \, e^{\rho/\nu} \tag{72}$$

where

$$a = \frac{\beta\Gamma(l+1-\nu)}{\pi\nu^l} \left(\frac{2}{\nu}\right)^\nu, \tag{73}$$

$$b = \{1 + \beta[A \cot \pi\nu + G]\} \frac{\nu^{l+1}(2/\nu)^\nu}{\Gamma(l+1-\nu)}. \tag{74}$$

Since $\mathscr{F}_n$ is an eigenfunction,

$$\mathscr{F}_n \sim a_n \rho^\nu{}_u e^{-\rho/\nu_n} \tag{75}$$

where

$$a_n = \beta(\epsilon_n) \frac{\Gamma(l+1-\nu_n)}{\pi\nu_n^l} \left(\frac{2}{\nu_n}\right)^{\nu_n}. \tag{76}$$

Substitution of (72) and (75) in (71) gives

$$\Delta_n = -\nu_n^2 a_n \left.\frac{\partial b}{\partial \nu}\right|_{\nu=\nu_n}. \tag{77}$$

Using (74) and (51) one obtains

$$\left.\frac{\partial b}{\partial \nu}\right|_{\nu=\nu_n} = -\frac{\pi A(\nu_n, l)\beta(\epsilon_n)\nu_n^{l+1}(2/\nu_n)^{-\nu_n}\zeta(\nu_n)}{\sin^2(\pi\mu(\epsilon_n))\Gamma(l+1-\nu_n)} \tag{78}$$

where

$$\zeta(\nu) = 1 + \frac{\partial \mu}{\partial \nu}. \tag{79}$$

From (76), (77) and (78) the normalization integral is

$$\Delta_n = \frac{\nu_n^3 \beta^2(\epsilon_n) A(\nu_n, l)\zeta(\nu_n)}{\sin^2(\pi\mu(\epsilon_n))}. \tag{80}$$

The normalized radial function $P_{nl}(r)$ is taken to satisfy

$$\int_0^\infty P_{nl}^2(r) \, dr = 1 \tag{81}$$

where the integration is over r and not $\rho = zr$. Making a convenient phase choice we put

$$P_{nl} = \frac{z^{1/2}(-1)^{n-l-1}\mathscr{F}_n}{\Delta_n^{1/2}}. \tag{82}$$

* Ham (11) quotes an expression similar to (71) and states that it can be evaluated using the Quantum Defect Method. He does not, however, give the explicit expression for the normalization factor as obtained in the present paper.

From (47) and (50),

$$\mathscr{F}_n = \frac{\beta(\epsilon_n)\Gamma(l+1-\nu_n)}{\pi \nu_n{}^l} y_5(\nu_n, l; \rho) \qquad (\rho > \rho_1) \tag{83}$$

and therefore

$$P_{nl} = \mathrm{K}(\nu_n, l) y_5(\nu_n, l; \rho) \qquad (\rho > \rho_1) \tag{84}$$

where

$$\mathrm{K}(\nu_n, l) = z^{1/2}[\zeta(\nu_n)\nu_n{}^2\Gamma(\nu_n + l + 1)\Gamma(\nu_n - l)]^{-1/2}. \tag{85}$$

If all the eigenvalues are known $\partial\mu/\partial\nu$ may be calculated and the exact normalized eigenfunctions determined for $\rho > \rho_1$. It may be noted that

$$\zeta(\nu) = 1 + \frac{\partial \mu}{\partial \nu} = 1 + \frac{2}{\nu^3}\frac{\partial \mu}{\partial \epsilon}. \tag{86}$$

For large ν, $\partial\mu/\partial\epsilon$ remains finite, and therefore $\zeta(\nu)$ tends to unity. The normalization factor then becomes

$$\mathrm{K}(\nu_n, l) = z^{1/2}[\nu_n{}^2\Gamma(\nu_n + l + 1)\Gamma(\nu_n - l)]^{-1/2} \qquad (\nu_n \text{ large}). \tag{87}$$

With $\nu_n = n$ this is the exact normalization factor for the pure Coulomb field (for which $\mu = 0$ and $\zeta = 1$). Hartree (**12**) has shown that the factor (87) satisfies a necessary recurrence relation for values of ν_n differing by integers. This is consistent with the exact result since the ν_n may be considered to differ by integers only when $\partial\mu/\partial\nu$ is negligible.

Numerical results for the bound-state normalization factor.—Hartree (**12**) and Bates and Damgaard (**5**) have shown that the normalization factor with $\zeta = 1$ is a good approximation when ν_n is not too small. In illustration of cases for

TABLE I

The normalization factor K *for 2p radial functions of atomic oxygen ions*

Ion	State	ν_2	K: Eqn. (85) with $\zeta = 1$	K: Eqn. (85) with $\zeta = \frac{(\nu_2 - 1)(\nu_2 + 2)}{\nu_2(\nu_2 + 1)}$	K: Exact value
O^0	$2p^4$ ^3P	0·890	imaginary	0·61	0·78
O^+	$2p^3$ ^2P	1·296	0·38	0·67	0·67
O^{+2}	$2p^2$ ^3P	1·483	0·48	0·71	0·71
O^{+3}	$2p$ ^2P	1·672	0·51	0·69	0·65

Note that the values of ν_2 are obtained from the energy parameters in the radial equations and not from measured energies.

which this approximation is not sufficient we consider the $2p$ radial functions for atomic oxygen ions calculated by Hartree, Hartree and Swirles (**20**). Fig. 2 shows the quantum defects for the $2p$ and $3p$ states of O^{+3}. It is seen that μ goes to unity for $\nu = 1$ and that $\mu(\epsilon)$ is a nearly linear function of ϵ for ν between 1 and ν_2. Assuming an exactly linear function we obtain

$$\zeta(\nu_2) = \frac{(\nu_2 - 1)(\nu_2 + 2)}{\nu_2(\nu_2 + 1)}. \tag{88}$$

Table I gives the normalization factors as calculated with $\zeta = 1$ and as calculated using (88) and also the exact values determined from numerical evaluation of the normalization integral (81).

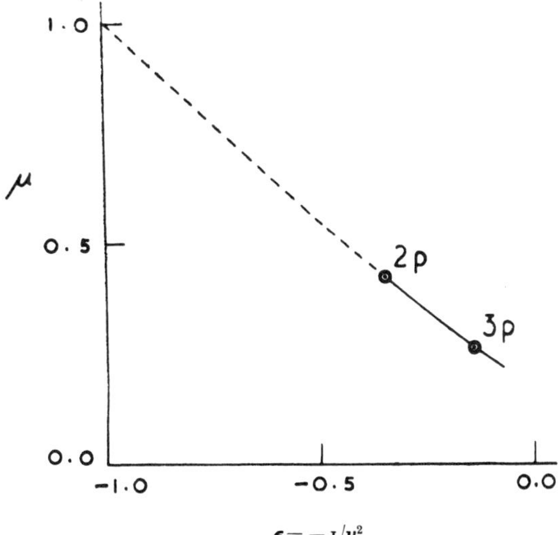

FIG. 2.—*Quantum defects for* O^{+3} *np.*

5. *Discussion of the Quantum Defect Method.*—The mathematical theory considered so far is concerned with central fields which tend to the attractive Coulomb form at large distances. A different approach would be needed in order to develop an exact theory applicable to many-electron atoms. Some of the results of the present theory may, however, be used directly.

Consider an atomic spectral series for which the ion is left in the ground state at the series limit. From the observed energy levels quantum defects may be obtained and extrapolated to positive energies. Just beyond the series limit the free electron will have a kinetic energy which will be too small for collisions to produce excitation of the ion. A single real phase parameter will then suffice to describe the asymptotic form of wave functions corresponding to definite angular momenta. In such cases there can be little doubt that accurate phases can be determined from the extrapolated quantum defects.

For higher energies the situation is more complicated. The asymptotic form of the wave function may be described in terms of a scattering matrix **S**, the off-diagonal elements of this matrix determining the probability of the free electron producing transitions between states of the ion. If all atomic energy levels were considered, including those corresponding to excitation of two or more electrons, it may be expected that information about the complete scattering matrix could in principle be deduced. The more complete theory required has not yet been developed. We consider some applications of the present form of the Quantum Defect Method.

Extrapolation procedures.—For the modified Coulomb field we have shown that the mixture parameter $\beta(\epsilon)$ can be expressed as a convergent power series in ϵ and that the quantum defect μ may be defined for all ϵ as a function of β. We have not shown that the quantum defect itself may be expressed as a convergent expansion in powers of ϵ but it may be shown (**12**) that μ may be expressed as an asymptotic expansion in powers of ϵ.

In practical applications information concerning the eigenvalues is always incomplete. Quantum defects obtained from observed energy levels may usually be represented by polynomials in ϵ containing only a few terms. These polynomials may be used for extrapolation to positive energies. The range of extrapolation is limited by incomplete knowledge of the exact eigenvalues and also by the fact that the simple theory is invalid for positive energies such that inelastic collisions may occur.

It may be convenient to extrapolate the η-defect when η varies more slowly than μ.

Comparison of calculated and empirical phases.—Such comparisons have been made in previous papers (**6, 7, 8, 9, 28**). In favourable cases, phases obtained from solutions of the continuous state Hartree–Fock equations (**21**) are in very good agreement with extrapolated quantum defects. The reason is that the Hartree–Fock equations for continuous states may be derived from a variational principle for the phase just as the equations for bound states may be derived from a variational principle for the energy. In cases for which the Hartree–Fock phases are less satisfactory it may be expected that configuration interaction is important; an example of such a case is the continuum corresponding to extrapolation of the 4*snp* series in Ca (**7**).

Calculation of elastic scattering cross-sections.—We consider ions with non-degenerate ground states. For elastic scattering of slow electrons the appropriate theory is then that for a modified Coulomb field. For scattering through an angle θ the cross-section per unit solid angle is (**22**)

$$I(\theta)=|\mathbf{a}_0 f(\theta)|^2 \tag{89}$$

where $\mathbf{a}_0$ is the Bohr radius and where

$$f(\theta)=\frac{1}{2ik}\sum_{l=0}^{\infty}(2l+1)P_l(\cos\theta)[e^{2i(\phi_l+\delta_l)}-1] \tag{90}$$

and*

$$\phi_l(\gamma)=\arg\Gamma(l+1-i\gamma). \tag{91}$$

It is convenient to put

$$I(\theta)=I_c(\theta)\mathscr{I}(\theta) \tag{92}$$

where $I_c(\theta)$ is the cross-section for the unmodified Coulomb field. The latter is given by

$$I_c(\theta)=|\mathbf{a}_0 f_c(\theta)|^2 \tag{93}$$

where $f_c(\theta)$ is obtained on putting $\delta_l=0$ in (90):

$$f_c(\theta)=\frac{1}{2ik}\sum_l(2l+1)P_l(\cos\theta)[e^{2i\phi_l}-1]. \tag{94}$$

This series may be summed (**22**) to give

$$f_c(\theta)=\frac{\gamma}{2k\sin^2(\theta/2)}\exp[2i(\phi_0+\gamma\ln\sin\tfrac{1}{2}\theta)] \tag{95}$$

and

$$I_c(\theta)=\left[\frac{z\mathbf{a}_0}{2k^2\sin^2(\theta/2)}\right]^2 \tag{96}$$

* We recall that $\delta_l(\gamma)$ is the phase for the l-wave due to the non-Coulomb part of the potential and that $\gamma=z/k$.

which is the Rutherford formula. From (90) and (94) we obtain

$$f(\theta)=f_c(\theta)+\frac{1}{2ik}\sum_l(2l+1)P_l\,e^{2i\phi_l}[e^{2i\delta_l}-1] \tag{97}$$

and, using (95),

$$\frac{f(\theta)}{f_c(\theta)}=1+\frac{2\sin^2(\theta/2)}{\gamma}\sum_l(2l+1)P_l(\cos\theta)e^{ip_l}\sin\delta_l \tag{98}$$

with

$$p_l(\theta,\gamma)=2(\phi_l-\phi_0)-2\gamma\ln\sin\theta/2+\delta_l. \tag{99}$$

It may be noted that

$$\begin{aligned}e^{2i(\phi_l-\phi_0)}&=\frac{\Gamma(l+1-i\gamma)\Gamma(1+i\gamma)}{\Gamma(l+1+i\gamma)\Gamma(1-i\gamma)}\\&=\frac{(1-i\gamma)(2-i\gamma)\dots(l-i\gamma)}{(1+i\gamma)(2+i\gamma)\dots(l+i\gamma)}\end{aligned} \tag{100}$$

and therefore

$$(\phi_l-\phi_0)=-\sum_{m=1}^{l}\arctan(\gamma/m). \tag{101}$$

From (99) we obtain finally

$$\mathscr{I}(\theta)=X^2+Y^2 \tag{102}$$

with

$$\left.\begin{aligned}X&=1+\frac{2\sin^2(\theta/2)}{\gamma}\sum_l(2l+1)P_l(\cos\theta)\cos p_l\sin\delta_l\\Y&=\frac{2\sin^2(\theta/2)}{\gamma}\sum_l(2l+1)P_l(\cos\theta)\sin p_l\sin\delta_l\end{aligned}\right\}. \tag{103}$$

It follows from these expressions that $\mathscr{I}(0)=1$ for all k^2 and that, for $k^2=0$, $\mathscr{I}(\theta)=1$ for all θ.

Fig. 3 shows results obtained for scattering of electrons by Na^+ ions. The phases* have been obtained using formulae given by Ham (**11**) for extrapolation of the η-defects. The cross-sections are seen to be significantly different from the corresponding Coulomb cross-sections. Complicated interference effects are particularly marked for small values of k.

Inelastic collisions.—For ions with degenerate ground states the Quantum Defect Method may be used to obtain accurate cross-sections for electron-induced transitions between ground-state fine structure components. The method has been applied (**10**) to the transitions C^+ $2p_{1/2}$–$2p_{3/2}$ and Si^+ $3p_{1/2}$–$3p_{3/2}$; these are of importance for the cooling of the interstellar gas.

For other transitions in ions produced by electron impact it has not yet proved to be possible to deduce cross-sections from quantum defect data alone. One may, however, make calculations in which the various parameters are so adjusted as to be consistent with extrapolated quantum defects (**10**).

The Quantum Defect Method may be used to obtain continuum wave functions having asymptotic forms which are virtually exact. Such functions have been used for the calculations of photo-ionization cross-sections (**13**).

* Only s- and p-phases have been taken into account. The higher order phases are small and will have little effect on the cross-section.

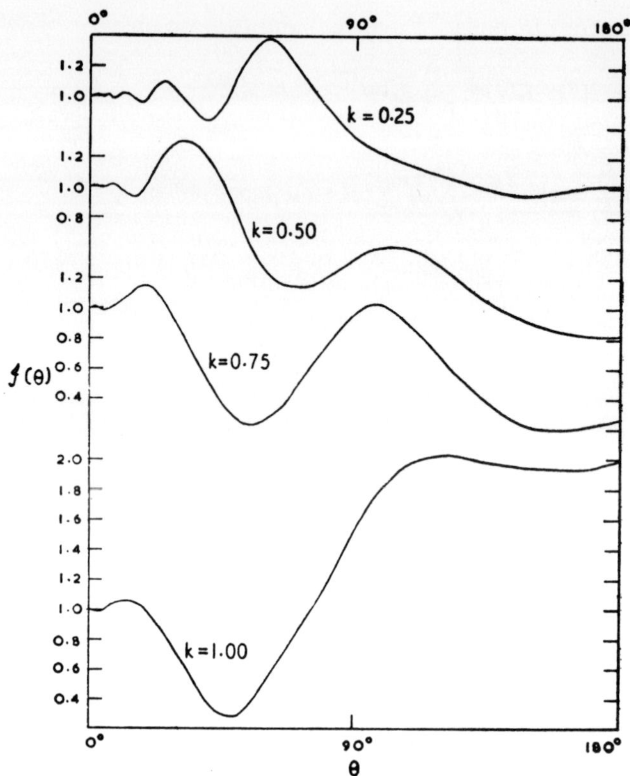

FIG. 3.—*Elastic electron scattering by* Na^+ *ions.* $\mathscr{I}(\theta)$ *is the factor by which the Coulomb cross-section must be multiplied,* θ *being the scattering angle.* k^2 *is the kinetic energy in Rydbergs* (13·60 *eV*). *Note that* $\mathscr{I}(\theta) = 1$ *for* $k = 0$.

APPENDIX

Expansions of the Coulomb functions in powers of the energy.—The expansion of the regular Coulomb function y_1 has been discussed in a number of papers (see (**23, 24, 25, 26, 27**)). One may first obtain

$$y_1(\kappa, l; \rho) = \sum_{p=0}^{\infty} a_p(\kappa, l)(2\rho)^{\frac{1}{2}(p+1)} J_{2l+1+p}(\sqrt{(8\rho)}) \tag{A 1}$$

where (**26**)

$$\left.\begin{aligned} &a_0 = 1, \qquad a_1 = 0, \qquad a_2 = (l+1)/4\kappa^2, \\ &a_p = \frac{1}{4\kappa^2 p}\{(2l+p)a_{p-2} - a_{p-3}\} \qquad (p \geqslant 3) \end{aligned}\right\}. \tag{A 2}$$

Putting

$$a_p(\kappa, l) = \sum_q a_{q,p}(l)\kappa^{-2q} \tag{A 3}$$

one obtains

$$\left.\begin{aligned} &a_{q,p} = 0 \quad \text{unless} \quad 2q \leqslant p \leqslant 3q, \\ &a_{0,0} = 1, \qquad a_{1,2} = (l+1)/4, \qquad a_{1,3} = -\frac{1}{12}, \\ &a_{q,p} = \frac{1}{4p}\{(2l+p)a_{q-1,p-2} - a_{q-1,p-3}\} \qquad (q \geqslant 1, \quad p \geqslant 3) \end{aligned}\right\}. \tag{A 4}$$

The expansion for y_1 is

$$y_1(\kappa, l; \rho) = \sum_{q=0}^{\infty} \kappa^{-2q} \sum_{p=2q}^{3q} a_{q,p}(l)(2\rho)^{\frac{1}{2}(p+1)} J_{2l+1+p}(\sqrt{(8\rho)}). \quad \text{(A 5)}$$

Various expansions for irregular Coulomb functions have been proposed. It has been shown (see Section 2) that the function y_3 cannot be represented by a convergent expansion in powers of the energy. It may, however, be represented by the *asymptotic* expansion

$$y_3(\kappa, l; \rho) = A(\kappa, l) \sum_{q=0}^{Q} \kappa^{-2q} \sum_{p=2q}^{3q} a_{q,p}(l)(2\rho)^{\frac{1}{2}(p+1)} Y_{2l+1+p}(\sqrt{(8\rho)}) + O(\kappa^{-2Q-2}), \quad \text{(A 6)}$$

(**25, 26, 27**). On expressing $A(\kappa, l)$ as a polynomial in κ^{-2}, (A 6) gives an asymptotic expansion in powers of κ^{-2}. The function $G(\kappa, l)$ (defined by equation (14)) may be represented by an asymptotic expansion in powers of κ^{-2} (**12, 27**). Substitution of the asymptotic expansions for y_3 and for Gy_1 in $y_4 = y_3 - Gy_1$ gives a convergent expansion for y_4 (Ham (**27**)). For numerical calculations the asymptotic expansion (A 6) is often convenient.

Department of Physics,
University College, London,
W.C.1:
1958 *May* 26.

References

(1) W. Prokofjew, *Z. Phys.*, **58**, 255, 1929.
(2) B. Trumpy, *Z. Phys.*, **61**, 54 and **66**, 720, 1929.
(3) L. Biermann, *Nachr. Akad. Wiss. Göttingen, Math-Phys., Kl.*, **2**, 116, 1946.
(4) E. A. Hylleraas, *Arch. Math. Naturv.* B, **48** No. 4, 1945.
(5) D. R. Bates and A. Damgaard, *Phil. Trans.*, A **242**, 101, 1949.
(6) M. J. Seaton, *Comptes Rendus*, **240**, 1317, 1955.
(7) M. J. Seaton, *Ann. d'Astrophys.*, **18**, 206, 1955.
(8) M. J. Seaton, *The airglow and the aurorae* (ed. E. B. Armstrong and A. Dalgarno), p. 289, Pergamon, 1955.
(9) M. J. Seaton and D. E. Osterbrock, *Ap. J.*, **125**, 66, 1957.
(10) M. J. Seaton, *Rev. Mod. Phys.*, **30**, 979, 1958.
(11) F. S. Ham, *Solid State Physics* (ed. F. Seitz and D. Turnbull), vol. **1**, p. 127, Academic Press, 1955.
(12) D. R. Hartree, *Proc. Camb. Phil. Soc.*, **24**, 89 and 426, 1927; *ibid.*, **25**, 310, 1929.
(13) A. Burgess and M. J. Seaton, *Rev. Mod. Phys.*, **30**, 992, 1958.
(14) E. T. Whittaker and G. N. Watson, *A course of modern analysis*, 4th ed., Ch. XVI, Cambridge, 1946.
(15) G. H. Wannier, *Phys. Rev.*, **64**, 358, 1943.
(16) Bateman Manuscript Project, *Higher Transcendental functions*, vol. **1**, p. 278, McGraw-Hill, 1953.
(17) P. Weiss, *Phys. Rev.*, **87**, 226, 1952.
(18) E. Fues, *Ann. d. Phys.*, **87**, 281, 1926.
(19) J. Hargreaves, *Proc. Camb. Phil. Soc.*, **25**, 75, 1929.
(20) D. R. Hartree, W. Hartree and B. Swirles, *Phil. Trans.*, A, **238**, 229, 1939.
(21) M. J. Seaton, *Phil. Trans.*, A, **245**, 469, 1953.
(22) W. Gordon, *Z. Phys.*, **48**, 180, 1928.
(23) F. L. Yost, J. A. Wheeler and G. Breit, *Phys. Rev.*, **49**, 174, 1936.
(24) J. G. Beckerley, *Phys. Rev.*, **67**, 11, 1945.
(25) T. S. Kuhn, *Quart. Appl. Math.*, **9**, 1, 1951.
(26) M. Abramowitz, *Jour. Math. and Phys.* (M.I.T.), **33**, 111, 1954.
(27) F. S. Ham, *Tables for the calculation of Coulomb Wave functions*, Tech. Rep. No. 204 Cruft Lab., Harvard, 1955.
(28) M. J. Seaton, *Proc. Phys. Soc.* A, **70**, 620, 1957.

1962 *Proc. Phys. Soc.* **80** 1073–7

Application of the Multichannel Effective Range Theory to Electron–Hydrogen Scattering

By R. DAMBURG and R. PETERKOP

Institute of Physics of the Latvian Academy of Sciences, Riga, U.S.S.R.

MS. received 4*th July* 1962

Abstract. The behaviour of cross sections in the neighbourhood of the 2s excitation threshold is investigated using **M** matrix theory. Calculations allowing for 1s–2s strong coupling are carried out for zero angular momentum in singlet, triplet and no-exchange cases. Threshold effects in all cases have the form of a cusp. A narrow resonance has been found below the threshold in the singlet case.

§1. INTRODUCTION

As was pointed out by Ross and Shaw (1961), the behaviour of cross sections in the neighbourhood of the threshold of 'new' channels is conveniently investigated by using **M** matrix theory. **M** is essentially the inverse of the reaction matrix **K**:

$$M_{ij} = k_i^{l_i+\frac{1}{2}}(\mathbf{K}^{-1})_{ij}k_j^{l_j+\frac{1}{2}} \tag{1}$$

where k_i and l_i are momentum and angular momentum of the ith channel.

If there is only one channel, we have $M = k^{2l+1}\cot\delta$, δ being the phase shift.

The main advantage of using the **M** matrix is that in the neighbourhood of the threshold of a 'new' channel, where cross sections change rapidly, elements of the **M** matrix are smooth functions of energy. Knowing the **M** matrix on one side of a reaction threshold, we can find cross sections in the neighbourhood of the threshold on the other side of it.

§2. CALCULATIONS

We calculated elements of the **M** matrix for the electron–hydrogen collision with zero total angular momentum taking into account two channels: 1s and 2s. Calculations were made above the threshold of excitation of the 2s state in the singlet (+), triplet (−) and no-exchange (0) cases.

The equations may be written in the following short form:

$$\begin{aligned} \mathscr{L}_{11}f_1+\mathscr{L}_{12}f_2 &= 0 \\ \mathscr{L}_{21}f_1+\mathscr{L}_{22}f_2 &= 0 \end{aligned} \tag{2}$$

where the $\mathscr{L}_{ij}$ are integro-differential operators. The explicit form of the equations is given by Marriott (1958).

The boundary conditions are

$$\begin{aligned} f_j^{(i)}(0) &= 0 \\ f_j^{(i)}(\infty) &\sim k_j^{-1}M_{ji}\sin k_jr+\delta_{ij}\cos k_jr. \end{aligned} \tag{3}$$

In (3) the **M** matrix has been transposed to that defined by Ross and Shaw. Index ij corresponds to transition from state j to state i.

Effective cross sections are expressed by the **M** matrix in the following way:

$$Q(1s\text{–}1s) = \frac{4\pi}{k_1^2+\gamma^2}; \qquad (k_1^2 \leqslant 0{\cdot}75 \text{ A.U.}) \tag{4}$$

$$Q(1s\text{–}1s) = \frac{4\pi(M_{22}^2+k_2^2)}{|D|^2}; \qquad (k_1^2 \geqslant 0{\cdot}75 \text{ A.U.}) \tag{5}$$

$$Q(1s\text{–}2s) = \frac{4\pi k_2 M_{21}^2}{k_1|D|^2}; \qquad Q(2s\text{–}2s) = \frac{4\pi(M_{11}^2+k_1^2)}{|D|^2} \tag{6}$$

$$\gamma = M_{11} - \frac{M_{12}M_{21}}{M_{22}+|k_2|}; \qquad k_2^2 = k_1^2 - 0{\cdot}75 \tag{7}$$

$$D = (M_{11}-ik_1)(M_{22}-ik_2)-M_{12}M_{21}. \tag{8}$$

The system of integro-differential equations (2) was solved by a non-iterative method used by Marriott (1958). Numerical integration was carried out by us using the XIth method of Milne (1953). Calculations were made by two methods by allowing for strong coupling ($\mathscr{L}_{12} = \mathscr{L}_{21}^{+}$) and by using the distorted wave approximation in which $\mathscr{L}_{12} = 0$. Our strong coupling cross sections at $k_1 = 0{\cdot}9$ A.U. are in good agreement with those obtained by Marriott (1958) and Smith (1960).

Table 1

Cross Sections at $k_1 = 0{\cdot}9$ A.U. (in units of πa_0^2)

	Q(1s–1s) +	Q(1s–1s) −	Q(1s–2s) +	Q(1s–2s) −
Marriott	1·7236	4·9109	0·153	0·00054
Smith	1·7432	4·9164	0·1536	0·0
Our results	1·7256	4·9108	0·1526	0·000545

From results obtained in the vicinity of the threshold ($k_2^2 \leqslant 0{\cdot}001$ A.U.) we found, by means of extrapolation, M_{ij} and their energy derivatives R_{ij} (effective radii) at the threshold point ($k_2 = 0$). They are given in table 2 in atomic units.

Table 2

The M Matrix and Effective Radii at the 2s Excitation Threshold

	Strong coupling calculation			Distorted wave calculation		
	+	−	0	+	−	0
M_{11}	1·1300	0·0301	0·9373	1·2518	0·0320	0·6273
M_{21}	−0·0629	−0·0017	−0·3097	−0·0638	−0·0017	−0·2516
M_{22}	−0·0356	−0·1206	0·2042	−0·0364	−0·1206	0·1534
R_{11}	4·82	1·20	7·64	3·58	1·20	1·2
R_{21}	−4·32	−0·06	−10·68	−4·38	−0·06	−7·8
R_{22}	11·54	5·14	19·2	11·44	5·14	16·0

As may be seen from table 2, in our case (except the triplet variant) the assumption of Ross and Shaw that non-diagonal effective radii R_{ij} ($i \neq j$) ought to be much smaller than the diagonal R_{ii} does not hold true. The inclusion of exchange decreases the difference between strong coupling and distorted wave calculations. Practically full agreement between strong coupling and distorted wave results is observed in the triplet case.

In the strong coupling calculation $M_{12} = M_{21}$, but in the distorted wave approximation $M_{12} = 0$. In the distorted wave approximation $M_{ii} = k_i \cot \delta_1$. One can approximately allow for strong coupling in the distorted wave method by putting $M_{12} = M_{21}$. One can also put in the **K** matrix $K_{12} = K_{21}$. The latter was applied by Burke and Seaton (1961) in the Born approximation. At the threshold point, $K_{21} = 0$ and therefore the allowance for strong 1s–2s coupling with the help of the **K** matrix leads to no change in the Q(1s–1s) at this point.

The effective range approximation assumes that

$$M_{ij}(E) = M_{ij}(E_0) + R_{ij}(E_0)(E - E_0). \tag{9}$$

We took E_0 to be equal to the threshold energy.

In figures 1, 2 and 3, cross sections obtained by the effective range approximation

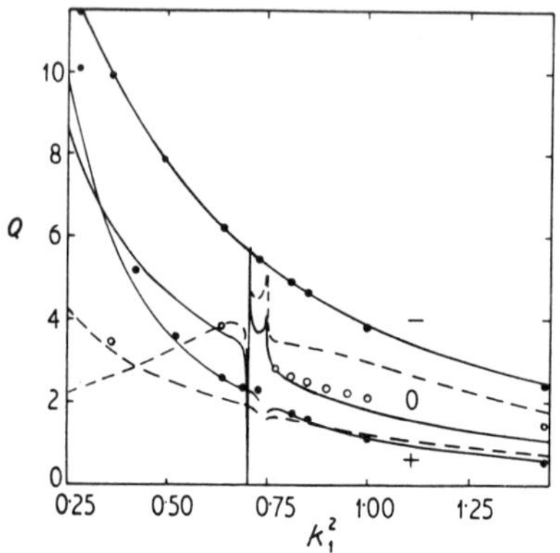

Figure 1. Q(1s–1s) in units of πa_0^2.

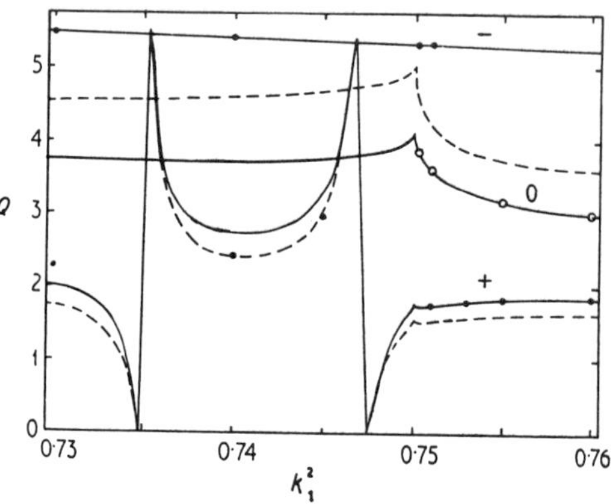

Figure 2. Q(1s–1s) in units of πa_0^2 in the vicinity of the 2s excitation threshold.

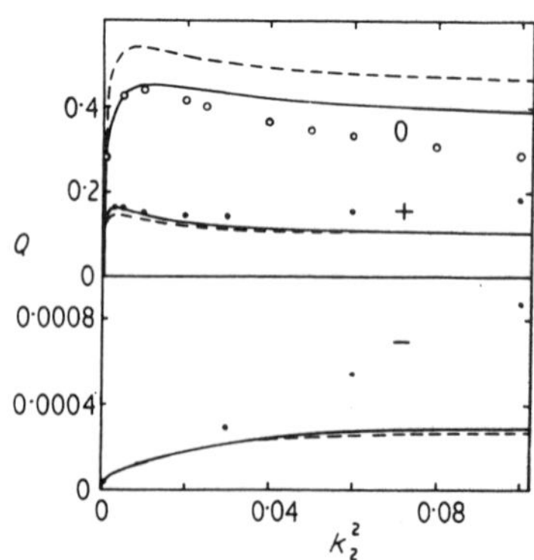

Figure 3. Q(1s–2s) in units of πa_0^2.

Solid curves—effective range approximation, strong coupling calculation; broken curves—effective range approximation, distorted wave calculation ($M_{12} = M_{21}$); circles—exact cross section values allowing for 1s–2s strong coupling. In the triplet case both effective range approximation variants are indistinguishable in figures 1 and 2.

are compared with exact values calculated by allowing for 1s–2s strong coupling: values obtained below the threshold by Smith and Burke (1961) and by Smith, McEachran and Fraser (1962) and above the threshold by Smith, Miller and Mumford (1960), by Marriott (1958) and by our calculations.

§ 3. DISCUSSION

Detailed analysis of the applicability of the effective range approximation is given by Ross and Shaw (1961). According to their theory, only one 'narrow' resonance can be described in the effective range approximation formalism and it can occur only below the threshold of channel two. This resonance corresponds to a bound state in the 'new' channel only.

Resonance effects are mainly determined by the element M_{22}. Consider the single channel two. The presence of bound states is determined by the following equation:

$$M_{22}(E_0)+\tfrac{1}{2}R_{22}(E_0)k_2^2-ik_2 = 0 \tag{10}$$

which has two solutions.

In the triplet case both zeros lie in the upper half of the k_2 plane but not on the imaginary k_2 axis. This case corresponds to 'case b' according to the Appendix D of Ross and Shaw (1961). Bound states are not present in this case.

In the singlet and no-exchange cases both zeros lie on the imaginary k_2 axis. This is the 'case a' where the upper zero is artificial according to Ross and Shaw. In the no-exchange case the lower zero lies just below the real axis and represents a 'virtual' bound state. In the singlet case the lower zero lies between 0 and iR_{22}^{-1} and represents an actual bound state (if we do not allow for strong coupling).

Thus from resonances shown in figures 1 and 2, the real resonance will be the one close to the threshold in the singlet case. In the region of other resonances higher terms should be taken into account in the expansion of the **M** matrix. The threshold effect is defined by the sign of the value of $M_{11}(E_0)-M_{21}^2(E_0)/M_{22}(E_0)$. In all cases it is positive and because of that a cusp is observed at all thresholds.

For the triplet case the range of applicability of the effective range approximation is large. The cusp in this case is extremely small.

In the singlet and no-exchange cases the effective range approximation has a small range of applicability. It may be extended if one takes into account higher terms in the expansion of the **M** matrix:

$$M_{ij} = M_{ij}(E_0)+\tfrac{1}{2}R_{ij}k_2^2+F_{ij}k_2^4+G_{ij}k_2^6+ \dots \tag{11}$$

We calculated F_{ij} and G_{ij} for the singlet case (with $\pm 10\%$ accuracy):

F_{11}	F_{21}	F_{22}	G_{11}	G_{21}	G_{22}
20	−34	67	500	−1600	2000

For $k_1^2 = 0{\cdot}74$ and $k_1^2 = 0{\cdot}745$ we obtained respectively

$$Q_1(1s\text{–}1s) = 2{\cdot}73, \quad Q_2(1s\text{–}1s) = 3{\cdot}30$$

(in units of πa_0^2) considering two terms of the expansion of the **M** matrix, and $Q_1 = 2{\cdot}45$ and $Q_2 = 3{\cdot}04$ considering four expansion terms. These values are in satisfactory agreement with those of Smith *et al.* (1962): $Q_1 = 2{\cdot}41$ and $Q_2 = 2{\cdot}95$ (the latter value Smith *et al.* regarded as provisional because of the slow convergence of iterations for $k_1^2 = 0{\cdot}745$).

Thus the application of **M** matrix formalism enables us to connect the cross sections above and below threshold for singlet and triplet cases. In the no-exchange case there are no exact values just below the threshold.

The distorted wave method allowing for strong coupling ($M_{12} = M_{21}$) in the region above the threshold is no more precise than the usual distorted wave method. In the vicinity of the threshold and below, it gives the same form for threshold and resonance effects as the strong coupling calculation.

ACKNOWLEDGMENTS

We wish to express our thanks to M. Gailitis for helpful discussion.

REFERENCES

BURKE, V. M., and SEATON, M. J., 1961, *Proc. Phys. Soc.*, **77**, 199.
MARRIOTT, R., 1958, *Proc. Phys. Soc.*, **72**, 121.
MILNE, W. E., 1953, *Numerical Solution of Differential Equations* (New York, London: Chapman and Hall), § 41.
ROSS, M. H., and SHAW, G. L., 1961, *Ann. Phys., N.Y.*, **13**, 147.
SMITH, K., 1960, *Phys. Rev.*, **120**, 845.
SMITH, K., MILLER, W. F., and MUMFORD, A. J. P., 1960, *Proc. Phys. Soc.*, **76**, 559.
SMITH, K., and BURKE, P. G., 1961, *Phys. Rev.*, **123**, 174.
SMITH, K., MCEACHRAN, R. P., and FRASER, P. A., 1962, *Phys. Rev.*, **125**, 553.

1963 *Sov. Phys.-JETP* **17** 1328–32

BEHAVIOR OF CROSS SECTIONS NEAR THRESHOLD OF A NEW REACTION IN THE CASE OF A COULOMB ATTRACTION FIELD

M. GAĬLITIS

Institute of Physics, Academy of Sciences, Latvian S. S. R.

Submitted to JETP editor December 15, 1962

J. Exptl. Theoret. Phys. (U.S.S.R.) 44, 1974-1981 (June, 1963)

The multichannel effective-radius theory is extended to the case of a Coulomb attractive field. It is shown that this theory leads to the presence of resonances in the cross sections below the threshold of the new reaction. Expressions are obtained for the averaged cross sections, widths, and shifts of the resonances.

INTRODUCTION

BAZ' has considered [1] the behavior of the elastic cross section near the threshold of the first inelastic reaction, and established the existence of resonances in the elastic scattering cross section. The resonances are the result of the fact that at definite energies below the thresholds the particles are able to excite a transition and remain themselves bound on one of the Bohr orbits in the Coulomb field. Such a state is unstable and the particles return to the ground states and move apart. To each of the unstable states there corresponds a resonance maximum (and a minimum) on the cross section curve. The distance between neighboring maxima is approximately equal to the distance between the corresponding Bohr levels. Near the threshold, where this distance is very small, only the cross section averaged over the resonances is of interest. Baz' has shown [1] that the average value of the elastic cross section below the first threshold is equal to the total cross section above the threshold. Fonda and Newton [2-4] investigated the behavior of the cross sections near higher thresholds. They expressed the scattering matrix below threshold in terms of its values above threshold and also demonstrated the continuity of the averaged total cross sections. In the present paper expressions are derived for the discontinuities in the averaged partial cross sections, widths, and shifts of the resonances. It is shown that these expressions are also applicable to the excitation of hydrogenlike ions by electrons, where polarization forces of the type r^{-3} act between the scattered electron and the ion in the excited state at large distances, in addition to the Coulomb forces.

We are considering d parallel nonrelativistic reactions $X(a, b_f) Y_f$, f = 1, . . . , d near the threshold of a new reaction $X(a, b_t) Y_t$. (Several channels, differing in the momenta of the particles b_t and Y_t can open immediately at this threshold. Accidental coincidence of thresholds of two essentially different reactions are disregarded.) We assume that at these energies there are no reactions in which three or more particles are produced. The particles b_t and Y_t are assumed oppositely charged. The remaining particles may or may not be charged.

We choose r_0 such that only Coulomb forces act between them when $r \geq r_0$. The wave function of the system of particles participating in the reaction is in this region a linear combination of single-particle wave functions in a Coulomb field. The cross section is expressed in terms of the coefficients of these functions, determined by smoothly joining the wave function of the system when $r \geq r_0$ with the same wave function for $r \leq r_0$. In place of the explicit solution for the latter, we shall use the smooth energy dependence of the logarithmic derivative. This, together with the known energy dependence of the Coulomb wave functions, is sufficient to determine the behavior of the cross sections below threshold and their connection with the cross sections above threshold.

1. FUNDAMENTAL EXPRESSIONS

We denote by l, J, m, r, v, and k the diagonal matrices whose elements are the orbital and total momenta, the reduced mass, the distance, the relative velocity, and the wave number of the channel. F(kr) and G(kr) are the regular and irregular Coulomb wave functions [5],

$$\eta = Z_b Z_Y/\hbar v, \qquad \sigma_l = \arg\Gamma(l+1+i\eta), \qquad \omega_l = \sigma_l - \sigma_0,$$

s is the spin of the channel.

The wave functions of the system for $r \geq r_0$ can be written in the form of a square matrix[6]

$$\psi = v^{-1/2}(F(kr) + G(kr)K). \quad (1)$$

Different columns in (1) correspond to different linearly independent solutions. Different rows correspond to the wave-function components for different channels. Only components of open channels and of those closed channels whose threshold region is under consideration are included in (1). The wave-function components for the remaining channels attenuate rapidly with increasing r, and when r_0 is sufficiently large their contribution is negligibly small. The channels that are open only above threshold will be called new, and summation over them will be denoted by Σ'. The remaining channels will be called old and the summation denoted by Σ''.

The symmetrical matrix K is connected with the cross sections in the following manner:

$$Q^J(i-f) = \frac{\pi}{k_i^2}\frac{2J+1}{2s_i+1}|T_{fi}|^2, \quad (2)$$

$$T = -2i(K^{-1}-i)^{-1}, \quad (3)$$

$$T = e^{-i\omega}[e^{2i\omega} - U]e^{-i\omega}. \quad (4)$$

We have chosen for the matrix T a definition (4) which differs from that used in [6] by the factors $e^{-i\omega}$, so as to simplify the formulas derived later.

The matrix K, as a function of the energy E, has a branch point at the threshold E_t of the new channel. We must separate in the formula for K the terms with the branching from the part that is analytic in the vicinity of the threshold. To this end we use the method employed by Teichmann for elastic scattering[7]. We define the R-matrix[8] by the relation

$$\frac{1}{\sqrt{m}}\psi(r_0) = R\frac{1}{\sqrt{m}}r_0\frac{d}{dr}\psi(r)\Big|_{r=r_0} \quad (5)$$

and, substituting (1) in (5), express K in terms of R,

$$K^{-1} = -\frac{G}{F} + \frac{1}{F\dot{F}} - \frac{1}{\rho^{1/2}\dot{F}}\left[\rho^{-1}\frac{F}{\dot{F}} - R\right]^{-1}\frac{1}{\dot{F}\rho^{1/2}},$$

$$\rho = kr_0, \quad F = F(\rho), \quad G = G(\rho), \quad \dot{F} = \frac{d}{d\rho}F(\rho). \quad (6)$$

It follows from the solution of the Schrödinger equation in the region $r \leq r_0$ (see [6]) that the R-matrix is analytic in the energy and has only simple poles on the real axis. The branching in (6) is due only to the Coulomb functions, which we therefore express in terms of the functions Ψ and Φ, which are regular in E when r_0 = const[5]:

$$F_l(\rho) = C_l\rho^{l+1}\Phi_l(\rho),$$

$$G_l(\rho) = \frac{\rho^{-l}}{(2l+1)C_l}\Big[\Psi_l(\rho) + \rho^{2l+1}p_l(\eta)\left(\ln 2\rho + \frac{q_l(\eta)}{p_l(\eta)}\right)\Phi_l(\rho)\Big],$$

$$C_l = \frac{2^l C_0}{(2l+1)!}\prod_{s=1}^{l}(s^2+\eta^2)^{1/2}, \qquad C_0 = \left(\frac{2\pi\eta}{e^{2\pi\eta}-1}\right)^{1/2},$$

$$p_l = 2\eta(2l+1)C_l^2/C_0^2, \qquad f(\eta) = {}^1/_2[\psi(i\eta) + \psi(-i\eta)]. \quad (7)$$

$q_l(\eta)/p_l(\eta) - f(\eta)$ is a rational function of η^2, which tends to a constant as $|\eta|^2 \to \infty$.

Substituting (7) in (6) we obtain

$$k^{l+1/2}(2l+1)!!\,C_lK^{-1}C_l(2l+1)!!\,k^{l+1/2} = M - \frac{(2l+1)!!^2}{(2l+1)}k^{2l+1}p_l(\ln k + f(\eta)), \quad (8)$$

where

$$M = \frac{(2l+1)!!}{r_0^{l+1/2}}\Big\{-\frac{\Psi}{(2l+1)\Phi} - \frac{k^{2l+1}p_l}{2l+1}r_0^{2l+1}\left(\ln 2r_0 + \frac{q_l}{p_l} - f(\eta)\right) + \frac{1}{\Phi\bar{\Phi}} - \frac{1}{\bar{\Phi}}\left[\frac{\Phi}{\bar{\Phi}} - R\right]^{-1}\frac{1}{\bar{\Phi}}\Big\}\frac{(2l+1)!!}{r_0^{l+1/2}},$$

$$\bar{\Phi}_l(\rho) = (l+1)\Phi_l(\rho) + \rho\frac{d}{d\rho}\Phi_l(\rho). \quad (9)$$

The terms $k^{2l+1}p_l$ and $q_l(\eta)/p_l(\eta) - f(\eta)$ contained in (9) are rational functions of k^2 and finite when $k^2 = 0$. The matrix M, which is symmetrical and real on the real axis, is therefore analytic in the vicinity of the threshold.

It follows from (3) and (8) that

$$T = k^{l+1/2}(2l+1)!!\,C_l\frac{-2i}{M-(2l+1)!!^2p_lk^{2l+1}\tau/(2l+1)} \times C_l(2l+1)!!\,k^{l+1/2}, \quad (10)$$

$$\tau = \ln k + f(\eta) + i\pi/(e^{2\pi\eta}-1). \quad (11)$$

The factor $(2l+1)!!$ has been singled out because $(2l+1)!!\,C = 1$ in the absence of a Coulomb field, when (10) goes over into formula (22) of Ross and Shaw[8].

The only term with a branch point in (10) is the diagonal matrix τ. The only elements of this matrix not analytic near the threshold are those corresponding to the new channels. They are all the the same and in the direct vicinity of the threshold they are equal to ($a|k_t| \ll 1$)

$$\tau_{tt}^a = \ln\frac{1}{a} - i\pi \quad (E > E_t), \quad \tau_{tt}^b = \ln\frac{1}{a} + \pi y \quad (E < E_t),$$

$$a = \frac{\hbar^2}{me^2|Z_{b_t}Z_{Y_t}|}, \quad y = \text{ctg}\frac{\pi}{a\varkappa}, \quad k_t = i\varkappa \text{ for } E < E_t. \quad (12)^*$$

*ctg = cot.

2. BEHAVIOR OF CROSS SECTIONS NEAR THE THRESHOLD OF ONE NEW CHANNEL

In this section and in the following we consider the region of energies near threshold, where M can be regarded as constant and equal to M_t, the value at threshold, and where formula (12) can be used for τ_{tt}. Eliminating M_t from (10), we relate T below threshold (T^b) with T above threshold (T^a):

$$T^b = [T^{a-1} + i\eta C_0^{-2}(\tau^a - \tau^b)]^{-1}. \tag{13}$$

All the elements of the matrix $\tau^a - \tau^b$ are proportional to $E - E_t$, except $\tau^a_{tt} - \tau^b_{tt} = -\pi(y + i)$. Neglecting the former and assuming that $\lim_{E\to E_t} 2\eta_t C_0^{-2}(\eta_t) = -1/\pi$, we obtain

$$T^b = (T^{a-1} + Y)^{-1}. \tag{14}$$

The matrix Y contains only one nonvanishing element $Y_{tt} = i(y + i)/2$, enabling us to simplify (14):

$$T^b_{fi} = T^a_{fi} - \frac{T^a_{ft}T^a_{ti}}{T^a_{tt}} + \frac{T^a_{ft}T^a_{ti}}{(T^a_{tt})^2}\frac{-2i}{y + i - 2i(T^a_{tt})^{-1}}. \tag{15}$$

The quantity y is a periodic function of $1/a = \hbar/a\sqrt{2m_t(E_t - E)}$. Therefore T^b and all the cross sections oscillate near threshold, assuming identical values at those energies for which $1/a\kappa$ differ by an integer. The form of the cross-section curve depends on T^a. If $\mathrm{Re}\, T^a_{tt} \ll 1$, then the values of T^b_{fi} are constant and approximately equal to T^a_{fi} at energies for which y differs noticeably from $-2\,\mathrm{Im}(T^a_{tt})^{-1}$. The cross sections are then equal to the corresponding cross sections above threshold. When $y \approx -2\,\mathrm{Im}(T^a_{tt})^{-1}$, the cross sections vary rapidly. The width of the resonance Γ and its shift Δ (if the latter is small) with respect to the value of $E_t - E = \hbar^2(2m_t a^2 n^2)^{-1}$ are given by the expressions

$$\frac{\Gamma}{D} = \frac{1}{2\pi}[2\,\mathrm{Re}\,T^a_{tt} - |T^a_{tt}|^2] = \frac{1}{2\pi}\sum_j{}'' |T^a_{tj}|^2,$$

$$\frac{\Delta}{D} = \frac{1}{2\pi}\,\mathrm{Im}\,T^a_{tt}, \tag{16}$$

where $D = 2\sqrt{2m_t}\,a\,(E_t - E)^{3/2}/\hbar$ is the distance between resonances.

The resonances are very close together near threshold, so that interest attaches to the cross sections averaged over the resonances

$$\overline{Q} = \frac{1}{D}\int_{E-D/2}^{E+D/2} Q(E)\,dE = \frac{1}{\pi}\int_{-\infty}^{+\infty} Q(y)\frac{dy}{1+y^2}. \tag{17}$$

Formula (16) expresses T^b_{fi} in the form $\alpha + \beta/(y - \gamma)$ and (17) reduces to the integral

$$\frac{1}{\pi}\int_{-\infty}^{+\infty}\left(\alpha + \frac{\beta}{y-\gamma}\right)\left(\alpha^* + \frac{\beta^*}{y-\gamma^*}\right)\frac{dy}{1+y^2}$$

$$= \frac{1}{\mathrm{Im}\,\gamma}\frac{|\beta|^2}{|\gamma + i|^2} + \left|\alpha - \frac{\beta}{\gamma + i}\right|^2,$$

$$\mathrm{Im}\,\gamma = 2|T^a_{tt}|^{-2}\,\mathrm{Re}\,T^a_{tt} - 1 = |T^a_{tt}|^{-2}\sum_j{}'' |T^a_{tj}|^2 > 0. \tag{18}$$

Using (2), (15), and (18), we obtain

$$\overline{Q^{bJ}(i-f)} = Q^{aJ}(i-f) + \frac{\pi}{k_i^2}\frac{2J+1}{2s_i+1}\frac{|T^a_{ft}|^2|T^a_{ti}|^2}{\sum_j'' |T^a_{tj}|^2} \tag{19}$$

$$= Q^{aJ}(i-f) + \frac{Q^{aJ}(i-t)\,Q^{aJ}(t-f)}{\sum_j'' Q^{aJ}(t-j)}.$$

Expression (19) shows that all averaged cross sections decrease abruptly at the threshold with increasing energy. The distribution of the jumps between the cross sections does not depend on the initial state and is equal to the distribution of the inelastic jumps in the averaged cross sections in scattering, where the new channel is the initial one. The behavior of the jumps in the averaged cross sections is analogous to their behavior in the compound-nucleus model. The sum of all the jumps is equal to the cross section of the new channel, i.e., the total cross section is continuous on the threshold. The latter was proved by Baz'[1] and by Fonda and Newton[2-4] by averaging the imaginary part of the scattering amplitude.

3. BEHAVIOR OF CROSS SECTIONS NEAR THE THRESHOLD OF SEVERAL NEW CHANNELS

If the particles b_t and Y_t have proper momenta, then, for the same threshold several (n) channels with different particle momentum for a constant total momentum J open up immediately. This case differs but little from the case considered in Sec. 2. Formula (14) remains in force, but the diagonal matrix Y now contains n equal non-zero elements $i(y + i)/2$. Expression (15) assumes the form

$$T^b_{fi} = T^a_{fi} - \sum_{mm'}{}' T^a_{fm}[\hat{T}^{-1}]_{mm'}T^a_{m'i}$$

$$+ \sum_{mm'}{}' T^a_{fm}\left[\frac{-2i}{(y + i - 2i\hat{T}^{-1})\hat{T}^2}\right]_{mm'} T^a_{m'i}\,. \tag{20}$$

Here $\hat{T}$ is a symmetrical matrix of rank n, containing the T^a-matrix elements relating the new channels only. Using the orthogonal matrix B that diagonalizes $\hat{T}$

$$B^{-1}\hat{T}B = t \equiv (t_k\delta_{kj}), \quad B^{-1} = B^T, \tag{21}$$

we transform (20) into

$$T^b_{fi} = T^a_{fi} - \sum_{mm'}' T^a_{fm} \{\hat{T}^{-1}\}_{mm'} T^a_{m'i} + \sum_k' \Theta_{ik}\Theta_{fk} \frac{-2i}{(y+i-2it_k^{-1})\, t_k^2};$$

$$\Theta_{jk} = \sum_m' T^a_{jm} B_{mk}. \qquad (22)$$

The consequences of (22) are similar to those of (15): near the threshold the cross sections oscillate and assume identical values when $1/a\kappa$ differ by integers. If Re $t_k \ll 1$, then for energy values when y differs appreciably from all the $-2\,\mathrm{Im}t_k^{-1}$ the cross sections are approximately equal to their values above threshold. When $y \approx -2\,\mathrm{Im}\, t_k^{-1}$ resonances are observed. Their widths and shift are

$$\frac{\Gamma_k}{D} = \frac{1}{2\pi}[2\,\mathrm{Re}\, t_k - |t_k|^2] \approx \frac{1}{2\pi}\sum_j'' |\Theta_{jk}|^2, \quad \frac{\Delta_k}{D} = \frac{1}{2\pi}\mathrm{Im}\, t_k. \qquad (23)$$

In analogy with (19) we obtain

$$\overline{Q^{bJ}(i-f)} - Q^{aJ}(i-f) = \frac{\pi}{k_i^2}\frac{2J+1}{2s_i+1}\sum_{k,\varkappa}' \frac{\Theta_{ik}\Theta^*_{i\varkappa}\Theta_{fk}\Theta^*_{f\varkappa}}{t_k - t_k t^*_\varkappa + t^*_\varkappa}$$

$$= \frac{\pi}{k_i^2}\frac{2J+1}{2s_i+1}\sum_{k,\varkappa}' \frac{\Theta_{ik}\Theta^*_{i\varkappa}\Theta_{fk}\Theta^*_{f\varkappa}}{\sum_j'' \Theta_{jk}\Theta^*_{j\varkappa}} \sum_m' B_{mk}B^*_{m\varkappa}. \qquad (24)$$

As in the case of one new channel, all the averaged cross sections decrease abruptly on the threshold with increasing energy, inasmuch as the right half of (24) is always positive. This can be readily verified by representing it in the form

$$\sum_{k,\varkappa}' \frac{a_k a^*_\varkappa}{1 - S_k S^*_\varkappa}, \quad |S_k| < 1$$

and expanding the denominator in a series. The sum of all the jumps is equal to the sum of the cross sections of the new channels. The form of (24) is more complicated than that of (19) because of the superposition of the neighboring resonances. In the important case when the sections of the new channels are small, i.e., the elements T^a_{jk} relating the old and the new channels are small and, in accord with (23), the resonances are narrow, Eq. (24) simplifies greatly. The terms with $k \neq \kappa$ in (24) are then much smaller than the terms with $k = \kappa$, and

$$\overline{Q^{bJ}(i-f)} - Q^{aJ}(i-f) = \frac{\pi}{k_i^2}\frac{2J+1}{2s_i+1}\sum_k' \frac{|\Theta_{ik}|^2 |\Theta_{fk}|^2}{\sum_j'' |\Theta_{jk}|^2}. \qquad (25)$$

4. THRESHOLD BEHAVIOR OF THE EXCITATION CROSS SECTIONS OF HYDROGENLIKE IONS BY ELECTRONS

Owing to the degeneracy of the levels of the hydrogenlike ion with different momenta, a strong polarization interaction exists between the electron and the ion in the excited state. This interaction decreases as $1/r^2$ with increasing distance between the electron and the ion.[1)] It is therefore impossible to choose r_0 such that the electron-ion interaction reduces to a Coulomb interaction when $r > r_0$, and the derivation presented above for the threshold behavior does not apply. We shall show that an account of the polarization interaction does not change the threshold behavior of the ion excitation cross sections, in contrast with the case of the excitation of hydrogen atoms, where the threshold behavior depends essentially on the polarization.

In analogy with [9] we choose r_0 such that Eq. (4) of [9] reduces when $r \geq r_0$ to

$$\left(\frac{d^2}{dr^2} - \frac{l(l+1)+\alpha}{r^2} + \frac{2Z}{r} + k^2\right)\psi = 0, \qquad (26)$$

where Z is the ion charge. The solution of (26) has the form $\psi = A\varphi$, with two systems of independent solutions existing for φ:

$$I_\lambda = -ie^{-\pi\eta/2}e^{-i\rho}(2\rho)^{\lambda+1}\, U_2(\lambda+1-i\eta,\, 2\lambda+2,\, 2i\rho) \underset{\rho\to\infty}{\sim} \exp\{-i(\rho - \pi\lambda/2 - \eta\ln 2\rho)\},$$

$$O_\lambda = ie^{-\pi\eta/2}e^{-i\rho}(2\rho)^{\lambda+1}\, U_1(\lambda+1-i\eta,\, 2\lambda+2,\, 2i\rho) \underset{\rho\to\infty}{\sim} \exp\{i(\rho - \pi\lambda/2 - \eta\ln 2\rho)\}. \qquad (27)$$

We use for the hypergeometric functions the notation of Morse and Feshbach [10] and the expansions of [5,11]

$$F(\lambda+1-i\eta,\, 2\lambda+2,\, 2i\rho) = (-2\eta)^{-2\lambda-1}\rho^{-2\lambda-1}e^{i\rho}\Gamma(2\lambda+2)\, y_1,$$

$$F(-\lambda-i\eta,\, -2\lambda,\, 2i\rho) = \Gamma(-2\lambda)\, e^{i\rho} y_2,$$

$$y_1 = \sum_{p=0}^{\infty} a_p (2Zr)^{(p+2\lambda+1)/2} J_{p+2\lambda+1}(2\sqrt{2Zr}),$$

$$y_2 = \sum_{p=0}^{\infty} b_p (2Zr)^{(p+2\lambda+1)/2} J_{p-2\lambda-1}(2\sqrt{2Zr}),$$

$$a_p = b_p = 0 \ \text{ if } \ p < 0, \qquad a_0 = b_0 = 1,$$

$$a_p = -[(p+2\lambda)\, a_{p-2} - a_{p-3}]/(2\eta)^2 p,$$

$$b_p = -[(p-2\lambda-2)\, b_{p-2} - b_{p-3}]/(2\eta)^2 p. \qquad (28)$$

If we represent ψ for $r \geq r_0$ in the form

$$\psi = k^{-1/2} A\, (I - OU'), \qquad (29)$$

we obtain for the matrix U an expression similar to Eq. (12) of [9]:

$$U = e^{i\pi l/2} A e^{-i(\pi\lambda/2+\sigma_\lambda)} U' e^{-i(\pi\lambda/2+\sigma_\lambda)} A^{-1} e^{i\pi l/2}. \qquad (30)$$

We express, in accord with (5), (27), and (28),

[1)] For the formulation of the problem and for the definition of the quantities a, A, and λ used here see [9].

the matrix U′ in terms of R and U_1, and U_2 from (27) in terms of Y_1 and y_2 from (28), obtaining the expression

$$U' = -\frac{\Gamma(\lambda+1+i\eta)}{\Gamma(\lambda+1-i\eta)} - 2i\frac{\Gamma(\lambda+1+i\eta)}{\Gamma(\lambda+1)e^{\pi\eta/2}}k^{\lambda+1/2}$$

$$\times\left[M - \frac{2\pi}{\sin 2\pi\lambda}\frac{\Gamma(\lambda+1+i\eta)e^{-i\pi(\lambda+1/2)}k^{2\lambda+1}}{\Gamma(-\lambda+i\eta)\Gamma^2(\lambda+1)}\right]^{-1}$$

$$\times\frac{\Gamma(\lambda+1+i\eta)k^{\lambda+1/2}}{\Gamma(\lambda+1)e^{\pi\eta/2}}, \quad (31)$$

which replaces (10)[2)], where M [which differs from that given by (9)] depends on E only through R and y_1 and y_2, so that it is analytic near the threshold E_t. In the region near the threshold, where M can be regarded as constant and the asymptotic expansion can be used for $\Gamma(\lambda_t + 1 + i\eta_t)$, it follows from (30) and (31) that

$$U^b_{fi} = U^a_{fi} - \sum_{mm'}{}' U^a_{fm}\left[\frac{1}{\hat{U} - \exp\{i(\pi l - 2\pi/a\varkappa)\}}\right]_{mm'} U^a_{m'i}, \quad (32)$$

where $\hat{U}$ is a symmetrical matrix of rank n, containing the matrix elements of U^a connecting new channels only. Upon substitution of (4) Eq. (32) coincides with (20), so that the results of Sec. 3 are applicable also in the case of excitation of hydrogenlike ions by electrons.

2)For diagonal elements with integral λ it is necessary to go in the square brackets to the limit λ → integer, since M also contains the term $(\sin 2\pi\lambda)^{-1}$. We then obtain the expression in the denominator of (10).

[1] A. I. Baz', JETP 36, 1762 (1959), Soviet Phys. JETP 9, 1256 (1959).

[2] L. Fonda and R. Newton, Ann. Phys. 7, 133 (1959).

[3] R. Newton and L. Fonda, Ann. Phys. 9, 416 (1960).

[4] L. Fonda, Suppl. Nuovo cimento 20, 116 (1961).

[5] M. M. Hull and G. Breit, Handb. Physik, XLI/I, Berlin, 1959, p. 408. C. E. Froberg, Revs. Modern Phys. 27, 399 (1955).

[6] A. M. Lane and R. G. Thomas, Revs. Modern Phys. 30, 257 (1958).

[7] T. Teichmann, Phys. Rev. 83, 141 (1951).

[8] M. H. Ross, and G. L. Shaw, Ann. Phys. 13, 147 (1961).

[9] M. Gailitis and R. Damburg, JETP 44, 1644 (1963), Soviet Phys. JETP 17, 1107 (1963).

[10] P. M. Morse and H. Feshbach, Methods of Mathematical Physics, vol. I., McGraw-Hill, N. Y. 1953.

[11] T. S. Kuhn, Quart. Appl. Mat. 9, 1 (1951). M. Abramowitz, J. Math, Phys. 33, 111 (1954). F. S. Ham, Quart. Appl. Math. 15, 31, (1957); Solid State Physics v. 1, 190, Academic Press, N. Y., 1955.

Translated by J. G. Adashko

312

1983 *Rep. Prog. Phys.* **46** 167–257

Quantum defect theory

M J Seaton

Department of Physics and Astronomy, University College London, Gower Street, London WC1E 6BT, UK

Abstract

Quantum defect theory (QDT) is concerned with the properties of an electron in the field of a positive ion and, in particular, with expressing those properties in terms of analytical functions of the energy. It provides a unified theory of bound states, including series perturbations, autoionisation and electron–ion scattering, both elastic and inelastic.

The main emphasis of the review is on the foundations of the theory. Properties of Coulomb functions are discussed in some detail and outline sketches are given of relevant topics in collision theory and radiative theory. One-channel and many-channel QDT are discussed separately. Applications to the following problems are considered: resonances, atomic collision calculations, systems with two energy levels of the ion core, helium, other rare gases, alkaline earths and other atomic systems, molecular hydrogen, dielectronic recombination.

This review was received in July 1982.

0034–4885/83/020167 + 90$09.00

Contents

1. Introduction

The non-relativistic energy for a hydrogenic ion is

$$E_n = \frac{-me^4}{2\hbar^2}\frac{Z^2}{n^2} \tag{1.1}$$

where Z is the charge on the nucleus, n is the principal quantum number and $m = m_e(1+m_e/M)^{-1}$ is the reduced mass, m_e being the electron mass and M the mass of the nucleus. Consider a series of spectral lines due to transitions $n \rightarrow n_0$ with n_0 fixed. Let the wavelengths be λ_n and the wavenumbers be $T_n = 1/\lambda_n$. Then

$$T_n = (E_n - E_{n_0})/hc = T_\infty - RZ^2/n^2 \tag{1.2}$$

where $R = me^4(4\pi\hbar^3 c)^{-1}$ is known as the *Rydberg constant* and $T_\infty = RZ^2/n_0^2$ is known as the *series limit.* We may put $R = R(\infty)(1+m_e/M)^{-1}$ where $R(\infty)$ is the value of R for infinite nuclear mass. Since $R(\infty)$ can be determined from spectroscopic measurements, it is one of the most accurately known of physical constants. Amin *et al* (1981) obtain 109 737.315 21 ± 0.000 11 cm^{-1}.

The discovery of series formulae was of great importance in the early history of spectroscopy. A formula equivalent to (1.2) was discovered empirically by Balmer (1885) for the series in hydrogen with $n_0 = 2$. The empirical formula of Rydberg (1889) for series in the spectra of alkali atoms is

$$T_n = T_\infty - RZ^2(n-\mu)^{-2} \tag{1.3}$$

where μ is the *quantum defect* and $Z = 1$ for neutral atoms. Subsequent work showed that accurate fits to experimental term values could be obtained on replacing μ by μ_n with μ_n a slowly varying function of T_n. These results suggest that alkali-like atoms have energy levels

$$E_n = \frac{-me^4}{2\hbar^2}\frac{Z^2}{(n-\mu_n)^2}. \tag{1.4}$$

The *effective quantum number* ν_n is defined by

$$\nu_n = n - \mu_n \tag{1.5}$$

(the notation $\nu_n = n^*$ is often used).

Some results for s, p, d and f series in Na are shown in figure 1. Quantum defects are plotted against the variable

$$\epsilon_n = \frac{T_n - T_\infty}{RZ^2} = \frac{-1}{\nu_n^2}. \tag{1.6}$$

The experimental data and the analysis are from Risberg (1956) who obtains $T_\infty =$ 41 449.44 cm^{-1} and fits the quantum defects to polynomials in ϵ_n of order not higher than cubics. If $\mu(\epsilon)$ is the continuous function obtained from such a fit we have $\mu_n = \mu(\epsilon_n)$. The results of figure 1 show that the quantum defects μ_{nl}, where l is the orbital angular momentum quantum number, decrease rapidly as l increases. They

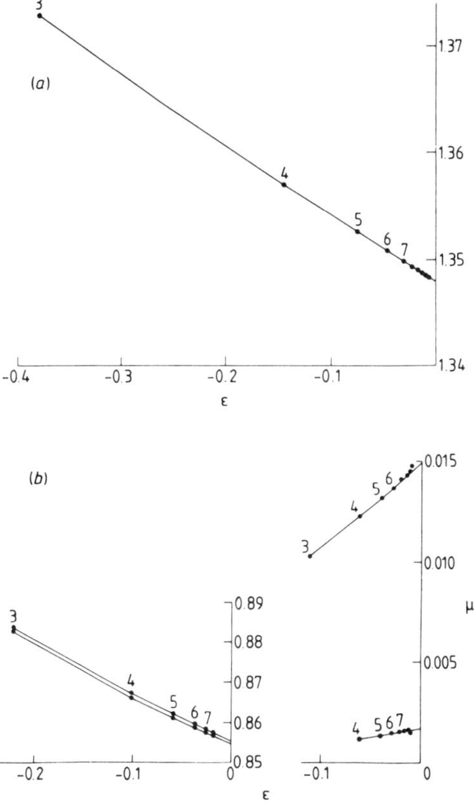

Figure 1. Quantum defects for Na I (from Risberg 1956). (*a*) s states, (*b*) left: p states; right, (top) d states, (bottom) f states.

are plotted on expanded scales and it must be noted that the quantum defects for the higher states (n large) are very sensitive to small errors in T_n, since a double cancellation occurs; firstly in calculating $(T_n - T_\infty)$ and secondly in calculating $(n - \nu_n) = \mu_n$. Much recent work has been concerned with making precision measurements for highly excited states.

Quantum defect formulae are of great importance for practical spectroscopy, for the following reasons: (*a*) fitting to the formulae gives values of T_∞ and hence provides ionisation potentials to spectroscopic accuracy; (*b*) values for quantum defects often aid identifications; (*c*) once quantum defects have been fitted to analytical formulae a lot of information is summarised concisely; (*d*) estimates can be obtained of wavelengths of lines which have not been observed; (*e*) more recent developments, which will be described in the present review, provide a unified treatment of diverse phenomena, such as series perturbations, autoionisation and resonances in electron–ion scattering. It is curious that, despite the importance of the subject, the theory of quantum defects is not discussed in most spectroscopy textbooks, which quote the basic formulae as though they had none other than a purely empirical justification.

Yet much of the theory is not new. It was discussed by Sommerfeld (1916, 1920) from the standpoint of the old quantum theory and a modern discussion, from the same standpoint, is given by Jaffé and Reinhardt (1977). For an electron moving in a pure Coulomb potential, classical mechanics gives the orbit to be an ellipse; and application of the Bohr–Sommerfeld quantum conditions gives equation (1.1) for the energy levels. For an alkali atom, Sommerfeld assumed that one electron can be excited easily, leaving the other electrons in a much more tightly bound atomic core. The excited electron is assumed to move in a potential due to the nucleus and to the core electrons, and to spend much of its time in an outer region where the nucleus is screened by the core electrons and where the potential is of Coulomb form, and some fraction of its time penetrating the region of the core. In this region there is a stronger attractive potential which causes the orbit to precess (there is no precession for the closed elliptical orbit for a pure Coulomb potential). Applying the quantum conditions, Sommerfeld obtained the expression (1.4) for the energy of the outer electron, with a quantum defect μ_n related to the rate of orbital precession. For states of large orbital angular momentum, there is little core penetration and μ_{nl} is small. In order to obtain agreement with experimental quantum defects for larger values of l one must allow for perturbation of the core by the outer electron, which gives rise to polarisation potentials behaving like α/r^4. This was first realised by Bohr (1923) and perturbation analyses were developed by Born and Heisenberg (1924) using the old quantum mechanics and by Waller (1926) using the new theory.

Hartree (1928) laid the foundations for the modern theory of quantum defects. He considered the radial Schrödinger equation

$$\left[-\frac{1}{2}\left(\frac{\mathrm{d}^2}{\mathrm{d}r^2}-\frac{l(l+1)}{r^2}\right)+[V(r)-E]\right]F(E,r)=0 \tag{1.7}$$

in Hartree units ($e=m=\hbar=1$), where $V(r)$ is the potential due to a nucleus of charge Z_0 screened by the spherically averaged charge distribution of N core electrons. For r sufficiently large there is complete screening and $V(r)=-Z/r$ where $Z=Z_0-N$. We put

$$V(r)=-Z/r \qquad \text{for} \qquad r\geqslant r_0. \tag{1.8}$$

Equation (1.7) has solutions $F_{\rm I}(E,r)$ and $F_{\rm II}(E,r)$ which are such that

$$F_{\rm I}(E,r)=0 \qquad \text{for} \qquad r\to 0 \tag{1.9}$$

$$F_{\rm II}(E,r)=0 \qquad \text{for} \qquad r\to\infty. \tag{1.10}$$

At the energy eigenvalues, $E=E_n$, $F_{\rm I}$ and $F_{\rm II}$ can be joined smoothly at $r=r_0$. This requires continuity of the logarithmic derivatives:

$$\frac{\mathrm{d}}{\mathrm{d}r}\ln F_{\rm I}=\frac{\mathrm{d}}{\mathrm{d}r}\ln F_{\rm II} \qquad \text{at } r=r_0 \quad \text{and} \quad E=E_n. \tag{1.11}$$

For values of E such that

$$|E|\ll|V(r)| \qquad \text{for} \qquad r\leqslant r_0 \tag{1.12}$$

$F_{\rm I}$ can be normalised in such a way that it is insensitive to E for $r<r_0$. For $r>r_0$, $V(r)$ is given by (1.8) and the Schrödinger equation (1.7) can be solved analytically. The solutions are known as *Coulomb functions* and the whole of *quantum defect theory* (QDT) hinges on a knowledge of their mathematical properties. This subject will be discussed in § 2. Since (1.7) is a second-order equation it has two linearly independent solutions. In § 2.7.3 we shall define the solutions $s(E,r)$ and $c(E,r)$.

They are insensitive to E at $r=r_0$ if (1.12) is satisfied. For $E<0$, s and c increase exponentially for $r\to\infty$ but $F_{\rm II}$ can be taken to be a linear combination of s and c such that (1.10) is satisfied. Hartree showed that a suitable linear combination is

$$F_{\rm II}=-\cos(\pi\nu)s+\sin(\pi\nu)c \tag{1.13}$$

where $E=-Z^2(2\nu^2)^{-1}$. It follows that all rapid variations of $F_{\rm II}$ as a function of E are due to the factors $\cos(\pi\nu)$ and $\sin(\pi\nu)$ in (1.13). Let us put $\nu=\nu_n=(n-\mu_n)$, implying $E=E_n=-Z^2(2\nu_n^2)^{-1}$, to obtain

$$F_{\rm II}=-(-1)^n[\cos(\pi\mu_n)s+\sin(\pi\mu_n)c]. \tag{1.14}$$

We may now take μ_n to be such that (1.11) is satisfied. At $r=r_0$, $F_{\rm I}$, s and c are slowly varying functions of E and it follows that $\mu_n=\mu(E_n)$ is also slowly varying. This completes Hartree's derivation of the quantum defect formula. The bound-state solutions of (1.7) may be normalised in such a way that $F=\mathscr{F}$ where

$$\mathscr{F}(E_n,r)=\cos(\pi\mu_n)s(E_n,r)+\sin(\pi\mu_n)c(E_n,r) \qquad \text{for } r\geqslant r_0. \tag{1.15}$$

Further interest in the theory of quantum defects was stimulated by the work of Bates and Damgaard (1949) whose Coulomb approximation provides a powerful method for the computation of bound–bound oscillator strengths for simple atomic systems. For $r>r_0$ they use Coulomb wavefunctions satisfying the boundary condition at $r\to\infty$ and with energies E_n determined spectroscopically. For transitions between excited states there is only a small contribution to the oscillator strengths from $r<r_0$ and a simple cut-off procedure was used. The theory was extended by Burgess and Seaton (1960) to calculate bound–free oscillator strengths (photoionisation). For this purpose it was necessary to obtain a relation between quantum defects for bound states and phases of wavefunctions for continuum states.

For $E>0$, s and c have asymptotic forms for $r\to\infty$:

$$s\sim(\pi k)^{-1/2}\sin(\zeta) \qquad c\sim(\pi k)^{-1/2}\cos(\zeta) \tag{1.16}$$

where

$$\zeta=kZr-\frac{1}{2}l\pi+\frac{1}{k}\ln(2kZr)+\arg\Gamma(l+1-\mathrm{i}/k) \tag{1.17}$$

and $k^2=2E/Z^2$. Put $\mu_n=\mu(E_n)$ and let $\delta(E)$ be the analytic continuation of $\pi\mu(E)$. Then for $E>0$ the continuation of the function (1.15) has asymptotic form

$$\mathscr{F}(E,r)\sim(\pi k)^{-1/2}\sin[\zeta+\delta(E)] \tag{1.18}$$

which was the result obtained by Seaton (1955, 1958a).

Meanwhile, an interest in quantum defect theory had arisen in connection with problems in solid-state physics (Kuhn and Van Vleck 1950) which led to developments in the mathematical theory, described in a review article by Ham (1955). The assumption made is that one can use a one-electron Hamiltonian for the calculation of the wavefunctions in valence and conduction bands, and that near any of the ions in the lattice this Hamiltonian is the same as that for a valence electron in the corresponding free atom. For applications to problems of the solid state it is necessary to consider, for finite values of r, the wavefunctions $\mathscr{F}(E,r)$ at negative energies E other than the eigenenergies E_n of the free atom. For $r>r_0$ these functions are

$$\mathscr{F}(E,r)=\cos(\pi\mu(E))s+\sin(\pi\mu(E))c \tag{1.19}$$

where $\mu(E)$ is interpolated from the experimental quantum defects $\mu(E_n)$.

It has been seen that, for simpler systems such as alkali atoms, the slow variation of quantum defects as functions of energy can be understood by considering the motion of an electron in the central potential of an ion core. It was first shown by Langer (1930), Shenstone (1931) and Shenstone and Russell (1932) that more rapid variations occur in systems, such as the alkaline earths, in which the core itself can be excited rather easily. Let us consider the case of Ca I. The Ca^+ ion core is an alkali-like system with a ground state $1s^2 2s^2 2p^6 3s^2 3p^6 4s$ and a number of low-lying excited states of which the first is $\ldots 3p^6 3d$. The complications which arise when an electron is added to such a core are illustrated by considering the $^1P^o$ series of Ca I. The energy levels, in order of increasing energy, are usually labelled

$$4s4p, 4s5p, 4s6p, 3d4p, 4s7p, 4s8p, \ldots \tag{1.20}$$

and 3d4p is referred to as a perturber level. Figure 2 shows quantum defects for this series obtained using wavelengths measured by Garton and Codling (1965) and, in equation (1.3), taking $n = 4, 5, 6, 7, 8, 9, \ldots,$ for the levels (1.20), so that $n = 7$ is used for the level labelled 3d4p and $n = 8, 9, \ldots,$ for those labelled 4s7p, 4s8p, ... It is seen that μ can be fitted to a smooth function of the energy and that this function increases by an amount approximately equal to unity in the vicinity of the perturbation. This treatment, in which the levels are numbered consecutively and no distinction is made between the 'perturber level' and the 'perturbed series', differs from the treatment of Langer, who used perturbation theory and obtained an expression for $\mu(E)$ having a pole at the position of the perturbing level.

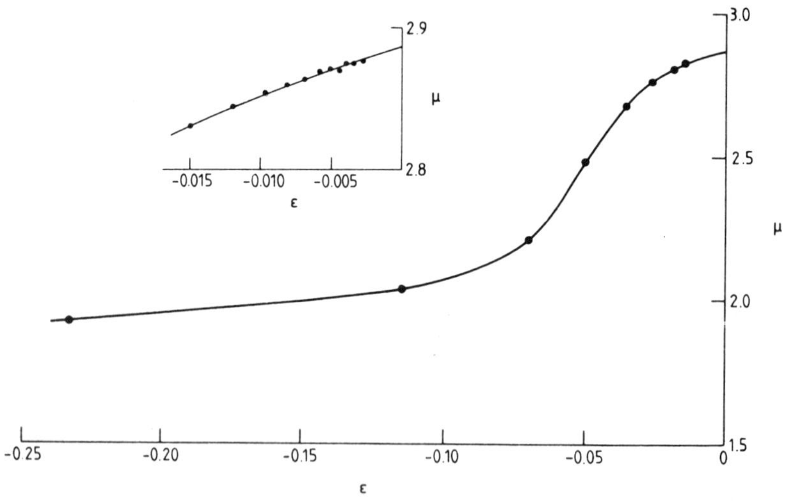

Figure 2. Quantum defects for Ca I $^1P^o$ (from Garton and Codling 1965). The labels usually used for the lowest six states are given by (1.20).

The 3d4p state of Ca I provides an example of a doubly excited state which lies below a first ionisation limit and causes a perturbation in a series converging to that limit. The work of Shenstone and Russell showed that there are many more doubly excited states of Ca I and that most of them lie above the first ionisation limit. The interaction between such states and continuum states (Ca^+ 4s + electron) produces the phenomenon of autoionisation, which was first discussed by Shenstone (1931) following a suggestion of Wigner. That the interactions which produce series perturbations are related to those which produce autoionisation was recognised by Shenstone. At higher energies one has interactions between continua, such as (Ca^+ 4s + electron) and (Ca^+ 3d + electron), which produce inelastic collisions.

Many-channel quantum defect theory was developed by Gailitis (1963) who considered resonances in electron–ion scattering cross sections and by Bely *et al* (1963) who considered perturbed series, autoionisation and inelastic scattering. In the many-channel theory one considers a number of core states and the differential equation (1.13) is replaced by a system of coupled equations of a type familiar in atomic collision theory.

A series of papers on quantum defect theory has been published by the group at University College London†.

A review by Fano (1975) is mainly concerned with work done by his group at Chicago. Fano (1970) first applied the theory to the analysis of molecular spectra, and gives a formulation of the theory in which use is made of *eigen quantum defects*.

Lu and Fano (1970) introduced graphical methods which can be used for systems in which the ion core can be considered to have two energy levels. Further generalisations of the theory are discussed by Greene *et al* (1979), who consider problems with: no long-range potentials; potentials behaving like $1/r^2$; Coulomb potentials. Applications to molecular problems are discussed further by, among others, Jungen and Atabek (1977).

Relations between quantum defect and configuration-interaction theories are discussed by Fano (1978) and Mies (1979). The relativistic theory has been developed by Johnson and Cheng (1979) and Lee and Johnson (1980).

The editors of these reviews instruct contributors to write in a style intended to be comprehensible to the non-specialist, and not to exceed a prescribed length. Compliance with these instructions usually requires that compromises should be made. The present contributor puts the main emphasis on the foundations of the theory at the expense of giving briefer accounts of some of its applications.

2. Coulomb functions

The theory of Coulomb functions is discussed by Whittaker and Watson (1927), Erdélyi *et al* (1953), Buchholz (1953) and Slater (1960).

2.1. Scaled variables

We use Z-scaled variables

$$\rho = \left(\frac{me^2}{\hbar^2}\right) Zr \qquad \epsilon = \left(\frac{\hbar^2}{me^4}\right)\frac{2E}{Z^2} \tag{2.1}$$

† I, General formulation (Seaton 1966a); II, illustrative one-channel and two-channel problems (Seaton 1966b); III, electron scattering by He^+ (Bely 1966); IV, the absorption of radiation by calcium atoms (Moores 1966); V, autoionising and bound states of the neutral beryllium atom (Moores 1967); VI, extrapolations along isoelectronic sequences (Doughty *et al* 1968); VII, analysis of resonance structures (Seaton 1969); VIII, resonances in the collision strengths for O^+ $2p^3$ $^2D_{3/2}$–$^2D_{5/2}$ (Martins and Seaton 1969); IX, complex quantum defects for the $e + Be^+$ system (Norcross and Seaton 1970); X, photoionisation (Dubau and Wells 1973); XI, clarification of some aspects of the theory (Seaton 1978); XII, complex quantum defects for the $He^+ + e^-$ system (Dubau 1978); XIII, further discussion of photoionisation (Dubau and Seaton 1983); XIV, dielectronic recombination (Bell and Seaton 1983).

and we put

$$\epsilon = k^2 \quad \text{for } \epsilon > 0 \qquad \epsilon = -1/\nu^2 \quad \text{for } \epsilon < 0. \tag{2.2}$$

In Hartree atomic units the radial variable is $r = \rho/Z$, the energy is $E = \epsilon Z^2/2$ and the electron wavenumber is kZ. The energy in Rydberg units is ϵZ^2. The effective quantum number ν is as defined in § 1.

Using (2.1), the Coulomb radial equation is

$$\left(\frac{d^2}{d\rho^2} - \frac{l(l+1)}{\rho^2} + \frac{2}{\rho} + \epsilon\right) F = 0. \tag{2.3}$$

2.2. *Descriptive properties*

2.2.1. Behaviour at the origin. Equation (2.3) has solutions $f(\epsilon, l; \rho)$ and $g(\epsilon, l; \rho)$ such that, in the limit of ρ small, f behaves like ρ^{l+1} and g like ρ^{-l}. It may be shown, by substitution in (2.3), that $f(\epsilon, l; \rho)$ has an expansion in powers of ρ. In order to expand $g(\epsilon, l; \rho)$, it is necessary to introduce terms in $\ln \rho$.

The functions $f(\epsilon, l; \rho)$ and $g(\epsilon, l; \rho)$ can be defined in such a way (§ 2.5.3) that they are analytic functions of ϵ.

2.2.2. Oscillatory solutions. From (2.3),

$$\left(\frac{d^2}{d\rho^2} + w\right) F = 0 \tag{2.4}$$

where

$$w = \frac{-l(l+1)}{\rho^2} + \frac{2}{\rho} + \epsilon. \tag{2.5}$$

The solutions F are oscillatory for values of ρ such that $w > 0$ and have points of inflection at values of ρ such that $w = 0$. The latter values are ρ_a and ρ_b where

$$\rho_a = \{-1 + [1 + \epsilon l(l+1)]^{1/2}\}/\epsilon \qquad \rho_b = \{-1 - [1 + \epsilon l(l+1)]^{1/2}\}/\epsilon. \tag{2.6}$$

For ϵ negative (but $\epsilon > -[l(l+1)]^{-1}$), ρ_a and ρ_b are both positive and (2.4) has oscillatory solutions for $\rho_a < \rho < \rho_b$.

For $\epsilon \geq 0$, ρ_b is negative and (2.4) has oscillatory solutions for $\rho \geq \rho_a$.

2.2.3. Asymptotic forms.

(*a*) $\epsilon < 0$. For $\epsilon < 0$ and $\rho > \rho_b$, w is negative. The JWBK method gives approximate solutions of (2.4) to be $\exp(\pm u)$ where

$$u = \int^{\rho} (-w)^{1/2} \, d\rho. \tag{2.7}$$

For $\rho \gg \rho_b$, $(-w)^{1/2} = (1/\nu) - (\nu/\rho) + O(\rho^{-2})$ and $u = (\rho/\nu) - \nu \ln(\rho) + O(\rho^0)$. Equation (2.4) therefore has solutions behaving like $\rho^{\nu} \exp(-\rho/\nu)$ and $\rho^{-\nu} \exp(+\rho/\nu)$ in the limit of ρ large.

(*b*) $\epsilon \geq 0$. For $\epsilon \geq 0$ and $\rho > \rho_a$, w is positive. The JWBK method gives approximate solutions of (2.4) to be $w^{-1/4} \sin(v)$ and $w^{-1/4} \cos(v)$ where

$$v = \int^{\rho} w^{1/2} \, d\rho. \tag{2.8}$$

Expressions for v are given by Seaton and Peach (1962) and an improved approximation is discussed by Burgess (1963). For $\epsilon=0$ and $\rho \gg \rho_a$ one obtains $w^{1/2}=(2/\rho)^{1/2}+O(\rho^{-3/2})$ and $v=(8\rho)^{1/2}+O(\rho^0)$ while for $\epsilon>0$ and $\rho\gg\rho_a$ one obtains $w^{1/2}=k+1/(k\rho)+O(\rho^{-2})$ and $v=kr+(1/k)\ln(\rho)+O(\rho^0)$.

2.2.4. Need for more powerful analytical methods. More powerful analytical methods are required in order to obtain all of the properties of Coulomb functions used in QDT. It will be found that special mathematical difficulties arise for the cases of physical interest, integer values of the variable l and real values of the variable ϵ. It will therefore be convenient to develop the mathematical theory for more general values of these variables and to obtain the physical solutions as limiting cases.

2.3. Series solutions

2.3.1. First form of series solutions. We consider the equation

$$\left(\frac{d^2}{d\rho^2}-\frac{\lambda^2-\frac{1}{4}}{\rho^2}+\frac{2}{\rho}+\epsilon\right)F=0 \tag{2.9}$$

which reduces to (2.3) if we take

$$\lambda=\pm(l+\tfrac{1}{2}). \tag{2.10}$$

We assume that (2.9) has solutions of the form

$$F(\epsilon,\lambda;\rho)=\sum_{n=0}^{\infty} a_n(\epsilon,\lambda)\rho^{n+\lambda+\frac{1}{2}} \tag{2.11}$$

with $a_0\neq 0$. Substitution in (2.9) gives

$$a_1=-2a_0(2\lambda+1)^{-1} \tag{2.12}$$

and, for $n\geqslant 2$,

$$a_n=-(2a_{n-1}+\epsilon a_{n-2})[n(n+2\lambda)]^{-1}. \tag{2.13}$$

Since λ enters (2.9) only through λ^2, $F(\epsilon,-\lambda;\rho)$ will be a solution if $F(\epsilon,\lambda;\rho)$ is a solution. A special difficulty with $\lambda=-(l+\frac{1}{2})$ is that (2.12) and (2.13) give a_n to be infinite for $n=(2l+1)$. The theory is therefore developed for arbitrary λ and (2.10) is considered later as a special case.

2.3.2. Second form of series solution. We put

$$\epsilon=-1/\kappa^2 \tag{2.14}$$

and

$$z=2\rho/\kappa \tag{2.15}$$

and consider solutions for complex z. Direct substitution shows that $F(\epsilon,\lambda;\rho)=y(\kappa,\lambda;z)$ is a solution of (2.9) where

$$y(\kappa,\lambda;z)=\frac{(\kappa z)^{\lambda+\frac{1}{2}}\exp(-z/2)}{\Gamma(\lambda+\frac{1}{2}-\kappa)}\sum_{n=0}^{\infty}\frac{\Gamma(\lambda+\frac{1}{2}-\kappa+n)z^n}{\Gamma(2\lambda+1+n)n!}. \tag{2.16}$$

2.3.3. Solutions analytic in ϵ. The series in (2.16) is readily shown to be convergent (by which we mean absolutely and uniformly convergent), as is the series for $\exp(-z/2)$.

We can therefore re-arrange the product of these two series to give a series of the form (2.11), and we may conclude that (2.11) is convergent. Comparing leading terms, one obtains $F(\epsilon, \lambda; \rho) = y(\kappa, \lambda; z)$ if one takes $a_0 = 2^{\lambda+\frac{1}{2}}[\Gamma(2\lambda+1)]^{-1}$.

The recurrence relation (2.13) is linear in ϵ and it follows that $a_n(\epsilon, \lambda)$ is a polynomial in ϵ and that (2.11) can be re-arranged as a convergent series in powers of ϵ. We conclude that the solutions $y(\kappa, \lambda; z)$ are analytic in ϵ.

2.4. *Contour integral representations*

2.4.1. The contour integral for $y(\kappa, \lambda; z)$. The reciprocal of the Γ function has a contour integral representation

$$\frac{1}{\Gamma(x)} = \frac{1}{2\pi i} \int_C \exp(t) t^{-x} \, dt \tag{2.17}$$

where the contour C is shown in figure 3. Using (2.17) for $[\Gamma(2\lambda+1+n)]^{-1}$ in (2.16)

$$y = \frac{(\kappa z)^{\lambda+\frac{1}{2}} \exp(-z/2)}{2\pi i} \int_C \exp(t) t^{-2\lambda-1} \left(\sum_{n=0}^{\infty} \frac{\Gamma(\lambda+\frac{1}{2}-\kappa+n)(z/t)^n}{\Gamma(\lambda+\frac{1}{2}-\kappa)n!} \right) dt. \tag{2.18}$$

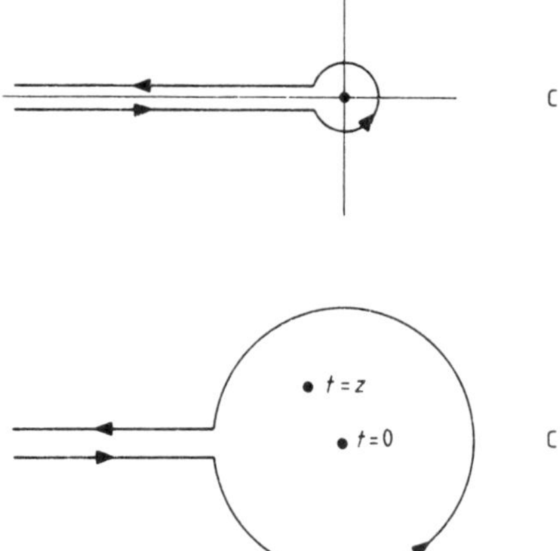

Figure 3. Contours in the complex t plane.

The binomial theorem is

$$(1-u)^{-\alpha} = \sum_{n=0}^{\infty} \frac{\Gamma(\alpha+n)u^n}{\Gamma(\alpha)n!} \tag{2.19}$$

and is valid for $|u| < 1$. The summation in (2.18) is therefore equal to $(1-z/t)^{-\lambda-\frac{1}{2}+\kappa}$ if $|z/t| < 1$, a condition which is satisfied if the contour C is deformed to the contour C′ of figure 3. Hence

$$y(\kappa, \lambda; z) = \frac{(\kappa z)^{\lambda+\frac{1}{2}} \exp(-z/2)}{2\pi i} \int_{C'} \exp(t) t^{-2\lambda-1} \left(1 - \frac{z}{t}\right)^{-\lambda-\frac{1}{2}+\kappa} dt. \tag{2.20}$$

2.4.2. The Whittaker function. Equation (2.9) has other solutions in terms of integrals with integrands similar to that of (2.20). The Whittaker function $W_{\kappa,\lambda}(z)$ may be defined by

$$W_{\kappa,\lambda}(z)=\frac{\Gamma(\kappa+\frac{1}{2}+\lambda)z^{\kappa}\exp(-z/2)}{2\pi\mathrm{i}}\int_{\mathrm{C}}\exp(t)t^{-\lambda-\frac{1}{2}-\kappa}\left(1-\frac{t}{z}\right)^{-\lambda-\frac{1}{2}+\kappa}\mathrm{d}t \quad (2.21)$$

for $|\arg(z)|<\pi$. It may be shown by direct substitution and sufficient effort that (2.21) is a solution of (2.9). Since κ enters (2.9) only through $\epsilon=-1/\kappa^2$, and since $z=2\rho/\kappa$, $W_{-\kappa,\lambda}(-z)$ must also be a solution. Care is needed in defining $-z$. We put

$$-z=\exp(\mathrm{i}\pi D)z \quad (2.22)$$

with

$$D=-1 \text{ for } \arg(z)>0 \qquad D=+1 \text{ for } \arg(z)>0. \quad (2.23)$$

Then $|\arg(-z)|<\pi$ if $|\arg(z)|<\pi$.

2.4.3. Relation between y and Whittaker functions. On deforming the contour C′ used in (2.20) and making a change of integration variable, we can express y in terms of two integrals over contours of type C. This gives

$$y(\kappa,\lambda;z)=\kappa^{\lambda+\frac{1}{2}}\exp(\mathrm{i}\pi\kappa D)\left(\frac{\exp[-\mathrm{i}\pi(\lambda+\frac{1}{2})D]W_{\kappa,\lambda}(z)}{\Gamma(\kappa+\frac{1}{2}+\lambda)}+\frac{W_{-\kappa,\lambda}[\exp(\mathrm{i}\pi D)z]}{\Gamma(-\kappa+\frac{1}{2}+\lambda)}\right) \quad (2.24)$$

which may be used for $0<\arg(z)<\pi$ with $D=-1$ or for $-\pi<\arg(z)<0$ with $D=+1$.

2.4.4. Asymptotic expansions of Whittaker functions. The integrand of (2.21) contains a factor $(1+t/z)^{-\lambda-\frac{1}{2}+\kappa}$ which, by (2.19), has the expansion

$$\left(1-\frac{t}{z}\right)^{-\lambda-\frac{1}{2}+\kappa}=\sum_{n=0}^{\infty}\frac{\Gamma(\lambda+\frac{1}{2}-\kappa+n)(t/z)^n}{\Gamma(\lambda+\frac{1}{2}-\kappa)n!} \quad (2.25)$$

valid for $|t/z|<1$. The expansion (2.25) is not convergent at all points on the contour C but, in the limit of $|z|$ large, one obtains a good approximation to $W_{\kappa,\lambda}$ on retaining only a finite number of terms in (2.25). This gives

$$W_{\kappa,\lambda}(z)=z^{\kappa}\exp(-z/2)\left(1-\frac{(\lambda+\frac{1}{2}-\kappa)(-\lambda+\frac{1}{2}-\kappa)}{1!}\frac{1}{z}+\ldots\right). \quad (2.26)$$

Since $W_{\kappa,\lambda}(z)$ is a solution of (2.9), $W_{\kappa,-\lambda}(z)$ must also be a solution. These two solutions have the same asymptotic form and it may be shown to follow that

$$W_{\kappa,\lambda}(z)=W_{\kappa,-\lambda}(z). \quad (2.27)$$

From (2.24), (2.26) and (2.27) one obtains asymptotic expansions for $y(\kappa,\lambda;z)$ and $y(\kappa,-\lambda;z)$.

2.5. Coulomb functions for $\lambda=l+\frac{1}{2}$

2.5.1. The functions $y(\kappa,l+\frac{1}{2};z)$ and $y(\kappa,-l-\frac{1}{2};z)$. For $\lambda\neq(l+\frac{1}{2})$, $y(\kappa,\lambda;z)$ and $y(\kappa,-\lambda;z)$ are linearly independent Coulomb functions. These two functions are not linearly independent for $\lambda=(l+\frac{1}{2})$; in the expansion (2.16) for $y(\kappa,-l-\frac{1}{2};z)$, the first $(2l+1)$ terms vanish and examination of subsequent terms shows that

$$y(\kappa,-l-\tfrac{1}{2};z)=-\mathscr{A}(\kappa,l+\tfrac{1}{2})y(\kappa,l+\tfrac{1}{2};z) \quad (2.28)$$

where

$$\mathscr{A}(\kappa, \lambda) = \frac{\Gamma(\kappa + \lambda + \frac{1}{2})}{\kappa^{2\lambda}\Gamma(\kappa - \lambda + \frac{1}{2})}. \tag{2.29}$$

We put

$$A(\epsilon, l) = \mathscr{A}(\kappa, l + \tfrac{1}{2}) = \frac{\Gamma(\kappa + l + 1)}{\kappa^{2l+1}\Gamma(\kappa - l)} \tag{2.30}$$

and, using $\Gamma(1+x) = x\Gamma(x)$, obtain

$$A(\epsilon, l) = \prod_{p=0}^{l} (1 + p^2\epsilon). \tag{2.31}$$

2.5.2. The function $\eta(\kappa, \lambda; z)$. For all λ, two linearly independent Coulomb functions are given by $y(\kappa, \lambda; z)$ and $\eta(\kappa, \lambda; z)$, where

$$\eta(\kappa, \lambda; z) = \frac{\mathscr{A}(\kappa, \lambda)\cos(2\pi\lambda) y(\kappa, \lambda; z) - y(\kappa, -\lambda; z)}{\sin(2\pi\lambda)}. \tag{2.32}$$

Taking the limit $\lambda \to (l + \frac{1}{2})$ we obtain

$$\eta(\kappa, l + \tfrac{1}{2}; z) = \frac{1}{2\pi}\frac{\mathrm{d}}{\mathrm{d}\lambda}\{A(\epsilon, l) y(\kappa, \lambda; z) + y(\kappa, -\lambda; z)\}|_{\lambda = l + \frac{1}{2}} + G(\kappa, l) y(\kappa, l + \tfrac{1}{2}; z) \tag{2.33}$$

where

$$G(\kappa, l) = \frac{1}{2\pi}\frac{\mathrm{d}}{\mathrm{d}\lambda}\mathscr{A}(\kappa, \lambda)|_{\lambda = l + \frac{1}{2}}. \tag{2.34}$$

Using (2.29)

$$G(\kappa, l) = \frac{A(\epsilon, l)}{2\pi}\{\psi(\kappa + l + 1) + \psi(\kappa - l) - 2\ln(\kappa)\} \tag{2.35}$$

where $\psi(x) = \mathrm{d}\ln\Gamma(x)/\mathrm{d}x$.

An alternative expression for η is obtained on using (2.24) and (2.27) to obtain an expression for $y(\kappa, -\lambda; z)$ in terms of $y(\kappa, \lambda; z)$ and $W_{\kappa,\lambda}(z)$. This gives

$$\eta(\kappa, l + \tfrac{1}{2}; z) = \cot[\pi(l + 1 - \kappa)] A y(\kappa, l + \tfrac{1}{2}; z) - \frac{\Gamma(l + 1 - \kappa)}{\pi\kappa^{l}} W_{\kappa, l + \frac{1}{2}}(z). \tag{2.36}$$

2.5.3. Solutions analytic in ϵ. We may now define the two solutions analytic in ϵ, $f(\epsilon, l; \rho)$ and $g(\epsilon, l; \rho)$.

The functions $y(\kappa, \lambda; z)$ and $y(\kappa, -\lambda; z)$ are analytic in ϵ for all λ. We take

$$f(\epsilon, l; \rho) = y(\kappa, l + \tfrac{1}{2}; z). \tag{2.37}$$

In the expression (2.33) for $\eta(\kappa, l + \frac{1}{2}; \rho)$, $G(\kappa, l)$ is not analytic in ϵ. The function $g(\epsilon, l; \rho)$ is taken to be

$$g(\epsilon, l; \rho) = \eta(\kappa, l + \tfrac{1}{2}; \rho) - G(\kappa, l) y(\kappa, l + \tfrac{1}{2}; z). \tag{2.38}$$

The two solutions analytic in ϵ are independent of ϵ in the limit of $\rho \to 0$ ($\lim_{\rho\to 0}(\rho^{-l-1} f) = 2^{l+1}[(2l+1)!]^{-1}$ and $\lim_{\rho\to 0}(\rho^{l} g) = (2l)![(\pi 2^{l})]^{-1}$).

Practical procedures for the computation of f and g are described by Seaton (1982a).

2.5.4. The function $G(\kappa, l)$. We consider ϵ real and put

$$G = \mathcal{G} + \mathrm{i}\mathcal{H} \tag{2.39}$$

where $\mathcal{G}$ and $\mathcal{H}$ are real. It follows from (2.34) that

$$\mathcal{H} = 0 \text{ for } \epsilon < 0 \qquad \mathcal{H} = \frac{A}{\exp(2\pi/k) - 1} \text{ for } \epsilon > 0. \tag{2.40}$$

Using Stirling's formula for $|\epsilon|$ small, and hence $|\kappa|$ large, one obtains the asymptotic expansion

$$\mathcal{G} = \frac{\epsilon A}{\pi}\left[\sum_{p=0}^{l} \frac{p}{1+p^2\epsilon} + \frac{1}{12}\left(1 + \frac{\epsilon}{10} + \frac{\epsilon^2}{21} + \frac{\epsilon^3}{20} + \ldots\right)\right] \tag{2.41}$$

(the coefficients can be expressed in terms of Bernoulli numbers). It follows from (2.31) that $A\ \Sigma_{p=0}^{l}\, p(1+p^2\epsilon)^{-1}$ is a polynomial in ϵ, and hence that $\mathcal{G}$ has an asymptotic expansion in powers of ϵ.

2.5.5. The function $h(\epsilon, l; \rho)$. The function h defined by

$$h(\epsilon, l; \rho) = -[\eta(\kappa, l+\tfrac{1}{2}; \rho) - \mathrm{i}\mathcal{H}y(\kappa, l+\tfrac{1}{2}; \rho)] = -(g + \mathcal{G}f) \tag{2.42}$$

has an asymptotic expansion in powers of ϵ. The asymptotic forms of h are simpler than those of g.

2.6. Asymptotic forms

The asymptotic forms of Coulomb functions are obtained using (2.26), together with (2.24) and (2.36).

2.6.1. Asymptotic forms for $\epsilon > 0$. For ϵ real and positive we put $\kappa = \mathrm{i}/k$ with $k > 0$ giving $z = \exp(-\mathrm{i}\pi/2)2\rho\kappa$, $\arg(z) = -\pi/2$ and $D = +1$ (the same final result is obtained with $\kappa = -\mathrm{i}/k$ and $D = -1$). After some manipulation, which involves use of the relation $\Gamma(x)\Gamma(1-x) = \pi/\sin(\pi x)$, we obtain

$$f \underset{\rho\to\infty}{\sim} \left(\frac{2}{\pi k}\right)^{1/2} \left(\frac{1 - \exp(-2\pi/k)}{A(k^2, l)}\right)^{1/2} \sin(\zeta) \tag{2.43}$$

and

$$h \underset{\rho\to\infty}{\sim} \left(\frac{2}{\pi k}\right)^{1/2} \left(\frac{A(k^2, l)}{1 - \exp(-2\pi/k)}\right)^{1/2} \cos(\zeta) \tag{2.44}$$

where

$$\zeta = k\rho - \frac{1}{2}l\pi + \frac{1}{k}\ln(2k\rho) + \arg\Gamma(l + 1 - \mathrm{i}z/k). \tag{2.45}$$

2.6.2. Functions with simple asymptotic forms for $\epsilon < 0$. For ϵ real and negative we put $\kappa = \nu$. The function $W_{\nu, l+\frac{1}{2}}(2\rho/\nu)$ is real and, for large ρ, is exponentially decreasing. The functions $W_{-\nu, l+\frac{1}{2}}[\exp(\mathrm{i}\pi D)2\rho/\nu]$ with $D = +1$ or $D = -1$ are complex and, for ρ large, are exponentially increasing. We define the real functions

$$\theta(\nu, l; \rho) = W_{\nu, l+\frac{1}{2}}\left(\frac{2\rho}{\nu}\right) \tag{2.46}$$

$$\xi(\nu, l; \rho) = \frac{1}{2}\left[\exp(\mathrm{i}\pi\nu) W_{-\nu, l+\frac{1}{2}}\left(\exp(\mathrm{i}\pi)\frac{2\rho}{\nu}\right) + \exp(-\mathrm{i}\pi\nu) W_{-\nu, l+\frac{1}{2}}\left(\exp(-\mathrm{i}\pi)\frac{2\rho}{\nu}\right)\right] \tag{2.47}$$

which have asymptotic forms

$$\xi \underset{\rho\to\infty}{\sim} \left(\frac{2\rho}{\nu}\right)^{-\nu} \exp\left(+\frac{\rho}{\nu}\right) \qquad \theta \underset{\rho\to\infty}{\sim} \left(\frac{2\rho}{\nu}\right)^{\nu} \exp\left(-\frac{\rho}{\nu}\right). \tag{2.48}$$

2.6.3. Relations for $\epsilon < 0$. Coulomb functions for $\epsilon < 0$ can be expressed as linear combinations of θ and ξ. For f and h we obtain

$$f = (-1)^l \nu^{l+1}\left(\frac{\sin(\pi\nu)\Gamma(\nu - l)}{\pi}\xi - \frac{\cos(\pi\nu)}{\Gamma(\nu + l + 1)}\theta\right) \tag{2.49}$$

and

$$h = (-1)^l \nu^{l+1} A\left(\frac{\cos(\pi\nu)\Gamma(\nu - l)}{\pi}\xi + \frac{\sin(\pi\nu)}{\Gamma(\nu + l + 1)}\theta\right). \tag{2.50}$$

2.7. Coulomb functions used in QDT

2.7.1. Restrictions on the range of QDT. QDT is concerned with states of atomic systems which have one outer electron, with Z-scaled energy ϵ, outside of a core of bound electrons. If ϵ is negative, the outer electron is bound but can only be considered to be 'outside' of the core if its binding energy is less than that of the core electrons. This restricts the theory to values of ϵ such that $\epsilon \gtrsim -1$. An electron with ϵ large and positive can ionise the core and hence give states with two electrons outside of the core, which contradicts our assumption that there is only one. The condition that it cannot ionise the core restricts ϵ to be such that $\epsilon \lesssim 1$. For these reasons we need to concern ourselves only with values of ϵ such that $|\epsilon| \lesssim 1$.

2.7.2. The functions f, g and h. The functions f and g are analytic in ϵ and hence have convergent expansions of the form

$$f(\epsilon, l; \rho) = \sum_{n=0}^{\infty} \epsilon^n f_{(n)}(l; \rho) \qquad g(\epsilon, l; \rho) = \sum_{n=0}^{\infty} \epsilon^n g_{(n)}(l; \rho). \tag{2.51}$$

Ham (1955) shows that the functions $f_{(n)}(l; \rho)$ and $g_{(n)}(l; \rho)$ can be expressed in terms of Bessel functions.

The function h is defined by $h = -(g + \mathcal{G}f)$ where $\mathcal{G}$ is zero for $\epsilon = 0$ and has an asymptotic expansion in powers of ϵ. For ϵ small a good approximation to $\mathcal{G}$ is obtained on retaining only a finite number of terms in its expansion. In this sense we may say that $\mathcal{G}$, and hence h, are 'nearly analytic' functions of ϵ. For most applications in QDT we may treat h as though it were an analytic function of ϵ.

Equations (2.43) and (2.44) contain factors $[1 - \exp(-2\pi/k)]$ which are close to unity for $\epsilon \lesssim 1$.

2.7.3. The functions s and c. We define

$$B(\epsilon, l) = \begin{cases} A(\epsilon, l)[1-\exp(-2\pi/k)]^{-1} & \text{for } \epsilon > 0 \\ A(\epsilon, l) & \text{for } \epsilon < 0 \end{cases} \tag{2.52}$$

and note that for most practical purposes in QDT we may put $B = A$.

The functions $s(\epsilon, l; \rho)$ and $c(\epsilon, l; \rho)$ are defined by

$$s = \frac{B^{1/2}}{2^{1/2}} f \qquad c = \frac{1}{2^{1/2} B^{1/2}} h. \tag{2.53}$$

These functions have simple asymptotic forms for $\epsilon > 0$,

$$s \sim (\pi k)^{-1/2} \sin(\zeta) \qquad c \sim (\pi k)^{-1/2} \cos(\zeta) \qquad \text{for } \epsilon > 0. \tag{2.54}$$

For $\epsilon < 0$,

$$\begin{aligned} s &= (-1)^l \left[\frac{\sin(\pi\nu)}{(2\nu)^{1/2}\pi K} \xi - \cos(\pi\nu) \left(\frac{\nu^3}{2}\right)^{1/2} K\theta \right] \qquad \epsilon < 0 \\ c &= (-1)^l \left[\frac{\cos(\pi\nu)}{(2\nu)^{1/2}\pi K} \xi + \sin(\pi\nu) \left(\frac{\nu^3}{2}\right)^{1/2} K\theta \right] \qquad \epsilon < 0 \end{aligned} \tag{2.55}$$

where

$$K(\nu, l) = [\nu^2 \Gamma(\nu + l + 1)\Gamma(\nu - l)]^{-1/2}. \tag{2.56}$$

The significance of K as a normalising factor will be discussed in § 2.11.3.

The equations (2.53) defining s and c contain factors $A^{1/2}$ and $A^{-1/2}$. It follows from (2.31) that these factors can be expanded in powers of ϵ if $|\epsilon| < 1/l^2$. For $l > 0$ and $\epsilon < 0$ (2.31) gives

$$A = \left(1 - \frac{1^2}{\nu^2}\right)\left(1 - \frac{2^2}{\nu^2}\right) \ldots \left(1 - \frac{l^2}{\nu^2}\right) \tag{2.57}$$

and it follows that $A = 0$ for $\nu = 1, 2, \ldots, l$. At these values of ν, (2.53) gives s to be zero and c to be infinite. The use of s and c is best restricted to energies such that $\epsilon > -1/l^2$.

2.7.4. The functions $\varphi^{\pm}$. Defining

$$\varphi^{\pm} = c \pm \mathrm{i}s \tag{2.58}$$

we have

$$\varphi^{\pm} \sim (\pi k)^{-1/2} \exp(\pm \mathrm{i}\zeta) \qquad \text{for } \epsilon > 0 \tag{2.59}$$

and

$$\varphi^{\pm} = (-1)^l \exp(\pm \mathrm{i}\nu) \left[\frac{\xi}{(2\nu)^{1/2}\pi K} \mp \mathrm{i} \left(\frac{\nu^3}{2}\right)^{1/2} K\theta \right] \qquad \text{for } \epsilon < 0. \tag{2.60}$$

2.8. Wronskians

For any two functions $a(\rho)$ and $b(\rho)$ the Wronskian is defined by

$$W(a, b) = a \frac{\mathrm{d}b}{\mathrm{d}\rho} - \left(\frac{\mathrm{d}a}{\mathrm{d}\rho}\right) b. \tag{2.61}$$

If a and b are solutions of an equation of the form (2.4), $W(a, b)$ is independent of ρ. The values of W for Coulomb functions can be deduced from leading terms in series expansions or from asymptotic forms. We obtain

$$\begin{aligned} W(h, f) &= -W(g, f) = 2/\pi \\ W(\theta, \xi) &= 2/\nu \\ W(c, s) &= 1/\pi \\ W(\varphi^-, \varphi^+) &= 2\mathrm{i}/\pi. \end{aligned} \tag{2.62}$$

2.9. Notation

Table 1 gives relations between the notations used here and those used elsewhere (we disregard trivial differences in normalising factors).

Table 1. Relations between present notation and notations used elsewhere.

Present notation	Notations used elsewhere
f	y_1 [a], $f^{(0)}$ [e]
η	y_3 [d]
g	y_4 [a], $g^{(0)}$ [e]
h	$\bar{g}$ [e]
s	s [b], $\mathscr{f}$ [c], f [e]
c	c [b], $\mathscr{g}$ [c], g [e]
$\varphi^{\pm}$	$\varphi_{\pm}$ [d]
$\mathscr{G}$	$\mathscr{G}$ [e]

[a] Seaton (1958a).
[b] Eissner *et al* (1969).
[c] Fano (1970).
[d] Seaton (1978).
[e] Greene *et al* (1979).

2.10. Bound states

The function $s(\epsilon, l; \rho)$ is zero for $\rho = 0$ and tends to zero for $\rho \to \infty$ if the coefficient of ξ vanishes in (2.50), i.e. if $\sin(\pi\nu)\Gamma(\nu - l) = 0$. This condition is satisfied if $\nu = n$ where $n \geqslant (l+1)$, which is the familiar result for the energy levels of a hydrogenic ion.

2.11. Normalisation

Put $s = s(\epsilon, l; \rho)$ and $s' = s(\epsilon', l; \rho)$. Writing (2.3) as $(\mathscr{L} + \epsilon)F = 0$ we have $(\mathscr{L} + \epsilon)s = 0$, and $(\mathscr{L} + \epsilon')s' = 0$ and

$$\int_0^\rho [s(\mathscr{L} + \epsilon')s' - s'(\mathscr{L} + \epsilon)s]\,\mathrm{d}\rho = 0 \tag{2.63}$$

giving, on integrating by parts,

$$\int_0^\rho ss'\,\mathrm{d}\rho = \frac{W(s, s')}{\epsilon - \epsilon'} \tag{2.64}$$

where $W(s, s')$ is evaluated at ρ. We consider the limit of $\rho \to \infty$.

2.11.1. The case of $\epsilon > 0$. For $\epsilon > 0$ and $\rho \to \infty$, $s \sim (\pi k)^{-1/2} \sin(\zeta)$ and $(\mathrm{d}s/\mathrm{d}\rho) \sim (k/\pi)^{1/2} \cos(\zeta)$. Considering the limit of $\epsilon' \to \epsilon$ we obtain $W(s, s') \sim (1/\pi) \sin(\zeta - \zeta')$ and $(\epsilon - \epsilon') = (k + k')(k - k') \simeq 2k(k - k')$, giving $W(s, s')/(\epsilon - \epsilon') \sim \sin(\zeta - \zeta') \times [2k(k - k')]^{-1}$. From (2.45), for $\rho \to \infty$, $(\zeta - \zeta') \sim (k - k')[\rho + (1/k^2) \ln(\rho)]$ and it follows that $\sin(\zeta - \zeta')[\pi(k - k')]^{-1} \sim \delta(k - k')$, the Dirac delta function. Since $\epsilon = k^2$, $\mathrm{d}\epsilon = 2k\, \mathrm{d}k$ and $\delta(k - k')/(2k) = \delta(\epsilon - \epsilon')$. We therefore obtain

$$\int_0^\infty s(\epsilon, l; \rho) s(\epsilon', l; \rho)\, \mathrm{d}\rho = \delta(\epsilon - \epsilon') \tag{2.65}$$

and may say that $s(\epsilon, l; \rho)$ is normalised per unit energy.

2.11.2. The case of $\epsilon = -1/n^2$, $\epsilon' = -1/n'^2$ with $n \neq n'$. For $\epsilon = -1/n^2$ and $\epsilon' = -1/n'^2$ we obtain

$$\int_0^\infty s(\epsilon_n, l; \rho) s(\epsilon_{n'}, l; \rho)\, \mathrm{d}\rho = 0 \qquad \text{for } n \neq n'. \tag{2.66}$$

2.11.3. The case of $\epsilon = \epsilon' = -1/n^2$. For $\epsilon < 0$ we put $\epsilon = -1/\nu^2$ and

$$s = \xi A_+(\nu) + \theta A_-(\nu) \tag{2.67}$$

where, from (2.54),

$$A_+(\nu) = (-1)^l \frac{\sin(\pi\nu)}{(2\nu)^{1/2} \pi K} \tag{2.68}$$

$$A_-(\nu) = -(-1)^l \cos(\pi\nu)(\nu^3/2)^{1/2} K. \tag{2.69}$$

We put $\epsilon = \epsilon_n = -1/n^2$ and $s = s_n$. Since $A_+(n) = 0$, we have

$$s_n = A_-(n)\theta. \tag{2.70}$$

We consider the limit $\epsilon' \to \epsilon_n$ and put

$$s' = s_n + (\epsilon' - \epsilon_n)(\mathrm{d}s/\mathrm{d}\epsilon)|_{\epsilon_n}. \tag{2.71}$$

In the limit of $\rho \to \infty$ we have $\theta \to 0$, $s_n \to 0$ and

$$s' \sim (\epsilon' - \epsilon_n)\xi \left(\frac{\mathrm{d}}{\mathrm{d}\epsilon} A_+\right)\bigg|_{\nu = n}. \tag{2.72}$$

From (2.70) and (2.72) we have, for $\epsilon' \to \epsilon_n$ and $\rho \to \infty$,

$$W(s_n, s') = -(\epsilon_n - \epsilon')\left[A_- \left(\frac{2}{\nu}\right) \frac{\mathrm{d}}{\mathrm{d}\epsilon} A_+\right]\bigg|_{\nu = n} \tag{2.73}$$

where we have used, from (2.62), $W(\theta, \xi) = (2/\nu)$. From (2.64) and (2.73) the normalisation integral is

$$\int_0^\infty s_n s_n\, \mathrm{d}\rho = -\left[A_- \left(\frac{2}{\nu}\right) \frac{\mathrm{d}}{\mathrm{d}\epsilon} A_+\right]\bigg|_{\nu = n}. \tag{2.74}$$

We put $(\mathrm{d}/\mathrm{d}\epsilon) = (\nu^3/2)(\mathrm{d}/\mathrm{d}\nu)$ and, using (2.67) and (2.68), obtain finally

$$\int_0^\infty s_n s_n\, \mathrm{d}\rho = n^3/2. \tag{2.75}$$

Combining (2.65) and (2.74)

$$\int_0^\infty s(\epsilon_n, l; \rho) s(\epsilon_{n'}, l; \rho)\, d\rho = (n^3/2)\delta(n, n'). \tag{2.76}$$

Putting $\delta\epsilon_n = (\epsilon_{n+1} - \epsilon_n)$ we have $\delta\epsilon_n \simeq 2/n^3$ for n large. The bound-state functions $s(\epsilon_n, l; \rho)$ are therefore normalised per unit energy in the limit of n large. Defining

$$P_{nl}(\rho) = -(-1)^{n+l}\left(\frac{2}{n^3}\right)^{1/2} s(\epsilon_n, l; \rho) = K(n, l)\theta(n, l; \rho) \tag{2.77}$$

we have

$$\int_0^\infty P_{nl}(\rho) P_{n'l}(\rho)\, d\rho = \delta(n, n'). \tag{2.78}$$

2.12. Graphs of Coulomb functions

Our discussion of Coulomb functions is concluded with some illustrative graphs.

Figure 4 illustrates the behaviour of s and c for $\epsilon = 0.0$ and $l = 0$ and 1. In the limit of $\rho \to 0$, s behaves like ρ^{l+1} and c like ρ^{-l}. For $\epsilon = 0$ the functions have one point of inflection, at $\rho_a = l(l+1)/2$; for $\rho > \rho_a$, s and c are oscillatory and differ in phase by $\pi/2$.

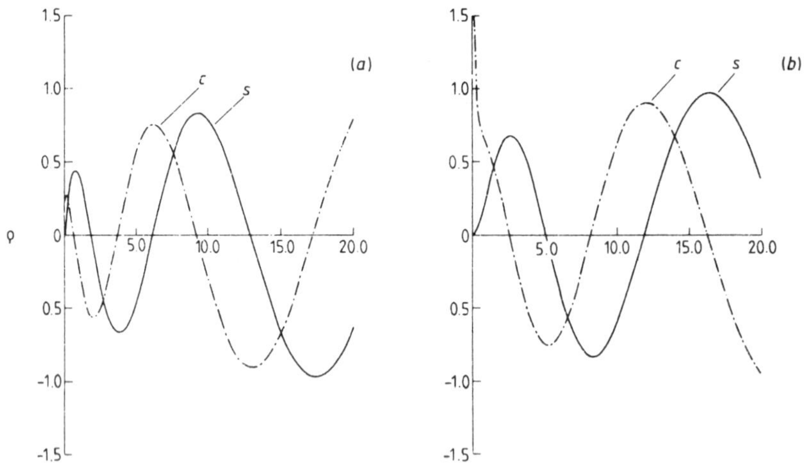

Figure 4. s and c for $\epsilon = 0.0$, and $l = 0$ (a) and 1 (b), plotted against ρ.

Figure 5 illustrates the behaviour of s and c for $\epsilon = -0.1$, 0.0 and +0.1 and $l = 0$ and 1. It is seen that s and c are insensitive to ϵ for smaller values of ρ. For $\epsilon < 0$, the functions have a second point of inflection at $\rho = \rho_b$; for $\epsilon = -0.1$ equation (2.9) gives $\rho_b = 20.0$ for $l = 0$ and $\rho_b = 18.94\ldots$ for $l = 1$. From figure 5 we see that the behaviour of s and c for $\epsilon = -0.1$ and $\rho \ll \rho_b$ is not essentially different from that of s and c for $\epsilon = 0.0$ and 0.1. The functions s and c have a quite different behaviour for $\epsilon < 0$ and $\rho > \rho_b$.

Figure 6 illustrates the behaviour of s and c for $\epsilon = 0.0$, 0.1, 0.2 and 0.3 and $l = 0$. For ρ large, these functions oscillate more rapidly for larger values of ϵ.

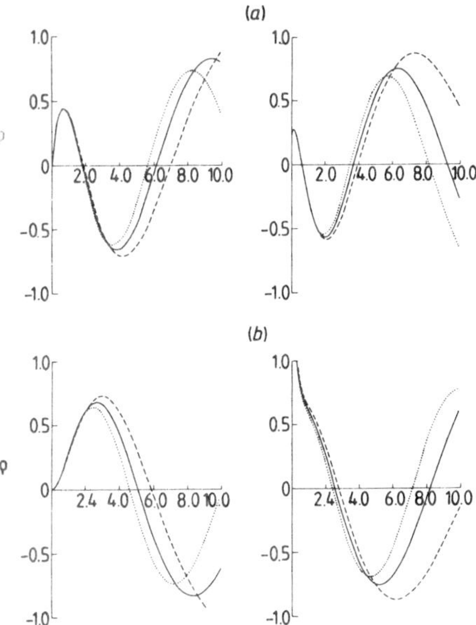

Figure 5. s (left) and c (right) for $\epsilon = -0.1$ (- - -), 0.0 (——) and +0.1 (. . .), and $l = 0$ (a) and 1 (b), plotted against ρ.

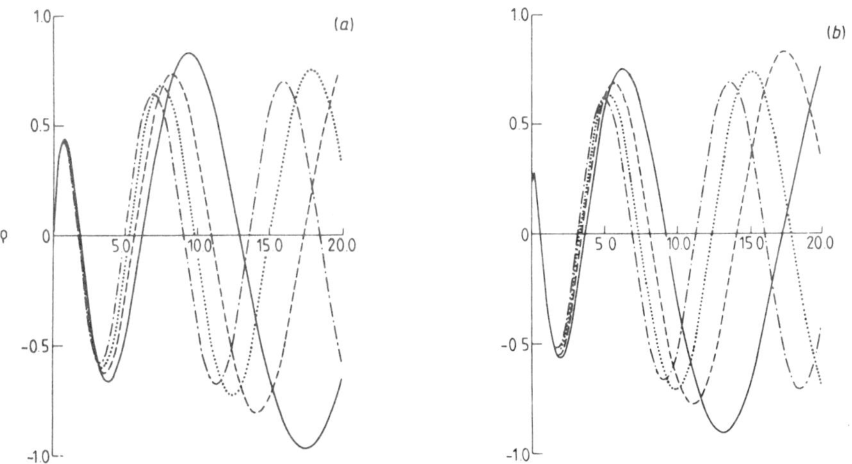

Figure 6. s (a) and c (b) for $\epsilon = 0.0$ (——), 0.1 (- - -), 0.2 (. . .) and 0.3 (- · -), and $l = 0$, plotted against ρ.

Figure 7 illustrates the behaviour of s and c for $\epsilon < 0$. Results are given for $\nu = 2.0$, 2.5, 3.0 and 3.5, and $l = 0$ and 1. For $\rho > \rho_b$, s and c are in general both exponentially increasing, but exponential decreases occur for s when $\nu = n$ (n being an integer) and for c when $\nu = n + \frac{1}{2}$.

Figure 8 illustrates the behaviour of θ for $\nu = 2.0$, 2.5, 3.0 and 3.5. For all ν, θ decreases exponentially in the limit of ρ large. The behaviour of θ for small ρ is shown on an inset figure; for $\rho \to 0$, $\theta \to 0$ only for $\nu = n$ where n is an integer.

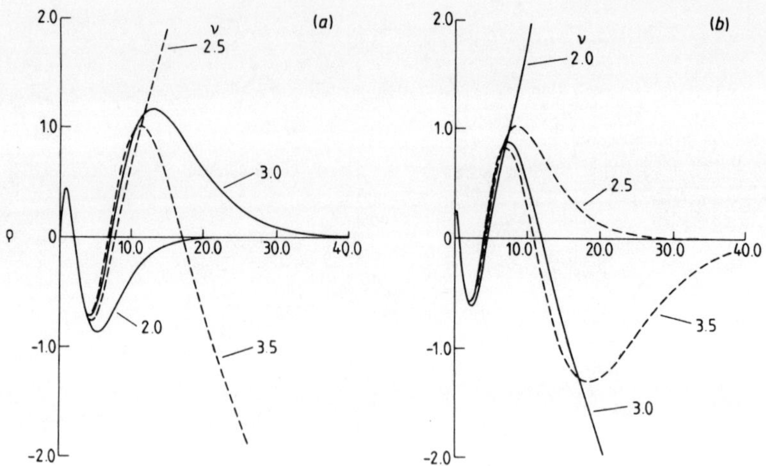

Figure 7. s (a) and c (b) for $\epsilon < 0$, and $l = 0$ (a) and 1 (b), plotted against ρ.

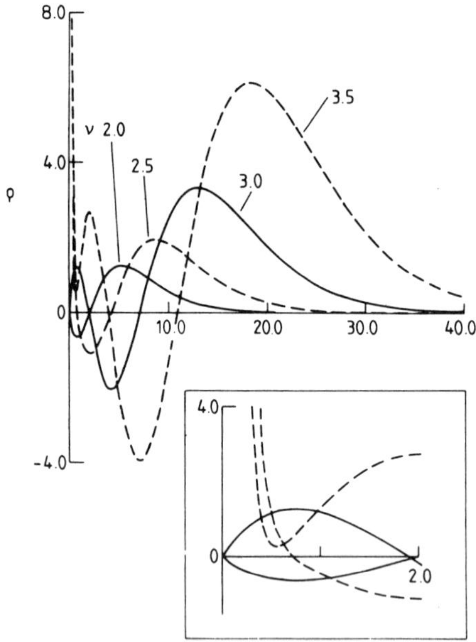

Figure 8. θ for $\nu = 2.0$, 2.5, 3.0 and 3.5, and $l = 0$, plotted against ρ.

3. Sketch of atomic collision theory

Before discussing the foundations of quantum defect theory we give brief sketches of atomic collision theory (in the present section) and of the theory of radiative transitions (in § 4).

3.1. The coupled equations of collision theory

We consider LS coupling but note that other coupling schemes may also be used. Let $\psi(\Gamma_i S_i L_i M_{S_i} M_{L_i})$ be the wavefunction for an N-electron target ion with energy

$E_i = E(\Gamma_i S_i L_i)$, and let $\chi(\frac{1}{2}l_i m_{s_i} m_{l_i})$ be the wavefunction for the spin and angle coordinates of a colliding electron. Let $\psi_i = \psi(\Gamma_i S_i L_i l_i SLM_S M_L)$ be the vector-coupled product

$$\psi_i = \sum_{\substack{M_{S_i} m_{s_i} \\ M_{L_i} m_{l_i}}} \psi(\Gamma_i S_i L_i M_{S_i} M_{L_i}) \chi(\tfrac{1}{2} l_i m_{s_i} m_{l_i})(M_{S_i} m_{s_i}|SM_S)(M_{L_i} m_{l_i}|LM_L) \tag{3.1}$$

where $(m_1 m_2|jm)$ is the vector-coupling coefficient $(j_1 j_2 m_1 m_2|j_1 j_2 jm)$. We refer to i as a channel index; it specifies a state of the target $(\Gamma_i S_i L_i)$ and the angular momentum (l_i) of the colliding electron. A function ψ_i has definite total angular momentum (SL) and parity (π). These functions are such that

$$(\psi_i|\psi_{i'}) = \delta(i, i') \qquad (\psi_i|H_N|\psi_{i'}) = E_i \delta(i, i') \tag{3.2}$$

where H_N is the Hamiltonian for the target.

For the $(N+1)$-electron system of target plus colliding electron we consider the use of expansions†

$$\Psi = \sum_{i=1}^{I} \psi_i \frac{1}{r} F_i(r) \tag{3.3}$$

where $F_i(r)$ is a radial function for the colliding electron and where all functions ψ_i in (3.3) have the same value of $SL\pi$. The equations satisfied by the functions F_i are obtained from

$$(\psi_i|H - E|\Psi) = 0 \tag{3.4}$$

where H is the total Hamiltonian, E the total energy, and where the integration in (3.4) is over all coordinates in ψ_i but not over the radial coordinate r. From (3.4) one obtains, in Hartree units,

$$\left[-\frac{1}{2}\left(\frac{d^2}{dr^2} - \frac{l_i(l_i+1)}{r^2}\right) - (E - E_i)\right] F_i + \sum_{i'=1}^{I} V_{ii'} F_{i'} = 0 \tag{3.5}$$

where

$$V_{ii'}(r) = \left(\psi_i \left| \sum_{n=1}^{N} |\boldsymbol{r} - \boldsymbol{r}_n|^{-1} - \frac{Z_0}{r} \right| \psi_{i'}\right) \tag{3.6}$$

Z_0 being the charge on the nucleus. We use

$$|\boldsymbol{r} - \boldsymbol{r}_n|^{-1} = \sum_{\lambda \geq 0} P_\lambda(\boldsymbol{r} \cdot \boldsymbol{r}_n) \gamma_\lambda(r, r_n) \tag{3.7}$$

where $\gamma_\lambda(r, r_n) = r^\lambda / r_n^{\lambda+1}$ for $r < r_n$ and $r_n^\lambda / r^{\lambda+1}$ for $r > r_n$, to obtain

$$V_{ii'}(r) \underset{r\to\infty}{\sim} -Z\delta(i, i')/r \tag{3.8}$$

with $Z = (Z_0 - N)$.

The energy of the colliding electron is $(E - E_i)$ and its Z-scaled energy is $\epsilon_i = 2(E - E_i)/Z^2$. Defining the scaled energies

$$\mathscr{E} = 2E/Z^2 \qquad \text{and} \qquad \mathscr{E}_i = 2E_i/Z^2 \tag{3.9}$$

we have

$$\mathscr{E} = \mathscr{E}_i + \epsilon_i. \tag{3.10}$$

† In practice additional 'bound channel' functions are often included. This is discussed in appendix A.

Defining $\rho = Zr$ and

$$U_{ii'}(\rho) = -2V_{ii'}(r)/Z^2 \tag{3.11}$$

equations (3.5) become

$$\left(\frac{d^2}{d\rho^2} - \frac{l_i(l_i+1)}{\rho^2} + \epsilon_i\right)F_i + \sum_{i'=1}^{I} U_{ii'}F_{i'} = 0 \tag{3.12}$$

or, in compact matrix notation,

$$(\mathscr{L} + \epsilon)\boldsymbol{F} = 0 \tag{3.13}$$

where

$$\mathscr{L} = \frac{d^2}{d\rho^2} - \frac{l(l+1)}{\rho^2} + \boldsymbol{U}(\rho). \tag{3.14}$$

Our convention in matrix equations is that quantities not in boldface type and without subscripts are diagonal matrices; thus ϵ and l in (3.13) and (3.14) are diagonal matrices with diagonal elements ϵ_i and l_i.

It follows from (3.6) and (3.11) that $\boldsymbol{U}$ is Hermitian, $\boldsymbol{U} = \boldsymbol{U}^\dagger$. With standard phase conventions, which we shall assume, $\boldsymbol{U}$ is real. We then have $\boldsymbol{U} = \boldsymbol{U}^* = \boldsymbol{U}^{\mathrm{T}}$ where superscript T denotes transpose.

We require Ψ to be bounded at the origin and hence $\boldsymbol{F}(\rho = 0) = 0$. The equations (3.12) have I linearly independent (LI) solutions satisfying this condition. We therefore take $\boldsymbol{F}$ to be a matrix with elements $F_{ij}(\rho)$ where $i = 1$ to I specifies a channel and $j = 1$ to I specifies a particular set of solutions.

If $\boldsymbol{F}$ and $\boldsymbol{F}'$ are solutions of $(\mathscr{L} + \epsilon)\boldsymbol{F} = 0$ and $(\mathscr{L} + \epsilon')\boldsymbol{F}' = 0$, we have

$$\int_0^{\rho_1} \{\boldsymbol{F}^\dagger(\mathscr{L} + \epsilon')\boldsymbol{F}' - [(\mathscr{L} + \epsilon)\boldsymbol{F}]^\dagger \boldsymbol{F}'\}\, d\rho = 0 \tag{3.15}$$

and integration by parts gives

$$(\mathscr{E} - \mathscr{E}')\int_0^{\rho_1} \boldsymbol{F}^\dagger \boldsymbol{F}'\, d\rho = \left[\boldsymbol{F}^\dagger \frac{d}{d\rho}\boldsymbol{F}' - \left(\frac{d}{d\rho}\boldsymbol{F}^\dagger\right)\boldsymbol{F}'\right]_{\rho=\rho_1}. \tag{3.16}$$

3.2. Electron exchange

In practice, it is advantageous to use functions which are explicitly antisymmetrised. Equation (3.3) is replaced by

$$\Psi = \mathscr{A}_S \sum_{i=1}^{I} \psi_i \frac{1}{r} F_i(r) \tag{3.17}$$

where $\mathscr{A}_S$ is an antisymmetrisation operator. One obtains equations of the form (3.13) with an operator $\mathscr{L}$ which contains non-local integral operators $\boldsymbol{X}_{ii'}$,

$$\mathscr{L} = \frac{d^2}{d\rho^2} - \frac{l(l+1)}{\rho^2} + \boldsymbol{U} + \boldsymbol{X}. \tag{3.18}$$

A full discussion is given by Eissner and Seaton (1972). Equation (3.16) remains valid if $\boldsymbol{X}$ is Hermitian in the domain $0 \leqslant \rho \leqslant \rho_1$.

3.3. Open and closed channels

Let the channels be ordered in such a way that $E_1 \leqslant E_2 \leqslant E_3 \ldots$, giving $\epsilon_1 \geqslant \epsilon_2 \geqslant \epsilon_3 \ldots$. *Open channels* have $\epsilon_i \geqslant 0$, and *closed channels* have $\epsilon_i < 0$. Let there be I_0 open channels. Then $\epsilon_i \geqslant 0$ for $i = 1$ to I_0 and $\epsilon_i < 0$ for $i = (I_0 + 1)$ to I.

If i is a closed channel, the radial functions $F_{ij}(\rho)$ will, in general, be exponentially increasing in the limit of ρ large. The physical solutions are such that Ψ is everywhere bounded. They can be constructed using radial functions $G_{ij}(\rho)$ which are such that

$$G_{ij}(\rho) \underset{\rho \to \infty}{\sim} 0 \qquad \text{if channel } i \text{ is closed.} \tag{3.19}$$

The number of LI solutions satisfying (3.19) is equal to I_0, i.e. $j = 1$ to I_0. Such functions may be partitioned as

$$\boldsymbol{G} = \begin{pmatrix} \boldsymbol{G}_{\mathrm{oo}} \\ \boldsymbol{G}_{\mathrm{co}} \end{pmatrix} \tag{3.20}$$

where $\boldsymbol{G}_{\mathrm{oo}}$ is the 'open–open' part and has elements G_{ij} with $i = 1$ to I_0 and $j = 1$ to I_0; and $\boldsymbol{G}_{\mathrm{co}}$ is the 'closed–open' part and has elements G_{ij} with $i = (I_0 + 1)$ to I and $j = 1$ to I_0. Equation (3.19) is

$$\boldsymbol{G}_{\mathrm{co}}(\rho) \underset{\rho \to \infty}{\sim} 0. \tag{3.21}$$

3.4. The reactance matrix and the scattering matrix

It follows from (3.8) and (3.11) that

$$U_{ii'}(\rho) \underset{\rho \to \infty}{\sim} \frac{2}{\rho} \delta(i, i'). \tag{3.22}$$

If, for ρ large, we put $U_{ii'}(\rho) = (2/\rho)\delta(i, i')$, the radial functions may be equated to linear combination of Coulomb functions.

The *reactance matrix* $\boldsymbol{R}$ is defined in terms of functions $\boldsymbol{G}(\boldsymbol{R}; \rho)$ with asymptotic forms, for $j = 1$ to I_0,

$$\left.\begin{aligned} G_{ij}(\boldsymbol{R}; \rho) &\sim s_i \delta(i, j) + c_i R_{ij} && \text{for } i = 1 \text{ to } I_0 \\ G_{ij}(\boldsymbol{R}; \rho) &\sim 0 && \text{for } i = (I_0 + 1) \text{ to } I. \end{aligned}\right. \tag{3.23}$$

In matrix notation,

$$\boldsymbol{G}_{\mathrm{oo}}(\boldsymbol{R}; \rho) \sim \boldsymbol{s} + \boldsymbol{c}\boldsymbol{R} \qquad \boldsymbol{G}_{\mathrm{co}}(\boldsymbol{R}; \rho) \sim 0. \tag{3.24}$$

The *scattering matrix* $\boldsymbol{S}$ is defined in terms of functions $\boldsymbol{G}(\boldsymbol{S}; \rho)$ with asymptotic forms

$$\boldsymbol{G}_{\mathrm{oo}}(\boldsymbol{S}; \rho) \sim \boldsymbol{\varphi}^- - \boldsymbol{\varphi}^+ \boldsymbol{S} \qquad \boldsymbol{G}_{\mathrm{co}}(\boldsymbol{S}; \rho) \sim 0. \tag{3.25}$$

It follows from (2.57) that $\boldsymbol{G}(\boldsymbol{S}; \rho) = -2\mathrm{i}\boldsymbol{G}(\boldsymbol{R}; \rho)(1 - \mathrm{i}\boldsymbol{R})^{-1}$ and that

$$\boldsymbol{S} = (1 + \mathrm{i}\boldsymbol{R})(1 - \mathrm{i}\boldsymbol{R})^{-1}. \tag{3.26}$$

For $\boldsymbol{U}$ real, $\boldsymbol{R}$ is real, $\boldsymbol{R} = \boldsymbol{R}^*$.

From (3.16) with $\mathscr{E} = \mathscr{E}'$ one obtains $\boldsymbol{R} = \boldsymbol{R}^{\mathrm{T}}$. It follows from (3.26) that $\boldsymbol{S} = \boldsymbol{S}^{\mathrm{T}}$ and $\boldsymbol{S}^*\boldsymbol{S} = 1$.

Let $\boldsymbol{G}(\boldsymbol{S};\rho)$ be calculated for energy $\mathscr{E}$ and $\boldsymbol{G}'(\boldsymbol{S};\rho)$ for energy $\mathscr{E}'$. It follows from (3.26), using the method of § 2.11.1, that

$$\int_0^\infty \boldsymbol{G}^\dagger(\boldsymbol{S};\rho)\boldsymbol{G}'(\boldsymbol{S};\rho) = 4\delta(\mathscr{E}-\mathscr{E}'). \tag{3.27}$$

3.5. The collision strength and collision cross section

The normalisation of φ^- and φ^+ is such that unit incoming flux in channel i gives an outgoing flux of $|S_{ij}|^2$ in channel j.

The transmission matrix is $\boldsymbol{T}=1-\boldsymbol{S}$. The partial collision strength, for given $SL\pi$, is

$$\Omega_{SL\pi}(i,j)=\tfrac{1}{2}(2S+1)(2L+1)|T_{SL\pi}(i,j)|^2 \tag{3.28}$$

and the total collision strength is

$$\Omega(\Gamma_iS_iL_i,\Gamma_jS_jL_j)=\sum_{SL\pi}\sum_{l_il_j}\Omega_{SL\pi}(\Gamma_iS_iL_il_i,\Gamma_jS_jL_jl_j). \tag{3.29}$$

For inelastic collisions the cross section is

$$Q(\Gamma_iS_iL_i\to\Gamma_jS_jL_j)=\frac{\Omega(\Gamma_iS_iL_i,\Gamma_jS_jL_j)}{(2S_i+1)(2L_i+1)}\frac{\pi a_0^2}{k_iZ} \tag{3.30}$$

where a_0 is the Bohr radius and where k_iZ is the electron wavenumber in atomic units.

With a Coulomb field the total elastic cross section is divergent. The differential cross section for Coulomb and modified Coulomb potentials is discussed by Seaton (1958a).

3.6. Bound states

Consider energies $\mathscr{E}$ such that all channels are closed. The equations (3.13) then have solutions which are everywhere bounded only for certain discrete values of $\mathscr{E}$, $\mathscr{E}=\mathscr{E}_n$ say. These are the bound states of the $(N+1)$-electron system.

It follows from (3.2) and (3.3) that

$$(\Psi|\Psi)=\sum_{i=1}^{I}\int_0^\infty |F_i|^2\,\mathrm{d}r. \tag{3.31}$$

For bound states we use functions $F_i=P_i$ normalised in such a way that

$$\sum_{i=1}^{I}\int_0^\infty |P_i|^2\,\mathrm{d}\rho=1. \tag{3.32}$$

3.7. Autoionisation

Consider energies such that some channels are open and some closed. For the special case in which there is no coupling between open and closed channels, we can have bound states in the closed channels. When there is coupling between open and closed channels we can have 'quasi-bound' states which can autoionise, and when such phenomena occur, the scattering matrix $\boldsymbol{S}(\mathscr{E})$ has poles at complex energies,

$$\mathscr{E}(\text{pole}) = \mathscr{E}_0 - \frac{\mathrm{i}}{2}\gamma_{\mathrm{A}} \tag{3.33}$$

with ϵ_0 and γ_{A} real. The autoionisation probability in atomic units is $\Gamma_{\mathrm{A}} = (Z^2/2)\gamma_{\mathrm{A}}$.

4. Sketch of radiative theory

4.1. Transitions between bound states

For the radiative transition $a \to b$ between two quantum states Ψ_a and Ψ_b the oscillator strength is

$$f_{ba} = \tfrac{2}{3}(E_b - E_a)|(\Psi_b|\vec{d}|\Psi_a)|^2 \tag{4.1}$$

where

$$\vec{d} = \sum_n \vec{r}_n \tag{4.2}$$

and where all quantities are in atomic units and the sum in (4.2) is over all electrons. The radiative transition probability for $b \to a$ with $E_b > E_a$ is

$$\Gamma_{\mathrm{R}}(b \to a) = (2\alpha^3/\tau_0)(E_b - E_a)^2 f_{ba} \tag{4.3}$$

where $\alpha = e^2(\hbar c)^{-1} = (137.036\ldots)^{-1}$ and $\tau_0 = \hbar^3(me^4)^{-1} = 2.4189\ldots \times 10^{-17}$ s is the atomic unit of time.

4.2. Transitions to the continuum

For transitions $a \to jE$ where j is an open channel as defined in § 3.3, the differential oscillator strength is

$$\frac{\mathrm{d}}{\mathrm{d}E} f_{ja}(E) = \tfrac{2}{3}(E - E_a)|(\Psi_j(E)|\vec{d}|\Psi_a)|^2 \tag{4.4}$$

where $\Psi_j(E)$ is normalised to

$$(\Psi_j(E)|\Psi_{j'}(E')) = \delta(j, j')\delta(E - E') \tag{4.5}$$

and has an outgoing wave only in channel j. The photoionisation cross section is

$$\sigma_{ja}(E) = 2\pi\alpha \frac{\mathrm{d}}{\mathrm{d}E} f_{ja}(E)\pi a_0^2. \tag{4.6}$$

4.3. Use of collision wavefunctions

4.3.1. Transitions between bound states. Consider transitions between two bound states, Ψ_a and Ψ_b. Let Ψ_b have an expansion of the type (3.3) with radial functions F_{ib} having normalisation such that

$$(\Psi_b|\Psi_b) = \sum_i \int_0^\infty |F_{ib}|^2 \,\mathrm{d}r = 1. \tag{4.7}$$

Using (3.3) we obtain

$$(\Psi_b|\vec{d}|\Psi_a)=\sum_i\int_0^\infty\left(\frac{1}{r}\right)F^*_{ib}(\psi_i|\vec{d}|\Psi_a)r^2\,dr \tag{4.8}$$

where the integrations for $(\psi_i|\vec{d}|\Psi_a)$ are over all coordinates in ψ_i. The same result, equation (4.8), is obtained if the antisymmetrised expansion (3.17) is used.

Using Z-scaled variables we obtain

$$f_{ba}=\tfrac{1}{3}(\mathscr{E}_b-\mathscr{E}_a)|\vec{D}_{ba}|^2 \tag{4.9}$$

where

$$\vec{D}_{ba}=\sum_i\int_0^\infty P^*_{ib}\vec{g}_{ia}\,d\rho \tag{4.10}$$

and where

$$\vec{g}_{ia}=Z^{1/2}r(\psi_i|\vec{d}|\Psi_a). \tag{4.11}$$

The radial functions P_{ib} are normalised to

$$\sum_i\int_0^\infty|P_{ib}|^2\,d\rho=1. \tag{4.12}$$

One can consider in a similar way the case that Ψ_a and Ψ_b both have expansions of the type (3.3).

4.3.2. Transitions to the continuum. We obtain

$$\frac{d}{d\mathscr{E}}f_{ja}(\mathscr{E})=\tfrac{1}{3}(\mathscr{E}-\mathscr{E}_a)|\vec{D}_{ja}\,(\text{out},\mathscr{E})|^2 \tag{4.13}$$

where

$$\vec{D}_{ja}\,(\text{out},\mathscr{E})=\sum_i\int G^*_{ij}\,(\text{out},\mathscr{E})\vec{g}_{ia}\,d\rho \tag{4.14}$$

$$\sum_i\int G^*_{ij}\,(\text{out},\mathscr{E})G_{ij'}\,(\text{out},\mathscr{E}')\,d\rho=\delta(j,j')\delta(\mathscr{E}-\mathscr{E}') \tag{4.15}$$

and G_{ij} (out) has an outgoing wave φ_i^+ only in channel $i=j$. Since $df/dE=(2/Z^2)\,df/d\mathscr{E}$, $\sigma=(4\pi\alpha/Z^2)(df/d\mathscr{E})\pi a_0^2$.

The functions $\boldsymbol{G}(\boldsymbol{S};\rho)$ have asymptotic form $\boldsymbol{G}(\boldsymbol{S};\rho)\sim(\varphi^--\varphi^+\boldsymbol{S})$ and are normalised to $4\delta(\mathscr{E}-\mathscr{E}')$ (equation (3.27)). We put

$$\boldsymbol{G}\,(\text{out},\rho)=-\tfrac{1}{2}\boldsymbol{G}(\boldsymbol{S};\rho)\boldsymbol{S}^* \tag{4.16}$$

which is normalised to $\delta(\mathscr{E}-\mathscr{E}')$. Since $\boldsymbol{S}\boldsymbol{S}^*=1$, $\boldsymbol{G}\,(\text{out};\rho)\sim\tfrac{1}{2}(\varphi^+-\varphi^-\boldsymbol{S}^*)$.

4.4. *Reduced matrix elements*

The operator $\vec{d}$ defined by (4.2) is a vector with Cartesian components d_x, d_y, d_z and spherical tensor components d_μ where $d_{-1}=(d_x-\mathrm{i}d_y)/\sqrt{2}$, $d_0=d_z$, $d_{+1}=-(d_x+\mathrm{i}d_y)/\sqrt{2}$. The matrix $\vec{\boldsymbol{D}}$ has components $\boldsymbol{D}_\mu$. We use $\boldsymbol{D}$ for the reduced matrix. In a representation in which total angular momenta are specified we have, by the Wigner–Eckart theorem,

$$D_\mu(\Gamma JM,\Gamma'J'M')=D(\Gamma J,\Gamma'J')\frac{(J'1JM|J'M1\mu)}{(2J+1)^{1/2}} \tag{4.17}$$

where $(j_1j_2jm|j_1m_1j_2m_2)$ is a vector coupling coefficient.

5. The one-channel theory

We describe the one-channel theory, essentially as developed by Ham (1955) and Seaton (1958a).

5.1. The radial equation

We consider the expansion (3.17) truncated to one term,

$$\Psi = \mathscr{A}_S \psi \frac{1}{r} F. \tag{5.1}$$

The radial function F is a solution of

$$(\mathscr{L} + \epsilon)F = 0 \tag{5.2}$$

where

$$\mathscr{L} = \frac{d^2}{d\rho^2} - \frac{l(l+1)}{\rho^2} + U(\rho) + X. \tag{5.3}$$

We consider solutions of (5.2) satisfying $F(\rho = 0) = 0$.

We put $\epsilon = k^2$ for $\epsilon \geqslant 0$ and $\epsilon = -1/\nu^2$ for $\epsilon < 0$.

5.2. A simple model

QDT can be developed as a mathematically rigorous theory if the real physical problem is replaced by a simple model problem. For the model, exchange is neglected ($X = 0$ in (5.3)) and the potential $U(\rho)$ is assumed to be equal to the Coulomb potential beyond some finite value of ρ:

$$U(\rho) = \frac{2}{\rho} \qquad \text{for } \rho \geqslant \rho_0 \text{ with } \rho_0 < \infty. \tag{5.4}$$

Ham (1955) shows that, with $X = 0$ in (5.3) and for all potentials $U(\rho)$ likely to be of interest, (5.2) has solutions $F(\epsilon, l; \rho)$ which are analytic in ϵ for all finite ϵ and finite ρ.

Assuming (5.4), $F(\epsilon, l; \rho)$ can be equated to a linear combination of two Coulomb functions for $\rho \geqslant \rho_0$. Choosing f and g:

$$F(\epsilon, l; \rho) = f(\epsilon, l; \rho)I(\epsilon, l) + g(\epsilon, l; \rho)J(\epsilon, l) \qquad \text{for } \rho \geqslant \rho_0. \tag{5.5}$$

Using (2.61), $I = -(\pi/2)W(g, F)$ and $J = (\pi/2)W(f, F)$ where the Wronskians can be evaluated at $\rho = \rho_0$. If $F(\epsilon, l; \rho)$ is analytic in ϵ, $I(\epsilon, l)$ and $J(\epsilon, l)$ are analytic in ϵ and hence have convergent expansions:

$$I(\epsilon, l) = \sum_{n=0}^{\infty} \epsilon^n I_{(n)}(l) \qquad J(\epsilon, l) = \sum_{n=0}^{\infty} \epsilon^n J_{(n)}(l). \tag{5.6}$$

Other quantities considered in the theory can be expressed in terms of I and J, and their analytical properties rigorously defined.

5.3. Discussion of the simple model

5.3.1. Exchange. Let I_c be the smallest binding energy for an electron in the core state ψ, and $\epsilon_c = -2I_c/Z^2$ the corresponding Z-scaled energy. If $X = 0$ in (5.3), (5.2) has solutions $F(\epsilon, l; \rho)$ which are exponentially increasing for ρ large, and for $\epsilon<0$ and not equal to an energy eigenvalue. For $X \neq 0$, (5.2) cannot have such solutions for $\epsilon<\epsilon_c$, since they would give the exchange integrals to be divergent. It would appear that, with $X \neq 0$, (5.2) does not have any solutions for $\epsilon<\epsilon_c$.

When exchange is included we shall assume, without proof, that (5.2) has solutions $F(\epsilon, l; \rho)$ which, for $\epsilon>\epsilon_c$, can be expressed as power series in ϵ, either convergent series or asymptotic series. We may assume that, for smaller values of $\rho, F(\epsilon, l; \rho)$ will vary slowly as a function of ϵ.

5.3.2. Long-range non-Coulomb potentials. It follows from (3.6), (3.7) and (3.11) that, if ψ in (5.1) is spherically symmetric, $\rho U(\rho)$ tends to 2 exponentially for ρ large. It may be expected that the neglect of exponentially decaying terms in $\rho U(\rho)$ for $\rho \geqslant \rho_0$, with ρ_0 large, will not lead to serious errors. More serious may be the neglect of contributions to $U(\rho)$ which, for ρ large, behave like ρ^{-n} with $n \geqslant 2$. For states which are not spherically symmetric, such contributions are given by the multipole expansion (3.7). They arise naturally in the many-channel theory. In a more realistic one-channel theory one should use an effective potential which allows for polarisation of the core, and hence gives a contribution to $U(\rho)$ behaving like ρ^{-4} for ρ large.

If one has long-range non-Coulomb potentials, (5.4) can be replaced by

$$U(\rho) \underset{\rho\to\infty}{\sim} 2/\rho. \tag{5.7}$$

That this leads to difficulties in the formulation of the theory can be appreciated as follows. One could attempt to define I and J on replacing (5.5) by

$$F \underset{\rho\to\infty}{\sim} fI+gJ. \tag{5.8}$$

For $\epsilon \geqslant 0$, f and g remain linearly independent in the limit of $\rho\to\infty$ and (5.8) gives a satisfactory definition of I and J. But for $\epsilon<0$, f and g are not, in general, linearly independent in the limit of $\rho\to\infty$ (since they both increase exponentially) and (5.8) fails to define I and J for all values of ϵ. Progress can be made using (2.42), (2.48) and (2.49) to obtain, from (5.8),

$$F \underset{\rho\to\infty}{\sim} (-1)^l \nu^{l+1} \frac{\Gamma(\nu-l)}{\pi}[\sin(\pi\nu)(I-\mathscr{G}J)-A\cos(\pi\nu)J]\xi. \tag{5.9}$$

We note that, in the asymptotic form, the coefficient of the exponentially decreasing function θ is undefined so long as the coefficient of the increasing function ξ is non-zero. For a given function $F(\epsilon, l; \rho)$, equation (5.9) defines the value of

$$\Gamma(\nu-l)[\sin(\pi\nu)(I-\mathscr{G}J)-A\cos(\pi\nu)J]. \tag{5.10}$$

If (5.10) is known as a continuous function of ϵ, J and $(I-\mathscr{G}J)$ are known at certain discrete values of ϵ: at $\epsilon=-1/n^2$ with $n \geqslant (l+1)$ the coefficient of $(I-\mathscr{G}J)$ vanishes and (5.10) gives the value of J; and at $\epsilon=-(n+\frac{1}{2})^{-2}$ the coefficient of J vanishes and (5.10) gives the value of $(I-\mathscr{G}J)$. Interpolation can then be used to define J and $(I-\mathscr{G}J)$ as continuous functions of ϵ, consistent with (5.11) and with extrapolation from values known for $\epsilon \geqslant 0$.

We may also consider the energy eigenvalues, $\epsilon_n=-1/\nu_n^2$. At the eigenvalues the coefficient (5.10) of ξ vanishes. Hence if the values of ϵ_n are known we can deduce the values of $J(I-\mathscr{G}J)^{-1}$ at $\epsilon=\epsilon_n$, and fit $J(I-\mathscr{G}J)^{-1}$ to a continuous function of ϵ. This is the procedure used in empirical analyses of experimental data.

The above arguments indicate how the theory could be developed with allowance for long-range non-Coulomb potentials. Throughout most of the present review we simplify the presentation by assuming that such potentials can be neglected for $\rho \geqslant \rho_0$ with ρ_0 finite. In § 5.7 we shall show how such potentials can be allowed for using perturbation theory.

5.4. *The functions Y, $\mathscr{R}$ and χ*

We assume (5.4) and define solutions $F(Y;\rho)$, $F(\mathscr{R};\rho)$ and $F(\chi;\rho)$ of (5.2) which are such that

$$\left.\begin{aligned} F(Y;\rho) &= f + hY \qquad &(5.11)\\ F(\mathscr{R};\rho) &= s + c\mathscr{R} \qquad &(5.12)\\ F(\chi;\rho) &= \varphi^- - \varphi^+\chi \qquad &(5.13)\end{aligned}\right\} \text{ for } \rho \geqslant \rho_0.$$

It follows from (2.42), (5.5) and (5.11) that

$$Y = -J(I - \mathscr{G}J)^{-1}. \tag{5.14}$$

For practical applications we may assume that I, J and $\mathscr{G}$ can be approximated by polynomials in ϵ, and hence that $Y(\epsilon, l)$ can be expressed as the ratio of two polynomials:

$$Y(\epsilon, l) = (a_0 + \epsilon a_1 + \ldots + \epsilon^p a_p)(b_0 + \epsilon b_1 + \ldots + \epsilon^q b_q)^{-1}. \tag{5.15}$$

From (2.52), (5.11) and (5.12) we have

$$\mathscr{R}(\epsilon, l) = B(\epsilon, l) Y(\epsilon, l). \tag{5.16}$$

We recall that, from (2.51), $B(\epsilon, l) = A(\epsilon, l)[1 - \exp(-2\pi/k)]^{-1}$ for $\epsilon > 0$ and $B(\epsilon, l) = A(\epsilon, l)$ for $\epsilon < 0$. In all practical applications of the theory (see § 2.7.1) we may assume that $\epsilon \ll (2\pi)^2$ and hence that, for $\epsilon > 0$, $(2\pi/k) \gg 1$ giving $B = A$. Thus, for all practical purposes we can put $B = A$ in (5.16).

From (2.57), (5.12) and (5.13),

$$\chi = (1 + \mathrm{i}\mathscr{R})(1 - \mathrm{i}\mathscr{R})^{-1}. \tag{5.17}$$

For $\epsilon \geqslant 0$, $\mathscr{R} = R$ and $\chi = S$, with R and S as defined in § 3.4.

5.5. *The η defect, the phase shift and the quantum defect*

5.5.1. The η defect. The quantity $Y(\epsilon, l)$ generally varies slowly with ϵ, but may have isolated poles. Following Ham we define η by

$$Y = \tan(\pi\eta). \tag{5.18}$$

Then η varies slowly with ϵ (and is equal to a half-integer at a pole in Y).

5.5.2. Phase shifts for $\epsilon \geqslant 0$. We define the phase shift δ by

$$\tan(\delta) = \mathscr{R}. \tag{5.19}$$

For $\epsilon > 0$ we than have, from (2.53) and (5.12),

$$F(\mathscr{R};\rho) \underset{\rho\to\infty}{\sim} (\pi k)^{-1/2} \sin(\zeta + \delta)[\cos(\delta)]^{-1}. \tag{5.20}$$

5.5.3. Bound states. Bound states occur at energies such that the coefficient of the exponentially increasing function ξ vanishes. From (5.9) the condition is

$$\Gamma(\nu-l)[\sin(\pi\nu)(I-\mathcal{G}J)-A\cos(\pi\nu)J]=0. \tag{5.21}$$

For $\epsilon<0$, (2.30) gives $A=\Gamma(\nu+l+1)[\nu^{2l+1}\Gamma(\nu-l)]^{-1}$. With Y defined by (5.14) the condition (5.21) is

$$\frac{\tan(\pi\nu)}{A(\epsilon,l)}+Y(\epsilon,l)=0 \tag{5.22}$$

and from (5.16) this condition is

$$\tan(\pi\nu)+\mathcal{R}=0 \tag{5.23}$$

so long as $A(\epsilon,l)\neq 0$.

For the pure Coulomb potential, $U(\rho)=2/\rho$, we have $J=0$ and hence $Y=0$ and $\mathcal{R}=0$. Equation (5.22) then gives the bound states to be at $\nu=n$ with $n\geqslant(l+1)$, as in § 2.10. We note that $[\tan(\pi\nu)/A]\neq 0$ for $\nu=m$ where

$$m=1,2,\ldots,l. \tag{5.24}$$

Equation (5.23) is often used as the equation for bound states for the case of $\mathcal{R}\neq 0$ but caution is required, since this can give spurious roots at energies lower than those of the lowest true bound state.

5.5.4. The quantum defect. The quantum defect μ is defined, for $\epsilon<0$, by

$$\mathcal{R}=\tan(\pi\mu). \tag{5.25}$$

Comparison with (5.19) shows that $\delta(\epsilon)$ is the analytic continuation of $\pi\mu(\epsilon)$ (a result strictly correct only for $(2\pi/k)\gg 1$).

Equation (5.23) has solutions at $\nu=n-\mu$, giving

$$\epsilon_n=-(n-\mu(\epsilon_n))^{-2}. \tag{5.26}$$

Given the function $\mu(\epsilon)$, (5.26) can be solved by iteration; a first estimate $\epsilon_n^{(0)}$ gives an improved estimate

$$\epsilon_n^{(1)}=-(n-\mu(\epsilon_n^{(0)}))^{-2}. \tag{5.27}$$

5.5.5. Use of the η defect. Since $A(\epsilon,l)=1$ for $l=0$, μ differs from η only for $l>0$.

For $l>0$, (5.16) constrains $\mathcal{R}(\epsilon,l)$ to be such that $\mathcal{R}=0$ for $\nu=m$ with m defined by (5.24), and μ extrapolated to $\nu=m$ must be equal to an integer. This explains the trend of the variation of μ with ϵ in many observed spectral series (Ham 1955, Seaton 1958a). The variation of $\eta(\epsilon)$ is generally slower than that of $\mu(\epsilon)$. This is illustrated in figure 9 for the Na I p series. Sheorey (1969) considers values of $Y=\tan(\pi\eta)$ for alkali-like systems and shows that the number of parameters required to fit $Y(\epsilon)$ to (5.15) is generally smaller than the number required to fit $\mu(\epsilon)$ to a polynomial in ϵ.

5.6. Normalisation

We consider the normalisation of the function

$$\mathcal{F}(\epsilon,l;\rho)=F(\mathcal{R};\rho)\cos(\pi\mu). \tag{5.28}$$

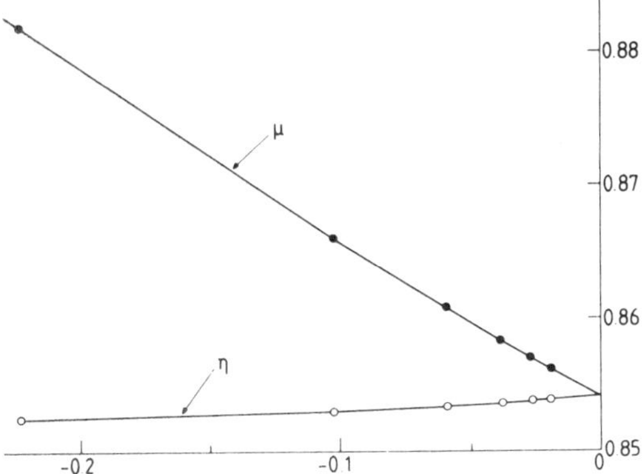

Figure 9. Quantum defects and η defects for Na I ^{2}P$^\circ$ (using the centres of gravity for the two components of the doublets).

For $\epsilon > 0$ we have, from (5.19) and (5.20),

$$\mathscr{F}(\epsilon, l; \rho) \underset{\rho\to\infty}{\sim} (\pi k)^{-1/2} \sin(\xi + \delta) \tag{5.29}$$

and obtain, by the method of § 2.11.1,

$$\int_0^\infty \mathscr{F}(\epsilon, l; \rho)\mathscr{F}(\epsilon', l; \rho)\, \mathrm{d}\rho = \delta(\epsilon - \epsilon'). \tag{5.30}$$

For $\epsilon < 0$ and $\rho \geqslant \rho_0$ we have, from (5.12) and (5.25), $\mathscr{F} = s\cos(\pi\mu) + c\sin(\pi\mu)$ and hence, using (2.54), $\mathscr{F} = A_+(\nu)\xi + A_-(\nu)\theta$ where

$$A_+(\nu) = \frac{(-1)^l \sin[\pi(\nu+\mu)]}{(2\nu)^{1/2}\pi K} \qquad A_-(\nu) = -(-1)^l \cos[\pi(\nu+\mu)]\left(\frac{\nu^3}{2}\right)^{1/2} K. \tag{5.31}$$

These expressions for A_+ and A_- differ from those of § 2.11.3 only in that $\sin(\pi\nu)$ and $\cos(\pi\nu)$ are replaced by $\sin[\pi(\nu+\mu)]$ and $\cos[\pi(\nu+\mu)]$. Proceeding as in § 2.11.3 we obtain

$$\int_0^\infty \mathscr{F}(\epsilon_n, l; \rho)\mathscr{F}(\epsilon_{n'}, l; \rho) = \delta(n, n')\tfrac{1}{2}\nu_n^3(1 + C_n) \tag{5.32}$$

where $\epsilon_n = -1/\nu_n^2$, $\nu_n = n - \mu(\epsilon_n)$ and $C_n = C(\nu_n)$, where

$$C(\nu) = \mathrm{d}\mu/\mathrm{d}\nu. \tag{5.33}$$

The normalisation integral for $\mathscr{F}(\epsilon_n, l; \rho)$ is seen to differ from that for $s(\epsilon_n, l; \rho)$ (equation (2.75)) by the factor $(1 + C_n)$. Since $\mathrm{d}\epsilon = (2/\nu^3)\,\mathrm{d}\nu$, $\mathrm{d}\mu/\mathrm{d}\nu = (2/\nu^3)\,\mathrm{d}\mu/\mathrm{d}\epsilon$. For ν large, $\mathrm{d}\mu/\mathrm{d}\epsilon$ tends to a finite limit and $[1 + C(\nu)]$ tends to unity. The significance of this factor can be understood by considering, as in § 2.11.3, the separations of the energy levels, $\delta\epsilon_n = (\epsilon_{n+1} - \epsilon_n)$. With $\epsilon_n = -(n - \mu_n)^{-2}$ we have, for n large, $\delta\epsilon_n \simeq (2/\nu_n^3)(1 - \mu_{n+1} + \mu_n) \simeq (2/\nu_n^3)(1 - C_n)$ and hence $1/(\delta\epsilon_n) \simeq (\nu_n^3/2)(1 + C_n)$. In the limit of n large the normalising factor in (5.32) is therefore equal to $1/(\delta\epsilon_n)$. The normalising factor $(1 + \mathrm{d}\mu/\mathrm{d}\nu)$ was given by Seaton (1958a). A much earlier derivation, for s states, was given by Fermi and Segré (1933) and is discussed by Foldy (1958).

For $\nu=\nu_n$ (5.31) gives $A_-=-(-1)^{n+l}(\nu_n^3/2)^{1/2}K$. The function defined by

$$P_{nl}(\rho)=-(-1)^{n+l}[(2/\nu_n^3)(1+C_n)^{-1}]^{1/2}\mathscr{F}(\epsilon_n,l;\rho) \tag{5.34}$$

is normalised to $(P_{nl}|P_{n'l})=\delta(n,n')$ and is equal to $(1+C_n)^{-1/2}K\theta$ for $\rho\geqslant\rho_0$.

5.7. Long-range potentials

In order to obtain correct quantum defects for states with larger values of l it is, as has already been noted in § 1, necessary to take account of long-range potentials.

The polarisation potential behaves, for r large, like $-\frac{1}{2}\alpha(\text{au})/r^4$ where $\alpha(\text{au})$ is the polarisability in atomic units, and hence gives a contribution to $U(\rho)$ behaving like α/ρ^4 where $\alpha=Z^2\alpha(\text{au})$. We put

$$U(\rho)=U_1(\rho)+U_4(\rho) \tag{5.35}$$

where

$$U_1(\rho)=\frac{2}{\rho} \quad\text{and}\quad U_4(\rho)=\frac{\alpha}{\rho^4} \quad \text{for } \rho\geqslant\rho_0. \tag{5.36}$$

We treat the polarisation potential U_4 as a perturbation. Let $\mathscr{F}^{(1)}$ be the radial function obtained with $U_4=0$, and with the normalisation adopted in § 5.6. For $\epsilon>0$, we take $\mathscr{F}^{(1)}$ to have a phase shift $\delta^{(1)}$, $\mathscr{F}^{(1)}\sim(\pi k)^{-1/2}\sin(\zeta+\delta^{(1)})$. Let $\delta=\delta^{(1)}+\Delta\delta$ be the phase shift for the potential (5.36). Using a variational principle for the phase shift (Percival 1960) we obtain

$$\Delta\delta=\pi(\mathscr{F}^{(1)}|U_4|\mathscr{F}^{(1)}). \tag{5.37}$$

For $\epsilon<0$, let $\mu^{(1)}$ be the quantum defect obtained using U_1 and put $\epsilon_n^{(1)}=-(n-\mu_n^{(1)})^{-2}$. Use of (5.35) gives $\epsilon_n=\epsilon_n^{(1)}+\Delta\epsilon$. The normalised function $P_{nl}^{(1)}$ is defined using (5.34), and bound-state perturbation theory gives

$$\Delta\epsilon=(P_{nl}^{(1)}|U_4|P_{nl}^{(1)}). \tag{5.38}$$

For $\nu_n^{(1)}$ large, $\Delta\epsilon=2\Delta\mu(\nu_n^{(1)})^{-3}$ and hence, using (5.34),

$$\Delta\mu=\left.\frac{(\mathscr{F}^{(1)}|U_4|\mathscr{F}^{(1)})}{1+C(\nu)}\right|_{\nu=\nu_n}. \tag{5.39}$$

Comparison with (5.37) shows that $\Delta\delta$ is obtained on extrapolating $\pi\delta\mu$.

Waller (1926) assumed that quantum defects for l large are determined entirely by the polarisation potential, α/ρ^4. He therefore put $U_1=2/\rho$ and $U_4=\alpha/\rho^4$ for all values of ρ and obtained

$$\mu_n=\left(s(\epsilon_n,l;\rho)\left|\frac{\alpha}{\rho^4}\right|s(\epsilon_n,l;\rho)\right). \tag{5.40}$$

For l large, the Coulomb functions $s(\epsilon_n,l;\rho)$ are small for smaller values of ρ and little error results from making the approximation of putting $U_4(\rho)=\alpha/\rho^4$ for all ρ. The integral in (5.40) may be evaluated to give

$$\mu_n=\alpha\frac{2[3-l(l+1)/n^2]}{l(l+1)(2l-1)(2l+1)(2l+3)} \tag{5.41}$$

and hence to give

$$\mu(\epsilon)=\alpha\frac{2[3+l(l+1)\epsilon]}{l(l+1)(2l-1)(2l+1)(2l+3)} \tag{5.42}$$

as an interpolated continuous function of ϵ.

Values of α are known exactly only for hydrogenic ions, for which $\alpha(\text{au})=9(2Z_0^4)^{-1}$. For He I d states, (5.42) gives $\mu(\epsilon)=0.002\,679+0.005\,36\epsilon$. Fitting of He I experimental data gives $\mu(\epsilon)=0.002\,094+0.003\,01\epsilon$ for ^{1}D and $\mu(\epsilon)=0.002\,869+0.006\,22\epsilon$ for ^{3}D. The difference between ^{1}D and ^{3}D shows that exchange effects are not negligible. They are eliminated to first order using $\bar{\mu}=(\mu[^1\text{D}]+3\mu[^3\text{D}])/4$ which gives $\bar{\mu}=0.002\,675+0.005\,42\epsilon$ in remarkable agreement with the polarisation formula.

5.8. *The relativistic one-channel theory*

In the introduction to an important paper on one-channel relativistic QDT, Johnson and Cheng (1979) remark that '... we must devote some space to an important but tedious discussion of the properties of relativistic Coulomb wavefunctions. To spare the casual reader some of the formal details, we summarise here the main lines of the argument presented ...'. The present review contains an account of the properties of the non-relativistic functions which, it is feared, some readers may find tedious. Constraints on space oblige us to advise the non-casual reader, interested in the relativistic theory, to consult the clear account given by Johnson and Cheng.

The Z-scaled energy for a bound state of a hydrogenic ion, obtained from solving the Dirac equation, is

$$\epsilon_{n\kappa}=\frac{2}{\beta^2}[(1+\beta^2/\nu_{n\kappa}^2)^{-1/2}-1] \tag{5.43}$$

where $\beta=\alpha Z$, $\alpha=e^2(\hbar c)^{-1}=1/137.036\ldots$ is the fine-structure constant;

$$\nu_{n\kappa}=n+(\kappa^2-\beta^2)^{1/2}-|\kappa| \tag{5.44}$$

and $\kappa=-(l+1)$ for $j=l+\frac{1}{2}$ and $\kappa=+l$ for $j=l-\frac{1}{2}$. The non-relativistic expression, $\epsilon_n=-1/n^2$, is recovered on taking the limit of $\beta\to 0$.

For alkali-like ions one can take the Z-scaled energy to be given by (5.43) with

$$\nu_{n\kappa}=n+(\kappa^2-\beta^2)^{1/2}-|\kappa|-\mu_{n\kappa} \tag{5.45}$$

where $\mu_{n\kappa}$ is the quantum defect which arises from interactions with the ion core. Putting $\mu_{n\kappa}=\mu(\epsilon_{n\kappa})$, Johnson and Cheng show that $\mu(\epsilon)$ can be expressed in terms of analytic functions of the energy. They derive the relativistic counterparts of all of the relations which have been derived for the non-relativistic theory.

We consider further the fine-structure splitting. In using (5.44) one allows for contributions to fine structure due to the potential, in atomic units, $V(r)=-Z/r$. The additional contributions due to the shorter-range ion-core potential are then included in the quantum defects $\mu_{n\kappa}$ in (5.45). In the non-relativistic theory all contributions to the fine-structure energy are included in the quantum defect. It follows that the differences between the quantum defects for the two components of a doublet are larger in the non-relativistic theory than in the relativistic theory. This is illustrated in table 2 which gives results for d states of N V, a lithium-like ion with $Z=5$.

Table 2. Quantum defects for d states of N v (from Johnson and Cheng 1979).

	Non-relativistic		Relativistic	
	$j=3/2$	5/2	3/2	5/2
n				
3	0.001490	0.001380	0.001324	0.001324
4	0.001778	0.001668	0.001570	0.001571
5	0.001876	0.001765	0.001643	0.001643
6	0.001879	0.001768	0.001629	0.001630
7	0.001817	0.001705	0.001556	0.001554
8	0.001704	0.001592	0.001434	0.001433

6. The many-channel theory

We describe, first, the many-channel theory, essentially as developed by Seaton (1966a, 1969, 1978).

6.1. Solutions of the coupled equations

We consider the expansions† (3.3) or (3.17) and the equations (3.13) with solutions $F_{ij}(\rho)$ where i is a channel index, $i=1$ to I, and j specifies some particular set of solutions. We shall be concerned with those solutions for which

$$F_{ij}(\rho)=0 \qquad \text{for } \rho=0 \tag{6.1}$$

and hence for which Ψ is bounded at the origin. In § 3 we imposed the further condition that Ψ should remain bounded for $\rho\to\infty$, giving the number of LI solutions to be equal to I_0, the number of open channels. In QDT we consider solutions for which no such further condition is imposed. The number of LI solutions is then equal to I, the total number of channels, and a complete set of solutions is given by a square matrix $\boldsymbol{F}(\rho)$ with elements $F_{ij}(\rho)$, $i=1$ to I and $j=1$ to I. If $\boldsymbol{F}_A(\rho)$ is some particular complete set, the columns of any other set of solutions $\boldsymbol{F}_B(\rho)$ can be expressed as linear combinations of the columns of $\boldsymbol{F}_A(\rho)$; that is to say $\boldsymbol{F}_B(\rho)=\boldsymbol{F}_A(\rho)\boldsymbol{C}$ with $\boldsymbol{C}$ independent of ρ.

We assume, as in the one-channel case, the existence of solutions $\boldsymbol{F}(\mathscr{E};\rho)$ which can be represented by expansions in powers of $\mathscr{E}$, and that all non-Coulomb potentials can be neglected beyond some finite value of ρ, giving

$$\boldsymbol{U}(\rho)=\frac{2}{\rho} \qquad \text{for } \rho\geqslant\rho_0 \text{ with } \rho_0<\infty. \tag{6.2}$$

We may then put

$$\boldsymbol{F}(\mathscr{E};\rho)=f(\epsilon,l;\rho)\boldsymbol{I}(\mathscr{E})+g(\epsilon,l;\rho)\boldsymbol{J}(\mathscr{E}) \tag{6.3}$$

where $\boldsymbol{I}(\mathscr{E})=-(\pi/2)\boldsymbol{W}(g,\boldsymbol{F})$ and $\boldsymbol{J}(\mathscr{E})=(\pi/2)\boldsymbol{W}(f,\boldsymbol{F})$ can be expanded in powers of ϵ.

† The effects of including 'bound channels' in the expansions is discussed in appendix A.

6.2. *The matrices* $\mathbf{Y}$, $\mathcal{R}$ *and* χ

We define the solutions $\boldsymbol{F}(\boldsymbol{Y};\rho)$, $\boldsymbol{F}(\mathcal{R};\rho)$ and $\boldsymbol{F}(\chi;\rho)$ to be such that

$$\left.\begin{aligned} \boldsymbol{F}(\boldsymbol{Y};\rho)&=f+h\boldsymbol{Y} \\ \boldsymbol{F}(\mathcal{R};\rho)&=s+c\mathcal{R} \\ \boldsymbol{F}(\chi;\rho)&=\varphi^{-}-\varphi^{+}\chi \end{aligned}\right\} \text{ for } \rho\geqslant\rho_0. \qquad \begin{aligned}(6.4)\\(6.5)\\(6.6)\end{aligned}$$

From (2.42) and (6.3) and (6.4),

$$\boldsymbol{Y}=-\boldsymbol{J}(\boldsymbol{I}-\mathcal{G}\boldsymbol{J})^{-1}. \tag{6.7}$$

We assume that $\boldsymbol{Y}$ can be approximated by

$$\boldsymbol{Y}(\mathscr{E})=(\boldsymbol{a}_0+\mathscr{E}\boldsymbol{a}_1+\ldots+\mathscr{E}^p\boldsymbol{a}_p)(\boldsymbol{b}_0+\mathscr{E}\boldsymbol{b}_1+\ldots+\mathscr{E}^q\boldsymbol{b}_q)^{-1}. \tag{6.8}$$

From (2.52), (6.4) and (6.5),

$$\mathcal{R}(\mathscr{E})=A^{1/2}\boldsymbol{Y}(\mathscr{E})A^{1/2} \tag{6.9}$$

where we put $B=A$ for $\epsilon\geqslant 0$. From (2.57), (6.5) and (6.6)

$$\chi=(1+\mathrm{i}\mathcal{R})(1-\mathrm{i}\mathcal{R})^{-1}. \tag{6.10}$$

Assuming $\boldsymbol{U}$ to be real we have, from (3.16),

$$\boldsymbol{Y}=\boldsymbol{Y}^{\mathrm{T}}=\boldsymbol{Y}^{*} \tag{6.11}$$

$$\mathcal{R}=\mathcal{R}^{\mathrm{T}}=\mathcal{R}^{*} \tag{6.12}$$

$$\chi=\chi^{T} \qquad \chi^{*}\chi=1. \tag{6.13}$$

6.3. *Diagonalisation of* $\mathbf{Y}$, $\mathcal{R}$ *and* χ

Let $\mathcal{R}$ have eigenvalues $\tan(\pi\bar{\mu}_j)$. Then

$$\mathcal{R}\boldsymbol{X}=\boldsymbol{X}\tan(\pi\bar{\mu}) \tag{6.14}$$

where $\tan(\pi\bar{\mu})$ is diagonal and where $\boldsymbol{X}$ can be taken to be real and normalised to

$$\boldsymbol{X}^{\mathrm{T}}\boldsymbol{X}=1. \tag{6.15}$$

It follows from (6.10) and (6.14) that $\chi\boldsymbol{X}=\boldsymbol{X}\exp(2\pi\mathrm{i}\bar{\mu})$. The quantities $\bar{\mu}_j$ are referred to as *eigen quantum defects* and $\bar{\delta}_j=\bar{\pi}\mu_j$ as *eigenphases*. Defining

$$\boldsymbol{F}(\bar{\mu};\rho)=\boldsymbol{F}(\mathcal{R};\rho)\boldsymbol{X}\cos(\pi\bar{\mu}) \tag{6.16}$$

we have

$$\boldsymbol{F}(\bar{\mu};\rho)=s\boldsymbol{X}\cos(\pi\mu)+c\boldsymbol{X}\sin(\pi\bar{\mu}) \qquad \text{for } \rho\geqslant\rho_0. \tag{6.17}$$

For the one-channel case, $F(\bar{\mu};\rho)=\mathscr{F}(\epsilon,l;\rho)$ of § 5.6.

The eigenvalues of Y are denoted by $\tan(\pi\eta_i)$.

6.4. *The matrices* $\boldsymbol{R}$ *and* $\boldsymbol{S}$

The reactance matrix $\boldsymbol{R}$ and the scattering matrix $\boldsymbol{S}$ are defined in § 3.4.

6.4.1. All channels open. When all channels are open we have $\epsilon_i = k_i^2 \geq 0$ for all i. It follows from the definitions (3.24) and (3.25) of $\boldsymbol{R}$ and $\boldsymbol{S}$ and (6.5) and (6.6) of $\boldsymbol{\mathcal{R}}$ and $\boldsymbol{\chi}$ that, for all channels open, $\boldsymbol{R} = \boldsymbol{\mathcal{R}}$ and $\boldsymbol{S} = \boldsymbol{\chi}$.

6.4.2. Some channels closed. When I_0 channels are open, with $I_0 < I$, we have $\epsilon_i = k_i^2 \geq 0$ for $i = 1$ to I_0 and $\epsilon_i = -1/\nu_i^2 < 0$ for $i = (I_0+1)$ to I. The function matrices $\boldsymbol{G}(\boldsymbol{R}; \rho)$ and $\boldsymbol{G}(\boldsymbol{S}; \rho)$ were introduced in § 3.4. The number of columns in each of these matrices is equal to I_0. We put

$$\boldsymbol{G}(\boldsymbol{R}; \rho) = \boldsymbol{F}(\boldsymbol{\mathcal{R}}; \rho)\boldsymbol{L} \qquad \text{and} \qquad \boldsymbol{G}(\boldsymbol{S}; \rho) = \boldsymbol{F}(\boldsymbol{\chi}; \rho)\boldsymbol{M} \tag{6.18}$$

where the matrices $\boldsymbol{L}$ and $\boldsymbol{M}$ each have I rows and I_0 columns.

For $\boldsymbol{F}(\boldsymbol{\mathcal{R}}; \rho)$ and $\boldsymbol{F}(\boldsymbol{\chi}; \rho)$ we use the partitioning

$$\boldsymbol{F} = \begin{pmatrix} \boldsymbol{F}_{oo} & \boldsymbol{F}_{oc} \\ \boldsymbol{F}_{co} & \boldsymbol{F}_{cc} \end{pmatrix} \tag{6.19}$$

with, as in § 3.3, subscripts o for 'open' and c for 'closed'. The elements F_{ij} are: $i = 1$ to I_0, $j = 1$ to I_0 for $\boldsymbol{F}_{oo}$; $i = 1$ to I_0, $j = (I_0+1)$ to I for $\boldsymbol{F}_{oc}$; $i = (I_0+1)$ to I, $j = 1$ to I_0 for $\boldsymbol{F}_{co}$; and $i = (I_0+1)$ to I, $j = (I_0+1)$ to I for $\boldsymbol{F}_{cc}$. The partitioning for ϵ is

$$\epsilon = \begin{pmatrix} k^2 & 0 \\ 0 & -1/\nu^2 \end{pmatrix} \tag{6.20}$$

and the partitioning for $\boldsymbol{L}$ and $\boldsymbol{M}$,

$$\boldsymbol{L} = \begin{pmatrix} \boldsymbol{L}_{oo} \\ \boldsymbol{L}_{co} \end{pmatrix} \qquad \boldsymbol{M} = \begin{pmatrix} \boldsymbol{M}_{oo} \\ \boldsymbol{M}_{co} \end{pmatrix} \tag{6.21}$$

is the same as that for $\boldsymbol{G}(\boldsymbol{\mathcal{R}})$ and $\boldsymbol{G}(\boldsymbol{S})$, equation (3.20).

Consider first the determination of $\boldsymbol{L}$. From (6.18), (6.19) and (6.21) we have

$$\boldsymbol{G}(\boldsymbol{R}) = \begin{pmatrix} \boldsymbol{F}_{oo}(\boldsymbol{\mathcal{R}})\boldsymbol{L}_{oo} + \boldsymbol{F}_{oc}(\boldsymbol{\mathcal{R}})\boldsymbol{L}_{co} \\ \boldsymbol{F}_{co}(\boldsymbol{\mathcal{R}})\boldsymbol{L}_{oo} + \boldsymbol{F}_{cc}(\boldsymbol{\mathcal{R}})\boldsymbol{L}_{co} \end{pmatrix} \tag{6.22}$$

and, using (6.5), obtain for $\rho \geq \rho_0$

$$\begin{aligned} \boldsymbol{G}_{oo}(\boldsymbol{R}) &= s_o\boldsymbol{L}_{oo} + c_o(\boldsymbol{\mathcal{R}}_{oo}\boldsymbol{L}_{oo} + \boldsymbol{\mathcal{R}}_{oc}\boldsymbol{L}_{co}) \\ \boldsymbol{G}_{co}(\boldsymbol{R}) &= s_c\boldsymbol{L}_{oo} + c_c(\boldsymbol{\mathcal{R}}_{co}\boldsymbol{L}_{oo} + \boldsymbol{\mathcal{R}}_{cc}\boldsymbol{L}_{co}). \end{aligned} \tag{6.23}$$

We must now take $\boldsymbol{L}$ to be such that (3.24) is satisfied. The first equation of (3.24) is satisfied if

$$\boldsymbol{L}_{oo} = 1 \tag{6.24}$$

and

$$\boldsymbol{R} = \boldsymbol{\mathcal{R}}_{oo} + \boldsymbol{\mathcal{R}}_{oc}\boldsymbol{L}_{co}. \tag{6.25}$$

The second equation of (3.24) is satisfied on eliminating the exponentially increasing functions ξ in the second equation of (6.23). From (2.54) the coefficients of ξ in s_c and c_c are proportional to $\sin(\pi\nu)$ and $\cos(\pi\nu)$ and ξ is eliminated in (6.22) on taking

$$\sin(\pi\nu)\boldsymbol{L}_{co} + \cos(\pi\nu)(\boldsymbol{\mathcal{R}}_{co}\boldsymbol{L}_{oo} + \boldsymbol{\mathcal{R}}_{cc}\boldsymbol{L}_{co}) = 0. \tag{6.26}$$

Using (6.24) this gives

$$\cos(\pi\nu)\boldsymbol{\mathcal{R}}_{co} + [\sin(\pi\nu) + \cos(\pi\nu)\boldsymbol{\mathcal{R}}_{cc}]\boldsymbol{L}_{co} = 0 \tag{6.27}$$

and hence

$$\boldsymbol{L}_{\text{co}} = -[\tan(\pi\nu) + \boldsymbol{\mathcal{R}}_{\text{cc}}]^{-1}\boldsymbol{\mathcal{R}}_{\text{co}}. \tag{6.28}$$

Substitution in (6.25) gives

$$\boldsymbol{R} = \boldsymbol{\mathcal{R}}_{\text{oo}} - \boldsymbol{\mathcal{R}}_{\text{oc}}[\tan(\pi\nu) + \boldsymbol{\mathcal{R}}_{\text{cc}}]^{-1}\boldsymbol{\mathcal{R}}_{\text{co}} \tag{6.29}$$

and, using (6.9),

$$\boldsymbol{R} = A_{\text{o}}^{1/2}\left[\boldsymbol{Y}_{\text{oo}} - \boldsymbol{Y}_{\text{oc}}\left(\frac{\tan(\pi\nu)}{A_{\text{c}}} + \boldsymbol{Y}_{\text{cc}}\right)^{-1}\boldsymbol{Y}_{\text{co}}\right]A_{\text{o}}^{1/2}. \tag{6.30}$$

The relations between $\boldsymbol{S}$ and χ is obtained in a similar way. From (6.18) and the first of equations (3.25)

$$\boldsymbol{M}_{\text{oo}} = 1 \qquad \boldsymbol{S} = \chi_{\text{oo}} + \chi_{\text{oc}}\boldsymbol{M}_{\text{co}}. \tag{6.31}$$

To eliminate the exponentially increasing solution we use (2.59) to obtain

$$\boldsymbol{M}_{\text{co}} = -[\chi_{\text{cc}} - \exp(-2\pi\mathrm{i}\nu)]^{-1}\chi_{\text{co}} \tag{6.32}$$

and hence

$$\boldsymbol{S} = \chi_{\text{oo}} - \chi_{\text{oc}}[\chi_{\text{cc}} - \exp(-2\pi\mathrm{i}\nu)]^{-1}\chi_{\text{co}}. \tag{6.33}$$

6.5. Bound states

When all channels are closed, $\epsilon_i = -1/\nu_i^2 < 0$ for all i, the equations (3.13) have solutions which are everywhere bounded only for certain discrete eigenvalues of the energy $\mathscr{E}$. Such solutions are denoted by column vectors $\boldsymbol{P}(\rho)$ with elements $P_i(\rho)$.

6.5.1. Energy eigenvalues. We put

$$\boldsymbol{P}(\rho) = \boldsymbol{F}(\boldsymbol{Y};\rho)\boldsymbol{M} \tag{6.34}$$

where $\boldsymbol{M}$ is a column vector. When all channels are closed we have, for $\rho \geqslant \rho_0$,

$$\boldsymbol{F}(\boldsymbol{Y};\rho) = (-1)^l\nu^{l+1}\left(\xi\frac{\Gamma(\nu-l)}{\pi}[\sin(\pi\nu) + \boldsymbol{A}\cos(\pi\nu)\boldsymbol{Y}] - \frac{\theta}{\Gamma(\nu+l+1)}[\cos(\pi\nu) - \boldsymbol{A}\sin(\pi\nu)\boldsymbol{Y}]\right). \tag{6.35}$$

The exponentially increasing solutions ξ are eliminated in (6.34) if

$$\Gamma(\nu-l)[\sin(\pi\nu) + \boldsymbol{A}\cos(\pi\nu)\boldsymbol{Y}]\boldsymbol{M} = 0. \tag{6.36}$$

Since $\boldsymbol{A} = \Gamma(\nu+l+1)[\nu^{2l+1}\Gamma(\nu-l)]^{-1}$, (6.36) is equivalent to

$$\left(\frac{\tan(\pi\nu)}{\boldsymbol{A}} + \boldsymbol{Y}\right)\boldsymbol{M} = 0 \tag{6.37}$$

which is the many-channel generalisation of (5.22). The condition for the determination of the energy eigenvalues is

$$\det\left(\frac{\tan(\pi\nu)}{\boldsymbol{A}} + \boldsymbol{Y}\right) = 0. \tag{6.38}$$

We can proceed in a similar way using $\boldsymbol{F}(\boldsymbol{\mathcal{R}};\rho)$. Putting

$$\boldsymbol{P}(\rho)=\boldsymbol{F}(\boldsymbol{\mathcal{R}};\rho)\boldsymbol{N} \tag{6.39}$$

the condition for the elimination of exponentially increasing solutions is

$$[\sin(\pi\nu)+\cos(\pi\nu)\boldsymbol{\mathcal{R}}]\boldsymbol{N}=0 \tag{6.40}$$

which can be satisfied if

$$\det[\tan(\pi\nu)+\boldsymbol{\mathcal{R}}]=0 \tag{6.41}$$

which is the many-channel generalisation of (5.23). It was noted in § 5.5.3 that caution is needed in using (5.23), which can give spurious roots. Similar caution is required in using (6.41). No such problems arise in using (6.38).

6.5.2. Normalisation. For all channels closed and $\mathscr{E}$ not necessarily an energy eigenvalue we obtain from (2.54), (6.5) and (6.39)

$$P(\rho)=\xi\boldsymbol{A}_{+}+\theta\boldsymbol{A}_{-} \qquad \text{for } \rho\geqslant\rho_0 \tag{6.42}$$

where

$$\begin{aligned}\boldsymbol{A}_{+}&=\frac{(-1)^l}{(2\nu)^{1/2}\pi\boldsymbol{K}}[\sin(\pi\nu)+\cos(\pi\nu)\boldsymbol{\mathcal{R}}]\boldsymbol{N}\\ \boldsymbol{A}_{-}&=-(-1)^l\left(\frac{\nu^3}{2}\right)^{1/2}\boldsymbol{K}[\cos(\pi\nu)-\sin(\pi\nu)\boldsymbol{\mathcal{R}}]\boldsymbol{N}.\end{aligned} \tag{6.43}$$

The normalisation condition for $\boldsymbol{P}(\rho)$ is obtained using the method of §§ 2.11 and 5.6. From (3.16) and (6.42) we obtain, at the eigenvalues,

$$\int_0^\infty \boldsymbol{P}^\dagger(\rho)\boldsymbol{P}(\rho)\,\mathrm{d}\rho=-\boldsymbol{A}^\dagger\left(\frac{2}{\nu}\right)\frac{\mathrm{d}}{\mathrm{d}\epsilon}\boldsymbol{A}_{+}. \tag{6.44}$$

At the eigenvalues we have $\boldsymbol{P}=\theta\boldsymbol{A}_{-}$. Using (6.40) to eliminate $\boldsymbol{\mathcal{R}}$ in the expression (6.43) for $\boldsymbol{A}_{-}$ we obtain

$$\boldsymbol{A}_{-}=-(-1)^l(\nu^3/2)^{1/2}[\cos(\pi\nu)]^{-1}\boldsymbol{K}\boldsymbol{N} \tag{6.45}$$

and we may therefore put

$$\boldsymbol{P}=\boldsymbol{K}\theta\boldsymbol{Z} \tag{6.46}$$

where

$$\boldsymbol{Z}=-(-1)^l(\nu^3/2)^{1/2}[\cos(\pi\nu)]^{-1}\boldsymbol{N}. \tag{6.47}$$

We take $\boldsymbol{P}(\rho)$ to be normalised to

$$\int_0^\infty \boldsymbol{P}^\dagger(\rho)\boldsymbol{P}(\rho)\,\mathrm{d}\rho=1. \tag{6.48}$$

Using (6.40), (6.43), (6.44) and (6.47) we obtain

$$\boldsymbol{Z}^\dagger[1+\boldsymbol{C}(\mathscr{E})]\boldsymbol{Z}=1 \tag{6.49}$$

where

$$\boldsymbol{C}(\mathscr{E})=q\left(\frac{\mathrm{d}}{\mathrm{d}\mathscr{E}}\boldsymbol{\mathcal{R}}\right)q \tag{6.50}$$

and

$$q = (-1)^l \left(\frac{2}{\pi \nu^3}\right)^{1/2} \cos(\pi \nu). \tag{6.51}$$

For the one-channel case, $\boldsymbol{C}(\mathscr{E})$ reduces to $C(\nu) = \mathrm{d}\mu/\mathrm{d}\nu$ as used in § 5.6.

We consider further the significance of the normalisation condition for highly excited states, such that ν_i is large in all channels. For such states the main contribution to (6.48) comes from $\rho \geqslant \rho_0$. It was shown in § 2.11.3 that the normalised bound-state wavefunctions of a hydrogenic system are $P_{nl}(\rho) = K(n, l)\theta(n, l; \rho)$ and it may be shown in a similar way that for ν large

$$\int_{\rho_0}^{\infty} [K(\nu, l)\theta(\nu, l; \rho)]^2 \, \mathrm{d}\rho \simeq 1. \tag{6.52}$$

For ν large it follows from (6.51) that q is small and that the condition (6.49) reduces to

$$\sum_i |Z_i|^2 \simeq 1 \tag{6.53}$$

which is seen to be consistent with (6.46) and (6.52). Defining

$$W_i = |Z_i|^2 \left(\sum_{i'} |Z_{i'}|^2\right)^{-1} \tag{6.54}$$

we may say that, for a highly excited bound state, W_i is the probability that the outer electron is in channel i.

6.6. Radiative transitions

The reduced dipole matrix is

$$\boldsymbol{D} = \int_0^{\infty} \boldsymbol{P}^{\dagger} \boldsymbol{g} \, \mathrm{d}\rho \tag{6.55}$$

for transitions to bound states and

$$\boldsymbol{D}(\text{out}) = \int_0^{\infty} \boldsymbol{G}^{\dagger}(\text{out}) \boldsymbol{g} \, \mathrm{d}\rho \tag{6.56}$$

for transitions to the continuum. The elements D_{ba} and $D_{ja}(\text{out})$ are defined in § 4. The radial functions $\boldsymbol{P}(\rho)$ in (6.55) have normalisation (6.49), and the functions $\boldsymbol{G}(\text{out})$ are defined by

$$\boldsymbol{G}(\text{out}) = -\tfrac{1}{2}\boldsymbol{G}(\boldsymbol{S})\boldsymbol{S}^*. \tag{6.57}$$

We define

$$\boldsymbol{F}(\boldsymbol{\chi}, \text{out}) = -\tfrac{1}{2}\boldsymbol{F}(\boldsymbol{\chi})\boldsymbol{\chi}^*. \tag{6.58}$$

It follows from (6.6) and (6.13) that

$$\boldsymbol{F}(\boldsymbol{\chi}, \text{out}) = \tfrac{1}{2}(\varphi^+ - \varphi^- \boldsymbol{\chi}^*) \qquad \text{for } \rho \geqslant \rho_0. \tag{6.59}$$

When all channels are open, $\boldsymbol{G}(\text{out}) = \boldsymbol{F}(\boldsymbol{\chi}, \text{out})$. When some channels are closed the relations between $\boldsymbol{G}(\text{out})$ and $\boldsymbol{F}(\boldsymbol{\chi}, \text{out})$ is similar to that between $\boldsymbol{G}(\boldsymbol{S})$ and $\boldsymbol{F}(\boldsymbol{\chi})$.

Using the method of § 6.4.2,

$$\boldsymbol{G}(\text{out}) = \boldsymbol{F}(\boldsymbol{\chi}, \text{out})\begin{pmatrix} 1_{\text{oo}} \\ -[\boldsymbol{\chi}^*_{\text{cc}} - \exp(+2\pi \mathrm{i}\nu)]^{-1}\boldsymbol{\chi}^*_{\text{co}} \end{pmatrix}. \tag{6.60}$$

We put

$$\boldsymbol{D}(\boldsymbol{\chi}, \text{out}) = \int_0^\infty \boldsymbol{F}^\dagger(\boldsymbol{\chi}, \text{out})\boldsymbol{g}\, \mathrm{d}\rho. \tag{6.61}$$

When some channels are closed we partition $\boldsymbol{F}(\boldsymbol{\chi}, \text{out})$ to obtain

$$\boldsymbol{D}(\boldsymbol{\chi}, \text{out}) = \begin{pmatrix} \boldsymbol{D}_{\text{o}}(\boldsymbol{\chi}, \text{out}) \\ \boldsymbol{D}_{\text{c}}(\boldsymbol{\chi}, \text{out}) \end{pmatrix} \tag{6.62}$$

where

$$\begin{aligned} \boldsymbol{D}_{\text{o}}(\boldsymbol{\chi}, \text{out}) &= \int_0^\infty [\boldsymbol{F}^\dagger_{\text{oo}}(\boldsymbol{\chi}, \text{out})\boldsymbol{g}_{\text{o}} + \boldsymbol{F}^\dagger_{\text{co}}(\boldsymbol{\chi}, \text{out})\boldsymbol{g}_{\text{c}}]\, \mathrm{d}\rho \\ \boldsymbol{D}_{\text{c}}(\boldsymbol{\chi}, \text{out}) &= \int_0^\infty [\boldsymbol{F}^\dagger_{\text{oc}}(\boldsymbol{\chi}, \text{out})\boldsymbol{g}_{\text{o}} + \boldsymbol{F}^\dagger_{\text{cc}}(\boldsymbol{\chi}, \text{out})\boldsymbol{g}_{\text{c}}]\, \mathrm{d}\rho. \end{aligned} \tag{6.63}$$

It follows from (6.60) that

$$\boldsymbol{D}(\text{out}) = \boldsymbol{D}_{\text{o}}(\boldsymbol{\chi}, \text{out}) - \boldsymbol{\chi}_{\text{oc}}[\boldsymbol{\chi}_{\text{cc}} - \exp(-2\pi \mathrm{i}\nu)]^{-1}\boldsymbol{D}_{\text{c}}(\boldsymbol{\chi}, \text{out}). \tag{6.64}$$

An expression similar to (6.64) is given by Dubau and Wells (1973).

An alternative approach is to use $\boldsymbol{G}(\boldsymbol{R})$ and $\boldsymbol{F}(\boldsymbol{\mathcal{R}})$ (Dubau and Seaton 1983). Relations between $\boldsymbol{S}$ and $\boldsymbol{R}$ and $\boldsymbol{G}(\boldsymbol{S})$ and $\boldsymbol{G}(\boldsymbol{R})$ are given in § 3.4. Using (6.57) we obtain

$$\boldsymbol{G}(\text{out}) = \frac{\mathrm{i}}{2}\boldsymbol{G}(\boldsymbol{R})(1 + \boldsymbol{S}^*). \tag{6.65}$$

Defining

$$\boldsymbol{D}(\boldsymbol{R}) = \int_0^\infty \boldsymbol{G}^\dagger(\boldsymbol{R})\boldsymbol{g}\, \mathrm{d}\rho \tag{6.66}$$

we have

$$\boldsymbol{D}(\text{out}) = -\frac{\mathrm{i}}{2}(1 + \boldsymbol{S})\boldsymbol{D}(\boldsymbol{R}). \tag{6.67}$$

We define

$$\boldsymbol{D}(\boldsymbol{\mathcal{R}}) = \int_0^\infty \boldsymbol{F}(\boldsymbol{\mathcal{R}})\boldsymbol{g}\, \mathrm{d}\rho \tag{6.68}$$

and using the relation (6.18) between $\boldsymbol{G}(\boldsymbol{R})$ and $\boldsymbol{F}(\boldsymbol{\mathcal{R}})$ obtain

$$\boldsymbol{D}(\boldsymbol{R}) = \boldsymbol{D}_{\text{o}}(\boldsymbol{\mathcal{R}}) - \boldsymbol{\mathcal{R}}_{\text{oc}}[\tan(\pi\nu) + \boldsymbol{\mathcal{R}}_{\text{cc}}]^{-1}\boldsymbol{D}_{\text{c}}(\boldsymbol{\mathcal{R}}). \tag{6.69}$$

The advantage of this approach is that all quantities in (6.69) are real. Complex quantities are introduced only in the final step of using (6.67) to obtain $\boldsymbol{D}(\text{out})$.

6.7. Formulation of the theory using eigen quantum defects

Formulations of the theory using eigen quantum defects are given by Fano (1970) and Lee and Lu (1973). In § 6.3 we defined a diagonal matrix $\bar{\mu}$ and a transformation matrix $\boldsymbol{X}$ such that $\boldsymbol{\mathcal{R}}=\boldsymbol{X}\tan(\pi\bar{\mu})\boldsymbol{X}^{\mathrm{T}}$, and functions $\boldsymbol{F}(\bar{\mu};\rho)=\boldsymbol{F}(\boldsymbol{\mathcal{R}};\rho)\boldsymbol{X}\cos(\pi\bar{\mu})$ such that $\boldsymbol{F}(\bar{\mu};\rho)=s\boldsymbol{X}\cos(\pi\bar{\mu})+c\boldsymbol{X}\sin(\pi\bar{\mu})$ for $\rho>\rho_0$.

6.7.1. All channels open. When all channels are open we have $\boldsymbol{R}=\boldsymbol{\mathcal{R}}=\boldsymbol{X}\tan(\pi\bar{\mu})\boldsymbol{X}^{\mathrm{T}}$ and $\boldsymbol{S}=\chi=\boldsymbol{X}\exp(2\pi\mathrm{i}\bar{\mu})\boldsymbol{X}^{\mathrm{T}}$. The functions $\boldsymbol{G}(\text{out})$, with asymptotic form $\frac{1}{2}(\varphi^{+}-\varphi^{-}\boldsymbol{S}^{*})$, are given by $\boldsymbol{G}(\text{out})=\mathrm{i}\boldsymbol{F}(\bar{\mu})\exp(-\mathrm{i}\pi\bar{\mu})\boldsymbol{X}^{\mathrm{T}}$.

6.7.2. Some channels closed. Lee and Lu (1973) introduced $\bar{\tau}$ and $\boldsymbol{T}$ such that $\boldsymbol{R}=\boldsymbol{T}\tan(\pi\bar{\tau})\boldsymbol{T}^{\mathrm{T}}$, $\boldsymbol{S}=\boldsymbol{T}\exp(2\pi\mathrm{i}\bar{\tau})\boldsymbol{T}^{\mathrm{T}}$ for the case of some channels closed. Putting $\boldsymbol{G}(\bar{\tau};\rho)=\boldsymbol{G}(\boldsymbol{R};\rho)\boldsymbol{T}\cos(\pi\bar{\tau})$ one has

$$\boldsymbol{G}_{\mathrm{oo}}(\bar{\tau};\rho)=s\boldsymbol{T}\cos(\pi\bar{\tau})+c\boldsymbol{T}\sin(\pi\bar{\tau}) \qquad \text{for } \rho\geqslant\rho_0 \tag{6.70}$$

$$\boldsymbol{G}_{\mathrm{co}}(\bar{\tau};\rho)\to 0 \qquad \text{for } \rho\to\infty \tag{6.71}$$

and $\boldsymbol{G}(\text{out})=\mathrm{i}\boldsymbol{G}(\bar{\tau})\exp(-\mathrm{i}\pi\bar{\tau})\boldsymbol{T}^{\mathrm{T}}$.

We may put

$$\boldsymbol{G}(\bar{\tau};\rho)=\boldsymbol{F}(\bar{\mu};\rho)\mathcal{N} \tag{6.72}$$

where $\mathcal{N}$ has I rows and I_0 columns.

Using the partitioning

$$\boldsymbol{X}=\begin{pmatrix}\boldsymbol{X}_{\mathrm{o}}\\ \boldsymbol{X}_{\mathrm{c}}\end{pmatrix} \tag{6.73}$$

(the elements of $\boldsymbol{X}_{\mathrm{o}}$ are X_{ij} with $i=1$ to I_0 and $j=1$ to I and those of $\boldsymbol{X}_{\mathrm{c}}$ are X_{ij} with $i=(I_0+1)$ to I and $j=1$ to I) we have, for $\rho>\rho_0$,

$$\boldsymbol{G}_{\mathrm{oo}}(\bar{\tau})=[s_{\mathrm{o}}\boldsymbol{X}_{\mathrm{o}}\cos(\pi\bar{\mu})+c_{\mathrm{o}}\boldsymbol{X}_{\mathrm{o}}\sin(\pi\bar{\mu})]\mathcal{N} \tag{6.74}$$

$$\boldsymbol{G}_{\mathrm{co}}(\bar{\tau})=[s_{\mathrm{c}}\boldsymbol{X}_{\mathrm{c}}\cos(\pi\bar{\mu})+c_{\mathrm{c}}\boldsymbol{X}_{\mathrm{c}}\sin(\pi\bar{\mu})]\mathcal{N}. \tag{6.75}$$

Equation (6.70) for $\boldsymbol{G}_{\mathrm{oo}}(\bar{\tau})$ is satisfied if

$$\begin{aligned}\boldsymbol{X}_{\mathrm{o}}\cos(\pi\bar{\mu})\mathcal{N}&=\boldsymbol{T}\cos(\pi\bar{\tau}).\\ \boldsymbol{X}_{\mathrm{o}}\sin(\pi\bar{\mu})\mathcal{N}&=\boldsymbol{T}\sin(\pi\tau).\end{aligned} \tag{6.76}$$

Solving for $\boldsymbol{T}$,

$$\boldsymbol{T}=\boldsymbol{X}_{\mathrm{o}}[\cos(\pi\bar{\mu})\mathcal{N}\cos(\pi\bar{\tau})+\sin(\pi\bar{\mu})\mathcal{N}\sin(\pi\bar{\tau})] \tag{6.77}$$

and eliminating $\boldsymbol{T}$,

$$\boldsymbol{X}_{\mathrm{o}}[\cos(\pi\bar{\mu})\mathcal{N}\sin(\pi\bar{\tau})-\sin(\pi\bar{\mu})\mathcal{N}\cos(\pi\bar{\tau})]=0. \tag{6.78}$$

Equation (6.71) for $\boldsymbol{G}_{\mathrm{co}}(\bar{\tau})$ is satisfied if

$$[\sin(\pi\nu)\boldsymbol{X}_{\mathrm{c}}\cos(\pi\bar{\mu})+\cos(\pi\nu)\boldsymbol{X}_{\mathrm{c}}\sin(\pi\bar{\mu})]\mathcal{N}=0. \tag{6.79}$$

The equations (6.78) and (6.79) may be written

$$\begin{aligned}&\sum_{\alpha=1}^{I}X_{\mathrm{i}\alpha}\sin[\pi(\bar{\mu}\alpha-\bar{\tau}_\beta)]\mathcal{N}_{\alpha\beta}=0 &&\text{for } i=1 \text{ to } I_0\\ &\sum_{\alpha=1}^{I}X_{\mathrm{i}\alpha}\sin[\pi(\bar{\mu}\alpha+\nu_i)]\mathcal{N}_{\alpha\beta}=0 &&\text{for } i=(I_0+1) \text{ to } I.\end{aligned} \tag{6.80}$$

For each value of β, we have I equations for $\mathcal{N}_{\alpha\beta}$, $\alpha = 1$ to I. These equations can be solved only if the determinant of the coefficients vanishes, which gives an eigenvalue problem for $\bar{\tau}_\beta$.

6.7.3. All channels closed. With all channels closed we put

$$\boldsymbol{P}(\rho) = \boldsymbol{F}(\bar{\mu}\,;\rho)\mathcal{N} \tag{6.81}$$

where $\mathcal{N}$ is a column vector. The condition $\boldsymbol{P}(\rho) \to 0$ for $\rho \to \infty$ gives the equations

$$[\sin(\pi\nu)\boldsymbol{X}\cos(\pi\bar{\mu}) + \cos(\pi\nu)\boldsymbol{X}\sin(\pi\bar{\mu})]\mathcal{N} = 0 \tag{6.82}$$

which may be written

$$\sum_{\alpha=1}^{I} X_{i\alpha} \sin[\pi(\nu_i + \bar{\mu}_\alpha)]\mathcal{N}_\alpha = 0 \tag{6.83}$$

and which give an eigenvalue problem for the determination of the bound-state energies.

Using the method of § 6.5.2, Lee and Lu (1973) show that the normalisation condition (6.48) is satisfied if

$$\begin{aligned} &\sum_i \left(\frac{\nu_i^3}{2}\right)\left(\sum_\alpha X_{i\alpha}\cos[\pi(\nu_i+\bar{\mu}_\alpha)]\mathcal{N}_\alpha\right)^2 + \sum_\alpha \left(\frac{\mathrm{d}\bar{\mu}_\alpha}{\mathrm{d}\mathscr{E}}\right)\mathcal{N}_\alpha^2 \\ &\quad + \frac{1}{\pi}\sum_{i\alpha\beta}\left(\frac{\mathrm{d}X_{i\alpha}}{\mathrm{d}\mathscr{E}}\right)X_{i\beta}\sin[\pi(\bar{\mu}_\alpha - \bar{\mu}_\beta)]\mathcal{N}_\alpha\mathcal{N}_\beta = 1. \end{aligned} \tag{6.84}$$

6.7.4. Radiative transitions. Let $\boldsymbol{D}(\bar{\mu})$ be calculated using functions $\boldsymbol{F}(\bar{\mu}\,;\rho)$. We then have: (i) for all channels open, $\boldsymbol{D}(\text{out}) = -\mathrm{i}\boldsymbol{X}\exp(\mathrm{i}\pi\bar{\mu})\boldsymbol{D}(\bar{\mu})$; (ii) for some channels closed, $\boldsymbol{D}(\text{out}) = -\mathrm{i}\boldsymbol{T}\exp(\mathrm{i}\pi\bar{\tau})\boldsymbol{D}(\bar{\tau})$ with $\boldsymbol{T}$ normalised to $\boldsymbol{T}^{\mathrm{T}}\boldsymbol{T} = 1$ and $\boldsymbol{D}(\bar{\tau}) = \mathcal{N}^{\mathrm{T}}\boldsymbol{D}(\bar{\mu})$ (the normalisation for $\mathcal{N}$ follows from (6.77)); (iii) for all channels closed, $\boldsymbol{D}(\text{bound}) = \mathcal{N}^{\mathrm{T}}\boldsymbol{D}(\bar{\mu})$ where $\mathcal{N}$ satisfies the normalisation condition (6.84).

6.8. *Different formulations of* QDT

6.8.1. Use of $\mathscr{R}$ *and* χ. All essential information required for QDT is provided by the matrix $\mathscr{R}(\mathscr{E})$. Since $\mathscr{R}$ is real and symmetric, it contains $\frac{1}{2}I(I+1)$ real numbers. The matrix $\chi = (1+\mathrm{i}\mathscr{R})(1-\mathrm{i}\mathscr{R})^{-1}$ is complex and symmetric and contains $\frac{1}{2}I(I+1)$ complex numbers: the relation $\chi^*\chi = 1$ gives $\frac{1}{2}I(I+1)$ equations of constraint.

$\mathscr{R}$ may have isolated poles. It is sometimes convenient to fit $\mathscr{R}$ to $\mathscr{R} = \mathscr{P}\mathscr{Q}^{-1}$ where $\mathscr{P}$ and $\mathscr{Q}$ do not have poles; from (6.41) we then have

$$\det[\sin(\pi\nu)\mathscr{Q} + \cos(\pi\nu)\mathscr{P}] = 0. \tag{6.85}$$

Eissner *et al* (1969) use phase-shifted functions defined by

$$\begin{aligned} \sigma &= s\cos(\pi\tau) + c\sin(\pi\tau) \\ \gamma &= c\cos(\pi\tau) - s\sin(\pi\tau) \end{aligned} \tag{6.86}$$

with some convenient choice of phases τ (their τ should not be confused with $\bar{\tau}$ of Lee and Lu). Defining $\mathscr{R}(\tau)$ by

$$\boldsymbol{F}(\mathscr{R}(\tau);\rho) = \sigma + \gamma\mathscr{R}(\tau) \qquad \text{for } \rho \geqslant \rho_0 \tag{6.87}$$

one obtains

$$\mathscr{R}(\tau)=[\cos(\pi\tau)+\mathscr{R}(0)\sin(\pi\tau)]^{-1}[\mathscr{R}(0)\cos(\pi\tau)-\sin(\pi\tau)]. \tag{6.88}$$

If $\mathscr{R}(\tau)$ is used in place of $\mathscr{R}(0)$ in QDT formulae, $\sin(\pi\nu)$ and $\cos(\pi\nu)$ are replaced by $\sin[\pi(\nu+\tau)]$ and $\cos[\pi(\nu+\tau)]$; thus equation (6.41) is replaced by

$$\det\{\tan[\pi(\nu+\tau)]+\mathscr{R}(\tau)\}=0. \tag{6.89}$$

The phases τ can always be chosen in such a way that $\mathscr{R}(\tau)$ does not have poles.

6.8.2. Use of $\bar{\mu}$ and $\boldsymbol{X}$. The diagonal matrix $\bar{\mu}$ contains I real numbers and the transformation matrix $\boldsymbol{X}$ contains I^2 real numbers. The condition $\boldsymbol{X}^{\mathrm{T}}\boldsymbol{X}=1$ gives $\frac{1}{2}I(I+1)$ equation of constraint and $\boldsymbol{X}$ therefore contains $\frac{1}{2}I(I-1)$ linearly independent numbers which Lee and Lu express in terms of generalised Euler angles.

Although $\bar{\mu}$ and $\boldsymbol{X}$ have been used in many applications of QDT, the present reviewer is of the opinion that there are advantages in using $\mathscr{R}$ or quantities closely related to $\mathscr{R}$: thus the relation (6.29) between $\boldsymbol{R}$ and $\mathscr{R}$ is simpler than the relations between $(\tau, \boldsymbol{T})$ and $(\bar{\mu}, \boldsymbol{X})$ given in § 6.7.2; and the normalisation condition (6.49) is simpler than the condition (6.84).

6.8.3. Degenerate channels. All channels are degenerate if the core energies, $\mathscr{E}_i$, are all equal. In such a case there is a clear advantage in diagonalisation of $\mathscr{R}$, since (6.83) reduces to $\sin[\pi(\nu+\bar{\mu}_\alpha)]=0$ giving $\nu_n=n-\bar{\mu}_\alpha$.

When one has groups of degenerate channels there are advantages in diagonalising certain partitioned matrices. This was noted by Gailitis (1963) and will be discussed in § 7.

6.9. Comments on .he use of QDT

The structure of QDT may be summarised as follows.

(i) The Coulomb radial equation, in which the potential is $U(\rho)=2/\rho$, can be solved analytically. The solutions f and g are analytic in ϵ. For smaller values of ρ the functions f, g, h, s, c, φ^- and φ^+ vary slowly with ϵ, and for $\epsilon<0$ they can be expressed as linear combinations of the exponentially increasing function ξ and the decreasing function θ. The coefficients of ξ and θ are proportional to the trigonometric functions $\sin(\pi\nu)$, $\cos(\pi\nu)$ and $\exp(\pm i\pi\nu)$ where $\epsilon=-1/\nu^2$.

(ii) We are concerned with solutions $\boldsymbol{F}$ of coupled equations (3.13) which contain potentials $\boldsymbol{U}(\rho)$ such that $\boldsymbol{U}(\rho)\sim 2/\rho$ for $\rho\to\infty$. The theory is simplified on assuming that all non-Coulomb potentials are of finite range, $\boldsymbol{U}(\rho)=2/\rho$ for $\rho\geqslant\rho_0$ with $\rho_0<\infty$. Equations (3.13) have solutions $\boldsymbol{F}(\mathscr{E};\rho)$ which, for ρ small, vary slowly with $\mathscr{E}$. The matrices $\boldsymbol{I}(\mathscr{E})$ and $\boldsymbol{J}(\mathscr{E})$ are defined by $\boldsymbol{F}(\mathscr{E};\rho)=f\boldsymbol{I}(\mathscr{E})+g\boldsymbol{J}(\mathscr{E})$ for $\rho\geqslant\rho_0$. Other matrices used in the theory are $\boldsymbol{Y}=-\boldsymbol{J}(\boldsymbol{I}-\mathscr{G}\boldsymbol{J})^{-1}$, $\mathscr{R}=A^{1/2}\boldsymbol{Y}A^{1/2}$, $\chi=(1+i\mathscr{R})(1-i\mathscr{R})$ and $\bar{\mu}$ and $\boldsymbol{X}$ where $\bar{\mu}$ is diagonal and $\chi\boldsymbol{X}=\boldsymbol{X}\exp(2\pi i\bar{\mu})$. These matrices vary slowly with $\mathscr{E}$ (except for isolated poles in $\boldsymbol{Y}$ and $\mathscr{R}$).

(iii) The physical solutions of (3.13) are obtained on eliminating ξ. Equation (6.29) gives the reactance matrix $\boldsymbol{R}$ in terms of $\mathscr{R}$; (6.33) gives the scattering matrix $\boldsymbol{S}$ in terms of χ; (6.38) and (6.43) give conditions for bound states; (6.49) gives a normalisation condition; and (6.64) and (6.69) give expressions for photoionisation amplitudes. In these equations all rapid variations with energy are given by the trigonometric functions $\sin(\pi\nu)$, $\cos(\pi\nu)$ and $\exp(-2\pi i\nu)$. Alternative formulations using eigen quantum defects are given in § 6.7.

Section 5.3.2 gives a sketch of how the one-channel theory could be developed without assuming the non-Coulomb potentials to be of finite range, and Seaton (1978) gives a similar discussion for the many-channel theory.

QDT is used for the empirical fitting of experimental data and in conjunction with *ab initio* calculations in which the coupled equations (3.13) are solved numerically.

It might be expected that the procedure in numerical work would be to compute directly the matrices, such as $\mathcal{R}$ or χ, which vary slowly with $\mathscr{E}$. This is not usually the case, for the reason that it is usually not a good approximation to assume the non-Coulomb potentials to be of short range. One could, of course, obtain a good approximation on taking such potentials to be of long but finite range, i.e. zero only for $\rho \geqslant \rho_0$ with ρ_0 large, but this would lead to computational difficulties since, for $\epsilon < 0$, the Coulomb functions have poor numerical linear independence when ρ is large. The procedures described in § 5.3.2 and by Seaton (1978) are not convenient for numerical work, but further progress should be possible using perturbation methods to compute approximate values of contributions from long-range potentials (the one-channel case was discussed in § 5.7).

The procedure which has been used extensively is to obtain solutions of the coupled equations which are everywhere bounded and to use the QDT formulae for empirical fitting. Thus, for example, $\boldsymbol{R}$ may be calculated at a fairly small number of energies and fitted to formulae obtained from (6.29) which give $\boldsymbol{R}$ as a continuous function of $\mathscr{E}$.

6.10. The relativistic many-channel theory

The relativistic many-channel theory is discussed by Lee and Johnson (1980). The relation between Z-scaled energies ϵ and effective quantum numbers ν is

$$\epsilon = \frac{2}{\beta^2}[(1+\beta^2/\nu^2)^{-1/2} - 1] \tag{6.90}$$

in the notation of § 5.8 (Johnson and Cheng (1979) and Lee and Johnson (1980) include the electron rest mass in the definition of ϵ and use units with $\hbar = c = 1$). The relativistic theory, which is formulated using the two-component relativistic Coulomb functions discussed by Johnson and Cheng, has a structure which is similar to that of the non-relativistic theory, and is best used for treating heavier systems and systems which are more highly ionised. For further details the reader is advised to consult the paper by Lee and Johnson.

7. Resonances

The use of QDT for the analysis of resonance structures is discussed by Seaton (1969).

7.1. The scattering matrix

The behaviour of the scattering matrix, as a function of energy, is given by (6.33):

$$\boldsymbol{S} = \chi_{oo} - \chi_{oc}[\chi_{cc} - \exp(-2\pi i\nu)]^{-1}\chi_{co}. \tag{7.1}$$

We assume that χ varies slowly with $\mathscr{E}$ and that all rapid variations of $\boldsymbol{S}$ as a function of $\mathscr{E}$ are due to the variation of $\exp(-2\pi i\nu)$ in (7.1).

Resonances are due to poles in $\boldsymbol{S}$ and occur at complex energies $\mathscr{E}$(pole) which are such that

$$\det[\chi_{cc} - \exp(-2\pi i\nu)] = 0 \qquad (7.2)$$

(when all channels are closed, (7.2) is equivalent to the equation (6.41) for the existence of bound states).

Consider that a threshold, $\epsilon_i = 0$, is approached from below. As $\epsilon_i \to 0$, $\nu_i \to \infty$ and one has an infinite series of resonances, the pattern being repeated each time that ν_i changes by unity. In addition to the rapid variations due to changes in ν_i, $\boldsymbol{S}$ has slower variations due to changes in the effective quantum numbers $\nu_{i'}$ belonging to higher thresholds.

7.2. *Contracted matrices*

The matrix χ is of dimension $I \times I$ and $\boldsymbol{S}$ is of dimension $I_o \times I_o$ where I_o is the number of open channels and $I_o < I$ if some channels are closed. We may say that $\boldsymbol{S}$ is a contraction of χ, the contraction being such as to eliminate exponentially increasing solutions. Such contractions can be carried out in several stages.

Consider two groups of channels, a lower group (a) which may be open or closed and an upper group (b) which are all closed. We partition χ,

$$\chi = \begin{pmatrix} \chi_{aa} & \chi_{ab} \\ \chi_{ba} & \chi_{bb} \end{pmatrix} \qquad (7.3)$$

and we define the contracted matrix

$$\chi(\text{contr}) = \chi_{aa} - \chi_{ab}[\chi_{bb} - \exp(-2\pi i\nu_b)]^{-1}\chi_{ba}. \qquad (7.4)$$

If all channels in a are open, $\boldsymbol{S} = \chi(\text{contr})$; and if some channels in a are closed we partition $\chi(\text{contr})$ to obtain

$$\boldsymbol{S} = \chi_{oo}(\text{contr}) - \chi_{oc}(\text{contr})[\chi_{cc}(\text{contr}) - \exp(-2\pi i\nu_c)]^{-1}\chi_{co}(\text{contr}) \qquad (7.5)$$

where ν_c is the matrix of effective quantum numbers for the closed channels in group a. Then $\boldsymbol{S}$ will have rapid variations due to ν_c in (7.5) and slower variations due to ν_b in (7.4).

The use of contracted matrices is implied in most empirical fitting of experimental data using QDT formulae. A large number of channels may be required to fit all data over an extended energy range, or for making accurate *ab initio* calculations, but all structures observed in a restricted range of lower energies may be accounted for using a small contracted matrix.

7.3. *Complex quantum defects*

We consider the region just below a new threshold. If the new threshold is not the highest threshold we assume that higher thresholds are eliminated by using a contracted χ matrix. With this understood, it may be assumed that the closed channels in (7.1) or (7.5) are all degenerate, i.e. all belong to the new threshold.

We diagonalise χ_{cc}:

$$\chi_{cc}\boldsymbol{X} = \boldsymbol{X}\chi'_{cc} \qquad (7.6)$$

with χ'_{cc} diagonal, and we put $\chi'_{cc} = \exp(2\pi i\mu_c)$ where μ_c is the *complex quantum defect* (Seaton 1969, Norcross and Seaton 1970). Let

$$\mu_c = \alpha_c + i\beta_c \tag{7.7}$$

with α_c and β_c real. It follows from (6.13) that

$$\chi^*_{co}\chi_{oc} + \chi^*_{cc}\chi_{cc} = 1 \tag{7.8}$$

and hence that $\chi^*_{cc}\chi_{cc} < 1$ and $\beta_c > 0$.

Since χ_{cc} is symmetric, $\boldsymbol{X}$ in (7.6) may be normalised to $\boldsymbol{X}^T\boldsymbol{X} = 1$. Since the closed channels are degenerate, $\exp(-2\pi i\nu)$ is a multiple of the unit matrix and commutes with $\boldsymbol{X}$. We therefore have

$$[\chi_{cc} - \exp(-2\pi i\nu)] = \boldsymbol{X}[\chi'_{cc} - \exp(-2\pi i\nu)]\boldsymbol{X}^T. \tag{7.9}$$

Defining $\chi'_{oc} = \chi_{oc}\boldsymbol{X}$ and $\chi'_{co} = \boldsymbol{X}^T\chi_{co}$ we obtain

$$\boldsymbol{S} = \chi_{oo} - \chi'_{oc}[\chi'_{cc} - \exp(-2\pi i\nu)]^{-1}\chi'_{co} \tag{7.10}$$

that is to say

$$S_{ij} = \chi_{ij} - \sum_p \frac{\chi'_{ip}\chi'_{pj}}{\exp(2\pi i\mu_p) - \exp(-2\pi i\nu)} \tag{7.11}$$

where i and j are open channels and p is summed over the degenerate closed channels.

The poles in $\boldsymbol{S}$ occur at complex effective quantum numbers $\nu(\text{pole}) = n - \mu_p$. Let the closed channels belong to a level of the core with energy $\mathscr{E}_c$. The energy of the pole is $\mathscr{E}(\text{pole}) = \mathscr{E}_c - [\nu(\text{pole})]^{-2}$. Putting $\mu_p = \alpha_p + i\beta_p$ and dropping subscripts p,

$$\mathscr{E}(\text{pole}) = \mathscr{E}_c - \frac{[(n-\alpha)^2 - \beta^2] + i[2(n-\alpha)\beta]}{[(n-\alpha)^2 + \beta^2]^2} \tag{7.12}$$

and hence, for $(n-\alpha) \gg \beta$,

$$\mathscr{E}(\text{pole}) \simeq \mathscr{E}_c - \frac{1}{(n-\alpha)^2} - \frac{i}{2}\gamma \tag{7.13}$$

where $\gamma = 4\beta(n-\alpha)^{-3}$ is the autoionisation probability as defined in § 3.6. Care is needed in interpreting γ as a width, using the usual formula $\Delta\mathscr{E}(\text{width}) = \gamma$, since in a Rydberg series the widths cannot be larger than the separations, $\Delta\mathscr{E}(\text{separations}) = 2(n-\alpha)^{-3}$. One obtains $\Delta\mathscr{E}(\text{width}) < \Delta\mathscr{E}(\text{separations})$ if $2\beta < 1$.

From (7.10) we have

$$\sum_i |\chi_{pi}|^2 = 1 - \exp(-4\pi\beta_p) \tag{7.14}$$

since $|\chi'_{pp}|^2 = \exp(-4\pi\beta_p)$. The resonances are very broad when $|\chi_{pi}|^2$ is large, i.e. when there is strong coupling between open and closed channels.

7.4. *The Gailitis average*

In all practical applications one requires collision cross sections averaged over some distribution function $f(v)$ of electron velocities. In the region just below a threshold the variation in the cross section due to an infinite series of resonances will be more rapid than the variation in $f(v)$. We may therefore compute cross sections averaged over resonances before integrating over $f(v)$.

We consider a region just below a new threshold and let ν be the effective quantum number for channels which become open at that threshold. We use

$$S_{ij}=\chi_{ij}-\sum_{p}\frac{\chi_{ip}\chi_{pj}}{\chi_{pp}-\exp(-2\pi i\nu)} \tag{7.15}$$

it being understood that, if necessary, $\boldsymbol{\chi}$ has been contracted and that χ_{cc} has been diagonalised. We define

$$\langle|S_{ij}|^2\rangle=\int_{\nu_0}^{\nu_0+1}|S_{ij}|^2\,d\nu \tag{7.16}$$

and evaluate the integral neglecting the energy variation of χ. Changing to an integration variable $z=\exp(2\pi i\nu)$, the integration is around the unit circle in the z plane. Using the fact that $|\chi_{pp}|<1$, one obtains

$$\langle|S_{ij}|^2\rangle=|\chi_{ij}|^2+\sum_{p,q}\frac{\chi_{ip}\chi_{pj}\chi^*_{iq}\chi^*_{qj}}{1-\chi_{pp}\chi^*_{qq}}. \tag{7.17}$$

Equation (7.17), first given by Gailitis (1963), has a simple interpretation for the case in which there is only one closed channel. It reduces to

$$\langle|S_{ij}|^2\rangle=|\chi_{ij}|^2+\frac{|\chi_{ip}|^2|\chi_{pj}|^2}{\sum_k|\chi_{pk}|^2} \tag{7.18}$$

where k is summed over all open channels and where the relation $1-|\chi_{pp}|^2=\sum_k|\chi_{pk}|^2$ follows from (7.8).

We interpret $|\chi_{ip}|^2$ as the probability of forming the resonance state p from i and $|\chi_{pj}|^2(\sum_k|\chi_{pk}|^2)^{-1}$ as the probability that break-up of the resonance gives j.

A similar analysis applies to the photoionisation amplitudes. Using (6.68) we obtain

$$\langle|D_i(\text{out})|^2\rangle=|D_i(\boldsymbol{\chi},\text{out})|^2+\sum_{p,q}\frac{\chi_{ip}\chi^*_{iq}D_p(\boldsymbol{\chi},\text{out})D^*_q(\boldsymbol{\chi},\text{out})}{1-\chi_{pp}\chi^*_{qq}} \tag{7.19}$$

and, using the unitarity of $\boldsymbol{\chi}$, it follows that

$$\sum_i\langle|D_i(\text{out})|^2\rangle=\sum_i|D_i(\boldsymbol{\chi},\text{out})|^2+\sum_p|D_p(\boldsymbol{\chi},\text{out})|^2. \tag{7.20}$$

The summation of i is over channels open below the new threshold, and of p is over those channels which become open at the new threshold. It follows from (7.20) that the total photoionisation cross section, averaged over resonances, is continuous across the threshold.

7.5. Isolated resonances

Consider resonances which are narrow compared with their separations, $\Delta\mathscr{E}(\text{width})\ll\Delta\mathscr{E}(\text{separation})$. Let there be a resonance at an energy close to $\mathscr{E}^{(0)}$. For closed channels $\mathscr{E}^{(0)}=\mathscr{E}_i-[\nu_i^{(0)}]^{-2}$. For $(\nu-\nu^{(0)})$ small, $\tan(\pi\nu)=\tan(\pi\nu_0)+(\nu-\nu_0)\times[\cos(\pi\mu^{(0)})]^{-2}$ and $(\mathscr{E}-\mathscr{E}^{(0)})=2(\nu-\nu_0)[\nu^{(0)}]^{-3}$. Using these formulae, (6.29) becomes

$$\boldsymbol{R}=\boldsymbol{\mathcal{R}}_{oo}-\boldsymbol{a}_{oc}(\mathscr{E}-\boldsymbol{\mathcal{B}})^{-1}\boldsymbol{a}_{co} \tag{7.21}$$

where $\boldsymbol{a}_{oc}=\boldsymbol{\mathcal{R}}_{oc}q$, $\boldsymbol{a}_{co}=q\boldsymbol{\mathcal{R}}_{co}$, $q=q(\nu^{(0)})$ is defined by (6.48), and

$$\boldsymbol{\mathcal{B}}=\mathscr{E}^{(0)}-q[\tan(\pi\nu^{(0)})+\boldsymbol{\mathcal{R}}_{cc}]q. \tag{7.22}$$

Diagonalisation of $\mathcal{B}$ gives

$$R_{ij} = \mathcal{R}_{ij} - \sum_p a'_{ip}(\mathcal{E} - \mathcal{B}'_p)^{-1} a'_{pj}. \tag{7.23}$$

Resonance formulae of the type (7.21) and (7.23) are obtained if the expansion (3.17) is replaced by

$$\Psi = \mathcal{A}_s \sum_{i\,\text{open}} \psi_i \frac{1}{r} F_i(r) + \sum_p \Phi_p C_p \tag{7.24}$$

where the functions Φ_p have the form of bound states for the $(N+1)$-electron system. This is discussed further in appendix A.

8. Use of QDT in atomic collision calculations

We illustrate the use of QDT in atomic collision calculations by considering excitation of ions with configurations $2s^2 2p^q$. These configurations give terms $S_c L_c$ for the ion core, where $S_c L_c = {}^2P^o$ for $q = 1$ and 5; $S_c L_c = {}^3P$, ${}^1D$ and ${}^1S$ for $q = 2$ and 4; and $S_c L_c = {}^4S^o$, ${}^2D^o$ and ${}^2P^o$ for $q = 3$. For $q = 2$ and 4 the collision strengths $\Omega({}^3P, {}^1D)$ contain Rydberg series of resonances at energies below the threshold for excitation of ${}^1S$, and for $q = 3$ there are similar resonances in $\Omega({}^4S^o, {}^2D^o)$.

Saraph *et al* (1969) obtained approximate solutions of the coupled equations (3.13) at energies such that all channels are open and with neglect of core states with configurations other than $2s^2\,2p^q$. At such energies the calculated matrix $\boldsymbol{R}$ is equal to $\mathcal{R}$ and one can put $\boldsymbol{Y} = A^{-1/2} \boldsymbol{R} A^{-1/2}$ (see (6.9)). The matrices $\boldsymbol{Y}$ can be fitted to (6.8) and extrapolated to the region in which some channels are closed. Saraph *et al* calculated collision strengths $\bar{\Omega}$, averaged over resonances, using methods described in § 7.4.

The core states with fine structure are $2s^2 2p^q S_c L_c J_c$. Martins and Seaton (1969) calculated collision strengths $\Omega({}^2D^o_{3/2}, {}^2D^o_{5/2})$ for O^+ using $\boldsymbol{Y}$ matrices from Saraph *et al* transformed to *jj* coupling. The number of *jj* channels is much larger than the number of *SL* channels and, at energies below the threshold for excitation of ${}^2P^o$, (6.33) was found to give complicated patterns of overlapping resonances.

Saraph and Seaton (1971) extrapolated the $\boldsymbol{Y}$ matrices of Saraph *et al* to the region of all channels closed and calculated positions of bound states from numerical solutions of an equation equivalent to (6.38). They used a pair coupling representation and experimental values for the energy separations of the core states, $2s^2 2p^q S_c L_c J_c$. The calculated fine-structure separations for $2s^2 2p^q nl$ states were found to be in good agreement with experimental results. Many series have perturbations due to the core having more than one term. Thus the calculations gave the dominant states in the first eight members of the O II ${}^2P^o$ series to be $({}^3P)3p$, $({}^1D)3'p$, $({}^3P)4p$, $({}^1S)3''p$, $({}^3P)5p$, $({}^1D)4'p$, $({}^3P)6p$ and $({}^3P)7p$, but not all of these have been observed. Saraph and Seaton obtained improved agreement between calculated and observed level positions using adjusted $\boldsymbol{Y}$ matrices (essentially adjustment of Slater parameters) and the adjusted matrices were used to obtain improved values for the collision strengths, particularly $\Omega({}^2P^o_{1/2}, {}^2P^o_{3/2})$ for C^+ and Ne^+.

The $2s^2 2p^2$ isoelectronic sequence was discussed by Eissner *et al* (1969). It is necessary to consider the *complex* of configurations with the same set of principal quantum numbers, $2s^2 2p^2$, $2s2p^3$ and $2p^4$. For Z large the separations between levels

Table 3. Channels included in calculations of Eissner and Seaton (1974) for $e+O^{2+}$, $SL\pi={}^2P^o$.

i	Channel	$\mathscr{E}_i$	i	Channel	$\mathscr{E}_i$	i	Channel	$\mathscr{E}_i$
1	$2s^22p^2(^3P)p$	0.000	4	$2s2p^3(^3D^o)d$	0.273	7	$2s2p^3(^1D^o)d$	0.426
2	$2s^22p^2(^1D)p$	0.046	5	$2s2p^3(^3P^o)d$	0.324	8	$2s2p^3(^1P^o)d$	0.479
3	$2s^22p^2(^1S)p$	0.098	6	$2s2p^3(^3P^o)s$	0.324	9	$2s2p^3(^1P^o)s$	0.479

belonging to the same complex are proportional to Z and the binding energy of an added electron, or the position of a resonance below a threshold, is proportional to Z^2. It follows that, as Z increases, the resonances belonging to higher configurations in the complex occur at lower values of the Z-scaled energy, $\mathscr{E}=2E/Z^2$. Eissner *et al* made calculations for O^{2+} using perturbation theory and QDT methods and showed that resonances of the type $2s2p^33s$ occur at energies just above the thresholds for excitation of the $2s^22p^2$ terms. Accurate solutions of the coupled equations were subsequently obtained by Eissner and Seaton (1974) with inclusion of channels belonging to $2s2p^3$ states of the O^{2+} target. They introduced methods of resonance analysis which have been used in a number of later papers. We consider, by way of illustration, their results for total angular momenta and parity $SL\pi={}^2P^o$. The nine channels, and target energies $\mathscr{E}_i$, are given in table 3. Figure 10 shows the computed collision strength $\Omega(1,2)=\Omega(^3Pp, {}^1Pp)$ for $0.00\leqslant\epsilon_2\leqslant0.13$; there is a Rydberg series of resonances converging to the threshold $\epsilon_3=0$, and two resonances in the region $\epsilon_3>0$ which are due to the channels $i=6$ and $i=4$.

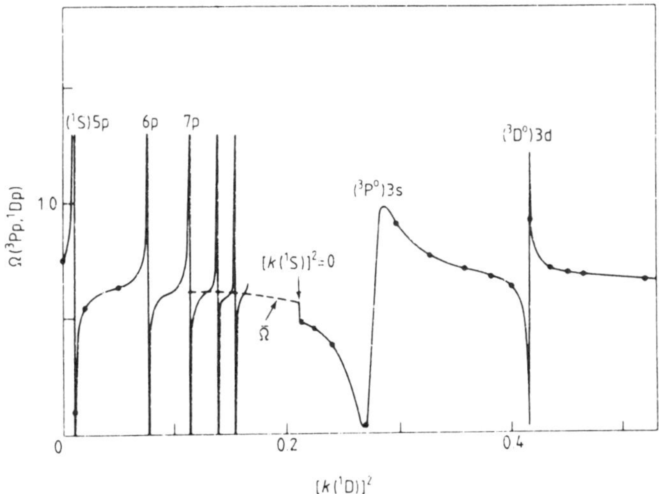

Figure 10. The collision strength $\Omega(^3Pp, {}^1Dp)$, $SL\pi={}^2P^o$, for the transition $2p^2\,{}^3P$–$2p^2\,{}^1D$ in O^{2+} (from Eissner and Seaton 1974); filled circles are computed points; the full lines are from fitting to QDT formulae; the broken curve is the Gailitis average.

Long-range non-Coulomb potentials are included in solving the coupled equations and, for reasons discussed in § 6.7, it would not be easy to compute matrices $\boldsymbol{Y}$ or $\mathscr{R}$ at energies such that some channels are closed. There is no difficulty in computing $\boldsymbol{Y}$ or $\mathscr{R}$ in the region of all channels open but, when the $2s2p^3$ channels are included, this region occurs at much higher energies ($\epsilon_1 > 0.479$, see table 3) and it is not practicable to extrapolate $\boldsymbol{Y}$ or $\boldsymbol{\mathcal{R}}$ to the lower energies considered in figure 10.

The two resonances of figure 10 in the region of $\epsilon_3 > 0$ are fairly well isolated and Eissner and Seaton found that the computed $\boldsymbol{R}$ matrix could be fitted accurately using (7.23). The computed points are shown as filled circles and results from the fit as smooth curves. The fitted $\boldsymbol{R}$ matrix from (7.23) can be used as a contracted $\boldsymbol{\mathcal{R}}$ matrix (§ 7.2) and used to calculate the collision strength in the region of $\epsilon_3 < 0$. Figure 10 shows the averaged collision strength $\bar{\Omega}(1, 2)$ and the structures of the first few resonances in the Rydberg series, computed using (6.29). A few check points, from solutions of the coupled equations in the near-threshold region, confirm the accuracy of these fitting procedures.

9. Systems with two energy levels of the ion core

9.1. *The* $^1P^o$ *series in Ca* I

We consider the $^1P^o$ series in Ca I, for which the quantum defects are shown in figure 2, as an illustrative two-channel problem. It is discussed further by Moores (1966), Carter *et al* (1971), Wynne and Armstrong (1979b) and Geiger (1979). We must note, however, that although the bound-state experimental data can be analysed using a two-channel theory, Mendoza (1979) shows that many more channels are required in order to obtain good accuracy in *ab initio* calculations.

The two channels, $4s\epsilon_1 p$ and $3d\epsilon_2 p$, have an energy difference

$$\Delta\mathscr{E} = \mathscr{E}_{3d} - \mathscr{E}_{4s} = \epsilon_1 - \epsilon_2 = 0.124\,943. \tag{9.1}$$

The condition for bound states is, from (6.41),

$$[\mathscr{R}_{11} + \tan(\pi\nu_1)][\mathscr{R}_{22} + \tan(\pi\nu_2)] - \mathscr{R}_{12}^2 = 0. \tag{9.2}$$

Moores (1966) shows that the observed levels can be fitted using a 2×2 matrix $\boldsymbol{Y}$ which varies slowly with $\mathscr{E}$, and that $\boldsymbol{\mathcal{R}} = \boldsymbol{A}^{1/2}\boldsymbol{Y}\boldsymbol{A}^{1/2}$ varies more rapidly.

Putting $\nu_1 = (n - \mu_1)$ we obtain from (9.2)

$$\tan(\pi\mu_1) = \mathscr{R}_{11} - \mathscr{R}_{12}^2[\mathscr{R}_{22} + \tan(\pi\nu_2)]^{-1} \tag{9.3}$$

which defines $\tan(\pi\mu_1)$ for all $\mathscr{E}$ such that $\epsilon_2 < 0$. If μ_1 is taken to be a continuous function of $\mathscr{E}$, it increases by an amount approximately equal to unity each time that ν_2 increases by unity. It is often more convenient to use μ_1 (modulo 1) $= (\mu_1 - m)$ with m_1 such that $0 \leqslant \mu_1$ (modulo 1) < 1. To avoid confusion we use the notation

$$\tilde{\mu}_1 = \mu_1 \text{ (modulo 1)}. \tag{9.4}$$

Figure 11 shows $\tilde{\mu}_1$ for Ca I $^1P^o$ computed using the $\boldsymbol{Y}$ matrix of Moores. It is seen that $\tilde{\mu}_1$ goes from 0 to 1 in a number of steps. There are an infinite number of steps below the threshold $\epsilon_2 = 0$ and the best part of one complete step in the region of bound states, $\epsilon_1 < 0$. In the general case, the number of steps for $\epsilon_1 < 0$ is determined by the value of $\nu_2(\infty)$, the value of ν_2 for $\nu_1 = \infty$. From (9.1), $\nu_2(\infty) = (\Delta\mathscr{E})^{-1/2}$ giving $\nu_2(\infty) = 2.8291$ for Ca I $^1P^o$.

Equations (6.40) and (6.47) are used to calculate Z_1 and Z_2. The bound-state radial functions are $P_1 \sim (K_1\theta_1)Z_1$, $P_2 = (K_2\theta_2)Z$ and the normalisation condition

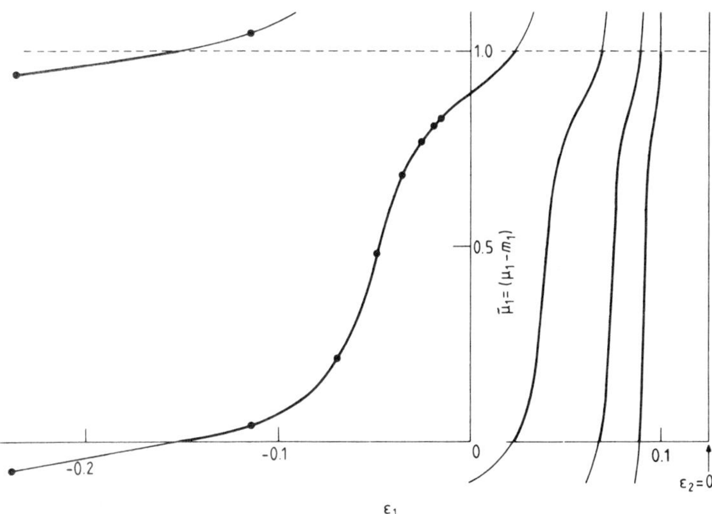

Figure 11. Quantum defects $\tilde{\mu}_1 = \mu_1$ (modulo 1) for Ca I $^1P^o$ from two-channel $\boldsymbol{Y}$ matrix of Moores (1966).

(6.49) is satisfied approximately if $|Z_1|^2 + |Z_2|^2 = 1$. Table 4 gives values of $W_1 = |Z_1|^2(|Z_1|^2 + |Z_2|^2)^{-1}$ and $W_2 = |Z_2|^2(|Z_1|^2 + |Z_2|^2)^{-1}$ which provides an indication of the contributions from each channel to the total wavefunction. It is seen that for all bound states channel 1 gives the dominant contribution and that the largest value of W_2 occurs for the state labelled '3d4p'. In the vicinity of this state the sum of the values of W_2 is close to unity; the state '3d4p' is, in effect, distributed among a number of $^1P^o$ states.

The resonances for Ca I $^1P^o$ are rather broad. The imaginary part of the complex quantum defect is $\beta = 0.1046$.

9.2. The Lu–Fano plot

The method introduced by Lu and Fano (1970) has been used extensively for the analysis of systems for which the ion core can be considered to have two energy levels.

The quantum defect μ_1 calculated using (9.3) has rapid variations with $\mathscr{E}$ due to variations of ν_2, where $\mathscr{E} = \mathscr{E}_2 - 1/\nu_2^2$, and slower variations due to variations with $\mathscr{E}$

Table 4. Mixing of Ca I $^1P^o$ states. Channels 1 and 2 are 4s ϵ_1 p and 3d ϵ_2 p. Results for two-channel model.

Spectroscopic label	W_1	W_2
4s 4p	0.773	0.227
4s 5p	0.855	0.145
4s 6p	0.753	0.247
3d 4p	0.720	0.280
4s 7p	0.871	0.129
4s 8p	0.941	0.059
4s 9p	0.970	0.030
4s 10p	0.980	0.020

of the elements of $\mathcal{R}$. In the method of Lu and Fano ν_2 is taken to be the independent variable. Equation (9.2) may be written

$$\Phi(\mu_1, \nu_2) = 0 \tag{9.5}$$

where

$$\Phi(\mu_1, \nu_2) = \det \begin{pmatrix} \mathcal{R}_{11} - \tan(\pi\mu_1) & \mathcal{R}_{12} \\ \mathcal{R}_{21} & \mathcal{R}_{22} + \tan(\pi\nu_2) \end{pmatrix}. \tag{9.6}$$

Let us define

$$\begin{aligned} \tilde{\mu}_1 &= \mu_1(\text{modulo } 1) = \mu_1 - m_1 \\ \tilde{\nu}_2 &= \nu_2\,(\text{modulo} 1) = \nu_2 - m_2 \end{aligned} \tag{9.7}$$

with m_1 and m_2 such that $0 \leqslant \tilde{\mu}_1 < 1$ and $0 \leqslant \tilde{\nu}_2 < 1$. Since $\mathscr{E} = \mathscr{E}_2 - 1/\nu_2^2$, a definite value of the integer m_2 corresponds to a definite range of energy $\mathscr{E}$. Let us put

$$\Phi_{m_2}(\tilde{\mu}_1, \tilde{\nu}_2) = \Phi(\mu_1, \nu_2 + m_2). \tag{9.8}$$

If $\mathcal{R}$ is independent of $\mathscr{E}$, Φ_{m_2} is independent of m_2. For bound states we have

$$\Phi_{m_2}(\tilde{\mu}_1, \tilde{\nu}_2) = 0. \tag{9.9}$$

We have $\mathscr{E} = \mathscr{E}_1 + \epsilon_1 = \mathscr{E}_2 + \epsilon_2$ giving

$$\Delta E - \epsilon_1 + \epsilon_2 = 0 \tag{9.10}$$

where $\Delta\mathscr{E} = (\mathscr{E}_2 - \mathscr{E}_1)$. Taking n_1 to be such that $\nu_1 = (n_1 - \tilde{\mu}_1)$, equation (9.10) may be written

$$\Theta_{n_1}(\tilde{\mu}_1, \tilde{\nu}_2) = 0 \tag{9.11}$$

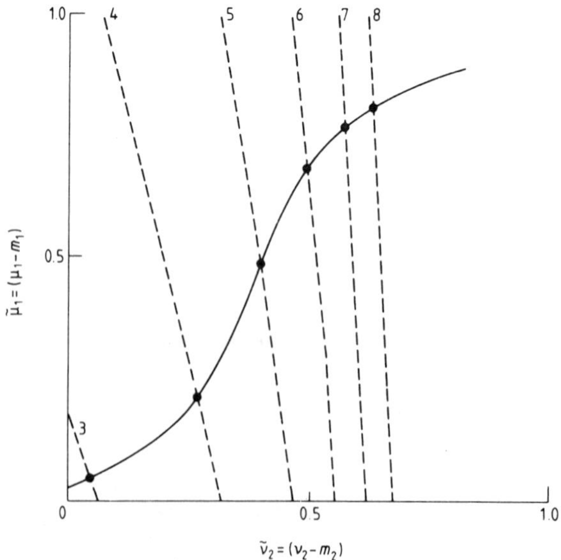

Figure 12. Graphical solutions for energy levels of Ca I ^{1}P^o. The full curve is the function defined by (9.8). The broken curves are functions defined by $\Theta_{n_1} = 0$ for various indicated values of n_1.

where

$$\Theta_{n_1}(\tilde{\mu}_1, \tilde{\nu}_2) = \Delta\mathscr{E} + \frac{1}{(n_1-\mu_1)^2} - \frac{1}{(\tilde{\nu}_2+m_2)^2}. \tag{9.12}$$

The energy levels are obtained from simultaneous solutions of (9.9) and (9.11). This is illustrated graphically in figure 12 for the Ca I $^1\mathrm{P}^\mathrm{o}$ series.

In the method of Lu and Fano, $\tilde{\mu}_1$ is plotted against $\tilde{\nu}_2$. For the two-channel case with $\mathscr{R}$ independent of $\mathscr{E}$, all points lie on a smooth curve defined by $\Phi = 0$. If $\mathscr{R}$ varies with $\mathscr{E}$ we have a set of curves, $\Phi_{m_2} = 0$. Figure 13 shows plots for Ca I $^1\mathrm{P}^\mathrm{o}$ obtained using the $\boldsymbol{Y}$ matrix of Moores. There is a segment of a curve for $m_2 = 1$ and curves are plotted for $m_2 = 2, 3$ and ∞.

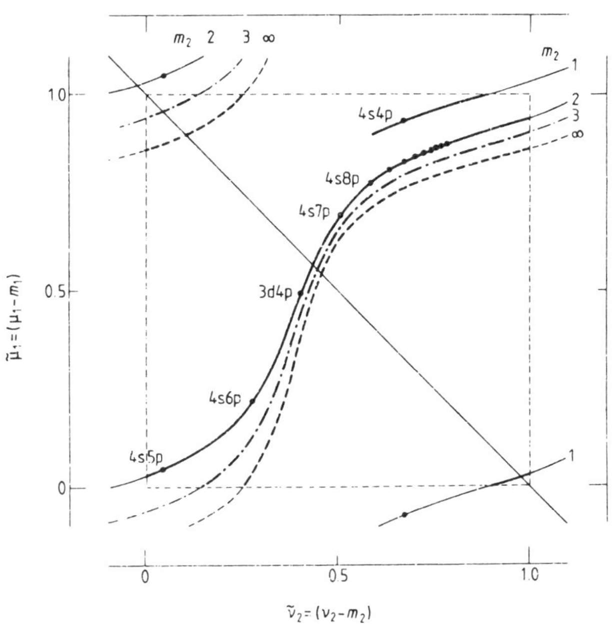

Figure 13. The Lu–Fano plot for Ca I $^1\mathrm{P}^\mathrm{o}$. The curves are obtained from $\Phi_{m_2} = 0$ using the $\boldsymbol{Y}$ matrix of Moores (1966).

The diagonal of the plot is defined by the equation

$$\tilde{\mu}_1 = 1 - \tilde{\nu}_2. \tag{9.13}$$

For each value of m_2, the diagonal has two intercepts with the curve $\Phi_{m_2} = 0$, and at these intercepts $\tilde{\mu}_1 = \bar{\mu}_1$ or $\bar{\mu}_2$ where $\bar{\mu}_1$ and $\bar{\mu}_2$ are the eigen quantum defects. This is proved as follows. If r is an eigenvalue of $\mathscr{R}$, det $(\mathscr{R} - r) = 0$ and (9.5) has this form if $r = \tan(\pi\mu_1) = -\tan(\pi\nu_2)$ which is satisfied when (9.13) is satisfied.

The use of Lu-Fano plots is restricted to systems with two core levels, $\mathscr{E}_1$ and $\mathscr{E}_2$, but not necessarily two channels. If there are more than two channels they may be assigned to two groups, group a belonging to $\mathscr{E}_1$ and group b to $\mathscr{E}_2$, and Φ may be defined by

$$\Phi = \det\begin{pmatrix} \mathcal{R}_{aa} - \tan(\pi\mu_1) & \mathcal{R}_{ab} \\ \mathcal{R}_{ba} & \mathcal{R}_{bb} + \tan(\pi\nu_2) \end{pmatrix}. \tag{9.14}$$

The equation $\Phi = 0$ then gives μ_1 as a multi-valued function of ν_2.

The plots are most useful for systems with small values of $\Delta\mathscr{E}$, which have a lot of interesting structure in the region of $|\epsilon_1|$ small in which $\boldsymbol{\mathcal{R}}$ is approximately constant.

9.3. *Photoionisation for two-channel systems*

Fano (1961) has given a formula for profiles of resonances in photoionisation cross sections for one open channel and an isolated pole in S at an energy $E(\text{pole}) = E_0 - \mathrm{i}\Gamma_A/2$; the formula is

$$|D|^2 = C(q+x)^2(1+x^2)^{-1} \tag{9.15}$$

where

$$x = (E - E_0)(\Gamma_A/2)^{-1} \tag{9.16}$$

C is the value of $|D|^2$ for the background continuum and q depends on the strength of the transition to the resonance state relative to that to the continuum.

Dubau and Seaton (1983) have given a QDT formula for two channels, one open and one closed. Using (6.68) and putting $\boldsymbol{D}(\chi, \text{out}) = -(\mathrm{i}/2)(1+\chi)\boldsymbol{D}(\boldsymbol{\mathcal{R}})$, we obtain

$$2\mathrm{i}\boldsymbol{D}(\text{out}) = (1, -\chi_{12}[\chi_{22} - \exp(-2\pi\mathrm{i}\nu_2)]^{-1})(1+\chi)\begin{pmatrix} D_1(\boldsymbol{\mathcal{R}}) \\ D_2(\boldsymbol{\mathcal{R}}) \end{pmatrix}. \tag{9.17}$$

Consistent with unitarity conditions we may put

$$\begin{gathered} \chi_{11} = \exp[2\pi\mathrm{i}(\delta + \mathrm{i}\beta)] \qquad \chi_{12} = \mathrm{i}[1 - \exp(-4\pi\beta)]^{1/2}\exp[\mathrm{i}\pi(\alpha+\delta)] \\ \chi_{22} = \exp[2\pi\mathrm{i}(\alpha + \mathrm{i}\beta)] \end{gathered} \tag{9.18}$$

where $\mu_c = \alpha + \mathrm{i}\beta$ is the complex quantum defect. After some algebra one obtains a formula of the form (9.15) where x, c and q are now defined by

$$x = \tan[\pi(\nu_2 + \alpha)][\tanh(\pi\beta)]^{-1} \tag{9.19}$$

$C = |P|^2$ and $q = Q/P$ where

$$\begin{aligned} Q &= D_1(\boldsymbol{\mathcal{R}})\sin(\pi\delta) - D_2(\boldsymbol{\mathcal{R}})[\tanh(\pi\beta)]^{-1/2}\cos(\pi\alpha) \\ P &= D_1(\boldsymbol{\mathcal{R}})\cos(\pi\delta) - D_2(\boldsymbol{\mathcal{R}})[\tanh(\pi\beta)]^{-1/2}\sin(\pi\alpha). \end{aligned} \tag{9.20}$$

It may be shown that the QDT expression (9.19) for x reduces to the expression (9.16) when the resonances are very narrow, $\beta \ll 1$.

The expression (9.19) gives $x = 0$ for $\nu_2 = (n - \alpha)$ and the range $-\infty < x < +\infty$ corresponds to $(n - \frac{1}{2} - \alpha) < \nu_2 < (n + \frac{1}{2} - \alpha)$. The pattern is repeated for each value of n. Figure 14 shows the profile factor $(q+x)^2(1+x^2)^{-1}$ plotted against $(\nu_2 + \alpha - n)$ for various values of β and q. Figure 15 shows photoionisation cross sections for Be calculated by Dubau and Wells (1973) and for Al calculated by Le Dourneuf *et al* (1975). The computed profiles are in satisfactory agreement with experimental results of Mehlman-Ballofet and Esteva (1969) and Esteva (1974), and Dubau and Seaton (1983) show that they can be fitted to the two-channel formula with all parameters linear and slowly varying functions of $\mathscr{E}$.

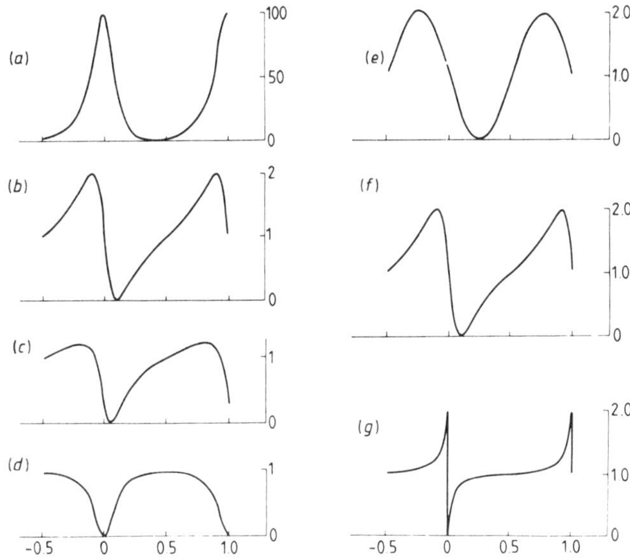

Figure 14. Photoionisation profiles for two channels, one open and one closed. The profile factor $(q+x)^2(1+x^2)^{-1}$ with $x=\tan[\pi(\nu_2+\alpha)][\tanh(\pi\beta)]^{-1}$ is plotted against $(\nu_2+\alpha-n)$. Curves on the left for $\beta=0.1$, those on the right for $q=-1.0$. (*a*) $q=-10.0$, (*b*) $q=-1.0$, (*c*) $q=-0.5$, (*d*) $q=0.0$, (*e*) $\beta=1.0$, (*f*) $\beta=0.1$, (*g*) $\beta=0.01$.

10. Helium

10.1. The one-channel theory

Seaton (1966b) analysed experimental He I quantum defects using the one-channel theory. In fitting $\mu_l(\epsilon)$ to polynomials in ϵ, or $Y(\epsilon, l)$ to expansions of the form (5.15), the method of least squares was used and the series limit T_∞ was treated as one of the adjustable parameters. A value of improved accuracy obtained for T_∞ ($T_\infty=$ $198\,310.76\pm0.01\ \mathrm{cm}^{-1}$) is of interest in connection with the determination of the He Lamb shift. Plots given for $\mu_l(\epsilon)$ illustrated the trend that μ_l tends to an integer as ϵ tends to $-1/l^2$ (see § 5.5.3).

Microwave measurements by Lamb *et al* (1973) show that positions of high levels obtained from the fitted parameters are more accurate than those from individual optical measurements; this is because the fitting implies a smoothing of the optical data. The elastic scattering phases obtained from the fitted quantum defects were found to be in satisfactory agreement with those computed by Sloan (1964) using the method of polarised orbitals.

10.2. The many-channel theory

For a hydrogenic ion of nuclear charge Z_N, the difference in energy between the ground state and the first excited state is $(E_2-E_1)=\frac{3}{4}(Z_N)^2/2$ atomic units. For an electron in the field of such an ion the asymptotic charge is $Z=(Z_N-1)$ and the Z-scaled energy difference is $(\mathscr{E}_2-\mathscr{E}_1)=(2/Z^2)(E_2-E_1)=\frac{3}{4}(Z+1)^2Z^{-2}$. For He^+, $(\mathscr{E}_2-\mathscr{E}_1)=3.0$. It is because this difference is sufficiently large that the one-channel

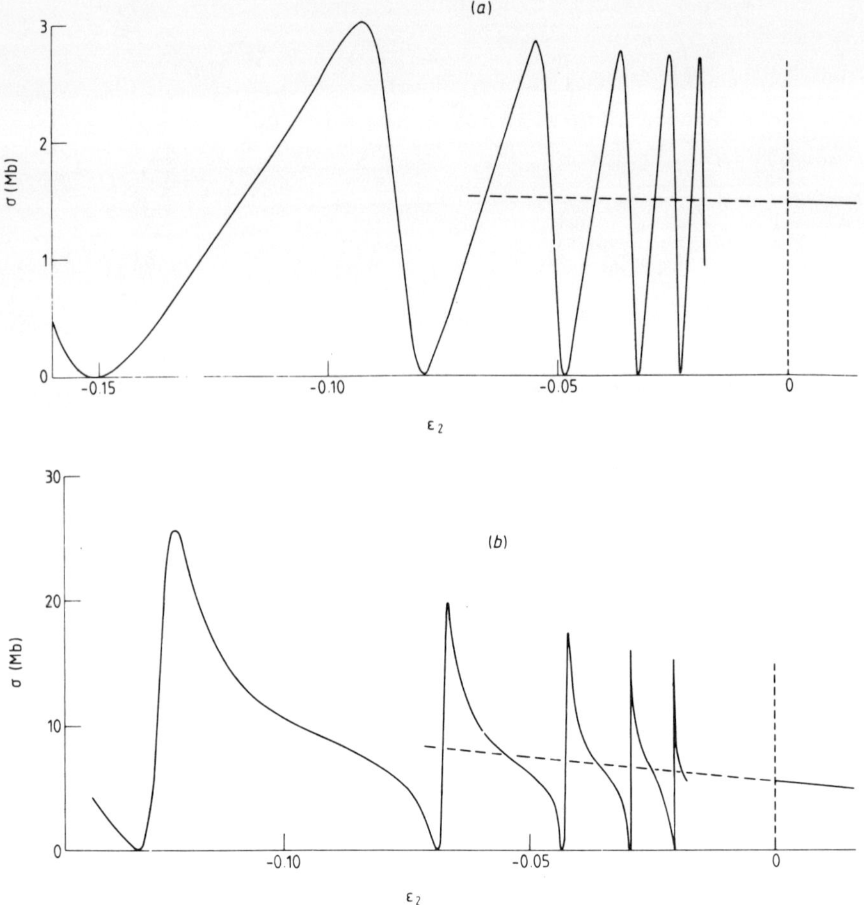

Figure 15. Photoionisation cross sections for Be (Dubau and Wells 1973) and Al (Le Dourneuf *et al* 1975). The computed cross sections can be fitted to the parametric formulae of § 9.3, with all parameters linear and slowly varying functions of the energy

theory can be used in discussing the bound states of He and the low-energy e–He^+ elastic scattering phase shifts. We now consider the region of inelastic scattering above the $n = 2$ threshold and the region of resonances below that threshold.

A unique feature of hydrogenic ions, in the non-relativistic approximation, is the degeneracy of the states nl for fixed n and variable l. One obtains long-range non-Coulomb potentials of which the most important are the dipole potentials, behaving asymptotically like ρ^{-2}, connecting degenerate states nl and $n(l \pm 1)$. That the effect of such potentials is larger for the degenerate case than for the non-degenerate case can be understood by considering the perturbation correction for a potential behaving like ρ^{-2} for $\rho \geqslant \rho_0$. This involves evaluation of integrals of the type

$$\int_{\rho_0}^{\infty} F_i \rho^{-2} F_j \, d\rho \tag{10.1}$$

where F_i and F_j are linear combinations of Coulomb functions for channels i and j.

Such integrals are largest when i and j are degenerate since, when they are non-degenerate, the magnitudes of the integrals are reduced by interference effects. In using QDT for systems with long-range non-Coulomb potentials, the matrices such as $\mathcal{R}$ and χ can be defined for energies such that all channels are open (see § 5.3.2). It is generally assumed that such matrices vary slowly as functions of the energy and can be extrapolated to energies such that some channels are closed. Numerical computations show such an assumption to be justified only for systems which do not have dipole potentials connecting degenerate states.

The theory for e–He$^+$ has been developed by Bely (1966) and Dubau (1978) who consider the states 1s, 2s and 2p giving channels

$$\begin{array}{lcccc} i= & 1 & 2 & 3 & 4 \\ & 1\mathrm{s}\epsilon_1 l & 2\mathrm{s}\epsilon_2 l & 2\mathrm{p}\epsilon_2(l+1) & 2\mathrm{p}\epsilon_2(l-1). \end{array} \tag{10.2}$$

If we neglect coupling to channel 1, and potentials decaying faster than ρ^{-2}, the equations for channels 2, 3 and 4 are

$$\left(\frac{\mathrm{d}^2}{\mathrm{d}\rho^2}+\frac{2}{\rho}-\frac{\boldsymbol{C}}{\rho^2}+\epsilon_2\right)\boldsymbol{F}=0 \tag{10.3}$$

in the limit of ρ large, where

$$\boldsymbol{C}=\begin{pmatrix} l(l+1) & C_{23} & C_{24} \\ C_{32} & (l+1)(l+2) & 0 \\ C_{42} & 0 & (l-1)(l) \end{pmatrix}. \tag{10.4}$$

The equations (10.3) can be solved on diagonalising $\boldsymbol{C}$, $\boldsymbol{CX}=\boldsymbol{X}c$. We put $c=\lambda^2-\frac{1}{4}$. Since $\boldsymbol{C}$ is real and symmetric, c is real and the elements of λ are either real or pure imaginary. Putting $\boldsymbol{F}=\boldsymbol{X}\bar{y}$ we obtain

$$\left(\frac{\mathrm{d}^2}{\mathrm{d}\rho^2}+\frac{2}{\rho}-\frac{\lambda^2-\frac{1}{4}}{\rho^2}+\epsilon_2\right)\bar{y}=0 \tag{10.5}$$

which is the equation discussed in § 2. Bely (1966) and Dubau (1978) consider solutions of (10.5) for values of λ which result from diagonalisation of $\boldsymbol{C}$. The functions of Bely have poor numerical linear independence, a difficulty which is avoided by Dubau on using functions defined by

$$\begin{aligned} \bar{s}&=\frac{\mathcal{A}y(\kappa,\lambda;\rho)-y(\kappa,-\lambda;\rho)}{2(2\mathcal{A})^{1/2}\sin(\pi\lambda)} \\ \bar{c}&=\frac{\mathcal{A}y(\kappa,\lambda;\rho)+y(\kappa,-\lambda;\rho)}{2(2\mathcal{A})^{1/2}\cos(\pi\lambda)} \end{aligned} \tag{10.6}$$

with $\epsilon=\epsilon_2$ and $\mathcal{A}=\mathcal{A}(\kappa,\lambda)$. These functions are such that, for $\lambda=l+\frac{1}{2}$, $\bar{s}=(-1)^l(A/2)^{1/2}y(\kappa,l+\frac{1}{2};\rho)$ and $\bar{c}=(-1)^{l+1}(2A)^{-1/2}\eta(\kappa,l+\frac{1}{2};\rho)$ where η is defined in § 2.5.2.

The asymptotic forms of $\bar{s}$ and $\bar{c}$, for $\epsilon>0$ (but $\epsilon\ll 4\pi^2$), are

$$\bar{s}\sim(\pi k)^{-1/2}\sin(u+\bar{p}) \qquad \bar{c}\sim(\pi k)^{-1/2}\cos(u+\bar{p}) \tag{10.7}$$

where

$$u=k\rho+(1/k)\ln(2k\rho) \tag{10.8}$$

and

$$\bar{p}=\begin{cases} -\dfrac{\pi}{4}+\lambda\dfrac{\pi}{2}+\arg\Gamma(\lambda+\frac{1}{2}-\mathrm{i}/k) & \text{for } \lambda \text{ real} \\ -\dfrac{\pi}{4}+\dfrac{1}{2}[\arg\Gamma(\lambda+\frac{1}{2}-\mathrm{i}/k)+\arg\Gamma(-\lambda+\frac{1}{2}-\mathrm{i}/k)] & \text{for } \lambda \text{ imaginary.} \end{cases} \tag{10.9}$$

These may be compared with the asymptotic forms of s and c, which may be written

$$s\sim(\pi k)^{-1/2}\sin(u+p) \qquad c\sim(\pi k)^{-1/2}\cos(u+p) \tag{10.10}$$

where

$$p=-l\frac{\pi}{2}+\arg\Gamma(l+1-\mathrm{i}/k). \tag{10.11}$$

The functions $\bar{s}$ and $\bar{c}$ are such that, for $\epsilon<0$,

$$\begin{aligned} \bar{s}&=\frac{\sin(\pi\nu)}{(2\nu)^{1/2}\pi K}\bar{\xi}-\cos(\pi\nu)\left(\frac{\nu^3}{2}\right)^{1/2}K\bar{\theta} \\ \bar{c}&=\frac{\cos(\pi\nu)}{(2\nu)^{1/2}\pi K}\bar{\xi}+\sin(\pi\nu)\left(\frac{\nu^3}{2}\right)^{1/2}K\bar{\theta} \end{aligned} \tag{10.12}$$

where $K=[\nu^2\Gamma(\nu+\lambda+\frac{1}{2})\Gamma(\nu-\lambda+\frac{1}{2})]^{-1/2}$ and where $\bar{\xi}$ and $\bar{\theta}$ are defined on replacing $(l+\frac{1}{2})$ by λ in the definitions (2.46) and (2.47) of θ and ξ.

Let us return to discussion of the four channels (10.2). When all channels are open we can define χ and $\bar{\chi}$ by using functions with asymptotic forms

$$\begin{aligned} \boldsymbol{F}(\chi;\rho)&\sim\varphi^- -\varphi^+\chi \\ \boldsymbol{F}(\bar{\chi};\rho)&\sim\bar{\varphi}^- -\bar{\varphi}^+\bar{\chi} \end{aligned} \tag{10.13}$$

where $\bar{\varphi}^{\pm}=\bar{c}\pm\mathrm{i}\bar{s}$. Using (10.7) and (10.10) Dubau shows that

$$\chi=\exp(-\mathrm{i}p)\boldsymbol{X}\exp(\mathrm{i}\bar{p})\bar{\chi}\exp(\mathrm{i}\bar{p})\boldsymbol{X}^{\mathrm{T}}\exp(-\mathrm{i}p). \tag{10.14}$$

The matrix $\bar{\chi}$ varies slowly with energy and can be extrapolated to the region of some channels closed. This is not true of χ. Using Stirling's formula one obtains

$$\arg\Gamma(\lambda_1+\tfrac{1}{2}-\mathrm{i}/k)-\arg\Gamma(\lambda_2+\tfrac{1}{2}-\mathrm{i}/k)=-(\lambda_1-\lambda_2)\frac{\pi}{2}+\frac{1}{2}(\lambda_1^2-\lambda_2^2)k+\mathrm{O}(k^2) \tag{10.15}$$

for λ_1 and λ_2 real.

It follows from (10.9), (10.11) and (10.14) that, for ϵ_2 small,

$$\chi=\boldsymbol{A}+\boldsymbol{B}\epsilon_2^{1/2}+\mathrm{O}(\epsilon_2) \tag{10.16}$$

where $\boldsymbol{A}$ and $\boldsymbol{B}$ are constant matrices.

When only channel 1 is open, it follows from (10.12) that the scattering matrix is given by

$$\boldsymbol{S}=\bar{\chi}_{\mathrm{oo}}-\bar{\chi}_{\mathrm{oc}}[\bar{\chi}_{\mathrm{cc}}-\exp(-2\pi\mathrm{i}\nu_2)]^{-1}\bar{\chi}_{\mathrm{co}} \tag{10.17}$$

and has poles at $\nu_2=n-\bar{\mu}$ where $\bar{\mu}$ is the complex quantum defect, obtained on diagonalising $\bar{\chi}_{\mathrm{cc}}$. Figure 16 shows results of Dubau for $SL\pi={}^1\mathrm{S}$ for which one has three channels, $1\mathrm{s}\epsilon_1\mathrm{s}$, $2\mathrm{s}\epsilon_2\mathrm{s}$ and $2\mathrm{p}\epsilon_2\mathrm{p}$. The full curves for $\epsilon_2>0$ show $\bar{\mu}$ (real and imaginary parts) obtained by Dubau from $\bar{\chi}_{cc}$. The dots and crosses show $\bar{\mu}$ obtained from positions and widths of resonances calculated by Burke and McVicar (1965) for $\epsilon_2 < 0$. The results for $\epsilon_2 > 0$ are seen to extrapolate smoothly to those for $\epsilon_2 < 0$. The broken curves of figure 16 show the complex quantum defect μ calculated from χ_{cc}; for $\epsilon_2 > 0$, μ behaves like $a + b\epsilon_2^{1/2}$ and μ cannot be extrapolated to $\epsilon_2 < 0$.

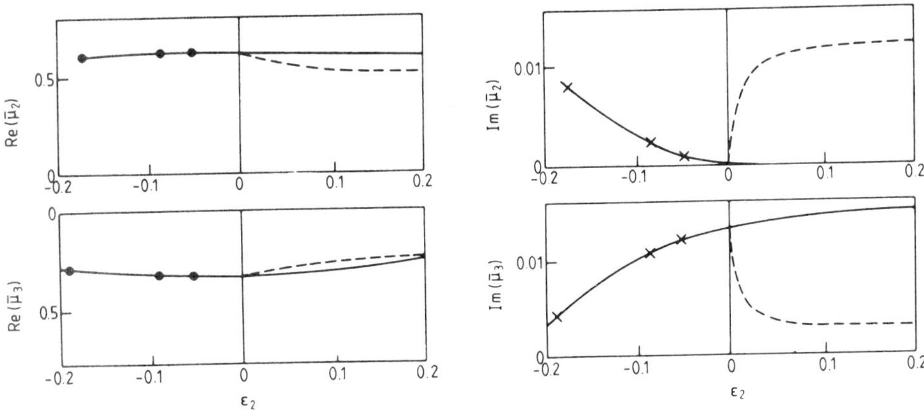

Figure 16. Complex quantum defects $\bar{\mu}$ for the e–He$^+$ system, $SL\pi = {}^1S$. Full curves from calculations of Dubau (1978) using the matrix $\bar{\chi}$ defined by (10.13). Filled circles and crosses from positions and widths of resonances, calculated by Burke and McVicar (1965). Broken curves from calculations using the matrix χ.

The theory has been used by Dubau (1973) to obtain autoionisation probabilities for the higher resonances and hence to calculate dielectronic recombination rates for He$^+$.

11. Rare gases other than helium

11.1. General discussion

Helium differs from the other rare gases in that its ion core has a ground configuration, He$^+$1s, without fine structure, and the one-channel theory can therefore be used in discussing the bound states and the low-energy elastic scattering. The rare gases other than helium have ion cores with ground configurations p^5 giving $^2P^o$ terms with fine-structure levels $J_c = \frac{3}{2}$ and $\frac{1}{2}$. There has, for the following reasons, been a lot of interest in applying QDT to the analysis of rare-gas spectra.

(*a*) Because the ion cores can be considered to have only two levels, the analysis of spectra for rare gases is relatively simple, and a lot of progress can be made in purely empirical analyses of experimental data. It is of interest to compare rare-gas spectra with the more complicated spectra of ions in the nitrogen sequence, for which the ion cores have configuration 2p^2, giving a ground term ^3P with fine structure $J_c = 0$, 1 and 2, and two excited terms, ^1D and ^1S; thus, for the nitrogen sequence one must consider at least five levels in the ion cores, and purely empirical analyses would be much more difficult.

(*b*) For systems with two core levels the graphical methods of Lu and Fano can be used.

(*c*) Experimental work on rare gases is easier than that for most other atomic systems and accurate data are therefore available for: energy levels, including high levels; f values; photoionisation, both in the autoionisation region between the two thresholds and in the region above the second threshold; more detailed studies of photoionisation giving angular distributions, spin polarisation parameters and, above the second threshold, branching ratios for leaving the ion in each of the fine-structure levels.

(*d*) Transitions between the fine-structure levels of Ne^+ and Ar^+, produced by collisions with slow electrons, are responsible for excitation and de-excitation of the radiative transitions $J_c=\frac{1}{2}\rightarrow J_c=\frac{3}{2}$ observed in infrared spectra of many astronomical objects.

The following contribute to the total energy for an excited state of a rare-gas atom, or for a slow collision between an electron and a rare-gas ion:

(*a*) the energy of the core, including fine structure;

(*b*) the energy of the outer electron in the central potential of the core, the quadrupole potential of the core and the core polarised by the outer electron (giving a potential behaving like α/ρ^4);

(*c*) exchange interactions between outer and core electrons;

(*d*) the spin–orbit energy of the outer electron.

If the magnetic interactions are neglected one may use the LS coupling scheme of § 3.1. For a given $SL\pi$ there are, in general, states of two types: (i) p^5l with $l=L$ giving a one-channel problem and (ii) p^5l and p^5l' with $l=(L-1)$ and $l'=(L+1)$, giving a two-channel problem.

With the exception of a few lower states, the separations between the states for LS coupling are smaller than the separations of the two core states and it is essential to use a representation in which J_c is specified. For given $J_cJ\pi$ one obtains, in general, a number of degenerate channels and hence an arbitrary choice of representation. Choices commonly used are pair coupling $(S_cL_cJ_c, lK, \frac{1}{2}J)$ or jj coupling $(S_cL_cJ_c, l\frac{1}{2}j, J)$. Table 5 gives the LS, pair and jj coupled states for p^5l, $l=$ s, p and d. An advantage of pair coupling is that, for given J_c and l, the quadrupole energy is diagonal with respect to K.

Table 5. p^5l states for LS coupling, pair coupling and jj coupling.

l	SL	J	J_c	K	J	J_c	j	J
s	$^1P^o$	1	$\frac{1}{2}$	$\frac{1}{2}$	0 1	$\frac{1}{2}$	$\frac{1}{2}$	0 1
	$^3P^o$	0 1 2	$\frac{3}{2}$	$\frac{3}{2}$	1 2	$\frac{3}{2}$	$\frac{1}{2}$	1 2
p	^1S	0	$\frac{1}{2}$	$\frac{1}{2}$	0 1	$\frac{1}{2}$	$\frac{1}{2}$	0 1
	^3S	1		$\frac{3}{2}$	1 2		$\frac{3}{2}$	1 2
	^1P	1	$\frac{3}{2}$	$\frac{1}{2}$	0 1	$\frac{3}{2}$	$\frac{1}{2}$	1 2 3
	^3P	0 1 2		$\frac{3}{2}$	1 2		$\frac{3}{2}$	0 1 2 3
	^1D	2		$\frac{5}{2}$	2 3			
	^3D	1 2 3						
d	$^1P^o$	1	$\frac{1}{2}$	$\frac{3}{2}$	1 2	$\frac{1}{2}$	$\frac{3}{2}$	1 2
	$^3P^o$	0 1 2		$\frac{5}{2}$	2 3		$\frac{5}{2}$	2 3
	$^1D^o$	2	$\frac{3}{2}$	$\frac{1}{2}$	0 1	$\frac{3}{2}$	$\frac{3}{2}$	0 1 2 3
	$^3D^o$	1 2 3		$\frac{3}{2}$	1 2		$\frac{5}{2}$	1 2 3 4
	$^1F^o$	3		$\frac{5}{2}$	2 3			
	$^3F^o$	2 3 4		$\frac{7}{2}$	3 4			

We consider transformations between these coupling schemes. Let α stand for the LS states $\{S_cL_c, \frac{1}{2}l, SLJ\}$ and i stand for the states $\{S_cL_cJ_c, lK, \frac{1}{2}J\}$ or $\{S_cL_cJ_c, l\frac{1}{2}j, J\}$. The states ψ_α, ψ_i depend on the coordinates of the core and the spin and angular coordinates of the added electron and are related by known algebraic transformations:

$$\psi_i = \sum_\alpha \psi_\alpha X_{\alpha i} \tag{11.1}$$

$$\psi_\alpha = \sum_i \psi_i X_{i\alpha}. \tag{11.2}$$

We consider states Ψ_α, Ψ_i which are such that (see (3.3))

$$\Psi_\alpha = \sum_{\alpha'} \psi_{\alpha'} \frac{1}{\rho} F_{\alpha'\alpha}(\rho) \tag{11.3}$$

$$\Psi_i = \sum_{i'} \psi_{i'} \frac{1}{\rho} F_{i'i}(\rho) \tag{11.4}$$

where

$$\left.\begin{aligned} F_{\alpha'\alpha} &= s_{\alpha'}\delta_{\alpha'\alpha} + c_{\alpha'}\mathscr{R}_{\alpha'\alpha} & (11.5)\\ F_{i'i} &= s_{i'}\delta_{i'i} + c_{i'}\mathscr{R}_{i'i} & (11.6)\end{aligned}\right\} \text{ for } \rho \geqslant \rho_0.$$

The Coulomb functions s_α, c_α depend on l_α and ϵ_α. In LS coupling core fine structure is neglected and all channels are degenerate; we may therefore put $\epsilon_\alpha = \epsilon$ independent of α. The functions s_i, c_i depend on l_i and ϵ_i and, since core fine structure is now included, ϵ_i depends on J_c for channel i; if $J_c = \frac{1}{2}$ we put $\epsilon_i = \epsilon(\frac{1}{2})$ and if $J_c = \frac{3}{2}$ we put $\epsilon_i = \epsilon(\frac{3}{2})$.

We now compare the functions Ψ_α and Ψ_i. The energies used for the calculation of these functions may be taken to be such that

$$\epsilon_\alpha = \epsilon = [\epsilon(\tfrac{1}{2}) + 2\epsilon(\tfrac{3}{2})]/3. \tag{11.7}$$

We consider the validity of using the expansion

$$\Psi_i = \sum_\alpha \Psi_\alpha X_{\alpha i} \tag{11.8}$$

which gives for $\rho \geqslant \rho_0$

$$\psi_i s_i + \sum_{i'} \psi_{i'} c_{i'} \mathscr{R}_{i'i} = \sum_\alpha \left(\psi_\alpha s_\alpha + \sum_{\alpha'} \psi_{\alpha'} c_{\alpha'} \mathscr{R}_{\alpha'\alpha} \right) X_{\alpha i}. \tag{11.9}$$

If $X_{\alpha i} \neq 0$ we have $l_\alpha = l_i$ but we do not have $\epsilon_\alpha = \epsilon_i$. In the limit of ρ large the Coulomb functions will be sensitive to the small energy differences, $\epsilon - \epsilon(\frac{1}{2})$ and $\epsilon - \epsilon(\frac{3}{2})$, and it follows that (11.8) cannot be satisfied for all values of ρ. Equation (11.8) can be satisfied in the inner region, $\rho \leqslant \rho_0$, if, at $\rho = \rho_0$, we can satisfy (11.9) and the derivative of (11.9) with respect to ρ. We may expect that, for ρ not much larger than ρ_0, the Coulomb functions will be oscillatory and insensitive to small energy differences. We assume that such differences can be neglected at $\rho = \rho_0$. We then have

$$\left.\begin{aligned} s_\alpha &= s_i & (11.10)\\ c_\alpha &= c_i & (11.11)\end{aligned}\right\} \text{ at } \rho = \rho_0 \text{ (for } l_\alpha = l_i)$$

(we recall that $X_{\alpha i} = 0$ unless $l_\alpha = l_i$) and similar equations for the derivatives. Using

(11.2) and (11.10), equation (11.9) gives

$$\sum_{i'} \psi_{i'} c_{i'} \mathcal{R}_{i'i} = \sum_{\alpha'} \psi_{\alpha'} c_{\alpha'} \sum_{\alpha} \mathcal{R}_{\alpha'\alpha} X_{\alpha i} \tag{11.12}$$

and using (11.2) and (11.11) this is seen to be satisfied if

$$\mathcal{R}_{i'i} = \sum_{\alpha\alpha'} X_{i'\alpha'} \mathcal{R}_{\alpha'\alpha} X_{\alpha i}. \tag{11.13}$$

We conclude that: (i) (11.8) can be used for $\rho \leq \rho_0$ so long as (11.10) and (11.11) can be assumed; (ii) (11.13) gives the relation between $\mathcal{R}_{i'i}$ and $\mathcal{R}_{\alpha'\alpha}$; (iii) Ψ_i must be used to obtain correct asymptotic forms for $\rho \to \infty$.

We use $\boldsymbol{\mathcal{R}}$ for the matrix with elements $\mathcal{R}_{i'i}$, $\boldsymbol{\mathcal{R}}(LS)$ for that with elements $\mathcal{R}_{\alpha'\alpha}$ and $\boldsymbol{X}_I$ for the known transformation matrix with elements $X_{\alpha i}$. Equation (11.13) can then be written

$$\boldsymbol{\mathcal{R}} = \boldsymbol{X}_I \, \boldsymbol{\mathcal{R}}(LS) \boldsymbol{X}_I^{\mathrm{T}}. \tag{11.14}$$

One approach in *ab initio* work is to start by making calculations in LS coupling, which gives 1×1 matrices for $l = L$ and 2×2 matrices for $l = (L-1)$, $l' = (L+1)$ (the coupling between l and l' is generally weak and the off-diagonal elements of the 2×2 matrices are generally small). The calculated matrices $\boldsymbol{\mathcal{R}}(LS)$ are transformed to matrices $\boldsymbol{\mathcal{R}}$ using (11.14). The asymptotic forms of the wavefunctions are obtained using experimental values for the separation of the core fine-structure states, $|\mathscr{E}(\frac{1}{2}) - \mathscr{E}(\frac{3}{2})|$. The effective quantum numbers, $\nu(\frac{1}{2})$, for channels with $J_c = \frac{1}{2}$ and $\nu(\frac{3}{2})$ for those with $J_c = \frac{3}{2}$ are such that

$$\frac{1}{\nu(\frac{1}{2})^2} - \frac{1}{\nu(\frac{3}{2})^2} = \mathscr{E}(\tfrac{1}{2}) - \mathscr{E}(\tfrac{3}{2}). \tag{11.15}$$

A further refinement would be to allow for the spin–orbit energy of the outer electron. An alternative *ab initio* approach is to use, from the outset, a fully relativistic theory.

In empirical approaches to the fitting of experimental data for the rare gases, one adopts a coupling scheme (pair or jj) and determines $\boldsymbol{\mathcal{R}}$ empirically. In many papers a diagonalisation of $\boldsymbol{\mathcal{R}}$ is used:

$$\boldsymbol{\mathcal{R}} = \boldsymbol{X} \tan(\pi \bar{\mu}) \boldsymbol{X}^{\mathrm{T}} \tag{11.16}$$

where $\bar{\mu}$ is the eigen quantum defect. One may put

$$\boldsymbol{X} = \boldsymbol{X}_{\mathrm{I}} \boldsymbol{X}_{\mathrm{II}} \tag{11.17}$$

where $\boldsymbol{X}_{\mathrm{I}}$ is the known transformation to LS coupling. The matrix $\boldsymbol{X}_{\mathrm{II}}$ differs from the unit matrix because of (i) coupling between $l = (L-1)$ and $l' = (L+1)$, and (ii) departures from LS coupling in the inner region, $\rho < \rho_0$.

11.2. Neon

The *ab initio* calculations of Saraph and Seaton (1971) have been described in § 8. All observed patterns of perturbations were reproduced for Ne $2p^5 l$ states with $l =$ s, p and d, including some perturbations due to interactions between s and d states. The results for d states showed systematic errors due to neglect of core polarisation. The weights W_i defined by (6.54) showed that pair coupling assignments are meaningful so long as there is no strong interaction between states belonging to different levels of the core.

The methods introduced by Seaton (1958b) and Saraph and Seaton (1971) for the determination of accurate collision strengths $\Omega(^2P_{1/2}, ^2P_{3/2})$ using empirically adjusted $\mathcal{R}$ matrices could be developed further and applied to other systems.

Starace (1973) has made studies of p^5s, and of $p^5p\ J = 0$. He uses a transformation matrix $\boldsymbol{U}$ and a diagonal matrix ξ defined by

$$(-\boldsymbol{J}\boldsymbol{I}^{-1})\boldsymbol{U} = \boldsymbol{U} \tan(\pi\xi) \tag{11.18}$$

($\boldsymbol{U}$ and ξ in his notation, $\boldsymbol{I}$ and $\boldsymbol{J}$ in that of § 6.1) and finds it possible to fit the experimental data with $\boldsymbol{U}$ independent of energy and ξ a linear function of energy.

For $J = 0$ and 2, the p^5s states give one-channel problems and, using LS coupling notation, the values of $\xi(^3P^o_0)$ and $\xi(^3P^o_2)$. For $J = 1$, p^5s gives a two-channel problem with eigenvalues $\xi(^3P^o_1)$ and $\xi(^1P^o_1)$. Starace finds the values of $\xi(^3P^o_J)$ to be nearly equal for the three values of J (0, 1 and 2). This is to be expected, since there is no spin–orbit energy for the outer electron in an s state.

For $J = 0$, p^5p gives a two-channel problem. For both $p^5s\ J = 1$ and $p^5p\ J = 0$, the transformation matrices $\boldsymbol{U}$ obtained empirically were found to be close to those for jj coupling (which, for these cases, are the same as those for pair coupling). This result is consistent with the neglect of coupling between p^5s and p^5d, and between p^5p and p^5f. Figure 17 shows the energy-dependent Lu–Fano plot obtained for p^5p $J = 0$. The levels are labelled np or n'p depending on whether the dominant parent states are $p^5\ ^2P^o_{3/2}$ or $p^5\ ^2P^o_{1/2}$.

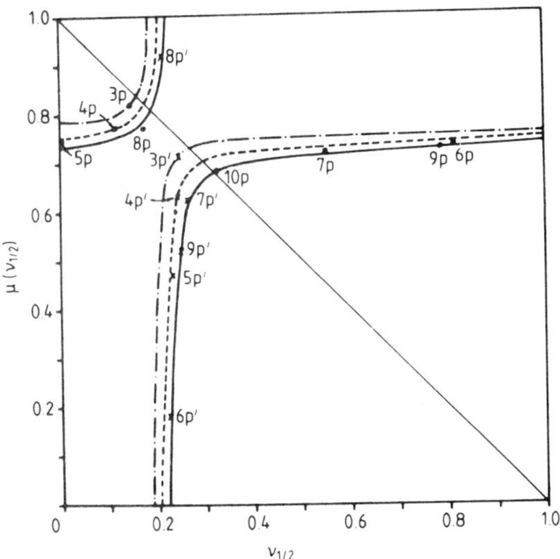

Figure 17. The Lu–Fano plot for Ne, $p^5p\ J = 0$ (from Starace 1973). – · –, $2.0 \leqslant \nu_{1/2} \leqslant 3.0$; – – –, $3.0 \leqslant \nu_{1/2} \leqslant 4.0$; ——, $\nu_{1/2}$ (modulo 1).

Starace computes oscillator strengths for $p^5n_1s\ J = 1 \rightarrow p^5n_2p\ J = 0$ with $(n_1, n_2) = (3, 3), (3, 4), (4, 3)$ and $(4, 4)$. QDT functions are used for initial and final states, and radial integrals are computed using the method of Bates and Damgaard (1949). The agreement with experiment is generally good.

Lu and Mansfield (1975) have made calculations for p^5s in the Ne isoelectronic sequence and shown that the coupling between $(^2P^o_{3/2})s$ and $(^2P^o_{1/2})s$ becomes less important for the highly ionised members.

11.3. Argon

Photoabsorption from the $^1S_0$ ground state of a rare gas leads to population of states with $J = 1$ and odd parity. These states give a five-channel problem (three channels belonging to $J_c = \frac{3}{2}$, two belonging to $J_c = \frac{1}{2}$). The $J = 1$ odd states of argon have been studied by the Chicago group: Lee and Lu (1973) consider empirical fitting of experimental data and Fano and Lee (1973) and Lee (1974a) make *ab initio* calculations in which a derivative matrix method is used to solve the coupled equations. Their analysis uses (i) a transformation matrix $\boldsymbol{X} = \boldsymbol{X}_{\mathrm{I}}\boldsymbol{X}_{\mathrm{II}}$ as discussed in § 11.1; (ii) the five eigen quantum defects $\bar{\mu}_\alpha$; (iii) the elements D_α of the dipole matrix $\boldsymbol{D}(\bar{\mu})$ calculated using functions $F(\bar{\mu}; \rho)$. The analysis of Lee and Lu involves fitting to the following experimental data: (i) energy levels analysed using Lu–Fano plots; (ii) the photoionisation cross section in the region of resonances below the $J_c = \frac{1}{2}$ threshold; (iii) the branching ratio, above the $J_c = \frac{1}{2}$ threshold, for photoionisation leaving the core in the $J_c = \frac{3}{2}$ and $J_c = \frac{1}{2}$ levels. Data obtained from the fits were used to calculate oscillator strengths for transitions to bound states.

In the Chicago work the ordering of the channels is, for LS coupling,

$$\begin{array}{llllll} \alpha = & 1 & 2 & 3 & 4 & 5 \\ lSL\pi = & \mathrm{d}\,^3\mathrm{D^o} & \mathrm{d}\,^1\mathrm{P^o} & \mathrm{d}\,^3\mathrm{P^o} & \mathrm{s}\,^3\mathrm{P^o} & \mathrm{s}\,^1\mathrm{P^o} \end{array} \tag{11.19}$$

and for $[J_c]lj$ with $J = 1$ odd,

$$\begin{array}{llllll} i = 1 & 2 & 3 & 4 & 5 \\ [J_c]lj = [\frac{3}{2}]\mathrm{d}\frac{5}{2} & [\frac{3}{2}]\mathrm{d}\frac{3}{2} & [\frac{3}{2}]\mathrm{s}\frac{1}{2} & [\frac{1}{2}]\mathrm{d}\frac{3}{2} & [\frac{1}{2}]\mathrm{s}\frac{1}{2}. \end{array} \tag{11.20}$$

Table 6 compares empirical values of $\bar{\mu}_\alpha$ and D_α from Lee and Lu with *ab initio* values from Lee. The agreement between experimental oscillator strengths and results from the empirical and *ab initio* calculations was generally good. Lee obtained

$$X_{\mathrm{II}} = \begin{pmatrix} 1.00 & 0.02 & -0.01 & 0.00 & 0.00 \\ -0.02 & 1.00 & 0.00 & 0.00 & 0.02 \\ 0.01 & 0.00 & 0.98 & 0.18 & 0.00 \\ 0.00 & 0.00 & -0.18 & 0.98 & 0.02 \\ 0.00 & -0.02 & 0.00 & -0.02 & 1.00 \end{pmatrix}. \tag{11.21}$$

The coupling between d $^1\mathrm{P^o}$ and s $^1\mathrm{P^o}$ gives off-diagonal elements $X_{\mathrm{II}}(2, 5)$ and $X_{\mathrm{II}}(5, 2)$ and that between d $^3\mathrm{P^o}$ and s $^3\mathrm{P^o}$ gives $X_{\mathrm{II}}(3, 4)$ and $X_{\mathrm{II}}(4, 3)$. The latter are seen to be the largest off-diagonal elements. Other off-diagonal elements are due to departures from LS coupling in the inner region.

Table 6. Eigen quantum defects $\bar{\mu}_\alpha$ and dipole matrix elements D_α for argon. Empirical values from Lee and Lu (1973), *ab initio* values from Lee (1974).

α	$\bar{\mu}_\alpha$		D_α	
	Empirical	*Ab initio*	Empirical	*Ab initio*
1	0.22	0.20	+0.03	+0.03
2	0.07	0.01	−2.12	−2.10
3	0.50	0.51	+0.12	−0.01
4	0.15	0.12	+0.03	+0.03
5	0.11	0.07	−1.40	−1.00

11.4. Krypton

Recent advances in spectroscopic studies of krypton have been made by Kaufman and Humphreys (1969), who obtain accurate positions of lower states using interferometric methods, Dunning and Stebbings (1974), who studied high even-parity states by laser excitation from a metastable state, Yoshino and Tanaka (1979), who studied high $J = 1$ odd-parity states by absorption from the ground state and Delsart *et al* (1981), who studied high $J = 0$, 1, 2 and 3 odd-parity states by two-step laser excitation from a metastable state. Oscillator strengths for krypton and xenon have been measured by Geiger (1977) using methods of high-energy electron spectroscopy.

The most complete QDT analysis for Kr is that of Aymar *et al* (1981a, b) who use graphical methods to obtain initial values for QDT parameters, and least squares fitting to obtain final values, including an improved value for the series limit. All data for higher odd-parity levels can be fitted using transformation matrices independent of energy and eigen quantum defects linear in energy. Results for $J = 1$ odd are compared with those from an earlier empirical study by Geiger (1977) and those from *ab initio* calculations by Johnson *et al* (1980). The work of Aymar *et al* is of particular interest because of the precision obtained and because a number of values of J are considered. The computer program used has been published (Robaux and Aymar 1982).

11.5. Xenon

QDT methods were applied to the analysis of the xenon spectrum by Lu (1971) in a paper which introduced methods of analysis used in much subsequent work. The variation with energy of the QDT parameters was neglected and no attempt was made to fit energy levels to highly numerical accuracy. Lu–Fano plots were given for odd-parity states with $J = 0$, 1, 2 and 3. The analysis of autoionisation and oscillator strengths was similar to that subsequently made by Lee and Lu (1973) for argon (see § 10.4). A more recent detailed study of xenon has been made by Geiger (1977).

11.6. Branching ratios, angular distributions and spin polarisations

In order to make the most stringent tests of *ab initio* calculations for photoionisation, or to obtain a maximum amount of information in empirical analyses, it is necessary to consider all of the detailed information which can be measured in experiments. Let $\sigma(^2P^o)$ be the total cross section for photoionisation of a rare-gas atom. Above the second threshold we can put $\sigma(^2P^o) = \sigma(^2P^o_{3/2}) + \sigma(^2P^o_{1/2})$ where $\sigma(^2P^o_{3/2})$ and $\sigma(^2P^o_{1/2})$ can be determined separately by measuring the energy of the photoelectrons. The *branching ratio* is $\sigma(^2P^o_{3/2})/\sigma(^2P^o_{1/2})$. If $I(\theta)$ is the differential cross section for photoionisation, where θ is the angle between the momenta of the incident photon and the ejected electron, the total cross section is

$$\sigma = 2\pi \int_{-1}^{+1} I(\theta) \, \mathrm{d} \cos(\theta). \tag{11.22}$$

It may be shown (see, for example, Lipsky 1967, Cooper and Zare 1968, Burke 1976) that, for unpolarised photons,

$$I(\theta) = \frac{\sigma}{4\pi} \{1 - \tfrac{1}{2}\beta P_2[\cos(\theta)]\} \tag{11.23}$$

where P_2 is a Legendre polynomial and β is known as the *asymmetry parameter.* For incident radiation with electric vector in the direction $\hat{\epsilon}$ the photoionisation cross section is proportional to $|\hat{\epsilon} \cdot \vec{\boldsymbol{D}}|^2$. Use of polarised radiation gives *spin polarisation* of the ejected electrons. The spin polarisation parameters enter expressions for the spin polarisation in terms of the momenta of the incident photon and ejected electron and the direction $\hat{\epsilon}$ for the photon polarisation. They are calculated on transforming from a representation in which J_c and J are specified to one in which the quantum number m_s of the ejected electron is specified. The derivation of expressions for angular distribution and spin polarisation parameters is discussed by Lee (1974b).

Johnson *et al* (1980) consider circularly polarised radiation in the region below the $J_c = \frac{1}{2}$ threshold and give expressions for β and spin polarisation parameters in terms of the amplitudes D_i(out) with $i = 1$, 2 and 3 in the notation of (11.20). They make calculations, for argon, krypton and xenon, using the relativistic random phase approximation and the relativistic QDT of Lee and Johnson (1980). Similar calculations for neon are made by Johnson and Le Dorneuf (1980). For Ar, Kr and Xe the calculations show that $\bar{\mu}_\alpha$ and $D_\alpha(\bar{\mu})$ have significant energy variations in the near-threshold regions. Plots are given for cross sections σ, angular distribution parameters β and spin polarisation parameters in the autoionisation regions. For Xe there is good agreement with the results of Eland (see Berkowitz (1979)) for σ, of Samson and Gardner (1973) for β and of Heinzmann *et al* (1979) for δ, the spin polarisation of the total flux. Comparisons are also made with previous calculations of β by Dill (1973) and Geiger (1977) and of δ by Lee (1974b).

Further experimental and theoretical work on Xe is described by Heinzmann (1980) who considers a parameter ξ which gives the angular distribution of spin polarisation for unpolarised radiation (measured using radiation in He and Ne resonance lines) and δ, the total spin polarisation for circularly polarised radiation (which was obtained using radiation from a synchrotron out of the synchrotron plane). Heinzmann gives expression for these parameters in terms of the amplitudes D_i(out). His work is of interest in that it gives results for energies, up to 41 eV, which are higher than those for transitions between bound states or to the autoionisation region.

12. Alkaline earths and other atomic systems

12.1. Beryllium

Insights into the use of QDT methods can be gained by considering *ab initio* calculations for simple systems. It might be thought that, for this purpose, the simplest systems would be those containing two electrons, but for these special problems arise: for $(e + He^+)$, discussed in § 10.2, they are a consequence of the degeneracy of the He^+ excited states; and for $(e + H_2^+)$, discussed in § 13, a consequence of the presence of two nuclei.

In the present subsection we consider the four-electron system $(e + Be^+)$ for which one can obtain two-electron model problems. The Be^+ ion has a ground state $1s^2 2s$, a first excited state $1s^2 2p$ and other excited states at considerably higher energies. The simplest model treats the $1s^2$ electrons as an inert closed shell. Close-coupling calculations† were made by Moores (1967) with inclusion of channels $2s\epsilon_1 l$, $2p\epsilon_2(l-1)$ and $2p\epsilon_2(l+1)$, denoted by $i=1$, 2 and 3. The coupling between 2s and 2p gives dipole potentials U_{12} and U_{13} behaving, for ρ large, like ρ^{-2}; and couplings between the 2p states give quadrupole potentials behaving like ρ^{-3}. There is no difficulty in calculating the matrix $\boldsymbol{Y}$ in the region of all channels open and, since the 2s and 2p states are non-degenerate, this matrix can be extrapolated smoothly to the region of some channels closed without making transformations of the type discussed in § 10.2. It would be difficult, however, for reasons discussed in §§ 5.3 and 6.9, to compute $\boldsymbol{Y}$ directly at energies such that some channels are closed.

The procedure of Moores was to calculate $\boldsymbol{Y}$ in the region of some channels open, fit to expressions of the type (6.8), extrapolate to the region of some channels closed and calculate $\boldsymbol{R}=\tan(\pi\mu_1)$ using (6.30). The quantum defects $\mu_1(\epsilon_1)$ calculated by this extrapolation procedure were compared with values obtained on solving the coupled equations in the region of some channels closed, with imposition of the boundary condition of elimination of the exponentially increasing solutions. The extrapolated results were found to be accurate at energies close to the 2p threshold but less accurate at lower energies and sensitive to the extrapolation formula used.

Further analysis of the series of resonances below the 2p threshold were made by Norcross and Seaton (1970). They calculated complex quantum defects μ_c by two methods: (i) by extrapolation from the region of all channels open (see § 7.3); and (ii) from analysis of the resonance structures in $\tan(\pi\mu_1)$ obtained from solutions of the coupled equations in the region of some channels closed. For all but the lowest resonances the two methods are in good agreement. By combining methods (i) and (ii) it was possible to obtain, by interpolation, accurate values of μ_c for all values of the energy.

The simple model used by Moores (1967) and Norcross and Seaton (1970) does not give results in close agreement with experiment. One deficiency of such a model is its failure to give accurate energies and wavefunctions for Be^+. Much more accurate calculations were made by Norcross and Seaton (1976), who allowed for polarisation of the $1s^2$ core. The first step was to obtain a polarisation potential with parameters adjusted so as to obtain close agreement with experiment for all Be^+ quantum defects. In treating $(e+Be^+)$ as a two-electron system, Norcross and Seaton allowed for polarisation of the core by both of the outer electrons, which gives a potential behaving like

$$-\alpha\left(\frac{1}{r_1^4}+\frac{1}{r_2^4}+\frac{2\hat{r}_1\cdot\hat{r}_2}{r_1^2r_2^2}\right) \tag{12.1}$$

for r_1 and r_2 both large. In solving the two-electron problem they include Be^+ states 2s, 2p and $\bar{3}d$ where $\bar{3}d$ makes partial allowance for the quadrupole polarisation of 2s and the dipole polarisation of 2p. The equations were solved subject to the imposition of physical boundary conditions and the results obtained were fitted to QDT formulae. The only series in Be with an obvious perturbation is ^1D, for which

† By 'close coupling' we mean exact solution of the coupled equations of § 3. This usage, which is employed extensively in the literature of atomic collision theory, was introduced to distinguish between exact solutions of the equations and earlier work in which weak coupling was assumed and perturbation methods used. A different usage is employed in parts of the QDT literature. There 'close coupling' refers to a representation of $\mathcal{R}$ diagonal (see § 6.7), and 'loose coupling' to one in which it is not (in some cases this leads to the usage of 'close coupling' for *LS* coupling and 'loose coupling' for *jj* coupling). The reasons for these alternative usages are not apparent to the present reviewer, who is of the opinion that they are best avoided.

the lowest member is designated '$2p^2$' and the higher members 2s*n*d. More careful analysis shows all series to have significant perturbations due to resonances above the 2s limit. It was found that the calculated quantum defects could be fitted to isolated-resonance formulae (see § 7.5 and appendix A):

$$\tan[\pi\mu(\epsilon_1)] = \sum_{i=1}^{I} \epsilon_1^{i-1} p_i - \frac{q}{\epsilon_1 - \Delta}. \tag{12.2}$$

For bound states one has $\epsilon_1 < 0$ in (12.2) and, for perturbing levels lying about Be^+2s, one has $\Delta > 0$. The effect of such perturbations are largest in the regions of the high states but it is difficult to establish their existence from a purely empirical analysis of the available spectroscopic data since the quantum defects for the high states are sensitive to the value assumed for the series limit, T_∞. An improved value of T_∞ for Be was obtained by Seaton (1976) on taking the quantum defects to be $\mu = \mu(\text{calc}) + \Delta\mu$ where $\Delta\mu$ is a small empirical correction to the calculated quantum defects of Norcross and Seaton. With this revised value for T_∞, the quantum defects for Be 2s*n*s ^{1}S show a comparatively rapid variation with energy for *n* large, due to perturbation by $2p^2$ ^{1}S which lies just above the 2s limit. Previous attempts to locate $2p^2$ ^{1}S are discussed by Lu (1974) in the course of a systematic study of perturbations by np^2 ^{1}S and np^2 ^{1}D in the spectra of alkaline-earth isoelectronic sequences.

12.2. *Other alkaline earths*

Calculations for $(e + Mg^+)$ have been made by Mendoza (1979, 1981a, b), using methods similar to those of Norcross and Seaton for $(e + Be^+)$, and give good results for energy levels, transition probabilities, photoionisation and collisional excitation.

The attempt to use similar methods for $(e + Ca^+)$ gives results which are not quite so satisfactory (Mendoza 1979). The attempt to treat $(e + Ca^{2+})$ as a system with one electron outside a polarisable core does not give close agreement with experiment for all of the levels, particularly the d levels. Mendoza finds that improved results are obtained on including channels belonging to excited states of the core, or polarised pseudostates. In order to make accurate calculations for $(e + Ca^+)$ it would probably be necessary to allow for core excitations as well as excitations of the valence electron.

Pandey *et al* (1980) and Armstrong *et al* (1981) have considered much simpler models for alkaline earths, in which core polarisation is neglected and simple approximations are used for the exchange interactions between the two outer electrons. It would be of interest to compare, in detail, the results from such a model with, for example, those obtained by Mendoza for $(e + Mg^+)$.

We turn to consideration of work on the empirical analysis of experimental data for the alkaline earths. For Ca^+ we recall that the three lowest levels are 4s, 3d and 4p, in that order. The Ca I $^1P^o$ series has already been discussed in §§ 1 and 9.1. Attempts to fit radiative data for $4s^2$–4s*n*p and $4s^2$–4sϵ_1p have been made in a number of papers (Moores 1966, Friedrich and Trefftz 1969, Carter *et al* 1971, Wells 1974, Geiger 1979). Some difficulties arise in attempting to make consistent, and correct, normalisation of experimental oscillator strength data for different spectral regions. In order to account for the full complexity of the observed photoionisation cross section (Ditchburn and Hudson 1960) it would be necessary to include several Ca^+ excited states and to allow for the breakdown of *LS* coupling, since intercombination features are observed. Improved high-dispersion spectra have been obtained by Brown and Ginter (1980), who give a QDT analysis.

The even-parity series of Ca I have been studied by Esherick *et al* (1976), Armstrong *et al* (1977) and Wynne and Armstrong (1979a) who give new experimental data for the high states. The 4s*n*s ^{1}S series has a single perturbation which Armstrong *et al* (1977) consider to be due to interaction with 4p^2 ^{1}S. They fit the data using two channels, 4sϵ_1s and 4pϵ_3p and take one eigen quantum defect $\bar{\mu}_1$ to be linear in energy and other parameters independent of energy. It was shown by Wynne and Armstrong that an equally good fit can be obtained using the two channels 4sϵ_1s and 3dϵ_2d and considering the perturbation to be due to 3d^2. *Ab initio* calculations have been made by Froese Fischer and Hansen (1981), in the multi-configuration Hartree–Fock approximation, who find that the perturbing state contains comparable admixtures of 4p^2 and 3d^2. The ^{3}D$_2$ series, considered by Esherick *et al* (1976), contains one perturbation, due to 3d5s, but the ^{1}D series contains three, 4p^2, 3d5s and 3d^2, and is analysed by Armstrong *et al* (1977) using four channels, 4sϵ_1d, 3dϵ_2s, 3dϵ_2d and 4pϵ_2p. The symmetric 4×4 matrix $\mathcal{R}$ contains 10 linearly independent parameters and Armstrong *et al* make a fit for ^{1}D with 11 adjustable parameters (one element of $\bar{\mu}$ linear in energy, other parameters independent of energy).

The different perturbations in the ^{3}D$_2$ and ^{1}D$_2$ series produce apparent crossings of the two curves of $\mu(\mathscr{E})$ against $\mathscr{E}$. If one were to allow for the mutual perturbations of these two series, due to breakdown of *LS* coupling, these crossings would be avoided but such effects in Ca appear to be small and the accuracy of the experimental data is such that they are not easily apparent.

Consideration of the even-parity series in Ca I has prompted a further study of *bound channels* in QDT, which is discussed in appendix A and by Seaton (1982). The expansion (A1.3) includes *free-channel* functions, containing radial functions $F_i(r)$ which are freely varied, and *bound-channel* functions which have the form of bound-state functions multiplied by coefficients C_j which are freely varied. The inclusion of bound channels does not alter the formal structure of the theory but may give rise to additional poles in $\boldsymbol{Y}$.

We consider the analysis of bound states in the alkaline earths in the approximation of including one free channel and a number of bound channels. This gives expressions of the form (5.15) for $Y(\epsilon, l)$ (the case of (e+Be$^+$) has already been discussed in § 12.1). The experimental data for Ca I ^{3}D$_2$ gives values of Y which can be fitted to a cubic plus one pole:

$$Y=\sum_{i=0}^{3}\alpha_i\epsilon^i-\beta(\epsilon-\Delta)^{-1} \tag{12.3}$$

and the quantum defect is given by $\tan(\pi\mu)=AY$ where, for $l=2$, $A=(1+\epsilon)(1+4\epsilon)$. For $\epsilon=-0.25$ we have $A=0$ and μ equal to an integer. Figure 18 shows experimental quantum defects for Ca I ^{3}D$_2$ and the smooth curve obtained from (12.3). This series is of interest in that the ground state has $\epsilon<-0.25$ and the smooth curve is seen to be such that $\mu=1$ at $\epsilon=-0.25$. Seaton (1982) shows that the data for ^{1}S can be fitted using one bound channel and those for ^{1}D using three bound channels.

In fitting Ca I data using a number of free channels and no bound channels, Armstrong *et al* (1977) exclude the lowest states in each series. This procedure is consistent in that, when antisymmetrised functions are used and exchange operators included, the coupled equations will not have any solutions in the region of such states, other than, of course, at the energy eigenvalues themselves (see § 5.3.1). The problem is that the binding energy of the 'added' electron becomes larger than that

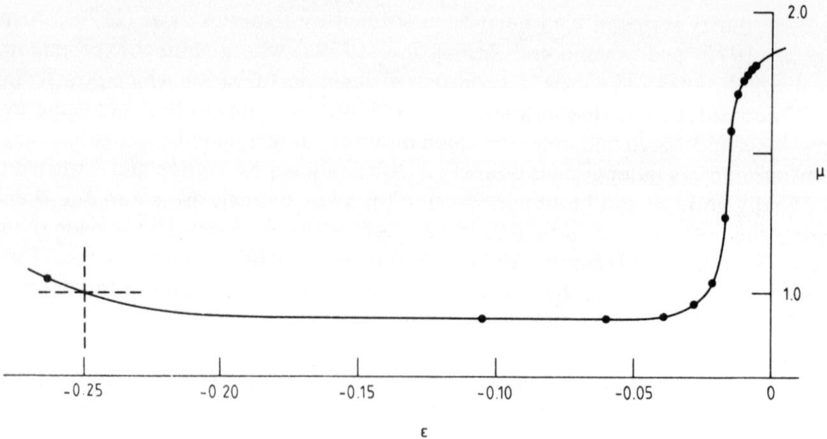

Figure 18. Quantum defects for Ca I $^3D_2$. Experimental points from Esherick *et al* (1976), full curve from equation (12.3). Note that $\mu(\epsilon) = 1$ for $\epsilon = -0.25$.

of an electron in an excited state of the 'target'. No such difficulty arises if one uses only one free channel and allows for correlations by inclusion of bound channels.

Different approaches to the use of free channels and bound channels can be compared by considering the different problems posed by the rare gases and the alkaline earths. The former have small energy separations between the target states of importance (they are due to fine-structure splitting), patterns of perturbations may be repeated many times in the region of bound states, and analysis is best made using a number of free channels. The alkaline earths, on the other hand, have larger separations between the target states of interest. Introduction of a free channel belonging to an excited state of the target will generally lead to, at most, one perturbation in the region of bound states; in this situation analysis using bound channels may be more convenient.

The analysis of extensive new data for the even-parity states of Sr I is discussed by Esherick (1977). Results for the ^1S series, which is perturbed by $5p^2$, confirm those obtained earlier by Lu (1974). The $^1D_2$ and $^3D_2$ series are of interest in showing clearly the mutual perturbations and avoided crossing which were difficult to detect in Ca I. The $J = 2$ even-parity series are analysed by Esherick using five channels and a *jj* coupling representation. The calculation of *g* factors for Sr is discussed by Wynne *et al* (1977), and Lu (1977) discusses near-threshold photoabsorption using a two-channel theory.

Turning to Ba, experimental results for high even-parity levels are given by Aymar *et al* (1978), who also give a theoretical analysis for $J = 0$. The analysis for $J = 2$, given by Aymar and Robaux (1979), shows that there are large departures from *LS* coupling. Lifetimes in the $^1D_2$ and $^3D_2$ series are discussed by Aymar *et al* (1981a, b), who include functions of bound-channel type. The absorption spectrum, for transitions from the ground state, $6s^2\ ^1S$, to states lying below the 6p limit, is discussed by Brown and Ginter (1978a).

The $^3P^o$ series in Ca, Sr and Ba are discussed by Armstrong *et al* (1979). In a review of the application of QDT of alkaline earths, Wynne and Armstrong (1979b) note the remarkably similar behaviour of quantum defects $\mu(\epsilon)$ for the $^1P^o$ series of Ca, Sr and Ba. Applications to inner-shell excitations are discussed by Newsom (1971) for Mg, Mansfield and Newsom (1977) for Ca, Mansfield and Newsom (1981) for Sr and Connerade *et al* (1979) for Ba.

12.3. Other atomic systems

Table 7 gives reference to papers, not cited elsewhere in the present review, concerned with applications of QDT to the analysis of the spectra of complex atomic systems.

Table 7. References to papers on applications of QDT for heavier atoms.

Atom	Reference
Mn	Brown and Ginter (1978b)
Ge	Brown *et al* (1977a)
Ag	Brown and Ginter (1977)
Sn	Brown *et al* (1977b)
Yb	Aymar *et al* (1980)
Au	Brown and Ginter (1978c)
Pb	Brown *et al* (1977c)
Ra	Armstrong *et al* (1980)

13. Molecular hydrogen

13.1. Introduction to applications of QDT to H_2

Fano (1970) first applied QDT to molecular problems. His work was concerned with perturbations in high states of H_2 produced by rotation of the H_2^+ core and it was stimulated by the precision spectroscopic measurements of Herzberg (1969) and Takezawa (1970), for levels with n up to 40. Further measurements were reported by Herzberg and Jungen (1972) and their analysis showed that perturbations are also produced by vibration of the core. The extension of the theory to include vibration was described, in a conference abstract, by Dill *et al* (1973) and applications were made by Atabek *et al* (1974) and Atabek and Jungen (1976). A detailed account of the theory is given by Jungen and Atabek (1977) and applications to autoionisation are discussed by Jungen and Dill (1980). The theory can also be applied to autodissociation, as was first noted by Dill *et al* (1973).

There are a number of analogies between QDT for the rare gases and H_2, since in both cases one is concerned with interactions due to small energy differences in the ion core. The theory rests on assumptions described in § 11.1: that there exists a value ρ_0 of ρ (the coordinate of the outer electron) such that the wavefunction is separable for $\rho \geqslant \rho_0$; that for $\rho \geqslant \rho_0$ the radial function for the outer electron is a linear combination of Coulomb functions; and that, for $\rho = \rho_0$, the Coulomb functions can be calculated neglecting the small differences in energy associated with the core states.

13.2. Sketch of the theory of H_2 states

The ground electronic state of H_2^+ is $1s\sigma\ {}^2\Sigma_g^+$ and the complete wavefunction is $\psi_{Nv}(\boldsymbol{R}, \boldsymbol{r})$ where $\boldsymbol{R}$ is the internuclear separation, $\boldsymbol{r}$ the coordinate of the electron, and N and v are quantum numbers for total angular momentum (excluding spin) and for vibration. The total energy is $E_+(N, v)$. The approximate expression

$$E_+(N, v) = E_0 + BN(N+1) + \omega(v + \tfrac{1}{2}) \tag{13.1}$$

where $B = 1.36 \times 10^{-4}$ and $\omega = 1.05 \times 10^{-2}$ (both in atomic units) shows that the spacing between rotational levels is much smaller than that between vibrational levels.

We consider two representations for those states of H_2 which can be reached by photoabsorption from the ground state, $H_2(1s\sigma)^2\,{}^1\Sigma_g^+$.

13.2.1. The Born–Oppenheimer states. In the Born–Oppenheimer (BO) approximation it is assumed that the speed of the electrons is large compared with that of the nuclei and the electronic wavefunction is calculated assuming the nuclei to be at rest. The states which can be reached by photoabsorption from the ground state are

$$\begin{aligned} &1s\sigma\,\epsilon p\sigma\ {}^1\Sigma_u^+ \\ &1s\sigma\,\epsilon p\pi\ {}^1\Pi_u^+,\quad {}^1\Pi_u^- \end{aligned} \tag{13.2}$$

and will be denoted by a channel index α. The BO electronic wavefunction, $\psi_e(R; r_1, r_2)$, depends parametrically on the inter-nuclear separation R (since we are concerned with singlet states, specification of spin coordinates is omitted).

The states ${}^1\Pi_u^+$ and ${}^1\Pi_u^-$ have equal energies in the BO approximation but different symmetry properties for reflection of electron coordinates in a plane containing the nuclei.

13.2.2. The collision states. In the BO approximation the electron angular momenta $\boldsymbol{l}$ are coupled to the inter-nuclear axis and only the components λ of $\boldsymbol{l}$ are defined. For the $(e + H_2^+)$ collision problem, and for the highly excited states of H_2, an outer electron may be at a large distance, ρ, from the H_2^+ core. For such states the speed of the outer electron may be much less than the speed of nuclear motion, the BO approximation is not valid and one must specify the quantum numbers l for the outer electron and N and v for the core. Using a channel index $i = \{N, v, l\}$, the wavefunction is

$$\Psi = \sum_i \psi_i \frac{1}{\rho} F_i(\rho) \qquad \text{for } \rho \geqslant \rho_0 \tag{13.3}$$

where ψ_i is an H_2^+ function multiplied by a function of the angle coordinates of the outer electron. The total angular momentum is $\boldsymbol{J} = \boldsymbol{N} + \boldsymbol{l}$.

Photoabsorption from the ground state gives states with $l = 1$. Table 8 gives, for $J = 0$, 1 and 2 and parity $\pi = +1$ and -1, the BO states which can be reached and

Table 8. States which can be reached by photoabsorption from $H_2\,{}^1\Sigma_g^+$.

J	π	BO	N
0	+1	${}^1\Sigma_u^+$	1
1	−1	${}^1\Sigma_u^+$, ${}^1\Pi_u^+$	0, 2
1	+1	${}^1\Pi_u^-$	1
2	+1	${}^1\Sigma_u^+$, ${}^1\Pi_u^+$	1, 3
2	−1	${}^1\Pi_u^-$	2

the allowed values of N; these results are obtained using rules for coupling of angular momenta and for deduction of parity (Herzberg 1950).

13.3. Rotational interactions

The presentation of the theory is simplified by considering, firstly, the case in which the H_2^+ core is in its ground vibrational state. The BO states can then be calculated with $R = R_0$, the H_2^+ equilibrium value. For $\rho \geqslant \rho_0$ the BO states are

$$\Psi_\alpha = \psi_\alpha \frac{1}{\rho} F_\alpha(\rho) \tag{13.4}$$

and the collision states are

$$\Psi_i = \sum_{i'} \psi_{i'} \frac{1}{\rho} F_{i'i}(\rho). \tag{13.5}$$

For reasons discussed in § 11.1 and at the end of § 13.1 we assume that, for ρ not much larger than ρ_0, the Coulomb functions are independent of the channel index, and hence that

$$\begin{aligned} F_\alpha &= s + c\mathcal{R}_\alpha \\ F_{i'i} &= s\delta_{i'i} + c\mathcal{R}_{i'i}. \end{aligned} \tag{13.6}$$

The relation between Ψ_α and Ψ_i is then

$$\Psi_i = \sum_\alpha \Psi_\alpha X_{\alpha i} \tag{13.7}$$

where $X_{\alpha i}$ is determined by the transformation between the states ψ_i and ψ_α. The transformation for $\boldsymbol{\mathcal{R}}$ is

$$\mathcal{R}_{i'i} = \sum_\alpha X_{i'\alpha} \mathcal{R}_\alpha X_{\alpha i}. \tag{13.8}$$

From table 8 it is seen that there are cases of two types, one-channel cases for $(J, \pi) = (0, +), (1, +1), (2, -1), \ldots,$ and two-channel cases for $(J, \pi) = (1, -1), (2, +1), \ldots$. For the latter, Fano (1970) shows that

$$\begin{array}{ccccc} & & \Lambda = \Sigma & \Pi & N = \\ \boldsymbol{X} = \dfrac{1}{\sqrt{(2J+1)}} & \Bigg(& \begin{array}{c} \sqrt{J} \\ -\sqrt{(J+1)} \end{array} & \begin{array}{c} \sqrt{(J+1)} \\ \sqrt{J} \end{array} \Bigg) & \begin{array}{c} (J-1) \\ (J+1). \end{array} \end{array} \tag{13.9}$$

The transformation (13.8) may be written

$$\boldsymbol{\mathcal{R}} = \boldsymbol{X} \mathcal{R}(\mathrm{BO}) \boldsymbol{X}^{\mathrm{T}} \tag{13.10}$$

where $\mathcal{R}(\mathrm{BO})$ is diagonal and has diagonal elements $\mathcal{R}_\alpha$. Defining ν_i by

$$E = E_+(i) - \frac{1}{2\nu_i^2} \tag{13.11}$$

(E in atomic units), we obtain the usual equation for bound states:

$$\det[\boldsymbol{\mathcal{R}} + \tan(\pi\nu)] = 0. \tag{13.12}$$

13.4. *Vibrational interactions*

Our simplified presentation of the theory is continued by considering, secondly, those values of (J, π) which give one-channel cases when vibration is neglected. Allowing for the dependence of the BO function on R we have

$$\Psi_\Lambda = \psi_\Lambda \frac{1}{\rho} F(R;\rho) \tag{13.13}$$

where $\Lambda = \Sigma$ or Π and

$$F(R;\rho) = s + c\mathscr{R}(R) \qquad \text{for } \rho > \rho_0. \tag{13.14}$$

The collision function is

$$\Psi_{Nv} = \psi_N \sum_{v'} \chi_{v'}(R) \frac{1}{\rho} F_{v'v}(\rho) \tag{13.15}$$

where $\chi_{v'}(R)$ is a vibrational function for H_2^+ and

$$F_{v'v}(\rho) = s\delta_{v'v} + c\mathscr{R}_{v'v} \qquad \text{for } \rho > \rho_0. \tag{13.16}$$

For the simple cases considered here we have $\psi_\Lambda = \psi_N$. Comparison of the coefficients of s shows that

$$\Psi_{Nv} = \Psi_\Lambda \chi_v(R) \tag{13.17}$$

and comparing the coefficients of c we obtain

$$\mathscr{R}(R)\chi_v(R) = \sum_{v'} \chi_{v'}(R)\mathscr{R}_{v'v}. \tag{13.18}$$

From the orthonormality of the functions $\chi_v(R)$ it follows that

$$\mathscr{R}_{v'v} = \int \chi_{v'}^*(R)\mathscr{R}(R)\chi_v(R)\,\mathrm{d}R. \tag{13.19}$$

13.5. *Rotation and vibration*

The BO states depend on $\{\alpha, R\}$, the collision states on $i = \{N, v, l\}$. The transformations are

$$\mathscr{R}_{i'i} = \sum_\alpha \int \mathrm{d}R\, X_{i'\alpha}\chi_{v'}^*(R)\mathscr{R}_\alpha(R)\chi_v(R)X_{\alpha i}. \tag{13.20}$$

The bound states are obtained on solving (13.12) with ν_i defined by (13.11).

Equation (13.20) gives an example of a *frame transformation* (Chang and Fano 1972), i.e. a transformation from the frame of fixed nuclei to the laboratory frame of rotating and vibrating nuclei.

13.6. *Some applications of the theory*

Fano (1970) considers excitation from the H_2 ground state (with $J' = 0$, $v' = 0$, $\pi' = +1$) to final states with $J = 1$, $v = 0$, $\pi = -1$, giving, with neglect of vibrational interactions, a two-channel problem. A satisfactory account of the observed rotational perturbations is obtained using the transformation (13.9). Fano also considers the intensities for transitions to bound and continuum states in terms of the amplitudes D_Σ and D_Π.

Since H_2 contains only two electrons, very accurate BO functions can be computed, particularly for the lower bound states. If $E_n(\boldsymbol{R})$ is the computed electronic energy for $1s\sigma np\sigma$ or $1s\sigma np\pi$, and $E_+(\boldsymbol{R})$ that for H_2^+, we have

$$E_n(\boldsymbol{R}) = E_+(\boldsymbol{R}) - \frac{1}{2[n - \mu_n(\boldsymbol{R})]^2}. \tag{13.21}$$

Atabek *et al* (1974) consider the state $1s\sigma 2p\pi\, c\, {}^1\Pi_u^-$ which, neglecting vibration, gives a one-channel problem. Values of $\mu_2(\boldsymbol{R})$ are obtained using (13.21), with $E_2(\boldsymbol{R})$ from Kołos and Wolniewicz (1965) and $E_+(\boldsymbol{R})$ from Wind (1965). The matrix $\mathscr{R}_{v'v}$ is calculated on putting $\mathscr{R}(\boldsymbol{R}) = \tan[\pi\mu_2(\boldsymbol{R})]$ and using (13.20).

The term values for the $c\, {}^1\Pi_u^-\ J = 1$ vibrational levels were calculated on solving (13.14) and using experimental values of Beckel *et al* (1970) for the H_2^+ vibrational levels. Comparison with observed term values showed the QDT results to be much more accurate than those obtained by Kołos and Wolniewicz (1968) in the BO approximation. Atabek *et al* note that both calculations use the same BO electronic energy for $c\, {}^1\Pi_u^-$; they differ in that the QDT calculations use vibrational functions for H_2^+ in (13.20), and equation (13.14), while the BO calculations use vibrational energies for H_2 obtained using the $c\, {}^1\Pi_u^-$ electronic energy.

Atabek *et al* also consider the high states, for ${}^1\Pi_u^-\ J = 1$, which lie between the $v = 0$ and $v = 1$ thresholds. The levels with approximate description $np\pi^-\ v = 1$ are perturbed by those with description $np\pi^-\ v = 2$. Good agreement is obtained with experiment despite the fact that, in using $\mathscr{R}_{v'v}$ from $c\, {}^1\Pi_u^-$, the energy variation of $\mathscr{R}_{v'v}$ is neglected.

The ${}^1\Sigma_u^+\ J = 0$ states considered by Atabek and Jungen (1976) also give one-channel problems when vibrational interactions are neglected. The BO quantum defects for $p\pi$ and $p\sigma$ are shown in figure 19. The curve for $p\pi$ is from the work of Atabek *et al* for $1s\sigma 2p\pi\ {}^1\Pi_u^-$. The curve for $p\pi$ is seen to be a different character, with μ small for $\boldsymbol{R}$ small and μ equal to unity for $\boldsymbol{R}$ large. This is a consequence of $np\sigma$ being a 'promoted' orbital. It follows from (13.19) that $\mathscr{R}_{v'v}$ is diagonal if $\mu(\boldsymbol{R})$ is independent of $\boldsymbol{R}$, and hence that vibrational interactions will be most important when $\mu(\boldsymbol{R})$ has a rapid variation with $\boldsymbol{R}$. This agrees with the experimental data which show perturbations for ${}^1\Sigma_u^+\ J = 0$ to be much more important than those for ${}^1\Pi_u^-$. The results for $p\sigma$ shown in figure 19 were obtained by Atabek and Jungen by fitting to the observed levels.

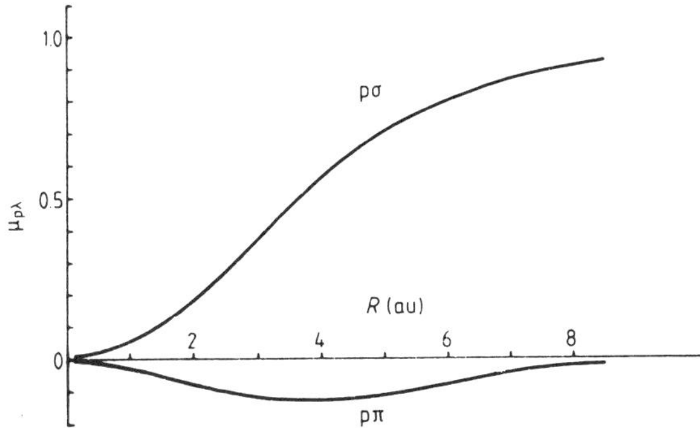

Figure 19. Quantum defects for H_2 $1s\sigma 2p\sigma\ {}^1\Sigma_u^+$ and $1s\sigma 2p\pi\ {}^1\Pi_u^-$ (from Atabek *et al* 1974).

Many refinements are included in the work of Jungen and Atabek (1977). They consider states of the five lowest configurations which can be reached by photoabsorption, $1s\sigma np\sigma$ with $n = 2$, 3 and 4 and $1s\sigma np\pi$ with $n = 2$ and 3. The quantum defects are obtained from *ab initio* calculations and, with the exception of $2p\sigma$, are found to vary slowly with n. Results are obtained for many values of v and J, for both H_2 and D_2. Agreement with experiment is generally very good and represents a considerable improvement over previous BO calculations.

Jungen and Dill (1980) consider the photoionisation of H_2 from its ground vibrational and rotational state to the region between the ionisation threshold ($N = 0$, $v = 0$) to the $N = 2$ threshold. The photoionisation amplitudes are estimated from absolute oscillator strength measurements. Figure 20 shows results of calculations for (i) direct ionisation; (ii) inclusion of rotational autoionisation due to the series $np\, N = 2$; (iii) inclusion of rotational and vibrational autoionisation (the important vibrational autoionisation states are $5p\pi\ v = 2$ and $7p\pi\ v = 1$). The calculated results are in good agreement with the experimental results of Dehmer and Chupka (1976).

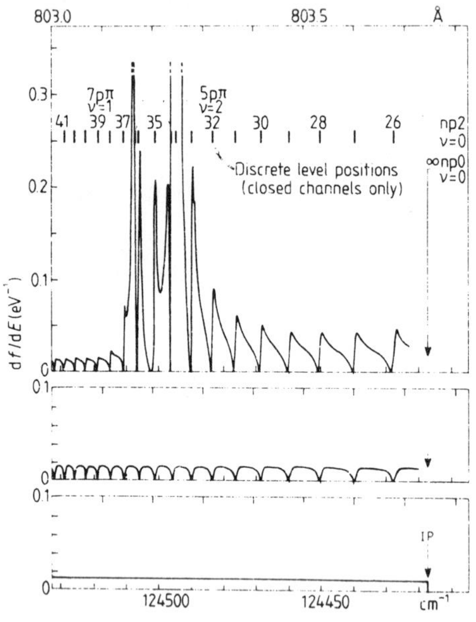

Figure 20. The cross section for photoionisation of H_2 in the region just above the first threshold (from Jungen and Dill 1980). The lower figure shows results of calculations neglecting rotational and vibrational interactions, the middle figure gives results including rotation and the top figure shows final results obtained with both rotational and vibrational interactions.

13.7. *Other recent applications to molecular problems*

There has been much recent work on applications of QDT to molecular problems but space permits only brief descriptions. It can be expected that further important advances will be made.

Raoult *et al* (1980) consider angular distributions in H_2 photoionisation, for which the theory was first formulated by Dill (1972) and Fano and Dill (1972). The work of Jungen and Dill (1980), described in § 13.6, has been extended by Raoult and

Jungen (1981) who consider resonances belonging to higher vibrational states, and asymmetry parameters β and branching ratios. There is, again, good agreement with the experimental results of Dehmer and Chupka (1976). Further discussions of the theory and its applications are given by Jungen and Raoult (1981) and Jungen (1982) who discusses both H_2 and N_2.

The process of molecular photoionisation:

$$h\nu + AB \rightarrow AB^+ + e \tag{13.22}$$

differs from atomic photoionisation only in that additional complexity results from the molecular systems having vibrational and rotational energies. The process of photodissociation:

$$h\nu + AB \rightarrow A + B \tag{13.23}$$

has no atomic counterpart. A third molecular process is dissociative recombination:

$$e + AB^+ \rightarrow A + B. \tag{13.24}$$

Since molecular dissociation energies, D, are generally smaller than ionisation energies, I, photodissociation can occur at energies below the threshold for photoionisation; above that threshold photodissociation and photoionisation are competing processes and both must be included in a complete theoretical treatment. Assuming $I > D$, the dissociative recombination process has a finite cross section in the limit of zero energy of the incident electron.

All of the above processes can occur via doubly excited states, AB^{**}:

$$h\nu + AB \leftrightarrow AB^{**} \begin{array}{l} \nearrow\!\!\!\swarrow\ AB^+ + e \\ \nwarrow\!\!\!\searrow\ A + B. \end{array}$$

This gives rise to structures in cross sections which can be analysed using an extension of QDT in which additional dissociation channels are included. This is discussed by Giusti (1980), Giusti-Suzor (1982) and by Jungen and Giusti-Suzor (1982) who consider photoionisation and photodissociation for the contrasting cases of H_2 and NO.

14. Dielectronic recombination

In the present section we use atomic units and not the Z-scaled units introduced in §§ 2 and 3.

The process of dielectronic recombination:

$$e + X^+ \rightarrow X^{**} \rightarrow X^* + h\nu \tag{14.1}$$

(where X^{**} is a doubly excited resonance state and X^* is a true bound-state) was first discussed by Massey and Bates (1942) and Bates and Massey (1943). The total rate of recombination in a plasma is obtained on summing the contributions from all doubly excited states and Burgess (1964) showed that the process can be particularly important in high-temperature plasmas since whole Rydberg series of doubly excited states can contribute.

We define the probability for a collision process as follows. From (3.28) and (3.30) the partial cross section is

$$Q_{SL\pi}(i\to j)=\frac{(2S+1)(2L+1)P_{SL\pi}(i,j)}{2\omega_i k_i^2}\pi a_0^2 \tag{14.2}$$

where ω_i is the initial statistical weight of the target, k_i is the initial electron wavenumber in atomic units, and

$$P_{SL\pi}(i,j)=|T_{SL\pi}(i,j)|^2 \tag{14.3}$$

where $\boldsymbol{T}$ is the transmission matrix. The total cross section is obtained on summing (14.2) over total angular momenta and parity, $SL\pi$. We interpret $P_{SL\pi}(i,j)$ as the probability that, for a given $SL\pi$, an incoming electron produces the transition $i\to j$. The capture process (14.1) has a probability $P_C(E)$ which varies rapidly with E in the vicinity of a resonance. Using detailed balancing arguments, Bates and Massey show that the probability integrated over the resonance profile is

$$\int_{\text{res}} P_C(E)\,\mathrm{d}E=2\pi\frac{\Gamma_A\Gamma_R}{\Gamma_A+\Gamma_R} \tag{14.4}$$

where Γ_A is the probability for the autoionisation process $X^{**}\to X^+ + e$ and Γ_R that for the radiative process $X^{**}\to X^* + h\nu$. The separations between resonances in a Rydberg series are given by

$$\delta E=Z^2(n-\alpha)^{-3} \tag{14.5}$$

where α is the real part of the complex quantum defect and a mean value of P_C is defined by

$$\bar{P}_C=\frac{1}{\delta E}\int_{\text{res}} P_C(E)\,\mathrm{d}E. \tag{14.6}$$

For high resonances in a Rydberg series we may assume that the captured electron acts as a 'spectator' in the radiative process, and that Γ_R is the probability for a radiative transition in the ion core and is independent of n. Since Γ_A behaves like n^{-3} for n large we have† $\Gamma_A \ll \Gamma_R$ for n sufficiently large and hence

$$\int_{\text{res}} P_C(E)\,\mathrm{d}E=2\pi\Gamma_A \qquad \text{for } n \text{ sufficiently large.} \tag{14.7}$$

Burgess considers the probability $P_{h\nu}$ for emission of a photon, both above and below an excitation threshold. If collisional excitation:

$$e+X_1\to e+X_2 \tag{14.8}$$

is followed by photon emission:

$$X_2\to X_1+h\nu \tag{14.9}$$

we have, above threshold,

$$P_{h\nu}=|\chi_{12}|^2 \tag{14.10}$$

† It follows from (4.3) that Γ_R is proportional to the cube of the fine-structure constant and hence that Γ_R is typically of order 10^{-6} atomic units. For smaller values of n one normally has $\Gamma_A \gg \Gamma_R$.

where we put $\boldsymbol{S} = \chi$. Below threshold the averaged probability of photon emission is

$$\bar{P}_{h\nu} = \bar{P}_C. \tag{14.11}$$

Burgess assumes that $P_{h\nu}$ above threshold joins smoothly to $\bar{P}_{h\nu}$ just below threshold, so that one may put

$$\bar{P}_C = |\chi_{12}|^2 b \tag{14.12}$$

where

$$b = \Gamma_R(\Gamma_R + \Gamma_A)^{-1} \tag{14.13}$$

is a branching ratio. Just below threshold (14.13) gives $b = 1$ and hence the required continuity at threshold between $\bar{P}_C$ below and $P_{h\nu}$ above. From (14.4), (14.6) and (14.12) one obtains

$$\Gamma_A = |\chi_{12}|^2 \delta E/2\pi. \tag{14.14}$$

These are the formulae which have been used in most calculations but one may doubt whether they are rigorously correct since (14.4) and (14.11) are both obtained by intuitive arguments. Such doubts are increased when one considers that for the very high states one has $\Gamma_R \gg \delta E$ and hence all resonance structure will be 'smeared out' by interaction with the radiation field. A more complete quantum-mechanical theory is also required in order to allow for overlapping resonances and possible interference effects.

The attempt to develop such a theory, as an extension of radiation damping theory, has been made by Davies and Seaton (1969). Interaction with the radiation field may be considered to give additional channels in the collision problem. It is shown by Seaton and Storey (1976) that the theory of Davies and Seaton is not entirely satisfactory, particularly in the region for which one has $\Gamma_R \gg \delta E$, and a fresh attack on the problem has been made by Bell and Seaton (1983) (see also Bell 1979).

Let $f_i(E, t)$ be the amplitude of the state with no photon and an incoming electron in channel i, and $g_{\beta\mu}(\omega, t)$ that for the final state in which a (μ, ω) photon has been emitted and the atom is in a bound state Ψ_β. The equations of Davies and Seaton may be written

$$\begin{aligned} \frac{\mathrm{d}}{\mathrm{d}t} f_i(E, t) &= -\mathrm{i} \exp(\mathrm{i}Et) \sum_{\beta\mu} \int \mathrm{d}\omega \, C_{i,\beta\mu}(E) \exp[-\mathrm{i}(E_\beta + \omega)t] g_{\beta\mu}(\omega, t) \\ \frac{\mathrm{d}}{\mathrm{d}t} g_{\beta\mu}(\omega, t) &= -\mathrm{i} \exp[\mathrm{i}(E_\beta + \omega)t] \sum_i \int \mathrm{d}E \, C_{\beta\mu,i}(E) \exp(-\mathrm{i}Et) f_i(E, t) \end{aligned} \tag{14.15}$$

where

$$C_{i,\beta\mu} = \left(\frac{2\omega^3\alpha^3}{3\pi}\right)^{1/2} (\Psi_\beta | d_\mu | \Psi_i(E)). \tag{14.16}$$

In (14.13), α is the fine-structure constant and $\Psi_i(E)$ has $\boldsymbol{S}$-matrix asymptotic form and is normalised to $(\Psi_i(E)|\Psi_{i'}(E')) = \delta(i, i')\delta(E - E')$. The solutions obtained by Davies and Seaton are in error due to an incorrect interchange of the order of limit processes. Bell and Seaton consider high Rydberg resonances and take $\Psi_i(E)$ to have a form given by QDT.

Space does not permit a detailed summary of the theory. For the two-channel case, Bell and Seaton obtain

$$b = \frac{[\exp(2\pi\Gamma_R/\delta E) - 1]}{[\exp(2\pi\Gamma_R/\delta E) - 1] + (2\pi\Gamma_A/\delta E)} \tag{14.17}$$

where Γ_A is defined by (14.14). It is seen that Γ_A and Γ_R enter (14.17) only through the ratios $(\Gamma_R/\delta E)$ and $(\Gamma_A/\delta E)$, whereas δE does not enter (14.13) explicitly. In the limit of $(2\pi\Gamma_R/\delta E) \ll 1$, (14.17) reduces to (14.13) and in the limit of $(2\pi\Gamma_R/\delta E) \gg 1$ (14.17) gives $b = 1$. Thus (14.13) and (14.17) agree in the limits of both smaller values of n and of $n \to \infty$. Equation (14.17) is consistent with the intuitive arguments leading to (14.13) but those arguments do not give correct results for all values of $(\Gamma_R/\delta E)$. In practice there is little difference in recombination rates calculated using (14.13) and (14.17).

The definition (14.14) of Γ_A requires further discussion. For two channels we have $\chi_{22} = \exp[2\pi i(\alpha + i\beta)]$ where $(\alpha + i\beta)$ is the complex quantum defect. From unitarity, $|\chi_{12}|^2 = 1 - |\chi_{22}|^2$ and hence (14.14) gives

$$\Gamma_A = [1 - \exp(-4\pi\beta)]\,\delta E/2\pi. \tag{14.18}$$

The poles in the scattering matrix occur at energies

$$E(\text{pole}) = E_2 - \frac{Z^2}{2(n - \alpha - i\beta)^2} \tag{14.19}$$

and if we define Γ_A to be such that the imaginary part of E(pole) is $-\frac{1}{2}\Gamma_A$ we obtain, for n large,

$$\Gamma_A = 2\beta\,\delta E. \tag{14.20}$$

The definitions (14.18) and (14.20) agree in the limit of $4\pi\beta \ll 1$. For larger values of β, (14.20) can give $(\Gamma_A/\delta E) > 1$, i.e. resonance widths greater than resonance separations. No such difficulty arises with (14.18).

A further effect of dielectronic recombination is to give a loss of flux in electron scattering channels. This was first discussed by Presnyakov and Urnov (1975) and has been discussed recently by Pradhan (1981). In the Gailitis expression (7.17) for $\langle |S_{ij}|^2 \rangle$ the terms with $p \neq q$ correspond to interference effects and, in practice, are found to be fairly small.

Neglecting these terms and using the unitarity of $\boldsymbol{\chi}$, (7.17) gives

$$\langle |S_{ij}|^2 \rangle = |\chi_{ij}|^2 + \sum_p \left(\frac{|\chi_{pi}|^2 |\chi_{pj}|^2}{\sum_k |\chi_{pk}|^2} \right) \tag{14.21}$$

where i, j and k are open-channel indices and p is a closed-channel index. We put

$$\Gamma_A(p, k) = \left(\frac{Z^2}{2\pi\nu_p^3} \right) |\chi_{pk}|^2 \tag{14.22}$$

and

$$\Gamma_A(p) = \sum_k \Gamma_A(p, k). \tag{14.23}$$

$\Gamma_A(p)$ is the autoionisation probability for resonances in channel p, $\Gamma_A(p, k)$ the probability that autoionisation gives an electron in channel k. Equation (14.21) can be written

$$\langle |S_{ij}|^2 \rangle = |\chi_{ij}|^2 + \sum_p \left(\frac{2\pi\nu_p^3}{Z^2}\right) \frac{\Gamma_A(p,i)\Gamma_A(p,j)}{\Gamma_A(p)}. \tag{14.24}$$

Let $\Gamma_R(p)$ be the probability for emission of radiation, for the transition from a resonance in channel p to a bound state. Allowing for the loss of flux due to this process of radiative stabilisation, (14.24) is replaced by

$$\langle |S_{ij}|^2 \rangle = |\chi_{ij}|^2 + \sum_p \left(\frac{2\pi\nu_p^3}{Z^2}\right) \left(\frac{\Gamma_A(p,i)\Gamma_A(p,j)}{\Gamma_A(p)+\Gamma_R(p)}\right). \tag{14.25}$$

The same result is obtained on replacing $\Sigma_k |\chi_{pk}|^2$ by

$$\sum_k |\chi_{pk}|^2 + \left(\frac{2\pi\nu_p^3}{Z^2}\right)\Gamma_R$$

in (14.21).

The averaged collision strength below threshold, $\langle \Omega(i,j) \rangle$, is proportional to $\langle |S_{ij}|^2 \rangle$ and the collision strength $\Omega(i,j)$ above threshold is proportional to $|\chi_{ij}|^2$. Using (14.21) we have that $\langle \Omega(1,2) \rangle$ below is larger than $\Omega(1,2)$ above. This is often known as the 'Gailitis jump'. When radiative channels are included the 'jump' disappears.

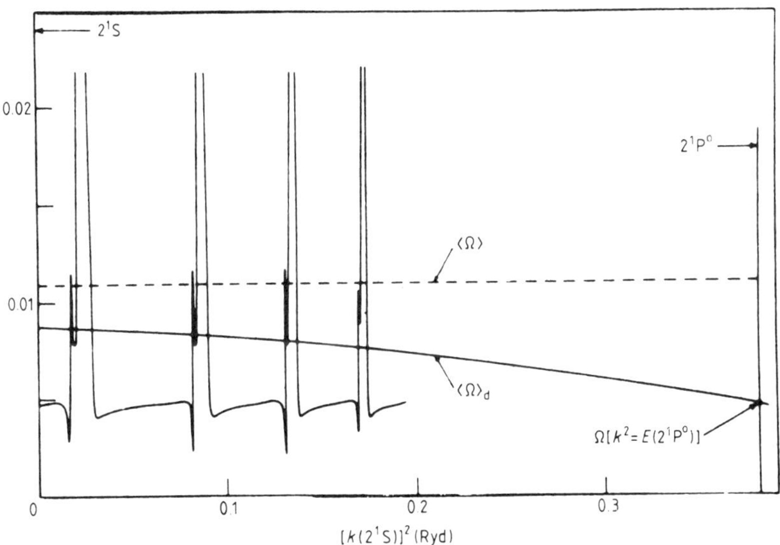

Figure 21. The collision strength for $1\,^1S$–$2\,^1S$ in O^{6+} (from Pradhan 1981). The average collision strength Ω is calculated without allowing for radiative decays; $\langle\Omega\rangle_d$ is calculated allowing for such decays. Note that, neglecting radiative decays, $\langle\Omega\rangle$ is the averaged collision strength below the $2\,^1P^o$ threshold, and that the collision strength above threshold is the continuation of the curve $\langle\Omega\rangle_d$; the difference between $\langle\Omega\rangle$ and $\langle\Omega\rangle_d$ at the threshold is the 'Gailitis jump'. When radiative decays are included there is no jump, $\langle\Omega\rangle_d$ below threshold being continuous with Ω above.

Figure 21 shows results obtained by Pradhan for the transition $1^1S \rightarrow 2^1S$ in the helium-like ion O^{6+}. The collision strength contains resonances in series converging to 2^1P. The matrix χ is calculated in the region above the 2^1P threshold and extrapolated to the region below. In figure 21 the following results are shown: the collision strength below the 2^1P threshold, with resonance structure; the averaged collision $\langle\Omega\rangle$ calculated using (14.21); the collision strength $\langle\Omega\rangle_d$ calculated allowing for radiative decays. It is seen that allowance for radiative decays can produce important changes for calculated excitation rate coefficients.

For highly ionised ions in an isoelectronic sequence the autoionisation probabilities Γ_A are approximately independent of Z but Γ_R is proportional to Z^4 (for transitions involving a change of principal quantum number). Radiative decays therefore become more important as Z increases.

Acknowledgments

The graphs of Coulomb functions, given in figures 4–8, were prepared using the SERC Interactive Computing Facility and FR 80 plotter. The author is much indebted to Dr Martin Crees who wrote the computer programs used.

Note added in proof

The author thanks Professor U Fano and Dr Ch Jungen for critical readings of this review. Professor Fano notes that a paper by Greene *et al* (1982), which is a sequel to Greene *et al* (1979), is now in press. Dr Jungen notes that most recent papers use N^+ (rather than N) for the angular momentum quantum number for an ion core; N can then be used, following the usual convention, for total angular momentum excluding spin.

Appendix A. Inclusions of bound channels

The expansion used in § 3 may be written

$$\Psi = \sum_i \Theta_i \tag{A1.1}$$

where

$$\Theta_i = \psi_i \frac{1}{r} F_i(r). \tag{A1.2}$$

Eissner and Seaton (1972) refer to the Θ_i as *free* channels since, in applying variational principles, the radial functions $F_i(r)$ are freely varied. It should be noted that a free channel may be open or closed, depending on the value of the energy. In practice, additional functions Φ_j are usually included in the expansion for Ψ:

$$\Psi = \sum_i \Theta_i + \sum_i \Phi_j C_j. \tag{A1.3}$$

The Φ_j have the form of bound-state functions for the $(N+1)$-electron system. Eissner and Seaton refer to them as *bound channels*. They are often referred to as *correlation functions*. In applying variational principles, the coefficients C_j in (A1.3) are freely varied. The radial functions F_i are constrained to be such that

$$(\Phi_j | \Theta_i) = 0. \tag{A1.4}$$

A compact notation is

$$\Psi = \Theta + \boldsymbol{\Phi}\boldsymbol{C} \tag{A1.5}$$

where

$$\Theta = \boldsymbol{\psi}\frac{1}{r}\boldsymbol{F} \tag{A1.6}$$

and

$$\boldsymbol{\psi} = (\psi_1, \psi_2, \ldots) \qquad \boldsymbol{\Phi} = (\Phi_1, \Phi_2, \ldots)$$
$$\boldsymbol{F} = \begin{pmatrix} F_1 \\ F_2 \\ \vdots \end{pmatrix} \qquad \boldsymbol{C} = \begin{pmatrix} C_1 \\ C_2 \\ \vdots \end{pmatrix}. \tag{A1.7}$$

The equations for the determination of $\boldsymbol{F}$ and $\boldsymbol{C}$ are

$$(\boldsymbol{\psi}|H - E|\Psi) = 0 \tag{A1.8}$$
$$(\boldsymbol{\Phi}|H - E|\Psi) = 0 \tag{A1.9}$$

(see, for example, Burke and Seaton (1971) or Eissner and Seaton (1972)), where the integrations in $(\boldsymbol{\psi}|H - E|\Psi)$ are over all coordinates except r. For simplicity in presentation we omit the Lagrange multipliers which should be included in (A1.8) in order to satisfy (A1.4).

Substitution of (A1.5) in (A1.8) and (A1.9) gives, using (A1.4),

$$\left[-\frac{1}{2}\left(\frac{\mathrm{d}^2}{\mathrm{d}r^2} - \frac{l(l+1)}{r^2}\right) - (E - E_{\mathrm{T}})\right]\boldsymbol{F} + \boldsymbol{V}\boldsymbol{F} + r(\boldsymbol{\psi}|H|\boldsymbol{\Phi})\boldsymbol{C} = 0 \tag{A1.10}$$

and

$$(\boldsymbol{\Phi}|H|\Theta) + \{(\boldsymbol{\Phi}|H - E|\boldsymbol{\Phi}) - E\}\boldsymbol{C} = 0 \tag{A1.11}$$

where E_{T} is the diagonal matrix of target energies. We note that

$$\begin{aligned}(\boldsymbol{\Phi}|H|\Theta) &= \int (\boldsymbol{\Phi}|H|\boldsymbol{\psi})\frac{1}{r}\boldsymbol{F}r^2\,\mathrm{d}r \\ &= \int r(\boldsymbol{\Phi}|H|\boldsymbol{\psi})\boldsymbol{F}\,\mathrm{d}r.\end{aligned} \tag{A1.12}$$

In terms of the scaled variables of § 3, the equations for $\boldsymbol{F}$ and $\boldsymbol{C}$ are

$$(\mathcal{L} + \epsilon)\boldsymbol{F} + \boldsymbol{W}\boldsymbol{C} = 0 \tag{A1.13}$$
$$-(\boldsymbol{W}|\boldsymbol{F}) + (\mathcal{H} - \mathcal{E})\boldsymbol{C} = 0 \tag{A1.14}$$

where

$$\boldsymbol{W}(\rho) = -\frac{2}{Z^2}r(\boldsymbol{\psi}|H|\boldsymbol{\Phi}) \tag{A1.15}$$

$$\mathcal{H} = \frac{2}{Z^2}(\boldsymbol{\Phi}|H|\boldsymbol{\Phi}). \tag{A1.16}$$

The formulation of QDT, with inclusion of bound channels, is discussed by Seaton (1982b). Let $\boldsymbol{U}$ and $\boldsymbol{W}$ be such that

$$\left.\begin{array}{l}\boldsymbol{U}(\rho)=2/\rho \\ \boldsymbol{W}(\rho)=0\end{array}\right\} \quad \text{for } \rho \geqslant \rho_0. \tag{A1.17}$$

We take $\boldsymbol{F}^{(0)}$ and $\boldsymbol{F}^{(1)}$ to be solutions of

$$(\mathscr{L}+\epsilon)\boldsymbol{F}^{(0)}=0 \tag{A1.18}$$

$$(\mathscr{L}+\epsilon)\boldsymbol{F}^{(1)}+\boldsymbol{W}=0 \tag{A1.19}$$

and to be such that

$$\boldsymbol{F}^{(0)}=f+h\boldsymbol{Y}^{(0)} \tag{A1.20}$$

$$\boldsymbol{F}^{(1)}=h\boldsymbol{Y}^{(1)} \quad \text{for } \rho \geqslant \rho_0. \tag{A1.21}$$

From (A1.19)

$$(\boldsymbol{F}^{(0)}|\mathscr{L}+\epsilon|\boldsymbol{F}^{(1)})+(\boldsymbol{F}^{(0)}|\boldsymbol{W})=0. \tag{A1.22}$$

Integrating by parts, and using (A1.14) and the Wronskian in (2.62), we obtain

$$\boldsymbol{Y}^{(1)}=\frac{\pi}{2}(\boldsymbol{F}^{(0)}|\boldsymbol{W}). \tag{A1.23}$$

The solution of (A1.13) is

$$\boldsymbol{F}=\boldsymbol{F}^{(0)}+\boldsymbol{F}^{(1)}\boldsymbol{C} \tag{A1.24}$$

and is such that

$$\boldsymbol{F}=f+h\boldsymbol{Y} \qquad \text{for } \rho \geqslant \rho_0 \tag{A1.25}$$

where

$$\boldsymbol{Y}=\boldsymbol{Y}^{(0)}+\boldsymbol{Y}^{(1)}\boldsymbol{C}. \tag{A1.26}$$

The coefficients $\boldsymbol{C}$ are obtained on substituting (A1.24) in (A1.14), which gives

$$-(\boldsymbol{W}|\boldsymbol{F}^{(0)})-(\boldsymbol{W}|\boldsymbol{F}^{(1)})\boldsymbol{C}+(\mathscr{H}-\mathscr{E})\boldsymbol{C}=0 \tag{A1.27}$$

and hence

$$\boldsymbol{C}=-(\mathscr{E}-\boldsymbol{D})^{-1}(\boldsymbol{W}|\boldsymbol{F}^{(0)}) \tag{A1.28}$$

where

$$\boldsymbol{D}=\mathscr{H}-(\boldsymbol{W}|\boldsymbol{F}^{(1)}). \tag{A1.29}$$

From (A1.23), (A1.26) and (A1.28),

$$\boldsymbol{Y}=\boldsymbol{Y}^{(0)}-\boldsymbol{B}^{\mathrm{T}}(\mathscr{E}-\boldsymbol{D})^{-1}\boldsymbol{B} \tag{A1.30}$$

where

$$\boldsymbol{B}=(\pi/2)^{1/2}(\boldsymbol{W}|\boldsymbol{F}^{(0)}). \tag{A1.31}$$

We may diagonalise $\boldsymbol{D}$:

$$\boldsymbol{D}\boldsymbol{G}=\boldsymbol{G}d \tag{A1.32}$$

to obtain

$$\boldsymbol{Y} = \boldsymbol{Y}^{(0)} - \boldsymbol{B}'^{\mathrm{T}}(\mathscr{E} - d)^{-1}\boldsymbol{B}' \tag{A1.33}$$

where $\boldsymbol{B}' = \boldsymbol{G}^{\mathrm{T}}\boldsymbol{B}$. From (A1.33),

$$Y_{ii'} = Y^{(0)}_{ii'} - \sum_j B'_{ji}(\mathscr{E} - d_j)^{-1}B'_{ji'}. \tag{A1.34}$$

The inclusion of bound channels in (A1.3) does not alter the formal structure of QDT. Thus if bound channels are included and some free channels are closed it is necessary, as before, to eliminate the exponentially increasing solutions in order to construct the physical solutions. The effect of including bound channels is to produce rapid variations in $\boldsymbol{Y}$ as a function of $\mathscr{E}$. Diagonalising $\boldsymbol{Y}$,

$$\boldsymbol{YX} = \boldsymbol{X} \tan(\pi\bar{\eta})$$

$\bar{\eta}$ varies slowly with $\mathscr{E}$ when bound channels are omitted but has rapid variations when they are included.

The relation between resonances due to bound channels and isolated resonances due to free channels is discussed in § 7.5. The normalisation formulae of § 6.5.2 remain valid when bound channels are included. In the vicinity of the energy of a bound channel, $\mathscr{E} \simeq d_j$ in (A1.34), the quantity $C(E) = q(\mathrm{d}\boldsymbol{\mathscr{R}}/\mathrm{d}\mathscr{E})q$ in (6.49) becomes large; it is shown by Seaton (1982b) that this accounts for the contribution of the bound channels to the normalisation integral.

References

Amin S R, Caldwell C D and Lichten W 1981 *Phys. Rev. Lett.* **47** 1234

Armstrong J A, Esherick P and Wynne J J 1977 *Phys. Rev.* A **15** 180

Armstrong J A, Jha S S and Pandey K C 1981 *Phys. Rev.* A **23** 2761

Armstrong J A, Wynne J J and Esherick P 1979 *J. Opt. Soc. Am.* **69** 211

Armstrong J A, Wynne J J and Tomkins F S 1980 *J. Phys. B: Atom. Molec. Phys.* **13** L133

Atabek O, Dill D and Jungen Ch 1974 *Phys. Rev. Lett.* **33** 123

Atabek O and Jungen Ch 1976 *Electron and photon interactions with atoms* ed H Kleinpoppen and M R C McDowell (New York: Plenum) p 163

Aymar M, Camus P, Dieulin M and Morrillon C 1978 *Phys. Rev.* A **18** 2173

Aymar M, Champeau R-J, Delsart C and Keller J-C 1981a *J. Phys. B: Atom. Molec. Phys.* **14** 4489

Aymar M, Débarre A and Robaux O 1980 *J. Phys. B: Atom. Molec. Phys.* **13** 1089

Aymar and Robaux O 1979 *J. Phys. B: Atom. Molec. Phys.* **12** 531

Aymar M, Robaux O and Thomas C 1981b *J. Phys. B: Atom. Molec. Phys.* **14** 4225

Balmer J J 1885 *Verhandl. Naturf. Gessel. Basel* **7** 548

Bates D R and Damgaard A 1949 *Phil. Trans. R. Soc.* A **242** 101

Bates D R and Massey H S W 1943 *Phil. Trans. R. Soc.* A **239** 269

Beckel C, Hansen B and Peek J 1970 *J. Chem. Phys.* **53** 3681

Bell R 1979 *PhD Thesis* London

Bell R and Seaton M J 1983 *J. Phys. B: Atom. Molec. Phys.* to be submitted

Bely O 1966 *Proc. Phys. Soc.* **88** 833

Bely O, Moores D L and Seaton M J 1963 *Atomic collision processes* ed M R C McDowell (Amsterdam: North-Holland) p 304

Berkowitz J 1979 *Photo-absorption, Photo-ionization and Photo-electron Spectroscopy* (New York: Academic) pp 177, 181

Bohr N 1923 *Ann. Phys., Lpz.* **71** 262

Born H and Heisenberg W 1924 *Z. Phys.* **23** 388

Brown C M and Ginter M L 1977 *J. Opt. Soc. Am.* **67** 1323

Brown C M and Ginter M L 1978a *J. Opt. Soc. Am.* **68** 817
—— 1978b *J. Opt. Soc. Am.* **68** 1541
—— 1978c *J. Opt. Soc. Am.* **68** 243
—— 1980 *J. Opt. Soc. Am.* **70** 87
Brown C M, Tilford S G and Ginter M L 1977a *J. Opt. Soc. Am.* **67** 584
—— 1977b *J. Opt. Soc. Am.* **67** 607
—— 1977c *J. Opt. Soc. Am.* **67** 1240
Buchholz H 1953 *Die Konfluente Hypergeometrische Funktion* (Berlin: Springer-Verlag) (Engl. transl. 1969 (Berlin: Springer-Verlag))
Burgess A 1963 *Proc. Phys. Soc.* **81** 442
—— 1964 *Astrophys. J.* **139** 776
Burgess A and Seaton M J 1960 *Mon. Not. R. Astron. Soc.* **120** 121
Burke P G 1976 *Atomic Processes and Applications* ed P G Burke and B L Moiseiwitsch (Amsterdam: North-Holland) p199
Burke P G and McVicar D D 1965 *Proc. Phys. Soc.* **86** 989
Burke P G and Seaton M J 1971 *Meth. Comput. Phys.* **10** 1
Carter V L, Hudson R D and Breig E L 1971 *Phys. Rev.* A **4** 821
Chang E S and Fano U 1972 *Phys. Rev.* A **6** 173
Connerade J P, Mansfield M W D, Newsom G H, Tracy D H, Baig M A and Thimm K 1979 *Phil. Trans. R. Soc.* A **290** 327
Cooper J and Zare R N 1968 *J. Chem. Phys.* **48** 942
Davies P C W and Seaton M J 1969 *J. Phys. B: Atom. Molec. Phys.* **2** 757
Dehmer P M and Chupka W A 1976 *J. Chem. Phys.* **65** 2243
Delsart C, Keller J C and Thomas C 1981 *J. Phys. B: Atom. Molec. Phys.* **14** 3355, 4241
Dill D 1972 *Phys. Rev.* A **6** 160
—— 1973 *Phys. Rev.* A **7** 1976
Dill D, Chang E S and Fano U 1973 *Electronic and Atomic Collisions* ed C Cobic and M V Kurepa (Belgrade: Institute of Physics) p 536
Ditchburn R W and Hudson R D 1960 *Proc. R. Soc.* A **256** 53
Doughty N A, Seaton M J and Sheorey V B 1968 *J. Phys. B: Atom. Molec. Phys.* **1** 802
Dubau J 1973 *Thesis* University of London
—— 1978 *J. Phys. B: Atom. Molec. Phys.* **11** 4095
Dubau J and Seaton M J 1983 *J. Phys. B: Atom. Molec. Phys.* to be submitted
Dubau J and Wells J 1973 *J. Phys. B: Atom. Molec. Phys.* **6** 1452
Dunning F B and Stebbings R F 1974 *Phys. Rev.* A **9** 2378
Eissner W, Nussbaumer H, Saraph H E and Seaton M J 1969 *J. Phys. B: Atom. Molec. Phys.* **2** 341
Eissner W and Seaton M J 1972 *J. Phys. B: Atom. Molec. Phys.* **5** 2187
—— 1974 *J. Phys. B: Atom. Molec. Phys.* **7** 2533
Erdélyi A, Magnus W, Oberhettinger F and Tricomi F G 1953 *Higher Transcendental Functions* vol 1 (New York: McGraw-Hill)
Esherick P 1977 *Phys. Rev.* A **15** 1920
Esherick P, Armstrong J A, Dreyfus R W and Wynne J J 1976 *Phys. Rev. Lett.* **36** 1296
Esteva J M 1974 *Thesis* CNRS **A 0** 9976
Fano U 1961 *Phys. Rev.* **124** 1866
—— 1970 *Phys. Rev.* A **2** 353
—— 1975 *J. Opt. Soc. Am.* **65** 979
—— 1978 *Phys. Rev.* A **17** 93
Fano U and Dill D 1972 *Phys. Rev.* A **6** 185
Fano U and Lee C-M 1973 *Phys. Rev. Lett.* **31** 1573
Fermi E and Segré E 1933 *Z. Phys.* **82** 729
Foldy L L 1958 *Phys. Rev.* **111** 1093
Friedrich H and Trefftz E 1969 *J. Quant. Spectrosc. Radiat. Transfer* **9** 333
Froese Fischer C and Hansen J E 1981 *Phys. Rev.* A **24** 631
Gailitis M 1963 *Sov. Phys.–JETP* **17** 1328
Garton W R S and Codling K 1965 *Proc. Phys. Soc.* **86** 1067
Geiger J 1977 *Z. Phys.* A **282** 129
—— 1979 *J. Phys. B: Atom. Molec. Phys.* **12** 2277
Giusti A 1980 *J. Phys. B: Atom. Molec. Phys.* **13** 3867

Giusti-Suzor A 1982 *Physics of Electronic and Atomic Collisions* ed S Datz (Amsterdam: North-Holland) p 381
Greene C, Fano U and Strinati G 1979 *Phys. Rev.* A **19** 1485
Green C H, Rau A R P and Fano U 1982 *Phys. Rev.* A in press
Ham F S 1955 *Solid St. Phys.* **1** 217
Hartree D R 1928 *Proc. Camb. Phil. Soc.* **24** 426
Heinzmann U 1980 *J. Phys. B: Atom. Molec. Phys.* **13** 4353, 4367
Heinzmann U, Schäfers F, Thimm K, Wolcke A and Kessler J 1979 *J. Phys. B: Atom. Molec. Phys.* **12** L679
Herzberg G 1950 *Molecular Spectra and Molecular Structure I. Spectra of Di-atomic Molecules* (New York: Van Nostrand) 2nd edn
—— 1969 *Phys. Rev. Lett.* **23** 1081
Herzberg G and Jungen Ch 1972 *J. Molec. Spectrosc.* **41** 425
Jaffé C and Reinhardt R P 1977 *J. Chem. Phys.* **66** 1285
Johnson W R and Cheng K T 1979 *J. Phys. B: Atom. Molec. Phys.* **12** 863
Johnson W R, Cheng K T, Huang K N and Le Dourneuf M 1980 *Phys. Rev.* A **22** 989
Johnson W R and Le Dourneuf M 1980 *J. Phys. B: Atom. Molec. Phys.* **13** L13
Jungen Ch 1982 *Physics of Electronic and Atomic Collisions* ed S Datz (Amsterdam: North-Holland) p 455
Jungen Ch and Atabek O 1977 *J. Chem. Phys.* **66** 5584
Jungen Ch and Dill D 1980 *J. Chem. Phys.* **73** 1
Jungen Ch and Giusti-Suzor A 1982 *Recent developments in electron–atom and electron–molecule collision processes* ed W Eissner *Daresbury Laboratory Rep.* DL/SC/R18
Jungen Ch and Raoult M 1981 *Faraday Symp.* **71** 253
Kaufman V and Humphreys C J 1969 *J. Opt. Soc. Am.* **59** 1614
Kołos W and Wolniewicz L 1965 *J. Chem. Phys.* **43** 2429
—— 1968 *J. Chem. Phys.* **48** 3672
Kuhn T S and Van Vleck J A 1950 *Phys. Rev.* **79** 382
Lamb W E, Mader D L and Wing W H 1973 *Proc. Estfahan Symp. on Fundamental and Applied Laser Physics* (New York: Wiley) p 523
Langer R M 1930 *Phys. Rev.* **35** 649
Le Dourneuf M, Vo Ky Lan, Burke P G and Taylor K T 1975 *J. Phys. B: Atom. Molec. Phys.* **8** 2640
Lee C M 1974a *Phys. Rev.* A **10** 584
—— 1974b *Phys. Rev.* A **10** 1598
Lee C M and Johnson W R 1980 *Phys. Rev.* A **22** 979
Lee C M and Lu K T 1973 *Phys. Rev.* A **8** 1241
Lipsky L 1967 *Proc. 5th Int. Conf. on Physics of Electronic and Atomic Collisions* (Leningrad: Nauka) Abstracts, p 617
Lu K T 1971 *Phys. Rev.* A **4** 579
—— 1974 *J. Opt. Soc. Am.* **64** 706
—— 1977 *Proc. R. Soc.* A **353** 431
Lu K T and Fano U 1970 *Phys. Rev.* A **2** 81
Lu K T and Mansfield M W D 1975 *Electron and photon interactions with atoms* ed H Kleinpoppen and M R C McDowell (New York: Plenum) p 627
Mansfield M W D and Newsom G H 1977 *Proc. R. Soc.* A **357** 77
—— 1981 *Proc. R. Soc.* A **377** 431
Martins P de A P and Seaton M J 1969 *J. Phys. B: Atom. Molec. Phys.* **2** 333
Massey H S W and Bates D R 1942 *Rep. Prog. Phys.* **9** 62
Mehlman-Ballofet G and Esteva J M 1969 *Astrophys. J.* **157** 945
Mendoza C 1979 *PhD Thesis* University of London
—— 1981a *J. Phys. B: Atom. Molec. Phys.* **14** 397
—— 1981b *J. Phys. B: Atom. Molec. Phys.* **14** 2465
Mies F H 1979 *Phys. Rev.* A **20** 1773
Moores D L 1966 *Proc. Phys. Soc.* **88** 843
—— 1967 *Proc. Phys. Soc.* **91** 830
Newsom G H 1971 *Astrophys. J.* **166** 243
Norcross D W and Seaton M J 1970 *J. Phys. B: Atom. Molec. Phys.* **3** 579
—— 1976 *J. Phys. B: Atom. Molec. Phys.* **9** 2983
Pandey K C, Jha S S and Armstrong J A 1980 *Phys. Rev. Lett.* **44** 1583
Percival I C 1960 *Proc. Phys. Soc.* **76** 206

Pradhan A K 1981 *Phys. Rev. Lett.* **47** 79
Presnyakov L P and Urnov A M 1975 *J. Phys. B: Atom. Molec. Phys.* **8** 1280
Raoult M and Jungen Ch 1981 *J. Chem. Phys.* **74** 3388
Raoult M, Jungen Ch and Dill D 1980 *J. Chem. Phys.* **77** 599
Risberg P 1956 *Ark. Fys.* **10** 583
Robaux O and Aymar M 1982 *Comp. Phys. Commun.* **25** 223
Rydberg J R 1889 *K. Svenska Vetensk. Akad. Handl.* **23** 11
Samson J A R and Gardner J L 1973 *Phys. Rev. Lett.* **31** 1327
Saraph H E and Seaton M J 1971 *Phil. Trans. R. Soc.* A **271** 1
Saraph H E, Seaton M J and Shemming J 1969 *Phil. Trans. R. Soc.* A **264** 77
Seaton M J 1955 *C. R. Acad. Sci., Paris* **240** 1317
—— 1958a *Mon. Not. R. Astron. Soc.* **118** 504
—— 1958b *Rev. Mod. Phys.* **30** 992
—— 1966a *Proc. Phys. Soc.* **88** 801
—— 1966b *Proc. Phys. Soc.* **88** 815
—— 1969 *J. Phys. B: Atom. Molec. Phys.* **2** 5
—— 1976 *J. Phys. B: Atom. Molec. Phys.* **9** 3001
—— 1978 *J. Phys. B: Atom. Molec. Phys.* **11** 4067
—— 1982a *Comp. Phys. Commun.* **25** 87
—— 1982b *J. Phys. B: Atom. Molec. Phys.* **15** 3899
Seaton M J and Peach G 1962 *Proc. Phys. Soc.* **79** 1296
Seaton M J and Storey P J 1976 *Atomic Processes and Applications* ed P G Burke and B L Moiseiwitsch (Amsterdam: North-Holland) p 133
Shenstone A G 1931 *Phys. Rev.* **38** 873
Shenstone A G and Russell H N 1932 *Phys. Rev.* **39** 415
Sheorey V B 1969 *J. Phys. B: Atom. Molec. Phys.* **2** 442
Slater L J 1960 *Confluent Hypergeometric Functions* (Cambridge: Cambridge University Press)
Sloan I H 1964 *Proc. R. Soc.* A **281** 151
Sommerfeld A 1916 *Ann. Phys., Lpz.* **51** 1
—— 1920 *Atombau und Spectrallinien* (Engl. trans. 1931 *Atomic Structure and Spectral Lines* (London: Methuen) 5th edn)
Starace A F 1973 *J. Phys. B: Atom. Molec. Phys.* **6** 76
Takezawa S 1970 *J. Chem. Phys.* **52** 2575, 5793
Waller I 1926 *Z. Phys.* **38** 635
Wells J 1974 *PhD Thesis* London
Whittaker E T and Watson G N 1927 *A Course of Modern Analysis* (Cambridge: Cambridge University Press)4th edn
Wind H 1965 *J. Chem. Phys.* **42** 2371
Wynne J J, Armstrong J A and Esherick P 1977 *Phys. Rev. Lett.* **39** 1520
Wynne J J and Armstrong J A 1979a *IBM J. Res. Dev.* **23** 490
—— 1979b *Comm. Atom. Molec. Phys.* **8** 155
Yoshino K and Tanaka Y 1979 *J. Opt. Soc. Am.* **69** 159

1985 *J. Phys. B: At. Mol. Phys.* **18** 1589–629

Dielectronic recombination: I. General theory

R H Bell† and M J Seaton

Department of Physics and Astronomy, University College London, Gower Street, London WC1E 6BT, England

Received 10 August 1984, in final form 10 December 1984

Abstract. In the process of dielectronic recombination (DR), radiative capture of an electron by a positive ion occurs via compound resonance states of the electron + ion system (states with two or more electrons excited). These states can decay by autoionisation, with probability A, or by radiative stabilisation, with probability R. In general, these are competing processes: one may have $A>R$ or $A<R$.

It is assumed that wavefunctions have been obtained for the electron + ion system, allowing for resonances but neglecting interactions with the radiation field. The interactions are then handled using the general formulation of the problem given by Davies and Seaton. A scattering matrix $\mathscr{S}$ is calculated allowing for radiative channels. It has submatrices $\mathscr{S}_{ee}$, which describes electron-electron scattering allowing for radiative decays, and $\mathscr{S}_{pe}$, which describes photon emission following electron capture. Using the unitarity of $\mathscr{S}$, the total DR rate can be expressed in terms of $\mathscr{S}_{ee}$.

The first case to be considered is that for which the electron + ion wavefunction can be described in terms of a set of open channels and a set of square integrable functions ('bound channels'). Formulae are obtained, of Breit–Wigner type, which are similar to but more general than those given in a number of previous papers.

The main new results are for resonances in Rydberg series converging to excited states of the recombining ion. The energies of the resonances are $E_n=-\frac{1}{2}z^2/(n-\mu)^2$ relative to the excited-ion state (z is the ion's charge, n an integer, μ a quantum defect and atomic units are used) and the resonance separations are $\delta E_n=(E_n-E_{n-1})=z^2/(n-\mu)^3$. The radiative transition occurs in the ion with probability R independent of n. For smaller values of n one has $A_n \gg R$ but A_n is proportional to δE_n and, as n increases, eventually becomes smaller than R. For very large n the separations δE_n become small compared with the radiative width and all resonance structure is wiped out.

Formulae derived from *ab initio* theory are compared with those deduced previously using intuitive arguments. There are some differences between our formulae and those used by Burgess but little difference in calculated total DR rates for plasma conditions.

1. Introduction

An electron can be captured by a positive ion into a doubly-excited resonance state,

$$X^+ + e \rightarrow X^{**} \tag{1.1}$$

and capture can be followed by the inverse process of autoionisation,

$$X^{**} \rightarrow X^+ + e \tag{1.2}$$

† Present address: 50 Fairfield Crescent, Edgware, Middlesex HA8 9AH, England.

0022-3700/85/081589+41$02.25

or by a stabilising radiative transition to a true bound state,

$$X^{**} \rightarrow X^{*} + h\nu \tag{1.3}$$

in which case dielectronic recombination (DR) is said to have occurred.

DR was first discussed by Massey and Bates (1942) and Bates and Massey (1943). Their concern was with recombination in plasmas at fairly low electron temperatures (such as occur in the earth's ionosphere) and they therefore considered low-lying doubly-excited states, with energies $E(X^{**})$ just above the energy $E(X^{+})$ of the ion ground state. Recent work has shown that DR involving such states is of importance in various astronomical objects (Harrington *et al* 1981, Stickland *et al* 1981, Clavel *et al* 1981, Storey 1981). For plasmas at higher electron temperatures (such as occur in the solar corona) Burgess (1964) showed that DR can be particularly important because it can take place *via* complete Rydberg series of resonances converging to the energy $E(X^{+*})$ of an excited state of the recombining ion.

Reviews of the theory of DR and of the calculation of DR rates of importance for laboratory and astronomical plasmas have been published by Seaton and Storey (1976) and by Dubau and Volonté (1980) and recent interest in the process has been heightened by experimental work in several groups (Mitchell *et al* 1983, Belić *et al* 1983, Dittner *et al* 1983). Presnyakov and Urnov (1974) have shown that the DR process can lead to significant reductions in rate coefficients for collisional excitation of positive ions, particularly for more highly ionised systems, and this is discussed further by Pradhan (1981). Expressions for the calculation of DR rates given by Bates and Massey (1943) and Burgess (1964) were obtained using intuitive arguments and although a number of attempts have been made to provide more rigorous foundations for the theory (Shore 1967, Trefftz 1967, 1969, 1970, Davies and Seaton 1969, Seaton and Storey 1976, Armstrong *et al* 1978, Rozman 1981, Hickman 1984) a completely satisfactory solution to the problem of deriving all of the required expressions from first principles of quantum mechanics has proved to be somewhat elusive. Davies and Seaton (1969) gave a rather general formulation, essentially a generalisation of radiation damping theory (Heitler 1954) but did not discuss detailed applications. Seaton and Storey encountered difficulties in attempts to use the theory of Davies and Seaton for the case of Rydberg series of resonances and it was thought by one of us (Seaton 1983a) that the theory was in error; subsequent work, to be discussed in the present paper, has shown that not to be the case.

The present paper gives a theory of DR which, we believe, overcomes the difficulties encountered in earlier work, and later papers in the present series discuss further applications of the theory. We make extensive use of formulae from quantum defect theory (QDT), which is reviewed by Seaton (1983a, hereafter referred to as S83). The contents of the present paper are summarised as follows. In § 2 we discuss structures of wavefunctions for electron–ion collisions neglecting interactions with the radiation field, and particularly those structures which describe the capture and autoionisation processes (1.1) and (1.2); § 3 summarises the theory used for interaction with the radiation field, essentially the theory of Davies and Seaton; § 4 describes some simpler illustrative applications (one open channel and either no resonances or just one resonance); and § 5 gives the theory for a finite number of resonances, which may be overlapping. The theory for infinite series of resonances, forming Rydberg series, becomes more intricate; in § 6 we consider the simplest such case, of one open channel and one closed channel giving a single Rydberg series of resonances; and in § 7 we consider many Rydberg series all converging to the same energy level of the ion core.

Further complications which arise when the ion core has fine structure will be discussed in a later paper of the present series.

Although some of our derivations become quite involved, our final results are fairly simple. Section 8 gives a summary of results and comparisons with those which have been obtained earlier using intuitive arguments (readers who are not interested in the details of derivation of formulae may prefer to turn directly to § 8). Although our results differ from the earlier ones, the differences are not such as to give important differences in final values for DR rate coefficients.

2. Structure of wavefunctions for the electron–ion collision problem

2.1. General form of the wavefunction

We use Hartree atomic units, $e = m = \hbar = 1$.

For the general formulation of the theory we assume that the Schrödinger equation for the electron–ion collision problem

$$(H - E)\Psi = 0 \tag{2.1}$$

has been solved exactly, neglecting interactions with the radiation field. We assume, further, that a good approximation to the structure of the solutions is obtained using an expansion similar to that employed in close-coupling theory with inclusion of correlation functions:

$$\Psi_{\alpha'}(\boldsymbol{R}, \boldsymbol{r}) = \sum_{\alpha} \psi_{\alpha}(\boldsymbol{R}, \hat{\boldsymbol{r}}) \frac{1}{r} F_{\alpha\alpha'}(r) + \sum_{j} \Phi_j(\boldsymbol{R}, \boldsymbol{r}) C_{j\alpha'} \tag{2.2}$$

where we use $\boldsymbol{R}$ for the coordinates of all electrons in the ion and $\boldsymbol{r}$ for those of the colliding electron. In (2.2), $\psi_{\alpha}(\boldsymbol{R}, \boldsymbol{r})$ is a vector-coupled product of a wavefunction for the ion times a function of the angular variables for the colliding electron. The $F_{\alpha\alpha'}(r)$ are radial functions for the colliding electron: α specifies a *collision channel* and α' a boundary condition. We use $\boldsymbol{F}(r)$ for the matrix with elements $F_{\alpha\alpha'}(r)$. The $\Phi_j(\boldsymbol{R}, \boldsymbol{r})$ are functions of bound-state type for the whole system (correlation functions) but should not be confused with true bound states, which are denoted by Ψ_{β}. We are mainly interested in functions Φ_j which represent states with two electrons excited. We refer to the Φ_j as *bound channels.* We use $\mathbf{C}$ for the matrix with elements $C_{j\alpha'}$.

In practice one must include electron spin and use antisymmetrised expansions. In applying the theory we use truncated antisymmetrised expansions of a form similar to (2.2) and calculate the radial functions $\boldsymbol{F}(r)$ and coefficients $\boldsymbol{C}$ on solving the usual coupled integro-differential equations of close-coupling theory. This gives a good approximation to the solutions of (2.1). In discussing the general theory of DR we find it convenient to use the unsymmetrised expansion (2.2); all results obtained remain valid when antisymmetrised functions are used.

The total energy is E and we put

$$E = E_{\alpha} + \tfrac{1}{2}k_{\alpha}^2 \tag{2.3}$$

where E_{α} is the energy of the ion state in α. We have $k_{\alpha}^2 > 0$ for *open channels* and $k_{\alpha}^2 < 0$ for *closed channels.* For the latter we put

$$\tfrac{1}{2}k_{\alpha}^2 = -\tfrac{1}{2}z^2/\nu_{\alpha}^2 \tag{2.4}$$

where z is the charge on the ion and ν_{α} is an *effective quantum number.*

We assume that, for $r > r_0$ with r_0 finite, the radial functions $\boldsymbol{F}(r)$ may be equated to linear combinations of Coulomb functions. Regular and irregular Coulomb functions commonly used in QDT (see S83) are $s_\alpha(r) = s(\varepsilon_\alpha, l_\alpha; zr)$ and $c_\alpha(r) = c(\varepsilon_\alpha, l_\alpha; zr)$ where $\varepsilon_\alpha = k_\alpha^2/z^2$ is the electron energy in z-scaled Rydberg units. In the present paper we use Coulomb functions $\phi_\alpha^+(r)$ and $\phi_\alpha^-(r)$ defined by

$$\phi_\alpha^\pm(r) = (c_\alpha(r) \pm \mathrm{i}\, s_\alpha(r))(2z)^{-1/2}. \tag{2.5}$$

These differ from the functions $\phi_\alpha^\pm$ used in S83 by a factor of $(2z)^{-1/2}$; this difference is a consequence of our use of Hartree energy units.

We consider mathematical solutions for the collision problem, of a type commonly used in QDT. For the mathematical solutions the radial functions $F_{\alpha\alpha'}$ are taken to be such that

$$F_{\alpha\alpha'}(r) = \phi_\alpha^-(r)\delta_{\alpha\alpha'} - \phi_\alpha^+(r)\chi_{\alpha\alpha'} \qquad (r \geqslant r_0) \tag{2.6}$$

or, more compactly,

$$\mathbf{F} = \boldsymbol{\phi}^- - \boldsymbol{\phi}^+\boldsymbol{\chi} \qquad (r \geqslant r_0) \tag{2.7}$$

where $\boldsymbol{\phi}^-$, $\boldsymbol{\phi}^+$ are diagonal matrices with diagonal elements ϕ_α^-, ϕ_α^+. When standard phase conventions are adopted, the matrix $\boldsymbol{\chi}$ is such that

$$\boldsymbol{\chi} = \boldsymbol{\chi}^{\mathrm{T}} \qquad \boldsymbol{\chi}^*\boldsymbol{\chi} = 1 \tag{2.8}$$

where superscript T denotes transpose.

For open channels the functions $\phi^\pm$ have asymptotic forms

$$\phi^\pm(r) \underset{r\to\infty}{\sim} (2\pi k)^{-1/2} \exp(\pm \mathrm{i}\zeta) \tag{2.9}$$

where

$$\zeta = kr - \tfrac{1}{2}l\pi + (z/k)\ln(2kr) + \arg\Gamma(l + 1 - \mathrm{i}z/k). \tag{2.10}$$

When all channels are open, $\boldsymbol{\chi}$ is equal to the scattering matrix $\mathbf{S}$ and the functions (2.2) are normalised to

$$(\Psi_\alpha(E)|\Psi_{\alpha'}(E')) = \delta_{\alpha\alpha'}\delta(E - E'). \tag{2.11}$$

The functions $F_{\alpha\alpha'}$ for closed channels are, in general, exponentially increasing in the limit of r large. The elimination of exponentially increasing solutions for the case of some channels closed will be discussed in § 2.3.

2.2. Bound channels

We are concerned with the resonance structure in the wavefunction for the electron–ion collision problem. In the present sub-section we consider structures which result from inclusion of bound-channel functions Φ_j in (2.2) and in § 2.3 we consider structures due to inclusion of closed collision channels.

When bound channels are omitted in (2.2), $\boldsymbol{\chi}$ is a slowly varying function of E. The effect of including bound channels is discussed by Seaton (1982, 1983b) and in appendix 1 of the present paper where the following results are obtained:

$$\chi_{\alpha\alpha'}(E) = X_{\alpha\alpha'} - \mathrm{i}\sum_{jj'} Y_{\alpha j}[(E - \mathbf{Z})^{-1}]_{jj'} Y_{\alpha'j'} \tag{2.12}$$

and

$$C_{j\alpha'}(E) = -\mathrm{i}(2\pi)^{-1/2}\sum_{j'}[(E - \mathbf{Z})^{-1}]_{jj'} Y_{\alpha'j'} \tag{2.13}$$

or, more compactly

$$\boldsymbol{\chi}(E)=\mathbf{X}-\mathrm{i}\mathbf{Y}(E-\mathbf{Z})^{-1}\mathbf{Y}^{\mathrm{T}} \tag{2.14}$$

$$\mathbf{C}(E)=-\mathrm{i}(2\pi)^{-1/2}(E-\mathbf{Z})^{-1}\mathbf{Y}^{\mathrm{T}}. \tag{2.15}$$

The matrices $\mathbf{X}$, $\mathbf{Y}$ and $\mathbf{Z}$ vary slowly with E and are such that

$$\mathbf{X}=\mathbf{X}^{\mathrm{T}} \qquad \mathbf{X}^*\mathbf{X}=1 \tag{2.16}$$

$$\mathbf{Z}=\mathbf{Z}^{\mathrm{T}} \tag{2.17}$$

$$\mathbf{Y}^{\dagger}\mathbf{Y}=\mathrm{i}(\mathbf{Z}-\mathbf{Z}^*) \tag{2.18}$$

$$\mathbf{Y}^{\dagger}\mathbf{X}=\mathbf{Y}^{\mathrm{T}}. \tag{2.19}$$

For the case of only one bound channel we may put

$$Z=E_0-\tfrac{1}{2}\mathrm{i}A \tag{2.20}$$

where E_0 and A are real. If all collision channels are open we have $\mathbf{S}=\boldsymbol{\chi}$ and $\mathbf{S}$ has a pole at $E=E_0-\frac{1}{2}\mathrm{i}A$. We interpret E_0 as the position of a resonance and A as its autoionisation probability. It follows from (2.18) that

$$\sum_{\alpha}|Y_{\alpha}|^2=A \tag{2.21}$$

and hence that $A>0$. Putting

$$A_{\alpha}=|Y_{\alpha}|^2 \qquad \text{giving} \qquad A=\sum_{\alpha}A_{\alpha} \tag{2.22}$$

we may interpret A_{α} as the probability of autoionisation to channel α. For the case of many closed channels and all collision channels open, the positions of resonances and their autoionisation probabilities are obtained on diagonalising $\mathbf{Z}$.

The treatment considered here differs from that of Trefftz (1970) in that she uses methods of perturbation theory to calculate the interactions between collision channels and bound channels.

2.3. *Closed channels*

When some collision channels are closed we use formulae from QDT. Let α' be an open channel and let the radial functions $G_{\alpha\alpha'}(r)$ have $\mathbf{S}$-matrix normalisation,

$$G_{\alpha\alpha'}(r) \underset{r\to\infty}{\sim} \begin{cases} \phi_{\alpha}^{-}\delta_{\alpha\alpha'}-\phi_{\alpha}^{+}S_{\alpha\alpha'} & \text{for } \alpha \text{ open} \\ 0 & \text{for } \alpha \text{ closed} \end{cases} \tag{2.23}$$

(we recall that $\mathbf{S}=\boldsymbol{\chi}$ when all channels are open). We partition G into open–open and closed–open sub-matrices,

$$\mathbf{G}=\begin{pmatrix}\mathbf{G}_{\mathrm{oo}}\\ \mathbf{G}_{\mathrm{co}}\end{pmatrix} \tag{2.24}$$

and may then write (2.23) as

$$\begin{aligned} G_{\mathrm{oo}}&\sim\phi^{-}-\phi^{+}S \\ G_{\mathrm{co}}&\sim 0. \end{aligned} \tag{2.25}$$

Partitioning χ into open-open, open-closed, closed-open and closed-closed sub-matrices

$$\chi = \begin{pmatrix} \chi_{oo} & \chi_{oc} \\ \chi_{co} & \chi_{cc} \end{pmatrix} \tag{2.26}$$

two results from QDT are

$$\mathbf{S} = \chi_{oo} - \chi_{oc}[\chi_{cc} - \exp(-2\pi i\nu)]^{-1}\chi_{co} \tag{2.27}$$

and

$$\mathbf{G}_{co}(r) = P(r)\mathrm{i}(\nu^3/z^2)^{1/2}\exp(-\mathrm{i}\pi\nu)[\chi_{cc} - \exp(-2\pi\mathrm{i}\nu)]^{-1}\chi_{co}. \tag{2.28}$$

In these equations ν is the diagonal matrix with diagonal elements ν_γ, defined by (2.4) (for closed channels we use γ in place of α for the channel index) and $P(r)$ is a diagonal matrix with diagonal elements $P_\gamma(r) = P(\nu_\gamma, l_\gamma; r)$ where

$$P(\nu, l; r) = -(-1)^l z^{1/2} K(\nu, l)\, W_{\nu, l+\frac{1}{2}}(2zr/\nu) \tag{2.29}$$

$$K(\nu, l) = [\nu^2\Gamma(\nu + l + 1)\Gamma(\nu - l)]^{-1/2} \tag{2.30}$$

and $W_{\nu,l+\frac{1}{2}}(2zr/\nu)$ is a Whittaker function, with asymptotic form $(2zr/\nu)^\nu \exp(-zr/\nu)$. For $\nu = n$ where n is an integer and $n \geqslant (l+1)$, $P(n, l; r)$ defined by (2.29) is equal to the hydrogenic radial function, normalised to unity, usually denoted by $P_{nl}(r)$.

If $P(\nu, l; r)$ is defined by (2.29) for all values of r, for $\nu \neq n$ it is divergent at the origin, $r = 0$. We adopt a different definition: that $P(r)$ is defined by (2.28) for all values of r. Equation (2.29) is then satisfied for $r \geqslant r_0$. The normalisation of the functions $P(\nu, l; r)$ is discussed by Dubau (1973), independently by Bhatti *et al* (1981), and in appendix 2 of the present paper. Using the notation $P(\nu, l; r) = P_\nu$ the normalisation condition is

$$(P_\nu | P_{\nu'}) = \frac{\sin[\pi(\nu - \nu')]}{\pi(\nu - \nu')}. \tag{2.31}$$

A completeness relation,

$$\sum_n (P_\nu | P_{n-b})(P_{n-b} | P_{\nu'}) = (P_\nu | P_{\nu'}) \tag{2.32}$$

is obtained in appendix 3.

The functions $\exp(-2\pi i\nu)$ in (2.27), (2.28) give rise to rapid variations in $\mathbf{S}$ and in $\mathbf{G}$ as functions of E. Poles occur to complex energies such that

$$\det[\chi_{cc} - \exp(-2\pi i\nu)] = 0. \tag{2.33}$$

2.4. Summary on resonance structures

Inclusion of bound channels in (2.2) gives rise to resonance structure in the matrix χ. In practice we consider such structures only for the case of all collision channels open, giving $\mathbf{S} = \chi$. Resonances arising from inclusion of bound channels will be referred to as *bound-channel* resonances. Within any finite range of energy the number of such resonances is finite.

Above a new threshold a closed channel becomes an open channel. As a new threshold is approached from below, one has $\nu \to \infty$ in (2.27) and hence an infinite series of resonances which will be referred to as *Rydberg resonances.* We shall be concerned with resonances high in Rydberg series, in the region immediately below a new threshold. In such a region we may neglect structures in χ which result from inclusion of bound channels.

2.5. *Rates for collisional excitation*

We quote some standard results from collision theory. The element $S_{\alpha\alpha'}$ of $\mathbf{S}$ gives an amplitude for the transition

$$(X^+ + e)_{\alpha'} \rightarrow (X^+ + e)_\alpha \tag{2.34}$$

with entrance channel α' and exit channel α. The dimensionless probability for the transition is

$$\mathscr{P}_{\alpha\alpha'} = |S_{\alpha\alpha'}|^2 \tag{2.35}$$

and the partial cross section is

$$Q(\alpha' \rightarrow \alpha) = \pi \mathscr{P}_{\alpha\alpha'} / k_{\alpha'}^2 \tag{2.36}$$

where, in any units, $k_\alpha = mv_\alpha/\hbar$ (v_α is the speed of the incident electron and the electron wavenumber k_α has dimensions of reciprocal length).

In order to define the total cross section it is necessary to consider sums over degenerate states. For the case of LS coupling we put

$$\alpha = (aS_aL_al_\alpha SLM_SM_L) \tag{2.37}$$

where a specifies a level of the ion core; S_aL_a are the angular momentum quantum numbers for level a; l_α is the orbital angular momentum of the colliding electron; and SLM_SM_L are the total angular momenta for the whole system. The scattering matrix is diagonal in SLM_SM_L and independent of M_SM_L. The partial collision strength is

$$\Omega_{SL}(al_\alpha, a'l_{\alpha'}) = \tfrac{1}{2}(2S+1)(2L+1)\mathscr{P}_{\alpha\alpha'} \tag{2.38}$$

and the total collision strength is

$$\Omega(a, a') = \sum_{SL} \sum_{l_\alpha l_{\alpha'}} \Omega_{SL}(al_\alpha, a'l_{\alpha'}). \tag{2.39}$$

The total cross section is

$$Q(a' \rightarrow a) = \pi\Omega(a, a')/(g_{a'}k_{a'}^2) \tag{2.40}$$

where $g_a = (2S_a+1)(2L_a+1)$ is the statistical weight for level a of the ion.

3. Interaction with the radiation field

3.1. *The time-dependent equations*

We use a formulation similar to that of Davies and Seaton (1969).

The electron–ion system has a Hamiltonian H, continuum-state functions $\Psi_\alpha(E)$ and bound-state functions Ψ_β; these functions are solutions of

$$(H-E)\Psi_\alpha(E) = 0 \qquad (H-E_\beta)\Psi_\beta = 0. \tag{3.1}$$

The functions $\Psi_\alpha(E)$ have $\mathbf{S}$ matrix normalisation, as discussed in § 2.3 and satisfy the normalisation condition (2.11).

The radiation field has states $\chi(0)$ with no photon and states $\chi(1, \mu, \omega)$ with one photon having polarisation μ and angular frequency ω.

We consider transitions between states

$$\Psi_\alpha(0, E, t) = \chi(0)\Psi_\alpha(E) \exp(-\mathrm{i}Et) \tag{3.2}$$

with no photon and one electron in the continuum, and states

$$\Psi_{\beta\mu}(1, \omega, t) = \chi(1, \mu, \omega)\Psi_\beta \exp(-\mathrm{i}E_\beta t) \tag{3.3}$$

with one photon and all electrons bound. The time-dependent state is

$$\Psi(t) = \sum_\alpha \int \mathrm{d}E\, \Psi_\alpha(0, E, t) f_\alpha(E, t) + \sum_{\beta\mu} \int \mathrm{d}\omega\, \Psi_{\beta\mu}(1, \omega, t) g_{\beta\mu}(\omega, t). \tag{3.4}$$

The interaction Hamiltonian is

$$H_{\mathrm{int}} = \sum_\mu \int \mathrm{d}\omega [D_\mu^* \exp(\mathrm{i}\omega t) C(\mu, \omega) + D_\mu \exp(-\mathrm{i}\omega t) A(\mu, \omega)] \tag{3.5}$$

where $C(\mu, \omega)$ is a creation operator,

$$C(\mu, \omega)\chi(0) = \chi(1, \mu, \omega) \tag{3.6}$$

and $A(\mu, \omega)$ an annihilation operator,

$$A(\mu, \omega)\chi(1, \mu', \omega') = \delta_{\mu\mu'}\delta(\omega - \omega')\chi(0). \tag{3.7}$$

We neglect terms

$$C(\mu, \omega)\chi(1, \mu', \omega') = \chi(2, \mu', \omega', \mu, \omega) \tag{3.8}$$

which lead to creation of a second photon. In (3.5), D_μ is the dipole operator:

$$D_\mu = \left(\frac{2\omega^3 \alpha(\mathrm{FSC})^3}{3\pi} \right)^{1/2} (R_\mu + r_\mu) \tag{3.9}$$

where $\alpha(\mathrm{FSC})$ is the fine-structure constant ($\alpha(\mathrm{FSC}) \simeq \frac{1}{137}$) and R_μ and r_μ are spherical-tensor components of $\mathbf{R}$ and $\mathbf{r}$. The usual density-of-states factors are included in the definition of D_μ.

We require the amplitudes $f_\alpha(E, t)$ and $g_{\beta\mu}(\omega, t)$ to be such that (3.4) is a solution of

$$(H + H_{\mathrm{int}})\Psi(t) = \mathrm{i} \frac{\partial}{\partial t} \Psi(t). \tag{3.10}$$

This gives the time-dependent equations

$$\frac{\mathrm{d}}{\mathrm{d}t} f_\alpha(E, t) = -\mathrm{i} \exp(\mathrm{i}Et) \sum_{\beta\mu} \int \mathrm{d}\omega\, D_{\alpha,\beta\mu}(E) \exp[-\mathrm{i}(E_\beta + \omega)t] g_{\beta\mu}(\omega, t) \tag{3.11}$$

$$\frac{\mathrm{d}}{\mathrm{d}t} g_{\beta\mu}(\omega, t) = -\mathrm{i} \exp[\mathrm{i}(E_\beta + \omega)t] \sum_\alpha \int \mathrm{d}E\, D_{\alpha,\beta\mu}^*(E) \exp(-\mathrm{i}Et) f_\alpha(E, t) \tag{3.12}$$

where

$$D_{\alpha,\beta\mu}(E) = (\Psi_\alpha(E)|D_\mu|\Psi_\beta). \tag{3.13}$$

We use an index b for (β, μ), take $\mathbf{D}(E)$ to be the matrix with elements (3.13) and $\boldsymbol{\Omega}$ to be the diagonal matrix with diagonal elements $(E_\beta + \omega)$. The equations (3.11),

(3.12) can then be written

$$\frac{\mathrm{d}}{\mathrm{d}t}\mathbf{f}(E, t) = -\mathrm{i}\exp(\mathrm{i}Et)\int \mathrm{d}\omega\, \mathbf{D}(E)\exp(-\mathrm{i}\Omega t)\mathbf{g}(\Omega, t) \tag{3.14}$$

$$\frac{\mathrm{d}}{\mathrm{d}t}\mathbf{g}(\Omega, t) = -\mathrm{i}\exp(\mathrm{i}\Omega T)\int \mathrm{d}E\, \mathbf{D}^{\dagger}(E)\exp(-\mathrm{i}Et)\mathbf{f}(E, t). \tag{3.15}$$

3.2. Solutions of the equations

We have considered two methods of solving the equations (3.14), (3.15). The first is that of Davies and Seaton (1969) who, using Laplace transforms, obtain expressions for **f** and **g** at $t=+\infty$ in terms of **f** and **g** at $t=0$. We have examined the method carefully and believe it to be correct. The point concerning the validity of an interchange of limits processes, which was questioned by Seaton (1983a), can be justified rigorously using methods described by Titchmarsh (1948). In the second method, which is described in detail by Bell (1979), we seek solutions which can be described as normal decay modes and which are such that

$$\frac{\mathrm{d}}{\mathrm{d}t}\mathbf{g} = \exp(\mathrm{i}\Omega t)\mathbf{J}\exp(-\lambda t) \qquad \frac{\mathrm{d}}{\mathrm{d}t}\mathbf{f} = \exp(\mathrm{i}Et)\mathbf{K}\exp(-\lambda t) \tag{3.16}$$

where λ is a diagonal matrix and where **J** and **K** are matrices which do not depend on t. We find that results obtained by the second method agree with those from the first. The first method, that of Davies and Seaton, gives more general solutions for the quantities of interest and is adopted here.

The slow variation of the factor $\omega^{3/2}$ in (3.9) is neglected (an equation, (3.20), is obtained later which gives the value of ω to be used in (3.9)). The initial conditions are taken to be such that no photon is present at time $t=0$, $\mathbf{g}(\Omega, 0)=0$, and that $\Psi(t=0)$ is a state with an incoming wavepacket in one entrance channel, say $\alpha=\alpha'$, and no outgoing wavepackets. The form of the wavepacket is determined by $f(E, 0)$. The packet is taken to be centred at $r=r_1$ where r_1 is large compared with the dimensions of the atom and with the positional width of the packet. These conditions imply that no interactions have taken place at $t=0$.

The solutions for $\mathbf{f}(E,\infty)$ and $\mathbf{g}(\Omega,\infty)$ obtained by Davies and Seaton are

$$\mathbf{f}(E,\infty) = [1-2\pi^2\mathbf{D}(E)(1+\mathbf{L}(E))^{-1}\mathbf{D}^{\dagger}(E)]\mathbf{f}(E, 0) \tag{3.17}$$

$$\mathbf{g}(\Omega,\infty) = -2\pi\mathrm{i}(1+\mathbf{L}(\Omega))^{-1}\mathbf{D}^{\dagger}(\Omega)\mathbf{f}(\Omega, 0) \tag{3.18}$$

where

$$\mathbf{L}(E) = -\mathrm{i}\pi\int \mathrm{d}E'\frac{\mathbf{D}^{\dagger}(E')\mathbf{D}(E')}{(E'-E-\mathrm{i}\varepsilon)} \tag{3.19}$$

and where it is to be understood that one takes the limit $\varepsilon\to 0$ after evaluating the integral (3.19) with $\varepsilon>0$.

It is seen that $\mathbf{g}(\Omega,\infty)$ is proportional to $\mathbf{f}(\Omega, 0)$ and hence that $g_{\beta\mu}(\omega,\infty)$ is proportional to $f_{\alpha'}(E, 0)$; this implies that the energy conservation condition

$$E_\beta + \omega = E \tag{3.20}$$

is satisfied. With this condition understood, equation (3.18) can be written

$$\mathbf{g}(E,\infty) = -2\pi\mathrm{i}(1+\mathbf{L}(E))^{-1}\mathbf{D}^{\dagger}(E)\mathbf{f}(E, 0). \tag{3.21}$$

3.3. *The scattering matrix with inclusion of radiative channels*

We express the solutions for $\mathbf{f}$ and $\mathbf{g}$ in terms of a scattering matrix $\mathscr{S}$ with partitioning

$$\mathscr{S}=\begin{pmatrix}\mathscr{S}_{\mathrm{ee}} & \mathscr{S}_{\mathrm{ep}}\\ \mathscr{S}_{\mathrm{pe}} & \mathscr{S}_{\mathrm{pp}}\end{pmatrix} \tag{3.22}$$

where $\mathscr{S}_{\mathrm{ee}}$ is the matrix for electron–electron scattering allowing for radiative decays ($\mathscr{S}_{\mathrm{ee}}$ has elements $\mathscr{S}_{\alpha\alpha'}$ and $\mathscr{S}_{\mathrm{ee}}=\mathbf{S}$ if the decays are neglected); $\mathscr{S}_{\mathrm{pe}}$ is the matrix for electron capture with photon emission (the elements are $\mathscr{S}_{\beta\mu,\alpha'}$); $\mathscr{S}_{\mathrm{ep}}$ that for the inverse process of photoionisation; and $\mathscr{S}_{\mathrm{pp}}$ (which we do not consider further) that for photon–photon scattering.

When interactions with the radiation field are neglected we use functions $\mathbf{\Psi}(E)$ with unit amplitudes for incoming waves and amplitudes $\mathbf{S}$ for outgoing waves. When the interactions are included we use functions $\mathbf{\Psi}(t)$ which have amplitudes $\mathbf{f}(E,0)$ for the incoming electron waves at time $t=0$, and amplitudes $\mathbf{Sf}(E,\infty)$ for the outgoing electron waves at time $t=\infty$. The matrices $\mathscr{S}_{\mathrm{ee}}$ and $\mathscr{S}_{\mathrm{pe}}$ are defined by

$$\mathbf{Sf}(E,\infty)=\mathscr{S}_{\mathrm{ee}}\mathbf{f}(E,0) \tag{3.23}$$

and

$$\mathbf{g}(E,\infty)=\mathscr{S}_{\mathrm{pe}}\mathbf{f}(E,0). \tag{3.24}$$

Using (3.17), (3.21) we obtain

$$\mathscr{S}_{\mathrm{ee}}=\mathbf{S}[1-2\pi^2\mathbf{D}(1+\mathbf{L})^{-1}\mathbf{D}^\dagger] \tag{3.25}$$

and

$$\mathscr{S}_{\mathrm{pe}}=-2\pi\mathrm{i}(1+\mathbf{L})^{-1}\mathbf{D}^\dagger. \tag{3.26}$$

It follows from the definition (3.19) of $\mathbf{L}$ that

$$\mathbf{L}(E)+\mathbf{L}^\dagger(E)=2\pi\int \mathrm{d}E'\,\frac{\mathbf{D}^\dagger(E')\mathbf{D}(E')\varepsilon}{(E'-E)^2+\varepsilon^2}. \tag{3.27}$$

Since

$$\lim_{\varepsilon\to 0}\frac{\varepsilon}{(E'-E)^2+\varepsilon^2}=\pi\delta(E-E') \tag{3.28}$$

we obtain

$$\mathbf{L}+\mathbf{L}^\dagger=2\pi^2\mathbf{D}^\dagger\mathbf{D}. \tag{3.29}$$

Using (3.25) and the relation $\mathbf{S}^\dagger\mathbf{S}=1$ we obtain

$$\mathscr{S}^\dagger_{\mathrm{ee}}\mathscr{S}_{\mathrm{ee}}=1-2\pi^2\mathbf{D}(1+\mathbf{L}^\dagger)^{-1}[(1+\mathbf{L})+(1+\mathbf{L}^\dagger)-2\pi^2\mathbf{D}\mathbf{D}^\dagger](1+\mathbf{L})^{-1}\mathbf{D}^\dagger \tag{3.30}$$

and hence, using (3.29)

$$\mathscr{S}^\dagger_{\mathrm{ee}}\mathscr{S}_{\mathrm{ee}}=1-4\pi^2\mathbf{D}(1+\mathbf{L}^\dagger)^{-1}(1+\mathbf{L})^{-1}\mathbf{D}^\dagger. \tag{3.31}$$

Comparison with (3.26) gives the conservation condition

$$\mathscr{S}^\dagger_{\mathrm{ee}}\mathscr{S}_{\mathrm{ee}}+\mathscr{S}^\dagger_{\mathrm{pe}}\mathscr{S}_{\mathrm{pe}}=1. \tag{3.32}$$

3.4. *Probabilities and cross sections*

For entrance channel α' and exit channel α the probability for electron–electron scattering allowing for radiative decays is

$$\mathscr{P}_{\alpha\alpha'} = |\mathscr{S}_{\alpha\alpha'}|^2. \tag{3.33}$$

The probability for recombination to a final state $\beta\mu$ is

$$\mathscr{P}_{\beta\mu,\alpha'} = |\mathscr{S}_{\beta\mu,\alpha'}|^2. \tag{3.34}$$

We usually require the total rate of recombination for a given entrance channel,

$$\mathscr{P}_{\alpha'}(\text{DR}) = \sum_{\beta\mu} |\mathscr{S}_{\beta\mu,\alpha'}|^2. \tag{3.35}$$

Using (3.32) we obtain

$$\mathscr{P}_{\alpha'}(\text{DR}) = (1 - \boldsymbol{\mathscr{S}}^{\dagger}_{\text{ee}}\boldsymbol{\mathscr{S}}_{\text{ee}})_{\alpha'\alpha'}. \tag{3.36}$$

The partial recombination cross section is

$$Q_{\alpha'}(\text{rec}) = \pi\mathscr{P}_{\alpha'}(\text{DR})/k^2_{\alpha'} \tag{3.37}$$

and the total recombination cross section is defined using formulae given in § 2.5.

4. Some simpler illustrative applications

In the usual first-order theory for the calculation of photoionisation and recombination rates one has

$$\boldsymbol{\mathscr{S}}_{\text{pe}} = -2\pi\text{i}\mathbf{D}^{\dagger}. \tag{4.1}$$

In the present section we consider two simple cases, and compare the first-order theory with the more exact unitary theory presented in § 3, for which $\boldsymbol{\mathscr{S}}_{\text{pe}}$ is given by (3.26).

For the first case, that of one open collision channel and no resonances, the first-order theory is shown to give a very good approximation. The second case, that of one open collision channel and one bound channel, corresponds to that considered by Fano (1961) in his discussion of resonance profiles in photoionisation. Using the unitary theory we obtain a formula which is similar to that of Fano and which reduces to his formula if the autoionisation probability, A, is much larger than the radiative probability, R.

It is shown in the present section that, whenever the continuum contributions to $\mathbf{D}$ are important, the first-order theory gives a satisfactory approximation. In subsequent sections we discuss the unitary theory neglecting the continuum contributions to $\mathbf{D}$.

4.1. *One open channel and no resonances*

For the case of one open channel and no resonances, the continuum wavefunction is

$$\Psi(E) = \psi\frac{1}{r}F \tag{4.2}$$

and the scattering matrix is of the form $S = \exp(2\text{i}\eta)$ with η real. For a final state

(β, μ) the matrix **D** has one element,

$$D=(\Psi(E)|D_\mu|\Psi_\beta). \tag{4.3}$$

Since D has no poles, we obtain from (3.19) $L=\pi^2|D|^2$ and from (3.25), (3.26)

$$\mathcal{S}_{ee}=S(1-\pi^2|D|^2)/(1+\pi^2|D|^2) \tag{4.4}$$

$$S_{pe}=-2\pi iD^*/(1+\pi^2|D|^2). \tag{4.5}$$

These results are consistent with the conservation condition

$$|\mathcal{S}_{ee}|^2+|\mathcal{S}_{pe}|^2=1. \tag{4.6}$$

In the expression (3.9) for D_μ there is a factor $\alpha(\text{FSC})^{3/2}\simeq 10^{-3}$. All other quantities in the expression (4.3) for D are of order unity (or smaller if cancellation effects occur). It follows that

$$|D|^2\lesssim 10^{-6} \tag{4.7}$$

and hence that, in the absence of resonances, the usual first-order expression (4.1) gives $|\mathcal{S}_{pe}|^2$ correct to about one part in 10^6.

4.2. One open channel and one bound channel

The next simplest case to consider is that of one open channel and one bound channel, giving

$$\Psi(E)=\psi\frac{1}{r}F+\Phi C(E). \tag{4.8}$$

From appendix 1 we have

$$\begin{aligned}
&F=F^{(0)}+F^{(1)}C\\
&F^{(0)}=\phi^- -\phi^+X \qquad \text{for } r\geqslant r_0\\
&F^{(1)}=-(2\pi)^{1/2}\phi^+Y \qquad \text{for } r\geqslant r_0\\
&C=-\mathrm{i}(2\pi)^{-1/2}(E-Z)^{-1}Y.
\end{aligned} \tag{4.9}$$

Consistent with equations (2.16) to (2.19) we may put

$$X=\exp(2\mathrm{i}\eta) \qquad Y=\gamma\exp(\mathrm{i}\eta) \qquad Z=E_0-\tfrac{1}{2}\mathrm{i}\gamma^2. \tag{4.10}$$

The autoionisation probability is

$$A=\gamma^2. \tag{4.11}$$

We introduce real functions $f(r)$ and $g(r)$ which are such that, for $r\geqslant r_0$,

$$f(r)=[\phi^+\exp(\mathrm{i}\eta)-\phi^-\exp(-\mathrm{i}\eta)]/(2\mathrm{i}) \tag{4.12}$$

and

$$g(r)=[\phi^+\exp(\mathrm{i}\eta)+\phi^-\exp(-\mathrm{i}\eta)]/2. \tag{4.13}$$

These functions have asymptotic forms

$$f\sim(2\pi k)^{-1/2}\sin(\zeta+\eta) \qquad g\sim(2\pi k)^{-1/2}\cos(\zeta+\eta). \tag{4.14}$$

Using (4.9) we obtain

$$F=-2\mathrm{i}\exp(\mathrm{i}\eta)(E-Z)^{-1}[(E-E_0)f-\tfrac{1}{2}\gamma^2 g] \tag{4.15}$$

and

$$\Psi=-2\mathrm{i}\exp(\mathrm{i}\eta)(E-Z)^{-1}[(E-E_0)\Psi^{(0)}+(8\pi)^{-1/2}\gamma\Phi'] \tag{4.16}$$

where

$$\Psi^{(0)}=\psi\frac{1}{r}f \tag{4.17}$$

and

$$\Phi'=\Phi-(2\pi)^{1/2}\gamma\psi\frac{1}{r}g. \tag{4.18}$$

We interpret Φ' as a modified bound-channel function, calculated allowing for interactions with the continuum.

For a final state (β,μ) we have

$$D=2\mathrm{i}\exp(-\mathrm{i}\eta)(E-Z^*)^{-1}[(E-E_0)P+Q] \tag{4.19}$$

where

$$P=(\Psi^{(0)}|D_\mu|\Psi_\beta) \tag{4.20}$$

is the continuum contribution and

$$Q=(8\pi)^{-1/2}\gamma(\Phi'|D_\mu|\Psi_\beta) \tag{4.21}$$

the resonance contribution.

We put

$$\delta=(2\pi)^{1/2}(\Phi'|D_\mu|\Psi_\beta) \tag{4.22}$$

and

$$R=\delta^2. \tag{4.23}$$

We interpret R as the radiative probability from the state Φ'. From (4.21) we have $Q=\delta\gamma/(4\pi)$.

In evaluating L we make a negligible error in neglecting the continuum term P in (4.19). With this approximation we obtain

$$L=\tfrac{1}{2}\mathrm{i}(E-Z)^{-1}\delta^2 \tag{4.24}$$

and

$$(1+L)^{-1}=(E-Z)/(E-Z') \tag{4.25}$$

where

$$Z'=Z-\tfrac{1}{2}\mathrm{i}\delta^2=E_0-\tfrac{1}{2}\mathrm{i}(A+R). \tag{4.26}$$

From (3.26) we obtain

$$\mathscr{S}_{\mathrm{pe}}=-4\pi\exp(\mathrm{i}\eta)(E-Z')^{-1}[(E-E_0)P+Q]. \tag{4.27}$$

Defining

$$x=(E-E_0)/[\tfrac{1}{2}\mathrm{i}(A+R)] \tag{4.28}$$

and

$$q=\frac{\delta}{2\pi\gamma P}\frac{A}{(A+R)} \tag{4.29}$$

we obtain

$$\mathscr{S}_{\mathrm{pe}}=-4\pi P\exp(\mathrm{i}\eta)\left(\frac{x+q}{x+i}\right) \tag{4.30}$$

which gives the familiar profile formula of Fano (1961), $|\mathscr{S}_{\mathrm{pe}}|^2$ proportional to $(x+q)^2/(1+x^2)$. However, our definitions of x and q differ from those of Fano in that we have used a unitary theory, whereas he uses a first-order theory for the interaction with the radiation field; our expressions agree with those of Fano for the case of $R \ll A$.

The maximum value of $(x+q)^2/(1+x^2)$ occurs for $x=1/q$ and is equal to $(1+q^2)$. From (4.11), (4.23) and (4.29) we obtain $q^2=AR/[2\pi P(A+R)]^2$. From (4.30), the maximum value of $|\mathscr{S}_{\mathrm{pe}}|^2$, as a function of E, is therefore

$$(|\mathscr{S}_{\mathrm{pe}}|^2)_{\max}=16\pi^2P^2+\frac{AR}{[\frac{1}{2}(A+R)]^2}. \tag{4.31}$$

The continuum contribution to (4.31), $16\pi^2P^2$, is small (since P^2 is proportional to $\alpha(\text{FSC})^3$). Neglecting the continuum term we have

$$(|\mathscr{S}_{\mathrm{pe}}|^2)_{\max}=\frac{AR}{[\frac{1}{2}(A+R)]^2}. \tag{4.32}$$

For all values of A and R this gives $(|\mathscr{S}_{\mathrm{pe}}|^2)_{\max}\leqslant 1$ and for $A=R$ it gives

$$(|\mathscr{S}_{\mathrm{pe}}|^2)_{\max}=1 \qquad (A=R). \tag{4.33}$$

Since R is of order 10^{-6}, this case can occur only for very narrow resonances. For such resonances, however, the first-order theory is clearly invalid.

4.3. *Neglect of continuum contributions*

In the present paper our main interest is in narrow resonances for which radiative transitions and autoionisation are competing processes, and for which $|\mathscr{S}_{\mathrm{pe}}|^2$ may be of order unity. We shall therefore neglect continuum processes, which give $|\mathscr{S}_{\mathrm{pe}}|^2 \ll 1$. For the case considered in § 4.2, we put $P=0$ and obtain (on noting that $S=\exp(2\mathrm{i}\eta)(E-E_0-\frac{1}{2}\mathrm{i}A)/(E-E_0+\frac{1}{2}\mathrm{i}A)$)

$$\mathscr{S}_{\mathrm{pe}}=-\exp(\mathrm{i}\eta)\frac{\gamma\delta}{[E-E_0+\frac{1}{2}\mathrm{i}(A+R)]} \tag{4.34}$$

$$\mathscr{S}_{\mathrm{ee}}=\exp(2\mathrm{i}\eta)\frac{[E-E_0-\frac{1}{2}\mathrm{i}(A-R)]}{[E-E_0+\frac{1}{2}\mathrm{i}(A+R)]}. \tag{4.35}$$

From (4.34) we obtain the Breit–Wigner formula for the DR probability:

$$\mathscr{P}(\text{DR})=|\mathscr{S}_{\mathrm{pe}}|^2=\frac{AR}{(E-E_0)^2+[\frac{1}{2}(A+R)]^2}. \tag{4.36}$$

A further approximation may also be noted: since we are interested in resonances for which γ is small we may use Φ in place of the modified function Φ' defined by (4.18).

5. Bound-channel resonances

In the present section we consider any number of open channels and any (finite) number of bound channels Φ_j, which may radiate to bound states Ψ_β.

5.1. Structure of the resonances

Let $\boldsymbol{\Psi}$ and $\boldsymbol{\Phi}$ be row vectors with elements ψ_α and Φ_j. Equation (2.2) may be written

$$\boldsymbol{\Psi}(E) = \boldsymbol{\psi}\frac{1}{r}\mathbf{F} + \boldsymbol{\Phi}\mathbf{C}. \tag{5.1}$$

We assume all collision channels to be open, giving $\boldsymbol{\chi} = \mathbf{S}$ and

$$\mathbf{F} = \boldsymbol{\phi}^- - \boldsymbol{\phi}^+\mathbf{S} \qquad \text{for } r > r_0. \tag{5.2}$$

From equations (2.14), (2.15)

$$\mathbf{S} = \mathbf{X} - \mathrm{i}\mathbf{Y}(E - \mathbf{Z})^{-1}\mathbf{Y}^{\mathrm{T}} \tag{5.3}$$

$$\mathbf{C}(E) = -\mathrm{i}(2\pi)^{-1/2}(E - \mathbf{Z})^{-1}\mathbf{Y}^{\mathrm{T}}. \tag{5.4}$$

The matrices $\mathbf{X}$, $\mathbf{Y}$ and $\mathbf{Z}$ vary slowly with E and satisfy the equations (2.16) to (2.19). It can be shown directly (Seaton 1983b) using (5.3) and (2.16) to (2.19) that

$$\mathbf{S}^{\dagger}\mathbf{S} = 1. \tag{5.5}$$

With all collision channels open, all resonance structure is due to inclusion of bound channels in (5.1). The positions and widths of the resonances are determined by the matrix $\mathbf{Z}$. Diagonalising $\mathbf{Z}$ we have

$$\mathbf{ZN} = \mathbf{N}z \tag{5.6}$$

with z diagonal. Since $\mathbf{Z}$ is symmetric we can take $\mathbf{N}$ to be normalised to $\mathbf{N}^{\mathrm{T}}\mathbf{N} = 1$. We may put

$$z_j = E_{0j} - \tfrac{1}{2}\mathrm{i}A_j \tag{5.7}$$

where E_{0j} is a resonance position and A_j an autoionisation width. Poles in $\mathbf{S}$ occur at complex energies $E = z_j$.

5.2. The radiative matrix elements

Our interest is in very narrow resonances, with autoionisation widths comparable with radiative widths. The radial functions $\mathbf{F}$ are always of order unity (this follows from the unitarity condition (5.5) for $\mathbf{S}$) but the coefficients $\mathbf{C}(E)$ of the bound-channel functions can be large at energies in the vicinity of a resonance, $E \simeq E_{0j}$. In calculating the radiative matrix elements we therefore neglect the continuum functions and put $\boldsymbol{\Psi}(E) = \boldsymbol{\Phi}\mathbf{C}(E)$, to obtain

$$\mathbf{D}(E) = \mathbf{C}^{\dagger}(E)\mathcal{D} \tag{5.8}$$

where $\mathcal{D}$ has elements

$$\mathcal{D}_{j,\beta\mu} = (\Phi_j|D_\mu|\Psi_\beta). \tag{5.9}$$

For one state Φ_j and many final states (β, μ) we have a total radiative probability (or width)

$$R_{(j)} = 2\pi \sum_{\beta\mu} |\mathscr{D}_{j,\beta\mu}|^2. \tag{5.10}$$

With many states Φ_j we introduce a radiative probability matrix,

$$\mathbf{R} = 2\pi \mathscr{D}\mathscr{D}^{\dagger}. \tag{5.11}$$

5.3. Evaluation of L

With $\mathbf{L}$ defined by (3.19) and $\mathbf{D}$ given by (5.8) we have

$$\mathbf{L} = \mathscr{D}^{\dagger}\mathbf{I}\mathscr{D} \tag{5.12}$$

where

$$\mathbf{I}(E) = -\mathrm{i}\pi \int \mathrm{d}E' \frac{\mathbf{C}(E')\mathbf{C}^{\dagger}(E')}{(E' - E - \mathrm{i}\varepsilon)} \tag{5.13}$$

Using (5.4),

$$\mathbf{I} = -\tfrac{1}{2}\mathrm{i} \int \mathrm{d}E' \frac{(E' - \mathbf{Z})^{-1}\mathbf{Y}^{\mathrm{T}}\mathbf{Y}^*(E' - \mathbf{Z}^*)^{-1}}{(E' - E - \mathrm{i}\varepsilon)} \tag{5.14}$$

and using (2.18)

$$\mathbf{I} = \tfrac{1}{2} \int \mathrm{d}E' \frac{(E' - \mathbf{Z})^{-1}(\mathbf{Z} - \mathbf{Z}^*)(E' - \mathbf{Z}^*)^{-1}}{(E' - E - \mathrm{i}\varepsilon)}. \tag{5.15}$$

Using the diagonalisation (5.6) we obtain

$$\mathbf{I} = \tfrac{1}{2}\mathbf{N}\mathbf{J}\mathbf{N}^{\dagger} \tag{5.16}$$

where

$$\mathbf{J}(E) = \int \mathrm{d}E' \frac{(E' - z)^{-1}(z\mathbf{N}^{\mathrm{T}}\mathbf{N}^* - \mathbf{N}^{\mathrm{T}}\mathbf{N}^* z^*)(E' - z^*)^{-1}}{(E' - E - \mathrm{i}\varepsilon)} \tag{5.17}$$

has elements

$$J_{jj'} = \int \mathrm{d}E' \frac{(z_j - z_{j'}^*)(\mathbf{N}^{\mathrm{T}}\mathbf{N}^*)_{jj'}}{(E' - z_j)(E' - E - \mathrm{i}\varepsilon)(E' - z_{j'}^*)}. \tag{5.18}$$

Completing the integration contour with a semicircle in the lower half of the complex plane, the only pole enclosed is at $E' = z_j$. We obtain

$$\mathbf{J}(E) = 2\pi\mathrm{i}(E - z)^{-1}\mathbf{N}^{\mathrm{T}}\mathbf{N}^* \tag{5.19}$$

and hence, using (5.6) and (5.16),

$$\mathbf{I} = \pi\mathrm{i}(E - \mathbf{Z})^{-1}. \tag{5.20}$$

5.4. The electron–electron scattering matrix

The electron–electron scattering matrix is, from (3.25),

$$\mathbf{S}_{\mathrm{ee}} = \mathbf{S}[1 - 2\pi^2\mathbf{D}(1 + \mathbf{L})^{-1}\mathbf{D}^{\dagger}]. \tag{5.21}$$

Using (5.8) and (5.12) we have

$$\mathbf{D}(1+\mathbf{L})^{-1}\mathbf{D}^{\dagger}=\mathbf{C}^{\dagger}\mathscr{D}(1+\mathscr{D}^{\dagger}\mathbf{I}\mathscr{D})^{-1}\mathscr{D}^{\dagger}\mathbf{C}. \tag{5.22}$$

Using the identity

$$(1+\mathscr{D}^{\dagger}\mathbf{I}\mathscr{D})\mathscr{D}^{\dagger}=\mathscr{D}^{\dagger}(1+\mathbf{I}\mathscr{D}\mathscr{D}^{\dagger}) \tag{5.23}$$

and equation (5.11) we obtain

$$\mathbf{D}(1+\mathbf{L})^{-1}\mathbf{D}^{\dagger}=(2\pi)^{-1}\mathbf{C}^{\dagger}\mathbf{R}[1+(2\pi)^{-1}\mathbf{I}\mathbf{R}]^{-1}\mathbf{C} \tag{5.24}$$

and using (5.20) we obtain

$$[1+(2\pi)^{-1}\mathbf{I}\mathbf{R}]=(E-\mathbf{Z})^{-1}(E-\mathbf{Z}') \tag{5.25}$$

where the matrix $\mathbf{Z}'$ is defined by

$$\mathbf{Z}'=\mathbf{Z}-\tfrac{1}{2}\mathrm{i}\mathbf{R}. \tag{5.26}$$

Substituting (5.25) in (5.24) and using (5.4) we obtain

$$\mathbf{D}(1+\mathbf{L})^{-1}\mathbf{D}^{\dagger}=(2\pi)^{-2}\mathbf{Y}^{*}(E-\mathbf{Z}^{*})^{-1}\mathbf{R}(E-\mathbf{Z}')^{-1}\mathbf{Y}^{\mathrm{T}} \tag{5.27}$$

and, substituting (5.27) in (5.21),

$$\mathscr{S}_{\mathrm{ee}}=\mathbf{S}[1-\mathbf{Y}^{*}(E-\mathbf{Z}^{*})^{-1}\tfrac{1}{2}\mathbf{R}(E-\mathbf{Z}')^{-1}\mathbf{Y}^{\mathrm{T}}]. \tag{5.28}$$

Using (5.3) we have

$$\mathbf{S}\mathbf{Y}^{*}=[\mathbf{X}-\mathrm{i}\mathbf{Y}(E-\mathbf{Z})^{-1}\mathbf{Y}^{\mathrm{T}}]\mathbf{Y}^{*}. \tag{5.29}$$

From (2.16) to (2.19), $\mathbf{X}\mathbf{Y}^{*}=\mathbf{Y}$ and $\mathbf{Y}^{\mathrm{T}}\mathbf{Y}^{*}=\mathrm{i}(\mathbf{Z}-\mathbf{Z}^{*})$ giving

$$\mathbf{S}\mathbf{Y}^{*}=\mathbf{Y}[1+(E-\mathbf{Z})^{-1}(\mathbf{Z}-\mathbf{Z}^{*})]=\mathbf{Y}(E-\mathbf{Z})^{-1}(E-\mathbf{Z}^{*}) \tag{5.30}$$

and hence, from (5.28)

$$\mathscr{S}_{\mathrm{ee}}=\mathbf{S}-\mathbf{Y}(E-\mathbf{Z})^{-1}\tfrac{1}{2}\mathbf{R}(E-\mathbf{Z}')^{-1}\mathbf{Y}^{\mathrm{T}}. \tag{5.31}$$

Again using (5.3),

$$\begin{aligned}\mathscr{S}_{\mathrm{ee}}&=\mathbf{X}-\mathrm{i}\mathbf{Y}(E-\mathbf{Z})^{-1}[1-\tfrac{1}{2}\mathrm{i}\mathbf{R}(E-\mathbf{Z}')^{-1}]\mathbf{Y}^{\mathrm{T}}\\&=\mathbf{X}-\mathrm{i}\mathbf{Y}(E-\mathbf{Z})^{-1}[E-\mathbf{Z}'-\tfrac{1}{2}\mathrm{i}\mathbf{R}](E-\mathbf{Z}')^{-1}\mathbf{Y}^{\mathrm{T}}.\end{aligned} \tag{5.32}$$

From (5.26), $(E-\mathbf{Z}'-\tfrac{1}{2}\mathrm{i}\mathbf{R})=(E-\mathbf{Z})$ and hence

$$\mathscr{S}_{\mathrm{ee}}=\mathbf{X}-\mathrm{i}\mathbf{Y}(E-\mathbf{Z}')^{-1}\mathbf{Y}^{\mathrm{T}}. \tag{5.33}$$

Equation (5.3) gives an expression for the scattering matrix neglecting radiative decays and equation (5.33) an expression for the electron–electron scattering matrix allowing for radiative decays *via* bound channels. These two expressions are seen to differ only in that $\mathbf{Z}$ in (5.3) is replaced by $\mathbf{Z}'$ in (5.33). The matrix $\mathbf{Z}'$ is defined by (5.26). Diagonalising $\mathbf{Z}'$ we obtain eigenvalues

$$z_j'=E_{0j}'-\tfrac{1}{2}\mathrm{i}\Gamma_j \tag{5.34}$$

where E_{0j}' is a resonance position shifted due to interaction with the radiation field and Γ_j is the *total* probability for decay of the resonance, due to both autoionisation and emission of radiation.

5.5. The DR probability

The DR probability is given by the diagonal elements of the matrix

$$\mathcal{P}(\mathrm{DR}) = 1 - \mathcal{S}^{\dagger}_{\mathrm{ee}}\mathcal{S}_{\mathrm{ee}}. \tag{5.35}$$

Using (5.33) and doing some algebra similar to that involved in deriving (5.33) from (5.28) we obtain

$$\mathcal{P}(\mathrm{DR}) = \mathbf{Y}^*(E - \mathbf{Z}'^*)^{-1}\mathbf{R}(E - \mathbf{Z}')^{-1}\mathbf{Y}^{\mathrm{T}}. \tag{5.36}$$

For the case of one collision channel and one bound channel, this equation reduces to the Breit–Wigner formula. Expressions for rates integrated over energy can be obtained from (5.36) on diagonalising $\mathbf{Z}'$.

6. A Rydberg series of resonances due to one closed channel

We consider in the present section and in § 7, the theory of DR for the case of resonances high in Rydberg series. Radiative decays are assumed to be due to transitions in the ion core. Since the theory for Rydberg resonances is somewhat intricate we start by considering, in the present section, the simplest case, for which there are two collision channels, one open and one closed, and one Rydberg series of bound states. We use indices α for the open channel, γ for the closed channel and β for the series of bound states. The more general Rydberg case will be considered in § 7.

We shall see that the theory for Rydberg resonances has many similarities to the theory for bound-channel resonances, described in § 5. The main difference for the Rydberg case is that, in order to obtain correct results for any one energy, it is necessary to evaluate summations for the complete series of autoionisation resonances (this will be discussed in § 6.5.2). The treatment of Seaton and Storey (1976) was unsatisfactory in that they attempted to consider separately the resonances belonging to definite values of the principal quantum number.

6.1. The collision wavefunctions

With one open channel α and one closed channel γ, the collision wavefunction is

$$\Psi_\alpha = \psi_\alpha \frac{1}{r} G_{\alpha\alpha}(r) + \psi_\gamma \frac{1}{r} G_{\gamma\alpha}(r) \tag{6.1}$$

and the total energy is

$$E = E_\gamma - \frac{z^2}{2\nu^2} \tag{6.2}$$

where ν is the effective quantum number in the closed channel. The open-channel function is

$$G_{\alpha\alpha} = \phi^-_\alpha - \phi^+_\alpha S_{\alpha\alpha} \tag{6.3}$$

where the scattering matrix is given by (see (2.27))

$$S_{\alpha\alpha} = \chi_{\alpha\alpha} - \chi_{\alpha\gamma}[\chi_{\gamma\gamma} - \exp(-2\pi\mathrm{i}\nu)]^{-1}\chi_{\gamma\alpha}. \tag{6.4}$$

The matrix

$$\chi = \begin{pmatrix} \chi_{\alpha\alpha} & \chi_{\alpha\gamma} \\ \chi_{\gamma\alpha} & \chi_{\gamma\gamma} \end{pmatrix} \tag{6.5}$$

is symmetric, $\chi = \chi^{\mathrm{T}}$, and unitary, $\chi^*\chi = \mathbf{1}$. It follows that $\chi_{\alpha\gamma} = \chi_{\gamma\alpha}$ and that

$$\begin{aligned} &|\chi_{\alpha\gamma}|^2 + |\chi_{\alpha\alpha}|^2 = 1 \\ &|\chi_{\alpha\gamma}|^2 + |\chi_{\gamma\gamma}|^2 = 1 \\ &\chi^*_{\alpha\alpha}\chi_{\gamma\alpha} + \chi^*_{\alpha\gamma}\chi_{\gamma\gamma} = 0. \end{aligned} \tag{6.6}$$

From equation (2.28) the closed-channel function is

$$G_{\gamma\alpha}(r) = P_\gamma(\nu; r) B_{\gamma\alpha}(\nu) \tag{6.7}$$

where $P_\gamma(\nu; r) = P(\nu, l_\gamma; r)$ is the radial function discussed in § 2.3 (for $r \geq r_0$, (2.29) gives P_γ in terms of the Whittaker function);

$$B_{\gamma\alpha}(\nu) = \mathrm{i}(\nu^3/z^2)t(\nu)[\chi_{\gamma\gamma} - \exp(-2\pi\mathrm{i}\nu)]^{-1}\chi_{\gamma\alpha} \tag{6.8}$$

and

$$t(\nu) = (z^2/\nu^3)^{1/2}\exp(-\mathrm{i}\pi\nu). \tag{6.9}$$

We note that

$$t^*(\nu) = (z^2/\nu^3)^{1/2}\exp(+\mathrm{i}\pi\nu) = (z^2/\nu^3)/t(\nu) \tag{6.10}$$

from which it follows that

$$B^*_{\gamma\alpha}(\nu) = -\frac{\mathrm{i}}{t(\nu)}[\chi^*_{\gamma\gamma} - \exp(+2\pi\mathrm{i}\nu)]^{-1}\chi^*_{\gamma\alpha}. \tag{6.11}$$

6.1.1. Poles and residues. We put

$$\chi_{\gamma\gamma} = \exp(2\pi\mathrm{i}\mu_\gamma) \tag{6.12}$$

where

$$\mu_\gamma = p + \mathrm{i}q \tag{6.13}$$

is the complex quantum defect, with $q > 0$ (see § 7.3 of S83). The scattering matrix has a pole at $\nu = \nu_n$, such that $\chi_{\gamma\gamma} = \exp(-2\pi\mathrm{i}\nu_n)$. We obtain

$$\nu_n = n - \mu_\gamma. \tag{6.14}$$

The energy at the pole is

$$E_n = E_\gamma - \frac{z^2}{(2\nu_n^2)} \tag{6.15}$$

In the limit of $\nu \to \nu_n$ we have

$$[\chi_{\gamma\gamma} - \exp(-2\pi\mathrm{i}\nu)] \to \chi_{\gamma\gamma}\{[1 - \exp[-2\pi\mathrm{i}(\nu - \nu_n)]\} \to 2\pi\mathrm{i}\chi_{\gamma\gamma}(\nu - \nu_n) \tag{6.16}$$

and

$$(E - E_n) \to (z^2/\nu_n^3)(\nu - \nu_n) \tag{6.17}$$

giving

$$[\chi_{\gamma\gamma} - \exp(-2\pi\mathrm{i}\nu)] \to 2\pi\mathrm{i}\chi_{\gamma\gamma}(\nu_n^3/z^2)(E - E_n) \tag{6.18}$$

and

$$B_{\gamma\alpha}(\nu) \to \frac{1}{2\pi}(E - E_n)^{-1} t(\nu_n) \chi_{\gamma\gamma}^{-1} \chi_{\gamma\alpha}. \tag{6.19}$$

6.1.2. Expansion for $G_{\gamma\alpha}(r)$. We expand $G_{\gamma\alpha}(r)$ using its poles and residues:

$$G_{\gamma\alpha}(r) = \frac{1}{2\pi} \sum_n P_\gamma(\nu_n\,;r)(E - E_n)^{-1} t(\nu_n) \chi_{\gamma\gamma}^{-1} \chi_{\gamma\alpha}. \tag{6.20}$$

A convenient matrix notation is to use $\mathbf{P}_\gamma$ for the row vector with element $P_\gamma(\nu_n\,;r)$, Z for the diagonal matrix with elements E_n, and $\mathbf{t}$ for the column vector with elements $t(\nu_n)$: equation (6.20) may then be written

$$G_{\gamma\alpha}(r) = \frac{1}{2\pi} \mathbf{P}_\gamma(r)(E - \mathbf{Z})^{-1} \mathbf{t} \chi_{\gamma\gamma}^{-1} \chi_{\gamma\alpha}. \tag{6.21}$$

6.2. The bound states

The bound states belong to a Rydberg series with principal quantum number which we denote by m. We put

$$\Psi_{m\beta} = \psi_\beta r^{-1} P_{m\beta}(r) \tag{6.22}$$

with radial functions

$$P_{m\beta}(r) = P_\beta(\nu_{m\beta}\,;r) \tag{6.23}$$

and effective quantum numbers

$$\nu_{m\beta} = m - \mu_\beta \tag{6.24}$$

where μ_β is a real quantum defect. We take $\mathbf{P}_\beta$ to be a column vector with elements $P_{m\beta}$ and the bound-state functions to form a column vector

$$\boldsymbol{\Psi}_\beta = \psi_\beta r^{-1} \mathbf{P}_\beta(r). \tag{6.25}$$

6.3. The radiative matrix elements

We assume that radiative transitions take place only *via* the closed channel, and hence that the radiative matrix element is

$$D_{\alpha,\mu m\beta} = \left(\psi_\gamma \frac{1}{r} G_{\gamma\alpha} | D_\mu | \psi_\beta \frac{1}{r} P_{m\beta} \right). \tag{6.26}$$

We further assume that the radiative transitions occur only in the ion core and we therefore replace $(R_\mu + r_\mu)$ in (3.9) by R_μ. With this approximation we have

$$D_{\alpha,\mu m\beta} = (G_{\gamma\alpha} | P_{m\beta}) a_{\gamma,\mu\beta} \tag{6.27}$$

where

$$a_{\gamma,\mu\beta} = (\psi_\gamma | D_\mu | \psi_\beta) \tag{6.28}$$

is the matrix element for the radiative transition in the core. Let $\mathbf{a}$ be the matrix with elements $a_{\gamma,m\beta}$ and $\mathbf{D}$ that with elements $D_{\alpha,\mu m\beta}$. We write (6.27) as

$$\mathbf{D} = (G_{\gamma\alpha} | \mathbf{P}_\beta) \mathbf{a} \tag{6.29}$$

and we put

$$\mathbf{D}^{\dagger} = \mathbf{a}^{\dagger}(\mathbf{P}_{\beta}|G_{\gamma\alpha}). \tag{6.30}$$

6.4. *The matrix* $\mathbf{L}$

The matrix $\mathbf{L}$ with elements $L_{\mu m\beta,\mu' m'\beta'}$, is, from (3.19),

$$\mathbf{L}(E) = -\mathrm{i}\pi \int \mathrm{d}E' \frac{\mathbf{D}^{\dagger}(E')\mathbf{D}(E')}{(E'-E-\mathrm{i}\varepsilon)}. \tag{6.31}$$

In the integrand, $\mathbf{D}(E')$ and $(E'-E-\mathrm{i}\varepsilon)^{-1}$ have poles above the real axis while $\mathbf{D}^{\dagger}(E')$ has poles below. We complete the integration contour with a semicircle below the real axis. All poles enclosed are then due to $\mathbf{D}^{\dagger}(E')$. We use

$$\mathbf{D}^{\dagger}(E') = (2\pi)^{-1}\mathbf{a}^{\dagger}(\mathbf{P}_{\beta}|\mathbf{P}_{\gamma})(E'-\mathbf{Z})^{-1}\mathbf{t}\chi_{\gamma\gamma}^{-1}\chi_{\gamma\alpha} \tag{6.32}$$

and

$$\mathbf{D}(E') = B^{*}_{\gamma\alpha}(\nu')(P_{\gamma}(\nu';r)|\mathbf{P}_{\beta})\mathbf{a} \tag{6.33}$$

to obtain

$$\mathbf{L} = \pi\mathbf{a}^{\dagger}\left(\sum_{n}(\mathbf{P}_{\beta}|P_{\gamma}(\nu_n))t(\nu_n)\chi_{\gamma\gamma}^{-1}\chi_{\gamma\alpha}(E-E_n)^{-1}B^{*}_{\gamma\alpha}(\nu_n)(P^{*}_{\gamma}(\nu_n)|\mathbf{P}_{\beta})\right)\mathbf{a} \tag{6.34}$$

(the reason for using $(P^{*}_{\gamma}(\nu_n)|\mathbf{P}_{\beta})$, and not $(P_{\gamma}(\nu_n)|\mathbf{P}_{\beta})$, is discussed at the end of appendix 2).

From (6.11),

$$B^{*}_{\gamma\alpha}(\nu_n) = -\frac{\mathrm{i}}{t(\nu_n)}\chi_{\gamma\gamma}(|\chi_{\gamma\gamma}|^2-1)^{-1}\chi^{*}_{\gamma\alpha} \tag{6.35}$$

and we obtain, making use of (6.6),

$$\mathbf{L} = \mathrm{i}\pi\mathbf{a}^{\dagger}(\mathbf{P}_{\beta}|\mathbf{P}_{\gamma})(E-Z)^{-1}(\mathbf{P}^{*}_{\gamma}|\mathbf{P}_{\beta})\mathbf{a}. \tag{6.36}$$

6.5. *The electron–electron scattering matrix*

We write equation (3.25) as

$$\mathscr{S}_{\alpha\alpha} = S_{\alpha\alpha}K \tag{6.37}$$

where K is defined by

$$K = 1 - 2\pi^2\mathbf{D}(1+\mathbf{L})^{-1}\mathbf{D}^{\dagger}. \tag{6.38}$$

6.5.1. Evaluation of K. In order to obtain an expression for $(1+\mathbf{L})^{-1}D^{\dagger}$ we use the expansion (6.32) for $D^{\dagger}$. We start by considering

$$\mathbf{L}\mathbf{a}^{\dagger}(\mathbf{P}_{\beta}|\mathbf{P}_{\gamma}). \tag{6.39}$$

From the completeness of the functions $\mathbf{P}_{\beta}$ (see appendix 3) we have

$$(\mathbf{P}^{*}_{\gamma}|\mathbf{P}_{\beta})(\mathbf{P}_{\beta}|\mathbf{P}_{\gamma}) = 1. \tag{6.40}$$

The total probability for a radiative transition in the ion core is

$$R = 2\pi\mathbf{a}\mathbf{a}^{\dagger} \tag{6.41}$$

where

$$\mathbf{a}\mathbf{a}^{\dagger} = \sum_{\mu} |a_{\gamma,\mu\beta}|^2. \tag{6.42}$$

Using these relations, and equation (6.36), we obtain

$$\mathbf{L}\mathbf{a}^{\dagger}(\mathbf{P}_{\beta}|\mathbf{P}_{\gamma}) = \mathbf{a}^{\dagger}(\mathbf{P}_{\beta}|\mathbf{P}_{\gamma})(E-Z)^{-1}\tfrac{1}{2}\mathrm{i}R \tag{6.43}$$

and hence

$$\begin{aligned}(1+\mathbf{L})\mathbf{a}^{\dagger}(\mathbf{P}_{\beta}|\mathbf{P}_{\gamma}) &= \mathbf{a}^{\dagger}(\mathbf{P}_{\beta}|\mathbf{P}_{\gamma})[1+(E-Z)^{-1}\tfrac{1}{2}\mathrm{i}R] \\ &= \mathbf{a}^{\dagger}(\mathbf{P}_{\beta}|\mathbf{P}_{\gamma})(E-Z')(E-Z)^{-1}\end{aligned} \tag{6.44}$$

where Z' is defined by

$$Z' = Z - \tfrac{1}{2}\mathrm{i}R. \tag{6.45}$$

From (6.44)

$$(1+\mathbf{L})^{-1}\mathbf{a}^{\dagger}(\mathbf{P}_{\beta}|\mathbf{P}_{\gamma}) = \mathbf{a}^{\dagger}(\mathbf{P}_{\beta}|\mathbf{P}_{\gamma})(E-Z)(E-Z')^{-1} \tag{6.46}$$

and using the expression (6.32) for $\mathbf{D}^{\dagger}$ we obtain

$$(1+\mathbf{L})^{-1}\mathbf{D}^{\dagger} = (2\pi)^{-1}\mathbf{a}^{\dagger}(\mathbf{P}_{\beta}|\mathbf{P}_{\gamma})(E-Z')^{-1}\mathbf{t}\chi_{\gamma\gamma}^{-1}\chi_{\gamma\alpha}. \tag{6.47}$$

We recall that $\mathbf{P}_{\beta}$ is a column vector with elements $P_{m\beta}$, $\mathbf{P}_{\gamma}$ a row vector with elements $P_{\gamma}(\nu_n)$ and $\mathbf{t}$ a column vector with elements $t(\nu_n)$. We now consider $\mathbf{D}(1+\mathbf{L})^{-1}\mathbf{D}^{\dagger}$ with $\mathbf{D}$ given by (6.33) and $(1+\mathbf{L})^{-1}\mathbf{D}^{\dagger}$ by (6.47). We use the definition (6.41) of R and the completeness relation (6.40) to obtain

$$\begin{aligned}\mathbf{D}(1+\mathbf{L})^{-1}\mathbf{D}^{\dagger} = &-\frac{\mathrm{i}R}{(2\pi)^2 t(\nu)}\chi_{\alpha\gamma}^{*}[\chi_{\gamma\gamma}^{*} - \exp(+2\pi\mathrm{i}\nu)]^{-1} \\ &\times (P_{\gamma}(\nu)|\mathbf{P}_{\gamma})(E-Z')^{-1}\mathbf{t}\chi_{\gamma\gamma}^{-1}\chi_{\gamma\alpha}.\end{aligned} \tag{6.48}$$

Substitution in (6.38), which defines K, gives

$$K = 1 + \tfrac{1}{2}\mathrm{i}R\chi_{\alpha\gamma}^{*}[\chi_{\gamma\gamma}^{*} - \exp(2\pi\mathrm{i}\nu)]^{-1}T\chi_{\gamma\gamma}^{-1}\chi_{\gamma\alpha} \tag{6.49}$$

where

$$\begin{aligned}T &= (P_{\gamma}(\nu)|\mathbf{P}_{\gamma})(E-Z')^{-1}t/t(\nu) \\ &= \sum_{n}(P_{\gamma}(\nu)|P_{\gamma}(\nu_n))(E-E_n+\tfrac{1}{2}\mathrm{i}R)^{-1}t(\nu_n)/t(\nu).\end{aligned} \tag{6.50}$$

6.5.2. Evaluation of T. In the present subsection we omit the subscript γ on P_{γ}. From the definition (6.9) of $t(\nu)$ we have

$$\frac{t(\nu_n)}{t(\nu)} = (\nu/\nu_n)^{3/2}\exp[\mathrm{i}\pi(\nu-\nu_n)]. \tag{6.51}$$

In (6.50), $(P(\nu)|P(\nu_n))$ is small unless

$$|\nu-\nu_n| \ll \nu. \tag{6.52}$$

When this condition is satisfied we have

$$\frac{t(\nu_n)}{t(\nu)} \simeq \exp[\mathrm{i}\pi(\nu-\nu_n)] \tag{6.53}$$

$$(E - E_n) \simeq (z^2/\nu^3)(\nu - \nu_n) \tag{6.54}$$

and

$$(E - E_n + \tfrac{1}{2}\mathrm{i}R) = (z^2/\nu^3)(\nu - \nu_n + \mathrm{i}\Delta(\nu)) \tag{6.55}$$

where $\Delta(\nu)$ is defined by

$$\Delta(\nu) = \tfrac{1}{2}R/(z^2/\nu^3) = \tfrac{1}{2}\nu^3 R/z^2. \tag{6.56}$$

The separation between adjacent resonances is $\delta E = (z^2/\nu^3)$ and we therefore have $2\Delta = R/\delta E$, the radiative width divided by the resonance separation. Since

$$\nu - \nu_n + \mathrm{i}\Delta(\nu) = \nu - n + \mu_\gamma + \mathrm{i}\Delta(\nu) \tag{6.57}$$

we see that iΔ can be considered to be a contribution to the complex quantum defect accounting for radiative decays. We can define a complex quantum defect, allowing for radiative decays, as

$$\mu'_\gamma = \mu_\gamma + \mathrm{i}\Delta(\nu) = p + \mathrm{i}(q + \Delta(\nu)). \tag{6.58}$$

Using equations (6.53) to (6.55) we obtain

$$T = \left(\frac{\nu^3}{z^2}\right) \sum_n \frac{(P(\nu)|P(\nu_n)) \exp[\mathrm{i}\pi(\nu - \nu_n)]}{(\nu - \nu_n + \mathrm{i}\Delta)}. \tag{6.59}$$

Introducing the radial function

$$P(\nu + \mathrm{i}\Delta) = P(\nu + \mathrm{i}\Delta;\, r) \tag{6.60}$$

we have, from (A2.10) and (A2.14) of appendix 2,

$$(P^*(\nu_n)|P(\nu + \mathrm{i}\Delta)) = \frac{\sin[\pi(\nu_n - \nu - \mathrm{i}\Delta)]}{\pi(\nu_n - \nu - \mathrm{i}\Delta)} \tag{6.61}$$

giving

$$\frac{\exp[\mathrm{i}\pi(\nu - \nu_n)]}{(\nu_n - \nu - \mathrm{i}\Delta)} = \frac{2\pi\mathrm{i}\exp(\pi\Delta)}{\{1 - \exp[-2\pi\mathrm{i}(\nu + \mu'_\gamma)]\}}(P^*(\nu_n)|P(\nu + \mathrm{i}\Delta)). \tag{6.62}$$

Substituting this result in (6.59) and using the completeness of the functions P_{ν_n} we obtain

$$T = \left(\frac{\nu^3}{z^2}\right) \frac{2\pi\mathrm{i}\exp(\pi\Delta)}{\{1 - \exp[-2\pi\mathrm{i}(\nu + \mu'_\gamma)]\}}(P(\nu)|P(\nu + \mathrm{i}\Delta)) \tag{6.63}$$

where, from (A2.10)

$$(P(\nu)|P(\nu + \mathrm{i}\Delta)) = \frac{\sin(\pi\mathrm{i}\Delta)}{\pi\mathrm{i}\Delta}. \tag{6.64}$$

Using this result we obtain, finally, for T:

$$T = \frac{2\mathrm{i}}{R} \frac{(g(\nu) - 1)\chi_{\gamma\gamma}}{[\chi_{\gamma\gamma} - g(\nu)\exp(-2\pi\mathrm{i}\nu)]} \tag{6.65}$$

where g is defined by

$$g(\nu) = \exp(2\pi\Delta) = \exp(\pi\nu^3 R/z^2). \tag{6.66}$$

6.5.3. Expression for $\mathcal{S}_{\alpha\alpha}$. Using (6.49) and (6.65) we have

$$K = 1 - \chi^*_{\alpha\gamma}[\chi^*_{\gamma\gamma} - \exp(+2\pi i\nu)]^{-1}(g-1)[\chi_{\gamma\gamma} - g\exp(-2\pi i\nu)]^{-1}\chi_{\gamma\alpha}. \tag{6.67}$$

We now calculate $\mathcal{S}_{ee}$ using (6.37). From (6.4) and (6.6) we obtain

$$S_{\alpha\alpha}\chi^*_{\alpha\gamma} = \chi_{\alpha\gamma}[\chi_{\gamma\gamma} - \exp(-2\pi i\nu)]^{-1}[\chi^*_{\gamma\gamma} - \exp(+2\pi i\nu)]\exp(-2\pi i\nu) \tag{6.68}$$

and hence

$$\mathcal{S}_{\alpha\alpha} = S_{\alpha\alpha} - \chi_{\alpha\gamma}[\chi_{\gamma\gamma} - \exp(-2\pi i\nu)]^{-1}\exp(-2\pi i\nu)(g-1) \times[\chi_{\gamma\gamma} - g\exp(-2\pi i\nu)]^{-1}\chi_{\gamma\alpha}. \tag{6.69}$$

Again using (6.4), this gives

$$\begin{aligned}\mathcal{S}_{\alpha\alpha} &= \chi_{\alpha\alpha} - \chi_{\gamma\alpha}[\chi_{\gamma\gamma} - \exp(-2\pi i\nu)]^{-1} \\ &\quad\times\{1 + \exp(-2\pi i\nu)(g-1)[\chi_{\gamma\gamma} - g\exp(-2\pi i\nu)]^{-1}\}\chi_{\gamma\alpha} \\ &= \chi_{\alpha\alpha} - \chi_{\alpha\gamma}[\chi_{\gamma\gamma} - \exp(-2\pi i\nu)]^{-1}\{[\chi_{\gamma\gamma} - g\exp(-2\pi i\nu)] \\ &\quad+ \exp(-2\pi i\nu)(g-1)\}[\chi_{\gamma\gamma} - g\exp(-2\pi i\nu)]^{-1}\chi_{\gamma\alpha}\end{aligned} \tag{6.70}$$

and it follows that

$$\mathcal{S}_{\alpha\alpha} = \chi_{\alpha\alpha} - \chi_{\alpha\gamma}[\chi_{\gamma\gamma} - g(\nu)\exp(-2\pi i\nu)]^{-1}\chi_{\gamma\alpha}. \tag{6.71}$$

6.6. The DR probability

6.6.1. The capture probability $\mathcal{P}_\alpha$ (DR). The probability of radiative capture is, from § 3,

$$\mathcal{P}_\alpha(\text{DR}) = \sum_{\mu m} |\mathcal{S}_{\mu m\beta,\alpha}|^2 = 1 - |\mathcal{S}_{\alpha\alpha}|^2. \tag{6.72}$$

For convenience of notation, in the present section we put

$$\mathcal{P}(\nu) = \mathcal{P}_\alpha(\text{DR}). \tag{6.73}$$

We obtain from (6.71), after some algebra, or directly using (3.26),

$$\mathcal{P}(\nu) = \frac{|\chi_{\alpha\gamma}|^2 G(\nu)}{[\chi^*_{\gamma\gamma} - g(\nu)\exp(+2\pi i\nu)][\chi_{\gamma\gamma} - g(\nu)\exp(-2\pi i\nu)]} \tag{6.74}$$

where

$$G(\nu) = g(\nu)^2 - 1 = \exp(4\pi\Delta(\nu)) - 1. \tag{6.75}$$

From (6.12), (6.13) we have $\chi_{\gamma\gamma} = \exp[2\pi i(p + iq)]$ with p and q real and from the unitarity conditions (6.6) we have $q > 0$ and $|\chi_{\alpha\gamma}|^2 = 1 - \exp(-4\pi q)$. Using these results we obtain from (6.74), after some more algebra,

$$\mathcal{P}(\nu) = \frac{\sinh(2\pi q)\sinh(2\pi\Delta(\nu))}{\sinh^2[\pi(q + \Delta(\nu))] + \sin^2[\pi(\nu + p)]}. \tag{6.76}$$

We recall that

$$\Delta(\nu) = \tfrac{1}{2}\nu^3 R/z^2 = \tfrac{1}{2}R/\delta E \tag{6.77}$$

where $\delta E = z^2/\nu^3$.

6.6.2. The mean capture probability. We define the mean capture probability to be

$$\langle\mathscr{P}(\nu)\rangle=\int_{\nu-1/2}^{\nu+1/2}\mathscr{P}(\nu')\,\mathrm{d}\nu'. \tag{6.78}$$

Using the integration variable $x=\exp(2\pi\mathrm{i}\nu)$ we obtain

$$\langle\mathscr{P}(\nu)\rangle=\oint\frac{\mathscr{P}(\nu')\,\mathrm{d}x}{2\pi\mathrm{i}x} \tag{6.79}$$

where the integration is counter-clockwise around the unit circle. We evaluate the integral neglecting the variation of $g(\nu)$; from (6.74), (6.78) we have

$$\langle\mathscr{P}(\nu)\rangle=\frac{|\chi_{\alpha\gamma}|^2G(\nu)}{2\pi\mathrm{i}}\oint\frac{\mathrm{d}x}{(\chi^*_{\gamma\gamma}-g(\nu)x)(x\chi_{\gamma\gamma}-g(\nu))}. \tag{6.80}$$

Since the only pole enclosed is at $x=\chi^*_{\gamma\gamma}/g(\nu)$ we obtain

$$\langle\mathscr{P}(\nu)\rangle=\frac{|\chi_{\alpha\gamma}|^2G(\nu)}{|\chi_{\alpha\gamma}|^2+G(\nu)} \tag{6.81}$$

where we have used the relation $|\chi_{\alpha\gamma}|^2=1-|\chi_{\gamma\gamma}|^2$, and where we recall that $|\chi_{\alpha\gamma}|^2=1-\exp(-4\pi q)$.

6.6.3. Discussion

6.6.3.1. The behaviour of the function $\mathscr{P}(\nu)$. Equation (6.76) gives an expression for the capture probability $\mathscr{P}(\nu)$ for the case of a complete Rydberg series of resonances; it is seen to have a form similar to that of the Breit–Wigner formula.

The expression (6.76) involves three parameters: the real part of the complex quantum defect, p, which determines the positions of the resonances; the imaginary part of the complex quantum defects, q, which is such that $q=\frac{1}{2}A/\delta E$ where A is the autoionisation width and $\delta E=z^2/\nu^3$ the separation of adjacent resonances (the relation $q=\frac{1}{2}A/\delta E$ follows from (6.13) to (6.15) on noting that $-\frac{1}{2}\mathrm{i}A_n$ is the imaginary part of the pole energy E_n); and $\Delta(\nu)$ which is such that $\Delta(\nu)=\frac{1}{2}R/\delta E$ where R is the radiative width (which is assumed to be a constant independent of ν).

For neutral atoms, $z=1$, we have $R\sim10^{-6}$ and $\Delta(\nu)\sim1$ for $\nu\sim100$. For $\Delta(\nu)\ll1$ (radiative width small compared with separations), $P(\nu)$ has resonance structure due to the variations of $\sin[\pi(\nu+p)]$ in (6.76). Maxima in $\mathscr{P}(\nu)$ occur at $\nu=n-p$ where n is an integer,

$$\mathscr{P}(n-p)=\frac{\sinh(2\pi q)\sinh[2\pi\Delta(n-p)]}{\sinh^2\{\pi[q+\Delta(n-p)]\}}. \tag{6.82}$$

The largest maximum, $\mathscr{P}(n-p)=1$, occurs when $q=\Delta(n-p)$, that is to say when the autoionisation probability is equal to the radiative probability.

For $\Delta(\nu)\gg1$ (radiative probability large compared with separations) the interactions with the radiation field cause a complete smearing out of the resonance structure. For $2\pi\Delta$ large we have $\sinh(2\pi\Delta)\simeq\frac{1}{2}\exp(2\pi\Delta)$, $\sinh^2[\pi(q+\Delta)]\simeq\frac{1}{4}\exp[2\pi(q+\Delta)]$ and hence

$$\mathscr{P}(\nu)\simeq1-\exp(-4\pi q)=|\chi_{\alpha\gamma}|^2. \tag{6.83}$$

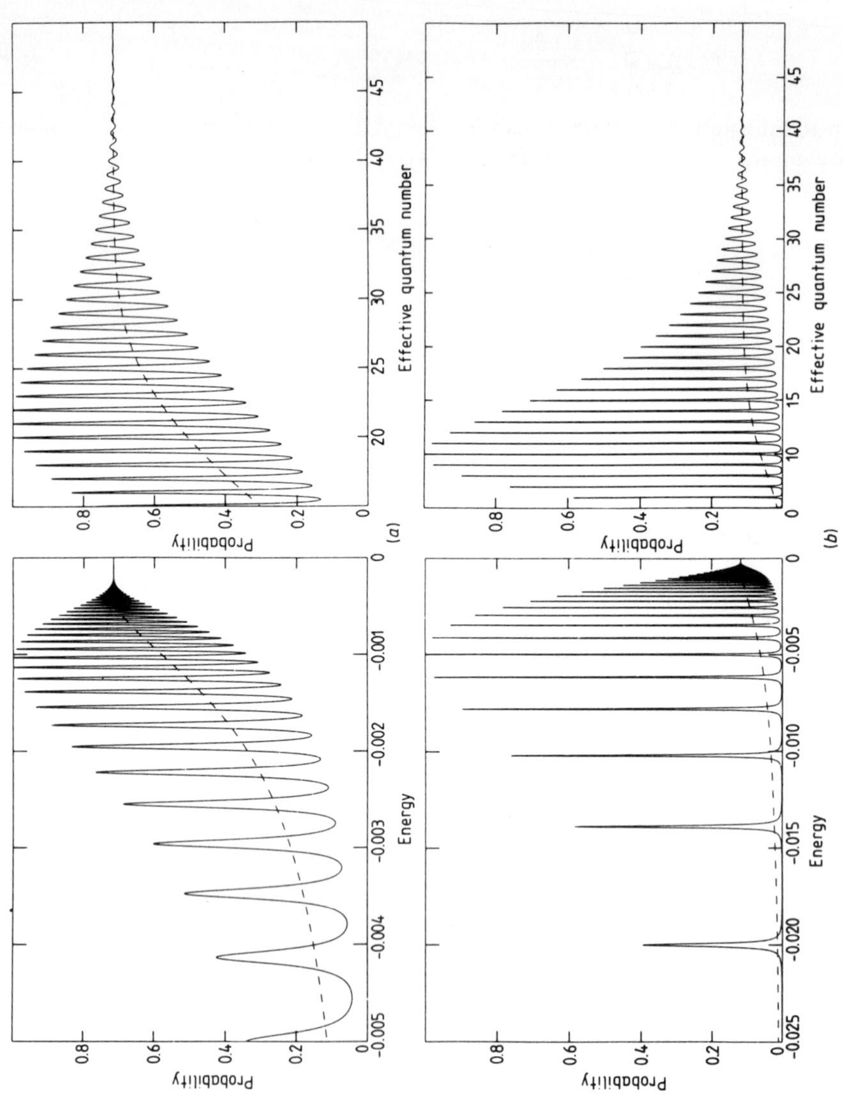

Probability
Energy
Effective quantum number
(a)
(b)

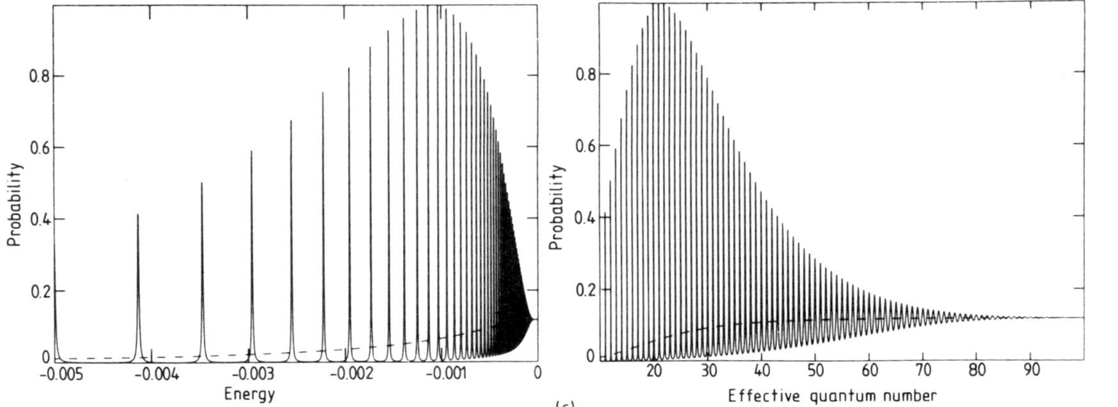

Figure 1. (a) Probability of DR for two-channel case. Full curve, probability $\mathscr{P}(\nu)$ calculated using (6.76); broken curve, mean probability $\langle\mathscr{P}(\nu)\rangle$ calculated using (6.81). Probabilities are plotted against energy, $-1/(2\nu^2)$, and effective quantum number ν for the closed channel. Results for $\delta=10^{-5}$ and $q=0.1$ where $\delta=\frac{1}{2}R/z^s$ (R is the radiative transition probability, z the charge on the ion), and where q is the imaginary part of the complex quantum defect. (b) As for (a) but with $\delta=10^{-5}$ and $q=0.01$. (c) As for (a) but with $\delta=10^{-6}$ and $q=0.01$.

Figure 1 (*a*), (*b*) and (*c*) shows some illustrative results. We plot $\mathscr{P}(\nu)$ against the energy variable $\varepsilon=-1/\nu^2$ and against ν for larger values of ν. We take $p=0$ and put

$$\Delta(\nu)=\nu^3\delta \tag{6.84}$$

where $\delta=\frac{1}{2}R/z^2$. The plots depend on two parameters, q for the autoionisation width and δ for the radiative width.

6.6.3.2. Figure 1(*a*) gives results for $q=0.1$ and $\delta=10^{-5}$. For this case we have fairly large autoionisation widths and a large radiative width (such a case may not occur in practice, since one can have a large radiative width for more highly-ionised systems but such systems usually have small autoionisation widths). The maximum value of $\mathscr{P}(\nu)$ occurs at $q=\Delta$, giving $\nu^3=10^4$, $\nu=21$. The radiative width is equal to the separation, $R=\delta E$, when $\Delta=0.5$ giving $\nu=37$. It is seen that the resonances are 'smeared out' for $\nu\gtrsim 37$.

6.6.3.3. Figure 1(*b*) gives results for $q=0.01$ and $\delta=10^{-5}$, that is to say autoionisation widths smaller than for figure 1(*a*) and the same radiative width. The condition $q=\Delta$ is satisfied for $\nu=10$ and the condition $\Delta=0.5$ for $\nu=37$.

6.6.3.4. Figure 1(*c*) gives results for fairly narrow resonances, $q=0.01$, and a smaller radiative width, $\delta=10^{-6}$. For this case we have $q=\Delta$ at $\nu=21$ and $\Delta=0.5$ at $\nu=79$.

7. Rydberg resonances for the many-channel case

We consider Rydberg resonances for the many-channel case. Our interest is in high Rydberg states, belonging to series converging to a new threshold and we therefore make the simplifying assumption that all closed channels belong to the same energy level of the ion core. The contracted $\boldsymbol{\chi}$ matrix is equal to the **S** matrix at energies just above new thresholds. If closed channels belonging to higher thresholds are included in the expansion (2.1) we use a contracted $\boldsymbol{\chi}$ matrix as described by Seaton (1978) and in S83.

An assumption implicit in this treatment is that the levels of the ion core are well separated. This will generally be true for *LS* coupling but not when account is taken of fine structure. Fine-structure effects will be discussed in a later part of the present series. In the present section *LS* coupling is assumed.

We use a notation of partitioned matrices as in § 2.3: o is used for open channels (α for individual open channels); c for closed channels (γ for individual closed channels); and b for bound states (β for individual channels giving bound states). Let $\boldsymbol{\psi}_{\mathrm{o}}$ be the row vector for the open-channel core functions, $\boldsymbol{\psi}_{\mathrm{c}}$ that for the closed-channel core functions: the expansion (2.1) can be written (neglecting bound channels $\boldsymbol{\Phi}$),

$$\Psi=\boldsymbol{\psi}_{\mathrm{o}}r^{-1}\mathbf{G}_{\mathrm{oo}}(r)+\boldsymbol{\psi}_{\mathrm{c}}r^{-1}\mathbf{G}_{\mathrm{co}}(r). \tag{7.1}$$

The total energy is

$$E=E_{\mathrm{c}}-\frac{z^2}{2\nu^2} \tag{7.2}$$

where, for the closed channels, E_{c} is the energy of the ion and ν the effective quantum number. The $\mathbf{S}$ matrix is given by (2.26). We diagonalise χ_{cc},

$$\chi_{cc}\mathbf{N} = \mathbf{N}\bar{\chi}_{cc} \tag{7.3}$$

where $\bar{\chi}_{cc}$ is diagonal and where $\mathbf{N}$ is normalised to

$$\mathbf{N}^{T}\mathbf{N} = 1. \tag{7.4}$$

Since the closed channels are degenerate, $\exp(-2\pi i\nu)$ in (2.26) is a multiple of the unit matrix and hence commutes with $\mathbf{N}$; from (2.26) we therefore obtain

$$\mathbf{S} = \chi_{oo} - \bar{\chi}_{oc}[\bar{\chi}_{cc} - \exp(-2\pi i\nu)]^{-1}\bar{\chi}_{co} \tag{7.5}$$

where

$$\bar{\chi}_{oc} = \chi_{oc}\mathbf{N} \qquad \bar{\chi}_{co} = \mathbf{N}^{T}\chi_{co}. \tag{7.6}$$

Similarly from (2.27) we obtain

$$\mathbf{G}_{co}(r) = \mathbf{P}_{c}(r)\mathbf{B}_{co} = \bar{\mathbf{P}}_{c}(r)\bar{\mathbf{B}}_{co} \tag{7.7}$$

where $\mathbf{P}_{c}(r)$ is the diagonal matrix $\mathbf{P}(r)$ used in (2.27), the definition of $\mathbf{B}_{co}$ follows from (2.27),

$$\bar{\mathbf{P}}_{c}(r) = \mathbf{P}_{c}(r)\mathbf{N} \tag{7.8}$$

and

$$\bar{\mathbf{B}}_{co} = i\left(\frac{\nu^{3}}{2z^{2}}\right)^{1/2} t(\nu)[\bar{\chi}_{cc} - \exp(-2\pi i\nu)]^{-1}\bar{\chi}_{co}. \tag{7.9}$$

The diagonal matrix $\bar{\chi}_{cc}$ has elements $\bar{\chi}_{\gamma\gamma} = \exp(2\pi i\mu_{\gamma})$ and $\mathbf{S}$ and $\mathbf{G}_{co}$ have poles at complex energies $E_{n\gamma}$ such that $\nu = \nu_{n\gamma} = n - \mu_{\gamma}$ and

$$E_{n\gamma} = E_{c} - \frac{z^{2}}{2\nu_{n\gamma}^{2}}. \tag{7.10}$$

Expanding $\mathbf{G}_{co}$ in terms of its poles and residues we obtain, as a generalisation of the results of § 6.1.2,

$$\mathbf{G}_{co}(r) = \frac{1}{2\pi}\bar{\bar{\mathbf{P}}}(r)(E - Z)^{-1}t\bar{\chi}_{cc}^{-1}\bar{\chi}_{co} \tag{7.11}$$

where $\bar{\bar{\mathbf{P}}}(r)$ is the matrix with elements

$$\bar{\bar{P}}_{\gamma,n\gamma'}(r) = P(\nu_{n\gamma'}, l_{\gamma}; r)N_{\gamma\gamma'} \tag{7.12}$$

and Z and t are diagonal matrices with diagonal elements $E_{n\gamma}$ and $t(\nu_{n\gamma})$ respectively.

The wavefunctions for the bound states are taken to be of the form

$$\boldsymbol{\Psi}_{b} = \sum_{\beta} \psi_{\beta} r^{-1} G_{\beta}(r) \equiv \boldsymbol{\psi}_{b} r^{-1}\mathbf{G}_{b}(r). \tag{7.13}$$

Since we are considering DR from resonances high in Rydberg series we assume that the radiative transitions are to highly-excited bound states in series converging to the ground level of the ion. We therefore include in (7.13) only those states ψ_{β} which belong to the lowest core level and we take the total energy for the bound states to be

$$E = E^{+} - \frac{z^{2}}{2\nu_{b}^{2}} \tag{7.14}$$

where E^+ is the energy for the ground level and ν_b the effective quantum number for the bound states. Letting $\boldsymbol{\chi}_{bb}$ be the $\boldsymbol{\chi}$ matrix for the bound states, the effective quantum number ν_b for bound states must be such that

$$\det[\chi_{bb} - \exp(-2\pi i \nu_b)] = 0. \tag{7.15}$$

We diagonalise $\boldsymbol{\chi}_{bb}$ to obtain eigenvalues $\bar{\boldsymbol{\chi}}_{\beta\beta} = \exp(2\pi i \mu_\beta)$. The wavefunction for a particular bound state is then

$$\Psi_{m\beta'} = \sum_\beta \psi_\beta r^{-1} G_{\beta,m\beta'}(r) \tag{7.16}$$

where

$$G_{\beta,m\beta'}(r) = P(\nu_{m\beta'}, l_\beta\,; r) M_{\beta\beta'} \tag{7.17}$$

$\nu_{m\beta} = m - \mu_\beta$ and M is real and such that

$$M^{\mathrm{T}} M = 1 \tag{7.18}$$

The matrix of bound-state functions $\boldsymbol{\Psi}_n$ has elements given by (7.16).

The radiative matrix element is

$$D_{\alpha,\mu m\beta} = (\Psi_\alpha | D_\mu | \Psi_{m\beta}) \tag{7.19}$$

where Ψ_α is an element of the collision function (7.1). As in § 6 we replace Ψ in the expression for $\mathbf{D}$ by $\boldsymbol{\psi}_c(1/r)\mathbf{G}_{co}$ and we take D_μ to operate only on the coordinates of the core electrons. We then obtain

$$\mathbf{D} = (G_{co} | \mathbf{a} | G_b) \tag{7.20}$$

where $\mathbf{a}$ has elements

$$a_{\gamma,\mu\beta} = (\psi_\gamma | D_\mu | \psi_\beta). \tag{7.21}$$

We consider the matrix defined by

$$\mathbf{R} = 2\pi \mathbf{a}\mathbf{a}^\dagger \tag{7.22}$$

and hence having elements

$$R_{\gamma\gamma'} = 2\pi \sum_{\mu m\beta} a_{\gamma,\mu m\beta}\, a^*_{\gamma',\mu m\beta} \tag{7.23}$$

(the quantum number m specifies individual bound states, as in § 6.2). We can take the index γ to stand for

$$\gamma = C_c l_\gamma S L M_S M_L \tag{7.24}$$

and the index β for

$$\beta = C_b l_\beta S' L' M'_S M'_L \tag{7.25}$$

where C_c specifies the excited level of the core and C_b the ground level; C_c is the same for all states γ and C_b the same for all states β. Since D_μ does not operate on $\hat{\mathbf{r}}$, $a_{\gamma,\mu m\beta}$ is zero unless $l_\gamma = l_\beta$. It follows that $\mathbf{R}$ is diagonal. The diagonal elements $R_{\gamma\gamma}$ give the total radiative decay probability for the excited core level and are independent of γ. We may therefore replace $\mathbf{R}$ by R, the total decay probability.

The further analysis for the many-channel case follows closely the analyses given in §§ 5 and 6. The final results are as follows. The electron–electron scattering matrix, allowing for radiative decay, is

$$\mathcal{S}_{ee}=\chi_{oo}-\chi_{oc}[\chi_{cc}-g(\nu)\exp(-2\pi i\nu)]^{-1}\chi_{co} \tag{7.26}$$

where

$$g(\nu)=\exp(\pi\nu^3R/z^2). \tag{7.27}$$

The DR probability is given by the diagonal elements of the matrix

$$1-\mathcal{S}^{\dagger}_{ee}\mathcal{S}_{ee}=G\chi_{oc}[\chi_{cc}-g\exp(-2\pi i\nu)]^{-1}[\chi^*_{cc}-g\exp(+2\pi i\nu)]^{-1}\chi^*_{co}. \tag{7.28}$$

For the many-channel case we consider the reduction in the probability of inelastic scattering due to radiative decays. The probability for inelastic scattering is $|\mathcal{S}_{\alpha\alpha'}|^2$ where $\mathcal{S}_{\alpha\alpha'}$ is an element of (7.26). For the probability averaged over resonances we obtain, using methods similar to those of § 6.6.2,

$$\langle|\mathcal{S}_{\alpha\alpha}|^2\rangle=|\chi_{\alpha\alpha'}|^2+\sum_{\gamma\gamma'}\frac{\bar{\chi}_{\alpha\gamma}\bar{\chi}_{\gamma\alpha'}\bar{\chi}^*_{\alpha\gamma'}\bar{\chi}^*_{\gamma'\alpha'}}{(G+1-\bar{\chi}_{\gamma\gamma}\bar{\chi}^*_{\gamma'\gamma'})} \tag{7.29}$$

where

$$G(\nu)=g(\nu)^2-1=\exp\left(\frac{2\pi\nu^3R}{z^2}\right)-1. \tag{7.30}$$

If radiative decays are neglected we have $G=0$ and (7.29) reduces to the familiar formula of Gailitis (1963, see also S83). The approximate formulae used by Presnyakov and Urnov (1974) and by Pradhan (1981) are of the form (7.29) with G defined by $G=2\pi\nu^3R/z^2$ and with terms $\gamma\neq\gamma'$ omitted; they are therefore similar to the DR formula of Burgess, which will be discussed in § 8.5.

The DR probability for entrance channel α is

$$\mathcal{P}_\alpha(\mathrm{DR})=1-\sum_{\alpha'}|\mathcal{S}_{\alpha'\alpha}|^2. \tag{7.31}$$

Using (7.29) and the unitarity of χ we obtain for the averaged probability

$$\langle\mathcal{P}_\alpha(\mathrm{DR})\rangle=G\sum_{\gamma\gamma'}\frac{\bar{\chi}_{\alpha\gamma}\bar{\chi}^*_{\gamma'\alpha}(N^{\mathrm{T}}N^*)_{\gamma\gamma'}}{(G+1-\bar{\chi}_{\gamma\gamma}\bar{\chi}^*_{\gamma'\gamma'})} \tag{7.32}$$

but, since $\mathbf{N}$ is complex, the condition (7.4) does not ensure that $\mathbf{N}^{\mathrm{T}}\mathbf{N}^*=\mathbf{1}$.

8. Summary and discussion

8.1. Probabilities and cross sections

We use $\mathcal{P}$(DR) for the dimensionless probability that DR occurs. Relations between probabilities $\mathcal{P}$, collision strengths Ω and collision cross sections Q are given in § 2.5. We use A and R, respectively, for the probabilities per unit time of the *autoionisation* process (1.2) and the *radiative* process (1.3). Since we use atomic units, with $\hbar=1$, the probability per unit time for any process is equal numerically to the associated energy width.

8.2. DR *for one open channel and one resonance*

For one channel open and one resonance we have, from (4.35),

$$\mathscr{P}(\mathrm{DR})=\frac{AR}{(E-E_0)^2+[\frac{1}{2}(A+R)]^2} \tag{8.1}$$

where E is the total energy of the system and E_0 the energy of the resonance state X^{**}. Integration over E gives

$$\int \mathscr{P}(\mathrm{DR})\,\mathrm{d}E=2\pi\frac{AR}{(A+R)} \tag{8.2}$$

which is the expression obtained by Bates and Massey using the intuitive argument that, by detailed balance, A is equal to the probability for the initial capture process (1.1) and that $R/(A+R)$ the branching ratio for (1.1) to be followed by (1.3).

8.3. *Many open channels and a finite number of resonances*

For many open channels and a finite number of resonances the electron–electron scattering matrix, neglecting interactions with the radiation field, has the form

$$\mathbf{S}=\mathbf{X}-\mathrm{i}\mathbf{Y}(E-\mathbf{Z})^{-1}\mathbf{Y}^{\mathrm{T}} \tag{8.3}$$

where $\mathbf{X}$, $\mathbf{Y}$ and $\mathbf{Z}$ vary slowly with E and satisfy the relations (2.16) to (2.19) which ensure that $\mathbf{S}$ is unitary, $\mathbf{S}^{\dagger}\mathbf{S}=\mathbf{1}$. The matrix $\mathbf{Z}$ has eigenvalues $E_{\mathrm{o}j}-\frac{1}{2}\mathrm{i}A_j$ where the $E_{\mathrm{o}j}$ and A_j are resonance positions and autoionisation widths. Allowing for radiative decays of the 'bound channel' function Φ_j in (2.2), the expression (8.3) for the electron–electron scattering matrix is replaced by

$$\mathscr{S}_{\mathrm{ee}}=\mathbf{X}-\mathrm{i}\mathbf{Y}(E-\mathbf{Z}')^{-1}\mathbf{Y}^{\mathrm{T}} \tag{8.4}$$

where

$$\mathbf{Z}'=\mathbf{Z}-\mathrm{i}\mathbf{R} \tag{8.5}$$

and $\mathbf{R}$ is a matrix for radiative decays defined by (5.9) and (5.11). Amplitudes for DR are given by the elements of the photon–electron scattering matrix, $\mathscr{S}_{\mathrm{pe}}$. The conservation condition is

$$\mathscr{S}^{\dagger}_{\mathrm{ee}}\mathscr{S}_{\mathrm{ee}}+\mathscr{S}^{\dagger}_{\mathrm{pe}}\mathscr{S}_{\mathrm{pe}}=1 \tag{8.6}$$

and the total DR probability for entrance channel α is given by the diagonal element $\mathscr{P}_{\alpha\alpha}(\mathrm{DR})$ of the matrix

$$\boldsymbol{\mathscr{P}}(\mathrm{DR})=1-\mathscr{S}^{\dagger}_{\mathrm{ee}}\mathscr{S}_{\mathrm{ee}}. \tag{8.7}$$

The positions and widths of the resonances, allowing for radiative decays, are given by the eigenvalues of $\mathbf{Z}'$. If the separations are larger than the widths, the DR probability is given by a sum of expressions of the type (8.1). The case of overlapping resonances can be handled by diagonalising $\mathbf{Z}'$. The results obtained in § 5 are similar to those obtained in a number of previous papers (for example, Trefftz 1970) but are somewhat more general.

8.4. *One open channel and one closed channel*

We consider two 'collision channels' in (2.2), with channel 1 open and channel 2 closed. The electron–electron scattering matrix has only one element, $S = S_{11}$, and from QDT

$$S = \chi_{11} - \chi_{12}[\chi_{22} - \exp(-2\pi i\nu)]^{-1}\chi_{21} \tag{8.8}$$

where ν is the effective quantum for channel 2 (the total energy is $E = E_2 - z^2/(2\nu^2)$) and where the matrix χ is unitary and is a slowly varying function of E. The complex quantum defect, μ_2, for the closed channel is defined by $\chi_{22} = \exp(2\pi i\mu_2)$. We put μ

$$\mu_2 = p + iq \tag{8.9}$$

with p and q real and $q > 0$. It follows from (8.8) that poles in S occur at complex energies.

$$E_n(\text{pole}) = E_2 - \frac{z^2}{2(n-p-iq)^2} \tag{8.10}$$

giving, for $(n-p) \gg q$,

$$E_n(\text{pole}) = E_2 - \frac{z^2}{2(n-p)^2} - \frac{iqz^2}{(n-p)^3}. \tag{8.11}$$

The separation between adjacent resonances is, from (8.11),

$$\delta E = z^2/(n-p)^3. \tag{8.12}$$

We can interpret the imaginary part of (8.11) as an autoionisation width, giving

$$A_n = 2qz^2/(n-p)^3 \tag{8.13}$$

so long as this gives widths much smaller than separations, $A_n \ll \delta E_n$.

We consider radiative decays to be due to $2 \to 1$ transitions in the ion core, with probability R independent of n. After some effort (§ 6) we obtain the expression

$$\mathcal{S}_{ee} = \chi_{11} - \chi_{12}[\chi_{22} - g(\nu)\exp(-\pi i\nu)]^{-1}\chi_{21} \tag{8.14}$$

for the electron–electron scattering matrix, where

$$g(\nu) = \exp(2\pi\Delta) \tag{8.15}$$

and

$$\Delta = \frac{\nu^3 R}{2z^2} = \tfrac{1}{2}R/\delta E \tag{8.16}$$

with $\delta E = z^2/\nu^3$. The DR probability is $\mathcal{P}(\text{DR}) \equiv \mathcal{P}(\nu)$ where, from (6.74),

$$\mathcal{P}(\nu) = \frac{|\chi_{12}|^2[g^2(\nu) - 1]}{[\chi_{22}^* - g(\nu)\exp(+2\pi i\nu)][\chi_{22} - g(\nu)\exp(-2\pi i\nu)]}$$

$$= \frac{\sinh(2\pi q)\sinh(2\pi\Delta)}{\sinh^2[\pi(q+\Delta)] + \sin^2[\pi(\nu+p)]}. \tag{8.17}$$

Some graphs for this quantity are given, and discussed, in § 6.6.3. The DR probability averaged over resonances is

$$\langle\mathcal{P}\rangle = (1/\delta E)\int_{\delta E} \mathcal{P}\, dE. \tag{8.18}$$

Using the first of equations (8.17) we obtain

$$\langle \mathscr{P} \rangle = \frac{|\chi_{12}|^2 G(\nu)}{|\chi_{12}|^2 + G(\nu)} \tag{8.19}$$

where

$$G(\nu) = \exp(4\pi\Delta(\nu)) - 1. \tag{8.20}$$

8.5. Comparison with the formula of Burgess

We summarise the method used by Burgess (1964). Following Bates and Massey (1943), he assumed (8.2) giving a mean probability, averaged over resonances, of

$$\langle \mathscr{P}_n(\text{Burgess}) \rangle = \frac{2\pi}{\delta E} \frac{A_n R}{A_n + R} \tag{8.21}$$

where A_n is the autoionisation probability for a resonance with principal quantum number n. Assuming A_n proportional to δE one has $A_n \propto n^{-3}$ and hence

$$\lim_{n\to\infty} \langle \mathscr{P}_n \rangle = (2\pi A_n/\delta E). \tag{8.22}$$

In order to obtain an expression for A_n, Burgess assumed that the photon production rate below threshold, averaged over resonances, joins smoothly to the photon production rate just above threshold, due to collisional excitation (CE). This assumption is justified by our work. Since all excitations are followed by photon emission, the probability for photon emission just above threshold is $\mathscr{P}(\text{CE}) = |\chi_{12}|^2$. Using (8.22) this gives $(2\pi A_n/\delta E) = |\chi_{12}|^2$ and substitution in (8.21) gives

$$\langle \mathscr{P}_n(\text{Burgess}) \rangle = \frac{|\chi_{12}|^2 (2\pi R/\delta E)}{|\chi_{12}|^2 + (2\pi R/\delta E)} \tag{8.23}$$

which differs from our formula (8.19) in having $(2\pi R/\delta E)$ in place of G. From (8.20) and (8.16),

$$G = \exp(2\pi R/\delta E) - 1 \tag{8.24}$$

which is equal to $(2\pi R/\delta E)$ for $(2\pi R/\delta E) \ll 1$, i.e. for radiative widths much smaller than resonance separations. Although our expression for $\langle \mathscr{P} \rangle$ differs from that of Burgess, the two are in agreement in the limits of both n small and n very large. Figure 2 gives comparisons of results from the two formulae for the three cases, (a), (b) and (c) of figure 1; it is seen that the differences are not large.

That the intuitive arguments of Burgess are not entirely correct is a consequence of his use of (8.2) which is valid for isolated resonances but not for the very high states for which the separations between the resonances become smaller than the radiative widths. A number of other workers have made the additional approximation of taking the autoionisation widths to be given by the imaginary parts of the complex energies at which the S matrix has poles, but for Rydberg series of resonances this is valid only if the autoionisation widths are smaller than the separations, a condition which is satisfied only when the coupling is weak, $|\chi_{12}|^2 \ll 1$.

Formulae for DR derived by Hickman (1984) using a complex potential model are discussed by Seaton (1984) who shows that our equation (8.14) can be deduced using an intuitive argument which is simple but not rigorous.

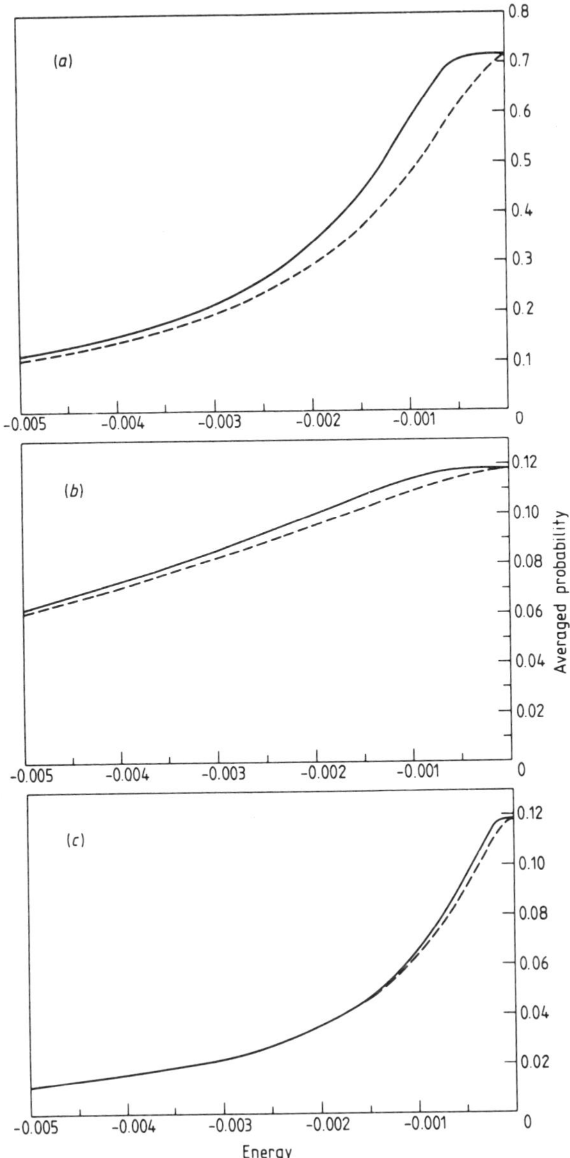

Figure 2. Mean probabilities for DR, $\langle \mathscr{P} \rangle$, calculated using our expression (8.19) (full curves) and the expression (8.23) of Burgess (broken curves). Cases (*a*), (*b*) and (*c*) as for figure 1(*a*), (*b*) and (*c*).

8.6. The many-channel case

In § 7 we considered any number of open channels and any number of closed channels all converging to the same threshold, and obtained equation (7.26),

$$\mathscr{S}_{\text{ee}} = \boldsymbol{\chi}_{\text{oo}} - \boldsymbol{\chi}_{\text{oc}}[\boldsymbol{\chi}_{\text{cc}} - g(\nu)\exp(-2\pi \mathrm{i}\nu)]^{-1}\boldsymbol{\chi}_{\text{co}}. \tag{8.25}$$

The averaged DR rate for entrance channel α is

$$\langle \mathscr{P}_\alpha(\text{DR})\rangle = G\sum_{\gamma\gamma'} \frac{\bar{\chi}_{\alpha\gamma}\bar{\chi}^*_{\gamma'\alpha}(N^{\text{T}}N^*)_{\gamma\gamma'}}{(G+1-\bar{\chi}_{\gamma\gamma}\bar{\chi}^*_{\gamma'\gamma'})} \tag{8.26}$$

where $\chi_{cc}\mathbf{N} = \mathbf{N}\bar{\chi}_{cc}$ with $\bar{\chi}_{cc}$ diagonal, $N^{\text{T}}N = 1$ and $\bar{\chi}_{oc} = \chi_{oc}\mathbf{N}$, $\bar{\chi}_{co} = \mathbf{N}^{\text{T}}\chi_{co}$. The expression (8.26) involves interference terms, $\gamma \neq \gamma'$. The formula of Burgess is recovered on neglecting the off-diagonal elements of χ_{cc} and replacing $G = [\exp(2\pi R/\delta E)-1]$ by $(2\pi R/\delta E)$. Calculations have been made of the DR rate for Mg^+, assuming LS coupling (Bell 1979). The total rate calculated using (8.26), summed over all contributing angular momenta, differs by only 3 per cent from the rate calculated using the formula of Burgess.

8.7. Inelastic scattering allowing for radiative decays

Equation (8.25) gives an expression for the scattering matrix allowing for radiative decays, from (7.29),

$$\langle |\mathscr{S}_{\alpha\alpha'}|^2\rangle = |\chi_{\alpha\alpha'}|^2 + \sum_{\gamma\gamma'} \frac{\bar{\chi}_{\alpha\gamma}\bar{\chi}_{\gamma\alpha'}\bar{\chi}^*_{\alpha\gamma'}\bar{\chi}^*_{\gamma'\alpha'}}{(G+1-\bar{\chi}_{\gamma\gamma}\bar{\chi}^*_{\gamma'\gamma'})}. \tag{8.27}$$

Putting $G = (2\pi R/\delta E)$, as in the method of Burgess, and neglecting terms with $\gamma \neq \gamma'$, this reduces to the formula of Presnyakov and Urnov (1974).

Acknowledgments

Figure 1 (*a*) (*b*), (*c*) was drawn using the FR 80 plotter of the SERC Central Computing Facility. We thank Dr P J Storey for writing the programs required for plotting these graphs.

We are indebted to Dr D L Moores and Mr K T Butler for careful readings of the typescript and to Dr E Trefftz for her care in refereeing the paper and making a number of suggestions for its improvement.

Appendix 1. Structure of the matrix χ due to inclusion of bound channels in the wavefunction expansion

Assuming an expansion of the type (2.2), the functions $\mathbf{F}(r)$ and coefficients $\mathbf{C}$ satisfy equations

$$(\mathbf{h}-\tfrac{1}{2}k^2)\mathbf{F}+\mathbf{UC}=0 \tag{A.1.1}$$

$$(\mathbf{U}|\mathbf{F})+(\mathscr{E}-E)\mathbf{C}=0 \tag{A.1.2}$$

(Eissner and Seaton 1974) where

$$\mathbf{h} = -\frac{1}{2}\left(\frac{\mathrm{d}^2}{\mathrm{d}r^2}-\frac{l(l+1)}{r^2}\right)+\mathbf{v} \tag{A.1.3}$$

$\mathbf{U}(r)$ is such that $(\mathbf{U}|\mathbf{F}) = (\boldsymbol{\Phi}|H|\psi r^{-1}\mathbf{F})$ and $\mathscr{E} = (\boldsymbol{\Phi}|H|\boldsymbol{\Phi})$. We assume standard phase conventions, which give $\mathbf{v}=\mathbf{v}^*=\mathbf{v}^{\text{T}}$, $\mathbf{U}=\mathbf{U}^*$, and $\mathscr{E}=\mathscr{E}^*=\mathscr{E}^{\text{T}}$.

We consider functions $\mathbf{F}^{(0)}$ and $\mathbf{F}^{(1)}$, solutions of

$$(\mathbf{h}-\tfrac{1}{2}k^2)\mathbf{F}^{(0)}=0 \tag{A.1.4}$$

$$(\mathbf{h}-\tfrac{1}{2}k^2)\mathbf{F}^{(1)}+\mathbf{U}=0 \tag{A.1.5}$$

and such that: for $r=0$, $\mathbf{F}^{(0)}=\mathbf{F}^{(1)}=0$; for $r\to\infty$,

$$\mathbf{F}^{(0)}\sim\phi^- - \phi^+\mathbf{X} \tag{A.1.6}$$

$$\mathbf{F}^{(1)}\sim -\sqrt{2\pi}\,\phi^+\mathbf{Y}. \tag{A.1.7}$$

Then

$$\mathbf{F}=\mathbf{F}^{(0)}+\mathbf{F}^{(1)}\mathbf{C} \tag{A.1.8}$$

is a solution of (A.1.1) and, from the definition (2.7) of $\mathbf{X}$

$$\chi=\mathbf{X}+\sqrt{2\pi}\,\mathbf{YC}. \tag{A.1.9}$$

Substitution of (A.1.8) in (A.1.2) gives

$$(\mathbf{U}|\mathbf{F}^{(0)})+[(\mathbf{U}|\mathbf{F}^{(1)})+\mathscr{E}-E]\mathbf{C}=0 \tag{A.1.10}$$

and hence

$$\mathbf{C}=(E-\mathbf{Z})^{-1}(\mathbf{U}|\mathbf{F}^{(0)}) \tag{A.1.11}$$

where

$$\mathbf{Z}=\mathscr{E}+(\mathbf{U}|\mathbf{F}^{(1)}). \tag{A.1.12}$$

We now consider functions $\mathbf{P}(r)$ and $\mathbf{Q}(r)$ such that: for $r=0$, $\mathbf{P}=\mathbf{Q}=0$; for $r\to\infty$,

$$\begin{aligned}\mathbf{P}(r)&\sim\phi^-(r)\mathbf{p}_- - \phi^+(r)\mathbf{p}_+\\ \mathbf{Q}(r)&\sim\phi^-(r)\mathbf{q}_- - \phi^+(r)\mathbf{q}_+.\end{aligned} \tag{A.1.13}$$

Defining

$$\mathbf{L}(\mathbf{P},\mathbf{Q})=(\mathbf{P}|(\mathbf{h}-\tfrac{1}{2}k^2)\mathbf{Q})-((\mathbf{h}-\tfrac{1}{2}k^2)\mathbf{P}|\mathbf{Q}) \tag{A.1.14}$$

integration by parts gives, using the Wronskian relation

$$\phi^-\frac{\mathrm{d}}{\mathrm{d}r}\phi^+-\left(\frac{\mathrm{d}}{\mathrm{d}r}\phi^-\right)\phi^+=\mathrm{i}/\pi \tag{A.1.15}$$

$$\mathbf{L}(\mathbf{P},\mathbf{Q})=\frac{\mathrm{i}}{2\pi}(\mathbf{p}_-^\dagger\mathbf{q}_- - \mathbf{p}_+^\dagger\mathbf{q}_+). \tag{A.1.16}$$

With different choices of $\mathbf{P}$ and $\mathbf{Q}$ we obtain the relations:

$\mathbf{P}$	$\mathbf{Q}$	Relation	
$\mathbf{F}^{(0)}$	$\mathbf{F}^{(0)}$	$\mathbf{X}^\dagger\mathbf{X}=\mathbf{1}$	(A.1.17)
$\mathbf{F}^{(0)*}$	$\mathbf{F}^{(0)}$	$\mathbf{X}^\mathrm{T}=\mathbf{X}$	(A.1.18)
$\mathbf{F}^{(1)}$	$\mathbf{F}^{(1)}$	$\mathbf{Y}^\dagger\mathbf{Y}=\mathrm{i}[(\mathbf{U}|\mathbf{F}^{(1)})-(\mathbf{F}^{(1)}|\mathbf{U})]$	(A.1.19)
$\mathbf{F}^{(1)*}$	$\mathbf{F}^{(1)}$	$(\mathbf{F}^{(1)}|\mathbf{U})^*=(\mathbf{U}|\mathbf{F}^{(1)})$	(A.1.20)
$\mathbf{F}^{(1)}$	$\mathbf{F}^{(0)}$	$\mathbf{Y}^\dagger\mathbf{X}=\mathrm{i}\sqrt{2\pi}(\mathbf{U}|\mathbf{F}^{(0)})$	(A.1.21)
$\mathbf{F}^{(0)*}$	$\mathbf{F}^{(1)}$	$\mathbf{Y}=\mathrm{i}\sqrt{2\pi}(\mathbf{F}^{(0)}|\mathbf{U})^*$.	(A.1.22)

We may now obtain equations (2.14) to (2.19). Since $\mathbf{U}=\mathbf{U}^*$, (A.1.22) gives $\mathbf{Y}^{\mathrm{T}}=\mathrm{i}\sqrt{2\pi}(\mathbf{U}|\mathbf{F}^{(0)})$; substitution in (A.1.11) gives (2.15); and substitution of (2.15) in (A.1.9) gives (2.14). Equations (A.1.17), (A.1.18) give (2.16). Using (A.1.12), (2.17) and (2.18) follow from (A.1.19) and (A.1.20). Equation (2.19) follows from (A.1.21) and (A.1.22).

Appendix 2. Overlap integrals

We use the notation $P_\nu(r)$ for the functions $P(\nu, l; r)$ introduced in § 2. These functions are defined by (2.28) for all r (which ensures that they are convergent at the origin) and are given by (2.29) for $r > r_0$ where r_0 is some finite number larger than the mean atomic radius. It is required to evaluate

$$(P_\nu | P_{\nu'}) = \int_0^\infty P_\nu P_{\nu'}\,\mathrm{d}r. \tag{A.2.1}$$

We consider that ν and ν' are both large and that the contribution to (A.2.1) from $r < r_0$ can be neglected, giving

$$(P_\nu | P_{\nu'}) = \int_{r_0}^\infty P_\nu P_{\nu'}\,\mathrm{d}r. \tag{A.2.2}$$

For $r > r_0$, (2.29) can be written, in the notation of S83, as

$$P_\nu(r) = -(-1)^l z^{1/2} K\theta \tag{A.2.3}$$

where θ is the Whittaker function. Putting

$$h = -\frac{1}{2}\left(\frac{\mathrm{d}^2}{\mathrm{d}r^2} - \frac{l(l+1)}{r^2}\right) - \frac{z}{r} \tag{A.2.4}$$

and

$$E_\nu = -z^2/(2\nu^2) \tag{A.2.5}$$

the functions P_ν are solutions of

$$(h - E_\nu)P_\nu = 0. \tag{A.2.6}$$

From the identity

$$\int_{r_0}^\infty [P_\nu(h - E_{\nu'})P_{\nu'} - P_{\nu'}(h - E_\nu)P_\nu]\,\mathrm{d}r = 0 \tag{A.2.7}$$

we obtain, integrating by parts,

$$\int_{r_0}^\infty P_\nu P_{\nu'}\,\mathrm{d}r = -\frac{1}{2(E_\nu - E_{\nu'})}\left(P_\nu \frac{\mathrm{d}}{\mathrm{d}r} P_{\nu'} - P_{\nu'} \frac{\mathrm{d}}{\mathrm{d}r} P_\nu\right)\bigg|_{r=r_0}. \tag{A.2.8}$$

For ν and ν' both large we have $(E_\nu - E_{\nu'}) \simeq -z^2(\nu' - \nu)/\nu^3$.

We express P_ν in terms of the regular and irregular Coulomb functions s and c using equation (2.55) of S83, to obtain

$$P_\nu = -z^{1/2}(2/\nu^3)^{1/2}[c \sin(\pi\nu) - s \cos(\pi\nu)]. \tag{A.2.9}$$

Assuming that r_0 is not large, the functions s and c at $r = r_0$ vary slowly with the energy and we may put

$$P_{\nu'} = -z^{1/2}(2/\nu^3)^{1/2}[c \sin(\pi\nu') - s \cos(\pi\nu')] \tag{A.2.10}$$

that is to say we take P_ν and $P_{\nu'}$ to differ only in the trigonometric functions of $\pi\nu$ and $\pi\nu'$. Using (A.2.9) and the Wronskian relation

$$c\frac{\mathrm{d}}{\mathrm{d}r}s - s\frac{\mathrm{d}}{\mathrm{d}r}c = z/\pi \tag{A.2.11}$$

we obtain

$$(P_\nu | P_{\nu'}) = \frac{\sin[\pi(\nu - \nu')]}{\pi(\nu - \nu')}. \tag{A.2.12}$$

A further point should be noted. The definition (A.2.1) of $(P_\nu | P_{\nu'})$ applies for functions which are real. For complex functions, (A.2.1) is replaced by

$$(P_\nu | P_{\nu'}) = \int_0^\infty P_\nu^* P_{\nu'} \, \mathrm{d}r \tag{A.2.13}$$

giving

$$(P_\nu^* | P_{\nu'}) = \int_0^\infty P_\nu P_{\nu'} \, \mathrm{d}r \tag{A.2.14}$$

which is the quantity used in § 6.5.2.

Appendix 3. A completeness relation

Defining

$$I(\nu, \nu') = \sum_n (P_\nu | P_{n-b})(P_{n-b} | P_{\nu'}) \tag{A.3.1}$$

we show that

$$I(\nu, \nu') = (P_\nu | P_{\nu'}). \tag{A.3.2}$$

Using (A.3.1) and (A.2.12),

$$I(\nu, \nu') = \sum_n \frac{\sin[\pi(\nu - n + b) \sin[\pi(\nu' - n + b)]}{\pi^2(\nu - n + b)(\nu' - n + b)} \tag{A.3.3}$$

and hence

$$I = \frac{\sin(\pi\alpha)\sin(\pi\alpha')}{\pi^2} J \tag{A.3.4}$$

where $\alpha = (\nu + b)$, $\alpha' = (\nu' + b)$ and

$$J = \sum_n \frac{1}{(n-\alpha)(n-\alpha')}$$

$$= \frac{1}{(\alpha - \alpha')} \sum_n \left(\frac{1}{(n-\alpha)} - \frac{1}{(n-\alpha')} \right). \tag{A.3.5}$$

Identifying the poles and residues of $\cot(\pi x)$ we obtain the expansion

$$\pi \cot(\pi x) = \sum_n \frac{1}{(n+x)} \tag{A.3.6}$$

and it follows that

$$J = \frac{\pi}{(\alpha - \alpha')}[-\cot(\pi\alpha) + \cot(\pi\alpha')] \tag{A.3.7}$$

and hence that

$$J = \frac{\pi \sin[\pi(\alpha - \alpha')]}{(\alpha - \alpha') \sin(\pi\alpha) \sin(\pi\alpha')}. \tag{A.3.8}$$

Substitution in (A.3.4) gives

$$I = \frac{\sin[\pi(\alpha - \alpha')]}{\pi(\alpha - \alpha')} = \frac{\sin[\pi(\nu - \nu')]}{\pi(\nu - \nu')}$$

$$= (P_\nu | P_{\nu'}). \tag{A.3.9}$$

References

Armstrong L, Theodosiou C E and Wall M J 1978 *Phys. Rev.* A **18** 2538
Bates D R and Massey H S W 1943 *Phil. Trans. R. Soc.* A **239** 269
Belić D S, Dunn G H, Morgan T J, Mueller D W and Trimmer C 1983 *Phys. Rev. Lett.* **50** 339
Bell R H 1979 *PhD Thesis* University of London
Bhatti S A, Cromer C L and Cooke W E 1981 *Phys. Rev.* A **24** 161
Burgess A 1964 *Astrophys. J.* **139** 776
Clavel J, Flower D R and Seaton M J 1981 *Mon. Not. R. Astron. Soc.* **197** 301
Davies P C W and Seaton M J 1969 *J. Phys. B: At. Mol. Phys.* **2** 757
Dittner P F, Datz S, Miller D P, Moak C D, Stelson P H, Bottcher C, Dress W B, Alton G D, Nešković N and Four C M 1983 *Phys. Rev. Lett.* **51** 31
Dubau J 1973 *PhD Thesis* University of London
Dubau J and Volonté S 1980 *Rep. Prog. Phys.* **43** 199
Eissner W and Seaton M J 1974 *J. Phys. B: At. Mol. Phys.* **5** 2187
Fano U 1961 *Phys. Rev.* **124** 1866
Gailitis M 1963 *Sov. Phys.-JETP* **17** 1328
Harrington J P, Lutz J H and Seaton M J 1981 *Mon. Not. R. Astron. Soc.* **195** 2P
Heitler W 1954 *Quantum Theory of Radiation* 3rd edn (Oxford: Clarendon) p 182
Hickman A P 1984 *J. Phys. B: At. Mol. Phys.* **17** L101
Massey H S W and Bates D R 1942 *Rep. Prog. Phys.* **9** 62
Mitchell J B A, Ng C T, Forand J L, Levac D P, Mitchell R E, Sen A, Miko D B and McGowan J W 1983 *Phys. Rev. Lett.* **50** 335
Pradhan A K 1981 *Phys. Rev. Lett.* **47** 79
Presnyakov L P and Urnov A M 1974 *J. Phys. B: At. Mol. Phys.* **8** 1280
Rozman L J 1981 *Physics of Electronic and Atomic Collisions* ed S Datz (Amsterdam: North-Holland) p 641
Seaton M J 1978 *J. Phys. B: At. Mol. Phys.* **11** 4067
—— 1982 *J. Phys. B: At. Mol. Phys.* **15** 3899
—— 1983a *Rep. Prog. Phys.* **46** 167
—— 1983b *Comment. At. Mol. Phys.* **13** 171
—— 1984 *J. Phys. B: At. Mol. Phys.* **17** L531
Seaton M J and Storey P J 1976 *Atomic Processes and Applications* ed P G Burke and B L Moiseiwitsch (Amsterdam: North-Holland) p 133
Shore B W 1967 *Rev. Mod. Phys.* **39** 439

Stickland D J, Penn C J, Seaton M J, Snijders M A J and Storey P J 1981 *Mon. Not. R. Astron. Soc.* **197** 107
Storey P J 1981 *Mon. Not. R. Astron. Soc.* **195** 27P
Titchmarsh E C 1948 *Introduction to the Theory of Fourier Integrals* (Oxford: Clarendon) pp 13, 42
Trefftz E 1967 *Z. Astrophys.* **65** 299
—— 1969 *Physics of the One- and Two-Electron Atoms* ed F Bopp and H Kleinpoppen (Amsterdam: North-Holland) p 839
—— 1970 *J. Phys. B: At. Mol. Phys.* **6** 763

1985 *J. Phys. B: At. Mol. Phys.* **18** 1631–5

Dielectronic recombination: II. Effects on cross sections for inelastic collisions between electrons and atomic ions

A K Pradhan† and M J Seaton‡

† Joint Institute for Laboratory Astrophysics, University of Colorado and National Bureau of Standards, Boulder, Colorado 80309, USA

‡ Department of Physics and Astronomy, University College London, Gower St, London WC1E 6BT, England

Received 20 September 1984

Abstract. Formulae obtained by Bell and Seaton in the first paper of the present series are used to calculate collision strengths for the 1^1S–2^1S transitions in O^{6+} at energies below the 2^1P threshold. Radiative decays lead to modifications of autoionising resonance profiles and substantial reductions in collision strengths averaged over resonances.

Cross sections for electron impact excitation of positive ions contain resonance structures forming Rydberg series converging to higher excitation thresholds. Radiative decays of the resonances, leading to capture of the colliding electron, were discussed by Presnyakov and Urnov (1975). They used intuitive arguments to obtain an expression for cross sections, calculated allowing for radiative decays and averaged over resonances. It can be shown that their expression is consistent with that obtained by Burgess (1964) for the process of dielectronic recombination (DR). The importance of the radiative decays depends on the ratio R/A where R is the probability for the radiative transition and A that for autoionisation. Along an isoelectronic sequence, R/A behaves like z^4, where z is the charge on the ion, and the radiative decays are therefore more important for more highly ionised systems.

Pradhan (1981) has considered excitation of the He-like ion O^{6+} with five levels, 1^1S, 2^3S, 2^3P, 2^1S and 2^1P in order of increasing excitation energy. Above the 2^1P threshold the collision strength $\Omega(1^1S, 2^1P)$ for the optically allowed $1^1S\rightarrow 2^1P$ transition is much larger than $\Omega(1^1S, 2^1S)$ for the optically forbidden $1^1S\rightarrow 2^1S$ transition. Below the 2^1P threshold, $\Omega(1^1S, 2^1S)$ contains a series of resonances converging to the 2^1P threshold. One can say that the electron incident on the ion in the 1^1S state can produce excitation of 2^1P below the 2^1P threshold but in doing so is captured into an autoionising state, and that subsequent autoionisation can leave the ion in either the 1^1S or 2^1S states. The probability R for radiative decay of the autoionising state is essentially that for the $2^1P\rightarrow 1^1S$ radiative transition in the ion core. The work of Pradhan showed that allowance for radiative decays leads to a substantial reduction in the 1^1S–2^1S collision strength averaged over resonances, $\bar{\Omega}(1^1S, 2^1S)$, throughout the entire energy range between the 2^1S and 2^1P thresholds.

A more rigorous theory has been developed by Bell and Seaton (1985) who obtain expressions for detailed resonance profiles allowing for radiative decays as well as for collision strengths averaged over resonances. We give a summary of the theory for

0022-3700/85/081631+05\$02.25

the case of O^{6+} with five levels. Let χ be the scattering matrix for energies such that all channels are open. This matrix varies slowly with energy and may be extrapolated to the region of some channels closed. Partitioning χ into open-open, open-closed, closed-open and closed-closed submatrices, quantum defect theory (see Seaton 1983) gives for the scattering matrix, when some channels are closed,

$$\mathbf{S} = \chi_{oo} - \chi_{oc}[\chi_{cc} - \exp(-2\pi i \nu)]^{-1}\chi_{co} \tag{1}$$

where ν is the diagonal matrix of effective quantum numbers in the closed channels. For energies between the 2^1S and 2^1P thresholds all closed channels belong to 2^1P and ν is a multiple of the unit matrix (if E is the total energy, in au, we have $E = E(2^1\text{P}) - z^2/(2\nu^2)$ with $z = 6$). We use a representation with χ_{cc} diagonal. If α and β are open channels, from (1) we obtain for $|S_{\alpha\beta}|^2$, averaged over resonances, the formula of Gailitis (1963),

$$\langle |S_{\alpha\beta}|^2\rangle = |\chi_{\alpha\beta}|^2 + \sum_{\gamma\gamma'} \frac{\chi_{\alpha\gamma}\chi_{\gamma\beta}\chi^*_{\alpha\gamma'}\chi^*_{\gamma'\beta}}{(1-\chi_{\gamma\gamma}\chi^*_{\gamma'\gamma'})} \tag{2}$$

where γ, γ' are closed channels. Presnyakov and Urnov neglect terms in (2) with $\gamma \neq \gamma'$. Using the unitarity of χ one then obtains

$$\langle |S_{\alpha\beta}|^2\rangle = |\chi_{\alpha\beta}|^2 + \sum_{\gamma}\left[|\chi_{\alpha\gamma}|^2|\chi_{\gamma\beta}|^2\left(\sum_{\alpha'}|\chi_{\gamma\alpha'}|^2\right)^{-1}\right] \tag{3}$$

where α' is summed over all open channels. We use $\mathscr{S}$ for the scattering matrix allowing for radiative decays. Using an intuitive argument, Presnyakov and Urnov modify (3) to allow for radiative decays and obtain an expression which may be written, in our notation, as

$$\langle |\mathscr{S}_{\alpha\beta}|^2\rangle = |\chi_{\alpha\beta}|^2 + \sum_{\gamma}\left[|\chi_{\alpha\gamma}|^2|\chi_{\gamma\beta}|^2\left(B + \sum_{\alpha'}|\chi_{\gamma\alpha'}|^2\right)^{-1}\right] \tag{4}$$

where

$$B = 2\pi\nu^3 R/z^2$$

and R is the probability, in au, for radiative decays. For entrance channel α, the total probability that radiative decays occur is $1 - \Sigma_\beta |\mathscr{S}_{\alpha\beta}|^2$ and is equal to the averaged DR probability, $\langle P_\alpha(\text{DR})\rangle$. Using the expression (4) of Presnyakov and Urnov, one obtains the expression of Burgess for $\langle P_\alpha(\text{DR})\rangle$.

In the more rigorous theory of Bell and Seaton the following results are obtained:

$$\mathscr{S} = \chi_{oo} - \chi_{oc}[\chi_{cc} - g(\nu)\exp(-2\pi i\nu)]^{-1}\chi_{co} \tag{5}$$

and

$$\langle |\mathscr{S}_{\alpha\beta}|^2\rangle = |\chi_{\alpha\beta}|^2 + \sum_{\gamma\gamma'} \frac{\chi_{\alpha\gamma}\chi_{\gamma\beta}\chi^*_{\alpha\gamma'}\chi_{\gamma'\beta}}{(G+1-\chi_{\gamma\gamma}\chi^*_{\gamma'\gamma'})} \tag{6}$$

where

$$g(\nu) = \exp(\tfrac{1}{2}B) = \exp(\pi\nu^3 R/z^2) \tag{7}$$

and

$$G = \exp(B) - 1 = \exp(2\pi\nu^3 R/z^2) - 1. \tag{8}$$

The resonance separations are $\delta E = z^2/\nu^3$ giving $B = 2\pi\nu^3 R/z^2 = 2\pi R/\delta E$. For ν sufficiently small we have $B \ll 1$ (radiative width much smaller than resonance separations) and $G \simeq B$. Neglecting terms with $\gamma \neq \gamma'$, (6) then agrees with (4).

We have made calculations for $\Omega(1^1S, 2^1S)$ in O^{6+} using (5) and (6). The matrix χ is calculated using methods described by Pradhan *et al* (1981) and Pradhan (1981). The important contributing states of total angular momentum and parity, $SL\pi$, are ^2S, $^2P^o$, ^2D, $^2F^o$ and ^2G. Figure 1(a) shows resonance structure in the total collision

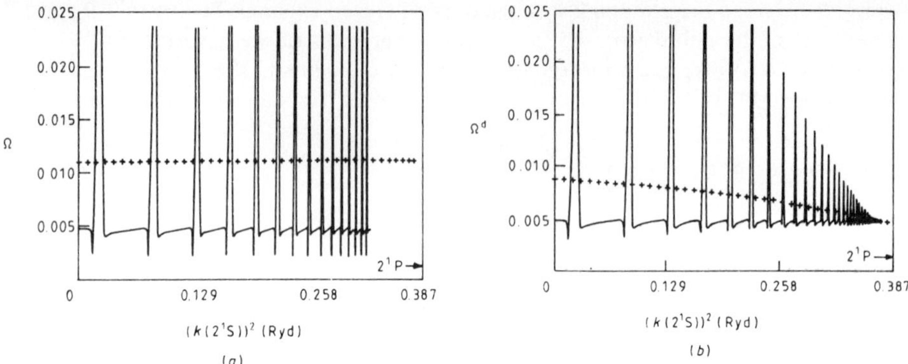

Figure 1. (a) The O^{6+} collision strength $\Omega(1^1S, 2^1S)$ calculated using equation (1) (full curve) and the averaged collision strength $\bar{\Omega}(1^1S, 2^1S)$ calculated using (2)(+++ curve). (b) The collision strengths $\Omega^d(1^1S, 2^1S)$ and $\bar{\Omega}^d(1^1S, 2^1S)$ calculated allowing for radiative decays, using (5) and (6).

strength, $\Omega(1^1S, 2^1S)$, for the region between the 2^1S and 2^1P thresholds, calculated using (1). The Gailitis-average collision strength, $\bar{\Omega}$, calculated using (2), is also shown. The effect of the resonances is such that $\bar{\Omega}$ is larger than the non-resonant background contribution to Ω by a factor of about 2.4. Figure 1(b) shows the collision strengths $\Omega^d(1^1S, 2^1S)$ and $\bar{\Omega}^d(1^1S, 2^1S)$ calculated using (5) and (6), i.e., allowing for radiative decays (dielectronic recombination). The effect of decays is to make $\bar{\Omega}^d$ smaller than $\bar{\Omega}$ throughout the entire range. As the 2^1P threshold is approached the resonance separations become small compared with the radiative width and all resonance structure is smeared out. At the 2^1P threshold, $\bar{\Omega}^d$ tends to the non-resonant background collision strength.

Figures 2(a) and 2(b) show, on an expanded scale, the first two groups of resonances in the total cross section, having principal quantum numbers $n = 10$ and 11. The resonances, in the order in which they appear, are $2^1Pns\ ^2P^o$, $np\ ^2S$ and ^2D, $nd\ ^2P^o$ and $^2F^o$, $nf\ ^2D$ and ^2G and $ng\ ^2F^o$. The results for Ω of figure 2(a) are calculated using (1) and those of figure 2(b) for Ω^d using (5). It is seen that radiative decays modify even the lowest resonances in the series. The modifications are greatest for resonances nl with l large (these states have smaller autoionisation probabilities A and hence larger values of R/A).

Figures 3(a) and 3(b) show results for a single value of angular momentum and parity, $SL\pi = {}^2D$.

Finally, we note that there are no large differences between results for $\bar{\Omega}^d$ obtained in the present paper, using equation (6) from Bell and Seaton, and the results for $\bar{\Omega}^d$ obtained by Pradhan (1981) using equation (4) from Presnyakov and Urnov. Bell and

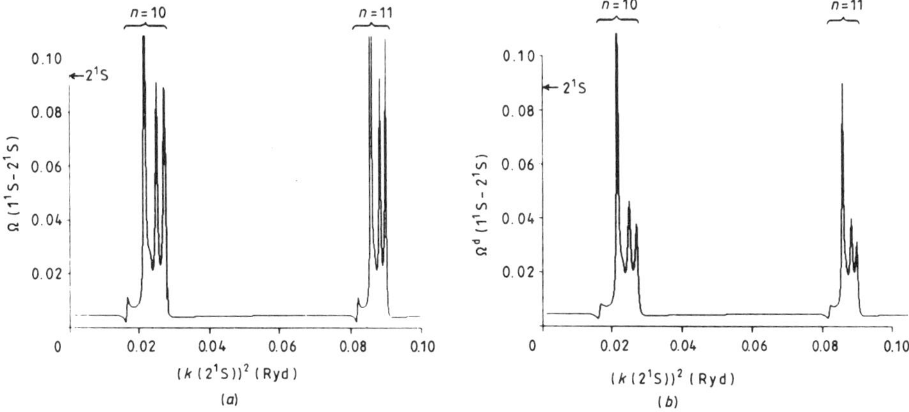

Figure 2. (a) The collision strength $\Omega(1^1S, 2^1S)$, and (b) the collision strength $\Omega^d(1^1S, 2^1S)$ allowing for radiative collisions, showing in greater detail the resonances with principal quantum numbers $n = 10$ and 11.

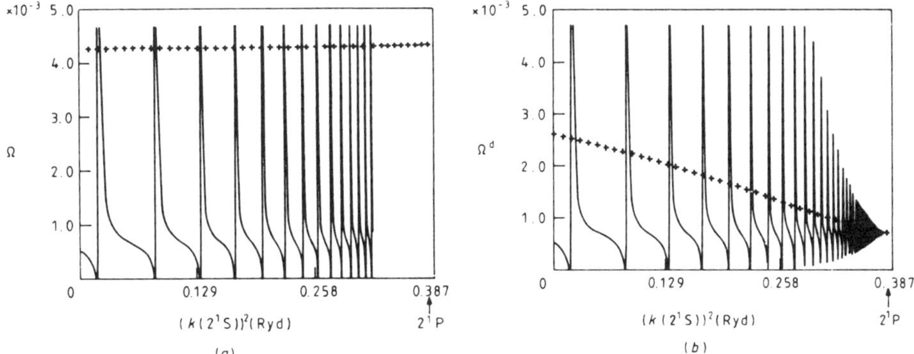

Figure 3. Contributions to 1^1S-2^1S collision strengths from states with $SL\pi = {}^2D$. (a) Ω and $\bar{\Omega}$,(b) Ω^d and $\bar{\Omega}^d$.

Seaton reach similar conclusions on comparing their formulae for DR rates with those of Burgess.

Acknowledgment

This work was supported by the US Department of Energy, Office of Fusion Energy (AKP) and by the Science and Engineering Research Council of the UK (MJS).

References

Bell R H and Seaton M J 1985 *J. Phys. B: At. Mol. Phys.* **18** 1589
Burgess A 1964 *Astrophys. J.* **139** 776
Gailitis M 1963 *Sov. Phys.-JETP* **17** 1328

Pradhan A K 1981 *Phys. Rev. Lett.* **47** 79
Pradhan A K, Norcross D W and Hummer D G 1981 *Phys. Rev.* A **23** 619
Presnyakov L P and Urnov A M 1975 *J. Phys. B: At. Mol. Phys.* **8** 1280
Seaton M J 1983 *Rep. Prog. Phys.* **46** 167

1979 *J. Phys. B: At. Mol. Phys.* **12** 863–79

Quantum defects for highly stripped ions†

W R Johnson‡‖ and K T Cheng§

‡Physics Department, University of Notre Dame, Notre Dame, Indiana 46556, USA
§Argonne National Laboratory, Argonne, Illinois 60439, USA

Received 26 June 1978

Abstract. A version of quantum-defect theory appropriate to the analysis of the spectra of highly ionised atoms is obtained from the single-electron Dirac equation. Quantum defects μ_n are defined as the principal quantum number n increases along a Rydberg series using Sommerfeld's relativistic one-electron energy level formula. Considering the analytic properties of solutions to the Dirac equation, we establish that μ_n can be extended smoothly away from the bound-state energies just as in the non-relativistic theory. Moreover, the relativistic $\mu(\epsilon)$ can be analytically continued beyond the threshold $\epsilon = mc^2$ and related to the non-Coulomb scattering phaseshift $\delta(\epsilon)$. At the threshold we establish the result $\delta(mc^2) = \pi\mu(mc^2)$ well known in non-relativistic quantum-defect theory. Several examples of the relativistic quantum-defect theory are given for highly stripped Na-like and Li-like ions.

1. Introduction

In the discussion to follow, we describe a version of the single-channel quantum-defect theory (QDT) based on the one-electron Dirac equation. Our study is directed toward an understanding of the structure of highly stripped ions such as those occurring in astrophysical and laboratory plasmas, or those produced in beam–foil stripping experiments. For such ions, the spectrum is dominated by the ionic Coulomb field, and the relativistic fine-structure effects become pronounced.

To narrow our arguments somewhat, we concentrate on the doublet spectra of highly stripped alkali-like ions. For such ions, it is natural to describe the interaction of the valence electron with the core in terms of a quantum defect. In order to distinguish clearly short-range core effects from the strong ionic Coulomb-field effects (which may include large fine-structure contributions), we are led to define the quantum defect μ_n using Sommerfeld's relativistic version of Rydberg's formula

$$T_{n\kappa} = mc^2\left(\frac{1}{[1+(\alpha Z/n^*)^2]^{1/2}} - 1\right) \tag{1}$$

where $T_{n\kappa}$ is the term energy, Z is the ionic charge, α is the fine-structure constant, and mc^2 is the electron rest energy. (In atomic units, $mc^2 = \alpha^{-2}$. We shall employ natural units $\hbar = c = 1$ in the following.) The subscripts n and κ of $T_{n\kappa}$ are the principal and angular quantum numbers, respectively; in particular, $\kappa = \mp(j+\frac{1}{2})$ for $j = l \pm \frac{1}{2}$, where j

† Work performed in part under the auspices of the US Department of Energy and by the National Science Foundation Grant No PHY77-15141.
‖ Argonne Universities Association Faculty Participant.

0022-3700/79/060863+17\$01.00

and l are the total and orbital angular momenta of the electron. Thus, the two term values $T_{n,1}$ and $T_{n,-2}$ describe the two level systems $np_{1/2}$ and $np_{3/2}$ associated with individual fine-structure components along the p series. The quantity n^* in equation (1) is an effective quantum number.

For a single electron in the nuclear Coulomb field $V(r) = -\alpha Z/r$, the eigenvalues of the Dirac equation lead to equation (1) with $n^*_{\text{Coul}} = n + \gamma - k$, where $k = |\kappa|$ and $\gamma = [k^2 - (\alpha Z)^2]^{1/2}$. In our discussion of alkali-like ions, we modify n^*_{Coul} to include core effects by subtracting a quantum defect $\mu_{n\kappa}$, so that

$$n^* = n + \gamma - k - \mu_{n\kappa}. \tag{2}$$

The quantum defects $\mu_{n\kappa}$ are determined so that equation (1) describes individual fine-structure components along a Rydberg series.

The non-relativistic limit is found from equation (1) by retaining only the lowest non-vanishing term in an expansion in powers of $(\alpha Z)^2$. Expanding equation (1), one finds

$$T_{n\kappa} \simeq -\frac{Z^2}{2n^{*2}} \quad \text{(au)} \tag{3}$$

with $n^* = n - \mu_{n\kappa}$. One important point is that the non-relativistic formula, equation (3), ascribes the fine-structure entirely to the parameters $\mu_{n\kappa}$ which describe the core interactions. Using the relativistic expression from equation (1) to analyse spectroscopic data, on the other hand, allows a clean separation of the fine-structure effects into core and ionic Coulomb contributions. As an alternative, for low charge Z, one can separate the core and ionic Coulomb contribution using a perturbative treatment (Edlén 1964).

One consequence of the analysis of actual spectroscopic data using the relativistic formula is that—over a wide range of ionic charges—the alkali fine structure is primarily due to the ionic Coulomb potential so that relativistic quantum defects are nearly identical for the two fine-structure components $\kappa = l, -l-1$ of a spectral line. By contrast, the non-relativistic quantum defects derived from equation (3) are strongly spin dependent.

We illustrate this fact in tables 1 and 2. First, in table 1, we give the quantum defects along the $nd_{3/2}$ and $nd_{5/2}$ series from the observed C IV and N V spectra (Bashkin and Stoner 1975); the non-relativistic quantum defects derived from equation (3) are compared with the relativistic ones from equation (1). While the order of magnitude of the non-relativistic quantum defects are the same for the fine-structure partners, the actual numerical values are noticeably different. By contrast, the relativistic quantum defect for the fine-structure components are numerically very close, illustrating that the ionic Coulomb potential is mainly responsible for the fine structure of these lines. This same feature characterises spectra of atoms of much higher ionisation, as illustrated in table 2, where non-relativistic and relativistic quantum defects for the d and f series of Na-like Mo (Mo XXXII) deduced from Dirac–Hartree–Fock term energies† are compared. Again, the non-relativistic quantum defects are different for the fine-structure components, while the relativistic ones are very close, illustrating the spin independence of the valence–core interaction for this very highly stripped ion.

Over the years, QDT (Ham 1955, Seaton 1958, Fano 1970) has been applied successfully to describe a wide range of atomic phenomena, including electron scattering (Seaton 1958), photoionisation (Dubau 1976), and electron capture (Lee 1977). To

† Calculated from a computer code by Desclaux (1975).

Table 1. The quantum defects along the $nd_{3/2}$ and $nd_{5/2}$ series in the observed spectra of C IV abd N V (Bashkin and Stoner 1975). The quantities μ_{NR} and μ_R are quantum defects obtained from the non-relativistic formula equation (3) and those from the relativistic one equation (1), respectively.

Ion	n	μ_{NR}		μ_R	
		$nd_{3/2}$	$nd_{5/2}$	$nd_{3/2}$	$nd_{5/2}$
C IV	3	0·001614	0·001533	0·001507	0·001498
	4	0·001912	0·001839	0·001779	0·001777
	5	0·002075	0·002015	0·001926	0·001936
	6	0·002211	0·002137	0·002051	0·002049
	7	0·002520	0·002461	0·002353	0·002365
N V	3	0·001490	0·001380	0·001324	0·001324
	4	0·001778	0·001668	0·001570	0·001571
	5	0·001876	0·001765	0·001643	0·001643
	6	0·001879	0·001768	0·001629	0·001630
	7	0·001817	0·001705	0·001556	0·001554
	8	0·001704	0·001592	0·001434	0·001433

Table 2. The DHF values of the quantum defects for nd and nf series in the spectra of Mo XXXII. The quantities μ_{NR} and μ_R are non-relativistic and relativistic quantum defects, respectively.

n	μ_{NR}				μ_R			
	$nd_{3/2}$	$nd_{5/2}$	$nf_{5/2}$	$nf_{7/2}$	$nd_{3/2}$	$nd_{5/2}$	$nf_{5/2}$	$nf_{7/2}$
3	0·0499	0·0449	—	—	0·0431	0·0427	—	—
4	0·0527	0·0474	0·00756	0·00531	0·0442	0·0434	0·00358	0·00361
5	0·0540	0·0486	0·01000	0·00775	0·0445	0·0436	0·00499	0·00502
6	0·0549	0·0494	0·01139	0·00914	0·0446	0·0437	0·00570	0·00573
7	0·0554	0·0500	0·01230	0·01004	0·0447	0·0438	0·00612	0·00614
8	0·0559	0·0504	0·01295	0·01068	0·0448	0·0439	0·00641	0·00642

study such problems, one requires a smooth function $\mu(\epsilon)$ which extends the quantum defects $\mu_{n\kappa}$ to energies ϵ other than the bound states. Ham (1955) showed how to construct such a function in the non-relativistic case and applied the resulting quantum-defect function $\mu(\epsilon)$ to problems in solid-state physics. Ham's method was extended in Seaton's mathematical theory of quantum defects (Seaton 1958). The QDT was further extended to include multi-channel situations and applied to various special problems in a series of papers by Seaton and his collaborators. A discussion of this QDT series can be found in Seaton (1970). Fano and co-workers, pursuing a parallel course, have also developed methods for understanding complex atomic spectra employing the multi-channel quantum-defect theory (Fano 1970, Lee and Lu 1973, Lee 1974).

In the present paper where we concentrate on the one-channel relativistic version of QDT, we employ analytic techniques similar to those developed in early papers of Ham (1955), Seaton (1958) and Fano (1970) to extend relativistic quantum defects away from the bound-state energies. In the non-relativistic theory, this was done using an analytic function $\beta(\epsilon)$ defined in terms of a pair of linearly independent analytic Coulomb wavefunctions. Ham (1955), and later Seaton (1958), employed a detailed analysis of the non-relativistic Coulomb wavefunctions to construct $\beta(\epsilon)$.

To carry out a parallel analysis in the relativistic case, we must first discuss several important details concerning relativistic Coulomb wavefunctions. Unfortunately, there is no sufficiently complete study of the relevant functions available, so we must devote some space to an important but tedious discussion of the properties of relativistic Coulomb wavefunctions. To spare the casual reader some of the formal details, we summarise here the main lines of the argument presented in the following section. We first establish that, under rather general conditions, relativistic electron theory leads to the existence of a function $\beta(\epsilon)$ analytic in the ϵ plane in terms of which a smooth quantum defect function $\mu(\epsilon)$ can be defined; of course, $\mu_n = \mu(\epsilon_n)$ at the bound states. Since $\beta(\epsilon)$ is analytic, it can be used to define an analytic continuation of $\mu(\epsilon)$ to the continuum region where $\epsilon > m$. In the continuum, $\mu(\epsilon)$ is shown to be related to $\delta(\epsilon)$, the short-range non-Coulomb part of the scattering phaseshift by

$$\cot \delta(\epsilon) = (1 + \cos \pi b \mathrm{e}^{-2\pi\nu}) \cot \pi\mu(\epsilon) + \sin \pi b \mathrm{e}^{-2\pi\nu}. \tag{4}$$

In equation (4) we have used $b = 2\gamma + 1$, where $\gamma = [k^2 - (\alpha Z)^2]^{1/2}$ (see equation (2)), and $\nu = \alpha Z \epsilon / p$, where $p = (\epsilon^2 - m^2)^{1/2}$ is the electron momentum. In the non-relativistic limit $\gamma \to k$, so that b becomes an odd integer, and equation (4) reduces to the form given in Seaton's (1958) paper, namely,

$$\tan \delta(\epsilon) = \tan \pi\mu(\epsilon) / (1 - \mathrm{e}^{-2\pi\nu}). \tag{5}$$

From either the relativistic or the non-relativistic formula, one extracts exactly the same threshold relation†

$$\tan \delta(\epsilon = m) = \tan \pi\mu(\epsilon = m) \tag{6}$$

which can be used to remove the arbitrariness in $\delta(\epsilon)$ by requiring

$$\delta(\epsilon = m) = \pi\mu(\epsilon = m). \tag{7}$$

Following our discussion of the existence of $\mu(\epsilon)$, we discuss the relation connecting Coulomb scattering wavefunctions above the threshold with the exponentially increasing and decreasing wavefunctions below the threshold. These important connection formulae are used in conjunction with equation (4) in the practical solution to photoionisation (Lee and Lu 1973, Dubau 1976) and scattering problems (Seaton 1966) using QDT.

Next we define and discuss a relativistic generalisation of Ham's η defect (Ham 1955). The η defect is expected to be somewhat more suitable for extrapolation over a large energy range than $\mu(\epsilon)$, and in the non-relativistic theory, $\eta(\epsilon)$ is used by Ham (1955) and by Seaton (1958) in preference to $\mu(\epsilon)$ to extrapolate or interpolate, particularly where high precision is required.

Finally, in the theory section, we determine wavefunction normalisation factors below and above threshold in terms of quantum defects. The resulting normalisation factors are important for a complete understanding of the energy dependence of matrix elements of physical observables.

† This relation is also obtained from the Dirac equation by Zitilis (1977).

2. Theory

Consider an N-electron alkali atom in which the valence electron sees an effective spherically symmetrical potential $V(r)$ which behaves as $-\alpha Z_{\rm nuc}/r$ near the origin and which behaves as $-\alpha Z/r$ asymptotically, with $Z = Z_{\rm nuc} - N + 1$. The Dirac equation for the valence electron is

$$H_0 \Psi(\boldsymbol{r}) = \epsilon \Psi(\boldsymbol{r}). \tag{8}$$

Here, $H_0 = \boldsymbol{\alpha} \cdot \boldsymbol{p} + \beta m + V(r)$ is the single-particle Dirac Hamiltonian, $\epsilon = m + T$ is the total energy of the electron with the term energy T, and $\boldsymbol{\alpha}$ and β are the usual Dirac matrices. We can write the four-component Dirac wavefunction $\Psi(\boldsymbol{r})$ as

$$\Psi_{n\kappa m}(\boldsymbol{r}) = \frac{1}{r}\begin{pmatrix} \mathrm{i} G_{n\kappa}(r)\Omega_{\kappa m}(\hat{r}) \\ F_{n\kappa}(r)\Omega_{-\kappa m}(\hat{r}) \end{pmatrix} \tag{9}$$

where, as in the introduction, $\kappa = \mp(j+\frac{1}{2})$ for $j = l \pm \frac{1}{2}$, $G_{n\kappa}$ and $F_{n\kappa}$ are the large and small component radial functions, and $\Omega_{\kappa m}$ is a spin–orbit eigenfunction. The functions $G_{n\kappa}$ and $F_{n\kappa}$ satisfy radial Dirac equations

$$\left(\frac{\mathrm{d}}{\mathrm{d}r} - \frac{\kappa}{r}\right) F + (m - \epsilon + V) G = 0$$

$$\left(\frac{\mathrm{d}}{\mathrm{d}r} + \frac{\kappa}{r}\right) G + (m + \epsilon - V) F = 0. \tag{10}$$

For convenience, we designate with bold sans serif letters column matrices with large and small radial components; e.g., $\mathbf{y} = \binom{G}{F}$ is a solution to equation (10).

2.1. *Dirac–Coulomb wavefunctions*

When $V(r) = -\alpha Z/r$, equation (10) becomes the Dirac equation for an electron moving in a Coulomb potential. If we define

$$k = |\kappa| \qquad \gamma = [k^2 - (\alpha Z)^2]^{1/2}$$

$$\lambda = (m^2 - \epsilon^2)^{1/2} \qquad z = 2\lambda r$$

$$\sigma = \alpha Z \epsilon/\lambda \qquad \sigma' = \alpha Z m/\lambda$$

solutions to equation (10) may be written

$$\mathbf{y} = \begin{pmatrix} (m+\epsilon)^{1/2} \\ (m-\epsilon)^{1/2} \end{pmatrix} z^{\gamma}\, \mathrm{e}^{-z/2}(-\kappa + \sigma')(v(z) \pm w(z)). \tag{11}$$

Here, the function $v(z)$ satisfies Kummer's equation

$$\mathrm{d}^2 v/\mathrm{d}z^2 + (b - z)\,\mathrm{d}v/\mathrm{d}z - av = 0 \tag{12}$$

with $a = \gamma - \sigma$ and $b = 2\gamma + 1$. The function $w(z)$ is given by

$$w(z) = \frac{1}{-\kappa + \sigma'} z^{1-a} \frac{\mathrm{d}}{\mathrm{d}z}(z^a v). \tag{13}$$

Following Slater (1960), four important solutions to equation (12) are

$$v_1(z) = {}_1F_1(a, b, z) \tag{14a}$$

$$v_2(z)=z^{1-b}{}_1F_1(1+a-b,2-b,z) \tag{14b}$$

$$v_5(z)=U(a,b,z) \tag{14c}$$

$$v_7(z)=\mathrm{e}^zU(b-a,b,-z) \tag{14d}$$

where ${}_1F_1$ and U are confluent hypergeometric functions. These particular solutions are singled out because v_1 and v_2 have simple behaviour near the origin, while v_5 and v_7 have simple behaviour for large $|z|$. From these v, we derive the following solutions to the Dirac–Coulomb equation:

$$\mathbf{y}_1=(m\pm\epsilon)^{1/2}z^{\gamma}\,\mathrm{e}^{-z/2}[(-\kappa+\sigma'){}_1F_1(a,b,z)\pm a{}_1F_1(a+1,b,z)] \tag{15a}$$

$$\mathbf{y}_2=(m\pm\epsilon)^{1/2}z^{-\gamma}\,\mathrm{e}^{-z/2}[(-\kappa+\sigma'){}_1F_1(1+a-b,2-b,z) \pm(1+a-b){}_1F_1(2+a-b,2-b,z)] \tag{15b}$$

$$\mathbf{y}_5=(m\pm\epsilon)^{1/2}z^{\gamma}\,\mathrm{e}^{-z/2}(-\kappa+\sigma')[U(a,b,z)\pm(\kappa+\sigma')U(a+1,b,z)] \tag{15c}$$

$$\mathbf{y}_7=(m\pm\epsilon)^{1/2}z^{\gamma}\,\mathrm{e}^{z/2}[(-\kappa+\sigma')U(b-a,b,-z)\mp U(b-a-1,b,-z)]. \tag{15d}$$

Here, upper and lower signs refer to the large and small components of $\mathbf{y}$, respectively. The solutions $\mathbf{y}_1$ and $\mathbf{y}_2$ are linearly independent and have a non-zero Wronskian

$$W(\mathbf{y}_1,\mathbf{y}_2)=G_1F_2-G_2F_1=4\gamma\lambda(-\kappa+\sigma'). \tag{16}$$

Two important relations between the four $\mathbf{y}$ functions are:

$$\mathbf{y}_1=\frac{\Gamma(b)}{\Gamma(b-a)}\mathrm{e}^{\varepsilon i\pi a}\mathbf{y}_5+\frac{\Gamma(b)}{\Gamma(a)}\mathrm{e}^{\varepsilon i\pi(a-b)}\mathbf{y}_7. \tag{17a}$$

$$\mathbf{y}_2=-\frac{\Gamma(2-b)}{\Gamma(1-a)}\mathrm{e}^{\varepsilon i\pi(a-b)}\mathbf{y}_5+\frac{\Gamma(2-b)}{\Gamma(1+a-b)}\mathrm{e}^{\varepsilon i\pi(a-b)}\mathbf{y}_7. \tag{17b}$$

Again, following Slater (1960), we use the sign convention

$$\varepsilon=\begin{cases}1 & \text{if } \arg(z)>0\\ -1 & \text{otherwise.}\end{cases} \tag{18}$$

In particular, we use $-z=\mathrm{e}^{-\varepsilon i\pi}z$ (z complex) to avoid any possible confusion.

2.1.1. Entire analytic functions. A function $F(\epsilon,\rho)$ is said to be an entire analytic function of ϵ if for all finite values of ϵ and ρ (a set of parameters), it can be represented by an expansion in powers of ϵ:

$$F(\epsilon,\rho)=\sum_{p=0}^{\infty}\epsilon^pf_p(\rho) \tag{19}$$

which is absolutely and uniformly convergent. In the appendix we show that there are two linearly independent solutions, $\mathbf{y}_R$ and $\mathbf{y}_I$, to the Dirac–Coulomb equation which are entire analytic functions of ϵ. In particular, $\mathbf{y}_R$ is a solution regular at the origin, while $\mathbf{y}_I$ is one singular at $r=0$. The analytic solutions $\mathbf{y}_R$ and $\mathbf{y}_I$ are related to $\mathbf{y}_1$ and $\mathbf{y}_2$ by:

$$\mathbf{y}_1=c_1\mathbf{y}_R \tag{20a}$$

$$\mathbf{y}_2=c_2\mathbf{y}_I \tag{20b}$$

where the constants c_1 and c_2 are given by

$$c_1 = (m+\epsilon)^{1/2}(2\lambda)^{\gamma}(-\kappa+\gamma+\sigma'-\sigma) \tag{21a}$$

$$c_2 = (m+\epsilon)^{1/2}(2\lambda)^{-\gamma}(-\kappa-\gamma+\sigma'-\sigma). \tag{21b}$$

Two essential points concerning $\mathbf{y}_\mathrm{R}$ and $\mathbf{y}_\mathrm{I}$ are that they are real functions of ϵ and that they change smoothly across the ionisation threshold. Consequently, $\mathbf{y}_\mathrm{R}$ and $\mathbf{y}_\mathrm{I}$ provide the necessary ingredients to relate the continuous spectrum to the discrete one.

2.1.2. Continuum wavefunctions. When $\epsilon > m$ (above threshold), we make the following replacements:

$$(m-\epsilon)^{1/2} \to -\mathrm{i}(\epsilon-m)^{1/2}$$

$$\lambda \to \mathrm{i}p \qquad p = (\epsilon^2 - m^2)^{1/2}$$

$$\sigma \to \mathrm{i}\nu \qquad \nu = \alpha Z\epsilon/p \qquad \sigma' = \mathrm{i}\nu' \qquad \nu' = \alpha Z m/p.$$

From $\mathbf{y}_5$ and $\mathbf{y}_7$ we can construct two Coulomb functions $\mathbf{f}(r)$ and $\mathbf{g}(r)$ which behave asymptotically as:

$$\mathbf{f}(r) \to \begin{pmatrix} [(\epsilon+m)/p]^{1/2} \cos(pr + \nu \ln 2pr + \delta_\kappa) \\ [(\epsilon-m)/p]^{1/2} \sin(pr + \nu \ln 2pr + \delta_\kappa) \end{pmatrix} \tag{22a}$$

$$\mathbf{g}(r) \to \begin{pmatrix} -[(\epsilon+m)/p]^{1/2} \sin(pr + \nu \ln 2pr + \delta_\kappa) \\ [(\epsilon-m)/p]^{1/2} \cos(pr + \nu \ln 2pr + \delta_\kappa) \end{pmatrix}. \tag{22b}$$

Here, δ_κ is the relativistic Coulomb phaseshift:

$$\delta_\kappa = \rho - \pi\gamma/2 - \arg\Gamma(\gamma + \mathrm{i}\nu) \tag{23}$$

with

$$\mathrm{e}^{2\mathrm{i}\rho} = (-\kappa + \mathrm{i}\nu')/(\gamma + \mathrm{i}\nu). \tag{24}$$

In terms of $\mathbf{y}_\mathrm{R}$ and $\mathbf{y}_\mathrm{I}$ we have

$$\mathbf{f}(r) = d_1 \mathbf{y}_\mathrm{R} \tag{25a}$$

$$\mathbf{g}(r) = e_1 \mathbf{y}_\mathrm{R} + e_2 \mathbf{y}_\mathrm{I} \tag{25b}$$

where

$$d_1 = c_1/N_1 \tag{26a}$$

$$e_1 = -(\operatorname{cosec} \pi b \mathrm{e}^{-2\pi\nu} + \cot \pi b)\, d_1 \tag{26b}$$

$$e_2 = \frac{\mathrm{i}\,\mathrm{e}^{-\mathrm{i}\pi a}\Gamma^2(b)}{\gamma(\gamma+\mathrm{i}\nu)|\Gamma(\gamma+\mathrm{i}\nu)|^2} \frac{c_2}{N_1} \tag{26c}$$

and where

$$N_1 = \frac{2p^{1/2}\Gamma(b)}{|\Gamma(\gamma+\mathrm{i}\nu)|} \exp(-\pi\nu/2 - \mathrm{i}\pi\gamma/2 + \mathrm{i}\rho). \tag{26d}$$

Note that d_1, e_1 and e_2 are all real and approach finite limits at threshold as $\epsilon \to m$.

2.2. Relations between the quantum defects and phaseshifts

Assuming that the valence electron sees an effective potential $V(r)$ which changes smoothly from $-\alpha Z_{\text{nuc}}/r$ at the origin to $-\alpha Z/r$ at large distances, there exists a solution $\mathbf{y}(r)$ to equation (10) such that (i) $\mathbf{y}(r)$ is regular at the origin, i.e., $\mathbf{y}(0)=0$, and (ii) $\mathbf{y}(r)$ is an entire analytic function of energy ϵ. Asymptotically, $\mathbf{y}$ can be expressed in terms of the two independent Coulomb wavefunctions $\mathbf{y}_{\mathrm{R}}$ and $\mathbf{y}_{\mathrm{I}}$ as

$$\mathbf{y}(r) \rightarrow a(\epsilon)\mathbf{y}_{\mathrm{R}}(r)+b(\epsilon)\mathbf{y}_{\mathrm{I}}(r). \tag{27}$$

Following the argument given in Ham (1955), we find that the function $\beta(\epsilon) = -b(\epsilon)/a(\epsilon)$ is itself analytic. In the following, we shall make use of the analyticity of $\beta(\epsilon)$ to define a quantum-defect function $\mu(\epsilon)$ which can then be related to the non-Coulomb phaseshifts $\delta(\epsilon)$ of the continuum wavefunctions.

2.2.1. Below threshold ($\epsilon < m$)—relation between $\beta(\epsilon)$ and the quantum defect. From equations (17), (20) and (27), we have, up to some normalisation constants,

$$\mathbf{y} \sim \left(1-\frac{c_1}{c_2}\frac{\Gamma(a)\Gamma(2-b)}{\Gamma(b)\Gamma(1+a-b)}\beta(\epsilon)\right)\mathbf{y}_1-\frac{c_1}{c_2}\frac{\Gamma(a)}{\Gamma(b-1)}\beta(\epsilon)\mathbf{y}_5. \tag{28}$$

At large distances, $\mathbf{y}_1 \rightarrow \mathrm{e}^{\lambda r}$ while $\mathbf{y}_5 \rightarrow \mathrm{e}^{-\lambda r}$. To prevent the wavefunction from diverging at bound states, we require the coefficient of $\mathbf{y}_1$ in the above formula to vanish. As a result, $\beta(\epsilon)$ is given at the bound states ($\epsilon=\epsilon_n$) as

$$\beta(\epsilon_n)=\left(\frac{c_2}{c_1}\right)\frac{\Gamma(b)\Gamma(1+a_n-b)}{\Gamma(2-b)\Gamma(a_n)}. \tag{29}$$

We define the quantum defect by $\mu_n=a_n+n-k$ at discrete eigenstates; this is equivalent to the expression given in equation (2). Utilising the relation

$$\Gamma(z)\Gamma(1-z)=\pi \operatorname{cosec} \pi z \tag{30}$$

$\beta(\epsilon)$ can then be expressed in terms of μ_n as

$$\beta(\epsilon_n)=-\left(\frac{c_2}{c_1}\right)\frac{\Gamma^2(b)\Gamma(1+\sigma_n-\gamma)}{2\pi\gamma\Gamma(1+\sigma_n+\gamma)}\left(\frac{1}{\cot \pi\mu_n-\cot \pi b}\right). \tag{31}$$

Since $\beta(\epsilon)$ is given by equation (31) at every bound state in a Rydberg series $\{\epsilon_n\}$ which converges to an accumulation point $\epsilon_\infty = m$ at the spectral head, $\beta(\epsilon)$ is defined everywhere in the complex ϵ plane by virtue of its analyticity. On the other hand, once $\beta(\epsilon)$ is uniquely determined by the discrete spectrum, we can introduce a quantum-defect function $\mu(\epsilon)$ to replace μ_n in equation (31) for energies other than ϵ_n. In this way, $\mu(\epsilon)$ is found to be real and continuous below threshold, and $\mu(\epsilon_n)=\mu_n$ at discrete eigenstates. The remaining question is to continue $\mu(\epsilon)$ analytically into the continuum region. To this end, we define two functions R and $\mathscr{B}$ by

$$R(z, z')=\left(\frac{\gamma+z}{z}\right)\left(\frac{-\kappa+\gamma+z'-z}{-\kappa-\gamma+z'-z}\right) \tag{32}$$

$$\mathscr{B}(z, \gamma)=\frac{\Gamma(z+\gamma)}{z^{2\gamma-1}\Gamma(1+z-\gamma)}. \tag{33}$$

Here we have $z=\sigma$, $z'=\sigma'$ for $\epsilon<m$ and $z=\mathrm{i}\nu$, $z'=\mathrm{i}\nu'$ for $\epsilon>m$. In general, we can write $\mathscr{B}(z,\gamma)$ as a complex function

$$\mathscr{B}(z,\gamma)=B(z,\gamma)+\mathrm{i}C(z,\gamma). \tag{34}$$

(i) *Below threshold* $\epsilon<m$

$$B(\sigma,\gamma)=\mathscr{B}(\sigma,\gamma) \tag{35a}$$

$$C(\sigma,\gamma)=0. \tag{35b}$$

(ii) *Above threshold* $\epsilon>m$

$$B(\mathrm{i}\nu,\gamma)=(1+\cos\pi b\mathrm{e}^{-2\pi\nu})A(\nu,\gamma) \tag{36a}$$

$$C(\mathrm{i}\nu,\gamma)=-\sin\pi b\mathrm{e}^{-2\pi\nu}A(\nu,\gamma) \tag{36b}$$

with

$$A(\nu,\gamma)=\frac{|\Gamma(\gamma+\mathrm{i}\nu)|^2\mathrm{e}^{\pi\nu}}{2\pi\nu^{2\gamma-1}}. \tag{36c}$$

It can be shown that $R(z,z')$ and $B(z,\gamma)$ (the real part of $\mathscr{B}(z,\gamma)$) defined above are real and continuous even at threshold. In terms of these functions, we can rewrite equation (31) as

$$\beta(\epsilon)=\frac{-\Gamma^2(b)}{2\pi\gamma(2\alpha Z\epsilon)^{2\gamma}R(z,z')B(z,\gamma)}\left(\frac{1}{\cot\pi\mu(\epsilon)-\cot\pi b}\right). \tag{37}$$

The function $\mu(\epsilon)$ thus defined is real and continuous in the entire spectrum and is a natural extension of the quantum defect μ_n.

2.2.2. Above threshold ($\epsilon>m$)*—relation between* $\beta(\epsilon)$ *and the phaseshifts.* For $\epsilon>m$, the regular solution to the Dirac equation $\mathbf{y}(r)$ defined in equation (27) behaves asymptotically as

$$\mathbf{y}(r)\to\begin{pmatrix}[(\epsilon+m)/p]^{1/2}\cos(pr+\nu\ln 2pr+\delta_\kappa+\delta(\epsilon))\\ [(\epsilon-m)/p]^{1/2}\sin(pr+\nu\ln 2pr+\delta_\kappa+\delta(\epsilon))\end{pmatrix} \tag{38}$$

where $\delta(\epsilon)$ is the short-range non-Coulomb phaseshift. In terms of Coulomb scattering functions $\mathbf{f}(r)$ and $\mathbf{g}(r)$, we have

$$\mathbf{y}(r)\to\cos\delta(\epsilon)\,\mathbf{f}(r)+\sin\delta(\epsilon)\,\mathbf{g}(r). \tag{39}$$

Up to a normalisation constant, $\mathbf{y}(r)$ can be expressed in terms of $\mathbf{y}_\mathrm{R}$ and $\mathbf{y}_\mathrm{I}$ as

$$\mathbf{y}(r)\to\mathbf{y}_\mathrm{R}+\left[\left(\frac{d_1}{e_2}\right)\cot\delta(\epsilon)+\frac{e_1}{e_2}\right]^{-1}\mathbf{y}_\mathrm{I}. \tag{40}$$

Comparing equations (27) and (40), we have

$$\beta(\epsilon)=-\left[\left(\frac{d_1}{e_2}\right)\cot\delta(\epsilon)+\frac{e_1}{e_2}\right]^{-1}. \tag{41}$$

After simplification, we arrive at the relation between $\beta(\epsilon)$ and the phaseshift $\delta(\epsilon)$

$$\beta(\epsilon)=\frac{-\Gamma^2(b)}{2\pi\gamma(2\alpha Z\epsilon)^{2\gamma}R(\mathrm{i}\nu,\mathrm{i}\nu')A(\nu,\gamma)}\left(\frac{1}{\cot\delta(\epsilon)-\operatorname{cosec}\pi b\mathrm{e}^{-2\pi\nu}-\cot\pi b}\right) \tag{42}$$

where $R(i\nu, i\nu')$ and $A(\nu, \gamma)$ have been defined in equations (32) and (36c). Comparing equations (37) and (42), we obtain the relation between the quantum defect $\mu(\epsilon)$ and phaseshift $\delta(\epsilon)$ given previously in equation (4):

$$\cot \delta(\epsilon) = (1 + \cos \pi b e^{-2\pi\nu}) \cot \pi\mu(\epsilon) + \sin \pi b e^{-2\pi\nu}.$$

As pointed out in the introduction, the above relation reduces to Seaton's non-relativistic formula, equation (5), in the limit $\alpha Z \ll 1$. Furthermore, we have the exact relation $\delta(\epsilon = m) = \pi\mu(\epsilon = m)$ at the threshold. Following Seaton's argument (Seaton 1958), we may regard $\mu(m)$ as the increase in the number of bound states which results from the short-range non-Coulomb part of the potential.

In figure 1 we present an example of a model potential calculation of the quantum defects μ_n ($n = 3$–8) in the nd series of the C^{3+} ion. The non-Coulomb phaseshift $\delta(\epsilon)$ of the d partial wave is also calculated from the same potential at a few energies above threshold. As one can see, the values of $\pi\mu_n$ indeed join smoothly with those of $\delta(\epsilon)$. A similar example on the N^{4+} ion is given in figure 2.

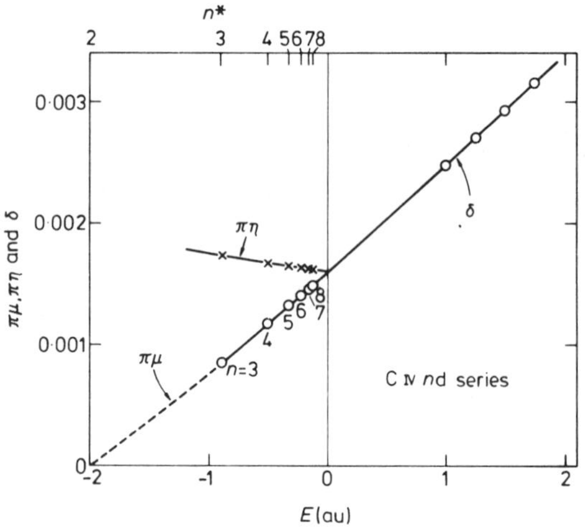

Figure 1. Quantum defects μ, η defects and phaseshifts δ in the nd series of C IV as obtained from a model potential calculation. Here, E is the term energy and n^* is the effective quantum number defined in equation (2).

2.2.3. Matching condition. In applying QDT to practical problems such as photoionisation and electron scattering, it is convenient to make use of a matching condition between wavefunctions above and below threshold based on their asymptotic behaviour. For this purpose, we introduce two Coulomb functions $\mathbf{y}_+$ and $\mathbf{y}_-$ in terms of $\mathbf{y}_5$ and $\mathbf{y}_7$ as:

$$\mathbf{y}_- = \left(\frac{2\lambda}{\pi}(\sigma' - \kappa)\Gamma(\sigma + \gamma + 1)\Gamma(\sigma - \gamma + 1)\right)^{-1/2} \mathbf{y}_5 \tag{43a}$$

$$\mathbf{y}_+ = -\left(\frac{\Gamma(\sigma + \gamma + 1)\Gamma(\sigma - \gamma + 1)}{2\pi\lambda(\sigma' - \kappa)}\right)^{1/2} e^{i\pi(b-a)} \mathbf{y}_7. \tag{43b}$$

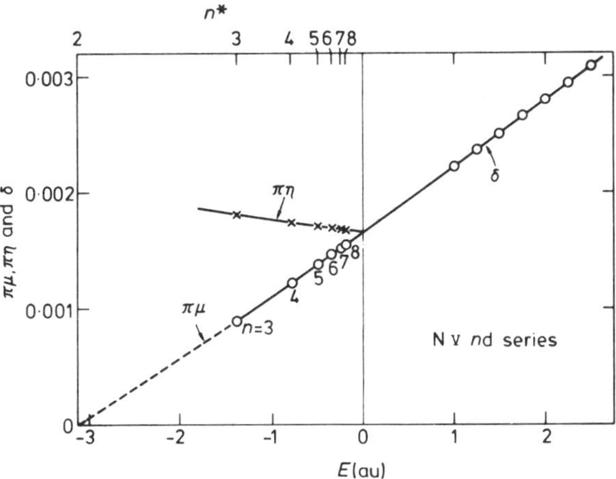

Figure 2. Quantum defects μ, η defects and phaseshifts δ in the nd series of N v as obtained from a model potential calculation. Here, E is the term energy and n^* is the effective quantum number defined in equation (2).

Asymptotically, we have $\mathbf{y}_\pm \to e^{\pm\lambda r}$. Furthermore, $\mathbf{y}_+$ and $\mathbf{y}_-$ are linearly independent with Wronskian $W(\mathbf{y}_-, \mathbf{y}_+) = 1$. Below the threshold we can express $\mathbf{y}_{\mathrm{R}}$ and $\mathbf{y}_{\mathrm{I}}$ in terms of $\mathbf{y}_\pm$ as:

$$\mathbf{y}_{\mathrm{R}} = \frac{\Gamma(b)}{s_+(\sigma, \sigma')(2\alpha Z\epsilon)^{\gamma-\frac{1}{2}}}\left[\frac{1}{\pi B(\sigma, \gamma)}\left(\frac{\sigma'-\kappa}{\sigma+\gamma}\right)\right]^{1/2}(e^{-i\pi a}\mathbf{y}_- - \sin\pi a\,\mathbf{y}_+) \tag{44a}$$

$$\mathbf{y}_{\mathrm{I}} = \frac{(2\alpha Z\epsilon)^{\gamma+\frac{1}{2}}}{s_-(\sigma, \sigma')\Gamma(b-1)}\left(\frac{\sigma+\gamma}{\sigma}\right)$$
$$\times\left[\pi B(\sigma, \gamma)\left(\frac{\sigma'-\kappa}{\sigma+\gamma}\right)\right]^{1/2}\operatorname{cosec}\pi b[e^{i\pi(b-a)}\mathbf{y}_- + \sin\pi(b-a)\,\mathbf{y}_+] \tag{44b}$$

where

$$s_\pm(z, z') = (-\kappa \pm \gamma + z' - z)(\epsilon + m)^{1/2} \tag{45}$$

with $z = \sigma$, $z' = \sigma'$ for $\epsilon < m$ and $z = i\nu$, $z' = i\nu'$ for $\epsilon > m$ as before.

Above the threshold, we have an analogous expression for $\mathbf{y}_{\mathrm{R}}$ and $\mathbf{y}_{\mathrm{I}}$ involving the Coulomb scattering functions **f** and **g**:

$$\mathbf{y}_{\mathrm{R}} = \frac{\Gamma(b)}{s_+(i\nu, i\nu')(2\alpha Z\epsilon)^{\gamma-\frac{1}{2}}}\left[\frac{1}{\pi A(\nu, \gamma)}\left(\frac{i\nu'-\kappa}{i\nu+\gamma}\right)\right]^{1/2}\mathbf{f}(r) \tag{46a}$$

$$\mathbf{y}_{\mathrm{I}} = \frac{(2\alpha Z\epsilon)^{\gamma+\frac{1}{2}}}{s_-(i\nu, i\nu')\Gamma(b-1)}\left(\frac{i\nu+\gamma}{i\nu}\right)\left[\pi A(\nu, \gamma)\left(\frac{i\nu'-\kappa}{i\nu+\gamma}\right)\right]^{1/2}$$
$$\times[\mathbf{g}(r) + (\operatorname{cosec}\pi b e^{-2\pi\nu} + \cot\pi b)\mathbf{f}(r)]. \tag{46b}$$

Note that the coefficients on the right-hand sides of the above equations are real, even though they appear complex. Since $\mathbf{y}_{\mathrm{R}}$ and $\mathbf{y}_{\mathrm{I}}$ are analytic functions, by comparing equations (44) and (46), we arrive at the matching condition (after some simplification):

$$\begin{array}{ll} \epsilon<m & \epsilon>m \\ \mathbf{y}_\alpha=\mathrm{e}^{-\mathrm{i}\pi a}\mathbf{y}_- - \sin\pi a\,\mathbf{y}_+ & \leftrightarrow N\mathbf{f} \end{array} \tag{47a}$$

$$\mathbf{y}_\beta=\mathrm{i}\,\mathrm{e}^{-\pi a}\mathbf{y}_- + \cos\pi a\,\mathbf{y}_+ \leftrightarrow \frac{1}{N}(\mathbf{g}+\sin\pi b\mathrm{e}^{-2\pi\omega}\,\mathbf{f}) \tag{47b}$$

with

$$N=(1+\cos\pi b\mathrm{e}^{-2\pi\nu})^{1/2}. \tag{48}$$

Now we introduce a real and continuous function $\mu(\epsilon)$ such that the general solution $\mathbf{y}$ to the Dirac equation is weighted asymptotically as

$$\mathbf{y}\rightarrow\cos\pi\mu\;\mathbf{y}_\alpha+\sin\pi\mu\;\mathbf{y}_\beta \tag{49a}$$

$$=\mathrm{e}^{-\mathrm{i}\pi(a-\mu)}\mathbf{y}-\sin\pi(a-\mu)\,\mathbf{y}_+ \qquad \epsilon<m \tag{49b}$$

$$=N\cos\pi\mu\;\mathbf{f}+\frac{1}{N}\sin\pi\mu(\mathbf{g}+\sin\pi b\mathrm{e}^{-2\pi\nu}\,\mathbf{f}) \qquad \epsilon>m. \tag{49c}$$

(i) Below threshold, for bound states to exist, the coefficient of $\mathbf{y}_+$ in equation (49*b*) must vanish, leading to the eigenvalue condition

$$\sin\pi(a-\mu)=0. \tag{50}$$

The function $\mu(\epsilon)$ thus has the physical interpretation of a quantum defect, since it satisfies the relation $\mu(\epsilon_n)=a_n+$integer at discrete eigenstates.

(ii) Above threshold, comparing equations (39) with (49*c*), we arrive at the relation between $\delta(\epsilon)$ and $\mu(\epsilon)$ as given in equation (4). We thus see in this way that the matching conditions given in equations (47*a*) and (47*b*) are sufficient to define the quantum-defect function and to relate it to the non-Coulomb phaseshift. The non-relativistic analogues of these matching conditions are given in the works of Seaton (1966) and Fano (1970); they have proved to be very useful in applying the QDT to practical problems.

2.2.4. The η defect. Non-relativistically, Ham (1955) has shown that the quantum defect $\mu(\epsilon)$ is subject to certain constraints in the non-physical region below the first bound state—namely $\mu(\epsilon)$ must be an integer when the effective quantum number $n^*\rightarrow 0, 1, \ldots, l$, with l being the angular momentum. Relativistically, similar constraints exist. To show this, let us rewrite equation (37) as

$$\beta(\epsilon)=\frac{-\Gamma^2(b)}{2\pi\gamma(2\alpha Zm)^{2\gamma}}\left(\frac{\kappa-\gamma}{\kappa+\gamma}\right)\frac{1}{\mathscr{A}(\sigma,\kappa)(\cot\pi\mu-\cot\pi b)} \tag{51}$$

where

$$\mathscr{A}(\sigma,\kappa)=\frac{(\sigma+\sigma'+\kappa+\gamma)(\sigma+\sigma'-\kappa-\gamma)}{2\sigma(\sigma+\sigma')}\left(\frac{\sigma}{\sigma'}\right)^{2\gamma}B(\sigma,\gamma). \tag{52}$$

In this way, all the energy-dependent terms in $\beta(\epsilon)$ are absorbed in $\mathscr{A}(\sigma,\kappa)$ and in $\mu(\epsilon)$. It can be shown that

$$\mathscr{A}(\sigma,\kappa)=0 \qquad \text{if } \sigma=\begin{cases}\gamma-1,\gamma-2,\ldots\geqslant 0 & \text{for } \kappa=-l-1<0\\ \gamma,\gamma-1,\ldots\geqslant 0 & \text{for } \kappa=l>0.\end{cases}$$

Since $\beta(\epsilon)$ does not necessarily diverge as $\mathscr{A}(\sigma, \kappa)$ vanishes, $\mu(\epsilon)$ must become an integer at these energies. This is the constraint on $\mu(\epsilon)$ in the relativistic theory which corresponds to the non-relativistic one mentioned above.

Following Ham (1955), we can define an η defect as

$$\cot \pi\eta = \mathscr{A}(\sigma, \kappa) \cot \pi\mu. \tag{53}$$

In the limit where $\alpha Z \ll 1$, we have

$$\mathscr{A}(\sigma, \kappa) \to \frac{\Gamma(\sigma + l + 1)}{\sigma^{2l+1}\Gamma(\sigma - l)} \tag{54}$$

for $\kappa = -l - 1$ or l. In this case, equation (53) reduces to its non-relativistic counterpart as given in Ham (1955) or Seaton (1958). Furthermore, since $\mathscr{A}(\sigma, \kappa) \to 1$ as $\epsilon \to m$, the η defect is identical to the μ defect at threshold—the same result as is obtained in the non-relativistic QDT. Since $\eta(\epsilon)$ is not restricted to pass through integer values at specific energies, it is expected to vary more slowly than $\mu(\epsilon)$ and is more suitable for numerical interpolation and extrapolation purposes. In the examples given in figures 1 and 2, the broken curve shows the continuation of the quantum defect to its first constraint position at $n^* = 2$. The curves for $\eta(\epsilon)$ are also plotted in these figures for comparison purposes. It is seen that $\eta(\epsilon)$ varies more slowly than $\mu(\epsilon)$ in these examples, and that η and μ are equal at threshold.

2.2.5. Normalisation of the radial functions. Let $\mathbf{y}_\kappa(\epsilon, r)$ and $\mathbf{y}_\kappa(\epsilon', r)$ be two regular solutions to the radial Dirac equation corresponding to energies ϵ and ϵ', respectively. We have, from equation (10),

$$\int_0^r \mathbf{y}_\kappa^\dagger(\epsilon, x)\mathbf{y}_\kappa(\epsilon', x)\, dx = \frac{G_\kappa(\epsilon', r)F_\kappa(\epsilon, r) - F_\kappa(\epsilon', r)G_\kappa(\epsilon, r)}{\epsilon - \epsilon'}. \tag{55}$$

In the limit $r \to \infty$, the above expression enables one to extract normalisation conditions for the radial wavefunctions simply in terms of their asymptotic behaviour. In the following, we are interested in those analytic functions $\mathbf{y}(\epsilon, r)$ which satisfy

$$\mathbf{y}(\epsilon, r) = \mathbf{y}_{\mathrm{R}}(\epsilon, r) - \beta(\epsilon)\mathbf{y}_{\mathrm{I}}(\epsilon, r) \qquad \text{for } r > r_0 \tag{56}$$

where r_0 is the radius beyond which the electron sees a pure Coulomb potential.

(i) *Continuum wavefunctions*

For $\epsilon > m$, the asymptotic form of the wavefunction $\mathbf{y}(\epsilon, r)$ is given in terms of a phaseshift δ as

$$\mathbf{y}(\epsilon, r) = C(\epsilon)\begin{pmatrix} [(\epsilon + m)/p]^{1/2} \cos(pr + \nu \ln 2pr + \delta_\kappa + \delta) \\ [(\epsilon - m)/p]^{1/2} \sin(pr + \nu \ln 2pr + \delta_\kappa + \delta) \end{pmatrix} \qquad \text{for } r > r_0. \tag{57}$$

The constant $C(\epsilon)$ is obtained from equations (42), (46*a*) and (46*b*) as

$$C(\epsilon) = \frac{\Gamma(b)}{s_+(\mathrm{i}\nu, \mathrm{i}\nu')(2\alpha Z\epsilon)^{\gamma - \frac{1}{2}}}\left[\frac{1}{\pi A(\nu, \gamma)}\left(\frac{\mathrm{i}\nu' - \kappa}{\mathrm{i}\nu + \gamma}\right)\right]^{1/2}$$
$$\times [\sin\delta(\cot\delta - \operatorname{cosec} \pi b e^{-2\pi\nu} - \cot \pi b)]^{-1}. \tag{58}$$

Utilising equation (55), we obtain the orthonormality condition for continuum wavefunctions

$$\int_0^\infty \mathbf{y}^\dagger(\epsilon, r)\mathbf{y}(\epsilon', r)\,\mathrm{d}r = \pi C^2(\epsilon)\delta(\epsilon-\epsilon') \tag{59}$$

where $\delta(\epsilon-\epsilon')$ is the Dirac δ function.

(ii) *Bound-state wavefunctions*

Below threshold, the analytic function $\mathbf{y}(\epsilon, r)$ is given in terms of the exponentially increasing and decreasing functions $\mathbf{y}_+$ and $\mathbf{y}_-$, as

$$\mathbf{y}(\epsilon, r) = \frac{N \sin \pi b}{\sin \pi(b-\mu)}[\mathrm{e}^{\mathrm{i}\pi(\mu-a)}\mathbf{y}_- + \sin \pi(\mu-a)\,\mathbf{y}_+] \qquad \text{for } r > r_0 \tag{60}$$

where the constant N is given by

$$N = \frac{\Gamma(b)}{s_+(\sigma, \sigma')(2\alpha Z\epsilon)^{\gamma-\frac{1}{2}}}\left[\frac{1}{\pi B(\sigma, \gamma)}\left(\frac{\sigma'-\kappa}{\sigma+\gamma}\right)\right]^{1/2}. \tag{61}$$

When $\epsilon' - \epsilon_n$, with ϵ_n being an eigen-energy, we put

$$\mathbf{y}_n = \mathbf{y}(\epsilon_n, r). \tag{62}$$

Since $\mathbf{y}_n$ vanishes at infinity, we have

$$\mathbf{y}_n = \frac{N_n \sin \pi b}{\sin \pi(b-\mu_n)}\mathrm{e}^{\mathrm{i}\pi(\mu_n-a_n)}\mathbf{y}_{n-} \qquad \text{for } r > r_0. \tag{63}$$

From equation (55), the norm of $\mathbf{y}_n$ is given by

$$\Delta_n = \int_0^\infty \mathbf{y}_n^\dagger \mathbf{y}_n\,\mathrm{d}r = \lim_{r\to\infty}\lim_{\epsilon-\epsilon_n}\frac{G_nF - F_nG}{\epsilon-\epsilon_n}. \tag{64}$$

Substituting equations (60) and (63) into the above relation, we arrive at

$$\Delta_n = \frac{\pi\sigma'^3 N_n^2 \zeta(\sigma_n)}{m(\alpha Z)^2}\,\frac{\sin^2 \pi b}{\sin^2 \pi(b-\mu_n)} \tag{65}$$

where the function $\zeta(\sigma)$ is given by

$$\zeta(\sigma) = 1 + \mathrm{d}\mu/\mathrm{d}\sigma. \tag{66}$$

Using equation (65), we can define a normalised radial function $\mathscr{F}_n(r)$ as

$$\mathscr{F}_n(r) = (-1)^{n-k}\Delta_n^{-1/2}\mathbf{y}_n(r). \tag{67}$$

Asymptotically, $\mathscr{F}_n$ can be expressed in terms of the Coulomb function $\mathbf{y}_5$ as

$$\mathscr{F}_n(r) = K_n \mathbf{y}_5(\epsilon_n, r) \qquad \text{for } r > r_0. \tag{68}$$

The constant K_n is then given by

$$K_n = (-1)^{n-k}\{\alpha Z/[2(\sigma_n' - \kappa)\sigma_n'^2\zeta(\sigma_n)\Gamma(\sigma_n+\gamma+1)\Gamma(\sigma_n-\gamma+1)]\}^{1/2}. \tag{69}$$

For the case of a pure Coulomb potential, we have $\mu = 0$ and $\zeta = 1$; and we find that the constant K_n in equation (69) reduces to the usual Coulomb normalisation factor. In table 3, we compare the wavefunction normalisation factors K_n determined from

Table 3. The normalisation factors K_n for $n\mathrm{p}_{3/2}$ radial wavefunctions of C IV and N V. The exact values of K_n and the values of σ_n are obtained from numerical solutions of the radial Dirac equation.

Ion	n	σ_n	K_n (au)	
			Analytic value (equation (69))	Exact value
C IV	2	1·98696	0·07320	0·07310
	3	2·98550	0·01967	0·01966
	4	3·98503	0·003904	0·003903
N V	2	1·98774	0·08178	0·08168
	3	2·98646	0·02196	0·02195
	4	3·98606	0·004357	0·004356

equation (69) with those obtained numerically in model potential studies of $n\mathrm{p}_{3/2}$ states in C IV and N V. The close agreement in these cases is not surprising in view of the fact that the analytical normalisation factors are determined from eigen-energies of the same model potential.

3. Conclusion

The relativistic analysis of quantum defects presented above is designed to aid in systematic studies of energy levels as Z increases along an isoelectronic sequence. Even for low degrees of ionisation there are advantages to the relativistic theory. For example, consider the elastic scattering of an electron by a closed-shell ion. Recall that the non-Coulomb part of the relativistic scattering phaseshift is obtained by analytic continuation of the quantum defect. Since the phaseshifts are relativistic, we have a tool for calculating the scattering asymmetry parameters, as well as the scattering cross section. Indeed, if we designate the non-Coulomb phaseshift above threshold by $\delta_\kappa^{\mathrm{NC}}$, we obtain the following scattering amplitudes (Akhiezer and Berestetskii 1965)

$$\text{non-spin-flip} \qquad a(\theta)=\frac{1}{2\mathrm{i}p}\sum_{l=0}^{\infty}[(l+1)a_{-l-1}+la_l]P_l(\cos\theta) \tag{70}$$

$$\text{spin-flip} \qquad b(\theta)=\frac{1}{2\mathrm{i}p}\sum_{l=0}^{\infty}(a_l-a_{-l-1})P_l^{(1)}(\cos\theta) \tag{71}$$

with

$$a_\kappa=\exp[2\mathrm{i}(\delta_\kappa+\delta_\kappa^{\mathrm{NC}})]-1 \tag{72}$$

where δ_κ is the Coulomb phaseshift defined in equation (23), and θ is the scattering angle. The differential scattering cross section for non-polarised electrons is given by

$$\mathrm{d}\sigma/\mathrm{d}\Omega=|a|^2+|b|^2 \tag{73}$$

while the left–right asymmetry parameter is

$$P(\theta)=2\,\mathrm{Im}(a^*b)/(|a|^2+|b|^2). \tag{74}$$

Another area in which the relativistic analysis may prove useful is the study of oscillator strengths along spectral series. The analytic wavefunctions considered in the previous section can be used to determine bound-state oscillator strengths. These analytic oscillator strengths can then be continued across the threshold $\epsilon = m$. They are related to the usual bound-state oscillator strengths f_n below threshold, and to $\mathrm{d}f/\mathrm{d}\epsilon$ above threshold. The quantity $\mathrm{d}f/\mathrm{d}\epsilon$ is just the photoionising cross section (up to a constant factor), so that a study of oscillator strengths for bound states determines the photoionisation cross section.

In many situations relativistic (spin–orbit) interactions play an important role in determining the size of the photoionisation cross section or angular distribution parameters (Johnson and Cheng 1978). Since electron spin is automatically included in the relativistic QDT, it provides an appropriate tool for investigating cases where spin–orbit effects are large.

It should be mentioned in closing that the extension of relativistic single-channel QDT to a multi-channel situation poses no particular mathematical problem beyond those already encountered in the non-relativistic case (Seaton 1966).

Acknowledgments

The authors are grateful to Professor U Fano and Dr C D Lin for helpful discussions.

Appendix

We require two independent solutions $\mathbf{y}_\mathrm{R}$ and $\mathbf{y}_\mathrm{I}$ to the Dirac–Coulomb equation which are regular and irregular functions at the origin, respectively, and which are entire analytic functions of energy. Using the series expansion method, we find:

(i) *Regular solution* $\mathbf{y}_\mathrm{R}(r)$.

$$\mathbf{y}_\mathrm{R}(r) = r^{\gamma}\begin{pmatrix} a_0 + a_1 r + \cdots \\ b_0 + b_1 r + \cdots \end{pmatrix} \tag{A.1}$$

where $a_0 = 1$, $b_0 = \alpha Z/(\gamma - \kappa)$ and where

$$a_n = \frac{-\alpha Z(\epsilon - m)a_{n-1} - (\gamma + n - \kappa)(\epsilon + m)b_{n-1}}{n(2\gamma + n)} \tag{A.2}$$

$$b_n = \frac{(\gamma + n + \kappa)(\epsilon - m)a_{n-1} - \alpha Z(\epsilon + m)b_{n-1}}{n(2\gamma + n)} \tag{A.3}$$

for $n = 1, 2, 3 \ldots$, etc.

(ii) *Irregular solution* $\mathbf{y}_\mathrm{I}(r)$

$$\mathbf{y}_\mathrm{I}(r) = r^{-\gamma}\begin{pmatrix} a_0 + a_1 r + \ldots \\ b_0 + b_1 r + \cdots \end{pmatrix} \tag{A.4}$$

where $a_0 = 1$ and $b_0 = -\alpha Z/(\gamma + \kappa)$. The recurrence relation is the same as equations (A.2) and (A.3), with $-\gamma$ replacing γ everywhere.

Since a_n and b_n are polynomials in ϵ, $\mathbf{y}_R$ and $\mathbf{y}_I$ must be analytic. Furthermore, it is easy to show that the series in equations (A.1) and (A.4) converge absolutely and uniformly for all finite values of ϵ. Thus, $\mathbf{y}_R$ and $\mathbf{y}_I$ are entire analytic functions. Finally, $\mathbf{y}_R$ and $\mathbf{y}_I$ are linearly independent, with Wronskian $W(\mathbf{y}_R, \mathbf{y}_I) = 2\gamma/\alpha Z$.

References

Akhiezer A I and Berestetskii V B 1965 *Quantum Electrodynamics* (New York: Interscience)
Bashkin S and Stoner J O Jr 1975 *Atomic Energy Levels and Grotrian Diagrams* vol 1 (Amsterdam: North-Holland)
Desclaux J P 1975 *Comput. Phys. Commun.* **9** 31
Dubau J 1976 *Electron and Photon Interactions with Atoms* ed H Kleinpoppen and M R C McDowell (New York: Plenum)
Edlén B 1964 *Encyclopedia of Physics* vol 27 (Heidelberg: Springer-Verlag)
Fano U 1970 *Phys. Rev.* A **2** 353
Ham F S 1955 *Solid State Physics* vol 1, ed F Seitz and D Turnbull (New York: Academic Press) p 127
Johnson W R and Cheng K T 1978 *Phys. Rev. Lett.* **40** 1167
Lee C M 1974 *Phys. Rev.* A **10** 584
——1977 *Phys. Rev.* A **16** 109
Lee C M and Lu K T 1973 *Phys. Rev.* A **8** 1241
Seaton M J 1958 *Mon. Not. R. Astron. Soc.* **118** 504
—— 1966 *Proc. Phys. Soc.* **88** 801
—— 1970 *Comments on Atomic and Molecular Physics* **2** 37
Slater L J 1960 *Confluent Hypergeometric Functions* (Cambridge: Cambridge University Press)
Zilitis V A 1977 *Opt. Spectrosk.* **43** 603

1980 *Phys. Rev.* A **22** 979–88
Reprinted with permission from the American Physical Society

Scattering and spectroscopy: Relativistic multichannel quantum-defect theory

C. M. Lee*
Laboratory for Laser Energetics, University of Rochester, Rochester, New York 14627

W. R. Johnson
Department of Physics, University of Notre Dame, Notre Dame, Indiana 46556
(Received 10 March 1980)

A formulation of relativistic multichannel quantum-defect theory is presented. The relativistic random-phase approximation is applied to calculate the eigenchannel parameters in the relativistic multichannel quantum-defect theory. The resonances due to inner-shell ($1s$) excitations below the K threshold for ions of the Be isoelectronic sequence are studied as illustrative examples.

I. INTRODUCTION

During the past decade there has been an increasing interest in the spectroscopy of atoms and ions of high nuclear charge. This interest arises because of the importance of highly charged ions in the physics of solar flares,[1] tokamak plasmas,[2] and laser-produced laboratory plasmas.[3] Laboratory measurements of the spectra and transition probabilities of highly stripped ions are made using accelerator beam-foil spectroscopy,[4] while neutral atoms of high nuclear charge are studied using synchrotron radiation.[5]

The theoretical methods that have been developed to understand and correlate the atomic data for highly charged atoms and ions have been based on the Dirac equation, since relativistic effects are important for systems in which αZ (Z being the nuclear charge) is not small. For complex atoms electron-electron correlation is also important, and a proper theoretical description of complex high-Z atoms must include both relativity and correlations.

Many of the features of complex atomic spectra can be understood in terms of a few important dynamical parameters using formal scattering theory.[6] One modern approach to scattering theory especially well suited to the description of single-electron ionization is the multichannel quantum-defect theory (MQDT).[7] In the recent past MQDT has been applied to investigate the complex spectra of atoms ranging from He to Ba,[8] and to describe photoelectron angular distributions[9] and spin polarization.[10] The MQDT has also been used to study molecular photoabsorption[11] and to estimate dissociative recombination cross sections between electrons and molecular ions.[12] In view of the many successes of MQDT in correlating atomic data for nonrelativistic systems, it is appropriate to consider a relativistic generalization designed to treat highly charged systems. The purpose of the present paper is to introduce such a generalization and to illustrate how the resulting theory is applied by giving a practical example.

The MQDT provides a detailed description of photoexcitation processes in terms of a relatively small number of parameters. Although these parameters may be obtained empirically, in the present paper we use MQDT in conjunction with a specific dynamical theory to obtain *ab initio* values of the MQDT parameters. In the present study we employ the relativistic random-phase approximation[13] (RRPA) to work out the dynamics. The RRPA has been applied successfully to study photoexcitation and photoionization of highly charged atoms. Applications have been made to the excitation of levels in closed-shell atoms of various isoelectronic sequences.[14] Oscillator strengths for transitions from the ground states to excited states of ions in these sequences are in excellent agreement with more sophisticated many-body calculations.[15]

Cross sections and angular-distribution parameters for photoionization of the closed-shell atoms of He, Be (Ref. 16), Ne, Ar, Kr, and Xe (Ref. 17) have been determined using RRPA.[18] For the lighter elements the RRPA calculations are in close agreement with nonrelativistic RPA calculations[19] and with other nonrelativistic many-body calculations.[20] For the heavier elements Kr and Xe, significant relativistic effects on angular distributions,[21] branching ratios,[22] and spin-polarization parameters[23] are found experimentally; these relativistic effects are explained by RRPA calculations[18,24]. Such applications have been concerned with the average behavior of photoionization cross sections; resonances have been either treated approximately[13] or completely neglected. The practical difficulties which the RRPA shares with other many-body theories are often connected with techniques for extracting information concerning highly excited bound states and resonances; the MQDT provides the proper theoretical framework for treating such questions.

In Sec. II we discuss in detail the relativistic

version of MQDT in which the basic quantum-defect parameters are defined and related to the photoabsorption spectrum. The theory presented in Sec. II is similar to the nonrelativistic MQDT, so we concentrate our attention on those details where the relativistic and nonrelativistic versions differ.

Section III is devoted to a brief outline of the RRPA. To extract the dynamical MQDT parameters, it is particularly convenient to approach RRPA from the time-dependent Hartree-Fock (TDHF) point of view. We discuss the RRPA in terms of TDHF equations for perturbed electronic orbitals, and we show how to determine the MQDT parameters from solutions to the TDHF equations. A reader interested only in the relativistic MQDT, and not in the dynamical model employed here, may skip Sec. III without loss of continuity.

We turn to an illustrative example in Sec. IV. This example is a detailed study of the location and shape of inner-shell $1s\to np$ resonances along the Be isoelectronic sequence. In neutral Be and in ions of low nuclear charge, the $J^{\pi}=1^{-}$ spectrum consists of a series of $(1s\,2s^2np)\,{}^1P$ resonances; the corresponding ${}^3P$ resonances do not show up in the absorption spectrum because of spin selection rules. By contrast, for Be-like Mo^{38+}, where relativistic effects are important, both the ${}^1P$ and ${}^3P$ resonances are prominent. Since we use a relativistic description throughout, we are able to follow the appearance and development of the forbidden excitations as Z increases along the Be isoelectronic sequence. To put our work in perspective we make comparisons with experimental measurements of inner-shell absorption spectra in the case of neutral Be. We obtain fair agreement between the theoretical and experimental determination of the very narrow resonance lines. For the higher-Z ions considered in Sec. IV (Ne^{6+} and Mo^{38+}), no experimental values are available; however, we expect that, since correlations are less important for higher-Z ions, the predicted spectra of the highly charged ions are at least as accurate as those for neutral Be.

In a subsequent paper[26] these relativistic MQDT techniques are applied to study Beutler-Fano autoionization resonances in the rare gases Ar, Kr, and Xe, and good agreement with recent experimental determinations[27] of the line profiles is obtained. An application of the relativistic MQDT to the autoionization resonances[28] in Ne near 575 Å has already appeared.[29]

II. RELATIVISTIC MULTICHANNEL QUANTUM-DEFECT THEORY

Let us consider an excited atomic system having angular momentum J and parity π with energy below the double-ionization threshold. This excited system consists of a probing (excited bound or excited continuum) electron and a residual ion of charge ζe; the ion itself may of course have various degrees of excitation.

The character of interaction between the probing electron and the ion varies over a range of distances. At large distances the interaction is governed by the static Coulomb potential $-\zeta e^2/r$. Stationary states are represented by linear combinations of electron-ion states combined to give angular momentum J, parity π. Each combination identifies a possible mode of dissociation and is called a dissociation channel, labeled by an index i. Asymptotically, the probing electron is described by a single Dirac orbital in the relativistic theory, viz,

$$u_i(\mathbf{r})=\frac{1}{r}\begin{pmatrix} iG_i(r)\Omega_{\kappa_i m_i}(\hat{r}) \\ F_i(r)\Omega_{-\kappa_i m_i}(\hat{r})\end{pmatrix}, \tag{1}$$

where κ_i and m_i are angular-momentum quantum numbers; $\kappa_i=\mp(j_i+\frac{1}{2})$ for $j_i=l_i\pm\frac{1}{2}$ ($\kappa_i=-1,1,-2,2,\ldots$, corresponds to the spectroscopic notation $s_{1/2}, p_{1/2}, p_{3/2}, d_{3/2},\ldots$). The symbol $\Omega_{\kappa_i m_i}$ designates a spherical spinor, while G_i and F_i are large and small component radial functions. In the sequel we use the abbreviated notation

$$y_i=\begin{pmatrix} G_i \\ F_i \end{pmatrix}$$

for the radial Dirac functions.

At small distances the probing electron and the atomic ion form a complex through which energy and angular momentum can be exchanged. Complicated interactions take place between the electron and ion, requiring an elaborate many-electron theory; in Sec. III we apply RRPA to solve the many-electron problem.

Outside the reaction zone, the radial wave function y of the excited electron orbital for a dissociation channel satisfies a radial Dirac equation

$$Hy=\epsilon y,$$

with

$$H=\begin{bmatrix} m+\dfrac{\alpha\zeta}{r} & \dfrac{d}{dr}-\dfrac{\kappa}{r} \\ -\dfrac{d}{dr}-\dfrac{\kappa}{r} & -m+\dfrac{\alpha\zeta}{r}\end{bmatrix}, \tag{2}$$

where ϵ is the orbital energy, including rest energy m. We use natural units ($\hbar=c=1$) unless otherwise noted. The radial wave function y is expressed as a linear combination of independent relativistic Coulomb wave functions[30] (f,g),

$$y=af+bg, \tag{3}$$

where a and b are as yet undetermined constants. These relativistic Coulomb wave functions have the following properties:

(1) The pair (f,g) are continuous functions (C^∞ functions) of energy $\epsilon - m$ across the threshold $\epsilon = m$.

(2) At large distances (f,g) have the asymptotic behavior:

(i) For $\epsilon > m$,

$$f \xrightarrow[r\to\infty]{} \begin{bmatrix} \left(\frac{\epsilon+m}{\pi p}\right)^{1/2} \cos\left(pr + \frac{\alpha\zeta\epsilon}{p}\ln 2pr - \frac{\pi}{2}(l+1) + \sigma_\kappa^c\right) \\ \left(\frac{\epsilon-m}{\pi p}\right)^{1/2} \sin\left(pr + \frac{\alpha\zeta\epsilon}{p}\ln 2pr - \frac{\pi}{2}(l+1) + \sigma_\kappa^c\right) \end{bmatrix}, \tag{4}$$

and

$$g \xrightarrow[r\to\infty]{} \begin{bmatrix} -\left(\frac{\epsilon+m}{\pi p}\right)^{1/2} \sin\left(pr + \frac{\alpha\zeta\epsilon}{p}\ln 2pr - \frac{\pi}{2}(l+1) + \sigma_\kappa^c\right) \\ \left(\frac{\epsilon-m}{\pi p}\right)^{1/2} \cos\left(pr + \frac{\alpha\zeta\epsilon}{p}\ln 2pr - \frac{\pi}{2}(l+1) + \sigma_\kappa^c\right) \end{bmatrix}, \tag{5}$$

with

$$\sigma_\kappa^c = \arg\Gamma(\gamma - i\alpha\zeta\epsilon/p) - \tfrac{1}{2}\pi(\gamma - l - 1) + \eta - \pi s\,, \tag{6}$$

where

$$s = \tfrac{1}{2}\left(\frac{\kappa}{|\kappa|} + 1\right), \quad \gamma = [\kappa^2 - (\alpha\zeta)^2]^{1/2},$$

$$e^{2i\eta} = \frac{-\kappa + i\alpha\zeta m/p}{\gamma + i\alpha\zeta\epsilon/p}\,,$$

and l is the orbital angular momentum.

(ii) For $\epsilon < m$,

$$f = e^{-i\pi a} y^- - \sin\pi a y^+\,, \tag{7}$$

$$g = e^{-i\pi(a-1/2)} y^- + \cos\pi a y^+\,, \tag{8}$$

with

$$a = \gamma + s - \nu\,, \tag{9}$$

$$\epsilon = \frac{m}{[1 + (\alpha\zeta)^2/\nu^2]^{1/2}}\,. \tag{10}$$

The parameter ν in Eq. (10) is an "effective" quantum number. Asymptotically,

$$y^- \xrightarrow[r\to\infty]{} \left(\frac{1}{2\lambda}\right)^{1/2} \left(\frac{(\nu' - \kappa)}{\Gamma(\nu+\gamma+1)\Gamma(\nu-\gamma+1)}\right)^{1/2} \left(\frac{\sqrt{m+\epsilon}}{\sqrt{m-\epsilon}}\right) (2\lambda r)^{\nu} e^{-\lambda r}\,, \tag{11}$$

$$y^+ \xrightarrow[r\to\infty]{} \frac{1}{\pi}\left(\frac{1}{2\lambda}\right)^{1/2} \left(\frac{\Gamma(\nu+\gamma+1)\Gamma(\nu-\gamma+1)}{(\nu' - \kappa)}\right)^{1/2} \left(\frac{-\sqrt{m+\epsilon}}{\sqrt{m-\epsilon}}\right) (2\lambda r)^{-\nu} e^{\lambda r}\,, \tag{12}$$

where

$$\lambda = (m^2 - \epsilon^2)^{1/2} \quad \text{and} \quad \nu' = \alpha\zeta m/\lambda\,.$$

Each dissociation channel is described in terms of a perturbed radial orbital function in the RRPA, and the dynamics is given by a set of coupled radial differential equations relating the various orbitals. For an M-channel problem there are $2M$ RRPA equations, as shown in Sec. III. The M "negative-frequency" RRPA orbitals (which account for ground-state correlations) are exponentially damped at large distances, while the remaining M "positive-frequency" orbitals (which describe the interchannel interaction including final-state correlations) are required to satisfy stationary-wave asymptotic boundary conditions:

$$y_j^{(i)} \xrightarrow[r\to\infty]{} f_j \delta_{ji} + g_j R_{ji}, \quad i,j = 1,\ldots,M\,. \tag{13}$$

The index j labels the dissociation channels, while the index i distinguishes between the independent solutions to the coupled dynamical equations. The number of independent solutions is just equal to the number of channels. The energy-dependent R matrix is symmetric; it is determined by the solution to the dynamical equations.

Let $U_{i\alpha}$ be the orthogonal matrix of eigenvectors of the matrix R_{ij},

$$\sum_j R_{ij} U_{j\alpha} = \lambda_\alpha U_{i\alpha}, \quad i = 1,\ldots,M \tag{14}$$

and introduce the eigen-quantum defects μ_α by $\lambda_\alpha = \tan\pi\mu_\alpha$. The quantities $\pi\mu_\alpha$ are just the scattering phase shifts when all channels are open. We construct eigenchannel orbitals as linear combinations of the stationary solutions (13):

$$z_j^\alpha = \sum_i U_{i\alpha} \cos\pi\mu_\alpha y_j^{(i)}\,,$$
$$\xrightarrow[r\to\infty]{} U_{j\alpha}(f_j \cos\pi\mu_\alpha + g_j \sin\pi\mu_\alpha), \quad j = 1,\ldots,M\,. \tag{15}$$

The eigenchannel solutions from Eqs. (15), the matrix $U_{i\alpha}$, and the eigen-quantum defects μ_α are only weakly energy dependent. The final physical boundary conditions are expressed in terms of these eigenchannel orbitals.

Physically acceptable solutions to the dynamical

equations are given by a superposition of eigenchannel orbitals:

$$y_j = \sum_\alpha A_\alpha z_j^\alpha, \quad j=1,\ldots,M \tag{16}$$

with mixing coefficients A_α determined by the boundary conditions. These boundary conditions differ for different ranges of the spectrum, namely, the discrete spectrum, the autoionizing spectrum, and the continuum. The mixing coefficients A_α are functions of the eigenchannel parameters $U_{i\alpha}$ and μ_α and of the energy of the excited system. The transition amplitude in the αth eigenchannel is represented by a reduced matrix element between the initial-state orbitals and those of the αth eigenchannel. Here, since we focus our attention on dipole excitations, we must consider the reduced matrix elements of the dipole operator D_α as additional eigenchannel parameters.

A. Discrete spectrum

The total energy of the excited atomic system is defined with respect to the ground-state energy of the ion by

$$\begin{aligned} E &= E_i + \epsilon_i - m \\ &= E_i + \frac{m}{[1+(\alpha\zeta)^2/\nu_i^2]^{1/2}} - m, \quad i=1,\ldots,M \end{aligned} \tag{17}$$

where E_i is the excitation energy of the ion in the ith dissociation channel. For $\alpha\zeta \ll 1$, Eq. (17) reduces to the familiar Rydberg formula. Each E corresponds to a set of numbers ν_i; the number of different ν_i equals the number K of nondegenerate states of the atomic ion. Equations (17) represent a set of $K-1$ independent equations of K unknowns ν_i.

In the discrete spectrum, all dissociation channels are closed, and the boundary conditions that $y_i \to 0$ as $r\to\infty$ for all $j=1,\ldots,M$ lead to the following set of M equations:

$$\sum_\alpha F_{i\alpha} A_\alpha = 0, \quad i=1,\ldots,M \tag{18}$$

with

$$F_{i\alpha} = U_{i\alpha} \sin\pi(a_i - \mu_\alpha). \tag{19}$$

The system of linear equations [Eq. (19)] has the compatibility condition

$$F(\{\nu\}) = \det|F_{i\alpha}| = 0 \tag{20}$$

and the nontrival solutions

$$A_\alpha = C_{i\alpha} \Big/ \Big(\sum_\alpha C_{i\alpha}^2\Big)^{1/2}, \tag{21}$$

where $C_{i\alpha}$ is the cofactor of the element $F_{i\alpha}$ of the determinant $|F_{i\alpha}|$, and the choice of the index i is arbitrary. Solutions to Eqs. (17) and (20), i.e., discrete sets of the numbers $\{\nu_i\}$, give discrete energy levels of the perturbed Rydberg series. For the nth state $\{\nu_i^n\}$, the oscillator strength is expressed as

$$f_n = \tfrac{2}{3}\omega \Big|\sum_\alpha A_\alpha^n D_\alpha\Big|^2 \Big/ N_n^2, \tag{22}$$

where ω is the photon energy in atomic units and D_α is the reduced dipole matrix element. The normalization factor (in atomic units) is

$$\begin{aligned} N_n^2 &= \sum_i \int_0^\infty dr\, y_i^\dagger y_i (\alpha\zeta)^2 \\ &= \frac{(\alpha\zeta)^2}{\pi} \sum_i \left[\frac{d}{dE}\Big(-\sum_\alpha U_{i\alpha}\sin\pi(a_i-\mu_\alpha)A_\alpha\Big)\right]_{E=E_n} \Big(\sum_\alpha U_{i\alpha}\cos\pi(a_i-\mu_\alpha)A_\alpha^n\Big) \\ &= \sum_i \nu_i^3 \left[1+\Big(\frac{\alpha\zeta}{\nu_i}\Big)^2\right]^{3/2} \Big(\sum_\alpha U_{i\alpha}\cos\pi(a_i-\mu_\alpha)A_\alpha^n\Big)^2 \\ &\quad + \sum_\alpha \Big(\frac{d\mu_\alpha}{dE}\Big)_{E_n} (A_\alpha^n)^2 + \sum_i\sum_\alpha\sum_{\alpha'} \Big(\frac{dU_{i\alpha}}{dE}\Big)_{E_n} U_{i\alpha'} \sin\pi(\mu_\alpha-\mu_{\alpha'}) A_\alpha^n A_{\alpha'}^n . \end{aligned} \tag{23}$$

B. Autoionization spectrum

In the autoionization spectrum, some dissociation channels are closed and other channels are open. Let us denote Q as the set of the closed dissociation channels and P as the set of the open dissociation channels. Further, let us define P_k as the set of the open dissociation channels pertaining to the kth state of the residual atomic ion. The union of the sets P_k equals the set of P. The number of members of the set P is the number of the open dissociation channels N_P, and the number of members at Q is the number of the closed dissociation channels $N_Q (N_P + N_Q = N)$, where N is the total number of relevant dissociation channels.

The asymptotic boundary conditions describing stationary-wave functions for autoionizing states are that the components of the wave function, Eq. (16), in the closed channels $i \in Q$ vanish exponentially as $r\to\infty$, and that the components in the open channel $i \in P$ consist of relativistic Coulomb standing waves with a common phase shift $\pi\tau$. Such conditions lead to

$$\sum_\alpha F_{i\alpha} A_\alpha = 0, \ \forall i \tag{24}$$

with

$$F_{i\alpha} = \begin{cases} U_{i\alpha} \sin\pi(a_i - \mu_\alpha), & i \in Q \\ U_{i\alpha} \sin\pi(\tau - \mu_\alpha), & i \in P. \end{cases} \tag{25}$$

Existence of nontrival solutions of the system of linear equations, Eq. (24) requires

$$F(\{\nu_i; i \in Q\}, \tau) = \det |F_{i\alpha}| = 0. \tag{26}$$

At each energy in the autoionization spectrum, Eq. (26) has roots τ denoted by $\{\tau_\rho; \rho = 1, \ldots, N_\rho\}$. The index ρ identifies a collisional eigenchannel, and the mixing coefficients A^ρ_α are given by an equation similar to Eq. (21), viz.,

$$A^\rho_\alpha = C_{i\alpha}(\{\nu_i, i \in Q\}, \tau_\rho) \Big/ \Big(\sum_\alpha [C_{i\alpha}(\{\nu_i, i \in Q\}, \tau_\rho)]^2\Big)^{1/2}. \tag{27}$$

The resulting stationary orbitals behave asymptotically as

$$y^\rho_i = \sum_\alpha A^\rho_\alpha z^\alpha_i \xrightarrow[r\to\infty]{} C^\rho_i(r), \tag{28}$$

where

$$C^\rho_i(r) = \begin{cases} T^\rho_i \begin{bmatrix} \left(\dfrac{\epsilon_i + m}{\pi p_i}\right)^{1/2} \cos\left(p_i r + \dfrac{\alpha\zeta\epsilon_i}{p_i} \ln 2p_i r - \dfrac{\pi}{2}(l_i + 1) + \sigma^c_i + \pi\tau_\rho\right) \\ \left(\dfrac{\epsilon_i - m}{\pi p_i}\right)^{1/2} \sin\left(p_i r + \dfrac{\alpha\zeta\epsilon_i}{p_i} \ln 2p_i r - \dfrac{\pi}{2}(l_i + 1) + \sigma^c_i + \pi\tau_\rho\right) \end{bmatrix} & \text{for } i \in P \\ \left(\sum_\alpha U_{i\alpha} \cos\pi(a_i - \mu_\alpha) A^\rho_\alpha\right) y^-_i \quad \text{for } i \in Q, \end{cases} \tag{29}$$

with

$$T^\rho_i = \sum_\alpha U_{i\alpha} \cos\pi(\tau_\rho - \mu_\alpha) A^\rho_\alpha. \tag{30}$$

The normalization for the collisonal channel ρ is given by

$$\bar{y}^\rho_i = y^\rho_i / N_\rho, \tag{31}$$

with

$$N_\rho = \Big(\sum_{i\in P} (T^\rho_i)^2\Big)^{1/2}. \tag{32}$$

The normalized matrix

$$\bar{T}^\rho_i = T^\rho_i / N_\rho \tag{33}$$

is an $N_P \times N_P$ orthogonal matrix. The reduced dipole matrix elements D_ρ in the collisional eigenchannels ρ are

$$\bar{D}_\rho = \Big(\sum_\alpha A^\rho_\alpha D_\alpha\Big) \Big/ N_\rho. \tag{34}$$

In order to obtain the probability of ejection of a photoelectron into any specific open dissociation channel, we then construct a traveling wave function as a superposition of the collisional eigenchannel functions

$$y^{(j-)}_i = \sum_\rho \bar{y}^\rho_i \bar{A}^{(j-)}_\rho, \tag{35}$$

which satisfies the "incoming-wave" boundary condition at infinity that the amplitude of the outgoing-wave vanish in all open channels $i \neq j$. The desired coefficients $\bar{A}^{(j-)}_\rho$ are

$$\bar{A}^{(j-)}_\rho = \bar{T}^\rho_i e^{-i\pi\tau_\rho}. \tag{36}$$

Thus, the oscillator strength density of the photoelectron group pertaining to the kth state of the residual atomic ion is

$$\begin{aligned} \frac{df_k}{dE} &= \tfrac{2}{3}\omega \sum_{i\in P_k} \Big|\sum_\rho \bar{A}^{(i-)}_\rho \bar{D}_\rho\Big|^2 \\ &= \tfrac{2}{3}\omega \sum_{i\in P_k} \sum_\rho \sum_{\rho'} \bar{T}^\rho_i \bar{T}^{\rho'}_i \cos\pi(\tau_\rho - \tau_{\rho'}) \bar{D}_\rho \bar{D}_{\rho'}. \end{aligned} \tag{37}$$

The total oscillator strength density is then expressed as

$$\frac{df}{dE} = \sum_k \frac{df_k}{dE} = \tfrac{2}{3}\omega \sum_\rho \bar{D}^2_\rho = \sum_\rho \frac{df^\rho}{dE}, \tag{38}$$

with

$$\frac{df^\rho}{dE} = \tfrac{2}{3}\omega \Big(\sum_\alpha A^\rho_\alpha D_\alpha\Big)^2 \Big/ N^2_\rho. \tag{39}$$

For applications to electron-atomic ion collisions in the resonance region, we may construct a traveling wave function as a superposition of the collisional eigenchannel wave functions

$$y_i^{(j+)} = \sum_\rho \bar{y}_i^\rho \bar{A}_\rho^{(j+)} , \tag{40}$$

which satisfies the "outgoing-wave" boundary conditions at $r = \infty$ that the amplitude of the incoming-wave vanish in all open channels $i \neq j$. The desired coefficients are

$$\bar{A}_\rho^{(j+)} = \bar{T}_j^\rho e^{i\pi\tau_\rho} . \tag{41}$$

Thus, the short-range scattering matrix element S_{ij}, for total angular momentum J and parity π, is

$$S_{ij} = \sum_\rho \bar{T}_i^\rho e^{i2\pi\tau_\rho} \bar{T}_j^\rho . \tag{42}$$

The full scattering matrix for J^π symmetry is

$$S_{ij} = \exp(i\sigma_i^c)\left(\sum_\rho \bar{T}_i^\rho e^{i2\pi\tau_\rho} \bar{T}_j^\rho\right)\exp(i\sigma_j^c) . \tag{43}$$

C. Continuous spectrum

In this energy range, all dissociation channels are open. In order to obtain the probability of ejection of a photoelectron into any specific channel j, the mixing coefficients in Eq. (16) are determined by requiring that the amplitude of the outgoing wave vanish in all channels $i \neq j$, namely,

$$A_\alpha^{(j-)} = U_{j\alpha} e^{-i\pi\mu_\alpha} . \tag{44}$$

The oscillator strength density of the photoelectron group pertaining to the kth state of the atomic ion is

$$\frac{df_k}{dE} = \frac{2}{3}\omega \sum_{i\in P_k} \left| \sum_\alpha A_\alpha^{(j-)} D_\alpha \right|^2$$
$$= \frac{2}{3}\omega \sum_{i\in P_k} \sum_\alpha \sum_{\alpha'} U_{i\alpha} U_{i\alpha'} \cos\pi(\mu_\alpha - \mu'_\alpha) D_\alpha D'_\alpha . \tag{45}$$

The total oscillator strength density is then

$$\frac{df}{dE} = \sum_k \frac{df_k}{dE}$$
$$= \frac{2}{3}\omega \sum_\alpha D_\alpha^2 . \tag{46}$$

In the context of electron-ion collision processes, the full scattering matrix for J^π symmetry can be written as

$$S_{ij} = \exp(i\sigma_i^c)\left(\sum_\alpha U_{i\alpha} e^{i2\pi\mu_\alpha} U_{j\alpha}\right)\exp(i\sigma_j^c) . \tag{47}$$

III. RELATIVISTIC RANDOM-PHASE APPROXIMATION

The problem outlined in the Introduction requires a solution to the many-body problem to determine the quantum-defect parameters. In this section we describe the RRPA which is an approximate relativistic many-body theory used to study photoexcitation and photoionization of highly charged atoms and ions.

For our present purposes it is convenient to formulate the RRPA in terms of perturbed single-particle orbitals from the TDHF theory.[31] The TDHF equations describe the response of a closed-shell atom to a time-dependent external field, which we take to be the field of photon incident on the atom. Let us suppose that the external time-dependent perturbation is given by the one-electron operator

$$V = \sum_i v_i(t) . \tag{48}$$

For a photon field of frequency ω and multipolarity J, M, λ (J and M are the photon angular-momentum quantum numbers and $\lambda = 0$ or 1 for magnetic or electric multipoles, respectively), we may write

$$v(t) = v_+ e^{-i\omega t} + v_- e^{i\omega t} , \tag{49}$$

with

$$v_+ = \vec{\alpha} \cdot \vec{a}_{JM}^\lambda , \quad v_- = v_+^\dagger , \tag{50}$$

where $\vec{a}_{JM}^\lambda$ is the photon multipole vector potential and $\vec{\alpha}$ is the usual Dirac matrix.

The TDHF equations for a closed-shell atom can be written in terms of single-particle orbitals ϕ_j as

$$i\frac{\partial\phi_j}{\partial t} = [h_0 + v(t)]\phi_j + V_{HF}\phi_j , \quad j = 1, \ldots, N . \tag{51}$$

Here, V_{HF} is the Hartree-Fock potential

$$V_{HF}\phi = \sum_j e^2 \int \frac{d^3r'}{R} [(\phi_j^\dagger \phi_j)'\phi - (\phi_j^\dagger \phi)'\phi_j] , \tag{52}$$

and

$$h_0 = \vec{\alpha} \cdot \vec{p} + \beta m - e^2 Z/r \tag{53}$$

is a one-electron Dirac Hamiltonian ($\vec{\alpha}$ and β are Dirac matrices). The RRPA equations are obtained from the TDHF approximation by linearizing in powers of the external field.

We let $u_i(\vec{r})$, $i = 1, \ldots, N$ be stationary orbitals for the Dirac-Fock (DF) ground state of an N-electron closed-shell atom. To carry out the linearization we set

$$\phi_j(\vec{r}, t) = u_i(\vec{r})e^{-i\epsilon_i t} + w_{i+}(\vec{r})e^{-i(\epsilon_i + \omega)t} + w_{i-}(\vec{r})e^{-i(\epsilon_i - \omega)t} . \tag{54}$$

In Eq. (54) ϵ_i is the DF eigenvalue for the orbital $u_i(\vec{r})$, and the functions $w_{i\pm}(\vec{r})$ describe the response of the orbital $u_i(\vec{r})$ to the perturbation $v_\pm(t)$. Substituting Eq. (54) into the TDHF equation

(51) and carrying out the linearization procedure, we obtain the inhomogeneous equations

$$(h_0 + v_{\mathrm{HF}} + \epsilon_i \pm \omega) w_{i\pm} + V_{\pm}^{(1)} u_i = v_{\pm} u_i , \quad i = 1, \ldots, N . \tag{55}$$

Here,

$$V_{\pm}^{(1)} u = \sum_j e^2 \int \frac{d^3 r'}{R} \{[(u_j^{\dagger} w_{j\pm})' + (w_{j\mp}^{\dagger} u_j)']u - (u_j^{\dagger} u)' w_{j\pm} - (w_{j\mp}^{\dagger} u)' u_j\} \tag{56}$$

describes the correlations due to the distortion of the DF potential.

A standard method for solving inhomogeneous equations such as those given in Eqs. (55) is to expand in terms of the complete set of solutions to the corresponding homogeneous system. Let us therefore consider the homogeneous equations

$$(h_0 + V_{\mathrm{HF}} + \epsilon_i \pm \omega) w_{i\pm} + V_{\pm}^{(1)} u_i = 0 , \tag{57}$$

along with the orthonormality constraint

$$\int d^3 r [(w_{i\pm}^{\dagger} u_j) + (u_i^{\dagger} w_{j\mp})] = 0 . \tag{58}$$

The solutions of interest to Eqs. (57) are those which have the angular symmetry dictated by the external field $v_{\pm}$.

Let us specialize to the case of an applied electric dipole field with $J = 1$ and $\lambda = 1$. Bearing in mind that the index i of the orbital $u_i(\vec{r})$ refers to a set of central-field quantum numbers $i = (n, \kappa, m)$, one readily establishes that the perturbed orbitals $w_{i\pm}$ have angular quantum numbers $\bar{\kappa} = -\kappa$, $\kappa \pm 1$. Thus, for example, an unperturbed $d_{3/2}$ orbital with $\kappa = 2$ will be excited to states with angular-momentum quantum numbers $\bar{\kappa} = 1$, -2, and 3, corresponding to $p_{1/2}$, $p_{3/2}$, and $f_{5/2}$ angular symmetries by the applied dipole field. In our example in Sec. IV the $s_{1/2}$ orbitals of Be-like ions which have $\kappa = -1$ are excited to $p_{1/2}$ and $p_{3/2}$ orbitals with $\bar{\kappa} = 1$ and -2, respectively.

Using the standard methods of Racah algebra, we may reduce Eqs. (57) to a set of coupled radial differential equations suitable for computer solution. We let $w_{k\pm}$ be the perturbed orbital (of frequency $\pm\omega$) associated with the excitation k: $(n, \kappa, m) \to (\epsilon, \bar{\kappa}, \bar{m})$. In the notation of the previous section, k corresponds to a dissociation channel in which the ion remains in the state $(n, \kappa, -m)$, while the electron is excited to a continuum state $(\epsilon, \bar{\kappa}, \bar{m})$. The electron orbital $w_{k\pm}$ can be written

$$w_{k\pm} = \frac{1}{r} \begin{bmatrix} iG_k(r) \Omega_{\bar{\kappa}\bar{m}}(\hat{r}) \\ F_k(r) \Omega_{-\bar{\kappa}\bar{m}}(\hat{r}) \end{bmatrix} . \tag{59}$$

We write, following the notation of the previous section,

$$y_{k\pm} = \begin{bmatrix} G_{k\pm}(r) \\ F_{k\pm}(r) \end{bmatrix} . \tag{60}$$

To describe the radial functions, we introduce

$$H_{\bar{\kappa}} = \begin{bmatrix} m + V_k(r) & \dfrac{d}{dr} - \dfrac{\bar{\kappa}_k}{r} \\ -\dfrac{d}{dr} - \dfrac{\bar{\kappa}_k}{r} & -m + V_k(r) \end{bmatrix} , \tag{61}$$

where $V_k(r)$ is the ion potential. We may then rewrite Eq. (57) as a set of coupled radial equations

$$(H_{\bar{\kappa}} - \epsilon_k \mp \omega) y_{k\pm} = O_{k\pm} , \tag{62}$$

where $O_{k\pm}$ is the coupling due to the correlation potential $V_{\pm}^{(1)}$. The radial equations (62) are written out in detail in Ref. 16.

Solutions to Eqs. (62) are either oscillatory or exponential at large r, depending on whether $\epsilon_k \pm \omega$ is larger or smaller than the electron mass m. The radial function y_{k+} represents the excited electron orbital, and at large distances (where V_k takes on its asymptotic behavior as an ionic Coulomb potential) the orbital y_{k+} is just a linear combination of the Coulomb wave functions given in Eq. (3). The negative-frequency perturbations y_{k-} describe the effects of correlation on the atomic ground state; these negative-frequency perturbed orbitals are always exponentially damped at large r.

The index k ranges over the various possible values of n, k, $\bar{\kappa}$ corresponding to the dipole excitation of an electron to state $\bar{\kappa}$, leaving an ion with a hole in the state n, κ. This is just the set of M dissociation channels of the atom discussed in Sec. II.

The number of equations in the system (62) is $2M$, twice the number of dissociation channels. For the M-negative-frequency orbitals $y_{k-}(r)$, the physically relevant boundary conditions are that the solutions be regular at the origin and exponentially damped at large r [see Eq. (64) below]. The remaining freedom in the solutions to Eqs. (62) is resolved by requiring that the solutions $y_{k+}(r)$ be regular at $r = 0$ and satisfy the boundary conditions for perturbed orbitals given in Eq. (13). Thus,

$$y_{k+}^{(i)} \xrightarrow[r\to\infty]{} f_k \delta_{ki} + g_k R_{ki} , \quad i, k = 1, \ldots, M \tag{63}$$

$$y_{k-}^{(i)} \xrightarrow[r\to\infty]{} 0 , \quad i, k = 1, \ldots, M . \tag{64}$$

Each value of the index i defines a stationary-wave solution to Eqs. (62), and the symmetric reaction matrix R_{ik} is determined as a function of energy from the solution. The normalization constant for the solutions to Eqs. (62) is given by

$$N^2 = \int dr \sum_k (y_{k+}^2 - y_{k-}^2), \tag{65}$$

as discussed, for example, in Ref. 16. Given a normalized solution to Eqs. (62) one finds the reduced dipole matrix element between a state described by $y_{k\pm}^{(i)}$ and the atomic ground state as

$$D^{(i)} = \sum_k (R_{k+}^{(i)} + R_{k-}^{(i)}), \tag{66}$$

where the radial integrals $R_k^{(i)}$ are given by

$$R_{k\pm}^{(i)} = C(k, \bar{k}) \int dr\, r(G_k G_{k\pm}^{(i)} + F_k F_{k\pm}^{(i)}). \tag{67}$$

In Eq. (67) the symbols G_k, F_k describes the ground-state DF orbitals, while $G_k^{(i)}$ and $F_k^{(i)}$ are the large and small components of the perturbed-orbital solution to Eqs. (62). The angular-momentum factor $C(k, \bar{k})$ in Eq. (67) is

$$C(k,\bar{k}) = \begin{cases} (-1)^{j_k+1/2}[(2j_k+1)(2\bar{j}_k+1)]^{1/2} \begin{pmatrix} j_k & j_k & 1 \\ -\frac{1}{2} & \frac{1}{2} & 0 \end{pmatrix}, & l_k + \bar{l}_k \text{ odd} \\ 0, \quad l_k + \bar{l}_k \text{ even} & \end{cases} \tag{68}$$

where the large round bracket designates a 3-j symbol. The RRPA equations give dipole matrix elements which are identical in length and velocity forms. In the velocity form Eq. (67) is replaced by

$$V_{k\pm}^{(i)} = \pm \frac{1}{\omega} C(k, \bar{k}) \int dr [(\kappa_k - \bar{\kappa}_k)(F_k G_{k\pm}^{(i)} + G_k F_{k\pm}^{(i)}) + (F_k G_{k\pm}^{(i)} - G_k F_{k\pm}^{(i)})]. \tag{69}$$

Dipole matrix elements D_α employed in Sec. II are obtained from Eq. (66) by forming the combinations of $D^{(i)}$ implied by Eq. (15), viz.,

$$D_\alpha = \sum_i D^{(i)} U_{i\alpha} \cos \pi \mu_\alpha. \tag{70}$$

The solution to the problem posed in the Introduction is now completely specified. We first solve Eqs. (62) subject to boundary conditions (63) and (64). The R matrix is then diagonalized to give the quantum-defect parameters μ_α and $U_{i\alpha}$, which are used together with the dipole matrix elements D_α to describe bound states, resonances, and the continuum, according to the method outlined in Sec. II.

IV. APPLICATIONS

As specific examples of the methods developed in the previous sections we consider the $1s \to np$ autoionization resonances for neutral beryllium and for several ions of the beryllium isoelectronic sequence. In these examples we ignore channels such as $1s\{(2s2p)^{1,3}P\}\binom{ns}{nd}$ associated with two electron excitations, since consideration of these channels requires a dynamical scheme beyond the RPA.[32]

In Table I we compare the calculated resonance positions below the K-shell threshold with the experimental data of Mehlman and Esteva.[25] These experimental data, which were based on synchronous-arc plasma-absorption measurements contain complicated structures due to mixtures of Be atoms, metastable Be atoms, and Be^+ ions. Mehlman and Esteva were able to determine the positions of the inner-shell excited states with high accuracy (±25 meV, i.e., ±0.02 Å). For the first resonance the width can be determined theoretically as $(\Delta\nu)_{1s\to 2p} = 4.0 \times 10^{-5}$, while the experimental measurement is limited by the spectral resolution (±0.02 Å). Part of the disagreement between the experimental and theoretical values shown in Table I is due to the omission of two electron excitations in the theory; another part, we believe, is due to the difficulty in determining the experimental threshold and the consequent loss of significance in determining the different between resonance positions and threshold. The theoretically predicted location of the autoionizing resonances is in agreement with a previous nonrelativistic TDHF calculation by Stewart *et al.*,[33] as is to be expected, since relativistic effects are insignificant for neutral beryllium.

The Be autoionizing sequence is illustrated in Fig. 1, where the predicted oscillator strength distribution is plotted against wavelength. In this figure we use the experimental value of the K threshold given by Mehlman and Esteva, instead of the theoretical RRPA value. The locations of the first three experimental lines are marked on Fig. 1 for comparison.

The evolution of the resonance profiles as Z increases along the isoelectronic sequence is shown in Fig. 2, where we plot theoretical absorption cross

TABLE I. Beryllium autoionizing states $[1s(2s)^2np]^1P$.

State	Expt. energy[a] (cm^{-1})	ν_{expt}[b]	ν_{theor}
$1s(2s)^2 2p$	931 300	1.31	1.27
$1s(2s)^2 3p$	979 200	2.6	2.40
$1s(2s)^2 4p$	988 200	4	3.42

[a] Mehlman and Esteva, Ref. 25.
[b] Based on an experimental threshold of 994 900 cm^{-1}.

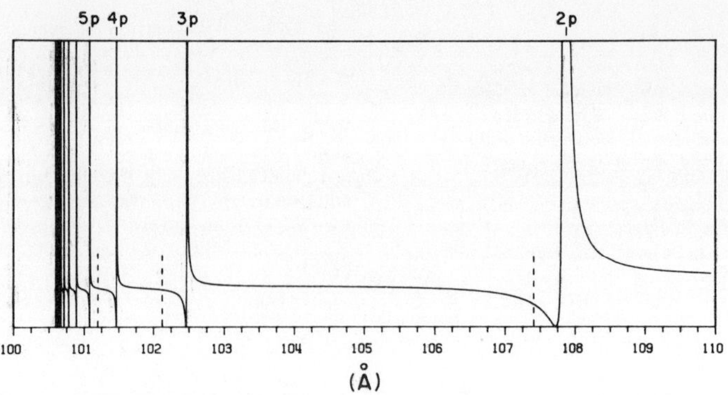

FIG. 1. Theoretical oscillator strength distribution for Be including $1s \to np$ autoionization resonances. The vertical dashed lines mark the location of experimental resonances (Mehlman and Esteva, Ref. 25). The experimental threshold at 994 900 cm^{-1} is used.

functions along with the short-range electron-ion scattering phase shift $\pi\tau$. For neutral beryllium, the phase shifts corresponding to the two open channels show resonances $[1s(2s)^2 2p]^3P$ and $[1s(2s)^2 2p]^1P$, $\nu \sim 1.15$ and $\nu \sim 1.27$, respectively. The character of the resonant states, i.e., ^1P and ^3P, is identified by the character of the eigenchannels which represent the dynamics at short range. Because of electron-electron interaction, the ^3P resonance occurs at lower energy than the ^1P resonance. As a consequence of electric dipole selection rules, the photoionization cross section shows only one resonance for neutral Be at the ^1P position.

For the Be-like neon ion Ne^{6+}, the character of the eigenchannels is no longer described by a pure LS coupling scheme. The smaller resonance near $\nu = 1.78$ is dominated by a ^3P state with a small admixture of ^1P character, while the larger resonance near $\nu = 1.81$ is dominantly ^1P with a small mixture of ^3P.

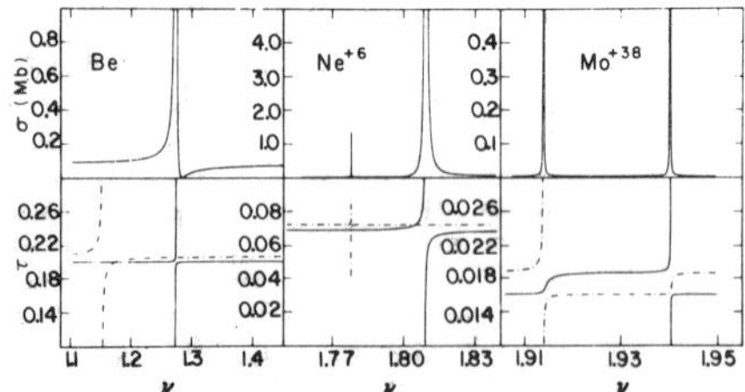

FIG. 2. Absorption cross sections (Mb) and open-channel phases parameters τ (phase $= \pi\tau$) are given in terms of the effective quantum number ν near the $1s \to 2p$ autoionizing resonances in the 4-electron ions Be, Ne^{6+}, and Mo^{38+}. The broken lines in the lower panels correspond to ^3P states, while the solid lines are for ^1P states.

As Z increases the coupling scheme changes from an almost pure LS scheme at neutral Be to a jj scheme for high Z. Indeed, for Be-like Mo^{38+} the eigenchannels are characterized by almost pure jj coupling. The positions of the first two Mo^{38+} resonances differ by $\Delta\nu \sim 0.03$, which is principally due to the spin-orbit difference $\Delta\nu_{so} = \frac{1}{4}(\alpha Z)^2 \sim 0.02$. Furthermore, the strengths of the two resonances are comparable, as a result of jj coupling.

In plotting the open-channel phases in Fig. 2, we have connected the individual phases smoothly through crossing points in the cases of Be and Ne^{6+}; however, our numerical analysis shows that the points of intersection in the figures are actually "anticrossings," as illustrated in the diagram for the Mo^{38+} phase shift.

Finally, we wish to stress that the eigenchannel parameters μ_α, $U_{i\alpha}$, and D_α in the relativistic MQDT can be obtained, as in the nonrelativistic MQDT, either from *ab initio* calculations to treat many-electron dynamics, or from a semiempirical analysis to fit available spectroscopic data. The two approaches complement each other; they may be pursued individually or by a judicious combination of the two approaches, which may prove to be a more satisfactory alternative. To treat open-shell atoms the RRPA dynamics outlined in the previous section is not adequate, and a more sophisticated dynamical scheme is required for the *ab initio* determination of the relativistic MQDT parameters.

ACKNOWLEDGMENTS

The authors wish to express their gratitude to Laboratoire pour l'Utilisation du Rayonnement Electromagnétique and Centre Européan De Calcul Atomique et Moleculaire for the arrangement of the workshop on inner-shell excitations of atoms, molecules, and solids, where parts of this work were completed. Thanks are due to Vo Ky Lan at the Observatoire de Paris, Meudon, for his hospitality and to M. LeDourneuf for her help at the Centre Inter-Regional de Calcul Electronique computing facility. The work of W. R. J. was supported in part by the U. S. National Science Foundation. The work of C. M. Lee was partially supported by the Laser Fusion Feasibility Project, which is sponsored by Exxon Research and Engineering Company, General Electric Company, Northeast Utilities Company, Empire State Electric Research Company, and the New York State Energy Research and Development Administration.

*Permanent address: Institute of Physics, Academy of Science, Peking, China.

[1]C. Jordan, in *Progress in Atomic Spectroscopy*, Part B, edited by W. Hanle and H. Kleinpoppen (Plenum, New York, 1979), pp. 1453–1483; A. K. Dupree, Adv. At. Mol. Phys. 14, 393 (1978).

[2]E. Hinnov, Phys. Rev. A 14, 1533 (1976); T. F. R. Group, J. Phys. (Paris) C4, 86 (1978).

[3]E. V. Aglitskii, V. A. Boiko, O. N. Krohkin, S. A. Pikuz, and A. Ya. Faenov, Sov. J. Quantum Electron. 4, 1152 (1975); P. G. Burkhalter, J. Reader, and R. D. Cowan, J. Opt. Soc. Am. 67, 1521 (1977).

[4]I. Martinson, in *Beam-Foil Spectroscopy*, edited by S. Bashkin (Springer, Berlin, 1976), pp. 31–61. H. G. Berry, Rep. Prog. Phys. 40, 155 (1977).

[5]K. Codling, in *Synchrotron Radiation, Techniques and Applications*, edited by C. Kunz (Springer, Berlin, 1979), pp. 231–265.

[6]M. L. Goldberger and K. M. Watson, *Collision Theory* (Wiley, New York, 1967), pp. 352 and 375.

[7]M. J. Seaton, Proc. Phys. Soc. London 88, 801 (1966); U. Fano, Phys. Rev. A 2, 353 (1970).

[8]U. Fano, J. Opt. Soc. Am. 65, 979 (1975), a review of work of Fano's group; J. A. Armstrong, P. Esherick, and J. J. Wynne, Phys. Rev. A 15, 180 (1977), a representative work from the IBM group; J. Dubau and J. Wells, J. Phys. B 6, 1452 (1973), a work from Seaton's group devoted to the Be atoms.

[9]D. Dill, Phys. Rev. A 7, 1976 (1973); J. Geiger, Z. Phys. A276, 219 (1976).

[10]C. M. Lee, Phys. Rev. A 10, 1598 (1974).

[11]Ch. Jungen and O. Atabek, J. Chem. Phys. 66, 5584 (1977).

[12]C. M. Lee, Phys. Rev. A 16, 109 (1977).

[13]M. Ya. Amusia and N. A. Cherepkov, Case Stud. At. Phys. 5, 47 (1975), nonrelativistic theory; G. Wendin, in *Vacuum Ultraviolet Radiation Physics*, edited by E. E. Koch, R. Haensel, and C. Kunz (Pergamon, Braunschweig, 1974), p. 225 (nonrelativistic theory); W. R. Johnson, C. D. Lin, K. T. Cheng, and C. M. Lee, Phys. Scr. 21, 409 (1980) (relativistic theory).

[14]W. R. Johnson and C. D. Lin, Phys. Rev. A 14, 565 (1976); C. D. Lin, W. R. Johnson, and A. Dalgarno, *ibid.* 15, 154 (1977); P. Shorer, C. D. Lin, and W. R. Johnson, *ibid.* 16, 1109 (1977); P. Shorer and A. Dalgarno, *ibid.* 16, 1502 (1977); P. Shorer, *ibid.* 18, 1060 (1978).

[15]Y. Accad, C. L. Pekeris, and B. Schiff, Phys. Rev. A 4, 885 (1971); C. M. Moser, R. K. Nesbet, and M. N. Gupta, *ibid.* 13, 17 (1976).

[16]W. R. Johnson and C. D. Lin, J. Phys. B 10, L331 (1077); C. D. Lin and W. R. Johnson, *ibid.* 12, 1677 (1979).

[17]W. R. Johnson and K. T. Cheng, Phys. Rev. A 20, 978 (1979).

[18]W. R. Johnson and C. D. Lin, Phys. Rev. A 20, 964 (1979).

[19]T. N. Chang, Phys. Rev. A 15, 2392 (1977).

[20]H. P. Kelly and R. L. Simons, Phys. Rev. Lett. 30, 529 (1973); P. G. Burke and K. T. Taylor, J. Phys. B 8, 2620 (1975); J. R. Swanson and L. Armstrong, Jr., Phys. Rev. A 16, 1117 (1977).

[21]J. L. Dehmer and D. Dill, Phys. Rev. Lett. 37, 1049 (1976); J. L. Dehmer, W. A. Chupka, J. Berkowitz, and W. T. Jiverey, Phys. Rev. A 12, 1966 (1975); D. L. Miller, J. D. Dow, R. G. Houlgate, G. V. Marr, and J. B. West, J. Phys. B 10, 3205 (1977).

[22]F. Wuilleumier, M. Y. Adam, P. Dhez, N. Sandner, V. Schmidt, and W. Melhorn, Phys. Rev. A 16, 646 (1977).

[23]U. Heinzmann, G. Schönhense, and J. Kessler, Phys. Rev. Lett. 42, 1603 (1979).

[24]K.-N. Huang, W. R. Johnson, and K. T. Cheng, Phys. Rev. Lett. 43, 1658 (1979).

[25]G. Mehlman and J. M. Esteva, Astrophys. J. 188, 191 (1974).

[26]W. R. Johnson, K. T. Cheng, and K.-N. Huang, Phys. Rev. A 22, 989 (1980).

[27]K. Radler and J. Berkowitz, J. Chem. Phys. 70, 221 (1979).

[28]K. Radler and J. Berkowitz, J. Chem. Phys. 70, 216 (1979).

[29]W. R. Johnson and M. LeDourneuf, J. Phys. B 13, L13 (1980).

[30]W. R. Johnson and K. T. Cheng, J. Phys. B 12, 863 (1979).

[31]A. Dalgarno and G. A. Victor, Proc. R. Soc. London A297, 291 (1966).

[32]F. Bely-Dubau, J. Dubau, and D. Petrini, J. Phys. B 10, 1613 (1977).

[33]R. F. Stewart, D. K. Watson, and A. Dalgarno, J. Chem. Phys. 63, 3222 (1975).

1980 *Phys. Rev.* A **22** 989–97
Reprinted with permission from the American Physical Society

Analysis of Beutler-Fano autoionizing resonances in the rare-gas atoms using the relativistic multichannel quantum-defect theory

W. R. Johnson
Physics Department, University of Notre Dame, Notre Dame, Indiana 46556

K. T. Cheng
Argonne National Laboratory, Argonne, Illinois 60439

K.-N. Huang
Physics Department, University of Notre Dame, Notre Dame, Indiana 46556

M. Le Dourneuf
Observatorie de Paris, Section d'Astrophysique, 92190 Meudon, France
(Received 31 March 1980)

The Beutler-Fano autoionizing resonances in the rare-gas atoms argon, krypton, and xenon are studied using the relativistic multichannel quantum-defect theory (MQDT). Dynamical parameters for the MQDT analyses are obtained from an *ab initio* relativistic-random-phase-approximation calculation. The position and profile of these resonances are in good agreement with experimental measurements. Angular distribution and spin polarization of photoelectrons in the resonance regions are also studied and are in close agreement with recent measurements.

I. INTRODUCTION

In this paper we report results of *ab initio* calculations of the Beutler-Fano autoionization resonances in argon, krypton, and xenon and compare our results with recent experimental measurements. The dipole-allowed single-electron excitation spectrum of the noble gases consists of five interacting Rydberg series. Three of these series arise from excitation of an outer $p_{3/2}$ electron to $ns_{1/2}$, $nd_{3/2}$, or $nd_{5/2}$ states; these three series converge to the $^2P^o_{3/2}$ ground state of the ion. The remaining two series arise from excitation of an outer $p_{1/2}$ electron to $ns_{1/2}$ or $nd_{3/2}$ states (termed ns' and nd' states in the sequel) and converge to the lowest excited $^2P^o_{1/2}$ state of the ion. The latter two series have members which lie in the continuum above the $^2P^o_{3/2}$ ionization threshold and are therefore subject to autoionization. The photoionization cross section in the energy interval between $^2P^o_{3/2}$ and $^2P^o_{1/2}$ thresholds exhibits two resonant series; a sharp series associated with excitations to the ns' states and a diffuse series associated with excitations to nd' states.

Autoionization resonances between the $^2P^o_{3/2}$ and $^2P^o_{1/2}$ thresholds in argon, krypton, and xenon were first observed by Beutler[1] and analyzed by Fano.[2] For argon, wavelengths of members of the ns' and nd' series and widths of the nd' resonances have been determined experimentally by Yoshino[3] and by Radler and Berkowitz.[4] Measurements of the absolute photoionization cross section for argon in the resonance region have been carried out by a number of workers[5-8] and are summarized by Hudson and Kiefer.[9]

Experimental measurements of the wavelengths of the ns' and nd' resonances for krypton have been reported by Yoshino and Tanaka,[10] and by Radler and Berkowitz,[4] while absolute cross sections for krypton have been determined by Metzger and Cook[7] and by Huffman, Tanaka, and Larrabee.[11] Locations of a number of ns' and nd' resonances are listed by Radler and Berkowitz[4] and by Moore[12] for xenon, and absolute photonionization cross sections for xenon in the resonance region have been reported by various groups.[5,7,13]

The more recent theoretical treatments of Beutler-Fano resonances have been based on multichannel quantum-defect theory (MQDT).[14] Discrete states and autoionization resonances in xenon were studied by Lu,[15] who employed empirically determined MQDT parameters. These studies were extended to argon by Lee and Lu.[16] The MQDT parameters for argon were also determined from an *ab initio* calculation by Lee,[17] who solved the many-electron Schrödinger equation in a limited spherical region. Further studies of autoionization in xenon and krypton using empirical MQDT parameters have been carried out by Geiger.[18,19]

Measurements of the angular-distribution asymmetry parameter β for xenon have been reported by Samson and Gardner.[20] The experimental values of β have been compared with theoretical MQDT values by Dill[21] and by Geiger.[18,19] The angular-distribution parameters depend on relative phases as well as magnitudes of photonioization amplitudes, so that comparisons of MQDT predictions for angular distributions with experiment place additional constraints on the MQDT parameters beyond those already included in cross-section comparisons.

Studies of spin polarization of photoelectrons also provide sensitive tests of MQDT parameters. Calculations of the spin polarization for xenon and argon in the MQDT formalism have been given by Lee.[22] These calculations have recently been subjected to experimental tests by Heinzmann *et al.*,[23] who measure the total spin polarization of xenon in the autoionizing region of the spectrum. In the following paragraphs we present theoretical studies of the Beutler-Fano resonances based on *ab initio* calculations of the MQDT parameters using the relativistic random-phase approximation (RRPA). A detailed account of the theory behind the present calculations is given in a previous paper.[24]

The Beutler-Fano resonances in Ne have already been considered using the present technique.[25] Here, we concentrate on the applications of the theory to argon, krypton, and xenon. In Sec. II we summarize the formulas required to calculate the photoabsorption cross section in the autoionization region. Our calculated MQDT parameters are then presented, along with angular distributions and spin polarization in Sec. III.

These studies should provide an aid to the understanding of systematic features of the Beutler-Fano resonances. Further experimental studies of the resonances, especially measurements of angular distributions and spin polarization, should provide sensitive tests of the theoretical MQDT parameters.

II. THEORETICAL CONSIDERATIONS

The theory of photoionization of atoms by polarized photons has been discussed by several authors.[22,26-30] We restrict our attention here to photoionization by circularly polarized incident radiation. To describe photoionization by unpolarized incident radiation, one must average the results presented below over the two states of circular polarization. For low-energy photoionization, where the electric dipole approximation is valid, the photoelectron angular distribution is given by

$$\frac{d\sigma}{d\Omega}=\frac{\sigma}{4\pi}[1-\tfrac{1}{2}\beta P_2(\cos\theta)]\,, \tag{1}$$

where θ is the angle between the photon momentum $\vec{k}$ and the electron momentum $\vec{p}$. In Eq. (1), σ is the photoionization cross section and β is a parameter which measures the asymmetry in the angular distribution. The polarization vector $\vec{P}$ of the photoelectron is conveniently expressed in a coordinate system with z axis along the photoelectron momentum vector $\vec{p}$, with y axis normal to the production plane (in the direction $\vec{k}\times\vec{p}$), and with x axis in the production plane, but perpendicular to $\vec{p}$ [in the direction $(\vec{k}\times\vec{p})\times\vec{p}$]. The components of $\vec{P}$ in this coordinate system are[30]

$$P_x=\pm\frac{\xi\sin\theta}{1-\frac{1}{2}\beta P_2(\cos\theta)}\,, \tag{2}$$

$$P_y=\frac{\eta\sin\theta\cos\theta}{1-\frac{1}{2}\beta P_2(\cos\theta)}\,, \tag{3}$$

$$P_z=\pm\frac{\zeta\cos\theta}{1-\frac{1}{2}\beta P_2(\cos\theta)}\,, \tag{4}$$

where the $\pm$ signs refer to incident photons of positive or negative helicity, respectively. The dynamical parameters σ, β, ξ, η, and ζ, which will be discussed explicitly below, are given in terms of reduced matrix elements of the dipole operator. Although we have restricted our discussion to the case of circularly polarized incident radiation, it is worth mentioning that the five dynamical parameters σ, β, ξ, η, and ζ suffice to describe low-energy photoionization for incident radiation of arbitrary polarization.[30]

For unpolarized incident radiation, only the component P_y is nonvanishing. If the incident radiation is circularly polarized, the spin polarization of the total electron flux (in the direction of incident radiation) is given by

$$P_{\rm tot}=\pm\delta\,, \tag{5}$$

where

$$\delta=\tfrac{1}{3}(\zeta-2\xi)\,. \tag{5a}$$

For the present purposes we consider photoionization of rare gases by photons with energy above the ${}^2P^o_{3/2}$ threshold, but below the ${}^2P^o_{1/2}$ threshold. In this energy region there are three open channels corresponding to the excitation of an outer $p_{3/2}$ electron to $s_{1/2}$, $d_{3/2}$, or $d_{5/2}$ continuum states. We label the reduced matrix element of the dipole operator in these channels by the photoelectron angular momentum $j=\frac{1}{2}$, $\frac{3}{2}$, and $\frac{5}{2}$. These dipole amplitudes D_j may be easily obtained from the MQDT. We write[24]

$$D_j=\sum_{\rho=1}^{3}\overline{T}^\rho_j e^{i\pi\tau_\rho}\overline{D}_\rho\,, \tag{6}$$

where $\pi\tau_\rho$ are the short-range eigenphases of the three open eigenchannels, $\overline{D}_\rho$ are the corresponding eigenamplitudes, and $\overline{T}^\rho_j$ is the orthogonal transformation matrix between the eigenchannels ($\rho=1,2,3$) and the open channels ($j=\frac{1}{2},\frac{3}{2},\frac{5}{2}$). The amplitudes D_j are determined from solutions to the RRPA equations, which satisfy outgoing-wave boundary conditions in the open channels. The quantities τ_ρ, $\overline{D}_\rho$, and $\overline{T}^\rho_j$ all may be obtained from the multichannel quantum-defect parameters μ_α, D_α, and $U_{i\alpha}(i,\alpha=1,\ldots,5)$, using the procedures described in detail in Ref. 24.

We may express the dynamical parameters σ, β, ξ, η, and ζ in terms of the amplitudes D_j (atomic units are used) by

$$\sigma=2\pi^2\alpha\frac{df}{dE}=\frac{4\pi^2\alpha}{3}\omega\bar{\sigma}\,, \tag{7}$$

$$\bar{\sigma}=|D_{1/2}|^2+|D_{3/2}|^2+|D_{5/2}|^2\,, \tag{8}$$

$$\beta=[-\tfrac{4}{5}|D_{3/2}|^2+\tfrac{4}{5}|D_{5/2}|^2-(1/\sqrt{5})(D_{1/2}D^*_{3/2}+\text{c.c.}) - (3/\sqrt{5})(D_{1/2}D^*_{5/2}+\text{c.c.})+\tfrac{3}{5}(D_{3/2}D^*_{5/2}+\text{c.c.})]\bar{\sigma}^{-1}\,, \tag{9}$$

$$\xi=[\tfrac{1}{2}|D_{1/2}|^2+\tfrac{2}{5}|D_{3/2}|^2-\tfrac{9}{10}|D_{5/2}|^2 - \tfrac{1}{4}\sqrt{5}(D_{1/2}D^*_{3/2}+\text{c.c.})+\tfrac{9}{20}(D_{3/2}D^*_{5/2}+\text{c.c.})\bar{\sigma}^{-1}\,, \tag{10}$$

$$\eta=i[(3/4\sqrt{5})(D_{1/2}D^*_{3/2}-\text{c.c.}) -(3/2\sqrt{5})(D_{1/2}D^*_{5/2}-\text{c.c.}) +\tfrac{3}{4}(D_{3/2}D^*_{5/2}-\text{c.c.})]\bar{\sigma}^{-1}\,, \tag{11}$$

$$\zeta=[-\tfrac{1}{2}|D_{1/2}|^2+\tfrac{1}{5}|D_{3/2}|^2+\tfrac{3}{10}|D_{5/2}|^2 -\tfrac{1}{2}\sqrt{5}(D_{1/2}D^*_{3/2}+\text{c.c.})-\tfrac{9}{10}(D_{3/2}D^*_{5/2}+\text{c.c.})]\bar{\sigma}^{-1}\,. \tag{12}$$

Our calculation of the angular distribution and spin polarization of photoelectrons within the resonance region involves several computational steps. First, we solve the many-electron problem describing photoexcitation using the RRPA to find the smooth MQDT parameters μ_α, D_α, and $U_{i\alpha}$ at a number of energies in the resonance region. For practical reasons we must limit these RRPA calculations to include only a few important intershell correlations. In the case of argon, we include the correlations within and between the 3*s* and 3*p* shells, and we treat the inner shells in a frozen-core approximation. The justification for this truncation has been discussed before.[31] For krypton, we include the 4*s*, 4*p*, and 3*d* shells in the dynamical calculation, while for xenon we include 5*s*, 5*p*, and 4*d* shells. The importance of including inner *d*-shell correlations in the RPA calculations has been discussed by Amusia[32]; we illustrate the effect of 4*d* correlations on the Beutler-Fano resonances of Xe in the following section.

Once the MQDT parameters have been determined from the RRPA calculations, we carry out the analysis described in Ref. 24 to determine the eigenphases and eigenamplitudes in the open channels, $\bar{D}_\rho$ and τ_ρ, and the transformation matrix $\bar{T}^\rho_j$. The production amplitudes D_j and the five parameters σ, β, ξ, η, and ζ describing photoionization are then easily obtained using Eqs. (7) through (12).

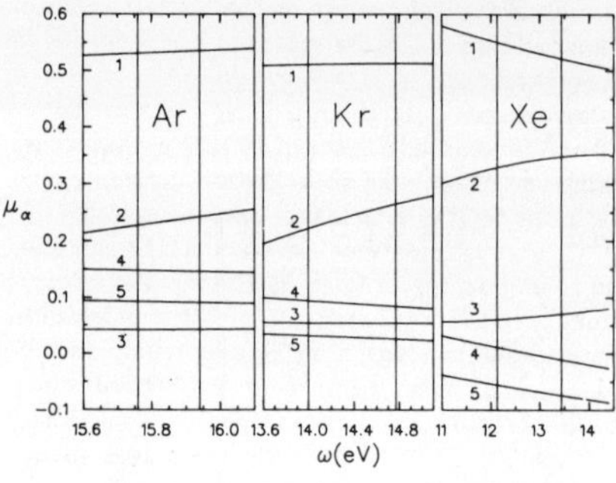

FIG. 1. Eigen-quantum defects μ_α plotted as functions of photon energy ω across the first two thresholds of the rare-gas atoms argon, krypton, and xenon.

In RRPA calculations the theoretical photoionization thresholds are given by the eigenvalues of the Dirac-Fock equations for the atomic ground state. Since these theoretical thresholds are only in fair agreement with experimental thresholds, we adopt in the following section the procedure of aligning the theoretical and experimental profiles at the second thresholds when presenting our comparisons with experimental data.

III. RESULTS AND DISCUSSIONS

As discussed in previous papers,[14-16,24] MQDT allows one to characterize the dynamics of photoionization processes in terms of a few parameters: the eigen-quantum defects μ_α, the eigen-dipole amplitudes D_α, and the transformation matrices $U_{i\alpha}$. These parameters are slowly varying functions of energy near the thresholds, so that they can be interpolated or extrapolated in the autoion-

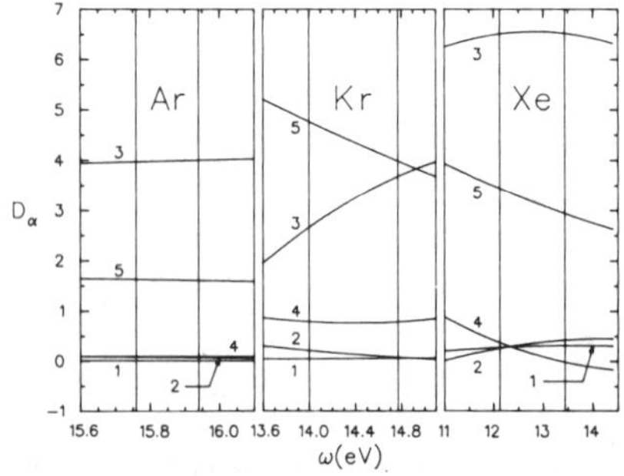

FIG. 2. Eigen-dipole amplitudes D_α plotted as functions of photon energy ω across the first two thresholds of the rare-gas atoms argon, krypton, and xenon.

TABLE I. Transformation matrices $U_{i\alpha}$, eigen-quantum defects μ_α, and eigen-dipole amplitudes D_α (in atomic units) at the $^2P^0_{1/2}$ thresholds of the rare-gas atoms argon, krypton, and xenon.

		Ar					Kr					Xe				
	i/α	1	2	3	4	5	1	2	3	4	5	1	2	3	4	5
$U_{i\alpha}$	1	0.056	0.001	0.041	−0.555	0.829	−0.086	−0.003	0.350	0.500	−0.787	0.074	−0.010	−0.146	0.346	0.924
	2	0.734	0.625	0.255	0.073	−0.014	−0.737	−0.615	−0.223	−0.136	−0.103	0.720	−0.633	0.279	−0.050	−0.002
	3	−0.546	0.333	0.767	−0.045	−0.032	0.546	−0.349	−0.671	−0.007	−0.360	−0.563	−0.323	0.743	0.121	0.113
	4	0.079	−0.003	−0.014	−0.826	−0.558	−0.119	0.006	−0.326	0.846	0.405	0.080	0.003	−0.034	0.929	−0.360
	5	0.392	−0.706	0.587	0.046	−0.024	−0.371	0.707	−0.521	−0.126	−0.274	0.391	0.704	0.590	0.011	0.065
μ_α		0.536	0.242	0.042	0.145	0.090	0.513	0.260	0.053	0.085	0.026	0.521	0.348	0.070	−0.011	−0.074
D_α		0.006	0.067	4.005	0.095	1.605	0.061	0.092	3.572	0.778	4.084	0.314	0.424	6.517	−0.021	2.938

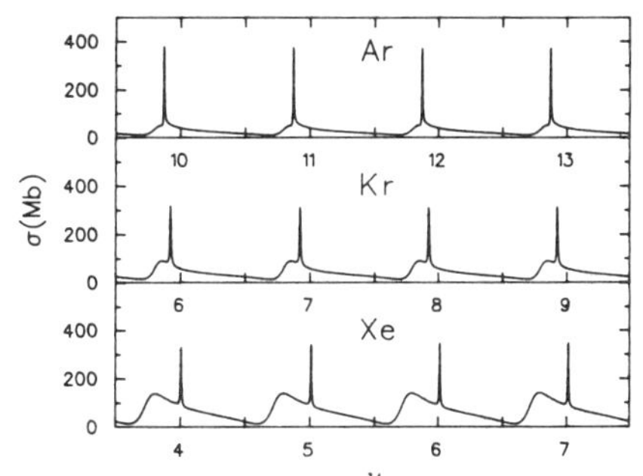

FIG. 3. Photoionization cross sections σ plotted against the effective quantum number ν.

ization region. To show the energy dependence of the MQDT parameters between the $^2P^o_{3/2}$ and $^2P^o_{1/2}$ thresholds of the rare gases argon, krypton, and xenon, we plot in Figs. 1 and 2 the values of μ_α and D_α obtained from the RRPA as functions of photon energies. As one can see from these figures, μ_α and D_α vary smoothly across the first two thresholds and are nearly linear functions of photon energies.

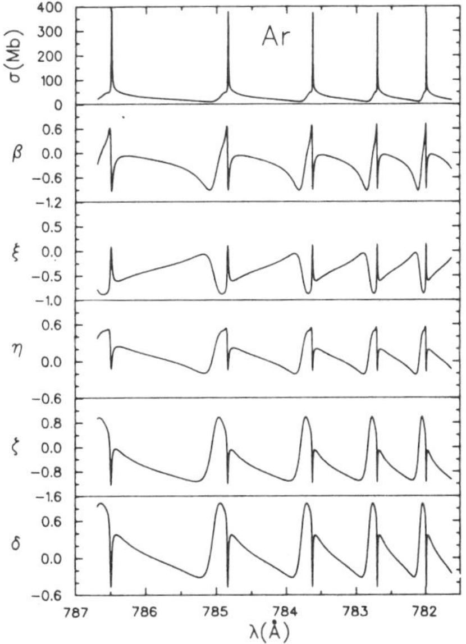

FIG. 4. Photoionization cross sections σ, angular-distribution asymmetry parameters β, and spin-polarization parameters ξ, η, ζ, and δ plotted as functions of photon wavelength λ in the autoionization region of argon.

For low-energy photoionization, the LS coupling scheme provides a good approximate description of the eigenchannels. In Fig. 2 the two strong channels ($\alpha = 3, 5$) with sizeable dipole amplitudes D_α correspond roughly to the dipole-allowed excitations from the ^1S ground state to the $(p^5d)\,^1P^o$ and $(p^5s)\,^1P^o$ states, respectively, while the three remaining weak channels ($\alpha = 1, 2, 4$) are associated with the forbidden excitations to the $(p^5d)\,^3P^o$, $(p^5d)\,^3D^o$, and $(p^5s)\,^3P^o$ states, respectively. A general trend here is the steady increase in the size of the corresponding eigenamplitudes D_α from argon to krypton to xenon, resulting in the increase in cross sections among these atoms, as shown below. The rapid increase in the magnitude of the three weak channels reflects the increasing importance of spin-orbit interactions in heavy atoms. The assignment of the jj-coupled channel indices i and the eigenchannel indices α are given by

$i, \alpha =$	1	2	3	4	5
i	$(^2P^o_{3/2})s_{1/2}$	$(^2P^o_{3/2})d_{3/2}$	$(^2P^o_{3/2})d_{5/2}$	$(^2P^o_{1/2})s_{1/2}$	$(^2P^o_{1/2})d_{3/2}$
α	$(p^5d)\,^3P^o$	$(p^5d)\,^3D^o$	$(p^5d)\,^1P^o$	$(p^5s)\,^3P^o$	$(p^5s)\,^1P^o$

Numerical values of μ_α, D_α, and $U_{i\alpha}$ at the $^2P^o_{1/2}$ thresholds of argon, krypton, and xenon are given in Table I.

In Fig. 3, the cross sections in the autoionizing regions of argon, krypton, and xenon are plotted as functions of the effective quantum number ν defined in the preceding paper.[24] (Nonrelativistically, ν is defined in terms of the energy ϵ relative to the $^2P^o_{1/2}$ threshold as $\epsilon = -\frac{1}{2}\nu^{-2}$.) One sees in these plots that the Beutler-Fano profiles are practically the same along the Rydberg series. This is not surprising in view of the weak energy dependence of the quantum-defect parameters. The implication of these regularities is that physical observables such as cross sections, asymmetry param-

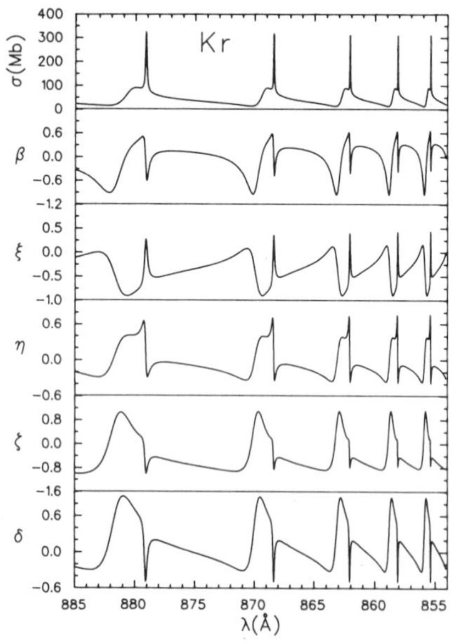

FIG. 5. Photoionization cross sections σ, angular-distribution asymmetry parameters β, and spin-polarization parameters ξ, η, ζ, and δ plotted as functions of photon wavelength λ in the autoionization region of krypton.

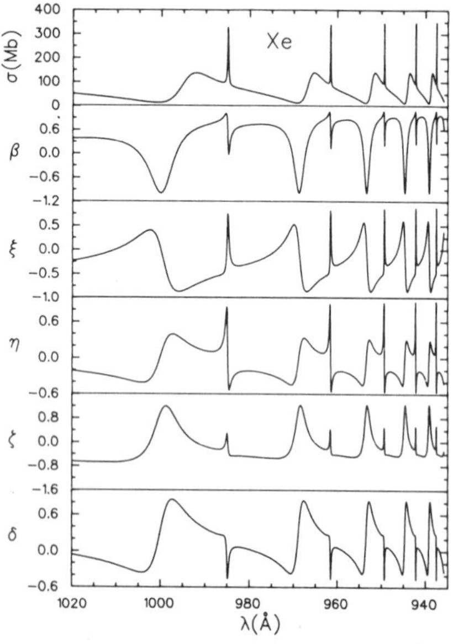

FIG. 6. Photoionization cross sections σ, angular-distribution asymmetry parameter β, and spin-polarization parameters ξ, η, ζ, and δ plotted as functions of photon wavelength λ in the autoionization region of xenon.

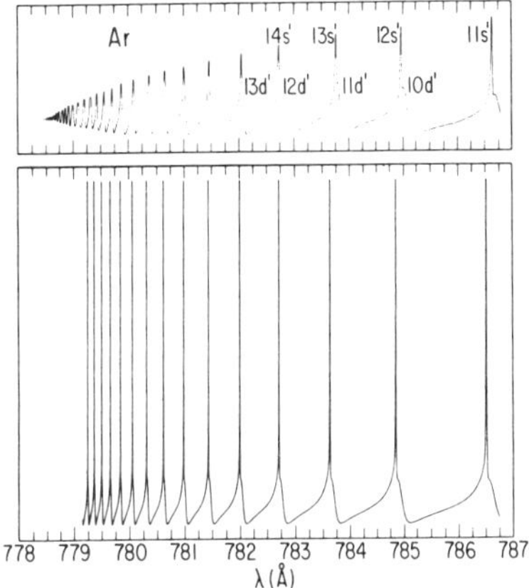

FIG. 7. Photoionization spectra of argon between the $^2P^o_{3/2}$ and $^2P^o_{1/2}$ thresholds. The upper graph shows experimental data by Radler and Berkowitz[4] with a photon resolution width of 0.02 Å, and the lower graph shows results of this calculation.

eters, and spin polarizations in the autoionizing region are characterized by their respective profiles at the first few resonances.

For argon, the sharp ns' resonances almost coincide with the broad nd' ones. The two resonances depart from each other in krypton, and they are well separated in xenon. From argon to krypton to xenon, one also sees that the magnitudes of the broad nd' resonances are increasing, while those of the ns' series remain fairly constant.

In Fig. 4 results of our present calculations on cross sections σ, angular asymmetry parameters β, and spin-polarization parameters ξ, η, ζ, and δ for argon are plotted as functions of photon wavelength λ. Similar plots for krypton and xenon are shown in Figs. 5 and 6, respectively. Because of the repetition of these parameters along the Rydberg series, only those results for the first few resonances are presented. These parameters, σ, β, ξ, η, ζ, and δ show two distinct patterns: sharp changes at the locations of ns' resonances, superimposed on slower changes arising from the nd' series.

In Figs. 7, 8, and 9, we compare our calculated resonance profiles with recent experimental measurements by Radler and Berkowitz[4] for argon,

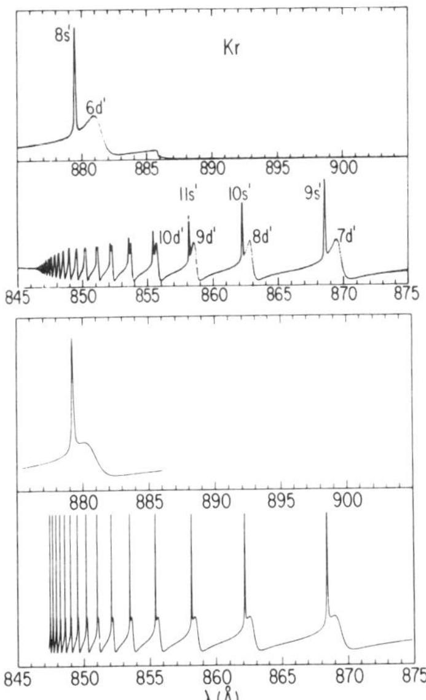

FIG. 8. Photoionization spectra of krypton between the $^2P^o_{3/2}$ and $^2P^o_{1/2}$ thresholds. The upper graph shows experimental data obtained by Berkowitz (Ref. 33) with a photon resolution width of 0.07 Å, and the lower graph shows results of this calculation.

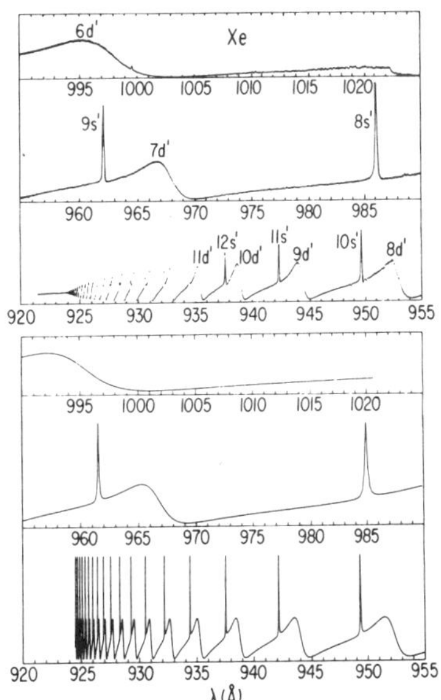

FIG. 9. Photoionization spectra of xenon between the $^2P^o_{3/2}$ and $^2P^o_{1/2}$ thresholds. The upper graph shows experimental data obtained by Eland (Ref. 33) with a photon resolution width of 0.07 Å, and the lower graph shows results of this calculation.

and by Berkowitz[33] for krypton and xenon. As can be seen from these figures, the general features of the autoionization profiles are well represented by the present calculations; however, there are minor discrepancies which should be noted. First, the positions of the resonances are slightly shifted. This shift is most pronounced near the $^2P^o_{3/2}$ threshold, and the theoretical resonances are located at shorter wavelengths than the experimental ones. Second, there is a rapid decrease in the size of the *ns′* peaks in the experimental spectrum which does not occur in the theoretical spectrum. For higher members of the Rydberg series, it is clear that the overall decrease in experimental resonance amplitudes is due to finite instrumental resolution; however, it is not clear whether the apparent decrease in the relative size of the experimental peaks for the first few resonances is a real effect or is due to instrumentation. From the theoretical point of view, it is difficult to understand a real decrease of the size along the Rydberg series, since such a decrease would imply a rapid change in the associated quantum-defect parameters in a relatively small energy region.

A third problem which can be noted is that the relative separation of the *ns′* and *nd′* resonances is somewhat different in the theoretical and experimental spectra. We believe that this problem is due, at least in part, to the limited correlation included in the present RRPA calculation. To illustrate the influence of inter-shell correlations on the resonance profiles, we present in Fig. 10 a comparison of two theoretical results in the (9*d′*, 11*s′*) region of the xenon spectrum. The first of these theoretical curves is a portion of the spectrum presented before in Fig. 9, which includes correlations from the 5*p*, 5*s*, and 4*d* shells. In the second curve, the 4*d* and 5*s* correlations are omitted. We find only an insignificant shift in the location of the theoretical resonance positions (less than 0.1 Å in this region), and for comparison purposes we simply line up the 11*s′* peaks in Fig. 10. As is apparent from the figure, the inclusion of 4*d* correlation increases the magnitude of the 9*d′* resonance and sharpens its profile. To compare with experiment, we superimpose on Fig. 10 the observed spectrum[33] in this region by first lining up the 11*s′* peak and then by scaling the experimental data to agree with theory on the shorter wavelength side of the 9*d′* resonance. The point here is that inter-shell correlation from 4*d* (and to a lesser extent, 5*s*) brings the theoretical profile of the 9*d′* resonances into closer agreement with observation. Also, because of the sharpening of the theoretical 9*d′* profile, there is an apparent increase in the separation of 9*d′* and 11*s′* resonances. This increased separation on including inter-shell correlations is in the right direction, but is still insufficient to account entirely for the observed separation.

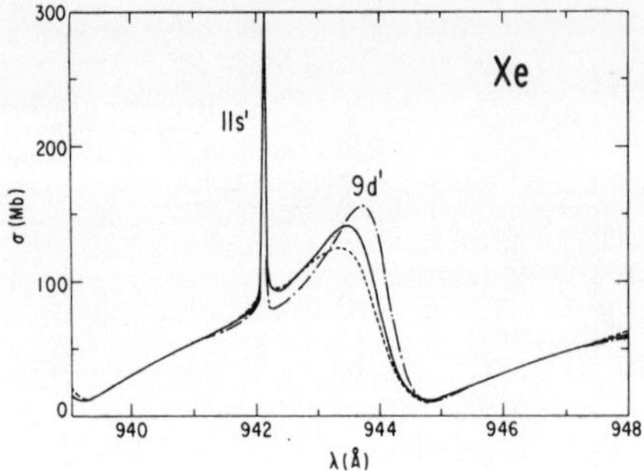

FIG. 10. Photoionization cross sections σ plotted against photon wavelength λ in the autoionization region of xenon. The solid curve is the result of this calculation, including correlations between the 5*p*, 5*s*, and 4*d* shells. The dashed curve is the result of this calculation with the 5*p*-shell correlations only. The dash-dot curve is the experimental data shown in Fig. 9 scaled to match theoretical curves on the shorter-wavelength side of the 9*d′* resonance. All curves are lined up at the 11*s′* resonance for comparison purposes.

In Fig. 11 we compare our theoretical predictions of the angular-distribution β parameters for xenon near the (6*d′*, 8*s′*) resonances with the measurements of Samson and Gardner.[20] To compare the profile of the β parameter, we align the theoretical and experimental 8*s′* resonances. The general agreement with experiment is good, although the data are not sufficiently detailed to reveal the structure of β near the 8*s′* resonance. We also

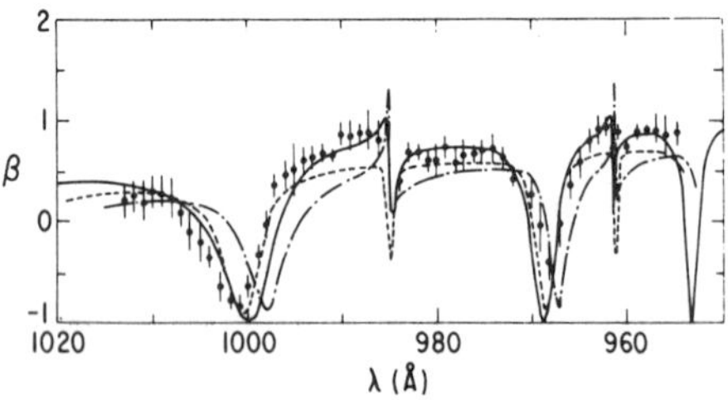

FIG. 11. Angular-distribution asymmetry parameters β plotted against photon wavelength λ in the autoionization region of xenon. Dots are experimental data by Samson and Gardner (Ref. 20). The solid curve is the result of this calculation. The dash-dot and the dashed curves are MQDT calculations by Dill (Ref. 21) and Geiger (Ref. 19), respectively. All curves are lined up at the 8*s′* resonances for comparison purposes.

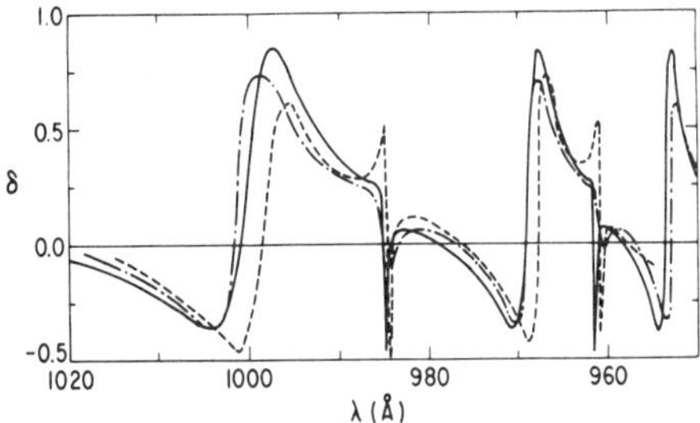

FIG. 12. Total spin-polarization parameters δ plotted against photon wavelengths λ in the autoionization region of xenon. The dash-dot curve is the experimental data by Heinzmann *et al*. (Ref. 23). The solid curve is the result of the present calculation. The dashed curve is the MQDT calculation of Lee (Ref. 22). All curves are lined up at the 8s′ resonances for comparison purposes.

show in the same graph two MQDT calculations by Dill[21] and by Geiger,[19] based on empirical quantum-defect parameters. Once again, the overall agreement is good, except for the detailed features near the 8s′ resonances.

In Fig. 12 our results on the total spin-polarization parameter δ in xenon near the (6d′, 8s′) resonances are compared with recent experimental measurements of Heinzmann *et al*.[23] In the same figure, we give the results of a MQDT calculation by C. M. Lee.[22] Again, as in the case of the β parameter, the general agreement between theoretical and experimental profiles is good. The two theoretical profiles differ somewhat at the 8s′ resonance. In this case, the experimental profile at the 8s′ resonance has been given in detail and is in somewhat better agreement with the present calculation.

In the above paragraphs, we have presented our MQDT studies of the Beutler-Fano profiles for the rare gases argon, krypton, and xenon using dynamical parameters determined from the RPPA. We have found that the systematic features of the profiles are in good agreement with existing experimental data. These comparisons show the utility of the MQDT–RRPA approach in predicting complex spectra involving strongly interacting channels. For further comparisons, it would be desirable to have measurements of β and δ parameters for krypton and argon. Measurements of the η parameter, such as those made at higher photon energies by Heinzmann *et al*.,[34] would also be useful since they would compliment the existing experimental data in the resonance region.

ACKNOWLEDGMENTS

The authors would like to thank Dr. J. Berkowitz for communicating unpublished data and for helpful discussions. We are also grateful to Dr. K. Yoshino and Dr. U. Heinzmann for helpful correspondence. The work of W. R. J. and K.-N. Huang are supported in part by the National Science Foundation under Grant No. PHY79-09229. The work of K. T. C. is supported by the U. S. Department of Energy. One of us (W.R.J.) thanks the Argonne National Laboratory for an appointment.

---

[1]H. Beutler, Z. Phys. 93, 177 (1935).
[2]U. Fano, Nuovo Cimento 12, 154 (1935).
[3]Kouichi Yoshino, J. Opt. Soc. Am. 60, 1220 (1970).
[4]K. Radler and J. Berkowitz, J. Chem. Phys. 70, 216 (1979); 70, 221 (1979).
[5]R. E. Huffman, Y. Tanaka, and J. C. Larrabee, J. Chem. Phys. 39, 902 (1963).
[6]R. D. Hudson and V. L. Carter, J. Opt. Soc. Am. 58, 227 (1968).
[7]P. H. Metzger and G. R. Cook, J. Opt. Soc. Am. 55, 516 (1965).
[8]Po Lee and G. L. Weisler, Phys. Rev. 99, 540 (1955).
[9]R. D. Hudson and L. J. Kiefer, At. Data 2, 205 (1971).
[10]K. Yoshino and Y. Tanaka, J. Opt. Soc. Am. 69, 159 (1979).
[11]R. E. Huffman, Y. Tanaka, and J. C. Larrabee, Appl. Opt. 15, 947 (1963).
[12]C. E. Moore, *Atomic Energy Levels NSRDS-NBS* (U.S. Government Printing Office, Washington, D.C., 1971), Vol. III, pp. 113–117.
[13]F. M. Matsunaga, K. Watanabe, and R. S. Jackson, J. Quant. Spectrosc. Radiat. Transfer 5, 329 (1965).
[14]M. J. Seaton, Proc. Phys. Soc. London 88, 801 (1966); U. Fano, Phys. Rev. A 2, 353 (1970).
[15]K. T. Lu, Phys. Rev. A 4, 579 (1971).
[16]Chia-Ming Lee and K. T. Lu, Phys. Rev. A 8, 1241 (1973).
[17]C. M. Lee, Phys. Rev. A 10, 584 (1974).
[18]J. Geiger, Z. Phys. A 276, 219 (1976).
[19]J. Geiger, Z. Phys. A 282, 129 (1977).
[20]J. A. R. Samson and J. L. Gardner, Phys. Rev. Lett. 31, 1327 (1973).
[21]D. Dill, Phys. Rev. A 7, 1976 (1973).
[22]C. M. Lee, Phys. Rev. A 10, 1598 (1974).
[23]U. Heinzmann, F. Schäfers, K. Thimm, A. Wolcke, and J. Kessler, J. Phys. B 12, L679 (1979).
[24]C. M. Lee and W. R. Johnson, Phys. Rev. A 22, 979 (1980).
[25]W. R. Johnson and M. Le Dourneuf, J. Phys. B 13, L13 (1980).
[26]B. Brehm, Z. Phys. 242, 195 (1971).
[27]V. L. Jacobs, J. Phys. B 5, 2257 (1972).
[28]N. A. Cherepkov, Zh. Eksp. Teor. Fiz. 38, 933 (1974) [Sov. Phys.—JETP 38, 463 (1974)]; J. Phys. B 12,

1279 (1979).
[29]R. H. Pratt, A. Ron, and H. K. Tseng, Rev. Mod. Phys. 45, 273 (1973).
[30]K.-N. Huang, Phys. Lett. 77A, 133 (1980); Phys. Rev. A 22, 223 (1980).
[31]W. R. Johnson and K. T. Cheng, Phys. Rev. A 20, 978 (1979).
[32]M. Ya. Amusia and N. A. Cherepkov, Case Stud. At. Phys. 5, 47 (1975).
[33]See J. Berkowitz, *Photoabsorption, Photoionization, and Photoelectron Spectroscopy* (Academic, New York, 1979), pp. 177 and 181.
[34]U. Heinzmann, G. Schönhense, and J. Kessler, Phys. Rev. Lett. 42, 1603 (1979).

1977 *Phys. Rev.* A **15** 817
Reprinted with permission from the American Physical Society

Erratum

Erratum: Quantum defect theory of *l* uncoupling in H_2 as an example of channel-interaction treatment [Phys. Rev. A 2, 353 (1970)]

U. Fano

On p. 357, Eq. (19′) should read

$$\nu_0^2 = \nu_2^2/(1 - 12B\nu_2^2) = \cdots .$$

On p. 359, Eq. (38) should read

$$\tan\eta = \frac{\sin^2\frac{1}{2}\pi\delta \sin^2 2\alpha}{1 - \sin^2\frac{1}{2}\pi\delta \sin^2 2\alpha} .$$

On p. 360, Eq. (48) should read in part

$$q = -\cot(\Delta_p - \Delta_s) ,$$

i.e., a sign should be reversed.

The asymptotic form of Coulomb functions in Sec. II, derived from Ref. 2, should be revised to reflect amendments introduced by W. Eissner *et al.* [J. Phys. B 2, 342 (1969)]. The revision, which removes stray imaginary terms from otherwise real expressions, has no effect on following portions of this paper but is relevant to other applications. I am indebted to Dr. J. Dubau for bringing this matter to my attention.

On p. 356, Eq. (10), the factor $e^{i\pi\nu}$ should read $\cos\pi\nu$, and the factor $e^{i\pi(\nu+1/2)}$ should read $(-\sin\pi\nu)$. In Eq. (15), the factor $e^{i\pi(\nu+\mu_\Lambda)}$ should read $\cos\pi(\nu+\mu_\Lambda)$. Recall that these equations must also be reinterpreted, unless $\nu \gg 1$, as stated in the second paragraph of Sec. II.

1970 *Phys. Rev.* A **2** 353–65
Reprinted with permission from the American Physical Society

Quantum Defect Theory of *l* Uncoupling in H_2 as an Example of Channel-Interaction Treatment*

U. Fano
Department of Physics, The University of Chicago, Chicago, Illinois 60637
(Received 6 March 1970)

Formulas are developed which account for the details of the high-resolution photoabsorption spectrum of H_2 near threshold observed by Herzberg. The treatment illustrates the application of Seaton's multichannel quantum defect method under schematic conditions, and might serve as a basis for further developments.

I. INTRODUCTION

The photoabsorption spectrum of $H_2\,{}^1\Sigma_g^+(v=0, J=0)$ has been observed recently[1] under high resolution near the two thresholds for ionization to $H_2^+\,|^2\Sigma_g^+$ $(v=0, N=0, 2,$ where N is the angular momentum apart from spin). Two long Rydberg series of lines appear, whose lower members correspond, respectively, to the standard H_2 classifications $1s\sigma\, np\sigma\,{}^1\Sigma_u^+$ and $1s\sigma\, np\pi\,{}^1\Pi_u^+$. As n increases the series change character progressively owing to l uncoupling and converge to the alternative rotational levels of H_2^+. In the course of this evolution, they perturb each other strongly with alternations of intensity and level density. The intensity alternations extend into the continuum between the thresholds where they take the form of Beutler lines with broad asymmetric profiles.

This paper develops formulas that predict the line positions and the intensity distribution throughout the region of intermediate coupling near the thresholds, in terms of a few experimentally accessible parameters. The development relies on Seaton's multichannel quantum defect method[2] (QDM) which has thus far received only limited application.

The essential feature of our problem is that the wave function of a highly excited electron with low orbital angular momentum extends over small-r and large-r regions where entirely different physical situations prevail.

(A) For small r values (e.g., $\lesssim 5$ Å), the usual circumstances of molecular physics prevail. The kinetic and potential energies of the excited electron exceed by far the rotational and vibrational energies of the nuclei, and the Born-Oppenheimer factorization of the molecular wave function applies. The electron's motion is rigidly coupled to the molecular axis $\hat{R}$ and the squared-orbital momentum projection $\Lambda^2 = (\vec{l}\cdot\hat{R})^2$ – equal to 0 or 1 for σ and π states – is a constant of the motion. The total angular momentum $\vec{J}$ is also constant, whereas the squared momentum of molecular ion rotation $\vec{N}^2$ does not have a definite value.

(B) For large-r values (e.g., $\gtrsim 50$ Å), the electron is effectively coupled to the H_2^+ core only by the isotropic Coulomb attraction and can exchange with this core hardly any energy or angular momentum. The squared angular momenta of the electron $\vec{l}^2$ and of the core $\vec{N}^2$ are constants of the motion, as is $\vec{J}=\vec{l}+\vec{N}$. (The electron-and nuclear-spin pairs are coupled into singlets and will accordingly be disregarded.) The electron's energy equals the total energy, less the core's ground-state energy, less the rotational core energy $B\vec{N}^2$. Accordingly, it has two possible values depending on the value of N, which may equal 0 or 2 as discussed below. Alternative vibrational states of the core do occur in photoabsorption but only $v=0$ will be considered here.

The complexity of the observed spectrum stems from the occurrence of these different situations, but can be unraveled owing to several helpful circumstances.

(i) The photoabsorption process takes place within the region occupied by the initial H_2 ground state, that is, in the innermost portion of region A. Its probability amplitude can be represented in terms of two dipole matrix elements D_σ and D_π, which we take to be normalized per unit atomic energy range and which are practically energy independent over the spectral range of interest to us, namely, over a few hundred cm^{-1} ($\lesssim 0.1$ eV). Details of level positions and of other final state characteristics influence the photoabsorption spectrum only through coefficients of a linear combination $D_\sigma A_\sigma + D_\pi A_\pi$.

(ii) The excited electron's interaction with the H_2^+ core is complicated only within region A. These complexities influence the electron's wave functions in the outer region only through the specification of parity and angular momentum quantum numbers and through boundary conditions on a surface surrounding the core. These boundary conditons can be represented by quantum defect parameters μ_σ and μ_π which are practically constant under the same conditions as D_σ and D_π are. The anisotropy of the interaction between the excited electron and the rest of the molecule is represented solely by these four parameters, because a central field prevails outside the boundary on which μ_σ and μ_π are defined.

(iii) The structure of the high resolution spectrum observed near threshold is determined by the properties of the excited electron's wave function in the region *outside* the H_2^+ core, where the electron experiences only a Coulomb attraction. This region includes the outer portion of region A and all of region B as well as the transition range between A and B. The study of wave functions outside the H_2^+ core constitutes the main subject of the QDM.

(iv) Wave functions appropriate to region B depend on the separation $6B$ of the first two rotational levels of the H_2^+ core. The parameter $B = 1.32 \times 10^{-4}$ a.u. completes, together with μ_σ, μ_π, D_σ and D_π, the set of *five parameters* on which the spectrum depends.

(v) Because a central Coulomb field prevails from the outer portion of region A all the way to infinity, the radial factors of the wave function are known analytically. Therefore wave functions appropriate to regions A and B differ only in their angular factors. The respective angular wave function factors for the two regions are also known analytically; they are related to one another by a 2×2 orthogonal transformation which connects the relevant eigenstates of $\Lambda^2 (\Lambda = \sigma$ or $\pi)$ to those of $\vec{N}^2 (N = 0$ or $2)$.

Conservation of angular momentum and parity in the photoabsorption by the even $J = 0$ ground state limits relevant excited-state quantum numbers to $J = 1$ and to negative ungerade parity. The "para" character of the homonuclear molecule limits the rotational quantum number N to even values, whereby the orbital quantum number l must be odd and the wave functions with Λ quantum numbers must have + character. In the one-center expansion of our Rydberg-state wave functions, the l quantum number is further restricted, for $r \gtrsim 2$ Å, to $l = 1$,[3] whereby $N \lesssim 2$.

On the basis of these considerations one may picture the state of a highly excited electron as follows. Most of the electron's probability distribution lies in region B, where the electron may belong to either the $N=0$ or the $N=2$ channel. Transitions between these channels occur insofar as the electron penetrates in the (comparatively small) region A where its interaction with the H_2^+ core becomes anisotropic. Hence, these transitions may be said to result from a short-range scattering process. For example, an electron bound in the $N=2$ channel, but with sufficient energy to escape by autoionization into the $N=0$ channel, may be said to autoionize by scattering on the H_2^+ core. (Indeed, such a picture of autoionization by scattering applies rather generally.)

In the past, the theory of molecular spectra has centered on states with low-to-moderate excitation, whose conditions correspond essentially to those of region A. The rotational energy is then given, in first approximation, by the expectation value of $B\vec{N}^2$ in the relevant eigenstate of Λ^2. From this point of view, the existence of off-diagonal elements of the matrix $(\Lambda | B\vec{N}^2 | \Lambda')$ disturbs the coupling of $\vec{l}$ to $\hat{R}$ so as to cause a certain amount of "l uncoupling"; the uncoupling has been generally treated by a perturbation expansion. Extension of accurate spectral analysis to the proximity of the ionization threshold requires now an "exact" treatment of the rotational energy; this task is quite manageable under the circumstances considered in this paper. The formulation of this treatment utilizing the QDM makes its extension beyond the ionization thresholds straightforward. (Indeed, the treatment for the continuum range is simplest and will be carried out first.) Further extensions, primarily to include interaction with the vibrational motions[4] and to make contact with the recent extensive and detailed experimental work on photoionization[5] seem quite possible but would require additional work.

The multichannel QDM is usually formulated in terms of the reaction matrix of collision theory. In our two-channel problem, not only is the reaction matrix of minimal (2×2) dimension, but its eigenvalues and eigenvectors are readily identified. These circumstances permit a particularly explicit treatment of multichannel theory, bringing out features that might otherwise have escaped attention. For this reason, Sec. V of this paper has been developed in greater detail than required by interpretation of the H_2 spectrum. Several concepts utilized there should be extended in a broader context. In this limited application, the matrix formalism of collision theory has appeared unnecessary. However, many equations of this paper can be rewritten in matrix language; some of them are well known in matrix form and remain valid for an indefinite number of channels, others seem to require further study.

Section II describes the transformation between eigenstates of Λ^2 and of $\vec{N}^2$; Sec. III, the treatment of radial wave functions which connects regions A and B. The remainder of the paper establishes four analytical results directly applicable to the H_2

spectrum, namely (i) Eq. (51) which determines the position of each line of the discrete spectrum near the lowest ionization limit, (ii) expression (59) for the intensity of each of these lines, (iii) expression (43) for the Beutler absorption spectrum between the $N=0$ and $N=2$ ionization limits, and (iv) expression (25) for the separate probabilities of photoionization into the two channels above the $N=2$ limit. The input data to these equations, namely, the 5 parameters μ_σ, μ_π, D_σ, D_π and B mentioned above and the angle α identified in Sec. II, may be regarded as known from other experiments or from theory. Alternatively, they may be fitted to the experimental results on H_2 photoabsorption near threshold as indicated in Sec. VII.

II. TRANSFORMATION BETWEEN EIGENSTATES OF Λ^2 AND $\vec{N}^2$

The eigenvectors of Λ^2 and $\vec{N}^2$ relevant to us constitute two pairs corresponding, respectively, to $\Lambda=0$ or 1 and to $N=0$ or 2. Therefore, each pair is related to the other by a 2×2 unitary transformation matrix $U_{N\Lambda}$. This matrix could be calculated by considerations of vector coupling, parity, etc. or by calculating and diagonalizing the matrix of Λ^2, $(N|(\vec{l}\cdot\hat{R})^2|N')$. Here we proceed, in essence, by constructing explicit representations of both sets of eigenvectors, $|N)$ and $|\Lambda)$ and calculating $U_{N\Lambda}=(N|\Lambda)$ as overlap integrals. These representations are wave functions of the internuclear axis orientation $\hat{R}\equiv(\theta,\phi)$ and of the electron direction vector $\hat{r}$. The vector $\hat{r}$ is represented by (ϑ,φ) or by (ϑ',φ') in the laboratory or molecular coordinate systems, respectively. The molecular system, with polar axis $\hat{R}$, is obtained from the laboratory system by three successive rotations with Euler angles (χ,θ,ϕ); we set $\chi=0$.

For reasons stated in Sec. I, the relevant wave functions of both sets have the total angular momentum quantum number $J=1$ and the orbital quantum number $l=1$. The eigenfunctions of $\vec{N}^2$ are obtained by simple angular momentum addition of rotational and electron wave functions,

$$\Phi_M^{(J,N)}(\hat{r},\hat{R})=\sum_m Y_{lm}(\vartheta,\varphi)Y_{N,M-m}(\theta,\phi)\times(lm,NM-m|lNJM)\,, \qquad (1)$$

where the Y are ordinary spherical harmonics, the $(\ldots|\ldots)$ represents a Wigner coefficient, and M is the total magnetic quantum number. The relevant (i.e., +) eigenfunctions of Λ^2 are

$$X_M^{(J,\Lambda=0)}=Y_{l0}(\vartheta',\varphi')D_{0M}^{(J)}(0,\theta,\phi)[(2J+1)/4\pi]^{1/2}\,, \qquad (2)$$

$$X_M^{(J,\Lambda=1,+)}=[Y_{l1}(\vartheta',\varphi')D_{1M}^{(J)}(0,\theta,\phi)+Y_{l,-1}(\vartheta',\varphi')D_{-1M}^{(J)}(0,\theta,\phi)][(2J+1)/8\pi]^{1/2}\,, \qquad (3)$$

where $D^{(J)}$ indicates a wave function of the symmetric rotor. The wave function for $\Lambda=0$ has automatically + character, whereas the + function with $\Lambda=1$ involves the linear combination in (3) with + sign. [A function $(\Lambda=1,-)$ would be obtained by reversing the sign in the brackets, but it does not occur for parahydrogen in our problem.]

The X and Φ wave functions are connected by the reciprocal transformation formulas

$$X_M^{(J,\Lambda+)}=\sum_{N=0,2}\Phi_M^{(J,N)}U_{N\Lambda}\,, \qquad (4)$$

$$\Phi_M^{(J,N)}=\sum_{\Lambda=0,1}X_M^{(J,\Lambda+)}\tilde{U}_{\Lambda N}\,, \qquad (5)$$

where the tilde indicates transposition. The explicit form of $U_{N\Lambda}$ is obtained by substituting in the left-hand side of (4), as given by (2) or (3), the inverse transformation

$$Y_{l\Lambda}(\vartheta',\varphi')=\sum_m Y_{lm}(\vartheta,\varphi)D_{m\Lambda}^{(l)}(-\pi-\phi,\theta,\pi) \qquad (6)$$

corresponding to the change of coordinates from the molecular axis $\hat{R}$ to the laboratory axis $\hat{z}$. A reduction of the resulting formula, described in Appendix A, yields

$$U_{N\Lambda}=(lJN0|l-\Lambda,J\Lambda)[2-\delta_{\Lambda0}]^{1/2}(-1)^{J+\Lambda}$$

$$\begin{array}{c|cc} (N)\backslash(\Lambda) & 0 & 1 \\ \hline 0 & \sqrt{\tfrac13} & \sqrt{\tfrac23} \\ 2 & -\sqrt{\tfrac23} & \sqrt{\tfrac13} \end{array} = \begin{vmatrix} \cos\alpha & \sin\alpha \\ -\sin\alpha & \cos\alpha \end{vmatrix}\,. \qquad (7)$$

Note, incidentally, that the Wigner coefficient in (7) corresponds to the vector subtraction formula $\vec{N}=-\vec{l}+\vec{J}$ rather than to $\vec{J}=\vec{l}+\vec{N}$.[6]

For purposes of comparison we utilize the $U_{N\Lambda}$ matrix to calculate the rotational energy matrix in the Λ representation, namely,

$$(\Lambda|B\vec{N}^2|\Lambda')=\sum_N \tilde{U}_{\Lambda N}BN(N+1)U_{N\Lambda'}$$

$$=B\begin{vmatrix} 4 & -\sqrt{8} \\ -\sqrt{8} & 2 \end{vmatrix}\,. \qquad (8)$$

The diagonal terms of (8) represent the rotational energy expectation values for the σ and π states and coincide with the terms linear in B calculated by Chiu.[7] The occurrence of off-diagonal terms in (8) may be regarded as the cause of breakdown of the (σ,π) classification of spectral levels, as noted in Sec. I.

The states represented by wave functions $\sum_\Lambda X_M^{(J\Lambda+)}A_\Lambda$ with arbitrary A_Λ form a set whose elements can be placed into one-to-one correspondence with the – real or complex – vectors A_Λ of a plane

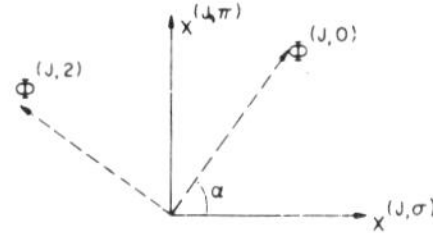

FIG. 1. Graphical relation between eigenvectors of Λ^2 and N^2.

or with states of light polarization or spin-$\frac{1}{2}$ orientation. Figure 1 shows a diagram, in which the pairs of states $X_M^{(J\Lambda+)}$ and $\Phi_M^{(JN)}$ are represented by alternative pairs of orthogonal unit vectors rotated with respect to one another by the angle $\alpha = \arccos\sqrt{\frac{1}{3}}$, in accordance with (7).

III. COULOMB FIELD WAVE FUNCTIONS.

Our problem concerns the motion of an electron *outside* a H_2^+ core, combined with the rotation of this core. The wave function for this combined system will be represented by superposing products of the X or Φ angular functions of Sec. II and of Coulomb field radial functions of the electron. Continuation of this wave function within the core is made unnecessary, in essence, by the use of boundary conditions on a spherical surface surrounding the core. Similarly, it is unnecessary to include explicit wave function factors representing the core vibration and the motion of the core electron or to symmetrize the core- and outer-electron coordinates explicitly.

The Coulomb field wave functions will be represented as linear combinations of two independent standard solutions $f(\nu, r)$ and $g(\nu, r)$ of the radial Schrödinger equation. Here ν is the effective quantum number of the electron, related to its energy ϵ in a. u. by $\epsilon = -1/(2\nu^2)$; in the continuum we have $\epsilon > 0$ and $\nu = i\gamma$. We utilize the pair of f and g functions described by Seaton in Eqs. (2.22), (2.24), and (2.25) of Ref. 2. An essential feature of these functions is their behavior for small r, respectively, as r^{l+1} and r^{-l} with *energy-independent coefficients*. It follows that f and g are practically independent of energy over the combined range $-\bar{\epsilon} < \epsilon < \bar{\epsilon}$, $r \ll 1/(2\bar{\epsilon})$, where the kinetic and potential energies are far larger than $|\bar{\epsilon}|$. Because of our interest in a narrow energy range near the ionization threshold, we assume $\epsilon \ll 1$ and simplify Seaton's formulas by setting $A = B = 1$ and $\mathcal{G} = 0$ in Ref. 2. (Corrections obtained by avoiding this approximation are given in Appendix B.)

The determination of energy levels and of intensities of discrete and continuous spectra depends primarily on the asymptotic behavior of $f(\nu, r)$ and $g(\nu, r)$ for large r. For $\epsilon > 0$, $\gamma = -i\nu \gg l$, $r \gg \gamma^2$ we have the familiar formulas

$$\begin{aligned} f(i\gamma, r) &\to (2\gamma/\pi)^{1/2}\sin[r/\gamma - \tfrac{1}{2}\pi l + \gamma\ln(2r/\gamma) \\ &\quad + \arg\Gamma(l+1-i\gamma)], \quad \text{as } r\to\infty \\ g(i\gamma, r) &\to -(2\gamma/\pi)^{1/2}\cos[r/\gamma - \tfrac{1}{2}\pi l + \gamma\ln(2r/\gamma) \\ &\quad + \arg\Gamma(l+1-i\gamma)], \quad \text{as } r\to\infty . \end{aligned} \tag{9}$$

Separate expressions for the ingoing and outgoing wave components are easily obtained from (9). For $\epsilon < 0$ we utilize exponentially rising and decreasing components $u(\nu, r)$ and $v(\nu, r)$ and write

$$\begin{aligned} f(\nu, r) &\to u(\nu, r)\sin\pi\nu - v(\nu, r)e^{i\pi\nu}, \quad \text{as } r\to\infty \\ g(\nu, r) &\to -u(\nu, r)\cos\pi\nu + v(\nu, r)e^{i\pi(\nu+1/2)}, \quad \text{as } r\to\infty , \end{aligned} \tag{10}$$

where the functions

$$\begin{aligned} u(\nu, r) &= (-1)^l \nu^{1/2}\pi^{-1}(2r/\nu)^{-\nu}\exp(r/\nu) \\ &\quad \times[\Gamma(\nu-l)\Gamma(\nu+l+1)]^{1/2}, \\ v(\nu, r) &= (-1)^l\nu^{1/2}(2r/\nu)^{\nu}\exp(-r/\nu) \\ &\quad \times[\Gamma(\nu-l)\Gamma(\nu+l+1)]^{-1/2}, \end{aligned} \tag{11}$$

have the r-independent product

$$u(\nu, r)v(\nu, r) = \nu/\pi . \tag{12}$$

Utilizing the radial functions f and g we now set out to construct wave functions $\Psi(\hat{R}, \hat{r}, r)$ of the core + electron system, for electron positions outside the core. We do this initially for situation A, by considering formally the unrealistic limit where the rotational energy parameter B vanishes and Λ is a good quantum number. We write then, for either $\Lambda = 0$ or 1,

$$\begin{aligned} \Psi_\Lambda(\hat{R}, \hat{r}, r) &= X_M^{(J\Lambda+)}(\hat{r}, \hat{R}) \\ &\quad \times[f(\nu, r)\cos\pi\mu_\Lambda - g(\nu, r)\sin\pi\mu_\Lambda], \end{aligned} \tag{13}$$

where X is given by (2) or (3) and μ_Λ is the quantum defect, regarding as a semiempirical parameter. [In a calculation from first principles one would obtain a wave function valid throughout the core and determine μ_Λ by matching (13) to that function at the core's edge.] For $\epsilon > 0$ the large-r Eq. (9) reduces (13) to

$$\begin{aligned} \Psi_\Lambda &\to X_M^{(J\Lambda+)}(2\gamma/\pi)^{1/2}\sin[r/\gamma - \tfrac{1}{2}l\pi + \gamma\ln(2r/\gamma) \\ &\quad + \arg\Gamma(l+1-i\gamma) + \pi\mu_\Lambda], \end{aligned} \tag{14}$$

showing that $\pi\mu_\Lambda$ equals the elastic scattering phase shift. For $\epsilon < 0$, substitution of (10) into (13) yields

$$\Psi_\Lambda = X_M^{(J\Lambda+)}[u(\nu, r)\sin\pi(\nu+\mu_\Lambda) - v(\nu, r)\exp i\pi(\nu+\mu_\Lambda)]. \tag{15}$$

The energy eigenvalues $\epsilon_{n\Lambda}$ are determined by the condition that Ψ_Λ remain finite as $r \to \infty$, i.e., that the coefficient of the rising exponential $u(\nu, r)$ vanish. Thereby, we obtain the Rydberg series formulas

$$\sin\pi(\nu + \mu_\Lambda) = 0, \quad \nu_{n\Lambda} = n - \mu_\Lambda, \quad \epsilon_{n\Lambda} = -1/2(n - \mu_\Lambda)^2. \tag{16}$$

For values of $n \sim 5$, the rotational energy is practically negligible, i.e., situation A prevails throughout the range of r where $\Psi_{n\Lambda} \neq 0$. Hence, approximate values of μ_Λ for our further calculation can be obtained by entering in (16) experimental values of $\epsilon_{n\Lambda}$ with $n \sim 5$ and applying, if desirable, the corrections from Appendix B.

Under the circumstances of situation B, i.e., when N is a good quantum number, a wave function Ψ_N contains the factor $\Phi_M^{(J,N)}(\hat{r}, \hat{R})$ given by (1). Accordingly, it has the form

$$\Psi_N(\hat{R}, \hat{r}, r) = \Phi_M^{(JN)}(\hat{r}, \hat{R})[f(\nu_N, r)c_N - g(\nu_N, r)d_N], \tag{17}$$

where the coefficients c and d remain undetermined pending specification of boundary conditions at the inner edge of region B. Here we have introduced a subscript N to the effective quantum number ν because ν_0 and ν_2 relate to the electron energy measured, respectively, from the photoionization thresholds $\epsilon = 0$ and $\epsilon = 6B$ and are accordingly defined by

$$\epsilon = -1/(2\nu_0^2) = 6B - 1/(2\nu_2^2). \tag{18}$$

This equation implies the explicit relations between ν_2 and ν_0:

$$\nu_2^2 = \nu_0^2/(1 + 12B\nu_0^2) = \nu_0^2 - 12B\nu_0^4 + \cdots, \tag{19}$$

$$\nu_0^2 = \nu_2^2/(1 - 12B\nu_2^2) = \nu_2^2 + 12B\nu_2^4 + \cdots. \tag{19'}$$

We are now finally able to write an expression of $\Psi(\hat{R}, \hat{r}, r)$ valid for *all r outside the core*, namely,

$$\Psi(\hat{R}, \hat{r}, r) = \sum_{\Lambda=0,1} \sum_{N=0,2} \Phi_M^{(JN)}(\hat{r}, \hat{R}) \times [f(\nu_N, r)U_{N\Lambda}\cos\pi\mu_\Lambda - g(\nu_N, r) \times U_{N\Lambda}\sin\pi\mu_\Lambda]A_\Lambda, \tag{20}$$

where $U_{N\Lambda}$ is given by (7) and the pair of coefficients A_Λ remain to be determined by boundary conditions at $r \to \infty$. Equation (20) reduces in region A to a superposition of the Ψ_Λ (13) with $\Lambda = 0$ and 1, because here one can set $\nu_2 \sim \nu_0$, i.e., $\nu_N = \nu$ after which the $\sum_N$ yields $X_M^{(J\Lambda+)}$ in accordance with (4). On the other hand, Eq. (20) can also be regarded as a superposition of the Ψ_N (17) with $N = 0, 2$ and with $c_N = \sum_\Lambda U_{N\Lambda}\cos\pi\mu_\Lambda A_\Lambda$, $d_N = \sum_\Lambda U_{N\Lambda}\sin\pi\mu_\Lambda A_\Lambda$.

The boundary conditions at large r will be applied separately in the following sections as appropriate to different spectral regions.

IV. PHOTOIONIZATION INTENSITY IN ALTERNATIVE OPEN CHANNELS

Absorption of a photon with energy $\hbar\omega$ in excess of the second ($N = 2$) ionization threshold yields two groups of photoelectrons with alternative energies $\hbar\omega - I_0$ and $\hbar\omega - I_2$, where I_0 and I_2 are the two threshold energies and $I_2 - I_0 = 6B$. To calculate the cross section for photoemission of each group $\bar{N}(= 0, 2)$, we determine sets of coefficients $A_\Lambda^{(\bar{N}-)}$ such that the outgoing portion of the wave function (20) vanishes in the channel with $N \neq \bar{N}$. [The label on the coefficients A_Λ indicates that the $\bar{N}$ channel selection is applied to the outgoing wave. The eigenfunction (20) constructed with these coefficients is indicated by $\Psi^{(\bar{N}-)}$ and is said to be normalized by an "ingoing wave boundary condition." Alternative sets of coefficients $\bar{A}^{(\bar{N}+)}$ and wave functions $\Psi^{(\bar{N}+)}$, obtained by applying a selection to the ingoing portion of the wave function, would be relevant to the process of radiative recombination which is the inverse of photoionization.]

In our energy range $\hbar\omega > I_2$, both ν_0 and ν_2 are imaginary, so that $\nu_N = i\gamma_N$ and $f(i\gamma_N, r)$ and $g(i\gamma_N, r)$ are given by (9). The outgoing wave portion of (20) is then represented for large r by

$$\Psi^{(\text{out})}(\hat{R}, \hat{r}, r) \to \sum_N \Phi_M^{(JN)}(\hat{R}, \hat{r}) \times (2\gamma_N/\pi)^{1/2}(1/2i)\exp i(r/\gamma_N + \cdots)\sum_\Lambda U_{N\Lambda} \times (\cos\pi\mu_\Lambda + i\sin\pi\mu_\Lambda)A_\Lambda, \quad \text{as } r \to \infty. \tag{21}$$

The channel selection condition is, then,

$$\sum_\Lambda U_{N\Lambda}\exp(i\pi\mu_\Lambda)A_\Lambda^{(\bar{N}-)} = \delta_{N\bar{N}}. \tag{22}$$

The coefficients obeying this condition are

$$A_\Lambda^{(\bar{N}-)} = \exp(-i\pi\mu_\Lambda)\tilde{U}_{\Lambda\bar{N}}. \tag{23}$$

The intrinsic transition amplitudes of the photoabsorption process depend on circumstances prevailing within the molecular core, i.e., in the situation called A in Sec. I. According to Sec. I, these amplitudes will be represented by the pair of coefficients $D_\Lambda (= D_\sigma, D_\pi)$ whose evaluation is discussed in Sec. VII. The intensity of photoelectric emission in the energy group $\bar{N}$ is, then,

$$I^{(\bar{N})} = |\sum_\Lambda A_\Lambda^{(\bar{N}-)*}D_\Lambda|^2. \tag{24}$$

Utilizing (23) and the explicit form (7) of $U_{N\Lambda}$, this gives

$$I^{(0)} = \cos^2\alpha\, D_\sigma^2 + \sin^2\alpha\, D_\pi^2 + 2\sin\alpha\cos\alpha\, D_\sigma D_\pi\cos\pi\delta, \tag{25}$$

$$I^{(2)} = \sin^2\alpha\, D_\sigma^2 + \cos^2\alpha\, D_\pi^2 - 2\sin\alpha\cos\alpha\, D_\sigma D_\pi\cos\pi\delta,$$

where we have $\alpha = \arccos\sqrt{\tfrac{1}{3}}$ according to (7) and

$$\delta = \mu_\sigma - \mu_\pi. \tag{26}$$

V. PHOTOIONIZATION INTENSITY IN THE BEUTLER SPECTRUM

The treatment of the spectral region between the two ionization thresholds involves many ramifications. The alternations of intensity in this part of the spectrum will be described in a manner slightly different from that of Beutler,[8] that is, we shall not introduce at the outset the occurrence, in the $N=2$ channel, of a series of discrete lines broadened by autoionization. However, the two-channel wave function of Sec. III will be shown to contain already the earlier results,[9] especially in the limit of well-separated lines.

Our treatment of this continuous but perturbed spectrum will be conducted on the one hand so that it can be adapted quite simply, in Sec. VI, to describe the discrete perturbed spectrum below the lowest ionization threshold. On the other hand it will display the transition amplitude to the single ($N=0$) channel open in this region as a superposition of the transition amplitudes calculated in Sec. IV for the two separate channels which are open above the second threshold I_2. A key quantity, emphasized in this paper and elsewhere[10] is the effective phase shift in the open channel $N=0$. Its "resonance" behavior reflects the existence of discrete structures (autoionizing levels) in the closed channel $N=2$.

A. The $N=0$ Phase Shift

In the continuum between the $N=0$ and $N=2$ thresholds, we have $\nu_0 = i\gamma_0$ imaginary and ν_2 real. Accordingly, we substitute in (20) the asymptotic expressions (9) for $f(i\gamma_0, r)$ and $g(i\gamma_0, r)$ and the expressions (10) for $f(\nu_2, r)$ and $g(\nu_2, r)$ to yield

$$\begin{aligned}\Psi(\hat{R}, \hat{r}, r) \to\ & \Phi_M^{(J0)}(\hat{r}, \hat{R})(2\gamma_0/\pi)^{1/2}\\ & \times[\sin(r/\gamma_0 + \cdots)\textstyle\sum_\Lambda U_{0\Lambda}\cos\pi\mu_\Lambda A_\Lambda\\ & + \cos(r/\gamma_0+\cdots)\textstyle\sum_\Lambda U_{0\Lambda}\sin\pi\mu_\Lambda A_\Lambda]\\ & + \Phi_M^{(J2)}(\hat{r}, \hat{R})[u(\nu_2, r)\textstyle\sum_\Lambda U_{2\Lambda}\sin\pi(\nu_2+\mu_\Lambda)A_\Lambda\\ & - v(\nu_2, r)\textstyle\sum_\Lambda U_{2\Lambda}\exp[i\pi(\nu_2+\mu_\Lambda)]A_\Lambda], \quad \text{as } r\to\infty. \end{aligned} \tag{27}$$

This expression enables us to formulate the boundary conditions at large r which determine the coefficients A_Λ.

Because the function $u(\nu_2, r)$ diverges at large r, its coefficient must vanish, i.e., we must have

$$\textstyle\sum_\Lambda U_{2\Lambda}\sin\pi(\nu_2+\mu_\Lambda)A_\Lambda = 0 \ . \tag{28}$$

In addition, normalization of the wave function at large r per unit energy range, as a standing wave with a real coefficient, requires the expression in the first brackets of (27) to equal $\sin(r/\gamma_0+\cdots+\Delta)$, where $\Delta=\pi\mu_0$ indicates a phase shift and μ_0 a quantum defect for the $N=0$ channel. This condition establishes the pair of equations

$$\begin{aligned}\textstyle\sum_\Lambda U_{0\Lambda}\cos\pi\mu_\Lambda A_\Lambda &= \cos\Delta \ ,\\ \textstyle\sum_\Lambda U_{0\Lambda}\sin\pi\mu_\Lambda A_\Lambda &= \sin\Delta \ .\end{aligned} \tag{29}$$

Alternative linear combinations of these equations are

$$\textstyle\sum_\Lambda U_{0\Lambda}\sin(\pi\mu_\Lambda-\Delta)A_\Lambda = 0 \ , \tag{30a}$$

$$\textstyle\sum_\Lambda U_{0\Lambda}\cos(\pi\mu_\Lambda-\Delta)A_\Lambda = 1 \ . \tag{30b}$$

The three equations (28), (30a), and (30b) determine A_σ, A_π, and the phase shift Δ as functions of ν_2. Specifically, $\Delta(\nu_2)$ is the eigenvalue of the homogeneous system consisting of (28) and (30a), whereas, (30b) normalizes A_Λ. These three equations can also be expressed in a form related to (22), namely,

$$\mathrm{Im}[\textstyle\sum_\Lambda U_{2\Lambda}\exp(i\pi\mu_\Lambda)\exp(i\pi\nu_2)A_\Lambda]=0, \tag{28'}$$

$$\textstyle\sum_\Lambda U_{0\Lambda}\exp(i\pi\mu_\Lambda)A_\Lambda = \exp i\Delta \ . \tag{30'}$$

The compatibility requirement of the homogeneous system consisting of (28) and (30a) is expressed, utilizing the values (7) of $U_{N\Lambda}$, by

$$\begin{vmatrix} -\sin\alpha\sin\pi(\mu_\sigma+\nu_2) & \cos\alpha\sin\pi(\mu_\pi+\nu_2)\\ \cos\alpha\sin(\pi\mu_\sigma-\Delta) & \sin\alpha\sin(\pi\mu_\pi-\Delta)\end{vmatrix} = 0 \ . \tag{31}$$

Among the alternative forms of this equation are

$$f(-\Delta)=\frac{\sin(\pi\mu_\sigma-\Delta)}{\sin(\pi\mu_\pi-\Delta)} = -\tan^2\alpha\frac{\sin\pi(\mu_\sigma+\nu_2)}{\sin\pi(\mu_\pi+\nu_2)} = 2f(\pi\nu_2) \ , \tag{32}$$

which lends itself to graphical solution,

$$\begin{aligned}F(\Delta,\nu_2) = {} & \sin^2\alpha\sin\pi(\mu_\sigma+\nu_2)\sin(\pi\mu_\pi-\Delta)\\ & + \cos^2\alpha\sin\pi(\mu_\pi+\nu_2)\sin(\pi\mu_\sigma-\Delta) = 0 \ ,\end{aligned} \tag{33}$$

which defines $\Delta(\nu_2)$ as an implicit function with period 1, and

$$\begin{aligned}&\tan(\Delta-\pi\bar{\mu}) - \tan(\Delta_p-\pi\bar{\mu})\\ &\quad = -\frac{\tan^2\frac{1}{2}\pi\delta\sin^2 2\alpha}{\tan\pi(\nu_2+\bar{\mu})-\tan(\Delta_p-\pi\bar{\mu})} \ ,\end{aligned} \tag{34}$$

which gives Δ explicitly as a function of ν_2, or ν_2 as an analogous function of Δ. In (34) we have introduced the notations

$$\bar{\mu} = \tfrac{1}{2}(\mu_\sigma+\mu_\pi) \ , \tag{35}$$

$$\tan(\Delta_p-\pi\bar{\mu}) = \tan\tfrac{1}{2}\pi\delta\cos 2\alpha \ , \tag{36}$$

where Δ_p indicates a "plateau value" of Δ as discussed below. The function $\Delta(\nu_2)$ is plotted in Fig. 2 utilizing the values $\mu_\sigma = 0.22$, $\mu_\pi = -0.06$ determined in Ref. 1 by fitting experimental data.

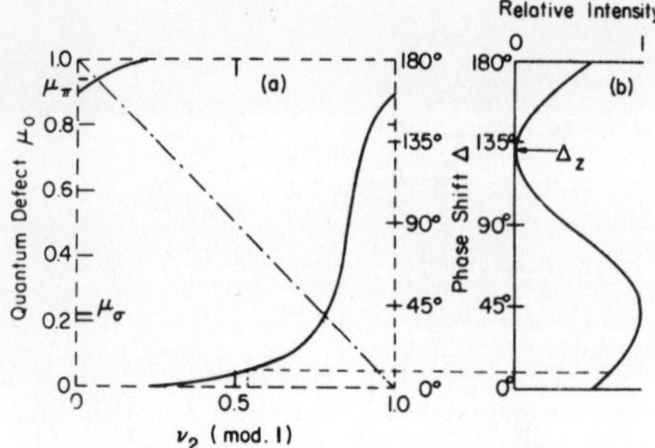

FIG. 2. (a) Solid line is the $N=0$ quantum defect function [Eq. (34)], and the dot-dashed line is the $\mu_0(\nu_2)+\nu_2=1$. (b) Plot of relative intensity $I(\nu_2)$ $(\sin^2\pi\delta\sin^2\alpha\cos^2\alpha)/I^{(2)}$ [Eq. (43)]. (The value of $\Delta_z\sim 133°$ is drawn from experimental data.) Dashed line indicates how to find the relative intensity (0.75) which corresponds to a spectral energy with $\nu_2=0.54$ (*mod.* 1).

Equation (34) shows that Δ would vary by a sharp step in the limit $\tan\frac{1}{2}\pi\delta\sin 2\alpha\to 0$, which corresponds to "vanishing interaction" between the two channels $N=0$ and 2. For weak interaction Δ would remain approximately equal to its plateau value Δ_p for all values of ν_2 except $\nu_2\sim\Delta_p/\pi-2\bar{\mu}$. At this special value of ν_2, Δ would rise stepwise by π. This vanishing interaction limit could occur for either of *two* reasons: (a) $\delta\sim 0$ implying that the H_2^+ core is effectively isotropic and does not cause any scattering between the channels $N=0$ and $N=2$, or (b) $\alpha\sim 0$ (*mod.* $\frac{1}{2}\pi$) implying that the eigenvectors of $\vec{N}_2$ and Λ^2 coincide. Neither reason obtains in our problem.

Note that μ_σ and μ_π coincide, *modulo* 1, with the values of $\mu_0=\Delta/\pi$ at the intersections of the plot of $\mu_0(\nu_2)$ with the plot's diagonal $\mu_0+\nu_2=1$. The values of μ_σ and μ_π are thus determined conveniently from an experimental plot of $\mu_0=\Delta/\pi$ against ν_2, as suggested elsewhere.[10]

The slope of this plot, which will be utilized below, is nonnegative. It can be expressed in alternative forms, e.g., as a function of ν_2 only or as a more symmetric function of μ_0 and ν_2. From (34) we obtain

$$\frac{d\mu_0}{d\nu_2}=\frac{1+\tan^2\pi(\nu_2+\bar{\mu})}{1+\tan^2\pi(\mu_0-\bar{\mu})}\,\frac{d\tan\pi(\mu_0-\bar{\mu})}{d\tan\pi(\nu_2+\bar{\mu})}$$
$$=\frac{\cos^2\pi(\mu_0-\bar{\mu})}{\cos^2\pi(\nu_2+\bar{\mu})}\,\frac{\tan^2\frac{1}{2}\pi\delta\sin^2 2\alpha}{[\tan\pi(\nu_2+\bar{\mu})-\tan(\Delta_p-\pi\bar{\mu})]^2}$$
$$=\frac{[1+\tan^2\pi(\nu_2+\bar{\mu})]\tan\eta}{[\tan\pi(\nu_2+\bar{\mu})-\tan(\Delta_p-\pi\bar{\mu})(1+\tan\eta)]^2+\tan^2\eta}\,, \tag{37}$$

where $\quad\sin\eta=\sin^2\frac{1}{2}\pi\delta\sin^2 2\alpha \quad (38)$

is an index of the interaction strength between the $N=0$ and $N=2$ channels. The following formulas are also noted:

$$\left(\frac{d\mu_0}{d\nu_2}\right)^{1/2}=-\cot\alpha\,\frac{\sin\pi(\mu_0-\mu_\sigma)}{\sin\pi(\nu_2+\mu_\sigma)}=\tan\alpha\,\frac{\sin\pi(\mu_0-\mu_\pi)}{\sin\pi(\nu_2+\mu_\pi)}\,, \tag{39}$$

$$\left(\frac{d\mu_0}{d\nu_2}\right)^{1/2}_{\nu_2=-\mu_\sigma}=\tan\alpha\,,\quad\left(\frac{d\mu_0}{d\nu_2}\right)^{1/2}_{\nu_2=-\mu_\pi}=\cot\alpha\,. \tag{39'}$$

B. The Wave Function

Compatibility of (30a) with (28) being ensured by relating Δ to ν_2, the inhomogeneous system consisting of (30a) and (30b) is solved by

$$A_\sigma(\nu_2)=-\frac{\sin(\pi\mu_\pi-\Delta)}{\sin\pi\delta\cos\alpha}=\frac{\sin(\pi\mu_\pi-\Delta)\bar{U}_{\sigma 2}}{\sin\pi\delta\sin\alpha\cos\alpha}\,, \tag{40}$$
$$A_\pi(\nu_2)=\frac{\sin(\pi\mu_\sigma-\Delta)}{\sin\pi\delta\sin\alpha}=\frac{\sin(\pi\mu_\sigma-\Delta)\bar{U}_{\pi 2}}{\sin\pi\delta\sin\alpha\cos\alpha}\,.$$

[These solutions would fail in the special case of vanishing interaction between the N channels, but in this case Eqs. (28), (30a), and (30b) are solved by inspection, the wave functions lying entirely in one channel or in the other.]

The vector $\vec{A}(\nu_2)$ with the components (40) can also be represented as a superposition of the $\bar{N}$-channel vectors $\vec{A}^{(0-)}$ and $\vec{A}^{(2-)}$ given by (23). To obtain this representation one may note that $\vec{A}^{(0-)*}\cdot\vec{A}(\nu_2)$ is given by (30′) and that $\vec{A}^{(2-)*}\cdot\vec{A}(\nu_2)$ is proportional to

$$\left(\frac{\partial F}{\partial\nu_2}+iF(\Delta,\nu_2)\right)\exp(-i\pi\nu_2)\,.$$

The result can be expressed as

$$\vec{A}(\nu_2)=\vec{A}^{(0-)}e^{i\Delta}+\vec{A}^{(2-)}e^{-i\pi\nu_2}(d\mu_0/d\nu_2)^{1/2}\,. \tag{41}$$

The factor $\exp(i\Delta)$ in (42) serves to adjust the "ingoing wave" normalization of $\vec{A}^{(0-)}$ to the "standing wave" normalization specified in (27). The last factor of (41) shows that the amplitude of the $N=2$ channel component is proportional to $(d\mu_0/d\nu_2)^{1/2}$.

Indeed, substitution of (29) and of the value of $\vec{A}^{(2-)*}\cdot\vec{A}$ into (27) reduces this wave function to the form

$$\Psi(\hat{R},\hat{r},r)_{r\to\infty}\Phi_M^{(J0)}(\hat{r},\hat{R})(2\gamma_0/\pi)^{1/2}$$
$$\times\sin[r/\gamma_0+\cdots+\Delta(\nu_2)]$$
$$+\Phi_M^{(J2)}(\hat{r},\hat{R})v(\nu_2,r)(d\mu_0/d\nu_2)^{1/2}\,. \tag{42}$$

It is well known from the theory of isolated autoionizing states that the admixture of a discrete wave function into a continuum has a squared amplitude proportional to $d\Delta/d\epsilon$, where ϵ is the energy of the state. Equation (42) extends the result to the whole continuum, irrespective of the sharpness of the autoionizing lines.

C. Intensity Distribution

The photoabsorption spectrum is described in the Beutler region by equivalent alternative formulas which are analogous to (24) but utilize the state amplitudes (40) or (41), respectively. The first of these formulas is

$$I(\nu_2) = [D_\sigma A_\sigma(\nu_2) + D_\pi A_\pi(\nu_2)]^2$$

$$= \frac{[-D_\sigma \sin(\pi\mu_\pi - \Delta)\sin\alpha + D_\pi \sin(\pi\mu_\sigma - \Delta)\cos\alpha]^2}{\sin^2\pi\delta\,\sin^2\alpha\,\cos^2\alpha}$$

$$= I^{(2)}\sin^2[\Delta(\nu_2) - \Delta_z]/\sin^2\pi\delta\,\sin^2\alpha\,\cos^2\alpha\,, \quad (43)$$

where $I^{(2)}$ is given by (25) and Δ_z indicates the value of Δ at which $I(\nu_2)$ vanishes. The zero-intensity value Δ_z is related to other parameters of our problem by

$$\frac{\sin(\Delta_z - \pi\mu_\sigma)}{\sin(\Delta_z - \pi\mu_\pi)} = f(-\Delta_z) = -\tan^2\alpha\frac{\sin\pi(\nu_{2z} + \mu_\sigma)}{\sin\pi(\nu_{2z} + \mu_\pi)} = \frac{D_\sigma}{D_\pi}\tan\alpha \quad (44)$$

and

$$\Delta_z = \arg[\vec{D}\cdot\vec{A}^{(2-)}] + \pi(\mu_\sigma + \mu_\pi)\,. \quad (44')$$

Equation (43) represents the alternation of intensity along the spectrum as a sinusoidal function of the grossly nonlinear energy scale Δ (Fig. 2). Part of the nonlinearity, due to the relationship between ν_2 and the energy ϵ, is characteristic of any Rydberg series; the part represented by $\Delta(\nu_2)$ is characteristic of the channel interaction. Two special values of $I(\nu_2)$ are noted:

$$I(-\mu_\sigma) = D_\sigma^2/\cos^2\alpha\,, \quad I(-\mu_\pi) = D_\pi^2/\sin^2\alpha\,. \quad (43')$$

Starting from (41) we have the alternative formula

$$I(\nu_2) = |\vec{D}\cdot\vec{A}(\nu_2)|^2 = \left|\vec{D}\cdot\left[\vec{A}^{(0-)}e^{i\Delta} + \vec{A}^{(2-)}e^{-i\pi\nu_2}\left(\frac{d\mu_0}{d\nu_2}\right)^{1/2}\right]\right|^2 = \left|\vec{D}\cdot\vec{A}^{(0-)} + \vec{D}\cdot\vec{A}^{(2-)}e^{-i(\pi\nu_2+\Delta)}\left(\frac{d\mu_0}{d\nu_2}\right)^{1/2}\right|^2$$

$$= \left|\vec{D}\cdot\vec{A}^{(0-)} + \vec{D}\cdot\vec{A}^{(2-)}\frac{\exp i(\pi\bar{\mu} - \Delta_p)(\tan\eta)^{1/2}[1 - i\tan\pi(\nu_2 + \bar{\mu})]}{\tan\pi(\nu_2 + \bar{\mu}) - \tan(\Delta_p - \pi\bar{\mu})(1 + \tan\eta) - i\tan\eta}\right|^2. \quad (45)$$

The last expression has been obtained utilizing (37) and the expression (34) for $\Delta(\nu_2)$. Note the reasonance form of the coefficient of $\vec{D}\cdot\vec{A}^{(2-)}$, regarded as a function of $\tan\pi(\nu_2 + \bar{\mu})$. Specifically, $\tan\eta$ represents the resonance width, here as in (37); the resonance center lies at

$$\tan\pi(\nu_2 + \bar{\mu}) = \tan(\Delta_p - \pi\bar{\mu})(1 + \tan\eta)\,,$$

where $\tan\eta$ represents a shift from the position in the vanishing interaction limit.

When the intensity (45) is averaged over one unit range of ν_2, the net effect of interference between the two amplitude terms can be shown to vanish. Moreover, the squared modulus of the coefficient of $\vec{D}\cdot\vec{A}^{(2-)}$ averages to unity. Thereby, one verfies that

$$\langle I(\nu_2)\rangle = |\vec{D}\cdot\vec{A}^{(0-)}|^2 + |\vec{D}\cdot\vec{A}^{(2-)}|^2 = I^{(0)} + I^{(2)}\,, \quad (46)$$

in agreement with Gailitis's general result.[11]

The oscillations of $I(\nu_2)$ are concentrated into sharp resonance lines when the function $\Delta(\nu_2)$ increases by sharp steps of height π separated by intervals of nearly one unit of ν_2 over which Δ remains near its plateau value, $\Delta \sim \Delta_p$. In this event, $I(\nu_2)$ may conveniently be represented by the resonance profile formula[9]

$$I = I_0(q + \epsilon)^2/(1 + \epsilon^2)\,, \quad (47)$$

with $\epsilon = -\cot(\Delta - \Delta_p)$, $q = \cot(\Delta_p - \Delta_z)$, $I_0 = I^{(2)}\sin^2(\Delta_p - \Delta_z)/\sin^2\pi\delta\,\sin^2\alpha\,\cos^2\alpha$. (48)

VI. DISCRETE SPECTRUM

The determination of energy levels, wave functions, and line intensities of the discrete spectrum constitutes at this point essentially a corollary to the treatment of Sec. V. The determination of energy levels and wave functions will proceed from the knowledge of continuum wave functions with a known phase shift along the line followed in Eqs. (14)–(16) for a single channel. The normalization of discrete wave functions, omitted in Sec. III, will be carried out in some detail as an illustration of the basic QDM procedure. Thereafter, the line intensity determination is straightforward.

A. Energy Levels and Wave Functions

To deal with the spectral region below the lowest ionization threshold, we substitute in (20) the asymptotic expressions (10) of f and g for both $N=2$ and $N=0$, which yeilds

$$\Psi(\hat{R},\hat{r},r) \to \sum_{N=0,2} \Phi_M^{(JN)}\{u(\nu_N,r)\sum_\Lambda U_{N\Lambda}\sin\pi(\nu_N+\mu_\Lambda)A_\Lambda - v(\nu_N,r)\sum_\Lambda U_{N\Lambda}\exp[i\pi(\nu_N+\mu_\Lambda)]A_\Lambda\}, \text{ as } r\to\infty. \quad (49)$$

The coefficient of $u(\nu_N, r)$ in this equation must be set to zero for both values of N, because $u(\nu_N, r)$ diverges for large r.

We proceed by applying this condition in two steps, firstly for $N=2$, so as to utilize the results of Sec. V and to display the relationship between the spectra in the discrete and in the Beutler region, and then for $N=0$. To this end we perform in (49) the same substitutions [i.e., (28), (29), etc.] which have reduced (27) to its form (42) and utilize the quantum defect symbol $\mu_0=\Delta(\nu_2)/\pi$. The result is

$$\Psi(\hat{R},\hat{r},r)=\sum_N \Phi_M^{(JN)}(\hat{r},\hat{R})F^{(N)}(r) \to \Phi_M^{(J0)}[u(\nu_0,r)\sin\pi(\nu_0+\mu_0)-v(\nu_0,r)\exp i\pi(\nu_0+\mu_0)]+\Phi_M^{(J2)}v(\nu_2,r)(d\mu_0/d\nu_2)^{1/2},$$

$$\text{as } r\to\infty. \quad (50)$$

The radial function $F^{(2)}(r)$ in (50) vanishes for large r and requires no further attention. The radial wave function $F_0(r)$, represented explicitly in the brackets of (50), has the same form as the radial factor of the wave function (15) which pertains to Born-Oppenheimer factorization, except for the replacement of ν and μ_Λ by ν_0 and $\mu_0(\nu_2)$, respectively. Therefore, the condition that Ψ remains finite as $r\to\infty$ determines the discrete energy levels by the *Rydberg-like formulas*, akin to (16),

$$\sin\pi[\nu_0+\mu_0(\nu_2)]=0, \quad \nu_{0n}=m-\mu_0(\nu_{2n}), \quad \epsilon_n=-1/(2\nu_{0n}^2)\,, \quad (51)$$

where n is a label for identifying each level and m is an integer, shown in Fig. 3, which may have the same value for two levels.

This equation represents all discrete levels as part of a *single* Rydberg series, with variable quantum defect $\mu_0(\nu_2)$.[10] It suggests the following procedure for determining the level positions: (i) Regard $-\nu_0$ and μ_0 as functions of ν_2 defined, respectively, by (19′) and (34); (ii) plot these functions *modulo* 1 as shown in Fig. 3; (iii) identify the abscissa of each intersection of the plots as one root ν_{2n} of $-\nu_0(\nu_2)=\mu_0(\nu_2)$ *mod*. 1 and its ordinate as $-\nu_{0n}$. In the spectral range 2–3 eV below the ionization threshold, we have $\nu_0\sim\nu_2$; since $\mu_0(\nu_2)$ has two branches, one finds two roots per every unit range of ν_2. Indeed, as long as $\nu_0\sim\nu_2$, the plot of $-\nu_0(\nu_2)$ *modulo* 1 coincides with the diagonal shown in Fig. 2 and the roots coincide with μ_σ and μ_π. In the spectral range near the ionization threshold, on the other hand, ν_0 varies by several units per every unit range of ν_2 so that a large number of roots is found per unit ν_2. This number diverges at the threshold I_0, that is, at $\nu_2=(12B)^{-1/2}$; here the Rydberg series converges with the limiting value of the quantum de-

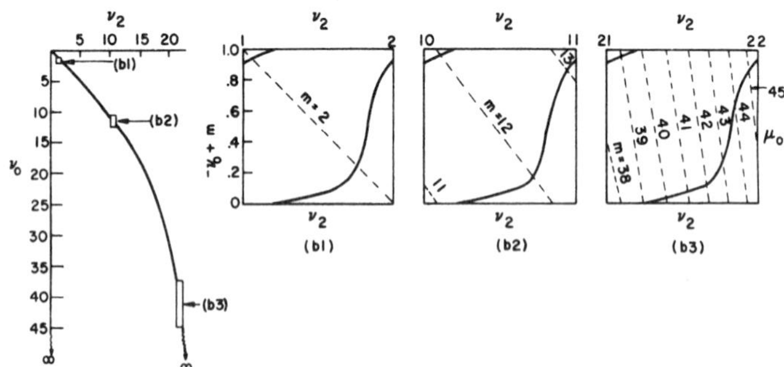

FIG. 3. Sample graphical determination of discrete level positions. (a) Small-scale plot of $-\nu_0(\nu_2)$. (b) Large scale-plots of $-\nu_0$ (dashed line) and μ_0 (solid line) for three sample ranges of ν_2: (b1) 1–2; (b2) 10–11; (b3) 21–22. Numerical labels on the dashed line curves indicate values of integer m. The value ν_{0n} of ν_0 corresponding to the ordinate of each intersection between the dashed line and the solid line represents the effective quantum number of level $\epsilon_n=-1/(2\nu_{0n}^2)$ a.u.

fect, $\mu_{01\mathrm{im}}=0.01$, given by (34) with $\nu_2=25.32$.

B. Normalization

Let us proceed now to the normalization of the wave function (50) for a level ϵ_n determined by (51). The QDM evaluates the integral over the squared wave function by a method[12] originally devised for continuum wave functions. This method considers the overlap integral of wave functions Ψ and Ψ_n calculated, respectively, for two slightly different energies ϵ and ϵ_n, the integral being extended initially only over the volume of a sphere with finite radius r. This volume integral is reduced to an integral over the surface of the sphere through multiplication by $\epsilon-\epsilon_n$ followed by use of the Schrödinger equation. The surface integral itself depends only on the asymptotic values of Ψ and Ψ_n. The calculation is set up as follows:

$$\int^{\infty}\Psi_n^2 d\tau=\lim_{\epsilon=\epsilon_n}\lim_{r=\infty}\int^{r}\Psi\Psi_n d\tau=\lim_{\epsilon=\epsilon_n}\lim_{r=\infty}(\epsilon-\epsilon_n)\int^{r}\frac{\Psi\Psi_n d\tau}{\epsilon-\epsilon_n}=\lim_{\epsilon=\epsilon_n}\lim_{r=\infty}\frac{1}{2}\sum_N\frac{-F_n^{(N)}dF^{(N)}/dr+F^{(N)}dF_n^{(N)}/dr}{\epsilon-\epsilon_n}\,. \quad (52)$$

The volume integration over $d\tau$ is understood to include all variables of the problem but the surface integral depends only on $F^{(N)}(r)$ and $dF^{(N)}/dr$. The key points here are: (a) The wave functions Ψ and Ψ_n are understood to be exact eigenfunctions of the complete problem; (b) irrespectively of their complication, the surface integral is expressed only in terms of $F^{(N)}$'s dependence *on the single variable* r that effectively ranges to infinity.

The evaluation of (52) now simplifies in many respects owing to the behavior of the $F^{(N)}$ for large r. First, since $F^{(2)}$ vanishes for large r, the $\sum_N$ reduces to the single term $N=0$. Second, we note from (11) that, in the large-r approximation,

$$\frac{du}{dr}\to\frac{u}{\nu_0}, \quad\text{and}\quad \frac{dv}{dr}\to\frac{-v}{\nu_0}, \quad\text{as}\quad r\to\infty\,. \quad (53)$$

Third, the factor $\sin\pi(\nu_0+\mu_0)$ vanishes at $\epsilon=\epsilon_n$ so that

$$F_n^{(0)}\to -v(\nu_{0n},r),\quad dF_n^{(0)}/dr=v(\nu_{0n},r)/\nu_{0n}\,. \quad (54)$$

Fourth, note that the factor $u(\nu_0,r)$ of the function $F^{(0)}$ is multiplied by $\sin\pi(\nu_0+\mu_0)$ which vanishes in the limit $\epsilon\to\epsilon_n$; accordingly, if the dependence of $u(\nu_0,r)$ on ν_0 is expanded in powers of $\nu_0-\nu_{0n}$, only the zero-order term with $\nu_0=\nu_{0n}$ need be considered. This yields

$$F^{(0)}\to u(\nu_{0n},r)\sin\pi(\nu_0+\mu_0)-v(\nu_0,r)\,,$$
$$\frac{dF^{(0)}}{dr}\to\frac{u(\nu_{0n},r)\sin\pi(\nu_0+\mu_0)}{\nu_{0n}}+\frac{v(\nu_0,r)}{\nu_0}\,. \quad (55)$$

Finally, only the first terms on the right-hand side of (55) will contribute in the large-r limit, because $v(\nu_0,r)$ vanishes.

Substitution of (54) and (55) into (52), taking into account the preceding considerations and the value (12) of the product uv, yields

$$\frac{1}{2}\left(-F_n^{(0)}\frac{dF^{(0)}}{dr}+F^{(0)}\frac{dF_n^{(0)}}{dr}\right)/(\epsilon-\epsilon_n)$$
$$\to\tfrac{1}{2}[v(\nu_{0n},r)u(\nu_{0n},r)\sin\pi(\nu_0+\mu_0)/\nu_{0n}$$
$$+u(\nu_{0n},r)\sin\pi(\nu_0+\mu_0)v(\nu_{0n},r)/\nu_{0n}]/(\epsilon-\epsilon_n)$$
$$=\sin\pi(\nu_0+\mu_0)/\pi(\epsilon-\epsilon_n)\,,\quad\text{as}\quad r\to\infty\,, \quad (56)$$

that is, a result independent of r. There remains to be taken the limit for $\epsilon\to\epsilon_n$. Utilizing (18), this limit yields

$$\int^{\infty}\Psi_n^2 d\tau=\left(\frac{d(\nu_0+\mu_0)}{d\epsilon}\right)_{\epsilon=\epsilon_n}=\nu_{0n}^3+\left(\frac{d\mu_0}{d\nu_2}\right)_{\nu_{2n}}\nu_{2n}^3\,. \quad (57)$$

The important result (57) has been known in various forms. For an ordinary Rydberg series with constant quantum defect μ, it is well known that the probability of transition to each level is proportional to the normalization factor ν_n^{-3}, where ν_n is the level's effective quantum number. The normalization formulas developed by Seaton[2] can be reduced to the form $[d(\nu_0+\mu_0)/d\epsilon]_{\epsilon_n}$. It has been suggested[13] that normalization coefficients be obtained under ordinary circumstances from the slope of a curve drawn through the points of a plot in which the position of each experimental line is entered at the coordinates (ϵ_n,n). Here we point out that this utilization of experimental data applies also to a strongly perturbed series; it suffices to draw a curve through the points (ϵ_n,n) – so as to obtain $d(\nu_0+\mu_0)/d\epsilon$ directly – or even only through a plot of points $[\mu_0(\nu_{2n}),\ \nu_{2n}]$ so as to obtain a value $d\mu_0/d\nu_2$ to be entered into the right-hand side of (57).

Since a correctly normalized wave function $\bar{\Psi}_n$ of the discrete spectrum must yield unity when substituted for Ψ_n in (57), the relationship of $\bar{\Psi}_n$

to the function $[\Psi(\hat{R}, \hat{r}, r)]_{\nu_0=\nu_{0n}}$ given by (50) is

$$\bar{\Psi}_n = [\Psi(\hat{R}, \hat{r}, r)]_{\nu_0=\nu_{0n}}\left(\frac{d\epsilon}{d(\nu_0+\mu_0)_n}\right)^{1/2}$$

$$= [\Psi(\hat{R}, \hat{r}, r)]_{\nu_0=\nu_{0n}}\left[\nu_{0n}^3 + \left(\frac{d\mu_0}{d\nu_2}\right)_n \nu_{2n}^3\right]^{-1/2}, \quad (58)$$

where $(d\mu_0/d\nu_2)_n$ indicates the value of (37) for $\nu_2 = \nu_{2n}$.

C. Line Intensities

The intensity of light absorption and of photoelectric emission has been represented in Secs. IV and V in the form $|\vec{D}\cdot\vec{A}|^2$. In the discrete spectrum, we deal with the corresponding absorption integrated over a line profile. The $\vec{A}_n$ coefficients to be entered into $|\vec{D}\cdot\vec{A}|^2$ to obtain the desired line intensity differ from the $\vec{A}(\nu_2)$ of Sec. V by the same normalization factor which relates the wave function $\bar{\Psi}_n$ to $[\Psi(\hat{R}, \hat{r}, r)]_{\nu_0=\nu_{0n}}$ in Eq. (58). Accordingly, each line intensity I_n of the discrete spectrum equals the continuum intensity $I(\nu_2)$ Eq. (43) multiplied by a normalization coefficient representing the width of a spectral band equivalent in strength to the discrete line,

$$I_n = I(\nu_{2n})\left(\frac{d\epsilon}{d(\nu_0+\mu_0)}\right)_n = I(\nu_{2n})\Big/\left[\nu_{0n}^3 + \left(\frac{d\mu_0}{d\nu_2}\right)_n \nu_{2n}^3\right]. \quad (59)$$

As noted at the end of Sec. VI A, in the range of $\nu_2 \lesssim 5$ where $\nu_0 \sim \nu_2$, the levels lie at $\nu_{0n} \sim \nu_{2n} \sim -\mu_\sigma$ and $\nu_{0n} \sim \nu_{2n} \sim -\mu_\pi$ (*mod.* 1). Substitution of (43′) and (39′) into (59) yields then

$$I_n \sim D_\sigma^2/\cos^2\alpha\,\nu_{0n}^3(1+\tan^2\alpha) = D_\sigma^2/\nu_{0n}^3, \text{ for } \nu_{0n} \sim -\mu_\sigma$$

$$(60)$$

$$I_n \sim D_\pi^2/\sin^2\alpha\,\nu_{0n}^3(1+\cot^2\alpha) = D_\pi^2/\nu_{0n}^3, \text{ for } \nu_{0n} \sim -\mu_\pi.$$

These results correspond to the initial definition of D_σ and D_π in Sec. I, as representing the intensities of the low lines of the two Rydberg series with normalization per unit energy range. The factors ν_{0n}^{-3} in (60) serve in fact to convert the normalization of the various I_n to a common basis, as noted below Eq. (57).

VII. PARAMETER FITTING FROM EXPERIMENTAL DATA

Section I took the point of view that the five input parameters of the theory, namely, μ_σ, μ_π, D_σ, D_π, and B are known initially from observations of the unperturbed lines of the Rydberg series or, in the case of B, from the spectrum of H_2^+. The value of α has been determined theoretically in Sec. II. This point of view has been preserved in essence throughout the paper, but one may also take the opposite view, of determining the parameters entirely from observations within the narrow spectral range of strong perturbation. This approach might become necessary in applications to similar problems where independent input is not available.

Following the procedure of Ref. 10, one would then determine the effective quantum number $\mu_{0n} = m - \nu_{0n}$ from the measured value of $\nu_{0n} = [2(I_0 - \epsilon_n)]^{1/2}$ of each discrete line, the integer m being so chosen that $0 \leqslant \mu_{0n} < 1$. All the values μ_{0n} so obtained are then plotted against the values of $\nu_{2n} = [2(I_2 - \epsilon_n)]^{1/2}$ (*mod.* 1) to form the main curve of Fig. 2. [The values of I_0 and I_2 are identified as convergence points of the line series.] The intersections of the plot $\mu_0(\nu_2)$ with the diagonal in Fig. 2 determine μ_σ and μ_π. The slope of the plot at these points determines α according to (39′). The absorption intensities in the Beutler continuum at such points, namely $I(-\mu_\sigma)$ and $I(-\mu)$, determine the magnitudes of D_σ and D_π according to (43′). The relative sign of D_σ and D_π is determined by (44′) utilizing the experimental determination of ν_{2z}; note that this sign depends on the conventions on α implied by Fig. 1.

ACKNOWLEDGMENTS

I am indebted to Professor G. Herzberg for communication of his results in advance of publication and for extensive correspondence and discussion. Thanks are also due to many colleagues, here and elsewhere, for discussions and suggestions, and particularly to K. T. Lu for preparation of the illustrations.

APPENDIX A: DERIVATION OF EQUATION (7).

The coordinate rotation matrix in (6) has the symmetry property[14]

$$D_{m\Lambda}^{(l)}(-\pi-\phi, \theta, \pi) = (-1)^{\Lambda-m} D_{-\Lambda,-m}^{(l)}(0, \theta, \phi) \quad . \quad (A1)$$

Substitution of this formula into (6) and of (6) into (2) and (3) yields

$$X_M^{(J,\Lambda=0)} = \sum_m Y_{lm}(\vartheta, \varphi)(-1)^{\Lambda-m} D_{0,-m}^{(l)}(0, \theta, \phi) \times D_{0,M}^{(J)}(0, \theta, \phi)[(2J+1)/4\pi]^{1/2}, \quad (A2)$$

$$X_M^{(J,\Lambda=1,+)} = \sum_m Y_{lm}(\vartheta, \varphi)(-1)^{\Lambda-m} \times [D_{-1,-m}^{(l)}(0, \theta, \phi) D_{1M}^{(J)}(0, \theta, \phi) + D_{1,-m}^{(l)}(0, \theta, \phi) \times D_{-1M}^{(J)}(0, \theta, \phi)][(2J+1)/8\pi]^{1/2}. \quad (A3)$$

The products of the two D factors in (A2) – one of which originates as a transformation matrix and the other as a rotor wave function – can be expanded into superpositions of

$$D_{0,M-m}^{(N)} = Y_{N,M-m}[4\pi/(2N+1)]^{1/2}$$

according to

$$D_{-\Lambda,-m}^{(l)} D_{\Lambda M}^{(J)} = \sum_N (l-\Lambda, J\Lambda \mid lJN0) \times D_{0,M-m}^{(N)}(lJN\, M-m \mid l-m, JM). \quad (A4)$$

For $\Lambda \neq 0$, (A3) contains two expansions (A4) which differ only by reversal of the sign of Λ. This reversal multiplies the expansion by $(-1)^{l+J-N}$, so that the sum vanishes when $l+J-N=2-N$ is odd (parity conservation rule) and consists of identical terms when it is even. Hence, substitution of (A4) into (A2) and (A3) yields

$$X_M^{(J,\Lambda+)} = \sum_m Y_{lm}(\vartheta, \varphi)(-1)^{\Lambda-m} \times \sum_{N=0,2} [2-\delta_{\Lambda 0}]^{1/2}(l-\Lambda, J\Lambda \,|\, lJN0) \times Y_{N,M-m}(\theta,\phi)\left(\frac{2J+1}{2N+1}\right)^{1/2}(lJN\,M-m\,|\,l-m, JM). \tag{A5}$$

Since the product of the last two factors of (A5) equals $(-1)^{J-N+m}(lm, NM-m\,|\,lNJM)$ and N is even, this formula is equivalent to

$$X_M^{(J,\Lambda+)} = \sum_N \Phi_M^{(J,N)}(2-\delta_{\Lambda 0})^{1/2}(l-\Lambda, J\Lambda\,|\,lJN0)(-1)^{J+\Lambda}, \tag{A6}$$

whence (7) follows by comparison with (4). [Alternatively, one could have multiplied (A5) by $\Phi_{M'}^{(J,N')*}$ and utilized the orthonormality of spherical harmonics and Wigner coefficients.]

APPENDIX B: EXTENSION OF THEORY TO A BROADER SPECTRAL RANGE.

The treatment of Coulomb wave functions in Sec. III assumes limiting values near threshold ($\epsilon \sim 0$) of the two parameters[2]

$$B(\nu, l) = \prod_{p=0}^{l} \frac{1+2\epsilon p^2}{1-\mathrm{St}(\epsilon)\exp(-2\pi\gamma)}, \tag{B1}$$

$$\mathcal{G}(\nu, l) = -(2\pi)^{-1}\prod_{p=0}^{l}(1+2\epsilon p^2) \times\left[\sum_{p=0}^{l} 4\epsilon q/(1+2\epsilon q^2) + \tfrac{1}{3}\epsilon - \tfrac{1}{15}\epsilon^2 + \tfrac{4}{63}\epsilon^3 - \cdots\right]. \tag{B2}$$

Here we have utilized the definitions $\epsilon = -1/(2\nu^2) = 1/(2\nu^2)$ and the step function $\mathrm{St}(\epsilon)$, which vanishes for $\epsilon < 0$, to avoid explicit use of the symbol[2] $A(\nu, l)$. The definitions (B1) and (B2) show that

$$\lim_{\epsilon=0} B(\nu, l) = 1 \quad \text{and} \quad \lim_{\epsilon=0} \mathcal{G}(\nu, l) = 0 \;.$$

When $B \neq 1$ and $\mathcal{G} \neq 0$, the Coulomb wave functions with the large-r forms (9) and (10) no longer coincide with the functions f and g which are energy independent at very small r. We shall indicate here the functions with the large-r forms (9) and (10) by German letters $\mathfrak{f}$ and $\mathfrak{g}$. Their relationship to f and g, which arises from the connection between the expansions of confluent hypergeometric functions into powers of $1/r$ and of r, is

$$\mathfrak{f}(\nu, r) = [B(\nu, r)]^{1/2} f(\nu, r) \;, \tag{B3}$$

$$\mathfrak{g}(\nu, r) = [B(\nu, r)]^{-1/2}[g(\nu, r) + \mathcal{G}(\nu, r) f(\nu, r)]. \tag{B4}$$

Equation (14), which is properly normalized per unit energy range and involves the standard value of the quantum defect μ_Λ, no longer represents the large-r form of (13) when $B \neq 1$ and $\mathcal{G} \neq 0$. To maintain the correct connection between (13) and (14), we replace μ_Λ in (13) by an adjusted – and more nearly constant – quantum defect ξ_Λ such that[2]

$$\cot\pi\xi_\Lambda = B(\nu, l)\cot\pi\mu_\Lambda - \mathcal{G}(\nu, l) \;. \tag{B5}$$

The right-hand side of (13) must also be renormalized through multiplication by

$$[B(\nu, l)]^{-1/2}\sin\pi\mu_\Lambda/\sin\pi\xi_\Lambda = [B^{-1}(\cos\pi\xi_\Lambda + \mathcal{G}\sin\pi\xi_\Lambda)^2 + B\sin^2\pi\xi_\Lambda]^{-1/2}. \tag{B6}$$

Equation (16) remains unchanged and the right-hand side of (17) must be multiplied by the factor (B6), with $\nu = \nu_N$ of course.

Equation (20), which underlies the main calculations of this paper, takes now two alternative forms appropriate to small and large r, namely,

$$\Psi(\hat{R}, \hat{r}, r) = \sum_\Lambda \sum_N \Phi_M^{(JN)}(\hat{r}, \hat{R}) \times [f(\nu_N, r)U_{N\Lambda}\cos\pi\xi_\Lambda - g(\nu_N, r)U_{N\Lambda}\sin\pi\xi_\Lambda]A_\Lambda = \sum_\Lambda \sum_N \Phi_M^{(JN)}(\hat{r}, \hat{R})[\mathfrak{f}(\nu_N, r)U_{N\Lambda}\cos\pi\mu_\Lambda - \mathfrak{g}(\nu_N, r)U_{N\Lambda}\sin\pi\mu_\Lambda]\mathcal{A}_\Lambda \;. \tag{B7}$$

Here, the normalization factor (B6) has been incorporated into the coefficients A_Λ and new coefficients $\mathcal{A}_\Lambda$ have been introduced for the large-r form. The connecting formulas (B3) and (B4) imply that the coefficients and quantum defects in (B7) are related by

$$\sum_\Lambda [B(\nu_N, l)]^{-1/2}[U_{N\Lambda}\cos\pi\xi_\Lambda + \mathcal{G}(\nu_N, l)U_{N\Lambda}\sin\pi\xi_\Lambda]A_\Lambda = \sum_\Lambda U_{N\Lambda}\cos\pi\mu_\Lambda\mathcal{A}_\Lambda, \tag{B8}$$

$$\sum_\Lambda [B(\nu_N, l)]^{1/2}U_{N\Lambda}\sin\pi\xi_\Lambda A_\Lambda = \sum_\Lambda U_{N\Lambda}\sin\pi\mu_\Lambda\mathcal{A}_\Lambda \;.$$

This system (which consists of four equations since $N=0$, 2) determines either of the sets of parameters (A_Λ, ξ_Λ) and $(\mathcal{A}_\Lambda, \mu_\Lambda)$ as a function of the other. Equation (B5) does *not hold* in general for our two-channel problem, unless suitably generalized to a matrix notation formula.[2]

All those considerations in Secs. IV–VI, which rely on the study of the large-r form of the wave functions, pertain actually to the calculation of the coefficients $\mathcal{A}_\Lambda$. On the other hand, all the *intensity* formulas are based on the definition $I = |\sum_\Lambda D_\Lambda A_\Lambda|^2$ which involves the A_Λ; hence these formulas should be recalculated, utilizing the implicit relations (B8) between (A_σ, A_π) and $(\mathcal{A}_\sigma, \mathcal{A}_\pi)$, unless the difference between these coefficients appears negligible.

However, a major simplification occurs in our problem owing to the smallness of the separation $6B$ between the two ionization thresholds. Either ϵ is so small that the approximations $B(\nu, l) \sim 1$ and $\mathcal{G} \sim 0$ obtain or it is sufficiently large for the approximation $\nu_2 \sim \nu_0 \sim \nu$ to obtain. In the latter event the subscript N can be removed from ν_N in (B8). The system (B8) resolves then into two separate identical systems pertaining, respectively, to the two values of Λ. Each system yields the relation (B5) between ξ_Λ and μ_Λ and the value (B6) for the ratio $\mathcal{a}_\Lambda / A_\Lambda$. Therefore, the intensities calculated in Secs. IV–VI should be simply multiplied by the square of (B6).

---

*Work supported in part by the U. S. Atomic Energy Commission, under Contract No. COO-1674-25.

[1]G. Herzberg, Phys. Rev. Letters 23, 1081 (1969). Similar results have been obtained by S. Takezawa, J. Chem. Phys. 52, 5793 (1970).

[2]M. J. Seaton, Proc. Phys. Soc. (London), 88, 801 (1966); also Monthly Notices Roy. Astron. Soc. 118, 504 (1958).

[3]See, e.g., U. Fano, Comments At. Mol. Phys. 1, 140 (1969).

[4]R. S. Berry and S. E. Nielsen, Phys. Rev. A 1, 383 (1970); 1, 395 (1970).

[5]W. A. Chupka and J. Berkowitz, J. Chem. Phys. 51, 4244 (1969); J. Berkowitz and W. A. Chupka, *ibid.* 51, 2341 (1969); these papers contain further references.

[6]See, particularly, J. H. Van Vleck, Rev. Mod. Phys. 23, 213 (1951).

[7]Y. N. Chiu, J. Chem. Phys. 41, 3235 (1964), Eq. (30). Note that these values differ from the familiar expression $B[J(J+1) - \Lambda^2]$ which disregards the energy term called $B(\vec{P}_\xi^2 + P_\eta^2)$ by Chiu. Chiu's symbol N corresponds to our J.

[8]See, e.g., H. Beutler, Z. Physik 93, 177 (1935).

[9]See, e.g., U. Fano, Phys. Rev. 124, 1866 (1961).

[10]K. T. Lu and U. Fano Phys. Rev. A 2, 81 (1790); see also M. J. Seaton, J. Phys. B2, 5 (1969).

[11]M. Gailitis, Zh. Eksperim. i Teor. Fiz. 44, 1974 (1963) [Soviet Phys. JETP 17, 1328 (1963)].

[12]See, e.g., A. Sommerfeld, *Atombau und Spektralinien* (Vieweg, Braunschweig, 1939), Vol. 2, p. 754.

[13]U. Fano and J. W. Cooper, Rev. Mod. Phys. 40, 441 (1968), Sec. 2.4.

[14]See, e.g., U. Fano and G. Racah, *Irreducible Tensorial Sets* (Academic, New York, 1959), Eqs. D.4 and D.24.

1973 *Phys. Rev.* A **8** 1241–57
Reprinted with permission from the American Physical Society

Spectroscopy and Collision Theory. II. The Ar Absorption Spectrum*

Chia-Ming Lee and K. T. Lu†
Department of Physics, The University of Chicago, Chicago, Illinois 60637
(Received 12 March 1973)

A collision-type formalism for spectral theory, described by Lu and applied originally to Xe, is improved and is applied to analyze the strongly perturbed absorption spectrum of Ar. The analysis expresses experimental data in terms of three sets of empirical parameters: five eigenquantum defects μ_α, one 5×5 orthogonal transformation matrix $\mathfrak{U}_{i\alpha}$, and five dipole matrix elements D_α. The energy dependence of the parameters is studied and is found to be approximately linear for the μ_α and insignificant for $\mathfrak{U}_{i\alpha}$ and D_α. With these parameters thus determined by fitting to the experimental data, the mixing coefficients of the Rydberg levels have been determined and their oscillator strength have been predicted and compared with available experimental data.

I. INTRODUCTION

A collision-theory analysis of highly perturbed spectra has been developed by Fano,[1] in a paper (to be called FH) which relies on Seaton's[2] multichannel quantum-defect method (QDM). FH expresses photoabsorption data obtained below, between, and above different ionization limits in terms of three sets of theoretical parameters. These parameters pertain to collisions of an electron with an ion core (atomic or molecular) and characterize the close-coupling eigenchannels of this system. A related paper by Lu[3] (to be called LX) extended FH to a multichannel case and applied it to analyze Xe spectra which involve five strongly perturbed series.

Although the treatment of LX was successful in describing a vast amount of photoabsorption data in a unified way, it had a number of limitations. First, all parameters were treated as if they were independent of energy, although an energy dependence was apparent in the imperfect fitting of the lower-lying levels. Second, LX did not determine uniquely all elements of the transformation matrix $\mathfrak{U}_{i\alpha}$ which connects the channels i of the dissociated system (electron + ion core) with the eigenchannels α of the same system in a close-coupling situation. Finally, the classification of levels provided by LX remained rather qualitative owing to insufficient knowledge of the matrix $\mathfrak{U}_{i\alpha}$.

The present work started with the modest aim of applying the method of LX to analyze the Ar spectra for which very extensive experimental data are available, primarily due to Yoshino.[4] However, in the course of the work, we succeeded in removing various limitations, thus extending the range of application of the formalism.

The method employed in this paper is basically the same as that of LX; moreover, the Ar spectrum has the same structure as that of Xe. Specifically, photoabsorption by Ar in its ground state leads to the same configurations as one finds in Xe, namely, to np^5d or np^5s, $J=1$, odd-parity states belonging to five series, of which three converge to the first ionization limit $I_{3/2}$ and two converge to the second ionization limit $I_{1/2}$.

In treating the energy dependence of the parameters, we found the procedures involving Seaton's B and $\mathcal{G}$ functions[1,2] to be rather unsuitable for Ar, whereas they had proven valuable for Ne.[5] We treated the energy dependence of the parameters by expanding them directly as linear functions of energy without prior elimination of the Coulomb field effects represented by B and $\mathcal{G}$. This aspect of the problem is discussed in Sec. III.

As in LX,[3] we chose the dissociation channels i to be jj coupled. On the other hand, the α channels are much more nearly LS coupled in Ar than in Xe. This permitted us to fit the $\mathfrak{U}_{i\alpha}$ matrix by minimizing the departure of these channels from LS coupling (Sec. IV).

Complete knowledge of the matrix $\mathfrak{U}_{i\alpha}$ has made it possible to characterize each discrete level of the absorption spectrum quantitatively as a superposition of dissociation channels i with mixing coefficients Z_i and alternatively as a superposition of close-coupling eigenchannels α with mixing coefficients $\mathfrak{A}_\alpha$. These coefficients are given in Table II for 20 excited levels.

As in LX we fitted intensity parameters to reproduce the observed intensity profile of the autoionization spectrum. The parameters were then utilized to predict the line intensities in the discrete spectrum. These predictions will be discussed in Sec. V with oscillator strengths determined by electron collision experiments[6] and with optical relative intensities estimated by Yoshino.[4]

Analysis of the absorption spectra of the Ne, Ar, Kr, Xe group by our method has now reached the following stage: The Ne $2p^5ns$ (and also the $2p^5np$) levels have been studied by Starace.[5] Ar is reported in this paper. For Kr the experimental data are too fragmentary to permit a detailed analysis. For Xe we have tried to improve the fit obtained in LX and to interpret it in greater detail but without sufficient success to warrant a new report.

One obvious application of the present treatment of noble gases would be to analyze the carbon group. In particular, the perturbations noticed by Andrew and Meissner[7] in the series $4s^2\,4p\,(^2P)nd$ of GeI are evidently similar to those occurring in the noble gases and should be amenable to a similar treatment. Other spectra (Al, Ba, and Hg) have been analyzed to some extent in this laboratory.

Sec. II will be a brief summary of the necessary analytic formulas. Numerical fitting will be treated in Sec. IV.

II. SUMMARY OF FORMULAS

This section summarized the formulas of LX and extends them as required for our purposes.

The atoms of interest have two ionization limits $I_{1/2}$ and $I_{3/2}$ corresponding to the doublet core states $p^5\,{}^2P_{1/2}$ and ${}^2P_{3/2}$, respectively. Photoionization of the ground state ($p^6\,{}^1S_0$) yields odd-parity states with $J=1$ consisting of the doublet ion and of a continuum electron in $s_{1/2}$, $d_{3/2}$, or $d_{5/2}$ orbits. These states belong to five *dissociation channels* i which will be labeled in the jj-coupling scheme by

$$\begin{array}{llllll} i= & 1 & 2 & 3 & 4 & 5 \\ \text{label}= & (^2P_{1/2})d_{3/2}, & (^2P_{3/2})d_{5/2}, & (^2P_{3/2})d_{3/2}, & (^2P_{1/2})s_{1/2}, & (^2P_{3/2})s_{1/2} \end{array} \tag{2.1}$$

(these labels are different from those used in LX). The states of the discrete absorption spectrum are often classified into five perturbed Rydberg series, but each state is actually a superposition of all five dissociation channels. The energy E of each state can be separated into the energy I of the ion and the energy ϵ of the excited or ionized electron. It is important that the energy separation differs for different channels. It is written in a.u. as

$$E = I_i + \epsilon_i = I_i - \frac{1}{2\nu_i^2} = \begin{cases} I_{3/2} - \dfrac{1}{2\nu_{3/2}{}^2} & (i=2,3,5) \\ I_{1/2} - \dfrac{1}{2\nu_{1/2}{}^2} & (i=1,4), \end{cases} \tag{2.2}$$

where $I_{3/2} < I_{1/2}$ and ν_i is imaginary or real according to whether $\epsilon_i > 0$ or $\epsilon_i < 0$. There are three spectral regions: *discrete, auto-ionization,* and *open continuum*, corresponding to $E < I_{3/2}$, $I_{3/2} < E < I_{1/2}$, and $I_{1/2} < E$, respectively.

The QDM[2] relies on the existence of a distance r_0 between the excited electron and the residual ion such that the interaction is purely Coulomb for $r > r_0$. For $r \geq r_0$, Eq. (2.12) of LX represents the wave function of the excited or ionized atom as a superposition of the wave functions of the five dissociation channels,

$$\Psi = \sum_i \phi_i \Big[\mathfrak{f}(\nu_i, l_i; r) \sum_\alpha \mathfrak{U}_{i\alpha} \cos\pi\mu_\alpha \mathfrak{A}_\alpha - \mathfrak{g}(\nu_i, l_i; r) \sum_\alpha \mathfrak{U}_{i\alpha} \sin\pi\mu_\alpha \mathfrak{A}_\alpha \Big], \qquad r \geq r_0 \tag{2.3}$$

where ϕ_i denotes the wave functions of the residual ion core, of spins, and of the angular part of the excited electron in the ith channel; $\mathfrak{f}(\nu_i, l_i; r)$ and $\mathfrak{g}(\nu_i, l_i; r)$ are regular and irregular Coulomb wave functions[8] for the ith channel. Symmetrization of the coordinate r of the excited electron with those of other electrons included in ϕ is implied in (2.3), though not indicated explicitly. The parameters of (2.3)—five *eigenquantum defects* μ_α and a 5×5 orthogonal *transformation matrix* $\mathfrak{U}_{i\alpha}$—represent boundary conditions on the wave function at $r=r_0$. The index α pertains to the close-coupling eigenchannels which characterize the effect of short-range interactions between the excited electron and the residual ion. These interactions, involving exchange and other electron correlations, prevail in the region $r<r_0$. These *eigenchannels* α are to be identified by fitting the parameters $\mathfrak{U}_{i\alpha}$ and μ_α to the experimental data. The coefficients $\mathfrak{A}_\alpha$ are to be determined by the boundary conditions at $r=\infty$. These conditions are different in the different ranges of the spectrum.

A. Discrete Spectrum, $E<I_{3/2}$

To each discrete state with an energy E corresponds a pair of values $(\nu_{1/2}, \nu_{3/2})$ determined by Eq. (2.2), from the value of E. The main boundary condition on discrete states, that $\Psi \to 0$ as $r \to \infty$, leads to relation (2.18) of LX,

$$\sum_\alpha F_{i\alpha} \mathfrak{A}_\alpha = 0 \quad \text{for all } i, \tag{2.4a}$$

where

$$F_{i\alpha} = \mathfrak{U}_{i\alpha} \sin\pi(\nu_i + \mu_\alpha). \tag{2.4b}$$

Equation (2.4a) has the compatibility condition

$$F(\nu_{1/2}, \nu_{3/2}) = \det|F_{i\alpha}| = 0 \tag{2.5}$$

and the solution

$$\mathfrak{A}_\alpha = C_{i\alpha}(\nu_{1/2}, \nu_{3/2}) / [\sum_\alpha C_{i\alpha}^2(\nu_{1/2}, \nu_{3/2})]^{1/2}, \quad (2.6)$$

where the index i can be chosen arbitrarily for convenience, and the $C_{i\alpha}(\nu_{1/2}, \nu_{3/2})$ is the cofactor of the element of the ith row and αth column of the determinant $|F_{i\alpha}|$.

The pair of values of $\nu_{1/2}$ and $\nu_{3/2}$ corresponding to each energy level must satisfy (2.2) and (2.5) simultaneously. For the nth particular pair $(\nu_{1/2,n}, \nu_{3/2,n})$, the coefficients $\mathfrak{A}_\alpha^{(n)}$ can be obtained according to Eq. (2.6). Since the wave function in Eq. (2.3) is normalized per unit energy in a.u., the normalization integral needs to be worked out. The result given by (3.13) of LX must be generalized when the energy dependence of the parameters cannot be neglected. The more general result, worked out in Appendix A, is

$$N_n^2 = \int |\Psi_n|^2 d\tau = \nu_{3/2,n}{}^3 N_{3/2,n} + \nu_{1/2,n}^3 N_{1/2,n} + \sum_\alpha \frac{d\mu_\alpha}{dE} (\mathfrak{A}_\alpha^{(n)})^2 + \sum_i \sum_\alpha \sum_\beta \frac{d\mathfrak{U}_{i\alpha}}{dE} \mathfrak{U}_{i\beta} \times \sin\pi(\mu_\alpha - \mu_\beta) \mathfrak{A}_\alpha^{(n)} \mathfrak{A}_\beta^{(n)}, \quad (2.7)$$

where

$$N_{1/2,n} = \sum_{i=1,4} [\sum_\alpha \mathfrak{U}_{i\alpha} \cos\pi(\nu_{i,n} + \mu_\alpha) \mathfrak{A}_\alpha^{(n)}]^2,$$
$$N_{3/2,n} = \sum_{i=2,3,5} [\sum_\alpha \mathfrak{U}_{i\alpha} \cos\pi(\nu_{i,n} + \mu_\alpha) \mathfrak{A}_\alpha^{(n)}]^2. \quad (2.8)$$

Therefore, the normalized wave function for $r > r_0$ can be represented as a superposition of the five dissociation channels in the form

$$\bar{\Psi}_n = \Psi_n / N_n = \sum_i \phi_i P_i^{(n)} Z_i^{(n)}, \quad (2.9)$$

where

$$P_i^{(n)} = [\nu_{i,n}^2 \tau(l_i + \nu_{i,n} + 1) \tau(\nu_{i,n} - l_i)]^{-1/2} \times (2r/\nu_{i,n})^{\nu_{i,n}} e^{-r/\nu_{i,n}} \quad \text{for } r \to \infty, \quad (2.10)$$

and

$$Z_i^{(n)} = (-1)^{l_i+1} \sum_\alpha \mathfrak{U}_{i\alpha} \cos\pi(\nu_{i,n} + \mu_\alpha) \mathfrak{A}_\alpha^{(n)} / N_n. \quad (2.11)$$

These coefficients Z_i satisfy the following normalization condition[2]

$$\sum_{ij} Z_i \zeta_{ij} Z_j = 1, \quad (2.12a)$$

where

$$\zeta_{ij} = \delta_{ij} + [(-1)^{l_i} (2/\pi\nu_i^3)^{1/2} \cos\pi\nu_i] \frac{d}{dE} \times (\sum_\alpha \mathfrak{U}_{i\alpha} \tan\pi\mu_\alpha \mathfrak{U}_{j\alpha}) [(-1)^{l_j} (2/\pi\nu_j^3)^{1/2} \cos\pi\nu_j]. \quad (2.12b)$$

This equation establishes a relationship between the five coefficients $Z_i^{(n)}$, which measure the mixing among the *five dissociation channels* (i channels) for the nth state in the region $r > r_0$, and the five coefficients $\mathfrak{A}_\alpha$, which measure the mixing among the *five close-coupling eigenchannels* (α channels). Through this equation, we can connect the present collisional approach to the traditional spectroscopic interpretation in terms of configuration interaction, as will be discussed in Sec. VI.

The oscillator strength, for the nth state, is written as

$$f_n = \frac{2(E_n - E_0)|\sum_\alpha D_\alpha \mathfrak{A}_\alpha^{(n)}|^2}{N_n^2} = \frac{2(E_n - E_0)|\sum_\alpha D_\alpha \mathfrak{A}_\alpha^{(n)}|^2}{N_{3/2,n}} \times \left[\nu_{3/2,n}^3 + \frac{N_{1/2,n}}{N_{3/2,n}} \nu_{1/2,n}^3 + \sum_\alpha \frac{d\mu_\alpha}{dE} (\mathfrak{A}_\alpha^{(n)})^2 + \sum_i \sum_\alpha \sum_\beta \frac{d\mathfrak{U}_{i\alpha}}{dE} \mathfrak{U}_{i\beta} \sin\pi(\mu_\alpha - \mu_\beta) \mathfrak{A}_\alpha^{(n)} \mathfrak{A}_\beta^{(n)}\right]^{-1} \quad (2.13)$$

according to Eq. (3.12) in LX, with the *five dipole-matrix elements* D_α defined in LX. These five dipole matrix elements D_α are regarded in this paper as the energy-independent parameters to be determined by fitting the Beutler-Fano profiles in the auto-ionization region.

B. Auto-ionization Spectrum, $I_{3/2} < E < I_{1/2}$

Following Sec. III B in LX, there are three open dissociation channels, $i = 2, 3, 5$ in the auto-ionization spectrum. Therefore, there shall be three *collision eigenstates*, $\rho = 1, 2, 3$, each collision eigenstate ρ being a superposition of the standing waves of the *three open dissociation channels* with the same *eigenphase shift* $\pi\tau_\rho$.

For each value of $\nu_{1/2}$ corresponding to an energy E, $I_{3/2} < E < I_{1/2}$; the boundary condition for the wave function in Eq. (2.3) at $r = \infty$ will lead to the following relations [see (3.23) of LX]:

$$\sum_\alpha \mathfrak{U}_{i\alpha} \sin\pi(\nu_{1/2} + \mu_\alpha) \mathfrak{A}_\alpha^{(\rho)} = 0 \quad \text{for the closed channels, } i = 1, 4$$
$$\sum_\alpha \mathfrak{U}_{i\alpha} \sin\pi(-\tau_\rho + \mu_\alpha) \mathfrak{A}_\alpha^{(\rho)} = 0 \quad \text{for the open channels, } i = 2, 3, 5, \quad (2.14)$$

for each of the three *collision eigenstates* ρ. The compatibility condition of Eq. (2.14) gives a relationship between τ_ρ and $\nu_{1/2}$, having the same form as Eq. (2.5) with $\nu_{3/2}$ replaced by $-\tau_\rho$, namely,

$$F(-\tau_\rho, \nu_{1/2}) = 0 \quad \text{for } \rho = 1, 2, 3. \quad (2.15)$$

[Note that (2.2) would require $\nu_{3/2}$ itself to be imaginary in the autoionization spectrum.] The solution of (2.14) is then

$$\mathfrak{A}^{\rho}_{\alpha} = C_{i\alpha}(-\tau_{\rho}, \nu_{1/2}) / [\sum_{\alpha} C^2_{i\alpha}(-\tau_{\rho}, \nu_{1/2})]^{1/2} \quad \text{for } \rho = 1, 2, 3. \tag{2.16}$$

For a given value of $\nu_{1/2}$, Eq. (2.15) is satisfied by three pairs $(-\tau_{\rho}, \nu_{1/2})$. For each pair there is a set of coefficients $\mathfrak{A}^{\rho}_{\alpha}$. Therefore, the density of oscillator strength in the auto-ionization spectrum can be represented as the sum of contributions corresponding to photoionization into the three collision eigenstates ρ, according to Eq. (3.25) in LX:

$$\frac{df}{dE} = \sum_{\rho=1}^{3} \frac{df^{(\rho)}}{dE}, \tag{2.17}$$

where

$$\frac{df^{(\rho)}}{dE} = \frac{2(E - E_0)|\sum_{\alpha} D_{\alpha} \mathfrak{A}^{\rho}_{\alpha}|^2}{N_{\rho}} \tag{2.18}$$

and

$$N_{\rho} = \sum_{i=2,3,5} |\sum_{\alpha} U_{i\alpha} \cos(-\tau_{\rho} + \mu_{\alpha}) \mathfrak{A}^{\rho}_{\alpha}|^2.$$

The dipole-matrix elements D_{α} can be determined by fitting the profiles in the auto-ionization spectrum according to Eqs. (2.17) and (2.18).

C. Open-Continuum Spectrum, $I_{1/2} < E$

In this energy range we use the coefficients $\mathfrak{A}_{\alpha}$, which satisfy the ingoing wave boundary condition at ∞ [Eq. (3.28) in LX]. The oscillator-strength densities of the two photoelectron groups can thus be expressed as

$$\begin{aligned} df_{3/2}/dE &= 2(E - E_0) \sum_{\alpha,\beta} \sum_{i=2,3,5} U^{\dagger}_{\alpha i} U_{i\beta} \times \cos\pi(\mu_{\alpha} - \mu_{\beta}) D_{\alpha} D_{\beta}, \\ df_{1/2}/dE &= 2(E - E_0) \sum_{\alpha,\beta} \sum_{i=1,4} U^{\dagger}_{\alpha i} U_{i\beta} \times \cos\pi(\mu_{\alpha} - \mu_{\beta}) D_{\alpha} D_{\beta}, \end{aligned} \tag{2.19}$$

according to Eqs. (3.29) and (3.30) in LX. Therefore, the total oscillator-strength density and the branching ratio of the two photoelectron groups can be written as

$$\frac{df_{\text{tot}}}{dE} = \frac{df_{3/2}}{dE} + \frac{df_{1/2}}{dE} = 2(E - E_0) \sum_{\alpha} D^2_{\alpha}, \tag{2.20}$$

and the ratio is

$$\frac{df_{3/2}/dE}{df_{1/2}/dE} = \frac{\sum_{i=2,3,5} \sum_{\alpha,\beta} U^{\dagger}_{\alpha i} U_{i\beta} \cos\pi(\mu_{\alpha} - \mu_{\beta}) D_{\alpha} D_{\beta}}{\sum_{i=1,4} \sum_{\alpha,\beta} U^{\dagger}_{\alpha i} U_{i\beta} \cos\pi(\mu_{\alpha} - \mu_{\beta}) D_{\alpha} D_{\beta}}. \tag{2.21}$$

From these formulas the Ar experimental uv photoabsorption data on level positions, line intensities, intensity profiles in the auto-ionization spectrum, the total photoabsorption cross section in the open-continuum spectrum, and the branching ratio of the two photoelectron groups can be expressed in terms of three sets of parameters, $(U_{i\alpha}, \mu_{\alpha}, D_{\alpha})$, which are slowly-varying functions of energy within the neighborhood of the ionization limits. The energy dependence of these parameters will be discussed in Sec. III.

III. ENERGY DEPENDENCE OF PARAMETERS

All parameters of QDM[2] should be considered as a slowly-varying function of energy within the neighborhood of ionization limits. In Fano's treatment of the H_2 spectrum the energy range of interest was about 4×10^{-3} a.u., near the ionization limits, hence these parameters were regarded to be energy independent. The possible extension to a broader spectral range has been discussed by Fano in Appendix B of FH.[1] Applying this method, Starace[5] has analyzed successfully the spectrum of Ne over the energy range of 1.8×10^{-1} a.u. near the ionization limits. In Lu's treatment[3] of the Xe spectrum, the energy dependence of these parameters has been disregarded over the energy range of 1.8×10^{-1} a.u. near the ionization limits. Nevertheless, only limited evidence has emerged of gross errors due to disregarding the energy dependence of these parameters for Xe. In the present problem of Ar, the energy range is about 1.6×10^{-1} a.u., near the ionization limits. The evidence for the energy dependence of these parameters is quite apparent as will be discussed in Sec. IV. Therefore, the energy dependence has to be taken into account and is going to be discussed in the following.

In the range $r > r_0$, the long-range Coulomb interaction is taken into account by representing the wave function of the excited electron as a linear superposition of analytically known Coulomb functions. The coefficients of this superposition have been related by Seaton[2] to elements of a short-range electron-ion scattering matrix. For this purpose, Seaton used two alternative pairs of Coulomb functions, each of which can serve as a basis for describing an arbitrary state of the electron in Coulomb field. These two pairs of

Coulomb functions denoted by $(\mathfrak{f}, \mathfrak{g})$ and (f, g) are related by a linear transformation[8] and coincide at the ionization limit.

The pair of Coulomb functions $(\mathfrak{f}, \mathfrak{g})$ forms a convenient basis for representing the wave function in the asymptotic region $r \to \infty$ over an energy range extending across the ionization limit. With this choice of Coulomb functions, the wave function in the region $r > r_0$ can be represented by Eq. (2.3) with the transformation matrix $\mathfrak{U}_{i\alpha}$ and the eigenquantum defects μ_α as parameters. Fano[1] and Lu[3] have emphasized that these parameters correspond to the eigenvalues and the eigenvectors of a short-range electron-ion scattering matrix (i.e., $S_{ij} = \sum_\alpha \mathfrak{U}_{i\alpha} e^{2i\pi\mu_\alpha} \mathfrak{U}^\dagger_{\alpha j}$), and that they represent boundary conditions on the wave functions at $r = r_0$, which characterize the dynamics of the excited electron in the region $r < r_0$. These parameters are expected to depend on energy, since they involve the energy dependence of the boundary conditions of the wave functions and the values of Coulomb function $(\mathfrak{f}, \mathfrak{g})$ at $r = r_0$.

The other pair of Coulomb functions (f, g) considered by Seaton is normalized to be energy independent at $r = 0$. In fact, these functions are approximately energy independent over the combined range (see Sec. III of Ref. 8):

$$-\bar{\epsilon} < \epsilon < \bar{\epsilon}, \quad r \ll 1/\bar{\epsilon}, \tag{3.1}$$

where $\bar{\epsilon}$ is much smaller than the potential and kinetic energies. As stressed by Seaton, f and g are analytic functions of energy ϵ, while $\mathfrak{f}$ and $\mathfrak{g}$ are continuous but nonanalytic functions of ϵ. The representation (2.3) of the wave function in the region $r > r_0$ can be replaced by a similar linear superposition of the Coulomb functions (f, g) with the parameters ξ_β and $U_{i\beta}$. The connection between $(\mathfrak{f}, \mathfrak{g})$ and (f, g) establishes an energy-dependent transformation between the two sets of parameters $(\mu_\alpha, \mathfrak{U}_{i\alpha})$ and $(\xi_\beta, U_{i\beta})$; the transformation identifies the index β.[5] The parameters ξ_β and $U_{i\beta}$ correspond to the eigenvalues and eigenvectors of the symmetric matrix IJ^{-1} of Seaton.[2] For practical calculations Seaton introduces the Y matrix which differs from matrix IJ^{-1} by a known function $\mathcal{G}$, $Y = (\mathcal{G} - IJ^{-1})^{-1}$.[2] In his second paper Seaton[9] introduced the concept of representing Y by asymptotic energy expansion together with simple poles. The parameters of the expression of Y as a function of energy were fitted by Moores[10] to experimental discrete levels of the Ca spectrum. Our approach differs from that of Seaton and co-workers[2,9,10] primarily because we fit in effect the eigenvalues and eigenvectors of a scattering matrix, which are smoother functions of energy than the separate elements of the Y matrix. Returning to our study of energy dependence, it has been surmised in FH and in the early phase of this work that the set of parameters $(\xi_\beta, U_{i\beta})$ would vary with energy more slowly than the set $(\mu_\alpha, \mathfrak{U}_{i\alpha})$, because a degree of energy dependence due to motion in the Coulomb field would have been removed by the transformation from $(\mu_\alpha, \mathfrak{U}_{i\alpha})$ to $(\xi_\beta, U_{i\beta})$. In fact, this might be the reason why Starace has analyzed Ne data only for s and p electrons successfully by fitting the parameters $(\xi_\beta, U_{i\beta})$ rather than $(\mu_\alpha, \mathfrak{U}_{i\alpha})$. Note that the difference between (f, g) and $(\mathfrak{f}, \mathfrak{g})$ becomes larger as the angular momentum l increases; for $l = 0$, the pair (f, g) almost coincide with the pair $(\mathfrak{f}, \mathfrak{g})$.[8]

In order to check the above surmise, detailed numerical experiments have been performed on the spectra of Ar and Xe. Our results indicate that in case of Ar, no advantage is achieved by fitting the parameters $(\xi_\beta, U_{i\beta})$ rather than parameters $(\mu_\alpha, \mathfrak{U}_{i\alpha})$, while in case of Xe, the parameters $(\xi_\beta, U_{i\beta})$ are more strongly energy dependent than the parameters $(\mu_\alpha, \mathfrak{U}_{i\alpha})$. This apparent contradiction can be explained as follows.

The values of the cutoff radius r_0, beyond which the interaction between the excited electron and the residual ion is purely Coulombian, can be taken from the Herman-Skillman potential model[11] for each atom, keeping in mind that they are probably underestimated. These values can be entered into Eq. (3.1) to verify whether the Coulomb functions (f, g) are still approximately energy independent at r_0 throughout the energy range of interest, extending from the lowest discrete level to the second ionization limit. The relevant quantities for Ne, Ar, and Xe are shown in Table I.

From the values of the ratios $(r_0\bar{\epsilon})$, the criterion

TABLE I. The relevant quantities of noble gases.

	Ne	Ar	Xe
r_0 (a.u.) Herman–Skillman	1.7	2.6	3.6
Energy range of interest $(\bar{\epsilon})$ (a.u.)	0.18	0.16	0.18
Combined region $(r \ll 1/\bar{\epsilon})$	$r \ll 5.4$	$r \ll 6.3$	$r \ll 5.4$
$r_0\bar{\epsilon}$	0.31	0.41	0.67

for the parameters $(\xi_\beta, U_{i\beta})$ to exhibit minimum energy dependence (i.e., $r_0\bar{\epsilon} \ll 1$) is not well satisfied in any case—Ne, Ar, or Xe, especially—if r_0 is underestimated. Thus we have no definite reason to choose the set of parameters $(\xi_\beta, U_{i\beta})$ to fit the data. We prefer the set of parameters $(\mu_\alpha, \mathfrak{U}_{i\alpha})$ corresponding to the Coulomb functions $(\mathfrak{f}, \mathfrak{g})$, which form a convenient basis for representing the wave function in the form of Eq. (2.3) in the asymptotic region $r \gg r_0$, and for applying the boundary conditions at $r = \infty$.

These parameters—transformation matrix $\mathfrak{U}_{i\alpha}$ and the eigenquantum defects μ_α—will then be treated as slowly-varying functions of energy within the neighborhood of ionization limits. The 5×5 orthogonal matrix $\mathfrak{U}_{i\alpha}$ is conveniently expressed in terms of ten angles, $\theta_k = 1, \ldots, 10$ as detailed in Appendix B. The energy dependence is then taken into account by expanding the eigenquantum defects μ_α and the angles θ_k into powers of the energy ϵ measured from the lowest ionization limit. In the case of Ar, the energy range of interest, extending from the lowest discrete level up to the second ionization limit, is about 0.16 a.u. We use the linear expansion

$$\mu_\alpha = u_\alpha^0 + \epsilon\mu_\alpha^1, \quad \alpha = 1, \ldots, 5,$$
$$\theta_k = \theta_k^0 + \epsilon\theta_k^1, \quad k = 1, \ldots, 10. \tag{3.2}$$

With regard to the dipole-matrix elements D_α, the fitting was not sufficiently accurate to justify the introduction of a nonzero linear coefficient D_α^1.

IV. NUMERICAL FITTING

Experiments on the UV photoabsorption of Ar have determined the following data: (i) the level positions of five Rydberg series belonging to $(3p^5)nd$ or $(3p^5)ns$, $J=1$, odd-parity states with n up to 58[4]; (ii) the photoabsorption cross section in the autoionization spectrum and in the continuous spectrum beyond the second ionization limit[12]; (iii) the branching ratio of photoelectron groups.[13]

In Sec. II, these quantities have been expressed in terms of three sets of theoretical parameters—eigenquantum defect μ_α, transformation matrix $\mathfrak{U}_{i\alpha}$, and dipole-matrix elements D_α. These parameters will now be determined by fitting the experimental data.

As a preliminary, we recall the surmise[14] that the close-coupling eigenchannels would be LS coupled, since the Coulomb interactions between excited and core electrons are stronger than spin-orbit coupling for $r < r_0$. The experience of LX for Xe confirmed that the α channels are approximately LS coupled, and also showed that the quadrupole coupling between s and d channels is rather weak. Accordingly we classify each α channel by approximate quantum numbers (i.e., by spectroscopy symbols):

$$\begin{array}{llllll} \alpha = & 1 & 2 & 3 & 4 & 5 \\ \text{Sym} = & (p^5d)\,^3D & (p^5d)\,^1P & (p^5d)\,^3P & (p^5s)\,^3P & (p^5s)\,^1P. \end{array} \tag{4.1}$$

It will be convenient to introduce an intermediate basis of channels $\bar{\alpha}$, for which the quantum numbers l ($\equiv s$ and d), L and S are *exact*, and to represent the connection between actual α channels and the $\bar{\alpha}$ channels by an orthogonal matrix $V_{\bar{\alpha}\alpha}$ which should differ only a little from unity. Thus we factor the matrix $\mathfrak{U}_{i\alpha}$ in the form

$$\mathfrak{U}_{i\alpha} = \sum_{\bar{\alpha}} \mathfrak{U}_{i\bar{\alpha}} V_{\bar{\alpha}\alpha}, \tag{4.2}$$

where $\mathfrak{U}_{i\bar{\alpha}}$ tranforms the jj-coupled channels i into the LS-coupled channels $\bar{\alpha}$. This transformation $\mathfrak{U}_{i\bar{\alpha}}$, which does *not* couple s and d channels, is known analytically[15] and is given by

$$\mathfrak{U}_{i\bar{\alpha}} = [(2j_c+1)(2j_e+1)(2L_{\bar{\alpha}}+1)(2S_{\bar{\alpha}}+1)]^{1/2} \begin{Bmatrix} 1 & \frac{1}{2} & j_c \\ l & \frac{1}{2} & j_e \\ L_{\bar{\alpha}} & S_{\bar{\alpha}} & 1 \end{Bmatrix} = \begin{array}{c|ccccc|} & \alpha=1 & 2 & 3 & 4 & 5 \\ i=1 & \sqrt{\frac{1}{2}} & \sqrt{\frac{1}{3}} & \sqrt{\frac{1}{6}} & 0 & 0 \\ 2 & -\sqrt{\frac{1}{10}} & \sqrt{\frac{3}{5}} & -\sqrt{\frac{3}{10}} & 0 & 0 \\ 3 & -\sqrt{\frac{2}{5}} & \sqrt{\frac{1}{15}} & \sqrt{\frac{8}{15}} & 0 & 0 \\ 4 & 0 & 0 & 0 & \sqrt{\frac{2}{3}} & -\sqrt{\frac{1}{3}} \\ 5 & 0 & 0 & 0 & \sqrt{\frac{1}{3}} & \sqrt{\frac{2}{3}} \end{array}, \tag{4.3}$$

where j_c and j_e pertain to the core and to the photoelectron in the i channels. The $V_{\bar{\alpha}\alpha}$ matrix will be fitted numerically.

The matrix factorization (4.2) has an important further advantage due to representing the dipole-matrix elements D_α in the form,

$$D_\alpha = \sum_{\alpha} V_{\alpha\bar{\alpha}}^\dagger D_{\bar{\alpha}}. \tag{4.4}$$

Owing to the selection rule that prohibits intercombination lines from the singlet ground state to triplet excited states, we know in advance that the three $D_{\bar{\alpha}}$ pertaining to triplet states ($\bar{\alpha} = 1, 3, 4$)

must vanish. Accordingly only two of the $D_{\bar{\alpha}}$, pertaining to two ^1P states *remain to be fitted* numerically.

a. Graphical representation.[16] The level position of each discrete line can be represented by a pair of numbers $(\nu_{1/2,n}, \nu_{3/2,n})$ from Eq. (2.2). However, levels with $\nu_{3/2,n} > 30$ have been discarded because the determination of the value of $\nu_{3/2,n}$ (modulo 1) becomes inaccurate near the threshold. As an important part of our procedure, each pair $(\nu_{1/2,n}, \nu_{3/2,n})$ is represented graphically by one point in the plot of $-\nu_{3/2}$ (mod 1) vs $\nu_{1/2}$ (mod 1) shown in Fig. 1. Since each pair $(\nu_{1/2,n}, \nu_{3/2,n})$ must satisfy Eqs. (2.2) and (2.5) simultaneously, each point of the plot must lie at the intersection of lines representing these equations. Equation (2.2) is represented in Fig. 1 by a family of nearly straight lines which cross the figure almost diagonally with varying obliquity; Fig. 1 shows only some fragments of these lines. A main objective of the fitting is to determine parameters $(\mu_\alpha, U_{i\alpha})$ such that the curve representing Eq. (2.5), $F(\nu_{1/2}, \nu_{3/2}) = 0$, passes as close as possible to all points of Fig. 1.

A look at the figure shows that the points for low-lying discrete levels do not lie on the same curve $F = 0$ as the others; these departures will have to be corrected by introducing the energy dependence of the parameters. The curve $F = 0$ shown in Fig. 1 pertains to the values of the parameters at the threshold $I_{3/2}$ which we call $(\mu^0_\alpha, U^0_{i\alpha})$.

Important applications of the plot "$-\nu_{3/2}$(mod 1) vs $\nu_{1/2}$(mod 1)" have been stressed in LX.

(i) The intersections of the curve $F = 0$ with the diagonal $\nu_{1/2} - \nu_{3/2} = 0$ have the ordinates $-\nu_{3/2}$ (mod 1) $= \mu_\alpha$; therefore their experimental determination yields the values of μ_α directly.

(ii) The slope $\mathcal{S}$ of the curve $F = 0$ at each of these intersections is

$$\mathcal{S}_\alpha = \sum_{i=1,4} U^2_{i\alpha} \Big/ \sum_{i=2,3,5} U^2_{i\alpha} . \tag{4.5}$$

In addition, should the α and $\bar{\alpha}$ channels coincide exactly, the following would also hold.

(a) The slopes (4.5) would be given by (4.2) as

$$\begin{array}{ll} \alpha: & 1\ 2\ 3\ 4\ 5 \\ \mathcal{S}: & 1\ \tfrac{1}{2}\ \tfrac{1}{5}\ 2\ \tfrac{1}{2}, \end{array} \tag{4.6}$$

(b) The separation of the matrix (4.2) into two submatrices for s and d states would cause the Eq. (2.5), $F = 0$, to factor into separate equations:

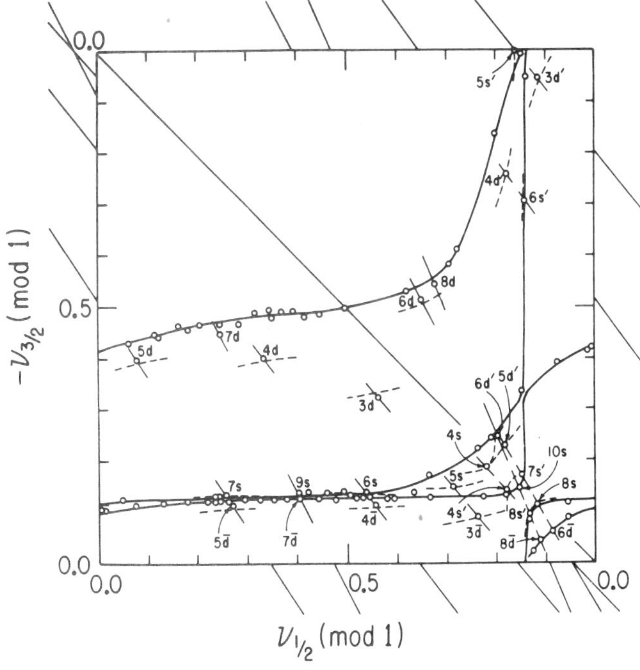

FIG. 1. $-\nu_{3/2}$ (mod 1) vs $\nu_{1/2}$ (mod 1): open circles are observed level positions. Solid curves represent Eq. (2.5), $F = 0$, with parameters fitted near the threshold $I_{3/2}$. The diagonal line represents $\nu_{1/2} - \nu_{3/2} = 0$. The function $-\nu_{3/2}(\nu_{1/2})$ defined by Eq. (2.2) is represented by sections of almost straight oblique lines. Each of the dotted curves represents a section of the curve $F = 0$ fitted to the energy of one of the low-lying levels.

$$F = F_s F_d = 0. \tag{4.7}$$

The curve representing $F = 0$ would then split into two branches that could cross freely. Figure 1 shows that three crossings are narrowly avoided near (0.86, 0.00), (0.86, 0.33), and (0.40, 0.13). The narrowness of the gaps at these quasicrossings indicates that the matrix $V_{\bar{\alpha}\alpha}$ is indeed near unity. This conclusion is confirmed by the fact that the slopes at the intersection with the diagonal are well represented by the values (4.6). Finally, the assumption $V_{\alpha\bar{\alpha}} = \delta_{\alpha\bar{\alpha}}$ leads to a value of 2 for the branching ratio of photoelectron groups (see Sec. VI C of LX); the experimental value of this ratio is indeed 1.98 for Ar, though it was 1.6 for Xe. See however the discussion in the note added in manuscript.

b. Determination of the parameters (μ_α^0, $U_{i\alpha}^0$, D_α) *at first ionization limit* $I_{3/2}$. We perform here an initial fitting of the parameters using only experimental data within ~0.18 eV of $I_{3/2}$. Values of μ_α^0 can be abtained from the intersection points between the diagonal $\nu_{1/2} - \nu_{3/2} = 0$ and a curve, representing $F(\nu_{1/2}, \nu_{3/2}) = 0$, drawn through the 54 experimental points with $10 < \nu_{3/2} < 30$. The correspondence between each of these intersection points and each of the α labels (4.1) is established using a combination of information from Moore's table,[17] from Hund's rules, and with the help of plots $-\nu_{3/2}$(mod 1) vs $\nu_{1/2}$(mod 1) for other values of J, namely, $J = 0, 2, 3, 4$. These plots are used because the character of the LS-coupled eigenchannels should be the same, irrespective of the J values, as discussed in the Appendix of LX. The values of μ_α^0 have been read off a graph similar to Fig. 1 to an accuracy of ~0.01 and have been assigned to the α channels as follows:

$$\begin{array}{llllll} \alpha = & 1 & 2 & 3 & 4 & 5 \\ \mu_\alpha^0 = & 0.22 & 0.07 & 0.50 & 0.15 & 0.11. \end{array} \tag{4.8}$$

Starting from these values of μ_α^0 and from the trial assumption that $V_{\bar{\alpha}\alpha} = \delta_{\bar{\alpha}\alpha}$, improved values of μ_α^0 and of $V_{\bar{\alpha}\alpha}$ were determined by a least-squares technique with aid of trial values. The quantity to be minimized is

$$\sum_n^{10 < \nu_{3/2,n} < 30} [F(\nu_{3/2,n}, \nu_{1/2,n})]^2, \tag{4.9}$$

with the F function given by (2.5). The numerical work was carried to three significant figures and yielded the more accurate values of μ_α^0, that is

$$\begin{array}{llllll} \alpha = & 1 & 2 & 3 & 4 & 5 \\ \mu_\alpha^0 = & 0.214 & 0.070 & 0.500 & 0.154 & 0.109. \end{array} \tag{4.10}$$

However, it yielded no significant departure of $V_{\bar{\alpha}\alpha}$ from $\delta_{\bar{\alpha}\alpha}$.

We proceed now to the dipole-matrix elements D_α, of which only D_2 and D_5 are nonzero owing to the intercombination selection rule and to $V_{\bar{\alpha}\alpha} \sim \delta_{\bar{\alpha}\alpha}$, as noted above. The parameters D_2 and D_5 were fitted to the intensity profile of the first autoionization line pair above the threshold $I_{3/2}$. This profile, extending over the range $8.64 < \nu_{1/2} < 9.64$, is shown in Fig. 2. The experimental photoabsorption data, from Hudson and Kieffer,[11] are of limited accuracy for the purpose of determining the total oscillator strength owing to intensity saturation in the incompletely resolved very sharp peak. These data were then complemented by the measured total oscillator strength in the smooth continuum just above $I_{1/2}$. [Recall that a series of absorption profiles identical to that of Fig. 2 are observed in each unit range of $\nu_{1/2}$ above 9.64, but that these profiles are increasingly sharp and difficult to resolve as the entire range from $I_{3/2}$ to $I_{1/2}$ extends only over 0.18 eV (1431.4 cm^{-1}).]

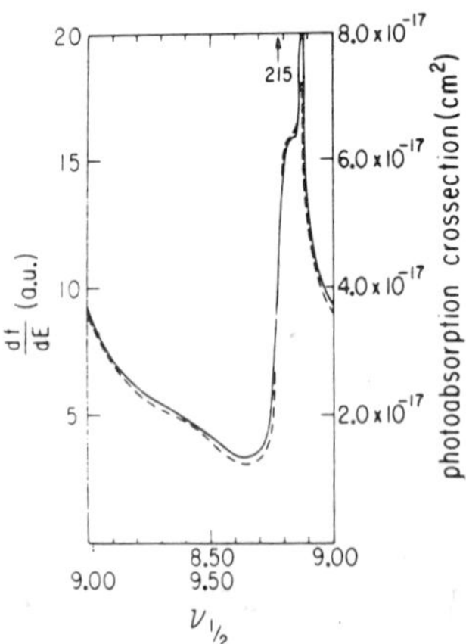

FIG. 2. Argon auto-ionization profile for $8.64 \le \nu_{1/2} \le 9.64$ (note abscissa scale mod 1). Dashed lines represent the experimental data from Ref. 12 and solid lines the theoretical fit with D_α listed in (4.12).

The qualitative features of the intensity profile of Fig. 2 are determined primarily by the values of the parameters μ_α^0 and $U_{i\alpha}^0$ and only to a lesser extent by the ratio of the nonzero dipole parameters D_2 and D_5. This ratio has a small influence because an increase of the amplitude D_5 of $p \to s$ excitation would raise the relative intensity of the sharp peak which is poorly resolved, and hence might escape detection. Actually, the values of μ_α^0 and $U_{i\alpha}^0$ determined by fitting the discrete spectrum account rather well for the qualitative features of the profile in Fig. 2. The value of D_2 is then determined primarily by the value of df/dE at any point of the profile, in Fig. 2, far from the sharp peak. The value of D_5 is determined by reconciling the observed height of the sharp peak with the value $df/dE \approx 7.4$ a.u. above the second ionization limit $I_{1/2}$. The D_α parameters thus adopted,

$$\begin{array}{lccccc} \alpha = & 1 & 2 & 3 & 4 & 5 \\ D_\alpha(\text{a.u.}) = & 0 & 2.1 & 0 & 0 & 1.4, \end{array}$$

determine through Eqs. (2.17) and (2.18) the theoretical curve shown in Fig. 2.

As a further step to evaluate the accuracy of fitting, we assumed the value 1.98 observed for the branching ratio (2.21) of photoelectron groups to be significantly different from the value 2. We then again fitted the matrix $V_{\bar{\alpha}\alpha}$, under the assumption of minimum departure of α channels from $\bar{\alpha}$ channels, so that Eq. (2.21) would yield 1.98 without increasing significantly the least-squares-sum value of (4.9). This procedure yielded

$$V_{\bar{\alpha}\alpha} = \begin{vmatrix} 1.000 & 0.001 & 0.000 & 0.000 & 0.001 \\ -0.001 & 1.000 & -0.005 & 0.000 & 0.000 \\ 0.000 & 0.005 & 1.000 & 0.000 & 0.003 \\ 0.000 & 0.000 & 0.000 & 1.000 & 0.000 \\ -0.001 & 0.000 & -0.003 & 0.000 & 1.000 \end{vmatrix},$$

$$\begin{array}{lccccc} \alpha = & 1 & 2 & 3 & 4 & 5 \\ D_\alpha(\text{a.u.}) = & -3.0\times10^{-3} & 2.1 & -2.0\times10^{-2} & 1.0\times10^{-4} & 1.4, \end{array} \quad (4.12)$$

and left the values (4.10) of μ^0_α unchanged. In Fig. 1, the curve $F(\nu_{1/2}, \nu_{3/2}) = 0$ calculated by using the adopted parameters $(\mu^0_\alpha, U^0_{i\alpha})$ shows a fair fitting with the experimental points. As shown in Fig. 2, the theoretical autoionization profile, determined by the parameters D_α(4.11), shows no significant difference from the theoretical profile determined by the parameters D_α(4.12). See, however, the discussion in the note added in manuscript.

c. Energy dependence of parameters $(\mu_\alpha, U_{i\alpha})$. In Sec. III, the energy dependence of the parameters has been discussed and represented by the linear expansion (3.2). After obtaining the parameters $(\mu^0_\alpha, U^0_{i\alpha})$ at the threshold $I_{3/2}$ from the previous procedures, we now proceed to determine the linear expansion coefficients $(\mu^1_\alpha, \theta^1_k)$.

As noted in Sec. IV a, the five parameters μ_α are determined by the five intersection points between the curve $F = 0$ and the diagonal line $\nu_{1/2} - \nu_{3/2} = 0$ in the plot $-\nu_{3/2}$(mod 1) vs $\nu_{1/2}$(mod 1). The slope of the curve $F = 0$ at each of these five intersections relates to the matrix $U_{i\alpha}$ from Eq. (4.5). If now the parameters $(\mu_\alpha, U_{i\alpha})$ are energy dependent, the equation $F(\nu_{1/2}, \nu_{3/2}) = 0$ will no longer be represented by a single curve on the $-\nu_{3/2}$(mod 1) vs $\nu_{1/2}$(mod 1) graph of Fig. 1. However, it will be represented by a single curve on a plot of $-\nu_{3/2}$(mod 1) vs $\nu_{1/2}$ itself rather than vs $\nu_{1/2}$(mod 1); such a plot is shown in Fig. 3, where the curve $F(\nu_{1/2}, \nu_{3/2}) = 0$ is simply drawn through the experimental points. As shown in Fig. 3, the point $[\nu_{1/2}, -\nu_{3/2}(\text{mod } 1)]$ representing each discrete level must lie at an intersection of the curve $F(\nu_{1/2}, \nu_{3/2}) = 0$ and of one of the almost straight

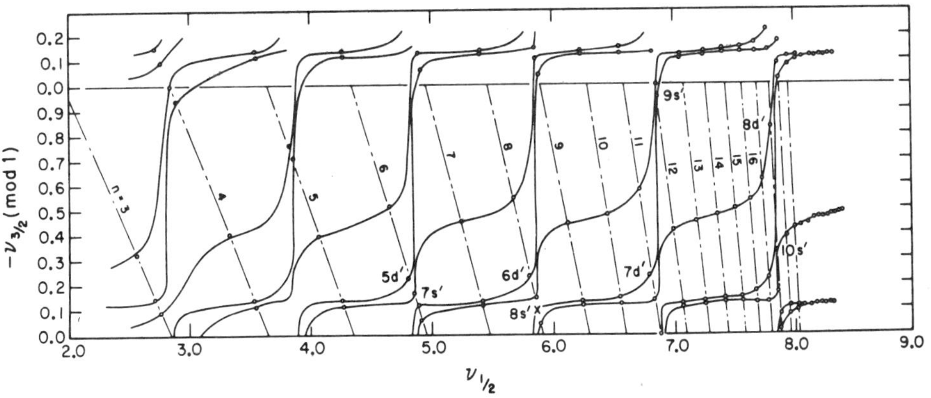

FIG. 3. Quantum defect $-\nu_{3/2}$ (mod 1) vs $\nu_{1/2}$ plot of Ar. Open circles are level positions. The full curves indicate the curve $F = 0$. The relation $-\nu_{3/2}(\nu_{1/2})$ defined by Eq. (2.2) is shown by dot-dashed lines.

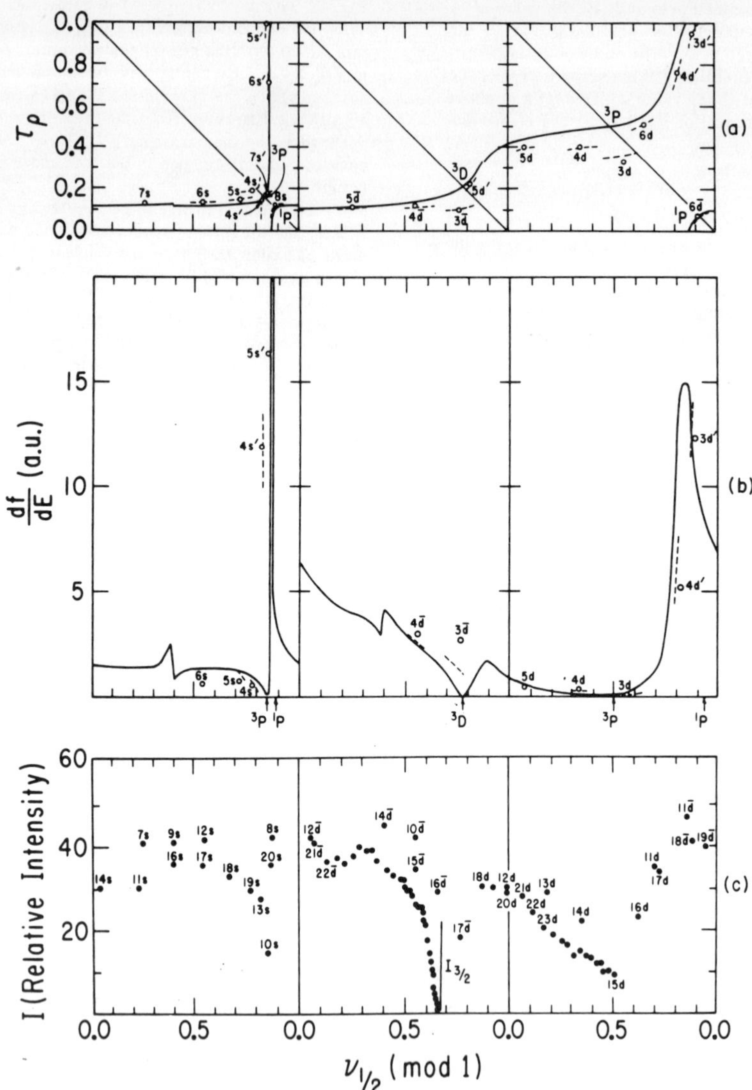

FIG. 4. Eigenphases τ_ρ and oscillator-strength densities $df^{(\rho)}/dE$ for separate eigenchannels $\rho = 1, 2, 3$. (a) Solid lines represent $\tau_\rho(\nu_{1/2})$ for parameters fitted near $I_{3/2}$, dashed lines represent $\tau_\rho(\nu_{1/2})$ for parameters fitted at the $\nu_{1/2}$ value of low-lying lines, and open circles represent experimental data $[\tau_\rho = -\nu_{3/2,n} \text{ (mod 1)}]$. (b) Solid lines represent $df^{(\rho)}/dE$ for parameters fitted near $I_{3/2}$, dashed lines represent $df^{(\rho)}/dE$ for parameters fitted at $\nu_{1/2}$ values of low-lying levels, and open circles represent experimental data from Ref. 6 renormalized by Eq. (5.1). (c) Closed circles represent experimental data (relative intensities) from Ref. 4.

lines (2.2) indicated by dashed oblique lines.

To fit the linear expansion coefficients $(\mu^1_\alpha, \theta^1_k)$, we first estimate a set of μ^1_α from the Fig. 3 and tentatively assign small values to θ^1_k as trial values. These parameters $(\mu^1_\alpha, \theta^1_k)$ thus can be determined by adjusting these parameters so that the curve $F(\nu_{1/2}, \nu_{3/2}) = 0$, for each of low-lying discrete levels, passes as close as possible to the corresponding experimental point.

The adopted values for the five μ^1_α are

$$\begin{array}{lccccc} \alpha = & 1 & 2 & 3 & 4 & 5 \\ \mu^1_\alpha(\text{a.u.})^{-1} = & 1.22 & 0.66 & 1.84 & -0.28 & -0.40. \end{array} \quad (4.13)$$

While the ordinates μ_α of the intersection points were thus found to vary appreciably as $\nu_{1/2}$ varies as shown in Fig. 1, the slopes of $F = 0$ at these points remained effectively constant. That is, the values of the parameters θ^1_k did not depart from zero significantly; in other words, the transformation matrix $U_{i\alpha}$ remains approximately the same as the matrix $U_{i\alpha}$ at $I_{3/2}$ throughout the range of interest. At this point we may return to the graph of Fig. 1 with $\nu_{1/2}$(mod 1) as the abscissa by folding again the curve $F(\nu_{1/2}, \nu_{3/2}) = 0$ with the values of μ_α given by (4.10) and (4.13); a segment of a curve $F(\nu_{1/2}, \nu_{3/2}) = 0$ for each of low-lying discrete levels, which is calculated by using the adopted parameters, is plotted as a dotted curve in Fig. 1. These curves show a fair fitting with the experimental points except the 3*d* line.

The implications of the parameters $(\mu_\alpha, U_{i\alpha}, D_\alpha)$ thus adopted over the whole spectral range of interest will be discussed in Secs. V and VI.

V. ANALYSIS OF OSCILLATOR STRENGTH

Equations (2.13), (2.17), and (2.18) establish the connection between the oscillator strength density df/dE in the auto-ionization spectrum and the oscillator strength f_n of each discrete line. Using the theoretical parameters $(\mu_\alpha, U_{i\alpha}, D_\alpha)$ obtained in Sec. IV, and recalling that the D_α have been fitted to the auto-ionization spectrum, we can now predict the values of f_n.

The quantity fitted in the auto-ionization spectrum is the total oscillator-strength density, since the measurements do not separate out the contributions of photoionization to the three collision eigenchannels, $\rho = 1, 2, 3$, Eq. (2.17). However, for the calculation of the f_n we must consider the separate terms of Eq. (2.17). Accordingly, Fig. 4 (b) shows the separate terms $df^{(\rho)}/dE$ for $\rho = 1, 2, 3$, plotted on adjacent graphs; the abscissa of each graph represents one unit range of $\nu_{1/2}$ (mod 1). (The sum of the ordinates of the three graphs coincides with the value of df/dE plotted in Fig. 2.) For purposes of identification, Fig. 4(a) shows the values of τ_ρ for $\rho = 1, 2, 3$, plotted against the same abscissas as $df^{(\rho)}/dE$.

In principle, each curve representing one of the τ_ρ in Fig. 4(a) should coincide with one of the branches of the plot of $-\nu_{3/2}$(mod 1) vs $\nu_{1/2}$(mod 1) in Fig. 1, since the τ_ρ are defined as roots of $F(\nu_{1/2}, -\tau_\rho) = 0$, and the plot of Fig. 1 represents $F(\nu_{1/2}, \nu_{3/2}) = 0$. However we have found it expedient to assign the labels $\rho = 1, 2, 3$ in such a way that the τ_ρ curves skip from one to another branch at the three "near crossing" points with coordinates (0.86, 0.00), (0.86, 0.33), and (0.40, 0.13). This artifice is analogous to the frequent practice of drawing energy-level diagrams of diatomic molecules according to a "diabatic" rather than "adiabatic" approximation. Because of this artifice, the plot of τ_1 represents, in effect, a root of the approximate Eq. (4.7), $F_s(\nu_{1/2}, -\tau_\rho) = 0$, while τ_2 and τ_3 represent roots of $F_d(\nu_{1/2}, -\tau_\rho) = 0$; that is, $\rho = 1$ represents an *s* branch and $\rho = 2, 3$ represent *d* branches. The artifice is justified, in our problem, by the weakness of *sd* coupling discussed in Sec. III. The main result of this artifice of plotting is to smooth out the plots of $df^{(\rho)}/dE$ in Fig. 4(b); a sizeable jag remains for $\rho = 1$ and 2 at the near crossing abscissa $\nu_{1/2} = 0.40$(mod 1), but no visible one appears at the position of the other near crossings.

To calculate the value of f_n for a line of the discrete spectrum with quantum numbers $(\nu_{1/2,n}, \nu_{3/2,n})$ one should take the value of $df^{(\rho)}/dE$ at the abscissa $\nu_{1/2} = \nu_{1/2,n}$ and with such ρ that $\tau_\rho = -\nu_{3/2,n}$ (mod 1) and then divide this value by the coefficient shown in (2.18). The results could then be compared with the experimental values of Ref. 6. Instead we have followed the procedure of LX of renormalizing each experimental value into an effective oscillator-strength density defined by

$$\left(\frac{df}{dE}\right)_n = f_n \left[\nu^3_{3/2,n} + \nu^3_{1/2,n}\left(\frac{N_{1/2,n}}{N_{3/2,n}}\right) + \sum_\alpha \frac{d\mu_\alpha}{dE}\left(\mathfrak{A}^{(n)}_\alpha\right)^2 + \sum_i \sum_\alpha \sum_\beta \frac{dU_{i\alpha}}{dE} U_{i\beta} \sin\pi(\mu_\alpha - \mu_\beta)\mathfrak{A}^{(n)}_\alpha \mathfrak{A}^{(n)}_\beta\right]. \quad (5.1)$$

Each of these renormalized data is plotted in Fig. 4(b) (as marked by circles) at the abscissa for which $\nu_{1/2} = \nu_{1/2,n}$ (mod 1) and $\tau_\rho = -\nu_{3/2,n}$(mod 1).

Also plotted in Fig. 4(b) are theoretical values of the oscillator-strength density adjusted to the low values of $\nu_{1/2} = \nu_{1/2,n}$ pertaining to low-lying discrete levels. The adjustment was made using Eq. (2.18) with the same dipole-matrix elements D_α fitted to the auto-ionization spectrum but taking $\mu_\alpha = \mu^0_\alpha + \mu^1_\alpha \epsilon_n$, with $\epsilon_n = -0.5/\nu^2_{3/2,n}$. The calculated values are listed in Table II.

TABLE II. Mixing coefficients $\mathfrak{U}_{\alpha}$ and Z_i and Ar oscillator strengths.

	$\mathfrak{U}_{\alpha}$					Z_i							
Moore's designation	1 $(d)\,^3D$	2 $(d)\,^1P$	3 $(d)\,^3P$	4 $(s)\,^3P$	5 $(s)\,^1P$	1 $d(\frac{1}{2},\frac{3}{2})$	2 $d(\frac{3}{2},\frac{5}{2})$	3 $d(\frac{3}{2},\frac{3}{2})$	4 $s(\frac{1}{2},\frac{1}{2})$	5 $s(\frac{3}{2},\frac{1}{2})$	f_n(expt.) [a]	f_n(calc.) [b]	$\left(\frac{df}{dE}\right)_n$(calc.) [c]
$4s[1\frac{1}{2}]^\circ$ $4s$	0.000	0.000	0.003	0.852	0.523	0.001	−0.003	0.003	0.396	0.948	$(7.0 \pm 0.7) \times 10^{-2}$	8.0×10^{-2}	0.6
$4s'[\frac{1}{2}]^\circ$ $4s'$	0.000	0.000	0.001	0.537	−0.844	−0.001	0.001	−0.001	0.946	−0.402	$(2.78 \pm 0.28) \times 10^{-1}$	2.1×10^{-1}	8.8
$3d[\frac{1}{2}]^\circ$ $3d$	−0.126	−0.078	0.989	−0.001	0.000	0.293	−0.526	0.743	0.001	−0.003	$(1.0 \pm 0.1) \times 10^{-3}$	1.6×10^{-3}	0.03
$5s[1\frac{1}{2}]^\circ$ $5s$	−0.030	−0.002	0.001	0.686	0.727	−0.017	0.007	0.020	0.142	0.997	$(2.8 \pm 0.3) \times 10^{-2}$	4.5×10^{-2}	1.1
$3d[1\frac{1}{2}]^\circ$ $3\bar{d}$	−0.866	0.491	−0.089	−0.001	−0.002	−0.357	0.667	0.623	0.000	−0.001	$(9.2 \pm 0.9) \times 10^{-2}$	4.5×10^{-2}	1.3
$5s'[\frac{1}{2}]^\circ$ $5s'$	0.001	0.002	0.001	0.742	−0.670	0.001	−0.002	0.001	0.995	0.146	$(1.24 \pm 0.4) \times 10^{-2}$	3.9×10^{-2}	51.2
$3d'[\frac{1}{2}]^\circ$ $3d'$	0.510	0.840	0.184	−0.006	0.007	0.849	−0.503	0.029	−0.008	−0.002	$(1.10 \pm 0.11) \times 10^{-1}$	1.28×10^{-1}	14.2
$4d[\frac{1}{2}]^\circ$ $4d$	−0.256	−0.154	0.954	−0.002	0.000	0.278	−0.512	0.790	0.001	−0.003	$(4 \pm 1) \times 10^{-3}$	2.6×10^{-3}	0.1
$6s[1\frac{1}{2}]^\circ$ $6s$	−0.001	0.001	0.000	0.612	0.791	0.000	−0.001	0.003	0.069	1.001	$(0.94 \pm 0.7) \times 10^{-2}$	2.3×10^{-2}	1.3
$4d[1\frac{1}{2}]^\circ$ $4\bar{d}$	−0.716	0.690	−0.101	−0.002	−0.003	−0.216	0.774	0.582	0.000	−0.003	$(4.8 \pm 1.4) \times 10^{-2}$	3.9×10^{-2}	2.4
$4d'[\frac{1}{2}]^\circ$ $4d'$	0.639	0.631	0.440	0.006	−0.003	0.875	−0.428	0.200	0.007	−0.001	$(1.5 \pm 0.4) \times 10^{-2}$	3.2×10^{-2}	10.7
$6s'[\frac{1}{2}]^\circ$ $6s'$	−0.005	−0.005	−0.002	0.807	−0.590	−0.007	0.002	0.000	0.998	0.081	$(2.2 \pm 0.4) \times 10^{-2}$	1.3×10^{-2}	161.0
$5d[\frac{1}{2}]^\circ$ $5d$	−0.501	−0.281	0.819	−0.003	0.001	0.359	−0.431	0.818	0.002	−0.003	$(3.2 \pm 0.3) \times 10^{-3}$	$4.3 \quad 10^{-3}$	0.5
$7s[1\frac{1}{2}]^\circ$ $7s$	−0.002	0.001	0.000	0.558	0.830	−0.001	0.000	0.003	0.058	1.000	$\left\{ (3.7 \pm 1.0) \times 10^{-2} \right.$	1.3×10^{-2}	1.5
$5d[1\frac{1}{2}]^\circ$ $5\bar{d}$	−0.550	0.830	−0.094	−0.002	−0.003	−0.185	0.828	0.523	0.000	−0.003		3.0×10^{-2}	3.6
$6d[\frac{1}{2}]^\circ$ $6d$	0.228	0.157	0.961	0.002	−0.001	0.465	−0.580	0.661	0.003	−0.002		7.5×10^{-4}	0.2
$5d'[\frac{1}{2}]^\circ$ $5d'$	0.990	0.125	−0.059	0.000	0.001	0.657	−0.235	−0.717	0.000	0.000		5.1×10^{-4}	0.18
$7s'[\frac{1}{2}]^\circ$ $7s'$	0.006	−0.001	0.000	0.978	−0.208	0.003	−0.003	−0.006	0.868	0.497		7.4×10^{-4}	0.6
$8s[1\frac{1}{2}]^\circ$ $8s$	0.000	0.000	0.000	−0.105	0.995	0.001	−0.002	0.002	−0.554	0.834		1.3×10^{-2}	3.8
$6d[1\frac{1}{2}]^\circ$ $6\bar{d}$	0.096	0.995	0.025	−0.002	0.003	0.534	0.811	0.232	−0.003	0.002		2.9×10^{-2}	8.5

[a] Experimental oscillator strength from Ref. 6. [b] Calculated oscillator strength from Eq. (2.13). [c] Calculated oscillator-strength density from Eq. (5.1).

The experimental points and the calculated value for low-lying lines appear to follow the main features of the main theoretical curve which was based on fitting to the auto-ionization spectrum.

As a further comparison with experimental evidence, we have also entered in Fig. 4(c) Yoshino's[4] experimental "relative line intensities" of highly excited levels of the three Rydberg series converging to $I_{3/2}$. This plot relies on the following considerations.

(i) Each line is unresolved in the spectrogram; accordingly its apparent strength is proportional not to f_n itself, but roughly to the product of f_n and the unresolved width and hence to $f_n \nu_{3/2,n}^3$.

(ii) For large values of $\nu_{3/2,n}$, the last three terms in the brackets of (5.1) are negligible as compared to the first one. Therefore, the relative intensity values given by Yoshino are approximately proportional to $(df/dE)_n$ and are accordingly plotted directly as ordinates in Fig. 4(c).

Interpretation. As shown in Fig. 4(b), the theoretical plot of $df^{(\rho)}/dE$ reproduces the main features of the variations of $(df/dE)_n$. In particular, it reproduces the points of near-zero intensity which correspond to the three points with "triplet" character among the five intersections between the diagonal $\nu_{1/2} - \nu_{3/2} = 0$ and the curve $F(\nu_{1/2}, \nu_{3/2}) = 0$ as marked in Fig. 4(a). (These zero points of the Ar spectrum were already noted in LX.) Thus the relative intensity for each of the 10s, 15d, 17$\bar{d}$ lines in Fig. 4(c), is very low because these levels are almost pure close-coupling eigenstates with triplet character. The occurrence of these low minima is consistent with the near coincidence of α channels with $\bar{\alpha}$ channels for Ar. We will return to this point with complementary evidence in Sec. VI.

VI. DISCUSSION

Several relationships have been established in this paper between spectral properties of Ar in the discrete and continuum regions. The plot of the equation $F(\nu_{1/2}, \nu_{3/2}) = 0$, in Fig. 1, determines, on the one hand, the position of discrete levels of strongly perturbed series and, on the other hand, the resonant behavior of collision eigenphases $\pi\tau_\rho$ in the auto-ionization region. Figure 4(b) provides the connection between the oscillator strengths in the discrete and the continuum, on the same scale. In all these regards the series of discrete levels and their adjoining continua can be indeed treated as a single unit.

The analytical treatment of this paper involves the parameters μ_α, D_α, and $U_{i\alpha}$. Values of these parameters have been obtained by fitting the discrete-level positions, the profile of auto-ionization lines, and the branching ratio in the open continuum, and are given in Eqs. (4.10), (4.12), and (4.13). The energy dependence of these parameters has been represented by expansion as linear functions of energy. Only the eigenquantum defect μ_α appears to be appreciably energy dependent over our spectral range. The over-all fitting is satisfactory except for that of the 3d level in Fig. 1. An expansion including a quadratic energy dependence of μ_α may be required to take care of this low-lying level.

Notice that the eigenquantum defect μ_α decreases with increasing energy in the s series. This downward drift also prevails in other noble gases, e.g., Ne, Kr, and Xe,[5,16,19] and appears to follow the normal trend of scattering phase shifts. That is, the phase shift due to an *attractive* potential (deeper than hydrogenic) is normally a decreasing function of energy except in the vicinity of thresholds or of resonances. By contrast, the presence of a centrifugal potential appears to reverse the situation for the d channels, whose eigenquantum defect μ_α increases with energy.[18,19] This upward drift is most apparent in Ar,[16] less so in Kr,[16] and more or less absent in Xe.[3] This may be correlated with the change in character of the balance between the electrostatic and centrifugal potential.[20] In Ar,[20] this balance creates a well defined potential barrier, which separates two valleys. This barrier confines a 3d electron near the bottom of the outer valley, where the potential is nearly hydrogenic. Increasing excitation permits a d electron to achieve some penetration of the centrifugal barrier toward the inner well where the attraction is much greater than hydrogenic. Thus the eigenquantum defect is larger for high d levels. References 19 and 20, complemented by additional estimates, suggest that the potential barrier is lower in Kr than in Ar and may even disappear for Xe. Accordingly the increase of μ_α with increasing energy, for d electrons, should be reduced along the sequence Ar, Kr, Xe. In fact for Xe all the d levels have approximately the same eigenquantum defect. This might be the reason why a rather satisfactory fitting was obtained for the d levels of Xe even though μ_α was treated as independent of energy. For Ne, the potential for d electrons is almost hydrogenic and therefore the eigenquantum defect is almost zero.

Observable properties have been related in this paper to the five close-coupling eigenchannels α of the complex $e + Ar^+$. With regard to the eigenchannels α, the electron's orbital momentum is 99.999% $l = 2$, and 0.001% $l = 0$ for three of them, and 99.999% $l = 0$, and 0.001% $l = 2$ for the other two. The s-d interference effects are quite small, being of order ~0.002, as characterized by the smallness of the nonzero elements $U_{i\alpha}$ with $i=1,2,3$, $\alpha=5$, which are of the order of 0.2%. Yet the nonzero value of these elements is manifested in the occurrence of avoided crossings between the curves of Fig. 1.

One of the three eigenchannels with predominantly d character, namely the one labeled by $\alpha=2$, is identified as predominantly singlet by the large value of the dipole-matrix parameter D_2 and the large magnitude of the oscillator-strength density at $\nu_{1/2}(\text{mod } 1)=-\mu_2(\text{mod } 1)=0.93$ in Fig. 4. The other singlet is assigned to one of the two eigenchannels with predominantly s character, $\alpha=5$, by the large value of D_5 and large magnitude of oscillator-strength density at $\nu_{1/2}(\text{mod } 1)=-\mu_5(\text{mod } 1)=0.891$. The other three eigenchannels α are identified as triplets as described in Sec. V.

Our data also provide a further characterization of discrete levels through a tabulation of mixing coefficients. Table I gives the components of the eigenvectors $\mathfrak{A}_\alpha$ and Z_i defined in Eq. (2.6) and (2.11). The entries pertain to 20 selected levels listed in order of increasing energy. The coefficients $\mathfrak{A}_\alpha$ and Z_i represent the mixing of α channels and i channels in the given level, respectively. However, they are so normalized according to Eq. [2.12(a)] that $\sum_i Z_i^2$ is not equal to unity for the lower levels. There are five levels, labeled $4s$, $5s$, $6s$, $7s$, and $8s$, with mixing coefficient $|Z_5|\geq 0.834$; these levels thus belong predominantly to the $i=5$ channel, with a small admixture from the other four i channels, the next more important admixture being from $i=4$. This set of levels represents the $3p^5(^2P_{3/2})ns\,[1\frac{1}{2}]^\circ$ series according to Moore's assignment. In three of them, $n=5,6,7$, the character is more than 0.99, $i=5$. However, $8s$ has 0.83 in $i=5$ and a substantial contribution from $i=4$, that is, $8s$ belongs neither purely to $i=5$ nor to $i=4$. These behaviors reveal themselves in Fig. 4(a), where the ns points with $n=5,6,7$ lie on the flat part of a curve with the same quantum defect and are almost unperturbed. In contrast, $8s$ lies near the intersection between the diagonal line $\nu_{1/2}-\nu_{3/2}=0$ and the curve $F(\nu_{1/2}-\tau_\rho)=0$. Table I shows that $8s$ has the character of $\alpha=5$, i.e., of a singlet state, even though it is *more highly excited* than the ns levels with $n=5,6,7$. Indeed the intensity plot in Fig. 4(b) gives $8s$ a large oscillator-strength density. For the d levels, we have followed Moore's tables by calling nd, with $n=3,4,5,6$, the levels classified as $3p^5(^2P_{3/2})nd[\frac{1}{2}]^\circ$, and $n\bar{d}$, with $n=3,4,5,6$, classified as $3p^5(^2P_{3/2})nd[1\frac{1}{2}]^\circ$. In fact, all of the mixing coefficients Z_i of these levels for $i=1,2,3$ are substantially large. This strong mixing of the i channels constitutes a clear departure from Moore's classification. Indeed most of these levels lie on a rising portion of the curves in Fig. 4(a) rather than on a flat portion. Again, $6\bar{d}$ has the character of a singlet, $\alpha=2$, as one can see both from the table and figure. Thus the combined utilization of Fig. 4 and Table II provides a qualitative and quantitative analysis of strongly perturbed Rydberg spectra of our type.

In the auto-ionization region, the connection between the collisional approach and the traditional interpretation of auto-ionization in terms of configuration interaction,[21] has been developed by Fano[1] for the two-channel problem of H_2. The traditional approach holds only when the resonance profiles are isolated from one another. Otherwise, the collisional approach followed in this paper appears preferable. According to this approach, autoionization of highly excited electrons in the "closed" channels $i=1$ and 4 results from their scattering into the "open" channels $i=2,3,5$ by close range interaction with the Ar^+ core; this scattering is inelastic because it drops the core from its excited doublet level $^2P_{1/2}$ to the ground level $^2P_{3/2}$. The mixing of *different* eigenchannels α into each state indicates the probability of the scattering from one i channel to another.

The complete fitting of the transformation matrix $U_{i\alpha}$ based on minimizing the departure of the α channel from LS coupling, constitutes a novelty of this paper. As a further check of this approach, we examine the relative strength of spin-orbit (so) coupling and electrostatic interaction. In our collision-type method, interaction strengths are represented by dimensionless numerical parameters equal to the shifts of quantum defect $\Delta\mu_{so}=0.0045$.[22] On the other hand, the eigenquantum-defect differences $\mu_\alpha-\mu_\beta$ of the close-coupling eigenchannels should arise primarily from electrostatic interaction. The smallest difference among the five μ_α in Ar is $\Delta\mu_{2,5}=0.035$, and is thus much larger than $\Delta\mu_{so}\cong 0.0045$. Also, as we have shown above, the quadrupole coupling between s and d channels is very weak. Therefore the LS characterization for the five α channels is fulfilled. This condition is satisfied even better in Ne; as a matter of fact Starace[5] has assumed complete decoupling of s and d channels. On the other hand, this condition is much less fulfilled for heavier noble gases, e.g., Xe.[3] The procedure for fitting $U_{i\alpha}$ developed here can still be applied to situations like that of Xe.

The graphical procedures used in this paper, as well as in LX and FH, apply to spectra with only *two* ionization thresholds in the range of interest. However, the determinant form (2.5) of the $F=0$ equation applies equally to spectra with more than two thresholds, and should be amenable to numerical solution even when the simple graphical approach fails.

Note added in proof. After this work had been completed, we received revised values of the branching ratios of the rare gases (Ne, Ar, Kr, Xe), measured by collecting electrons at the "magic angle" of 54°44′ by Samson.[13] The new branching ratio for Ar is 1.87 ±0.06, instead of the value 1.98 which was utilized by us. This change of the experimental data changes the matrix V [Eq. (4.12)] and the dipole-matrix elements D_α into

$$V_{\bar{\alpha}\alpha} = \begin{vmatrix} 1.000 & 0.009 & 0.000 & 0.000 & 0.006 \\ -0.009 & 0.999 & -0.041 & 0.000 & -0.001 \\ 0.000 & 0.041 & 0.999 & 0.000 & 0.027 \\ 0.000 & 0.000 & 0.000 & 1.000 & -0.001 \\ -0.006 & 0.000 & -0.027 & 0.001 & 1.000 \end{vmatrix},$$

$$\begin{array}{llllll} \alpha = & 1 & 2 & 3 & 4 & 5 \\ D_\alpha = & -2.8\times10^{-2} & 2.1 & -1.2\times10^{-1} & 1.4\times10^{-3} & 1.4. \end{array}$$

The numerical values on the s-d mixing in the α channels given in Sec. VI are thus changed. However no significant changes result in (i) the $-\nu_{3/2}$ (mod 1) vs $\nu_{1/2}$(mod 1) plot of Fig. 1, up to the second decimal digit included, (ii) the df/dE vs $\nu_{1/2}$(mod 1) plot of Fig. 2, and (iii) the mixing coefficients, $\mathfrak{A}_\alpha$ and Z_i, for s and s' series; changes of the order of ±0.01 do occur in the d, $\bar{d}$, and d' series.

The over-all characterization of the discrete levels discussed in this paper remains thus unaffected. On the other hand, observable quantities which depend sensitively on interferences between different channels are affected by the greater departure of the α channels from the $\bar{\alpha}$ channels. For example, according to Eq. (2.13), the oscillator strengths f_n are proportional to $|\sum_\alpha \mathfrak{A}_\alpha^{(n)} D_\alpha|^2$; the increase of the dipole-matrix element D_3 to reach the order of 10^{-1} a.u. modifies appreciably the small oscillator strengths for the d series given in Table II (see Table III). Other oscillator strengths are not changed significantly.

TABLE III. The corrected oscillator strengths.

	Table II	Corrected
$3d$	1.6×10^{-3}	4.7×10^{-3}
$4d$	2.6×10^{-3}	4.9×10^{-3}
$5d$	4.3×10^{-3}	6.2×10^{-3}
$6d$	7.5×10^{-3}	4.1×10^{-3}

ACKNOWLEDGMENTS

The authors wish to express their deep gratitude to Professor U. Fano for suggesting this problem and for continuous guidance and support. We also wish to thank Dr. A. F. Starace for helpful discussions, and Dr. Dan Dill for the critical reading of the manuscript.

APPENDIX A: NORMALIZATION INTEGRAL

According to the same procedure as in Sec. VI of FH,[1] the normalization integral of the wave function (2.3) can be written as follows:

$$\int^\infty \Psi_n^2 d\tau = \lim_{E\to E_n}\lim_{r\to\infty}\int^r \psi\psi_n\, d\tau = \lim_{E\to E_n}\lim_{r\to\infty}(E-E_n)\int^r (\psi\psi_n/E-E_n)\,d\tau$$

$$= \frac{1}{\pi}\sum_i\left(\frac{d}{dE}\left[\sum_\alpha \mathfrak{U}_{i\alpha}\sin\pi(\nu_i+\mu_\alpha)\mathfrak{A}_\alpha\right]_{E=E_n}\right)\left[\sum_\alpha \mathfrak{U}_{i\alpha}\cos\pi(\nu_{i,n}+\mu_\alpha)\mathfrak{A}_\alpha^{(n)}\right]. \tag{A1}$$

For convenience, the normalization integral is decomposed into three contributions with derivatives $d\mathfrak{U}_{i\alpha}/dE$, $d[\sin\pi(\nu_i+\mu_\alpha)]/dE$, and $d\mathfrak{A}_\alpha/dE$, respectively. These three contributions are

$$N_1 = \frac{1}{\pi}\sum_i\left(\sum_\alpha \frac{d\mathfrak{U}_{i\alpha}}{dE}\sin\pi(\nu_{i,n}+\mu_\alpha)\mathfrak{A}_\alpha^{(n)}\right)\left[\sum_\beta \mathfrak{U}_{i\beta}\sin\pi(\nu_{i,n}+\mu_\beta)\mathfrak{A}_\beta^{(n)}\right]$$

$$= \frac{1}{\pi}\sum_i\left(\sum_\alpha\sum_\beta \frac{d\mathfrak{U}_{i\alpha}}{dE}\mathfrak{U}_{i\beta}\sin\pi(\nu_{i,n}+\mu_\alpha)\cos\pi(\nu_{i,n}+\mu_\beta)\mathfrak{A}_\alpha^{(n)}\mathfrak{A}_\beta^{(n)}\right.$$

$$\left. -\sum_\alpha\sum_\beta \frac{d\mathfrak{U}_{i\alpha}}{dE}\mathfrak{U}_{i\beta}\cos\pi(\nu_{i,n}+\mu_\alpha)\sin(\nu_{i,n}+\mu_\beta)\mathfrak{A}_\alpha^{(n)}\mathfrak{A}_\beta^{(n)}\right)$$

$$= \frac{1}{\pi}\sum_i\sum_\alpha\sum_\beta \frac{d\mathfrak{U}_{i\alpha}}{dE}\mathfrak{U}_{i\beta}\sin\pi(\mu_\alpha-\mu_\beta)\mathfrak{A}_\alpha^{(n)}\mathfrak{A}_\beta^{(n)}, \tag{A2}$$

since

$$\sum_{\alpha}\mathfrak{U}_{i\alpha}\sin\pi(\nu_{i,n}+\mu_{\alpha})\mathfrak{A}_{\alpha}^{(n)}=0;$$

$$\begin{aligned}
N_2&=\frac{1}{\pi}\sum_i\left[\sum_{\alpha}\mathfrak{U}_{i\alpha}\left(\frac{d}{dE}\sin\pi(\nu_{i,n}+\mu_{\alpha})\right)_{E=E_n}\mathfrak{A}_{\alpha}^{(n)}\right]\left[\sum_{\beta}\mathfrak{U}_{i\beta}\cos\pi(\nu_{i,n}+\mu_{\beta})\mathfrak{A}_{\beta}^{(n)}\right]\\
&=\sum_i\nu_{i,n}^3\left[\sum_{\alpha}\mathfrak{U}_{i\alpha}\cos\pi(\nu_{i,n}+\mu_{\alpha})\mathfrak{A}_{\alpha}^{(n)}\right]^2_{E=E_n}+\sum_i\sum_{\alpha}\sum_{\beta}\left(\frac{d\mu_{\alpha}}{dE}\right)_n\mathfrak{U}_{i\alpha}\mathfrak{U}_{i\beta}\cos\pi(\nu_{i,n}+\mu_{\alpha})\cos\pi(\nu_{i,n}+\mu_{\beta})\mathfrak{A}_{\alpha}^{(n)}\mathfrak{A}_{\beta}^{(n)}\\
&=\sum_i\nu_i^3\left[\sum_{\alpha}\mathfrak{U}_{i\alpha}\cos\pi(\nu_{i,n}+\mu_{\alpha})\mathfrak{A}_{\alpha}^{(n)}\right]^2+\sum_i\sum_{\alpha}\sum_{\beta}\frac{d\mu_{\alpha}}{dE}\mathfrak{U}_{i\alpha}\mathfrak{U}_{i\beta}\cos\pi(\mu_{\alpha}-\mu_{\beta})\mathfrak{A}_{\alpha}^{(n)}\mathfrak{A}_{\beta}^{(n)}\\
&=\sum_i\nu_i^3\left[\sum_{\alpha}\mathfrak{U}_{i\alpha}\cos\pi(\nu_{i,n}+\mu_{\alpha})\mathfrak{A}_{\alpha}^{(n)}\right]^2+\sum_{\alpha}\frac{d\mu_{\alpha}}{dE}(\mathfrak{A}_{\alpha}^{(n)})^2,
\end{aligned}\tag{A3}$$

since

$$\sum_i\mathfrak{U}_{i\alpha}\mathfrak{U}_{i\beta}=\delta_{\alpha\beta};$$

$$\begin{aligned}
N_3&=\frac{1}{\pi}\sum_i\left[\sum_{\alpha}\mathfrak{U}_{i\alpha}\sin\pi(\nu_{i,n}+\mu_{\alpha})\left(\frac{d\mathfrak{A}_{\alpha}}{dE}\right)_{E=E_n}\right]\left[\sum_{\beta}\mathfrak{U}_{i\beta}\cos\pi(\nu_{i,n}+\mu_{\beta})\mathfrak{A}_{\beta}^{(n)}\right]\\
&=\frac{1}{\pi}\sum_i\sum_{\alpha}\sum_{\beta}\mathfrak{U}_{i\alpha}\mathfrak{U}_{i\beta}\sin\pi(\mu_{\alpha}-\mu_{\beta})\left(\frac{d\mathfrak{A}_{\alpha}}{dE}\right)_{E=E_n}\mathfrak{A}_{\beta}^{(n)}\\
&=\frac{1}{\pi}\sum_{\alpha}\sum_{\beta}\delta_{\alpha\beta}\sin\pi(\mu_{\alpha}-\mu_{\beta})\left(\frac{d\mathfrak{A}_{\alpha}}{dE}\right)_{E=E_n}\mathfrak{A}_{\beta}^{(n)}=0.
\end{aligned}\tag{A4}$$

Therefore,

$$\begin{aligned}
N_n^2&=\int^{\infty}|\Psi_n|^2d\tau\\
&=N_1+N_2+N_3\\
&=\sum_i\nu_{i,n}^3\left[\sum_{\alpha}\mathfrak{U}_{i\alpha}\cos\pi(\nu_{i,n}+\mu_{\alpha})\mathfrak{A}_{\alpha}^{(n)}\right]^2\\
&\quad+\sum_{\alpha}\frac{d\mu_{\alpha}}{dE}\mathfrak{A}_{\alpha}^{(n)2}+\frac{1}{\pi}\sum_i\sum_{\alpha}\sum_{\beta}\frac{d\mathfrak{U}_{i\alpha}}{dE}\mathfrak{U}_{i\beta}\\
&\quad\times\sin\pi(\mu_{\alpha}-\mu_{\beta})\mathfrak{A}_{\alpha}^{(n)}\mathfrak{A}_{\beta}^{(n)}.
\end{aligned}\tag{A5}$$

APPENDIX B: REPRESENTATION OF $n\times n$ ORTHOGONAL MATRIX $\mathfrak{U}$

Geometrically, an $n\times n$ orthogonal matrix describes a rotation in n-dimensional space. Since a rotation in n-dimensional space can be decomposed into a combination of the $\binom{n}{c}$ elementary rotations, $\mathfrak{U}$ can be represented as a vector with $\binom{n}{2}$ components θ_α with the following convention for the subscripts:

$$\begin{array}{lccccccc}\alpha= & 1 & 2 & \cdots & n-1 & n & \cdots & \tfrac{1}{2}n(n-1)\\ (i,j)= & (1,2) & (1,3) & \cdots & (1,n) & (2,3) & \cdots & (n-1,n).\end{array}$$

The component θ_α of the vector represents a finite angle of rotation in the (i,j) plane. The explicit expression of $n\times n$ matrix $\mathfrak{U}$ is

$$\mathfrak{U}=\prod_{\alpha=1}^{\frac{1}{2}n(n-1)}R^{\alpha}(\theta_\alpha),\tag{B1}$$

where

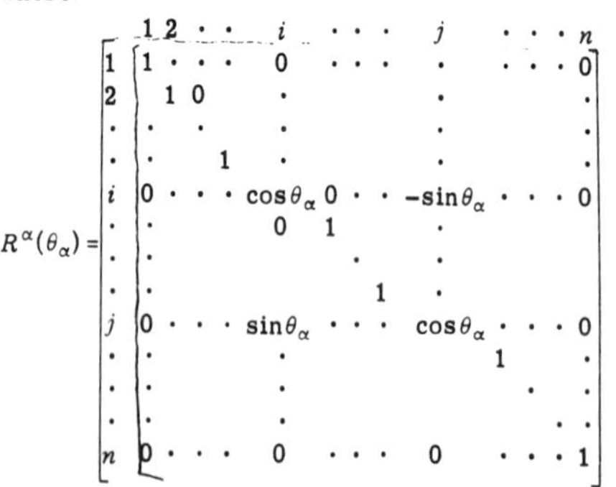

*Work supported by U.S. Atomic Energy Commission under Contract No. C00-1674-77.

†K. T. Lu participated only in the later stages of this work.

[1]U. Fano, Phys. Rev. A **2**, 353 (1970).

[2]M. J. Seaton, Proc. Phys. Soc. Lond. **88**, 801 (1966).

[3]K. T. Lu, Phys. Rev. A **4**, 579 (1971).

[4]K. Yoshino, J. Opt. Soc. Am. **60**, 1220 (1970).

[5]A. F. Starace, J. Phys. B **6**, 76 (1973).

[6]S. Natali, C. E. Kuyatt, and S. R. Mielczarek (unpublished). The line intensities are measured by electron spectroscopy. (We

express our appreciation to Dr. Kuyatt for allowing us use of the data before publication.)
[7]K. L. Andrew and K. W. Meissner, J. Opt. Soc. Am. **49**, 146 (1959).
[8]J. L. Dehmer and U. Fano, Phys. Rev. A **2**, 304 (1970).
[9]M. J. Seaton, Proc. Phys. Soc. Lond. **88**, 815 (1966).
[10]D. L. Moores, Proc. Phys. Soc. Lond. **88**, 843 (1966).
[11]F. Herman and S. Skillman, *Atomic Structure Calculations* (Prentice-Hall, Englewood Cliffs, N.J., 1963).
[12]R. D. Hudson and L. J. Kieffer, At. Data **2**, 205 (1971).
[13]J. A. R. Samson and R. B. Cairns, Phys. Rev. **173**, 80 (1968). We thank Professor Samson for communicating to us his new values of the branching ratio of the rare gases, measured at the "magic angle." The new ratios are Ne = 1.98, Ar = 1.87 ± 0.06, Kr = 1.64, and Xe = 1.44.
[14]U. Fano, Bull. Am. Phys. Soc. **13**, 37 (1968).
[15]U. Fano and G. Racah, *Irreducible Tensorial Sets* (Academic, New York, 1959), p. 68.
[16]K. T. Lu and U. Fano, Phys. Rev. A **2**, 81 (1970).
[17]C. E. Moore, *Atomic Energy Levels*, Natl. Bur. Std. Circ. No. 467 (U.S. GPO, Washington, D.C., 1958), Vol. III.
[18]B. Edlén, *Encyclopedia of Physics* (Springer-Verlag, Berlin, 1964), Vol. 27, p. 80.
[19]E. U. Condon and G. H. Shortley, *Theory of Atomic Spectra* (Cambridge U.P., New York, 1964), p. 143.
[20]A. R. P. Rau and U. Fano, Phys. Rev. **167**, 7 (1968).
[21]U. Fano, Phys. Rev. **124**, 1866 (1961).
[22]U. Fano, Comments At. Mol. Phys. **2** (1), 30 (1970).

1973 *Phys. Rev.* A **7** 1976–87
Reprinted with permission from the American Physical Society

Resonances in Photoelectron Angular Distributions*

Dan Dill
Department of Physics, The University of Chicago, Chicago, Illinois 60637
(Received 29 January 1973)

Photoelectron angular distributions should show pronounced variations with energy across autoionization resonances. This prediction applies quite generally to both atomic and molecular autoionization. Examples illustrate both the magnitude of the spectral variation and the inability of the Cooper-Zare model to account for the phenomenon. Calculations are reported for autoionization in xenon between the fine-structure levels $5p^5\ {}^2P^o_{3/2}$ and $5p^5\ {}^2P^o_{1/2}$ of the ion ground-state doublet. An analysis is given of the recent measurements by Niehaus and Ruf on autoionizing levels of the mercury Rydberg series $5d^96s^2(^2D)np$ and $5d^96s^2(^2D)n'f$ below the $Hg^+\ 5d^96s^2\ {}^2D_{5/2}$ threshold.

I. INTRODUCTION

The determination of the spectral variation of photoelectron angular distributions through autoionization resonances is a new and essentially untapped resource for photoelectron spectroscopy. Here this class of spectroscopic measurements is theoretically analyzed. The analysis predicts quite generally not only sharp spectral variations of the angular distributions across resonance features, but more importantly, angular distributions that should depart markedly from those predicted by direct (nonresonant) ionization models, such as the Copper–Zare model.[1] Deviations from direct ionization predictions arise owing to the enhancement by the autoionization process of the effects of just those forces that are often sufficiently weak as to go undetected in nonresonant photoionization. Accordingly, these resonances in photoelectron angular distributions are a sensitive new probe of photoejection dynamics.

This study rests on the angular-momentum-transfer formulation of angular correlations, given recently by Dill and Fano.[2] It also draws on extensive experience in analyzing the dynamical origin and significance of the various angular momentum transfers allowed in any given ionization process. This paper reports the most important implications and results for autoionization of the general dynamical analysis, whose full description is deferred to a separate report.[3]

Consider the schematic ionization process

$$A(J_0\pi_0)+\gamma(j_\gamma=1,\ \pi_\gamma=-1)$$
$$\rightarrow A^+(J_c\pi_c)+e[ls\,j,\ \pi_e=(-1)^l]\ , \qquad (1)$$

in which photoelectrons are ejected by electric dipole interaction from a generic unpolarized (atomic or molecular) target A. Dill and Fano (DF)[2] obtain a general expression representing the photoelectron angular distribution of process (1) in terms of separate components characterized by alternative magnitudes of

$$\vec{j}_t=\vec{J}_c+\vec{s}-\vec{J}_0=\vec{j}_\gamma-\vec{l}\ , \qquad (2)$$

the angular momentum transferred in the ionization. The allowed values of $\vec{j}_t$ are those consistent with the balance of total angular momentum $\vec{J}$ and parity π,

$$\vec{J}=\vec{J}_0+\vec{j}_\gamma=\vec{J}_c+\vec{s}+\vec{l}\ , \qquad (3)$$

$$\pi=\pi_0\pi_\gamma=\pi_c\pi_e$$
$$=-\pi_0=(-1)^l\pi_c\ . \qquad (4)$$

For any given ionization process, a range of j_t will be allowed, with ionization dynamics determining the relative contribution of different j_t components. However, each angular-momentum-transfer component has a characteristic angular distribution with an analytical structure independent of the dynamical details of the ionization. The predicted sharp variation of the angular distribution across resonance features follows from (a) the general structure of the distribution for each j_t, and (b) the spectral variation of the weights of distributions with different j_t. The deviations of the resulting angular distributions from predictions of direct ionization models, such as the Cooper–Zare formula,[1] follow from autoionization enhancing angular momentum transfers that can be unimportant in nonresonant ionization.[2,3]

Resonances in photoelectron angular distributions have already been calculated for the example of rotational autoionization in H_2 and the results show clearly the effects predicted here.[4] Additional examples will be given in this report. The specific examples have been chosen for their amenability to experimental verification: The resonances are wide and the spectral variations of the angular distributions are quite pronounced. Further, for both examples energy analysis of the photoelectrons is not required since only one group of photoelectrons is produced. Calculations are given for the angular distribution of xenon photoelectrons ejected into the autoionization region between the ionization thresholds of the $5p^5\,{}^2P^\circ_{3/2}$ and $5p^5\,{}^2P^\circ_{1/2}$ levels of the Xe^+ ground-state doublet. The results illustrate both a pronounced resonance in the angular distribution as well as substantial deviations from the predictions of the Cooper–Zare model.[1] The photoionization of mercury is analyzed in the resonant region between the $5d^{10}6s\,{}^2S_{1/2}$ level of ion ground state and the $5d^9 6s^2\,{}^2D_{5/2}$ level of the first excited state of Hg^+. This example is particularly striking: Since the ionization proceeds by an $s\rightarrow p$ transition, one might expect the photoelectrons to carry away the full anisotropy of the photon-target interaction, i.e., throughout the spectral region one might expect a fixed $\cos^2\theta$ angular distribution, peaking along the electric vector of the light. Actually, owing to the enhancement of spin-orbit effects in the resonant region, the analysis here predicts strong variations with energy which can extend over the full range from $\cos^2\theta$ to $\sin^2\theta$. Evidence of this sharp energy dependence of the Hg photoelectron angular distributions is seen in the recent measurements of Niehaus and Ruf.[5]

It is to be emphasized that a variety of dynamical interactions can result in multiple angular momentum transfers. In particular, *the forces are not restricted to just those of magnetic origin.*[2,3] Thus, in rotational autoionization of H_2, it is the torque exerted by the anisotropic *electric* field of the molecular ion that leads to parity-unfavored angular molecular transfer. For both the Xe and Hg examples, as we shall see, the anisotropy is provided by a combination of the *electrostatic* (exchange) interaction which separates different *LS* terms of the electron-ion complex and the spin-orbit interaction which results in a breakdown of *LS* coupling. Effects of spin-orbit interaction alone on photoelectron angular distributions have been theoretically analyzed recently by Walker and Waber [T. E. H. Walker and J. T. Waber, Phys. Rev. Letters 30, 307 (1973)]. As these authors point out, their results are an example of the angular-momentum-transfer formulation and, in fact, can be obtained directly from the general expression for the angular distribution given below [Eq. (8)]. Nonetheless, it is important to realize that spin-orbit interactions are but one mechanism for the production of multiple angular momentum transfers.

II. ANGULAR-MOMENTUM-TRANSFER FORMULATION OF ANGULAR CORRELATIONS

The differential cross section for the ionization (1) is given by

$$\frac{d\sigma}{d\Omega}=\frac{\sigma}{4\pi}\,[1+\beta P_2(\cos\theta)], \qquad (5)$$

where σ is the integrated cross section and the asymmetry parameter β pertains to light linearly polarized along the z axis. The resolution of the differential cross section into components with different angular momentum transfers proceeds as follows[2]: The allowed j_t values are determined from Eqs. (2)–(4). Then one determines an asymmetry parameter β for each value of j_t according to whether the parity change in the target is $\pm(-1)^{j_t}$: Values of j_t for which $\pi_0\pi_c = (-1)^{j_t}$ are said to be parity favored and have an asymmetry parameter

$$\beta(j_t)_{\text{fav}} = \frac{(j_t+2)|\bar{S}_+(j_t)|^2 + (j_t-1)|\bar{S}_-(j_t)|^2 - 3[j_t(j_t+1)]^{1/2}[\bar{S}_+(j_t)\bar{S}_-(j_t)^\dagger + \bar{S}_+(j_t)^\dagger S_-(j_t)]}{(2j_t+1)[|\bar{S}_+(j_t)|^2 + |\bar{S}_-(j_t)|^2]} , \quad (6)$$

where $S_\pm(j_t)$ denotes the photoionization amplitude[6] for a given j_t and for $l = j_t \pm 1$; values of j_t for which $\pi_0\pi_c = -(-1)^{j_t}$ are called parity unfavored and have a $\sin^2\theta$ angular distribution, i.e.,

$$\beta(j_t)_{\text{unf}} = -1 , \quad (7)$$

independently of dynamics. The observed asymmetry parameter is an average over the asymmetry parameters $\beta(j_t)$, weighted by the corresponding partial integrated cross sections $\sigma(j_t)$,

$$\beta = \left(\sum_{j_t}^{\text{fav}}{}' \sigma(j_t)\beta(j_t) - \sum_{j_t}^{\text{unf}}{}' \sigma(j_t) \right) \Big/ \sum_{j_t} \sigma(j_t) . \quad (8)$$

In (8) the primed sums are over only the parity-favored and parity-unfavored j_t values, respectively. The integrated cross sections have the structure

$$\sigma(j_t)_{\text{fav}} = \pi\lambda\!\!\!^{-2}[(2j_t+1)/(2J_0+1)] \times [|\bar{S}_+(j_t)|^2 + |\bar{S}_-(j_t)|^2] , \quad (9)$$

$$\sigma(j_t)_{\text{unf}} = \pi\lambda\!\!\!^{-2}[(2j_t+1)/(2J_0+1)]|\bar{S}_0(j_t)|^2 , \quad (10)$$

where $\bar{S}_0(j_t)$ is the photoionization amplitude[6] for the value of j_t equal to the escape orbital momentum l; $\lambda\!\!\!^-$ is the photon wavelength (divided by 2π).

The expression (8) for β is the essential tool of our analysis. It is a general result, following only from the specification of the total angular momentum and parity of each element of the ionization reaction.[2] No assumptions are required about angular-momentum coupling schemes or indeed in regard to any details of internal structure of the various reactants (e.g., atom or molecule, charged or neutral). Accordingly, (8) embodies a very general conceptual framework with which to analyze effects of autoionization on photoelectron angular distributions.

That photoelectron angular distributions should vary markedly with energy across a resonance in the integrated cross section σ follows from the analytical structure of this asymmetry parameter β: σ and β are merely two sides of the same coin, namely, alternative manifestations of the resonant behavior of ionization amplitudes that is characteristic of autoionization. That is, a resonance in σ implies and is implied by a resonance in β. Note however that, whereas the integrated cross section depends only on the squared moduli of the ionization amplitudes, the asymmetry parameter depends as well on the moduli and phases of these amplitudes through the interference terms in the $\beta(j_t)_{\text{fav}}$. Therefore, the resonance in β can be an even more pronounced spectral feature than the resonance in σ.

That autoionization may enhance the role of otherwise weak terms of (8) can be seen as follows[3]: The initial stage of photoabsorption imparts $j_\gamma = 1$ units of orbital momentum to the photoelectron, which has initially orbital momentum $\vec{l}_0$, yielding an orbital momentum

$$\vec{l}' = \vec{j}_\gamma + \vec{l}_0 . \quad (11)$$

At this point the angular momentum transfer is

$$\vec{j}_t' = -\vec{l}_0 = \vec{j}_\gamma - \vec{l}' , \quad (12)$$

with the single value $j_t' = l_0$. Additional angular momentum transfers are in general allowed, arising from anisotropic interactions of the photoelectron with the rest of the target (including its own spin) during its subsequent escape. Such interactions—*both electrostatic* (*orbit-orbit*) *and magnetic* (*spin-orbit*)—can change the orientation and even the magnitude of the orbital momentum from $\vec{l}'$ to $\vec{l}$, thereby resulting in additional angular momentum transfers $\vec{j}_t - \vec{j}_t'$. Of course, the electron must interact with a nonzero angular momentum, for only in that case will it experience a torque and thereby be able to exchange angular momentum. [Thus, e.g., interaction with a $^1S_0$ core, even if polarized, cannot be lead to exchange of angular momentum.] Now autoionization consists of the decay of a bound state in which the electron experiences a protracted and generally anisotropic interaction with the core. This extended interaction should enhance the angular momentum transfers in addition to $j_t' = l_0$. In contrast, the Cooper–Zare model,[1] for example, implies complete absence of any anisotropic electron-core interaction and leads to a formula obtained from (8) by setting $j_t = l_0$ and

$$\bar{S}(l_0) = i^{-l} e^{i\delta_l}(l||C^{[1]}||l_0)R(l, l_0) , \quad (13)$$

where R is a reduced radial dipole matrix element and the other symbols have their usual meaning.[2,3]

Therefore, the measurement of resonances in photoelectron angular distributions should yield results that can deviate considerably from those predicted by the Cooper–Zare model.

III. EXAMPLES

We have then pointed out two key elements of the effect of autoionization on photoelectron angular distributions: One is the sharp spectral variation of β and the other is the enhanced importance of alternative angular momentum transfers. To illustrate these resonant features of photoelectron angular distributions we shall consider examples from the photoionization of xenon and mercury.

A. Autoionization in Xenon

Ionization of xenon leaves the ion in a doublet ground state,

$$\mathrm{Xe}(5p^6\ {}^1S_0) + \gamma(j_\gamma = 1,\ \pi_\gamma = -1)$$
$$\to \mathrm{Xe}^+(5p^5\ {}^2P^\circ_{3/2,1/2}) + e(l = 0,\ 2) \quad . \tag{14}$$

The ionization potential for production of the lower level ${}^2P^\circ_{3/2}$ of the ion doublet is $I_{3/2} = 12.127$ eV, and the doublet splitting is $I_{1/2} - I_{3/2} = 1.43$ eV. The spectral region between these fine-structure thresholds was found by Beutler[7] long ago to consist of a Rydberg series of autoionization resonances, with widths comparable to their level separations. The autoionization consists of the escape of an electron, initially bound to the ${}^2P^\circ_{3/2}$ core. It is represented schematically by

$$\left.\begin{matrix} 5p^5({}^2P^\circ_{1/2})nd \\ 5p^5({}^2P^\circ_{1/2})n's \end{matrix}\right\} \xrightarrow[J=1,\ \pi=-1]{\text{autoionization}} \left\{\begin{matrix} 5p^5({}^2P^\circ_{3/2})\epsilon d \\ 5p^5({}^2P^\circ_{3/2})\epsilon s \end{matrix}\right. .$$

Only the total angular momentum ($J = 1$) and parity ($\pi = -1$) are necessarily conserved in the autoionization. In particular, the orbital momentum $\vec{l}$ of the photoelectron can be changed by the autoionization process.

The Rydberg character of the series of autoionizing levels emerges when this spectral region is mapped onto the scale of the effective quantum number $\nu_{1/2}$ with respect to the upper threshold $I_{1/2}$ of the ion doublet. This quantum number is defined by

$$\hbar\omega = I_{1/2} - 1/2\nu^2_{1/2}\ , \tag{15}$$

where energies are in atomic units. The mapping reveals that each pair of Rydberg levels, $5p^5({}^2P^\circ_{1/2})nd$ and $5p^5({}^2P^\circ_{1/2})n's$, fits within one unit of $\nu_{1/2}$, and that the Beutler–Fano resonance profiles repeat unchanged with period one in $\nu_{1/2}$. On the other hand, owing to the nonlinear relationship between the energy $\hbar\omega$ and the effective quantum number $\nu_{1/2}$, the profile widths and separations decrease rapidly on the energy scale from about 0.44 eV ($3.5 \leq \nu_{1/2} \leq 4.5$) just above the ${}^2P^\circ_{3/2}$ threshold to zero ($\nu_{1/2} = \infty$) at the ${}^2P^\circ_{1/2}$ threshold.

The linearity and periodicity of the Rydberg series on the $\nu_{1/2}$ scale is a central element of the theoretical understanding of these resonances, i.e., the whole autoionization region can be understood from the properties of the spectrum over a single unit range of $\nu_{1/2}$. Here, then, we want to determine the angular distribution of photoelectrons ejected into this spectral region and we will do this for the single unit range $3.5 \leq \nu_{1/2} \leq 4.5$.

The first step is to analyze the ionization process (14) according to the scheme outlined in Sec. II. The angular momentum and parity quantum numbers are

$$\begin{matrix} J_0 = 0\ , & j_\gamma = 1\ , & J_c = \frac{3}{2}\ , & l = 0,\ 2, \\ \pi_0 = +1\ , & \pi_\gamma = -1\ , & \pi_c = -1\ , & \pi_l = +1\ . \end{matrix} \tag{16}$$

Note that $J_c = \frac{3}{2}$ since we consider only photon energies below $I_{1/2}$; $\pi_l = +1$ is determined by parity balance with $\pi = -1$; $l = 0,\ 2$ is then fixed by angular momentum balance with $J = 1$. The angular momentum transfers allowed are then determined by Eq. (2) to be

$$\begin{matrix} j_t = 1\ \text{(parity favored)} & \text{for } l = 0,\ 2, \\ j_t = 2\ \text{(parity unfavored)} & \text{for } l = 2\ . \end{matrix} \tag{17}$$

Note that odd j_t values are parity favored since $\pi_0\pi_c = -1$, and that s waves ($l = 0$) contribute only to the parity-favored component. For this case, the asymmetry parameter (8) is then

$$\beta = [\sigma(1)\beta(1) - \sigma(2)]/[\sigma(1) + \sigma(2)]\ . \tag{18}$$

Now from (6) we have

$$\beta(1) = \frac{3|\bar{S}_d(1)|^2 - 3(2)^{1/2}[\bar{S}_d(1)\bar{S}_s(1)^\dagger + \bar{S}_d(1)^\dagger\bar{S}_s(1)]}{3[|\bar{S}_d(1)|^2 + |\bar{S}_s(1)|^2]}\ , \tag{19}$$

and (9) and (10) give

$$\sigma(1) = 3\pi\lambda\!\!\!^{-2}[|\bar{S}_d(1)|^2 + |\bar{S}_s(1)|^2]\ , \tag{20}$$

$$\sigma(2) = 5\pi\lambda\!\!\!^{-2}|\bar{S}_d(2)|^2\ , \tag{21}$$

where the subscripts +, −, 0, on $\bar{S}(j_t)$ are replaced with s or d for $l = 0$ or $l = 2$. Therefore the asymmetry parameter is given by

$$\beta = \frac{|\bar{S}_d(1)|^2 - \frac{5}{3}|\bar{S}_d(2)|^2 - \sqrt{2}[\bar{S}_d(1)\bar{S}_s(1)^\dagger + \bar{S}_d(1)^\dagger\bar{S}_s(1)]}{|\bar{S}_d(1)|^2 + \frac{5}{3}|\bar{S}_d(2)|^2 + |\bar{S}_s(1)|^2}\ . \tag{22}$$

The parity-unfavored contribution to the asymmetry parameter (22) is given by the photoelectric intensity $|S_d(2)|^2$ for ejection of d waves into the continuum with the transfer of two units of angular momentum. The physical origin here of $j_t = 2$ is the spin-orbit final-state interaction which causes a breakdown of LS coupling as the electron escapes from the ion core. This may be seen by expressing Eq. (2) for $\vec{j}_t$ in terms of the approximately good LS quantum numbers of the ion core, i.e.,

$$\vec{j}_t = \vec{L}_c + \vec{S}_c + \vec{s}, \tag{23}$$

using the fact that $J_0 = 0$. Now the electrons are originally coupled into a singlet; i.e., before photoabsorption the photoelectron's spin was *antiparallel* to the net spin of the rest of the atom. In the absence of spin-orbit interaction in the final state the spins would remain antiparallel and therefore yield $\vec{S}_c + \vec{s} = 0$ in (23). Thereby j_t would be limited to the single value $j_t = L_c = 1$. Therefore, the contribution of the $j_t = 2$ component to (22) is an index of the effects of spin-orbit interaction on the photoionization process.

To determine the angular distribution quantitatively, the transition amplitudes $\bar{S}_l(j_t)$ must be evaluated. This is done in two steps. First the $\bar{S}_l(j_t)$ are expressed in terms of the more familiar reduced dipole transition amplitudes $P_l^{[1]}(J)$ characterized by the total angular momentum J. Then, the matrix elements $P_l^{[1]}(J)$ are determined semiempirically using the multichannel quantum-defect theory (MQDT).[8] The $P_l^{[1]}(J)$ are evaluated in terms of the interaction parameters of the MQDT, which have been determined[8(d)] by fitting theory to spectroscopic measurements.

Reduced Dipole Matrix Elements $P_l^{[1]}(J)$

The formulation of DF, summarized in Sec. II, expresses angular distribution in terms of rotationally invariant matrix elements of a scattering operator S, which connects the initial state (target + photon) to the final state (residual ion + electron) and which is characterized by the angular momentum transferred between unobserved elements of the "scattering process." These matrix elements are written

$$\bar{S}_l(j_t) \equiv ((J_c s)J_{cs} l|\bar{S}(j_t)|J_0 j_\gamma = 1). \tag{24}$$

The formulation in terms of S matrix elements derives from the original treatment[9] which is designed for a class of reactions more general than photoprocesses. The formulation in terms of j_t exploits the incoherence of the angular momentum transferred between unobserved reactants to represent the differential cross section by separate characteristic contributions for the various j_t. On the other hand, photoionization *calculations* are conveniently done in terms of the more familiar reduced dipole matrix elements, rather than S matrix elements, characterized by the total angular momentum J of the process, rather than the angular momentum transfer j_t. Two steps are required to establish the connection between the amplitudes $\bar{S}_l(j_t)$ and the reduced dipole transition amplitudes. First, the connection is given between the scattering amplitudes $\bar{S}_l(j_t)$ and S matrix elements[9]

$$S_l(J) \equiv ((J_c s)J_{cs} l|S(J)|J_0 j_\gamma = 1), \tag{25}$$

characterized by the total angular momentum J of the whole system. Then the scattering amplitudes $S_l(J)$ are expressed in terms of the usual reduced dipole matrix elements[10]

$$P_l^{[1]} \equiv ((J_c s)J_{cs} l, J- ||P^{[1]}||J_0), \tag{26}$$

normalized with the incoming-wave boundary conditions (denoted by the minus sign) that are appropriate to photoionization.

The amplitudes $\bar{S}_l(j_t)$ and $S_l(J)$ are shown in Ref. 9 to be related by the recoupling expansion

$$((J_c s)J_{cs} l|\bar{S}(j_t)|J_0 j_\gamma = 1) = \sum_J (-1)^{J_0 - J - 1}(2J+1) \times \begin{Bmatrix} l & J_{cs} & J \\ J_0 & 1 & j_t \end{Bmatrix} ((J_c s)J_{cs} l|S(J)|J_0 j_\gamma = 1). \tag{27}$$

The connection between the amplitudes $S_l(J)$ and $P_l^{[1]}(J)$ is found by comparing the alternative expressions[11] for the photoionization cross section given in terms of these amplitudes by

$$\sigma = 4\pi^2 \alpha \hbar\omega (2J_0+1)^{-1} \times \sum_{M_c \mu m M_0} \sum_l |(J_c M_c, s\mu, lm, -|P_{m_\gamma}^{[1]}|J_0 M_0)|^2, \tag{28a}$$

$$\sigma = 3\pi \lambda\!\!\!^{-2} (2J_0+1)^{-1} \times \sum_{M_c \mu m M_0} \sum_l |(J_c M_c, s\mu, lm|S|J_0 M_0, j_\gamma m_\gamma)|^2. \tag{28b}$$

In (28) the summations over m quantum numbers and the factor $(2J_0+1)^{-1}$ arise from averaging over the initial magnetic substates and summing over the final magnetic substates and spin polarizations. Using the expansions[12]

$$(J_c M_c, s\mu, lm, -|P_{m_\gamma}^{[1]}|J_0 M_0) = \sum_{J_{cs} J} (2J+1)^{-1/2} (J_c M_c, s\mu, lm|(J_c s)J_{cs} l, JM) \times ((J_c s)J_{cs} l, J- ||P^{[1]}||J_0)(JM|J_0 M_0, j_\gamma m_\gamma), \tag{29a}$$

$$(J_c M_c, s\mu, lm|S(J)|J_0 M_0, j_\gamma m_\gamma) = \sum_{J_{cs} J} (J_c M_c, s\mu, lm|(J_c s)J_{cs} l, JM) \times ((J_c s)J_{cs} l|S(J)|J_0 j_\gamma)(JM|J_0 M_0, j_\gamma m_\gamma), \tag{29b}$$

and the symmetry and orthonormality of the Wigner coefficients,[13] the summations over m quantum numbers can be carried out to give

$$\sigma = \tfrac{4}{3}\pi^2 \alpha \hbar\omega (2J_0+1)^{-1} \times \sum_{J_{cs} J l} |((J_c s)J_{cs} l, J- ||P^{[1]}||J_0)|^2, \tag{30a}$$

$$\sigma = \pi \lambda\!\!\!^{-2} (2J_0+1)^{-1}$$

$$\times \sum_{J_{cs}Jl} (2J+1) |((J_c s) J_{cs} l | S(J) | J_0 j_\gamma = 1)|^2 . \quad (30b)$$

Comparison of the two expressions (30) gives the desired connection,

$$((J_c s) J_{cs} l | S(J) | J_0 j_\gamma = 1)$$
$$= n(\lambda\!\!\!^-)(2J+1)^{-1/2}((J_c s) J_{cs} l, J - ||P^{[1]}|| J_c), \quad (31)$$

$$n(\lambda\!\!\!^-)^2 = 4\pi^2 \alpha \hbar\omega / 3\pi\lambda^2 . \quad (32)$$

For the xenon example we have $J_0 = 0$, and therefore J is restricted to the single value $J = j_\gamma = 1$ and $J_{cs} = j_t$. Accordingly, substitution of (31) into (27) gives

$$((J_c s) J_{cs} = j_t l | \bar{S}(j_t) | J_0 = 0, j_\gamma = 1)$$
$$= n(\lambda\!\!\!^-)(2j_t + 1)^{-1/2}(-1)^{1+j_t}$$
$$\times ((J_c s) J_{cs} = j_t l, J = 1 - ||P^{[1]}|| J_0 = 0), \quad (33)$$

for the connection between the amplitudes $\bar{S}_l(j_t)$ and $P_l^{[1]}(J=1)$.

Before proceeding to the evaluation of the amplitudes $P_l^{[1]}(J)$, it is important to note one new aspect of the expansions (29), not included in the corresponding expansion equation (5) or Ref. 9. This is the explicit account taken here of the photoelectron spin. This introduces an additional summation over the total *unobserved* final-state angular momentum J_{cs}.[14] With this additional feature, the whole analysis of Ref. 9 carries through as before with the following additional result. The final-state angular momentum J_{cs} is incoherent in both the differential and integrated cross section, just as j_t is incoherent, since the orientation of the core (M_c) and the spin polarization of the photoelectron (μ) are unobserved, and $d\sigma/d\Omega$ must be summed over all possible values of J_{cs} [compare (30)]. This means that the expressions given in Sec. II for σ and $\sigma\beta$ must in general be summed over J_{cs}. Often, however, as for the xenon example and, as we shall see, for the mercury example as well, only a single value of J_{cs} occurs.

Evaluation of $P_l^{[1]}(J)$ by MQDT

A fundamental concept of the MQDT is the resolution of the ionization process into an initial photon absorption and the photoelectron's subsequent interaction with, and escape from, the target. These two aspects of the process contribute different characteristic elements to the ionization dynamics. Photon absorption occurs inside the atom and is characterized by real transition amplitudes

$$D_\alpha = (\alpha J = 1 || P^{[1]} || J_0 = 0) \quad (34)$$

connecting the atomic ground state to close-coupling eigenstates $(\alpha|$ which depend on the electron-core interaction at short range. [These eigenstates $(\alpha|$, with *standing-wave* normalization, are discussed in Ref. 8(c).] Electron escape, on the other hand, probes the effects of long-range interaction, in particular, the temporary retention of the electron by the $^2P^\circ_{1/2}$ ion to form autoionizing states. The electron escape is characterized by the eigenchannels $(\rho|$ of elastic scattering of the electron by the xenon ion in the $^2P^\circ_{3/2}$ level in the range of electron kinetic energies $\epsilon < I_{1/2} - I_{3/2} = 1.43$ eV, with the whole system having angular momentum $J = 1$ and odd parity. Retention of the electron in an autoionizing state through excitation of the ion from the $^2P^\circ_{3/2}$ level to the $^2P^\circ_{1/2}$ level, causes the elastic scattering phase shifts $\pi\tau_\rho$ to experience a pronounced spectral variation that is the earmark of multichannel resonance. This energy dependence is described below and discussed in detail in Ref. 8(c). It constitutes a central aspect of the MQDT treatment [see especially Fig. 2 of Ref. 8(c)]. In essence, the sum of the phase shifts $\pi\tau_\rho$ for all channels $(\rho|$ undergoes a net increase of 2π for every unit range of $\nu_{1/2}$, i.e., every time the energy passes through the combined Beutler–Fano profile of the two $(ns, n'd) \rightarrow (\epsilon s, \epsilon d)$ autoionization resonances.

The dipole matrix elements $P_l^{[1]}$ are expressed in terms of the amplitudes D_α in two steps. First, the final states $((J_c s) J_{cs} l, J|$, with *standing-wave* normalization (i.e., without the "minus" quantum number) are expanded into the elastic-scattering channels $(\rho|$ and then re-expanded in terms of the close-coupling eigenchannels $(\alpha|$. This represents *real* dipole transition amplitudes as

$$((J_c s) J_{cs} l, J = 1 || P^{[1]} || J_0 = 0)$$
$$= \sum_{\rho\alpha} ((J_c s) J_{cs} l | \rho)(\rho | \alpha) D_\alpha , \quad (35)$$

in terms of *real* transformation coefficients $(J_{cs} l | \rho)$ and $(\rho | \alpha)$. Next, the change is made from standing-wave to incoming-wave normalization through multiplication by the phase factor

$$\exp[i(\sigma_{lJ_c} - \tfrac{1}{2} l\pi + \pi\tau_\rho)] = i^{-l} \exp[i(\sigma_{lJ_c} + \pi\tau_\rho)] . \quad (36)$$

In (36) σ_{lJ_c} is the Coulomb phase $\arg\Gamma(l + 1 - i/k)$, which is determined by the electronic orbital momentum l and kinetic energy $\epsilon(\text{a.u.}) = \frac{1}{2}k^2$ in the $J_c = \frac{3}{2}$ channel. The result is then

$$((J_c s) J_{cs} l, J = 1 - || P^{[1]} || J_0 = 0)$$
$$= i^{-l} e^{i\sigma_{lJ_c}} \sum_{\rho\alpha} ((J_c s) J_{cs} l | \rho) e^{i\pi\tau_\rho} (\rho | \alpha) D_\alpha . \quad (37)$$

In the Appendix it is shown how the various elements of this expansion—the transition amplitudes D_α, the transformation matrices $(\rho | \alpha)$ and $(J_{cs} l | \rho)$, and the phase shifts $\pi\tau_\rho$—are determined in terms of the interaction parameters of the MQDT. These interaction parameters in turn have been determined for our purposes by C. M. Lee[8(d)] and are tabulated in the Appendix. Also, in the Appendix a simple method is given for the evaluation of the Coulomb phase shift differences $\sigma_{j_t+1}-\sigma_{j_t-1}$ which occur in the interference term of $\beta(j_t)_{tav}$.

The spectral variation of the phase shifts $\pi\tau_\rho$ determine the resonance pattern of both the integrated cross section σ and the asymmetry parameter β. As outlined in the Appendix, in addition to the explicit dependence on τ_ρ in (37), both sets of transformation coefficients, $(J_{cs}l|\rho)$ and $(\rho|\alpha)$, depend on the τ_ρ through functions $\sin\pi\tau_\rho$ or $\cos\pi\tau_\rho$. The dipole transition amplitude (37) contains the additional multiplicative factor $e^{i\pi\tau_\rho}$, and is therefore invariant to the transformation $\tau_\rho \rightarrow \tau_\rho+1$. Now each τ_ρ is a strongly nonlinear but monotonically increasing function of $\nu_{1/2}$. This functional dependence is characteristic for each ρ. Under the transformation $\nu_{1/2}\rightarrow\nu_{1/2}+1$, the curves $\tau_\rho(\nu_{1/2})$ for different ρ permute among themselves such that the same set of curves occurs for each unit range of $\nu_{1/2}$. Two of the τ_ρ curves will be shifted upward by 1 in the permutation, corresponding to the net increase by 2π in the elastic-scattering phase shift. But each shift by 1 leaves (37) invariant. Further, since (37) is summed over all ρ, the permutation merely amounts to interchanging terms of the sum, and therefore the permutation also leaves (37) invariant. We therefore obtain the fundamental result that the dipole transition amplitude (37) is invariant under the transformation $\nu_{1/2}\rightarrow\nu_{1/2}+1$, except for the slight energy dependence of the Coulomb phase.

This invariance accounts for the periodicity of the Rydberg series of autoionizing levels. The periodicity is exact in the integrated cross section, for which the Coulomb phase effect vanishes altogether because of the incoherence of the transition amplitudes for alternative orbital momenta l. Further, the Coulomb phase effect is small in the asymmetry parameter for low $\nu_{1/2}$ and becomes rapidly negligible as $\nu_{1/2}$ increases, owing to the corresponding rapid decrease in the change of electronic kinetic energy per unit range of $\nu_{1/2}$. That is, the resonance profiles in β are themselves very nearly also periodic in $\nu_{1/2}$, with period 1.

The amplitudes for transitions into the channels $(\rho|$ superpose coherently in the angular distribution, and the resulting interference terms are proportional to $\cos(\sigma_d-\sigma_s+\pi\tau_\rho-\pi\tau_{\rho'})$. These interference terms vanish in the integrated cross section (see Appendix) but they cause the angular distribution to contain information not present in the integrated cross section. This circumstance is especially important to the theory for the following reason: The semiempirical fitting procedures used by Lee[8(d)] and earlier by Lu,[8(c)] based on photoabsorption (integrated cross section) measurements, have encountered difficulties which have prevented, so far, the unambiguous determination of all of the interaction parameters of the theory. Independent information contained in measured angular distributions should help in removing these ambiguities. Conversely, however, it should be kept in mind that the measured spectral variation of β may differ from the results given here, owing to these uncertainties in the values of the semiempirical interaction parameters utilized in this paper. Nonetheless, these differences should be minor. The gross features of the β resonance will remain.

Results for Xenon

The results are given graphically in Fig. 1 for one unit range of $\nu_{1/2}$ from $\nu_{1/2}=3.5$ to $\nu_{1/2}=4.5$, corresponding to the range of photoelectron kinetic energies $0.196 \leqslant \epsilon \leqslant 0.635$ eV. Figure 1(a) gives the integrated cross section σ. The broad maximum peaking at $\nu_{1/2}\approx 0.75$ (modulo 1) is due primarily to autoionization of $5d^5(^2P^\circ_{1/2})nd$. The autoionization of $5d^5(^2P^\circ_{1/2})ns$ is concentrated at the "s resonance" on the high-energy shoulder of the broad peak at $\nu_{1/2}\approx 0.975$. As an index of the strength of parity-unfavored ionization, the ratio

$$r=\sigma(1)/[\sigma(1)+\sigma(2)] \tag{38}$$

is plotted in Fig. 1(b). This ratio could range between 1 for fully parity-favored ($j_t=1$) ionization, and zero for fully unfavored ($j_t=2$) ionization. Since s waves contribute only to $\sigma(1)$, r is also an index of the relative strength of s- and d-wave ionization. Finally, in Fig. 1(c) the spectral variation of the asymmetry parameter β is plotted, showing a pronounced resonant structure in the photoelectron angular distribution.

The spectral variation of β shows two striking features as $\nu_{1/2}$ increases: There is first a broad dip (width ≈ 0.1 eV) toward $\beta=-1$ ($j_t=2$), and then a rapid oscillation (width ≈ 0.06 eV) across the s resonance in the integrated cross section. This behavior can be understood qualitatively when viewed with the corresponding spectral features of σ and r. The broad minimum in β at $\nu_{1/2}\approx 0.64$ occurs when the photoelectric current is a minimum. Further, s-wave ionization is negligible in this region, and the minimum of β is paralleled by a corresponding minimum of r. Then, as the photoelectric current rises and passes through the broad d-resonance maximum in σ, parity-favored d-wave ionization increases, resulting in an increase of r and a corresponding increase of β away from -1. In the region of the s resonance the coupling between the s- and d-wave ionization channels produces sharp variations in the relative contributions of s- and d-wave ionization. These variations are reflected in the rapid oscillation of β across the s resonance: First, immediately below the s resonance, s-wave ionization is depressed and d-wave ionization is enhanced, resulting in a sharp rise of

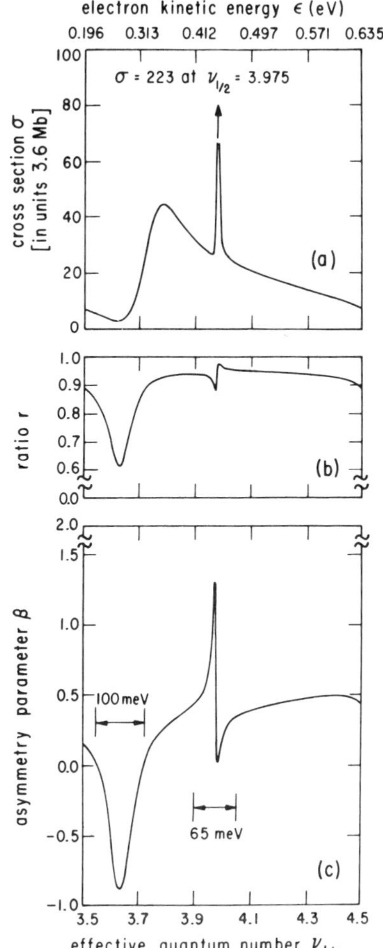

FIG. 1. Plot of xenon autoionization cross section and photoelectron asymmetry parameter resonance for one unit range of $\nu_{1/2}$. (a) Integrated cross section $\sigma=\sigma(1)+\sigma(2)$; (b) dimensionless cross-section ratio $\sigma(1)/\sigma$; (c) asymmetry parameter β. Note the breaks in the ordinate scales, and the nonlinear scale of electron kinetic energies

β away from the s-wave value $\beta=0$, and a corresponding dip of r. Then, s-wave ionization rapidly increases to a peak at the s-resonance maximum, sending β near 0 and r toward 1. Finally, on the high-energy side, s-wave ionization drops off once again, r drops and levels off, and β rises to a smooth plateau.

This cycle will repeat periodically in $\nu_{1/2}$, with period 1, up to the threshold of the $^2P^\circ_{1/2}$ level. The Coulomb phase interference term causes $\beta(\nu_{1/2}=3.5)$ to differ from $\beta(\nu_{1/2}=4.5)$ by 0.3.

An analogous β resonance occurs in all of the rare gases, and in fact the calculations also have been performed for argon, showing results similar to those for xenon. The xenon example is given here since xenon has the broadest resonances and therefore is a good candidate for experimental verification.

As pointed out in Sec. II, resonances in β are to be expected, in general, whenever autoionization occurs. Indeed, going to higher photon energies the rare gases provide another example of this effect, in the region of $nsnp^6(^2S_{1/2})n'p \rightarrow ns^2np^5(^2P^\circ)\times[\epsilon s, \epsilon d]$ resonances of the Rydberg series converging to the $nsnp^6\,{}^2S_{1/2}$ level of the rare-gas ion. These resonances have not been sufficiently well resolved by photoelectron spectroscopy to map the resonant structure of β, but evidence of the effect is already apparent in the measurements by Codling and Mitchell[15] for argon and in the recent results of Van der Wiel and Brion[16] on neon.

B. Autoionization in Mercury

As a further example consider the photoionization of mercury below the first excited state of the mercury ion. Production of the ion ground state,

$$\mathrm{Hg}(5d^{10}6s^2\ {}^1S_0)+\gamma(j_\gamma=1,\ \pi_\gamma=-1) \rightarrow \mathrm{Hg}^+(5d^{10}6s\ {}^2S_{1/2})+e(l=1), \quad (39)$$

has an ionization potential of 10.43 eV. Embedded in this ionization continuum are autoionizing Rydberg levels $5d^96s^2(^2D)np$ and $5d^96s^2(^2D)n'f$ converging to the first excited term $5d^96s^2(^2D)$ of the ion.[17] The lower level $^2D_{5/2}$ of the doublet has the ionization potential 14.83 eV, and the spin-orbit interaction raises the $^2D_{3/2}$ level 1.87 eV higher. Autoionization occurs between these fine structure ^2D thresholds, analogous to the xenon example, but here we consider rather autoionizing levels in the 4.43-eV spectral range between the $^2S_{1/2}$ and $^2D_{5/2}$ levels of different terms.

This first autoionization region provides a striking example of the theory. Owing to parity and angular momentum conservation, the electron must leave the atom with the single orbital momentum $l=1$. The Cooper–Zare formula[1] then predicts that the electron is ejected with full anisotropy, i.e., that the angular distribution is fixed at $\cos^2\theta$ ($\beta=2$), independently of dynamics, throughout the whole spectral range. Actually, this will be the case only if the spin of the photoelectron and the net spin of the rest of the atom *remain coupled into a singlet through the ionization process*, for only then must the electron necessarily carry away the full anisotropy of the photon-atom interaction. If ionization proceeds through autoionizing levels in which the spins are coupled into a triplet, then the angular distribution can be *depolarized by the unobserved nonzero spin*[18] and β can deviate from 2.

The angular-momentum-transfer analysis of the ionization reaction provides the quantitative basis for these ideas. Since $J_0=0$ and $J_c=S_c=\frac{1}{2}$, the angular momentum transfer (2) becomes for this case

$$\vec{j}_t=\vec{S}_c+\vec{s}\,. \tag{40}$$

Thus the allowed angular momentum transfers are

$$\begin{aligned} j_t&=0 \quad (\text{parity favored}, \quad \beta=2),\\ j_t&=1 \quad (\text{parity unfavored}, \quad \beta=-1). \end{aligned} \tag{41}$$

If the spins remain antiparallel, then only $j_t=0$ can occur and from (6), $\beta=2$ as predicted by the Cooper–Zare formula. On the other hand, $j_t=1$, with $\beta=-1$, arises because spin-orbit forces disturb this singlet coupling. The observed asymmetry parameter is [cf. (6) and (8)]

$$\beta=[2\sigma(0)-\sigma(1)]/[\sigma(0)+\sigma(1)], \tag{42}$$

which ranges from $\beta=2$ ($\cos^2\theta$) for fully-parity-favored ionization, to $\beta=-1$ ($\sin^2\theta$) for fully-parity-unfavored ionization. From (9) and (10) the integrated cross sections are given by

$$\begin{aligned} \sigma(0)&=\pi\lambda\!\!\!^{-2}|\bar{S}_p(0)|^2,\\ \sigma(1)&=3\pi\lambda\!\!\!^{-2}|\bar{S}_p(1)|^2. \end{aligned} \tag{43}$$

However, from (33), the matrix elements $\bar{S}_p$ are directly proportional to the corresponding reduced dipole matrix elements $P_p^{[1]}$, and the angular distribution depends only on the ratio of p-wave ionization with $j_t=0$ or $j_t=1$. This dependence is conveniently characterized by a strength parameter

$$s=\frac{|P_p^{[1]}(0)|^2-|P_p^{[1]}(1)|^2}{|P_p^{[1]}(0)|^2+|P_p^{[1]}(1)|^2}, \tag{44}$$

which ranges from $s=1$ for fully-parity-favored ionization, through $s=0$ for equal intensity with $j_t=0$ and $j_t=1$, to $s=-1$ for fully-parity-unfavored ionization. In terms of s the asymmetry parameter (42) becomes

$$\beta=\tfrac{1}{2}(3s+1), \tag{45}$$

i.e., $\beta=2$ for $s=1$, $\beta=\frac{1}{2}$ for $s=0$, and $\beta=-1$ for $s=-1$.

Niehaus and Ruf[5] have measured β at two closely spaced energies in the $^2S_{1/2}-{}^2D_{5/2}$ spectral region, using the ArI doublet resonance lines at 11.63 and 11.83 eV. The lower line falls on the high-energy shoulder of the autoionizing level $[5d^9(^2D_{3/2})6s^2]\times 6p_{3/2}\,{}^3D_1^\circ$ for which the electron and core spins are coupled into a *triplet*.[19] At this lower energy they find $\beta=1.25\pm0.1$, corresponding to $s=0.5$. At 11.83 eV they find $\beta=2.13\pm0.1$, corresponding to $s=1.0$. [For dipole ionization β can be no larger than 2 and therefore this latter value may reflect experimental uncertainty.] Over an energy range of 0.2 eV the strength parameter changes by a factor of 2. This indicates that indeed parity-unfavored ionization is substantial in this autoionization spectral region. It would be very desirable to fully map the spectral variation of β throughout this range.

IV. CONCLUSION

Resonant structure in photoelectron angular distributions is predicted to be a general feature of all atomic and molecular autoionization resonances. These β resonances are sensitive probes of atomic and molecular dynamics. They can determine the cumulative effects of forces that might go undetected in direct ionization. Owing to the interferences present in all angular distribution measurements, the measured β spectra contain new and independent dynamical information not available in integrated cross sections. The experimental investigation of these angular distribution resonances is an essentially untapped and rich resource of information for chemical physics.

ACKNOWLEDGMENT

I am grateful to Professor U. Fano for valuable discussions and for his encouragement.

APPENDIX

MQDT Interaction Parameters

The reduced dipole matrix elements (37) are evaluated for xenon in terms of the following interaction parameters, which have been obtained by Lee.[8(d)] (i) One set of parameters consists of five real dipole transition amplitudes D_α. The quantum number α of the close-coupling eigenchannels runs from 1 to 5 and corresponds approximately to the five LS channels of the electron–ion-core system at short range, with $J=1$ and $\pi=-1$:

$$\begin{array}{rccccc} \alpha= & 1, & 2, & 3, & 4, & 5;\\ LS\text{ channel}= & p^5d\,{}^3D_1^\circ, & p^5d\,{}^1P_1^\circ, & p^5d\,{}^3P_1^\circ, & p^5s\,{}^3P_1^\circ, & p^5s\,{}^1P_1^\circ. \end{array} \tag{A1}$$

The values of D_α, normalized per unit energy range in atomic units, are

$$\alpha = \quad 1 \; , \quad 2 \; , \quad 3 \; , \quad 4 \; , \quad 5 \; ;$$
$$D_\alpha/\sqrt{3} = -0.28462 \, , \; 4.09991 \, , \; -0.50779 \, , \; -0.03106 \, , \; 1.83506 \; . \tag{A2}$$

The factor $1/\sqrt{3}$ in (A2) is necessary because the D_α, as defined here by Eq. (34), are rotationally invariant matrix elements of r, whereas Lee[8(d)] and Lu[8(c)] define the D_α as matrix elements of z [see Eq. (2.4) of Ref. 8(c)]. The two definitions are made equivalent through multiplication by $\langle (z/r)^2 \rangle^{1/2} = 1/\sqrt{3}$. With this normalization, the unit of cross section $\pi ƛ^2 n(ƛ)^2$ [see Eqs. (20), (21), (32), and (33)] is determined by setting $\lambda^{-1} = I_{3/2}$, since the MQDT interaction parameters are assumed to be constant throughout the autoionization range from $I_{3/2}$ to $I_{1/2}$.[8(c)] This gives $\pi ƛ^2 n(ƛ_{3/2})^2 = 3.6$ Mb.
(ii) Another set of parameters consists of five close-coupling eigenphase shifts $\pi\mu_\alpha$, or the equivalent quantum defects

$$\alpha = \quad 1 \; , \quad 2 \; , \quad 3 \; , \quad 4 \; , \quad 5 \; ;$$
$$\mu_\alpha = 0.360 \, , \; 0.120 \, , \; 0.560 \, , \; 0.040 \, , \; -0.007 \, . \tag{A3}$$

(iii) The last set of parameters forms the 5×5 orthogonal transformation matrix

$$U_{i\alpha}^{(J=1)} \equiv (J_c(ls)j \,|\, \alpha)^{(J=1)} \, , \tag{A4}$$

connecting the close-coupling eigenchannels $(\alpha J|$ to the asymptotic jj-coupled dissociation channels $(J_c(ls)j, J|$. In (A4), the quantum number i of the dissociation channels is given by

$$i = \quad 1 \; , \quad 2 \; , \quad 3 \; , \quad 4 \; , \quad 5 \; ;$$
$$jj \text{ channel} = {}^2P^\circ_{3/2} d_{5/2} \, , \; {}^2P^\circ_{3/2} d_{3/2} \, , \; {}^2P^\circ_{3/2} s_{1/2} \, , \; {}^2P^\circ_{1/2} d_{3/2} \, , \; {}^2P^\circ_{1/2} s_{1/2} \, . \tag{A5}$$

The transformation matrix is

$$U = \begin{array}{c|ccccc|} \alpha = & 1 & 2 & 3 & 4 & 5 \\ i=4 & 0.67094 & 0.65800 & 0.34073 & -0.00514 & 0.02744 \\ 1 & -0.36223 & 0.69301 & -0.62240 & -0.00545 & -0.03342 \\ 2 & -0.64673 & 0.29448 & 0.70336 & -0.00230 & 0.01741 \\ 5 & 0.01115 & 0.00640 & 0.02454 & 0.81549 & -0.57810 \\ 3 & -0.01573 & 0.00451 & -0.03462 & 0.57871 & 0.81463 \end{array} \, . \tag{A6}$$

and is published here for the first time.[8(d)]

Evaluation of Dipole Transition Amplitudes

Using these interaction parameters, the elements of the expansion (37) are determined as follows: From (A5) we see that there are three elastic-scattering eigenchannels $(\rho|$ open below $I_{1/2}$. The corresponding phase shifts $\pi\tau_\rho$ are determined for each value of $\nu_{1/2}$ as roots of the cubic equation obtained by setting to zero the determinant of the linear system [Sec. III A of Ref. 8(c)]

$$\sum_{j=1}^{3} \left[-\cot\pi(\nu_{1/2} + \tau_\rho)\delta_{ij} + \left(\sum_{\alpha=1}^{5} U_{i\alpha} \cot\pi(\nu_{1/2} + \mu_\alpha) U_{\alpha j} \right) \right] C_j = 0 \, , \quad i = 1, 2, 3 \, . \tag{A7}$$

The transformation coefficients $(\rho|\alpha)$ which connect the elastic-scattering channels $(\rho|$ to the close-coupling channels $(\alpha|$ is expressed in terms of a vector $\vec{A}$ [defined in Ref. 8(c)] by

$$(\rho|\alpha) = A_\alpha / N_\rho \, , \tag{A8}$$

where N_ρ is a normalization factor given below. The vector $\vec{A}$ is determined from the system of equations [Eq. (3.23) of Ref. 8(c)]

$$\sum_\alpha U_{i\alpha} \sin\pi(\mu_\alpha - \tau_\rho) A_\alpha = 0 \, , \quad i = 1, 2, 3 \, . \tag{A9}$$

The transformation coefficients $((J_c s) J_{cs} l | \rho)$ are given by

$$((J_c s) J_{cs} = j_t l | \rho) = \sum_i ((J_c s) J_{cs} = j_t l | i = (J_c(ls)j))^{(J=1)} T_{i\rho} / N_\rho \, , \tag{A10}$$

in terms of the geometrical transformation coefficient

$$((J_c s)J_{cs}=j_t l\,|\,J_c(ls)j)^{(J=1)}$$

$$=(-1)^{J_c+j+1}(2j+1)^{1/2}(2j_t+1)^{1/2}\begin{Bmatrix}\frac{1}{2} & l & j\\ 1 & J_c & j_t\end{Bmatrix}, \quad \text{(A11)}$$

of the dynamical transformation $T_{i\rho}$, given by [see Eq. (3.21) of Ref. 8(c)]

$$T_{i\rho}=\sum_\alpha U_{i\alpha}\cos\pi(\mu_\alpha-\tau_\rho)A_\alpha\,,\quad i=1,2,3, \quad \text{(A12)}$$

and of the normalization factor N_ρ, given by

$$N_\rho^2=\sum_{i=1}^{3}|T_{i\rho}|^2. \quad \text{(A13)}$$

Incoherence of Channels ρ in the Integrated Cross Section

The integrated cross section is proportional to

$$\sum_{j_t l}|P_l(j_t)|^2. \quad \text{(A14)}$$

But, owing to the orthonormality property

$$\sum_{j_t l}(\rho'|J_{cs}=j_t l)(J_{cs}=j_t l|\rho')=\delta_{\rho\rho'}\,, \quad \text{(A15)}$$

(A14) is incoherent in the transition amplitudes for the alternative channels ρ, i.e.,

$$\sum_{j_t l}|P_l(j_t)|^2=\sum_{\rho=1}^{3}\left(\left|\sum_\alpha A_\alpha D_\alpha\right|^2\Big/N_\rho^2\right). \quad \text{(A16)}$$

This result corresponds to Eq. (3.25) of Ref. 8(c).

Evaluation of Coulomb Phase Shift Differences

The Coulomb phase difference $\sigma_d-\sigma_s$ which occurs in the interference term of the asymmetry parameter (22) can be conveniently evaluated using the following algebraic scheme which derives from Bethe.[1] The difference $\sigma_{j_t+1}-\sigma_{j_t-1}$ can be represented as the angle of a right triangle whose adjacent side A, opposite side O, and hypotenuse H are given by

$$A=j_t(j_t+1)-(Z/k)^2,$$
$$O=-(Z/k)(2j_t+1), \quad \text{(A17)}$$
$$H=\{[(j_t+1)^2+(Z/k)^2][j_t^2+(Z/k)^2]\}^{1/2},$$

where the ionic charge is Z and the photoelectron kinetic energy is $\epsilon\,(\text{a.u.})=\frac{1}{2}k^2$. These relations are exact and therefore valid for all values of k^2. Thus, e.g., $\cos(\sigma_d-\sigma_s)$ for the xenon example ($Z=1$) is given by

$$\cos(\sigma_d-\sigma_s)=(2k^2-1)/(4k^4+5k^2+1)^{1/2} \quad \text{(A18)}$$

and is seen to increase monotonically from -1 at $k=0$ to $+1$ at $k=\infty$. This treatment of Coulomb phase differences is simple and bypasses the numerical errors which may occur when the phase shifts are first calculated separately as arguments of the complex Γ function and then subtracted.

---

*Work supported by U. S. Atomic Energy Commission Contract No. C00-1674-74.

[1]J. Cooper and R. N. Zare, in *Lectures in Theoretical Physics: Atomic Collision Processes*, edited by S. Geltman, K. T. Mahanthappa, and W. E. Brittin (Gordon and Breach, New York, 1969), Vol. XI-C, pp. 317–337; see also H. Bethe, in *Handbuch der Physik*, edited by H. Geiger and K. Scheel (Springer, Berlin, 1933), Vol. 24, Pt. 1, pp. 482–484; P. Auger and F. Perrin, J. Phys. Rad. 8, 93 (1927).

[2]D. Dill and U. Fano, Phys. Rev. Letters 29, 1203 (1972).

[3]D. Dill (unpublished).

[4]D. Dill, Phys. Rev. A 6, 160 (1972).

[5]A. Niehaus and M. W. Ruf, Z. Physik 252, 84 (1972).

[6]A detailed definition of the amplitudes $\bar{S}(j_t)$, in terms of reduced dipole matrix elements, will be given in Sec. III.

[7]H. Beutler, Z. Physik 93, 177 (1935). For recent work see, e.g., K. Yoshino, J. Opt. Soc. Am. 60, 1220 (1970).

[8]The multichannel quantum-defect theory was developed originally by Seaton in (a) M. J. Seaton, Proc. Phys. Soc. (London) A88, 801 (1966). Fano has adapted and extended the theory to the present application in the basic paper (b) U. Fano, Phys. Rev. A 2, 353 (1970), which treats the photoabsorption spectrum of H_2 near threshold. Building on the methods and concepts of Fano's work, Lu developed and applied the MQDT to the semiempirical analysis of the xenon photoabsorption spectrum near the ionization threshold, in (c) K. T. Lu, Phys. Rev. A 4, 579 (1971). C. M. Lee has extended the fitting procedure of Ref. 8(c) in order to obtain the interaction parameters of Lu's analysis in the form required to evaluate the xenon photoelectron angular distribution. This extended fitting procedure will be published shortly, with application to the argon-threshold-region photoabsorption spectrum, in (d) C. M. Lee and K. T. Lu, Phys. Rev. A (to be published); and the xenon parameters are given here in the Appendix. I am indebted to C. M. Lee for providing these results and allowing me to use them prior to publication.

[9]U. Fano and D. Dill, Phys. Rev. A 6, 185 (1972).

[10]See, e.g., B. W. Shore and D. H. Menzel, *Principles of Atomic Spectra* (Wiley, New York, 1968), p. 442, Eq. (8.5).

[11]For the dipole expression in Eq. (28) see, e.g., H. A. Bethe and E. E. Salpeter, *Quantum Mechanics of One- and Two-Electron Atoms* (Springer-Verlag, Berlin, 1957), Eqs. (59.5) and (69.5); for the S-matrix expression see Ref. 8 of U. Fano and D. Dill, Phys. Rev. A 6, 185 (1972), and F. J. Blatt and V. Weisskopf, *Theoretical Nuclear Physics* (Wiley, New York, 1952), pp. 517–521.

[12]The factor $(2J+1)^{-1/2}$ in (29a) arises through application of the Wigner–Eckart theorem.

[13]The properties are the symmetry relation

$$|(JM|J_0M_0,\,j_\gamma m_\gamma)|^2=\frac{2J+1}{2j_\gamma+1}|(j_\gamma m_\gamma|JM,\,J_0\text{-}M_0)|^2,$$

and the orthonormality equations

$$\sum_{M_0}|(j_\gamma m_\gamma|JM,\,J_0\text{-}M_0)|^2=1,$$

$$\sum_{M_c \mu m} (J_{cs} l J M \mid J_c M_c, s\mu, lm) \times (J_c M_c, s\mu, lm \mid J'_{cs} l J' M) = \delta_{J_{cs} J'_{cs}} \delta_{JJ'}.$$

[14]In this regard see the first two paragraphs of Ref. 9, Sec. I.

[15]P. M. Mitchell and K. Codling, Phys. Letters 38A, 31 (1972).

[16]M. J. Van der Wiel and C. E. Brion, J. Electr. Spectrosc. 1, 739 (1973).

[17]The photoionization spectrum of Hg below the Hg^+ ($^2D_{5/2}$) threshold has been studied by several authors: (a) B. Brehm, Z. Naturforsch. 21a, 196 (1966); (b) J. Berkowitz and C. Lifshitz, J. Phys. B 1, 438 (1968); (c) R. B. Cairns, H. Harrison, and R. I. Schoen, J. Chem. Phys. 53, 96 (1970).

[18]Depolarization of angular correlations, owing to the lack of information about the orientation of unobserved angular momenta, is a concept that is central to the theory of angular correlation. See U. Fano and G. Racah, *Irreducible Tensorial Sets* (Academic, New York, 1959), Chap. 19, and especially p. 110, last paragraph.

[19]W. C. Martin, J. Sugar, and J. L. Tech, Phys. Rev. A 6, 2022 (1972) have verified the triplet character of this level using intermediate coupling theory with configuration interaction. They report that the $5d^9(^2D_{3/2})\,6s^26p_{3/2}$ ($J=1$, $\pi=-1$) level at 11.62 eV is composed of 63.3% $^3D_1^\circ$, and 29.5% $^3P_1^\circ$, the total singlet composition is less than 7%. These authors find substantial triplet character for many of the autoionizing levels of Hg [see also W. C. Martin, J. Sugar, and J. L. Tech, J. Opt. Soc. Am. 62, 1488 (1972)]. The deviation of β from 2 is evidence of this triplet character.

1974 *J. Phys. B: At. Mol. Phys.* **7** 2533–48

Electron impact excitation of metastable levels in O^{2+}

W Eissner and M J Seaton

Department of Physics and Astronomy, University College London

Received 28 May 1974

Abstract. Collision strengths have been calculated for electron-impact transitions between the O^{2+} ground configuration terms ^3P, ^1D and ^1S, for energies up to 0·5 Ryd above excitation thresholds. The wavefunction expansion for the $(e+O^{2+})$ system includes all O^{2+} terms for which the dominant configurations are $2s^22p^2$ and $2s2p^3$. Configuration-interaction wavefunctions are used for the target states. These functions give accurate oscillator strengths for all transitions of the type $2s^22p^2$–$2s2p^3$.

The collision strengths are calculated on solving systems of coupled integro-differential equations. The (^3P–^1D) collision strength contains resonances of the type $2s^22p^2(^1S)nl$, and all collision strengths contain resonances of the type $2s2p^3nl$.

The method used to obtain collision strengths for the $(e+O^{2+})$ system is also used to calculate the positions of bound states for O^+. It is found that inclusion of the $2s2p^3$ states of O^{2+} in the expansions gives a marked improvement in the agreement with experiment for the positions of the O^+ bound states. The calculated positions of the near-threshold resonance states, $2s2p^3(^3D^o)3s\ ^2D^o$ and $2s2p^3(^3P^o)3s\ ^2P^o$, are in satisfactory agreement with positions predicted from iso-electronic sequence extrapolations.

The collision strengths are integrated over Maxwell distributions for the velocity of the colliding electron, to obtain rate coefficients for use in astronomical applications. It is estimated that the errors in the computed coefficients should be less than 10%.

There are large differences between the present results and those obtained by Ormonde *et al* (1973). These differences are discussed and it is concluded that the results of Ormonde *et al* are in error.

1. Introduction

It is difficult to measure cross sections for electron-impact excitation of metastable levels, or cross sections for excitation of positive ions. So far as we are aware, no experimental work has been done on the excitation of metastable levels in positive ions (with the exception of the special case of the 1s–2s transition in He^+ studied by Dance *et al* (1966) and by Peart and Dolder (1973)). On the other hand, many of the spectrum lines observed in gaseous nebulae are produced by electron impact excitation of metastable levels in positive ions. It is now possible to measure the intensities of these lines to within an accuracy of a few per cent. In order to interpret these observations, excitation rate coefficients are required to a similar accuracy. The forbidden lines in O^{2+} are of particular astronomical interest and we have therefore attempted to make accurate calculations for this case.

We are concerned with transitions between the three lowest levels in O^{2+}, that is to say the terms ^3P, ^1D and ^1S in the $2s^22p^2$ configuration. In all earlier work these were the only O^{2+} terms considered in the problem (Hebb and Menzel 1940, Seaton 1953, 1955, Saraph *et al* 1966, 1969, Smith *et al* 1969, Henry *et al* 1969). It was shown by

Eissner *et al* (1969) that it is necessary to include O^{2+} terms belonging to the configuration $2s2p^3$, since they give rise to resonances of the type $2s2p^33s$ in the near-threshold collision strengths. It has also been shown that, in calculations of this type, it is necessary to use accurate target functions (Eissner 1972, Saraph 1973). In the present work we have included all O^{2+} target terms for which the dominant configurations are $2s^22p^2$ and $2s2p^3$ and we have used accurate configuration-interaction target functions. These functions give good results for the O^{2+} $2s^22p^2$–$2s2p^3$ oscillator strengths. The techniques which have been used to solve the $(e+O^{2+})$ collision problem have also been used to calculate the positions of bound states in O^+. Comparisons with positions determined experimentally provide a check on the accuracy of our calculations.

Preliminary results of our calculations were presented at the 1972 Liège conference on Planetary Nebulae (Eissner and Seaton 1973). Some further recent calculations on excitation of O^{2+} have been published by Ormonde *et al* (1973). We will show that these are good reasons for believing these calculations to be in error.

2. Wavefunctions for O^{2+}

Our formulation of the collision problem can be obtained from a variational principle and it can be shown that, if exact target wavefunctions are used, the errors in the calculated collision strengths will be of quadratic order in the error in the wavefunctions for the $(e+O^{2+})$ system. However, the use of approximate target functions will introduce first order errors in the collision strengths. It is therefore desirable to use accurate target functions. The accuracy of these functions can be judged by considering the calculation of energies and oscillator strengths.

The coupling between target states with dominant configurations $2s^22p^2$ and $2s2p^3$ is of importance in the collision problem. The coupling potentials are proportional, in the asymptotic region, to $\sqrt{f}$ where f is the corresponding oscillator strength. It is therefore particularly important that the wavefunctions used should give accurate results for these oscillator strengths.

We use *configuration-interaction* (CI) wavefunctions: $(2s^22p^2+2p^4+2s2p^23d)$ for the even parity terms, ^3P, ^1D and ^1S; and $(2s2p^3+2s^22p3d)$ for the odd parity terms, $^5S^o$, $^3D^o$, $^3P^o$, $^1D^o$, $^3S^o$ and $^1P^o$. These functions are calculated using the program of Eissner and Nussbaumer (1969). The radial functions are calculated using potentials of the form

$$V_l(r) = -\frac{2\zeta(r/\lambda_l)}{r} \tag{2.1}$$

where $\zeta(r)$ is the effective charge in the Fermi–Thomas–Dirac statistical model, and λ_l is a scale parameter. The same radial functions are used for all states of O^{2+}. The scale parameters λ_l were initially adjusted so as to minimize the energy sum $\{E(^3P)+E(^1D)+E(^1S)\}$. This procedure automatically gives a rather larger value of λ_d ($\lambda_d \simeq 2$) and hence a contracted 3d orbital (Nussbaumer 1971); in the notation of Weiss (1970) this orbital is of 'correlation' rather than 'spectroscopic' type. The energies are not sensitive to the exact value of λ_d and hence the variational principle does not provide a good method for the determination of this parameter. The value of λ_d finally adopted, $\lambda_d = 1{\cdot}9$, was chosen so as to obtain the best agreement with accurate oscillator strengths, as obtained from experiment and from large CI calculations (Smith and Wiese 1971). The other two scale parameters are $\lambda_s = 1{\cdot}269$ and $\lambda_p = 1{\cdot}210$.

Results for energies and oscillator strengths of O^{2+} are given in tables 1 and 2. Table 1 gives term energies calculated using *single-configuration* (SC) functions ($2s^22p^2$ and $2s2p^3$ only) and our CI functions, together with the experimental energies. It is seen that the SC functions give poor results for the intervals (^3P–^1S) and (^3P–$^5S^o$), the calculated position of $^5S^o$ being lower than that of ^1S. Improved results are obtained with the CI functions. Table 2 gives results for the $2s^22p^2$–$2s2p^3$ oscillator strengths. It is seen that the use of SC functions gives oscillator strengths in error by a factor of two or more.

Table 1. Term energies for O^{2+} in Rydberg units.

Dominant configuration	Term	Energy		
		Experiment	SC functions	CI functions
$2s2p^3$	$^1P^o$	1·916	1·985	2·052
	$^3S^o$	1·795	1·847	1·924
	$^1D^o$	1·703	1·785	1·836
	$^3P^o$	1·296	1·261	1·333
	$^3D^o$	1·092	1·061	1·122
	$^5S^o$	0·550	0·400	0·474
$2s^22p^2$	^1S	0·392	0·500	0·393
	^1D	0·183	0·200	0·203
	^3P	0·000	0·000	0·000

Table 2. Oscillator strengths for O^{2+}.

Transition	f		
	Smith and Wiese (1971)	SC functions	CI functions
$^3P \to {}^3D^o$	0·11	0·195	0·108
$\to {}^3P^o$	0·14	0·139	0·140
$\to {}^3S^o$	0·18	0·272	0·195
$^1D \to {}^1D^o$	0·30	0·525	0·315
$\to {}^1P^o$	0·23	0·197	0·230
$^1S \to {}^1P^o$	0·27	0·656	0·274

3. The collision problem

The $i \to i'$ collision cross section, in units of πa_0^2, is

$$Q(i \to i') = \frac{\Omega(i, i')}{\omega_i k_i^2} \tag{3.1}$$

where

$$\Omega(i, i') = \sum_{SL\pi} \sum_{ll'} \Omega_{SL\pi}(il, i'l') \tag{3.2}$$

and where: ω_i is the statistical weight of target level i; k_i^2 is the kinetic energy of the incident electron in Rydberg units; and $\Omega_{SL\pi}(il, i'l')$ is the contribution to the total collision strength $\Omega(i, i')$ from states with total angular momenta and parity $SL\pi$ and orbital angular momenta l, l' for the colliding electron.

Previous work has shown that the only significant contributions to the collision strengths $\Omega(^3\text{P}, ^1\text{D})$ and $\Omega(^3\text{P}, ^1\text{S})$ come from $l = 0$, 1 and 2 and $l' = 0$, 1 and 2. These are the only values of l, l' considered in the present work. Larger values of l, l' make more important contributions to $\Omega(^1\text{D}, ^1\text{S})$, but these contributions can easily be estimated from the results of previous work. We shall return to the question of the convergence of the partial wave expansions in §§ 4.4 and 6.

We use the ID approximation of Eissner and Seaton (1972), which involves the solution of coupled *integro-differential* equations. The equations are set up using methods discussed by Eissner (1975) and solved using methods described by Seaton (1974). The wavefunction for the $(e + O^{2+})$ system has the form

$$\Psi = \sum_{i=1}^{\text{NCHF}} \Theta_i + \sum_{j=1}^{\text{NCHB}} \Phi_j c_j \tag{3.3}$$

where the free-channel functions Θ_i contain radial functions $F_i(r)$ for the colliding electron. These radial functions, and the coefficients c_j of the bound channel functions Φ_j, are varied so as to satisfy the Kohn variational principle. This gives the ID equations. In these equations we use experimental energy differences, $(k_i^2 - k_{i'}^2)$, instead of calculated differences.

The use of i for a channel index in (3.3) should not be confused with the use of i for a state of the target system in (3.1): i in (3.3) corresponds to $ilSL\pi$ in (3.2).

Table 3 gives a list of the free channels and bound channels for the important cases of $SL\pi = {}^2\text{P}^\text{o}$ and ${}^2\text{D}^\text{o}$. It should be noted that the use of CI target functions increases the number of bound channels included and hence makes better allowance for correlation effects in the solution of the collision problem (Seaton 1974).

4. Results for the collision strengths

4.1. Analysis of resonances

The collision strengths contain many resonances within about 0·5 rydberg of the excitation thresholds, which is the energy range of interest for astronomical applications. In order to be economic in the use of computer time, we have fitted the computed reactance matrices to analytic functions of the energy. Let $\mathbf{F}(r)$ be a matrix with elements $F_{ii'}(r)$ where i specifies a channel and i' a boundary condition. When all channels are open we may put

$$\mathbf{F} \underset{r\to\infty}{\sim} k^{-1/2}\{(\sin\zeta) + (\cos\zeta)\mathscr{R}\}, \tag{4.1}$$

where ζ is the phase of the regular Coulomb function. We can now make an analytic continuation of $\mathscr{R}$ to the region in which some channels are closed. The methods used are described in a series of papers on quantum defect theory (see Seaton 1966 and 1969). When some channels are closed we may partition $\mathscr{R}$ according to the scheme

$$\mathscr{R} = \begin{pmatrix} \mathscr{R}_{\text{oo}} & \mathscr{R}_{\text{oc}} \\ \mathscr{R}_{\text{co}} & \mathscr{R}_{\text{cc}} \end{pmatrix} \tag{4.2}$$

Table 3. Free channels and bound channels for the $(e+O^{2+})$ system.

$SL\pi$	Free channels i	(S_iL_i)	Bound channels j	Configuration
$^2P^o$	1	(^3P) p	1	$2s2p^33d$
	2	(^1D) p	2	
	3	(^1S) p	3	
	4	$(^3D^o)$ d	4	
	5	$(^3P^o)$ d	5	$2s^22p3d^2$
	6	$(^3P^o)$ s	6	
	7	$(^1D^o)$ d	7	
	8	$(^1P^o)$ d	8	$2s^22p^3$
	9	$(^1P^o)$ s	9	$2p^5$
$^2D^o$	1	(^3P) p	1	$2s2p^33d$
	2	(^1D) p	2	
	3	$(^3D^o)$ d	3	
	4	$(^3D^o)$ s	4	
	5	$(^3P^o)$ d	5	
	6	$(^1D^o)$ d	6	$2s^22p\ 3d^2$
	7	$(^1D^o)$ s	7	
	8	$(^3S^o)$ d	8	
	9	$(^1P^o)$ d	9	$2s^2\ 2p^3$

Note. If n bound channels have the same configuration this indicates that, for the value of $SL\pi$ considered, this configuration gives n linearly independent bound channel functions.

(open–open, open–closed, closed–open and closed–closed). We may also partition **F** according to the scheme

$$\mathbf{F} = \begin{pmatrix} \mathbf{F}_{oo} \\ \mathbf{F}_{co} \end{pmatrix} \tag{4.3}$$

where

$$\mathbf{F}_{oo} \underset{r\to\infty}{\sim} k^{-1/2}\{(\sin\zeta)+(\cos\zeta)\mathbf{R}\} \tag{4.4}$$

and

$$\mathbf{F}_{co} \underset{r\to\infty}{\sim} 0. \tag{4.5}$$

We then have (Seaton (1969)

$$\mathbf{R} = \mathcal{R}_{oo} - \mathcal{R}_{oc}(\tan\pi\nu_c + \mathcal{R}_{cc})^{-1}\mathcal{R}_{co} \tag{4.6}$$

which involves the effective quantum numbers ν_c in the closed channels. These are defined by

$$E = E_i - \frac{z^2}{\nu^2} \tag{4.7}$$

where E is the total energy, E_i the target energy for channel i and z is the charge on the ion ($z = 2$ for O^{2+}).

The matrix $\mathscr{R}$ varies slowly as a function of the energy. As we go through a resonance the energy variation of $\mathbf{R}$ is mainly due to the variation of $\tan \pi\nu_c$ in (4.6). The scattering matrix is given by

$$\mathbf{S} = (\mathrm{i}-\mathbf{R})(\mathrm{i}+\mathbf{R})^{-1} \tag{4.8}$$

and the partial collision strength by

$$\Omega_{SL\pi}(il, i'l') = \tfrac{1}{2}(2S+1)(2L+1)|S_{SL\pi}(il, i'l')|^2. \tag{4.9}$$

In principle it should be possible to calculate $\mathscr{R}$ in the region of all-channels-open and, on making an analytic continuation, calculate $\mathbf{R}$ at all energies using (4.6). In practice this cannot be done, since the region of all-channels-open corresponds to comparatively high energies (see table 1) and extrapolations of $\mathscr{R}$ over an extended energy range become unreliable. Our procedure is therefore to use empirical fitting procedures based on (4.6).

Expanding about an energy $E^{(0)}$ we have

$$\tan \pi\nu = \tan \pi\nu^{(0)} + (E-E^{(0)})/q^2 \tag{4.10}$$

where

$$q^2 = \frac{2z^2}{\pi\nu^3}(\cos \pi\nu^{(0)})^2. \tag{4.11}$$

Substitution of (4.10) in (4.6) gives

$$\mathbf{R} = \mathscr{R}_{\mathrm{oo}} - \mathscr{R}_{\mathrm{oc}} q(E-\mathbf{B})^{-1} q \mathscr{R}_{\mathrm{co}} \tag{4.12}$$

where

$$\mathbf{B} = E^{(0)} - q(\tan \pi\nu_{\mathrm{c}}^{(0)} + \mathscr{R}_{\mathrm{cc}})q. \tag{4.13}$$

If we now diagonalize $\mathbf{B}$ we obtain

$$R_{ii'} = \mathscr{R}_{ii'} - \sum_{\lambda} a_{i\lambda}(E-E_\lambda)^{-1} a_{i'\lambda}. \tag{4.14}$$

In the region of an isolated pole in $\mathbf{R}$ we may fit to

$$R_{ii'} = \mathscr{R}_{ii'} - a_i(E-E_1)^{-1} a_{i'} \tag{4.15}$$

assuming $\mathscr{R}_{ii'}$ to be a polynomial in E and a_i, $a_{i'}$ and E_1 to be constants. We always check that the number of parameters which have to be adjusted to obtain a fit is smaller than the number of energies for which $\mathbf{R}$ has been calculated. In some cases we have overlapping resonances. If, in some energy range, we have n poles in R, we fit to (4.14) with $\lambda = 1$ to n.

For the case of a Rydberg series of resonances converging to the threshold for the opening of new channels, we may redefine $\mathscr{R}$ to be the matrix $\mathbf{R}$ calculated at energies just above the new threshold. On fitting to (4.14) or (4.15) at energies below the new threshold, we can deduce the values of the elements of $\mathscr{R}$ below the threshold. We can then interpolate $\mathscr{R}$ to obtain the complete series of resonances.

4.2. Results

4.2.1. Results for $\Omega(^3P, ^1D)$. Results for $\Omega_{SL\pi}(^3\mathrm{P}, ^1\mathrm{D})$ are shown in figure 1(*a*) for odd parity states ($SL\pi = {}^2\mathrm{P}^{\mathrm{o}}$ and ${}^2\mathrm{D}^{\mathrm{o}}$) and in figure 1(*b*) for even parity states ($SL\pi = {}^2\mathrm{P}^{\mathrm{e}}$, ${}^2\mathrm{D}^{\mathrm{e}}$ and ${}^2\mathrm{F}^{\mathrm{e}}$). For $SL\pi = {}^2\mathrm{P}^{\mathrm{o}}$ and ${}^2\mathrm{D}^{\mathrm{e}}$ we have series of resonances converging to the threshold for the opening of the $2p^2\ ^1S$ channels. For the other values of $SL\pi$ there is no coupling $2p^2\ ^1S$. The present results for the positions of the $2s2p^3(^3D^o)3s\ ^2D^o$ and $2s2p^3(^3P^o)3s\ ^2P^o$ resonances are in good agreement with the positions obtained by Eissner *et al* (1969), but the present calculations give a smaller width for the 3s $^2D^o$ resonance.

4.2.2. Results for $\Omega(^3P, ^1S)$. Results for $\Omega_{SL\pi}(^3P, ^1S)$ are shown in figure 2 for $SL\pi = {}^2P^o$ and $^2D^e$. For $^2P^o$ we obtain a complicated system of four overlapping resonances in the region of $[k(^1S)]^2$ between 0·4 and 0·6.

4.2.3. Results for $\Omega(^1D, ^1S)$. Results for $\Omega_{SL\pi}(^1D, ^1S)$ are shown in figure 3 for $SL\pi = {}^2P^o$, $^2S^e$ and $^2D^e$. For $SL\pi = {}^2P^o$ the background contribution is small and the resonances

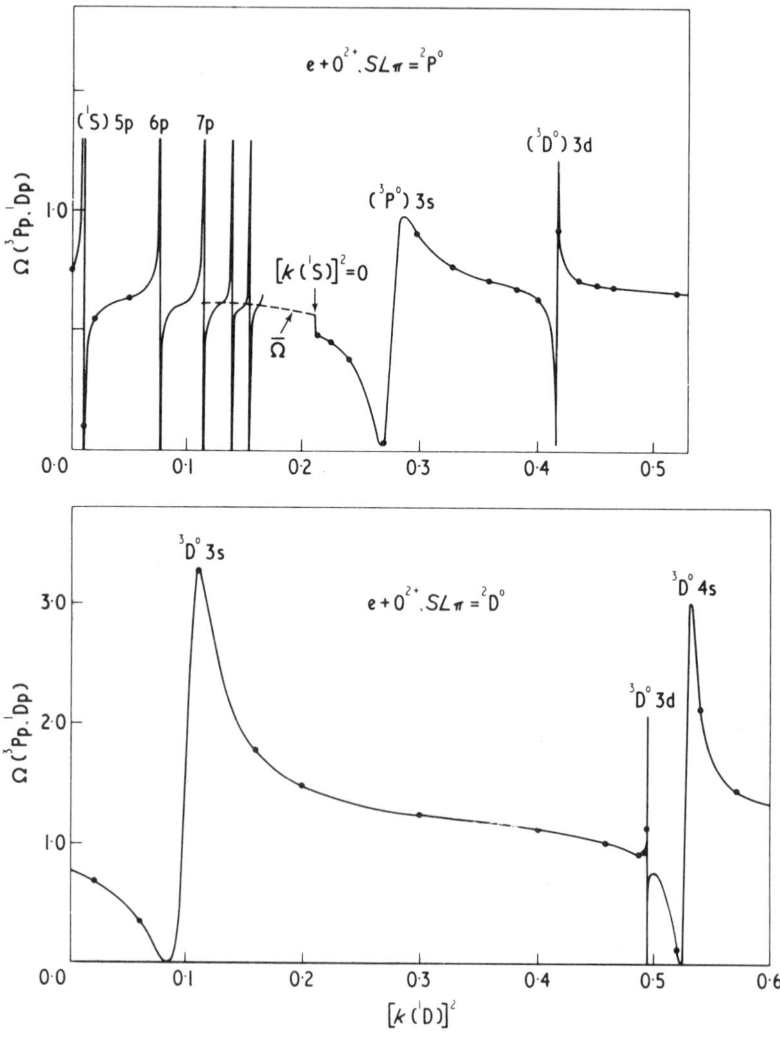

Figure 1(*a*). Results for $\Omega_{SL\pi}(^3P, ^1D)$, $SL\pi = {}^2P^o$ and $^2D^o$.

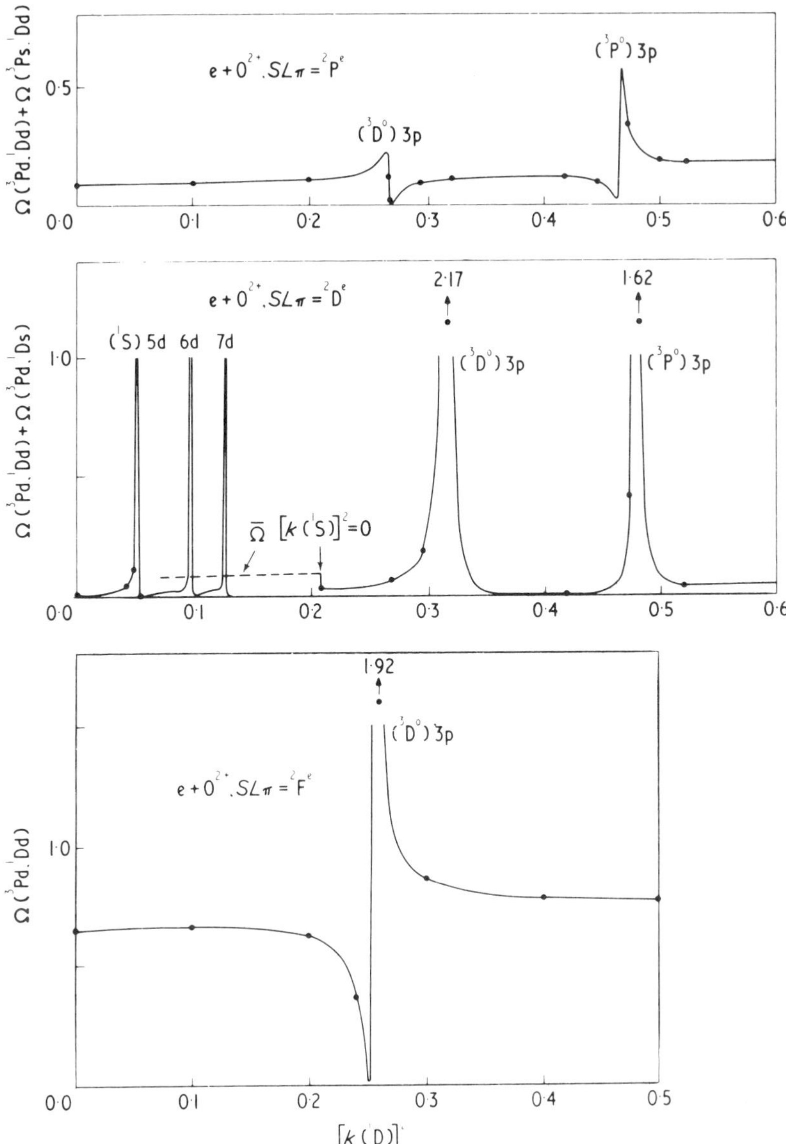

Figure 1(*b*). Results for $\Omega_{SL\pi}({}^3P, {}^1D)$, $SL\pi = {}^2P^e, {}^2D^e$ and ${}^2F^e$.

in the region of $[k({}^1S)]^2 = 0{\cdot}4$ to $0{\cdot}6$ are clearly resolved. From an examination of the dominant contributions to the wavefunctions these resonances can be identified as $2s2p^3({}^3D^o)4d$, $({}^3P^o)3d$, $({}^1P^o)3s$ and $({}^3P^o)4s$.

4.2.4. Results for $\Omega({}^3P, {}^5S^o)$. The collision strength $\Omega({}^3P, {}^5S^o)$ has been calculated by Jackson (1973) using the methods of the present paper.

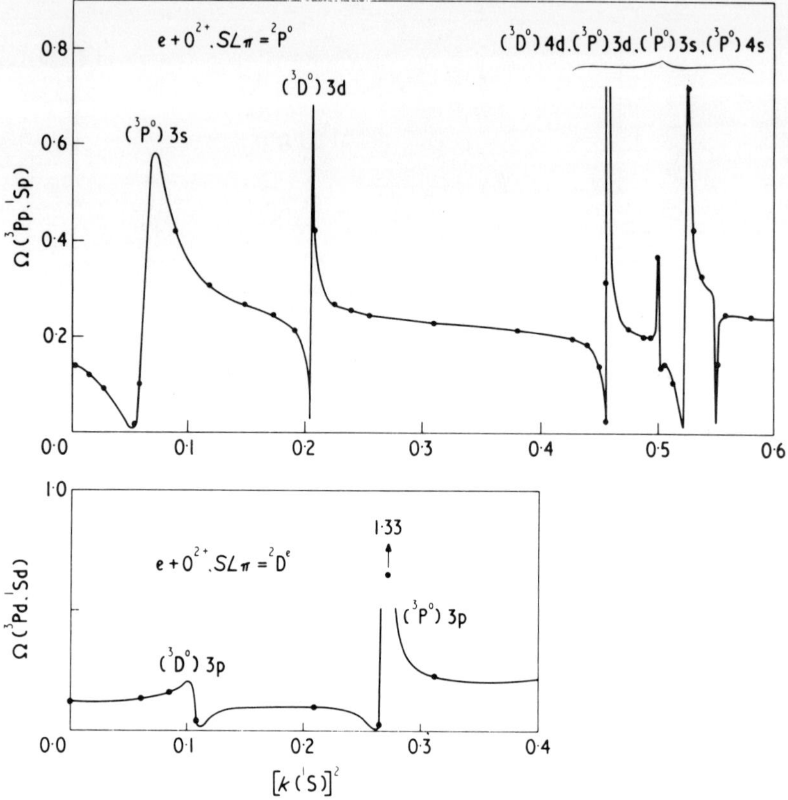

Figure 2. Results for $\Omega_{SL\pi}(^3\mathrm{P}, {}^1\mathrm{S})$, $SL\pi = {}^2\mathrm{P}^{\circ}$ and ${}^2\mathrm{D}^{\mathrm{e}}$.

4.3. *Comments on computational techniques*

In figures 1, 2 and 3 the points for which the coupled equations were solved have been marked as filled circles. The full lines are obtained using the fitting procedures described in § 4.1. The broken lines are collision strengths averaged over resonances (Gailitis 1963, Seaton 1969).

It was found that, with high closed channels, numerical integrations tended to be unstable in the asymptotic region. This difficulty was overcome using the Fox–Goodwin method described by Norcross and Seaton (1973).

4.4. *Reaction rates*

For applications in astronomy we require the reaction rates $q(i \to i') = \langle v_i Q(i \to i') \rangle$ where v_i is the velocity of the incident electron and the average is over a Maxwell distribution. For *de-excitation* ($E_i > E_{i'}$) we have

$$q(i \to i') = \frac{8{\cdot}63 \times 10^{-6} \Upsilon(i \to i')}{\omega_i T^{1/2}} \mathrm{cm}^3\,\mathrm{s}^{-1} \tag{4.16}$$

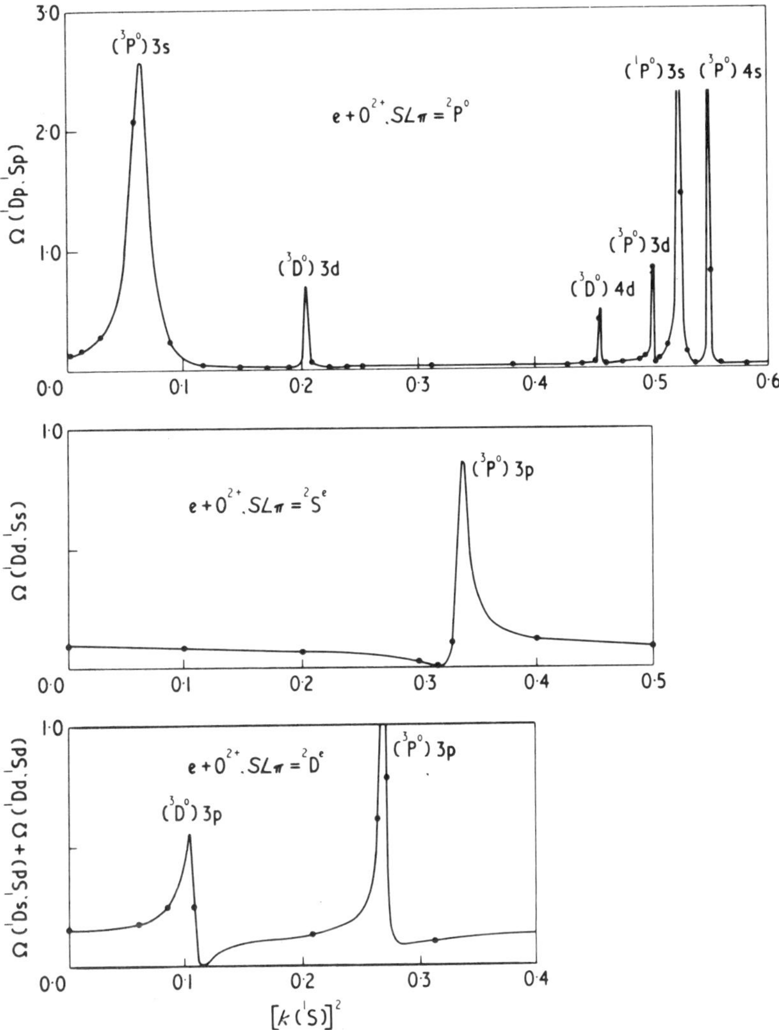

Figure 3. Results for $\Omega_{SL\pi}(^1\mathrm{D}, ^1\mathrm{S})$, $SL\pi = {}^2\mathrm{P}^{\mathrm{o}}$, ${}^2\mathrm{S}^{\mathrm{e}}$ and ${}^2\mathrm{D}^{\mathrm{e}}$.

where T is the temperature in K and where

$$\Upsilon(i \to i') = \int_0^\infty \Omega(i, i') \exp\left(-\frac{mv_i^2}{2kT}\right) \mathrm{d}\left(\frac{mv_i^2}{2kT}\right). \tag{4.17}$$

For *excitation* we have

$$q(i' \to i) = \frac{\omega_i}{\omega_{i'}} q(i \to i') \exp\left(-\frac{E_i - E_{i'}}{kT}\right). \tag{4.18}$$

Results for $\Upsilon(^1\mathrm{D} \to {}^3\mathrm{P})$, $\Upsilon(^1\mathrm{S} \to {}^3\mathrm{P})$ and $\Upsilon(^1\mathrm{S} \to {}^1\mathrm{D})$ are given in table 4, for values of T in the range 5000 K to 20000 K. This table includes: (i) contributions from the values of $SL\pi$ considered in the present work; (ii) contributions from l and l' greater

Table 4. The parameters Υ required for the calculation of rate coefficients (see § 4.4).

T(K)	Υ(^{1}D → ^{3}P)						
	$SL\pi$ = ^{2}P^o	^{2}D^o	^{2}P^e	^{2}D^e	^{2}F^e	$(l, l') > 2$	Total
5000	0·659	0·704	0·095	0·048	0·657	0·007	2·17
7500	0·647	0·789	0·097	0·059	0·659	0·007	2·26
10000	0·639	0·881	0·099	0·067	0·663	0·007	2·36
12500	0·634	0·953	0·100	0·074	0·668	0·007	2·44
15000	0·629	1·003	0·102	0·081	0·673	0·007	2·49
17500	0·626	1·034	0·103	0·086	0·675	0·007	2·53
20000	0·623	1·050	0·104	0·091	0·675	0·007	2·55
					Saraph *et al* (1969)		2·39

T(K)	Υ(^{1}S → ^{3}P)			
	$SL\pi$ = ^{2}P^o	^{2}D^e	$(l, l') > 2$	Total
5000	0·1472	0·1286	0·0002	0·276
7500	0·1757	0·1288	0·0002	0·305
10000	0·1955	0·1293	0·0002	0·325
12500	0·2083	0·1303	0·0002	0·339
15000	0·2162	0·1314	0·0002	0·348
17500	0·2208	0·1322	0·0002	0·353
20000	0·2229	0·1324	0·0002	0·356
			Saraph *et al* (1969)	0·335

T(K)	Υ(^{1}S → ^{1}D)				
	$SL\pi$ = ^{2}P^o	^{2}S^e	^{2}D^e	$(l, l') > 2$	Total
5000	0·4239	0·0972	0·1662	0·1197	0·807
7500	0·4738	0·0955	0·1702	0·1197	0·859
10000	0·4704	0·0941	0·1716	0·1197	0·856
12500	0·4491	0·0934	0·1718	0·1197	0·834
15000	0·4230	0·0930	0·1710	0·1197	0·807
17500	0·3968	0·0928	0·1694	0·1197	0·779
20000	0·3722	0·0926	0·1671	0·1197	0·752
				Saraph *et al* (1969)	0·310

Note. In the columns headed '$(l, l') > 2$' we include the contribution from $(l, l') = (1, 3)$ and $(3, 1)$.

than 2, from the work of Saraph *et al* (1969); (iii) total Υ's, obtained on summing (i) and (ii); (iv) total Υ's of Saraph *et al.*

For Υ(^{1}D → ^{3}P) and Υ(^{1}S → ^{3}P), the contributions from $(l, l') > 2$ are very small. This contribution is larger for Υ(^{1}S → ^{1}D), because this transition has an electric quadrupole moment. It is shown by Seaton (1955) that, for the larger values of (l, l'), the collision strengths for such a transition are proportional to the square of the quadrupole moment, S_2^2. The ratio of S_2^2 obtained from our wavefunctions to that obtained from the functions of Saraph *et al* is 1·023. We have multiplied the contributions to Υ(^{1}S → ^{1}D) from $(l, l') > 2$, calculated by Saraph *et al*, by this factor.

From table 4 it is seen that the present results for Υ(^{1}D → ^{3}P) and Υ(^{1}S → ^{3}P) are close to those of Saraph *et al*. That they are so close is largely fortuitous, since the present results contain a lot of structure not present in the earlier results. There is a much bigger difference in the results for $\Upsilon(^1S \rightarrow {}^1D)$, which can be explained as follows. Saraph *et al* noted that, in the calculation of the $SL\pi = {}^2P^o$ contribution to $\Omega(^1D, {}^1S)$ there is considerable cancellation between monopole exchange integrals and quadrupole potential integrals, which results in this contribution being small. The same effect is seen in our calculations away from resonances. The larger value of $\Upsilon(^1S \rightarrow {}^1D)$ obtained in the present work is a consequence of the $(^3P^o)3s\ {}^2P^o$ resonance, shown in figure 3.

5. The calculations of Ormonde *et al* (1973)

Calculations including $2s^22p^2$ and $2s2p^3$ target states have been made by Ormonde *et al* (1973). Their results are markedly different from ours. Our results should be the more accurate in that we use CI target states, whereas they use SC target states. We have, however, also made SC calculations and find that the results are not very different from the more accurate CI results. We can therefore conclude that the differences between our results and those of Ormonde *et al* are not due to the use of different target states.

The most important differences, so far as astrophysical applications are concerned, are in the positions of the resonances $2s2p^3(^3D^o)3s\ {}^2D^o$ and $(^3P^o)3s\ {}^2P^o$. Fortunately, information on the positions of these resonances can be inferred from experimental data.

Eissner *et al* (1969) consider the C iso-electronic sequence. For ions with residual charge $z = 2$, 3, 4, 5 and 6 (ie O^{2+}, F^{3+}, Ne^{4+}, Na^{5+} and Mg^{6+}), $2s2p^3(^5S^o)3s\ {}^4S$ is observed as a true bound state. Using the program of Eissner and Nussbaumer, Eissner *et al* calculated the position of the 3s $^4S^o$ level. They used a CI function $(2s2p^33s+2s^22p^3)$ and adjusted the scale parameters in their potentials so as to obtain exact agreement with the experimental quantum defects. These are defined to be such that the total energy is $E = E(^5S^o) - z^2/(n-\mu)^2$. The same radial functions were then used in calculations for the states: $(^3D^o)3s\ {}^2D^o$ using the CI function $(2s2p^3(^3D^o)3s+2s2p^3(^1D^o)3s+2s^22p^3)$; and $(^3P^o)3s\ {}^2P^o$ using the CI function $(2s2p^3(^3P^o)3s+2s2p^3(^1P)3s+2s^22p^3)$. The level 3s $^2D^o$ is observed, as a true bound state, for $z = 3$, 5 and 6. The differences between observed and calculated quantum defects were $\{\mu(\text{obs})-\mu(\text{calc})\} = 0{\cdot}016$, 0·016 and 0·017. For $z = 2$, Eissner *et al* neglected coupling to the continuum for the state $(^3D^o)3s\ {}^2D^o$ and obtained $\mu(\text{calc}) = 0{\cdot}788$. The estimate of μ for this state, from the iso-electronic sequence extrapolation, is therefore $\mu(\text{extrap}) = 0{\cdot}788+0{\cdot}016 = 0{\cdot}804$. The extrapolation can also be done in a purely empirical way, since $(1/\mu)$ is very nearly a linear function of z; this gives $\mu(\text{extrap}) = 0{\cdot}80 \pm 0{\cdot}01$.

The method used by Eissner *et al* gives $\mu((^3P^o)3s\ {}^2P^o) = \mu((^3D^o)3s\ {}^2P^o)$. The 3s $^2P^o$ level is observed only for $z = 6$; for this case $\mu(3s\ {}^2D^o)$ agrees with $\mu(3s\ {}^2P^o)$ to three significant figures. We may therefore take $\mu(\text{extrap}) = 0{\cdot}80$ for the 3s $^2P^o$ state in the $(e+O^{2+})$ system.

In the present work we allow for coupling to the continuum. The 3s $^2D^o$ and $^2P^o$ resonances are fairly broad and the calculated positions depend to some extent on the exact way in which they are defined. We take the positions to be defined by the real parts of the complex quantum defects (see Seaton 1969 and Norcross and Seaton 1970). We consider that the iso-electronic sequence extrapolations should give estimates for the real part of the complex quantum defect correct to within about 0·01 for $^2D^o$ and 0·02 for $^2P^o$.

The methods described in the present paper have been used to calculate the position of the bound state, O^+ $(^5S^o)3s\ {}^4S^o$. Table 5 gives quantum defects for the states $(^5S^o)3s\ {}^4S^o$, $(^3D^o)3s\ {}^2D^o$ and $(^3P^o)3s\ {}^2P^o$ in the $(e+O^{2+})$ system. The following results are included:

Table 5. Quantum defects for O^+ $2s2p^3(^5S^o)3s\ ^4S^o$ and real parts of complex quantum defects for $(e+O^{2+})$ $2s2p^3(^3D^o)3s\ ^2D^o$ and $(^3P^o)3s\ ^2P^o$.

State	Values of μ		
	(a)	(b)	(c)
$3s\ ^4S^o$	0·774	0·761	—
$3s\ ^2D^o$	0·80	0·82	0·60
$3s\ ^2P^o$	0·80	0·78	0·58

(a) Experiment for $^4S^o$, iso-electronic sequence extrapolation for $^2D^o$ and $^2P^o$.
(b) Results of present calculations.
(c) Results of Ormonde *et al* (1973).

(a) experiment for $^4S^o$ and iso-electronic sequence extrapolation for $^2D^o$ and $^2P^o$; (b) the present calculations; and (c) values obtained by Ormonde *et al* for $^2D^o$ and $^2P^o$. It is seen that the agreement between (a) and (b) is satisfactory but that the differences between (a) and (c) are much larger than the errors in the iso-electronic sequence extrapolations. We therefore conclude that the resonance positions obtained by Ormonde *et al* are in error.

The iso-electronic problem of electron scattering by N^+ has been discussed by Saraph and Seaton (1974), who obtain results which are very different from those of Ormonde *et al.* Independent calculations by Robb (1974) give results in good agreement with those of Saraph and Seaton.

6. Calculations for bound states

The methods used to solve the $(e+O^{2+})$ collision problem can also be used to calculate bound states for O^+ (Seaton 1974). We consider results for states for which the dominant configurations are $2s^22p^2np$ and $2s^22p^3$.

In table 6 the present results for states with dominant configuration $2s2p^2np$ are compared with experiment and with the results of Saraph and Seaton (1971), who considered only the $2s^22p^2np$ configurations. It is seen that the use of CI functions for the O^{2+} states, and the inclusion of O^{2+} states for which the dominant configuration is $2s2p^3$, gives a marked improvement in the agreement with experiment. As is to be expected, this improvement is most striking for $2s^22p^2(^3P)3p\ ^4S^o$, which is perturbed by $2s2p^3(^5S^o)3s\ ^4S^o$, which lies between 4p and 5p $^4S^o$ (the results for $(^5S^o)3s\ ^4S^o$ have already been given in table 5).

The calculated quantum defects show a systematic error in that they are always smaller than the experimental values. This must be due to our having neglected a large part of the polarizability of the O^{2+} core. We allow for a part of the polarizability of $2s^2$ in $2s^22p^2$, by including $2s2p^3$ states, but we make no allowance for the polarizability of $2p^2$.

Another source of error, for $SL\pi = {}^2P^o, {}^2D^o, {}^2F^o, {}^4P^o$ and $^4D^o$, arises from our neglect of f-waves. This error is likely to be largest for $^2F^o$. Without f-waves, the free channels for $^2F^o$ are $(^1D)p$, $(^3D^o)d$, $(^3P^o)d$, $(^1D^o)d$ and $(^1P^o)d$. When f-waves are included we have three additional channels, $(^3P)f$, $(^1D)f$ and $(^1S)f$. Calculations have been made for this case. It is found that the difference between calculated and experimental effective

Table 6. Effective quantum numbers for $2s^22p^2np$ states in O^+.

State	ν (exp)	ν (calc) Present work	ν (calc) − ν (exp) Present work	ν (calc) − ν (exp) Saraph and Seaton (1971)
$(^3P)3p\ ^2S^o$	2·353	2·357	0·004	0·043
$(^3P)3p\ ^2P^o$	2·518	2·524	0·006	0·032
$(^1D)3p\ ^2P^o$	2·482	2·489	0·007	0·038
$(^3P)4p\ ^2P^o$	3·542	3·551	0·009	0·036
$(^3P)3p\ ^2D^o$	2·472	2·489	0·017	0·033
$(^1D)3p\ ^2D^o$	2·439	2·458	0·019	0·045
$(^3P)4p\ ^2D^o$	3·526	3·545	0·019	0·036
$(^3P)5p\ ^2D^o$	4·502	4·518	0·016	0·033
$(^1D)3p\ ^2F^o$	2·419	2·438	0·019	0·025
$(^3P)3p\ ^4S^o$	2·481	2·494	0·013	0·065
$(^3P)4p\ ^4S^o$	—	3·458	—	—
$(^3P)5p\ ^4S^o$	—	4·626	—	—
$(^3P)3p\ ^4P^o$	2·418	2·428	0·010	0·028
$(^3P)4p\ ^4P^o$	—	3·449	—	—
$(^3P)5p\ ^4P^o$	4·450	4·456	0·006	0·025
$(^3P)3p\ ^4D^o$	2·394	2·403	0·009	0·030
$(^3P)4p\ ^4D^o$	3·420	3·428	0·008	0·028
$(^3P)5p\ ^4D^o$	4·429	4·436	0·007	0·027

Notes. (i) The effective quantum numbers ν given for states labelled $(S_iL_i)np\,SL$ are such that the total energy is $E = E(S_iL_i) - z^2/\nu^2$.

(ii) The experimental effective quantum numbers are taken from Saraph and Seaton (1971), who give references and describe the method used to eliminate fine structure.

quantum numbers is 0·019 without f-waves and 0·017 with f-waves. It can therefore be concluded that the error produced by neglect of f-waves is quite small.

Table 7 gives results for the O^+ ground configuration terms. The present results are compared with experiment and with the results of large multi-configuration Hartree–Fock (MCHF) calculations of Bagus *et al* (1971). We consider ionization energies, since this is the quantity given directly by our method. It is seen that our results are more accurate than the MCHF results, particularly for $SL\pi = {}^2D^o$ and ${}^2P^o$. The experimental energies may differ from exact solutions of the Schrödinger problem in the third figure after the decimal point, due to relativistic effects. Our calculated

Table 7. Ionization energies for O^+ ground configuration terms (Rydberg units).

Term	I (exp)	I (exp) − I (calc) Present work	I (exp) − I (calc) Bagus *et al*
^2P	2·214	+0·015	0·052
^2D	2·339	0·000	0·028
^4S	2·583	−0·031	0·035

results may also contain purely numerical errors in this figure, due to the use of a fairly coarse mesh for the tabulation of the radial functions and the use of a small (32 bit) word-length (further details are given by Seaton 1974).

7. Accuracy of the collision strength calculations

The calculated $2p^2np$ quantum defects agree with experiment to within 2 to 4 per cent. Scattering phase shifts are given by $\eta = \pi\mu$, where μ is the extrapolated quantum defect. The collision strengths behave, essentially, like squares of phase shifts. We therefore estimate that the collision strengths should be correct to within about 5 to 10 per cent.

A further estimate of accuracy can be made by considering the results for the bound states in the $^2D^o$ series (table 6). We have four observed levels, for which the dominant contributions to the wavefunctions are $2s^22p^2(^3P)\,3p$, $(^1D)\,3p$, $(^3P)\,4p$ and $(^3P)\,5p$. Let us consider the **R** matrix for the channels $(^3P)\,p$ and $(^1D)\,p$, denoted by 1 and 2. For bound states we have

$$(R_{11}+\tan\pi\nu_1)(R_{22}+\tan\pi\nu_2)-R_{12}^2 = 0 \tag{7.1}$$

(Seaton 1966) or, putting $\nu_1 = n-\mu_1$,

$$\tan\pi\mu_1 = R_{11}-\frac{R_{12}^2}{R_{22}+\tan\pi\nu_2}. \tag{7.2}$$

For the particular series we are considering, (7.2) is not convenient because μ_1 is close to 0·5 and $\tan\pi\mu_1$ is therefore very large. It is better to work with the matrix $\boldsymbol{\rho} = \mathbf{R}^{-1}$. In place of (7.2) we have

$$\cot\pi\mu_1 = \rho_{11}-\frac{\rho_{12}^2}{\rho_{22}+\cot\pi\nu_2}. \tag{7.3}$$

For an unperturbed series, such as $^4D^o$ in table 6, we find that it is a good approximation to take $\cot\pi\mu_1$ to be a linear function of the energy. In (7.3) we therefore assume that ρ_{11} is a linear function of the energy and that ρ_{22} and ρ_{12} are independent of the energy. The elements of $\boldsymbol{\rho}$ can then be deduced from the four observed levels in the $^2D^o$ series. We consider results at the energy of the $(^1D)\,3p$ level. From fitting to the experimental data we obtain

$$\boldsymbol{\rho}(\text{exp}) = \begin{pmatrix} 0{\cdot}003 & 0{\cdot}246 \\ 0{\cdot}246 & -0{\cdot}222 \end{pmatrix} \tag{7.4}$$

and from using the same procedure to fit to the calculated data we obtain

$$\boldsymbol{\rho}(\text{calc}) = \begin{pmatrix} 0{\cdot}062 & 0{\cdot}260 \\ 0{\cdot}260 & -0{\cdot}158 \end{pmatrix}. \tag{7.5}$$

We can now calculate the **S** matrices at the energy of the $(^1D)3p$ level, to obtain $|S_{12}(\text{exp})|^2 = 0{\cdot}206$ and $|S_{12}(\text{calc})|^2 = 0{\cdot}231$. This indicates an error of about 12% in the calculated collision strength. The contribution to $\Upsilon(^1D\text{–}^3P)$ from $SL\pi = {}^2D^o$ is about 37%, and an error of 12% in this contribution therefore gives an error of 4·5% in Υ. The results of table 6 indicate that the calculations for $SL\pi = {}^2P^o$ are more accurate than those for $SL\pi = {}^2D^o$.

We have made bound state calculations only for the odd parity states. For the even parity states, the contributions from d-waves in the initial and final channels are much more important than the contributions from s-waves. The results for the d-waves in the even parity states should be more accurate than those from the p-waves in the odd parity states. We can therefore feel rather confident that the errors in our final results for the Υ's are probably about 5% and in any case not larger than 10%.

References

Bagus P S, Hibbert A and Moser C 1971 *J. Phys.* B: *Atom. Molec. Phys.* **4** 1611–32

Dance O F, Harrison M F A and Smith A C H 1966 *Proc. R. Soc.* A **290** 74–93

Dolder K T and Peart B 1973 *J. Phys.* B: *Atom. Molec. Phys.* **6** 2415–26

Eissner W 1972 *Proc. Int. Conf. on Physics of Electronic and Atomic Collisions* (Amsterdam: North-Holland) pp 460–78

—— 1975 *J. Phys.* B: *Atom. Molec. Phys.* to be submitted

Eissner W and Nussbaumer H 1969 *J. Phys.* B: *Atom. Molec. Phys.* **2** 1028–43

Eissner W, Nussbaumer H, Saraph H E and Seaton M J 1969 *J. Phys.* B: *Atom. Molec. Phys.* **2** 341–55

Eissner W and Seaton M J 1972 *J. Phys.* B: *Atom. Molec. Phys.* **5** 2187–98

—— 1973 *Mem. Soc. R. Sci. Liège* **5** 203–8

Gailitis M 1963 *Sov. Phys.-JETP* **17** 1328–32

Hebb M H and Menzel D H 1940 *Astrophys. J.* **92** 408–23

Henry R J W, Burke P G and Sinfailam A L 1969 *Phys. Rev.* **178** 218–25

Jackson A R G 1973 *J. Phys.* B: *Atom. Molec. Phys.* **6** 2325–33

Norcross D W and Seaton M J 1970 *J. Phys.* B: *Atom. Molec. Phys.* **3** 579–84

—— 1973 *J. Phys.* B: *Atom. Molec. Phys.* **6** 614–21

Nussbaumer H 1971 *Astrophys. J.* **166** 411–22

Ormonde S, Smith K, Torres B W and Davies A R 1973 *Phys. Rev.* A **8** 262–95

Robb W D 1974 *Phys. Rev. Lett.* in press

Saraph H E 1973 *J. Phys.* B: *Atom. Molec. Phys.* **6** L243–6

Saraph H E and Seaton M J 1971 *Phil. Trans. R. Soc.* A **271** 1–39

—— 1974 *J. Phys.* B: *Atom. Molec. Phys.* **7** L36–40

Saraph H E, Seaton M J and Shemming J 1966 *Proc. Phys. Soc.* **89** 27–34

—— 1969 *Phil. Trans. R. Soc.* A **264** 77–105

Seaton M J 1953 *Proc. R. Soc.* A **218** 400–16

—— 1955 *Proc. R. Soc.* A **231** 37–52

—— 1966 *Proc. Phys. Soc.* **88** 801–14

—— 1969 *J. Phys.* B: *Atom. Molec. Phys.* **2** 5–11

—— 1974 *J. Phys.* B: *Atom. Molec. Phys.* **7** 1817–40

Smith K, Conneely M J and Morgan L A 1969 *Phys. Rev.* **177** 196–203

Smith M W and Wiese W L 1971 *Astrophys. J. Suppl.* **23** 103–92

Weiss A W 1970 *Nucl. Instrum. Meth.* **90** 121–31

1970 *Phys. Rev.* A **2** 81–6
Reprinted with permission from the American Physical Society

Graphic Analysis of Perturbed Rydberg Series*

K. T. Lu and U. Fano
Department of Physics, The University of Chicago, Chicago, Illinois 60637
(Received 19 January 1970)

A method of analysis of all level positions in multiple strongly perturbed series of levels is presented and illustrated by examples from the rare gas and Ba spectra. The method is suggested by Seaton's multichannel quantum defect theory but is presented here as an empirical approach. It emphasizes the dependence of a perturbation on the periodicity of the perturbing series and aims at extracting significant information from experimental data in compact form; this goal will be pursued in later works.

Treatments of the configuration interaction of two (or more) mutually perturbing series currently utilize the Shenston-Russell-Edlên formula,[1] which derives from a second-order perturbation theory by Langer.[2] They also utilize plots of the quantum defect $\mu = n - n^*$ against term value T_n. These plots show an irregularity, similar to an anomalous dispersion curve, wherever a perturbing level of another series occurs in the spectrum. This formalism maintains a clear distinction between a foreign perturbing level, or series of levels, and the perturbed series. If the foreign level were counted among those of the perturbed series, all higher levels of this series would be assigned a value of n one unit higher.

However, this distinction between the two series becomes somewhat artificial when configuration interaction cannot be adequately described by second-order perturbation effects and the wave function of a perturbing level becomes strongly admixed into many levels of the perturbed series. Moreover, plotting μ against T_n fails to emphasize the periodicity of the perturbing series and its convergence to a different limit, beyond that of the perturbed series.

These limitations of the current treatments are overcome, in principle, by Seaton's multichannel quantum defect theory.[3] Efforts have been underway for some time[4] to extend the applications of this theory to extract a maximum of information from experimental data. At this time, it seems useful to present here a graphical method of manipulating data which displays and utilizes simultaneously the Rydberg character of different series and will prove useful for further developments. These developments will be reported later,[5,6] but some of their results will be anticipated here. This paper presents an empirical approach for the analysis of experimental data, but postpones a full explanation of its theoretical basis.

Quantum defect theories express the eigenvalue equation for the discrete levels of a Rydberg series by an equation of the type

$$\sin\pi(\nu_n + \mu) = 0 . \qquad (1)$$

Here ν_n (often called n^*) is the effective quantum number of a level and μ is the quantum defect, which is approximately constant for a whole Rydberg series in the absence of perturbation. The solutions of (1) are, of course,

$$\nu_n = n - \mu . \qquad (2)$$

The Shenstone-Russell-Edlên formula makes μ apparently singular at the spectral location of each separate perturbing level. This paper presents evidence showing that μ is conveniently plotted as a monotonically increasing function of energy which exhibits the Rydberg periodicity of the perturbing series. Moreover, this periodicity is brought out most clearly by plotting together the μ values of all perturbed series which converge to the same limit.

It has been known for a long time[7] that the quantum defect of a Rydberg series extrapolates in the continuum beyond the series limit into the phase shift of the electron-ion scattering problem according to the formula

$$\delta = \pi\mu . \qquad (3)$$

The multichannel theory[3] extends this connection, δ being an eigenphase shift of the scattering problem. The plots of μ shown here and in earlier literature[8] exhibit the series perturbations in the form familiar from the theory of resonance scattering.[9]

As a first illustration we consider a "multichannel" system, that is, an atom with several series, some of which converge to the ground state and some to an excited state of an ion. Photoabsorption by Xe in its ground state leads to p^5d or p^5s, $J=1$, odd-parity states belonging to five series, of which three, called[10] $5p^5(^2P_{3/2})ns[1\frac{1}{2}]^\circ$, $5p^5(^2P_{3/2})nd[\frac{1}{2}]^\circ$, $5p^5(^2P_{3/2})nd[1\frac{1}{2}]^\circ$, converge to the first ionization potential $T_{1\infty}$ and two, $5p^5(^2P_{1/2})ns'[\frac{1}{2}]^\circ$, $5p^5(^2P_{1/2})nd'[1\frac{1}{2}]^\circ$, converge to a second ionization potential $T_{2\infty}$. It is usually said that the first three series are perturbed by members of the last two series. Here we do not distinguish initially between the "perturbed series" and the "perturbing level," or even between level series converging to the same limit. Instead, we consider equally all experimental levels T_m, with $J=1$ and odd parity, below the first limit $T_{1\infty}$; to each of them we assign *two* alternative effective quantum numbers ν_1 and ν_2 defined, respectively, by the equations

$$T_m = T_{1\infty} - R/\nu_{1m}^2, \tag{4a}$$

$$T_m = T_{2\infty} - R/\nu_{2m}^2, \tag{4b}$$

where R is the Rydberg constant. These equations imply the functional relation between ν_1 and ν_2,

$$\nu_1 = \nu_2/(1-\nu_2^2\,\Delta)^{1/2}, \quad \Delta = (T_{2\infty} - T_{1\infty})/R. \tag{5}$$

We also assign to each level T_m a single quantum defect

$$\mu_m = n - \nu_{1m}, \tag{6}$$

where n is an integer shown in Fig. 2(a), which may have the same value for two (or more) levels; we shall usually plot only the decimal part of μ, i.e., we plot μ *modulo* 1, except for some extrapolation to display periodicities.

Values of μ_m for the 28 levels available for Xe[10] are thus plotted in Fig. 1 against ν_{2m}. To emphasize the connection with usual practice, the 25 points (μ_m, ν_{2m}) usually assigned to series converging to $T_{1\infty}$ are marked by open circles, and the three points usually assigned to series converging to $T_{2\infty}$ are marked by ×. The solid and dashed lines joining the points are interpolated curves which represent all quantum defects as points lying on the plot of a *multivalued continuous function* $\mu(\nu_2)$. In terms of this continuous function, Eqs. (4a), (4b), and (2) imply together that the observed levels are determined as the roots of the equation, analogous to (1),

$$\sin\pi[\nu_1 + \mu(\nu_2)] = 0. \tag{7}$$

In this paper $\mu(\nu_2)$ is regarded as an empirical function. The connection of this equation with Seaton's theory is outlined in the Appendix. As shown in Fig. 2(a), Eq. (7) requires each point (μ_m, ν_{2m}) to lie at an intersection of the curve $\mu(\nu_2)$ and of a plot *modulo* 1 of the function $\nu_1(\nu_2)$ defined by (5).

Key properties of $\mu(\nu_2)$ are apparent by inspection of Fig. 1, but are brought out more clearly by constructing the corresponding graph for Ar [Fig. 2(a)]. This construction has recently been made possible by extensive new data.[11] Most of the available points are concentrated in the interval $8.0 < \nu_2 < 8.4$ near the $T_{1\infty}$ limit. The remaining

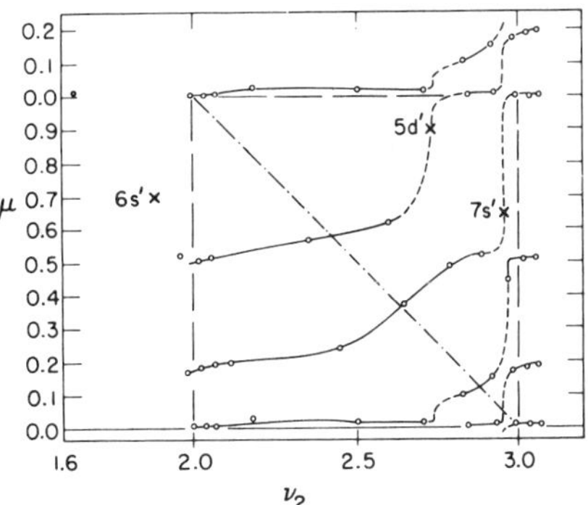

FIG. 1. The quantum defect μ-versus-ν_2 plot of Xe. Open circles (o) are level positions of the three series $p^5(^2P_{3/2})ns[1\frac{1}{2}]^\circ$, $p^5(^2P_{3/2})nd[\frac{1}{2}]^\circ$, and $p^5(^2P_{3/2})nd[1\frac{1}{2}]^\circ$. Crosses (×) labeled nl' correspond to level positions of the two series $p^5(^2P_{1/2})ns'[\frac{1}{2}]^\circ$ and $p^5(^2P_{1/2})nd'[1\frac{1}{2}]^\circ$. Solid and dashed lines are interpolated with varying degree of confidence. A diagonal dot-dash line represents the equation $\mu(\nu_2)+\nu_2=1$. The curves outside the basic unit square are the repeated portions of those within the square.

67 points might not quite suffice to draw the curves $\mu(\nu_2)$ with reasonable assurance, were it not for the clear pattern of periodicity in ν_2. (This periodicity is less clear in Fig. 1, where most data fall in a unit range of ν_2.) The periodicity is utilized in Fig. 2(b), where the data of Fig. 2(a) are replotted *modulo* 1, i.e., with successive unit intervals of ν_2 folded onto one another. The points corresponding to such different intervals are now seen to interpolate rather smoothly to define a single multibranched periodic curve $\mu(\nu_2)$.

Inspection of both Figs. 1 and 2(b) reveals now the key point of this paper, namely, that $\mu(\nu_2)$ is a periodic function of ν_2. One may also consider $\mu(\nu_2)$ as defined implicitly by an equation $F(\mu, \nu_2) = 0$, where F is a periodic function of both μ and ν_2 (see Appendix). Thereby each branch of the curve which exits from one margin of the basic unit square of the plot reappears at the corresponding point of the opposite margin. If one regards such corresponding points on opposite margins as the "same" point, all branches of the curve in each figure are seen to be parts of a single continuous curve. In our example, where three "series" converge to $T_{1\infty}$ and two to $T_{2\infty}$, any vertical line drawn across the basic unit square intersects $\mu(\nu_2)$ three times and any horizontal line intersects it two times.

The theory of unperturbed Rydberg series relies on the possibility of introducing a quantum defect μ which is nearly constant as the energy of succes-

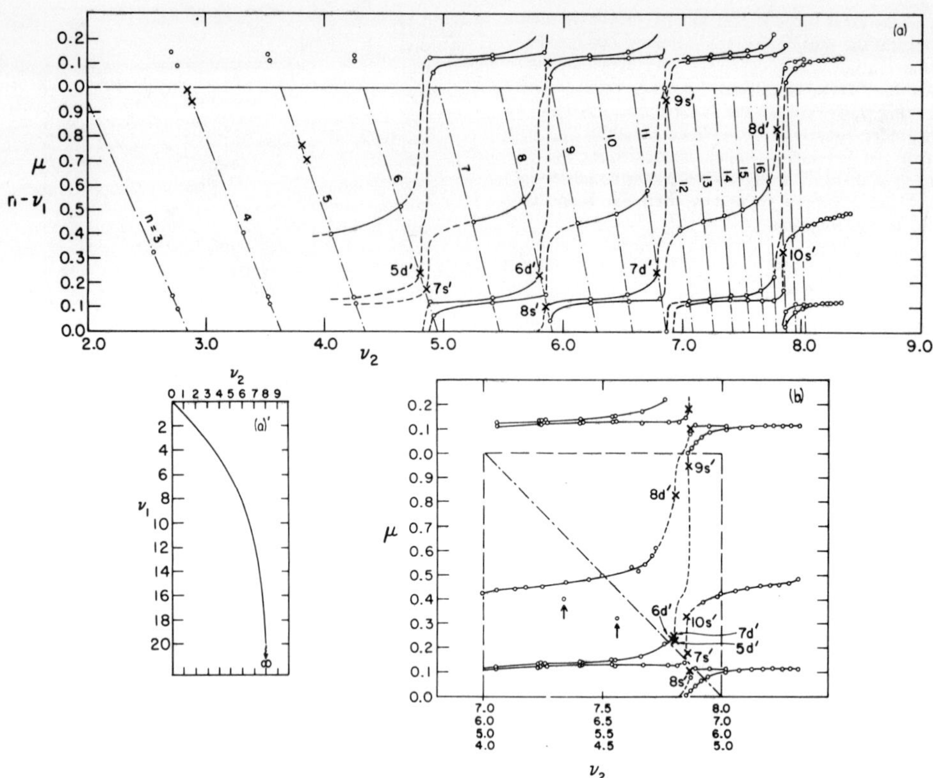

FIG. 2. (a) The quantum defect μ-versus-ν_2 plot of Ar. Open circles and cross marks as in Fig. 1. No interpolated curves are drawn for $\nu_2 < 4$ because points are too sparse. The function $-\nu_1(\nu_2)$ defined by Eq. (5) is shown in the box (a)′ and again, *modulo* 1, by a dot-dash line. Numerical labels on these curves indicate values of integer n of Eq. (6). (b) The folded quantum defect μ-versus-ν_2 plot of Ar. The data of (a) with $\nu_2 > 4.0$ are replotted with ν_2 scale *modulo* 1, i.e., with successive intervals of ν_2 folded onto one another. Some points with $\nu_2 < 4.0$ are included (↑) to show departure from the interpolated curves.

sive levels varies along the spectrum. (Minor variations of μ for the lower levels of a series are commonly observed and taken for granted.) Here, we have introduced the *function* $\mu(\nu_2)$ which is similarly understood to be nearly, but not quite, periodic along the spectrum. In Fig. 1, three points lie at $\nu_2 < 2$; shift of each of these points by one unit of the abscissas brings each of them close to – but not quite on – one branch of the curve $\mu(\nu_2)$. Similarly, Fig. 2(a) includes a number of points with $\nu_2 < 4$; some of these points are those that appear in Fig. 2(b) at substantial distance from the curves.

The crosses, which represent levels usually assigned to "perturbing" series lie on sharply rising portions of the curves. Indeed, if all interactions between series were very weak, all unperturbed levels of the "perturbed" series would be represented by circles lying on flat portions of the curves, i.e., with constant μ values. The flat portions would be separated by sharp steps and pairs of curves would nearly touch each other at the step corners. Some of the curves in Figs. 1 and 2 actually rise quite sharply, leaving small gaps at the corners, other steps are smoother; in some instances no reasonably flat, i.e., unperturbed, portion of curve is seen. The magnitude of the gaps between successive curves provides a visual estimate of the strength of the interaction between different series.

Proceeding now to a more detailed analysis of data, one notes in Fig. 1 a number of points with nearly the same quantum defect $\mu_1 \simeq 0.0$. This set of points represents the $5p^5(^2P_{3/2})ns[1\frac{1}{2}]^\circ$ channel according to Moore's assignment. Among the other points one can roughly single out two groups with quantum defects $\mu \cong 0.5$ and 0.18, respectively. These two sets of points represent levels labeled[10] $5p^5(^2P_{3/2})nd[\frac{1}{2}]^\circ$ and $5p^5(^2P_{3/2})nd[1\frac{1}{2}]^\circ$, respectively. However, they belong neither purely to $nd_{3/2}$ nor to $nd_{5/2}$; most of them being also appreciably perturbed. Their character remains to be determined in a separate paper.[6] Since the level usually labeled $p^5P_{1/2}5d'$ lies at $\nu_2 = 2.73$, the appearance of the curves indicates that this level interacts strongly with $P_{3/2}s$. The $p^5P_{1/2}7s'$ level, which lies at 2.95, interacts less strongly, mostly with $P_{3/2}d[\frac{1}{2}]$ and less so with $P_{3/2}d[1\frac{1}{2}]$ and $P_{3/2}s_{1/2}$. (Incidentally, it had been previously suggested[12] that s' levels would interact mostly with $P_{3/2}s_{1/2}$.) The pair of steps in the curves between $\nu_2 = 2.6$ and 3.0 resembles the steps that would be observed in scattering-phase plots in the region of two somewhat overlapping resonances. In the scattering problem the sum of the phases of all open channels increases by π at each resonance; here the sum of the three μ values increases by unity.[13]

The unperturbed part of the $nd[1\frac{1}{2}]^\circ$ series of Ar [Figs. 2(a) and 2(b)] is almost degenerate with the $ns[1\frac{1}{2}]^\circ$ series unlike the corresponding series of Xe. The levels of the "perturbing" series $p^5P_{1/2}nd'$ lie at $\nu_2 = 4.81$, 5.80, 6.79, and 7.8, corresponding to $n = 5$, 6, 7, and 8, respectively. The appearance of the curves indicates that this series interacts strongly with the series $P_{3/2}d[\frac{1}{2}]^\circ$, less strongly with $P_{3/2}d[1\frac{1}{2}]^\circ$, and very weakly with $P_{3/2}s[1\frac{1}{2}]^\circ$. The levels of the "perturbing" series $p^5P_{1/2}ns'$, which lie at $\nu_2 = 4.85$, 5.87, 6.86, and 7.85, corresponding to $n = 7$, 8, 9, and 10, respectively, interact less strongly with $P_{3/2}s[1\frac{1}{2}]^\circ$ and very weakly with $P_{3/2}d[\frac{1}{2}]^\circ$ and $P_{3/2}d[1\frac{1}{2}]^\circ$. The fact that the interaction between the d and s series is so weak was also noted by Yoshino.[11]

For Kr we have only a limited amount of data,[14] which are plotted for purposes of orientation and comparison in Figs. 3(a) and 3(b). The lower levels are not adequate to construct good interpolated curves in Fig. 3(a), nor do they fit well the folded plot in Fig. 3(b). The analysis may be improved when higher levels become available experimentally. The curve pattern of Kr shows characteristics intermediate between those of Ar and Xe. There is no degeneracy among series as in Ar. The curves also show the two stepwise jumps and an interaction between d and s series which is larger than in Ar and smaller than in Xe.

A final illustration, Fig. 4 shows a $\mu(\nu_2)$ plot for the single principal series $(6smp\,^1P_1^\circ)$ of Ba.[15] As $T_{1\infty}$ we take, of course, the series limit $6s\,^2S_{1/2}$, and as $T_{2\infty}$, we take somewhat arbitrarily the series limit $5d\,^2D_{5/2}$, to which the series $5dmp\,^1P_1^\circ$ converges. Here the interpolated curve of the principal series jumps, in essence, by one unit of μ whenever it passes through one of the perturbing levels $5d6p\,^1P_1^\circ$, $5d8p\,^1P_1^\circ$, $5d8p\,^3D_1^\circ$, and $5d8p\,^3P_1^\circ$, which are marked by $\times$ in the graph. Note how the points corresponding to $6smp\,^1P_1^\circ$ with $m = 6$, 7, 8, and 9 lie rather well, but not perfectly, along the line determined by higher m values. In fact the plot serves to call attention to possibly aberrant data. In particular, the point corresponding to the level called by Garton[15] $5d4f\,^1P_1^\circ$ (?) does not seem to be in the right place. The levels $5d8p\,^3D_1^\circ$ and $5d8p\,^3P_1^\circ$ actually belong to series converging to the limit $5d\,^2D_{3/2}$

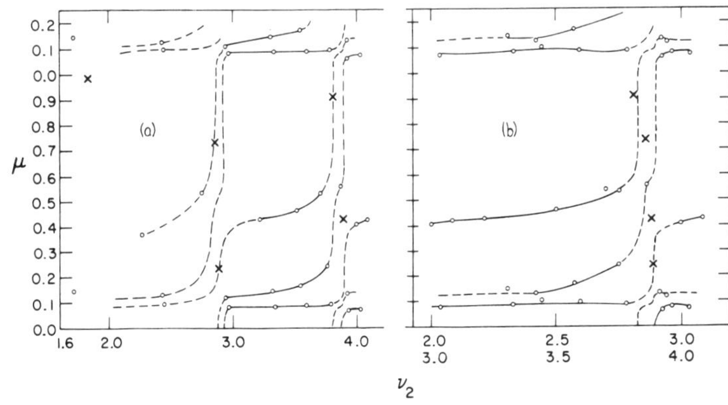

FIG. 3. (a) The quantum defect μ-versus-ν_2 plot of Kr. Because of the sparse distribution of the points the interpolated curves are sketched most tentatively. (b) The data of (a) with $\nu_2 > 2.0$ replotted *modulo* 1. The tentative interpolated curves are drawn through the points with $\nu_2 > 3.0$.

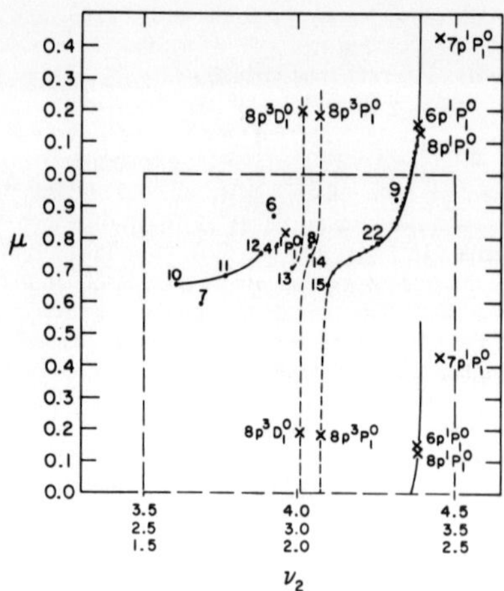

FIG. 4. The quantum defect μ-versus-ν_2 plot of Ba. Dot numbers indicate the n value of principal series $6snp\,^1P_1^\circ$ levels. Crosses labeled $nl^{1,3}L_1^\circ$ correspond to $5dnl^{1,3}L_1^\circ$.

of Ba$^+$, which lies somewhat below $T_{2\infty}$ ($5d^2D_{5/2}$ of Ba$^+$). The extension of the present paper, to take into account the occurrence of three or more series limits, remains to be explored.

We wish, however, to indicate some of the directions of the investigation now in progress.[6] Line strengths and g factors may be plotted as periodic functions of ν_2, showing systematic variations related to those of $\mu(\nu_2)$. The intensity plots extrapolate into the Beutler spectra of the continuum (see, e.g., Fig. 28 of Ref. 16). A point of particular interest concerns the intersections of $\mu(\nu_2)$ curves with the $\mu(\nu_2)+\nu_2 = 1$ diagonal lines shown in Figs. 1 and 2(b).[5] States of the atom represented by points close to these intersections are approximately eigenstates of the scattering problem of one electron colliding with a ground state or excited ion core with near-zero energy; the μ values at the intersections represent, to within a factor of π, the eigenphases of this scattering problem.

We emphasize in conclusion that the curves in the graphs of this paper have no claim to accuracy. They have been drawn somewhat sketchily for purposes of illustration; more accurate determinations should follow. Eventually the approach outlined here might be utilized by experimenters interested in specific spectra.

APPENDIX: CONNECTION WITH SEATON'S THEORY

The levels T_m of a discrete spectrum are determined in Ref. 3 through the roots of the determinant Eq. (7), namely,

$$|\tan\pi\nu + \underline{R}| = 0. \tag{A1}$$

Here $\underline{R}$ is a reaction matrix which varies slowly as a function of energy and ν is a matrix whose eigenvalues $\nu_\alpha = \nu_1, \nu_2, \ldots,$ are related to one another and to the energy E by the equation analogous to our Eq. (4):

$$E/hc = T_{1\infty} - R/\nu_1^2 = T_{2\infty} - R/\nu_2^2 \ldots = T_{\alpha\infty} - R/\nu_\alpha^2 = \ldots, \tag{A2}$$

the $T_{\alpha\infty}$ being levels of the ion core spectrum ($T_{1\infty} \leq T_{2\infty} \leq \cdots$). Equation (A1) is an eigenvalue equation because the ν_α are all related to one another. (In the main text of this paper, with reference to rare gases a single symbol ν_1 has been used to indicate three degenerate eigenvalues while ν_α ν_2 has been used for the other two values.)

To connect Seaton's work[17] with the empirical procedure of this paper we envisage the solution of (A1) by a two-step parametric procedure. In the first step we *replace* in (A1) the elements ν_1 of the diagonalized matrix ν by a parameter $-\mu$. The equation thus obtained then defines μ as an implicit function $\mu(\nu_2 \ldots \nu_\alpha \ldots)$ of all elements of ν with $\alpha > 1$; this definition holds for any value of the energy E. [This paper actually determines the same function μ by interpolation of experimental data rather than from implied knowledge of $\underline{R}$ and solution of the modified equation (A1).] The next step of the procedure consists of seeking those values of ν_1 related to $\nu_2 \cdots \nu_\alpha \cdots$ by (A2), which fulfill with $\mu(\nu_2 \cdots \nu_\alpha \cdots)$ the eigenvalue equation

$$\tan\pi[\nu_1 + \mu(\nu_2 \cdots \nu_\alpha \cdots)] = 0, \tag{A3}$$

equivalent to (7).

---

*Work supported by the U. S. Atomic Energy Commission, under Contract No. COO-1674-27.

[1] B. Edlén, *Encyclopedia of Physics* (Springer-Verlag, Berlin, 1964), Vol. 27, p. 80.

[2] R. Langer, Phys. Rev. 35, 649 (1930).

[3] M. J. Seaton, Proc. Phys. Soc. (London) 88, 801 (1966).

[4] K. T. Lu, Bull. Am. Phys. Soc. 13, 37 (1968).

[5] U. Fano (unpublished).

[6] K. T. Lu (unpublished).

[7] See, for example, M. J. Seaton, Monthly Notices Roy. Astron. Soc. 118, 504 (1958).

[8]See, for example, M. J. Seaton, Proc. Phys. Soc. (London) 88, 815 (1966); and D. L. Moores, *ibid.* 88, 843 (1966).

[9]P. G. Burke and H. M. Schey, Phys. Rev. 126, 147 (1962). This fact is also well known in nuclear and particle physics.

[10]C. E. Moore, *Atomic Energy Levels* (U. S. Natl. Bur. Std., Washington, D. C., 1958), Circ. No. 467, Vol. III.

[11]K. Yoshino, Optical Society of America 1969 Annual Meeting program No. WH12 (unpublished); K. Yoshino (unpublished). (We express our appreciation to Dr. Yoshino for allowing us full use of the data before formal publication.)

[12]H. Beutler, Z. Physik 93, 177 (1935); F. J. Comes and H. G. Sälzer, Phys. Rev. 152, 29 (1966).

[13]J. Macek, Phys. Rev. A 1, 618 (1970); and see, for example, P. G. Burke, J. W. Cooper, and S. Ormonde, *ibid.* 183, 245 (1969).

[14]Reference 10, Vol. II (1952).

[15]W. R. S. Garton and K. Codling, Proc. Phys. Soc. (London) 75, 87 (1960); W. R. S. Garton and F. S. Tomkins, Astrophys. J. 158, 1219 (1969).

[16]U. Fano and J. Cooper, Rev. Mod. Phys. 40, 441 (1968).

[17]M. J. Seaton, J. Phys. B2, 5 (1969).

1984 *J. Phys. B: At. Mol. Phys.* **17** 215–30

Alternative parameters of channel interactions: I. Symmetry analysis of the two-channel coupling

A Giusti-Suzor† and U Fano

Department of Physics, University of Chicago, Chicago, Illinois 60637, USA

Received 8 February 1983, in final form 26 July 1983

Abstract. Translations of the scales of the Lu–Fano plots are introduced phenomenologically to bring out the symmetry of two-channel coupling. These shifts in the (ν_i) space amount to a phase renormalisation of the Coulomb basis wavefunctions of ODT, which eliminates the diagonal elements of Seaton's reactance matrix. The off-diagonal element of the resulting matrix measures the effective coupling strength and the new origins of the Lu–Fano plot axes mark the extrema of channel admixture. In the continuous energy range, the parameters $(\tilde{\mu}_i, \xi)$ provide a compact expression of the cross section for any two-channel (one open and one closed) process, including the whole series of resonances due to the discrete states in the closed channel. A similar generalisation of the Beutler–Fano resonance formula has been previously achieved by Dubau and Seaton.

1. Introduction

Analysis of atomic and molecular optical spectra has increasingly utilised the multichannel quantum defect theory (MQDT) (Wynne and Armstrong 1979, Jungen and Dill 1980, Seaton 1983 and references therein). The original procedures (Seaton 1966, Fano 1970, 1975) reflected the context of their original applications. In particular the early semiempirical analyses of multichannel spectra (Fano 1970, Lu and Fano 1970, Lu 1971) dealt with noble gases or with H_2. In these cases, transitions of the ion core due to interaction with a Rydberg electron merely change its spin–orbit coupling scheme or its ro-vibrational state. However the same interaction produces far more drastic changes in the cores of group II–VII atoms with open valence shells or in molecules prone to dissociation. Therefore the recent application of MQDT to phenomena confined to high Rydberg states (Aymar and Robaux 1979, Gallagher *et al* 1980) or to molecular dissociation (Giusti 1980, Giusti-Suzor *et al* 1983, Giusti-Suzor and Jungen 1984) does not fit readily in the existing pattern: spectral analysis must sort out different effects of electron–core interaction and parametrise them separately before attempting a comprehensive description.

Adaptation of MQDT procedures to such varied tasks requires added flexibility. To this end we shall utilise in this paper a phase renormalisation of Coulomb-field radial functions, introduced a long time ago by Eissner *et al* (1969) (see also Seaton 1983, § 6.8). The standard base pair (f, g) is replaced by a phase-shifted pair:

$$(f, g) \rightarrow (f \cos \phi - g \sin \phi, g \cos \phi + f \sin \phi). \qquad (1)$$

† NSF-CNRS Fellow; permanent address: Laboratoire de Photophysique Moleculaire, Université Paris-Sud, Batiment 213, 91405 Orsay Cedex, France.

0022-3700/84/020215+16$02.25

This phase renormalisation induces a transformation of Seaton's reactance matrix R. Here we will utilise sets of phase parameters (one of them for each channel) which eliminate the diagonal elements of the transformed matrix—a reactance matrix of this form has been introduced heuristically by Giusti (1980) in a particular context of molecular channels, and more consciously by Cooke and Cromer (1983) for the analysis of atomic spectra. Its general significance, i.e. sorting out the intra- and interchannel effects of channel interaction, will be analysed in a separate paper (Giusti-Suzor and Fano 1984, paper II in this series) that derives MQDT formulae from a Hamiltonian formulation of channel interactions. Another possible role of the transformation (1) is to take into account indirectly and compactly those effects of the electron–core interactions which need not be specified for a limited purpose. An example of this process will be described in a further paper (III) in this series dealing with a localised perturbation of the Ba Rydberg series previously fitted within an extensive MODT analysis (Aymar and Robaux 1979).

In the present paper most of the developments will deal with two channels only, for the purpose of simplicity and because the consequences of the basis transformation (1) are more transparent in this case. The multichannel case is being examined by Cooke and Cromer (1983) with the goal of obtaining a more effective representation of channel functions, especially in the autoionisation range of energies. In order to be specific, our treatment will be further restricted to the original QDT framework of ion plus electron complexes, for which the pair of base functions (f, g) and its modification (1) are solutions of the radial wave equation for an electron in a Coulomb field. However, extension to complexes whose fragments interact at long range through other forces, in particular to dissociative molecular channels (atomic fragments) will often be indicated. The two pairs of Coulomb functions in equation (1) have merely to be replaced by the corresponding pairs, in accordance with Mies (1980) or with Greene *et al* (1979, 1982). We stress finally that although several points in this paper were familiar in different forms before, an essential purpose of the present formulation is to unify them, in particular by closely relating the analysis of bound and continuous spectra and by connecting the results of configuration mixing and quantum defect theories.

2. Alternative formulations of two-channel QDT

2.1. *The compatibility equation and the Lu–Fano plot*

We first give a brief review of basic MDQT formulae as an introduction of the following developments. In the single-channel QDT, the energies of the successive levels of a Rydberg series are identified as zeros of the coefficient of the divergent part of the Rydberg electron wavefunction,

$$\sin \pi(\nu+\mu)=0. \tag{2}$$

Here ν is the effective principal quantum number, related to the binding energy ε by

$$\varepsilon=E-I=-\frac{\text{Ryd}}{\nu^2} \qquad (\text{Ryd}=\text{Rydberg constant}) \tag{3}$$

and μ is the quantum defect ($\pi\mu$ extrapolates smoothly beyond the series limit I into the phaseshift of the electron–ion scattering). Note that the condition (2) for bound states in a closed channel holds for any kind of long-range potential, $\nu+\mu$ representing more generally the number of half wavelengths of the radial wavefunction between $r=0$ and $r=\infty$ and μ the contribution of the 'reaction zone' ($r<r_0$). The only change is in the relation between ν and the energy: equation (3) is specific to the Coulomb field and should be replaced by another relation, possibly numerical, for other long-range interactions (see § IV of Greene *et al* 1982).

Equation (2) has been extended in different ways to multichannel situations where several Rydberg series perturb each other as they converge to different thresholds. For a given energy E, one defines an effective quantum number ν_i for each threshold I_i by the relation:

$$E = I_i - \mathrm{Ryd}/\nu_i^2. \tag{4}$$

Seaton (1966) showed that each discrete level E corresponds to a set of ν_i values which simultaneously fulfill equation (4) and the determinantal equation that ensures compatible convergence of the radial wavefunction in all channels:

$$\det|\delta_{ij} \tan \pi\nu_i + R_{ij}| = 0. \tag{5}$$

Here R is a reactance matrix that represents the effect of the *short-range* electron–core interactions, whereas the influence of long-range interactions is incorporated in the radial wavefunctions (Coulomb wavefunctions in the present framework).

This formulation in the basis of the 'fragmentation' (or 'collision') channels fits into the conventional scattering theory and has been very fruitful for extrapolation across thresholds and for resonance predictions. Fano (1970, 1975) has introduced an alternative basis, the short-range interaction eigenchannels α. Equation (5) was thus cast in the equivalent form:

$$\det|U_{i\alpha} \sin \pi(\nu_i + \mu_\alpha)| = 0 \tag{6}$$

where $\tan \pi\mu_\alpha$ are the eigenvalues of Seaton's R matrix and $[U_{i\alpha}]$ is the orthogonal matrix of its eigenvectors. The eigenchannel formulation of MQDT is mainly relevant when the matrix U, which describes the change of coupling when going from the collision channels to the eigenchannels, is close to a simple frame transformation matrix known e.g. from angular momentum theory. This occurs in the prototype case of l uncoupling in H_2 (Fano 1970) or in the passage from L–S to j–j coupling in noble-gas atoms.

In the case of two channels equations (5) and (6) can be written explicitly

$$\begin{vmatrix} \tan \pi\nu_1 + R_{11} & R_{12} \\ R_{12} & \tan \pi\nu_2 + R_{22} \end{vmatrix} = (\tan \pi\nu_1 + R_{11})(\tan \pi\nu_2 + R_{22}) - R_{12}^2 = 0 \tag{5'}$$

and

$$\begin{vmatrix} \cos\theta \sin \pi(\nu_1 + \mu_{\alpha_1}) & \sin\theta \sin \pi(\nu_1 + \mu_{\alpha_2}) \\ -\sin\theta \sin \pi(\nu_2 + \mu_{\alpha_1}) & \cos\theta \sin \pi(\nu_2 + \mu_{\alpha_2}) \end{vmatrix} = 0 \tag{6'}$$

where θ is the rotation angle that identifies the 2×2 orthogonal matrix $[U_{i\alpha}]$. These two equations express the same functional relation between ν_1 and ν_2, represented by the same graph in the (ν_1, ν_2) plane, in terms of different sets of three independent short-range parameters.

Equations (5′) and (6′) are periodic (with unit period) in both ν_1 and ν_2, regarded as independent variables, to within any weak energy dependence of the short-range parameters, which is neglected here and commented upon in § 3.3. Their cyclic behaviour may thus be represented adequately within any unit square of the (ν_1, ν_2) plane, to be called a 'reduced zone' (Armstrong *et al* 1981). This zone has usually been chosen as $[0, 1]\times[0, 1]$, for reference to the hydrogenic case; a graph of the effective quantum defect $\mu = -\nu_1$ as a function of ν_2 (zone 1 of figure 1) constitutes then a Lu–Fano plot, which is generally constructed by fitting equation (6′) to observed discrete levels. The eigenchannel parameters of equation (6′) may be 'read' off this plot at its *intersection with the diagonal* $\nu_1 = \nu_2$, because one column of the determinant (6′) vanishes upon setting $\nu_1 = \nu_2 = -\mu_{\alpha_1}$ (or $-\mu_{\alpha_2}$). The ordinate of each intersection thus measures an eigenquantum defect μ_α. The slopes of the plot at those points equal $\tan^2\theta$ and $\cot^2\theta$ (equation (39′) of Fano 1970).

This familiar choice for the reduced zone obscures important aspects of the plot's general behaviour and its symmetry properties which appear clearly on the extended

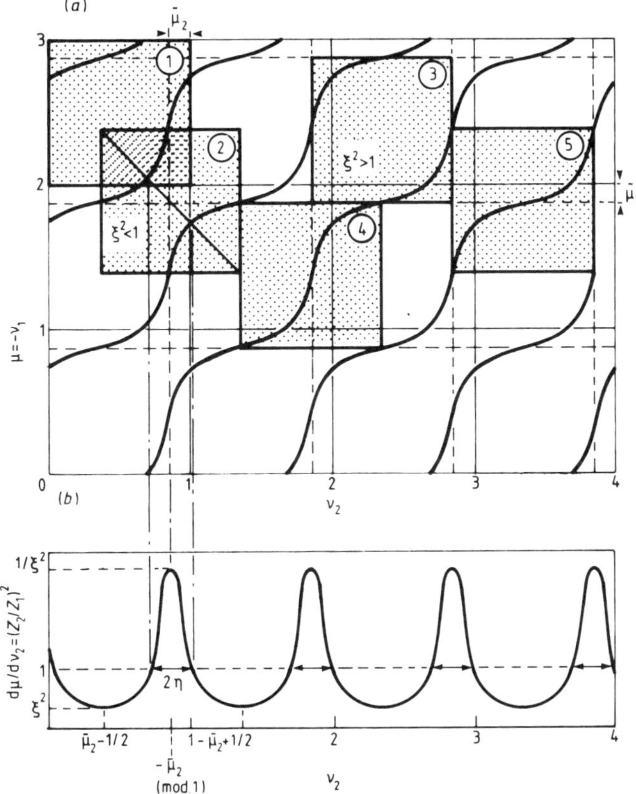

Figure 1. (*a*) Extended plot of the doubly periodic equations (5′), (6′) or (8) for two-channel interaction (adapted from figure 10 of Herzberg and Jungen 1972). The full line grid corresponds to the usual axes of the Lu–Fano plot, lying at integer values of ν_i as shown in the reduced zone 1. The broken line grid corresponds to $\nu_i = -\tilde{\mu}_i$ (modulo 1) and intersects the plot at its points of maximum or minimum slope. The reduced zones 2 to 5 are adapted to the symmetry of the plot as discussed in the text. (*b*) Slope $d\mu/d\nu_2$ of the curves in (*a*), equal to the squared ratio of channel amplitudes $(Z_2/Z_1)^2$ (equation (3)), as a function of ν_2. The arrows mark the points of equal admixture of both channels at the intersection of the plot with the diagonal $\nu_1 + \tilde{\mu}_1 = \nu_2 + \tilde{\mu}_2$.

plot of figure 1. This extended plot consists of equally spaced periodic wavy lines progressing in the general direction $-\nu_1/\nu_2 = 1$. The physical meaning of the inflection points where the lines attain their maximum or minimum slope (see the lower part of figure 1) is shown by the connection between the slope and the ratio of the amplitudes $Z_i (i = 1, 2)$ of the two fragmentation channels (Fano 1970, Herzberg and Jungen 1972):

$$\frac{d(-\nu_1)}{d\nu_2} = \frac{d\mu}{d\nu_2} = \left(\frac{Z_2}{Z_1}\right)^2. \tag{7}$$

Each inflection point thus marks a maximum contribution of one of the fragmentation channels in the total wavefunction. Note that the plots of different two-channel processes will differ only in the *amplitude* of oscillations of the wavy lines and in their *phase* which is identified by the coordinates of the inflection points. These quantities form the centrepiece of the further parametrisation introduced in this paper.

2.2. Symmetrised Lu–Fano plots

Different kinds of reduced zone adapted to the oscillatory behaviour of the graph are identified on figure 1 as zones 2 and 3, 4 and 5. We shall deal mainly with zone 2 where an inflection point lies in the centre of each side and the two branches of the plot come closest in the centre of the square. This zone emphasises the effect of channel mixing which is directly related to the closest approach of the two branches, as we shall see below.

The centre of the square 2 with coordinates $\tilde{\mu}_1$ and $1-\tilde{\mu}_2$ (modulo 1) is a centre of symmetry such that the plot's equation becomes symmetric in the new variables $\tilde{\nu}_i = \nu_i + \tilde{\mu}_i$. By analogy with equation (5′) which depicts an equilateral hyperbola in the coordinates $\tan \pi\nu_i$, we cast the compatibility equation in the form

$$\begin{vmatrix} \tan \pi(\nu_1 + \tilde{\mu}_1) & \xi \\ \xi & \tan \pi(\nu_2 + \tilde{\mu}_2) \end{vmatrix} = \tan \pi(\nu_1 + \tilde{\mu}_1) \tan \pi(\nu_2 + \tilde{\mu}_2) - \xi^2 = 0. \tag{8}$$

It is easily shown that the translations $(\nu_i \to \nu_i + \tilde{\mu}_i)$ in the ν_i space amount to a phase renormalisation of the radial basis wavefunctions in each channel, given by equation (1) with $\phi = \pi\tilde{\mu}_i$. Phase-shifted basis functions were first introduced by Eissner *et al* (1969) for a numerical purpose, namely to remove the isolated poles of the R matrix which arise if one of the eigenquantum defects μ_α is half integer. Here, the transformation incorporates part of the effect of short-range interactions in the basis functions such that the transformed reaction matrix is a purely coupling matrix without diagonal elements, according to equation (8):

$$\tilde{R} = \begin{bmatrix} 0 & \xi \\ \xi & 0 \end{bmatrix}. \tag{9}$$

The interconnections among the alternative sets of parameters in equations (5′), (6′), and (8) are displayed by expanding each determinant as a function of the $\tan \pi\nu_i$ and equating the coefficients of $\tan \pi\nu_1$, $\tan \pi\nu_2$ and $\tan \pi\nu_1 \tan \pi\nu_2$:

$$R_{11} = \cos^2\theta \tan \pi\mu_{\alpha_1} + \sin^2\theta \tan \pi\mu_{\alpha_2} = \frac{\tan \pi\tilde{\mu}_1 + \xi^2 \tan \pi\tilde{\mu}_2}{1 - \xi^2 \tan \pi\tilde{\mu}_1 \tan \pi\tilde{\mu}_2} \tag{10a}$$

$$R_{22} = \sin^2\theta \tan \pi\mu_{\alpha_1} + \cos^2\theta \tan \pi\mu_{\alpha_2} = \frac{\tan \pi\tilde{\mu}_2 + \xi^2 \tan \pi\tilde{\mu}_1}{1 - \xi^2 \tan \pi\tilde{\mu}_1 \tan \pi\tilde{\mu}_2} \tag{10b}$$

$$R_{12}=\tfrac{1}{2}\sin 2\theta(\tan \pi\mu_{\alpha_1}-\tan \pi\mu_{\alpha_2})=\frac{1}{\cos \pi\tilde{\mu}_1 \cos \pi\tilde{\mu}_2}\,\frac{\xi}{1-\xi^2 \tan \pi\tilde{\mu}_1 \tan \pi\tilde{\mu}_2}. \tag{10c}$$

Note that part of these relations may be obtained from the general relation (2.17) of Eissner *et al* (1969):

$$R=(\sin \pi\tilde{\mu}+\cos \pi\tilde{\mu}\tilde{R})(\cos \pi\tilde{\mu}-\sin \pi\tilde{\mu}\tilde{R})^{-1} \tag{11}$$

where e.g. $\cos \pi\tilde{\mu}$ denotes the diagonal matrix with elements $\cos \pi\tilde{\mu}_i$.

Inversion of equations (10) yields the expression of each parameter $\tilde{\mu}_i$, ξ in terms of the matrix elements R_{ij}:

$$\tan 2\pi\tilde{\mu}_1=\frac{2(R_{11}+R_{22}\Delta_R)}{1-\Delta_R^2-(R_{11}^2-R_{22}^2)} \tag{12a}$$

$$\tan 2\pi\tilde{\mu}_2=\frac{2(R_{22}+R_{11}\Delta_R)}{1-\Delta_R^2-(R_{22}^2-R_{11}^2)} \tag{12b}$$

$$\xi=\frac{2R_{12}}{[(R_{11}+R_{22})^2+(1-\Delta_R)^2]^{1/2}+[(R_{11}-R_{22})^2+(1+\Delta_R)^2]^{1/2}} \tag{12c}$$

with $\Delta_R=\det|R_{ij}|=R_{11}R_{22}-R_{12}^2$.

Equation (12c) implies $|\xi|\leqslant 1$. This restriction, adequate for the plot in the reduced zone 2, will be commented on in § 3.2. It has been imposed on the intermediate formula:

$$\xi^2=\tanh\left(\tfrac{1}{4}\ln\frac{(R_{11}+R_{22})^2+(1-\Delta_R)^2}{(R_{11}-R_{22})^2+(1+\Delta_R)^2}\right). \tag{12d}$$

Replacement of tanh by coth, with a corresponding shift of $\tilde{\mu}_1$ and $\tilde{\mu}_2$ by $\frac{1}{2}$, would yield $\xi^2>1$ as required for the plot in zone 3.

We show in the following section that the parameters $\{\tilde{\mu}_i, \xi\}$ represent characteristics of the plot and of the energy dependence of the system which are not readily expressed in terms of the R-matrix elements or of the eigenchannel parameters (μ_α, θ).

3. Interpretation of the parameters $(\tilde{\mu}_i, \xi)$

3.1. Evolution of the channel admixture

Differentiation of equation (8) yields an analytical expression of the slope (7) as a function of ν_2:

$$\frac{d\mu}{d\nu_2}=\left(\frac{Z_2}{Z_1}\right)^2=\frac{\xi^2[1+\tan^2 \pi(\nu_2+\tilde{\mu}_2)]}{\xi^4+\tan^2 \pi(\nu_2+\tilde{\mu}_2)}$$
$$=[\xi^2\cos^2 \pi(\nu_2+\tilde{\mu}_2)+\xi^{-2}\sin^2 \pi(\nu_2+\tilde{\mu}_2)]^{-1} \tag{13}$$

displayed in the lower part of figure 1. Its extrema lie at the middle of each side of the unit square, in accordance with our prescription, with a maximum at $\nu_2=-\mu_2$ (for $\xi^2\leqslant 1$). The ratio $(Z_2/Z_1)^2$ oscillates between its minimum value ξ^2 (at $-\nu_1=\tilde{\mu}_1$) and its maximum value $1/\xi^2$, never reaching zero nor infinity unless $\xi=0$. Both fragmentation channels are thus admixed in the actual wavefunction at all energies and no point in the plot corresponds to a 'pure' i channel. Although being far from new (Fano 1961, Herzberg and Jungen 1972) this result is more clearly displayed in the present approach.

Two extreme cases are noted: for $\xi = 0$ the plot consists of two crossing straight lines corresponding to independent Rydberg series; for $\xi = 1$, it consists of two parallel straight lines of slope one representing two strongly coupled channels with constant equal admixture. Thus the parameter ξ appears as an index for the strength of the channel coupling.

For $0 < \xi < 1$, equal admixture of both channels is reached at two points only, the orthogonal intersections of the plot with the diagonal of the unit square 2, that is, at $\tan \pi(\nu_1 + \tilde{\mu}_1) = \tan \pi(\nu_2 + \tilde{\mu}_2) = \pm \xi$. These points mark the closest approach of the two branches of the plot (Cooke and Cromer 1983) and their abscissae $\tilde{\mu}_2 \pm \eta$ provide a graphical determination of $\xi = \tan \pi\eta$ ($|\eta| < \frac{1}{4}$), directly related to the repulsion between the two branches at their avoided crossing.

In the discrete region of the spectrum the points of extreme or equal admixture of the fragmentation channels will seldom correspond to actually observed levels, that is, to intersections of the plot with the curve representing the energy constraint deduced from equation (4):

$$\nu_1 = \nu_2 \left(1 - \frac{I_2 - I_1}{\text{Ryd}} \nu_2^2\right)^{-1/2}. \tag{14}$$

Their role emerges more clearly from a study of the continuous energy spectrum between the two thresholds $I_1 < E < I_2$. Here the plot displays the variations of the *short-range phaseshift* in the open channel as a function of the energy $E = I_2 - \text{Ryd}/\nu_2^2$. Its slope is proportional to the *time delay* (Wigner 1955) (more precisely to the extra time delay due to non-hydrogenic interactions)

$$2\hbar \frac{\mathrm{d}(\pi\mu)}{\mathrm{d}E} = 2h \frac{\mathrm{d}\mu}{\mathrm{d}\nu_2} \frac{\mathrm{d}\nu_2}{\mathrm{d}E} = h \frac{\nu_2^3}{2\,\text{Ryd}} \frac{\mathrm{d}\mu}{\mathrm{d}\nu_2} \tag{15}$$

which has been shown by Smith (1960) to represent the lifetime of a resonant multichannel scattering process.

The *resonance centres* are defined as the energies at which the phaseshift varies most rapidly and the time delay is maximum, as is most clearly displayed by the reduced zone 4 in figure 1. These energies correspond to the points $\nu_2 = -\tilde{\mu}_2$ (modulo 1) on the plot and follow the Rydberg-like formula:

$$E_n = I_2 - \frac{\text{Ryd}}{(n - \tilde{\mu}_2)^2}. \tag{16}$$

Corresponding values of the time delay

$$2\hbar \left(\frac{\mathrm{d}(\pi\mu)}{\mathrm{d}E}\right)_n = \frac{\pi\hbar(n - \tilde{\mu}_2)^3}{2\,\text{Ryd}} \frac{1}{\xi^2} \tag{17}$$

are usually identified as $4T_n$, where T_n is the resonance lifetime reciprocal to the resonance width

$$\Gamma_n = \frac{2\,\text{Ryd}}{(n - \tilde{\mu}_2)^3} \frac{2\xi^2}{\pi}. \tag{18}$$

Note that a non-zero time delay, $\pi\hbar\nu_2^3\xi^2/2\,\text{Ryd}$, due to mixing with the closed channel, occurs even at the points $\nu_2 = -\tilde{\mu}_2 \pm \frac{1}{2}$ which correspond to energies farthest from any resonance. Another point to be noted is that the resonance energies E_n (equation

(16)) include the *shift* of the discrete levels in the closed channel, due to their interaction with the continuum.

Equations (16) to (18) show that the parameters $\tilde{\mu}_2$ and ξ characterise the resonant behaviour of the open-channel phaseshift, which governs the resonance pattern in the various cross sections. These results apply directly to non-Coulomb long-range interactions, provided equation (16) is replaced by the appropriate relation between ν_2 and the energy, and equation (18) by the more general expression

$$\Gamma_n = \left(\frac{\mathrm{d}E}{\mathrm{d}\nu_2}\right)_n \frac{2\xi^2}{\pi}. \tag{19}$$

3.2. Resonances in elastic and photoabsorption cross sections

The partial cross section for elastic scattering in an open channel with orbital momentum l and wavenumber k can be written as

$$\sigma_l(E) = \frac{4\pi}{k^2}(2l+1)\sin^2(\delta_l + \pi\mu) \tag{20}$$

where δ_l is the contribution of the long-range potential to the open-channel phaseshift. The short-range part $\pi\mu$ may be further expressed as the sum of a constant part $\pi\tilde{\mu}_1$ and a resonant part $\pi(\mu - \tilde{\mu}_1)$, which varies according to equation (8) (with $\nu_1 = -\mu$). The cross section (20) becomes:

$$\sigma_l(E) = \frac{4\pi}{k^2}(2l+1)\sin^2(\delta_l + \pi\tilde{\mu}_1)\frac{[\cot(\delta_l + \pi\tilde{\mu}_1) + \cot\pi(\mu - \tilde{\mu}_1)]^2}{1 + \cot^2\pi(\mu - \tilde{\mu}_1)} \tag{21a}$$

$$= \frac{4\pi}{k^2}(2l+1)\sin^2(\delta_l + \pi\tilde{\mu}_1)\frac{[-\cot(\delta_l + \pi\tilde{\mu}_1) + \tan\pi(\nu_2 + \tilde{\mu}_2)/\xi^2]^2}{1 + \tan^2\pi(\nu_2 + \tilde{\mu}_2)/\xi^4}. \tag{21b}$$

The last factor of equation (21b) has the familiar form (Fano 1961)

$$\frac{(q+\varepsilon)^2}{1+\varepsilon^2} \tag{22}$$

where $\varepsilon = \tan\pi(\nu_2 + \tilde{\mu}_2)/\xi^2$ represents a 'reduced energy' which vanishes at each resonance ($\nu_2 = n - \tilde{\mu}_2$) and runs from $-\infty$ to $+\infty$ between two successive resonances. The profile index

$$q = -\cot(\delta_l + \pi\tilde{\mu}_1) \tag{23}$$

determines the resonance shape, which is the same for a whole series of resonances.

For electron–ion collisions, δ_l is the Coulomb phaseshift and the total elastic cross section diverges. Equation (21b) then has a restricted interest but may nevertheless be useful for describing compactly a resonance pattern in e–atom or atom–atom collisions. For example, Colle (1981) studied in the two-channel QDT formalism a collision between two oxygen atoms colliding along the $^5\Pi_u$ repulsive potential and coupled by spin–orbit effect to the bound levels in the B $^3\Sigma_u^-$ state (the closed channel). Equation (92) of Colle's paper is akin to our equation (21a), but was not expressed explicitly as a function of the closed-channel wavenumber ν_2 as in our equation (21b) because the expression in terms of the eigenchannel parameters used by Colle would have been too complicated.

The corresponding analysis of the photoabsorption cross section requires additional short-range parameters that represent dipole transition amplitudes. In the present framework we define dipole matrix elements $\tilde{D}_i$ $(i=1,2)$ associated to the channel wavefunctions $\tilde{\psi}_i$ which are written outside the reaction zone:

$$\tilde{\psi}_i = \sum_j \phi_j(\tilde{f}_j\delta_{ji} - \tilde{g}_j\tilde{R}_{ji})\dagger \qquad (r > r_0). \tag{24}$$

Here ϕ_j combines the wavefunctions of the ion core and of the angular and spin part of the external electron in the jth channel, and $(\tilde{f}_i, \tilde{g}_i)$ denotes the phase-shifted Coulomb base pair (cf equation (1)):

$$\begin{aligned}\tilde{f}_i &= f_i \cos \pi\tilde{\mu}_i - g_i \sin \pi\tilde{\mu}_i \\ \tilde{g}_i &= g_i \cos \pi\tilde{\mu}_i + f_i \sin \pi\tilde{\mu}_i.\end{aligned} \tag{25}$$

Adopting the approach used by Seaton (1966, 1983) in the standard basis of Coulomb functions, we expand the energy eigenfunction Ψ as:

$$\Psi = \sum_i Z_i \cos \pi(\nu_i + \tilde{\mu}_i)\tilde{\psi}_i. \tag{26}$$

From equation (24) we deduce the matricial relation satisfied by the column vector Z of the channel components Z_i:

$$[\sin \pi(\nu + \tilde{\mu}) + \cos \pi(\nu + \tilde{\mu})\tilde{R}]Z = 0$$

or explicitly:

$$\begin{aligned}\sin \pi(\nu_1 + \tilde{\mu}_1)Z_1 + \xi \cos \pi(\nu_2 + \tilde{\mu}_2)Z_2 = 0 \\ \xi \cos \pi(\nu_1 + \tilde{\mu}_1)Z_1 + \sin \pi(\nu_2 + \tilde{\mu}_2)Z_2 = 0.\end{aligned} \tag{27}$$

The homogeneous system (27), whose compatibility condition is equation (8), yields the ratio:

$$Z_2/Z_1 = -\frac{\sin \pi(\nu_1 + \tilde{\mu}_1)}{\xi \cos \pi(\nu_2 + \tilde{\mu}_2)} \tag{28}$$

which is just another form of equation (13).

The expansion (26) leads to an expression of the absorption intensity I in terms of the parameters $\tilde{D}_i$:

$$I = \left| \sum_i Z_i \cos \pi(\nu_i + \tilde{\mu}_i)\tilde{D}_i \right|^2 \tag{29}$$

or, according to equation (28),

$$I = |Z_1|^2[\cos \pi(\nu_1 + \tilde{\mu}_1)\tilde{D}_1 - \sin \pi(\nu_1 + \tilde{\mu}_1)\tilde{D}_2/\xi]^2. \tag{30}$$

The factor $|Z_1|^2$ is determined by the normalisation of the total wavefunction in each spectral range. In the *bound spectrum* the amplitudes Z_{in} of the energy-normalised channel wavefunctions Ψ_i for a discrete state n must satisfy $\Sigma_i Z_{in}{}^2 \nu_{in}^3 = 1$ and the intensity is:

$$I_n = \frac{[\cos \pi(\nu_{1n} + \tilde{\mu}_1)\tilde{D}_1 - \sin \pi(\nu_{1n} + \tilde{\mu}_1)\tilde{D}_2/\xi]^2}{\nu_{1n}^3 + [\mathrm{d}(-\nu_1)/\mathrm{d}\nu_2]_n \nu_{2n}^3}. \tag{31}$$

The intensity I_n of a discrete level that happens to lie at an inflection point of the plot is directly related to one of the two quantities $\tilde{D}_i$ as shown in table 1. In the

† In equation (24), adapted from equation (3.23) of Seaton (1983), the minus sign replaces his plus sign because his functions c_i are equivalent to our $(-g_i)$.

Table 1. Photoabsorption at special points of the plot of figure 1.

$-\nu_1$† (mod. 1)	ν_2 (mod. 1)	Slope $d(-\nu_1)/d\nu_2=(Z_2/Z_1)^2$ (equation (14))	Discrete spectrum I_n (equation (31))	Autoionisation region $I(\nu_2)$ (equation (33))
$\tilde{\mu}_1$	$-\tilde{\mu}_2\pm\frac{1}{2}$	ξ^2	$\tilde{D}_1^2/(\nu_{1n}^3+\xi^2\nu_{2n}^3)$	$\tilde{D}_1^2$
$\tilde{\mu}_1\pm\frac{1}{2}$	$-\tilde{\mu}_2$	ξ^{-2}	$\tilde{D}_2^2/(\xi^2\nu_{1n}^3+\nu_{2n}^3)$	$\tilde{D}_2^2/\xi^2$

† $\mu=-\nu_1$ represents the effective quantum defect in the discrete spectrum, and $\pi\mu$ is the open-channel phaseshift in the autoionisation region.

autoionisation spectrum above the first threshold, energy normalisation requires unit amplitude in the only open channel ($|Z_1|^2=1$). The continuum intensity

$$I=|\cos\pi(\mu-\tilde{\mu}_1)\tilde{D}_1+\sin\pi(\mu-\tilde{\mu}_1)\tilde{D}_2/\xi|^2 \tag{32}$$

reaches periodically the characteristic values quoted in the last column of table 1; the parameters $\tilde{D}_i$ may thus easily be fitted using experimental autoionisation spectra. Since the inflection points $\nu_i=-\tilde{\mu}_i$ of the plot do not correspond to pure i channels (see § 3.1) we stress that the parameters $\tilde{D}_1$ and $-\tilde{D}_2/\xi$ represent respectively the absorption to the continuum including some admixture of the resonant discrete level, and the absorption to the discrete level including some admixture of the continuum, actually the minimum admixture in both cases.

The energy dependence of the photoabsorption (or photoionisation) cross section $\sigma(E)=4\pi^2e^2(\omega/c)I(\nu_2)$ in the autoionisation region is displayed by expressing equation (32) as a function of ν_2 only:

$$\begin{aligned} I(\nu_2)&=\tilde{D}_1^2\sin^2\pi(\mu-\tilde{\mu}_1)[\cot\pi(\mu-\tilde{\mu}_1)+\tilde{D}_2/\xi\tilde{D}_1]^2\\ &=\tilde{D}_1^2\frac{[\tan\pi(\nu_2+\tilde{\mu}_2)/\xi^2-\tilde{D}_2/\xi\tilde{D}_1]^2}{1+\tan^2\pi(\nu_2+\tilde{\mu}_2)/\xi^4}. \end{aligned} \tag{33}$$

Note that this expression applies without change to the case of *molecular photodissociation*, in a spectral region where bound vibrational levels are predissociated into a nuclear continuum.

Equation (33) exhibits a series of resonances similar to those of the elastic cross section (21*b*), with the same enegies and widths given by equations (16) and (18). On the other hand the profile index, specific to each excitation process, is for photoabsorption

$$q=-\tilde{D}_2/\xi\tilde{D}_1. \tag{34}$$

The value of q^2 can be fitted from the ratio of the intensities obtained at $\nu_2=-\tilde{\mu}_2$ and $-\tilde{\mu}_2\pm\frac{1}{2}$ (modulo 1), according to table 1.

An expression very close to equation (33) has been previously obtained by Dubau and Seaton (1984) for the photoionisation cross section (see Seaton 1983, § 9.3). Their derivation rests on a parametrisation of the S matrix rather than of the R matrix and involves a *complex* quantum defect $\mu=\alpha+i\beta$ for the closed channel. The imaginary part β accounts for the decay of the discrete levels in the closed channel due to the interaction with the open channel, and is related to our interaction parameter ξ by

$$\tanh\pi\beta=\xi^2 \tag{35}$$

the other parameters α and δ correspond to our $\tilde{\mu}_2$ and $\tilde{\mu}_1$, respectively. Note that equations (35) and (12*d*) yield directly the expression of β in terms of matrix elements R_{ij}:

$$\exp(-4\pi\beta)=[(R_{11}-R_{22})^2+(1+\Delta_R)^2][(R_{11}+R_{22})^2+(1-\Delta_R)^2]^{-1}. \tag{36}$$

The dipole moment amplitudes D_i used by Dubau and Seaton correspond to the standard channel wavefunctions

$$\psi_i=\sum_j \phi_j(f_i\delta_{ij}-g_jR_{ji}) \qquad (r>r_0) \tag{37}$$

which, according to equations (11) and (25), are related to the alternative channel wavefunctions $\tilde{\psi}_i$ (equation (24)) by:

$$[\tilde{\psi}_1, \tilde{\psi}_2]=[\psi_1, \psi_2](\cos\pi\tilde{\mu}-\sin\pi\tilde{\mu}\tilde{R}). \tag{38}$$

From equation (38) we deduce the relations

$$\begin{aligned}\tilde{D}_1&=D_1\cos\pi\tilde{\mu}_1-\xi D_2\sin\pi\tilde{\mu}_2\\ \tilde{D}_2&=-\xi D_1\sin\pi\tilde{\mu}_1+D_2\cos\pi\tilde{\mu}_2\end{aligned} \tag{39}$$

which identify the quantities $P=\tilde{D}_1$ and $Q=-\tilde{D}_2/\xi$ given by equation (9.20) of Seaton (1983), and demonstrate the equivalence between the two expressions

$$q=Q/P=-\tilde{D}_2/\xi\tilde{D}_1 \tag{40}$$

for the profile index. The two sets of parameters $(\delta, \alpha, \beta, D_1, D_2)$ and $(\tilde{\mu}_1, \tilde{\mu}_2, \xi, \tilde{D}_1, \tilde{D}_2)$ are thus very close although they derive from rather different approaches: the introduction of a complex quantum defect is specific to the autoionisation range of energies, whereas the derivation from the Lu–Fano plot refers primarily to the discrete region, with two closed channels, and is then extended to the continuous range in accordance with the QDT procedure.

The expressions (21*b*) and (33) for two-channel resonant cross sections, which have the common form

$$\sigma=\sigma_0\frac{[q+\tan\pi(\nu_2+\tilde{\mu}_2)/\xi^2]^2}{1+\tan^2\pi(\nu_2+\tilde{\mu}_2)/\xi^4} \tag{41}$$

deserve several comments. First, near each resonance E_n one may use the first-order expansion

$$\varepsilon=\frac{\tan\pi(\nu_2+\tilde{\mu}_2)}{\xi^2}\simeq\frac{\pi}{\xi^2}(\nu_2+\tilde{\mu}_2-n)\simeq\frac{\pi}{\xi^2}\left(\frac{\mathrm{d}\nu_2}{\mathrm{d}E}\right)_n(E-E_n)=\frac{E-E_n}{\Gamma_n/2}$$

where Γ_n is the width defined by equation (19). The expression (41) then coincides with the usual resonance formula (Fano 1961) in a limited energy range around the resonance centre. This energy range covers the entire resonance in the case of narrow widths ($\xi^2\ll 1$) but when the width increases the wings are more and more perturbed by the closest resonances on both sides and the cyclic expression (41) becomes necessary†. Actually the case of closely lying discrete levels coupled to the same continuum has been studied by Fano (1961) and in greater detail by Mies (1968), using the configuration mixing (CM) approach. For the successive levels of a single

† A quantitative comparison between profiles computed from equation (33) or from the original Fano formula has been made recently by Connerade (1983).

Rydberg series they obtained an expression similar to equation (41) with the definition of ε:

$$\frac{1}{\varepsilon}=\sum_n \frac{\Gamma_n}{2(E-E_n)}.$$

The theoretical connection between the present approach and the CM theory will be developed in paper II.

Another comment concerns the restriction $\xi^2 \leqslant 1$ stated in § 2 (equations (12c)–(12d)). The corresponding restriction $\Gamma_n \leqslant (2/\pi)\,(\mathrm{d}E/\mathrm{d}\nu_2)_n$ means roughly that the width may not exceed the resonance spacing. One sees easily that any two-channel resonance pattern may be parametrised within this restriction, provided one transforms equation (41) into the equivalent form, for $|\xi|>1$:

$$\sigma=\sigma_0 q^2 \frac{[q'+\tan \pi(\nu_2+\tilde{\mu}_2')/\xi'^2]^2}{1+\tan \pi^2(\nu_2+\tilde{\mu}_2')/\xi'^4} \tag{42}$$

with $q'=1/q$, $\tilde{\mu}_2'=\tilde{\mu}_2+\frac{1}{2}$ and $\xi'^2=1/\xi^2<1$†. Equation (42) describes a series of resonances whose energies and widths are (for the Coulomb field case):

$$E_n'=I_2-\frac{\text{Ryd}}{(n-\tilde{\mu}_2-\frac{1}{2})^2} \qquad \Gamma_n'=\frac{2\,\text{Ryd}}{(n-\tilde{\mu}_2-\frac{1}{2})^3}\,\frac{2}{\pi\xi^2}. \tag{43}$$

Hence the resonance centres are now located at intermediate energies with respect to the initial parametrisation, and the larger is the interaction parameter ξ, the narrower are the 'apparent' resonance widths Γ_n'. This result may be interpreted by means of the extended plot of figure 1: the resonance centres correspond to the inflection points where the time delay is maximum, that is $\nu_2=-\tilde{\mu}_2$ for $\xi^2<1$ (see the lower part of the figure); the other set of inflection points (at $\nu_2=-\tilde{\mu}_2-\frac{1}{2}$) correspond instead to a minimal time delay. If ξ increases and becomes larger than one, the roles of the two sets of inflection points exchange ($1/\xi^2<\xi^2$) and the resonance characteristics are given by equation (43). In the limiting case $\xi=1$ the time delay is constant and the cross section has an oscillatory behaviour without any definite resonances, the total wavefunction being always equally shared between the open and closed channels. This resonance narrowing effect in the case of very strong interaction has been described by Mies (1968) for autoionised Rydberg series (see his figure 1) and by Child (1974) for predissociated levels. The case of predissociation is most illustrative since it can be interpreted in terms of a change from the diabatic regime to the adiabatic one, when the electronic interaction between the diabatic molecular states becomes too large. The corresponding change from crossing potential curves to curves with a strongly avoided crossing induces a redefinition of the bound vibrational levels, which corresponds to the shift found above for the resonance positions. In both extreme cases (very weak or very large interaction), the resonance width, given by equations (88) or (90) of Child (1974), is very small.

Finally we indicate a straightforward generalisation of equation (41) to the case of $(N-1)$ non-interacting closed channels coupled to a single open channel. This situation is often encountered in molecular spectra, when several Rydberg series, whose limits are different vibrational levels of a same ion state, are electronically coupled to the continuum of a lower state (electronic continuum if it is a lower state of the ion, i.e.

† The corresponding change $\tilde{\mu}_1 \rightarrow \tilde{\mu}_1' = \tilde{\mu}_1+\frac{1}{2}$ preserves the compatibility equation (8), which is now written $\tan \pi(\nu_1+\tilde{\mu}_1')\tan \pi(\nu_2+\tilde{\mu}_2')=\xi'^2$.

for electronic autoionisation; nuclear continuum if it is a dissociating state of the neutral, i.e. for predissociation). The vibrational coupling between these Rydberg series may often be neglected at least in a first step, in so far as the quantum defect does not vary rapidly with the internuclear distance. The compatibility equation may thus be cast in a form equivalent to (8):

$$\begin{vmatrix} \tan\pi(\nu_1+\tilde{\mu}_1) & \xi_2 & \xi_3 \dots & \xi_N \\ \xi_2 & \tan\pi(\gamma_2+\tilde{\mu}_2) & & \\ \xi_3 & & \ddots & 0 \\ \vdots & 0 & & \\ \xi_N & & & \tan\pi(\nu_N+\tilde{\mu}_N) \end{vmatrix} = 0 \quad (44)$$

or

$$\tan\pi(\mu-\tilde{\mu}_1) = -\sum_{i=2}^{N} \xi_i^2 \cot\pi(\nu_i+\tilde{\mu}_i) \quad (45)$$

where $\pi\mu = -\pi\nu_1$ is as above the open-channel phase, and ξ_i is the effective coupling strength between the ith closed channel and the continuum. The cross section is written:

$$\sigma = \sigma_0\left(1+\sum_{i=2}^{N} q_i\xi_i^2 \cot\pi(\nu_i+\tilde{\mu}_i)\right)^2\left[1+\left(\sum_{i=2}^{N} \xi_i^2 \cot\pi(\nu_i+\tilde{\mu}_i)\right)^2\right]^{-1}. \quad (46)$$

For photoabsorption one has more specifically

$$\sigma_0 \propto \tilde{D}_1^2 \qquad q_i = -\tilde{D}_i/\xi_i\tilde{D}_1$$

with the same definition for the $\tilde{D}_i$ as in the two-channel case.

3.3. Energy dependence of the parameters

The alternative sets of short-range parameters, related by equations (10), (12) and (39), have been considered above as energy independent. Not much will change if a smooth energy dependence has to be introduced to fit the experimental data (Moores 1966, Starace 1973, Lee and Lu 1973). The extended plot of figure 1 is then no longer exactly periodic, in particular the inflection points lie no longer on a grid parallel to the coordinate axes. Nevertheless, these points usually still lie on almost straight lines as displayed, for example, in figure 2 which was obtained from a fit to experimental discrete levels of Kr. This observation means that the parameters $\tilde{\mu}_i$ depend linearly on the energy, with first derivative $d\tilde{\mu}_i/dE$ given by the slopes of the straight lines. In the autoionisation region Dubau and Seaton (1984) arrive at the same conclusion since they fit the photoionisation spectra of Be and Al with linear two-channel parameters, slowly varying with the energy. A stronger energy dependence, as encountered for example in the discrete spectrum of Be (Greene 1981) would alter the behaviour of a two-channel system more deeply. Note also that for predissociation, the interaction parameter ξ often has an oscillatory behaviour with the energy because it is roughly proportional to the overlap integrals between the vibrational wavefunction of the dissociative state and that of the successive bound levels in the potential well of the attractive curve (see figures 2–4 of Murrell and Taylor 1969). The resulting oscillations of the resonance widths (equation (19)) destroy the quasi-periodicity of the resonance pattern, for example in the elastic cross section for two colliding oxygen atoms (figure 2 of Colle 1981).

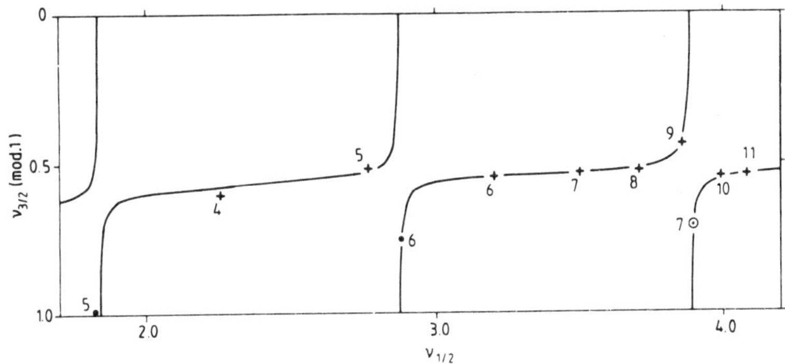

Figure 2. Lu–Fano plot for excited levels of two interacting Rydberg series of Kr with $J=0$ ($4p^5\ ^2P_{3/2}nd$ and $4p^5\ ^2P_{1/2}ns'$). Observed levels: + nd, · ns'; ⊙ predicted 7s'. Note slight departure from periodicity in $\nu_{1/2}$ reflecting energy dependence of parameters, mainly of $\tilde{\mu}_{3/2}$ (from Aymar *et al* 1981, Aymar 1982).

4. Summary and discussion

This paper is rooted in the realisation that the QDT formulations in terms of Seaton's reactance matrix R or of its eigenchannels (Fano 1970) are relevant only to the analysis of the *total* short-range electron–core interaction. Accordingly they need not provide the best guidelines for the analysis of resonance features or for spectral studies of limited scope.

We have used for each channel a flexible base pair of radial functions which incorporate the influence of the diagonal elements R_{ii} together with the effect of long-range forces embodied in the usual pair (f, g). The diagonal elements of Seaton's matrix equation (5) reduce then to $\tan \pi(\nu_i+\tilde{\mu}_i)$ as in equation (8). The MQDT equations thus transformed exhibit greater symmetry and so does the plot of equation (8) in zone 2 of figure 1, i.e., the plot with coordinates $\nu_1+\tilde{\mu}_1$ and $\nu_2+\tilde{\mu}_2$. The R matrix of a channel pair reduces to a single off-diagonal element ξ which measures the effective coupling strength between the two channels and can be restricted to $|\xi|\leq 1$. The parameters $\tilde{\mu}_i$ and ξ lead to simple expressions of the energies, widths and profile index of a whole series of resonances arising in the cross section for various two-channel processes. The analytical expression (41) in terms of the closed-channel wavenumber ν_2 applies to any kind of long-range forces, i.e., to any kind of collision partners or fragmentation products (the resulting resonance pattern as a function of the energy may instead differ considerably from one process to the other).

In the multichannel case with some of the channels open (N_{op}) and some closed, several MDQT works (Lu 1971, Lee and Lu 1973, Jungen and Dill 1980) utilised as asymptotic basis sets at each energy the N_{op} eigenchannels of the *scattering matrix*. Their wavefunctions include an admixture of all the closed and open channels and each eigenchannel ψ_ρ is characterised by having the same phase parameter τ_ρ in all its open channel components. Generally, a resonance in a given process does not correspond to the rapid change of a single eigenphase τ_ρ, but is spread out on several (often on all) S-matrix eigenchannels. As has been illustrated numerically by Burke *et al* (1969) for the case of He+e scattering, a more significant characterisation of resonance features is achieved by selecting eigenstates of the *time-delay operator* introduced analytically by Smith (1960) and discussed for two channels in § 3 of this paper. Extensive utilisation of the flexibility afforded for this purpose by the choice of the phases in equation (1) lies in the future.

Acknowledgments

Key elements of the present paper originated independently from one of us (AGS) and from Cooke and Cromer (1983) who centre on multichannel aspects. We are indebted to W Cooke for making his material available in advance of publication and for several stimulating discussions. We are also indebted to K T Lu and to Ch Jungen for helpful discussions. Our work was supported by the US Department of Energy, Office of Basic Energy Sciences and by the NSF-CNRS exchange program.

Note added in proof. A detailed MQDT analysis of diatomic predissociation and inelastic atomic collisions has just been achieved by F H Mies (1984 to be published) and F H Mies and P S Julienne (1984 to be published). Their results, especially for the case of two channels (Mies and Julienne 1984) are in good accordance with the main points of the present paper.

References

Armstrong J A, Sudhanshu S Jha and Pandey K C 1981 *Phys. Rev.* **23** 2761–75
Aymar M 1982 private communication
Aymar M and Robaux O 1979 *J. Phys. B: At. Mol. Phys.* **12** 531–46
Aymar M, Robaux O and Thomas C 1981 *J. Phys. B: At. Mol. Phys.* **14** 4255–70
Burke P G, Cooper J W and Ormonde S 1969 *Phys. Rev.* **183** 245–64
Child M S 1974 *J. Mol. Spectrosc.* **53** 280–301
Colle R 1981 *J. Chem. Phys.* **74** 2910–9
Connerade J P 1983 *J. Phys. B: At. Mol. Phys.* **16** L329–35
Cooke W E and Cromer C L 1983 *Phys. Rev.* A to be published
Dubau J and Seaton M J 1984 *J. Phys. B: At. Mol. Phys.* **17** 381–403
Eissner W, Nussbaumer H, Saraph H E and Seaton M J 1969 *J. Phys. B: At. Mol. Phys.* **2** 341–55
Fano U 1961 *Phys. Rev.* **124** 1866–78
—— 1970 *Phys. Rev.* A **2** 353–65
—— 1975 *J. Opt. Soc. Am.* **65** 979–87
Gallagher T F, Safinya K A and Cooke W E 1980 *Phys. Rev.* A **21** 148–56
Giusti A 1980 *J. Phys. B: At. Mol. Phys.* **13** 3867–94
Giusti-Suzor A, Bardsley J N and Derkits C 1983 *Phys. Rev.* A **28** 582–91
Giusti-Suzor A and Fano U 1984 *J. Phys. B: At. Mol. Phys.* to be submitted
Giusti-Suzor A and Jungen Ch 1984 *J. Chem. Phys.* to be published
Giusti-Suzor A and Lefebvre-Brion H 1980 *Chem. Phys. Lett.* **76** 132–5
Greene C H 1981 *Phys. Rev.* A **23** 661–78
Greene C H, Fano U and Strianti G 1979 *Phys. Rev.* A **19** 1485–509
Greene C H, Rau A R P and Fano U 1982 *Phys. Rev.* A **26** 2441–59
Herzberg G and Jungen Ch 1972 *J. Mol. Spectrosc.* **41** 425–86
Jungen Ch and Dill D 1980 *J. Chem. Phys.* **73** 3338–45
Lee C M and Lu K T 1973 *Phys. Rev.* A **8** 1241–57
Lu K T 1971 *Phys. Rev.* A **4** 579–96
Lu K T and Fano U 1970 *Phys. Rev.* A **2** 81–6
Mies F H 1968 *Phys. Rev.* **175** 164–75
—— 1980 *Mol. Phys.* **41** 953–72
Moores D L 1966 *Proc. Phys. Soc.* **88** 843

Murrell J N and Taylor J M 1969 *J. Mol. Spectrosc.* **16** 609–21
Seaton M J 1966 *Proc. Phys. Soc.* **88** 801–14
—— 1983 *Rep. Prog. Phys.* **46** 167–257
Smith F T 1960 *Phys. Rev.* **118** 349–56
Starace A F 1973 *J. Phys. B: At. Mol. Phys.* **6** 76–92
Wigner E P 1955 *Phys. Rev.* **98** 145–7
Wynne J J and Armstrong J A 1979 *Comment. At. Mol. Phys.* **8** 155–71

1987 *J. Phys. B: At. Mol. Phys.* **20** 3645–62

Parametrisation of resonance structures in the multichannel quantum defect theory

J M Lecomte

Laboratoire Aimé Cotton, CNRS II, Bâtiment 505, 91405 Orsay, France

Received 8 August 1986, in final form 26 February 1987

Abstract. The resonance structures corresponding to autoionising states interacting with many continua are parametrised using the multichannel quantum defect theory. All the involved continua are taken into account in an effective way, and collisional parameters which describe the position and width of the resonances are then introduced. The interactions between open and closed channels are parametrised in an original way in terms of vectors whose relative orientation allows stabilisation effects to be described. The introduction of a unitary transformation which rotates the 'degenerate' closed channels greatly simplifies the analysis. This theory is well adapted to the case of core-excited Rydberg states of barium, as is briefly discussed.

1. Introduction

In any formulation of the multichannel quantum defect theory (MQDT), the key point is that the short- and long-range interactions between the ionic core and the electron are separately analysed. The long-range effects are analytically described; the short-range effects are characterised by empirical parameters describing the behaviour of the wavefunction in the ionic core. The different formulations of the theory are related to the different definitions of these parameters: collisional parameters (Seaton 1983), eigenchannel parameters (Lu and Fano 1970) and collisional phaseshifted parameters (Giusti-Suzor and Fano 1984a, b, Cooke and Cromer 1985). Most of the applications describe the characteristics of complex series of atomic Rydberg states (Seaton 1983, Aymar 1984).

We are interested here with the use of this theory for the description of total photoionisation spectra in an energy range where resonance series appear. These resonances arise from autoionising states, for example, the so-called core-excited Rydberg states in alkaline-earth atoms. In such states the atom is described as consisting of an electron in a Rydberg orbit and an ionic core with an excited electron. Several spectra of such atoms have been studied in the past few years, the most recent studies including the works by Gounand *et al* (1983), Bloomfield *et al* (1984), Tran *et al* (1984), Mullins *et al* (1985), Cooke and Cromer (1985) and Camus *et al* (1985). The involved states in these cases are localised outside the core, therefore the collisional formulations of the MQDT defining the parameters on the exit or the collision channels are more suitable than the eigenchannel ones. Giusti-Suzor and Fano (1984a, b) and later Cooke and Cromer (1985) defined collisional phaseshifted parameters. The

0022-3700/87/153645+18$02.50

intrachannel effects due to the core are separated from the interchannel coupling. Moreover the analysis of autoionised states can be carried out by distinguishing the closed channels which correspond to the resonance series from the open channels, grouped into a small number of 'effective continua' (Cooke and Cromer 1985). The resonance structures can then be described in terms of a few parameters; this approach has been successfuly used to analyse the experimental spectra of Cooke and Cromer (1985).

The purpose of this paper is to rigorously deduce the minimal set of parameters required for the description of the resonance structures and the total photoionisation cross sections. Following Seaton (1983) the whole problem involving N closed channels coupled to M open channels is treated by considering a matrix defined in the closed-channel space only. However, instead of the scattering matrix χ_{cc} used by Seaton we introduce a new reactance matrix κ_{cc}.

The $N(N+1)$ elements of κ_{cc} are sufficient to describe the resonance whatever the number M of open channels is (M increases rapidly when the excitation of the core increases). However, it is not possible to identify these coefficients directly from the experimental spectra. Consequently it is necessary to transform the κ_{cc} reactance matrix by changing the Coulomb basis wavefunctions describing the channels, according to a procedure developed by Giusti-Suzor and Fano (1984a) and by Cooke and Cromer (1985). This method is adapted to the study of an imaginary matrix. The set of parameters useful in describing the resonances is then deduced. It includes the phaseshifted quantum defects previously employed by Giusti-Suzor and Fano (1984a) and Cooke and Cromer (1985). But in our work the effects of open channels are completely described without introducing 'effective continua'. In fact, the open channels are globally taken into account by considering the complex reactance matrix κ_{cc} or its transformed equivalent. All the interactions between open and closed channels are described in an original way through the introduction of a set of N vectors with a length related to the width of the resonance profile. In previous analyses only the N quantum defects and the N vector lengths have been fitted to experimental data. Such studies rest on the assumption that each closed channel is coupled to only one 'effective continuum'. In fact such treatment neglects some interactions and we prove that these neglected effects are described by $N(N-1)/2$ parameters. These parameters corresponding to the angles between the vectors measure the interference effects between the autoionisation paths. A correct description of these interference effects is of prime importance to explain the stabilisation of some excited levels observed experimentally (Giusti-Suzor and Lefebvre-Brion 1984, Neukammer *et al* 1985, Boulmer *et al* 1987).

In addition, we give a rigorous treatment of the situations involving 'degenerate' closed channels through the introduction of a rotation matrix in the transformation which changes the matrix κ_{cc}. In this way it is possible to perform an exact interchannel diagonalisation in the subspace of 'degenerate' closed channels (Giusti-Suzor and Fano 1984b, Mullins *et al* 1985). The main characteristics of the resonance structures are described by a 'density of states' which does not depend on this rotation matrix, while the oscillator strength density is tightly dependent on it.

A few basic principles of Seaton's theory are presented in § 2 and the complex reaction matrix is introduced. The frame transformation leading to the new set of collisional parameters and the interpretation of these parameters are given in § 3. Some spectra previously discussed are analysed in § 4 allowing us to illustrate by various examples the different points discussed in this formal work.

2. MQDT formalism

2.1. Reactance and scattering matrices

Quantum defect theory (Seaton 1983) considers the system from the collisional standpoint: an external electron interacts with the ionic core, which possesses a finite spatial extension. There exists a sphere with radius r_0 including all the core electrons. Beyond r_0 the electron is subjected to long-range interactions of a pure Coulomb potential; in this region the radial function describing the electron in a well defined collisional or dissociative channel can be analytically represented by a linear combination of Coulomb functions. Each collisional channel is related to a 'target state' (ionic state plus angular part of the electron function) and characterised by a set of quantum numbers defined in the *jj*-coupling scheme.

Non-Coulombic interactions at short range induce transitions of the ion plus electron system from one channel to another one. In Seaton's formalism (Seaton 1983) short-range electron–ion interactions are accounted for by the reactance matrix R (real and symmetric) or by the scattering matrix χ (unitary and symmetric). These matrices operating on the whole set of channels are related by

$$\chi=(1+\mathrm{i}R)(1-\mathrm{i}R)^{-1}. \tag{2.1}$$

The corresponding matrix elements can be considered as empirical parameters to be determined by fitting the experimental data. In all this work, any energy dependence of the parameters is neglected. In the autoionised energy range, the N closed channels i (set c) are coupled to the M open channels α (set o). Seaton (1983) has shown that the positions and widths of autoionising resonances are characterised by the matrix elements of the χ_{cc} matrix, i.e. the restriction of the χ matrix on the space c of closed channels. In this work we introduce a new matrix κ_{cc}, defined on the same space c. The relation between χ_{cc} and κ_{cc} is obtained by expressing κ_{cc} in terms of the elements of R by using equation (2.1):

$$\chi_{\mathrm{cc}}=(I_{\mathrm{cc}}+\mathrm{i}\kappa_{\mathrm{cc}})(I_{\mathrm{cc}}-\mathrm{i}\kappa_{\mathrm{cc}})^{-1} \tag{2.2}$$

where

$$\kappa_{\mathrm{cc}}=R_{\mathrm{cc}}-R_{\mathrm{co}}(\mathrm{i}I_{\mathrm{oo}}+R_{\mathrm{oo}})^{-1}R_{\mathrm{oc}} \tag{2.3}$$

and I_{cc} and I_{oo} are the unit $N\times N$ and $M\times M$ matrices.

From equation (2.2), κ_{cc} may be considered as the reactance matrix related to the scattering matrix χ_{cc}. Both χ_{cc} and κ_{cc} matrices operate in the space of closed channels only, however, the contributions due to the M open channels are globally taken into account in an effective way. Let us note that the χ_{cc} matrix is no longer a unitary one and that κ_{cc} now has complex elements.

The different scattering and reactance matrices defined on the complete set of channels and on the restricted spaces o or c are summarised in table 1.

2.2. Interpretation of the different terms of the κ_{cc} matrix

In this work we use the matrix κ_{cc} rather than the matrix χ_{cc}. In fact the elements of the former matrix have a more apparent physical meaning. From equation (2.3) κ_{cc} can be expanded as

$$\kappa_{\mathrm{cc}}=R_{\mathrm{cc}}-R_{\mathrm{co}}R_{\mathrm{oo}}(I_{\mathrm{oo}}+R_{\mathrm{oo}}^2)^{-1}R_{\mathrm{oc}}+\mathrm{i}R_{\mathrm{co}}(I_{\mathrm{oo}}+R_{\mathrm{oo}}^2)^{-1}R_{\mathrm{oc}}. \tag{2.4}$$

Table 1. Reactance and scattering matrices defined on the complete set of channels and on the restricted spaces o or c. All the matrices are symmetric.

Whole set of channels (c+o)	Restriction to open channels (o)	Restriction to closed channels (c)
Reaction matrix	K_{oo} real	κ_{cc} complex
R real	$K_{oo} = R_{oo} - R_{oc}(\mathrm{tg}\,\pi\nu_c + R_{cc})^{-1}R_{co}$ (Seaton 1983)	$\kappa_{cc} = R_{cc} - R_{co}(\mathrm{i}I_{oo} + R_{oo})^{-1}R_{oc}$ (equation (2.3))
Collision matrix	S_{oo} unitary (Seaton 1983)	χ_{cc} non-unitary
$\chi = (1+\mathrm{i}R)(1-\mathrm{i}R)^{-1}$	$S_{oo} = (I_{oo} + \mathrm{i}K_{oo})(I_{oo} - \mathrm{i}K_{oo})^{-1}$	$\chi_{cc} = (I_{cc} + \mathrm{i}\kappa_{cc})(I_{cc} - \mathrm{i}\kappa_{cc})^{-1}$ (equation (2.2))

The different parts of this matrix can be interpreted in the framework of the configuration interaction theory (Fano 1961). The imaginary part is due to the interactions between closed and open channels (R_{co} and R_{oc}). All the continuum–continuum interactions are included in the term $(I_{oo} + R_{oo}^2)^{-1}$, which was determined from the study of the coupling of a discrete state with several continua (Lefebvre and Beswick 1972). This imaginary contribution is responsible for the autoionisation of the states belonging to the closed channels and for the existence of resonance structures with a finite width. The real part of κ_{cc} consists of two terms. The leading term is the so-called direct coupling term R_{cc} between closed channels. It is modified by a second term $-R_{co}R_{oo}(I_{oo} + R_{oo}^2)^{-1}R_{oc}$, which describes the indirect coupling between closed channels through the continua. The diagonal matrix elements of this latter term correspond to the energy shift of the resonances arising from the coupling between the continuum states (Lefebvre and Beswick 1972). Furthermore, the non-diagonal matrix elements are also modified by the continuum–continuum state interaction (Lecomte and Luc-Koenig 1985).

2.3. Oscillator strength density and density of states

We obtain the expression for photoionisation cross sections from the coefficients obtained from the development of the wavefunctions describing the core plus electron system.

2.3.1. Amplitude of closed channels. A set of M independent physical solutions is constituted by the Φ_o^- functions behaving as a single outgoing wave in a well defined channel α. Following Bell and Seaton (1985) the asymptotic development may be written as

$$\Phi_o^- = \tfrac{1}{2}\psi_o(\varphi_o^+ - \varphi_o^- S_{oo}^*) + \psi_c P_c(r) A_{co}^- \tag{2.5}$$

where $\varphi^{\pm} = c \pm \mathrm{i}s$, c and s are the Coulomb functions normalised per unit energy range, $\psi_{o(c)}$ are the target state functions, S_{oo}^* is the complex conjugate (CC) of the collision matrix between the open channels (cf table 1), $P_c(r)$ is the set of exponential decreasing functions

$$P_i(r) = s_i \cos \pi\nu_i - c_i \sin \pi\nu_i \tag{2.6}$$

where ν_i is the effective quantum number of the closed channel i and A_{co}^- is the amplitude of the closed channels:

$$A_{co}^- = \mathrm{i}\exp(\mathrm{i}\pi\nu_c)[\exp(2\mathrm{i}\pi\nu_c) - \chi_{cc}^*]^{-1}\chi_{co}^*. \tag{2.7}$$

We can write a similar expression for the incoming wave Φ_o^+. Using equations (2.1) and (2.3) we can write $A_{co}^{\pm}$ in terms of κ_{cc}:

$$A_{co} = (\cos \pi\nu_c)^{-1}(H_{cc}^{\pm})^{-1}R_{co}(\pm \mathrm{i} I_{oo} + R_{oo})^{-1} \qquad (2.8)$$

where

$$H_{cc}^{+} = \tan \pi\nu_c + \kappa_{cc} \qquad (2.9)$$

and

$$H_{cc}^{-} = (H_{cc}^{+})^{*}$$

operate on the closed-channel space only.

2.3.2. Density of oscillator strength. Photoionisation from a given initial state φ_{in} may be described in terms of the set d_c of electric dipole transition elements between φ_{in} and the closed channels *i*. Each *di* is defined by

$$di = (\varphi_{in}|D|\psi_i P_i(r)) \qquad (2.10)$$

where D is the dipole moment operator. We focus our analysis on the spectra of the core-excited Rydberg states of alkaline-earth atoms obtained by photoionisation of Rydberg bound states. When the energy of the final state is not near to an ionisation threshold I_α, direct excitation to a continuum is negligible compared with the excitation to a closed channel (Tran *et al* 1984). Moreover, the elements *di* can be often calculated. In fact it can be assumed that during the transition process only the core is excited, while the outer electron is a spectator. This is the basis of the ICE (isolated core excitation) approximation (Cooke *et al* 1978). The initial and excited states can be described only by their asymptotic development, such as that in equation (2.5). Then the elements *di* are separated into an analytically calculated radial part and an angular part related to the angular coupling of the initial and final states (Bhatti and Cooke 1983). The partial cross section for the excitation of the continuum states Φ_α^- can be expressed as

$$s_\alpha = \sum_{i,j} A_{i\alpha}^{-*} \sigma_{ij} A_{j\alpha}^{-} \qquad (2.11)$$

where σ_{ij} denotes an element of the $N \times N$ matrix

$$\sigma_{cc} = {}^t d_c d_c. \qquad (2.12)$$

The total cross section is equal to

$$s = \sum_\alpha s_\alpha = \mathrm{Tr}({}^t A_{co}^{-*} \sigma_{cc} A_{co}^{-}) = \mathrm{Tr}((A_{co}^{\mp})({}^t A_{co}^{\mp *})\sigma_{cc}). \qquad (2.13)$$

With the help of equation (2.8) s can be written as

$$s = \mathrm{Tr}[\cos(\pi\nu_c)^{-1}(H_{cc}^{+*})^{-1}\Gamma_{cc}(H_{cc}^{+})\cos(\pi\nu_c)^{-1}\sigma_{cc}] \qquad (2.14)$$

where $\Gamma_{cc} = \mathrm{Im}(\kappa_{cc})$. In this expression the unique variable is the energy through the effective quantum numbers ν_c.

2.3.3. Density of states. From equation (2.13) we define the density of states

$$ds = \mathrm{Tr}({}^t A_{co}^{\pm *} A_{co}^{\pm}) = \mathrm{Tr}(A_{co}^{\pm}\, {}^t A_{co}^{\pm *}). \qquad (2.15)$$

This quantity does not depend on the excitation process, contrary to the cross section s.

However, the fast energy variations of both quantities are very similar since they are related to the same denominator $\Delta\Delta^*$ where $\Delta = \det(\cos \pi\nu_c H^+_{cc})$. Thus, the density of states reproduces the characteristics of the resonances related to the poles of Δ whatever the excitation process is.

3. Collisional parameters

We consider here only the cases where total photoionisation cross sections are known. The problem is to determine the collisional parameters so that we can reproduce the observed spectra. The $N(N+1)$ elements of the κ_{cc} (or χ_{cc}) matrix cannot be easily fitted to experimental data because they are not directly related to the spectral characteristics, position or width of the resonances. In particular, the position of an autoionised level is not given by R_{ii} or $\mathrm{Re}(\kappa_{ii})$ only (Giusti-Suzor and Fano 1984a). The aim of the following sections is to define from the κ_{cc}-(or χ_{cc}) matrix elements new alternative parameters which can be more easily determined; several sets of parameters will be introduced with different physical meanings.

3.1. *Transformation of reaction and scattering matrices*

Dealing with two channels only, Giusti-Suzor and Fano (1984a) have shown how to deduce from the R matrix new phaseshifted quantum defects connected with the collision channels. Their method is generalised to the case where the complex matrix κ_{cc} is used instead of the real matrix R. Moreover, it is extended by including an orthogonal matrix defined on particular subspace of channels—open channels or degenerate closed channels—as explained below.

3.1.1. New basis set. The principle of the method is to redefine the collision channels and to introduce a new basis set of Coulomb radial functions. For each new collision channel (b) one has

$$\theta_b = \sum_a W_{ab}(\psi_a s_a \cos \pi\delta_b + \psi_a c_a \sin \pi\delta_b) \tag{3.1}$$

$$\bar{\theta}_b = \sum_a W_{ab}(-\psi_a s_a \sin \pi\delta_b + \psi_a c_a \cos \pi\delta_b) \tag{3.2}$$

where a denotes all the channels open (o) or closed (c). The real rotation matrix W represents the changing of the collision channels. The transformation is chosen so that $W_{co} = 0$, namely W works separately inside the spaces of either the closed or open channels. Moreover, in the closed-channel space (c) only degenerate channels (identical core state and thus same effective quantum number) are superposed by W_{cc}.

3.1.2. Transformed matrices. The transformation represents changing the asymptotic development of the wavefunction $F \sim s + cR$ (Seaton 1983) to the new form

$$F' \sim \theta + \bar{\theta}R'. \tag{3.3}$$

This gives the transformed reactance matrix R':

$$R' = [\cos(\pi\delta)'WRW - \sin(\pi\delta)][\sin(\pi\delta)'WRW + \cos(\pi\delta)]^{-1} \tag{3.4}$$

with $\cos(\pi\delta)$ and $\sin(\pi\delta)$ representing diagonal matrices in the total space $\{(\mathrm{o})+(\mathrm{c})\}$. Eissner and Seaton (1969) as well as Cooke and Cromer (1985) have described a similar transformation of R, however, here an additional rotation is introduced. The elements of R_{cc}, R_{oo} and R_{co} are completely mixed by the transformation.

In the same way the χ matrix is transformed into

$$\chi' = \exp(-\mathrm{i}\pi\delta)' W\chi W \exp(-\mathrm{i}\pi\delta). \tag{3.5}$$

The restriction χ'_{cc} of the matrix χ' to the closed-channel space (c) depends only on the contracted matrix χ_{cc}, and on δ_c and W_{cc}, because one has $W_{co} = W_{oc} = 0$. Thus, using equation (2.2) we obtain the transformation of κ_{cc}:

$$\kappa'_{cc} = (\cos \pi\delta_c' W_{cc}\kappa_{cc} W_{cc} - \sin \pi\delta_c)(\sin \pi\delta_c' W_{cc}\kappa_{cc} W_{cc} + \cos \pi\delta_c)^{-1} \tag{3.6}$$

whose complex elements depends on the phaseshifts δ_c and on the matrix elements W_{cc} characterising the closed channels only.

The main conclusion of this section must be emphasised: the transformation of κ_{cc} does not involve any operation on the open channels. κ'_{cc} is completely independent on any transformation performed within the open-channel space (o). In particular, the open channels might remain unchanged by the transformations (3.1) and (3.2), i.e. we may choose $W_{oo} = 1$ and $\delta_o = 0$.

3.2. *Set of collisional parameters related to the closed channels*

3.2.1. Phaseshifts δ_c and matrix W_{cc}. When the system does not involve closed degenerate channels ($W_{cc} = 1$), the N phaseshifts δ_c characterising the closed channels are defined by the N equations

$$\mathrm{Re}(\kappa'_{ii}) = 0 \qquad i = 1, N \tag{3.7}$$

which require intrachannel diagonalisation. In equation (3.7) $\mathrm{Re}(\kappa'_{ii})$ differs from the direct coupling term R'_{ii} since it includes all the couplings induced by the continua (equation (2.4)). The solutions δ_i of equation (3.7) describe the core effects within the closed channels but depend on all the interactions between closed and open channels. Our approach can be compared to that introduced by Giusti-Suzor and Fano (1984a, b) and generalised by Cooke and Cromer (1985). In their approach the phaseshifts characterising the closed and open channels are determined by solving the system

$$\begin{cases} R'_{ii} = 0 \qquad i = 1, N & (3.8a) \\ R'_{oo} = 0. & (3.8b) \end{cases}$$

Solving equation (3.8*b*) involves diagonalising the matrix R'_{oo} and determining the phaseshifts δ_o of the open channels, i.e. performing inter- and intrachannel diagonalisation on the open-channel set. It must be emphasised that it is necessary to consider all the equations (3.8) simultaneously. Moreover, these operations cannot be done without introducing a rotation matrix $W_{oo} \neq I_{oo}$ in the transformations (3.1) and (3.2), and thus the W_{oo} matrix elements are solutions of the whole system (3.8) just as the phaseshifts δ_c and δ_o are. One has to solve a very complicated system of $N + M(M+1)/2$ equations. However, it can be demonstrated that the two systems (3.7) and (3.8) are equivalent (cf appendix). A main conclusion of our analysis is that, when concerned with total photoionisation spectra only, all the parameters related to the open channels are useless and in such situations the reduced system (N equations

(3.7)), which determines the phaseshifts δ_i only, is much more suitable than the complete system (3.8).

When the system involves some closed degenerate channels, supplementary conditions can be added to system (3.7), so that for example

$$\mathrm{Re}(\kappa'_{ij}) = 0 \tag{3.9}$$

where i and j pertain to the same block of degenerate channels. This condition is equivalent to an interchannel diagonalisation of $\mathrm{Re}(\kappa'_{cc})$.

By solving simultaneously both systems (3.7) and (3.9) one can determine all the phaseshifts δ_c and the W_{cc}-matrix elements; it must be noticed that the δ_c depend on the conditions chosen to build W_{cc}.

Once the phaseshifts δ_c and W_{cc} have been determined, the κ'_{cc} matrix elements are obtained from equation (3.6) and the real R'_{cc}-matrix elements from equation (3.4).

The part of the asymptotic development of $\Phi_o^{\pm}$ (equation (2.5)) restricted to the closed-channel space is written as $P'_c A'^{\pm}_{co}$. The set P'_c represents P'_i functions with form similar to $\psi_c P_c(r)$ (equation (2.6)):

$$P'_i = \theta_i \cos \pi(\nu_i + \delta_i) - \bar{\theta}_i \sin \pi(\nu_i + \delta_i). \tag{3.10}$$

It is easy to show that P'_i is an exponentially decreasing function along the r coordinate. Indeed, using equations (3.1) and (3.2), we obtain

$$P'_c = \psi_c P_c(r)\, W_{cc}.$$

By definition we have

$$A'^{\pm}_{co} = {}^{t}W_{cc} A^{\pm}_{co}. \tag{3.11a}$$

Using equations (2.7), (2.1), (2.3) and (3.5) and the commutation relation between the matrices W_{cc} and $\exp(\mathrm{i}\pi\nu_c)$ we obtain

$$A'^{\pm}_{co} = \cos \pi(\nu_c + \delta_c)^{-1} (H'^{\pm}_{cc})^{-1} R'_{co} (\pm \mathrm{i} I_{oo} + R'_{oo})^{-1} \exp(\mp \mathrm{i}\pi\delta_o)\, {}^{t}W_{oo} \tag{3.11b}$$

with

$$H'^{+}_{cc} = \tan \pi(\nu_c + \delta_c) + \kappa'_{cc} \text{ and } H'^{-}_{cc} = (H'^{+}_{cc})^{*}$$

where the real part of a diagonal element is $\tan \pi(\nu_i + \delta_i)$.

3.2.2. $\boldsymbol{R}'_i$ vectors. The total state density and the total oscillator strength density involve the matrix product

$$A'^{\pm}_{co}\, {}^{t}A'^{\pm *}_{co} = \cos \pi(\nu_c + \delta_c)^{-1} (H'^{\pm}_{cc})^{-1} \Gamma'_{cc} (H'^{\mp}_{cc})^{-1} \cos \pi(\nu_c + \delta_c)^{-1} \tag{3.12}$$

where $\Gamma'_{cc} = \mathrm{Im}(\kappa'_{cc})$.

The Γ'_{cc} matrix, which describes the coupling between the closed channels via the resonant continua, does not depend on any transformation performed in the open-channel subspace. This is a positive definite matrix of dimension N which can always be expressed as the product of two transposed matrices:

$$\Gamma'_{cc} = L\, {}^{t}L \tag{3.13a}$$

where L is a lower triangular matrix of dimension N. Each row i of L defines the components of a vector $\boldsymbol{R}'_i$. Thus each element Γ'_{ij} can be expressed as a scalar product:

$$\Gamma'_{ij} = \boldsymbol{R}'_i \cdot \boldsymbol{R}'_j. \tag{3.13b}$$

The N vectors $\boldsymbol{R}'_i$ describe the resonant coupling between closed and open channels. Each vector has, at most, N non-zero components, which means that only N continua are required for a complete description of total state densities. These continua are 'effective' continua (Fano 1961, Cooke and Cromer 1985). According to equation (3.13a), the first closed channel is coupled to the first continuum, the second closed channel to two continua (including the first one), etc. The relative order of the closed channels is arbitrary and the definition of the effective continua is not unique. The couplings between the N closed channels and the N effective continua are determined by at most N^2 coefficients. Let us note that in most of the previous analyses of autoionised states (Cooke and Cromer 1985, Mullins *et al* 1985) each closed channel is assumed to be coupled to only one effective continuum. Thus several parameters describing the coupling between closed and open channels are surmised, without any rigorous justification, to have zero values. The non-zero parameters are in fact associated with the N 'lengths' of vectors $\boldsymbol{R}'_i$. A main aim of our study is to demonstrate that only $N(N-1)/2$ additional parameters are necessary to complete the exact description of the couplings between closed and open channels. These parameters are the angles between the $\boldsymbol{R}'_i$ vectors.

3.2.3. Interpretation of the parameters. The $N(N+1)$ MQDT parameters may be classified in five distinct sets: N phaseshifts δ_i, N vector lengths $|\boldsymbol{R}'_i|$, $N(N-1)/2$ angles between the vectors $\boldsymbol{R}'_i$, the $\mathrm{Re}(\kappa'_{ij})$-matrix elements with i and j pertaining to distinct blocks of closed channels and the $\boldsymbol{W}_{\mathrm{cc}}$ matrix elements. The last two sets contain $N(N-1)/2$ parameters. In the following we successively examine each class of parameters and discuss their physical meaning.

3.2.3a. Isolated resonances: phaseshifts δ_i and vector lengths $|\boldsymbol{R}'_i|$. When the direct coupling between the closed channels is negligible or weak ($\mathrm{Re}(\kappa'_{ij}) \ll 1$) and when the resonances pertaining to a given channel are well separated from other resonances (no overlapping) the structure of the spectrum is completely described by the two sets of parameters δ_i and $|\boldsymbol{R}'_i|$. Each pair of these parameters thus describe an isolated channel (Giusti-Suzor and Fano 1984, Seaton 1983). When $|\boldsymbol{R}'_i|^2 \ll 1$, δ_i is a quantum defect which characterises the position of the resonances in the channel i and $|\boldsymbol{R}'_i|$ is related to the normalised autoionisation half-width (Cooke and Cromer 1985)

$$|\boldsymbol{R}'_i|^2 = (\pi\nu_i^3/2)\Gamma_i. \tag{3.14}$$

However, in the case $|\boldsymbol{R}'_i|^2 \simeq 1$ ($|\boldsymbol{R}'_i|$ cannot be greater than 1 (Seaton 1983)) it is no longer possible to interpret $|\boldsymbol{R}'_i|^2$ as a reduced width. The total state density has an oscillatory behaviour without any definite resonance.

3.2.3b. Relative positions of vectors $\boldsymbol{R}'_i$-stabilised states in the continuum. The crucial role played by the angles between the $\boldsymbol{R}'_i$ vectors may be analysed by studying the stabilisation phenomenon, namely the presence of bound states embedded in continua but completely uncoupled to the continua. Following Cooke and Cromer (1985) the wavefunction of a stabilised state consists of a linear superposition of functions F' (equation (3.3)) without any component in the open channels and with exponentially decaying components in the closed channels (Seaton 1983). If there exists a column vector z_{c} such that

$$\begin{cases} [\tan\pi(\nu_{\mathrm{c}}+\delta_{\mathrm{c}}) + R'_{\mathrm{cc}}]z_{\mathrm{c}} = 0 & (3.15) \\ R'_{\mathrm{oc}}z_{\mathrm{c}} = 0 & (3.16) \end{cases}$$

then z_c represents a stabilised state in the continuum. Equation (3.15) gives the position of the state. From equation (3.16) we have

$$\Gamma'_{cc} z_c = R'_{co}(1 + R'^2_{oo})^{-1} R'_{oc} z_c = 0$$

or equivalently

$$\boldsymbol{R}'_i \cdot \left(\sum_{j=1} z_j \boldsymbol{R}'_j \right) = 0. \qquad (3.17a)$$

This condition may be satisfied only if

$$\sum_{j=1}^{N} z_j \boldsymbol{R}'_j = 0. \qquad (3.17b)$$

Therefore a necessary but insufficient condition for the existence of a stabilised state is that the $\boldsymbol{R}'_i$ are not linearly independent. Therefore the maximum number of requisite effective continua is $N-1$.

It follows from equation (3.17*b*) that stabilisation is due to destructive interferences between the paths of autoionisation represented by the couplings between closed and open channels. The less favourable situation (no interference) occurs when the $\boldsymbol{R}'_i$ are orthogonal. The angles between the $\boldsymbol{R}'_i$ vectors are closely related to the occurrence of interferences between alternative autoionisation paths.

3.2.3c. Coupling between closed channels: $Re(\kappa'_{cc})$. Assuming that the resonance widths R'^2_i and the elements of the matrix Γ'_{cc} are much smaller than the separation between the resonances, the relation giving the poles of Δ (cf § 2.3.3) reduces to

$$\det[\tan \pi(\nu_c + \delta_c) + \mathrm{Re}(\kappa'_{cc})] = 0. \qquad (3.18)$$

When all channels are closed, equation (3.18) is identical to the condition for bound states in Rydberg series. The graph of this equation in the ν_c space constitutes a Lu–Fano plot, which allows us to fit the non-diagonal elements of $\mathrm{Re}(\kappa'_{cc})$.

However, it must be noticed that equation (3.18) has been obtained under some restrictive conditions. In many physical situations (broad or overlapping resonances) it is quite impossible to draw a Lu–Fano plot of the centres of *resonance* (Mullins *et al* 1985, Cooke and Cromer 1985).

3.2.3d. Elements of the rotation matrix W_{cc} coupling the degenerate closed channels. From equations (3.11*a*, *b*), giving the amplitudes of the closed channels, it is obvious that the state density (equation (2.15)) does not depend on any transformation performed within a given block of degenerate channels. However, the observed spectra depend not only on the state density but also on transition matrix elements. The determination of these latter elements requires the knowledge of the exact coupling scheme describing the target states. The rotation W_{cc} transforms the collisional closed channels which were initially defined in the *jj*-coupling scheme. In the case where all the channels are closed the κ'_{cc} matrix is real and reduces to the R' matrix. One has then the well known property that the energies of Rydberg states pertaining to degenerate channels do not depend on an orthogonal transformation of the channels; the complete determination of the wavefunctions, however, requires the knowledge of the exact coupling of the target states (Lu 1971, Aymar 1984).

3.3. Cross section in the transformed basis

Using the expressions (3.11) of the transformed amplitudes of closed channels, the total cross section in an ICE spectrum can be written as

$$s = \mathrm{Tr}(W_{cc}(A'^{-}_{co})('A'^{-*}_{co})'W_{cc}\sigma_{cc}) \tag{3.19}$$

where σ_{cc} is the matrix already defined in equation (2.12). Using the new functions (3.10) and (2.10), the cross section may be expressed in terms of $\sigma'_{cc} = 'd'_c d'_c$, where the transformed dipole matrix elements are such that $d'_c = d_c W_{cc}$.

Hence, from equations (3.11) and (3.19) we find that

$$s = \mathrm{Tr}[\cos\pi(\nu_c+\delta_c)^{-1}(H'^{+*}_{cc})^{-1}\Gamma'_{cc}(H'^{+}_{cc})^{-1}\cos\pi(\nu_c+\delta_c)^{-1}\sigma'_{cc}] \tag{3.20}$$

where

$$\Gamma'_{cc} = \mathrm{Im}(\kappa'_{cc}) \qquad \text{or} \qquad \Gamma'_{ij} = \boldsymbol{R}'_i \cdot \boldsymbol{R}'_j$$

and

$$H'^{+}_{cc} = \tan\pi(\nu_c+\delta_c) + \kappa'_{cc}$$

or

$$H'^{+}_{ij} = \tan\pi(\nu_i+\delta_i)\delta_{ij} + \mathrm{Re}(\kappa'_{ij}) + \mathrm{i}\boldsymbol{R}'_i \cdot \boldsymbol{R}'_j.$$

The parameters (W_{cc}, δ_i, $\mathrm{Re}(\kappa'_{ij})$, $\boldsymbol{R}'_i$) are adjustable parameters and are fitted on experimental spectra.

4. Application to systems involving two or three closed channels

In this section, the definition of the MQDT parameters will be illustrated by some particular examples corresponding to two or three closed channels ($N = 2$ or 3), the number of continua being arbitrary. In particular, we shall analyse the physical meaning of the new MQDT parameters introduced in our approach: the angles between the $\boldsymbol{R}'_i$ vectors and the rotation matrix W_{cc}.

4.1. Exact or approximate stabilisation ($N = 2$, $M = 1$ or 2)

The experimental data are expressed in terms of six parameters: the quantum defects of the closed channels, δ_1 and δ_2, the vector lengths R'_1 and R'_2, the angle between the $\boldsymbol{R}'_i$ vectors β, and the direct coupling $\mathrm{Re}(\kappa'_{12})$ noted r_{12}.

If $\beta = 0$ or π ($\boldsymbol{R}'_1$ parallel to $\boldsymbol{R}'_2$) the problem reduces to that previously analysed in detail by Giusti-Suzor and Lefebvre-Brion (1984), i.e. two closed channels coupled to a unique continuum. If $\beta = \pi/2$ ($\boldsymbol{R}'_1$ orthogonal to $\boldsymbol{R}'_2$), each closed channel is coupled to only one particular continuum. If, in addition, $r_{12} = 0$, the system gives two completely decoupled two-channel systems. With $\beta = \pi/2$, the interference rate between autoionisation paths is minimal (even null if $r_{12} = 0$) and stabilisation cannot occur. At the other extreme ($\beta = 0$) the interference rate is maximal and stabilised states can appear.

Stabilised $5d_{3/2}nd_{3/2}$ $J = 0$ states ($n \sim 60$) have been observed in barium by Neukammer *et al* (1985). The stabilisation is due to destructive interference between two alternative ionising paths: direct autoionisation to the $5s\varepsilon S\ {}^1S_0$ continuum and indirect autoionisation through the $5d_{5/2}nd_{5/2}$ $J = 0$ channel.

Figure 1 shows a typical representation of the 'scaled' widths of resonances when channel 2 converges to a lower limit than channel 1. As the energy is varied, ν_2 pass through several cycles while ν_1 is nearly constant. All the MQDT parameters are kept constant. The 'scaled' widths are calculated with an exact but complicated relation. We have checked that, in most cases, similar results are obtained by using a simpler and more comprehensive expression

$$\pi\tfrac{1}{2}\Gamma\nu_2^3=(t_1R_2'\cos\beta-r_{12}R_1')^2/(t_1^2+R_1'^4)+R_2'^2\sin^2\beta \tag{4.1}$$

where $t_1=\tan\pi(\nu_1+\delta_1)$.

When $\beta=0$ this relation reduces to that previously obtained by Cooke and Cromer (1985) for describing a similar situation involving only one continuum. When $\beta=\pi/2$ the plot is a symmetrical Lorentzian profile: there is no stabilisation. When $\beta\neq\pi/2$ it is an asymmetric Fano profile with Fano parameter $q=-r_{12}/R_1'R_2'\cos\beta$. When $\beta=0$ the minimum is zero and exact stabilisation may occur. With $\beta\neq 0$, $\pi/2$ there is always a non-zero minimum. The stabilisation can only be 'approximate'. The width of the resonances in the neighbourhood of the minimum (resonance numbered 20 or 21) increases when β increases from $\beta=0$. This relation clearly shows that the angle β measures the interference rate between alternative ionising paths.

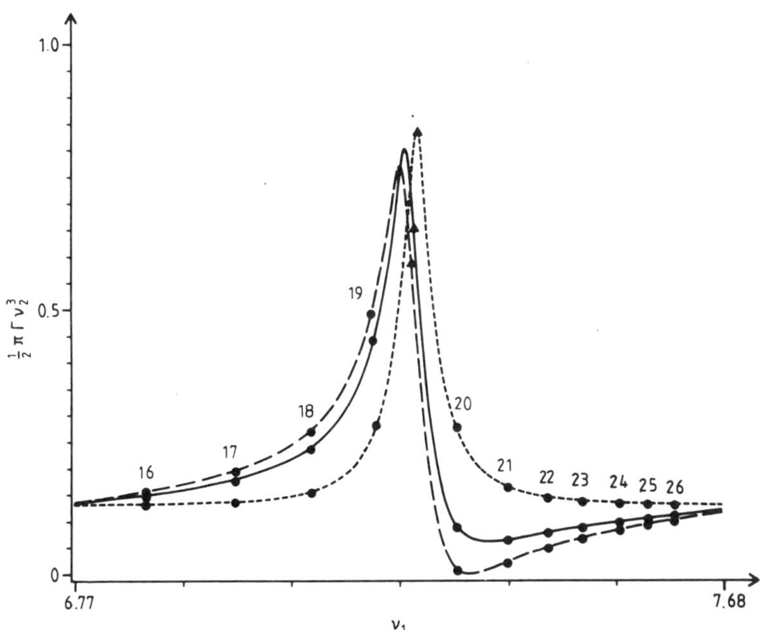

Figure 1. Calculated widths $\frac{1}{2}\pi\Gamma\nu_2^3$ plotted against ν_1. The studied case corresponds to $\nu_1\ll\nu_2$. I_1 and I_2 are chosen to be the energies of the Ba^+ $6p_{3/2}$ and $6p_{1/2}$ levels respectively. The widths are calculated using parameters extracted from Mullins *et al* (1985). They correspond to the system (denoted + by Mullins *et al* and B in the text, § 4.2.4) formed by the two channels $6p_{3/2}nd$ (+) $J=3$ and $6p_{1/2}nd$ $J=3$ coupled to open channels. $\delta_1=2.75$, $\delta_2=2.79$, $|\boldsymbol{R}_1'|=0.32$, $|\boldsymbol{R}_2'|=0.36$, $r_{12}=0.27$. ● represents the resonances $6p_{1/2}nd$ $(16\leq n\leq 26)$ and ▲ represents the perturber $6p_{3/2}10d$ (+). · · · · ·, $\beta=\pi/2$ (no stabilisation; - - -, $\beta=0$ (exact stabilisation); ——, $\beta=\pi/4$ (approximate stabilisation).

4.2. System involving two degenerate channels ($N=3$)

We consider a system with three closed channels, the number of continua being arbitrary. We assume that two closed channels are degenerate. The orthogonal transformation W_{cc} between these channels depends only on the rotation angle θ.

4.2.1. Definition of closed channels. There are several different ways to define the rotation angle each of which allows us to accurately define the target states.

In § 3.2 the angle θ defining the orthogonal transformation of the two degenerate channels (1 and 2) has been determined using the system (3.7) and the condition $\mathrm{Re}(\kappa'_{12})=0$ (equation 3.9)). The latter condition may be replaced by another one, acting for example on the imaginary part of κ'_{cc} which describes the coupling between the open and closed channels.

More precisely, the MQDT parameters may be determined from

$$\left.\begin{aligned} \mathrm{Re}(\kappa'_{ii}) &= 0 \quad (4.2a) \\ \mathrm{Im}(\kappa'_{13}) &= 0 \quad (4.2b) \end{aligned}\right\} \quad i=1,3.$$

Using equation (4.2b) the angle θ is chosen such that $\boldsymbol{R}'_1 \cdot \boldsymbol{R}'_3 = 0$, which means that the vectors associated to two non-degenerate channels are orthogonal. The relative position of the $\boldsymbol{R}'_i$ vectors, shown schematically in figure 2, depends on the two angles β and φ.

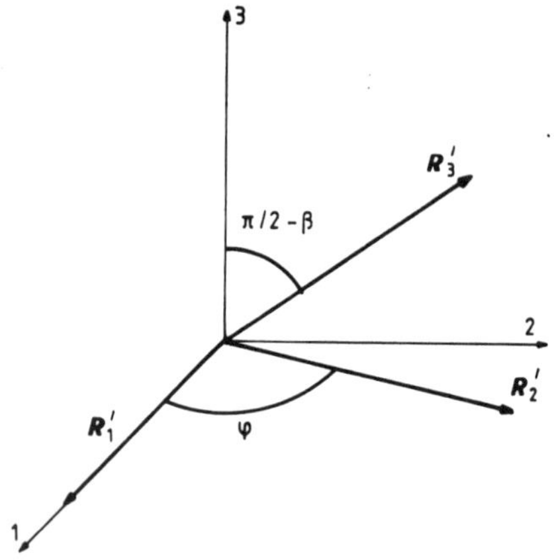

Figure 2. Geometrical representation of the $\boldsymbol{R}'_i$ vectors in the case of $N=3$ and two degenerate channels $\pi/2-\beta=(3, \boldsymbol{R}'_3)$ and $\varphi=(\boldsymbol{R}'_1, \boldsymbol{R}'_2)$. $\boldsymbol{R}'_3$ is in the plane 2-3.

The angle θ does not act on the $\boldsymbol{R}'_i$ vectors' space, but it is related to the definitions of the collision channels 1 and 2. Let us study the important case corresponding to two degenerate channels with the same orbital momentum of the outer electron ($l_1=l_2$). Through the transformation (3.1) we have transformed the initial target states ψ_1 and ψ_2 into $\Phi_A(\theta)$ and $\Phi_B(\theta)$:

$$\begin{aligned} \Phi_A(\theta) &= \psi_1 \cos\theta - \psi_2 \sin\theta \\ \Phi_B(\theta) &= \psi_1 \sin\theta + \psi_2 \cos\theta \end{aligned} \quad (4.3)$$

where ψ_1 and ψ_2 are given in the jj-coupling scheme. The rotation θ consists in changing the angular coupling of the target states.

4.2.2. Density of states: angle of stabilisation β. When the interference rate between the degenerate channels is minimal ($\varphi = \pi/2$ or equivalently $\boldsymbol{R}'_1 \cdot \boldsymbol{R}'_2 = 0$), the system can be separated into two subsystems A and B (see figure 3). The former subsystem consists of one closed channel characterised by a particular target state $\Phi_A(\theta)$ which depends on θ and on one open channel. The latter subsystem consists of two closed channels ($\Phi_B(\theta)$ and ψ_3) and two continua. These two subsystems are coupled through the off-diagonal elements $\mathrm{Re}(\kappa'_{12})$ and $\mathrm{Re}(\kappa'_{13})$. When these elements are negligible compared with the scaled half-widths $|\boldsymbol{R}'_i|^2$ (for example for highly excited core states) the total system gives two completely independent subsystems. The total state density, which depends on β and φ but not on θ, consists of the superposition of the densities characterising each individual system. The system A is formed by resonances characterised by one quantum effect δ_1 and one scaled half-width $|\boldsymbol{R}'_1|^2$. The system B is similar to the two-closed-channels system studied in § 4.1. Stabilisation phenomena may occur, according to the value of the angle β which measures the interference rate between alternative autoionisation paths.

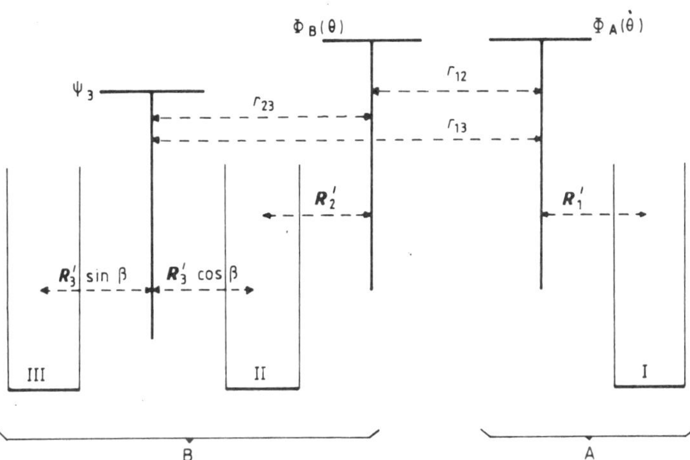

Figure 3. Effective continua and couplings between the channels in the case of $N = 3$ and two degenerate channels. The system is separated into two systems A and B. $\varphi = \pi/2$ (see figure 2). I, II and III are the effective continua; $\boldsymbol{R}'_1$, $\boldsymbol{R}'_2$ and $\boldsymbol{R}'_3$ are represented on figure 2. r_{12}, r_{13} and r_{23} are the direct couplings between closed channels (i.e. the off-diagonal elements of the matrix $\mathrm{Re}(\kappa'_{cc})$).

4.2.3. Density of oscillator strength: θ, angle of mixing of the two systems A and B. The angle θ plays a role only in the density oscillator strength through the dipole moment transition parameters. The transformation of the target states is equivalent to a transformation of the dipole moment d_1 and d_2 (cf § 3.3) giving

$$\begin{aligned} D_A &= d_1 \cos\theta - d_2 \cos\theta \\ D_B &= d_1 \sin\theta + d_2 \cos\theta. \end{aligned} \tag{4.4}$$

The relative importance of D_A and D_B strongly depends on the excitation process through the dipole moments d_1 and d_2 and the angle θ. It can happen that one parameter is negligible and only one subsystem, A or B, is observable.

4.2.4. Examples. Mullins *et al* (1985) have recently studied the excitation of (6p*n*d) $J=3$ autoionised states of Ba from 6s*n*d $^{1,3}D_2$ states. The autoionised spectrum consists of three closed channels coupled to several continua. The two degenerate channels $6p_{3/2}nd_{3/2}$ and $6p_{3/2}nd_{5/2}$ interact with the $6p_{1/2}nd_{5/2}$ channel. Mullins *et al* (1985) observed that the system A (noted −) is excited dominantly starting from 6s*n*d $^3D_2$ states. Mullins *et al* (1985) and Gounand *et al* (1983) noted that the system B (noted +) is observed when exciting 6s*n*d $^1D_2$ states. This system may be described by an MQDT model involving two closed channels coupled to continua. Stabilisation phenomena, caused by destructive interferences between ionising paths, may occur, according to the value of the angle β. In the vicinity of the $6p_{3/2}10d$ $J=3$ levels, complex structures correspond to the high member ($16 \leqslant n \leqslant 26$) of the $6p_{1/2}nd$ $J=3$ series perturbed by the $6p_{3/2}10d$ levels.

This system has been analysed by Mullins *et al* (1985) assuming a minimal interference rate ($\beta=\pi/2$). The main features of the experimental spectrum corresponding to system B have also been reproduced by Giusti-Suzor and Lefebvre-Brion (1984) with the opposite assumption, that is a maximal interference rate ($\beta=0$). The reduced widths of resonances pertaining to the system B are shown in figure 1. These widths have been calculated with parameters extracted from the MQDT model fitted by Mullins *et al* (1985) to the experimental data. Figure 1 clearly displays the strong β dependence

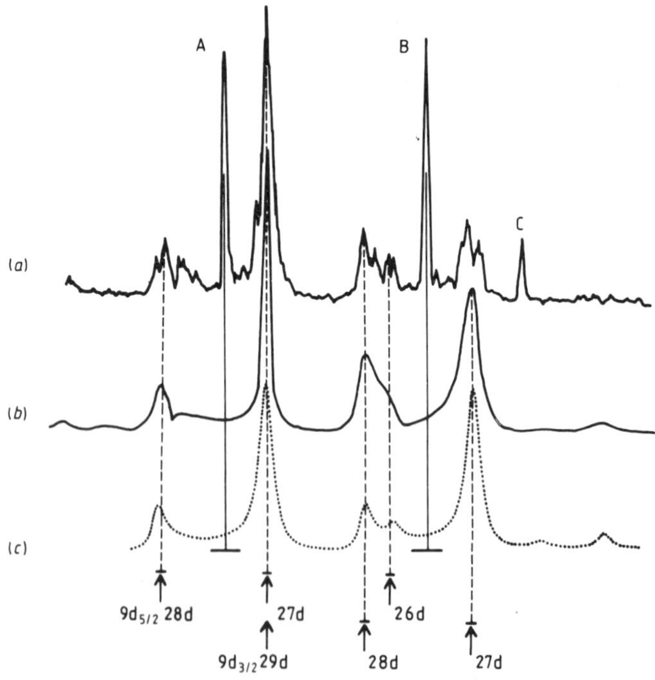

Figure 4. (*a*) Excitation spectrum 6s26d $^3D_2 \rightarrow$ (9d*n*d) $J=4$ recorded by Camus *et al* (1985) in neutral barium. (*b*) Preliminary fit of the observed profiles using our new formulation ($\beta \simeq \pi/3$). The narrowing resonance is related to an approximate stabilisation (Boulmer *et al* 1987). (*c*) Computed profiles previously published by Camus *et al* (1985). The resonance series are superposed ($\beta=\pi/2$). The line resonances A, B and C correspond to ionic transitions. Each resonance $9d_{5/2}nd$ corresponds in fact to the superposition of two degenerate levels.

of the widths of the $6p_{1/2}nd$ $J=3$ levels with $n=20$, 21. 'Approximate' stabilisation occurs for intermediate values of β. However, the precision of the observed spectra does not allow us to choose between the different interpretations.

The $9dnd$ $J=4$ spectrum of Ba has been recently observed by Camus *et al* (1985). This system is formed by three closed channels $9d_{5/2}nd_{3/2}$, $9d_{5/2}nd_{5/2}$ and $9d_{3/2}nd_{5/2}$ coupled to a large number of continua. The observations have been interpreted by Camus *et al* (1985) using the particular hypothesis ($\delta_1=\delta_2, |\boldsymbol{R}_1'|=|\boldsymbol{R}_2'|, \varphi=\pi/2$), which consists of considering the two degenerate channels as completely superposed.

In addition, they assumed that the system B is broken into two subsystems, each formed by one closed channel coupled to one continuum. This consists of putting $\mathrm{Re}(\kappa_{13}')=0$ and $\beta=\pi/2$.

The condition $\varphi=\pi/2$ is certainly satisfied. In fact $\varphi\neq\pi/2$ would result in interference between the channels 1 and 2; the degenerate channels should be characterised by narrow resonances (the smaller the width, the smaller φ) in contradiction with the observations.

In contrast, the justification of the conditions $\mathrm{Re}(\kappa_{13}')=0$ and $\beta=\pi/2$ is less apparent. The introduction of a non-zero angle β allows us to interpret some anomalies observed in the experimental spectrum: the narrowing of a resonance occurring in the vicinity of the crossing of the two $9d_{5/2}nd$ and $9d_{3/2}nd$ series might be due to the effect of interferences between autoionising paths. Figure 4 presents a comparison of calculated profiles with the experimental ones (figure 4(a)). Figure 4(b) shows a preliminary fit using a non-zero value of β, while figure 4(c) reproduces the previously mentioned calculations of Camus *et al* (1985). Additional more precise measurements are in progress. Such data might allow us to check to what extent destructive interferences between autoionisation paths still occur for systems involving a large number of continua ($M>30$ in the present case).

5. Conclusion

In this work, we have reduced the general MQDT problem involving M open and N closed channels to the simpler problem involving only the closed channels. We have suggested an alternative set of a small number of MQDT parameters that are useful in analysing photoionisation spectra. In this formulation only the characteristics of the closed channels and their mutual couplings are described. The presence of the open channels is taken into account by introducing a complex reactance matrix κ_{cc} instead of the usual real R matrix. The imaginary part of the matrix elements is written as the scalar product of vectors. Each of these vectors corresponds to a closed channel. The length of a vector globally represents the properties of the resonances observed in the corresponding channel. The stabilisation phenomenon is interpreted as due a particular relative orientation of these vectors. Furthermore, the coupling scheme between the ionic core and the external electron is not necessarily of the *jj* type, which greatly simplifies the analysis when degenerate channels are involved.

All the introduced parameters refer to the closed channels; therefore it is not necessary to explicitly construct the effective continua from the open channels defined initially. Indeed, the continua are globally taken into account.

The present study has been restricted to the description of total photoionisation spectra, which are the more experimentally investigated spectra. Taking into account the developments in experimental technology (photoelectron spectroscopy or energy analysis of the residual ion) it would be worthwhile to describe the continua less globally. The formalism we have introduced can be straightforwardly deduced by partitioning the open channels into several blocks, corresponding for example to different energy levels of the residual ion.

Acknowledgments

I am much indebted to Dr M Aymar and E Luc-Koenig for help and encouragement and for several stimulating discussions. Special thanks go to Professor P Camus, Dr P Pillet and Dr A Giusti-Suzor for critical reading of this paper.

Appendix. Equivalence of systems (3.7) and (3.8)

We show in this section that the solutions δ_i of equation (3.7) are solutions of equations (3.8); moreover, it is possible to construct from the δ_i a complete set of solutions of equations (3.8).

The transformations introduced in § 3.1 are in the whole space $\{(o)+(c)\}$. The transformation which is defined by the set of parameters δ_c, δ_o and W_{oo} is noted $T(\delta_c, \delta_o, W_{oo})$.

We start from the δ_i such that

$$\mathrm{Re}(\kappa'_{ii}) = 0 \tag{A.1}$$

and perform two successive transformations:

$$\begin{cases} R \\ \kappa_{cc} \end{cases} \xrightarrow{T(\delta_i, 0, I_{oo})} \begin{cases} R' \\ \kappa'_{cc} \end{cases} \xrightarrow{T(0, \delta_o, W_{oo})} \begin{cases} R'' \\ \kappa''_{cc} \end{cases}$$

we have $R'_{ii} \neq 0$ and $R'_{oo} \neq 0$ in general.

The parameters of the second transformation are defined through the diagonalisation of R'_{oo}:

$$R'_{oo} = W_{oo} \tan \delta_o \, {}'W_{oo}. \tag{A.2}$$

We can easily check that the R'' matrix is such that

$$R''_{oo} = 0 \tag{A.3}$$

and

$$R''_{co} = R'_{co} W_{oo} \cos \pi\delta_o. \tag{A.4}$$

Moreover, we have from equation (3.6)

$$\kappa''_{cc} = \kappa'_{cc}. \tag{A.5}$$

It follows from equation (A.3) and from the development (2.3) of κ''_{cc} that

$$\mathrm{Re}(\kappa''_{cc}) = \mathrm{Re}(\kappa'_{cc}) = R''_{cc} \tag{A.6}$$

and

$$R''_{ii} = 0. \tag{A.7}$$

Using the scattering matrices (on the whole set of channels $\{(o)+(c)\}$) χ, χ' and χ'', and the transformation (3.5) $\chi' = \exp(-i\pi\delta)\,'W\chi W\exp(-i\pi\delta)$ we obtain immediately

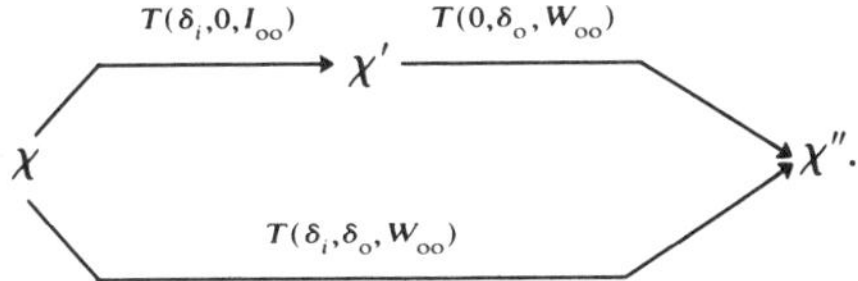

The same scheme holds for the reaction matrices: R'' is obtained from R through the transformation $T(\delta_i, \delta_o, W_{oo})$. Then the set $(\delta_i, \delta_o, W_{oo})$ is a solution of the system

$$\begin{cases} R^0_{ii} = 0 \\ R^0_{oo} = 0 \end{cases} \qquad i = 1, N$$

which is equivalent to equations (3.8) or the addition of equations (A.7) and (A.3).

References

Aymar M 1984 *Phys. Rep.* **110** 164–220
Bell R J and Seaton M J 1985 *J. Phys. B: At. Mol. Phys.* 1589–629
Bhatti S A and Cooke W E 1983 *Phys. Rev.* A **28** 756–9
Bloomfield L A, Freeman F R, Cooke W E and Bokor J 1984 *Phys. Rev. Lett.* **53** 2234–7
Boulmer J, Camus P and Pillet P 1987 *J. Opt. Soc. Am.* B to be published
Camus P, Pillet P and Boulmer J 1985 *J. Phys. B: At. Mol. Phys.* **18** 481–7
Cooke W E and Cromer C L 1985 *Phys. Rev.* A **32** 2725–38
Cooke W E, Gallagher T F, Edelstein S A and Hill R M 1978 *Phys. Rev. Lett.* **40** 178–81
Eissner W and Seaton M J 1969 *J. Phys. B: At. Mol. Phys.* **2** 341–55
Fano U 1961 *Phys. Rev.* **124** 1866–78
Giusti-Suzor A and Fano U 1984a *J. Phys. B: At. Mol. Phys.* **17** 215–29
—— 1984b *J. Phys. B: At. Mol. Phys.* **17** 4277–82
Giusti-Suzor A and Lefebvre-Brion R 1984 *Phys. Rev.* A **30** 3057–65
Gounand F, Gallagher T F, Sandner W, Safinya K A and Kachru R 1983 *Phys. Rev.* A **27** 1925–38
Lecomte J M and Luc-Koening E 1985 *J. Phys. B: At. Mol. Phys.* **18** 3139–47
Lefebvre R and Beswick J A 1972 *Mol. Phys.* **23** 1223–34
Lu K T 1971 *Phys. Rev.* A **4** 579–96
Lu K T and Fano U 1970 *Phys. Rev.* A **2** 281–6
Mullins D C, Zhu Y, Xu E Y and Gallagher T F 1984 *Phys. Rev.* A **32** 2234–41
Neukammer J, Rinneberg H, Jönnsson G, Cooke W E, Hieronymus H, Kõnig A and Vietzke K 1985 *Phys. Rev. Lett.* **55** 1979–82.
Seaton M J 1983 *Rep. Prog. Phys.* **46** 167–257
Tran N H, Pillet P, Kachru R and Gallagher T F 1984 *Phys. Rev.* A **29** 2640–50

1979 *IBM J. Res. Dev.* **23** 490–503
Reprinted with permission from International Business Machines Corporation; copyright 1979

Systematic Behavior in Alkaline Earth Spectra: A Multichannel Quantum Defect Analysis

J J Wynne and J A Armstrong

"Two-electron" atoms are more complex than "one-electron" atoms because of electron-electron interactions. This leads to spectra that are not well understood. Using multiple photon excitation and ionization detection, we have extensively studied Rydberg series of states in Ca, Sr, and Ba. The spectroscopic data were interpreted by using multichannel quantum defect theory (MQDT). In this context, systematic trends in the atom series Ca, Sr, and Ba are discussed.

Introduction

The revolution in optical spectroscopy, driven by the tunable laser, has spawned a renaissance in atomic spectroscopy. This is a field with a long and continuing history of achievements. However, the unique capabilities provided by tunable lasers have allowed new and exciting studies of the electronic structure of atoms. We have been studying the high Rydberg states in the alkaline earth atoms Ca, Sr, and Ba. These atoms have two valence electrons outside a closed, rare-gas-like electronic shell and are therefore classified as two-electron atoms. As such, they provide a "laboratory" for the study of the effects of electron correlation on atomic spectra and atomic structure. This "laboratory" is well matched to the capabilities of tunable dye lasers because it is possible to use a small number of visible or near-visible photons to selectively excite and study well-defined Rydberg series of states converging on the first and several higher ionization limits of these atoms.

Many aspects of the electronic structure of many-electron atoms are still not well understood. The simplest picture of such atoms is that each electron moves in a well-defined orbit, independent of the other electrons. However, spin-orbit coupling effects and Coulomb repulsion between electrons modify this picture, and lead to admixtures of configurations and splitting of terms that depend on the relative orientations of total orbital angular momentum L and total spin S. These effects hamper our ability to identify the states of many-electron atoms and to classify them with one-electron labels. When one looks at tabulations of atomic energy levels [1], there are obvious large gaps in the list of known members of many Rydberg series. Laser spectroscopy offers the possibility of enormously extending our knowledge of these series.

In a series of papers [2–4] we have reported on laser spectroscopic investigations of extensive Rydberg series of different types in Ca, Sr, and Ba. Similar studies have been reported by several European groups [5–8]. Previous to the laser-assisted work, studies of these atoms by absorption spectroscopy [9–11] had yielded extensive information on the $^1P_1^0$ Rydberg series. Multichannel quantum defect theory (MQDT), a method developed by Seaton [12] and by Fano and coworkers [13–16], has proven to be very useful in analyzing these spectroscopic data [2–4, 6, 8, 17].

In this paper we briefly review our laser spectroscopic methods but *our main point is a discussion of the systematic trends in the alkaline earths that our MQDT analyses have uncovered.* In particular, MQDT extracts from complicated spectra a small set of parameters that describes, in a compact manner and with surprising accuracy, the details of these spectra. What we have found in each case analyzed is that for a spectrum of given J and parity, all but one of the set of parameters are nearly the same for Ca, Sr, and Ba. This similarity goes well beyond what has been previously recognized from comparisons of state-

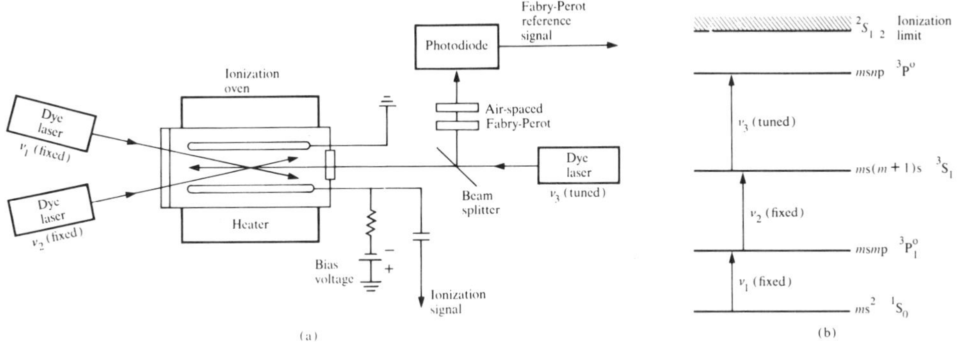

Figure 1 (a) Experimental setup for multiphoton ionization spectroscopy (MIS). (b) Energy level diagram of typical alkaline earth atom showing three-photon sequence that excites $^3P^o$ states.

by-state analyses of the spectra. An earlier analysis by Lu [18] briefly considered systematic trends in the alkaline earths.

Multiphoton ionization spectroscopy (MIS)

The tunable dye laser as a spectroscopic tool gives the experimenter the ability to start with atoms in a known state, *e.g.*, the ground state, and excite them along a carefully chosen excitation pathway. The pathway is determined by both the photon energies and the laser polarization. Of course, such selective excitation is possible with nonlaser light sources, but the vastly superior spectral brightness of laser light allows a large fraction of the excited atoms to end up in a single final state. This is true even when the excitation involves more than one photon. Such selective multiphoton excitations were, in effect, not observable before the advent of lasers. By using combinations of tunable lasers, Rydberg series that are not connected to the ground state by one-photon transitions may be observed and identified. For example, the use of an even number of photons leads to final states of the same parity as the ground state. Such states cannot be observed in absorption from the ground state by electric dipole transitions. In general, with tunable lasers we can excite and observe final states of a particular pre-selected symmetry.

In our work on the alkaline earths, we have used a simple experimental setup to detect excitation to high Rydberg states. Our technique is based on detection of ionization that follows population of a high Rydberg state by multiphoton excitation, hence the name multiphoton ionization spectroscopy (MIS). The essence of MIS is illustrated in Fig. 1. A pipe containing the atomic species of interest is heated to provide adequate vapor pressure ($\approx$10–100 Pa). For the alkaline earths Ca, Sr, and Ba, a typical operating temperature is 800°C, at which temperature the pipe is red-hot. One or more laser beams [three are shown in Fig. 1(a)] are aimed into the pipe to spatially overlap between a pair of parallel-plate ionization-detecting electrodes. The detection mechanism, which has been known for over 50 years [19], is based on neutralization of the space charge that is produced by thermionic emission of the heated electrodes. With a negative dc bias voltage applied to one of the electrodes, a dc current flow of thermionically emitted electrons occurs. If the bias voltage is small enough, a significant space charge is established, with its maximum density near the negative electrode. If positive ions are introduced into the gas, they partially neutraliz the space charge, allowing more electrons to flow between the electrodes. Due to the much greater drift velocity of the electrons than the ions (approximately inversely proportional to their respective masses), many extra electrons flow from the negative to the positive electrode while each positive ion is traveling from the positive to the negative electrode. In addition, the positive ions may become trapped near the maximum of the space charge density, prolonging their lifetime and allowing proportionally more electron current. The net result is a gain of $\approx 10^5$ in the extra charge flowing in the external circuit due to the introduction of positively charged ions into the vapor between the electrodes. Such a detector, called a space-charge-limited thermionic diode, provides us with a simple and convenient method for detecting

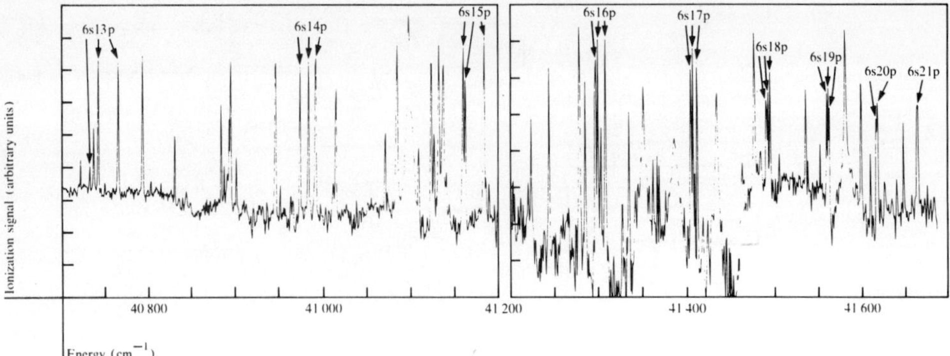

Figure 2 Multiphoton ionization spectrum of Ba. The labels correspond to the members of the 6snp series, with the arrows designating various terms (*e.g.*, $^3P^o_1$, $^3P^o_2$, $^1P^o_1$). The ionization limit is 42034.9 cm^{-1}.

ionization in the hot atomic vapor. The resulting signal is linear in the number of ions over a dynamic range of several orders of magnitude.

The mechanism by which the ions are produced requires some explanation. In our studies of bound Rydberg states, the laser photons excite the atoms to a final bound state. Once this final state is populated it must still be ionized to give a signal. Of the possible photo- or collisional ionization mechanisms, we believe that the predominant one is chemi-ionization, *i.e.*, the production of stable molecular ions upon collision of an excited atom with a ground state atom. This conclusion is based on optical spectroscopic observations of atomic ions showing that these ions last only $\approx 100\ \mu s$, whereas the conductive species responsible for the ionization signal lasts for ≈ 1 ms [20]. Furthermore, recent multiphoton ionization studies carried out in an atomic beam of Sr have detected the production of Sr_2^+, in support of the chemi-ionization mechanism [21].

To record a spectrum, one of the lasers is scanned and the ionization signal recorded. In the setup shown in Fig. 1(a), ν_1 and ν_2 are held fixed while ν_3 is scanned. Part of the beam at ν_3 is split off and passed through a Fabry-Perot interferometer, providing a reference signal. The transmission through the Fabry-Perot is periodic in ν_3, showing peaks at fixed frequency spacing of $c/2\ell$, where c is the speed of light and ℓ is the air-space thickness between the two plates of the interferometer. Once this spacing is calibrated against known peaks in the ionization signal, it provides a frequency marker against which the frequencies of unknown peaks in the ionization spectrum can be calibrated. The signals may be recorded on chart paper or stored digitally in a computer.

Figure 2 shows a spectrum of Ba that was recorded by using three lasers (as in Fig. 1). This spectrum is representative of many such spectra that we have recorded for Ca, Sr, and Ba. For the purposes of this article, the most important things to point out about Fig. 2 are that there is an excellent signal-to-noise ratio and that most of the peaks represent states of Ba that have not been previously identified or classified. The reader is referred to Refs. [2–4] for more details about the experimental methods and for presentations of the extensive new data.

Multichannel quantum defect theory (MQDT)

Traditional analyses of atomic spectra have been aimed at the determination of the energies of the initial and final states of one-photon transitions, the strengths of such transitions, and the classification of the states. When Rydberg series are noninteracting, as in alkali metal ("one-electron") atoms, this classification is relatively straightforward. However, when the series are strongly interacting, as in "many-electron" atoms, such classification becomes difficult. In many cases classifications of states by a pure configuration are wrong. Each state may be instead a mixture of several configurations. However, the mixture describing one state is *not* unrelated to the mixtures describing the other states of the same series. This important fact is recognized by MQDT, which analyzes the spectra in terms of interactions between entire series instead of between individual configurations.

A full exposition here of what MQDT is and why it works would take us too far afield. The important point for the reader of this article to appreciate and keep in mind is that there is a new way of analyzing complex spectra and that the result of this analysis is a small set of parameters that gives a highly accurate and compact description of the spectra. As noted earlier, all but one of the parameters describing a given spectral type are nearly invariant along the atom series Ca, Sr, and Ba. We believe that this near-invariance reflects dynamical symmetries in the electron-electron interaction of the two outermost electrons.

Because we wish to present results based on our MQDT analyses, we must define certain terms, especially the parameters of MQDT. However, on the basis of the following short definitions the reader is not expected to understand the details of MQDT; those wishing to pursue a deeper understanding are referred to Refs. [2, 13–16].

The first term we need to define is *channel*. A channel is a group of states that encompasses both a region of discrete energies and a continuum. One type of channel, known as a collision- or i-channel, describes a set of states with an outer electron of various energies, an electronic core in a definite energy level, specific angular momenta of the outer electron and the core, and their coupling. This type of channel is an appropriate description of the atom when the Rydberg electron is far from the core. As an example, in Ba the 6s*n*p $^1P_1^o$ discrete states and the 6sεp $^1P_1^o$ continuum together constitute an *L, S*-coupled collision channel. Another type of channel, known as an eigen- or α-channel, diagonalizes the noncentral part of the electron-electron interaction. This type of channel is an appropriate description of the atom when the Rydberg electron is close to or penetrating the core.

Interactions among channels are described by MQDT in terms of a small number of physically meaningful parameters whose values may be derived from experimental measurements. These parameters represent the effects of the electronic core on the wave function of the excited electron. They are: 1) the elements of a unitary matrix $U_{i\alpha}$ that specifies the transformation diagonalizing the matrix for noncentral scattering of the excited electron by the core; 2) the eigendefects μ_α that specify the eigenvalues of the scattering matrix, exp $(i2\pi\mu_\alpha)$; and 3) the electric dipole matrix elements D_α that describe transitions from a given discrete state, such as the ground state, to each eigenchannel. The *U*-matrix transforms a basis set of α-channels to a basis set of i-channels.

An important role is played in MQDT by effective quantum numbers ν_i, defined in terms of equations of the type

$$E = I_i - R/(\nu_i)^2, \tag{1}$$

where E is the measured energy of a state, I_i is the *i*th ionization threshold (or series limit) of the atom, and R is the Rydberg constant appropriate for the atom in question. The essence of MQDT is an analytical relationship, parametrized in terms of $U_{i\alpha}$ and μ_α, between the ν_i:

$$\det |U_{i\alpha} \sin \pi(\nu_i + \mu_\alpha)| = 0. \tag{2}$$

For an *N*-limit, *M*-channel problem, there are *N* equations of the type given by Eq. (1); $U_{i\alpha}$ is an $M \times M$ matrix and there are *M* eigendefects μ_α. The solution to Eq. (2) is an $(N-1)$-dimensional surface Σ in the *N*-dimensional ν_i-space. This surface is parametrized in terms of $U_{i\alpha}$ and μ_α.

Our work has dealt primarily with finding values for $U_{i\alpha}$ and μ_α such that Eqs. (1) and (2) fit our experimental data when expressed in terms of ν_i. We have carried out this fitting with the graphical aid of Lu-Fano plots [22]. These plots present a graph of ν_i *versus* ν_j for various pairs of effective quantum numbers. One finds as many values of ν_i as there are relevant series limits in the spectra of interest. This number is the same as the number of distinct core states in the configurations that are admixed in the spectrum.

Systematic behavior in the series Ca, Sr, and Ba

Up to this point we have been reviewing and summarizing our earlier work as an introduction to what we now present. We have found that the values of $U_{i\alpha}$ and μ_α that give good MQDT parametrizations of the experimental data on Rydberg series are intimately related to one another along the atom series Ca, Sr, and Ba. Thus, for example, if we consider the $^1P_1^o$ series of these three atoms, we find a common subset of these parameters that describes the channel interactions *in all three atoms* between the *m*s*n*p and $(m-1)$d*n*′p channels.

We note that Be and Mg are not included in this set of alkaline earth atoms. The difference between Be and Mg and the heavier alkaline earths is that the latter have low-lying empty d shells in the electronic core, allowing for the possibility of low-lying core-electron excitations to d states. This common feature seems to lie at the root of the systematic behavior we will be discussing. We believe that this observed systematic behavior also extends to the heaviest alkaline earth Ra; however, there are not yet sufficient data to be certain. (See the discussion in the next section.)

The common subset of parameters gives the *shape* of the surface Σ as distinct from its *position*. This subset consists of the *U*-matrix and a set of new combinations of eigendefects numbering one less than the original set. The "omitted" parameter is the average eigendefect

$$\bar{\mu} = \left(\sum_{\alpha=1}^{M} \mu_\alpha\right)/M, \tag{3}$$

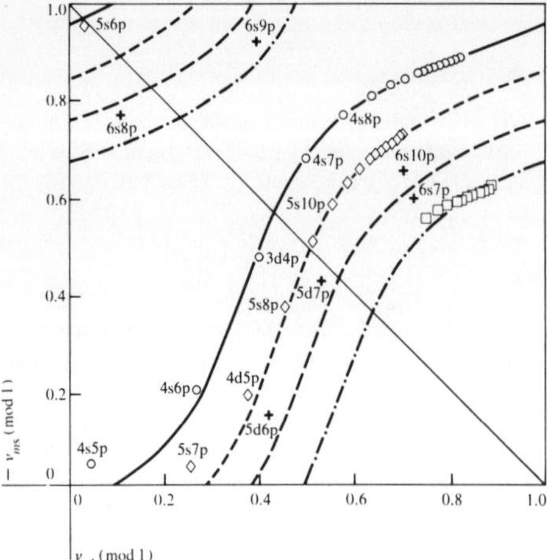

Figure 3 Lu-Fano plots for the bound $^1P_1^o$ spectra of Ca (———, ○) Sr (- - -, ◇), Ba (— —, +), and Ra (— • —, □). The curves are MQDT fits using the parameters given in Table 1. Intersections of the curves with the diagonal line give the μ_1 (upper left-hand curves) and μ_2 (lower curves) values. The integers m (see ordinate) are 4(Ca), 5(Sr), 6(Ba), and 7(Ra). The integers n (see abscissa) are 3(Ca), 4(Sr), 5(Ba), and 6(Ra).

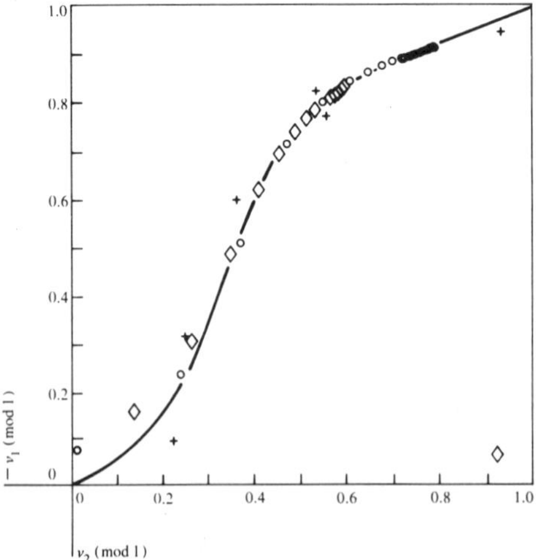

Figure 4 Composite Lu-Fano plot produced by displacing the curves and data of Fig. 3 along the diagonal so as to superimpose the curves. Data points are shown for Ca (○), Sr (◇), and Ba (+).

and it alone changes substantially from atom to atom. A representation of the new eigendefect parameters is found by taking $\mu_\alpha' = \mu_\alpha - \bar{\mu}$ and substituting this into Eq. (2), which now reads

$$\det |U_{i\alpha} \sin \pi(\nu_i + \mu_\alpha' + \bar{\mu})| = 0. \tag{4}$$

If we also take $\nu_i' = \nu_i + \bar{\mu}$ and substitute this into Eq. (4), we have

$$\det |U_{i\alpha} \sin \pi(\nu_i' + \mu_\alpha')| = 0. \tag{5}$$

What we have done in the transformation is displace Σ *along the N-space body diagonal* ($\nu_1 = \nu_2 = \cdots = \nu_N$) *a distance* $\bar{\mu}$. *This displacement does not change the shape of the surface* Σ.

We now assert that the shape of Σ corresponds to the interaction among channels and that this shape is nearly constant within a given spectrum along the series Ca, Sr, and Ba. Thus we will allow the surfaces fitting the data to be displaced along the body diagonal to see how well they may be superimposed for these three atoms. We believe that the constancy of the shape of Σ mirrors invariances in the dynamics of the correlated "two-electron" system along the series Ca, Sr, and Ba. We shall now show by a series of examples how this procedure works.

- *$^1P_1^o$ spectra*

As a first example we consider the spectral series seen in absorption from the ground state. The main series correspond to the $ms^2\ ^1S_0$-$msnp\ ^1P_1^o$ transitions. The spectra are complicated by interactions between the $msnp$ and $(m-1)dn'p$ channels. Thus the real states of the $^1P_1^o$ series are mixtures of these two types of configurations. From the MQDT point of view, this is a two-limit, two-channel problem. In the i-channel representation, we have channel 1 consisting of the electronic core in the ms state and the Rydberg electron in a p state. Channel 2 consists of the core in the $(m-1)$d state and the Rydberg electron in a p state. In both channels the electron angular momenta are coupled to produce a $^1P_1^o$ state. To treat these spectra we need experimental values for the energies E of the real states and values for the ionization limits I_1 and I_2. These may be found in the literature [1, 4, 9–11]. For I_2 we take the average of the two limits $I_{^2D_{3/2}}$ and $I_{^2D_{5/2}}$, corresponding to the lowest excited d states of the alkaline-earth ions. Thus we ignore spin-orbit effects. This is an approximation which depends on the fact that the spin-orbit splitting of the D core is small compared to the energy required to excite the core from an S to a D state, and our result is here applied only to the bound spectra below the I_S limit. Next we calculate values for ν_1 and ν_2 by using Eq. (1). For these simple two-channel cases, it is appropriate to present Lu-Fano plots in the form $-\nu_1$(mod 1) *versus* ν_2(mod 1), *i.e.*, to plot the

Table 1 Two-channel MQDT parameters for various states using I_S and I_D.

Parameter	$^1P_1^o$				$^3P_1^o$			$^3D_2$		
	Ca	*Sr*	*Ba*	*Ra*	*Ca*	*Sr*	*Ba*	*Ca*	*Sr*	*Ba*
Θ	0.60	0.60	0.60	0.60	0.30	0.30	0.30	0.25	0.25	0.25
μ_1	0.97	0.89	0.83	0.76	0.97	0.90	0.85	0.86	0.80	0.78
μ_2	0.57	0.49	0.43	0.36	0.87	0.80	0.75	0.35	0.29	0.27
$\Delta = [\mu_1 - \mu_2](\text{mod } 1)$	0.40	0.40	0.40	0.40	0.10	0.10	0.10	0.51	0.51	0.51
$\bar{\mu} = [\mu_1 - \Delta/2](\text{mod } 1)$	0.77	0.69	0.63	0.56	0.92	0.85	0.80	0.605	0.545	0.525

points on the unit square. The choice of signs for the plot follows the convention established by Lu and Fano [22]. The neglect of the integers in the ν_i for the purposes of such plots is a reflection of the fact that Eq. (2) is invariant to changes in the values of ν_i by integers. This form of the plot facilitates a fitting of the surface Σ (here a curved line) to the experimental data.

In Fig. 3 we present Lu-Fano plots for Ca, Sr, and Ba. The points correspond to the experimentally measured energies of the $^1P_1^o$ spectra with the following omissions. First, the lowest state in each atom is excluded because MQDT does not apply unless one electron has a large orbit. Next, the highest members of the series ($n > 25$) for Ca and Sr are left off the plots to avoid visual bunching as the ms ionization limits are approached. Note that only the lowest member of the perturbing series, 3d4p in Ca and 4d5p in Sr, is bound for these two atoms. For Ba the situation is more complicated. There are several perturbing configurations in the bound portion of the spectrum. The two lowest members of the 5dnp channel (5d6p and 5d7p) are completely bound, while the next member (5d8p) is mixed into the 6snp channel in both the bound and continuum regions of the spectrum. This last state is therefore excluded for our present purposes of comparison with Ca and Sr. Finally, the presence of 5d4f perturbers in Ba has no analogue in Ca and Sr, so we only consider Ba states that are not influenced by the 5d4f configuration. Thus, the plot for Ba shows data limited to the energy region below the 6s11p state in Ba.

In any two-channel problem, the unitary matrix $U_{i\alpha}$ has only one independent element. It is convenient to express it in terms of the parameter Θ, with $U_{12} = \sin \Theta$. Figure 3 shows curves that are solutions of Eq. (2) with values for the parameters Θ, μ_1, and μ_2 as given in Table 1 under the heading $^1P_1^o$. Note that the intersections of the upper-left to lower-right diagonal line ($\nu_1 = \nu_2$) with the curves occur at $\nu_1(\text{mod } 1) = \nu_2(\text{mod } 1) = -\mu_1$ and at $\nu_1(\text{mod } 1) = \nu_2(\text{mod } 1) = -\mu_2$. This is a consequence of Eq. (2), which is seen to have such special solutions. The values of Θ, μ_1, and μ_2 were chosen to give good fits to the data for all three atoms and to demonstrate the point made earlier that the curves (surfaces) Σ would have the same shape in all three atoms. Thus Θ is the same for all curves in Fig. 3, as is $\Delta = \mu_1 - \mu_2$. The only thing that changes in the parametrization is $\bar{\mu} = \mu_1 - \Delta/2$. This is further dramatized by displacing each plot along the diagonal so that the curves are superimposed, as shown in Fig. 4. (In this and succeeding plots of superimposed curves, the ν_i values are not transposed by $\bar{\mu}$, as Eq. (5) would suggest, but by $\bar{\mu} + k$, where k is a constant that displaces the curves into a position in the unit square chosen for its suitability in displaying the data.) Thus, one may compactly represent all of the data on the $^1P_1^o$ series in these atoms by the two common parameters, $\Theta = 0.6$ and $\Delta = 0.4$, and a third parameter $\bar{\mu}$ that is different for each atom. This remarkable result indicates that, to a good approximation, the effect of the configuration interaction on the $^1P_1^o$ spectra of these three atoms is identical when seen from the standpoint of MQDT. The channel interaction is expressed in terms of the two parameters Θ and Δ. The differing values of the average quantum defect $\bar{\mu}$ reflect the differences in size and, correspondingly, polarization and penetration of the core by the Rydberg electron as one goes from Ca to Sr to Ba. These changes in $\bar{\mu}$ are not unlike the changes in the quantum defect characterizing one-electron alkali-metal atoms seen along the series Li, Na, K, Rb, and Cs.

A test of our claim of near-invariance in the configuration interaction among the alkaline earths with low-lying d shell excitations is found by examining data for Ra. Using unpublished data provided by Tomkins [23], we get the Lu-Fano plot also shown in Fig. 3. Data are not available on the lower-lying levels needed to test the region where the theoretical curve is rising but we see that the existing data fit on a curve corresponding to the parameters given in Table 1. This curve is produced by displacing the composite Ca-Sr-Ba curve along the $\nu_1 = \nu_2$ diagonal. We hope to take more data on Ra to test the other portions of the curve; however, it is already clear that the Ra $^1P_1^o$ series substantially fits the picture we are presenting.

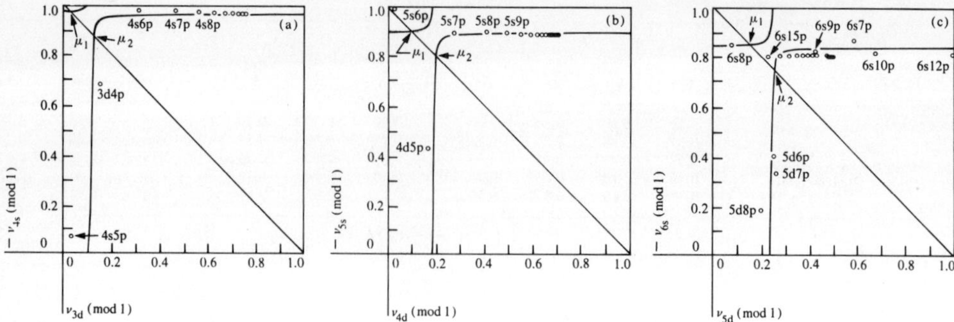

Figure 5 Lu-Fano plots for the bound $^3P_1^o$ spectra of (a) Ca, (b) Sr, and (c) Ba. The curves are MQDT fits using the parameters given in Table 1.

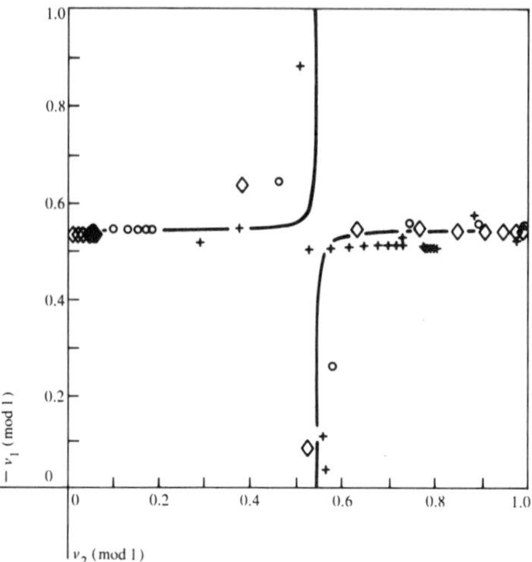

Figure 6 Composite Lu-Fano plot produced by displacing the curves and data of Fig. 5 along the diagonal so as to superimpose the curves. Data points are shown for Ca (○), Sr (◇), and Ba (+).

In fitting the data on the $^1P_1^o$ states, we have not allowed the MQDT parameters to have any energy dependence [15]. If we allow the eigendefect μ_1 to increase with decreasing energy, we can improve the MQDT fits to the experimental data. This additional dependence reflects the fact that low-lying states penetrate and polarize the core more than do high-lying states. For our present purposes, omission of such energy dependence causes the lowest energy states in our plots to deviate from the curves in the Lu-Fano plots in the direction of decreasing ν_1. This will be more obvious in the examples to be considered in the following sections.

- *$^3P_1^o$ spectra*

An examination of the odd-parity $J = 1$ spectra of the alkaline earths shows that the separation into singlet and triplet states is a very good approximation [4]. These series consist primarily of configurations having the core in an S state with no orbital angular momentum and consequently no spin-orbit splitting.

Having already treated the singlet spectra, we now consider the triplet spectra. As in the previous case, the i-channels are described as having the core in the *m*s state (channel 1) or in the ($m - 1$)d state (channel 2), with the Rydberg electron in a p state for both i-channels. But now the coupling between electrons produces a $^3P_1^o$ state. Following the same procedure as already described, we calculate ν_1 and ν_2 for each state of these spectra in Ca, Sr, and Ba. The results are plotted in Fig. 5. Again, the lowest state in each atom is excluded from the plots. For Ca and Sr, only the lowest member of the perturbing series is bound, while for Ba there are three bound members, 5d6p, 6d7p, and 5d8p. For Ba the 6s11p and 6s13p states are perturbed by 5d4f configurations, while the 6s14p state is perturbed by the 5d8p $^3D_1^o$ state. Since there are no analogous effects in Ca or Sr, these states are omitted from Fig. 5(c).

The solid curves in Fig. 5 are solutions to Eq. (2) obtained by using the values for the various parameters given in Table 1 for the $^3P_1^o$ state. As before, the only differences among the curves are the values chosen for $\bar{\mu}$. Note how the three perturbing states in Ba all lie near the

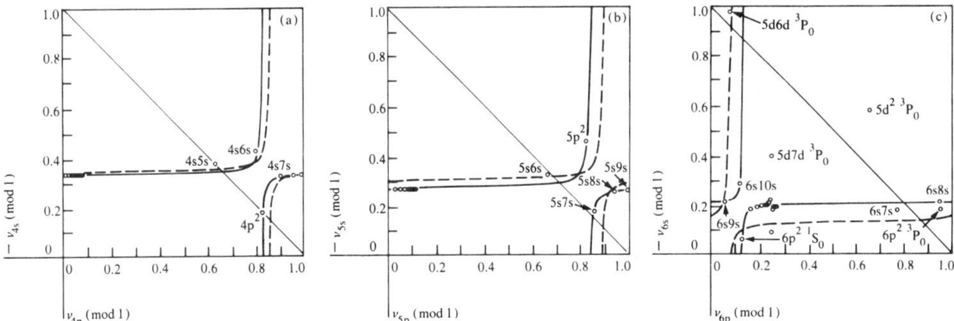

Figure 7 Lu-Fano plots for the bound $^1S_0$ spectra of (a) Ca, (b) Sr, and (c) Ba. The solid curves are MQDT fits using the parameters given in columns 2-4 of Table 2, while the dashed curves use the parameters of columns 5-7 of Table 2.

Table 2 Two-channel MQDT parameters for $^1S_0$ states.

Parameter	*Using I_S and I_P; fitting atoms:*						*Using I_S and I_D; fitting atoms:*					
	Individually			*With common Θ, Δ*			*Individually*			*With common Θ, Δ*		
	Ca	*Sr*	*Ba*	*Ca*	*Sr*	*Ba*	*Ca*	*Sr*	*Ba*	*Ca*	*Sr*	*Ba*
Θ	0.22	0.35	0.10	0.22	0.22	0.22	0.18	0.22	0.28	0.23	0.23	0.23
μ_1	0.35	0.295	0.21	0.36	0.32	0.14	0.335	0.28	0.185	0.345	0.355	0.115
μ_2	0.17	0.145	0.87	0.14	0.10	0.92	0.745	0.82	0.415	0.735	0.745	0.495
$\Delta = [\mu_1 - \mu_2](\text{mod } 1)$	0.18	0.15	0.34	0.22	0.22	0.22	0.59	0.46	0.77	0.61	0.61	0.61
$\bar{\mu} = [\mu_1 - \Delta/2](\text{mod } 1)$	0.26	0.22	0.04	0.25	0.21	0.03	0.04	0.05	0.80	0.04	0.05	0.80

vertical part of the curve in Fig. 5(c). If we displace each plot along the diagonal to superimpose the curves, we get the composite plot shown in Fig. 6. While the result is not as good a fit to the data as the $^1P_1^o$ case, it still shows how remarkably similar the effects of electron correlation are in all three atoms. Once again, the channel interaction is nearly the same, with only the average quantum defect $\bar{\mu}$ changing substantially.

An interesting observation is that the configuration interaction is much stronger in the singlet than in the triplet spectrum. In terms of the Lu-Fano plots, this appears as a much larger gap between the two branches of the curves at their distance of closest approach. We interpret this to be a consequence of the tendency of parallel-spin electrons (forming triplet states) to avoid one another, as compared to electrons with antiparallel spins (forming singlet states). If the outer electrons avoid one another, the intermixing of configurations due to Coulomb repulsion between these two electrons is diminished. Hence, a smaller configuration interaction results. Further discussion on this result may be found in Ref. [4]. We shall see that this result is also a feature of the even-parity $J = 2$ spectra.

The separation of the odd-parity Ba $J = 1$ spectrum into singlet and triplet spectra, followed by two-channel MQDT treatments, is an approximation that is made for our present purposes. Note, however, that we have successfully analyzed the complete Ba odd $J = 1$ spectrum with an eight-channel MQDT fit, which includes the effects of the 5d4f states [4]. A discussion of the detailed eight-channel fit is outside the scope of this paper.

• $^1S_0$ *spectra*

For this group of spectra the main series of interest are the *msns* states that have only one even-parity $J = 0$ term, the $^1S_0$. As we shall see, there is more than the usual uncertainty in labeling the perturbers actually seen in the bound $^1S_0$ spectra. The configurations mp^2 or $(m - 1)d^2$, the lowest members of the possible perturbing series, contribute to these perturbers.

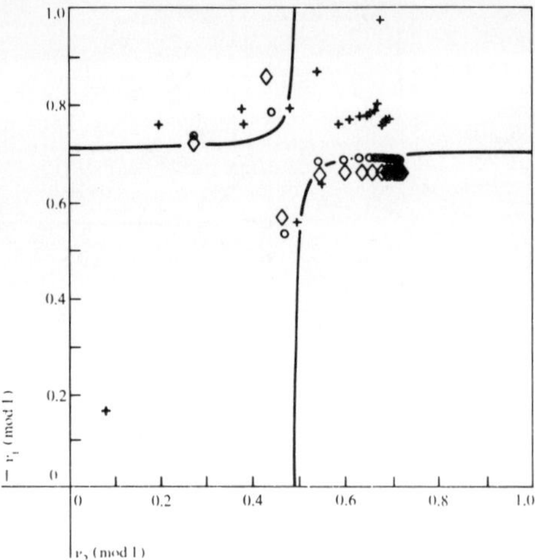

Figure 8 Composite Lu-Fano plot produced by displacing the dashed curves and data of Fig. 7 along the diagonal so as to superimpose the curves. Data points are shown for Ca (○), Sr (◇), and Ba (+).

The bound state region of Ca has been extensively studied by Armstrong *et al.* [2]; that of Sr, by Esherick [3] and Ewart and Purdie [6]; and that of Ba, by Rubbmark *et al.* [7] and Aymar *et al.* [8]. The data for Ca and Sr show a main series with relatively constant quantum defect and a single perturbing level that has been labeled mp^2 in the literature [1]. There is no evidence for another $^1S_0$ perturber in the bound spectra of Ca and Sr. Recent calculations by Nesbet and Jones [24] on Ca predict that $4s^2$, $4p^2$, and $3d^2$ are heavily admixed in the $^1S_0$ spectrum. Their results indicate that the $4s^2$ configuration dominates the ground state but that $4p^2$ and $3d^2$ occur with nearly equal weight in the two other states that have these labels. Their calculation predicts one such state with an energy in excellent agreement with the state traditionally labeled $4p^2$, while the other state ("$3d^2$") is found to lie well above the 4s ionization limit. This is consistent with the experimental results that show only a single perturber in the bound region of the spectrum.

We present Lu-Fano plots for the $^1S_0$ spectra of Ca and Sr in Figs. 7(a) and (b). The data are taken from Refs. [2] and [3]. In these plots channel 1 has an *m*s core and an s Rydberg electron, while channel 2 has an *m*p core and a p Rydberg electron; $I_{^2S_{1/2}}$ is chosen for I_1, while the average of $I_{^2P_{1/2}}$ and $I_{^2P_{3/2}}$ is chosen for I_2, where these labels refer to the ion ground state and the first excited state, respectively. The states so labeled are those that have the highest amount of mp^2 perturber when the data are treated with two-channel MQDT fittings.

The data for Ba are not so straightforward. There are several perturbers of the 6s*n*s series in the bound region and their identification is somewhat in question. The data are presented in a Lu-Fano plot in Fig. 7(c), where the average ^2P limit is used for I_2. This plot shows several points well removed from the horizontal line near $-\nu_{6s} = 0.21$, where most of the data lie. These points correspond to the perturbers, and we will shortly present a labeling scheme based on comparisons with Ca and Sr.

We first find two-channel, two-limit MQDT parametrizations for Ca and Sr. Solid curves corresponding to such parametrizations (see columns 2 and 3 in Table 2) are given in Figs. 7(a) and (b). The data on Ba [Fig. 7(c)] show a region near the lower left corner of the Lu-Fano plot where a gap, analogous to that of Ca and Sr, might occur between the two branches of a two-channel fit. This suggests that such a gap is due to the $6p^2$ configuration. With appropriate values for the parameters (column 4 in Table 2), we generate the solid curve shown in Fig. 7(c). The fit is seen to be reasonable if one ignores the three points labeled $5d^2$, 5d6d, and 5d7d $^3P_0$.

We see that all three parameters Θ, Δ, and $\bar{\mu}$ differ in the three atoms when one tries to fit each atom by itself (columns 2–4 in Table 2). If we look for values for Θ and Δ that are common for these atoms we get the results shown by the dashed curves in Fig. 7. The parameters are given in columns 5–7 of Table 2. Displacement of these curves and the data along the $\nu_1 = \nu_2$ diagonal to superimpose the curves produces the composite plot of Fig. 8. While this fit of all three atoms with a curve of the same shape leaves much to be desired, our comparison affords evidence for an identification of $6p^2$ $^1S_0$ for the state in Ba so labeled in Fig. 7(c). This state was labeled 5d6d $^1S_0$ by Aymar *et al.* [8] on the basis of their four-channel MQDT parametrization. However, they ended up without a $6p^2$ $^1S_0$ state. Our label is in basic agreement with Rubbmark *et al.* [7], although they reverse the 6s10s and $6p^2$ labels as compared to us. The state we label 6s8s, in agreement with Rubbmark *et al.*, has been labeled $6p^2$ in the pre-laser tabulations [1].

As an alternative to our choice of the $6p^2$ $^1S_0$ perturber, we consider the possibility that this state is the $5d^2$ $^1S_0$ and the corresponding perturbers in Ca and Sr are $3d^2$ and $4d^2$, respectively. Thus we plot the data using the average of the *D*-limits for I_2. The results are shown in Fig. 9, where the solid curves correspond to the two-channel parameters of Table 2 (columns 8–10), while the dashed curves

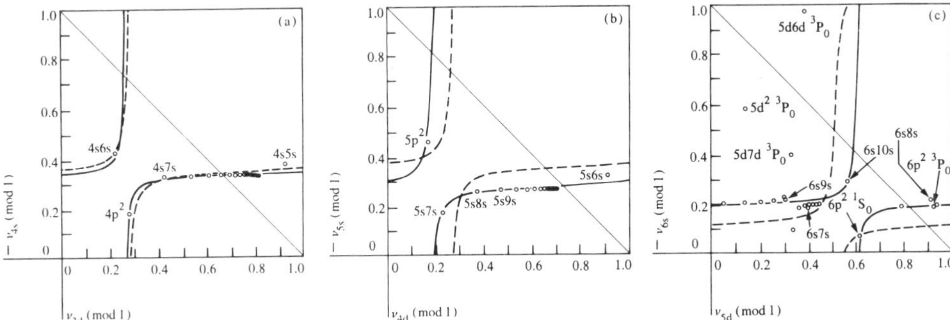

Figure 9 Lu-Fano plots for the bound $^1S_0$ spectra of (a) Ca, (b) Sr, and (c) Ba. The solid curves are MQDT fits using the parameters given in columns 8–10 of Table 2, while the dashed curves use the parameters given in columns 11–13 of Table 2.

have common values of Θ and Δ (see columns 11–13 in Table 2). (Note that the states are still labeled according to the scheme adopted for Fig. 7.) Displacement of the dashed curves and the data along the $\nu_1 = \nu_2$ diagonal produces the composite plot shown in Fig. 10. This composite fit appears to be marginally worse than that of Fig. 8.

The question arises as to which is the better choice of labels, mp^2 or $(m-1)d^2$. The heavy admixture of ms^2, mp^2, and $(m-1)d^2$ into three different $^1S_0$ states in each atom ought to depend strongly on the details of the wave functions of each configuration in the core. Since these configurations all correspond to tightly bound as opposed to Rydberg electrons, MQDT should be inadequate in describing the interactions and admixtures. In the absence of data that could give information about the interactions between these configurations, such as autoionizing spectra, we have insufficient information to make a choice between mp^2 and $(m-1)d^2$. The reader may draw his own conclusions from a comparison of Figs. 8 and 10. In general, the $^1S_0$ spectra show much less striking evidence for systematic near-invariance in the MQDT parameter set than do the $^1P^o_1$ and $^3P^o_1$ spectra. We believe this reflects a particular difficulty that MQDT encounters in describing electron-electron interactions between electrons *in the same orbit*.

As for the other perturbers, the labels given in Fig. 7(c) agree with those of Moore [1] and Aymar *et al.* [8]. Thus the $5d^2$, 5d6d, and 5d7d configurations are present as $^3P_0$ states in the bound region. Only the 5d7d $^3P^o_0$ state has an observable effect on the 6s*n*s $^1S_0$ series. Since no comparable effects are seen in Ca and Sr, we do not treat this state or the appropriate channels in our MQDT fits.

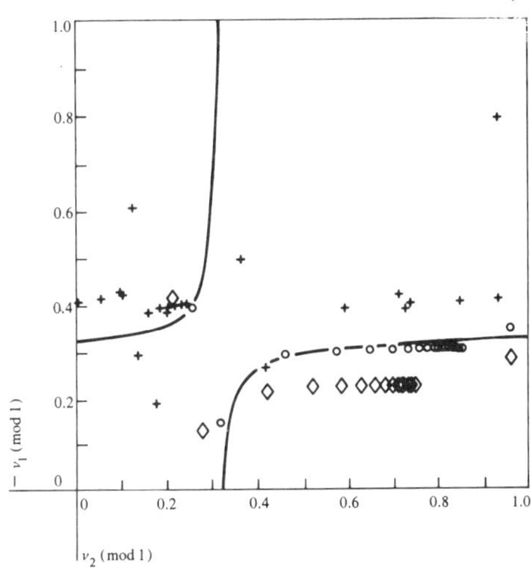

Figure 10 Composite Lu-Fano plot produced by displacing the dashed curves and data of Fig. 9 along the diagonal so as to superimpose the curves. Data points are shown for Ca (○), Sr (◇), and Ba (+).

• *$^3D_2$ spectra*

We next turn to the $^3D_2$ spectra, with the main series being the *msn*d states. Possible perturbing series are *mpnp*, $(m-1)$d*n*s, and $(m-1)$d*n*d. The bound state data for Ca are given up to 4s16d in Moore's tables [1]. We did not observe many $^3D_2$ states in our studies [2] because singlet-to-triplet transitions were very weak due to the small spin-orbit interaction in Ca. However, many more

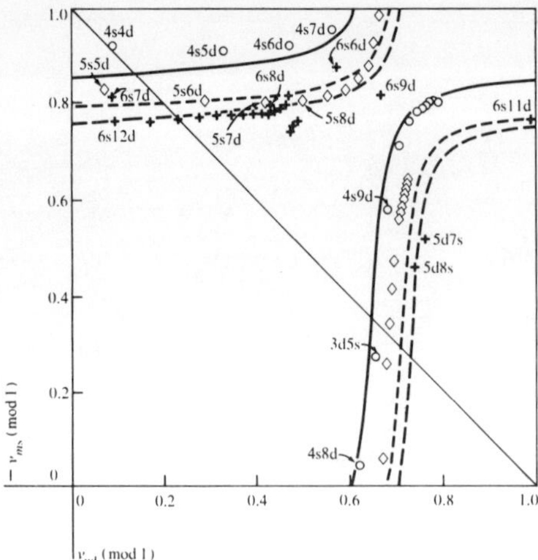

Figure 11 Lu-Fano plots for the bound $^3D_2$ spectra Ca (——, ○), Sr (- - -, ◇), and Ba (— —, +). The curves are MQDT fits using the parameters given in Table 1.

$^3D_2$ states in Sr were seen by Esherick [3] in our laboratory. Rubbmark *et al.* [7] and Aymar *et al.* [8] studied the $^3D_2$ states in Ba, and Aymar and Robaux [17] have published an MQDT analysis of the data of Aymar *et al.* [8].

The data for Ca and Sr show evidence for the presence of a single perturbing level whose influence is distributed over many members of the *msnd* series. In Ca, the singlet-triplet interaction is sufficiently weak that singlets and triplets may be treated in a completely decoupled manner. The MQDT analysis of Ca in Ref. [2] showed that while the Σ surfaces for the $^1D_2$ and $^3D_2$ series cross in several places, the real states near these crossing points show no evidence of these crossings. In contrast, the analysis of Sr in Ref. [3] shows clear evidence of an avoided crossing between the $^1D_2$ and $^3D_2$ surfaces, with several states having nearly equal amounts of $^1D_2$ and $^3D_2$ character. Despite this, over most of the spectrum the triplets and singlets may be distinguished in Sr.

We present Lu-Fano plots for the bound $^3D_2$ spectra of Ca and Sr in Fig. 11. In these plots channel 1 has an *m*s core and a d Rydberg electron, while channel 2 has an $(m-1)$d core and an s Rydberg electron. An average of $I_{^2D_{3/2}}$ and $I_{^2D_{5/2}}$ is chosen for I_2. The state so labeled for Ca is that with the highest amount of 3d5s perturber when the data are analyzed with MQDT. The analogous configuration, 4d6s in Sr, is so diluted throughout the 6s*n*d series that no state has more than eight percent 4d6s $^3D_2$ character. Thus, no individual state can be appropriately given this label [3]. The choice of the $(m-1)$d*n*s channel as the perturbing channel is based on the following reasoning: 1) there is only a single perturber and 2) it cannot be the lowest member of the *mpnp* or $(m-1)$d*n*d channels, since these are the mp^2 and $(m-1)d^2$ states, respectively, and neither can have a $^3D_2$ term by the Pauli exclusion principle. Note that in contrast to the $^1S_0$ spectra, and in analogy to the $^1P^o_1$ and $^3P^o_1$ spectra, neither the main channel *msnd* nor the perturbing channel $(m-1)$d*n*s has states with both electrons in the same orbit.

The data for Ba are, once again, more complicated. The analysis of Aymar and Robaux (17) identifies the following states as having significant perturbing effect on the 6s*n*d $^3D_2$ spectrum; 5d7s $^3D_2$, 5d8s $^3D_2$, 6p^2 $^1D_2$, 5d8s $^1D_2$, and 5d7d $^1D_2$. The analysis of Aymar and Robaux also gives labels to many other perturbers, but these others do not appear to significantly influence the 6s*n*d $^3D_2$ spectrum. The state labeled 6p^2 $^1D_2$ has only a weak influence so we omit it. Similarly, the influence of the state labeled 5d8s $^1D_2$ on the triplets is confined to the 6s10d $^3D_2$ state. To facilitate our comparison with Ca and Sr, we omit both of these states. The 5d7d $^1D_2$ influences the triplet spectrum through its perturbation of the 6s*n*d $^1D_2$ series, forcing it to cross the $^3D_2$ series. This leads to an easily observed avoided crossing between singlets and triplets. For present purposes, the 5d7d $^1D_2$ perturber is omitted. The balance of the $^3D_2$ spectrum of Ba is presented in a Lu-Fäno plot in Fig. 11, using the average of the ^{2}D limits for I_2. The two perturbers 5d7s and 5d8s fall away from the horizontal line near $-\nu_s = 0.78$, where most of the other data points lie. In all three plots of Fig. 11, the lowest members of the main series are, as usual, omitted.

Curves corresponding to two-channel, two-limit MQDT parametrizations for all three atoms (using the values given for the $^3D_2$ states in Table 1) are also presented in Fig. 11. While the curves for Ca and Sr nicely reproduce the pattern shown by the data, Ba requires additional explanation. Due to the influence of the state labeled 6p^2 $^1D_2$ by Aymar and Robaux, the 6s9d $^3D_2$ state is pushed to higher energy, causing it to lie off the curve. The higher *n* states show the effects of the 5d7d $^1D_2$ state mentioned earlier. Other than that, the two-channel curve reproduces the Ba data nicely, allowing for the usual deviation of the lowest-lying states (6s6d and 6s7d in this case) from the curve. In particular, both the 5d7s and 5d8s perturbers lie very close to the vertical portion of the curve.

The curves given in Fig. 11 have all used the same value for Θ and Δ, as in the previous cases. Displacement of the curves and the data along the $\nu_1 = \nu_2$ diagonal to superimpose the curves produces the composite plot of Fig. 12. Once again, the behavior of all three atoms is remarkably similar.

• *$^1D_2$ spectra*
For this group of spectra the main series are again the *msnd* states. However, this situation is much more complicated than the $^3D_2$ spectra because the *mpnp* and $(m-1)dnd$ channels *do* influence the $^1D_2$ spectra and because the perturbing channels include states with two electrons in identical orbits. The bound state regions of Ca and Sr were extensively studied in our laboratory [2, 3]. Rubbmark *et al.* [7] and Aymar *et al.* [8] studied Ba, with Aymar and Robaux [17] carrying out an MQDT analysis.

To treat the Ca and Sr data, we first neglect the spin-orbit splitting of the ^2D and ^2P core states, leaving a three-limit problem. This is consistent with the separation of the data into singlets and triplets mentioned earlier. In a three-limit case the MQDT approach yields three values of ν_i for each state. The corresponding Lu-Fano plots are three-dimensional [2]. We present a stereo view of a three-dimensional Lu-Fano plot of the $^1D_2$ data of Ca [2] and Sr [3] in Fig. 13. (For help in viewing stereo figures, see the figure caption.) For comparison to theory, the Sr data has been displaced -0.05 along the ν_i-space body diagonal. The points that are next-to-nearest to the right-hand face of the cube correspond to the states traditionally labeled mp^2 [1]. The other perturbers are so diluted throughout the *msnd* series that no individual states merit the $(m-1)d(m+1)s$ or md^2 labels. In the case of Sr the $4d^2$ perturber lies entirely in the continuum above the I_S limit and appears as an autoionizing state [3].

A surface Σ corresponding to four-channel, three-limit MQDT parametrizations, using the values for Ca in Table 3, is also presented in Fig. 13. In this MQDT analysis channel 1 has an *m*s core and a d Rydberg electron, channel 2 has an $(m-1)$d core and an s Rydberg electron, channel 3 has an $(m-1)$d core and a d Rydberg electron, and channel 4 has an *m*p core and a p Rydberg electron. Instead of showing the surface translated into the unit cube where it would have many sheets (branches) and look very confusing, we have displayed one continuous sheet covering a range for $-\nu_s$ of 0.72 to 3.72, while ν_d and ν_p vary from 0 to 1.0. Here, instead of displacing the surface Σ along the ν_i-space body diagonal a distance 0.05 in going from Ca to Sr, we have displaced the Sr data -0.05 to produce a composite plot.

This figure dramatically illustrates the fact that the surfaces Σ_{Ca} and Σ_{Sr}, in which the respective $^1D_2$ states lie, *differ only by translation along the body diagonal.* The shapes of the two surfaces are identical. This exhibits the invariance of the channel interactions from atom to atom within a spectral class for a many-channel, many-limit case.

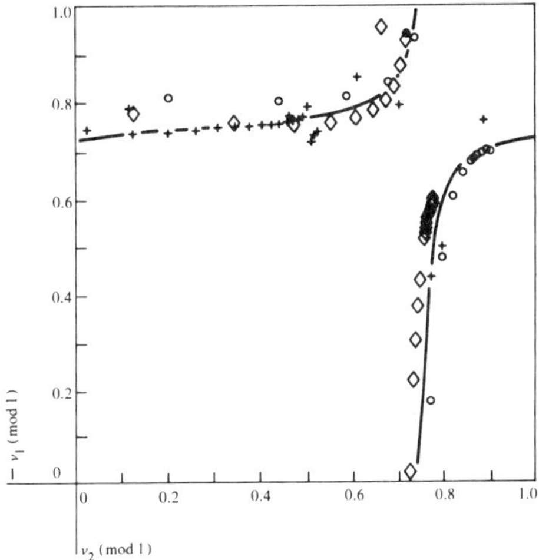

Figure 12 Composite Lu-Fano plot produced by displacing the curves and data of Fig. 11 along the diagonal so as to superimpose the curves. Data points are shown for Ca (○), Sr (◇), and Ba (+).

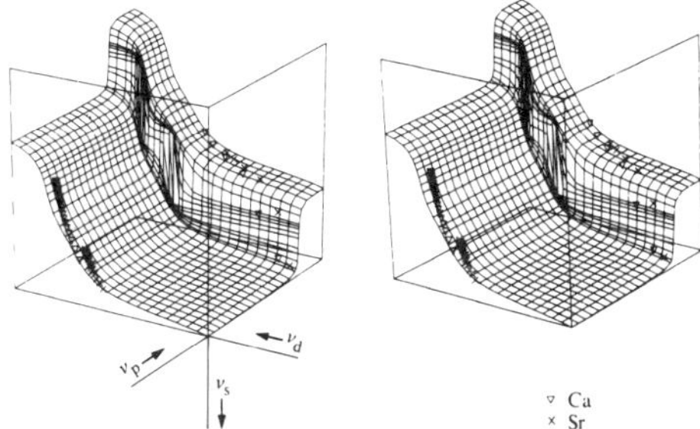

Figure 13 Stereo Lu-Fano plot for the bound $^1D_2$ spectra of Ca and Sr. The surface is a four-channel MQDT fit to Ca using the parameters given in Table 3. The Sr data points have been moved -0.05 along the body diagonal to lie in or near the surface. The three axes are dimensionless with $-\nu_s$ values running from 0.72 to 3.72, and ν_d and ν_p running from 0 to 1.0. If no stereo viewer is available, place the figure about one foot from your eyes, place an obstruction between your eyes so that each eye can see only one image, and then fuse the two images.

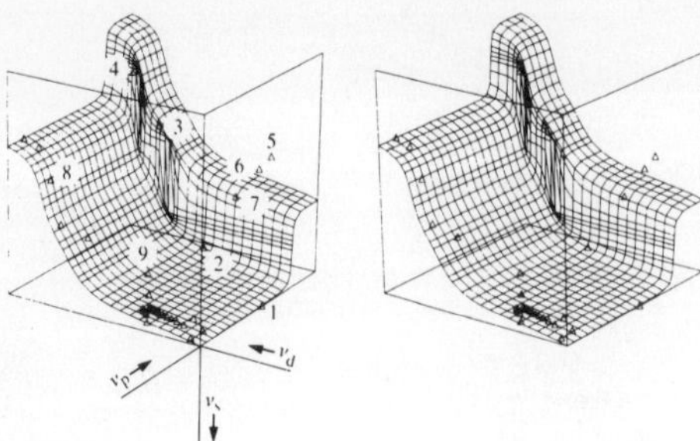

Figure 14 Stereo Lu-Fano plot for the bound $^1D_2$ spectrum of Ba. The surface corresponds to the Ca parameters of Table 3. The data points have been moved −0.12 along the body diagonal to lie near the surface.

Table 3 Four-channel MQDT parameters for $^1D_2$ states using I_S, I_D, and I_P.

Parameter				
i, α	1	2	3	4
$\|i\rangle$	$[ms]d$	$[(m-1)d]s$	$[(m-1)d]d$	$[mp]p$
I_i	I_S	I_D	I_D	I_P
$\mu'_\alpha = [\mu_\alpha - \bar{\mu}]^\dagger(\text{mod } 1)$	0.43	0.95	0.79	0.83
$U_{i\alpha}$; $i = 1$	0.837	−0.444	−0.301	−0.109
$i = 2$	0.415	0.896	−0.149	−0.054
$i = 3$	0.333	0	0.942	−0.044
$i = 4$	0.130	0	0	−0.992

$^\dagger\bar{\mu}_{Ca} = 0.39$, $\bar{\mu}_{Sr} = 0.34$, $\bar{\mu}_{Ba} = 0.27$.

Turning to Ba, we find once again a more complicated situation. First, Ba has several members of the perturbing series 5d*n*s and 5d*n*d occurring in the bound region of the spectrum. Next, the spin-orbit splitting of the D core is sizeable (800 cm^{-1}), making the separation into singlet and triplet states questionable. Perturbers with singlet and triplet labels are both found to perturb the 6s*n*d $^1D_2$ spectrum. Next, in the spectral region from the 6s6d to 6s11d $^1D_2$ states, there are ten perturbing levels according to the labeling scheme of Aymar *et al.* [8]. This confuses the spectrum and makes any labeling scheme quite speculative. Finally, the lowest members of some of the perturbing channels have configurations with identical electrons, *i.e.*, mp^2. Such states do not satisfy the MQDT criterion of having a highly excited Rydberg electron. This is a problem with Ca and Sr as well.

With these problems in mind we took the Ba data [8] and translated the corresponding points in three-dimensional ν_i-space along the body diagonal to see how nearly the data would lie in the surface Σ that contains the data points for Ca and Sr. The result is shown in Fig. 14, where the Ba data have been displaced −0.12 along the diagonal. The ν_s(mod 1) values have had the integers 1 or 2 added to them to move them close to that part (sheet) of Σ we have plotted. While the fit is not as good as for Ca or Sr, the majority of the points lie close to the same surface. In particular, the points along the left-most contours follow the shape of the surface remarkably well. The points indicated by the numbers 1 to 9 have been labeled by Aymar and Robaux [17] as $^1D_2$ states with the following configuration: 1-5d^2, 2-6s6d, 3-5d7s, 4-6s7d, 5-6s8d, 6-5d6d, 7-6p^2, 8-5d8s, and 9-5d7d (these are given in order of increasing energy). Of these, 4, 5, 6, and 7 appear to lie farthest away from the surface.

We cannot, on the basis of this three-limit, four-channel fit, propose an acceptable set of assignments for all the $^1D_2$ states. A more complicated fit, including the effects of the spin-orbit interaction and singlet-triplet mixing, is required. However, the comparison with Ca and Sr in the manner we have carried out in this paper casts some doubt on the assignments of Aymar and Robaux [17]. In particular, the 6p^2 state ought to lie near the sharply rising contour at $\approx\nu'_p = 0.78$ (the prime indicates the shift of the data by −0.12 in ν_i-space) rather than near $\nu'_p = 0.99$, as does point 7 in Fig. 14. Similarly, the 5d7s and 5d8s states ought to lie near the gradually rising contour at $\approx\nu'_d = 0.66$, whereas points 3 and 8 have $\nu'_d = 0.73$ and 0.82, respectively. Finally, the 5d^2, 5d6d, and 5d7d states ought to have $\nu'_d = 0.82$, whereas points 1, 6, and 9 have $\nu'_d = 0.01$, 0.28, and 0.36, respectively.

This example of the $^1D_2$ states in Ca, Sr, and Ba further highlights the extent to which spectral regularities and invariances are revealed by the point of view and apparatus of MQDT.

Conclusion

We have studied spectral regularities along the series Ca, Sr, Ba (and in one case, Ra). The series studied were $^1P^o_1$, $^3P^o_1$, $^1S_0$, $^3D_2$, and $^1D_2$. For each spectral type, except the $^1S_0$, most states in the observed series can be described quantitatively by a set of MQDT parameters of which all but one are the same for Ca, Sr, and Ba. Even in the series of Ba that have perturbing channels not found in the other atoms, the description appropriate to Ca and Sr is still found to provide an excellent "starting" description of the spectra for Ba. These points are exemplified in Figs. 4, 12, 13, and 14.

Multichannel quantum defect theory affords a framework with which to describe the channel interactions of two-electron spectra independently of particular values for series limits. This is the significance of the surface Σ, described by Eq. (2). The near-invariance of the surface Σ along the series Ca, Sr, Ba (and Ra?), appropriate to a given spectral type, implies that there are basis sets of two-electron wave functions; *i.e.*, these are basis channels that diagonalize the noncentral Coulomb scattering of the two-electron system, independent of whether one is dealing with Ca, Sr, or Ba. Similarly, there is a set of common eigenphase shifts (eigenquantum defects) μ'_α that belong to a given spectral type. It remains for future work to exhibit these invariances in the more usual non-MQDT treatment of configuration interactions and/or in a group theoretical description of the dynamical symmetries of "two-electron" atoms.

References

1. C. E. Moore, *Atomic Energy Levels*, National Bureau of Standards Circular, No. 467 (U.S. Government Printing Office, Washington, DC), Vol. 1 (1949), Vol. 2 (1952), Vol. 3 (1958).
2. J. A. Armstrong, P. Esherick, and J. J. Wynne, *Phys. Rev. A* **15**, 180 (1977).
3. P. Esherick, *Phys. Rev. A* **15**, 1920 (1977).
4. J. A. Armstrong, J. J. Wynne, and P. Esherick, *J. Opt. Soc. Amer.* **69**, 211 (1979).
5. D. J. Bradley, P. Ewart, J. V. Nicholas, and J. R. D. Shaw, *J. Phys. B* **6**, 1594 (1973).
6. P. Ewart and A. F. Purdie, *J. Phys. B* **9**, L437 (1976).
7. J. R. Rubbmark, S. A. Borgström, and K. Bockasten, *J. Phys. B* **10**, 421 (1977).
8. M. Aymar, P. Camus, M. Dieulin, and C. Morillon, *Phys. Rev. A* **18**, 2173 (1978).
9. W. R. S. Garton and K. Codling, *J. Phys. B* **1**, 106 (1968).
10. W. R. S. Garton and F. S. Tomkins, *Astrophys. J.* **158**, 1219 (1969).
11. C. M. Brown, S. G. Tilford, and M. L. Ginter, *J. Opt. Soc. Amer.* **63**, 1454 (1973).
12. M. J. Seaton, *Proc. Phys. Soc. (Lond.)* **88**, 801 (1966).
13. U. Fano, *Phys. Rev. A* **2**, 353 (1970).
14. K. T. Lu, *Phys. Rev. A* **4**, 579 (1971).
15. C. M. Lee and K. T. Lu, *Phys. Rev. A* **8**, 1241 (1973).
16. U. Fano, *J. Opt. Soc. Amer.* **65**, 979 (1975).
17. M. Aymar and O. Robaux, *J. Phys. B*, accepted for publication.
18. K. T. Lu, *J. Opt. Soc. Amer.* **64**, 706 (1974).
19. K. H. Kingdon, *Phys. Rev.* **21**, 408 (1923).
20. J. P. Hermann and J. J. Wynne, *J. Opt. Soc. Amer.* **68**, 1412 (1978).
21. E. F. Worden, J. A. Paisner, and J. G. Conway, *Opt. Lett.* **3**, 156 (1978).
22. K. T. Lu and U. Fano, *Phys. Rev. A* **2**, 81 (1970).
23. F. S. Tomkins, Chemistry Division, Argonne National Laboratory, Argonne, IL 60439, private communication.
24. R. K. Nesbet and J. W. Jones, *Phys. Rev. A* **16**, 1161 (1977).

Received January 18, 1979; revised March 17, 1979

The authors are located at the IBM Thomas J. Watson Research Center, Yorktown Heights, New York 10598.

1984 *J. Opt. Soc. Am.* **B 1** 239–45

Reprinted from **Journal of the Optical Society of America B,** Vol. *1*, page 239, April 1984.

Multichannel-quantum-defect theory wave functions of Ba tested or improved by laser measurements

M. Aymar

Laboratoire Aimé Cotton, Centre National de la Recherche Scientifique II, Bâtiment 505, 91405 Orsay Cedex, France

Received November 7, 1983; accepted December 12, 1983

Considerable efforts have been made recently to understand the detailed structure and properties of high Rydberg states of the bound even-parity spectrum of Ba. Interest in Ba is motived by the ability to study interacting Rydberg series. The perturbations of $6snl$ Rydberg series by $5d7d$ levels are reflected not only in the level structure but also in various observable quantities recently investigated by high-resolution laser spectroscopy. A wealth of data has been obtained on lifetimes, isotopes shifts, Landé factors, hyperfine structures, photoelectron angular distributions, etc. Previously the multichannel-quantum-defect theory (MQDT) has permitted one successfully to interpret the energies of perturbed series. The new possibilities offered by laser spectroscopy to analyze state mixing have provided an opportunity to probe the MQDT wave functions. This paper shows how the data provided by different measurements complement one another to check or even extend the MQDT models derived from energies. Various examples concerning the perturbed $6sns$ and $6snd$ Rydberg series are presented. For the $6snd$ $^{1,3}D_2$ series, energies are insufficient to get the full MQDT parameters needed to determine the wave functions, and g_J-factor and hyperfine-structure measurements have been essential to improve MQDT wave functions.

1. INTRODUCTION

The high-Rydberg states of Ba have received considerable attention during the last decade. The major motive for the interest in alkaline-earth atoms is the ability to use them to study interacting Rydberg series. By far, the perturbed Rydberg series of the bound even-parity spectrum of Ba have been the most thoroughly studied.

The perturbations of Rydberg series $6snl$ of Ba by doubly excited states pertaining to $5dn'l'$ and $6p^2$ configurations are reflected primarily in the level structure. Until 1981 most of experimental investigations on $6snl$ series have had term analysis as their main purpose. Laser spectroscopy has provided a wealth of new energy data on the highly excited states of the Rydberg series $6sns$ $^1S_0$, $^3S_1$, $6snd$ $^1D_2$, $^3D_{1,2,3}$, and $6sng$ and on perturbers of these series.[1,2] Simultaneously the spectroscopic data have been analyzed by the multichannel-quantum-defect theory (MQDT), which was introduced by Seaton.[3,4] The new formulation of this method presented by Lu,[5] Lee and Lu,[6] and Fano[7] has proven to be a powerful tool for studying perturbed Rydberg series. Semiempirical analyses of the observed spectra (for J values ranging from 0 to 5) have been successfully carried out by the use of MQDT.[1,8,9]

Recently high-resolution laser spectroscopy has been used to study the detailed structure and properties of Rydberg states of even-parity Ba.[10] The perturbations of Rydberg series are reflected in various observable quantities, such as lifetimes,[11–14] Landé factors,[15] hyperfine structures and isotope shifts,[16–24] photoelectron angular distributions,[25] and Stark[12,26] and diamagnetic shifts.[27] The new possibilities offered by laser spectroscopy to analyze state mixing and intermediate coupling of Rydberg states are of particular interest to probe MQDT wave functions derived from energy data.

The aim of this paper is to show how the data provided by different measurements complement one another to check and even extend MQDT models derived from energies. For this purpose numerous examples are presented and discussed; these illustrations concern the highly excited Rydberg series $6sns$ $^1S_0$, $^3S_1$, and $6snd$ $^{1,3}D_2$, which have been the most thoroughly studied. A crucial point emphasized in this paper is that energy values sometimes represent a data set that is too limited to get the full MQDT parameters needed to determine the wave functions; experimental information on quantities other than energies is then valuable input data for improved MQDT calculations. In this context the examples relevant to the highly excited $6snd$ $^{1,3}D_2$ levels are particularly instructive.

2. MULTICHANNEL-QUANTUM-DEFECT THEORY

Basic Definitions

Since MQDT analysis of bound spectra is now a well-known procedure,[1,5–9,28–31] here I only briefly recall some basic concepts and equations.

MQDT is an exact parameterization of the energies and wave functions of levels of interacting Rydberg series. For a spectrum of given J and parity, the interactions of different series are described in terms of two sets of parameters whose values can be derived from energy data by a graphical method based on the use of Lu–Fano curves.

The essence of MQDT lies in separating the effects of long- and short-range interactions between the electron and the ion core. When the electron is far from the core ($r > r_0$), the interaction is Coulombian and the ion–electron dissociated system is described by collision channels i identified in coupling. The effect of short-range non-Coulomb interactions are characterized by the close-coupling channels α. The MQDT parameters, the eigenquantum defect μ_α and the elements of the orthogonal transformation matrix $U_{i\alpha}$ represent boundary conditions of the wave functions at $r = r_0$.

An important role is played in MQDT by effective quantum number ν_i, defined for each ionization limit I_i by

$$E = I_i - R/\nu_i^2, \tag{1}$$

where E is the state energy and R is the mass-corrected Rydberg constant. Discrete levels correspond to a set of ν_i values that simultaneously fulfill Eq. (1) and an analytic relationship

$$\det |U_{i\alpha} \sin \pi(\nu_i + \mu_\alpha)| = 0, \tag{2}$$

which ensures correct asymptotic behavior of the wave functions. The graphical representation of MQDT (Lu–Fano curves) is in terms of plots of ν_i versus ν_j. Analysis of a spectrum thus consists of adjusting the MQDT parameters so that term values given by Eqs. (1) and (2) agree with experimental data.[28]

In addition to providing a means of fitting experimental energies, MQDT also yields detailed information on the wave functions of the eigenstates. From μ_α and $U_{i\alpha}$ parameters one can calculate the fractional admixture of channels (identified in any pure-coupling scheme) for each bound state.

Remarks on the $U_{i\alpha}$ Transformation Matrix

A crucial point emphasized in 1971 by Lu[5] is that experimental data on the energy levels are often insufficient to determine the complete $U_{i\alpha}$ matrix. When more than one channel converges on a given threshold, energies give only some combinations of $U_{i\alpha}$ elements that are invariant under orthogonal transformation of these channels. Experimental data that distinguish these channels are necessary to determine the individual matrix elements. When such data are not available, some assumptions have to be made to construct the $U_{i\alpha}$ matrix. In previous MQDT parameterizations of alkaline earths,[1,8,9,29–31] the close-coupling channels are assumed to be nearly LS coupled because the electrostatic interactions are expected to be stronger than spin–orbit-coupling effects for $r < r_0$. Accordingly one introduces an intermediate basis $\bar{\alpha}$ of pure LS-coupled channels[6] and factors the matrix $U_{i\alpha}$ in the form

$$U_{i\alpha} = \sum_{\bar{\alpha}} R_{i\bar{\alpha}} V_{\bar{\alpha}\alpha}, \tag{3}$$

where the $R_{i\bar{\alpha}}$ matrix corresponds to the standard jj–LS transformation. Moreover, the $V_{\bar{\alpha}\alpha}$ matrix can be generated by successive rotations through generalized Eulerian angles θ_{kl}.[6,28] The fitting of the matrix $V_{\bar{\alpha}\alpha}$ (or of the θ_{kl} angles) is then based on minimizing the departure of α channels from $\bar{\alpha}$ channels. The validity of this assumption in Ba is discussed later.

A last comment concerns the relative signs of the θ_{kl} mixing angles, which cannot be derived unambiguously from energies. It is shown later how to remove part of this indeterminacy.

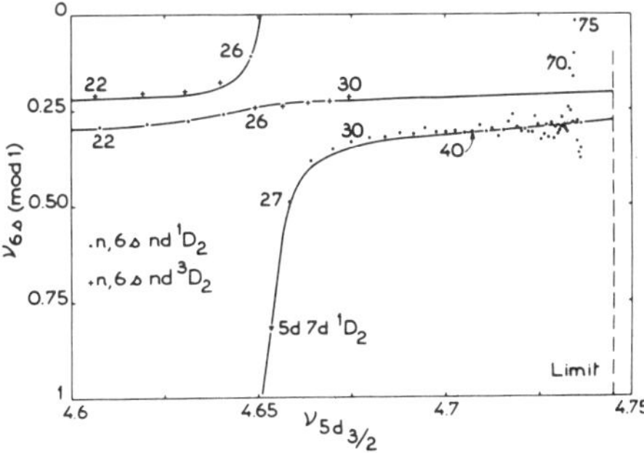

Fig. 1. Lu–Fano plot of the $J = 2$ high-lying levels of Ba calculated with the nine-channel MQDT model of Ref. 8. The experimental points are from Ref. 8.

3. MULTICHANNEL-QUANTUM-DEFECT THEORY TREATMENTS OF HIGHLY EXCITED STATES OF Ba

The MQDT models[1,8,9] developed to describe the entire $J = 0$, $J = 1$, and $J = 2$ bound even-parity spectra of Ba involve, respectively, nine, seven, and four interacting channels. However, the admixtures of several channels into the highly excited states are negligible. This paper deals with Rydberg levels $6sns\ ^1S_0$, $^3S_1$, and $6snd\ ^{1,3}D_2$ located in the spectral range where perturbations are due only to $5d7d$ levels. The wave functions of these levels can be written as expansions on a limited basis of pure LS-coupled channels as follows:

$6sns\ ^1S_0$ ($n \geq 14$) and $5d7d\ ^3P_0$ perturber:

$$\Psi_i = \alpha_i |6sns\ ^1S_0\rangle + \sum_{k={}^1S_0,{}^3P_0} \epsilon_i^k |5d7d\ k\rangle, \tag{4}$$

$6sns\ ^3S_1$ ($n \geq 14$) and $5d7d$ $J = 1$ perturbers:

$$\Psi_i = \alpha_i |6sns\ ^3S_1\rangle + \sum_{k={}^3D_1,{}^3S_1,{}^1P_1,{}^3P_1} \epsilon_i^k |5d7d\ k\rangle, \tag{5}$$

$6snd\ ^{1,3}D_2$ ($n \geq 17$) and $5d7d\ ^1D_2$ perturber:

$$\Psi_i = \alpha_i |6snd\ ^1D_2\rangle + \beta_i |6snd\ ^3D_2\rangle + \sum_{k={}^1D_2,{}^3D_2,{}^3F_2,{}^3P_2} \epsilon_i^k |5d7d\ k\rangle. \tag{6}$$

The mixing coefficients are identified with MQDT admixture coefficients obtained in the treatment of the entire spectra. The introduction of $5dnd$ interacting channels accounts for the intermediate coupling of the perturbers, which is never described well in any pure-coupling scheme.[1,8,9]

In the following, the Rydberg character of a level i will correspond to the α_i^2 (or $\alpha_i^2 + \beta_i^2$) quantity, whereas the total perturber character is measured by

$$\epsilon_i^2 = \sum_k |\epsilon_i^k|^2. \tag{7}$$

For the $J = 2$ spectrum, in addition to the nine-channel MQDT model,[8] a simpler three-channel model has been developed.[10,12,21] This model describes the highly excited spectrum [$6snd\ ^{1,3}D_2$ ($n \geq 17$), $5d7d\ ^1D_2$] well but does not allow one to define the intermediate coupling of the perturber. Equation (6) is reduced to

$$\Psi_i = \alpha_i|6snd\ ^1D_2\rangle + \beta_i|6snd\ ^3D_2\rangle + \epsilon_i|5d7d\ ^1D_2\rangle. \quad (8)$$

The influence of the $5d7d\ ^1D_2$ perturber on the energies of the $6snd\ ^{1,3}D_2$ levels is illustrated by the Lu–Fano plot in Fig. 1.[8] As explained in Ref. 8, the perturber induces a large mixing between the singlet and triplet Rydberg levels that depends strongly on n. On each branch of the Lu–Fano plot (continuous or broken), there is a reversal of the LS character for a particular n value. Moreover, MQDT[8] has shown that the $5d7d$ character occurs mainly in levels located on the broken branch of the Lu–Fano plot.

4. HOW ATOMIC OBSERVABLES OTHER THAN ENERGIES ALLOW ONE TO TEST OR IMPROVE MULTICHANNEL-QUANTUM-DEFECT THEORY WAVE FUNCTIONS

A. Lifetimes

Lifetimes of Rydberg states are sensitive to perturbation by doubly excited states because Rydberg levels are long lived, whereas doubly excited ones are rather short lived.

Three papers treat the measurement of lifetimes in the perturbed $6snd\ ^{1,3}D_2$ series.[11–13] The shortening of lifetimes of the $6snd$ levels, which interact strongly with $5d7d\ ^1D_2$, is reproduced well by a theoretical procedure based on MQDT,[11] as shown in Fig. 2. Under some assumptions described in Ref. 11, the radiative decay rate of any level i [$6snd\ ^{1,3}D_2$ ($n \geqslant 17$), $5d7d\ ^1D_2$] can be expressed as

$$\Gamma_i = \alpha_i^2\Gamma_1 + \beta_i^2\Gamma_3 + \epsilon_i^2\Gamma_{5d7d\ ^1D_2}, \quad (9a)$$

$$\Gamma_1 = \gamma_1/(n_i^*)^3, \quad \Gamma_3 = \gamma_3/(n_i^*)^3. \quad (9b)$$

Γ_1 and Γ_3 are the decay rates of pure singlet and triplet Rydberg levels, respectively, which are expected to have a $(n_i^*)^{-3}$ dependence (n^* is the effective quantum number associated with the 6s limit). $\Gamma_{5d7d\ ^1D_2}$ is the decay rate of a pure $5d7d\ ^1D_2$ perturber, i.e., involving no admixture of the $6snd$ channels. The α_i^2, β_i^2, and ϵ_i^2 MQDT admixture coefficients [Eqs. (6) and (7)] correspond to the nine-channel MQDT model[8] derived from energies. The γ_1,γ_3, and $\Gamma_{5d7d\ ^1D_2}$ quantities were fitted to experimental lifetimes.

The optimal values obtained for the lifetimes of pure Rydberg or perturber levels need comment. For the n^* of the perturber one has $1/\Gamma_1$ = 2.3 μsec, $1/\Gamma_3$ = 4.1 μsec, and $1/\Gamma_{5d7d\ ^1D_2}$ = 0.067 μsec. The lifetimes of unperturbed singlet or triplet levels are not much different, and both are much larger than the lifetime of the pure perturber. Therefore radiative decay rates of perturbed states, dominated largely by the perturber decay rate, are insensitive to the differences between Γ_1 and Γ_3. The shortening of lifetimes permits one to determine the total perturber character ϵ_i^2 [Eq. (7)] but not to specify the exact distribution of Rydberg character between singlet and triplet, i.e., the individual α_i^2 *and* β_i^2 coefficients.

Similar measurements and theoretical interpretation have been performed for the $6sns\ ^1S_0$ series.[14] The shortened lifetime observed for the $6s18s\ ^1S_0$ level, which strongly interacts with the $5d7d\ ^3P_0$ level, is described well by the theory based on MQDT[1] (see Fig. 3).

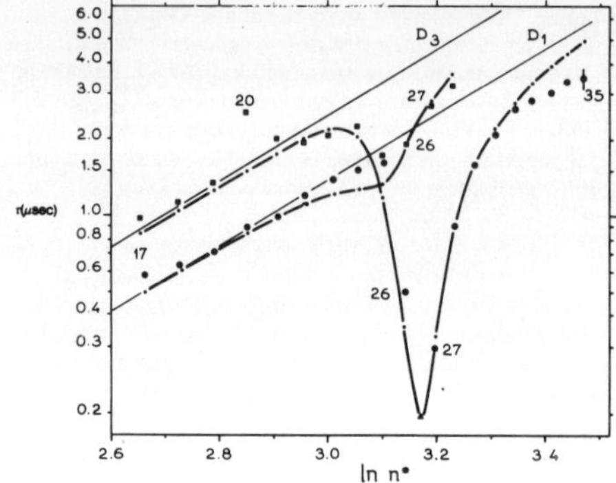

Fig. 2. Lifetimes of the $6snd\ ^{1,3}D_2$ perturbed series of Ba. Comparison of experimental data (●, $6snd\ ^1D_2$, ■, $6snd\ ^3D_2$; ▲, $5d7d\ ^1D_2$ perturber) with the computed lifetimes (•). For the $5d7d$ level, the theoretical and experimental values coincide. The D_1 and D_3 lines correspond, respectively, to $1/\Gamma_1$ and $1/\Gamma_3$ (from Ref. 11).

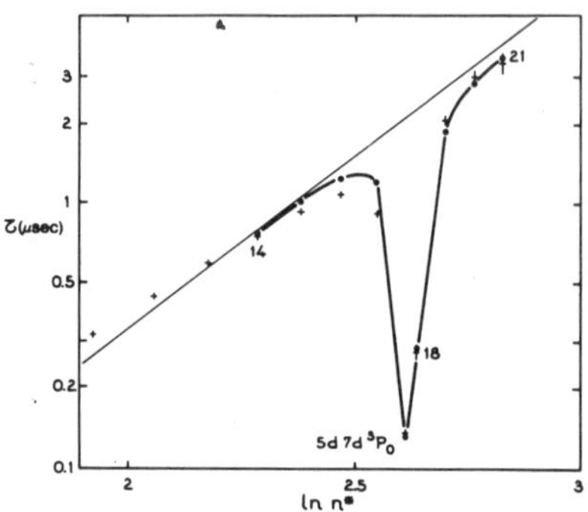

Fig. 3. Lifetimes of the $6sns\ ^1S_0$ perturbed series of Ba. Comparison of experimental data (+, $6sns\ ^1S_0$, $5d7d\ ^3P_0$) with theoretical lifetimes (●). The straight line corresponds to $1/\Gamma$, the lifetime of the pure $6sns\ ^1S_0$ level (from Ref. 14).

B. Isotope-Shift Measurements for Even Isotopes

Isotope shifts for transitions to $6sns\ ^1S_0$ and $6snd\ ^1D_2$ Rydberg levels have been investigated[16–19,22,23] for various even and odd isotopes of Ba. Isotope shifts of optical transitions are currently analyzed in terms of the contributions of the mass shift and of the finite nuclear size (volume or field shift). The separation of both contributions can be achieved by using a King diagram, as explained in Refs. 17, 22, and 23.

Here I consider only information provided by the determination of volume shifts in even isotopes. The volume shift is proportional to the change in electron density at the nucleus $\Delta\rho$ in the transition under study. Admixture of doubly excited states into a Rydberg level causes a change in $\Delta\rho$ from which the perturber character of the level can be derived. The decrease of electron density for the $6s18s\ ^1S_0$ level strongly mixed with the $5d7d\ ^3P_0$ level is evident in Fig. 4 (from Ref. 17).

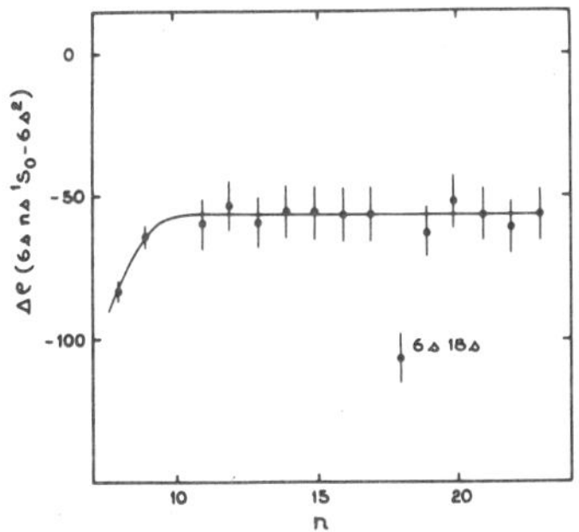

Fig. 4. Isotopic volume shift for the even isotopes of Ba. Change in the charge density at the nucleus between the ground state and the $6sns$ $^1S_0$ Rydberg state (from Ref. 17, courtesy of H. Rinneberg).

Table 1. Rydberg Character of Ba Levels Derived from Isotope Shifts and Predicted by MQDT

Level	Isotope Shift	MQDT
$6s18s$ $^1S_0$	0.67(7)[a]	0.71[b]
$5d7d$ $^3P_0$	0.35[c]	0.35[b]
$5d7d$ $^1D_2$	0.59(5)[d]	0.60[e]
$5d7d$ $^1D_2$	0.59(5)[d]	0.56[f]
$5d7d$ $^1D_2$	0.59(5)[d]	0.58[g]

[a] Refs. 17 and 23.
[b] Ref. 1.
[c] Ref. 17.
[d] Ref. 22.
[e] Ref. 8.
[f] Ref. 15.
[g] Ref. 10.

The change of charge density at the nucleus in the transition from a given level to the ith level of the $6snd$ $^1D_2$ series can be expressed as

$$\Delta\rho_i = (\alpha_i{}^2 + \beta_i{}^2)\Delta\rho_{6snd} + \epsilon_i{}^2\Delta\rho_{5d7d}, \qquad (10)$$

where the $\Delta\rho$ on the right-hand side correspond to pure Rydberg or perturber levels, as in Eq. (9a). A comparison of either Eqs. (9) and (10) or Figs. (3) and (4) shows that the types of information provided by lifetime and volume shift measurements, although similar, are not identical. Lifetimes are determined mainly by the outer part of the wave functions, whereas volume shifts are determined by the part close to the nucleus. Thus the decay rates depend on the Rydberg electron and on the intermediate coupling of levels, whereas the $\Delta\rho$ quantities are nearly independent of these quantities. The sensitivity of lifetimes and isotope shifts to detect and analyze the perturbations depends on the case considered, as discussed in Ref. 10. In any case, lifetimes and volume shifts give information only on the total Rydberg character or, equivalently, on the total perturber character. Rydberg characters derived from isotope shifts for several levels (perturbed Rydberg states or perturbers[17,22,23]) are in good agreement with MQDT predictions,[1,8,10,15] as shown in Table 1.

C. Landé Factors

Landé factors have been measured[15] in the perturbed $6snd$ $^{1,3}D_2$ series. These data have been valuable as input data for improved MQDT treatment of the highly excited $J = 2$ spectrum of Ba.[10,15] For the first time, observable quantities other than energies have been introduced in the MQDT analysis of an alkaline-earth spectrum.

Initially, the g_J factor of any level i near the $5d7d$ $^1D_2$ perturber has been calculated with the MQDT mixing coefficients [Eq. (6)] from Ref. 8 by using the expression

$$g_J{}^i = \alpha_i{}^2 g_J(^1D_2) + \beta_i{}^2 g_J(^3D_2) + \sum_k (\epsilon_i{}^k)^2 g_J(k). \qquad (11)$$

Theoretical predictions (long-dashed curve of Fig. 5) strongly disagree with experiment, whereas the same MQDT model[8] describes the lifetimes well[11] (see Section 4.A).

Afterward, two improved MQDT interpretations[10,15] of experimental data were obtained. Energies are fully independent of the θ angle describing the orthogonal transformation of the two $6snd$ $^{1,3}D_2$ close-coupling channels α. Therefore the previous MQDT model[8] fitted to energies was based on the assumption of pure LS-coupled α channels; i.e., the θ angle had a zero value. The g_J factors that differentiate the two $6snd$ channels permit a partial removal of the indeterminacy of the θ angle. An improved nine-channel MQDT model has been fitted to energies and g_J factors, which now involves a nonzero-value θ angle.[15] The agreement between theory (short-dashed curve of Fig. 5) and experiment is greatly improved. Let me note that the α_i and β_i mixing coefficients depend on θ, whereas the total Rydberg character $\alpha_i{}^2 + \beta_i{}^2$ is independent of θ. This explains why the previous MQDT model[8] was unsuited to reproduce well the singlet–triplet mixing reflected by g_J factors but was adequate to describe the lifetimes.

The main discrepancies between the short-dashed curve of Fig. 5 and experimental data concern the $5d7d$ $^1D_2$ perturber and the neighboring $6snd$ $^1D_2$ levels, which have the maximum $5d7d$ character. The cause of these discrepancies can be understood by analyzing g_J factors with a simple three channel model.[10] The expression for the g_J factor of a level i, in terms of the mixing coefficients of Eq. (8), is

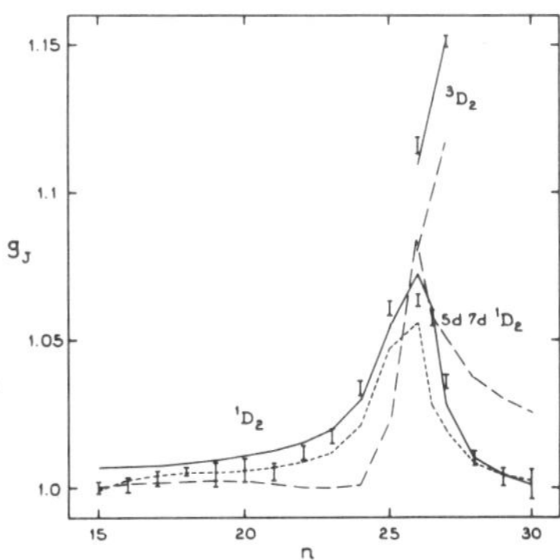

Fig. 5. g_J factors for $6snd$ $^{1,3}D_2$ and $5d7d$ $^1D_2$ levels of Ba. The bars correspond to experimental data of Gräfström *et al.*[15] The curves give MQDT results as follows: long-dashed curve, nine-channel model with $\theta = 0$ (Ref. 8); short-dashed curve, nine-channel model with $\theta = 0.35$ (Ref. 15); solid curve, three-channel model.[10] For $^3D_2$ levels the solid and short-dashed curves are superposed.

$$g_J{}^i = \alpha_i{}^2 g_J({}^1D_2) + \beta_i{}^2 g_J({}^3D_2) + \epsilon_i{}^2 \bar{g}. \quad (12)$$

The $\bar{g}$ parameter corresponds to the g_J factor of a pure $5d7d$ perturber, i.e., neglecting its mixing with Rydberg states. The MQDT parameters, as well as the $\bar{g}$ parameter, were fitted to the experimental energies and g_J factors, leading to a satisfactory agreement between theory (solid curve of Fig. 5) and experiment. The optimal value $\bar{g}$ = 1.105 is only slightly larger than the value

$$g = \left[\sum_k (\epsilon_i{}^k)^2 g_J(k)\right] \Big/ \sum_k (\epsilon_i{}^k)^2 = 1.03 \quad (13)$$

obtained with the nine-channel model.[15] Here again, the energies are insufficient to determine all the $U_{i\alpha}$ parameters that define the $\epsilon_i{}^k$ coefficients, i.e., the exact coupling of the $5d7d$ perturber.

It appears clear that g_J factors are sensitive to all the mixing coefficients involved in the MQDT wave functions, not only to the singlet–triplet mixing coefficients but also to the intermediate coupling of the $5d7d$ perturber. However, g_J factors, depending only on squared mixing coefficients, cannot completely remove the indeterminacy of the relative signs of the θ_{kl} MQDT angles.[10]

The theoretical interpretations of the g_J factors also gave evidence of the breakdown of the LS characterization of close-coupling channels relevant to the description of the $6snd$ and $5d7d$ levels. In fact, spin–orbit effects in Ba are found to play a greater role than previously surmised, as is discussed in much detail in Ref. 10.

D. Hyperfine-Structure Measurements

The hyperfine structure of perturbed Rydberg states of Ba has been investigated intensively[16,19–24] and has been found to be quite sensitive to state mixing. For Rydberg states the preponderant hyperfine interaction is the Fermi-contact interaction between the nuclear spin **I** and the electron spin **s** of the 6s electron:

$$H_{\text{hf}} = a_{6s}\,\mathbf{I}\cdot\mathbf{s}. \quad (14)$$

The hyperfine interaction can mix the fine structure of neighboring n states having the same total quantum number F when the states are separated by an energy interval comparable with or smaller than the hyperfine splitting. For details on the role of this induced hyperfine mixing, the reader is referred to the literature.[18,19,21–24] Here I consider only two examples, in which the interpretation of hyperfine structures in terms of splitting factors A (i.e., neglecting induced hyperfine mixing) has yielded quantitative information on state-mixing coefficients.

The first example concerns the perturbed $6sns\ {}^3S_1$ series recently studied by two groups.[19,23] The admixture of the $5d7d$ J = 1 perturber into several $6sns\ {}^3S_1$ states was studied[19,23] from observed hyperfine splitting between the $F = I + 1$ and $F = I - 1$ components. This splitting is directly related to the $\alpha_i{}^2$ Rydberg character of the perturbed $6sns\ {}^3S_1$ level by the relation

$$A_i = \tfrac{1}{4}(E_{I+1}{}^i - E_{I-1}{}^i) = \tfrac{1}{2}\alpha_i{}^2 a_{6s}. \quad (15)$$

The knowledge of a_{6s} permits the determination of $\alpha_i{}^2$ coefficients in good agreement with MQDT predictions (see Table 2). Moreover, it has been emphasized[19] in this particular example that hyperfine structure is a much more sensitive tool than isotope shift to detect a small admixture of the perturber (see Figs. 2 and 3 of Ref. 19).

Table 2. Rydberg Character of $6sns\ {}^3S_1$ Levels of Ba Derived from Hyperfine Structures and Predicted by MQDT

Level	Hyperfine Structure Ref. 23	Hyperfine Structure Ref. 19	MQDT (Ref. 9)
6s15s	0.946(1)	0.94(1)	0.974
6s16s	0.9970(4)	–	0.998
6s17s	0.9987(4)	–	0.998
6s18s	0.99625(5)	–	0.995
6s19s	–	0.96(1)	0.973
6s20s	0.9177(4)	0.91(1)	0.918

The second example concerns the perturbed $6snd\ {}^1D_2$ series coupled to the $6snd\ {}^3D_2$ series by spin–orbit interaction. Three independent investigations of hyperfine structures were performed.[16,20–22,24] The data obtained on states near the $5d7d\ {}^1D_2$ perturber have provided a stringent test of the MQDT wave functions derived from g_J factors.[10,15] From their observations, Rinneberg and Neukammer[20–22] and Eliel and Hogervorst[24] deduced the amount as well as the relative signs of the α_i and β_i singlet and triplet mixing coefficients involved in Eq. (6) or (8). For simplicity the hyperfine structures are discussed here in terms of splitting factors A, although such an analysis is only approximate for several $6snd\ {}^1D_2$ levels located near the $5d7d\ {}^1D_2$ level.[24] For details on the exact procedure used to analyze experimental data, the reader is referred to Refs. 10, 21, and 24. For any level i ($6snd\ {}^1D_2$, $5d7d\ {}^1D_2$), one has

$$A_i = a_{6s}(\beta_i{}^2 - 2\alpha_i\beta_i\sqrt{6})/12. \quad (16)$$

Derivation of the β_i mixing coefficients from the experimental A_i factors required additional information on the total Rydberg character $\alpha_i{}^2 + \beta_i{}^2$, and isotope shifts, lifetimes, or MQDT can yield this information.

The β_i data have been interpreted using MQDT first in Ref. 21 and then combined with the Landé factor in Refs. 10 and 24 using either a three- or a nine-channel model. The n dependence of the β mixing coefficient derived from hyperfine structures is reproduced well in various analyses. The best agreement between theory and experiment is obtained with the three-channel model[10,21] (solid curve of Fig. 6), which also leads to the best interpretation of the g_J factors (see Section 5.C). Hyperfine structures for $6snd\ {}^1D_2$ series have confirmed the breakdown of the LS characterization of $6snd\ {}^{1,3}D_2$ close-coupling channels pointed out by the g_J factors[15], i.e., the requirement of a nonzero value for the θ angle coupling these two channels. In addition, hyperfine structures, sensitive to the relative signs of the α_i and β_i mixing coefficients, allow one to determine the relative signs of the θ_{kl} MQDT mixing angles, which cannot be derived from energies and can only be partially deduced from g_J factors.[10] The short- and long-dashed curves plotted in Fig. 6 correspond to a θ angle with zero value and to two different choices of the relative signs of the mixing angles involved in the MQDT model[8] fitted to energies.

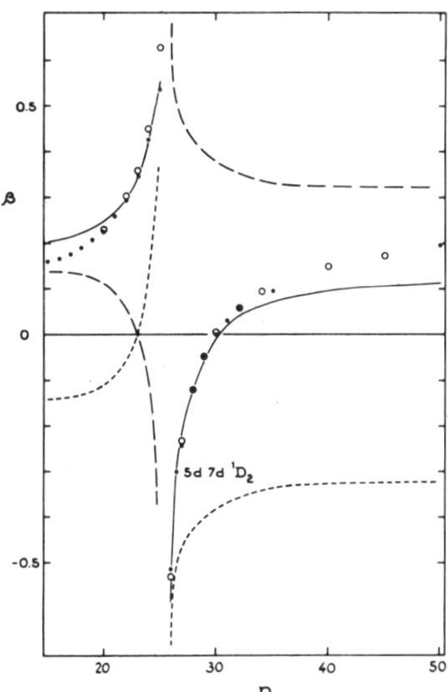

Fig. 6. β mixing coefficients for 6*snd* $^1D_2$ and 5*d*7*d* $^1D_2$ levels of Ba. The points are deduced from hyperfine structures: •, Refs. 21 and 22; o, Ref. 24. The curves correspond to MQDT calculations: solid curve, three-channel model[10,21]; long- and short-dashed curves, nine-channel model with $\theta = 0$.[8]

In conclusion, hyperfine structure appears to be the most sensitive tool to analyze singlet–triplet mixing in the 6*snd* $^{1,3}D_2$ series. However, hyperfine structures reflect only the spin density at the nucleus of the 6*s* electron; then additional information on the perturber character of level is needed to derive state-mixing coefficients for perturbed 6*snd* levels.

E. Some Other Tools for Probing Multichannel-Quantum-Defect Theory Wave Functions of Perturbed Levels

Angular distributions of photoelectrons have been measured after stepwise resonant multiphoton ionization of the ground state of Ba through 6*snd* $^{1,3}D_2$ Rydberg levels.[25] The angular distributions of 6*sϵl* photoelectrons are found to be quite sensitive to both the perturbation of the Rydberg series by the 5*d*7*d* $^1D_2$ level and the singlet–triplet mixing of 6*snd* Rydberg states. The quantitative interpretation of the experimental data[25] has given additional evidence of the validity of the improved MQDT models derived from g_J factors and hyperfine structures.[10,15,21]

Properties of Rydberg states in the presence of external fields can also be used to detect perturbation of Rydberg series. So, for example, large Stark and diamagnetic shifts are expected for Rydberg levels, while similar shifts are minuscule for the valence state perturber. Zero-field MQDT wave functions can be used to interpret such properties, provided that the external field is weak. For papers on Stark and diamagnetic shifts in which MQDT had served to interpret experimental data obtained in Ba, the reader is referred to Refs. 12, 26, and 27.

5. CONCLUSION

The main emphasis of this paper has been on the observable properties of highly excited states of Ba and the strong connection between laser spectroscopy and theoretical interpretation by the use of MQDT. The interplay of experiment and theory has greatly enlarged our understanding of the detailed structure of highly excited states of the even-parity spectrum of Ba. The examples presented throughout the paper clearly demonstrate how the various data provided by high-resolution laser spectroscopy complement one another to check and even extend MQDT models derived from energies. So the g_J factors and hyperfine structures have been essential to obtain reliable wave functions for the 6*snd* $^{1,3}D_2$ highly excited levels. In Sr the singlet–triplet mixing in the 5*snd* $^{1,3}D_2$ series as derived from g_J factors[32] and hyperfine structures[33] was found to be in perfect agreement with the predictions of the MQDT model[30] fitted to energies. Therefore, because of its complexity, the Ba spectrum, which is governed by the mutual influence of electrostatic interactions and spin–orbit coupling,[10] is much more attractive than the lighter alkaline earths. The breakdown of the *LS* characterization of the 6*snd* close-coupling channels in Ba stresses the large effect of spin–orbit-coupling interaction, in contrast to the situation with Sr. However, in Ba, the role of spin–orbit coupling and more generally of relativistic effects is not yet perfectly understood. Further theoretical and experimental investigations on Rydberg series of Ba are necessary to clarify the situation. For this purpose, additional experimental data on the detailed structure of bound Rydberg states of the odd-parity spectrum, mainly on the singlet–triplet mixing in 6*snp* $^{1,3}P_1$ series, are highly desirable.

The Laboratoire Aimé Cotton is affilated with the Université Paris-Sud.

REFERENCES

1. M. Aymar, P. Camus, M. Dieulin, and C. Morillon, "Two-photon spectroscopy of neutral barium: observations of the highly excited even levels and theoretical analysis of the $J = 0$ spectrum," Phys. Rev. A **18**, 2173–2183 (1978).
2. P. Camus, M. Dieulin, and A. El Himdy, "Two-step optogalvanic spectroscopy of neutral barium: observations and interpretation of the even levels below the 6*s* ionization limit with $J = 1, 3, 4$ and 5," Phys. Rev. A **26**, 379–390 (1982).
3. M. J. Seaton, "Quantum defect theory. I—General formulation," Proc. Phys. Soc. **88**, 801–814 (1966).
4. M. J. Seaton, "Quantum defect theory," Rep. Prog. Phys. **46**, 167–257 (1983).
5. K. T. Lu, "Spectroscopy and collision theory. The Xe absorption spectrum," Phys. Rev. A **4**, 579–596 (1971).
6. C. M. Lee and K. T. Lu, "Spectroscopy and collision theory. II—The Ar absorption spectrum," Phys. Rev. A **8**, 1241–1257 (1973).
7. U. Fano, "Unified treatment of perturbed series, continuous spectra, and collisions," J. Opt. Soc. Am. **65**, 979–987 (1975), and references therein.
8. M. Aymar and O. Robaux, "Multichannel quantum-defect analysis of the bound even-parity $J = 2$ spectrum of neutral barium," J. Phys. B **12**, 531–546 (1979).
9. M. Aymar and P. Camus, "Multichannel quantum-defect analysis of the bound even-parity spectrum of neutral barium," Phys. Rev. A **28**, 850–857 (1983).
10. M. Aymar, "Rydberg series of alkaline-earth atoms Ca through Ba. The interplay of laser spectroscopy and multichannel quantum defect analysis," Phys. Rep. (to be published).
11. M. Aymar, R. J. Champeau, C. Delsart, and J. C. Keller, "Life-

times of Rydberg levels in the perturbed 6*snd* $^{1,3}D_2$ series of barium I," J. Phys. B **14,** 4489–4496 (1981).
12. T. F. Gallagher, W. Sandner, and K. A. Safinya, "Probing configuration interaction of the Ba 5*d*7*d* $^1D_2$ state using radio-frequency spectroscopy and lifetime measurements," Phys. Rev. A **23,** 2969–2977 (1981).
13. K. Bathia, P. Grafström, C. Levinson, H. Lundberg, L. Nilsson, and S. Svanberg, "Natural radiative lifetimes in the perturbed 6*snd* $^1D_2$ sequence of barium," Z. Phys. A **303,** 1–5 (1981).
14. M. Aymar, P. Grafström, C. Levinson, H. Lundberg, and S. Svanberg, "Perturbation of Ba 6*sns* $^1S_0$ sequence by the 5*d*7*d* $^3P_0$ state, probed by lifetime measurements". J. Phys. B **15,** 877–882 (1982).
15. P. Grafström, C. Levinson, H. Lundberg, S. Svanberg, P. Grundevik, L. Nilsson, and M. Aymar, "Zeeman effect in the perturbed 6*snd* $^{1,3}D_2$ sequences of Ba I; test of MQDT wave functions," Z. Phys. A **308,** 95–101 (1982).
16. P. Grafström, J. Zhan-Kui, G. Jönsson, S. Kröll, C. Levinson, H. Lundberg, and S. Svanberg, "Hyperfine structure and isotope shift of highly excited barium-I states," Z. Phys. A **306,** 281–284 (1982).
17. J. Neukammer, E. Matthias, and H. Rinneberg, "Systematic investigation of electron densities at the nucleus for Rydberg states of barium," Phys. Rev. A **25,** 2426–2429 (1982).
18. H. Rinneberg, J. Neukammer, and E. Matthias, "Isotope shifts of perturbed 6*sns* $^1S_0$ and $^3S_1$ Rydberg states of odd barium isotopes," Z. Phys. A **306,** 11–18 (1982).
19. J. Neukammer and H. Rinneberg, "Hyperfine structure of perturbed 6*sns* $^3S_1$ Rydberg states of Ba," J. Phys. B **15,** L425–L429 (1982).
20. H. Rinneberg and J. Neukammer, "Resonance in singlet-triplet mixing in two-electrons systems caused by perturbing configurations," Phys. Rev. Lett. **49,** 124–127 (1982).
21. H. Rinneberg and J. Neukammer, "Hyperfine structure and three-channel quantum defect theory of 6*snd* $^1D_2$ Rydberg states of Ba," Phys. Rev. A **27,** 1779–1789 (1983).
22. H. Rinneberg and J. Neukammer, "Hyperfine structure and configuration interaction of the 5*d*7*d* $^1D_2$ perturbing state of barium," J. Phys. B **15,** L825–L829 (1982).
23. W. Hogervorst and E. R. Eliel, "High resolution spectroscopy on 6*sns* Rydberg states of Ba," Z. Phys. A **310,** 19–26 (1983).
24. R. E. Eliel and W. Hogervorst, "Hyperfine structure in 6*snd* Rydberg configuration in barium," J. Phys. B **16,** 1881–1893 (1983).
25. E. Matthias, P. Zoller, D. S. Elliott, N. D. Piltch, S. J. Smith, and G. Leuchs, "Influence of configuration mixing in intermediate states on resonant multiphoton ionization," Phys. Rev. Lett. **50,** 1914–1917 (1983), and references therein.
26. K. A. H. van Leeuwen, W. Hogervorst, and B. H. Post, "Stark effect in barium 6*snd* $^1D_2$ Rydberg states; evidence of strong perturbations in the $^1F_3$ series," Phys. Rev. A **28,** 1901–1908 (1983).
27. R. J. Fonck, F. L. Roesler, D. H. Tracy, K. T. Lu, F. S. Tomkins, and W. R. S. Garton, "Atomic diagmagnetism and diamagnetically induced configuration mixing in laser-excited barium," Phys. Rev. Lett. **39,** 1513–1516 (1977).
28. O. Robaux and M. Aymar, "A program for analyzing the Rydberg series of highly excited discrete spectra by MQDT," Comp. Phys. Commun. **25,** 223–236 (1982).
29. J. A. Armstrong, P. Esherick, and J. J. Wynne, "Bound even parity $J = 0$ and $J = 2$ spectra of Ca; a multichannel quantum-defect theory analysis," Phys. Rev. A **15,** 180–196 (1977).
30. P. Esherick, "Bound, even-parity $J = 0$ and $J = 2$ spectra of Sr," Phys. Rev. A **15,** 1920–1936 (1977).
31. J. A. Armstrong, J. J. Wynne, and P. Esherick, "Bound, odd-parity $J = 1$ spectra of the alkaline earths: Ca, Sr, and Ba," J. Opt. Soc. Am. **69,** 211–230 (1979).
32. J. J. Wynne, J. A. Armstrong, and P. Esherick, "Zeeman effect of $J = 2$ states of Sr: *g* factor variation for interacting Rydberg series," Phys. Rev. Lett. **39,** 1520–1523 (1977).
33. R. Beigang, E. Matthias, and A. Timmermann, "Influence of singlet–triplet mixing on the hyperfine structure of 5*snd* Rydberg states in ^{87}Sr," Phys. Rev. Lett. **47,** 326–329 (1981); **48,** 290 (1982).

1972 *J. Phys. B: At. Mol. Phys.* **5** 2187–98

Computer programs for the calculation of electron–atom collision cross sections
I. General formulation

W EISSNER and M J SEATON

Department of Physics and Astronomy, University College, London

MS received 2 June 1972

Abstract. The wavefunction for electron collisions with an N-electron atom is expanded in the form

$$\Psi = \sum_i \Theta_i + \sum_j \Phi_j c_j$$

where Θ_i is an antisymmetrized product of an atomic eigenfunction times an orbital function θ_i for the colliding electron; θ_i contains a radial function F_i; Φ_j has the form of a bound state function for the $(N+1)$-electron problem. The condition is imposed that $(P_y|F_i) = 0$ if $l_y = l_i$, where P_y is an atomic radial function and where l_y and l_i are orbital angular momenta associated with P_y and F_i; this condition does not imply a restriction on Ψ so long as a suitable set of states Φ_j is included.

The variational principle is discussed and two approximations are described: (i) The distorted wave (DW) approximation is valid when the coupling is not too strong. Particular attention is paid to the normalization of the DW functions F_i. The coefficients c_j are determined using the variational principle. (ii) The variational principle is used to obtain a set of coupled integro-differential (ID) equations for the determination of the functions F_i and the coefficients c_j.

1. Introduction

Eissner and Nussbaumer (1969) have described a computer program for the automatic calculation of atomic wavefunctions. In the present series of papers we describe some computer programs for the automatic calculation of electron–atom collision cross sections. Our interest is mainly in the excitation of neutral atoms and positive ions at near-threshold energies of importance for astrophysical applications.

We consider two approximations:

(i) The Distorted Wave approximation. This can be used when the coupling is not too strong. This condition is always satisfied for the contributions from sufficiently large total orbital angular momentum. For highly ionized systems (in practice more than about two or three times ionized) the method can be used for the calculation of all angular momentum contributions. We shall refer to this method as the DW approximation.

(ii) The exact solution of the coupled Integro Differential equations which are obtained on expanding the total wavefunction in terms of atomic eigenfunctions. This method may be used in cases for which the DW method fails. It has previously been referred to as the continuum-state Hartree–Fock method (Seaton 1953) or as the close-coupling method (Burke and Smith 1962). We shall refer to it as the ID approximation.

The present paper describes our general formulation of the problem. Later papers will give a more detailed algebraic formulation and a description of the numerical methods used in the DW approximation (Eissner 1973), and a description of the numerical methods used in the ID approximation (Seaton 1973). The computer programs will be submitted for publication in *Computer Physics Communications*. In the present series of papers we will, so far as possible, use notations similar to those used in the FORTRAN programs.

2. Formulation of the collision problem

2.1. Expansion of the wavefunction

We consider an ion with N electrons and nuclear charge Z. The charge on the ion is $z = (Z-N)$. Let the ion hamiltonian H_{I} have eigenfunctions χ_i,

$$H_{\mathrm{I}}\chi_i = E_i\chi_i. \tag{2.1}$$

These are simultaneous eigenfunctions for the total spin and orbital angular momentum for the ion, and we may therefore put

$$\chi_i = \chi(\alpha_i S_i L_i M_{S_i} M_{L_i}). \tag{2.2}$$

For the collision problem we have to consider a system with $(N+1)$ electrons. Let $\boldsymbol{x}_p = (\boldsymbol{r}_p, \sigma_p)$ be the space and spin co-ordinate of electron p. We put

$$\chi_i(\bar{\boldsymbol{x}}_p) = \chi_i(\boldsymbol{x}_1, \boldsymbol{x}_2, \ldots, \boldsymbol{x}_{p-1}, \boldsymbol{x}_{p+1}, \ldots, \boldsymbol{x}_{N+1}). \tag{2.3}$$

We introduce orbital functions for the colliding electron,

$$\theta_i(\boldsymbol{x}) = Y_{l_i m_{l_i}}(\hat{\boldsymbol{r}})\delta(m_{s_i}, \sigma)\frac{1}{r}F_i(r). \tag{2.4}$$

For the $(N+1)$ electron system we introduce vector-coupled anti-symmetric functions

$$\begin{aligned}\Theta_i &= \Theta(\alpha_i S_i L_i l_i SLM_S M_L)\\ &= (N+1)^{-1/2}\sum_{p=1}^{N+1}\sum_{M'_S m_s}\sum_{M'_L m_l}(-1)^{p-N-1}\\ &\quad\times C^{S_i\,1/2\,S}_{M'_S m_s\,M_S}C^{L_i\,l_i\,L}_{M'_L m_l M_L}\chi(\alpha_i S_i L_i M'_S M'_L|\bar{\boldsymbol{x}}_p)\theta_i(l_i m_s m_l)\chi_p).\end{aligned} \tag{2.5}$$

Let H be the total hamiltonian and let Ψ be an eigenfunction of H,

$$H\Psi = E\Psi \tag{2.6}$$

where E is the total energy. We represent Ψ as an expansion in the functions Θ_i,

$$\Psi = \sum_i \Theta_i. \tag{2.7}$$

The radial functions $F_i(r)$ are required to be such that (2.6) is satisfied. In §6 we give equations which the functions F_i should satisfy when a truncated expansion is used in place of (2.7).

2.2 *Boundary conditions*

We require that Ψ should be everywhere bounded.

2.2.1. Behaviour at the origin. The functions $F_i(r)$ must be such that $F_i(r) \to 0$ as $r \to 0$. It may be shown to follow that, in the limit of r small, $F_i(r)$ behaves like r^{l_i+1}. We may therefore put

$$\lim_{r\to 0} (r^{-l_i-1}F_i(r)) = A_i \tag{2.8}$$

where A_i is a constant.

2.2.2. Behaviour for r large. In the limit of r large the functions F_i are solutions of

$$\left[\frac{d^2}{dr^2} - \frac{l_i(l_i+1)}{r^2} + \frac{2z}{r} + k_i^2\right] F_i = 0 \tag{2.9}$$

where

$$k_i^2 = E - E_i. \tag{2.10}$$

We measure r in Bohr radii and use Rydberg units for energies.

For $k_i^2 > 0$ the functions F_i may be taken to have asymptotic form

$$F_i \underset{r\to\infty}{\sim} k_i^{-1/2}\{(\sin \xi_i)a_i + (\cos \xi_i)b_i\} \qquad (k_i^2 > 0) \tag{2.11}$$

where

$$\xi_i = \zeta_i + \tau_i \tag{2.12}$$

$$\zeta_i = k_i r - \tfrac{1}{2}l_i\pi + \frac{z}{k_i}\ln(2k_i r) + \arg\Gamma\left(l_i + 1 - \frac{iz}{k_i}\right). \tag{2.13}$$

It may be noted that the regular Coulomb function $f_i(r)$, which is a solution of (2.9) for all values of r and is such that $f_i(0) = 0$, has asymptotic form $f_i(r) \sim \text{constant} \times \sin(\zeta_i)$.

The choice of τ_i in (2.12) is essentially arbitrary. The quantities a_i and b_i in (2.11) will, of course, depend on the choice adopted for τ_i.

For $k_i^2 < 0$ we require that

$$F_i(r) \to 0 \text{ as } r \to \infty \qquad (k_i^2 < 0). \tag{2.14}$$

We refer to i an open channel if $k_i^2 \geqslant 0$ and a closed channel if $k_i^2 < 0$. Let the states i be arranged in order of increasing E_i; $E_1 \leqslant E_2 \leqslant E_3 \ldots$. Let the number of open channels be $NCHOP$: then

$$k_i^2 \geqslant 0 \text{ for } i = 1, \ NCHOP$$
$$k_i^2 < 0 \text{ for } i > NCHOP \tag{2.15}$$

We use the notation $i = L, M$ to mean $i = L, L+1, \ldots, M$.

2.3. *Linearly independent solutions*

Equation (2.6) has $NCHOP$ linearly independent solutions satisfying the boundary conditions (2.8), (2.11) and (2.14). Let $\Psi'_{i'}$ be a solution with boundary condition i' ($i' = 1, NCHOP$) and let this solution contain radial functions $F_{ii'}$. These functions are such that

$$\lim_{r\to 0}\{r^{-l_i-1}F_{ii'}(r)\} = A_{ii'} \qquad \text{for all } i\text{; and } i' = 1, NCHOP \tag{2.16}$$

$$F_{ii'}(r) \underset{r\to\infty}{\sim} k_i^{-1/2}\{(\sin \xi_i)a_{ii'} + (\cos \xi_i)b_{ii'}\} \qquad \begin{array}{l}\text{for } i = 1, NCHOP\\ \text{and } i' = 1, NCHOP\end{array} \tag{2.17}$$

$$F_{ii'}(r) \to 0 \text{ as } r \to \infty \qquad \text{for } i > NCHOP \text{ and } i' = 1, NCHOP \tag{2.18}$$

A particular solution, say $i' = i''$, may be defined on specifying $NCHOP$ numbers in the i'' columns of the three square arrays

$$\left.\begin{array}{l}A_{ii'}\\ a_{ii'}\\ b_{ii'}\end{array}\right\} \begin{array}{l} i = 1, NCHOP\\ i' = 1, NCHOP.\end{array} \tag{2.19}$$

A complete set of $NCHOP$ linearly independent solutions may be defined on making $NCHOP$ linearly independent specifications of $NCHOP$ numbers in the i' columns of the arrays (2.19). It is, of course, to be understood that the boundary condition (2.18) is always imposed.

One possible way of defining the $NCHOP$ linearly independent solutions would be to take

$$A_{ii'} = \delta_{ii'} \text{ for } i = 1, NCHOP \text{ and } i' = 1, NCHOP. \tag{2.20}$$

Other ways of specifying the linearly independent solutions will be discussed by Seaton (1973).

2.4. The collision problem

The equations (2.17) may be written in matrix notation

$$\mathbf{F} \sim k^{-1/2}\{(\sin \xi)\mathbf{a} + (\cos \xi)\mathbf{b}\} \tag{2.21}$$

where it is to be understood that quantities not in boldface type, and without subscripts, are diagonal matrices.

We may also consider functions $\mathbf{F}^{(\mathbf{R})}$ and $\mathbf{F}^{(\mathbf{S})}$ with $\mathbf{R}$ matrix or $\mathbf{S}$ matrix boundary conditions:

$$\mathbf{F}^{(\mathbf{R})} \sim k^{-1/2}\{(\sin \xi) + (\cos \xi)\mathbf{R}\} \tag{2.22}$$

$$\mathbf{F}^{(\mathbf{S})} \sim k^{-1/2}\{e^{-i\xi} - e^{+i\xi}\mathbf{S}\}. \tag{2.23}$$

The functions $\mathbf{F}^{(\mathbf{S})}$ are complex, while the functions $\mathbf{F}^{(\mathbf{R})}$ are real. The relations between the matrices are

$$\mathbf{R} = \mathbf{b}\mathbf{a}^{-1} \tag{2.24}$$

$$\mathbf{S} = (1 + i\mathbf{R})(1 - i\mathbf{R})^{-1}. \tag{2.25}$$

The matrices $\mathbf{R}$ and $\mathbf{S}$, as defined here, depend on the choice of τ_i in (2.12); it is readily shown that

$$\mathbf{R}(\tau) = \{\sin(\tau' - \tau) + \cos(\tau' - \tau)\mathbf{R}(\tau')\}\{\cos(\tau' - \tau) - \sin(\tau' - \tau)\mathbf{R}(\tau')\}^{-1} \tag{2.26}$$

$$\mathbf{S}(\tau) = e^{i(\tau' - \tau)}\mathbf{S}(\tau')\, e^{i(\tau' - \tau)}. \tag{2.27}$$

The probability of a transition from channel i to channel i' is $|S_{ii'}|^2$. This is independent of τ. The total collision strength for the transition $(\alpha_i S_i L_i, \alpha_{i'} S_{i'} L_{i'})$ is defined as

$$\Omega(\alpha_i S_i L_i, \alpha_{i'} S_{i'} L_{i'}) = \tfrac{1}{2} \sum_{SL} \sum_{ll'} (2S+1)(2L+1)|T(\alpha_i S_i L_i l SL, \alpha_{i'} S_{i'} L_{i'} l' SL)|^2 \qquad (2.28)$$

where $\mathbf{T} = \mathbf{S} - \exp(2i\tau)$.

The collision cross section is

$$Q(\alpha_i S_i L_i \to \alpha_{i'} S_{i'} L_{i'}) = \frac{\Omega(\alpha_i S_i L_i, \alpha_{i'} S_{i'} L_{i'})}{k_i^2 (2S_i+1)(2L_i+1)} \; (\pi a_0^2). \qquad (2.29)$$

For the case of $(\alpha_i S_i L_i) = (\alpha_{i'} S_{i'} L_{i'})$, (2.29) gives the total elastic cross section for scattering by neutral atoms. For positive ions the total elastic cross section is divergent.

2.5. The case of all channels closed

When all channels are closed the boundary conditions may be written

$$\left.\begin{array}{l} F_i(r) \to 0 \text{ as } r \to 0 \\ F_i(r) \to 0 \text{ as } r \to \infty \end{array}\right\} \text{all } i. \qquad (2.30)$$

These conditions can be satisfied only for certain discrete values of the energy E, corresponding to the eigenenergies for the $(N+1)$-electron problem.

3. The variation principle

Let $\mathbf{\Psi}$ be a vector with components Ψ_i which are solutions of (2.6) and let the radial functions $\mathbf{F}$ in $\mathbf{\Psi}$ have the $\mathbf{R}$ matrix asymptotic form (2.22).

Consider a trial function $\mathbf{\Psi}^t$ containing a radial function $\mathbf{F}^t$ with asymptotic form

$$\mathbf{F}^t \sim k^{-1/2}\{(\sin \xi) + (\cos \xi)\mathbf{R}^t\}. \qquad (3.1)$$

Put $\mathbf{\Psi}^t = \mathbf{\Psi} + \delta\mathbf{\Psi}$, $\mathbf{R}^t = \mathbf{R} + \delta\mathbf{R}$. It should be noted that the same phases τ are used in $\mathbf{\Psi}$ and in $\mathbf{\Psi}^t$.

Define $(\mathbf{\Psi}|H-E|\mathbf{\Psi})^t = (\mathbf{\Psi}^t|H-E|\mathbf{\Psi}^t)$ and

$$\delta(\mathbf{\Psi}|H-E|\mathbf{\Psi}) = (\mathbf{\Psi}|H-E|\mathbf{\Psi})^t - (\mathbf{\Psi}|H-E|\mathbf{\Psi}). \qquad (3.2)$$

Using Green's theorem it may be shown that

$$\delta\{(\mathbf{\Psi}|H-E|\mathbf{\Psi}) - \mathbf{R}\}$$
$$= (\delta\mathbf{\Psi}|H-E|\mathbf{\Psi}) + (\mathbf{\Psi}|H-E|\delta\mathbf{\Psi})^\dagger + (\delta\mathbf{\Psi}|H-E|\delta\mathbf{\Psi}). \qquad (3.3)$$

From (2.6) it follows that

$$(\delta\mathbf{\Psi}|H-E|\mathbf{\Psi}) = (\mathbf{\Psi}|H-E|\delta\mathbf{\Psi})^\dagger = 0 \qquad (3.4)$$

and (3.3) therefore reduces to

$$\delta\{(\mathbf{\Psi}|H-E|\mathbf{\Psi}) - \mathbf{R}\} = (\delta\mathbf{\Psi}|H-E|\delta\mathbf{\Psi}). \qquad (3.5)$$

We may say that $\delta\{(\mathbf{\Psi}|H - E|\mathbf{\Psi}) - \mathbf{R}\}$ vanishes to first order for small variations about the exact function $\mathbf{\Psi}$.

Since $(\Psi|H-E|\Psi) = 0$, equation (3.5) may be written

$$\mathbf{R} = \mathbf{R}^{\mathrm{t}} - (\Psi|H-E|\Psi)^{\mathrm{t}} + (\delta\Psi|H-E|\delta\Psi). \tag{3.6}$$

The Kohn corrected R matrix is defined by

$$\mathbf{R}^{\mathrm{K}} = \mathbf{R}^{\mathrm{t}} - (\Psi|H-E|\Psi)^{\mathrm{t}}. \tag{3.7}$$

It is seen that $\mathbf{R}^{\mathrm{K}}$ differs from $\mathbf{R}$ by a quantity of quadratic order in $\delta\Psi$.

We may now describe the DW and ID approximations in their simplest forms. More general forms will be considered in § 6.

In the DW approximation the radial functions are calculated neglecting coupling between channels. We then have

$$F_{ii'}^{\mathrm{DW}} = \mathscr{F}_i \delta_{ii'}. \tag{3.8}$$

The functions $\mathscr{F}_i$ are taken to have asymptotic form

$$\mathscr{F}_i \sim k_i^{-1/2} \sin \xi_i. \tag{3.9}$$

It should be noted that this defines τ_i and the normalization of $\mathscr{F}_i$. We then have $\mathbf{R}^{\mathrm{t}} = 0$. The DW approximation for $\mathbf{R}$ is obtained from (3.7),

$$\mathbf{R}^{\mathrm{DW}} = -(\Psi|H-E|\Psi)^{\mathrm{DW}}. \tag{3.10}$$

The error in the *corrected* matrix $\mathbf{R}^{\mathrm{DW}}$, is then of quadratic order in the error in the wavefunctions. The DW method is a good approximation if all elements of $\mathbf{R}^{\mathrm{DW}}$ are small (in practice it will generally give results of acceptable accuracy if $R^{\mathrm{DW}}(i, i') \lesssim 1/2$ for all i, i'). It should be noted that any choice of τ_i other than that which gives (3.9) will generally give larger values of $\mathbf{R}^{\mathrm{DW}}$.

In the ID approximation we use wavefunctions obtained on truncating (2.7),

$$\Psi_i^{\mathrm{ID}} = \sum_{i'=1}^{i_m} \Theta_{i'i}^{\mathrm{ID}}. \tag{3.11}$$

Let Ψ^{ID} contain radial functions $\mathbf{F}^{\mathrm{ID}}$ and let $\Delta\Psi^{\mathrm{ID}}$ be a variation in Ψ^{ID} due to variations $\Delta\mathbf{F}^{\mathrm{ID}}$ in $\mathbf{F}^{\mathrm{ID}}$. We impose the condition that

$$(\Delta\Psi|H-E|\Psi)^{\mathrm{ID}} = 0 \tag{3.12}$$

for all variations $\Delta\mathbf{F}^{\mathrm{ID}}$. It may be shown to follow from this condition that $(\Psi|H-E|\Psi)^{\mathrm{ID}} = 0$ and hence that

$$\mathbf{R} = \mathbf{R}^{\mathrm{ID}} - (\delta\Psi|H-E|\delta\Psi) \tag{3.13}$$

(see, eg Percival and Seaton 1957). In the ID approximation the error in the *trial* $\mathbf{R}$ matrix is of quadratic order in the error in the wavefunctions.

It should be noted that, in formulating the variational principle, it has been assumed that we have exact eigenfunctions χ_i for the target system. In practice, for all atoms with more than one electron, we use approximate target functions. This introduces errors into the calculated $\mathbf{R}$ matrices which are of first order in the errors in the target functions. It is clearly important to use the most accurate target functions that can conveniently be handled.

4. Target states

In practice we use some approximate set of target states which are not solutions of (2.1). We require these states to be such that

$$(\chi_i|\chi_{i'}) = \delta_{ii'}, \qquad \text{and } (\chi_i|H_1|\chi_{i'}) = E_i\delta_{ii'} \tag{4.1}$$

which defines E_i. With an infinite set of states, (4.1) is equivalent to (2.1) but in practice we use a finite set.

The states χ_i are taken to be linear combinations of Slater states, that is to say anti-symmetrized products of one-electron orbitals. They are set up as follows:

(i) We define a number $NRAD$ of bound-state radial functions,

$$P_\gamma(r);\ \gamma = 1, NRAD. \tag{4.2}$$

For each radial function we have quantum numbers $n_\gamma l_\gamma$ defined in the usual way. We require that the functions P_γ satisfy the relations

$$(P_\gamma|P_\gamma) = 1, \qquad (P_\gamma|P_{\gamma'}) = 0 \qquad \text{if } l_\gamma = l_{\gamma'} \text{ but } n_\gamma \neq n_{\gamma'}. \tag{4.3}$$

(ii) We define a set of configurations, which give rise to all of the terms of interest and which contain the radial functions P_γ; $\gamma = 1, NRAD$. The configurations may be listed as

$$C(q_1, q_2, \ldots, q_{NRAD}) = \prod_{\gamma=1}^{NRAD} (n_\gamma l_\gamma)^{q_\gamma} \tag{4.4}$$

where $\Sigma_\gamma q_\gamma = N$, the number of electrons in the target system. For each configuration C we set up Slater states and states $\chi(C\beta SLM_SM_L)$ as linear combinations of Slater states. The quantum number β is included to allow for the possibility that, for a given $CSLM_SM_L$, there may be more than one linearly independent state. We may now diagonalize the matrix

$$(C\beta SL|H_1|C'\beta' SL) \tag{4.5}$$

to obtain states $\chi(\alpha SL)$ which satisfy (4.1). In the collision problem we do not necessarily use all of the states obtained from the diagonalization of (4.5); we must therefore specify the number of states to be included.

In order to illustrate some of the problems which arise we consider two examples:

Example (i). Suppose that we wish to solve the collision problem with inclusion of the two lowest ^{1}S states for an ion in the He sequence. We could calculate radial functions for 1s^2 and 1s2s ^{1}S in the Hartree–Fock approximation. We obtain different 1s functions for 1s^2 and 1s2s. Furthermore, the 1s and 2s functions in 1s2s ^{1}S are not orthogonal (Seaton 1953). From the three linearly independent radial functions obtained in the Hartree–Fock approximation we can construct three orthogonal radial functions. These could be labelled 1s, 2s and 3s. Using these functions we have six possible configurations, 1s^2, 1s2s, 1s3s, 2s^2, 2s3s and 3s^2. We may now diagonalize the matrix of H_1 using these six configurations and select the two states of lowest energy for use in the collision problem. It should be noted that, using the original Hartree–Fock states, we do not obtain (1s^2 ^{1}S$|H_1|$1s2s ^{1}S) = 0. The states obtained using six configurations and diagonalizing the matrix of H_1 should be more accurate than the original Hartree–Fock states.

Example (ii). For systems with a larger number of electrons we generally use radial functions calculated using a scaled statistical model potential (Eissner and Nussbaumer

1969). For the radial function P_γ we use a potential $v_{l_\gamma}(r)$ which does not depend on n. The orthogonality relation in (4.1) is then satisfied automatically.

It should be emphasized that our target states differ from those frequently used in atomic structure work in that we use the same set of radial functions P_γ for all target states to be included in the collision problem. This greatly simplifies the algebraic formulation of the problem.

5. States for the collision problem

Using the states χ_i, as defined in § 4, we may set up the states Θ_i defined by (2.5). We refer to the states Θ_i as *free channels*, since they contain radial functions F_i which may be freely varied so as to satisfy (3.12). Let the number of free channels be $NCHF$. The wavefunction for the collision problem is then

$$\Psi = \sum_{i=1}^{NCHF} \Theta_i. \tag{5.1}$$

It should be noted that a free channel may be an open channel ($k_i^2 \geqslant 0$) or a closed channel ($k_i^2 < 0$).

The form of the expansion, and of the equations satisfied by the radial functions, are independent of the boundary conditions to be imposed. We therefore omit subscripts which define the boundary conditions. We now impose an orthogonality condition on the radial functions F_i,

$$(P_\gamma|F_i) = 0 \qquad \text{if } l_\gamma = l_i. \tag{5.2}$$

Imposition of this condition will not alter the function Ψ so long as we add to (5.1) a suitable linear combination of functions Φ_j which have the form of bound-state functions for the $(N+1)$ electron problem. We refer to the functions Φ_j as *bound channels*, and denote the number of such channels by $NCHB$. In place of (5.1) we then have

$$\Psi = \sum_{i=1}^{NCHF} \Theta_i + \sum_{j=1}^{NCHB} \Phi_j c_j. \tag{5.3}$$

The set of states Φ_j is determined as follows. Let the target state χ_i, as defined in § 4, contain a state $\chi(C\beta S_i L_i)$ and suppose that we impose the orthogonality condition $(P_\gamma|F_i) = 0$ for $l_\gamma = l_i$. We must then include in (5.3) a state

$$\Phi(C\beta S_i L_i n_\gamma l_\gamma SL). \tag{5.4}$$

The states (5.4) are taken to be fully anti-symmetric. Some of these states vanish when anti-symmetrized; in this case they are omitted in calculating $NCHB$. The functions Φ_j are taken to be such that

$$(\Phi_j|\Phi_{j'}) = \delta_{j,j'}. \tag{5.5}$$

There are several reasons for imposing the condition (5.2): (i) It greatly simplifies the algebraic formulation. (ii) When this condition is not imposed the functions F_i are not, in general, uniquely defined (Seaton 1953). This can lead to inaccuracies in numerical integrations (Norcross 1969). (iii) In the DW approximation the coefficients c_j in (5.3) can be treated as variational parameters. This improves the accuracy of the approximation. (iv) In the ID approximation we can include the minimum set of functions Φ_j

required when the orthogonality conditions (5.2) are imposed, plus some additional functions Φ_j which may improve the accuracy of the calculations. These additional functions are similar to the correlation functions as used, for example, by Burke and Taylor (1966).

A further point may be noted. In certain circumstances some of the orthogonality conditions (5.2) may be omitted, without this leading to difficulties in algebraic formulation or to lack of uniqueness in the radial functions F_i. This point will be discussed further by Eissner (1973). Let us define an array **ORT** with elements $ORT(\gamma, i) = 0$ or 1. This is such that $ORT(\gamma, i) = 0$ if $l_\gamma \neq l_i$. The orthogonality condition $(P_\gamma|F_i) = 0$ is imposed only when $ORT(\gamma, i) = 1$. In compact matrix notation the orthogonality conditions may be written

$$(\mathbf{P}|\mathbf{ORT}|\mathbf{F}) = 0. \tag{5.6}$$

6. Reduction of integrals

Using (5.3) we obtain

$$(\Psi|H-E|\Psi) = \sum_{ii'}(\Theta_i|H-E|\Theta_{i'}) + \sum_{ij'}(\Theta_i|H-E|\Phi_{j'})c_{j'} + \sum_{ji'}c_j^*(\Phi_j|H-E|\Theta_{i'}) + \sum_{jj'}c_j^*(\mathcal{H}_{jj'}-E)c_{j'} \tag{6.1}$$

where

$$\mathcal{H}_{jj'} = (\Phi_j|H|\Phi_{j'}). \tag{6.2}$$

The reduction of the integrals in (6.1) is discussed in the Appendix to the present paper for the case of electron–hydrogen collisions, and by Eissner (1973) for the general case of electron collisions with many-electron atoms. The following results are obtained:

$$(\Psi|H-E|\Psi) = \sum_{ii'}(F_i|(h_i-k_i^2)\delta_{ii'} + W_{ii'}|F_{i'}) + \sum_{ij'}(F_i|U_{ij'})c_{j'} + \sum_{ji'}c_j^*(U_{i'j}|F_{i'}) + \sum_{jj'}c_j^*(\mathcal{H}_{jj'}-E)c_{j'} \tag{6.3}$$

where

$$h_i = -\frac{\mathrm{d}^2}{\mathrm{d}r^2} + \frac{l_i(l_i+1)}{r^2} - \frac{2Z}{r}, \tag{6.4}$$

$$W_{ii'}F_{i'} = 2\sum_\nu \{f_{ii'}^{(\nu)}y_\lambda(P_\gamma P_{\gamma'})F_{i'} - g_{ii'}^{(\nu)}y_\lambda(P_\gamma F_{i'})P_{\gamma'}\}, \tag{6.5}$$

$$U_{ij} = \sum_\nu \{a_{ij}^{(\nu)}h_i P_\gamma \delta(l_i, l_\gamma) + 2b_{ij}^{(\nu)}y_\lambda(P_\gamma P_{\gamma''})P_{\gamma'}\}; \tag{6.6}$$

$f_{ii'}^{(\nu)}$, $g_{ii'}^{(\nu)}$, $a_{ij}^{(\nu)}$ and $b_{ij}^{(\nu)}$ are algebraic coefficients and

$$y_\lambda(AB|r) = r^{-\lambda-1}\int_0^r A(x)B(x)x^\lambda\,\mathrm{d}x + r^\lambda \int_r^\infty A(x)B(x)x^{-\lambda-1}\,\mathrm{d}x. \tag{6.7}$$

The sums over ν in (6.5), (6.6) contain some finite (usually small) number of terms. It is to be understood that λ, γ and γ' in (6.5) depend on i, i' and ν; and that λ, γ, γ' and γ'' in (6.6) depend on i, j and ν.

The equation (6.3) can be written in a compact matrix notation,

$$(\Psi|H-E|\Psi) = (\mathbf{F}|h-k^2+\mathbf{W}|\mathbf{F})+(\mathbf{F}|\mathbf{U})\mathbf{c}+\mathbf{c}^{\dagger}(\mathbf{U}|\mathbf{F})+\mathbf{c}^{\dagger}(\mathscr{H}-E)\mathbf{c}. \qquad (6.8)$$

We may now define our two approximations more precisely.

(i) *The* DW *approximation.* We choose some suitable central potentials V_i and calculate functions $\mathscr{F}_i$ which are solutions of

$$(h-k_i^2+V_i)\mathscr{F}_i = 0 \qquad (6.9)$$

and are such that

$$F_i(0) = 0 \qquad \mathscr{F}_i \sim k_i^{-1/2}\sin\xi_i. \qquad (6.10)$$

We next define F_i by

$$F_i = \mathscr{F}_i - \sum_{\gamma}\delta_{l_i l_\gamma}(P_\gamma|F_i)P_\gamma. \qquad (6.11)$$

These functions satisfy the condition (5.2). We require that (6.8) should be stationary for variations of the coefficients c_j: we obtain

$$(\mathbf{U}|\mathbf{F})+(\mathscr{H}-E)\mathbf{c} = 0. \qquad (6.12)$$

Substitution of (6.12) in (6.8) gives

$$(\Psi|H-E|\Psi) = (\mathbf{F}|h-k^2+\mathbf{W}-\mathbf{U}^{\dagger}(\mathscr{H}-E)^{-1}\mathbf{U}|\mathbf{F}). \qquad (6.13)$$

From (3.10) the final result for the **R** matrix in the DW approximation is

$$R_{ii'}^{\mathrm{DW}} = -(F_i|(h_i-k_i^2)\delta_{ii'}+\{\mathbf{W}-\mathbf{U}^{\dagger}(\mathscr{H}-E)^{-1}\mathbf{U}\}_{ii'}|F_{i'}). \qquad (6.14)$$

(ii) *The* ID *approximation.* We require that (3.12) should be satisfied for variations of the functions **F**, subject to the constraint (5.6). We therefore introduce Lagrange multipliers $\lambda_{i\gamma}$, which are such that $\lambda_{i\gamma} = 0$ if $ORT(\gamma, i) = 0$. We obtain

$$(h-k^2+\mathbf{W})\mathbf{F}+\mathbf{U}\mathbf{c}+\boldsymbol{\lambda}\mathbf{P} = 0. \qquad (6.15)$$

The equations to be satisfied in the ID approximation are (5.6), (6.12) and (6.15). It may be noted that **c** may be eliminated between (6.12) and (6.15), to give

$$(h-k^2+\mathbf{W})\mathbf{F}-\mathbf{U}(\mathscr{H}-E)^{-1}(\mathbf{U}|\mathbf{F})+\boldsymbol{\lambda}\mathbf{P} = 0. \qquad (6.16)$$

This, however, can lead to difficulties, since $(\mathscr{H}-E)$ can be singular for certain values of E.

Acknowledgments

This work has been supported by the Science Research Council. Part of the work has been carried out during a visit by M J Seaton to the Joint Institute for Laboratory Astrophysics, University of Colorado and National Bureau of Standards.

Work done at JILA was supported by the Advanced Research Projects Agency, The Department of Defense and was monitored by US Army Research Office—Durham, Box CM, Duke Station, Durham, North Carolina 27706, under Contract No DA-31-124-ARO-D-139.

Appendix. Electron–hydrogen collisions

Percival and Seaton (1957) have discussed the integro-differential equations for e–H collisions, without imposition of orthogonality constraints. They define algebraic coefficients

$$f_\lambda(l_1 l_2, l'_1 l'_2 ; L) = (l_1 l_2 L | P_\lambda(\hat{\boldsymbol{r}}_1 \cdot \hat{\boldsymbol{r}}_2) | l'_1 l'_2 L) \tag{A1}$$

and

$$g_\lambda(l_1 l_2, l'_1 l'_2 ; L) = (-1)^{l_1+l_2-L} f_\lambda(l_1 l_2, l'_2 l'_1 ; L) \tag{A2}$$

and give an expression for f_λ in terms of Racah coefficients.

We consider the equations for e–H collisions, with imposition of orthogonality constraints.

(i) *Notation.* Percival and Seaton use nl_1 for the atomic electron and l_2 for the colliding electron. We use $n_i L_i$ for the atomic electron and l_i for the colliding electron. We put

$$i = (n_i L_i l_i SL) \qquad \text{where } i = 1, NCHF. \tag{A3}$$

(ii) *The functions* Θ_i. We form vector-coupled functions

$$\theta_i(1, 2) = \theta(n_i L_i l_i SL | 1, 2) \tag{A4}$$

and anti-symmetric functions

$$\Theta_i(1, 2) = \frac{1}{\sqrt{2}}\{\theta_i(1, 2) - \theta_i(2, 1)\}. \tag{A5}$$

When orthogonality constraints are not imposed we have

$$\Psi = \sum_{i=1}^{NCHF} \Theta_i. \tag{A6}$$

(iii) *Orthogonality constraints.* Let i and i' be two states included in (A3), where we do not exclude the case of $i = i'$. If $l_i = L_{i'}$ we impose the orthogonality constraint

$$(P_{\gamma'} | F_i) = 0 \text{ where } \gamma' = (n_{i'} L_{i'}) \text{ and } L_{i'} = l_i. \tag{A7}$$

(iv) *The functions* Φ_j. When the constraints (A7) are imposed, the expansion (A6) is replaced by

$$\Psi = \sum_{i=1}^{NCHF} \Theta_i + \sum_{j=1}^{NCHF} \Phi_j c_j \tag{A8}$$

where

$$j = (n_j L_j n'_j L'_j SL). \tag{A9}$$

In (A8) we include all antisymmetric functions Φ_j which can be constructed using the set of bound state functions, $n_i L_i$, included in (A3). The functions Φ_j are constructed as follows. We first form vector-coupled functions

$$\phi_j(1, 2) = \phi(n_j L_j n'_j L'_j SL | 1, 2). \tag{A10}$$

We then have two cases. (a) If $(n_j L_j) \neq (n'_j L'_j)$ we form the normalized antisymmetric functions

$$\Phi_j(1, 2) = \frac{1}{\sqrt{2}}\{\phi_j(1, 2) - \phi_j(2, 1)\} \qquad \text{for } (n_j L_j) \neq (n'_j L'_j). \tag{A11}$$

(b) If $(n_jL_j) = (n'_jL'_j)$ the functions ϕ_j are symmetric if $(S+L)$ is odd and anti-symmetric if $(S+L)$ is even. For the case of $(S + L)$ even the normalized anti-symmetric functions are

$$\Phi_j = \phi_j \text{ for } (n_jL_j) = (n'_jL'_j) \text{ and } (S+L) \text{ even.} \tag{A12}$$

(v) *The operator* **W**. We obtain

$$W_{ii'}F_{i'} = 2\sum_{\lambda} \{f_\lambda(L_il_i, L_{i'}l_{i'}; L)y_\lambda(P_\gamma P_{\gamma'})F_{i'} - g_\lambda(L_il_i, L_{i'}l_{i'}; SL)y_\lambda(P_\gamma F_{i'})P_{\gamma'}\} \tag{A13}$$

where $\gamma = (n_iL_i)$, $\gamma' = (n_{i'}L_{i'})$ and

$$g_\lambda(l_1l_2, l'_1l'_2; SL) = (-1)^{1-S}g_\lambda(l_1l_2, l'_1l'_2; L). \tag{A14}$$

It may be noted that

$$f_0(l_1l_2, l'_1l'_2; L) = \delta_{l_1l'_1}\delta_{l_2l'_2}$$
$$g_0(l_1l_2, l'_1l'_2; SL) = (-1)^{l_1+l_2-L+1-S}\delta_{l_1l'_2}\delta_{l_2l'_1}. \tag{A15}$$

(vi) *The operator* **U**. We have two cases.

(a) For $(n_jL_j) \neq (n'_jL'_j)$,

$$U_{ij} = \{\delta_{n_in_j}f_0(L_il_i, L_jL'_j; L)h_i + 2\sum_\lambda f_\lambda(L_il_i, L_jL'_j; L)y_\lambda(P_\gamma P_{\gamma'})\}P_{\gamma''}$$
$$-\{\delta_{n_in'_j}g_0(L_il_i, L_jL'_j; SL)h_i + 2\sum_\lambda g_\lambda(L_il_i, L_jL'_j; SL)y_\lambda(P_\gamma P_{\gamma''})\}P_{\gamma'} \tag{A16}$$

where $\gamma = (n_iL_i)$, $\gamma' = (n_jL_j)$, $\gamma'' = (n'_jL'_j)$.

(b) For $(n_jL_j) = (n'_jL'_j)$,

$$U_{ij} = \sqrt{2}\{\delta_{n_in_j}f_0(L_il_i, L_jL_j; L)h_i + 2\sum_\lambda f_\lambda(L_il_i, L_jL_j; L)y_\lambda(P_\gamma P_{\gamma'})\}P_{\gamma'} \tag{A17}$$

where $\gamma = (n_iL_i)$, $\gamma' = (n_jL_j)$.

References

Burke P G and Smith K 1962 *Rev. mod. Phys.* **34** 458–502

Burke P G and Taylor A J 1966 *Proc. Phys. Soc.* **88** 549–62

Eissner W 1973 *J. Phys.* B: *Atom. molec. Phys.* to be submitted

Eissner W and Nussbaumer H 1969 *J. Phys.* B: *Atom. molec. Phys.* **2** 1028–43

Norcross D W 1969 *J. Phys.* B: *Atom. molec. Phys.* **2** 1300–3; and corrigendum 1971 *J. Phys.* B: *Atom. molec. Phys.* **4** 628

Percival I C and Seaton M J 1957 *Proc. Camb. Phil. Soc.* **53** 654

Seaton M J 1953 *Phil. Trans. R. Soc.* A **245** 469–99

—— 1973 *J. Phys.* B: *Atom. molec. Phys.* to be submitted

1983 *Phys. Rev.* A **28** 2209–16
Reprinted with permission from the American Physical Society

Atomic photoionization in a strong magnetic field

Chris H. Greene
Department of Physics and Astronomy, Louisiana State University, Baton Rouge, Louisiana 70803
(Received 22 April 1983)

The photoionization of hydrogen atoms in a strong magnetic field is formulated as a multichannel problem by representing the asymptotic electron wave function in cylindrical coordinates. Departures from cylindrical symmetry close to the nucleus are incorporated by an R-matrix treatment at short range, which then merges with standard quantum-defect procedures. The R-matrix calculation utilizes the eigenchannel approach, recast in noniterative form. At the field strength treated here, $B=4.7\times10^9$ G, the photoionization cross section displays narrow "autoionizing" resonances near the excited Landau thresholds.

I. INTRODUCTION

Magnetic white dwarfs exhibit field strengths approaching 10^9 G, and so the interpretation of white-dwarf spectra requires a detailed understanding of atomic properties in such fields.[1] Also, the quantum-mechanical motion of an electron in combined Coulomb and magnetic fields constitutes one of the simplest nonseparable problems in atomic physics; its properties have fundamental implications for the physics of all correlated systems.[2,3] While for certain atomic properties the Hamiltonian appears to be quasiseparable,[4,5] for other properties the nonseparability becomes paramount, e.g., the decay of quasi-Landau levels observed at smaller fields,[6] or the "autoionization" of states lying below excited Landau thresholds.[7,8]

In the discrete spectrum much attention has been given to the problem of calculating energy levels and oscillator strengths (see, e.g., Ref. 9). The most complete and accurate results, obtained over a wide range of field strengths, appear to be those of Wunner and Ruder.[10] Photoionization of hydrogen in strong fields was treated by Kara and McDowell,[11] though many aspects of their calculation are unrealistic because of an inadequate treatment of Coulomb field effects at the Landau thresholds. At much higher fields ($B\sim10^{12}$ G) than considered here, a realistic calculation of the photoionization cross section was obtained by Wunner *et al.*,[12] though the "autoionizing" resonances were not included. The energies and widths of several autoionizing states were calculated by Friedrich and Chu,[7] and scattering amplitudes were obtained by Onda,[8] though the photoionization cross section was not given.

The central purpose of this paper is the presentation of a multichannel R-matrix calculation of the photoionization cross section for hydrogen. For this prototype calculation only a single value of the magnetic field (4.7×10^9 G) and of the photon polarization ($\hat{\epsilon}\,||\,\vec{B}$) will be considered. The escape of the photoelectron to $z\to\pm\infty$ is conveniently represented here in terms of regular and irregular Coulomb wave functions, whereby the standard results of quantum-defect theory[13–17] suffice to translate the short-range R-matrix calculation into a photoionization cross section. Clark[18] has recently developed a closely related approach to treat negative-ion photodetachment in a magnetic field.

A second objective of this study is to test the feasibility of a noniterative reformulation of the eigenchannel R-matrix approach,[19] which amounts to a generalization of the variational expression derived by Kohn[20] for the logarithmic derivative of the wave function at a finite radius. This version of R-matrix theory is closely related to that described by Taylor,[21] Lane and Robson,[22] Purcell and Chatwin,[23] and Oberoi and Nesbet.[24] A new application of the same approach to molecular dynamics is described by Rouzo and Raseev.[25] The results of the calculation, given in Sec. III, show how the short-range reaction matrix varies strongly as the energy is increased from far below a threshold up to the near-threshold region. The reaction matrix, nonetheless, becomes a smooth function of the energy once E is within a few electron volts of an opening threshold. These results complement an earlier study of this energy dependence in a much different context.[26]

II. NONITERATIVE CALCULATION OF R-MATRIX EIGENSTATES

In the interest of keeping this paper self-contained, the noniterative eigenchannel R-matrix method will be derived in this section. This method retains the rapid convergence of the eigenchannel approach[19,27] while eliminating its main disadvantage, namely, the necessity for iteratively diagonalizing the Hamiltonian several times at each energy. It is closely related to the noniterative approach developed by Refs. 21–24, with the main difference being the present emphasis on R-matrix *eigenstates* rather than on the full R matrix. The motivation for focusing on these eigenstates stems from their having a simple physical interpretation in many problems.[14,28]

The common thread in all R-matrix methods[29] is their solution of the Schrödinger equation within a finite reaction volume Ω of configuration space. The scattering properties of a many-particle system are known once the normal logarithmic derivative $(\partial\psi/\partial n)\psi^{-1}$ is specified on the surface Σ enclosing the reaction volume. The goal of theory is to determine this information in the form of an R matrix.

Consider the Ritz variational expression for the Schrödinger energy eigenvalue

$$E=\frac{\int_\Omega \psi^*(-\frac{1}{2}\nabla^2\psi+V\psi)d\omega}{\int_\Omega \psi^*\psi d\omega}, \quad (1)$$

where $d\omega$ is the differential volume element of configuration space and the integrals extend only over the reaction volume Ω. (In a many-particle system, ∇^2 must be interpreted as $\sum_i \nabla_i^2/m_i$.) Application of Green's theorem transcribes Eq. (1) into

$$E=\frac{\int_\Omega(\frac{1}{2}\vec{\nabla}\psi^*\cdot\vec{\nabla}\psi+\psi^*V\psi)d\omega-\frac{1}{2}\int_\Sigma \psi^*(\partial\psi/\partial n)d\sigma}{\int_\Omega \psi^*\psi d\omega}, \quad (2)$$

in which an additional integral must now be evaluated over the surface Σ of the reaction volume. The differential area element is $d\sigma$, and the normal derivative $\partial\psi/\partial n$ on Σ can be expressed in terms of ψ and the constant b through

$$\frac{\partial\psi}{\partial n}+b\psi=0, \quad (3)$$

on Σ. [In a multichannel problem having several degenerate continuum solutions, Eqs. (2) and (3) are meant to hold for each independent solution ψ_β with its associated constant b_β. It is worth emphasizing that b in Eq. (3) is not necessarily a constant for any arbitrary state ψ. Yet we can always look for the set of R-matrix eigenstates ψ_β for which b_β *is* a constant on Σ, in which case b_β is interpreted as an eigenvalue of the R-matrix.] Consequently, the expression (2) can be written in a form more useful for continuum states at a given energy E, as an equation for the unknown $b(E)$:

$$b=\frac{\int_\Omega[-\vec{\nabla}\psi^*\cdot\vec{\nabla}\psi+2\psi^*(E-V)\psi]d\omega}{\int_\Sigma \psi^*\psi d\sigma}. \quad (4)$$

This expression for b is clearly real and, moreover, the first variation δb vanishes to first order in small deviations $\delta\psi$ of the wave function from the exact solution ψ. This stationary property follows upon evaluating δb using Eq. (4) and then applying Green's theorem. Importantly, *no constraint* needs to be imposed on the trial functions in showing that $\delta b=0$. In particular, the trial functions need not have any specified logarithmic derivative on the reaction surface Σ. In this sense Eq. (4) is a less restrictive variational expression than Eq. (1), since the *energy* functional is stationary only when all trial functions have the same normal logarithmic derivative on Σ as the exact solution.

A. Generalized eigenvalue problem for the R-matrix eigenstates

A particularly convenient method for using Eq. (4) to calculate the R-matrix follows from using trial functions which are a linear combination of basis functions y_k:

$$\psi=\sum_k c_k y_k. \quad (5)$$

In the following, the c_k, y_k, and ψ will be assumed real without loss of generality, as the Hamiltonian is real. It should be pointed out that the y_k need not be orthogonal, and also they need not have any particular logarithmic derivative on Σ; indeed their normal logarithmic derivatives should ideally span a range of values. Equation (4) then takes the form

$$b[c_k]=\sum_{k,l} c_k\Gamma_{kl}c_l \Big/ \sum_{k',l'} c_{k'}\Lambda_{k'l'}c_{l'}. \quad (6)$$

The real symmetric matrices $\underline{\Gamma}$ and $\underline{\Lambda}$ are defined by

$$\Gamma_{kl}=\int_\Omega[-\vec{\nabla}y_k\cdot\vec{\nabla}y_l+2y_k(E-V)y_l]d\omega, \quad (7)$$

$$\Lambda_{kl}=\int_\Sigma y_k y_l d\sigma. \quad (8)$$

The element Γ_{kl} can also be put in the more conventional form involving the matrix element of a non-Hermitian Hamiltonian and a surface term

$$\Gamma_{kl}=2\int_\Omega y_k(E-H)y_l d\omega-\int_\Sigma y_k\frac{\partial y_l}{\partial n}d\sigma. \quad (9)$$

A necessary condition for b to be stationary with respect to small variations of the c_k is that

$$\frac{\partial b}{\partial c_k}=0=\left[\sum_{k',l'} c_{k'}\Lambda_{k'l'}c_{l'}\right]^{-1}\left[\sum_l \Gamma_{kl}c_l-b\sum_l \Lambda_{kl}c_l\right]. \quad (10)$$

In matrix notation this amounts to the generalized eigenvalue equation

$$\underline{\Gamma}\vec{c}=b\underline{\Lambda}\vec{c}. \quad (11)$$

Properties of the eigensystem (11) are discussed in Ref. 30, and an algorithm for numerical solution which appears to be stable and efficient is presented in Refs. 31 and 32. Eigenvectors $\vec{c}_\beta$ and $\vec{c}_{\beta'}$ corresponding to distinct eigenvalues b_β and $b_{\beta'}$ are orthogonal over the reaction surface Σ:

$$\vec{c}_\beta\underline{\Lambda}\vec{c}_{\beta'}=0, \quad (12)$$

when $b_\beta\neq b_{\beta'}$. The number of nontrivial solutions to Eq. (11) depends on the problem being studied. On physical grounds the number of eigensolutions must equal the number of open and weakly closed channels, which amounts to those reaction channels having non-negligible amplitude on Σ. The remaining channels, which may contribute at short range but are exponentially small on Σ, are referred to as strongly closed.

The concept of reaction channels arises here by introducing a complete set of real orthonormal surface harmonics ϕ_n which span the surface Σ:

$$\int_\Sigma \phi_n\phi_{n'}d\sigma=\delta_{nn'}. \quad (13)$$

Each of the basis functions y_k can be expanded in terms of the ϕ_n over Σ according to

$$y_k=\sum_n a_{kn}\phi_n, \quad (14)$$

on Σ, and the matrix elements of $\underline{\Lambda}$ are then simply

$$\Lambda_{kl}=\sum_{n} a_{kn}a_{ln} \ . \tag{15}$$

If Γ and Λ are $n \times n$ matrices and Λ is nonsingular, then there are n nontrivial solutions to Eq. (11). In practice, however, Λ is highly singular because only a few terms (i.e., *channels)* contribute to the summation in (15). In the extreme case of a single-channel problem, the matrix Λ is factorable and thus has but one nonzero eigenvalue, whereby the system (11) has exactly one nontrivial solution at each energy, as expected. Similarly, if a total of N channels are required to span the reaction surface Σ at any given energy, then the number of relevant eigensolutions is also N. Besides the eigenvalues b_β, the R-matrix eigenvectors are also required to construct the matrix $\underline{R}$. In the present formulation these R-matrix eigenvectors are (within normalization) the projection of the β'th eigensolution

$$\psi_\beta=\sum_{k} y_k c_{k\beta} \tag{16}$$

onto the n'th surface harmonic

$$\begin{aligned} Z_{n\beta}&=\int_\Sigma \phi_n \psi_\beta d\sigma / N_\beta \\ &=\sum a_{kn} c_{k\beta}/N_\beta \ , \end{aligned} \tag{17}$$

with the normalization constant given by

$$N_\beta^2=\sum_{n} \left[\int_\Sigma \phi_n \psi_\beta d\sigma \right]^2 . \tag{18}$$

Finally, the R matrix is given by

$$R_{nn'}=\sum_{\beta} Z_{n\beta} b_\beta Z_{n'\beta} \ , \tag{19}$$

which is automatically symmetric.

Whereas testing the symmetry of the R matrix provides a useful check on convergence for some methods of calculation, it is of no value in the present approach since $\underline{R}$ in (19) is automatically symmetric. One useful test of convergence is the comparison of b_β obtained from (11) with the value obtained from the less-accurate expression

$$b_{\beta,\text{approx}}=-\frac{\int_\Sigma \psi_\beta (\partial\psi_\beta/\partial n) d\sigma}{\int_\Sigma \psi_\beta^2 d\sigma} \ , \tag{20}$$

calculated using the variational wave function ψ_β. Since Eq. (20) differs from the exact b_β in first order ($\delta\psi$), while Eq. (11) differs in second order ($\delta\psi^2$), their difference measures the convergence of the expansion (5).

B. Photoionization in a magnetic field

Next, the specific application of R-matrix theory to treat the photoionization of hydrogen in a strong field is addressed. In cylindrical coordinates the Hamiltonian is (in a.u.)

$$H=-\frac{1}{2}\frac{\partial^2}{\partial\rho^2}-\frac{1}{2\rho}\frac{\partial}{\partial\rho}-\frac{1}{2}\frac{\partial^2}{\partial z^2}+\frac{L_z^2}{2\rho^2}+\alpha L_z+\tfrac{1}{2}\alpha^2\rho^2 -(\rho^2+z^2)^{-1/2} \ , \tag{21}$$

where $\alpha \equiv eB/2m_e c$ is the Larmor frequency. As is well known, this Hamiltonian is approximately separable in spherical coordinates at short distances ($r \lesssim r_c$), but is nearly separable in cylindrical coordinates farther out ($r \gtrsim r_c$). If the electron kinetic energy is roughly T, then this critical radius lies in the neighborhood of $r_c \approx (2T)^{1/2}\alpha^{-1}$ bohr radii. The photoelectron can escape beyond this distance only by moving parallel to the magnetic field, and its kinetic energy of escape depends on the azimuthal quantum number m and on the asymptotic Landau quantum number n characterizing the motion in ρ. To account properly for this channel structure the asymptotic wave function in a photoionization calculation *must* be expressed in cylindrical coordinates even for relatively weak fields. Notice also that for large $|z| \gg r_c$, the potential energy assumes the form $-1/|z|$. Hence, each Landau component of ψ can be written as a linear combination of regular and irregular Coulomb functions (f,g) in z. This feature permits the use of multichannel quantum-defect theory[15,17] to express the photoionization cross section in terms of a short-range reaction matrix and dipole matrix elements, all of which vary smoothly with energy.

In the terminology of Sec. II A, then, the reaction surface Σ is cylindrical, with the ends of the cylinder taken to be

$$z=\pm z_0 \ . \tag{22}$$

The actual value of z_0 will depend on the energy range and on the magnetic field in general. The radius of the cylinder ρ_0 need not be specified here since the electron can escape only a finite distance in ρ at any given energy. Thus, ρ_0 will simply be imagined large enough to contain all possible excitations in ρ. The natural choice for the surface harmonics in this problem are the wave functions associated with the Landau levels of a free electron in a magnetic field:

$$\phi_n^{(m)}(\rho,\phi) \equiv N_{nm} e^{im\phi} e^{-\alpha\rho^2/2} (\alpha\rho^2)^{|m|/2} L_n^{(|m|)}(\alpha\rho^2) \ , \tag{23}$$

with an associated residual channel-energy level

$$E_n^{(m)}=\alpha(2n+|m|+m+1) \ . \tag{24}$$

Here N_{nm} is a normalization constant and $L_n^{(|m|)}$ is an associated Laguerre polynomial.

As quantum-defect procedures have been described elsewhere, relevant details will only be summarized here. The key point is that solutions to the Schrödinger equation at $|z| \geq z_0$ can be represented in terms of regular and irregular Coulomb wave functions f_n and g_n. In the present problem f_n and g_n are the $l=0$ energy-normalized base pair of Refs. 15–17 evaluated at the appropriate channel energy

$$\epsilon_n^{(m)}=E-E_n^{(m)} \ , \tag{25}$$

which can be positive or negative. The explicit form of the β'th independent solution of Eq. (11) can thus be written in terms of constants $I_{n\beta}$ and $J_{n\beta}$ as

$$\psi_\beta=\sum_{n} \phi_n^{(m)}(\rho,\phi)[f_n(z)I_{n\beta}-g_n(z)J_{n\beta}], \quad z \geq z_0 \ . \tag{26}$$

As each ψ_β is either even or odd about $z=0$, it suffices to consider the region $z \geq z_0$ only.[33] The coefficients $I_{n\beta}$ and $J_{n\beta}$ are found by projecting each ψ_β onto $\psi_n^{(m)}$, as well as its normal derivative

$$\frac{\partial\psi_\beta}{\partial z} = -b_\beta\psi_\beta = \sum_n \phi_n^{(m)}(\rho,\phi)[f_n'(z)I_{n\beta} - g_n'(z)J_{n\beta}], \quad z \geq z_0 \tag{27}$$

and solving the resulting linear system after setting $z=z_0$. The coefficients $I_{n\beta}$ and $J_{n\beta}$ define the reaction matrix $K_{nn'}$ through

$$\underline{K} = \underline{J}\,\underline{I}^{-1}, \tag{28}$$

though it is useful in practice to deal separately with the reaction matrix eigenvalues $\tan\pi\mu_\alpha$ and eigenvectors $U_{n\alpha}$ by solving

$$\underline{K}\,\underline{U} = \underline{U}\tan(\pi\underline{\mu}). \tag{29}$$

The eigenstates of the short-range reaction matrix will be denoted Ψ_α; these are linear combinations of the R-matrix eigenstates ψ_β:

$$\Psi_\alpha = \sum_{\beta n}\psi_\beta(\underline{I}^{-1})_{\beta n}U_{n\alpha}\cos(\pi\mu_\alpha). \tag{30}$$

The Ψ_α have the standard asymptotic form

$$\Psi_\alpha = \sum_n \phi_n^{(m)}(\rho,\phi)[f_n(z)U_{n\alpha}\cos(\pi\mu_\alpha) - g_n(z)U_{n\alpha}\sin(\pi\mu_\alpha)], \quad z \geq z_0. \tag{31}$$

Photoabsorption depends lastly on the dipole matrix elements between[33] Ψ_α and Ψ_0, the ground-state wave function

$$d_\alpha = \int_0^{2\pi} d\phi \int_0^\infty \rho\, d\rho \int_0^\infty dz\, \Psi_\alpha^*(\rho,\phi,z)\hat{\epsilon}\cdot\vec{r}\,\Psi_0(\rho,\phi,z). \tag{32}$$

The d_α, μ_α, and $U_{n\alpha}$ determine the total and partial photoionization cross sections, as detailed in Ref. 17. These parameters are smooth functions of energy (as seen in Sec. III), even through ionization thresholds. This fact allows them to be calculated on a coarse mesh of energies, even though the cross section itself changes rapidly with energy.

III. PROTOTYPE CALCULATION

The R-matrix procedure discussed above has been implemented at one value of the magnetic field, $B=4.7\times10^9$ G, corresponding to $\alpha=1$ in Eq. (21). It has also been limited to incident photons which are linearly polarized along the field axis $\hat{z}$, thereby selecting only the $m=0$, odd-parity final state. The main elements required for a realistic calculation can be discerned from Fig. 1, which shows the diagonal potential matrix elements for $m=0$ and $\alpha=1$:

$$V_{nn}(z) = E_n^{(0)} - \int_0^\infty \rho\, d\rho \int_0^{2\pi} d\phi\, |\phi_n^{(0)}(\rho,\phi)|^2 \times(\rho^2+z^2)^{-1/2}, \tag{33}$$

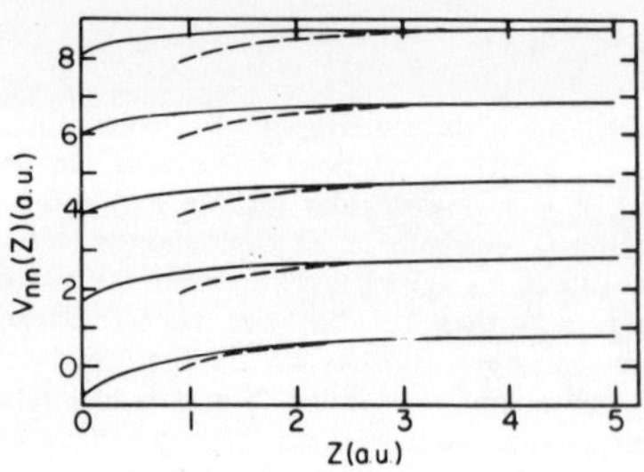

FIG. 1. Diagonal potential matrix elements which are given by Eq. (33), converging to the lowest five Landau levels as $|z|\to\infty$. These are the potentials relevant for $m=0$, at a magnetic field 4.7×10^9 G. Also shown as dashed curves are the purely Coulombic potentials $-1/|z|$ converging to each Landau level.

where $E_n^{(0)}=\alpha(2n+1)$ and $n=0$, 1, etc. These integrals were found numerically using Simpson's rule, except at $z=0$ where they could be evaluated simply in closed form. Of course, their convergence to the asymptotic form

$$V_{nn}(z) \xrightarrow[|z|\to\infty]{} \alpha(2n+1) - 1/|z| + O(|z|^{-3}) \tag{34}$$

is slower for high n because $\langle\rho\rangle$ is correspondingly larger and (34) is satisfied only for $|z| \gg \langle\rho\rangle$. Figure 1 demonstrates that for energies up to a few a.u. a choice for the R-matrix boundary of $z_0=5$ a.u. should easily suffice. Inspection of the off-diagonal terms of the potential matrix confirms that they are negligible at $z_0=5$ a.u.

The calculation of the $m=0$, even-parity ground state is perfomed at only one energy. Rather than using the R-matrix approach, it is therefore more convenient to diagonalize the Hamiltonian in an orthonormal basis set which vanishes at $|z| \geq z_0=5$ a.u. and has a vanishing z derivative at $z=0$. The basis set of 23 functions which was used is (for $|z|<z_0$)

$$y_{nj}(\rho,\phi,z) = \begin{cases} \phi_n^{(0)}(\rho,\phi)(2/z_0)^{1/2}\cos[\pi(j-\tfrac{1}{2})z/z_0], \\ \qquad n=0,1 \text{ and } j=1,\ldots,10 \\ \phi_n^0(\rho,\phi)(8\gamma/\pi)^{1/4}\exp(-\gamma z^2), \quad n=2,3,4. \end{cases} \tag{35}$$

Matrix elements of H were evaluated numerically using an efficient Filon scheme for integrals involving cosine functions and Simpson's rule for the others.[34] The parameter γ was optimized roughly at the value $\gamma=2.5$ and then held constant. The lowest energy eigenvalue so obtained is $E_0=-0.002\,38$ a.u., corresponding to a binding energy of 1.0024 a.u. This value is in satisfactory agreement with more sophisticated calculations, e.g., Wunner and Ruder[10] obtain a binding energy of 1.0222 a.u. Figure 2 shows the ground-state wave-function components in each of the Landau channels; it confirms that the higher channels ($n \geq 1$) contribute only at short range.

The $m=0$, odd-parity final state has a node at $z=0$. This fact diminshes the effect of the spherical volume

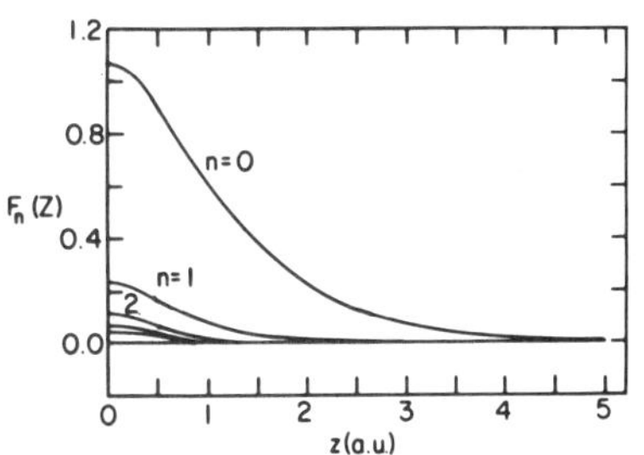

FIG. 2. Components of the even-parity ground-state wave function are shown in the lowest five Landau channels, for $m=0$.

near $r=0$, and implies that the cylindrical expansion should converge more rapidly than it did for the ground state; hence, only the lowest two Landau channels were retained. The variational basis used for the R-matrix calculation is

$$y_{nj}(\rho,\phi,z)=\phi_n^{(0)}(\rho,\phi)\sin(jz/2)\ , \tag{36}$$

with $n=0$ or 1 and $j=1,2,\ldots,j_{\max}$. The final calculation used $j_{\max}=15$, giving a total of 30 nonorthogonal basis functions. Some of the calculations were repeated with $j_{\max}=10$ to assure convergence of the expansion.

For an N-channel problem there are N independent solutions ψ_β obtained by solving the eigenvalue problem (11). It is useful to write these solutions in the form

$$\psi_\beta(\rho,\phi,z)=\sum_n \phi_n^{(0)}(\rho,\phi)F_{n\beta}(z),\quad |z|\geq z_0\ . \tag{37}$$

The R matrix can be calculated from values of $F_{n\beta}(z_0)$ and its derivative with respect to z, $F'_{n\beta}(z_0)$:

$$\underline{R}=-\underline{F}'(z_0)[\underline{F}(z_0)]^{-1}\ . \tag{38}$$

If the ψ_β are eigenstates of the R matrix, then $F'_{n\beta}(z_0)=-b_\beta F_{n\beta}(z_0)$, and $\underline{R}$ is given by

$$R_{nn'}=\sum_\beta F_{n\beta}(z_0)b_\beta[\underline{F}(z_0)^{-1}]_{\beta n'}\ , \tag{39}$$

which is equivalent to Eq. (19). But $\underline{F}'(z_0)$ can also be calculated by explicitly taking the derivative of the variational solution. This gives, in general, a less-accurate R matrix when (38) is evaluated. In Table I the two-channel R matrices calculated at $E=3.1$ a.u. for the $m=0$, odd-parity final state using these two methods are compared. The R matrix found using (39) is denoted "variational"; it is exactly symmetric at all levels of approximation. The less-accurate R matrix found from (38) is denoted "approximate"; its asymmetry $R_{10}-R_{01}$ provides one measure of the inaccuracy of the calculation. The last row of Table I gives the R matrix calculated by direct numerical integration of the two-channel close-coupling equations. Its good agreement with the best variational results (third row of Table I) confirms the accuracy of the calculation.

A. Single-channel regime

At energies below 2.4 a.u. the first excited-channel wave function $(n=1)$ has no appreciable amplitude at $|z|=z_0$. This fact reduces the problem to a one-channel situation, even though the excited channel(s) still contribute to Ψ *within* the reaction zone. Throughout this one-channel regime the results of the R-matrix calculation can be summarized by a single-channel quantum defect $\mu_0(E)$ and a single-dipole matrix element $d_0(E)$ connecting the ground state to an energy-normalized final state. In keeping with the quantum-defect point of view, these are calculated as a continuous function of energy even below the $n=0$ ionization threshold at $E_0^{(0)}=1$ a.u. The regular and irregular Coulomb wave functions needed at $z_0=5$ a.u. were evaluated by a power-series expansion. Figure 3 shows the quantum defect throughout this energy range. It varies slowly with energy in general, though below $E=0.8$ a.u., μ_0 begins to vary more rapidly with E, reflecting the presence of the lowest member of the Rydberg series at $E=0.7029$ a.u. The differential oscillator-strength distribution df/dE, calculated using both the length and velocity[35] forms of the dipole matrix element, appears in Fig. 4. This gives the photoionization cross section at $E>1$ a.u. Below that energy the photoabsorption spectrum is entirely discrete, despite the fact that μ_0 and df/dE are shown continuously. The energies of the bound levels are determined by the requirement that

$$\mu_0(E)+\nu_0(E)=P\ , \tag{40}$$

for P integer, where the effective quantum number is

$$\nu_0(E)=[2(E_0^{(0)}-E)]^{-1/2}\ . \tag{41}$$

At those energies satisfying (40) the discrete oscillator strengths are related to df/dE by

$$f_N=(df/dE)_{E_N}/(\nu_0^3+d\mu_0/dE)_{E_N}\ . \tag{42}$$

These bound-state energies and the oscillator strengths connecting to the ground state are compared in Table II

TABLE I. R matrices at $E=3.1$ a.u. as calculated by different methods.

Method	$j_{\max}$	R_{00}	R_{11}	R_{01}	R_{10}
Variational	10	0.969 98	0.117 72	0.165 78	0.165 78
Approximate	10	0.924 03	0.121 98	0.160 74	0.150 99
Variational	15	0.971 17	0.117 78	0.166 03	0.166 03
Approximate	15	0.973 87	0.119 27	0.167 21	0.169 33
Close coupling		0.971 26	0.117 78	0.166 04	0.165 98

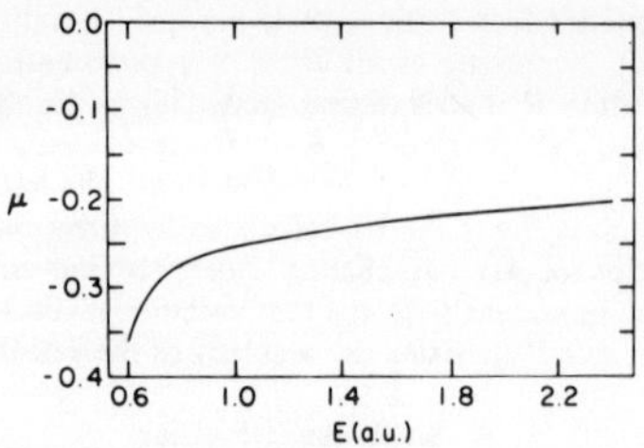

FIG. 3. Single-channel quantum defect ($m=0$, even parity) is shown as a smooth function of energy near the lowest ionization threshold $E_0^{(0)}=1$ a.u.

with the calculation of Refs. 10 and 36. While the agreement is reasonably good, there is a noticeable discrepancy between the length oscillator strengths and the velocity values for all levels but the lowest. The origin of this discrepancy is not fully understood. The oscillator-strength sum, which should be unity, is found to be $\sum f_N=0.954$ (length), $\sum f_N=0.950$ (velocity). These sums include the lowest seven discrete levels, an estimate of the remaining discrete levels, and the continuum up to $E=2.4$ a.u. The discrete spectrum contributes roughly 75% of this sum compared with 56.5% for hydrogen in zero external field.[37]

B. Two-channel regime

The energy range from $E=2.4$–4 a.u., which contains the first excited $m=0$ Landau threshold ($n=1$) at $E_1^{(0)}=3$ a.u., can be treated as a two-channel problem. The 2×2 orthogonal matrix $U_{n\alpha}$ can be parametrized in terms of a mixing angle θ; together with the eigenquantum defects μ_2 and μ_1, these determine the reaction matrix. The two dipole matrix elements d_α were calculated in the dipole length approximation using (32) and in the dipole velocity approximation using an analogous expres-

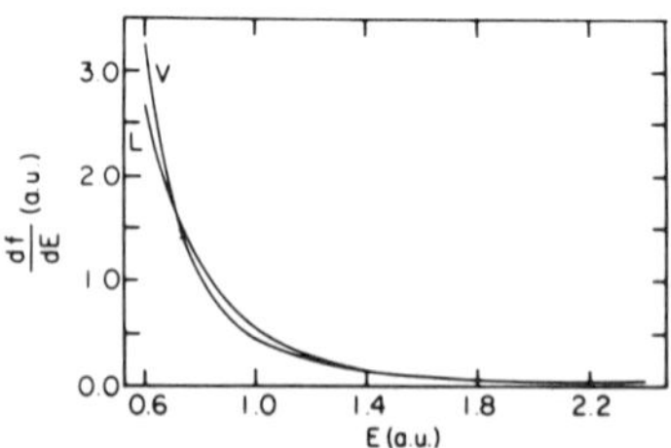

FIG. 4. Oscillator-strength distribution df/dE obtained when the hydrogen ground state is photoionized using photons linearly polarized along the field axis at $B=4.7\times10^9$ G. These are shown using both the length (L) and velocity (V) forms of the dipole matrix element. While the curves are shown continuously at $E<1$ a.u., in fact, the observed spectrum is then entirely discrete, with oscillator strengths given in Table II.

TABLE II. Binding energies (a.u.) and oscillator strengths of the $m=0$, odd-parity bound states.

$E_0^{(0)}-E_N$		f_N		
Present	Ref. 10	Length	Velocity	Ref. 36
0.2971	0.2977	0.664	0.679	0.687
0.09772	0.09685	0.0676	0.0587	0.0578
0.04712	0.04690	0.0189	0.0161	
0.02761	0.02752	0.00791	0.00671	

sion. These five smoothly varying short-range parameters are shown as a function of energy in Figs. 5–7. The increasing value of $|\theta/\pi|$ implies a gradual increase in the channel mixing as the energy increases from $E=2.4$ a.u. toward the $n=1$ threshold. This onset of channel mixing is more gradual than that observed in Ref. 26. It is best reflected in the squared off-diagonal scattering matrix element

$$|S_{01}|^2=\sin^2(2\theta)\sin^2[\pi(\mu_1-\mu_2)] ,$$

which (for $E>3$ a.u.) can be interpreted as the probability that an ($m=0$, odd-parity) electron incident in the lowest Landau channel will emerge in the first excited channel. Figures 5 and 6 show that $|S_{01}|^2$ increases from the small value 0.004 at $E=2.4$ up to the non-negligible value 0.16 at $E=3.2$. Nonetheless, this implies a relatively small channel mixing throughout this energy range, as expected because of the node at $z=0$ in the final state. In simpler terms, the small mixing suggests that the Schrödinger equation is nearly separable in cylindrical coordinates, at least for $m=0$, odd-parity states at this (and higher) magnetic field strengths.

The agreement of length and velocity dipole matrix elements in Fig. 7 provides some evidence that two Landau channels adequately describe the final state. Still, the 10–15% discrepancy between the $\alpha=2$ matrix elements is most likely traceable to the omission of the $n=3$ Landau channel near $z=0$. The parameters of Figs. 5–7 permit a calculation of differential oscillator-strength distribution, through, e.g., Eqs. (30)–(40) of Ref. 26. The resulting df/dE are shown in Fig. 8 with both the length

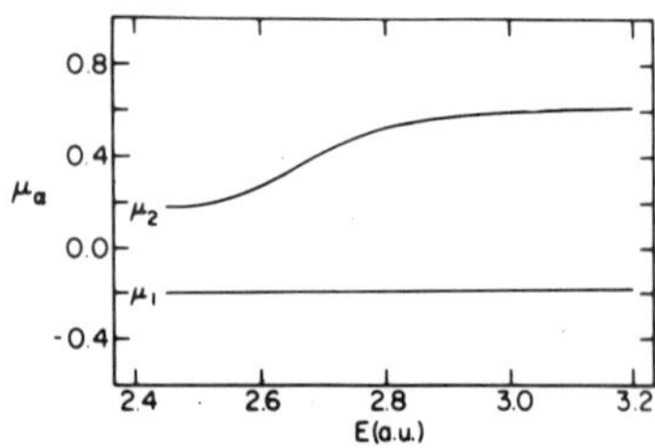

FIG. 5. Two eigenquantum defects μ_α are shown as a function of energy in the two-channel range near the first excited Landau threshold ($n=1$, $m=0$, odd parity) $E_1^{(0)}=3$ a.u. These are the eigenphaseshifts of the reaction matrix, divided by π.

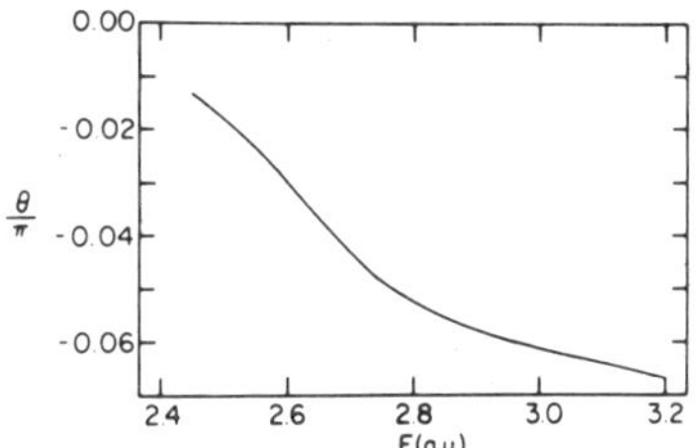

FIG. 6. Mixing angle θ/π for the final state is shown near $E_1^{(0)}=3$ a.u. This is the rotation angle needed to diagonalize the symmetric 2×2 reaction matrix. Notice the increasing importance of channel mixing as the threshold is approached from below.

and velocity results demonstrating the significant enhancement of oscillator strength which is produced by the autoionizing levels. These narrow states form a Rydberg series converging on the $n=1$ threshold at $E=3$ a.u.; the first four states are shown in Fig. 8. Above threshold, the oscillator strength becomes again a smooth continuum, and photoelectrons having two possible speeds v_z at $|z|\to\infty$ would be detected in an experiment. The data of Figs. 5 and 6 imply that near threshold the slower photoelectrons escape roughly four times as often as the faster photoelectrons.

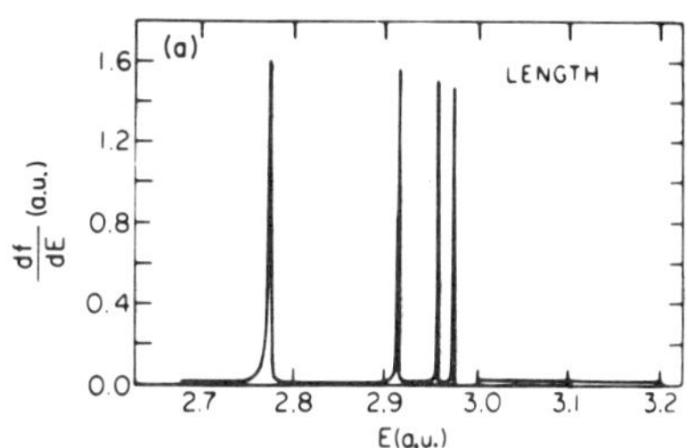

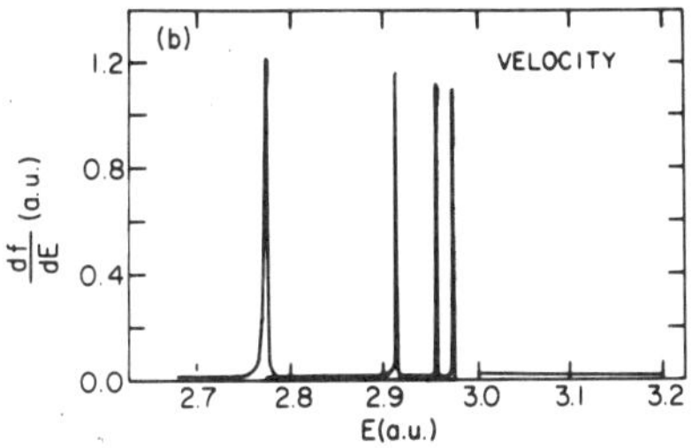

FIG. 8. Total photoionization oscillator strength of hydrogen at $B=4.7\times10^9$ G is given as a function of energy near the first excited Landau threshold $E_1^{(0)}=3$ a.u. Resonances form an infinite Rydberg series converging on this threshold, though only the lowest four are shown. Oscillator strength has a finite value at threshold and decreases slowly as the energy is increased further. It is shown using (a) length and (b) velocity forms of the dipole matrix element.

IV. CONCLUSIONS

The qualitative and quantitative aspects of hydrogen photoionization in a strong magnetic field can clearly be treated realistically by the approach outlined in Sec. II. While the calculation presented has neither attempted high accuracy nor has it considered a range of magnetic field strengths, it has overcome the main difficulties to be encountered in such studies. A main conclusion reached is that for $B\geq4.7\times10^9$ G, the coupling between cylindrical channels is so small as to be perturbative, at least for $m=0$ and odd parity. This is probably not true of the $m=0$, even-parity channels. In this regard it may be interesting to survey the photoionization cross section at field strengths 10 to 100 times weaker, as the resonances will be significantly broader as the channel mixing (i.e., nonseparability) increases. For these lower field strengths another consequence of the increased channel interactions will be strong-level perturbations, which should necessitate a multichannel quantum-defect treatment even for the calculation of bound states. In addition, the lower field strengths are probably more relevant to astrophysical problems than is the field $B=4.7\times10^9$ G.

A second major result of the calculation is to provide further evidence that in any multichannel calculation the mixing with excited channels becomes important at considerably lower energies than the actual location of the first resonances in those channels. Comparison of Figs. 6

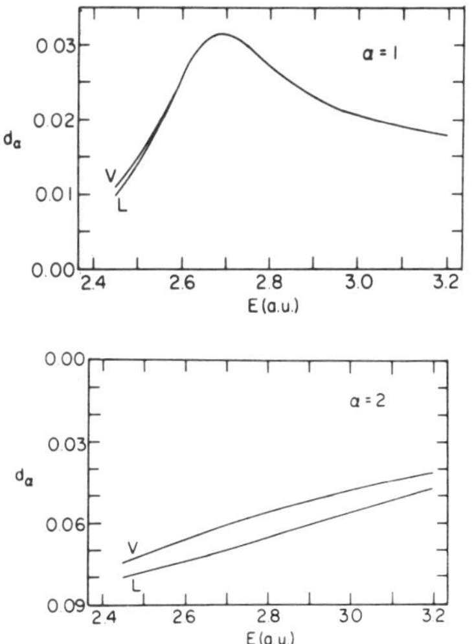

FIG. 7. Dipole matrix elements d_a calculated using either length (L) or velocity (V) forms.

and 8 shows the channel mixing begins to grow above $E=2.5$ a.u., though the lowest resonance lies at $E=2.75$ a.u. Nonetheless, the onset of channel mixing does occur more smoothly as a function of energy here than was observed in Ref. 26. The circumstances determining this aspect are not well understood and remain a topic for future studies.

ACKNOWLEDGMENTS

Conversations with A.R.P. Rau have been extremely useful. I thank G. Wunner for providing results prior to publication and R. K. Nesbet for comments on an earlier version of this manuscript. This work was supported in part by the National Science Foundation.

---

[1]J. R. P. Angel, E. P. Borra, and J. D. Landstreet, Astrophys. J. Suppl. Ser. 45, 359 (1981).

[2]U. Fano, Phys. Rev. A 22, 2660 (1980).

[3]U. Fano and C. D. Lin, in *Atomic Physics IV*, edited by G. zu Putlitz, E. W. Weber, and A. Winnacker (Plenum, New York, 1975) p. 47; A. R. P. Rau, J. Phys. (Paris) Colloq. 43, C2-211 (1982).

[4]C. W. Clark and K. T. Taylor, Nature 292, 437 (1981).

[5]E. A. Solov'ev, Zh. Eksp. Teor. Fiz. 82, 1762 (1982) [Sov. Phys.—JETP 55, 1017 (1982)]; C. W. Clark, Phys. Rev. A 24, 605 (1981); D. R. Herrick, *ibid.* 26, 323 (1982); C. J. Goebel and T. W. Kirkman (unpublished).

[6]W. R. S. Garton and F. S. Tomkins, Astrophys. J. 158, 839 (1969).

[7]H. Friedrich and M. Chu, Phys. Rev. A 28, 1423 (1983).

[8]K. Onda, J. Phys. Soc. Jpn. 45, 216 (1978).

[9]R. H. Garstang, Rep. Prog. Phys. 40, 105 (1977).

[10]G. Wunner and H. Ruder, J. Phys. (Paris) Colloq. 43, C2-137 (1982).

[11]S. M. Kara and M. R. C. McDowell, J. Phys. B 14, 1719 (1981).

[12]G. Wunner, H. Ruder, H. Herold, and W. Schmitt (unpublished).

[13]M. J. Seaton, Proc. Phys. Soc. London 88, 801 (1966); Rep. Prog. Phys. 46, 167 (1983).

[14]U. Fano, J. Opt. Soc. Am. 65, 979 (1975).

[15]C. M. Lee and K. T. Lu, Phys. Rev. A 8, 1241 (1973).

[16]C. H. Greene, A. R. P. Rau, and U. Fano, Phys. Rev. A 26, 2441 (1982).

[17]C. H. Greene, Phys. Rev. A 22, 149 (1980).

[18]C. W. Clark, Phys. Rev. A 28, 83 (1983).

[19]C. M. Lee, Phys. Rev. A 10, 584 (1974); U. Fano and C. M. Lee, Phys. Rev. Lett. 31, 1573 (1973).

[20]W. Kohn, Phys. Rev. 74, 1763 (1948).

[21]J. L. Jackson, Phys. Rev. 83, 301 (1951).

[22]A. M. Lane and D. Robson, Phys. Rev. 178, 1715 (1969).

[23]J. E. Purcell, Phys. Rev. 185, 1279 (1969); R. A. Chatwin, Phys. Rev. C 2, 1167 (1970).

[24]R. S. Oberoi and R. K. Nesbet, Phys. Rev. A 8, 215 (1973); 9, 2804 (1974); R. K. Nesbet, *Variational Methods in Electron-Atom Scattering Theory* (Plenum, New York, 1980).

[25]H. Le Rouzo and G. Raseev (unpublished).

[26]C. H. Greene, Phys. Rev. A 23, 661 (1981).

[27]Ch. Jungen, in *Invited Papers in the Twelfth International Conference on the Physics of Electronic and Atomic Collisions*, edited by S. Datz (North-Holland, Amsterdam, 1982), p. 455.

[28]See, e.g., E. S. Chang and U. Fano, Phys. Rev. A 6, 173 (1972); also Ch. Jungen and O. Atabek, J. Chem. Phys. 66, 5584 (1977).

[29]P. G. Burke and W. D. Robb, Adv. At. Mol. Phys. 11, 143 (1975).

[30]J. H. Wilkinson, *The Algebraic Eigenvalue Problem*, (Oxford University Press, London, 1965).

[31]C. B. Moler and G. W. Stewart, SIAM J. Numer. Anal. 10, 241 (1973).

[32]The calculations in this paper have used subroutine EIGZF, which is found in the IMSL Library. It is a product of IMSL, Inc., 7500 Bellaire Blvd., 6th Floor, NBC Building, Houston Texas 77036-5085.

[33]Integrations in the z coordinate are generally performed over the interval $-\infty < z < \infty$. Since all nonzero integrals needed here involve integrands which are symmetric about $z=0$, they have all been evaluated instead over the range $0 \le z < \infty$ only. The $l=0$ Coulomb wave functions (f,g) of Ref. 17 are energy normalized over $0 \le z < \infty$, which permits their use without rescaling by the factor $1/\sqrt{2}$, as would be needed if (f,g) were to be energy normalized over $-\infty < z < \infty$.

[34]F. Scheid, *Theory and Problems of Numerical Analysis, Schaum's Outline Series* (McGraw-Hill, New York, 1968).

[35]See, e.g., A. F. Starace, in *Handbuch der Physik* edited by W. Mehlhorn (Springer, Berlin, 1982), Vol. XXXI.

[36]G. Wunner (private communication).

[37]H. A. Bethe and E. E. Salpeter, *Quantum Mechanics of One- or Two-Electron Atoms* (Springer, Berlin, 1957), p. 265.

1991 *Phys. Rev.* A **44** 1773–90
Reprinted with permission from the American Physical Society

Spin-orbit effects in the heavy alkaline-earth atoms

Chris H. Greene
Department of Physics and Joint Institute for Laboratory Astrophysics, University of Colorado, Boulder, Colorado 80309-0440

Mireille Aymar
Laboratoire Aimé Cotton, Centre National de la Recherche Scientifique II, Bâtiment 505, 91405 Orsay CEDEX, France

(Received 28 November 1990)

Complex bound and autoionizing spectra of strontium, barium, and radium are treated at two different levels of approximation using multichannel-quantum-defect techniques (MQDT) and eigenchannel R-matrix calculations. In the first level, the R-matrix calculations are conducted entirely in LS coupling without explicitly including any spin-orbit terms in the Hamiltonian. A subsequent frame transformation to jj coupling is then carried out prior to the MQDT treatment. In the second level, the entire R-matrix calculation is conducted in jj coupling, including spin-orbit terms in the Hamiltonian explicitly within the R-matrix box. The two methods give nearly identical results even for atoms as heavy as barium, each showing good agreement with experiment, but significant differences begin to appear between the two levels of approximation for atomic radium.

I. INTRODUCTION

Recent years have seen a major enhancement in our ability to calculate energy levels, oscillator strengths, and autoionization observables for the alkaline-earth atoms. The combination of an eigenchannel R-matrix approach [1,2] with multichannel-quantum-defect theory [3–5] (MQDT) has now been shown in numerous calculations to describe experimental spectra to spectroscopic accuracy, even for atoms as heavy as strontium [6–8]. Some other progress has been achieved [9] without making use of the power of MQDT, but such methods have only limited applicability to the highly perturbed and overlapping Rydberg series which are the rule in all of these atoms.

In the studies performed to date [6–8], a variational R-matrix calculation is carried out in LS coupling neglecting all fine-structure effects. The interactions between all open and closed channels are then characterized by elements of a smooth reaction matrix K^{LS} and by corresponding dipole matrix elements d^{LS}. This LS-coupled information is then transformed into jj coupling using the well-known orthogonal transformation matrix [10] $\langle jj|LS\rangle$. The spin-orbit terms in the Hamiltonian are now used indirectly, in that the *experimental* j-dependent ionization threshold energies E_i are used when solving the MQDT equations for the continuum states on a finite-energy mesh, just as in earlier semiempirical studies [4,11,12]. This guarantees that Rydberg series will be present converging to each allowed ionization threshold, thus permitting a highly accurate description of spin-orbit effects in light atoms, even when those effects are strongly nonperturbative. In more physical terms, this scheme amounts in effect to neglecting the spin-orbit interaction, except to the extent that energy conservation affects the energy of the outermost electron at $r \geq r_0$.

The high accuracy of calculations based on the above scheme is now well established for Mg [13], for Ca [7], and for Sr [6]. However, in very heavy atoms such as Ba and Ra, this approach is more questionable since spin-orbit effects become much larger and are expected to affect the motion of the outer two valence electrons even at small radii $r \leq r_0$. In the first test of the jj-LS frame transformation in an atom as heavy as Ba, however, Aymar [14] recently showed it to give surprisingly good results, demonstrating largely the same good level of agreement with experiment as had been found previously for Sr [6(a)].

The present study examines in more detail the range of validity of the jj-LS frame transformation by calculating numerous spectra for Sr, Ba, and Ra and comparing the jj-LS calculation with a full jj-coupled calculation that includes spin-orbit terms explicitly in the Hamiltonian even within the R-matrix box. The latter calculations are carried out along the same lines developed in a recent treatment [15] of the alkali-metal negative ions Rb^-, Cs^-, and Fr^-. This fully jj-coupled approach is expected to give a much more realistic description of spin-orbit effects when they are important, and accordingly the comparison of the two types of calculation will help to clarify the range of validity of frame transformation calculations in general. This matter has some practical importance, since each jj-coupled calculation requires 5–10 times more computer time and memory than does an equivalent R-matrix calculation in LS coupling. (This is because for a given total angular momentum J there are typically three times as many jj-coupled channels as LS channels, which translates into roughly nine times as many matrix elements to reach an equivalent level of convergence in jj coupling.) For atoms with only two valence electrons, either approach is modest enough to be performed on a small computer workstation, but an order of magnitude could make an enormous difference in the feasibility of such calculations for open-shell atoms with more valence-shell electrons.

Accounting for detailed spectroscopy of these heavy atoms, much of it only recently acquired [16], is a second major goal of this paper. Theoretical results are shown to adequately describe photoionization spectra of ground-state strontium and also of the ground state and an excited state of barium. Less experimental information is available for comparison in the case of atomic radium [17–19], but most of the $J=1$, odd-parity bound spectrum which has been observed is in reasonable agreement with the calculations reported here. A single major discrepancy in the radium bound spectrum suggests either a gross misclassification in the early measurements of Rasmussen [17], or else it indicates a serious limitation of the present approach for the heaviest atoms in the Periodic Table. Deciding between these two alternative interpretations may require another experimental measurement of the radium spectrum.

Beyond the issues of agreement or disagreement between theory and experiment are several substantial questions concerning the methods of calculation needed to treat heavy atoms in the Periodic Table. One major question is whether the Schrödinger equation suffices to describe electronic wave functions even in atoms as heavy as radium. The answer expected from traditional multiconfiguration Hartree-Fock-style calculations (MCHF) would be negative, as the inner shell electrons are highly relativistic in radium. However, our present results apparently confirm those of Ref. [15] in suggesting that use of empirical information about the "one-valence-electron" system (e.g., Ra^+) allows one to bypass a full description based on the Dirac equation and to continue using the more convenient Schrödinger description. This and other related issues are addressed below in Sec. IV.

Section II outlines the *jj*-coupled *R*-matrix approach including spin-orbit effects within the reaction zone. Once the *jj*-coupled reaction matrix and dipole matrix elements have been obtained, the generation of the full spectrum follows standard MQDT formulations. On the other hand, the *LS*-coupled *R*-matrix calculations are not discussed at depth here since they have been presented in detail elsewhere [6–8,14]. Section III is instead devoted to analyzing the theoretical and experimental spectra of these heavy atoms, with their implications considered further in Sec. IV.

II. EIGENCHANNEL *R*-MATRIX CALCULATION IN *jj*-COUPLING

The model Hamiltonian used to describe the two valence electrons will be represented as

$$H=T_1+T_2+V(r_1)+V(r_2)+V_{\text{s.o.}}(1)+V_{\text{s.o.}}(2)+\frac{1}{r_{12}} \quad (1)$$

Here T_1 and T_2 are the usual Schrödinger kinetic-energy operators for each electron. The screening of the nuclear charge Z by the core electrons produces in the end a net electron-core effective potential represented by $V(r_i)$ for the ith electron. The potential $V(r)$ is chosen to have the following analytical form:

$$V(r)=-\frac{1}{r}[2+(Z-2)\exp(-\alpha_1^l r)+\alpha_2^l r\exp(-\alpha_3^l r)]-\frac{\alpha_d}{2r^4}\{1-\exp[-(r/r_c^l)^6]\}, \quad (2)$$

where Z is the nuclear charge and α_d the dipole polarizability of the doubly charged positive ion [14,15]. The empirical parameters α_i^l and r_c^l are adjusted to obtain optimum agreement between the energy eigenvalues E_{nlj} of the one-electron Schrödinger equation and the experimental energies of the alkaline-earth ion. The values of the parameters obtained for Sr^+, Ba^+, and Ra^+ are given in Table I. It is important to note that the potential $V(r)$ depends on the orbital angular-momentum quantum number l of the electron, which makes it formally a nonlocal potential but in a trivial way.

The spin-orbit interaction, e.g., between electron i and the screened nucleus, is represented by a potential having the form

$$V_{\text{s.o.}}(i)=\frac{\mathbf{s}_i\cdot\mathbf{l}_i}{2c^2}\frac{1}{r_i}\frac{\partial V}{\partial r_i}\left[1-\frac{V(r_i)}{2c^2}\right]^{-2}, \quad (3)$$

with $c=137.036$ the speed of light in atomic units. The last factor in Eq. (3) is suggested by the Dirac equation [20], but is typically omitted in perturbation treatments. Here it is included as in some previous studies in order to make solutions to the radial Schrödinger equation well defined near the origin $r_i\to 0$.

The first aim of the calculation is to determine the channel-mixing parameters of multichannel-quantum-defect theory, after which any desired observable can be rapidly calculated. These parameters are determined by solving the Schrödinger equation variationally within the finite reaction volume $\mathcal{V}$ in configuration space, defined by $\max(r_1,r_2)\le r_0$, while the reaction surface $\mathcal{S}$ is the set of all points for which $\max(r_1,r_2)=r_0$. This calculation is also referred to as an *eigenchannel* *R*-matrix calculation since it determines variationally the *eigenstates* of the *R*-matrix at some specified energy E, having a constant normal logarithmic derivative on the reaction surface $\mathcal{S}$.

The specifics of the general eigenchannel formulation are derived elsewhere (see, e.g., Ref. [8]), so we only summarize the formulas and the logic followed in their solution. The βth desired eigenstate of the Hamiltonian (and of the R matrix) is represented within $\mathcal{V}$ by a basis-set expansion,

$$|\Psi_\beta\rangle=\sum_k |y_k\rangle Z_{k\beta}, \quad (4)$$

in terms of constant coefficients $Z_{k\beta}$ and a two-electron *jj*-coupled basis set $|y_k\rangle$. The negative of the normal logarithmic derivative of $|\Psi_\beta\rangle$ on $\mathcal{S}$, denoted b_β, constant over $\mathcal{S}$, is determined as an eigenvalue of a generalized eigenvalue problem,

TABLE I. Semiempirical parameters in Eq. (2) describing the model potential experienced by the outermost (valence) electron in Sr^+, Ba^+, and Ra^+.

	l	α_1	α_2	α_3	r_c
Sr^+	0	3.4187	4.7332	1.5915	1.7965
$\alpha_d=7.5$	1	3.3235	2.2539	1.5712	1.3960
	2	3.2533	3.2330	1.5996	1.6820
	≥ 3	5.3540	7.9517	5.6624	1.0057
Ba^+	0	3.0751	2.6107	1.2026	2.6004
$\alpha_d=11.4$	1	3.2304	2.9561	1.1923	2.0497
	2	3.2961	3.0248	1.2943	1.8946
	≥ 3	3.6237	6.7416	2.0379	1.0473
Ra^+	0	3.7702	4.9928	1.5179	1.3691
$\alpha_d=18$	1	3.9430	5.0552	3.6770	1.0924
	2	3.7008	4.7748	1.4956	2.2784
	≥ 3	3.8125	5.0332	2.1016	1.2707

$$\underline{\Gamma}\mathbf{Z}=b\underline{\Lambda}\mathbf{Z} . \tag{5}$$

The matrices $\underline{\Gamma}$ and $\underline{\Lambda}$ are defined in terms of the Hamiltonian H in Eq. (1) and the matrix elements of the Bloch operator

$$L_{kk'}\equiv\tfrac{1}{2}\left\langle\!\!\left\langle y_k \left| \frac{\partial}{\partial n} \right| y_{k'}\right\rangle\!\!\right\rangle . \tag{6}$$

Here the double bracket matrix element $\langle\langle\ |\ |\ \rangle\rangle$ indicates an integral over the reaction surface $\mathcal{S}$, including a trace over the spin degrees of freedom of the electrons, while single brackets indicate a matrix element over the whole reaction volume. Then Eq. (5) contains a reaction volume overlap matrix element $O_{kk'}\equiv\langle y_k|y_{k'}\rangle$ in addition to the Hamiltonian and Bloch operator matrices,

$$\Gamma_{kk'}=2EO_{kk'}-2H_{kk'}-2L_{kk'} . \tag{7}$$

Equation (5) also contains a surface overlap matrix element

$$\Lambda_{kk'}=\langle\langle y_k|y_{k'}\rangle\rangle . \tag{8}$$

The basis set used here consists of nonorthogonal jj-coupled independent-electron numerical basis functions, eigenfunctions of the independent-electron Hamiltonian operator $H_0=H-1/r_{12}$. Note that in this basis, both H and H_0 are not Hermitian, but owing to the Bloch operator, Γ is Hermitian.

A. Design of the basis set

The two-electron basis set is composed of one-electron eigenfunctions, obeying

$$(T+V+V_{\text{s.o.}})|n(sl)jm\rangle=E_{nlj}|n(sl)jm\rangle , \tag{9}$$

with l and j denoting the one-electron orbital and total angular momenta, and n denoting a "principal quantum number" labeling different radial solutions having the same l and j. An unsymmetrized two-electron eigenfunction jj coupled to form a state of definite total angular momentum J will then be written in notation analogous to that of Biedenharn and Louck [21]

$$|n_1(sl_1)j_1,n_2(sl_2)j_2JM\rangle . \tag{10}$$

A fully antisymmetric two-electron basis function is labeled by additional curly brackets $\{\ \}$,

$$|\{n_1(sl_1)j_1,n_2(sl_2)j_2JM\}\rangle=\frac{1}{\sqrt{2(1+\delta_{n_1n_2}\delta_{l_1l_2}\delta_{j_1j_2})}}[\,|n_1(sl_1)j_1,n_2(sl_2)j_2JM\rangle - (-1)^{j_1+j_2-J}|n_2(sl_2)j_2,n_1(sl_1)j_1JM\rangle] . \tag{11}$$

Thus the basis-vector label k in Eq. (4) is meant as a short-hand notation for the quantum numbers $\{n_1l_1j_1n_2l_2j_2\}$.

In a basis consisting of such jj-coupled two-electron functions it is more or less straightforward to calculate all matrix elements in the variational expression (5) and to determine corresponding eigenvalues $-b_\beta$ of the R matrix. In practice it has proven convenient to specify the two-electron basis in a manner which guarantees that the number of such eigenvalues obtained at any given energy E coincides with the number of channels in the calculation. This can be accomplished using the terminology of "closed-type" and "open-type" two-electron basis vectors, $|y_k^c\rangle$ and $|y_k^o\rangle$, respectively, along the same lines discussed previously for the LS-coupled calculations [6–8,14]. To begin with, two different sets of *one-electron* orbitals are first defined which obey Eq. (9). The first set is composed of orbitals $|n(sl)j\rangle$ denoted closed-type because they are chosen to vanish at $r=r_0$. Because the closed-type one-electron orbitals each obey the same boundary condition at the origin and at r_0, and because they are eigenfunctions of a Hermitian operator, they are automatically orthogonal over the radial interval $0\le r\le r_0$. The second set of one-electron orbitals is denoted open-type because they will be included in the two-electron basis set only in channels to be treated as open (or weakly closed in the sense of MQDT), and will have the radial quantum number denoted by a bar, i.e, $|\bar{n}(sl)j\rangle$. Then each two-electron closed-type basis vector $|y_k^c\rangle$ is given by Eq. (11), and it consists of two closed-type radial orbitals.

Accordingly, $|y_k^c\rangle$ vanishes on the two-electron reaction surface $\mathcal{S}$. The open-type two-electron basis vector $|y_k^o\rangle$ is instead comprised of a closed-type orbital for n_1, but an open-type one-electron orbital $|\bar{n}_2(sl_2)j_2\rangle$ is used in place of the closed-type $|n_2(sl_2)j_2\rangle$ in Eq. (11). The logic of this choice is that one electron is thereby allowed to reach the reaction surface $\mathcal{S}$ in any channel for which at least one open-type two-electron basis vector is included, but in no channel are *both* electrons able to move beyond r_0 simultaneously. As in usual MQDT treatments, only one electron escape beyond the reaction surface is considered here. In practice two open-type two-electron basis vectors $|y_k^o\rangle$ defined in the above fashion are included for each MQDT channel, while many more closed-type $|y_k^c\rangle$ are included. (As was discussed in detail in Ref. [8(b)], the closed-type basis set vanishing at r_0 is complete by itself, in the ordinary sense of quantum mechanics. Yet extra flexibility is provided, for describing general bound and continuum wave functions that fail to vanish at $r=r_0$, by including these open-type basis vectors. But as a rule of thumb the open-type basis should be kept comparatively small so that the basis does not become linearly dependent to within the machine precision. Experimentation with one to three open-type basis vectors per channel has shown little sensitivity to this number, and most eigenchannel R-matrix calculations presently use two per channel.) If for a given set of $\{n_1l_1j_1l_2j_2\}$, only closed-type radial orbitals $u_{n_2l_2j_2}(r)$ are included in the full basis set, the channel $i\equiv\{n_1l_1j_1l_2j_2\}$ is called "strongly closed," while the presence of an open-type orbital $u_{\bar{n}_2l_2j_2}(r)$ in the full basis signifies that the corresponding channel i to be treated as open or weakly closed in the eventual MQDT calculation.

B. Streamlined solution of the generalized eigensystem

After the basis set is designed and all matrix elements calculated, Eq. (5) can be solved numerically at each desired energy E. The jj-coupled calculations presented in this paper typically involve 200 to 500 basis functions for any given J, depending on factors such as the value of r_0, the values of J and parity, the number of channels, and the degree of convergence desired. For such large basis sets the solution of the linear eigensystem (5) at many energies can become time consuming, and for this reason a much more efficient "streamlined" solution method of Greene and Kim [8(b)] is used.

For a modest box size r_0 less than about 25 a.u., good convergence is achieved using eight closed-type basis vectors per channel and two open-type vectors per channel. The calculations here involved typically five to 15 open or weakly closed channels, with an additional 100 to 300 strongly closed two-electron basis vectors included to improve the flexibility of the basis set to describe various polarization and correlation effects. Thus the total variational basis set breaks up into mostly closed-type basis vectors with perhaps an order of magnitude fewer open-type vectors. It is this fact that the streamlined formulation takes advantage of, by partitioning the matrices in Eqs. (5)–(8) into open and closed partitions, e.g.,

$$\underline{\Gamma}=\begin{bmatrix}\underline{\Gamma}^{cc} & \underline{\Gamma}^{co}\\ \underline{\Gamma}^{oc} & \underline{\Gamma}^{oo}\end{bmatrix}, \tag{12}$$

and

$$\underline{\Delta}=\begin{bmatrix}0 & 0\\ 0 & \underline{\Delta}^{oo}\end{bmatrix}. \tag{13}$$

This partitioning breaks Eq. (5) into two coupled matrix equations,

$$\underline{\Gamma}^{cc}\mathbf{Z}^c+\underline{\Gamma}^{co}\mathbf{Z}^o=0\ , \tag{14}$$

and

$$\underline{\Gamma}^{oc}\mathbf{Z}^c+\underline{\Gamma}^{oo}\mathbf{Z}^o=b\underline{\Delta}^{oo}\mathbf{Z}^o\ . \tag{15}$$

Using (14) to eliminate $\mathbf{Z}^c$ from (15) leads to a matrix equation of much smaller dimension for $\mathbf{Z}^o$, specifically

$$[\underline{\Gamma}^{oo}-\underline{\Gamma}^{oc}(\underline{\Gamma}^{cc})^{-1}\underline{\Gamma}^{co}]\mathbf{Z}^o=b\underline{\Delta}^{oo}\mathbf{Z}^o\ . \tag{16}$$

Inversion of the matrix $\underline{\Gamma}^{cc}$ at many energies E is efficiently accomplished by first transforming the closed-portion of the basis set into the energy-independent representation in which $\underline{H}^{cc}$ is diagonal, with eigenvalues E_λ and orthonormal eigenvectors $X_{k\lambda}$. (Here we have used the facts that $\underline{O}^{cc}$ is the unit matrix and that $\underline{L}^{cc}$ vanishes for our present choice of basis set.) Using $\underline{\Omega}$ to denote the matrix in square brackets in Eq. (16), its energy dependence is now given analytically

$$\Omega_{kk'}=2(EO^{oo}_{kk'}-H^{oo}_{kk'}-L^{oo}_{kk'})-2\sum_\lambda\frac{(EO^{oc'}_{k\lambda}-H^{oc'}_{k\lambda}-L^{oc'}_{k\lambda})(EO^{c'o}_{\lambda k'}-H^{c'o}_{\lambda k'}-L^{c'o}_{\lambda k'})}{E-E_\lambda}\ . \tag{17}$$

In Eq. (17) the notation c' implies that the closed portion of the two-electron basis set is now in the transformed representation, i.e., $O^{oc'}_{k\lambda}=\sum_{k'}O^{oc}_{kk'}X_{k'\lambda}$, $O^{c'o}_{\lambda k}=\sum_{k'}O^{co}_{k'k}X_{k'\lambda}$, etc. The semianalytic energy dependence of all matrices in (16), combined with the much smaller number of open-type than closed-type basis vectors, thus makes (16) much faster to solve on a fine-energy mesh than the original Eq. (5).

Recalling that the dimension of the reduced equation (16) is $2N_o$, where N_o is the number of open and weakly closed channels in the calculation, this generalized eigenvalue equation has N_o nontrivial eigenvalues b_β and corresponding eigenvectors $Z^o_{k\beta}$ at each energy E. The component $\psi_{i\beta}$ of the βth eigenstate in the ith channel can be written as

$$\psi_{i\beta}=\sum_{\bar{n}_2} Z^o_{i\bar{n}_2,\beta} u_{\bar{n}_2 l_{2i} j_{2i}}(r_0) , \tag{18}$$

where we use the notation $k\equiv\{i\bar{n}_2\}$ for the open portion of the variational two-electron basis set. The function $u_{\bar{n}_2 l_{2i} j_{2i}}(r)$ denotes the radial part of an open-type one-electron eigenfunction $|\bar{n}_2(sl_{2i})j_{2i}\rangle$ as described in Sec. II A above. The radial derivative of this same solution is then simply $\psi'_{i\beta}=-b_\beta\psi_{i\beta}$. By matching $\psi_{i\beta}$ and its derivative to a linear combination of regular and irregular Coulomb functions (f_i,g_i) in the ith channel, one obtains the usual "smooth" reaction matrix $K_{ii'}$ of MQDT, and the relevant dipole matrix elements needed for a photoionization calculation also (Eqs. (13)–(17) of Ref. [7(a)], for instance). In the calculations presented below, the value of r_0 typically was chosen in the range of 18 to 20 a.u.

III. RESULTS OF THE CALCULATIONS

A. Absorption spectra of Sr, Ba, and Ra

In this section we analyse the limitations of the approximations inherent in the jj-LS frame transformation by comparing the results previously obtained with LS-coupled R-matrix calculations and the jj-LS frame transformation for the absorption spectra of Sr [6(a)] and of Ba [14] with those obtained in the present work by performing R-matrix calculations in jj coupling. In addition, new R-matrix results obtained for Ra will be presented. These latter results deal with not only the photoionization spectrum, but also with the $J=1$ odd-parity bound spectrum.

R-matrix calculations performed either in LS coupling or jj coupling to investigate the absorption spectrum below the $m_0p_{3/2}$ threshold of Sr ($m_0=5$), Ba ($m_0=6$), and Ra ($m_0=7$) include 13 odd-parity $J=1$ channels, which are the jj-coupled ionization channels involved in the MQDT calculation of observables. [These channels are $m_0snp_{1/2}$, $m_0snp_{3/2}$, $(m_0-1)d_{3/2}np_{1/2}$, $(m_0-1)d_{3/2}np_{3/2}$, $(m_0-1)d_{3/2}nf_{5/2}$, $(m_0-1)d_{5/2}np_{3/2}$, $(m_0-1)d_{5/2}nf_{5/2}$, $(m_0-1)d_{5/2}nf_{7/2}$, $m_0p_{1/2}ns$, $m_0p_{1/2}nd_{3/2}$, $m_0p_{3/2}ns$, $m_0p_{3/2}nd_{3/2}$, $m_0p_{3/2}nd_{5/2}$.] Theoretical predictions for Sr and Ba will be compared with the most recent experimental data. In particular, we will consider the new experimental photoionization spectra obtained for Sr and Ba by Griesmann, Esser, and Hormes [16] using synchrotron radiation. The results obtained with R-matrix calculations in LS and jj coupling will be referenced throughout this paper as LS and jj results, respectively. However, the reader must keep in mind that R-matrix calculations performed in LS coupling are combined with MQDT calculations, including fine-structure effects through a frame transformation. Although all photoionization cross sections have been calculated using both the length and velocity forms, only velocity results are presented here, except in one case. In general, the agreement between length and velocity calculations is comparable to that shown later in Fig. 4, although it is somewhat poorer for the radium calculations.

1. Strontium

Figure 1 compares the experimental results [16] obtained in Sr below the $4d_{3/2}$ threshold with the jj results. The measurement of Griesmann, Esser, and Hormes [16] is not absolute. To compare it with our results, we have normalized the experimental curve by adjusting the peak photoionization cross section of the dominant resonance, namely the $4d6p\ ^1P$ resonance around 196 nm. Assignments of the other resonant states $4dnp$, $4dnf$, and $5p6s$ $J=1$ can be found in Refs. [6(a)] and [22]. The overall agreement between experiment and theory is better than that previously obtained [6(a)] by performing the variational calculation in LS coupling. However, the improvements are not due to the different treatment of the spin-orbit interaction but derive instead from three other differences. First, the one-electron model potential $V(r)$ used in Ref. [6(a)] had a very simple form involving only three parameters, while the new jj calculations are done with a more sophisticated model potential including l-dependent screening and polarization terms. As a result the theoretical energy levels of Sr^+ are more accurate here than in Ref. [6(a)]. Second, inaccuracies occur in MQDT calculations conducted following the LS-coupled R-matrix calculation in the low-energy range when experimental one-electron energies (including the fine-structure splitting) are used for the MQDT threshold energies. Difficulties arise when some channels are "strongly closed" but treated as "weakly closed" in the R-matrix calculation. In fact, during the matching procedure on

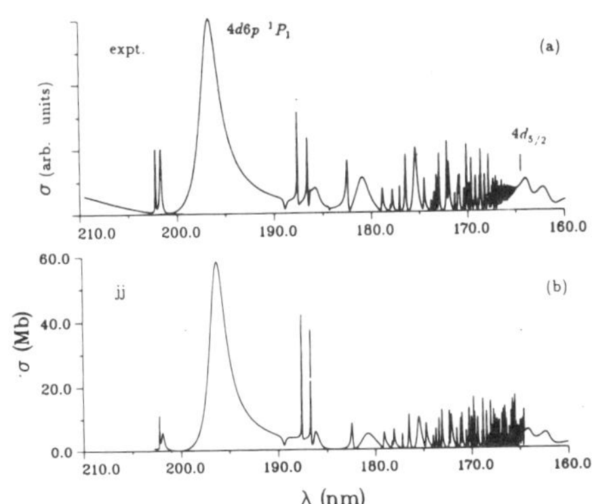

FIG. 1. Photoabsorption spectrum of Sr in the wavelength region from 210 to 160 nm, below the $4d_{5/2}$ threshold: total photoionization cross section. (a) Relative measurement of Griesmann, Esser, and Hormes [16] (wavelength resolution ~0.05 nm). (b) jj results.

the reaction surface which gives the K matrix, the Coulomb functions are evaluated with theoretical photoelectron energies, the fine structure of the thresholds being neglected at this point but accounted for later in the MQDT calculation. At the low-energy end in each channel, the MQDT parameters acquire a strong energy dependence and the MQDT results become very sensitive to the value of the threshold energy. This point has been discussed by Greene and Kim [7(a)], and will be documented later in the following sections. It was found [14] in Ba that these difficulties can be removed by using theoretical energies instead of experimental ones for the threshold energies associated with some strongly closed channels. In Ref. [6(a)], the use of experimental energies for the $5p_{1/2,3/2}$ thresholds led to inaccuracies in the energy range corresponding to the lowest end of the $5pns$, nd series, although use of experimental threshold energies definitely improves the calculation near those thresholds. It is to be stressed that these difficulties inherent in the frame-transformation approximation are almost absent when R-matrix calculations are done in jj coupling because then the experimental and theoretical threshold energies are very close. Finally, the previous R-matrix calculations [6(a)] were carried out with a small two-electron basis set, whereas the present jj-coupled R-matrix calculation introduces an enlarged basis set, improving the convergence of the variational calculation.

Figure 2 compares experimental [16] and theoretical photoionization cross sections of Sr at higher energies, between the $4d_{5/2}$ and $5p_{3/2}$ thresholds. In this energy range, the jj results, curve 2(b), are very similar to those previously obtained in the LS calculations [6(a)]. The comparison of experiments with the results obtained with LS-coupled R-matrix calculation is discussed further in Ref. [6(a)].

2. Barium

The eigenchannel R-matrix calculation performed in jj coupling for the absorption spectrum of Ba uses the same model potential $V(r)$ as in the previous calculations [14] done in LS coupling. Moreover, the basis sets used in both calculations are very similar. Thus, the comparison of theoretical predictions of the two different treatments furnishes a more stringent probe of the frame-transformation approximation.

Figure 3 compares the LS [14] and jj-photoionization cross sections, calculated between the $6s$ and $5d_{3/2}$ thresholds, with the observations of Griesmann, Esser, and Hormes [16]. It is evident from Fig. 3 that R-matrix calculations in either LS or jj coupling are almost identical. For the reasons outlined above, the spin-orbit-averaged theoretical energy was used for the $6p_{1/2,3/2}$ threshold in the MQDT calculation done with the K matrix deduced from the LS-coupled R-matrix calculation. Various minor discrepancies between the LS results and experiment, detailed in Ref. [14], have not been removed by performing the R-matrix calculation in jj coupling. The structure of the autoionizing odd-parity $J=1$ spectrum below the $5d_{3/2}$ threshold is due to the $5dnp$ and $5dnf$ resonances. This spectrum has been recently reinvestigated using laser spectroscopy by Gounand *et al.* [23], the $5dnlJ=1^o$ levels being excited from the $5d^2\,{}^1S_0$

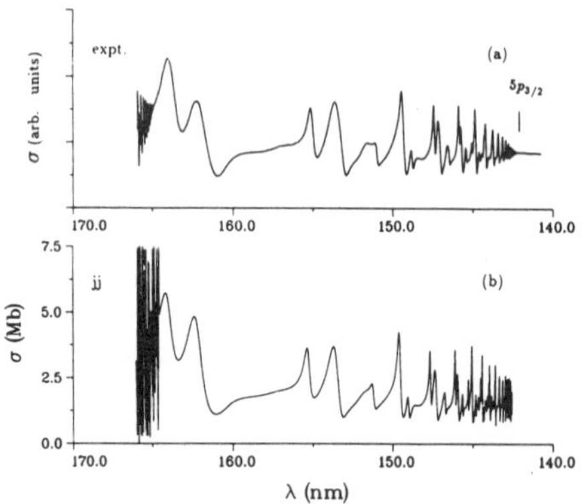

FIG. 2. Same as Fig. 1 in the wavelength region from 166 to 143 nm, between the $4d_{5/2}$ and $5p_{3/2}$ thresholds.

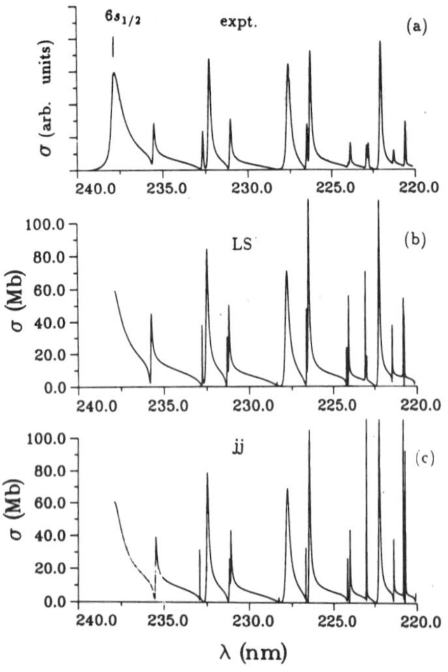

FIG. 3. Photoabsorption spectrum of Ba in the wavelength region from 240 to 220 nm, below the $5d_{3/2}$ threshold: total photoionization cross section. (a) Relative measurement of Griesmann, Esser, and Hormes [16] (wavelength resolution ~0.05 nm). (b) LS results. (c) jj results.

level; in addition, the assignments of the observed resonances deduced from the R-matrix calculation [14] are presented in Ref. [23].

The experimental results obtained in Ba by Brown and Ginter [24] just below the $6p_{1/2}$ threshold are compared with R-matrix results in Fig. 4. The LS and jj results are here again almost identical, both calculations reproducing the experimental spectrum accurately. Here curves 4(b) and 4(c) display both length and velocity results which are very close, giving some confidence in the convergence of the variational calculations.

3. Radium

Radium involves a much stronger spin-orbit interaction than Ba and thus the comparison of R-matrix results obtained in either LS or jj coupling is of particular interest for analyzing the limitations of the frame transformation for describing spin-orbit effects. The $J=1$ odd-parity spectrum of Ra is largely unknown experimentally, however, as the only available data besides that compiled by Moore [18] concerns the $7snp\ ^1P_1$ principal series observed by Tomkins and Ercoli and analyzed by Armstrong, Wynne, and Tomkins [19]. We first consider how these data relevant to the bound spectrum are described by R-matrix calculations, after which we will present R-matrix predictions for the photoionization spectrum.

(a) $J=1$ odd-parity bound spectrum of Ra. The $J=1$ odd-parity spectrum of Ra is much simpler than the homologous spectrum of Ba since only three doubly excited states appear, while in Ba 12 doubly excited levels lie below the first ionization limit. The three doubly excited levels perturbing the $7snp\ ^{1,3}P_1$ Rydberg series of Ra are the $6d7p\ ^3D_1$, $^3P_1$, and $^1P_1$ levels. The two former levels appear in the table of Moore[18] while the last one has not been observed. Two previous two-channel MQDT calculations have been carried out to analyze the perturbation of the $7snp\ ^1P$ series by the $6d7p\ ^1P$ level. The first analysis was the empirical treatment of Armstrong, Wynne, and Tomkins [19]. The second, performed by Kim and Greene [25], used an eigenchannel R-matrix calculation in LS-coupling and ignored fine-structure effects altogether. In both studies, the classification of low-lying levels was questioned.

We have performed two different R-matrix calculations, namely a five-channel treatment in LS coupling, including the $7snp$ and $6dnp$ $J=1$ channels, and a 13-channel treatment in jj coupling, introducing in addition the $6dnf$, $7pns$, and $7pnd$ $J=1$ channels. These latter channels do not support any bound level and have accordingly been treated as strongly closed in the LS calculation. Figure 5 displays the theoretical Lu-Fano plots showing $-\nu_{7s}$ (mod 1) vs $\nu_{6d_{3/2}}$ (ν_{7s} and $\nu_{6d_{3/2}}$ are the

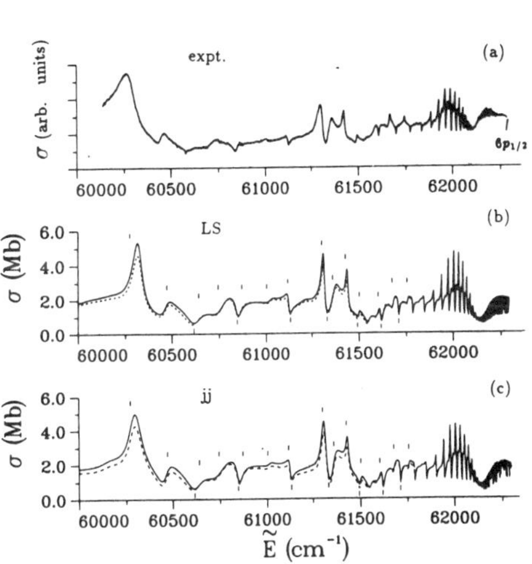

FIG. 4. Photoabsorption spectrum of Ba in the energy range from 60 000 cm^{-1} to the $6p_{1/2}$ threshold: total photoionization cross section. (a) Relative measurement of Brown and Ginter [24] (energy resolution ~0.11 cm^{-1}). (b) LS velocity (——) and length results (— — —); vertical bars indicate the position of the observed absorption peaks and minima. (c) jj velocity (——) and length results (— — —).

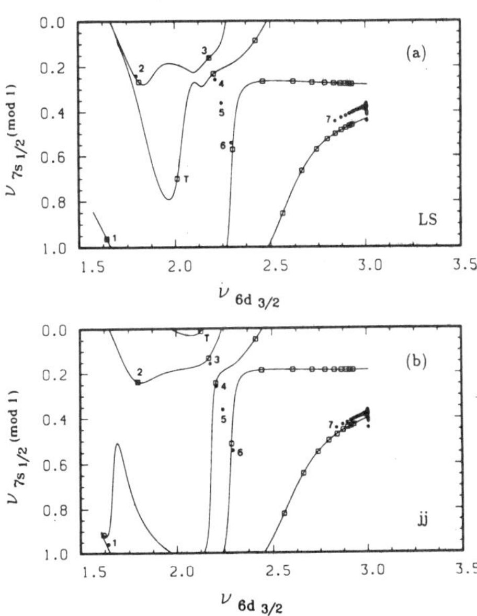

FIG. 5. Lu-Fano plot of the $J=1$ odd-parity bound levels of Ra in the $-\nu_{7s}$ (mod 1) against $\nu_{6d_{3/2}}$ plane. (a) ——, LS calculation; ●, experimental level positions [17–19]: (1) $7s7p\ ^3P_1$, (2) $7s7p\ ^1P_1$ (3)–(6) see the text, (7) $7s13p\ ^1P_1$; □, theoretical level positions [for (T) see the text]. (b) ——, jj calculation.

effective quantum numbers relative to the $7s$ and $6d_{3/2}$ thresholds). Curves 5(a) and 5(b) compare the LS and jj results, respectively, with the known levels [17–19]. The low-lying $7s7p\ ^{1,3}P$ levels [18] and the high-lying $7snp\ ^1P$ levels [19] ($n \geq 13$) agree well with our calculated values, a better description of high-lying levels being obtained with jj calculation. Discrepancies occur in the intermediate-energy range, however, with level number (5) being far from each theoretical curve. However, in this energy range the two theoretical curves are rather different and both support one theoretical energy level (labeled by T) which cannot be associated with any experimental level. The energy value predicted by the jj calculation is 30477 cm^{-1}. Four levels, namely the $7s8p\ ^{1,3}P$ levels and $6d7p\ ^3D$, ^3P levels, are expected to lie in the range $2.0 \leq \nu_{6d_{3/2}} \leq 2.5$ in agreement with our predictions. The four levels are so intermixed that it is not possible to unambiguously label them, except in the case of level number (6) for which the calculations seem to confirm the classification of Russell [17] adopted by Moore [18], namely $6d7p\ ^3P$. Additional experimental investigations would help greatly to sort out this problem and provide a decisive judgment on the accuracy of the present calculations. Our prediction for the $6d7p\ ^1P$ level, at 37 855 cm^{-1}, based on the jj calculation, is close to previous semiempirical predictions [19] (37 895–38 048 cm^{-1}).

Some more points should be discussed here. Our calculations show the oscillator strengths of the $7s^2$–$7snp\ ^3P$ series are considerably weaker than those of the $7s^2$-$7snp\ ^1P$ series, by a factor of about 800 for $n \geq 13$ in the jj calculation. It is surprising that the triplet states are so weakly excited from the ground state, considering the strength of the radium spin-orbit interaction. But this result is consistent with the fact that the triplet $7snp\ ^3P$ series has not been observed in absorption measurements from the ground state. It must be noted that the triplet $m_0snp\ ^3P$ series of the lighter alkaline earths have not been observed from the ground state either, but that is less surprising owing to the smaller spin-orbit interaction. Multistep laser investigation, as performed in Ca, Sr, and Ba [12], would be desirable to extend our knowledge of the Ra spectrum, although its radioactivity clearly causes additional experimental difficulties. Unlike Ba, oscillator strengths of the principal series of Ra behave very regularly in the high-energy range because no perturber occurs there.

One major feature in all the heavy alkaline-earth atoms is the occurrence of a very strong interaction between the m_0snp and $(m_0-1)dnp\ ^1P$ channels; just as for Ca [7(a)], Sr [6(b)], and Ba [14], the corresponding channels of Ra are almost equally mixed at the m_0s threshold, the channel-mixing angle around 0.2π being close to the theoretical maximum 0.25π. [The channel-mixing angle θ is defined for a two-channel system in MQDT as the rotation angle needed to transform the reaction matrix into its diagonal representation. An arbitrary mixing angle θ can be transformed into the range $-\pi/4 \leq \theta \leq \pi/4$ by using the fact that the transformation $\theta \rightarrow \theta + n\pi$ simply multiplies each eigenstate of the reaction matrix by $(-1)^n$, and also by using the fact that the transformation $\theta \rightarrow \pi/2 - \theta$ amounts to interchanging the arbitrary ordering of the eigenstates.] This near invariance of the channel mixing, discussed by Wynne and Armstrong [26], was used by Armstrong, Wynne, and Tomkins [19] to predict the locations of unobserved Ra levels.

(b) Absorption spectrum of Ra. Figure 6 compares cross sections for photoionization of ground state Ra obtained below the $6d_{3/2}$ threshold using R-matrix calculations in LS and jj coupling. Curves 6(a) and 6(b) display marked differences in the whole energy range. Both MQDT calculations introduce 13 channels. In addition, an eight-channel R-matrix calculation in LS coupling was carried out in which the $6pns, nd$ channels were treated as strongly closed and omitted from the MQDT calculation. The corresponding result is almost identical to curve (b) in Fig. 6, thus excluding the possibility that differences between curves 6(a) and 6(b) might result from inaccuracies in the treatment of the strongly closed channels converging to the $6p$ thresholds. The autoionizing structures in this energy range are due to $6dnp$ ($n \geq 8$) and $6dnf$ $J=1$ resonances and perhaps to $7p8s$ $J=1$ resonances which are likely to lie below the $6d_{3/2}$ threshold. The resonances appear to be so intermixed that we have not attempted to identify them.

The LS and jj results obtained for the photoionization cross sections between the $6d_{5/2}$ and $7p_{3/2}$ thresholds are compared in Fig. 7. In this energy range, clearly curves 7(a) and 7(b) bear a much closer resemblance than in the lower-energy range. However, differences, mainly near the strong resonant peak around 60 000 cm^{-1}, are much more visible than in Figs. 3 and 4 corresponding to Ba. Here again we have not attempted to identify the predicted structures due to transitions to the $7pns$ and $7pnd$ levels. Since the resonances are very broad and overlapping, it is almost certain that, just as for the homologous reso-

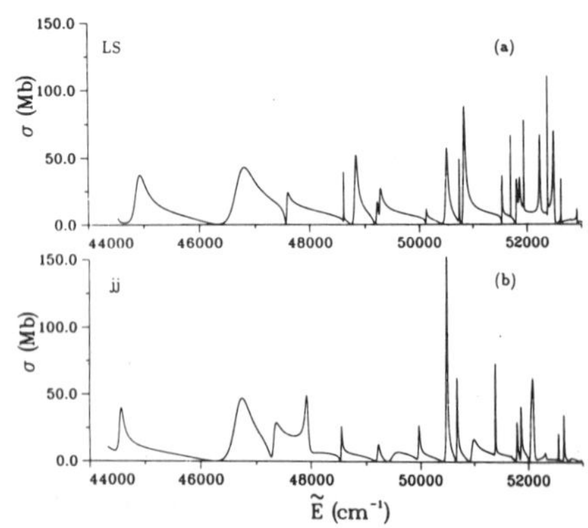

FIG. 6. Total photoionization cross section of ground-state Ra in the energy range from 44 000 to 53 000 cm^{-1}, below the $6d_{3/2}$ threshold. (a) LS results. (b) jj results.

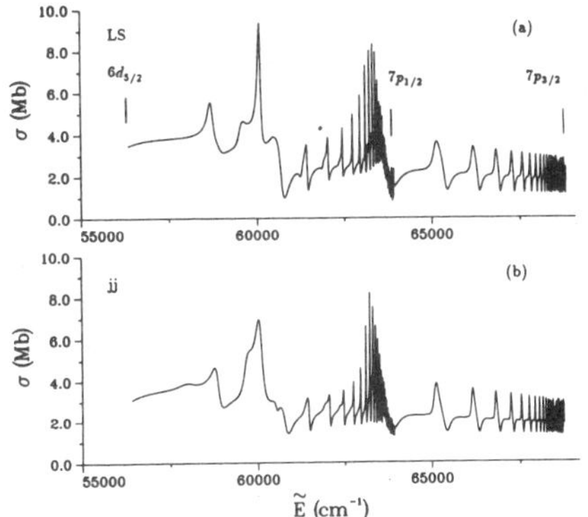

FIG. 7. Same as Fig. 6 in the energy range between the $6d_{5/2}$ and $7p_{3/2}$ thresholds.

nances of Ba [14], an independent particle label is meaningless for most of them.

It is worth noting that the autoionization pattern predicted for Ra above the $6d_{5/2}$ threshold presents many similarities with that observed for Ca, Sr, and Ba in the corresponding energy range. Below the $7p_{1/2}$ threshold, the structures with periodic enhancement due to low-lying $7p_{3/2}nl$ levels mixed with Rydberg levels $7p_{1/2}nl$ bear much resemblance to the structures observed in Ba (see Fig. 4) and in Sr (see Fig. 9 of Ref. [6(a)]). In the same way, the periodic pattern of asymmetrical absorption profiles predicted between the $7p_{1/2}$ and $7p_{3/2}$ thresholds is very similar to that observed near the m_0p thresholds of Sr [6(a)], Ba [14], and even Ca [7(a),7(b)]. These marked similarities throughout the heaviest alkaline earths reflect the systematic near invariance of the electronic channel mixings. Here, the strongest channel mixing in all the heavy alkaline-earth atoms corresponds to the $m_0p\epsilon d$-$(m_0-1)d\epsilon f$ mixing. From Figs. 1, 3, and 6, it is evident that below the $(m_0-1)d$ threshold there is much less evidence for near invariance in electronic channel mixing, although much of this difference in appearance is due to the variation of the threshold splitting $(m_0-1)d-m_0p$ in these atoms.

The large differences between curves (a) and (b) of Fig. 5 clearly suggest some limitations of the frame transformation to account for spin-orbit effects in Ra. To understand these limitations, recall that the term "frame transformation" in this paper implies the following procedure. First, *LS*-coupled reaction matrices are obtained variationally, neglecting all spin-orbit interactions. Second, these *LS*-coupled matrices are recoupled by a geometric *jj*-*LS* transformation to give a single *jj*-coupled reaction matrix. (It is easy to show that this recoupling, by itself, will have no effect on the spectrum whatsoever if the initial-state calculation ignores fine-structure interactions, and if *j*-independent thresholds are used in the MQDT calculation.) The third step is to perform the quantum-defect calculation on a fine-energy grid, but *including the experimental j-dependent spin-orbit splitting* of the inner electron threshold energies. This last step is the only one which uses ion-dependent information about the spin-orbit interaction, but it has a major qualitative effect at energies close to fine-structure-split thresholds. Accordingly, the frame transformation accounts mainly for the spin-orbit interaction of the inner electron, while it is neglected for the outer excited electron except through effects associated with energy conservation. A good measure of the core-electron spin-orbit interaction is the difference between the quantum defects of spin-orbit-split doublet-ion levels. For the m_0p ion core, $\Delta\mu=0.045$ and 0.12 for Ba^+ and Ra^+, respectively, while for the $(m_0-1)d$ ion core one has $\Delta\mu=0.012$ and 0.030. It is much more difficult to estimate the strength of the spin-orbit interaction for the outer electron, but it is clear that this is far stronger for the *np* electrons involved in $(m_0-1)dnp$ levels in Ra than in the lighter alkaline earths. Part of the difference between curves (a) and (b) in Fig. 5 is thus likely to result from the incomplete treatment of the spin-orbit interaction by the *jj*-*LS* frame transformation. Moreover, below the 6*d* threshold the energy dependence of the MQDT parameters is stronger than above, and thus the results are much more sensitive to the values of the threshold energies used in the MQDT calculation than are the results above the 6*d* threshold. The spin-orbit interaction strength of *nd* electrons in the 7*pnd* levels of Ra is likely to be of the same order of magnitude as that of *np* electrons in the 5*dnp* levels of Ba. But no difference occurs between curves (b) and (c) of Fig. 3, while some are visible in Fig. 6. These differences probably result from the fact that the fine-structure splitting of the 7*p* level of Ra^+ is three times larger than the splitting of the 5*d* level of Ba^+.

No experimental spectrum is available in this energy for Ra and thus the reliability of the predictions remains to be tested. In particular, one major open question is whether the present *jj*-coupling calculation adequately accounts for relativistic effects beyond the spin-orbit interaction. As argued elsewhere [15], these are described at least approximately through the use of a semiempirical model potential $V(r)$, but this approximation has not yet been adequately tested in an atom as heavy as radium.

B. $J=0$ and 2 even-parity spectra of Ba below the $5d_{3/2}$ threshold

This section deals with the $J=0$ and 2 even-parity spectra of Ba, which have been extensively investigated since the advent of laser spectroscopy. In addition to experimental investigations making a term analysis [27,28] high-resolution laser studies have been used to probe the detailed structures and properties of Rydberg levels and doubly excited autoionizing levels. Most of the pioneering investigations dealt with the perturbations of the $6sns\ ^1S_0$ and $6snd\ ^{1,3}D_2$ series by low-lying doubly excited states. Perturbations are reflected not only by level structure but also by lifetimes, hyperfine structure, Landé *g* factors, and various other observables (see Ref. [4] and references therein). Now the focus has shifted toward the autoionizing Rydberg states, experiments having revealed interesting properties such as the extreme stability [29] of some levels.

The MQDT has enjoyed remarkable success in describing the experimental data. However, previous analyses have utilized empirical MQDT parameters which are adjusted to agree with a particular set of measurements. The complexity of Ba is such that these empirical treatments encounter serious difficulties relating to the energy dependence of the parameters, to the occurrence of isolated perturbers, and to the sometimes overwhelming number of interacting channels [4]. The previous empirical analyses devoted to bound spectra [27,30] were achieved using the eigenchannel MQDT formalism [3(b)], whereas now the phase-shifted reaction-matrix approach [3(c)] is currently used in the autoionizing region, although this is purely a matter of convenience.

It is of particular interest to analyse how the eigenchannel R-matrix approach is able to handle the bound spectrum of Ba, in particular the complicated $J=2^e$ spectrum investigated so extensively in previous works. The empirical MQDT treatments showed large departures from the geometric jj-LS frame transformation. R-matrix calculations performed in the present work for the $J=0^e$, $J=2^e$ bound spectra, in both LS and jj coupling, should shed light on whether these departures are due to the spin-orbit interaction or to the approximations used to simplify the fitting of empirical MQDT parameters. Moreover, by overcoming the limitations of empirical MQDT, it should be possible to get new insight into the nature of channel mixing and to check the assignments of levels, especially in cases where the term designation has been controversial.

In the autoionizing region, the new R-matrix results presented below will concern the $5dnl$ $J=0^e$ and $J=2^e$ autoionizing Rydberg series and the $6p^2\,{}^1S_0$ level, excited from either the $6s6p$ or else the $5d6p\,{}^1P_1$ bound level. Here again calculations are conducted in both LS and jj coupling.

1. $J=0$ even-parity bound spectrum

The $J=0^e$ spectrum consists of the $6sns\,{}^1S_0$ Rydberg series and of six doubly excited levels, low-lying members of Rydberg series converging to the $5d_j$ or $6p_j$ thresholds. Five-channel R-matrix and MQDT calculations are carried out. The jj-coupled ionization channels are $6s_{1/2}ns_{1/2}$, $5d_{3/2}nd_{3/2}$, $5d_{5/2}nd_{5/2}$, $6p_{1/2}np_{1/2}$, and $6p_{3/2}np_{3/2}$. Identifications of the observed levels were based on an empirical four-channel MQDT treatment [27] involving only one $6pnp$ channel introduced to take care of the $6p^2\,{}^3P_0$ bound level.

Figure 8 compares the theoretical Lu-Fano plots to experiment, curves 8(a) and 8(b) showing the LS and jj results, respectively. To avoid inaccuracies at the low-energy end of the $6pnp$ channels, the theoretical $6p_{1/2,3/2}$ threshold energy is used in the MQDT calculation conducted following the R-matrix calculation in LS coupling. The dots correspond to experimental data [18,27,31]. For $n \geq 30$, the very precise energy values measured by Neukammer *et al.* [31] are plotted instead of those given in Ref. [27]. Note that a recent experiment [23] has confirmed that the $5d^2\,{}^1S_0$ level [32] lies at 25 873.83 cm^{-1}. The curves of Fig. 8 have to be compared with Fig. 4 of Ref. [27] which, however, does not include the low-energy levels for $\nu_{5d_{3/2}} \leq 2.9$. Both R-matrix calculations correctly reproduce the perturbations of the $6sns$ series by the $5d6d\,{}^3P_0$, $5d6d\,{}^1S_0$, $5d7d\,{}^3P_0$, and $6p^2\,{}^3P_0$ levels, the jj results being closer to experiment. The description of the low-lying $5d^2$ levels is less satisfying; note that theoretical curves 8(a) and 8(b) are quite different in the low-energy range, a point which is discussed in Sec. IV. One key result of the calculations concerns the assignment of the various levels: the present study completely confirms the identifications deduced from empirical MQDT analysis [27]. The classifications of some levels are still questioned [26,33] but we hope this study will close the controversy.

2. $J=2$ even-parity bound spectrum

The $J=2^e$ bound spectrum is much more complicated than the $J=0^e$ spectrum, since it involves 16 doubly excited levels pertaining to $5dnd$ ($n=5$–7), $5dns$ ($n=7,8$),

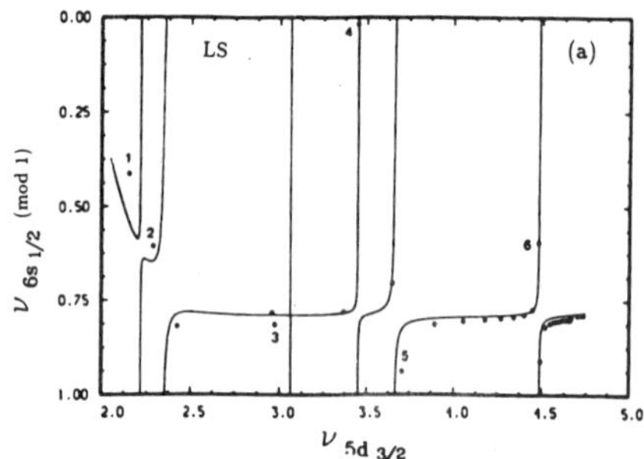

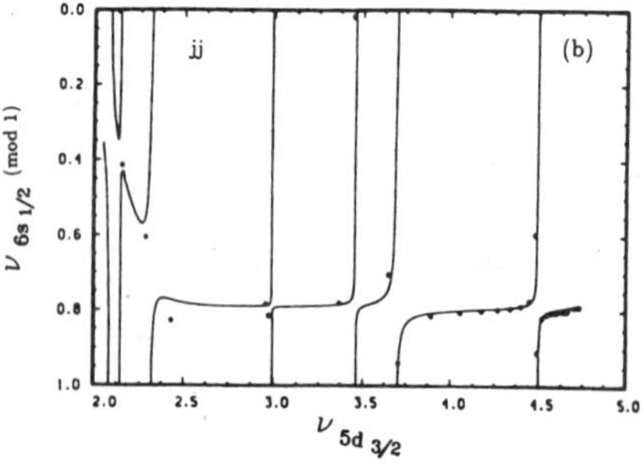

FIG. 8. Lu-Fano plot of the $J=0$ even-parity bound levels of Ba in the $-\nu_{6s}$ (mod 1) against the $\nu_{5d_{3/2}}$ plane. (a) ——, LS calculation; ●, experimental level positions [18,27,31]: (1) $5d^2\,{}^3P$, (2) $5d^2\,{}^1S$, (3) $6p^2\,{}^3P$, (4) $5d6d\,{}^3P$, (5) $5d6d\,{}^1S$, (6) $5d7d\,{}^3P$; (b) ——, jj calculation.

and $6p^2$ configurations which perturb in several places the $6snd\ ^1D_2$ and $^3D_2$ Rydberg series. The *R*-matrix and MQDT calculations are conducted using 11 channels. The *jj*-coupled channels are as follows: $6s_{1/2}nd_{3/2}$, $6s_{1/2}nd_{5/2}$, $5d_{3/2}ns_{1/2}$, $5d_{3/2}nd_{3/2}$, $5d_{3/2}nd_{5/2}$, $5d_{5/2}ns_{1/2}$, $5d_{5/2}nd_{3/2}$, $5d_{5/2}nd_{5/2}$, $6p_{1/2}np_{3/2}$, $6p_{3/2}np_{1/2}$, and $6p_{3/2}np_{3/2}$. This spectrum was analyzed previously [30] by an empirical nine-channel MQDT treatment which disregarded the $6p^2\ ^3P_2$ level and introduced only one $6pnp$ channel to account for the $6p^2\ ^1D_2$ level. Later, experimental data on Landé *g* factors and hyperfine structure have permitted improvements on the initial MQDT model which had been derived from energies [4] alone.

Figure 9 compares the theoretical Lu-Fano plots with experimental data [18,27,31,34]. Note that the low-lying $6s5d\ ^{1,3}D_2$ levels fall below the energy range considered in the present study. Here again curves 9(a) and 9(b) correspond, respectively, to the *LS* and *jj* results. As in the

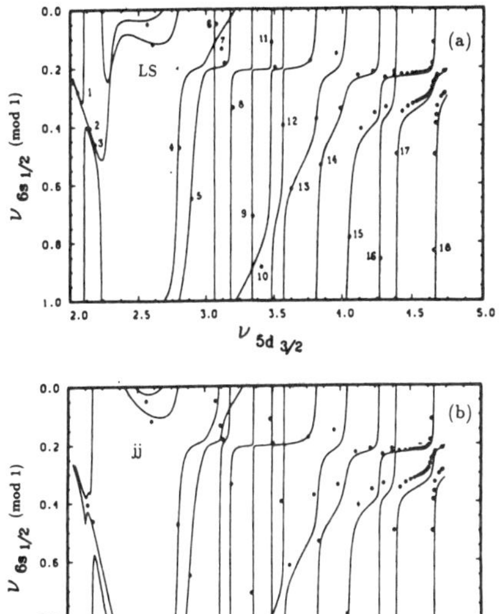

FIG. 9. Lu-Fano plot of the $J=2$ even-parity bound levels of Ba in the $-\nu_{6s}$ (mod 1) against the $\nu_{5d_{3/2}}$ plane. (a) ——, *LS* calculation; ●, experimental level positions [18,27,31,34]. Level assignments recommended by the present *jj* calculation are as follows: $5d^2\ ^3F_2$, $^1D_2$, $^3P_2$, (1,2,3); $5d7s\ ^3D_2$, $^1D_2$, (4,5); $6p^2\ ^3P_2$, $^1D_2$, (6,7); $5d6d\ ^3D_2$, $^3F_2$, (8,9); $6s7d\ ^1D_2$, (10); $5d6d\ ^1D_2$, $^3P_2$, (11,12); $6s8d\ ^1D_2$, (13); $5d8s\ ^3D_2$, $^1D_2$, (14,15); $5d7d\ ^3D_2$, $^3F_2$, $^1D_2$, (16,17,18). These classifications are discussed further in the text and in Table II. (b) ——, *jj* calculation.

$J=0^e$ case, both Lu-Fano plots differ in the low-energy range where the MQDT parameters have a strong energy dependence. These curves are to be compared with Fig. 1 of Ref. [30], which, however, does not consider the low-lying levels. There, the points were exactly on the curve, whereas some deviations are visible here. However, the agreement between theory and experiment is generally quite satisfying, accounting for the great complexity of the spectrum despite some noticeable discrepancies. We have numbered the doubly excited levels as well as some $6snd$ levels, whose assignment will be discussed later. The calculations correctly reproduce the perturbations due to the $5d^2\ ^3F_2$, $^1D_2$, $^3P_2$ (1,2,3), $5d6d\ ^3D_2$, $^3F_2$, $^1D_2$, $^3P_2$ (8,9,11,12), and $5d7d\ ^3D_2$, $^3F_2$, $^1D_2$ (16,17,18) levels. The perturbations due to the $5dns\ ^3D_2$, $^1D_2$ levels (4 and 5 for $n=7$; 14 and 15 for $n=8$) are also well described. The *R*-matrix calculations permit the previous assignment of levels to be checked. Only a few levels, as detailed below, were misclassified in Ref. [30]. The levels listed above were correctly assigned in Ref. [30]. In contrast, in the intermediate-energy range where there are more perturbers than there are $6snd$ levels, our calculations suggest a reclassification of levels 6, 10, and 13 as indicated in Table II. This new classification is close to the original one given by Moore [18]. It is worth noting that several experiments [33] have shown recently that levels 6 and 7 are predominantly $6p^2$ in character, in harmony with our new classification. One must keep in mind that no level corresponds actually to an independent-electron configuration: the $6p^2$ levels interact strongly with nearby levels, and in the higher-energy range several doubly excited levels are strongly mixed with dense Rydberg levels.

Here again, the MQDT calculation done with the *K* matrix deduced from *LS*-coupled *R*-matrix calculations utilizes theoretical $6p_{1/2,3/2}$ threshold energies. This was found *essential* for the *LS* calculation to give a correct description of the whole spectrum, in contrast with the $J=0^e$ case where only a small improvement is achieved in this fashion. From Figs. 9 and 7, it appears that neglect of the fine structure of the $Ba^+(6p)$ ionic level in the *LS* calculation reduces the accuracy compared to the *jj* calculation only for the $6p^2$ levels (6 and 7 on Fig. 9 and 3 on Fig. 8). For the other $J=2^e$ levels, the two calculations are comparable, or perhaps the *LS* results agree somewhat better with experiment overall.

We now turn to a comparison of the MQDT channel-mixing parameters obtained at the $6s$ threshold, either empirically or from *R*-matrix calculations. To better analyze the departures from the frame transformation, instead of the eigenvectors $U_{i\alpha}$ of the *jj*-coupled *K* matrix, we compare instead the eigenvectors $V_{\bar{\alpha}\alpha}$ of the matrix R^tKR, where *R* denotes the *jj*-*LS* geometric recoupling transformation. The $\bar{\alpha}$ channels correspond to pure *LS*-coupled channels. The α channels are the exact eigenchannels of the reaction matrix. If spin-orbit effects are negligible within the reaction volume, the matrix $V_{\bar{\alpha}\alpha}$ is expected to be block diagonal. The only MQDT parameters whose values are close in the three different treatments are the eigenquantum defects μ_α associated with

TABLE II. Positions of four experimental $J=2^e$ levels of Ba, showing the new classifications obtained in the present *jj*-coupled calculation.

Level number	Energy (cm^{-1})	Ref. [18]	Label Ref. [30]	*jj* result
6	35 344.42	$6p^2\,{}^1D_2$	$6s7d\,{}^1D_2$	$6p^2\,{}^3P_2{}^a$
7	35 616.94	$6p^2\,{}^3P_2$		$6p^2\,{}^1D_2{}^a$
10	37 434.95	$6s7d\,{}^1D_2$	$6s8d\,{}^1D_2$	$6s7d\,{}^1D_2$
13	38 556.18	$6s8d\,{}^1D_2$	$6p^2\,{}^1D_2$	$6s8d\,{}^1D_2$

[a]Levels 6 and 7 are so strongly mixed that these classifications should not be taken too seriously.

the 5*dns* and 5*dnd* channels and the $V_{\bar{\alpha}\alpha}$ matrix elements describing the mixing between the 6*snd* and 6*pnp* ${}^1D_2$ channels; the corresponding channel mixing, with a channel-mixing angle $\sim 0.2\pi$, clearly dominates the $J=2^e$ bound spectrum of Ba. Within the diagonal blocks corresponding to a given *LS* symmetry, the matrix elements deduced from the *jj*-coupled *R*-matrix calculation bear a close resemblance to those obtained by neglecting spin-orbit terms in the reaction volume. Both *R*-matrix calculations differ considerably from the matrices fitted to experiment. In particular, the mixing between perturbing channels, almost completely neglected in Ref. [30], is not negligible. The *R*-matrix calculation in *jj*-coupling shows strong departures from the *jj*-*LS* frame transformation, mainly within the three 6*pnp* channels and for other channels interacting with the 6*pnp* channels. Except for the 6*snd*-6*pnp* ${}^1D_2$ interaction, the empirical values introduced for several mixing angles are very different from those calculated with the *R*-matrix approach. In particular, the 6*snd* ${}^1D_2$ and ${}^3D_2$ eigenchannels are almost uncoupled (mixing angle ~ 0.02 rad) while the empirical value derived from hyperfine measurements was 0.42. Clearly, the neglect of the spin-orbit interaction for the 6*p* electron and the various other approximations involved in the empirical treatments affects the MQDT parameters considerably.

The preceding paragraph shows that quite different sets of MQDT parameters can give energy levels in good agreement. A good description of level positions is clearly necessary but not sufficient to demonstrate that a proper description of channel interactions has been achieved. Data on more sensitive observables are needed to probe the wave functions. This point has been largely documented in Ref. [4] for the even-parity $J=2$ spectrum of Ba. We have already mentioned that the new designations of levels 6 and 7 in Fig. 9 are supported by several experiments [33]. Now we consider another example in greater detail. It is known that hyperfine-structure measurements provide a very sensitive probe of the singlet-triplet mixing in high-lying 6*snd* levels. Figure 10 illustrates the evolution with *n* of the admixture coefficient β of the 6*snd* ${}^3D_2$ channel in the high-lying $5d7d$ and 6*snd* ${}^1D_2$ levels. The plus and cross symbols marked are derived from the hyperfine measurements of Eliel and Hogervorst [35], and those of Rinneberg and Neukammer [36], respectively. Note that β is an amplitude characterizing the singlet-triplet mixing, which is constrained to lie in the range $-1/\sqrt{2} \le \beta \le 1/\sqrt{2}$. The value $\beta=0$ is obtained only when S $(=0$ or $1)$ is a good quantum number, whereas the equal mixing of singlet and triplet states is obtained when β approaches one of the extrema just noted. For the explicit definition, see Refs. [35] and [36]. The dashed and solid curves are drawn through the theoretical *LS* and *jj* values, respectively. Compared to experiment, both theoretical curves are slightly shifted to the right, the predicted $5d7d\,{}^1D_2$ levels being a little too high. However, it is evident that the experimental behavior is better reproduced by the *R*-matrix calculation in *jj* coupling than by the one carried out in *LS* coupling.

3. The $J=0$ and 2 even-parity autoionizing spectrum of Ba

The even-parity spectrum of Ba below the $5d_{5/2}$ threshold has been investigated using multistep laser spectroscopy [28,29,37–39] the 5*dnl* levels being excited from several $5d6p$ bound levels. For $J=0^e$, these investigations led to the observation of the two $5d_{3/2}nd_{3/2}$ and $5d_{5/2}nd_{5/2}$ $J=0$ Rydberg series up to large-*n* values and to the identification [38] of the $6p^2\,{}^1S_0$ level with a broad autoionizing resonance at 48 000 cm^{-1}. For $J=2^e$, the eight expected series 5*dns*, 5*dnd*, and 5*dng* have been observed [28,39]. The autoionization widths observed for

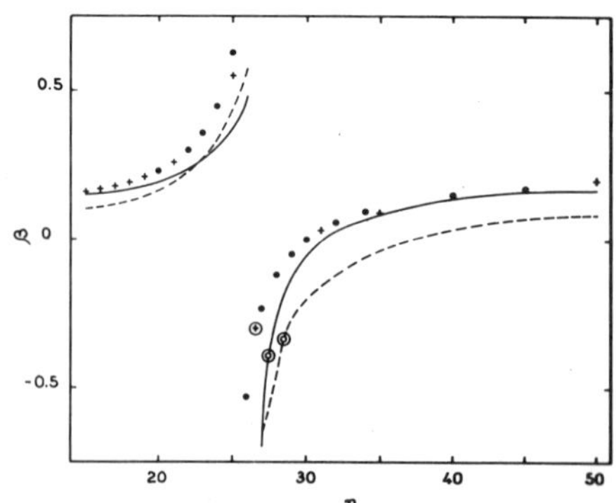

FIG. 10. Singlet-triplet β mixing coefficients for the high-lying 6*snd* ${}^1D_2$ and $5d7d\,{}^1D_2$ levels of Ba. The *LS* results (— - —) and *jj* results (——) are compared with values deduced from hyperfine structure: ● [35], + [36]. The circled symbols correspond to the $5d7d\,{}^1D_2$ level.

the 5*dnd* $J=0$ and 2 resonances are much smaller than the separation between adjacent peaks. Thus, level identifications were obtained by analyzing the experimental data with empirical MQDT models ignoring the coupling with the $6s\epsilon l$ continua [28,37]. Moreover, for $J=2^e$, the 5*dnd*-5*dng* interaction was neglected. These studies, restricted to the positions of autoionizing resonances only, were completed by additional empirical MQDT studies which account for the interactions of autoionizing levels with the continua. A first theoretical description [38] dealt with the profiles of the $6p^2\,{}^1S_0$ level and of the nearby 5*dnd* $J=0$ resonances. Later, the high-lying $5d_{3/2}nd_{3/2}$ levels were reinvestigated by Neukammer *et al.* [29] and dramatic deviations of the autoionization widths from the $\nu_{5d_{3/2}}^{-3}$ law (expected in the absence of channel interactions) have been observed, some levels being metastable against autoionization decay. This unusual suppression of the autoionization process for particular levels has been accurately described using MQDT [29,40]. Finally, Bente and Hogervorst [39] successfully used an MQDT parametrization to reproduce $J=2^e$ autoionizing features which they observed by photoionizing the $5d6p\,{}^1F_3$ bound level.

Most of the experimental photoionization measurements performed to date in the alkaline-earth atoms have been relative measurements which fail to provide an absolute scale to the cross section. Recently two absolute cross-section measurements were carried out for photoionization of the Ba $6s6p\,{}^1P_1^o$ state at energies near the 6*s* ionization threshold [41,42]. Two very recent calculations have obtained theoretical cross sections for comparison with these experiments, one a Wigner-Eisenbud *R*-matrix calculation by Bartschat and McLaughlin [43], and the other a *jj*-coupled *R*-matrix calculation by Greene and Theodosiou [44] using the same techniques described above. These two calculations are both in agreement that the measured cross sections of Kallenbach, Köck, and Zierer [41] are too large by a factor of approximately 5.5. On the other hand, the calculation of Ref. [44] suggests that the measured cross section of Burkhardt *et al.* [42] is correct in its absolute normalization to within a factor of 2.

Next we consider photoionization of excited barium 6*s*6*p* and $5d6p\,{}^1P_1$ levels into $J=0^e$ and $J=2^e$ final states. Channel-mixing parameters in the final $J=0^e$ and $J=2^e$ autoionizing energy range were obtained by performing variational calculations in either *LS* or *jj* coupling. The same *l*-dependent model potential $V(r)$ and two-electron basis set are used in both calculations (except that the spin-orbit interaction terms are, of course, omitted from the *LS* calculation). The wave functions of the initial state were obtained by diagonalizing, the two-electron Hamiltonian within the reaction volume. In the calculation done in *LS* coupling the starting level is assumed to be a pure ${}^1P_1$ level, this restriction being of course relaxed in the *jj* calculation. Also, the starting level is assumed to be isotropic in both calculations, ignoring the fact that it is typically aligned if prepared by laser excitation.

(a) The $J=0$ even-parity autoionizing spectrum of Ba. Figure 11 compares the *LS* and *jj* results obtained for the

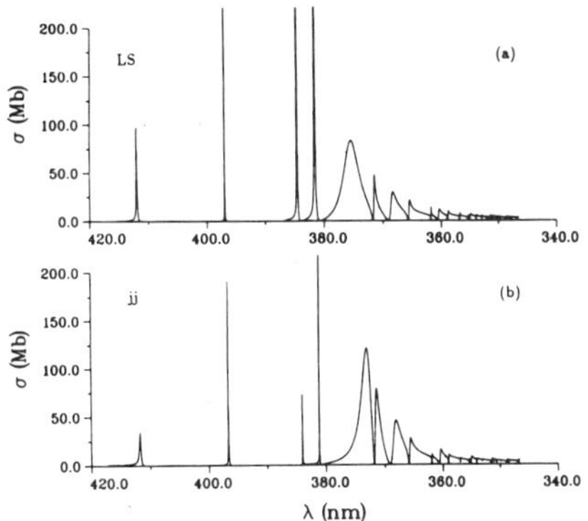

FIG. 11. Partial photoionization cross section for the $6s6p\,{}^1P_1 \to J=0^e$ symmetry of Ba in the wavelength region from 420 to 340 nm, below the $5d_{3/2}$ threshold. (a) *LS* results. (b) *jj* results.

$J=0^e$ partial photoionization cross section of the $6s6p\,{}^1P_1$ level of Ba, below the $5d_{3/2}$ threshold. As in the energy range below the 6*s* threshold, five channels are included in each calculation. Marked differences are visible in the 380–360-nm wavelength range where the $6p^2\,{}^1S_0$ level is expected to lie [38]. In the *LS* calculation this level is identified with the broad resonance around 375 nm, relatively well separated from the nearby $5d_{3/2}10d_{3/2}$ resonance in the right wing. In contrast, the *jj* spectrum shows a more complicated pattern of overlapping resonances. To achieve convergence in the $J=0^e$ *R*-matrix calculations it was found essential to include in the two-electron basis set strongly closed functions $nln'l$, involving large orbital momenta $l=3$ and 4. MQDT calculations performed with the *K* matrix deduced from the *LS*-coupled *R*-matrix calculation were performed using either experimental or theoretical threshold energies for the 6*pnp* channels; both calculations give almost identical results and thus the differences between curves 11(a) and 11(b) do not result from inaccuracies such as those which occur below the 6*s* threshold.

To better analyze the origin of the differences, we consider also the photoionization of the excited $5d6p\,{}^1P_1$ level, for which some observations are available. The *LS* and *jj* results obtained for the $5d6p\,{}^1P_1 \to J=0^e$ spectrum of Ba are compared in Fig. 12, in the same energy range as above. In addition, we indicate the positions of the resonance peaks observed by Camus *et al.* [28] and Aymar, Camus, and Hindy [38], the autoionizing levels being excited from the $5d6p\,{}^1P_1$ or ${}^3P_1$ level. Both curves 12(a) and 12(b), corresponding to *LS* or *jj* results respectively, correctly reproduce the positions of resonant peaks. Note that the differences between the calculated peak heights in curves 12(a) and 12(b) are not significant for the sharp resonances, curve 12(a) being obtained with a finer energy mesh than curve 12(b). However, near 45 000 cm^{-1}, the observed structures, i.e., the

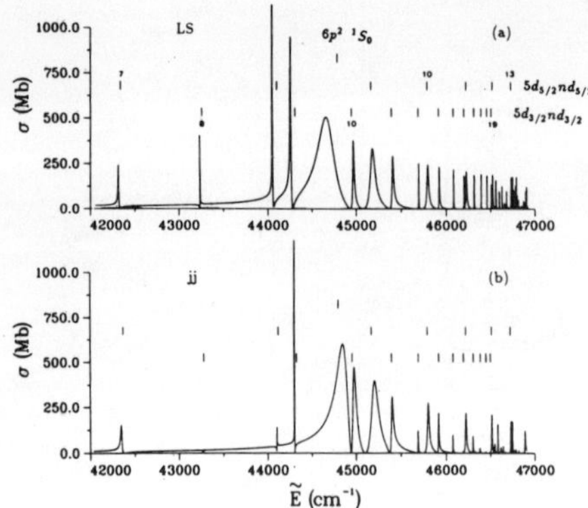

FIG. 12. Partial photoionization cross section for the $5d6p\ ^1P_1 \rightarrow J=0^e$ symmetry of Ba in the energy range 42 000 to 47 000 cm^{-1}, below the $5d_{3/2}$ thresold (the energies are relative to the ground state of Ba). (a) *LS* results; the vertical bars indicate the positions of observed autoionizing states [28,38]; some members of the $5d_{3/2}nd_{3/2}$ and $5d_{5/2}nd_{5/2}$ Rydberg series are labeled by their *n* value. (b) *jj* results.

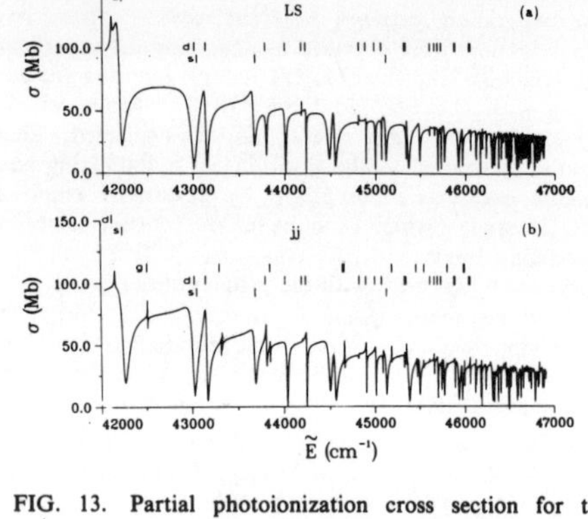

FIG. 13. Partial photoionization cross section for the $6s6p\ ^1P_1 \rightarrow J=2^e$ symmetry of Ba in the energy range 42 000 to 47 000 cm^{-1}, below the $5d_{3/2}$ threshold (the energies are relative to the ground state of Ba). (a) *LS* results (11-channel calculation); the vertical bars indicate the positions of autoionizing levels observed by Camus *et al.* [28]; the $5dns$ (*s*), $5dnd$ (*d*) and $5dng$ (*g*) are distinguished by the different ordinates used for the bars. (b) *jj* results (14-channel calculation).

broad and asymmetric resonance corresponding to the $6p^2\ ^1S_0$ level, and the nearby peak associated with the $5d_{3/2}10d_{3/2}$ level are better described by the *jj* calculation. The irregular variation of the autoionization widths along the $5d_{3/2}nd_{3/2}$ series, observed by Neukammer *et al.* [29], for $n \geq 14$, is already visible in Fig. 12 for smaller-*n* values. Because these observations have already been successfully interpreted we have not attempted to include enough energy points in our calculation to represent the spectrum just below the $5d_{3/2}$ threshold in greater detail.

The present calculations completely confirm the observations of Ref. [38] concerning the $6p^2$ autoionizing resonance. Note that the recent calculation of Bartschat and McLaughlin [43] fails to observe this level above the 6*s* threshold, in contrast to Ref. [44]. In addition, the small differences between curves (a) and (b) in Figs. 11 and 12 suggest a mild failure of the frame transformation for describing the $6p^2\ ^1S_0$ level. For this state, of course, the spin-orbit interaction must play the largest role among all members of the $6pnp\ ^1S_0$ Rydberg series.

(b) The $J=2$ even-parity autoionizing spectrum of Ba. Now we turn to the $J=2^e$ partial cross sections for photoionization of the $6s6p\ ^1P_1$ level of Ba, calculated below the $5d_{3/2}$ threshold. The 11 channels introduced in the *R*-matrix calculations done in *LS* coupling are identical to those used in calculating the bound spectra. In *jj* coupling, two different calculations were carried out, involving either 11 or 14 channels. The latter calculation treats the $5dng$ $J=2$ channels as "weakly closed," whereas the former calculation treats them as "strongly closed."

Figure 13 compares the 11-channel *LS* results with the 14-channel *jj* results. [The 11-channel *jj* calculation only differs from curve 13(b) by the absence of the sharp peaks associated with the $5dng$ resonances.] Curves 13(a) and 13(b) are in good agreement. In particular, the maxima of narrow peaks and minima are located at the same energies in both calculations. Especially in the lower-energy range, however, the shapes of the broad resonances show some noticeable differences. No experimental data are available for comparison except at particular energies [41,42] (see Fig. 3 of Ref. [44] for a comparison of theory and experiment).

To check the reliability of the calculations, positions of the resonances observed by Camus *et al.* [28] by photoionizing the $5d6p$ levels are marked by vertical bars. The predicted positions of the $5dng$ resonances [curve 13(b)] agree well with the observation. These sharp resonances are added to a more complex pattern of broad resonances separated by sharp windows. The detailed correspondence between these resonances and the observed $5dns$ and $5dnd$ levels is not obvious. To clarify the situation, we have calculated the $J=2^e$ partial photoionization cross section of the $5d6p\ ^1P_1$ level. The results of the 11-channel calculation done in *LS* coupling are compared on Fig. 14 with the same set of experimental level positions. Nearly every vertical bar can be associated without ambiguity to a calculated peak, the additional theoretical resonances being very weak. Moreover, we have verified that the previous assignments of levels are supported by our calculation. From Fig. 14, it is clear that the *R*-matrix description of the $J=2^e$ autoionizing resonance positions is accurate. The resonances displayed in Fig. 14 are relatively narrow, in agreement with the observations [28], but in apparent contradiction with Fig. 13.

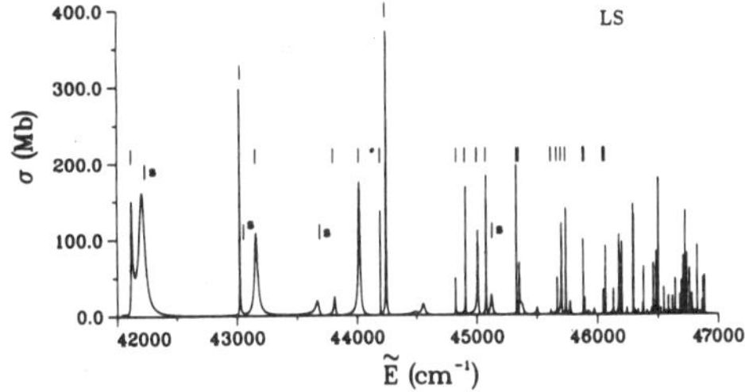

FIG. 14. Partial photoinozation cross section for the $5d6p\,^1P_1 \rightarrow J=2^e$ symmetry of Ba, below the $5d_{3/2}$ threshold. The calculated *LS* results are shown as a solid curve, while vertical bars indicate the positions of auotionizing levels observed by Camus *et al.* [28]. All the bars indicate $5dnd$ resonances except those labeled by (*s*), which are associated with $5dns$ levels.

We now address the striking differences between Figs. 13 and 14. Since in both cases, the spectral density of the $J=2^e$ autoionizing states is the same, the differences must derive from differences in the mode of excitation of the resonances. In photoionization of the $5d6p$ level, the excitation of the $6s\epsilon l$ continua is negligible compared to that of the autoionizing Rydberg series $5dns,nd$ while the opposite holds for photoionization of the $6s6p$ level. For overlapping resonances, the distribution of oscillator strength is strongly affected by interference effects between autoionizing paths and direct photoionization paths, and as explained by Mies [45] the apparent widths of resonances depend on the excitation dipole matrix elements. Studying the case where two Rydberg series autoionize to the same continua, Mies [45] found two extreme cases when changing the Fano-profile indices of the series while other characteristics of the series remain unchanged. In these two cases, the oscillator strength is either collected in narrow intervals, giving sharp peaks like those visible in Fig. 14, or else it is spread out over a larger energy range while the cross section exhibits windows in narrow intervals, but with the narrow peaks and the sharp windows being located at the same places. Similar behavior is found here, as Fig. 13 shows that the positions of most of the dips between the broad resonances coincide with the peak positions in Fig. 14. These structures are much more complicated than in the case studied by Mies [45], however, owing to the large number of interloping Rydberg series. Note that similar broadening and narrowing effects also occur in Figs. 11 and 12, respectively. The small differences between the resonance profiles in curves (a) and (b) of Fig. 13 probably result from differences between calculated dipole matrix elements owing to the different selection rules which have been imposed in either *LS* or *jj* coupling. Experimental observation of the Ba $6s6p\,^1P_1 \rightarrow J=2^e$ spectrum is desirable to probe the reliability of the calculations.

IV. DISCUSSION

This study demonstrates, along with the work discussed in Ref. [14], that an *R*-matrix calculation in *LS* coupling combined with a *jj*-*LS* MQDT frame transformation accurately describes channel interactions even in an atom as heavy as Ba. This conclusion is borne out by the good agreement between *LS*-coupling calculations and experiment, and also by the close agreement of the *LS*-coupled calculations with *jj*-coupled calculations, including spin-orbit terms in the short-range Hamiltonian explicitly. In atomic radium the agreement between these two methods of calculation deteriorates noticeably compared to barium, though more so in some spectral ranges than in others.

The complex and detailed spectral features and perturbations described in this paper give further evidence of the efficiency of the MQDT description at handling many interacting channels. Nevertheless, some of the difficulties encountered with the MQDT description should be pointed out. A first analytical complication occurs whenever the photoelectron energy in a channel i becomes less than $-1/2l_i^2$, at which point the "energy-normalized" Coulomb functions (f_i,g_i) become complex and therefore inconvenient. This complication is readily bypassed by matching to "analytic" Coulomb functions (f_i^0,g_i^0) in that particular channel [3(b)]. This modifies the linear MQDT equations in a standard manner (see, e.g., Eqs. (9)–(11) of Ref. [15]) and causes no further difficulties.

A more fundamental difficulty occurs whenever a channel is deeply enough closed for the exponential growth of its channel components to become "large" within the reaction zone $r \le r_0$. Our experience has shown that several difficulties then arise. (i) The short-range MQDT parameters then acquire an appreciable r_0 dependence, although normally the calculated observables such as energy-level positions and photoabsorption intensities remain independent of r_0 to a good approximation. (ii) The same MQDT parameters begin to vary much more rapidly with energy in this range, in stark contrast to their smooth behavior closer to an ionization threshold. (iii) The use of experimental thresholds in such a strongly-closed channel can adversely affect the MQDT calculation of observables, sometimes producing artificial resonances and at other times causing a physical resonance or bound state to disappear altogether; as discussed in Sec. III, this last problem can be corrected by using theoretical ionization threshold energies in such channels for consistency. Examples of (i) and (ii) can be seen in some of the Lu-Fano plots of this paper. For instance, in Fig. 5 the theoretical *LS*-coupled and *jj*-coupled curves look complicated and extremely different at low energies, in contrast to the actual predicted bound levels, most of which agree reasonably well in the two calculations. These difficulties do not invalidate the use of MQDT in any way, but they require some caution and it will be desirable to eventually understand them better.

Another major conclusion of this paper and of Ref. [14] is that eigenchannel *R*-matrix methods are now capable of predicting the short-range parameters of multichannel-quantum-defect theory for any atom having two valence electrons, to an accuracy which is spectroscopically useful. Reference [15] shows that this is true of the alkali-metal negative ions in addition to the neutral atoms treated in this paper. These computations are all comparatively modest, as the largest *jj*-coupled calculations have required only several hours of CPU time on a desktop computer workstation. To achieve spectroscopically useful results in the heavy alkaline-earth atoms like barium, the single most important element in the computational scheme is the use of a semiempirical potential $V(r)$ which generates one-electron levels (e.g., of Ba^+) in agreement with experiment. For the moment, such calculations have not been extended to the vicinity of thresholds much higher than the $6p$ thresholds of Ba^+. It should be possible to go as high in energy as the Ca calculations of Ref. [7(c)], without substantial modification of the present computer programs. However, the CPU time grows rapidly owing to the larger number of channels, and also owing to the larger value needed for the box radius r_0, which in turn demands more basis functions per channel to reach an equivalent level of convergence. Thus to go above the $8s$ threshold of Ba^+ would be somewhat more appropriate for high-speed supercomputers than for desktop workstations at their present level of technology.

A large number of previous MQDT analyses have been conducted semiempirically to interpret experimental data, especially in barium. Nearly all of these studies found that it is practically impossible to obtain enough experimental information to uniquely pin down the MQDT parameters in this fashion if the number of channels exceeds three of four. The success of small-scale calculations, like those of this paper, in calculating accurate short-range MQDT parameters should greatly help such analyses. Some discrepancies will always remain between theory and experiment, as in Fig. 9 for instance, and more generally whenever a perturber location is just slightly in error in the calculation. But even in such cases the calculated K matrix should serve as an excellent starting point for further semiempirical optimization.

It is familiar from many previous studies of atomic photoionization that observables relating to the *anisotropy* of the photoprocesses are more sensitive probes of wave-function accuracy than is the energy spectrum or the total photoionization cross section. A very recent study of Ba photoionization by Lange *et al.* [46] shows generally good agreement between measured and calculated photoelectron angular distribution asymmetry parameters β in several different channels. The calculations of Ref. [46] were conducted in *LS* coupling followed by a *jj-LS* frame transformation like the *LS* results of the present study. The origin of some remaining discrepancies between theoretical and experimental asymmetry parameters in Ref. [46] was further investigated by recalculating β using the present *jj*-coupled approach. While the *jj* and *LS* results are not completely identical, they both show comparable agreement with experiment, implying that the main limitations of these calculations do *not* stem from deficiencies of the *jj-LS* frame transformation, even in atomic barium.

Apparently these small deviations between theory and experiment in Ref. [46] derive instead from the model Hamiltonian used [Eq. (1)], which approximates the effect of the inner electrons by a semiempirical model potential which is (radially) local. Another possible source of error in some sensitive observables could be an incomplete convergence of the variational basis-set expansion, as this has not been exhaustively investigated. Most observables are relatively insensitive to the basis-set size, but there have been occasional indications (as in the unexpected importance of f and g states for describing the Ba $6p^2\,{}^1S_0$ resonance in Fig. 12) that a larger basis set might improve the convergence in some regions of the spectrum.

Despite the clear general success of the *jj-LS* frame transformation for barium, there are nevertheless a few specific instances where the fully *jj*-coupled calculations give improved agreement with the experimental spectrum. Examples visible from the present calculations include the low-lying $6p^2$ levels, specifically level number 3 in Fig. 8 which is the $6p^2\,{}^3P_0$ bound level, and the $6p^2\,{}^1S_0$ autoionizing level in Fig. 12. While the positions of these levels are accurately reproduced by the *jj*-coupled calculation, the *LS* calculation places the ${}^3P_0$ level too high and the ${}^1S_0$ level too low. This possibly indicates that the frame transformation underestimates the strength of the spin-orbit interaction for states in which the two electrons have a comparable degree of excitation. (This can be seen from the fact that this interaction tends to cause these two *LS*-coupled energy levels to repel each other.) Figure 10 also suggests that the singlet-triplet mixing is described somewhat more accurately by the *jj*-coupled calculation.

Finally, we point out again that the *jj*-coupled results are far more stable near the bottom end of each Rydberg series. In that energy range spurious results occasionally show up in the *LS*-coupled calculations using experimental threshold energies, as sometimes an unphysical bound (or quasibound) level may appear, or sometimes a physical bound level may disappear altogether. The origin of this instability is now well understood, relating to the exponential growth of strongly closed channel components. Use of the frame-transformation method in this energy range requires extra caution, and in some cases it is essential to use *theoretical* threshold energies for such channels [see item (iii) above], rather than experimental thresholds as are frequently used in MQDT.

A last question of some practical importance is whether the present calculations based entirely on a Schrödinger-level description of the valence electron dynamics can adequately describe relativistic effects. These are known to be important in heavy atoms such as barium and radium, yet all relativistic effects aside from the spin-orbit interaction are *apparently* neglected here. On this question we believe that, as discussed in Ref. [15], many relativistic effects on the two-electron spectrum such as the "relativistic mass correction" are indirectly incorporated through our use of a semiempirical one-electron model potential. In the course of adjusting the potential to reproduce experimental one-electron levels, we effectively guarantee that the phase of the outermost electron, e.g., in Ra^+, is correct at large distances $r \gtrsim 1$

a.u. This is the most important part of configuration space for determining the nature of valence electron-electron correlations, and the electron velocities are clearly small enough at these larger radii for the Schrödinger equation to suffice. Moreover, the dipole matrix elements used in calculating photoabsorption intensities are primarily sensitive to this region of configuration space beyond the inner shells. For these reasons we expect the present *jj*-coupled photoabsorption calculations, nonrelativistic except for the spin-orbit interaction, to be valid even for atomic radium. Less accurate results should be anticipated, on the other hand, for observables more sensitive to the wave function very close to the nucleus.

A better test of the accuracy of the present radium calculations is highly desirable, considering the sparse nature of available spectra. The Lu-Fano plots of Ra $J=1^{\circ}$ levels in Fig. 5 show generally good agreement between theory and experiment, as the calculated quantum defects μ_{7s} deviate by less than $\Delta\mu=0.05$ for all levels, except for two in Fig. 5. The exceptions include experimental level number 5 in Fig. 5, which apparently has no theoretical counterpart, and the theoretical level labeled *T*, having no experimentally observed counterpart. Judging from the accuracy of our calculations for Sr and Ba, and for most of the Ra levels, the most likely explanation of these major discrepancies is probably some type of experimental error or misclassification. Given our approximate treatment of relativistic effects, however, further theoretical and experimental effort is certainly needed before this conclusion can be confidently accepted.

ACKNOWLEDGMENTS

We thank M. Le Dourneuf, J. M. Launay, and P. G. Burke for their support and hospitality during a workshop in Meudon, France, where this work was initiated. We are grateful to U. Griesmann for providing us with cross-section photoionization data prior to publication. The work of one of us (C.H.G.) was supported in part by the National Science Foundation.

---

[1] C. H. Greene, Phys. Rev. A **28**, 2209 (1983); **32**, 1880 (1985); for the original iterative formulation, see also U. Fano and C. M. Lee, Phys. Rev. Lett. **31**, 1573 (1973).

[2] H. Le Rouzo and G. Raseev, Phys. Rev. A **29**, 1214 (1984).

[3] (a) M. J. Seaton, Rep. Prog. Phys. **46**, 97 (1983); (b) U. Fano and A. R. P. Rau, *Atomic Collisions and Spectra* (Academic, Orlando, 1986); (c) A. Giusti-Suzor and U. Fano, J. Phys. B **17**, 215 (1984); W. E. Cooke and C. L. Cromer, Phys. Rev. A **32**, 2725 (1985).

[4] M. Aymar, Phys. Rep. **110**, 163 (1984); M. Aymar, J. Opt. Soc. Am. B **1**, 239 (1984).

[5] C. H. Greene and Ch. Jungen, Adv. At. Mol. Phys. **21**, 51 (1985).

[6] (a) M. Aymar, J. Phys. B **20**, 6507 (1987); (b) M. Aymar, E. Luc-Koenig, and S. Watanabe, *ibid.* **20**, 4235 (1987); M. Aymar and J. M. Lecomte, *ibid.* **22**, 223 (1989).

[7] (a) C. H. Greene and L. Kim, Phys. Rev. A **36**, 2706 (1987); (b) L. Kim and C. H. Greene, *ibid* **36**, 4272 (1987); (c) **38**, 2361 (1988); (d) also V. Lange, U. Eichmann, and W. Sandner, J. Phys. B **22**, L361 (1989).

[8] (a) C. H. Greene, in *Fundamental Processes of Atomic Dynamics,* edited by J. Briggs, H. Kleinpoppen, and H. Lutz (Plenum, New York, 1988); (b) C. H. Greene and L. Kim, Phys. Rev. A **38**, 5953 (1988).

[9] See, e.g., T. N. Chang and Y. S. Kim, Phys. Rev. A **34**, 2609 (1986).

[10] I. I. Sobel'man, *An Introduction to the Theory of Atomic Spectra* (Pergamon, New York, 1972).

[11] C. M. Lee and K. T. Lu, Phys. Rev. A **8**, 1241 (1973); also K. T. Lu, J. Opt. Soc. Am. **64**, 706 (1974).

[12] J. A. Armstrong, J. J. Wynne, and P. Esherick, J. Opt. Soc. Am. **69**, 211 (1979).

[13] C. J. Dai, G. W. Schinn, and T. F. Gallagher, Phys. Rev. A **42**, 223 (1990); P. F. O'Mahony and C. H. Greene, *ibid.* **31**, 250 (1985); P. F. O'Mahony, *ibid.* **32**, 908 (1985).

[14] M. Aymar, J. Phys. B **23**, 2697 (1990).

[15] C. H. Greene, Phys. Rev. A **42**, 1405 (1990).

[16] U. Griesmann, B. Esser, and J. Hormes (unpublished).

[17] E. Rasmussen, Z. Phys. **87**, 607 (1934). For a relabeling of some levels see H. N. Russell, Phys. Rev. **46**, 989 (1934).

[18] C. E. Moore, *Atomic Energy Levels I–III,* Natl. Bur. Stand. (U.S.) Circ. No. 467 (U.S. GPO, Washington, DC, 1958).

[19] J. A. Armstrong, J. J. Wynne, and F. S. Tomkins, J. Phys. B **13**, L133 (1980); F. S. Tomkins and B. Ercoli, Appl. Opt. **6**, 1299 (1967).

[20] E. U. Condon and G. H. Shortley, *The Theory of Atomic Spectra* (Cambridge University Press, Cambridge, England, 1935).

[21] L. C. Biedenharn and J. D. Louck, *Angular Momentum in Quantum Physics* (Addision-Wesley, Reading, MA, 1981).

[22] R. D. Hudson, V. L. Carter, and P. A. Young, Phys. Rev. **180**, 77 (1969).

[23] F. Gounand, B. Carré, P. R. Fournier, P. d'Oliveira, and M. Aymar, J. Phys. B **24**, 1309 (1991).

[24] C. M. Brown and M. L. Ginter, J. Opt. Soc. Am. **68**, 817 (1978).

[25] L. Kim and C. H. Greene, J. Phys. B **22**, L175 (1989).

[26] J. J. Wynne and J. A. Armstrong, IBM J. Res. Dev. **23**, 490 (1979).

[27] M. Aymar, P. Camus, M. Dieulin, and C. Morillon, Phys. Rev. A **18**, 2173 (1978).

[28] P. Camus, M. Dieulin, and A. El Himdy, Phys. Rev. A **26**, 379 (1982); P. Camus, M. Dieulin, A. El Himdy, and M. Aymar, Phys. Scr. **27**, 125 (1983); M. Aymar, P. Camus, and A. El Himdy, *ibid.* **27**, 183 (1983).

[29] J. Neukammer, H. Rinneberg, G. Jönsson, W. E. Cooke, H. Hieronymus, A. König, K. Vietzke, and H. Springer Bolk, Phys. Rev. Lett. **55**, 1979 (1985).

[30] M. Aymar and O. Robaux, J. Phys. B **12**, 531 (1979); see also Ref. [4].

[31] J. Neukammer, G. Jönsson, A. König, K. Vietzke, H. Hieronymus, and H. Rinneberg, Phys. Rev. A **38**, 2804 (1988).

[32] J. Verges, unpublished result.

[33] J. E. Hunter III, J. S. Keller, and R. S. Berry, Phys. Rev.

A **33**, 3138 (1986); K. A. H. Leeuwen and W. Hogervorst, Z. Phys. A **316**, 149 (1984); W. H. King and M. Wilson, J. Phys. B **18**, 23 (1985).
[34] H. P. Palenius, Phys. Lett. A **56**, 451 (1976).
[35] E. R. Eliel and W. Hogervorst, J. Phys. B **16**, 1881 (1983).
[36] H. Rinneberg and J. Neukammer, Phys. Rev. A **27**, 1779 (1983); H. Rinneberg and J. Neukammer, J. Phys. B **15**, L825 (1982).
[37] J. J. Wynne and J. P. Hermann, Opt. Lett. **4**, 106 (1979).
[38] M. Aymar, P. Camus, and A. El Himdy, J. Phys. B **15**, L759 (1982).
[39] E. A. J. M Bente and W. Hogervorst, Z. Phys. D **14**, 119 (1989).
[40] M. Aymar, J. Phys. B **18**, L763 (1985).
[41] A. Kallenbach, M. Kock, and G. Zierer, Phys. Rev. A **38**, 2356 (1988).
[42] C. E. Burkhardt, J. L. Libbert, Jian Xu, J. J. Leventhal, and J. D. Kelley, Phys. Rev. A **38**, 5949 (1988).
[43] K. Bartschat and B. McLaughlin, J. Phys. B **23**, L439 (1990).
[44] C. H. Greene and C. E. Theodosiou, Phys. Rev. A **42**, 5773 (1990).
[45] F. Mies, Phys. Rev. **175**, 164 (1968).
[46] V. Lange, M. Aymar, U. Eichmann, and W. Sandner, J. Phys. B **24**, 91 (1991).

Part B

Generalizations

1978 *Phys. Rev.* A **17** 93–9
Reprinted with permission from the American Physical Society

Connection between configuration-mixing and quantum-defect treatments

U. Fano
Department of Physics, University of Chicago, Chicago, Illinois 60637
(Received 10 June 1977)

Alternative formulations of atomic problems are connected in terms of the Green's function for the radial motion of a single electron in an effective potential field. The quantum-defect theory (QDT) represents an electron outside an ion by a Coulomb function with a phase shift contributed by the additional interactions that prevail inside the ion core. We interpret the phase-shifted portion of this radial function as arising from the application of a Green's function propagator to the short-range operator that represents the coupling to other channels of the electron + ion system. The alternative representation of the same radial function, as a superposition of zero-order functions for configurations with different energies, differs only by having the same Green's function expanded in a set of zero-order eigenfunctions. This analysis of configuration mixing leads to a reformulation of reaction matrices which extends the initial QDT treatment of bound states.

I. INTRODUCTION

Quantum-defect treatments of atomic and molecular processes are finding increasing application, but their connection with more general and familiar treatments does not seem to have been described adequately. This circumstance has emerged particularly in a recent and extensive treatment of the excitation and ionization of the H_2 molecule, which also outlines an extension toward dissociation.[1] Reference 1 transforms the familiar formulation of the full H_2 Hamiltonian into an application of quantum-defect procedures, without describing the transformation in full detail.

Seaton's original formulation of the multichannel quantum-defect theory (MQDT)[2] rests on a truncated close-coupling expansion of the many-particle eigenfunction, and on extrapolation of a reaction matrix from the continuous to the discrete spectrum. Some of the later applications of MQDT have been frankly phenomenological.[3] Here we derive, instead, the MQDT from a configuration-mixing point of view, namely, by representing eigenfunctions of the full Hamiltonian as superpositions of eigenfunctions of an approximate Hamiltonian. The derivation is somewhat laborious but it may clarify concepts of the MQDT and remove the limitations indicated above. Section I outlines the background of the problem and the main results; details follow in separate sections for the continuous and discrete spectra.

One key idea of the MQDT is to regard a many-electron system as separated into a core and an escaping electron, when the radial distance r of an electron from the center of the core has attained a sufficient value r_0. Wave functions of core + electron are then represented in the range $r \geqslant r_0$ by a close-coupling expansion

$$\Psi = \sum_i [F(\epsilon_i, r)\cos\delta_i - G(\epsilon_i, r)\sin\delta_i]\Phi_i B_i \ . \qquad (1)$$

Here the expression in brackets is the radial wave function of the escaping electron, Φ_i represents a coupled wave function of the orbital and spin coordinates of this electron and of the core in its ith stationary state, and the coefficients B_i identify a particular superposition of channels. Different channel functions Φ_i may include the same stationary state of the core; hence energy parameters ϵ_i and ϵ_j of Eq. (1) with different indices may be equal. The variable of the radial function r indicates the distance of the electron *farthest* from the core's center; this specification removes the need for explicit antisymmetrization. The radial function in Eq. (1) is a general standing-wave solution of the radial equation for an electron moving in the field generated by the core at $r > r_0$, which is generally Coulombic and excludes exchange effects. This radial function is represented as the superposition of two particular solutions, F and G, with G lagging in phase by 90° with respect to F. The two functions (F, G) are usually defined also within the core, i.e., for $r_0 \geqslant r \geqslant 0$, as wave functions of a Coulomb field or, alternatively, of a more realistic atomic field which may include a nonlocal potential that represents exchange; in either case, F is the function regular at $r = 0$. The functions F and G are normalized per unit energy and ϵ_i represents the energy of the escaping electron, i.e., the difference of the total energy E of the system and of the energy level E_i of the core in the state Φ_i. The parameters δ_i are provided by a separate calculation of channel coupling at $r \leqslant r_0$ or by analysis of experimental data.

Seaton's original form of the MQDT assumes initial knowledge of a finite (i.e., truncated) set of functions Φ_i and of the corresponding core energies E_i. For total energies E in excess of all E_i, the Schrödinger equation of electron + core reduces then to a system of coupled equations for radial functions $f_i(\epsilon_i, r)$, regular at $r=0$, with $\epsilon_i = E - E_i$. [Here again, as in Eq. (1), energies ϵ_i and ϵ_j with different indices may be equal.] Each f_i is represented for $r \geq r_0$ as a superposition of Coulomb functions F and G. A general energy eigenfunction is then represented for $r \geq r_0$ by Eq. (1), in which the coefficients

$$B_i \cos\delta_i = C_i \tag{2}$$

remain free while

$$B_i \sin\delta_i = -\pi \sum_j K_{ij}(E) C_j\,, \tag{3}$$

with the reaction matrix K_{ij} determined by the solution of the coupled radial equations for $r \leq r_0$. The K matrix is then extrapolated to energies E lower than one or more of the core energies E_i, whereby some of the escape energies ϵ_i become negative, meaning that the corresponding channels are closed. The requirement that $f_i(\epsilon_i, r)$ remain finite at $r = \infty$ in the closed channels imposes linear restrictions on the coefficients C_i. When all channels are closed, these restrictions are compatible only for discrete energy eigenvalues E. Reference 3 has described applications of the MQDT from a more flexible point of view, which is compatible with Seaton's approach, or alternatively with empirical fitting of the core energies E_i and of the reaction matrix $K_{ij}(E)$, or with calculation of K_{ij} by the R-matrix procedures which are confined to the finite volume of the core.[4]

Configuration-mixing treatments follow instead the traditional approach of separating the Hamiltonian of the whole system into two terms, $H = H_0 + V$, and of constructing eigenfunctions of H as superpositions of a complete set of eigenfunctions of H_0. To maintain a degree of parallelism with Seaton's approach we indicate the eigenfunctions of H_0 by $\mathcal{A}\{f_{in}(r)\Phi_i\}$—where $\mathcal{A}$ stands for the antisymmetrization that is required for r values within the core—i.e., we set $(H_0 - E_i - \epsilon_{in})\mathcal{A}\{f_{in}\Phi_i\} = 0$. The constructional definition of f_{in} and Φ_i differs, however, here and in the MQDT. In the elementary form of configuration mixing, Φ_i is constructed from a complete orthonormal set of independent particle wave functions and $f_{in}(r)$ is itself the radial part of one of these functions; the set of Φ_i is also implicitly complete, rather than truncated. Correlations may be built initially in the Φ_i by superposition of Slater determinants in the case of configuration mixing, while the MQDT places fewer restrictions on the Φ_i. We write the equation for the radial functions f_{in} as

$$h_i(r) f_{in}(r) = \epsilon_{in} f_{in}(r)\,, \quad 0 \leq r \leq \infty\,. \tag{4}$$

An index i has been appended here to the Hamiltonian to allow for modifications of the elementary procedure, in which the optical potential—including exchange—depends on the core state[5]; any such modification should, however, preserve the orthonormality of the $\mathcal{A}\{f_{in}\Phi_i\}$. The eigenvalues ϵ_{in} of Eq. (4), and the corresponding eigenvalues $E_i + \epsilon_{in}$ of H_0, are of course unrestricted to allow the mixing of configurations of different energy. The desired eigenfunctions of H are then represented by

$$\Psi = \sum_i \sum_n D_{in} \mathcal{A}\{f_{in}(r)\Phi_i\}\,, \tag{5}$$

where each $\sum_n$ generally extends over a discrete and continuous spectrum and the coefficients D_{in} obey a system of equations

$$\epsilon_{in} D_{in} + \sum_j \sum_m (in|V|jm) D_{jm} = D_{in}\epsilon_i \tag{6}$$

with $\epsilon_i = E - E_i$. (We consider in this paper eigenfunctions of H_0 and of H that are eigenfunctions of the total angular momentum and are of the standing wave type.)

The main purpose of this paper is to derive the MQDT representation (1) of a wave function from the more general, configuration-mixing expansion (5). The derivation consists essentially of two remarks drawn from the Green's function treatment of scattering theory[6]:

(a) When Eq. (6) is solved in terms of a reaction matrix K, the summation over n in Eq. (5) can be expressed in terms of the eigenfunction expansion of the Green's function of the radial Hamiltonian $h_i(r)$,

$$\mathcal{G}_i(\epsilon_i; r, r') = \sum_n f_{in}(r)(\epsilon_i - \epsilon_{in})^{-1} f_{in}(r')\,. \tag{7}$$

The applications to atomic problems utilize the "two-potential" approach[6] of including a part of the interaction in the Hamiltonian $h_i(r)$ and the rest in the operator V of Eq. (6), and envisage the direct numerical solution of Eq. (4) and of the K matrix form of Eq. (6).

(b) The summation over configurations n of different energies is then carried out simply by replacing the expanded form (7) of the Green's function by its alternative form in terms of the regular and irregular solutions of Eq. (4) at the energy ϵ_i that are 90° out of phase,

$$\mathcal{G}_i(\epsilon_i; r, r') = \pi g_i(\epsilon_i, r) f_i(\epsilon_i, r')\,, \quad r > r'\,. \tag{8}$$

[In Eq. (8) the coefficient is set to π by the normalization of f and g per unit energy and by the phase of g lagging that of f; these conventions differ from those of Ref. 6.]

The application of these remarks is straightforward as long as one deals only with continuous spectra, i.e., with positive values of ϵ_i (Sec. II). In practice, however, the study of electron + core systems generally involves negative values of ϵ_i, corresponding to closed channels. This circumstance will lead us, in Sec. III, to modify the procedure for solving Eq. (6) in terms of a K matrix, in order to avoid the singular behavior of this matrix in the presence of closed channels. Seaton has avoided these singularities by extrapolating to negative ϵ_i the K matrix calculated for a truncated set of channels with all $\epsilon_i > 0$. He points out that extrapolation is in fact dependable over the most relevant range of 3–5 eV below a threshold. On the other hand it may be worthwhile to develop an alternative that avoids numerical extrapolations as well as truncations of the set of channels Φ_i which are realistic only when the spectrum of threshold energies E_i shows a major gap. Properties of the Green's functions for a discrete spectrum will in fact permit us to construct a "smoothed out" reaction matrix $K^{(s)}$ which is equivalent to Seaton's but can be calculated for closed channels without resorting to extrapolation. The matrix $K^{(s)}$ is substantially equivalent to an R matrix.

II. CONFIGURATION MIXING IN THE CONTINUUM

Consider initially energies E in excess of all E_i, i.e., such that all $\epsilon_i > 0$, as in Seaton's treatment. The linear system, Eq. (6), is then singular because the coefficient of D_{in}, namely, $\epsilon_i - \epsilon_{in}$, may vanish. There results a singularity of the solutions D_{in} which may be represented by the superposition of two terms. One term equals $\delta(\epsilon_i - \epsilon_{in})C_i$, where C_i is a coefficient discussed below; its contribution to the wave function (5) does not admix configurations of different energies,

$$\sum_n \mathcal{A}\{f_{in}(r)\Phi_i\}\delta(\epsilon_i - \epsilon_{in})C_i = \mathcal{A}\{f_i(\epsilon_i, r)\Phi_i\}C_i . \quad (9)$$

[We have introduced in Eq. (9), as in Eq. (8), the symbol $f_i(\epsilon_i, r)$ to indicate the $f_{in}(r)$ with $\epsilon_{in} = \epsilon_i$, i.e., "on the energy shell."] The other term represents the mixing of configurations of the same channel i but with energies other than ϵ_i (i.e., "off the energy shell"); the energy ϵ_i is *excluded* here from the mixing by restricting the superposition through the principal part integration symbol P. Solution of Eq. (6) for standing-wave boundary conditions have then the form[7]

$$D_{in} = \delta(\epsilon_i - \epsilon_{in})C_i + P(\epsilon_i - \epsilon_{in})^{-1} \times \sum_k (f_{in}\Phi_i|K(E)|f_k(\epsilon_k)\Phi_k)C_k . \quad (10)$$

The coefficients of the second—configuration mixing—term of Eq. (10) constitute a column of off-shell elements of the reaction matrix K; we have replaced here the row indices "*in*" of this matrix by the fuller description of a base function, $f_{in}\Phi_i$. The channel coefficients C_i, akin to those of Eqs. (2) and (3), are not determined by the Schrödinger equation (6) but serve to identify one eigenfunction of a degenerate manifold, usually in terms of boundary conditions; this paper does not concern itself with their determination. The K-matrix elements, instead, replace the D_{in} as the parameters to be determined by Eq. (6). When Eq. (10) is substituted into (6), multiplication of D_{in} by $\epsilon_i - \epsilon_{in}$ eliminates the δ-function term and cancels the pole of the second term. There results a linear combination of coefficients C_k set equal to zero, a condition which is met by setting to zero the coefficient of each C_k. Thus one obtains the Lippmann-Schwinger equation for the K matrix [Eq. (2.23), Ref. 6],

$$\begin{aligned}(f_{in}\Phi_i|K(E)|f_k(\epsilon_k)\Phi_k) &= (f_{in}\Phi_i|V|f_k(\epsilon_k)\Phi_k) \\ &+ \sum_j \sum_m (f_{in}\Phi_i|V|f_{jm}\Phi_j)P(\epsilon_j - \epsilon_{jm})^{-1} \\ &\times (f_{jm}\Phi_j|K(E)|f_k(\epsilon_k)\Phi_k) . \quad (11)\end{aligned}$$

This system of coupled integral equations can be reduced to a finite algebraic system by taking the $\sum_m$ over a discrete mesh and by truncating it at large ϵ_{jm}; numerical solutions of this system have been obtained for particular applications.[8]

What matters for us is that the mixing of configurations can be described in terms of a K matrix which is calculable, and that this description is obtained by substituting Eq. (10) into (6). The resulting wave function is

$$\Psi = \sum_i \Phi_i \Big(f_i(\epsilon_i, r)C_i + \sum_n f_{in}(r) \frac{P}{\epsilon_i - \epsilon_{in}} \times \sum_j (f_{in}\Phi_i|K(E)|f_j(\epsilon_j)\Phi_j)C_j \Big) . \quad (12)$$

The decisive step consists now of noticing that the $\sum_n$ in Eq. (12) operates on the same set of n-dependent factors as the $\sum_n$ on the right-hand side of Eq. (7), if we allow for the following differences of notation: The P symbol is generally implied but not explicitly included in the expansion (7) of a Green's function[6]; on the other hand the variable r' of the last factor of Eq. (7) is only implied—and integrated over—in the definition of the K-matrix element. The effect of *configuration mixing* in the ith channel can thus be represented in *terms of the Green's function* $\mathcal{G}_i$ of Eq. (7), evaluated for the single on-shell energy ϵ_i, and of the operator K, *without further resort* to explicit superposition of wave functions $f_{in}(r)$ for all different energies ϵ_{in}. This fact is familiar in analytic manipulations of scattering theory; we apply it here in the context of a Green's function of the single-particle Hamiltonian h_i with an unspecified optical potential.

Our application consists then of replacing, in Eq. (12), the expanded form (7) of the Green's function by the form (8) which involves only radial functions on the energy shell. The restriction on Eq. (8), namely $r > r'$, is particularly appropriate to our aim of studying the wave function for radial distances larger than the core radius r_0. This restriction provides us in fact with an operational definition of r_0; this core radius must exceed all radial distances r' at which the interaction V and the corresponding reaction operator K are nonzero. [Recall, however, that the optical potential of the core need not vanish at $r > r_0$, because this potential is included in $h_i(r)$, i.e., in the portion H_0 of the Hamiltonian. We are only assuming that the aggregate effect of the *short range* interactions V, represented by K, is confined to $r' < r_0$.] Replacing then in Eq. (12) the expression (7) of $\mathcal{G}_i$ by the expression (8), we have

$$\Psi = \sum_i \Phi_i\Big(f_i(\epsilon_i, r)C_i + \pi g_i(\epsilon_i, r) \times \sum_j (f_i(\epsilon_i)\Phi_i|K(E)|f_j(\epsilon_j)\Phi_j)C_j\Big), \quad r \geq r_0. \tag{13}$$

All the reaction matrix elements in this expression belong now to the submatrix pertaining to states on the energy shell; these elements can be indicated more briefly by $K_{ij}(E)$, as in Eq. (3), since the rows and columns of the submatrix are labeled adequately by the channel indices (i, j). Thereby Eq. (13) reduces to

$$\Psi = \sum_i \Phi_i\Big(f_i(\epsilon_i, r)C_i + \pi g_i(\epsilon_i, r)\sum_j K_{ij}(E)C_j\Big), \quad r \geq r_0. \tag{14}$$

The configuration-mixing wave function Ψ has thus been reduced to coincide, at least in essence, with the quantum-defect wave function (1), as complemented by Eqs. (2) and (3). More specifically, Eqs. (1) and (14) coincide altogether for $r > r_0$ if the pair of radial functions (F, G) of Eq. (1) obeys the equation $[h_i(r) - \epsilon_i]y(r) = 0$ not only for $r > r_0$, as specified by their definition, but also for $r < r_0$. If this equation is obeyed only for $r > r_0$, F and G must nevertheless be represented in this range as superpositions of (f_i, g_i); Eqs. (1) and (14) are then brought to coincide by appropriate adjustments of Eqs. (2) and (3).

In conclusion, we stress the critical role played in this section by the inhomogeneous Eq. (11). Its solution, combined with later use of Green's function properties, enables us to eliminate explicit consideration of the mixing of off-the-shell configurations, by embodying the effect of this mixing in the on-the-shell submatrix $K_{ij}(E)$. This simplification rests, of course, on the fact that we have confined our aim to a description of the wave function for $r > r_0$ only.

III. DISCRETE LEVELS AND DETACHMENT THRESHOLDS

Substantial adaptation of the procedure of Sec. II is required when the energy E falls below one or more of the threshold E_i. That procedure, namely, the treatment of configuration mixing by solving the Lippmann-Schwinger Eq. (11) followed by substitution of the Green's function expansion, remains valid—but only in principle—as long as E exceeds at least one threshold. Two practical difficulties arise which must be bypassed.

Firstly, when E drops below a channel threshold E_i, the energy parameter $\epsilon_i = E - E_i$ becomes negative. The δ function singularity in Eqs. (9) and (10) becomes then inoperative, except for isolated values of E at which ϵ_i coincides with one of the discrete negative levels ϵ_{in}. Thereby, the parameter C_i of that channel drops out, in accordance with the fact that the degeneracy of eigenstates equals the number of open channels (i.e., of channels with $E > E_i$). A real difficulty stems instead from the changed character of the $(\epsilon_i - \epsilon_{in})^{-1}$ singularity in Eqs. (10)–(12). The contribution of this singularity depends on ϵ_i smoothly only as long as the ϵ_{in} form a continuum and the principal part is taken in the $\sum_n$, but it oscillates sharply when the set of ϵ_{in} is discrete; indeed, the $\sum_n$ becomes very large whenever ϵ_i gets much closer to one particular level ϵ_{in} than to any other one. This singular behavior, familiar in the study of Green's functions of systems with discrete spectra (Ref. 6, p. 99), makes numerical treatment of Eq. (11) quite impractical and contributes to its solution singularities that reflect the presence of autoionizing levels in the closed channels with $E_i > E$. One might say that the discrete structure of the spectrum of ϵ_{in} renders inoperative the stipulation of principal part integration across the pole of $(\epsilon_i - \epsilon_{in})^{-1}$.

A second difficulty arises at energies E close to any of the thresholds E_i, owing to singularities of the continuum $f_{in}(r)$ as functions of ϵ_{in} at $\epsilon_{in} = 0$. This difficulty is particularly conspicuous in the absence of a Coulomb field, as $f_{in}(r)$ vanishes at the threshold for finite values of r, causing the matrix elements of V and K to vanish as well, with branch point singularities.

These difficulties are bypassed by the quantum-defect theory through the following approach. One remarks that both the discrete structure of the spectrum for negative ϵ_{in} and the singularities at $\epsilon_{in}=0$ result from boundary conditions imposed on the $f_{in}(r)$ at $r=\infty$. On the other hand the calculation of configuration mixing in terms of a K matrix, as formulated in Sec. II, deals exclusively with short-range interactions. Accordingly one should extrapolate freely the K-matrix calculation below any or all of the thresholds E_i by removing temporarily the boundary conditions at $r=\infty$ upon the eigenfunctions of the radial operators h_i. Thereby one will obtain a modified K matrix—to be called $K^{(s)}$—that depends smoothly on E. The discrete structure of spectra for negative ϵ_i will reemerge in a further step of calculation, when the boundary conditions at $r=\infty$ are imposed again.[3]

As anticipated in Sec. I, the results to be obtained in this manner, by taking into account the short-range interactions equally below any threshold as above it, will be substantially equivalent to those obtained by the R-matrix procedures which are manifestly independent of threshold effects and of boundary conditions at $r=\infty$.[4] In fact, the calculation of the $K^{(s)}$ matrix by solving the modified Lippmann-Schwinger equation, to be presented here, may be regarded as a configuration mixing technique for the solution of the core problem in R-matrix theory. This remark contributes to our goal of illustrating the connections between alternative theoretical approaches.

A. Radial functions for negative energy

The approach to be followed requires smooth, analytic if possible, extrapolation of continuum wave functions $f_i(\epsilon_i, r)$ into the range of negative ϵ_i. Seaton's MQDT (Ref. 2) and its extensive applications by his school also involve extrapolation of K matrices, or rather of equivalent R matrices, which are calculated numerically for $\epsilon_i>0$ but are also applied for negative values of ϵ_i. Here we introduce the concept of calculating the matrix $K_{ij}^{(s)}(E)$ independently for different values of E, larger or smaller than any of the threshold E_i.

Regarding the extrapolation of wave functions to negative energies, consider initially that construction of a regular solution $f_{in}(r)$ of Eq. (4), proceeding from $r=0$ to any finite r, develops in the same manner whether ϵ_{in} is positive or negative; indeed the zero point of ϵ_{in} is defined as the value of the potential in h_i at $r=\infty$, which is not relevant to the integration at lower r. Similarly, one can construct a second, independent solution of Eq. (4) at each ϵ_{in}. Our treatment in Sec. II requires, however, that we identify at each positive energy ϵ_i a *standard pair* of solutions of Eq. (4), normalized per unit energy, of which f_i is regular at $r=0$ and g_i lags in phase by 90° at large r. The extrapolation of this standard pair to negative values of ϵ_i—which need not coincide with eigenvalues ϵ_{in}—has been provided by Seaton[2] for Coulomb field functions. An extension of his results is discussed in a separate paper[9] and will only be outlined here.

At the outset, we can identify an alternative base pair of solutions of Eq. (4), $f_i^0(\epsilon_i, r)$ and $g_i^0(\epsilon_i, r)$, for any ϵ_i by imposing boundary conditions at $r=0$ rather than at large r. These conditions do not depend on ϵ_i and fix the Wronskian of the pair. This pair coincides with the standard pair of Sec. II at large r when $\epsilon_i=0$ and the field of $h_i(r)$ is Coulombic. Otherwise the base pair identified at $r=0$ is related to the standard pair identified for large r at $\epsilon_i>0$ by a unimodular (but not unitary) transformation which leaves the Wronskian invariant. Our procedure consists of *extrapolating this transformation* to negative ϵ_i; we thus generate a standard pair (f_i, g_i) normalized at large r, starting from the base pair identified at $r=0$ for any value of ϵ_i.[9] By "large r" we mean r values comparable to or larger than the core radius r_0. For negative ϵ_i, we shall also consider a third pair of base functions, obtained by orthogonal transformation of the standard pair,

$$u_i(\epsilon_i, r)=f_i(\epsilon_i, r)\sin\chi_i - g_i(\epsilon_i, r)\cos\chi_i\,, \qquad (15)$$
$$v_i(\epsilon_i, r)=-f_i(\epsilon_i, r)\cos\chi_i - g_i(\epsilon_i, r)\sin\chi_i\,.$$

This pair, and hence the angle $\chi_i(\epsilon_i)$, is identified by the requirement that $v_i(\epsilon_i, r)$ vanishes at $r=\infty$. The discrete eigenvalues $\epsilon_{in}<0$ of h_i are thus identified by the condition $\chi_i(\epsilon_{in})=n\pi$. (In the quantum-defect literature χ is indicated by $\pi\nu$ for Coulomb functions.) The functional relationship between the parameter χ_i and the energy ϵ_i also serves to represent the normalization ratio of a discrete wave function $f_{in}(r)$ to the standard function $f_i(\epsilon_{in}, r)$,

$$f_{in}(r)=N_{in}f_i(\epsilon_{in}, r)\,, \quad N_{in}^2=\pi(d\epsilon_i/d\chi_i)_{\epsilon_{in}}\,. \qquad (16)$$

[In Sec. II, i.e., for $\epsilon_{in}>0$, f_{in} was normalized per unit energy and hence identical to $f(\epsilon_{in}, r)$.]

B. Lippmann-Schwinger equation for negative energy

We return now to the problem of solving Eq. (6) for the coefficients D_{in} at energies E lower than the channel threshold E_i. Having removed temporarily the requirement that Ψ remain finite at $r=\infty$, we can allow Ψ to include the term represented by Eq. (9) even for $\epsilon_i<0$; later application of the boundary condition will restrict the coefficient C_i.[10] We might then represent D_{in} once again by Eq. (10), but an adjustment is required to make its pole term yield a smooth contribution in the discrete, i.e., to achieve the same effect as the principal part symbol does in the continuum. The adjustment consists of subtracting from D_{in} a term that just cancels its pole. Finally our notation must also be adjusted, because the $\sum_n$ does not generally include, for closed channels, a term with $\epsilon_{in}=\epsilon_i$. Thus we rewrite Eqs. (5) and (10) in a form that distinguishes the contributions of discrete and continuous spectra in terms of the step function $\Theta(x)$, which equals 1 for $x\geqslant 0$ and 0 for $x<0$,

$$\Psi=\sum_i\Big(f_i(\epsilon_i,r)\Phi_i d_i+\sum_n f_{in}(r)\Phi_i D_{in}\Big), \tag{17}$$

$$d_i=C_i-\Theta(-\epsilon_i)\pi\cot\chi_i(\epsilon_i)\times\sum_k(f_i(\epsilon_i)\Phi_i|K^{(s)}|f_k(\epsilon_k)\Phi_k)C_k, \tag{18}$$

$$D_{in}=[\Theta(\epsilon_i)P+\Theta(-\epsilon_i)](\epsilon_i-\epsilon_{in})^{-1}\times\sum_k(f_{in}\Phi_i|K^{(s)}|f_k(\epsilon_k)\Phi_k)C_k. \tag{19}$$

To verify that the contributions of the poles of Eqs. (18) and (19) cancel out, consider that $\pi\cot\chi_i$ has poles $\pi[\chi_i(\epsilon_i)-\chi_i(\epsilon_{in})]^{-1}$ and that the normalization ratio of $f_i(\epsilon_i)$ and of f_{in} is given by Eq. (16).

Upon substitution of Eq. (17) in the Schrödinger equation $(H_0+V-E)\Psi=0$, the operator H_0-E yields zero when applied to the term with the coefficient d_i and yields $\epsilon_{in}-\epsilon_i$ for the terms with D_{in}. The analog of Eq. (11) results then by projecting the equation on $f_{in}\Phi_i$ and setting to zero the coefficient of each C_k,

$$\begin{aligned}(f_{in}\Phi_i|K^{(s)}(E)|f_k(\epsilon_k)\Phi_k)&=(f_{in}\Phi_i|V|f_k(\epsilon_k)\Phi_k)\\&+\sum_j\sum_m(f_{in}\Phi_i|V|f_{jm}\Phi_j)[\Theta(\epsilon_j)P+\Theta(-\epsilon_j)](\epsilon_j-\epsilon_{jm})^{-1}(f_{jm}\Phi_j|K^{(s)}(E)|f_k(\epsilon_k)\Phi_k)\\&-\sum_j(f_{in}\Phi_i|V|f_j(\epsilon_j)\Phi_j)\Theta(-\epsilon_j)\pi\cot\chi_j(\epsilon_j)(f_j(\epsilon_j)\Phi_j|K^{(s)}(E)|f_k(\epsilon_k)\Phi_k).\end{aligned} \tag{20}$$

This equation differs from Eq. (11) mainly by the insertion of its last term, which effectively replaces the symbol P for negative values of ϵ_j. As contemplated above, this new term causes the $K^{(s)}$ matrix elements to remain smooth even for energies E corresponding to poles of $(\epsilon_j-\epsilon_{jm})^{-1}$ with negative values of ϵ_j. (The resonances observed experimentally at those energies will be represented instead by sharp variations of the coefficients C_j.)

Actual solution of Eq. (20) to yield rows of the $K^{(s)}$ matrix with negative values of ϵ_{in} extends the range of practical application of the Lippmann-Schwinger equations. It will actually be unnecessary to solve Eq. (20) for each of the infinitely numerous levels n of a Rydberg series. For negative values of ϵ_{in} one should rather divide Eq. (20) by the normalization coefficient N_n, and thereby replace the discrete wave function f_{in} in the row indices of $K^{(s)}$ by the corresponding continuum-normalized standard function $f_i(\epsilon_{in})$; solution of the equation for a rather coarse mesh of values of ϵ_{in} should prove adequate for negative as it does for positive values of ϵ_{in}. Similarly, evaluation of the $\sum_m$ over negative ϵ_{jm} may be simplified in Eq. (20) by factoring the normalization N_{jm} out of f_{jm} in the matrix elements, after which the matrix elements depend smoothly on ϵ_{jm} and the $\sum_m N^2_{jm}(\epsilon_j-\epsilon_{jm})^{-1}$ approximates $\pi\cot\chi_j(\epsilon_{jm})$.

C. Green's function for negative energy

We come thus eventually to reducing our wave function (17) to the form utilized by the quantum-defect theory. Here, as in Sec. II, the $\sum_n$ in Eq. (17) operates on the same set of n-dependent factors as the $\sum_n$ in Eq. (7) and can thus be replaced by an alternative expression of the Green's function $\mathcal{G}_i(\epsilon_i;r,r')$. However, for negative values of ϵ_i, the appropriate expression of $\mathcal{G}_i$ is no longer given by Eq. (8).

As discussed, e.g., on p. 99 of Ref. 6, the Green's function of an operator $h_i(r)$ has poles at the negative eigenvalues ϵ_{in} of its discrete spectrum. These poles are not represented by the simple expression $\pi g_i(\epsilon_i,r)f_i(\epsilon_i,r')$ on the right of Eq. (8). They are included by subtracting from that expression a term $f_i(\epsilon_i,r)f_i(\epsilon_i,r')$ with a suitable coefficient that diverges at $\epsilon_i=\epsilon_{in}$ and vanishes at or near the middle of the interval between two successive poles, ϵ_{in} and $\epsilon_{i,n+1}$;

comparison with Ref. 6 shows this coefficient to coincide with our function $\pi \cot\chi(\epsilon_i)$.

We conclude that the expression on the right of Eq. (8) remains indeed relevant for negative values of ϵ_i. It no longer serves to represent the contribution $\mathcal{G}_i(\epsilon_i; r, r')$ of the $\sum_n$ in Eq. (17), but the *combined contribution* of $\mathcal{G}_i$ and of the function $-f_i(\epsilon_i, r)\pi \cot\chi_i f_i(\epsilon_i, r')$ which arises from the term of Eq. (17) with the coefficient d_i. This expression might be indicated by $\mathcal{G}^{(s)}$ and described as a smoothed-out Green's function. Thereby Eq. (17) reduces, for $r > r_0$, to the final Eq. (14) of Sec. II. In other words, Eq. (14) *remains valid for negative values of* ϵ_i, provided we use the reaction matrix $K_{ij}^{(s)}(E)$ calculated from the singularity free Eq. (20).

D. Resonances and threshold effects

This result completes our derivation of the quantum-defect expression of a wave function outside the core from the configuration-mixing expression with coefficients determined by a Lippmann-Schwinger equation. As anticipated above, the discrete structures of spectra observed below detachment thresholds are introduced in quantum-defect treatments as a subsequent step, by enforcing the requirement that the radial functions in channels with negative ϵ_i remain finite at $r \to \infty$. This requirement is formulated by replacing in Eq. (14) the standard pair of functions (f_i, g_i) by the alternative base pair (u_i, v_i), defined by Eq. (15), and setting to zero the coefficient of the diverging function u_i. This procedure gives

$$\Theta(-\epsilon_i)\left(\sin\chi_i(\epsilon_i)C_i - \pi\cos\chi_i(\epsilon_i)\sum_j K_{ij}^{(s)}(E)C_j\right) = 0 . \tag{21}$$

Resonant behavior of the coefficients C_j occurs at near-zero minima of the determinant

$$\mathrm{Det}|\Theta(-\epsilon_i)[\sin\chi_i\delta_{ij} - \pi\cos\chi_i K_{ij}^{(s)}]\Theta(-\epsilon_j)| . \tag{22}$$

When all channels are closed, the system of equations (21) has nonzero solutions C_j only when its determinant, Eq. (22) vanishes; this condition fixes the energy levels of the bound states of core + electron.

The additional singularities, which are associated with threshold behavior, at $\epsilon_i = 0$, particularly in the absence of long-range forces, are removed instead by an additional procedure with which we do not concern ourselves here. This procedure involves a further smoothing out of the Eq. (20), beyond the insertion of the $\cot\chi_i$ term and the replacement of the discrete radial functions f_{in} by the $f_i(\epsilon_i)$; it replaces the $f_i(\epsilon_i)$ of the standard pair with the base functions normalized at $r = 0$, through the unimodular transformation mentioned above. This transformation is carried out explicitly in Seaton's work[2]; it replaces the calculation of the $K^{(s)}$ matrix through Eq. (20) by the calculation of a still smoother, but substantially equivalent, R matrix.

ACKNOWLEDGMENTS

I am indebted to Christian Jungen and especially to Dan Dill for stimulating discussions and suggestions. I also thank M. Inokuti for reviewing the manuscript. This work was supported by the U. S. Energy Research and Development Administration, Division of Physical Research, Contract No. COO-1674-130.

---

[1]C. Jungen and O. Atabek, J. Chem. Phys. 16, 5584 (1977).

[2]M. J. Seaton, Proc. Phys. Soc. Lond. 88, 801 (1966).

[3]U. Fano, J. Opt. Soc. Am. 65, 979 (1975).

[4]P. G. Burke and D. Robb, Adv. Atomic and Mol. Phys. 11, 144 (1975); U. Fano and C. M. Lee, Phys. Rev. Lett. 31, 1573 (1973); C. M. Lee, Phys. Rev. A 10, 584 (1974); see also "R-Matrix Symposium" in *Invited Papers to the Tenth ICPEAC, Electronic and Atomic Collisions*, edited by G. Watel (North-Holland, Amsterdam, 1977).

[5]This formulation might be extended to include long range, but *local*, potentials that couple different channels (i,j), as outlined, e.g., by E. S. Chang and U. Fano, Phys. Rev. A 6, 173 (1972).

[6]See, e.g., L. S. Rodberg and R. M. Thaler, *The Quantum Theory of Scattering* (Academic, New York, 1967), Chap. 4, especially Sec. 3.

[7]See, e.g, Ref. 6, pp. 239–240.

[8]See, e.g., P. L. Altick and E. N. Moore, Phys. Rev. 147, 59 (1966); A. F. Starace, Phys. Rev. A 2, 118 (1970).

[9]C. Greene *et al*. (unpublished).

[10]The traditional application of configuration mixing would represent Ψ at large r by a superposition of the decaying exponentials of many different wave functions f_{in}. The MQDT procedure replaces this superposition by the single function $v_i(\epsilon_i, r)$ with the value of ϵ_i at which $\chi_i(\epsilon_i) = 0$.

1979 *Phys. Rev.* A **20** 1773–83
Reprinted with permission from the American Physical Society

Perturbed Rydberg series: Relationship between quantum-defect and configuration-interaction theory

F. H. Mies
Molecular Spectroscopy Division, National Bureau of Standards, Washington, D.C. 20234
(Received 15 November 1978; revised manuscript received 24 May 1979)

In this paper the theory of a single Rydberg series perturbed by an interloping state is examined. General analytic expressions are presented which apply both to the bound-state regions and to the autoionizing regions above the Rydberg ionization limit. The parameters are expressed in terms of configuration-interaction matrix elements, which can be calculated *a priori*. There is no restriction to weak-perturbation theory, and the wave function and all its associated properties, such as transition moments, can be extracted. Explicit calculations are presented for the Cd$(5s\ nd)\,^1D_2$ series perturbed by the Cd$(5p^2)\,^1D_2$ valence state, and excellent results are obtained.

I. INTRODUCTION

The theoretical calculation of atomic and molecular wave functions by variational techniques, which utilize trial functions to describe the individual electron orbitals, has worked very well for the lowest-energy states of a given symmetry. This is particularly true for "valence states," i.e., states whose predominant configurations are constructed from valence orbitals. In principle, the methods can be used to obtain excited states of the same symmetry by promoting the electrons to Rydberg-type orbitals. However, at these higher energies other configurations become almost degenerate, and the calculations must be expanded to include multi-configuration interactions. In particular, the interaction between a Rydberg state and an *excited* valence configuration can be quite strong, and the entire concept of a valence state and Rydberg state becomes confused. The problem of maintaining orthogonality between succeeding eigenstates also becomes a chore, and no simple methods have been devised to extrapolate the results to high principal quantum numbers.

Our goal in this paper is to demonstrate that there is a wealth of information about excited electronic states buried in present day *ab initio* calculations that can be extracted simply be recognizing the analytic properties of Rydberg states. This can be accomplished by incorporating multi-channel quantum defect theory (QDT). into the framework of configuration interaction and variational theories for bound electronic states. The results of QDT are derived by utilizing the analytic properties of the Rydberg electron for *arbitrary* energy. However, we will show that the necessary parameters can be obtained from *ab initio* calculations at *discrete* energies, without any modification of the codes that are presently employed. The QDT, with parameters derived from low-energy variational calculations, allows us to describe the properties of electronic states at higher energies, including the continuum. This is particularly imperative for molecular Rydberg series, where experimental data are both imprecise and difficult to interpret.

The theory is developed for the simplest conceivable case of a single Rydberg series, originating from an $L=0$ ion core, that is perturbed by a single interloping state originating from a different configuration. There is no restriction on the strength of the perturbation, and the perturbing level may lie imbedded within the series, or equally well, lie within the ionization continuum, in which case autoionization is encountered.

The theory is applied to an analysis of *ab initio* calculations for the Cd $(^1D_2)$ Rydberg series which converges to the Cd$^+$ $(^2S_{\frac{1}{2}})$ ground-state ion. The experimental term values are shown in Fig. 1, and are measured in units of $e^2/2a_0$ with respect to the ionization limit where $\epsilon=0$. This system was very carefully chosen for its simplicity and conformity to the present theoretical model. The unperturbed series originates from the Cd $(5s\ nd)$ $^1D_2$ configuration. The unperturbed levels, $n=5$–10, calculated by Stevens[1] are shown in Fig. 1. In addition, a Cd $(5p^2)$ $^1D_2$ valence state calculated to exist at E_p, approximately 1 eV *above* the ionization limit, strongly interacts with the series. Aside from higher members of the Cd $(5p\,np)$ $^1D_2$ series, which can be incorporated into the theory with little modification, no other significant perturbations are expected.

In Sec. II we review the quantum-defect theory[2] for a single unperturbed series. The asymptotic behavior of the Rydberg orbital is expressed as an analytic function of the continuous energy variable,

$$\epsilon=-1/\nu^2\,. \tag{1}$$

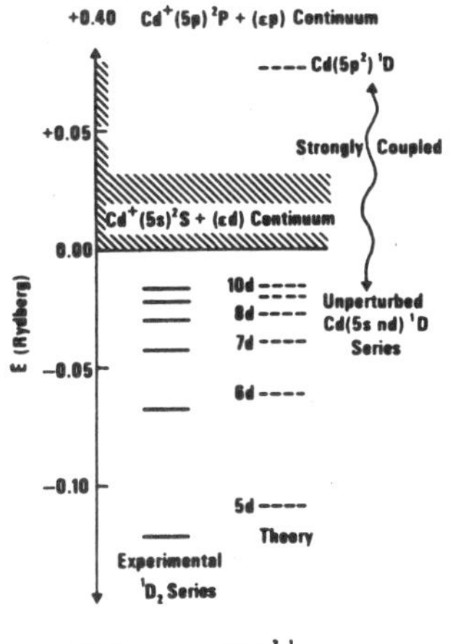

FIG. 1. Energy levels (Ry) of Cd $^1D_2$ series relative to the Cd⁺ $(5s)\,^2S_{1/2}$ ionization limit. The theoretical levels for the unperturbed Cd $(5s\,nd)\,^1D_2$ series and the perturbing Cd $(5p^2)\,^1D_2$ valence state were calculated by Stevens (Ref. 1). The matrix elements for these strongly coupled states are given in Table I.

At large distances, the orbital radial function must be a solution of the second-order Coulomb equation and can be uniquely represented as a linear combination of the regular and irregular Coulomb functions in terms of a single parameter $\tan\xi^0$. For arbitrary $\epsilon<0$ the Coulomb functions contain exponentially rising terms and the solution is not well behaved. The eigenvalues ϵ_n^0 are determined by the following condition which insures that the amplitude which multiplies these terms vanishes and the resultant orbital exponentially decays at large distances,

$$\tan\pi(\nu+\mu^0)=0\,. \tag{2}$$

This equation imposes the requirement that $\nu=n-\mu^0$, for integer values of $n\geq l+1$ and yields the usual expression for the Rydberg eigenvalues,

$$\epsilon_n^0=-1/(n-\mu^0)^2\,. \tag{3}$$

The defect, $\mu^0=\mu^0(\epsilon)$, is a slowly varying function of ϵ

$$\tan\pi\mu^0=A(\epsilon)/[\cot\pi\xi^0+\mathcal{G}(\epsilon)]\underset{\epsilon\to 0}{\sim}\tan\pi\xi^0\,. \tag{4}$$

A and $\mathcal{G}$ are monotonic functions of ϵ which approach 1 and 0, respectively, at the ionization limit and are known functions[2] defined by the orbital angular momentum l of the Rydberg orbital. The Rydberg series is uniquely defined by the parameter $\xi^0=\xi^0(\epsilon)$ which is a slowly varying analytic function of ϵ. The slowly varying nature of $\mu^0(\epsilon)$ and $\xi^0(\epsilon)$ is, of course, the essence of the QDT. The parameters may be determined from an analysis of the *ab initio* calculations for the unperturbed series.

In Sec. III the theory is developed for the perturbation of the series caused by a single interloping state. A very simple analytic representation of the perturbation of the eigenvalues is obtained, in which the quantum defect μ^0 in Eqs. (2) and (3) is replaced by μ_p,

$$\tan\pi\mu_p=A/[\cot\pi(\xi^0+\lambda_P)+\mathcal{G}]\underset{\epsilon\to 0}{\sim}\tan\pi(\xi^0+\lambda_P)\,. \tag{5}$$

The perturbation introduces an "energy-dependent" phase shift λ_p given by

$$\tan\pi\lambda_P=-\Gamma/2(\epsilon-\overline{E}_P) \tag{6}$$

which is added to the phase shift ξ^0 for the unperturbed series. $\overline{E}_P$ is the "position" of the perturbing level, possibly modified by a principal part "shift" which we shall ignore for present purposes. The "width" $\Gamma=\pi V_p^2$ is a measure of the strength of the interaction.

In Sec. IV we determine the normalization of the perturbed wave function and show how Γ can be evaluated from the usual configuration-interaction (CI) matrix elements between the perturbed and the unperturbed Rydberg states. Even without recourse to theoretical calculations, Eq. (5) is a very useful representation for analyzing a perturbed series. The parameter $\pi\lambda_P(\epsilon)$ introduced by the perturbing level varies from zero, as $\epsilon\to-\infty$, to $\frac{1}{2}\pi$ at $\epsilon=\overline{E}_P$, to π as $\epsilon\to+\infty$. The energy dependence of the unperturbed parameter ξ_0 is modified in such a way as to introduce an *additional* eigenvalue into the spectrum of the Rydberg levels. If the width $\Gamma=\pi V_P^2$ is very small, then the additional level will be well localized in analogy to the usual perturbation results. However, a large width will spread the contribution of the perturbing state over the entire Rydberg series and possibly into the ionization continuum, where it can influence the photoionization spectrum.

In Sec. V the theory is applied to the Cd$(^1D_2)$ series. Conclusions are summarized in Sec. VI.

II. THEORY OF UNPERTURBED RYDBERG SERIES

We shall assume LS coupling is valid for the given, *unperturbed* Rydberg series. Further, we only consider the simplest case of a series originating from an *ion*-core state $|L_c,M_{L_c};S_c,M_{S_c}\rangle=\phi_0(\vec{r}_c)$ with total orbital angular momentum $L_c=0$. This may be combined with the spin functions, $|s,m_s\rangle$, of the $s=\frac{1}{2}$ Rydberg electron to define a total spin $\vec{S}=\vec{S}_c+\vec{s}$, which is *assumed* to commute with the total Hamiltonian $H_T(\vec{r}_c,\vec{r})$.

$$|S,M_S\rangle = \sum_{M_{S_c},m_s} \phi_0(\vec{r}_c)\,|s,m_s\rangle \times C(S_c,s,S;M_{S_c},m_s,M_S)\,. \qquad (7)$$

The Rydberg orbital, $\psi_0(\vec{r}) = Y_{l,m_l}(\hat{r})\rho_0(r)/r$, with orbital angular momentum l defined by the Legendre function $Y_{l,m_l}(\hat{r})$, is described by a radial function $\rho_0(r)$ which must be well behaved at the origin, i.e.,

$$\rho_0(r) \underset{r\to 0}{\sim} (2r)^{l+1}/(2l+1)\,! \qquad (8)$$

We shall be particularly concerned about the long-range properties of $\rho_0(r)$.

The general function for the Rydberg state can be expressed as an antisymmetrized product,

$$|\alpha_0;\rho_0\rangle = \mathcal{Q}[|S,M_S\rangle\psi_0(\vec{r})] = \mathcal{Q}[|\alpha_0\rangle\rho_0(r)/r]\,, \qquad (9a)$$

where

$$|\alpha_0\rangle \equiv |S,M_S;l,m_l\rangle = |S,M_S\rangle Y_{l,m_l}(\hat{r}) \qquad (9b)$$

and the antisymmetrization operator $\mathcal{Q}$ is normalized such that

$$\langle\alpha_0,\rho_0|\alpha_0,\rho_0'\rangle = \int_0^\infty dr\,\rho_0^*(r)\rho_0'(r)\,. \qquad (9c)$$

We can construct a complete set of states $\{\alpha\}$ from the complete set of ion-core states, $\{\phi_c(\vec{r}_c)\} = \{\phi_0,\phi_1,\ldots\}$, and the various orbital angular momentum states $Y_{l,m_l}(\hat{r})$. These are referred to as channel states, and they play an important role in QDT and scattering theory. The channel states are simple product functions which already contain the orbital and spin angular momentum of the Rydberg electron. They span the entire space of the system, *with the exclusion of the Rydberg radial coordinate r*. The total wave function $|E\rangle$, at total energy E, is expanded in a truncated set of channel states

$$|E\rangle = \sum_i C_i\,|\alpha_i;\rho_i\rangle\,, \qquad (10)$$

and coupled integro-differential equations are generated for the radial functions $C_i\rho_i(E,r)/r$,

$$\langle\alpha_i|H_T(\vec{r}_c,\vec{r}) - E|E\rangle = 0 \quad \text{for all } i\,. \qquad (11)$$

In the limit of large r, where the Rydberg electron is ionized and its interaction with the core electrons vanishes, these channel states $|\alpha\rangle$ *must be uncoupled* by the total Hamiltonian $H_T(\vec{r}_c,\vec{r})$. (This is the *definition* of a channel state. The assumption concerning *LS*-coupling for the ion-core states can be removed by selecting appropriate combinations of "approximate" channel states which properly diagonalize the *asymptotic* spin-orbit interaction.) The asymptotic Hamiltonian is separable and the channel states define the following differential equation for the radial portion of the total wave function:

$$\langle\alpha_i|H_T - E|E\rangle \underset{r\to\infty}{\sim} C_i\left((E_{\alpha_i} - E) - \frac{1}{r}\frac{\partial^2}{\partial r^2}r + \frac{l_i(l_i+1)}{r^2} - \frac{2}{r}\right)\frac{\rho_i(E,r)}{r} = 0 \qquad (12a)$$

This is, of course, the Coulomb equation. *Each* α state defines an ionization limit, with the threshold energy E_α and an orbital angular momentum l for the escaping electron. Therefore, *each* $|\alpha\rangle$ *identifies an entire unperturbed Rydberg series.* QDT examines the analytic properties of the radial functions in each channel. Whenever E equals a proper bound-state eigenvalue E_n, the radial functions must vanish asymptotically,

$$\sum_i C_i\frac{\rho_i(E,r)}{r} \underset{r\to\infty}{\sim} 0\,, \quad E = E_n\,. \qquad (12b)$$

For the one-channel expansion $|E\rangle = C_0|\alpha_0;\rho_0\rangle$ defined in Eq. (9), ρ_0 must satisfy Eqs. (11) and (12) for $E_{\alpha_i} = E_{\alpha_0}$.

Seaton[2] has defined Coulomb functions $f(\epsilon,r)$ and $g(\epsilon,r)$ that are solutions of the Coulomb equation (12a) and are *analytic* functions of the asymptotic energy ϵ,

$$\epsilon = (E - E_\alpha)\,. \qquad (13)$$

Further, these functions are exact, independent solutions of Eq. (12a) for *all* ϵ and *all* r, with the Wronskian normalized as follows,

$$f\frac{\partial g}{\partial r} - g\frac{\partial f}{\partial r} = \frac{2}{\pi}\,. \qquad (14)$$

The leading terms in the power-series expansion of f and g are independent of ϵ; in fact, Eq. (8) is equivalent to the expansion of the solution f, which is regular as $r\to 0$, while g approaches $-(2l)!/\pi(2r)^l$ and is divergent at the origin. The radial function $\rho_0(\epsilon,r)$ is related to these functions as follows:

$$\rho_0 \underset{r\to 0}{\sim} f(\epsilon,r) \qquad (15)$$
$$\rho_0 \equiv \psi_R^0(\epsilon,r)\mathcal{C}$$

where

$$\psi_R^0 \underset{r\to\infty}{\sim} (f\cos\pi\xi^0 - g\sin\pi\xi^0)\,. \qquad (16)$$

The important parameter $\xi^0 = \xi^0(\epsilon)$ is a slowly varying analytic function of ϵ. The single-channel expansion $|E\rangle = C_0|\alpha_0;\rho_0\rangle$ defines a solution of Eq. (11) which is exact and has its asymptotic properties given by Eq. (16) for *all E*. However, for $E < E_{\alpha_0}$, well-behaved solutions, which are square integrable, are only obtained at *particular* eigenvalues $E = E_n^0 = \epsilon_n^0 + E_{\alpha_0}$, when ψ_R^0 vanishes asymptotically, i.e.,

$$\psi_R^0(\epsilon_n^0, r) \underset{r\to\infty}{\sim} 0 \, . \tag{17}$$

The asymptotic properties of f and g have been well described by Seaton, and, for $\epsilon < 0$, both functions contain exponentially rising functions Φ^-. The discrete eigenvalues defined by (17) occur when the amplitude associated with the Φ^- components in Eq. (16) vanishes. Seaton has shown this condition is satisfied when

$$[\cos\pi\xi^0 + \sin\pi\xi^0(\mathcal{G} + A\cot\pi\nu_n^0)]\frac{\Phi^-}{\Gamma(l+1-\nu_n^0)} = 0 \, , \tag{18}$$

where

$$\epsilon_n^0 = -1/(\nu_n^0)^2 \, . \tag{19}$$

Except for $\xi^0 = 0$ [which corresponds to the "pure" Coulomb case, where the eigenvalues occur at the poles of the γ function $\Gamma(l+1-\nu_n^0)$, i.e., $\nu_n^0 = l+1, l+2, \ldots$] the eigenvalues are defined by the condition

$$\cot\pi\xi^0 + \mathcal{G} + A\cot\pi\nu_n^0 = 0 \, . \tag{20}$$

Letting

$$\cot\pi\mu^0 = (\mathcal{G} + \cot\pi\xi^0)/A \, , \tag{21}$$

we obtain Eqs. (1)–(4) in which the eigenvalues occur at $\nu_n^0 = (n - \mu^0)$, with $n = l+1, l+2, \ldots$, such that

$$E_n^0 = E_{\alpha_0} - 1/(n-\mu^0)^2 \, . \tag{22}$$

$\mu^0(\epsilon_n^0)$ is the "quantum defect" for the state with eigenvalues E_n^0 and is related to the parameter ξ^0 through Eq. (21). The parameter A is a function of ϵ and l

$$A = \prod_{p=0}^{l}(1 + p^2\epsilon) \tag{23}$$

and, $A \to 1$, as $\epsilon \to 0$. The function $\mathcal{G}$ is not analytic in ϵ and is given in terms of the Ψ functions,[3] which are logarithmic derivatives of Γ functions,

$$\mathcal{G}(\nu, l) = \frac{A}{2\pi}[\psi(\nu + l + 1) + \psi(\nu - l) - 2\ln\nu] \, . \tag{24}$$

This function approaches zero as $\epsilon \to 0$. Only at threshold, $\epsilon = 0$, does $\mu^0 = \xi^0$. The *known* functions A and $\mathcal{G}$ introduce significant energy variation in the quantum defect μ^0, especially for $l \neq 0$; generally much more substantial extrapolation of the spectra can be obtained from an analysis based on ξ^0 held constant, with μ^0 varying according to Eq. (21).

The usual self-consistent-field (SCF) variational calculations construct a trial function similar to Eqs. (7) and (9), but with one important exception. The ion-core function $\phi_0(\vec{r}_c)$ in QDT is constrained to be the asymptotic ion-core state of the ionized system. The only variable in the wave function is the radial function $\rho_0(r)$ [*and* the number of ion-core channels included in the expansion of Eq. (10)]. SCF calculations, on the other hand, replace $\phi_0(\vec{r}_c)$ by a variational trial function ϕ_0^v which is not necessarily an eigenfunction of the ion-core Hamiltonian, and *both* ϕ_0^v and ρ_0^v are varied to obtain a best energy.

If we use the single configuration $|\alpha_0^v;\rho_0^v\rangle$ in a variational calculation, we will obtain a set of eigenvalues E_n^v, which represent the Rydberg levels of the *unperturbed* series originating from the $L_c = 0$ core and an l Rydberg electron. For an $l = 2$ electron, which corresponds to, say, the $\ldots 5s\,(^2S_{1/2})\, nd\, ^1D$ series of Cd, we might expect that the variationally calculated core function will closely approximate the true ion-core state. Certainly as n becomes large and the Rydberg electron density within the core diminishes (as n^{-3}), this will be true. For present purposes we will *assume* $\phi_0^v \equiv \phi_0$ and the radial function $\rho_0(r)$ is the only variational function. In this case, with no core polarization, both variational theory and QDT give identical results, i.e., $E_n^v = E_n^0$ in Eq. (22). Let

$$|E_n^0\rangle = N_n^0|\alpha_0;\rho_0(E_n^0)\rangle \, , \tag{25}$$

where ρ_0 is the variationally determined radial wave function with boundary conditions given by Eq. (8) in the frozen field of the $^2S_{1/2}$ core. Let

$$F_n^0(r) = N_n^0\rho_0(E_n^0, r) \equiv (N_n^0\mathcal{C})\psi_R^0(E_n^0, r) \tag{26}$$

such that

$$\int F_n^{0*}(r)F_{n'}^0(r)\,dr = \delta_{n,n'} \tag{27}$$

and

$$\langle E_n^0|H_T|E_{n'}^0\rangle = E_n^0\delta_{n,n'} \, . \tag{28}$$

Given a set of eigenvalues $\{E_n^0\}$ determined from the variational solution of Eq. (11), we obtain a "quantum defect" $\mu^0(E_n^0)$ from Eq. (22) and then use Eq. (21) to determine the parameter $\xi^0(E_n^0)$. *If* the amplitudes of the functions f and g in Eqs. (15) and (16) are determined primarily in the range of small r values where $f(\epsilon, r)$ and $g(\epsilon, r)$ are almost independent of ϵ, then we might expect $\xi^0(\epsilon)$ to be only a slowly varying function of ϵ. This requires that all interactions, except the asymptotic Coulomb and centrifugal terms in Eq. (12a) vanish rapidly with increasing r. The QDT is predicated on this fact that ξ^0 is an insensitive function of ϵ, particularly in the vicinity of the threshold. Calculated ξ^0 values of the Cd ($5s\,nd$) $^1D_2$ Rydberg series are given in Table I and exhi-

TABLE I. Calculated properties of CdI $(5s\ nd)^1D_2$.

Term	Unperturbed eigenvalues E_n^0 (Ry) (calculated)	(μ^0-2) (derived)	(ξ^0-2) (derived)	CI matrix element $\langle P\|H_T\|E_n^0\rangle$	Γ(Ry) (derived)
5s5d	−0.108 5049	−0.035 815	−0.071 404	−0.044 414	0.298 56
5s6d	−0.061 173 5	−0.043 135	−0.061 068	+0.027 650	0.224 23
5s7d	−0.039 266 6	−0.046 477	−0.057 541 [a]	−0.020 014	0.199 99 [a]
5s8d	−0.027 319 9	−0.050 075	−0.057 902	+0.015 557 2	0.194 52
5s9d	−0.020 084 4	−0.056 198	−0.062 444	−0.012 951 2	0.205 48
5s10d	−0.015 364 0	−0.067 663	−0.073 274	+0.010 899	0.211 81
$(5p)^2$	+0.076 938 5 [a]				

[a] Optimum parameters chosen: $(\xi^0-2)=-0.057\,541$; $\Gamma=0.199\,99$ Ry, $\bar{E}_p=0.076\,938\,5$ Ry.

bit this desired feature.

Note that by using the nondivergent SCF solutions $|E_n^0\rangle$, we have obtained an asymptotic analytic representation of the radial function $\psi_R^0(\epsilon,r)$ for *all* energies. This function is regular at the origin and, for $\epsilon>0$, is just the scattering, or continuum, wave function, in channel α_0. The asymptotic form of ψ_R^0 yields a scattering phase shift δ_0 in terms of the parameter ξ^0, i.e.,[4]

$$\psi_R^0(\epsilon,r)\underset{\substack{r\to\infty\\ \epsilon>0}}{\sim}\left(\frac{2\gamma}{\pi}\right)^{1/2}B^{1/2}\frac{\sin\pi\xi^0}{\sin\pi\delta^0}\sin(\omega+\delta^0)\,, \tag{29}$$

where the phase shift is given by

$$\cot\delta_0=[\cot\pi\xi^0+R_e(\mathcal{G})]/B\,,$$

where $\gamma^2=1/\epsilon$, $R_e(\mathcal{G})\equiv\frac{1}{2}(\mathcal{G}+\mathcal{G}^*)$, $B=A(\epsilon,l)/[1-\exp(-2\pi\gamma)]$ and

$$\omega=r/\gamma+\gamma\ln(2r/\gamma)-\tfrac{1}{2}\pi l-\arg(\Gamma(l+1+i\gamma))\,.$$

Near threshold, where $\gamma\to\infty$, $\epsilon\to0^+$, $B\approx A\approx1$, and $R_e(\mathcal{G})\approx0$, we obtain

$$\delta^0\approx\pi\mu^0\approx\pi\xi^0\quad\text{as }\epsilon\to0^+\,. \tag{30}$$

In addition to the function $\psi_R^0(\epsilon,r)$, which is regular at the origin, we can obtain an *independent* solution at energy ϵ, which is also analytic in ϵ and irregular at the origin. We define $\psi_I^0(\epsilon,r)$ by the following Wronskian condition,

$$\psi_R^0\frac{\partial\psi_I^0}{\partial r}-\psi_I^0\frac{\partial\psi_R^0}{\partial r}=2/\pi\,. \tag{31}$$

where $\psi_I^0\underset{r\to0}{\sim}g(\epsilon,r)\mathcal{C}$ and

$$\psi_I^0\underset{r\to\infty}{\sim}(f\sin\pi\xi^0+g\cos\pi\xi^0) \tag{32}$$

Both ψ_R^0 and ψ_I^0 are solutions of the equation

$$\langle\alpha_0|H_T-(E_{\alpha_0}+\epsilon)|\alpha_0,\psi^0(\epsilon)\rangle=0\,. \tag{33}$$

It should be recognized that, for $\epsilon>0$, the solution $\psi_R^0(\epsilon,r)$ corresponds to the distorted-wave approximation (DWA)[5] in channel $|\alpha_0\rangle$ in the absence of coupling to other channels.

III. PERTURBATION OF THE RYDBERG SERIES

Let us suppose that, in addition to the ground-state series $|E_n^0\rangle$ in Eq. (25), we have determined various, *excited*, variational channel state wave functions which have the following properties.

$$|P\rangle=\sum_{i\neq0}C_i^P|\alpha_i;\rho_i^P\rangle \tag{34a}$$

such that

$$\langle P|H_T|P'\rangle=E_P\delta_{P,P'} \tag{34b}$$

with the additional condition that the following integral vanish for *any* arbitrary radial function $F(R)$,

$$\langle P|\alpha_0;F\rangle=0\,. \tag{35}$$

The "perturbing" states $|P\rangle$ correspond to *well-behaved* bound-state solutions originating from *excited* ion-core states $|\alpha_i\rangle$ and are orthogonal to the unperturbed Rydberg series with $F=F_n^0$ given by Eq. (26). This will generally be true, especially if the set of channel states $|\alpha_i\rangle$ only contain core angular momenta L_c and Rydberg orbital angular momenta l which both differ from the values in $|\alpha_0\rangle$ [see Eqs. (7) and (9)]. If $|P\rangle$ is generated from a variational calculation which allows arbitrary variations in the core functions, then small nonorthogonalities *may* be introduced which invalidate Eq. (35), but only in very special cases, generally involving configurations with excited cores of the same symmetry as $|\alpha_0\rangle$. These complications are not present in the Cd $(^1D_2)$ calculations, and can be treated, if necessary, with minor variation of the present theory. We shall proceed with the assumption contained in Eq. (35).

If we consider just one perturbing state, and one Rydberg series, we obtain a particularly simple result. Let

$$|E\rangle = |\alpha_0;F_E\rangle + B_P(E)|P\rangle , \tag{36}$$

where we must choose the coefficient $B_P(E)$, and the radial function $F_E(r)$ to satisfy the condition

$$\langle E'|H_T - E|E\rangle = 0 \tag{37}$$

for *all* E and E' and such that $|E\rangle$ is well behaved.

Equation (36) may be regarded as a restricted version of the multichannel QDT expressed in Eqs. (10)–(12). The excited channel states, with $\alpha_i \neq \alpha_0$, have been collected together in Eq. (34a), and by some variational technique the radial functions $C_i^P\rho_i^P$ have been generated which decay asymptotically and satisfy Eq. (12b) when $E = E_P$. This set is "frozen" and introduced into Eq. (36) with one linear parameter $B_P(E)$ which must be adjusted to satisfy Eq. (37). Ideally the QDT would allow each $C_i\rho_i$ function to vary independently until Eqs. (11) and (12) are satisfied, and in this sense Eq. (36) is much more restrictive. However, since the variationally determined state $|P\rangle$ can indirectly encompass the *infinite* set of channel states, it offers compensations that are not easy to assess.

From another point of view, Eq. (36) is exactly equivalent to Fano's configuration interaction theory[6] (CIT) for a single resonance imbedded in a continuum. Fano simply expands the undetermined function F_E using the *complete* set of *well-behaved* solutions for the unperturbed series that were generated in Sec. II,

$$F_E^{\rm CIT}(r) = \sum_n A_n(E)F_n^0(r) + \int_0^\infty d\epsilon' C_{\epsilon'}(E)\psi_R^0(\epsilon',r) , \tag{38}$$

where F_n^0 is defined in Eq. (26) for the bound eigenstates and ψ_R^0 has the asymptotic properties in Eqs. (16) and (29) for $E' = E_{\alpha_0} + \epsilon'$ above the ionization limit. $F_E^{\rm CIT}(r)$ vanishes at the origin, and is always nondivergent for $r \to \infty$. The coefficients $A(E)$ and $C(E)$ are determined from Eq. (37). Below the ionization limit $E < E_{\alpha_0}$, solutions are obtained *only* at discrete energies $E = E_n$ which define the eigenvalues for the perturbed series.

In contrast, we shall expand $F_E(r)$ as follows:

$$F_E(r) = \psi_R^0(\epsilon,r) - \tan\pi\lambda_P G_I(\epsilon,r) , \tag{39a}$$

where G_I is required to be well behaved as $r \to 0$, and asymptotically

$$G_I(\epsilon,r) \underset{r\to\infty}{\sim} \psi_I^0(\epsilon,r) . \tag{39b}$$

Note that ψ_R^0 and F_E are not yet normalized. Also, the entire perturbed radial dependence has been incorporated into the function G_I. When the parameter $\lambda_P = 0$, for all ϵ, we retrieve the unperturbed solution ψ_R^0. Unlike $F_E^{\rm CIT}$ in Eq. (38) which is well behaved and only defines solutions at discrete energies E_n when E is below the ionization limit, F_E in Eq. (39) yields mathematically proper solutions at *all* energies using the unperturbed radial functions $\psi_R^0(\epsilon,r)$ and $\psi_I^0(\epsilon,r)$ *on* the energy shell $E = E_{\alpha_0} + \epsilon$. Of course these functions are asymptotically divergent when $\epsilon < 0$ and well-behaved solutions must be extracted, in analogy to Eq. (17), by imposing the condition $F_E(r) \sim 0$ as $r \to \infty$.

Let us define the following matrix elements:

$$V_P \equiv \langle P|H_T - E|\alpha_0;\psi_R^0(\epsilon)\rangle , \tag{40}$$

$$W_P \equiv \langle P|H_T - E|\alpha_0;G_I(\epsilon)\rangle , \tag{41}$$

and a shifted energy for the perturbing state $|P\rangle$,

$$\bar{E}_P = E_P + \tfrac{1}{2}\pi W_P V_P . \tag{42}$$

The term $\frac{1}{2}\pi W_P V_P$ in Eq. (42) corresponds to the principal-part shift one obtains in CIT.[6] Since we have no simple method of evaluating W_P with any certainty, we shall introduce it as the sole adjustable parameter in the present theory. (It is possible that multi-configuration calculations can be used to estimate the shift $\frac{1}{2}\pi W_P V_P$, or explicit numerical integrations can be employed using pseudopotential methods.) The Cd results are obtained *assuming* $\bar{E}_P = E_P$.

From Eqs. (36), (37), and (39) we obtain the algebraic equation

$$\langle P|H_T - E|E\rangle = B_P(E)(E_P - E) + V_P - \tan\pi\lambda_P W_P = 0 \tag{43}$$

and an integro-differential equation for the radial function $F_E(r)$,

$$\langle\alpha_0|H_T - E|\alpha_0;F_E\rangle + B_P(E)\langle\alpha_0|H_T - E|P\rangle = 0 . \tag{44}$$

The matrix elements in Eq. (43) involve an integration over the entire $(N+1)$-electron coordinate space, including the radial coordinate r of the Rydberg electron, in analogy to the usual CIT.[6] In Eq. (44) we only integrate over the coordinates of the channel state $|\alpha_0\rangle$, and this defines a second-order differential equation for the radial function $F_E(r)$ which has the asymptotic properties expressed in Eq. (12a). Using the homogeneous solutions of Eq. (44) we obtain the result[7]

$$-\tan\pi\lambda_P G_I(\epsilon,r') = \tfrac{1}{2}\pi B_P(E)\left(\psi_R^0(r')\int_{r'}^\infty dr\,\psi_I^0(r)\frac{r^2}{r}(\langle\alpha|H_T|P\rangle + O)(N+1)^{1/2} + \psi_I^0(r')\int_0^{r'} dr\,\psi_R^0(r)\frac{r^2}{r}(\langle\alpha|H_T|P\rangle + O)(N+1)^{1/2}\right) . \tag{45a}$$

Note that both $\langle\alpha|H_T|P\rangle$ and O are functions of the radial coordinate r. The operator O represents the contribution of the $(N+1)$ electron exchange terms to the solution of Eq. (44). These terms are, of course, implicit functions of F_B in Eq. (39). We shall assume that the *exchange* terms in Eq. (45a) are well approximated using $F_B \approx \psi_R^0$. In the limit, as $r \to \infty$, the first integral vanishes, and the second integral is equivalent to V_P in Eq. (40). This should be an excellent approximation in most cases, and is certainly consistent with the assumption that W_P in Eq. (41) can be neglected. Comparing the asymptotic form of Eq. (45a) to Eqs. (39a) and (39b), we obtain the result,

$$-\tan\pi\lambda_P \cong \tfrac{1}{2}\pi B_P(E) V_P \tag{45b}$$

and from Eq. (43) we find

$$B_P = V_P/(E - \bar{E}_P) \tag{46}$$

such that

$$-\tan\pi\lambda_P = \frac{\pi}{2}\frac{V_P^2}{E-\bar{E}_P} \equiv \frac{\Gamma}{2(E-\bar{E}_P)}\,. \tag{47}$$

If more than one perturbing level were included in Eq. (36), the right-hand side of Eq. (47) would consist of a sum of such resolvent terms.

Using these results in Eq. (36), we obtain

$$|E\rangle = |\alpha_0;\psi_R^0\rangle + [V_P/(E-\bar{E}_P)](\tfrac{1}{2}\pi V_P|\alpha_0;G_I\rangle + |P\rangle) \tag{48}$$

$$\underset{r\to\infty}{\sim} (1/\cos\pi\lambda_P) \times [|\alpha_0;f\rangle\cos\pi(\xi_0+\lambda_P) - |\alpha_0;g\rangle\sin\pi(\xi_0+\lambda_P)]\,. \tag{49}$$

From the asymptotic properties of Eq. (49), we derive a "quantum defect" $\mu_P(\epsilon)$ for the perturbed series,

$$\cot\pi\mu_P = [\cot\pi(\xi_0+\lambda_P) + \mathcal{G}]/A\,, \tag{50}$$

which yields Eq. (5). This is to be compared to the result, Eq. (25), for the unperturbed series. In order to evaluate the important parameter Γ in Eq. (47), we must discuss the proper normalization of the Rydberg functions.

IV. EVALUATION OF Γ AND NORMALIZATION OF THE PERTURBED RYDBERG SERIES

The "width" parameter $\Gamma = \pi V_P^2$ depends on the "reduced" CI matrix element $V_P(E)$ in Eq. (40) and is an analytic function of the total energy E. The *ab initio* calculations provide matrix elements, $\langle E_n^0|H_T|P\rangle$, between the normalized, *unperturbed*, Rydberg wave functions $|E_n^0\rangle$ given in Eqs. (25) and (26), and the normalized, perturbing valence state, $|P\rangle$, in Eq. (34). These are listed in column 5 of Table I for the Cd series. The reduced matrix element, at the specific energies E_n^0 can be determined by the relationship, $\langle E_n^0|H_T|P\rangle \equiv (N_n^0\mathcal{C})V_P(E_n^0)$. The normalization for the *unperturbed* wave function, i.e., $(N_n^0\mathcal{C})^2$, is obtained from Eq. (53) with $\nu_n = \nu_n^0$, and $\lambda_P = 0$. The final result is

$$\Gamma = \pi V_P^2 = \pi\langle E_n^0|H_T|P\rangle^2 \times\left[A\left(\frac{\sin^2\pi\xi^0}{\sin^2\pi\mu^0}\right)\left(\nu_n^{03} + 2\,\frac{\partial\mu^0}{\partial E_n^0}\right)\right]\,. \tag{51}$$

The normalization factor in brackets is often well approximated as $[\nu_n^3/A]$. We expect that $\Gamma(E_n^0)$ is only a slowly varying function of the total energy. This is demonstrated by the calculated widths obtained from Stevens' CI matrix elements[1] for Cd given in Table I.

The perturbed wave function in Eq. (48) is *not* normalized. Let $\psi_B = N_B|E\rangle$ such that $\langle\psi_B|\psi_{B'}\rangle = \delta(E-E')$, for $E > E_{\alpha_0}$, $\langle\psi_{B_n}|\psi_{B_{n'}}\rangle = \delta_{n,n'}$, for $E_n < E_{\alpha_0}$. The normalization can be evaluated as follows,[8]

$$\langle\psi_B|\psi_{B'}\rangle = N_B N_{B'} \lim_{r\to\infty}\left(\frac{F_B\frac{\partial}{\partial r}F_{B'} - F_{B'}\frac{\partial}{\partial r}F_B}{E-E'}\right)\,. \tag{52}$$

We find

$$N_B^2 = [\cos^2\pi\lambda_P \sin^2\pi\mu_P/\sin^2\pi(\xi^0+\lambda_P)2B]\theta_B^{-1}\,, \tag{53}$$

where $\theta_B = 1$, for $E > E_{\alpha_0}$, and $\theta_B = \tfrac{1}{2}\nu_n^3 + \partial\mu_P/\partial E_n$, for $E_n < E_{\alpha_0}$ and we obtain

$$\psi_B = \frac{1}{(2B\theta_B)^{1/2}}\frac{\sin\pi\mu_P}{\sin\pi(\xi^0+\lambda_P)}[\cos\pi\lambda_P|\alpha_0,\psi_R^0\rangle - \sin(\pi\lambda_P)(2/\pi V_P)(\tfrac{1}{2}\pi V_P|\alpha_0,G_I\rangle + |P\rangle)]$$

$$\equiv \frac{-1}{(2B\theta_B)^{1/2}}\frac{\sin\pi\mu_P}{\sin\pi(\xi^0+\lambda_P)}\left[\frac{(E-\bar{E}_P)|\alpha_0,\psi_R^0\rangle + V_P(\tfrac{1}{2}\pi V_P|\alpha_0,G_I\rangle + |P\rangle)}{[(E-\bar{E}_P)^2 + \tfrac{1}{4}\Gamma^2]^{1/2}}\right]. \tag{54}$$

From Eq. (54) we can derive simple expressions for properties such as the transition moments and evaluate the influence of the perturbing levels on the oscillator strength to both bound-state levels and the photoionization continuum.

V. DISCUSSION OF RESULTS OBTAINED FOR Cd (5s nd) ^1D

The accuracy of the theoretical results and the *simplicity* of their application can be demonstrated using *ab initio* pseudopotential calculations provided by Stevens[1] for Cd (5s *nd*) ^1D interacting with the Cd $(5p^2)$ ^1D valence states shown in Fig. 1. The results of the variational calculations are summarized in Table I. These can be used to obtain the three parameters, ξ^0, Γ, and $\bar{E}_P$ we require to characterize the entire system.

The term values for the first six members of the unperturbed series (5s *nd*) are given in column 2. Using Eq. (3) we obtain the unperturbed quantum defect μ_0 in column 3, and from these, using Eq. (4), in conjunction with (27) and (28), we extract the parameter ξ^0 listed in column 4. The "philosophy" of the QDT requires this parameter to become a slowly varying, almost constant, function as *nd* increases. The variational trial function used to calculate this series is most accurately optimized at small n, and the deviation in ξ^0 for 10*d*, from what appears to be a stabilized value for $n = 7, 8, \sim 9$ is not unexpected. In the application of the present theory we shall choose our parameters from the results for 5s 7*d* as the best compromise. Very similar results would be obtained if, say, 5s 8*d* had been chosen.

The second parameter Γ should also be an insensitive function of energy since it measures the interaction of the "tight," perturbing valence state with the inner, energy insensitive portion of the unperturbed Rydberg states. The calculated matrix elements $\langle 5s\ nd | H_T | 5p^2 \rangle$ are listed in column 5 and, of course, these decrease rapidly because of the normalization factor which scales the 5s *nd* functions. Using Eq. (51) to reduce these data, e.g., $\Gamma \approx \pi \nu_n^{03} \langle 5s\ nd | H_T | 5p^2 \rangle / A$, we obtain the relatively constant Γ values in column 6. Again, we shall employ the 7*d* value as the best compromise between increasing n and decreasing accuracy.

The third parameter $\bar{E}_P$ is given by Eq. (42). We are not prepared to evaluate the shift implied by the quantity $W_P V_P$, and the unshifted value $E_P = 0.07694$ calculated by Stevens will be employed. Various elaborate hand waving arguments can be used to rationalize this *assumption*, but the excellent results we obtain for Cd are, for the moment, the best justification. Obviously, if E_P were to occur, not *above* the ionization limit, but imbedded in the Rydberg series, the exact position $\bar{E}_P$ is more critical, and $\bar{E}_P$ might best be used as an adjustable parameter.

The results of the calculations are given in Table II. Column 2 lists the experimentally observed term values for the $^1D_2$ series.[9] Column 3 lists the predictions of the single configuration variational calculations obtained by Stevens, and column 4 gives the results of solving a (7×7) CI secular equation for the interaction of $5d - 10d$ with the $5p^2$ term located at $+8443\ \text{cm}^{-1}$. Obviously the CI calculations make a significant improvement in the eigenvalues.

The result of the present theory is shown in columns 5, 6, and 7. Given a value for Γ, $\bar{E}_P$, and ξ^0, we must evaluate $\lambda_P(E)$, $A(E)$, and $\mathcal{G}(\nu, l)$ from Eqs. (47), (23), and (24), respectively, at some value of E. These quantities are then used in Eq. (50) to arrive at an estimate for $\mu_P(E)$. Ideally this process should be iterated until E agrees with the predicted eigenvalue, E_n, i.e.,

$$E_n = [n - \mu_P(E_n)]^{-2} \tag{55}$$

As a good first approximation we may evaluate $\mu_P(E_n^0)$ at the "unperturbed" eigenvalue E_n^0. This result is given in column 5 using the parameters Γ and ξ^0 obtained for each individual *nd* state in Table I. Column 6 shows the results obtained from the 7*d* parameters which we have chosen as the "best" set based on the quality of the calculations employed by Stevens[1]. (The resultant quantum defects are shown by the dashed curve in Fig. 2.) Finally column 7 presents the results of the present theory when the experimental energies E_n in column 1, are employed to evaluate $\lambda_P(E_n)$, $A(E_n)$, and $\mathcal{G}(\nu_n, l)$. This is essentially just a lazy way to achieve the self-consistency required by Eq. (55). In the absence of experimental data essentially the same results would be achieved from a single iteration of Eq. (55) with E_n replaced by $[n - \mu_P(E_n^0)]^{-2}$.

The resultant quantum defects for each level are plotted in Fig. 2. The solid curve is drawn through the predicted values obtained from column 7, using the 7*d* parameters. The circles are the experimentally observed values, and these seem to converge on the theoretical curve. Some of the deviations may be due to "other" unknown perturbations which may interact weakly with the $^1D_2$ series.

The boxes in Fig. 2 indicate the quantum defects obtained from the (7×7) CI results in column 4. The serious deviations one obtains as n increases demonstrates the limitations of using a "truncated" basis for the unperturbed series. Effectively the theoretical expressions derived from QDT yield an *analytic* solution to the *infinite* CI

TABLE II. Comparison of experimental and calculated term values for Cd $(5s\ nd)\,^1D_2$.

	Expt. (cm^{-1}) (E_n)	Calc. unperturbed series (E_n^0)	Results of (7 × 7) CI	Results of QDT Using (nd) and E_n^0	Results of QDT Using (7d) and E_n^0	Results of QDT Using (7d) and E_n
5*d*	−13 320.3	−11 907	−13 004	−12 862	−12 636	−12 552
6*d*	−7 405.3	−6 713	−7 250	−7 301	−7 239	−7 205
7*d*	−4 701.7	−4 309	−4 592	−4 653	−4 653	−4 640
8*d*	−3 243.1	−2 998	−3 162	−3 220	−3 226	−3 221
9*d*	−2 370.6	−2 204	−2 308	−2 361	−2 362.0	−2 359.6
10*d*	−1 804.3	−1 686	−1 749	−1 798	−1 801.3	−1 800.2
11*d*	−1 419.3				−1 416.8	−1 417.3
12*d*	−1 145.4				−1 143.9	−1 144.2
13*d*	−943.5				−942.7	−942.8
14*d*	−790.6				−789.9	−790.1
15*d*	−671.9				−671.6	−671.6
16*d*	−578.18					−577.90
17*d*	−502.61					−502.46
18*d*	−440.95					−440.87
19*d*	−389.92					−389.94
20*d*	−347.38					−347.34
21*d*	−311.30					−311.35
...						
23*d*	−280.76					−280.68
24*d*	−231.45					−231.51
25*d*	−211.56					−211.63
26*d*	−194.12					−194.20
∞	0					
p^2		+8 443	+10 695			

secular equation, *including* the contribution of the Rydberg continuum states. The *ab initio* calculations should obviously be directed at obtaining a highly optimized Rydberg wave function at some modest n value in order to obtain the best ξ^0 and Γ values, rather than expanding calculational resources on obtaining a large number of less accurate Rydberg states.

There are, of course, possible contributions from higher-lying $(5p\ np)\,^1D_2$ Rydberg states, of which the so-called valence state $(5p^2)\,^1D_2$ is the lowest, and highly atypical, first member. These states add terms, $\sum_n \Gamma_{P,n}/2(E-E_{P,n})$, to the right-hand side of Eq. (47) which can be evaluated by applying QDT to estimate the widths and positions of the higher states from Γ and $\bar{E}_P$ for the lowest state. However, this rapidly leads us into the next level of theory that must be developed which is a complete generalization of the Seaton multichannel QDT and will be deferred to a later paper. Certainly with regard to the perturbation of the bound states these higher states can be neglected since they are distant, with smaller widths, and any errors introduced will be of the same order as approximating $\bar{E}_P$ by the unperturbed position of the valence state E_P.

We may usually approximate the pre-bracket factors in Eq. (54) simply by $(2\theta_E)^{-1/2}$; the "contribution" of the perturbing $|P\rangle$ state to each, normalized, perturbed Rydberg level is then

$$\frac{G(E)}{\theta_E}=\frac{2}{\pi}\,\frac{\sin^2\pi\lambda_p}{\Gamma\theta_E}\equiv\frac{\Gamma}{2\pi\left[(E-\bar{E}_P)^2+\frac{1}{4}\Gamma^2\right]}\,\frac{1}{\theta_E}\,. \qquad (56)$$

The quantity $G(E)$ is plotted in Fig. 3 for Cd $^1D_2$. If the entire contribution of the perturbing level were to lie above the ionization limit, where $\theta_E = 1$, and E were continuous, we would recognize $G(E)$ as the normalized Lorentzian line shape,

$$\int dE\, G(E) \simeq 1 \qquad (57)$$

or, equivalently, as the Fano-Beutler profile[6] associated with the "autoionizing" $|5p^2\rangle$ state.

In view of our neglect of the higher members of the $(5p\ np)\,^1D_2$ series this line shape is only approximate, particularly for $E > \bar{E}_P$, since this resonance will overlap the $(5p\ 6p)$ state at $E \approx 0.27$, which should have a width $\approx \frac{1}{4}\Gamma \approx 0.05$ Ry. The details of such overlapping effects are incidental to the purpose of this paper and can be easily treated in the more general, multichannel approach.

In the present case, the "autoionizing" width, i.e., $\Gamma = 2.722$ eV, is sufficiently large that it "extends" below the ionization limit and perturbs the Rydberg series. The contribution to each bound state is

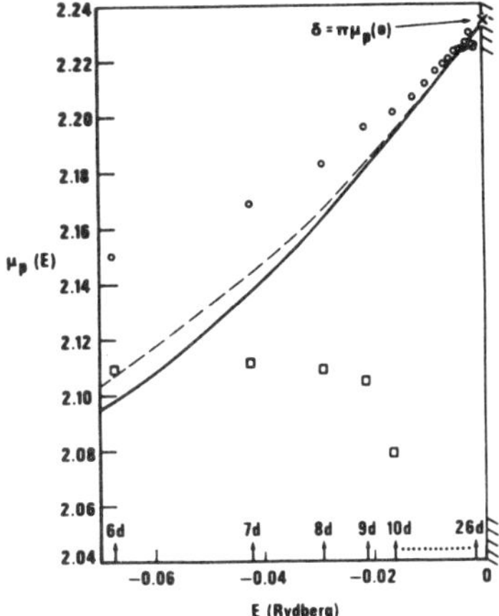

FIG. 2. Quantum defects for Cd $^1D_2$, i.e., $E(nd) = (n-\mu_p)^{-2}$, with $n \geq 5$. The circles represent the experimental values obtained by Brown, Tilford, and Ginter (Ref. 8). The dashed curve is the result of the present theory, utilizing the unperturbed eigenvalues E_n^0 as an *initial* guess for E in Eqs. (47), (50), (23), and (24). The solid curve is the result of utilizing the experimental eigenvalues to evaluate the parameters λ_p, A, and $\mathcal{G}$, and closely approximates the final results that would be obtained from an iteration of the theoretical expressions. The squares present the predictions of a (7 × 7) CI solution using the first six $(5s\ nd)\ ^1D$ wave functions and the $(5p^2)\ ^1D$ state obtained by Stevens (Ref. 1).

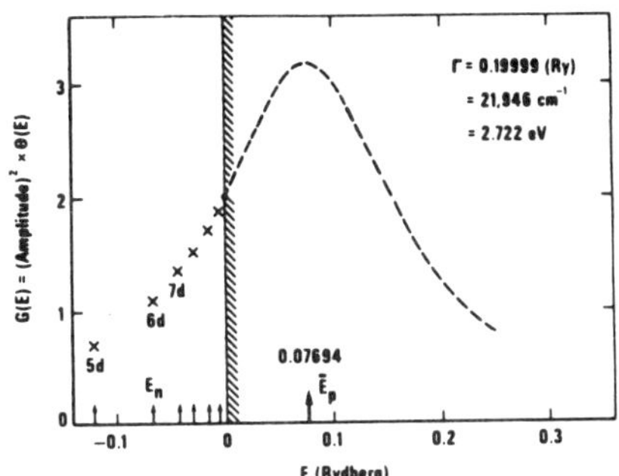

FIG. 3. Contribution of the perturbing Cd $(5p^2)\ ^1D$ valence state to the ^1D Rydberg series. $G(E) = (\text{amplitude})^2 \times \theta(E)$ measures the (amplitude)^2 of the $(5p^2)\ ^1D$ state to the total wave function at energy E. The quantity $\theta(E) = 1$ for $E \geq 0$, and the envelope of $G(E)$ represents the usual Fano-Beutler line shape for this autoionizing state. Below the ionization limit, $E < 0$, $\theta(E_n) = \frac{1}{2}\nu_n^3$ weights the contribution of the valence state to the discrete Rydberg states at $E_n = \nu_n^{-2}$.

$$G(E_n)/\theta_{E_n} \cong G(E_n)(2/\nu_n^3)\ .$$

If we may approximate the summation over bound states as follows, we simply recover the normalization condition in Eq. (57), i.e.,

$$\sum_{n=l+1}^{\infty} G(E_n)\frac{2}{\nu_n^3} \cong \int dn\, G(E_n)\frac{2}{\nu_n^3} \approx \int_{-\infty}^{0} dE\, G(E)\ . \quad (58)$$

It should be obvious that the present theory is completely equivalent to Fano's CI theory.[6] In fact, Fano has completely summarized the application of CI theory to perturbed Rydberg series when $\bar{E}_p$ lies ***below*** the ionization limit in Appendix B of his paper and has extracted most of the qualitative features we have discussed. Our contribution has been to generalize his results and obtain quantitative relationships which permit use of calculated CI matrix elements to obtain the pertinent parameters.

VI. SUMMARY AND CONCLUSIONS

We have developed theoretical expressions for the perturbation of a Rydberg series by a single interloping energy level. The interaction can be expressed in terms of a generalized "width" which is easily calculable by variational techniques and allows for simple extrapolation of bound-state calculations into the ionization continuum. The expressions are in complete analogy to Fano's CI theory, but apply equally well to autoionizing states and to configuration interaction in the Rydberg region, below the ionization limit. Even without access to theoretical calculations, Eq. (5) gives a very useful parametric form for analyzing perturbed Rydberg series without restriction to weak interactions and permits treatment of perturbing levels *above* the ionization limit.

The theory can easily be generalized to configuration interaction between many Rydberg series, and explicit relationships can be derived between CIT and multichannel QDT. In this way we can utilize the theoretical apparatus developed for bound-state variational calculations to obtain cross sections for photoionization and autoionization phenomena. Such calculations are particularly imperative for molecular systems where the experimental data is often unavailable or uninterpretable.

ACKNOWLEDGMENTS

I especially wish to thank Professor M. J. Seaton and the entire staff of the Physics Department at the University College London for their hospitality, and their many beneficial discussions during my tenure as visiting scientist. I am indebted to Dr. W. J. Stevens for his invaluable assistance in providing the theoretical parameters required in this paper. This research was supported in part by the Office of Laser Fusion, Department of Energy.

---

[1]W. J. Stevens, unpublished MCSCF calculations using the Cd effective core potential of H. Basch *et al.*, J. Chem. Phys. 68, 4005 (1978).

[2](a) M. J. Seaton, Proc. Phys. Soc. 88, 801, 815 (1966); (b) Comments At. Phys. 2, 37 (1970).

[3]*Handbook of Mathematical Functions*, Natl. Bur. Stand. Appl. Math. Ser. 55, edited by M. Abramowitz and I. Stegun (U.S. GPO, Washington, D. C., 1959), p. 258. Equation (24) is well approximated for $l = 2$, by $(A/2\pi)\,\epsilon\,(37/6 - 1079\epsilon/60 + \cdots)$, as $\epsilon \to 0$.

[4]J. Dehmer and U. Fano, Phys. Rev. A 2, 304 (1970).

[5]N. F. Mott, and H. S. W. Massey, *The Theory of Atomic Collisions*, 3rd ed. (Oxford University, London, 1965).

[6]U. Fano, Phys. Rev. 124, 1866 (1961).

[7]See Chap. IV of Ref. 5.

[8]This is discussed in Ref. 2(a), pp. 811–814.

[9]C. M. Brown, S. G. Tilford, and M. L. Ginter, J. Opt. Soc. Am. 65, 1404 (1975).

1981 *Phys. Rev.* A **24** 619–22
Reprinted with permission from the American Physical Society

Stark effect of nonhydrogenic Rydberg spectra

U. Fano
Department of Physics, University of Chicago, Chicago, Illinois 60637
(Received 17 February 1981)

Rydberg states of an atom in an electric field are represented in terms of hydrogen eigenfunctions for the same field but scattered by the ionic core of the actual atom. The effect of the Stark field on photoabsorption is expressed in terms of scattering and frame-transformation parameters.

Rau has singled out a set of diverse phenomena characterized by a wave function that extends over two regions of space, where fields of different symmetry prevail.[1] The Stark effect of Rydberg spectra, which exhibits remarkable features,[2] affords a prototype example amenable to complete treatment. The Rydberg electron's motion separates in parabolic coordinates in the Coulomb plus Stark field at large distances from an atom, but in polar coordinates near or within the atomic core where the Stark field is usually negligible.

Representing Rydberg states in terms of (a) hydrogenic wave functions at large distances, and (b) scattering of these wave functions by the ionic core of an atom (or molecule) is the trademark of the quantum-defect method (QDT).[3] Parameters of aspects (a) and (b) of a state are calculated separately and then combined analytically; the combination often involves a frame transformation, e.g., when *jj* and *LS* coupling are used in (a) and (b), respectively. This approach looks well suited to treating the Stark effect.[4] To this end it presupposes a full development of Stark-field wave functions for H (Ref. 5) and their adaptation to QDT,[6] which are now in hand. The necessary scattering parameters are common to other QDT applications. Combining these two sets of parameters requires a transformation from the frame of parabolic quantum numbers to the angular momentum frame which is normal for core interactions. This Communication shows how existing transformation formulas can in fact be extended, as required by the Stark problem, within a *limited region* of space. Thus it introduces the concept of a *local frame transformation*, which seems relevant to the treatment of any localized interaction.

The treatment developed here disregards quantities of order Fr_0 a.u., where the field strength $F = O(10^{-5}$ a.u.) in current experiments and the ionic radius $r_0 = O(1$ a.u.).

A complete set of electron eigenfunctions for the Coulomb plus Stark potential, $-e^2/r + eFz$, with $(e,F) > 0$, is represented by[5–7]

$$\psi_{nm}(F,\epsilon;\xi,\eta,\phi) = N_{n\epsilon}u_{1m}(F,\epsilon,\beta_n;\xi) \times u_{2m}(F,\epsilon,1-\beta_n;\eta) \times e^{\pm im\phi}(2\pi)^{-1/2} , \quad (1)$$

where $-\infty < \epsilon < \infty$ is the energy, $\xi = r + z = r(1+\cos\theta)$, $\eta = r(1-\cos\theta)$, $n = 0, 1, \ldots,$ is the number of nodes of u_{1m}, and $m \geq 0$. (The number n is indicated by n_1 in Refs. 5–7.) The functions u_{1m} and u_{2m} are Coulombic for $(\xi,\eta) \leq O(1)$ (in atomic units), are regular and normalized to $\xi^{m/2}$ or $\eta^{m/2}$ as $(\xi,\eta) \to 0$, and are renormalized by $N_{n\epsilon}$ to ensure the completeness

$$\tfrac{1}{4}(\xi+\eta)2\pi \sum_n \int_{-\infty}^{\infty} d\epsilon\, \psi_{nm}^*(F,\epsilon;\xi',\eta',\phi)\psi_{nm}(F,\epsilon;\xi,\eta,\phi) = \delta(\xi-\xi')\delta(\eta-\eta') .$$

The separation parameter $\beta_n(m,F,\epsilon)$ is determined by requiring u_{1m} to converge (after n nodes) at $\xi \to \infty$ where the potential rises steadily. The potential falls instead at $\eta \to \infty$ where u_{2m} oscillates after tunneling through a barrier if ϵ falls in the range of bound-state levels.[7]

Replacement of the bare nucleus of H by a "frozen" ionic core with spherical symmetry—such as occurs in alkali atoms—preserves the invariance of ϵ and m in the set (1) but couples wave functions with different values of n. The n quantum number serves here as the channel index in a scattering process, as the orbital l does in other phenomena. Exchanges of energy and/or angular momentum with the core would be treated by enlarging the set of channels, but we forego this extension for simplicity.

The scattering effect of the ionic core upon the electron states (1) stems from a potential V, the excess of the electron's interaction with the core (including exchange) over its interaction $-e^2/r$ with a bare H^+ ion. Eigenfunctions Ψ_{nm} for the complete potential $-e^2/r + V + eFz$ are constructed by adding to each ψ_{nm} a superposition of the set (1) extending over all channels and energies, but over a single m,

$$\Psi_{nm}(F,\epsilon;\xi,\eta,\phi)=\psi_{nm}(F,\epsilon;\xi,\eta,\phi)+\sum_{n'}\int_{-\infty}^{\infty}d\epsilon'\,\psi_{n'm}(F,\epsilon';\xi,\eta,\phi)\frac{\mathcal{P}}{\epsilon-\epsilon'}N^{*}_{n'\epsilon'}\langle n'\epsilon'|K_m^{(so)}(\epsilon)|n\epsilon\rangle N_{n\epsilon}\ . \quad (2)$$

Here $\mathcal{P}$ indicates principal part integration at $\epsilon'\sim\epsilon$, and the coefficients of the superposition are matrix elements of a reaction operator K,[8] from which $N^{*}_{n'\epsilon'}$ and $N_{n\epsilon}$ have been factored out. The ansatz (2) reduces the Schrödinger equation for Ψ_{nm} to an equation for K which we need not solve here. The label "*so*" means that $K^{(so)}$ operates on the eigenfunctions $u_{1m}u_{2m}$ of the outer field[8] normalized independently of ϵ at $(\xi,\eta)\rightarrow 0$. This specification makes $K^{(so)}$ "smooth and analytic"[9] in its dependence on ϵ as well as *independent of the Stark field F*, which is negligible wherever V is nonzero. The coefficients $N_{n\epsilon}$ depend instead on F; they are given explicitly in Ref. 6.

The integration over ϵ' in Eq. (2) may seem laborious but is in fact trivial just at $\eta\rightarrow\infty$ where (2) serves to determine the channel mixing and the normalization of Ψ_{nm}. Note that the second term of (2) may be viewed as resulting from the application to ψ_{nm} firstly of the operator $K^{(so)}$ and then of the standing-wave Green's function for motion in the Coulomb-Stark (CS) field[8]

$$G^{(CS)}(\vec r,\vec r\,')=\sum_{n'm}\int_{-\infty}^{\infty}d\epsilon'\,\psi_{n'm}(F,\epsilon';\xi,\eta,\phi)\frac{\mathcal{P}}{\epsilon-\epsilon'}\,\psi_{n'm}(F,\epsilon';\xi',\eta',\phi')\ . \quad (3)$$

At $\eta\rightarrow\infty$, ψ_{nm} oscillates so rapidly, as a function of ϵ', that its contributions to $G^{(CS)}$ cancel except at $\epsilon'\sim\epsilon$, where they yield[8]

$$G^{(CS)}(\vec r,\vec r\,')=\pi\sum_{n'm}\chi_{n'm}(F,\epsilon;\xi,\eta,\phi)\psi_{n'm}(F,\epsilon;\xi',\eta',\phi'),\quad \eta>\eta'\ . \quad (4)$$

Here χ_{nm} is identified as that solution of the equation for ψ_{nm} which oscillates at $\eta\rightarrow\infty$ with the same amplitude as ψ_{nm} but with a 90° phase lag; this identification suffices to define χ_{nm} at all η, but (4) remains qualified by $\eta>\eta'$. Substitution of (3) and (4) reduces Eq. (2) to

$$\Psi_{nm}(F,\epsilon;\xi,\eta,\phi)=\psi_{nm}(F,\epsilon;\xi,\eta,\phi)+\pi\sum_{n'}\chi_{n'm}(F,\epsilon;\xi,\eta,\phi)N^{*}_{n'\epsilon}\langle n'\epsilon|K_m^{(so)}(\epsilon)|n\epsilon\rangle N_{n\epsilon},\quad (\xi,\eta)>O(1\text{ a.u.})\ . \quad (5)$$

This reformulation relieves us of any need to consider matrix elements of $K^{(so)}$ off diagonal in (ϵ',ϵ). The diagonal elements are treated next. Note how the integration over ϵ' has served here to isolate relevant aspects of Ψ_{nm} at $\eta\rightarrow\infty$, while aspects relevant at $(\xi,\eta)=O(1)$ are considered below; the behavior of Ψ_{nm} at intermediate ranges of (ξ,η) is *not* relevant to us.

Invariance of the ionic core under rotations about its nucleus implies that V and $K^{(so)}$ are diagonal in the orbital quantum number l. Indeed, the Schrödinger equation with the potential $-e^2/r+V$ has solutions analogous to (2) but separable in polar coordinates,

$$F_l^0(\epsilon,r)Y_{lm}(\theta,\phi)=\left\{f_l^0(\epsilon,r)+\int_{-\infty}^{\infty}d\epsilon'\,f_l^0(\epsilon',r)N^{*}_{l\epsilon'}\frac{\mathcal{P}}{\epsilon-\epsilon'}N_{l\epsilon'}\langle l\epsilon'|K^{(so)}(\epsilon)|l\epsilon\rangle\right\}Y_{lm}(\theta,\phi)\ . \quad (6)$$

Here $f_l^0(\epsilon,r)$ is a hydrogen radial function normalized to r^l at $r\rightarrow 0$ and renormalized to unit energy (at $\epsilon \lesseqgtr 0$) by the separate factor $N_{l\epsilon}$.[9] (The index 0 means "normalized at $r\rightarrow 0$," as in Ref. 9.) Reference to the Coulomb (C) Green's function

$$\begin{aligned}G^{(C)}(\vec r,\vec r\,')&=\sum_{lm}Y_{lm}(\theta,\phi)\int_{-\infty}^{\infty}d\epsilon'\,f_l^0(\epsilon',r)N_{l\epsilon'}\frac{\mathcal{P}}{\epsilon-\epsilon'}N^{*}_{l\epsilon'}f_l^0(\epsilon',r')Y_{lm}(\theta',\phi')\\&=\pi\sum_{lm}Y_{lm}(\theta,\phi)\bar g_l(\epsilon,r)|N_{\epsilon l}|^2f_l^0(\epsilon,r')Y_{lm}(\theta',\phi'),\quad r>r'\ ,\end{aligned} \quad (7)$$

where $\bar g_l(\epsilon,r)$ (Ref. 9) lags 90° behind $f_l^0(\epsilon,r)$ at $r\rightarrow\infty$, yields the analog of Eq. (5)

$$F_l^0(\epsilon,r)=f_l^0(\epsilon,r)+\pi\bar g_l(\epsilon,r)|N_{l\epsilon}|^2\langle l\epsilon|K^{(so)}(\epsilon)|l\epsilon\rangle\ . \quad (8)$$

The reaction matrix element can thus be expressed as

$$\langle l\epsilon|K^{(so)}|l\epsilon\rangle=-\pi^{-1}\tan\delta_l^{(s)}(\epsilon)|N_{l\epsilon}|^{-2}\ , \qquad (9)$$

in terms of the phase shift $\delta_l^{(s)}$ of $F_l^0(\epsilon,r)$ with respect to $f_l^0(\epsilon,r)$. This phase shift can be obtained from the zero-field spectrum of bound-state levels of the atom, by quantum-defect analysis,[3,9] or it can be calculated *ab initio* for the ϵ of interest. Note that $\delta_l^{(s)}\neq 0$ for $l\leqslant 3$ only, as higher partial waves do not penetrate atomic cores.

We come now to our main task of transforming the matrix (9) to the frame of parabolic coordinates. The transformation can be introduced through the expansion of the parabolic ψ_{nm} into spherical harmonics with r-dependent coefficients,

$$\psi_{nm}(F,\epsilon;\xi,\eta,\phi)=N_{n\epsilon}\sum_l b_{nl}(F,\epsilon,m;r)\,Y_{lm}(\theta,\phi)\ . \qquad (10)$$

Since $\psi_{nm}/N_{n\epsilon}$ and $f_l^0 Y_{lm}$ obey equations that differ negligibly at $r\leqslant O(1\text{ a.u.})$, where $|eFz| << |-e^2/r|$, and have analogous normalization, the coefficients b_{nl} must be proportional to $f^0(r)$ at small r and do not depend on F explicitly. However, b_{nl} does depend on F implicitly through the separation parameter β_n of Eq. (1) which results from the quantization at $\xi\rightarrow\infty$. It is the restriction to $r\leqslant O(1)$ which makes the transformation (10) "local." Recalling that $u_{1m}u_{2m}\rightarrow(\xi\eta)^{m/2}=r^m\sin^m\theta$ as $(\xi,\eta\rightarrow 0)$ while $f_l^0(\epsilon,r)\rightarrow r^l$ as $r\rightarrow 0$, we write

$$b_{nl}(F,\epsilon,m;r)=a_{lm}(\beta_n,\nu)r^l[1+O(r)]\ . \qquad (11)$$

On the right-hand side, ϵ has been replaced by $\nu=(-\epsilon/13.6\text{ eV})^{-1/2}$ and n by β_n of Eq. (1). According to (10) and (11), the coefficients a_{lm} are obtained by projecting the confluent hypergeometric factors of ψ_{nm}, with $F=0$, onto the associated Legendre polynomials $P_l^m(\cos\theta)$,

$$a_{lm}(\beta_n,\nu)=r^{-l}\int_{-1}^{1}d(\cos\theta)\sin^{2m}\theta P_l^m(\cos\theta)$$
$$\times F\left[-\beta_n\nu+\tfrac{1}{2}(m+1),m+1,\frac{r(1+\cos\theta)}{\nu}\right]F\left[(\beta_n-1)\nu+\tfrac{1}{2}(m+1),m+1,\frac{r(1-\cos\theta)}{\nu}\right]\ . \qquad (12)$$

This integral over powers of $\cos\theta$ reduces manifestly to a hypergeometric polynominal.

Specifically, Eq. (12) is known to reduce to a Wigner coefficient,[10]

$$a_{lm}(\beta_n,\nu)=(-1)^l\langle lm\,|\,jm_1,jm_2\rangle\left[2\frac{n!\Gamma(\nu-n-m)\Gamma(\nu+l+1)}{(n+m)!\Gamma(\nu-n)\Gamma(\nu-l)}\right]^{1/2}\frac{2^l m!^2}{(2l+1)!\nu^{l-m}}\ ,$$
$$j=\tfrac{1}{2}(\nu-1),\quad m_1=(\tfrac{1}{2}-\beta_n)\nu+\tfrac{1}{2}m,\quad m_2=(\beta_n-\tfrac{1}{2})\nu+\tfrac{1}{2}m\ . \qquad (13)$$

The origin of this result is that the transformations among H wave functions with equal energy are generated by the orbital momentum $\vec{L}$ together with the Lenz vector $\vec{A}$. Each of the combined operators $\frac{1}{2}(\vec{L}\pm\vec{A})$ acts on the wave functions as a spin of magnitude $j=\frac{1}{2}(\nu-1)$, and these two spins add up to l as indicated by (13).[10] The *analyticity of* (12) shows that irrational values of the effective spin j, which become complex for $\epsilon>0$, are no obstacle to this limited extension of angular momentum algebra. The influence of the Stark field F on the coefficients a_{nl} is implied by the parameters β_n which are themselves determined by the wave-function convergence at $\xi\rightarrow\infty$. The reaction matrix elements in Eq. (2) are thus given by (9), (11), and (13) in the form

$$\langle n'\epsilon|K_m^{(so)}(\epsilon)|n\epsilon\rangle=-\pi^{-1}\sum_l a_{lm}^*(\beta_{n'},\nu)\tan\delta_l^{(s)}(\epsilon)|N_{l\epsilon}|^{-2}a_{lm}(\beta_n,\nu)\ . \qquad (14)$$

In this key result the sum extends only over the few terms with $\delta_l^{(s)}\neq 0$, thus making it unnecessary to consider higher terms of the expansion (10). It is this circumstance that justifies setting $F\rightarrow 0$ in the calculation of the transformation coefficients a_{lm} though not in the value of β_n.

Finally, the Stark effect of the photoabsorption spectrum is to be calculated by evaluating the transition dipole moment D_{nm} from the atomic ground state to the various states Ψ_{nm}. These states overlap with the ground state only at $(\xi,\eta)\leqslant O(1\text{ a.u.})$, where the first term of Eq. (2), ψ_{nm}, is expanded into radial functions $f_l^0(\epsilon,r)$ by (10) and (11). The second term of Ψ_{nm} depends on $\vec{r}$ through the function $G^{(CS)}(\vec{r},\vec{r}')$, Eq. (4), which *coincides locally* with $G^{(C)}(\vec{r},\vec{r}')$, Eq. (7), because $G^{(CS)}$ and $G^{(C)}$ obey the same inhomogeneous equation at $r\leqslant O(1)$. Accordingly, the entire Eq. (2) is expanded in terms of (6),

$$\Psi_{nm}(F,\epsilon;\xi,\eta,\phi)\underset{r\leqslant O(1)}{\rightarrow}\sum_l F_l^0(\epsilon,r)\,Y_{lm}(\theta,\phi)\times a_{lm}(\beta_n,\nu)N_{n\epsilon}\ . \qquad (15)$$

This expansion enables us to expand the desired dipoles D_{nm} in terms of the corresponding zero-field dipoles d^0_{lm} for transition to the eigenstates (6) of $\bar{l}^2$. In the alkali-type atoms considered here, with spherical ionic cores, the only transition allowed in photoabsorption leads to p states; hence only the $l=1$ term of (15) contributes to the spectrum. Thus we have

$$\begin{aligned} D_{nm}(\epsilon) &= \int d\bar{r}\, \Psi^*_{nm} r_m R(r) \\ &= N^*_{n\epsilon} a^*_{1m}(\beta_n, \nu) \\ &\quad \times \int_0^\infty dr\, F_1^{0*}(\epsilon;r) r R(r) \sqrt{1/3} \\ &= N^*_{n\epsilon} a^*_{1m}(\beta_n, \nu) d_1^0(\epsilon) \ , \end{aligned} \tag{16}$$

where r_m equals z for $m=0$ and $\sqrt{1/2}(x+iy)$ for $m=1$, and $R(r)$ is the wave function of the s ground state. The dipole $d_1^0(\epsilon)$ can be obtained from the intensities of zero-field spectral lines, much as the $\delta_l^{(s)}(\epsilon)$ are obtained from the levels, or it can be calculated *ab initio*.

Equation (16) merely distributes the zero-field dipole amplitude d_1^0 linearly among the parabolic channels to yield the D_{nm}. The effect of scattering on the Stark spectrum emerges when the Ψ_{nm} are orthonormalized. In the spherical frame, Eqs. (8) and (9) show F_l^0 to be normalized by the coefficient $\cos\delta_l^{(s)} = (1+tg^2\delta_l^{(s)})^{-1/2}$. In the parabolic frame the matrix element $\pi N^*_{n'\epsilon} \langle n'\epsilon | K_m^{(so)}(\epsilon) | n\epsilon \rangle N_{n\epsilon}$ plays the same role as

$$\pi |N_{l\epsilon}|^2 \langle l\epsilon | K^{(so)}(\epsilon) | l\epsilon \rangle = -\tan\delta_l^{(s)}(\epsilon)$$

does in Eq. (8). Accordingly, the set Ψ_{nm} is orthonormalized by the matrix $[1+(\pi N^*_\epsilon K_m^{(so)} N_\epsilon)^2]^{-1/2}$, and the photoabsorption spectrum is proportional to

$$\sum_{n'n} D^*_{n'm}(\epsilon)\, \{1+[\pi N^*_\epsilon K_m^{(so)}(\epsilon) N_\epsilon]^2\}^{-1}_{n'n}\, D_{nm}(\epsilon) \ , \tag{17}$$

with $m=0$ or 1 depending on the incident polarization. [A corrective factor $\mathcal{G}$, which appears in Eqs. (4.15) and (3.17) of Ref. 9 but is very small for Rydberg states, has been ignored here for simplicity but should be considered in more complete treatments.]

Application to the observed Stark spectra of alkali atoms is planned.

I am very indebted to J. H. Macek, D. A. Harmin, A. R. P. Rau, and other colleagues for contributions and criticism leading to the present approach. The work has been supported by the U.S. Department of Energy, Office of Basic Energy Sciences.

---

[1]A. R. P. Rau, J. Phys. B 12, L193 (1979).

[2]R. R. Freeman, N. P. Economou, G. C. Bjorklund, and K. T. Lu, Phys. Rev. Lett. 41, 1463 (1978).

[3]See, e.g., U. Fano, J. Opt. Soc. Am. 65, 979 (1975).

[4]U. Fano, Colloq. Int. CNRS 273, 130 (1977).

[5]E. Luc-Koenig and A. Bachelier, J. Phys. B 13, 1743,1769 (1980).

[6]D. A. Harmin (unpublished).

[7]See, e.g., H. A. Bethe and E. E. Salpeter, *Quantum Mechanics of One- and Two-Electron Atoms* (Springer, Berlin, 1957), Secs. 6 and 53; L. D. Landau and E. M. Lifshitz, *Quantum Mechanics*, 3rd ed. (Pergamon, Oxford, 1977), p. 287ff.

[8]See, e.g., L. S. Rodberg and L. M. Thaler, *Quantum Theory of Scattering* (Academic, New York, 1967), Sec. 4.4, p. 237ff.

[9]C. H. Greene, U. Fano, and G. Strinati, Phys. Rev. A 19, 1485 (1979), especially pp. 1499–1505.

[10]See, e.g., Landau and Lifshitz, Ref. 7, p. 132, and especially L. C. Biedenharn and P. J. Brussaard, *Coulomb Excitation* (Clarendon, Oxford, 1965), p. 107.

1986 *J. Phys. B: At. Mol. Phys.* **19** 3011–25

Multichannel quantum-defect theory of the Stark effect

K Sakimoto
Institute of Space and Astronautical Science, Komaba, Meguro-ku, Tokyo 153, Japan

Received 1 October 1985, in final form 23 January 1986

Abstract. A theoretical treatment of the Stark effect based on the quantum-defect theory and the WKB method, which was recently developed for alkali atoms by Harmin, is generalised to include ion cores of any state. Also investigated is the effect of an electric field on electron–ion resonance scattering. The present theory serves also as an extension of the multichannel quantum-defect theory developed by Seaton and his co-workers to the theory of an electron–ion system under the influence of an external electric field. Numerical calculations of some simple problems are performed.

1. Introduction

Recently, great interest has been concentrated on the subject of highly excited atoms in a uniform electric field (Kleppner *et al* 1983). Owing to the small energy difference between adjacent states with different principal quantum numbers *n*, highly excited states are easily mixed by a relatively weak electric field, and it is possible to investigate the higher-order Stark effect. Furthermore, the field ionisation method is a useful experimental technique to detect a highly excited atom (Dunning and Stebbings 1983). It is important to know the ionisation threshold and the Stark level structure. Autoionisation of the doubly excited state and dielectronic recombination are greatly influenced by an electric field (Freeman and Bjorklund 1978, Safinya *et al* 1980, Jaffe *et al* 1984, Dunn *et al* 1984). Theoretical developments of this problem are definitely required.

For hydrogenic atoms in an electric field, the product of the electric field vector and a generalised Runge–Lenz vector is exactly conserved (Redmond 1964). The system has a so-called dynamical symmetry. Owing to this fact, the Hamiltonian of the hydrogenic atom in an electric field is separable in parabolic coordinates (Bethe and Salpeter 1977). In the case of highly excited atoms which are not hydrogen-like, the excited electron feels a different field from the Coulomb plus electric fields in a core region. Because of this difference from the hydrogenic case, the Hamiltonian is not entirely separable. Zimmerman *et al* (1979) calculated the Stark levels of an alkali metal by using a method of spherical basis expansion, and succeeded in explaining the observed complex Stark level structure. However, this method requires very many terms of the expansion as *n* or the strength of the electric field increases.

An elegant way to avoid a tedious calculation was suggested by Fano (1981). In the core region, where the electric field is usually negligible, we can use the quantum-defect theory based on the spherical basis. At large distances from the core, only the Coulomb plus electric fields are present. The motion of the excited electron is separable in parabolic coordinates in this region. A proper frame transformation can connect the spherical and the parabolic bases. Although the motion in the Coulomb plus

0022-3700/86/193011+15$02.50

electric fields is not solved in closed form, a WKB approximation is useful to know the analytical property of the wavefunction. Harmin (1982) applied this idea to the photoabsorption of an alkali metal in an electric field. He calculated a continuous spectrum because all the Stark states have continuous energies. Furthermore, Harmin (1984) calculated the Stark energy level of the alkali metal.

Although Harmin's work is an excellent approach to the problem of the Stark effect, his theory is restricted only to an alkali metal with a closed-shell core. In the present paper, we extend Harmin's theory to the general case where an ion core is in an arbitrary state. As a result, we can easily investigate the Stark energy level of any Rydberg atoms and the effect of an electric field on the doubly excited autoionising state (or electron–ion resonance scattering). Thus, the present theory is also an extension of the multichannel quantum-defect theory developed by Seaton and his co-workers (Seaton 1983) to the theory of an electron–ion system under the influence of an external electric field.

In the presence of an electric field, all the discrete Stark states are quasibound states embedded in the continuum. In order to apply the multichannel quantum-defect theory to the present case, we must give a boundary condition for the quasibound Stark state. Harmin (1984) defined the quasibound energy level as the centre of the resonance peak in the continuous photoabsorption spectrum. However, this treatment is not convenient for the present extension. Here, we define the condition of the quasibound state by imposing an appropriate boundary condition in the barrier region that is made by the Coulomb plus electric fields. For this purpose, the generalised WKB method developed by Miller and Good (1953) is very useful. Then, the construction of the theory is easily performed along the lines of the treatment of the usual multichannel quantum-defect theory (Seaton 1983).

For convenience, Harmin's theory is reformulated in § 2 in a simpler manner and with a little modification. We consider the addition of an important normalisation factor, which was dismissed in Harmin's paper. Furthermore, the analytical behaviour of the wavefunction in the barrier region is derived in this paper for the first time. This is very important for the multichannel quantum-defect theory formulated in § 3. A general formula is obtained in § 3 for the calculation of the Stark energy level of any Rydberg atom. The influence of an electric field on the resonance scattering of electrons from ions is also discussed in § 3. We derive a general expression for a reactance matrix in which the Stark effect of the resonance state is explicitly considered. Numerical results of some simple problems are given in § 4. Stark energy levels of Li and K are calculated. A model calculation for the resonance scattering is performed.

2. Outline of Harmin's theory with some modifications

Because only the Stark effect of the bound state is taken into account, the negative energy of the electron is assumed. Atomic units are used unless otherwise stated.

2.1. WKB solutions

We consider electron motion in the Coulomb plus uniform electric fields, whose potential is

$$V = -1/r + Fz \tag{1}$$

where the direction of the electric field F is chosen to be the z axis. The Schrödinger equation is separable in parabolic coordinates (ξ, η, ϕ) which are related to the polar coordinates (r, θ, ϕ) by

$$\xi = r(1+\cos\theta) \qquad \eta = r(1-\cos\theta) \qquad \phi = \phi.$$

Writing the wavefunction as

$$\psi(F\varepsilon\beta m) = (2\pi\xi\eta)^{-1/2} f(\xi) g(\eta)\, \mathrm{e}^{\mathrm{i}m\phi} \tag{2}$$

we obtain the separated form of the differential equations for f and g (Bethe and Salpeter 1977)

$$\left(\frac{\mathrm{d}^2}{\mathrm{d}\xi^2} - \frac{m^2-1}{4\xi^2} + \frac{\beta}{\xi} + \tfrac{1}{2}\varepsilon - \tfrac{1}{4}F\xi\right) f(\xi) = 0 \tag{3a}$$

$$\left(\frac{\mathrm{d}^2}{\mathrm{d}\eta^2} - \frac{m^2-1}{4\eta^2} + \frac{1-\beta}{\eta} + \tfrac{1}{2}\varepsilon + \tfrac{1}{4}F\eta\right) g(\eta) = 0 \tag{3b}$$

where β is the separation constant, m the magnetic quantum number, and ε the energy of the electron. Although they are one-dimensional problems, equations (3) cannot be solved in closed form. In the present work, the WKB approximation is used to calculate the eigenvalue and to inspect the behaviour of the wavefunction in various regions.

Figure 1 shows a typical case where the quasibound state is possible. Classically allowed regions are restricted to $\xi_a < \xi < \xi_b$ for ξ motion and $\eta_a < \eta < \eta_b$, $\eta > \eta_c$ for η motion. At small ξ and η, the behaviour of the wavefunction is analytically known as (Bethe and Salpeter 1977)

$$f(\xi) = CN_\xi \xi^{(m+1)/2} \qquad \xi \sim 0 \tag{4a}$$

$$g(\eta) = N_\eta \eta^{(m+1)/2} \qquad \eta \sim 0 \tag{5a}$$

where C, N_ξ and N_η are constant factors. In the regions where $\xi_a \ll \xi \ll \xi_b$ and $\eta_a \ll \eta \ll \eta_b$, we employ the WKB wavefunctions

$$f(\xi) = C\left(\frac{2}{\pi p(\xi)}\right)^{1/2} \sin\left(\int_{\xi_a}^{\xi} p(\xi')\, \mathrm{d}\xi' + \pi/4\right) \qquad \xi_a \ll \xi \ll \xi_b \tag{4b}$$

where

$$p^2(\xi) = \tfrac{1}{2}\varepsilon - \frac{m^2}{4\xi^2} + \frac{\beta}{\xi} - \tfrac{1}{4}F\xi$$

and

$$g(\eta) = \left(\frac{2}{\pi q(\eta)}\right)^{1/2} \sin\left(\int_{\eta_a}^{\eta} q(\eta')\, d\eta' + \pi/4\right) \qquad \eta_a \ll \eta \ll \eta_b \tag{5b}$$

where

$$q^2(\eta) = \tfrac{1}{2}\varepsilon - \frac{m^2}{4\eta^2} + \frac{1-\beta}{\eta} + \tfrac{1}{4}F\eta.$$

Without loss of generality, m is taken to be non-negative. The factors N_ξ and N_η are uniquely determined by imposing the conditions that $f(\xi)$ and $g(\eta)$ have the forms (4b) and (5b), respectively. Harmin (1981) derived N_η in the form

$$N_\eta = (\nu^{m/2} m!)^{-1} \left(\frac{2\Gamma(\nu - \beta\nu + \frac{1}{2} + m/2)}{\Gamma(\nu - \beta\nu + \frac{1}{2} - m/2)}\right)^{1/2} \tag{6}$$

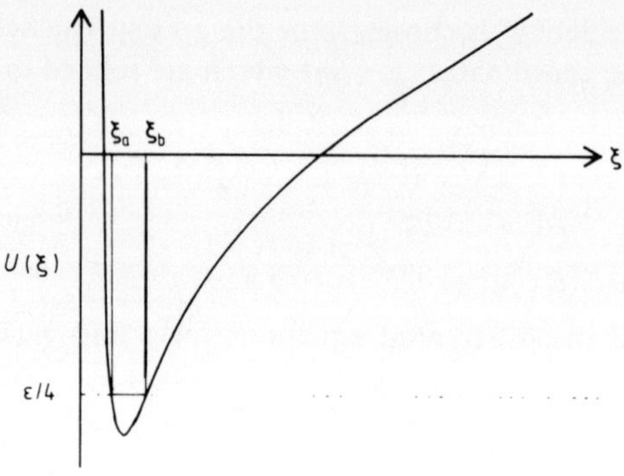

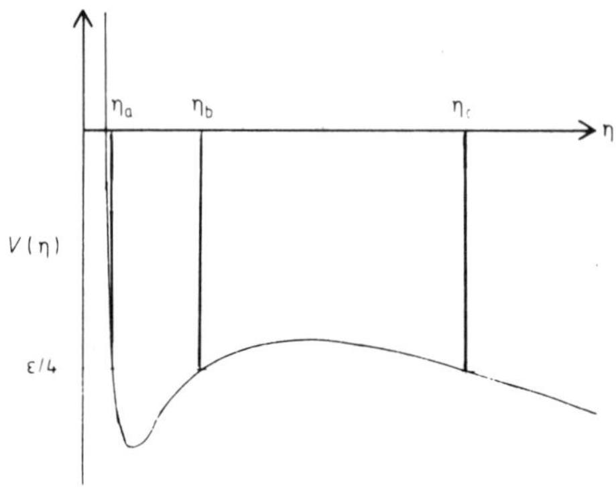

Figure 1. Effective potential energies

$$U(\xi)=\frac{m^2}{8\xi^2}-\frac{\beta}{2\xi}+\tfrac{1}{8}F\xi$$

and

$$V(\eta)=\frac{m^2}{8\eta^2}-\frac{1-\beta}{2\eta}-\tfrac{1}{8}F\eta$$

respectively, for ξ and η motion. In this example, $V(\eta)$ has a local maximum, and a quasibound motion is possible.

where ν is the effective quantum number defined by

$$\varepsilon=-\tfrac{1}{2}\nu^{-2} \tag{7}$$

and $\Gamma(x)$ is a gamma function. Harmin disregarded the factor N_ξ. However, this factor should be taken into account to define a frame transformation (see § 2.2). N_ξ is given by the replacement $\beta\to 1-\beta$ in (6). The factor C is the proper normalisation constant of the bound-type wavefunction $f(\xi)$. Luc-Koenig and Bachelier (1980) give

$$C=\left(\frac{1}{\pi}\int_{\xi_a}^{\xi_b}\frac{\mathrm{d}\xi}{p(\xi)\xi}\right)^{-1/2}. \tag{8}$$

We can analytically continue the solution $(5b)$ to the region $\eta\gg\eta_a$ (including also the turning points η_b, η_c and the tunnelling region $\eta_b<\eta<\eta_c$) by using the generalised

WKB method (Miller and Good 1953). The generalised WKB solution that has the expression (5b) at $\eta_a \ll \eta \ll \eta_b$ is (Rice and Good 1962, Harmin 1981)

$$g(\eta)=\left(\frac{2Q(\rho)}{\pi q(\eta)}\right)^{1/2}\left[\left(\frac{k}{2}\right)^{1/2}\sin\Delta\, W_1(\rho)+\left(\frac{1}{2k}\right)^{1/2}\cos\Delta\, W_2(\rho)\right] \qquad \eta \gg \eta_a \tag{9}$$

where the variable η is mapped into ρ by

$$\int_{-2\alpha}^{\rho} Q(\rho')\,\mathrm{d}\rho' = \int_{\eta_b}^{\eta} q(\eta')\,\mathrm{d}\eta' \tag{10}$$

with

$$Q^2(\rho)=\tfrac{1}{4}\rho^2-\alpha^2 \tag{11}$$

$$\alpha=\left(\frac{1}{\pi}\left|\int_{\eta_b}^{\eta_c} q(\eta)\,\mathrm{d}\eta\right|\right)^{1/2} \tag{12}$$

$$\Delta=\int_{\eta_a}^{\eta_b} q(\eta)\,\mathrm{d}\eta+\tfrac{1}{2}\arg\Gamma(\tfrac{1}{2}+\mathrm{i}\alpha^2)+\tfrac{1}{2}\alpha^2(1-\log\alpha^2) \tag{13}$$

$$k=[1+\exp(2\pi\alpha^2)]^{1/2}-\exp(\pi\alpha^2). \tag{14}$$

In equation (9), $W_1(\rho)$ and $W_2(\rho)$ are the parabolic cylinder functions (see appendix). The range $\eta_a \ll \eta \ll \eta_b$ corresponds to $\rho^2 \gg \alpha^2$. Inserting (A.2) into (9) gives equation (5b). In the region $\eta_b \ll \eta \ll \eta_c$ (i.e. $\rho^2 \ll \alpha^2$), we obtain from (A.3)

$$g(\eta)=W_1(0)\left(\frac{2Q(\rho)}{\pi q(\eta)}\right)^{1/2}\left[\left(\frac{k}{2}\right)^{1/2}\sin\Delta\,\mathrm{e}^{-\alpha\rho}+\left(\frac{1}{2k}\right)^{1/2}\cos\Delta\,\mathrm{e}^{+\alpha\rho}\right] \qquad \eta_b \ll \eta \ll \eta_c. \tag{15}$$

For later discussions, we need another independent (irregular) solution $\bar{g}(\eta)$ of (3b). The WKB approximation of $\bar{g}(\eta)$ is defined by a replacement of a sine function with a cosine function in (5b). The generalised WKB solution similar to (9) is

$$\bar{g}(\eta)=\left(\frac{2Q(\rho)}{\pi q(\eta)}\right)^{1/2}\left[\left(\frac{k}{2}\right)^{1/2}\cos\Delta\, W_1(\rho)-\left(\frac{1}{2k}\right)^{1/2}\sin\Delta\, W_2(\rho)\right] \qquad \eta \gg \eta_a. \tag{16}$$

The form at $\eta_b \ll \eta \ll \eta_c$ is

$$\bar{g}(\eta)=W_1(0)\left(\frac{2Q(\rho)}{\pi q(\eta)}\right)^{1/2}\left[\left(\frac{k}{2}\right)^{1/2}\cos\Delta\,\mathrm{e}^{-\alpha\rho}-\left(\frac{1}{2k}\right)^{1/2}\sin\Delta\,\mathrm{e}^{+\alpha\rho}\right] \qquad \eta_b \ll \eta \ll \eta_c. \tag{17}$$

The ξ motion is bound. The Bohr–Sommerfeld quantisation rule gives

$$\int_{\xi_a}^{\xi_b} p(\xi)\,\mathrm{d}\xi=\pi(n_1+\tfrac{1}{2}) \tag{18}$$

where n_1 is a non-negative integer. From (18), the separation constant β takes a discrete value at a fixed energy.

2.2. Frame transformation

From (2), (4a) and (5a), the wavefunction at small ξ and η is

$$\psi(F\varepsilon\beta m)=N(F\varepsilon\beta m)(\xi\eta)^{m/2}(2\pi)^{-1/2}\,\mathrm{e}^{\mathrm{i}m\phi} \tag{19}$$

where

$$N(F\varepsilon\beta m)=CN_\xi N_\eta. \tag{20}$$

At $r \ll F^{-1/2}$, where the Schrödinger equation does not depend on F, the wavefunction divided by $N(F\varepsilon\beta m)$ is independent of F, i.e.

$$\frac{\psi(F\varepsilon\beta m)}{N(F\varepsilon\beta m)}=\frac{\psi^0(\varepsilon\beta m)}{N^0(\varepsilon\beta m)} \qquad r \ll F^{-1/2} \tag{21}$$

where $\psi^0(\varepsilon\beta m)=\psi(F=0, \varepsilon\beta m)$ is the Coulomb function in parabolic coordinates and we put $N^0(\varepsilon\beta m)=N(F=0, \varepsilon\beta m)$. From Bethe and Salpeter (1977)

$$N^0(\varepsilon\beta m)=(\nu^{m+1/2}m!^2)^{-1}\left(\frac{2\Gamma(\beta\nu+\frac{1}{2}+m/2)\Gamma(\nu-\beta\nu+\frac{1}{2}+m/2)}{\Gamma(\beta\nu+\frac{1}{2}-m/2)\Gamma(\nu-\beta\nu+\frac{1}{2}-m/2)}\right)^{1/2}. \tag{22}$$

The wavefunction $\psi^0(\varepsilon\beta m)$ can be expanded in terms of the spherical Coulomb function $\psi^0(\varepsilon lm)=h_{\varepsilon l}(r)Y_{lm}(\theta,\phi)$, where the radial function $h_{\varepsilon l}(r)$ is chosen to satisfy the normalisation condition per unit energy range at $\nu=n$ (n being an integer). Thus, we have

$$\psi^0(\varepsilon\beta m)=\sum_l X_{\beta l}(\varepsilon m)\psi^0(\varepsilon lm). \tag{23}$$

In the special case of $\nu=n$, $X_{\beta l}(\varepsilon m)$ is reduced to (Hughes 1967)

$$X_{\beta l}(\varepsilon m)=(-1)^{n_2-n+1}(j\lambda_+ j\lambda_- | lm) \tag{24}$$

where $(j\lambda_+ j\lambda_- | lm)$ is the Clebsch-Gordan (CG) coefficient and

$$j=\tfrac{1}{2}(n-1) \qquad \lambda_\pm=\tfrac{1}{2}[m\pm(n_2-n_1)]$$

$$n_1=\beta n-\tfrac{1}{2}(m+1) \qquad n_2=(1-\beta)n-\tfrac{1}{2}(m+1).$$

For $\nu \neq n$ and an arbitrary β, Fano (1981) showed that $X_{\beta l}(\varepsilon m)$ can be given by an analytical continuation of the CG coefficient with real arguments. We define the frame transformation matrix $U_{\beta l}(F\varepsilon m)$ by

$$\psi(F\varepsilon\beta m)=\sum_l U_{\beta l}(F\varepsilon m)\psi^0(\varepsilon lm) \qquad r \ll F^{-1/2}. \tag{25}$$

From (6) and (20)–(23), we obtain

$$U_{\beta l}(F\varepsilon m)=(2\nu)^{1/2}CX_{\beta l}(\varepsilon m). \tag{26}$$

The outer boundary, $r_b=F^{-1/2}$, for equation (25) is very far below the usual experimental condition (e.g. $r_b \sim 2000$ au for $F=1$ kV cm^{-1}). Therefore, we can use the frame transformation (25) over a wide region. It should be noted that the frame transformation matrix (26) is not orthogonal except at $F=0$.

Next, we consider the irregular Coulomb function $\bar{\psi}^0(\varepsilon lm)$ and the parabolic irregular function $\bar{\psi}(F\varepsilon\beta m)$, where

$$\bar{\psi}(F\varepsilon\beta m)=(2\pi\xi\eta)^{-1/2}f(\xi)\bar{g}(\eta)\,e^{im\phi}. \tag{27}$$

As was discussed by Harmin (1982), we can show that the irregular functions are transformed as

$$\bar{\psi}^0(\varepsilon lm)=\sum_\beta U^t_{l\beta}(F\varepsilon m)\bar{\psi}(F\varepsilon\beta m) \tag{28}$$

where U^t is the transposed matrix of U.

Harmin (1982) imposed a different normalisation condition for $g(\eta)$. However, the difference is not essential. We can derive the same result obtained in the next section whether we use the present or Harmin's normalisation condition.

3. Multichannel quantum-defect theory

3.1. Quasibound Stark state

We develop a multichannel theory following Seaton (1983). We consider an electron moving around an ion core, which is not necessarily a point charge, under an applied field.

3.1.1. Frame transformation of R matrix. We write a solution of the Schrödinger equation for the system with total energy $E(<0)$ at $r \gg r_0$ (r_0 being the radius of the core) as

$$\Phi(\gamma\beta m) = \sum_{\gamma'\beta'm'} \Upsilon_{\gamma'} G(\gamma'\beta'm'; \gamma\beta m) \tag{29}$$

where Υ_γ denotes the wavefunction of the core in its state γ, and the electric field is assumed to have no effects on the core state. For simplicity, we omit the spin variable and the antisymmetrisation of electrons. At $r \gg r_0$, only the Coulomb and the electric fields are present. Thus, a solution G for the electron motion can be expressed as a linear combination of $\psi(F\varepsilon\beta m)$ and $\bar{\psi}(F\varepsilon\beta m)$ with $\varepsilon = E - E_\gamma$ (E_γ being the energy of the core with respect to the ground state):

$$G(\gamma'\beta'm'; \gamma\beta m) = \psi(F\varepsilon\beta m)\delta_{\gamma'\gamma}\delta_{\beta'\beta}\delta_{m'm} + \bar{\psi}(F\varepsilon'\beta'm')R_{\gamma'\beta'm',\gamma\beta m} \qquad r \gg r_0. \tag{30}$$

The coefficient R is determined by the smooth connection of G to an inner ($r < r_0$) solution at the core boundary. Generally, the electron-core interaction mixes the β or m channels. Therefore, R is non-diagonal with respect to β and m. When $r \ll F^{-1/2}$, we can express a solution for the present system also in terms of the spherical Coulomb functions:

$$\Phi^0(\gamma l m) = \sum_{\gamma'l'm'} \Upsilon_{\gamma'}[\psi^0(\varepsilon l m)\delta_{\gamma'\gamma}\delta_{l'l}\delta_{m'm} + \bar{\psi}^0(\varepsilon'l'm')R^0_{\gamma'l'm',\gamma l m}] \qquad r_0 \ll r \ll F^{-1/2}. \tag{31}$$

R^0 is the usual R matrix defined in the quantum-defect theory (Seaton 1983). The solution (31) can be expressed as a linear combination of $\Phi(\gamma\beta m)$. Multiplying $\Phi(\gamma\beta m)$ by $U^{-1}_{l\beta}$, summing over β, and using (25) and the inverse relation of (28), we obtain

$$\begin{aligned}\sum_{\beta} \Phi(\gamma\beta m) U^{-1}_{l\beta} &= \sum_{\gamma'l'm'} \Upsilon_{\gamma'}\Big(\psi^0(\varepsilon l m)\delta_{\gamma'\gamma}\delta_{l'l}\delta_{m'm} \\ &\quad + \bar{\psi}^0(\varepsilon'l'm') \sum_{\beta'\beta} U^{-1}_{l'\beta'} R_{\gamma'\beta'm',\gamma\beta m}(U^{-1})^{t}_{\beta l}\Big).\end{aligned} \tag{32}$$

The right-hand side of (32) has the same expression as (31). Thus, (32) is identical to $\Phi^0(\gamma\beta m)$, and we define the frame transformation of the R matrix by

$$R_{\gamma'\beta'm',\gamma\beta m} = \sum_{l'l} U_{\beta'l'} R^0_{\gamma'l'm',\gamma l m} U^{t}_{l\beta}. \tag{33}$$

It should be noticed that the present approach is applicable only at electric fields such that $F^{-1/2} \gg r_0$.

3.1.2. General formula for Stark level calculation. In order to impose the condition of the quasibound state, we take a linear combination of G to make new functions

$$H_i = \sum_j G(i;j) L_j \tag{34}$$

where i (and j) abbreviates γ, β and m, and L is a constant column vector. Notice that the functions $g(\eta)$ and $\bar{g}(\eta)$ have the forms (15) and (17) at $\eta_b \ll \eta \ll \eta_c$. We define the quasibound state by requiring that at $\eta_b \ll \eta \ll \eta_c$, terms proportional to $e^{+\alpha\rho}$ in H vanish. This condition gives the equation

$$\exp(+\alpha_i \rho_i) \sum_j [\cos \Delta_i\, \delta_{ij} - \sin \Delta_i\, R_{ij}] L_j = 0 \tag{35}$$

which can be satisfied if

$$\det|\cot \Delta_i\, \delta_{ij} - R_{ij}| = 0. \tag{36}$$

Condition (36) is the extension of the multichannel quantum-defect theory to the case where the electric field is present, and gives the general formula for the calculation of the Stark energy level of any Rydberg atom. When R_{ij} are all zero, (36) gives the WKB approximation of the hydrogenic Stark level.

3.1.3. $F \to 0$. When $F = 0$, we can easily show that

$$C = (2\nu)^{-1/2} \qquad \Delta = \pi(-\tfrac{1}{2}m + \nu - \nu\beta) \qquad n_1 = \nu\beta - \tfrac{1}{2}(m+1).$$

From these relations,

$$U_{\beta l}(F=0, \varepsilon m) = X_{\beta l}(\varepsilon m)$$

and

$$\det \left| \tan(\pi\nu_\gamma)\delta_{\gamma'\gamma}\delta_{\beta'\beta}\delta_{m'm} + \sum_{l'l} X_{\beta' l'} R^0_{\gamma' l' m', \gamma l m} X^{t}_{l\beta} \right| = 0. \tag{37}$$

Even if $F = 0$, X is not an orthogonal matrix for $\nu \neq n$. However, by using an approximation method for the calculation of CG coefficients discussed in appendix II of Brussaard and Tolhoek (1957), we can show that X nearly satisfies an orthogonality relation even for $\nu \neq n$ when ν is much larger than unity. Therefore, equation (37) is almost identical to

$$\det|\tan(\pi\nu_\gamma)\delta_{\gamma'\gamma}\delta_{l'l}\delta_{m'm} + R^0_{\gamma' l' m', \gamma l m}| = 0. \tag{38}$$

This is the condition of the bound state in the usual multichannel quantum-defect theory (Seaton 1983) in the spherical basis.

3.1.4. Stark level of alkali metals. For alkali metals, the R matrix can be expressed as

$$R^0_{\gamma' l' m', \gamma l m} = \tan(\pi\mu_l)\delta_{l'l}\delta_{\gamma'\gamma}\delta_{m'm} \tag{39}$$

where μ_l is the quantum defect, and only the ground state of the core is considered. Inserting (39) into (33), we obtain from (36)

$$\det \left| \cot \Delta_\beta\, \delta_{\beta'\beta} - \sum_l U_{\beta' l} \tan(\pi\mu_l) U^{t}_{l\beta} \right| = 0. \tag{40}$$

This equation gives the Stark levels of alkali metals.

Harmin (1984) also obtained an equation for the Stark level of alkali metals. He defined the quasibound Stark level as the centre of the resonance peak in the continuous photoabsorption spectrum. We can easily show that Harmin's result is identical to (40). Equation (40) can be rewritten as

$$\det\left|\sum_{\beta} U^{-1}_{l'\beta} \cot \Delta_{\beta} (U^{t})^{-1}_{\beta l} - \tan(\pi\mu_l)\delta_{l'l}\right| = 0. \tag{41}$$

With use of matrix inversion

$$[U^{-1} \cot \Delta (U^{t})^{-1}]^{-1} = U^{t} \tan \Delta\, U$$

we obtain from (41)

$$\det\left|\cos(\pi\mu_l)\delta_{l'l} - \sum_{\beta} U^{t}_{l'\beta} \tan \Delta_{\beta}\, U_{\beta l} \sin(\pi\mu_l)\right| = 0. \tag{42}$$

This is the same result as derived by Harmin. (Notice the phase difference of Δ_{n_1} in his paper.)

3.2. *Effects of electric field on resonance scattering*

We consider resonance scattering of electrons from ions. When the incident energy of the electron is somewhat below the excitation threshold of the ion, the electron may be trapped in a Rydberg state (nl) with an excited ion. This doubly excited state decays by re-emitting the electron in a finite lifetime. This process corresponds to resonance scattering, which is well described in terms of the multichannel quantum-defect theory (Seaton 1983). If the resonance state is a highly excited Rydberg state, resonance scattering is expected to be significantly influenced by the presence of an electric field. Here, we apply the present theory to electron–ion resonance scattering under the influence of an electric field.

Consider an experimental apparatus in which the collision region is centred between two parallel plates separated by a distance d. Then, the maximum potential of the electric field is Fd. Here, we assume that the collision energies of the incident and scattered electrons are much larger than $|Fd|$. Then, we can neglect the Stark effect for the free electron, and its motion can be well described in terms of the spherical basis. On the other hand, we fully take into account the Stark effect of a Rydberg electron in a resonance state. (As in § 3.1, the distortion of the ion state by the electric field is assumed to be negligible.)

We write a solution of the Schrödinger equation for electron–ion scattering in a form similar to (29). Here, both cases $\varepsilon < 0$ and $\varepsilon > 0$ are possible. The function G with $\varepsilon < 0$ can be given by (30) except that the total energy E is positive. For the continuum state ($\varepsilon > 0$) of the electron, we can have the following solutions (Seaton 1983): when $\varepsilon' < 0$ and $\varepsilon > 0$, we have

$$G(\gamma'\beta'm';\, \gamma lm) = \bar{\psi}(F\varepsilon'\beta'm')R'_{\gamma'\beta'm',\gamma lm} \tag{43a}$$

when $\varepsilon' > 0$ and $\varepsilon < 0$, we have

$$G(\gamma'l'm';\, \gamma\beta m) = \bar{\psi}^{0}(\varepsilon'l'm')(R')^{t}_{\gamma'l'm',\gamma\beta m} \tag{43b}$$

and when $\varepsilon' > 0$ and $\varepsilon > 0$, we have

$$G(\gamma'l'm';\, \gamma lm) = \psi^{0}(\varepsilon lm)\delta_{\gamma'\gamma}\delta_{l'l}\delta_{m'm} + \bar{\psi}^{0}(\varepsilon'l'm')R^{0}_{\gamma'l'm',\gamma lm}. \tag{43c}$$

For the continuum state, we adopt the polar-coordinate representation. The matrix R' is transformed from R^0 by

$$R'_{\gamma'\beta'm',\gamma lm} = \sum_{l'} U_{\beta' l'} R^0_{\gamma' l' m',\gamma lm}. \tag{44}$$

We take a linear combination of G in a similar way to (34). Then, following the method of Seaton (1983) that takes both the open and closed channels into account, we can obtain the expression for the reactance matrix $\mathcal{R}$:

$$\mathcal{R}_{\gamma' l' m',\gamma lm} = R^0_{\gamma' l' m',\gamma lm} + \sum_{\substack{\gamma_1\beta_1 m_1 \\ \gamma_2\beta_2 m_2}} (R')^{t}_{\gamma' l' m',\gamma_1\beta_1 m_1} [\cot\Delta - R]^{-1}_{\gamma_1\beta_1 m_1,\gamma_2\beta_2 m_2} R'_{\gamma_2\beta_2 m_2,\gamma lm}. \tag{45}$$

The second term of (45) represents the resonance contribution where the Stark effect is taken into account. The subscripts (γ_i, β_i, m_i) with $i = 1$ and 2 denote the closed channels. The quantity $\cot\Delta - R$ is the matrix defined in (36). Resonances occur at poles of the reactance matrix. This condition is determined by $\det|\cot\Delta - R| = 0$.

4. Numerical results of some simple problems

4.1. Stark levels of alkali metals

In order to see the validity of the present method, we calculate some Stark energy levels of alkali metals, and compare them with the accurate quantum-mechanical calculations by Zimmerman *et al* (1979). To evaluate the determinant (40), we must perform the integrals (8), (12), (13) and (18). These integrals can be expressed in terms of elliptic functions (Harmin 1981), and are easily evaluated. In figure 2, we show the left-hand side of (40) (matrix determinant) as a function of energy. Zero points of the matrix determinant correspond to the quasibound energies. Figure 3

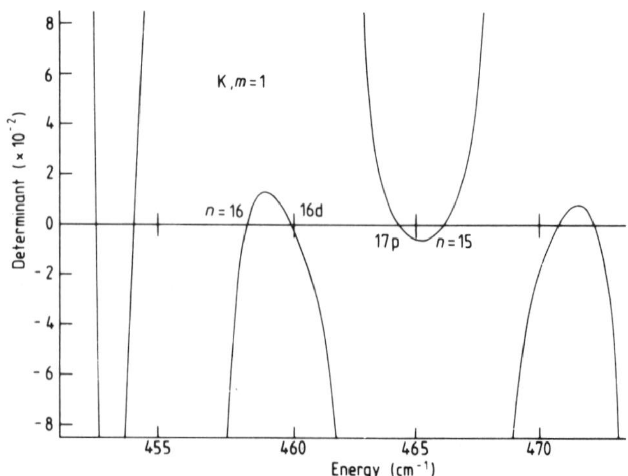

Figure 2. Determinants of the matrix $\cot\Delta_i\,\delta_{ij} - R_{ij}$ for potassium, $m = 1$, as a function of energy. The field strength is $F = 3\ \text{kV cm}^{-1}$. The energy indicates the negative energy measured from the ionisation threshold in the limit of $F = 0$. The level identification for some zero points is shown.

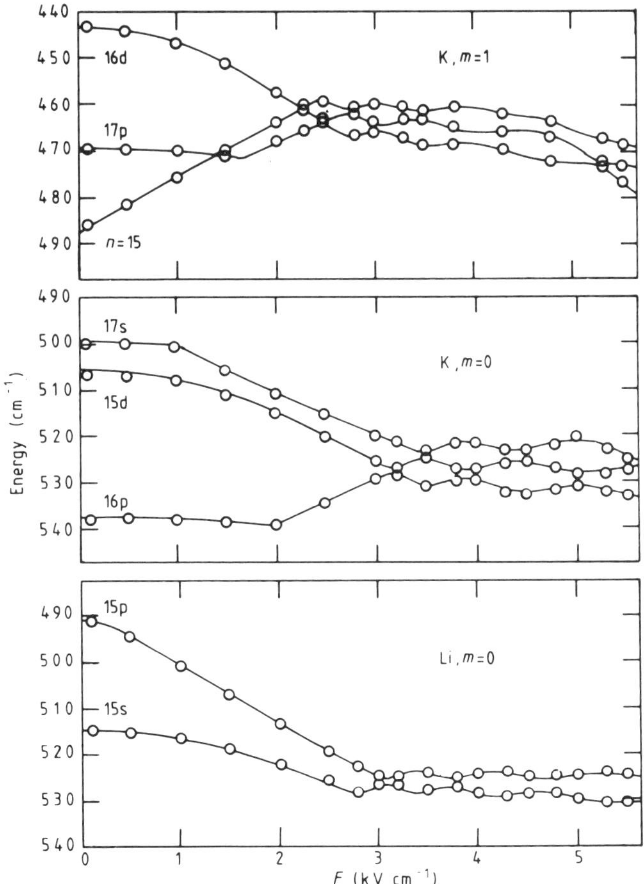

Figure 3. Stark level structures in the vicinity of $n = 15$ for lithium and potassium. The energy indicates the negative energy measured from the ionisation threshold in the limit of $F = 0$.

shows an example of the Stark energy levels of Li and K in the vicinity of $n = 15$. The quantum defects used in the calculations are cited from Zimmerman *et al* (1979): $\mu_0 = 0.399$, $\mu_1 = 0.053$, $\mu_2 = 0.002$ for Li; $\mu_0 = 2.178$, $\mu_1 = 1.712$, $\mu_2 = 0.267$, $\mu_3 = 0.010$ for K. Agreements with the results of Zimmerman *et al* (1979) are generally good. The Stark level structure of a Li atom was also calculated by Harmin (1984).

4.2. *Resonance scattering*

Next, we present a model calculation for resonance scattering. We consider two states ($\gamma = 1$ and 2) of the ion and the case where the incident energy of the electron is less than the threshold of the $\gamma = 1 \rightarrow 2$ excitation. Furthermore, for simplicity, we assume that only the s-wave ($l = 0$) electron can excite the ion and be trapped into the s state of the Rydberg atom. Then, the off-diagonal parts of the R^0 matrix with respect to γ vanish unless $l = 0$.

Because only elastic scattering occurs, the reactance matrix is expressed in terms of the phaseshift. We denote the s-wave phaseshift by κ. Then, the s-wave reactance matrix is equal to $\tan\kappa$, and equation (45) with $\gamma=\gamma'=1$ and $l=l'=0$ reduces to

$$\tan\kappa = R^0_{100,100} + \sum_{\beta\beta'} (R')^{t}_{100,2\beta 0}[\cot\Delta - R]^{-1}_{2\beta 0,2\beta' 0} R'_{2\beta' 0,100}. \tag{46}$$

In the present model, the scattering matrix is obtained from $\mathcal{S}=\exp(2i\kappa)$, and the (s-wave) elastic cross section is proportional to $|1-\mathcal{S}|^2$.

In the vanishing field limit, as is discussed in § 3.1.3, we can easily show that (46) reduces to

$$\tan\kappa = R^0_{100,100} - R^0_{100,200}[\tan(\pi\nu) + R^0_{200,200}]^{-1} R^0_{200,100}. \tag{47}$$

That is, in the case of $F=0$, resonances appear in the form of a simple Rydberg series. However, at $F\neq 0$, as is shown in (46), many channels (β) are mixed in a resonance state, and thus we have many more resonance peaks than in the case of $F=0$. (Notice that the resonance occurs at $\tan(\pi\nu)+R^0_{200,200}=0$ for $F=0$ and at $\det|\cot\Delta - R|=0$ for $F\neq 0$.) In the present model, the Rydberg state ($m=0$) with $l=1,\ldots,n-1$ of the trapped electron cannot contribute to the resonance at $F=0$. However, when the electric field is present, these $n-1$ states are coupled with the s state (the Stark mixing of the angular momentum). As a result, all the n states (including the s state) can have characteristics of the resonance s state. As the field strength increases from zero, first this l mixing is important. At a strong field, the Rydberg states with different n are also mixed. All these facts are represented as the mixing of the β channels in (46).

Figure 4 shows the resonance profile calculated by (46). The values of R^0 chosen here are: $R^0_{100,100}=0.316$, $R^0_{100,200}=R^0_{200,100}=0.579$ and $R^0_{200,200}=1.06$, which are the same as in a model calculation of Hickman (1984). The uppermost part of figure 4 shows the resonance near the incident energy of $\varepsilon = E_2 - E_1 - 2.3\times 10^{-3}$ au in the absence of an electric field. The peak shown corresponds to the resonance through the 15s Rydberg electron trapped in the excited ($\gamma=2$) ion. The lower two parts in figure 4 show the resonance profile at finite field strengths.

Because $R^0_{200,200}$ is not zero, the ns Rydberg state in the resonance has a non-zero quantum defect. Therefore, there is an energy difference between the ns and nl ($l\geq 1$) Rydberg levels with $\gamma=2$. (The energy of the nl levels is just equal to the hydrogenic one $\varepsilon = E_2 - E_1 - 1/2n^2$ because of their vanishing quantum defect.) Owing to this energy difference, the Stark mixing between the ns and nl ($l\geq 1$) states is not strong provided that the field strength is small. The middle part of figure 4 shows a case of relatively weak field strength ($F=1$ kV cm^{-1}). There, 14 other states within the $n=15$ manifold also have a slight characteristic of the resonance s state because of the weak Stark mixing, and appear as narrow sharp peaks. As the field strength increases, the degree of Stark mixing increases. In this case (the lowest part of figure 4 for $F=$ 2 kV cm^{-1}), we can no longer indicate which peak mainly possesses the intrinsic s-state character. In fact, the widths of all the resonance peaks are of nearly the same magnitude.

Figure 5 shows the halfwidth and the peak position of the intrinsic (s-wave) resonance as a function of field strength. The ns resonance position at $F=0$ lies on the lower-energy side of the hydrogenic nl ($l\geq 1$) levels. Thus, with increasing field strength, the peak of the intrinsic ns resonance is shifted to a lower energy as a result of Stark mixing. If the ns resonance at $F=0$ lies on the higher-energy side of the nl ($l\geq 1$) levels, then the peak of the ns resonance at $F\neq 0$ is shifted to a higher energy.

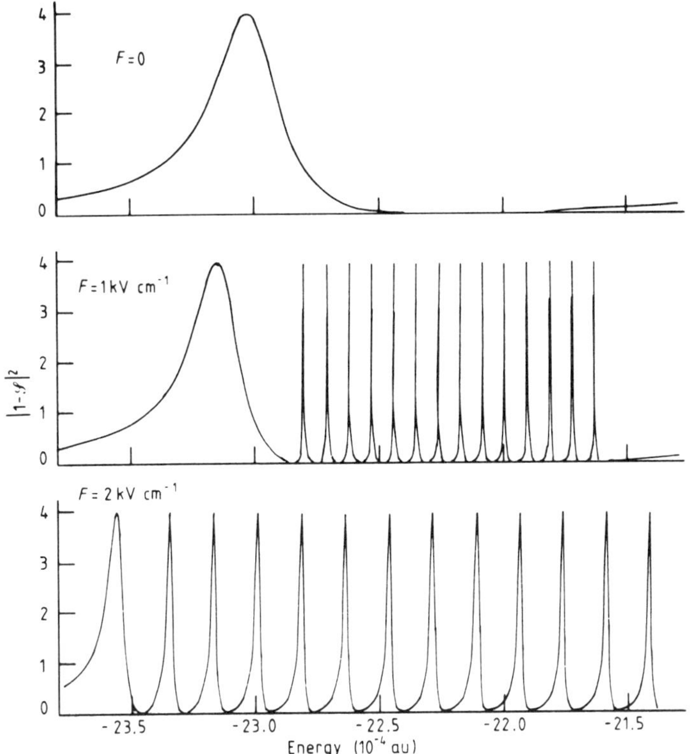

Figure 4. Resonance profiles corresponding to the $n = 15$ Rydberg state of the trapped electron at $F = 0$, 1 and $2\ \text{kV cm}^{-1}$. The energy is measured from the excitation threshold in the limit of $F = 0$.

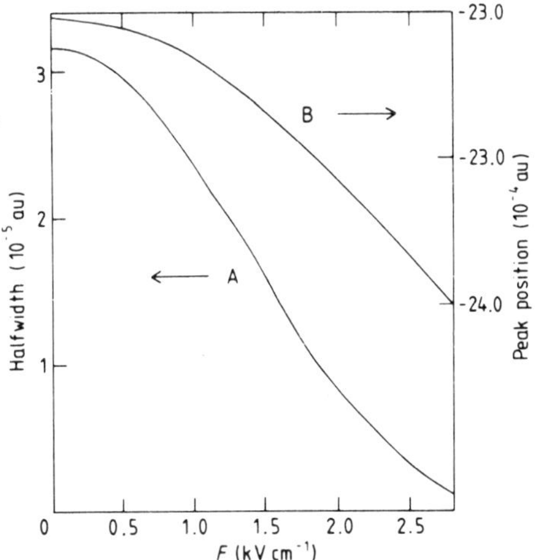

Figure 5. Halfwidths (curve A) and peak positions (curve B) of the intrinsic resonance corresponding to the 15s state as a function of F.

This behaviour is just the same as the Stark shift seen in the Stark effect of alkali metals (Zimmerman *et al* 1979).

The magnitude of the resonance width is related to the degree of s-state character that the resonance state possesses. Therefore, as the field strength becomes large, the widths of the *nl* ($l \geqslant 1$) resonances increases while that of the *ns* resonance decreases. In the weak-field region ($F < 1\,\mathrm{kV\,cm^{-1}}$), the width can be expressed in the form $A - BF^2$, where A and B are constants. This dependence on F is expected by perturbation theory, as shown by Hickman (1984). However, the dependence of the width on F is not simple at strong field strengths. The present theory can be used to investigate the behaviour of the resonance profile not only at weak but also at strong field strengths.

5. Summary and discussion

We formulate the theory of the Stark effect for an arbitrary Rydberg state using the multichannel quantum-defect theory. We applied the theory to some simple cases such as numerical examples. If we know the matrix R^0, a calculation for any complex system can be easily performed.

For resonance scattering, we have neglected the Stark effect on the incident and scattered electrons (and the continuum state is described in polar coordinates). In order to take this effect into account, we must treat the continuum state of the electron in parabolic coordinates. Furthermore, it is necessary to take into account an arbitrary direction of the incident electron in parabolic coordinates. (We choose the electric field direction as the z axis.) This treatment is not easy and remains to be done.

Recently, Bell and Seaton (1985) developed an *ab initio* theory of dielectronic recombination. Combining their and the present theories, we can deal with the effect of an electric field on dielectronic recombination. The present theory can be easily applied to photoabsorption to a doubly excited autoionising state under the influence of an electric field. These subjects will be studied in a subsequent paper.

Acknowledgments

The author would like to thank Professor Y Itikawa for discussions and his critical reading of the manuscript and the referee for his advice on revising the manuscript.

Appendix. Parabolic cylinder functions

We consider the differential equation,

$$\left(\frac{\mathrm{d}^2}{\mathrm{d}\rho^2}+\frac{1}{4}\rho^2-\alpha^2\right)W(\rho)=0 \tag{A.1}$$

where α is a positive constant. The solution of this equation is expressed in terms of the parabolic cylinder functions (§ 19.17 in Abramowitz and Stegun 1965). The two linearly indepent parabolic cylinder functions have the following analytical properties:

if $\rho \ll -\alpha$,

$$W_1(\rho)=\left(\frac{2}{kQ(\rho)}\right)^{1/2}\sin\left(\int_\rho^{-2\alpha} Q(\rho')\,\mathrm{d}\rho'+\sigma+\pi/4\right) \tag{A.2a}$$

$$W_2(\rho)=\left(\frac{2k}{Q(\rho)}\right)^{1/2}\cos\left(\int_\rho^{-2\alpha} Q(\rho')\,\mathrm{d}\rho'+\sigma+\pi/4\right) \tag{A.2b}$$

where

$$\sigma=\tfrac{1}{2}\arg\Gamma(\tfrac{1}{2}+\mathrm{i}\alpha^2)+\tfrac{1}{2}\alpha^2(1-\log\alpha^2)$$

and if $\rho^2 \ll \alpha^2$,

$$W_1(\rho)=W_1(0)\,\mathrm{e}^{-\alpha\rho} \tag{A.3a}$$

$$W_2(\rho)=W_1(0)\,\mathrm{e}^{+\alpha\rho} \tag{A.3b}$$

where $Q(\rho)$, Δ and k are defined in (11), (13) and (14), respectively.

References

Abramowitz M and Stegun I A 1965 *Handbook of Mathematical Functions* (Washington: National Bureau of Standards)

Bell R H and Seaton M J 1985 *J. Phys. B: At. Mol. Phys.* **18** 1589

Bethe H A and Salpeter E E 1977 *Quantum Mechanics of One- and Two-Electron Atoms* (New York: Plenum)

Brussaard P J and Tolhoek H A 1957 *Physica* **23** 955

Dunn G H, Belic D S, Djuric N and Mueller D W 1984 *Atomic Physics* vol 9, ed R S VanDyck and E N Fortson (Singapore: World Scientific) p 505

Dunning F B and Stebbings R F 1983 *Rydberg States of Atoms and Molecules* ed R F Stebbings and F B Dunning (Cambridge: Cambridge University Press) p 315

Fano U 1981 *Phys. Rev.* A **24** 619

Freeman R R and Bjorklund G C 1978 *Phys. Rev. Lett.* **40** 118

Harmin D A 1981 *Phys. Rev.* A **24** 2491

—— 1982 *Phys. Rev.* A **26** 2656

—— 1984 *Phys. Rev.* A **30** 2413

Hickman A P 1984 *J. Phys. B: At. Mol. Phys.* **17** L101

Hughes J W B 1967 *Proc. Phys. Soc.* **91** 2656

Jaffe S M, Kachru R, Tran N H, van Linden van den Heuvell H B and Gallagher T F 1984 *Phys. Rev.* A **30** 1828

Kleppner D, Littman M G and Zimmerman M L 1983 *Rydberg States of Atoms and Molecules* ed R F Stebbings and F B Dunning (Cambridge: Cambridge University Press) p 73

Luc-Koenig E and Bachelier A 1980 *J. Phys. B: At. Mol. Phys.* **13** 1743

Miller S C and Good R H 1953 *Phys. Rev.* **91** 174

Redmond P J 1964 *Phys. Rev.* **133** B1352

Rice M H and Good R H 1962 *J. Opt. Soc. Am.* **52** 239

Safinya K A, Delpech J F and Gallagher T F 1980 *Phys. Rev.* A **22** 1062

Seaton M J 1983 *Rep. Prog. Phys.* **46** 167

Zimmerman M L, Littman M G, Kash M M and Kleppner D 1979 *Phys. Rev.* A **20** 2251

1989 *J. Phys. B: At. Mol. Opt. Phys.* **22** 2727–39

Influence of electric fields on highly excited states of H_2: quantum-defect-theory approach

K Sakimoto

Institute of Space and Astronautical Science, Yoshinodai, Sagamihara, Kanagawa 229, Japan

Received 5 January 1989, in final form 7 April 1989

Abstract. A quantum-defect theory of the Stark effect given recently by Sakimoto is applied to study the effects of an electric field on highly excited states of molecular hydrogen. The Stark energy levels are calculated. Also investigated are field effects on photoionisation through rotational autoionisation.

1. Introduction

The Stark effect is a very old topic. However, it still raises an interesting and new problem. When we perturb a hydrogen atom in its ground state by applying an electric field, the field needed is about 10^9 V cm^{-1}. It is a fairly strong field, and can never be realised in present-day experiments. However, if we consider an atom in highly excited states, the required field can be much weaker. Highly excited states therefore play an important role in the study of the Stark effect. It is only in recent years that we have been able experimentally to control atoms and molecules in such highly excited states.

Processes involving highly excited states are well described by use of a quantum-defect theory (QDT). Fano (1970) and Jungen and Atabek (1977) developed a QDT for molecules. Later, the theory has been applied to interpret various kinds of processes concerning highly excited states of molecules (see Greene and Jungen 1985).

On the other hand, it was found that a QDT is also useful to investigate the Stark effect of highly excited states of atoms. A central idea was first introduced by Fano (1981). Harmin (1981) applied Fano's idea to the study of the Stark effect of alkaline atoms, which consist of a Rydberg electron and an ion core in its ground state. Recently, the present author has extended the theory to the case of an ion core in any state (Sakimoto 1986). An important result of these studies is that, once we know the field-free quantities appearing in the field-free QDT, the calculation for the Stark effect can be accomplished in a straightforward manner by use of frame transformation. This reveals that the QDT of the Stark effect can be directly extended to the molecular case.

Furthermore, it has recently become possible to study the Stark effect of molecules experimentally by use of high-resolution spectroscopy (Cooper *et al* 1983, Chevaleyre *et al* 1986, Janik *et al* 1987, Bordas *et al* 1988). Thus, at this time, it is very important to formulate a basic theory for the Stark effect of molecules.

The present paper gives a QDT formulation for studying effects of an electric field on molecules in highly excited states. The theory is constructed by combining the QDT

0953-4075/89/172727+13$02.50

of molecules and that of the Stark effect. A formulation is given for a hydrogen molecule whose ion core (H_2^+) is in the ground electronic state $X\,^2\Sigma_g^+$ but in arbitrary vibrational and rotational states. The theory can easily be extended to any molecules in any electronic states.

2. Quantum-defect theory

Choose an electric-field vector $\boldsymbol{F}$ oriented along the z axis. Then, the field potential acting on an electron is given by Fz. Atomic units are used unless otherwise stated. In order to emphasise the physical idea of the theory, we discuss the problem by dividing the distance r between an excited electron and a molecular ion H_2^+ into four regions.

Region I: $r_F < r < \infty$. At large distances, the electron motion is determined only by the Coulomb and electric-field potential, $V = -1/r + Fz$. The lower limit $r_F = F^{-1/2}$ is the distance at which the Coulomb force is comparable to the field. Parabolic coordinates are the most appropriate frame in this region (Sakimoto 1986).

Region II: $r_{BO} < r < r_F$. As the distance gets smaller, the Coulomb force dominates the electric field. This region always exists unless the field is extraordinarily strong. (For example, $r_F = 2000$ au for $F = 1$ kV cm^{-1}.) Since the electron motion is determined by the pure Coulomb potential, polar coordinates are a good frame as well as parabolic ones. The lower limit r_{BO} is the distance at which the velocities of nuclear and electron motions are comparable.

Region III: $r_c < r < r_{BO}$. The distance r_c has a magnitude comparable to the radius of the molecular ion. In this region, the short-range and exchange interactions are still negligible. The interaction is the pure Coulomb one as in region II. The physical importance of this region lies in the point that the Born-Oppenheimer (BO) approximation is well satisfied there. Jungen and Atabek (1977) discussed whether we can define this region. The existence of this region is critical for the efficiency of molecular QDT (Fano 1970, Jungen and Atabek 1977).

Region IV: $0 < r < r_c$. The electron motion is complicated here because the short-range and exchange interactions are present. However, the BO approximation is powerful to solve the problem. Physical quantities are well described in a molecular frame where the z axis is directed along the molecular axis.

We thus have three different frames depending on the distance r: the molecular (BO), spherical and parabolic ones. An important point is that there is always a common region where two of these frames apply equally well. First, we obtain a solution of the Schrödinger equation for the whole system independently in each region. Then, we can smoothly match the solutions in the common regions. An appropriate boundary condition is imposed in region I. In this way, we can completely determine the wavefunction over the entire space.

In this study, we neglect the Stark effect on H_2^+. This is well justified because molecular hydrogen does not have a permanent electric dipole moment. A rough estimate shows that (even if it is a polar molecule) the Stark effect is negligible unless both the dipole moment and the field strength are fairly strong or unless the principal quantum number of the excited electron is extraordinarily large. (The latter case should be noted because the change in the large principal quantum number is significant even if the energy of H_2^+ is only slightly changed owing to an electric field.)

2.1. *Solutions in $r_c < r < r_F$ (regions II and III)*

In regions II, III and IV, the situation is exactly the same as that for the field-free QDT of molecules. Therefore, a general solution of the Schrödinger equation for the whole system is just equal to a field-free one. In regions II and III, a solution can be specified by the total angular momentum $\boldsymbol{J} = \boldsymbol{N} + \boldsymbol{l}$, where $\boldsymbol{N}$ is the rotational angular momentum of H_2^+ and $\boldsymbol{l}$ is the orbital angular momentum of the excited electron. We can write a solution that is valid through regions II and III in the following form (Jungen and Atabek 1977):

$$\Phi_{vNl}^{JM}(\boldsymbol{K}^J) = \sum_{v'N'} \chi_{v'}^{N'}(R) Y_{N'l}^{JM}(\hat{\boldsymbol{R}}, \hat{\boldsymbol{r}}) G_{v'N',vN}^{Jl}(r) \tag{1}$$

where χ_v^N is a vibrational wavefunction of H_2^+, Y_{Nl}^{JM} is an eigenfunction of the total angular momentum (J, M), $\boldsymbol{R}$ is a vector of the internuclear distance of H_2^+, and $G_{v'N',vN}^{Jl}$ is a radial part of the solution for the excited electron. This form of expansion is familiar in scattering problems. Summation over v' and N' is necessary because these H_2^+ states are mixed by a short-range e–H_2^+ interaction through region IV. In the present study, we assume that l is conserved. This is a good approximation for H_2 because the molecule is nearly a spherically symmetric system. In fact, many experimental results can be explained by assuming l conserved (Greene and Jungen 1985). In the solution (1), channels are indicated by (v, J, N, l).

In regions II and III, the radial function G is reduced to a sum of regular $s_{\varepsilon l}(r)$ and irregular $c_{\varepsilon l}(r)$ Coulomb functions (ε being the energy of the excited electron). Hence, we have

$$G_{v'N',vN}^{Jl}(r) = s_{\varepsilon'(v'N')l}(r)\delta_{v'N',vN} + c_{\varepsilon'(v'N')l}(r) K_{v'N',vN}^{Jl}. \tag{2}$$

The summation in equation (1) must be taken so as to conserve the total energy $E = E_{vN} + \varepsilon$ (E_{vN} being the energy of H_2^+). The coefficient $K_{v'N',vN}^{Jl}$ is an element of the K matrix, which is determined by smoothly matching the solution to that given in region IV. The K matrix includes all the information on the complex electron motion in region IV.

Following Fano (1970) and Jungen and Atabek (1977), the K matrix can be expressed in terms of another K matrix, $K_{l\Lambda}^{\mathrm{BO}}(R)$, which is the essential quantity in the field-free QDT of molecules, namely

$$K_{v'N',vN}^{Jl} = \sum_{\Lambda} \langle N|\Lambda\rangle^{Jl} \langle \chi_{v'}^{N'}| K_{l\Lambda}^{\mathrm{BO}}(R) |\chi_v^N\rangle \langle \Lambda|N\rangle^{Jl} \tag{3}$$

where

$$\langle N|\Lambda\rangle^{Jl} = (-1)^{J+\Lambda-N}[2/(1+\delta_{\Lambda 0})]^{1/2}(l - \Lambda J \Lambda | N0). \tag{4}$$

In order to obtain the K^{BO} matrix, the Schrödinger equation for the total system is solved first under the assumption that the BO approximation is well satisfied at whole distances r. Then, by fitting the electronic part of the solution to a form similar to (2) at a large distance, we can obtain the K^{BO} matrix. Since the molecular frame is appropriate in this case, the calculation is done at each fixed internuclear distance $\boldsymbol{R}$.

The K^{BO} matrix is given as a function of R, the angular momentum of the excited electron l, and its projection onto the internuclear axis Λ. The presence of region III is very important to obtain the relation (3). In region III, we can match the BO solution to the other one (1) smoothly.

Since l is conserved in this case and thus the K^{BO} matrix is diagonal, K^{BO} is expressed in terms of a quantum defect $\mu_{l\Lambda}(R)$ by (Greene and Jungen 1985)

$$K^{BO}_{l\Lambda}(R) = \tan[\pi\mu_{l\Lambda}(R)]. \tag{5}$$

2.2. Solutions in $r_c < r < \infty$ (regions I, II and III)

We can write a general solution that is valid through regions I, II and III. Because an electric field is not negligible in region I, the angular momentum l cannot be a channel index in this solution. The electronic part of the solution is given in parabolic coordinates, $\xi = r + z$, $\eta = r - z$, $\phi = \tan^{-1}(y/x)$. Basis functions for the electron motion are the Stark ones defined as (Sakimoto 1986, 1987)

$$\psi_{\varepsilon\beta m}(r) = (2\pi\xi\eta)^{-1/2} f_{\varepsilon\beta m}(\xi) g_{\varepsilon\beta m}(\eta)\, e^{im\phi} \tag{6a}$$

$$\psi'_{\varepsilon\beta m}(r) = (2\pi\xi\eta)^{-1/2} f_{\varepsilon\beta m}(\xi) g'_{\varepsilon\beta m}(\eta)\, e^{im\phi}. \tag{6b}$$

These are the eigenfunctions of the Schrödinger equation for the electron motion in the potential $V = -1/r + Fz$, and are expressed as a product of the ξ, η and ϕ components. With the choice of the z axis along the field direction, the magnetic quantum number m is conserved. The quantity β is a separation constant. Since the ξ motion is always bound in the potential V, the separation constant β is quantised and has a discrete value at a given ε and m. In this study, the quantisation of β is done by using a WKB method (Sakimoto 1986). The functions $g(\eta)$ and $g'(\eta)$ are regular and irregular at the origin, respectively. These functions play the same role as the regular and irregular radial Coulomb functions. Hence, it is important to know the analytical behaviour at large η when we impose a physical boundary condition on a solution. The analytical behaviour of these functions is investigated by using a WKB method as well (Sakimoto 1986).

Using the Stark functions (6), we can express a general solution at $r > r_c$ by

$$\Phi^{F}_{vN\beta m}(\boldsymbol{K}^{F}) = \sum_{v'N'\beta'm'} \chi^{N'}_{v'}(R)\, Y_{N'M\text{-}m'}(\hat{\boldsymbol{R}})[\psi_{\varepsilon'(v'N')\beta'm'}(r)\delta_{v'N'\beta'm',vN\beta m} + \psi'_{\varepsilon'(v'N')\beta'm'}(r) K^{F}_{v'N'\beta'm',vN\beta m}] \tag{7}$$

where Y_{NM-m} is a spherical harmonic that is an eigenfunction of the rotation of H_2^+, and K^F is a K matrix defined for a finite field. This matrix has an element non-diagonal with respect to β and m because these quantities are not conserved in region IV. Channels are specified by (v, N, β, m) in this case. We must determine the K^F matrix. This can be done by use of frame transformation.

2.3. Frame transformation

When the distance r is much less than r_F, the electric field is negligible. There, the Stark functions (6) are also solutions of the Schrödinger equation for the pure Coulomb potential. At $r<r_F$, therefore, we can have a relation between the Stark and the spherical Coulomb functions in the form (Sakimoto 1986)

$$\psi_{\varepsilon\beta m}(\xi\eta\phi)=\sum_l U_{\beta l}(F\varepsilon m)Y_{lm}(\hat{r})s_{\varepsilon l}(r) \tag{8a}$$

$$\psi'_{\varepsilon\beta m}(\xi\eta\phi)=\sum_l (U^{-1})^{t}_{\beta l}(F\varepsilon m)Y_{lm}(\hat{r})c_{\varepsilon l}(r) \tag{8b}$$

where $U_{\beta l}(F\varepsilon m)$ is a matrix element of frame transformation between polar and parabolic coordinates. An explicit expression for this quantity is given in Sakimoto (1986).

By using the transformation (8), we can match the two solutions (1) and (7) in an analytical way. Hence, we have a relation between the field-free K and field-dependent K^{F} matrices, i.e.

$$K^{\mathrm{F}}_{v'N'\beta'm',vN\beta m}=\sum_l U_{\beta' l}(F\varepsilon' m')K^{l}_{v'N'm',vNm}U_{\beta l}(F\varepsilon m) \tag{9}$$

where

$$K^{l}_{v'N'm',vNm}=\sum_J (lm'N'M-m'|JM)K^{Jl}_{v'N',vN}(lmNM-m|JM). \tag{10}$$

The transformation (10) is necessary because m is a good channel index in the presence of an electric field, and J is not so. From the relations (9), (10), (3) and (5), we can completely determine the K^{F} matrix in terms of the field-free quantum defect.

2.4. Boundary condition

As was discussed in Sakimoto (1986), there is no bound state in atoms and molecules quantum mechanically when an electric field is applied. This is because the potential $V=-1/r+Fz$ becomes negatively infinite as $z\to-\infty$. However, the potential V has a local maximum so that a quasi-bound state can exist for energies less than $\varepsilon_0=-2F^{1/2}$. If the tunnelling probability through the barrier is very small, we can actually regard the quasi-bound state as a true bound state. This picture is exemplified by viewing the electron motion along the η coordinate.

In this section, we consider only a quasi-bound state. Processes concerning continuum states are treated in § 4. Following Sakimoto (1986), the behaviour of g and g' around the barrier is as follows:

$$g_{\varepsilon\beta m}(\eta)=C_{\varepsilon\beta m}(\eta)\{\sin[\Delta(F\varepsilon\beta m)]D_{\varepsilon\beta m}(\eta)+\cos[\Delta(F\varepsilon\beta m)]I_{\varepsilon\beta m}(\eta)\} \tag{11a}$$

$$g'_{\varepsilon\beta m}(\eta)=C_{\varepsilon\beta m}(\eta)\{\cos[\Delta(F\varepsilon\beta m)]D_{\varepsilon\beta m}(\eta)-\sin[\Delta(F\varepsilon\beta m)]I_{\varepsilon\beta m}(\eta)\} \tag{11b}$$

where Δ is a WKB phase integral of the quasi-bound motion defined by

$$\Delta(F\varepsilon\beta m)=\int_a^b q(\eta)\,\mathrm{d}\eta \tag{12}$$

with two turning points a and b, and the local momentum $q(\eta)$ given by

$$q^2(\eta)=\varepsilon/2-m^2/(4\eta^2)+(1-\beta)/\eta+F\eta/4.$$

In equation (11), the function $C(\eta)$ varies slowly with η. The function $D(\eta)$ ($I(\eta)$) exponentially decreases (increases) with η in the barrier region. The function $I(\eta)$ has a large amplitude outside the barrier. Hence, as was done in Sakimoto (1986), we define a bound state by setting the exponentially increasing component $I(\eta)$ to vanish identically.

3. Stark energy levels

In order to obtain the wavefunction of the bound state, we take a linear combination of the solution (7). We set

$$\Psi \equiv \sum_{vN\beta m} \Phi^{\mathrm{F}}_{vN\beta m}(\boldsymbol{K}^{\mathrm{F}}) L_{vN\beta m} \tag{13}$$

where $L_{vN\beta m}$ is a constant coefficient. Noticing the form (11), we impose the condition that the wavefunction Ψ does not have an increasing component $I(\eta)$ in the barrier region. This can be satisfied if we have

$$\det\left|\cot[\Delta(F\varepsilon\beta m)]\delta_{v'N'\beta'm',vN\beta m} - K^{\mathrm{F}}_{v'N'\beta'm',vN\beta m}\right| = 0. \tag{14}$$

Roots for the energy ε of equation (14) give the Stark energy levels of molecules. As is discussed in § 2.3, the molecular quantity required in the calculation of equation (14) is only the quantum defect defined under the assumption of the BO approximation. In particular for H_2, the BO quantum defects are already known both theoretically and experimentally (Greene and Jungen 1985).

4. Autoionisation

In this section, we study photoionisation from the ground state of H_2. We assume that the molecular ion is in the electronically ground state. Accordingly, we are interested in the field effect of rotational and vibrational autoionisation.

We assume that the electric field is so weak that its effect on the ejected electron is negligible. However, it does not mean that the field effect is always negligible on the photoionisation. If ionisation occurs via a resonance in which an electron is trapped into a high Rydberg state, then a very weak field can affect the ionisation process.

We can expect that neglect of the field effect on the ejected electron does not significantly alter the total photoionisation cross section for the ejected electron energy, $\varepsilon > 0$. This is because the total cross section is given by summing over all the final channels. Also at the energies $\varepsilon_0 < \varepsilon < 0$, we have open channels. In this energy region, however, we can never neglect the field effect on the ejected electron. (If we neglect the field, the states with $\varepsilon < 0$ no longer correspond to open channels.)

As in the work of Harmin (1981), it is possible to take the field effect on the ejected electron into account. In this study, however, we employ the formalism based on the scattering theory given by Seaton (1983), which is more suitable to study a scattering process like dissociative recombination. The problem in applying the theory to the present case is that the z axis is already chosen along the field direction, not along the incident or scattered direction of the electron. Consequently, we must take any direction of the incident or scattered electron in parabolic coordinates. However, this procedure is not so simple because the free motion of the electron in the field is not described in a single term of a plane wave. To avoid this complexity, we assume that ε is much larger (>0) than the energy gain that the field supplies to the ejected electron. The field effect can thereby always be negligible for the open channels, and accordingly the partial cross section is specified in terms of the spherical basis.

Since the field effect is negligible on the ejected electron, the oscillator strength for a bound-free transition is given in the usual form

$$\mathrm{d}f_i(E)/\mathrm{d}E = 2(E - E_0)|(\Psi_i(E)|z|\Psi_0)|^2 \tag{15}$$

where Ψ_0 is a wavefunction of H_2 in the electronically ground state (X $^1\Sigma_g^+$), E_0 is its energy and Ψ_i is a wavefunction for a final state with $i=(v, N, l, m)$.

The wavefunction Ψ_i is obtained in the same way as for the wavefunction (13). We set

$$\Psi_i = \sum_\alpha \Phi_\alpha^{\mathrm{F}}(\boldsymbol{x}^{\mathrm{F}}) L_{\alpha,i}. \tag{16}$$

As has been assumed above, the Stark effect is neglected for open channels ($\varepsilon > \varepsilon_0$) while it is fully taken into account for closed channels ($\varepsilon < \varepsilon_0$). The motion of the ejected electron is described in terms of the spherical Coulomb functions. Thus, in equation (16), the subscript α indicates (v, N, l, m) for open channels and (v, N, β, m) for closed channels. The function $\Phi_\alpha^{\mathrm{F}}(\boldsymbol{x}^{\mathrm{F}})$ has the following form:

$$\Phi_\alpha^{\mathrm{F}}(\boldsymbol{x}^{\mathrm{F}}) = \sum_{\alpha'} \Upsilon_{\alpha'}(\boldsymbol{R}) H_{\alpha',\alpha}(\boldsymbol{r}) \tag{17}$$

where

$$\Upsilon_\alpha(\boldsymbol{R}) = \chi_v^N(R) Y_{NM-m}(\hat{\boldsymbol{R}}) \tag{18}$$

$$H_{\alpha',\alpha}(\boldsymbol{r}) = \psi_{\alpha'}^{+}(\boldsymbol{r})\delta_{\alpha',\alpha} - \psi_{\alpha'}^{-}(\boldsymbol{r})(\chi_{\alpha',\alpha}^{\mathrm{F}})^* \tag{19}$$

with

$$\psi_\alpha^{\pm} = \tfrac{1}{2}(\psi'_\alpha \pm \mathrm{i}\psi_\alpha). \tag{20}$$

For closed channels, ψ' and ψ are the Stark functions (6); and for open channels, these are the spherical Coulomb functions given by

$$\psi_\alpha = Y_{lm}(\hat{\boldsymbol{r}}) s_{\varepsilon l}(r) \qquad \psi'_\alpha = Y_{lm}(\hat{\boldsymbol{r}}) c_{\varepsilon l}(r) \qquad \varepsilon > \varepsilon_0. \tag{21}$$

It should be noted that the normalisation condition in the electronic part of the solution (19) is different from that in the solution (7). Using the relation (20), we can easily verify (in matrix form)

$$\boldsymbol{\chi}^{\mathrm{F}} = (\mathbf{1} + \mathrm{i}\boldsymbol{K}^{\mathrm{F}})(\mathbf{1} - \mathrm{i}\boldsymbol{K}^{\mathrm{F}})^{-1}. \tag{22}$$

This matrix is just equal to an S matrix when all the channels are open. The matrix $\boldsymbol{K}^{\mathrm{F}}$ is the same as the one defined in equation (9) except that the frame transformation is done only for closed channels; i.e. $\boldsymbol{K}^{\mathrm{F}} = \boldsymbol{X}\boldsymbol{K}\boldsymbol{X}^{\mathrm{t}}$, where $\boldsymbol{X}_{\mathrm{oo}} = \mathbf{1}$, $\boldsymbol{X}_{\mathrm{oc}} = \boldsymbol{X}_{\mathrm{co}}^{\mathrm{t}} = \mathbf{0}$ and $\boldsymbol{X}_{\mathrm{cc}} = \boldsymbol{U}$. The functions (20) and the matrix (22) have been introduced in Sakimoto (1987) to discuss resonance scattering. The normalisation condition in the solution (19) is the one proper for photoionisation (Seaton 1983). Noticing again the form (11), we eliminate exponentially increasing components in the wavefunction (16). This determines the coefficient $L_{\alpha,i}$. We find in matrix form

$$\boldsymbol{L}_{\mathrm{oo}} = \mathbf{1} \tag{23a}$$

$$\boldsymbol{L}_{\mathrm{co}} = -[(\boldsymbol{\chi}_{\mathrm{cc}}^{\mathrm{F}})^* + \mathrm{e}^{2\mathrm{i}\boldsymbol{\Delta}}]^{-1}(\boldsymbol{\chi}_{\mathrm{co}}^{\mathrm{F}})^*. \tag{23b}$$

In these equations, matrices are partitioned into open (o) and closed (c) channel parts.

Defining a dipole matrix $D_i = (\Psi_i|z|\Psi_0)$, and using the coefficients (23), we can partition D_i into a direct (d) and a resonance (r) part, $D_i = D_i^{\mathrm{d}} + D_i^{\mathrm{r}}$, where

$$\underline{D}^{\mathrm{d}} = \underline{d}_{\mathrm{o}}^{\mathrm{F}}(\boldsymbol{\chi}^{\mathrm{F}}) \tag{24a}$$

$$\underline{D}^{\mathrm{r}} = -\boldsymbol{\chi}_{\mathrm{oc}}^{\mathrm{F}}[\boldsymbol{\chi}_{\mathrm{cc}}^{\mathrm{F}} + \mathrm{e}^{-2\mathrm{i}\boldsymbol{\Delta}}]^{-1}\underline{d}_{\mathrm{c}}^{\mathrm{F}}(\boldsymbol{\chi}^{\mathrm{F}}). \tag{24b}$$

We have used an under-bar notation to express a row vector, and

$$\underline{d}^{\mathrm{F}}(\boldsymbol{\chi}^{\mathrm{F}}) = \begin{pmatrix} \underline{d}_{\mathrm{o}}^{\mathrm{F}}(\boldsymbol{\chi}^{\mathrm{F}}) \\ \underline{d}_{\mathrm{c}}^{\mathrm{F}}(\boldsymbol{\chi}^{\mathrm{F}}) \end{pmatrix} = ([\boldsymbol{\Phi}^{\mathrm{F}}(\boldsymbol{\chi}^{\mathrm{F}})]^{\mathrm{t}}|z|\Psi_0). \tag{25}$$

Since the relation (8) is given between the eigenfunctions of the different Schrödinger equations, the frame-transformation matrix $\boldsymbol{X}$ (or $\boldsymbol{U}$) is not orthogonal. However, if the field is not so strong, we can assume that the matrix $\boldsymbol{X}$ is orthogonal (Sakimoto 1986). Then, we have

$$\underline{d}^{\mathrm{F}}(\boldsymbol{\chi}^{\mathrm{F}}) = \boldsymbol{X}\underline{d}(\boldsymbol{\chi}) \tag{26}$$

where $\underline{d}(\boldsymbol{\chi})$ is a field-free quantity defined by

$$\begin{aligned} d_{vNlm}(\boldsymbol{\chi}) &= \sum_J (lmNM-m|JM)(\Phi^{JM}_{vNl}(\boldsymbol{\chi}^J)|z|\Psi_0) \\ &\equiv \sum_J (lmNM-m|JM)\, d^{JM}_{vNl}(\boldsymbol{\chi}^J). \end{aligned} \tag{27}$$

The function $\Phi^{JM}_{vNl}(\boldsymbol{\chi}^J)$ is similar to the one given by equation (1) except in the $\boldsymbol{\chi}^J$ matrix normalisation, where $\boldsymbol{\chi}^J = (\mathbf{1}+\mathrm{i}\boldsymbol{K}^J)(\mathbf{1}-\mathrm{i}\boldsymbol{K}^J)^{-1}$ with $\boldsymbol{K}^J$ defined by equation (3). Expressing $\Phi^{JM}_{vNl}(\boldsymbol{\chi}^J)$ in terms of the function (1), we obtain (cf Seaton 1983)

$$\underline{d}^{JM}(\boldsymbol{\chi}^J) = -(\mathrm{i}/2)(\mathbf{1}+\boldsymbol{\chi}^J)\underline{d}^{JM} \tag{28}$$

with

$$d^{JM}_{vNl} = (\Phi^{JM}_{vNl}(\boldsymbol{K}^J)|z|\Psi_0). \tag{29}$$

The function $\Phi^{JM}_{vNl}(\boldsymbol{K}^J)$ is just the same as that given by equation (1). The dipole matrix (29) is usually introduced in the field-free QDT of molecules (Dill 1972, Greene and Jungen 1985). Dill (1972) and Jungen and Dill (1980) have shown that this quantity can be expressed in terms of a dipole matrix $d^{\mathrm{BO}}_{l\Lambda}(R)$ calculated under the assumption of the BO approximation by

$$d^{JM}_{vNl} = \sum_{\Lambda} \langle N|\Lambda\rangle^{Jl} \langle J_0|\Lambda\rangle^{Jl} \langle \chi^N_v|d^{\mathrm{BO}}_{l\Lambda}(R)|\chi^{J_0}_{v_0}\rangle (J_0M_010|JM) \tag{30}$$

where v_0, J_0 and M_0 are quantum numbers for H_2, and $\chi^{J_0}_{v_0}$ is a vibrational wavefunction of H_2. It should be noted that the BO dipole matrix given by Jungen and Dill (1980) corresponds to $\cos[\pi\mu_{l\Lambda}(R)]d^{\mathrm{BO}}_{l\Lambda}(R)$ owing to the difference in the normalisation condition.

The calculation of the oscillator strength (15) requires the quantum defects and the dipole matrix elements to be calculated under the assumption of the BO approximation. Thus, the calculation can be done in a straightforward manner as in the case of the Stark energy levels.

5. Application

In this study, we neglect the R dependence of the quantum defect $\mu_{l\Lambda}(R)$ for simplicity. This approximation is justified if we restrict the electron energy to some narrow range (see Greene and Jungen 1985). In fact, Herzberg and Jungen (1972) explained nicely their experimental results for energy levels of H_2 in terms of the R-independent quantum defect. Also, for other molecules, the R dependence of the quantum defect is sometimes neglected to analyse the experimental results (Greene and Jungen 1985). However, because of this approximation, we cannot discuss the effect due to the vibrational motion of H_2^+.

We take the quantum defects, the first ionisation threshold of H_2 and the rotational constant of H_2^+ from the paper by Herzberg and Jungen (1972): $\mu_{s\Sigma}=-0.0845$, $\mu_{p\Sigma}=0.185$ and $\mu_{p\Pi}=-0.082$. For simplicity, we assume that $\mu_{l\Lambda}=0$ for $l\geqslant 2$.

5.1. *Stark energy levels*

Figure 1 shows the calculated field-free energy levels of para H_2 below the first ionisation limit (H_2^+, $v=0$, $N=0$) in the energy range 123 000–123 500 cm^{-1}. The energy is measured from the ground state of H_2 ($X\,^1\Sigma_g^+$, $v_0=0$, $N_0=0$). By using the quantum defects given above, we can reproduce well the experimental values of Herzberg and Jungen (1972). Assignment of the non-hydrogenic levels is given by (lJ). In constrast to the atomic case, many more levels are crowded into a narrow energy range. This is because there are many Rydberg series, each of which converges to the energy of H_2^+ in a rotationally excited state. Rydberg states with the same value of the principal quantum number n are degenerate when they have $l\geqslant 2$. These hydrogenic levels are also shown in figure 1 and are labelled by (Nn). Since the level space is small, we can easily imagine that even a weak field can significantly perturb the level structure compared with the atomic case.

When a field is present, actually all the l and N values are mixed together. In the calculation of the Stark energy levels, we include the rotational states up to $N=6$. Mixing of l is automatically taken into account in the frame transformation (9). Figure

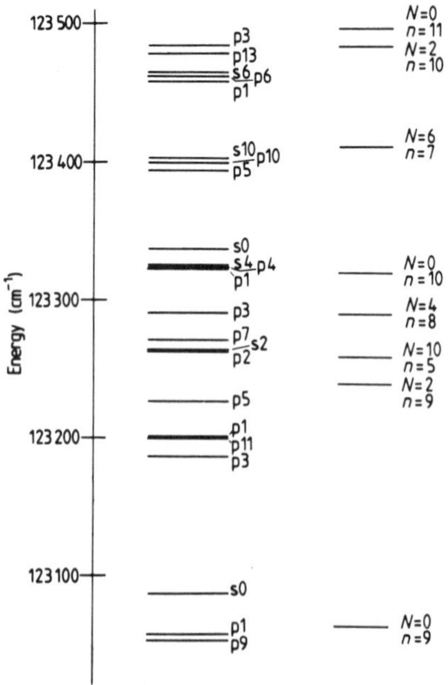

Figure 1. Calculated field-free energy levels (cm^{-1}) of para H_2. The energy is measured from the ground state of H_2 ($X\,^1\Sigma_g^+$, $v_0=0$, $N_0=0$). Assignments are given for non-hydrogenic levels by (lJ) and for hydrogenic levels by (Nn).

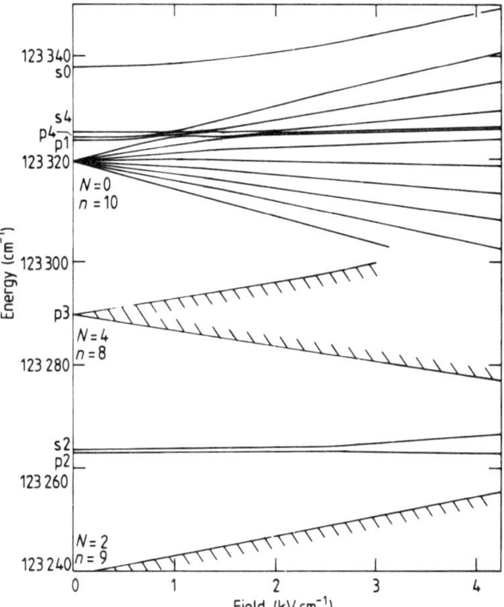

Figure 2. Calculated Stark energy levels (cm^{-1}) of H_2 in the range 123 240–123 350 cm^{-1}. The total magnetic quantum number is set to $M = 0$. The energy is measured from the ground state of H_2 ($X\,^1\Sigma_g^+$, $v_0 = 0$, $N_0 = 0$). The details of the ($N = 4$, $n = 8$) manifold are shown in figure 3.

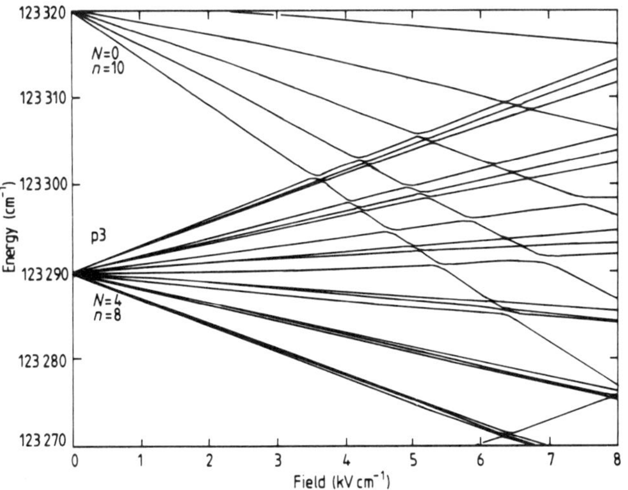

Figure 3. Calculated Stark energy levels (cm^{-1}) for the ($N = 4$, $n = 8$) manifold of H_2. The total magnetic quantum number is set to $M = 0$. The energy is measured from the ground state of H_2 ($X\,^1\Sigma_g^+$, $v_0 = 0$, $N_0 = 0$).

2 shows the Stark energy levels calculated in the range 123 240–123 350 cm^{-1}. Since the states with $N \geqslant 7$ are neglected, we do not have the field-free levels labelled by p7 and ($N=10$, $n=5$). In the hydrogenic manifolds with ($N=4$, $n=8$) and ($N=2$, $n=9$), there are too many levels to be shown separately.

Figure 3 shows the magnification of the hydrogenic ($N=4$, $n=8$) manifold. An interesting feature is that these degenerate levels are split into 18 levels although this manifold is composed of states with $l=2$–7. (It should be noted that since the p3 level is very near to the hydrogenic one, the manifold apparently seems to have split into 19 levels.) This is explained as follows. When a field is applied, the total magnetic quantum number $M=J_z=l_z+N_z$ is exactly conserved, but $m=l_z$ is not conserved. The mixing of m is due to the non-diagonal K matrix (9) or (10). In this study, the K matrix vanishes if $l \geqslant 2$, and consequently it vanishes unless $m=-1$, 0 or 1. This is the reason why the hydrogenic manifold with $l=2$–7 is split into 18 ($=3\times 6$) levels. (If the quantum defect with $l=2$ does not vanish, the number of splittings becomes $5\times 6=30$.)

A similar effect is also seen for alkaline atoms if we take the fine structure into account (Zimmerman *et al* 1979). In this case, not m but $M=l_z+s_z$ should be conserved, s_z being the z component of the spin angular momentum.

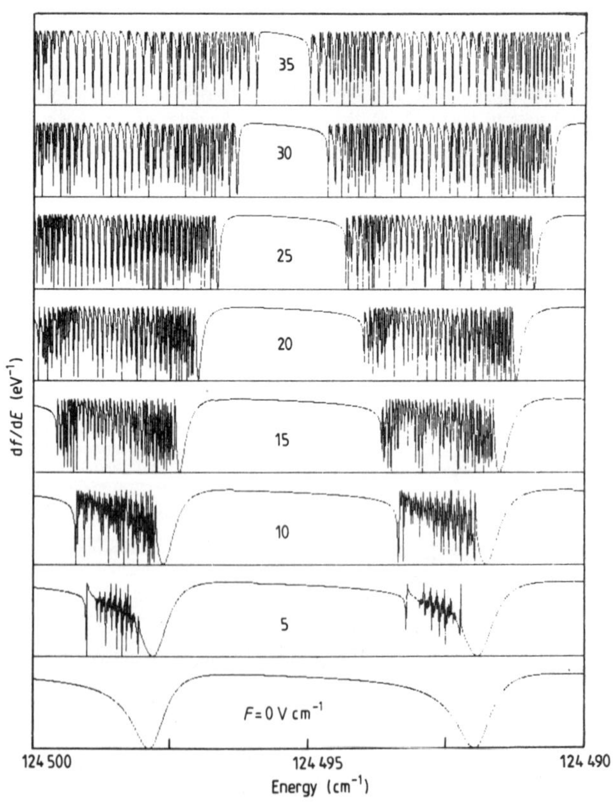

Figure 4. Calculation of photoabsorption of H_2 (X $^1\Sigma_g^+$, $v_0=0$, $N_0=0$) above the first ionisation threshold (124 417.2 cm^{-1}). The energy (cm^{-1}) is measured from the ground state of H_2 (X $^1\Sigma_g^+$, $v_0=0$, $N_0=0$).

The situation is different in the $N=0$ case at weak fields. Since $N=0$, m is conserved as long as N is not mixed. When a field is weak, N is also conserved. The hydrogenic manifold with $(N=0, n)$ is therefore split into only $n-2$ levels.

5.2. *Rotational autoionisation*

Above the first ionisation limit (124 417.2 cm^{-1}), oscillator strengths have a structure due to autoionisation. Jungen and Dill (1980) calculated field-free oscillator strengths in an energy range between the ionisation limits with $N=0$ and 2. The final state they considered is $(J=1, l=1, N=0, v=0)$. Since $J=1$ and $l=1$, only the $N=2$ state contributes to rotational autoionisation.

In this study, we investigate field effects on the rotational autoionisation. Jungen and Dill (1980) give the BO dipole matrices, which are assumed to be independent of R. The Franck–Condon factor needed in the calculation of (30) is taken from the paper by O'Neil and Reinhardt (1978). We include only the $N=0$ and 2 states.

Above the first ionisation limit, the principal quantum number of the Rydberg series converging to the $N=2$ limit is larger than 25. Therefore, we expect that a very weak field can perturb the rotational autoionisation. The results are shown in figure 4, where the field is varied from 0 to 35 V cm^{-1}. Jungen and Dill (1980) found a drastic effect due to vibrational autoionisation. However, since we neglect the vibrational motion, we cannot reproduce the same structure obtained by Jungen and Dill at $F=0$.

When a field is applied, l and J are no longer conserved, and all the Rydberg states are coupled with the field-free autoionising state (p1). We have Rydberg states labelled by s2, p2 and $(N=2, n)$, which are embedded in the continuum. The energy positions of these states are just above the autoionising p1 level (cf figure 1). However, these states can never autoionise in the absence of a field because we assume that the quantum defect is zero for $l \geqslant 2$. For a finite field, as is mentioned in § 5.1, the hydrogenic degenerate $(N=2, n)$ manifold is split into $3(n-2)$ levels. Each of these states comes to autoionise because of the coupling with the intrinsic autoionising state. Since the principal quantum number related to figure 4 is about 30, more than 90 states contribute to autoionisation in a narrow energy region. This makes the resonance structure very complex even if the applied field is very weak.

6. Summary and discussion

We have formulated a basic theory for the Stark effect of molecules in highly excited states by using the QDT. As in the case of the field-free QDT, the BO quantum defect $\mu_{l\Lambda}(R)$ contains the essential part of the physics. Once we know the BO quantum defects, we can solve the problem directly by introducing the three types of frame transformation (3), (9) and (10). Furthermore, the physical boundary condition can be imposed in an analytical way in parabolic coordinates, which are the most appropriate frame in the presence of an electric field. This is the main advantage in introducing a QDT.

We have applied the theory to calculate the Stark energy levels and to study the field effect on rotational autoionisation. In doing that, we have neglected the vibrational motion of molecules. Since vibrational autoionisation is a very important process in ionisation of H_2, we should include the vibrational motion in the next step of the study. However, even if we neglect the vibrational motion, the structure of the Stark energy levels or resonance profiles is very complicated compared with the atomic case.

When a molecule is in a highly excited state, there can be another channel, predissociation, that can compete with autoionisation. It is very interesting to study how a field affects such a competing process. Furthermore, the field effect on dissociative recombination should be investigated. This process is very similar to dielectronic recombination in collisions between an electron and an atomic ion. It is now evident that dielectronic recombination is significantly affected by a very weak field (Muller *et al* 1987, Sakimoto 1987). This fact is rather a severe problem in doing experiments because we can never set up experimental apparatus perfectly free from fields. No-one has succeeded in meaşuring the true field-free cross section for dielectronic recombination by beam experiments. We may have the same situation also in experiments for dissociative recombination. This should be critically examined.

Finally, if the field effect is drastically different for autoionisation and for predissociation in some molecule, an electric field would be a useful tool to investigate the dynamics of such a molecule in a highly excited state. By tuning the field strength, we may be able to adjust the degree of relative importance of autoionisation and predissociation.

Acknowledgments

The author would like to thank Dr H Takagi for valuable discussions on the field-free QDT of molecules and on dissociative recombination, and Professor Y Itikawa for his critical reading of the original manuscript. Thanks are also due to one of the referees for advice on revising the manuscript.

References

Bordas C, Brevet P F, Broyer M, Chevaleyre J, Labastie P and Perrot J P 1988 *Phys. Rev. Lett.* **60** 917
Chevaleyre J, Bordas C, Broyer M and Labastie P 1986 *Phys. Rev. Lett.* **57** 3027
Cooper J W, Saloman E B, Cole B E and Shardanand 1983 *Phys. Rev.* A **28** 1832
Dill D 1972 *Phys. Rev.* A **6** 160
Fano U 1970 *Phys. Rev.* A **2** 353
—— 1981 *Phys. Rev.* A **24** 619
Greene C H and Jungen C 1985 *Adv. At. Mol. Phys.* **21** 51
Harmin D A 1981 *Phys. Rev.* A **24** 2491
Herzberg G and Jungen C 1972 *J. Mol. Spectrosc.* **41** 425
Janik G R, Mullins O C, Mahon C R and Gallagher T F 1987 *Phys. Rev.* A **35** 2345
Jungen C and Atabek O 1977 *J. Chem. Phys.* **66** 5584
Jungen C and Dill D 1980 *J. Chem. Phys.* **73** 3338
Muller A, Belic D S, DePaola B D, Djuric N, Dunn G H, Mueller D W and Timmer C 1987 *Phys. Rev.* A **36** 599
O'Neil S V and Reinhardt W 1978 *J. Chem. Phys.* **69** 2126
Sakimoto K 1986 *J. Phys. B: At. Mol. Phys.* **19** 3011
——1987 *J. Phys. B: At. Mol. Phys.* **20** 807
Seaton M J 1983 *Rep. Prog. Phys.* **46** 167
Zimmerman M L, Littman M G, Kash M M and Kleppner D 1979 *Phys. Rev.* A **20** 2251

1986 *Phys. Rev. Lett.* **57** 2931–4
Reprinted with permission from the American Physical Society

Quadratic Zeeman Effect for Nonhydrogenic Systems: Application to the Sr and Ba Atoms

P. F. O'Mahony and K. T. Taylor
Royal Holloway and Bedford New College, University of London, Egham, Surrey TW20 0EX, England
(Received 2 September 1986)

A method has been developed to calculate the spectrum over the l- and n-mixing regimes of an arbitrary nonhydrogenic system in an external laboratory-strength magnetic field. The original experimental results on Sr and Ba are explained in detail for the first time. The importance of quadratic Zeeman spectroscopy in the quantitative study of atomic structure is made clear.

PACS numbers: 32.60.+i, 31.50.+w

Garton and Tomkins,[1] in 1968, performed the first experiments on atoms in highly excited Rydberg states in an externally applied magnetic field. These experiments have stimulated new developments in both theoretical and experimental atomic physics because an atom in an external magnetic field provides the simplest example of a system where two disparate symmetries compete with one another. These symmetries are the cylindrical symmetry of the applied field and the spherical one of the Coulomb field. This competition results in a Hamiltonian which is nonseparable in any coordinate system.

This Letter reports the first calculations on Rydberg states of nonhydrogenic atoms in a uniform magnetic field that permit a detailed comparison with experimental spectra over the l- and n-mixing regimes for any atom.

The approach followed has hinged on two physical realizations. The first is that for magnetic fields of laboratory strength (5–50 kG or $\beta = 10^{-6}$ to 10^{-5} in atomic units), the quadratic Zeeman potential $\frac{1}{8}\beta^2 r^2 \sin^2\theta$ can, up to some radius r_0 equal to several hundred bohrs about the nucleus, be completely ignored in comparison to the potentials arising from the much stronger electron-nucleus and electron-electron interactions. This reduces the description of this inner region in the problem to that appropriate to a conventional multielectron atom completely free of any externally imposed fields. The second realization is that in the outer region (which extends in radial coordinates from r_0 to ∞ and in which the quadratic Zeeman potential is crucially important), only one electron of the atom will ever be found. Thus the Schrödinger equation for the complex atom in this outer region reduces to the one-electron form already investigated for hydrogen. These two physically distinct regions will now each be discussed separately in some detail.[2]

Since the magnetic field potential is a negligible term in the Hamiltonian over the inner region $0 \leq r \leq r_0$, we can acquire wave-function solutions to the Schrödinger equation there such that each solution corresponds to a fixed specific value of total orbital angular momentum L and total spin S. To progress further, this inner region must be divided into two subregions. The innermost of these, besides encompassing all but one of the atomic electrons, must also be sufficiently large so that low-lying excited configurations of the remaining electrons are entirely contained within it. Over the remaining part of the inner region the single electron present must, even for a fixed L, be described by a multicomponent wave function reflecting the various possible levels of excitation of the residual electrons of the atom. However, by the boundary of this region all but those components linked to the ground state of the residual ion will be considered to have undergone exponential decay and died away. For Sr and Ba atoms, the ground state of the residual positive ion is ^2S and so for a given L there remains only a single component corresponding to a specific value l of the orbital angular momentum of the single electron.

Near the boundary $r = r_0$, where only the Coulomb potential has significance, this single component can be written as a linear combination of energy-normalized regular and irregular Coulomb functions, $f_l(r)$ and $g_l(r)$, respectively; viz.,

$$P_l(r) = f_l(r) + \tan(\pi\mu_l) g_l(r) \tag{1}$$

The quantity μ_l is the energy-dependent quantum defect and parametrizes the energy-dependent coupling between this component and those that die off exponentially before r_0 is reached. Generally for $l > 4$, μ_l is approximately zero.

It is appropriate to calculate solutions in the inner region for all values of L allowed by parity considerations, each time gaining a component of the form (1) on the outer boundary. Before long, however, there is some L_{max} with a corresponding l_{max} in (1) such that for $l > l_{max}$ the angular momentum barrier is so large that the regular Coulomb function $f_l(r)$ evaluated at $r = r_0$ is completely negligible. This provides a valuable restriction on the amount of inner-region calculation necessary.

The true inner-region wave function that matches on to the wave function in the outer region is composed from all these differing l contributions. A vector $\mathbf{F}$ of radial parts can thus be written such that

$$\mathbf{F} = \mathbf{P} \cdot \mathbf{A}, \tag{2}$$

where $\mathbf{P}$ is a diagonal matrix with elements $P_l(r)$ of Eq. (1) and $\mathbf{A}$ is a vector of numbers specifying the relative weighting of each $P_l(r)$ in the true wave function.

This weighting is governed by the direct action of the magnetic field potential in the outer region where the problem has been reduced to one electron moving under the combined influence of Coulomb and magnetic fields. Thus previous work on the hydrogen atom is relevant. The Hamiltonian over the region $r_0 \leq r \leq \infty$ takes a form identical to that for hydrogen which, in atomic units, is

$$H = -\tfrac{1}{2}\nabla^2 - r^{-1} + \beta l_z + \tfrac{1}{2}\beta^2 r^2 \sin^2\theta, \tag{3}$$

where $\beta = ehB/2mc$ and the magnetic field vector **B** is chosen to point along the z axis. This Hamiltonian is symmetric with respect to coordinate inversion and so parity is a good quantum number. In addition, since the potential does not depend on the azimuthal angle ϕ, it is also diagonal in the magnetic quantum number m.

In solving the corresponding Schrödinger equation in this outer region, we have chosen a method which is a variant of the R-matrix approach first employed in atomic physics by Burke, Hibbert, and Robb.[3] This rests on the ability of a finite number of suitably chosen basis functions to represent the wave function adequately over a limited region of space. In our work such basis functions have been obtained by diagonalization of the Hamiltonian (plus a surface term that arises because integration is confined to the outer region) over a set of linearly independent Sturmian functions $S^{\xi}_{nl}(r)$ satisfying *arbitrary* boundary conditions at $r = r_0$.

The basis functions have energies E_k and are denoted by ψ_k, where

$$\psi_k = \sum_{nl} C^k_{nl} S^{(\xi)}_{nl}(r) Y_{lm}(\theta,\phi) \tag{4a}$$

$$= \sum_l d_{lk}(r) Y_{lm}(\theta,\phi). \tag{4b}$$

The energies E_k are the eigenvalues and C^k_{nl} the eigenvector components resulting from the diagonalization. In turn the Sturmian functions, already exploited advantageously by Clark and Taylor[4] in the hydrogen-atom magnetic field problem, are such that for a given orbital angular momentum quantum number l each satisfies the differential equation

$$\left[\frac{d^2}{dr^2} - \frac{l(l+1)}{r^2} + \frac{\xi n}{r} - \frac{\xi^2}{4}\right] S^{(\xi)}_{nl}(r) = 0, \tag{5}$$

where n is an integer label with $n > l$ and ξ is a positive constant common to the basis.

A direct result of the R-matrix procedure (see Burke and Robb[5]) is that the R matrix, which relates the vector **F** of radial wave-function components given by Eq. (2) to the radial derivative of this vector on the boundary $r = r_0$, viz.,

$$\mathbf{F}(r_0) = \mathbf{R} \cdot \mathbf{F}'(r_0), \tag{6}$$

can be expressed as

$$R_{ll'} = \sum_k \frac{d_{lk}(r_0) d_{l'k}(r_0)}{E_k - E}, \tag{7}$$

where E is the total energy of the single last electron in the outer region and l_{max} provides the upper limit on both l and l'. It should be noted that Eq. (7), for a given field strength, is *independent* of the complexity of the atom in the inner region and once calculated can be used for an arbitrary atom.

This approach for the outer region has allowed us to use there—with minimal amendment—the computer programs written by Clark and Taylor[4] to examine the l- and n-mixing regimes in hydrogen. Some alterations result from the need to evaluate matrix elements of the Hamiltonian (3) over the limited radial range from r_0 to ∞ rather than from 0 to ∞. This breaks the selection rules on label n enjoyed by matrix elements of $1/r$ and r^2 in the Sturmian set. Nevertheless the new nonzero integrals are readily evaluated analytically and the overall banded structures of the Hamiltonian and overlap matrices remain, albeit with somewhat wider bandwidths than in the hydrogen problem. Moreover the limited radial range means that a smaller number of Sturmian functions is needed. Full details of the methods for these new integrals will be presented in a future publication.

Once calculations have been performed in the two regions there remains the problem of joining wave functions across the one-electron boundary. This is achieved by use of Eq. (6) above. Solution and derivative vectors $\mathbf{F}(r_0)$ and $\mathbf{F}'(r_0)$ are provided on the boundary from the inner-region side while expression (7) for the R matrix means that this depends entirely on the outer-region calculation. Thus Eq. (6) acts as a matching condition on the boundary between the regions and will be satisfied only at certain discrete energies E. On substituting (2) in (6) we obtain

$$(\mathbf{P} - \mathbf{R} \cdot \mathbf{P}') \cdot \mathbf{A} = 0. \tag{8}$$

The discrete energies are therefore those at which

$$\det(\mathbf{P} - \mathbf{R} \cdot \mathbf{P}') = 0, \tag{9}$$

and the vector **A** is also given by solution of (8) at each such energy.

In this matching of wave functions on the boundary $r = r_0$ we have been able to take full advantage of the efficient energy search algorithm due to Seaton[6] and the associated computer program. In Seaton's program, and indeed in all applications of the R-matrix method before this one, the R-matrix form (7) has resulted from calculations in the *inner* region surrounding the nucleus, and the vector (2) has been supplied by the *outer*-region calculation. The inversion of roles has necessitated some changes to Seaton's program but not of a fundamental nature.

For each wave function obtained and properly normalized, the coefficient A_1 gives the fraction of $L = 1$ character in it. The oscillator strength linking any such state to an $L = 0$ ground state is then simply proportional to $|A_1|^2$ times the field-free oscillator strength at the ener-

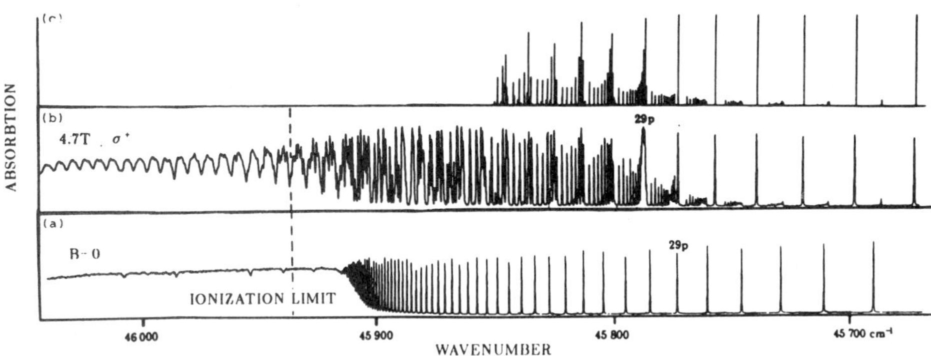

FIG. 1. Strontium absorption spectrum vs transition wave number from the ground state: (a) field-free spectrum; (b) experimental densitometer tracing for strontium in a magnetic field of 4.7 T; (c) theoretical photoabsorption spectrum in a field of 4.7 T. The theoretical results give the absolute oscillator strengths, but for n less than 29 the strongest lines have been reduced in size to facilitate comparison with the nonabsolute experimental measurements.

gy E of that state.

The theory outlined above has, in this first application, been directed towards the heavy complex atoms Sr and Ba for which magnetic field experimental data exist. We have used experimental energies[7] for field-free states in $^1P^\circ$ and $^1F^\circ$ symmetries to determine the relevant μ_l's in (1) and experimental values[8] for oscillator strengths of field-free discrete transitions from the ^1S-component ground states of these atoms. In general, and especially for light atoms, these data can be provided by *ab initio* calculations using standard methods. This is all that is required to determine the spectrum in a laboratory magnetic field of arbitrary strength.

The calculated results are compared with the experimental measurements of Lu, Tomkins, and Garton[9] on Sr and Ba in Figs. 1 and 2, respectively. The striking differences between the Sr and Ba spectra in the presence of a magnetic field has long been a source of puzzlement. [Their field-free spectra are quite similar in the energy range as evidenced by the frames of Figs. 1(a) and 2(a).] We find that the differences are due to the perturbers "$4d5p$" in Sr and "$5d8p$" in Ba. In zero field the quantum defects of the $5snp$ and $5snf$ series in Sr each rise through unity as a result of the $4d5p\,^1P^\circ,^1F^\circ$ resonances. However this rise occurs predominantly at low n and the quantum defects of the $5snp$ and $5snf$ series are slowly varying at $n \approx 29$, being approximately 0.8 and 0.1, respectively. This results in the $5snp$ and $5snf$ levels being separated from the higher l's in the n manifold. These levels, in the presence of a magnetic field, can be associated with the strongest lines in the spectrum of Fig. 1(b) for $n < 29$.

In contrast, the $5d8p$ perturber in Ba lies in the vicinity of the ionization threshold and causes the zero-field quantum defects of the $6snp$ and $6snf$ series to change with energy in this region. Surprisingly, however, these

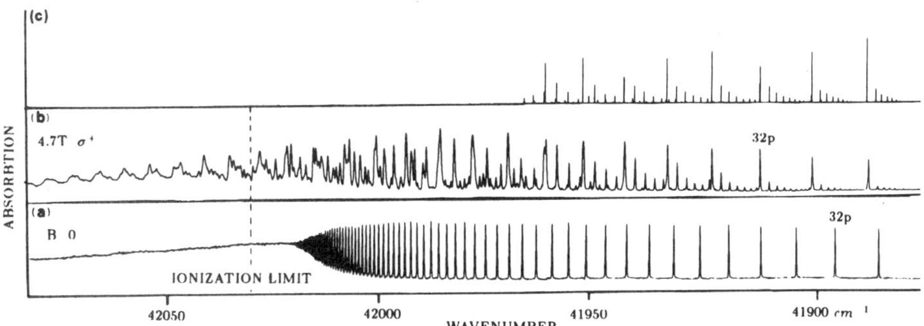

FIG. 2. Barium absorption spectrum vs transition wave number from the ground state: (a) field-free spectrum; (b) experimental densitometer tracing for barium in a magnetic field of 4.7 T; (c) theoretical photoabsorption spectrum in a field of 4.7 T.

quantum defects are both approximately zero (modulo one) in the energy range of interest around $n \sim 32$. This fortuitous coincidence explains the hydrogenic-type pattern that is observed in Ba in the presence of a magnetic field [Fig. 2(b)] in the inter-l- and inter-n-mixing regions.

The spectrum of Sr in the presence of a magnetic field should, however, be more typical of an arbitrary atom since the Ba spectrum results from an unusual coincidence in value of two important quantum defects.

In passing we point out that as $n = 29$ is approached in Sr, our calculations indicate a much more drastic redistribution of oscillator strength among levels than do the vapor experiments. This has important consequences for the use of magnetic rotation spectroscopy in measuring oscillator strengths.[10]

Some qualitative understanding of the spectra of Sr and Ba near $E = 0$ and above the ionization threshold exists but as yet no reliable theoretical approach has been developed to treat this region. The method worked out here for photoabsorption in nonhydrogenic systems should be readily applicable to photoionization in nonhydrogenic systems once a method has been found to handle the photoionization of hydrogen in a magnetic field.

In conclusion, we have described a general method to treat any nonhydrogenic system in laboratory magnetic fields of arbitrary strength. We have, for the first time, obtained excellent agreement with the original experiments of Garton and Tomkins on Sr and Ba. In doing so we have revealed in detail how the quadratic Zeeman spectroscopy of atoms can indirectly provide valuable information on symmetries (such as the $^1F^\circ$ in these cases) that are inaccessible in field-free single photoabsorption by ground-state atoms. We believe that our new theoretical understanding turns this spectroscopy into a quantative technique.

This work was supported in part by a research grant from the Science and Engineering Research Council, United Kingdom.

---

[1]W. R. S. Garton and F. S. Tomkins, Astrophys. J. **158**, 839 (1968).

[2]This approach is similar in spirit to that used by D. A. Harmin, Phys. Rev. Lett. **49**, 128 (1982), for atoms in external electric fields. However, he was able to utilize the separability of the Stark problem in parabolic and spherical coordinates. No comparable separability holds in the magnetic case.

[3]P. G. Burke, A. Hibbert, and W. D. Robb, J. Phys. B **4**, 153 (1971).

[4]C. W. Clark and K. T. Taylor, J. Phys. B **15**, 1175 (1982).

[5]P. G. Burke and W. D. Robb, Adv. Atom. Molec. Phys. **11**, 143 (1975).

[6]M. J. Seaton, J. Phys. B **18**, 2111 (1985).

[7]Sr energy levels: I. R. Rubbmark and S. A. Borgstrom, Phys. Scr. **18**, 196 (1978); M. A. Baig and J. P. Connerade, J. Phys. B **17**, L271 (1984). Ba energy levels: B. H. Post, W. Vassen, W. Hogervorst, M. Aymer, and O. Robaux, J. Phys. B **18**, 187 (1985); W. R. S. Garton and F. S. Tomkins, Astrophys. J. **158**, 1219 (1969).

[8]Sr oscillator strengths: W. R. S. Garton, J. P. Connerade, M. A. Baig, J. Hormes, and B. Alexa, J. Phys. B **16**, 389 (1983). Ba oscillator strengths: W. H. Parkinson, E. M. Reeves, and F. S. Tomkins, J. Phys. B **9**, 157 (1976).

[9]K. T. Lu, F. S. Tomkins, and W. R. S. Garton, Proc. Roy Soc. London, Ser. A **362**, 421 (1978).

[10]J. P. Connerade, private communication.

1987 *Phys. Rev.* A **36** 5467
Reprinted with permission from the American Physical Society

Erratum

Erratum: Interaction between a Rydberg atom and neutral perturbers [Phys. Rev. A 36, 971 (1987)]

Ning Yi Du and Chris H. Greene

In our paper Eqs. (9) and (10) were modified incorrectly. Thus, Eq. (9) should be replaced by

$$\underline{K}=[\cos(\pi\underline{\eta})\underline{\bar{K}}+\sin(\pi\underline{\eta})][\cos(\pi\underline{\eta})-\sin(\pi\underline{\eta})\underline{\bar{K}}]^{-1}\ , \tag{9}$$

and Eq. (10) should be replaced by

$$\underline{K}^{\text{eff}}=[\cos(\pi\underline{\eta})\underline{K}'+\sin(\pi\underline{\eta})][\cos(\pi\underline{\eta})-\sin(\pi\underline{\eta})\underline{K}']^{-1}\ . \tag{10}$$

All of our calculations were based on these correct equations.

1987 *Phys. Rev.* A **36** 971–4
Reprinted with permission from the American Physical Society

Interaction between a Rydberg atom and neutral perturbers

Ning Yi Du and Chris H. Greene
Department of Physics and Astronomy, Louisiana State University, Baton Rouge, Louisiana 70803
(Received 6 February 1987)

A simple combination of quantum defect theory with a nonperturbative Fermi-type analysis is shown to reliably determine a class of molecular Born-Oppenheimer potential curves. Applications to alkali-metal-atom–rare-gas-atom dimers verify that their potential curves are determined to semiquantitative accuracy using two pieces of information: the alkali-atom quantum defects and the electron–rare-gas scattering phase shifts.

In quantum mechanics one is often faced with a system composed of two or more simpler subsystems, each of which is separately understood. A fundamental question of some practical importance arises naturally: To what extent can the properties of the composite system be predicted using only the scattering properties of the simpler subsystems, i.e., without solving the full Schrödinger equation for all particles? In an early effort of this type, Fermi obtained a simple formula[1] for the energy levels of a Rydberg atom $A(nlm)$ in the presence of a neutral perturber B located on the quantization axis at $\mathbf{R}=R\hat{\mathbf{z}}$, namely,

$$E_{nlm}(R)=E_{nl}^{(A)}-2\pi L\,|\,\psi_{nlm}(\mathbf{R})\,|^2 \ . \tag{1}$$

Here $L=\lim_{k\to 0}(\tan\delta_0/k)$ is the s-wave e-B scattering length, and $E_{nl}^{(A)}$ and $\psi_{nlm}(\mathbf{R})$ are the unperturbed energy and wave function for the Rydberg electron moving in the field of the ion core A^+.

The simplicity of Fermi's treatment contrasts sharply with the best methods used currently.[2,3] Pascale's formulation[2] starts, for instance, by finding elaborate l-dependent pseudopotentials which can reproduce the e-A^+ quantum defects and the low-energy e-B scattering phase shifts as accurately as possible. After the pseudopotentials are determined, the full one-electron Hamiltonian is diagonalized at each R in some large variational basis set. While this approach is less transparent than Fermi's, it has accounted quantitatively for many experimental features observed in alkali-metal–rare-gas systems which cannot be explained by Eq. (1), including the alkali absorption spectrum in the presence of a perturbing buffer gas.[4]

The aim of this Rapid Communication is to show how a treatment approaching the simplicity of Fermi's describes molecular potential curves (or their equivalent adiabatic quantum defects) surprisingly well. It adopts much of the philosophy given in the preceding paragraph; the main differences are the following. (i) The pseudopotential representing the e-B interaction is simplified to such an extreme (Dirac δ function) that its Schrödinger equation is exactly solvable. (ii) By using quantum defect theory to represent the Rydberg electron wave function in the field of the ion A^+, the need for the e-A^+ pseudopotential is eliminated altogether. (iii) The final results are expressed in terms of a smooth R-dependent reaction matrix which directly describes scattering processes $l\to l'$ caused by the perturber.

The only *approximate* feature in this outline is the use (i) of an unrealistically oversimplified e-B pseudopotential. If highly accurate wave functions and molecular potential curves are needed for any particular application, it is straightforward to replace the approximation (i) by a variational R-matrix treatment[5] of a physically reasonable e-B interaction, such as that of Ref. 2. Nevertheless, we show below that the crude approximation (i), which is far simpler and faster to implement numerically, successfully describes most features of the potential curves.

In the following, we consider the simplest such problem: an alkali atom A interacting with a single rare-gas perturber B. Prior to imposing boundary conditions at large electron distances $r\to\infty$, a set of independent multichannel wave functions is characterized by a real, symmetric body-frame reaction matrix $K_{ll'}(R)$,

$$\psi_{l'}(\mathbf{r},R)=\sum_l Y_{l\lambda}(\hat{\mathbf{r}})[f_l(r)\delta_{ll'}-g_l(r)K_{ll'}(R)] \ . \tag{2}$$

In Eq. (2), (f_l,g_l) are the usual energy-normalized Coulomb wave functions[6] evaluated at the appropriate electronic energy $\varepsilon(R)$, while $\mathbf{r}$ is the electron position relative to the positive ion A^+. The perturber B is located on the body-frame quantization axis at $\mathbf{R}=R\hat{\mathbf{z}}$ relative to A^+, whereby λ is conserved. This expression is essentially exact for $r>R+\max(r_A,r_B)$, with r_B the radius of the perturber and r_A the radius of A^+. The reaction matrix depends on λ and weakly on $\varepsilon(R)$. Once it is known, the discrete electronic energies $\varepsilon(R)$ are the roots of a determinantal equation

$$\det[\tan(\pi\nu)\delta_{ll'}+K_{ll'}(R)]=0 \ ,$$

where $\nu\equiv[-2\varepsilon(R)]^{-1/2}$, *which ensures that the bound-state wave function decays exponentially at large r.* Denoting the eigenvalues of $K_{ll'}(R)$ by $\tan[\pi\mu_\alpha(R)]$, the allowed energies are thus quantized in terms of a positive integer n by a Rydberg formula for each eigenchannel (in a.u.):

$$\varepsilon_{n\alpha}(R)=-\frac{1}{2[n-\mu_\alpha(R)]^2} \ . \tag{3}$$

At this point, we introduce the ansatz that the e-B interaction potential can be approximated by

$$V_{e\text{-}B}(\mathbf{r})=V_0\delta(\mathbf{r}-\mathbf{R}) \ . \tag{4}$$

This potential represents only that part of the interaction which affects the s-wave component ($l_0=0$) of e-B scattering. (A small p-wave correction can be non-negligible at $R\lesssim 10$ a.u. and will be added perturbatively below.) Our derivation of the reaction matrix consists of two parts. First, the electronic Schrödinger equation will be solved in integral form for an electron moving in the joint potentials of A^+ and of B [Eq. (4)]. This can be solved in closed form after using a truncated partial-wave expansion $\psi_{l'}=\sum_l Y_{l\lambda}(\hat{\mathbf{r}})M_{ll'}(r)$, for the l'-th independent solution. The coupled integral equations satisfied by $M_{ll'}(r)$ have the form

$$M_{ll'}(r)=F_l(r)\delta_{ll'}+\sum_{l''}\int dr'\mathcal{G}_l(r,r')V_{ll''}(r')M_{l''l'}(r') , \quad (5)$$

with

$$V_{ll''}(r')=(V_0/R^2)Y_{l\lambda}^*(\hat{\mathbf{R}})Y_{l''\lambda}(\hat{\mathbf{R}})\delta(r'-R) . \quad (6)$$

The atomic (A) radial Green's function

$$\mathcal{G}_l(r,r')=\pi F_l(r_<)G_l(r_>)$$

is given in terms of quantum-defect-shifted Coulomb functions for $r>r_A$ which are, respectively, regular and irregular when continued back to the ionic origin,

$$F_l(r)=f_l(r)\cos(\pi\eta_l)-g_l(r)\sin(\pi\eta_l) ,$$
$$G_l(r)=f_l(r)\sin(\pi\eta_l)+g_l(r)\cos(\pi\eta_l) . \quad (7)$$

The experimental atomic quantum defects η_l are then taken from standard references.

The coupled equations (5) can be solved trivially, giving a simple reaction matrix $\bar{K}$ relative to the atomic radial solutions (F_l,G_l):

$$\bar{K}_{ll'}(R)=-\frac{CF_l(R)Y_{l\lambda}^*(\hat{\mathbf{R}})F_{l'}(R)Y_{l'\lambda}(\hat{\mathbf{R}})}{1-C\sum_{l''}|Y_{l''\lambda}(\hat{\mathbf{R}})|^2F_{l''}(R)G_{l''}(R)} , \quad (8)$$

where $C=\pi V_0/R^2$. This result is one of the key formulas of the present study. The numerator of Eq. (8) is just the first-order Born approximation to $\bar{K}_{ll'}$. Inclusion of higher-order scattering processes is thus seen to rescale every matrix element by the same constant (which depends on ε and R). The resulting matrix (8) is readily transformed into the more usual (f_l,g_l) representation of Eq. (2) through the relation

$$\underline{K}=[\underline{\bar{K}}+\tan(\pi\underline{\eta})][\underline{I}-\tan(\pi\underline{\eta})\underline{\bar{K}}]^{-1} , \quad (9)$$

in which $\underline{\eta}$ is a diagonal matrix containing the atomic quantum defects. The dimension of $\underline{K}$ is $(l_{\max}+1)\times(l_{\max}+1)$ $(\nu<l_{\max}\leq\nu+1)$. For any $l_{\max}\geq n_a$ (n_a being the number of nonzero atomic quantum defects), there are only n_a+1 nonzero eigenvalues $\tan(\pi\mu_a)$ of $\underline{K}$,[7] which can therefore be found by diagonalizing an effective reaction matrix $\underline{K}^{\text{eff}}$ of dimension $(n_a+1)\times(n_a+1)$:

$$\underline{K}^{\text{eff}}=[\underline{I}-\tan(\pi\underline{\eta})\underline{K}']^{-1}[\underline{K}'+\tan(\pi\underline{\eta})] . \quad (10)$$

Here an auxiliary matrix $\underline{K}'$ has been introduced such that $K'_{ll'}\equiv\bar{K}_{ll'}$ for $l,l'<n_a$, and where

$$K'_{n_a,l}=K'_{l,n_a}\equiv\left[\sum_{l'(\geq n_a)}|a_{l'}|^2\right]^{-1/2}\sum_{l''(\geq n_a)}a_{l''}\bar{K}_{l''l} ,\ (l<n_a) ,$$
$$K'_{n_a,n_a}\equiv-\text{sgn}(D)\sum_{l(\geq n_a)}|a_l|^2 . \quad (11)$$

The remaining quantities in Eq. (11) are

$$a_l=|D|^{1/2}F_l(R)Y_{l\lambda}(\hat{\mathbf{R}}) ,$$
$$D=(\pi V_0/R^2)\left[1-(\pi V_0/R^2)\sum_{l''}|Y_{l''\lambda}(\hat{\mathbf{R}})|^2F_{l''}(R)G_{l''}(R)\right]^{-1} . \quad (12)$$

The ε- and R-dependent constant V_0 is now determined by a consistency requirement. The net phase change induced in the wave function close to the perturber must equal the experimental s-wave phase shift $\delta_{l_0=0}$ experienced by a *free* electron which encounters $V_{eB}(\mathbf{r})$. The energy of this free electron is the kinetic energy at $\mathbf{R}$,

$$T_e\equiv\tfrac{1}{2}k^2=\varepsilon(R)+1/R . \quad (13)$$

The desired relation is found by solving Eq. (5) again, only using free-particle solutions,

$$f_l^{\circ}=(2k/\pi)^{1/2}rj_l(kr) ,$$
$$g_l^{\circ}=(2k/\pi)^{1/2}rn_l(kr) ,$$

in place of the atomic (F_l,G_l), and, of course, using the free-particle Green's function

$$\mathcal{G}_l(r,r')=\pi f_l^{\circ}(r_<)g_l^{\circ}(r_>) .$$

This procedure yields a free-electron reaction matrix having the structure (8). Its lone nonzero eigenvalue must coincide with the experimental $\tan\delta_{l_0=0}$ for e-B scattering,

$$\tan\delta_0=\frac{-C\sum_l|f_l^{\circ}(R)Y_{l\lambda}(\hat{\mathbf{R}})|^2}{1-C\sum_l|Y_{l\lambda}(\hat{\mathbf{R}})|^2f_l^{\circ}(R)g_l^{\circ}(R)} . \quad (14)$$

Equation (14) thus determines $V_0=CR^2/\pi$ analytically in terms of the known e-B phase shift for each choice of R and of $\varepsilon(R)$, without any matrix manipulations.

Several details of the above procedure deserve elaboration. First of all, it is essential that the partial wave expansions be truncated to the same $l_{\max}$ in the free-particle expression (14) as in the atomic expression (8). In fact, the summations in the denominators of Eqs. (8) and (14) are formally divergent if the summations are carried out to $l \rightarrow \infty$. This reflects the fact that a Dirac δ function is too highly singular to give an analytically solvable Schrödinger equation in *three* dimensions. Despite this singularity, stable and converged results are obtained when the same $l_{\max}$ is used in both calculations. Our treatment is in some sense a "renormalization" which can produce accurate reaction matrices even though it deals with a highly singular δ-function potential. A second point concerns the dominance of the $l_0 = 0$ partial wave in the e-B scattering amplitude. At low k values, the s-wave phase shift is expected to predominate, but the Coulombic contribution to the electron kinetic energy can be several volts at small R values, and the p-wave phase shift can be non-negligible. Accordingly, if $l_0 = 1$ (or higher partial waves) are important, they are easily handled in first-order perturbation theory.[8] This amounts to simply adding the following correction to the right-hand side of Eq. (8):

$$\Delta \bar{K}_{ll'} = \frac{6\pi}{k^3} \tan\delta_{l_0=1} \{\nabla[F_l(r) Y^*_{l\lambda}(\hat{\mathbf{r}})] \cdot \nabla[F_{l'}(r) Y_{l'\lambda}(\hat{\mathbf{r}})]\}_{\mathbf{r}=\mathbf{R}} \, . \qquad (15)$$

It should also be pointed out that whenever $R > -1/\varepsilon(R)$, the value of k becomes imaginary, since $T_e < 0$. This causes some difficulty, since the e-B scattering phase shifts are normally available only at positive energies. But since $k^{-1}\tan\delta_{l_0=0}$ is smooth near $T_e = 0$ it can be simply extrapolated to (or calculated *ab initio* at) negative energies. It should be stressed that high accuracy is not needed in this extrapolation, since each potential curve rapidly approaches its asymptotic atomic limit in this region anyway. The behavior of the p-wave correction (15) is slightly more problematic, since $k^{-2l_0-1}\tan\delta_{l_0}$ diverges as $T_e \rightarrow 0$ for $l_0 \geq 1$ because of the long-range polarization field.[9] This divergent piece is appreciable only in the immediate vicinity of $k = 0$ [i.e., where $R = -1/\varepsilon(R)$], and it can in practice be simply subtracted. The localized divergence is clearly unphysical since it stems from the most distant long-range contributions to the phase shift, whereas the actual electronic wave function decays exponentially at large distances.

Figure 1(a) compares some of the resulting ($\lambda = 0$) CsHe potential curves with Pascale's variational pseudopotential calculation, the best results obtained to date for this dimer. Note the correct representation of many critical features, especially the two avoided crossings between $5d\sigma$ and $7s\sigma$ and between $7p\sigma$ and $6d\sigma$. The avoided crossing between $5d\sigma$ and $7s\sigma$ is apparently particularly sensitive; it did not show up in the first (l-independent) pseudopotential calculations, but its existence has been experimentally confirmed (see Fig. 1 of Dubourg, Ferray, Visticot, and Sayer[4]). [We also find that this crossing disappears if the first-order Born approximation is used in place of Eqs. (8) and (11).] The potential curves of NaHe, shown in Fig. 1(b), are also informative. The calculations of Ref. 2 (l-dependent pseudopotential) and of Ref. 10 (l-independent model potential) agree with each other and with our results reasonably well. Similar comparisons have been obtained for other alkali–rare-gas dimers.

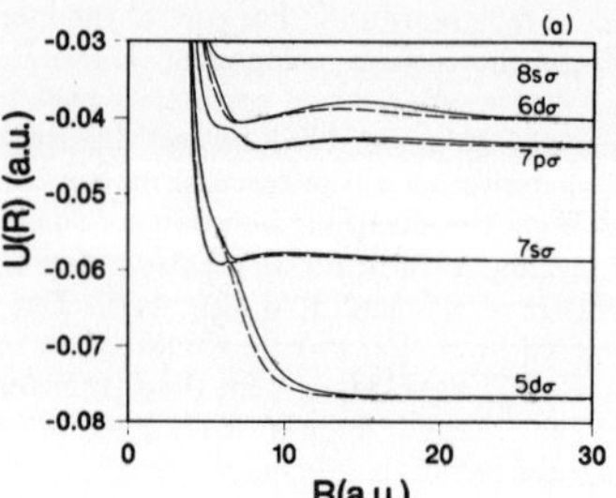

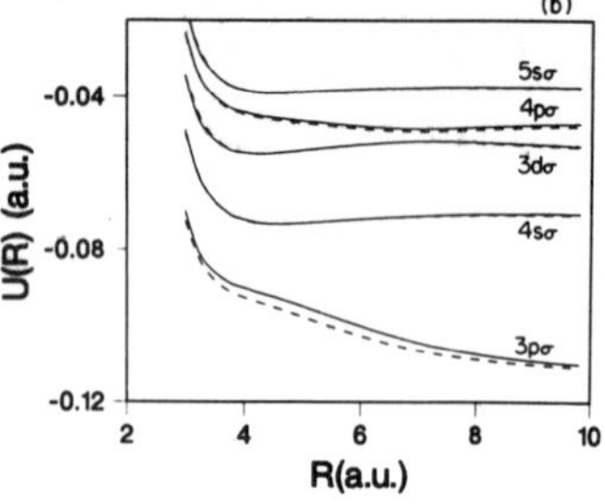

FIG. 1. Several Born-Oppenheimer potential curves resulting from our nonperturbative Fermi-type analysis (dashed curves) are compared to the best calculations of Pascale (smooth curves, from Ref. 2) for (a) CsHe and (b) NaHe. The curves shown include the potential for the alkali-ion–rare-gas system in addition to the energy $\varepsilon(R)$ of the valence electron.

The good agreement between elaborate potential curve calculations and this nonperturbative Fermi-type analysis suggests that two-body scattering properties can in fact be used to determine major features of the composite system. By expressing final results in terms of a body-frame reaction matrix $K_{ll'}(R)$, this treatment provides an immediate link to quantum-defect formulations,[6] thereby permitting a straightforward description of non-Born-Oppenheimer processes such as predissociation, l-changing collisions, and Penning ionization. This procedure outlined above can also be generalized without much difficulty to deal with a *multichannel* Rydberg atom (e.g., barium) which is perturbed by a rare-gas atom. The extension to handle *several* perturbing atoms simultaneously also appears to be clear-cut, and may prove useful for studying Rydberg states of clusters. On the other hand, treatment of *open-shell* perturbing atoms appears somewhat more difficult than the above description of rare-gas perturbers, and will require further development.

To date, numerous efforts to extend Fermi's treatment have been made (see, e.g., Refs. 3, 7, 9, and 11). The most relevant is the recent work of de Prunele,[7] which also determines the alkali-metal–rare-gas potential curves from the atomic quantum defects and the *e-B* scattering information. The theoretical formulation of de Prunele is substantially different, however, since in Ref. 7 the "smooth" Coulomb Green's function[6] is not used, and accordingly a "smooth" reaction matrix is not obtained. In our multichannel quantum defect theory treatment, the reaction matrix varies smoothly with energy (even across threshold), and it embodies all the information of channel interactions. Consequently, the calculation can be performed on a coarse energy mesh. By linking more directly to quantum-defect theory, processes such as dissociative recombination can also be treated using the *K* matrix obtained by the present method. Nevertheless, Ref. 7 has obtained results which seem very close to ours, and the two treatments should enjoy a comparable advantage over conventional calculations in simplicity and in ease of computation.

We acknowledge useful discussions with A. R. P. Rau. Access to the unpublished results of J. Pascale is also appreciated. This work was supported in part by the National Science Foundation and in part by the Alfred P. Sloan Foundation.

[1]E. Fermi, Nuovo Cimento **11**, 157 (1934).

[2]J. Pascale, Phys. Rev. A **28**, 632 (1983).

[3]M. Philippe, F. Masnou-Seeuws, and P. Valiron, J. Phys. B **12**, 2493 (1979); P. Valiron, A. L. Roche, F. Masnou-Seeuws, and M. E. Dolan, *ibid.* **17**, 2803 (1984).

[4]For a thorough review of research prior to 1982, see N. Allard and J. Kielkopf, Rev. Mod. Phys. **54**, 1103 (1982). Some other and more recent references are K. Ueda, J. Quant. Spectrosc. Radiat. Transfer **33**, 77 (1985); R. Kachru, T. Mossberg, and S. R. Hartmann, Phys. Rev. A **21**, 1124 (1980); I. Dubourg, M. Ferray, J. P. Visticot, and B. Sayer, J. Phys. B **19**, 1165 (1986).

[5]See, e.g., C. H. Greene, Phys. Rev. A **32**, 1880 (1985), and references therein.

[6]U. Fano and A. R. P. Rau, *Atomic Collisions and Spectra* (Academic, New York, 1986); M. J. Seaton, Rep. Prog. Phys. **46**, 167 (1983); C. H. Greene and Ch. Jungen, Adv. At. Mol. Phys. **21**, 51 (1985).

[7]E. de Prunelé, Phys. Rev. A **35**, 496 (1987).

[8]A. Omont, J. Phys. (Paris) **38**, 1343 (1977).

[9]T. F. O'Malley, L. Spruch, and L. Rosenberg, J. Math. Phys. **2**, 491 (1961).

[10]J. Hanssen, R. McCarroll, and P. Valiron, J. Phys. B **12**, 899 (1979).

[11]F. I. Dalidchik and G. K. Ivanov, Teor. Eksp. Khim. **8**, 6 (1972).

1984 *Phys. Rev.* A **30** 3321
Reprinted with permission from the American Physical Society

Errata

Erratum: General form of the quantum-defect theory. II [Phys. Rev. A 26, 2441 (1982)]

Chris H. Greene, A. R. P. Rau, and U. Fano

In the first line of Eq. (A2), $\exp\{\mp i\frac{1}{2}\pi[\nu \mp (\lambda_R+1)]\}$ should be replaced by $\exp\{i\frac{1}{2}\pi[\nu \mp (\lambda_R+1)]\}$.

In the second line of Eq. (A2), replace $e^{(1/2)\pi\alpha}$ by $e^{\pm(1/2)\pi\alpha}$.

The first factor on the right of Eq. (A7) should be $2^{-(\lambda_R+1)}$.

In the first line of Eq. (B4), replace tanh by coth.

We take this opportunity to record explicit formulas for $\mathcal{G}_\lambda$ and other QDT parameters in the presence of a dipole field, that is, when $\lambda = -\frac{1}{2}+i\alpha$. The procedure for this was given after Eq. (B10), but not worked out earlier. To form the QDT parameters $\mathcal{B}_{-\lambda-1}$ and $\delta_{-\lambda-1}$, one follows Appendix A except that in place of Eq. (3.9), one initially starts with the combination $[J(\lambda)-J(\lambda^*)]/(2i)$ of Jost functions. This finally results in certain changes in sign for the parameters relevant to $-\lambda-1$ from those for λ given in Tables I and II. For B and A in these tables, the sign of the terms involving a cosine is changed to minus. For the phase parameters η and β, an additional $\pi/2$ appears in the argument of the tangent function that stands at the end of the definitions of ϕ and $\bar{\phi}$. Finally, the parameter $\mathcal{G}$ as given by Eq. (B8) is

$$\mathcal{G} = -\frac{B\sin[z_\alpha - z_{-\alpha} - 2\alpha\ln 2k]}{(Z_\alpha^2 - Z_{-\alpha}^2)/2Z_\alpha Z_{-\alpha}}\ , \quad \epsilon > 0$$

$$= -\frac{A\sin 2(y-\alpha\ln 2\kappa)}{\sinh 2\pi\alpha}\ , \quad \epsilon < 0\ .$$

We would like to note two publications that have appeared since. One is a detailed application of QDT to diatomic predissociation which has considerable overlap with our treatment [Frederick H. Mies, J. Chem. Phys. **80**, 2514 (1984)]. Another is a comprehensive review of QDT [M. J. Seaton, Rep. Prog. Phys. **46**, 167 (1983)].

1982 *Phys. Rev.* A **26** 2441–59
Reprinted with permission from the American Physical Society

General form of the quantum-defect theory. II

Chris H. Greene and A. R. P. Rau
Department of Physics and Astronomy, Louisiana State University, Baton Rouge, Louisana 70803

U. Fano
Department of Physics, University of Chicago, Chicago, Illinois 60637
(Received 14 June 1982)

An earlier paper by the same title recast the treatment of an electron in the field of an ionic core [quantum-defect theory (QDT)] into a form largely independent of the character of the long-range field. Analytical complexities and limitations of that paper are eliminated here by further reformulation. Qualitative aspects are highlighted. A WKB analysis interprets the QDT parameters as phase integrals over certain ranges of the radial coordinate; the limitations of the WKB analysis are then removed through the Milne approach to wave equations. The QDT analysis is connected to Jost's formulation of scattering, extended to long-range fields by explicit treatment of the relevant singularities. The effects of interactions confined to an internal many-body core are represented by QDT parameters wholly independent of the external region.

I. INTRODUCTION

Quantum-defect theory[1,2] (QDT) served initially to exploit a characteristic feature of an electron excited into a discrete or continuum state of an atomic system: The electron moves prevalently *outside* the residual ionic core, under the simple influence of the core's Coulomb field. The motion in this external region, which determines most of the spectral and scattering observables of the excited system, can be described with reference to standard Coulomb eigenfunctions. The motion *within* the multiparticle ionic core is, instead, more complicated, but is confined within a small region of space of the order of a molecular diameter. Owing to this confinement, the mutual influence of the motions in the internal and external regions can be represented by *a few* QDT parameters only—akin to scattering lengths—which can be determined either by fitting to experimental data or by *ab initio* calculations restricted to the core. QDT predicts spectral and scattering data on the basis of (1) this connection at the core's boundary and (2) the connection between the amplitude and phase of the outer-field wave function at the core's edge and at $r\rightarrow\infty$.

These two connections differ characteristically in their dependence on the excitation energy: Since strong interactions prevail in the core, the connection parameters at the core's edge depend on energy on a rather coarse scale; that is, it is generally sufficient to evaluate them at intervals of ~ 1 eV. Far outside the core, on the other hand, the forces are weak and the wave-function behavior may depend very sensitively upon energy. The connection between the wave functions at $r\rightarrow\infty$ and at the core's edge has then to be studied analytically.

The success of the QDT program in spectroscopy and scattering suggests an extension of its approach, not only to the electron motion in non-Coulomb outer potentials but also to the motion of *any pair of fragments* released by the breakdown of a compound system. With this aim an earlier paper, to be referred to as I,[3] has developed the QDT analysis for long-range interactions proportional to $1/r^2$ ("dipole field") or constant ("zero field"). The former occurs notably in electron plus excited hydrogen and electron plus polar molecule combinations, the latter in electron plus neutral scattering or in photodetachment from negative ions. Paper I dealt briefly with more general potentials in its Sec. II E.

QDT procedures were then extended to the $1/r^4$ polarization potential and to the interatomic potential in molecular dissociation.[4,5] More recently a QDT procedure has been applied to photoabsorption in a Stark field whose long-range potential does not vanish at infinity.[6] We underscore this further evidence of the flexibility of QDT's procedure without treating it explicitly.

Paper I considered different long-range fields

separately. Its analysis and its choice of parameters remained rather cumbersome owing to limited appreciation of several relevant factors. It now seems worthwhile to return to the subject, redeveloping most of Secs. II and IV of I, while its Introduction and its analysis of relevant Green's-function properties (Sec. III) remain serviceable. Notation will follow historical precedents except where new considerations recommend departures to be noted explicitly.

The present paper intends to be self-contained except for occasional references to paper I. The structure of QDT and the problems faced in implementing it, particularly in the definition and construction of its parameters, are reviewed in Sec. II. The results of paper I for the QDT parameters of specific long-range interactions are rederived in Sec. III in a unified way and combined into Jost functions. A general treatment is then developed in Sec. IV, having in mind potentials that may be known only numerically. A WKB procedure serves to identify the QDT parameters in terms of phase integrals over classically allowed or forbidden ranges of r. A phase-amplitude treatment, free from approximations, affords then a constructive procedure for the parameters. Finally, Sec. V replaces Sec. IV of paper I, which combines the parameters of separate fragmentation channels in the external region into a single multichannel treatment. The combination involves the QDT parameters discussed in previous sections, as well as those that arise from the internal region and provide the connection at the core's boundary.

II. MAJOR ELEMENTS OF QDT

Our development rests on the following circumstances.

(a) The QDT parameters serve to interconnect different solutions of a Schrödinger equation, identified by their behavior at short and large radial distances, respectively. Scattering phase shifts η are thus defined at positive energies $\epsilon=\frac{1}{2}k^2$ by connecting a solution $f^0(r)$, regular at the origin, to a pair of solutions $f^{\pm}(r)$ having the basic structure $\exp(\pm ikr)$ at $r\rightarrow\infty$,

$$f^0(r)=B^{-1/2}(2/\pi k)^{1/2}(f^+e^{i\eta}-f^-e^{-i\eta})/2i$$
$$=B^{-1/2}f(r)\,. \qquad (2.1)$$

The function $f(r)$ is identified here as the real superposition of f^+ and f^- normalized per unit energy in atomic units and proportional to f^0.

The amplitude factor $B^{-1/2}$ arises as follows: The solution $f^0(r)$ of a single-particle Schrödinger equation is generally controlled at small r by the centrifugal term of the potential $l(l+1)/r^2$, which causes $f^0(r)\rightarrow r^{l+1}$ as $r\rightarrow 0$. Since the energy term of the equation is dwarfed here by $l(l+1)/r^2$, $f^0(r)$ is appropriately normalized independently of ϵ. The factor $B^{-1/2}$ of (2.1) serves thus to connect the normalizations of f^0 and of $f^{\pm}$—or, rather, of f—which are controlled by different considerations; B is generally a function of ϵ, as the phase η also is. As we will see, the two functions f^0 and f are each useful in its own particular context. The dependence on wave-function behavior at large distances, introduced into f through B, makes f^0 the more useful form when the focus is on dependences at small r. On the other hand, for an item in (d) and (e) below and in Appendix B, it is more convenient to work with f. [The specification of $f^0(r)$ at small r differs in the presence of an attractive potential $-a/r^p$ with $p\geq 2$, which dwarfs in turn the centrifugal potential and causes $f^0(r)$ to oscillate faster and faster as $r\rightarrow 0$. In this event $f^0(r)$ is identified adequately by an arbitrary, but energy-independent, standardization of its amplitude and phase at an appropriately small value of r (see, e.g., Ref. 4).]

(b) In the bound-state part of the spectrum we have $\epsilon<0$ and the corresponding wave number is imaginary, $k=i\kappa$, whereby $f^{\pm}$ has the essential form $e^{\mp\kappa r}$. The amplitude corresponding to $B^{-1/2}$ is indicated here by $A^{-1/2}$ and the complex phases $\exp(\pm i\eta)$ are replaced by real coefficients $(D\cos\beta,\ D^{-1}\sin\beta)$. The factors $D^{\pm 1}$ are monotonic functions of ϵ which adjust the scale of $f^{\pm}$, respectively, while the oscillating factors $(\cos\beta,\ \sin\beta)$ correspond more closely to $\exp(\pm i\eta)$ as we shall see. The counterpart of (2.1) is thus

$$f^0(r)=A^{-1/2}(\pi\kappa)^{-1/2}(\sin\beta\, D^{-1}f^- -\cos\beta\, Df^+)$$
$$=A^{-1/2}f(r)\,. \qquad (2.2)$$

Bound states occur at energies for which (2.2) remains finite at $r\rightarrow\infty$, as expressed by the condition that the coefficient of the rising exponential $f^-(r)$ vanishes,

$$\sin\beta=0\,. \qquad (2.3)$$

(c) In the presence of a multiparticle core confined to radial distances $r<r_0$, the radial function of an electron (or other fragment) excited out of the core is analogous to $f^0(r)$ at $r>r_0$. However, this function differs generally from f^0 by a phase shift determined by its normal logarithmic derivative at r_0. We represent the function with a phase shift δ^0 by

$$F^0(r) \propto f^0(r)\cos\delta^0 - g^0(r)\sin\delta^0 , \qquad (2.4)$$

where $g^0(r)$ is an independent solution of the same equation as $f^0(r)$, to be identified appropriately. The value of δ^0 is fitted to experimental data or calculated by solving the Schrödinger equation within the core and fitting (2.4) to the excited electron's wave function at the core's edge. The considerations that underlie the definition of $g^0(r)$ are central to the present paper.

(d) Briefly, the structure of Eq. (2.4) suggests that $g^0(r)$ should lag $f^0(r)$ in phase by 90°, in order that δ^0 combine additively with η in Eq. (2.1). However, neither f^0 nor g^0 oscillates in general as $r \to 0$. [This difficulty does not arise in the case noted at the end of (a) when a strong attractive field caused $f^0(r)$ to oscillate rapidly as $r \to 0$.] QDT traditionally identifies an oscillation 90° out of phase with reference to oscillatory behavior at $r \to \infty$ for $\epsilon > 0$. One might thus define $g^0(r)$ as $\propto B^{1/2}(f^+e^{i\eta}+f^-e^{-i\eta})$, where the + sign in the parentheses provides the desired phase lag and the exponent $\frac{1}{2}$ of B is adjusted to remove B from the Wronskian $W(f^0,g^0)$. However, the presence of the phase shift η in this tentative definition of $g^0(r)$ introduces an unwanted dependence of g^0 on phenomena occurring at large r and an equally unwanted sensitivity to the energy near the threshold $\epsilon = 0$. Accordingly one includes in $g^0(r)$ an additional term proportional to f^0, which does not contribute to $W(f^0,g^0)$, with a coefficient $\mathscr{G}(\epsilon)$ designed to remove the unwanted dependence on large-r and threshold effects, setting

$$\begin{aligned} g^0(r) &= -B^{1/2}(2/\pi k)^{1/2}[f^+(r)e^{i\eta}+f^-(r)e^{-i\eta}]/2 - \mathscr{G}f^0 = B^{1/2}g(r) - \mathscr{G}f^0 , \quad \epsilon > 0 \\ &= -A^{1/2}(\pi\kappa)^{-1/2}(\cos\beta\, D^{-1}f^- + \sin\beta\, Df^+) - \mathscr{G}f^0 = A^{1/2}g(r) - \mathscr{G}f^0 , \quad \epsilon < 0 . \end{aligned} \qquad (2.5)$$

[The function $g(r)$ is normalized like $f(r)$ in (2.1) and lags it by 90° at $r \to \infty$.]

It remains awkward to define a solution to be utilized at small r starting from its behavior at $r \to \infty$. There is indeed in general a second independent solution besides f^0 near the origin, but the linear independence fails at just the physical values of the orbital angular momentum as we shall see in Sec. III. This complication is well known in the theory of special functions but the standard remedy will be seen to lead back to Eq. (2.5).

(e) The definition of g^0, and the actual construction of the coefficient $\mathscr{G}$, are but a part of a broader problem that has emerged in the QDT development, namely, that an irregular solution of a wave equation is not readily identified because addition of a regular solution fails to modify it appreciably near the singular point. This difficulty of defining a function through boundary conditions at a point of singularity bears as well on the identification of f^- at $\epsilon < 0$. Inadequate appreciation of this fact has been troublesome in the past (see I, Appendix A); dealing with it occupies much of the present paper. Two different avenues will be pursued for this purpose in Secs. III and IV, namely, reliance on analytical forms of the initial Eqs. (2.1) and (2.2), where this form is available, and alternative approaches akin to the WKB analysis under more general circumstances.

(f) QDT treatments have been expressed in terms of the six parameters B, η, A, β, D, and $\mathscr{G}$, originally for the Coulomb field and in the extensions of paper I. More compactly, however, solutions specified at small r are connected to those specified at large r by Jost functions $J^{\pm}(\epsilon,l)$, which embody all those parameters,

$$f^0 = J^+f^+ + J^-f^- . \qquad (2.6)$$

Section III will concern itself particularly with $J^{\pm}$. The two Jost functions are related by the symmetry property[7]

$$J^-(k) = [J^+(k^*)]^* . \qquad (2.7)$$

For $\epsilon > 0$, i.e., k real, Eq. (2.7) amounts to the obvious complex conjugation symmetry of the two terms of (2.6), $J^- = J^{+*}$, but the meaning of (2.7) is nontrivial for $\epsilon < 0$.

The Jost functions pertaining to the wave function $F^0(r)$, Eq. (2.4), embody as well the additional parameter δ^0, which represents the effect of core interaction and is equivalent to the quantum defect $\bar{\mu}$, the namesake of QDT. The calculation of δ^0 is essential to QDT applications, of course, but is regarded as a problem of core dynamics to be pursued separately. The excitation of an electron (or other core fragment) by photoabsorption is similarly regarded as a core phenomenon to be studied separately; QDT itself takes into account the events at large radial distances subsequent to photoabsorption.

(g) We have dealt thus far with radial wave functions of a particle, or fragment, at $r > r_0$ pertaining to a *single* mode of excitation (i.e., to a "single channel"). In fact, QDT deals generally with multichannel situations. Paper I's approach to a multichannel process started from the close-coupling representation of a system's complete wave function $\Psi = \sum_i \{M_i(r)\Phi_i(\omega)\}$, and reduced its Schrödinger equation laboriously to Eq. (I.4.14) for a reaction matrix $K^{(s0)}$, which represents the effect of short-range interactions. It then directed attention to the eigenvalues $\tan\delta_\alpha^{(s0)}$ and eigenfunctions $U_{i\alpha}^{(s0)}$ of the matrix $K^{(s0)}$, which combine with the QDT parameters for the several excitation channels i to predict observable results. Viewing the problems of core dynamics as a separate problem, we shall indicate instead, in Sec. V, how an R-matrix treatment of the multiparticle interactions at $r \le r_0$ determines the parameters $\delta_\alpha^{(s0)}$ and $U_{i\alpha}^{(s0)}$ without reference to circumstances prevailing at large radial distances. These parameters serve then as input for QDT treatments of outer-region phenomena.

III. QDT PARAMETERS OF ANALYTIC WAVE FUNCTIONS

We consider here the QDT properties of solutions of the radial Schrödinger equation

$$\left[-\frac{1}{2}\frac{d^2}{dr^2} + \frac{\lambda(\lambda+1)}{2r^2} - \frac{b}{r} - \epsilon\right] f(\epsilon,\lambda,b,r) = 0 , \tag{3.1}$$

in atomic units. This equation applies to a free particle for $\{\lambda = l,\ b = 0\}$, to an electron in a Coulomb field for $\{\lambda = l,\ b = Z\}$, and to an electron in a combined Coulomb and dipole field for

$$\{\lambda(\lambda+1) = l(l+1) - a,\ b = Z\} .$$

Its analytic solutions thus contain all the elements developed in Sec. II of paper I, whose Eq. (I.2.1) is analogous to (3.1).

We depart from I, however, by normalizing the basic solution f^0 of (3.1) as

$$f^0(\epsilon,\lambda,b,r) \underset{r\to 0}{\longrightarrow} r^{\lambda+1} , \quad \lambda \ge -\tfrac{1}{2} \tag{3.2}$$

because the coefficient $a_0(\lambda)$ introduced in I proves unnecessary here. The normalization (3.2) does not apply for a strongly attractive dipole field, with $a > l(l+1) + \frac{1}{4}$, i.e., $\lambda(\lambda+1) < -\frac{1}{4}$, because λ becomes complex

$$\lambda = -\tfrac{1}{2} \pm [a - l(l+1)]^{1/2} = -\tfrac{1}{2} \pm i\alpha . \tag{3.3}$$

In this case the solutions of (3.1) oscillate at $r \to 0$, as noted in Sec. II at the end of (a); since $\lambda(\lambda+1) = \lambda^*(\lambda^*+1)$, we shall utilize the real standard solution

$$\tfrac{1}{2}[f^0(\epsilon,\lambda,b,r) + f^0(\epsilon,\lambda^*,b,r)] \underset{r\to 0}{\longrightarrow} r^{1/2}\cos(\alpha \ln r) . \tag{3.4}$$

[This convention differs from that of (I.2.5).]

The familiar solution of (3.1) with the boundary condition (3.2) is

$$f^0(\epsilon,\lambda,b,r) = r^{\lambda+1} e^{-br/\nu} F(\lambda+1-\nu, 2\lambda+2, 2br/\nu) , \tag{3.5}$$

for real values of λ, with

$$\nu = \begin{cases} ib/k , & \epsilon = \tfrac{1}{2}k^2 > 0 ; \quad (3.6a) \\ b/\kappa , & \epsilon = -\tfrac{1}{2}\kappa^2 < 0 . \quad (3.6b) \end{cases}$$

For complex values of λ, the combination (3.4) of solutions (3.5) applies.

The solutions $f^\pm$ introduced in Eq. (2.1) are characterized by their asymptotic form

$$f^\pm(\epsilon,\lambda,b,r) \underset{r\to\infty}{\longrightarrow} e^{\mp br/\nu} r^{\pm\nu} \tag{3.7}$$

with the notation (3.4) and for either sign of ϵ. The connection (2.6), or (2.1), between f^0 and $f^\pm$ takes now the explicit form

$$f^0(\epsilon,\lambda,b,r) = J^+(\epsilon,\lambda,b) f^+(\epsilon,\lambda,b,r) + J^-(\epsilon,\lambda,b) f^-(\epsilon,\lambda,b,r) ; \tag{3.8}$$

when λ is complex, $J^\pm$ is defined as

$$\tfrac{1}{2}[J^\pm(\epsilon,\lambda,b) + J^\pm(\epsilon,\lambda^*,b)] . \tag{3.9}$$

The Jost functions in (3.8), which are equivalent to the Wronskians utilized in paper I, have explicit expressions obtained by comparing Eq. (3.8) to the asymptotic representation at $r \to \infty$ of the confluent hypergeometric functions in (3.5),

$$J^{\pm}(\epsilon,l,b)=\pm W(f^{\mp},f^{0})/2ik$$

$$=\begin{cases}\dfrac{\Gamma(2\lambda+2)}{\Gamma(\lambda+1\pm\nu)}\left[\dfrac{1}{2k}\right]^{\lambda+1\mp\nu}e^{i(1/2)\pi[\nu\mp(\lambda+1)]}\,, & \epsilon>0 \quad (3.10a)\\[2ex] \dfrac{\Gamma(2\lambda+2)}{\Gamma(\lambda+1\pm\nu)}\left[\dfrac{1}{2\kappa}\right]^{\lambda+1\mp\nu}\begin{cases}\cos[\pi(\lambda+1-\nu)]\,,\\ 1\,,\end{cases} & \epsilon<0\,. \quad (3.10b)\end{cases}$$

Whereas (3.10a) explicitly satisfies the general analytic condition (2.7), it is clear that (3.10b), in general, does not. However, for all physically meaningful values of ν, i.e., at the bound-state eigenvalues when $\epsilon<0$, the relation (2.7) is indeed valid.

A. Connections of $J^{\pm}$ across the $\epsilon=0$ threshold

Equations (3.10) are connected, in the main, by analytical continuation of the various ϵ-dependent parameters, indicated by (3.6) for ν and by $k\rightarrow i\kappa$. Indeed the ratio of Γ functions maintains the same form as a function of ν. The factor $(1/2k)^{\lambda+1\mp\nu}$ spins off a coefficient $\exp[-i\frac{1}{2}\pi(\lambda+1\mp\nu)]$ as $k\rightarrow i\kappa$. This coefficient merges into the last factor on the right of Eq. (3.10), but the result of this merger is nontrivial for J^{+} and has given rise to misunderstandings in the literature as detailed in Appendix A of I. The key element here is that the function f^{0} should remain real. Accordingly the last factor of (3.10b), resulting from this merger, has been set at $\mathrm{Re}\{\exp[-i\pi(\lambda+1-\nu)]\}$ in the expression of J^{+} as it was in paper I. This nontrivial aspect of the connection from $\epsilon>0$ to $\epsilon<0$ relates to the branch point of $k=(2\epsilon)^{1/2}$ at $\epsilon=0$ and further to the essential singularity of the wave function at $\{\epsilon=0,r=\infty\}$. The values of J^{-} remain, instead, simply connected across $\epsilon=0$, because in this case the power of i spun off from $(1/2k)^{\lambda+1+\nu}$ simply cancels the last factor of Eq. (3.10a).

This connection of $J^{\pm}$ across the threshold relates to the broader question of the specification of J^{+} at $\epsilon<0$, which has been introduced in Sec. II (e) and will be discussed extensively in Sec. IV. For the limited purposes of the present section we have bypassed the broader question by following the procedure of paper I.

B. QDT parameters as functions of $J^{\pm}$

The parameters B, η, A, β, D introduced in Sec. II are expressed simply in terms of the Jost functions $J^{\pm}$, as one sees by comparing Eqs. (2.1)–(2.3) with (2.6) and (2.7). The phase shift contributed by the potential of Eq. (3.1) over the whole range $0\leq r\leq\infty$ is obtained from

$$e^{2i\eta}=-\frac{J^{+}(\epsilon,\lambda,b)}{J^{-}(\epsilon,\lambda,b)}\,,\quad \epsilon>0 \tag{3.11a}$$

where the $-$ sign originates from the sign difference in Eqs. (2.1) and (2.6). The phase η includes a term $(b/k)\ln(2k)$, which combines with the factor $r^{\pm\nu}$ of (3.7) to yield the diverging phase contribution of the Coulomb field (not included in the usual phase shift). The counterpart of (3.11a) for $\epsilon<0$ is

$$D^{2}\cot\beta=-\frac{J^{+}(\epsilon,\lambda,b)}{J^{-}(\epsilon,\lambda,b)}\,,\quad \epsilon<0\,. \tag{3.11b}$$

Recall that the ratio J^{+}/J^{-} must be replaced, when λ is complex, by its more general form

$$\frac{J^{+}}{J^{-}}\rightarrow\frac{J^{+}(\epsilon,\lambda,b)+J^{+}(\epsilon,\lambda^{*},b)}{J^{-}(\epsilon,\lambda,b)+J^{-}(\epsilon,\lambda^{*},b)}\,.$$

The separation of the factors D and $\cot\beta$ in (3.11b) requires an additional consideration relating to their respective physical interpretations which emerge in Sec. IV. We note here that $\beta=\pi(\nu-\lambda)$ when λ is real.

The amplitude coefficients are obtained, instead, from the products of J^{+} and J^{-}, utilizing their relation (2.7),

$$B=\frac{1}{J^{+}(\epsilon,\lambda,b)J^{-}(\epsilon,\lambda,b)}\,,\quad \epsilon>0 \tag{3.12a}$$

$$A\csc(2\beta)=\frac{1}{J^{+}(\epsilon,\lambda,b)J^{-}(\epsilon,\lambda,b)}\,,\quad \epsilon<0\,. \tag{3.12b}$$

The separation of the factors A and $\csc(2\beta)$ relates to that of D and $\cot\beta$ noted above. For complex values of λ one must again substitute

$$\frac{1}{J^+J^-} \to \frac{4}{[J^+(\epsilon,\lambda,b)+J^+(\epsilon,\lambda^*,b)][J^-(\epsilon,\lambda,b)+J^-(\epsilon,\lambda^*,b)]} . \tag{3.13}$$

Substitution of the explicit form of $J^{\pm}$, Eqs. (3.10), yields readily the well-known expressions of the amplitudes and phases in the particular cases of the Coulomb field and zero field. For complex values of λ, instead, the reduction of the results to manifestly real form is somewhat laborious. A procedure for this purpose is developed in Appendix A and the combined results are presented in Tables I and II.

Large-r expressions of the standing-wave radial function f^0 in terms of the out- and in-going functions $f^{\pm}$ and of the QDT parameters are

$$f^0(\epsilon,\lambda,b,r) \underset{r\to\infty}{\to} [B(k,\lambda)]^{-1/2}\left[\frac{2}{\pi k}\right]^{1/2} \sin[kr+(b/k)\ln r+\eta] , \quad \epsilon>0 \tag{3.14a}$$

$$f^0(\epsilon,\lambda,b,r) \underset{r\to\infty}{\to} [A(\kappa,\lambda)]^{-1/2}(\pi\kappa)^{-1/2}(D^{-1}\sin\beta\, e^{\kappa r}r^{-\nu}-D\cos\beta\, e^{-\kappa r}r^{\nu}) , \quad \epsilon<0 . \tag{3.14b}$$

Because of the similar roles of the parameters A and B and of β and η at $\epsilon<0$ and $\epsilon>0$, respectively, we shall occasionally indicate each of these pairs by a single symbol, namely,

$$\mathscr{B}_\lambda = \begin{cases} A(\kappa,\lambda) , & \epsilon<0 \\ B(k,\lambda) , & \epsilon>0 \end{cases} \tag{3.15}$$

or

$$\delta_\lambda = \begin{cases} \beta(\kappa,\lambda) , & \epsilon<0 \\ \eta(k,\lambda) , & \epsilon>0 . \end{cases} \tag{3.16}$$

The remaining QDT parameter $\mathscr{G}(\epsilon,\lambda,b)$ has been introduced in Eq. (2.5) to remove from $g^0(\epsilon,\lambda,b,r)$ all effects of propagation at large r. As indicated in Sec. II (d), a second solution of Eq. (3.1) identified by an energy-independent boundary condition at the origin, is available and is generally independent of $f^0(\epsilon,\lambda,b,r)$. Since Eq. (3.1) is invariant under the substitution $\lambda\to-(\lambda+1)$, the function $f^0(\epsilon,-\lambda-1,b,r)$ is such a solution. Its independence is verified by constructing the Wronskian $W(f^0_\lambda,f^0_{-\lambda-1})$, which is seen from Eq. (3.2) to equal $-(2\lambda+1)$. [We have introduced here the symbol f^0_λ for $f^0(\epsilon,\lambda,b,r)$.] The Wronskian vanishes when $\lambda=-\frac{1}{2}$. In addition, however, the power expansion of $f^0_{-\lambda-1}$ from Eq. (3.5) shows that the coefficients of all terms beyond the one in r^λ are divergent for critical values $\lambda=\lambda_c$. In general, the values of λ_c are integers or half-integers, but only the half-integers occur for $b=0$. This divergence was formally avoided in paper I by introducing a coefficient $a_0(\lambda)$ in the definition (3.2). This meant, however, that $a_0(\lambda)f^0_{-\lambda-1}\propto f^0_\lambda$ at $\lambda=\lambda_c$, whereby its linear independence was lost anyhow. In addition, the coefficients A and B, which reflect the dependence of f^0 on large r, appear automatically in the power expansion of $f^0_{-\lambda-1}$, a function identified by an energy-independent boundary condition at the origin. The definition of the irregular function g^0 with a smooth dependence both on ϵ and λ thus requires an appropriate subtraction from $f^0_{-\lambda-1}$ of its singularities in λ and of its nonanalytic dependences on ϵ at $\epsilon=0$. Details of this subtraction are given in Appendix B. The result takes the form given in Eq. (2.5) with

$$\mathscr{G}_\lambda = -\mathscr{B}_\lambda \cot(\delta_\lambda-\delta_{-\lambda-1}) , \tag{3.17}$$

except at the critical values $\lambda=\lambda_c$ where the above expression becomes singular. We refer to Appendix B for the result in those special cases and for the details of the derivation.

IV. GENERAL TREATMENT

The analytical representation of QDT parameters in Sec. III relates the amplitude and phase of wave functions in the asymptotic range, $r\to\infty$, to the corresponding values at shorter distances. Indeed we had noted in Sec. I that weak long-range interactions have great influence on the motion of slow particles and that the energy dependence of this influence has to be studied analytically. It was a main point of paper I that the long-range field of Eq. (3.1) encompasses all the characteristic asymptotic situations. A broader range of potentials that are relevant at shorter ranges was considered briefly in Sec. II E of I, namely, potentials that depart from the form (3.1) at intermediate ranges of r, $r_0<r<\infty$. Here we deal with QDT parameters of a general medium-range potential regardless of its very long-range behavior.

It should also be contemplated that different portions of radial range external to the core may be treated separately. For example, an electron at very large distances from a neutral atom can be regarded as moving in a zero, or purely centrifugal, potential. Yet, down to much smaller radii, the electron-atom interaction is well represented by a $1/r^4$ polarization potential. Consequently, the phase shifts referred to the zero-field basis (spherical Bessel functions) can change fairly rapidly over a small range of energies near a threshold, even after the standard k^{2l+1} dependence of the centrifugal potential is removed; phase shifts referred to the exact Mathieu-function solutions of the Schrödinger equation with an r^{-4} potential vary far more slowly and smoothly with the energy. This has been demonstrated clearly in both single-channel and multichannel problems.[4,8]

Indeed the QDT procedures apply even when the potential does not converge to zero at large distances. For example, in the theory of the Stark effect, these procedures have been applied with great success by expressing the wave functions (f^+,f^-) or (f,g) in terms of Airy functions.[6] We shall not concern ourselves here any further with this flexibility of QDT, but concentrate instead on refining and implementing the remarks of Sec. II on identifying, interpreting, and interrelating the QDT parameters.

A. WKB analysis

We take advantage here of the WKB approximation which is appropriate to the treatment of long-range fields including the Coulomb field, at least in the absence of sharp field variations. The breakdown of WKB near a nucleus can be bypassed by the Langer procedure[9]; one first avoids the singularity by replacing r by $\ln r$ and then shows that the correct phase is obtained in any case in the r variable by adding a mock-centrifugal potential $1/4r^2$ and then following the WKB routine.[10]

Note that the WKB method is generally applicable in the outer region of atomic fields, barring only sharp irregularities. In particular, the analytic condition for applicability of WKB is seen to hold for the potential (3.1), even as $\epsilon\to 0$, particularly for large r. Consider that the rate of change of the wavelength $\lambda\!\!\!^{-}$

$$\frac{d\lambda\!\!\!^{-}}{dr}=\frac{d}{dr}\left[2\epsilon+\frac{2b}{r}-\frac{\lambda(\lambda+1)}{r^2}\right]^{-1/2}$$

vanishes at sufficiently large r, irrespective of the value of ϵ. For $\lambda\neq 0$ any centrifugal barrier can also be treated within a WKB framework.

Accordingly we combine here a long-range potential of interest with the centrifugal term and with the "Langer correction" into the effective potential, setting

$$U(r)=(\lambda+\tfrac{1}{2})^2/2r^2+V(r) \tag{4.1}$$

in a.u. For definiteness λ will be considered real in this section, and $U(r)$ will be assumed to have a single potential minimum and to vanish as $r\to\infty$. Three regions of interest, I, II, and III, are separated by two classical turning points $r_1(\epsilon)$ and $r_2(\epsilon)$ with r_2 taken to be infinite for $\epsilon\geq 0$. (Additional turning points may be considered, when relevant.) The regular solution, with behavior near $r=0$ of the form $f^0\to r^{\lambda+1}$ can be found immediately within the WKB approximation.

Region I, $0<r\ll r_1$

$$f_{\rm I}^0=\left[\frac{\lambda+\frac{1}{2}}{\kappa_{\rm I}}\right]^{1/2}e^{\varphi_{\rm I}(r)}\ , \tag{4.2}$$

where

$$\varphi_{\rm I}(r)=\int^{r}dr'\kappa_{\rm I}(r')\ , \tag{4.3}$$

and as usual

$$\kappa_{\rm I}(r)=[2U(r)-2\epsilon]^{1/2}\ . \tag{4.4}$$

When the centrifugal potential dominates near $r=0$ we can be more explicit about $\varphi_{\rm I}(r)$,

$$\varphi_{\rm I}(r)=\lim_{s\to 0}\left[\int_s^r\kappa_{\rm I}(r')dr'+(\lambda+\tfrac{1}{2})\ln s\right]\ . \tag{4.5}$$

These results adequately describe $f^0(r)$ in the classically forbidden region near $r=0$ and well inside the first turning point $r\ll r_1$. For $r\approx r_1$, where $\kappa_{\rm I}(r)$ approaches zero, this WKB wave function is a very poor approximation. Here the radial solution is replaced by a linear combination of Airy functions which are exact solutions to the Schrödinger equation for a linear potential. The explicit form of $f^0(r)$ near $r=r_1$ is not required here but will be needed to connect the approximate solution (4.2) at $r\ll r_1$ to an analogous expression at $r\gg r_1$ in region II. These connection formulas are given in standard texts.[10] Their use requires a transcription of (4.2):

$$f_{\mathrm{I}}^{0}(r)=\left[\frac{\lambda+\frac{1}{2}}{\kappa_{\mathrm{I}}}\right]^{1/2} e^{\varphi_{\mathrm{I}}(r_1)}\exp\left[-\int_r^{r_1}\kappa_{\mathrm{I}}(r')dr'\right] . \tag{4.6}$$

Region II, $r_1 \ll r \ll r_2$

The continuation of (4.6) into the classically allowed range is

$$f_{\mathrm{II}}^{0}(r)=2\left[\frac{\lambda+\frac{1}{2}}{k_{\mathrm{II}}}\right]^{1/2} e^{\varphi_{\mathrm{I}}(r_1)}\sin\left[\int_{r_1}^{r}k_{\mathrm{II}}(r')dr'+\frac{\pi}{4}\right] , \tag{4.7}$$

where the wave vector is now

$$k_{\mathrm{II}}(r)=[2E-2U(r)]^{1/2} . \tag{4.8}$$

At positive energies the outer turning point r_2 goes to ∞ and the asymptotic form is given by Eq. (3.14a), with B and η given in the WKB approximation by

$$B^{-1}=\pi(2\lambda+1)\exp[2\varphi_{\mathrm{I}}(r_1)] , \tag{4.9}$$

$$\eta=\lim_{r\to\infty}\left[\int_{r_1}^{r}k_{\mathrm{II}}(r')dr'-kr-(b/k)\ln r\right]+\frac{\pi}{4} . \tag{4.10}$$

For negative energies the wave function is extended into region III through the second turning point r_2, again using the WKB connection formulas. First we rewrite Eq. (4.7) as

$$f_{\mathrm{II}}^{0}(r)=-2\left[\frac{\lambda+\frac{1}{2}}{k_{\mathrm{II}}}\right]^{1/2} e^{\varphi_{\mathrm{I}}(r_1)}$$

$$\times\sin\left[\int_{r}^{r_2}k_{\mathrm{II}}(r')dr'+\tfrac{1}{4}\pi-\beta\right] \tag{4.11}$$

with the phase shift

$$\beta(\kappa)=\int_{r_1}^{r_2}k_{\mathrm{II}}(r')dr'+\tfrac{1}{2}\pi . \tag{4.12}$$

Region III, $r_2 \ll r < \infty$ ($\epsilon<0$ only)

The connection formulas give, for r not too near r_2,

$$f_{\mathrm{III}}^{0}(r)=2\left[\frac{\lambda+\frac{1}{2}}{\kappa_{\mathrm{III}}}\right]^{1/2}$$

$$\times e^{\varphi_{\mathrm{I}}(r_1)}(\sin\beta\, e^{\varphi_{\mathrm{III}}(r)}-\tfrac{1}{2}\cos\beta e^{-\varphi_{\mathrm{III}}(r)}) . \tag{4.13}$$

The exponent $\varphi_{\mathrm{III}}(r)$ is now defined through

$$\varphi_{\mathrm{III}}(r)=\int_{r_2}^{r}\kappa_{\mathrm{III}}(r')dr' \tag{4.14}$$

and

$$\kappa_{\mathrm{III}}(r)=[2U(r)-2E]^{1/2} . \tag{4.15}$$

To extract the asymptotic form of $f^0(r)$ and the negative-energy Jost functions, we cast once again $\varphi_{\mathrm{III}}(r)$ into the form

$$\varphi_{\mathrm{III}}(r)=\lim_{s\to\infty}\left[\kappa s-(b/\kappa)\ln s-\int_r^s\kappa_{\mathrm{III}}(r')dr'\right]+w(\kappa) \tag{4.16}$$

with

$$w(\kappa)=-\lim_{s\to\infty}\left[\kappa s-(b/\kappa)\ln s-\int_{r_2}^{s}\kappa_{\mathrm{III}}(r')dr'\right] . \tag{4.17}$$

Most importantly for our purposes, the large-r limit of $\varphi_{\mathrm{III}}(r)$ is simply

$$\varphi_{\mathrm{III}}(r)\underset{r\to\infty}{\to}\kappa r-(b/\kappa)\ln r+w(\kappa) , \tag{4.18}$$

whereby the functions $\exp(\pm\varphi_{\mathrm{III}})$ in Eq. (4.13) are now identified with the exponential solutions $f^{\pm}$ of Eq. (3.7)

$$e^{\pm\varphi_{\mathrm{III}}(r)}=e^{\pm w(\kappa)}f^{\mp} . \tag{4.19}$$

The connection (2.2) between f^0 and $f^{\pm}$ accordingly takes the form given in Eq. (3.14b) with $\beta(\kappa)$ given in Eq. (4.12) and A coinciding with B in Eq. (4.9). Finally the rescaling amplitude D is given by

$$D=(\tfrac{1}{2})^{1/2}\exp[-w(\kappa)] . \tag{4.20}$$

These results represent the QDT parameters in terms of WKB phase or tunneling integrals. The essential physical origin of each parameter is thus identified, and, in particular, the range of integration that contributes to it, regardless of the accuracy of the WKB approximation. Accordingly we have gathered the parameters and will discuss each in turn.

$B(k)$. The amplitude coefficient $B^{-1/2}$ of the asymptotic form (4.13) of f^0 represents the ratio of this function, normalized as $r^{\lambda+1}$ at $r\to 0$, to the energy-normalized wave function $(2/\pi k)^{1/2}\sin(kr+\ldots)$ for $\epsilon>0$. For our potential with a single minimum, B involves only the tunneling integral from $r=0$ to $r=r_1$, Eq. (4.9). The centrifugal barrier at large r would introduce an additional factor near threshold in the absence of a countervailing long-range attraction. Tunneling under this second barrier would then add to B the additional factor $k^{2\lambda+1}$ which appears in Wigner's threshold law.

$\eta(k)$. The positive-energy phase measures the number of half-wavelengths of f^0 between $r=0$ and $r=\infty$. As this would be infinite at all $\epsilon \geq 0$, η is given by Eq. (4.10) as a finite complement to the infinite term represented by the phase $kr+(b/k)\ln r$ of f^+. Together with the usual phase shift, η included the centrifugal contribution $-\lambda\pi/2$.

$A(\kappa)$. This amplitude parameter is the negative-energy extension of B, which relates f^0 to an energy-normalized form in the classically allowed range. For our potential with a single minimum, A and B coincide, both of them representing the amplitude of tunneling under the centrifugal barrier from $r=0$ into the oscillatory region II. The WKB expression for A and B can be evaluated analytically for the potential of Eq. (3.1). The WKB result is not exact except in the limit of large λ, in contrast to Eq. (4.21).

$\beta(\kappa)$. This phase function measures the number of half-wavelengths of f^0 between $r=0$ and $r=\infty$ for $\epsilon<0$. Bound states are identified by the condition $\beta=(n+1)\pi$, $n=0,1,\ldots$. For a long-range Coulomb field, the WKB integral gives exactly the value obtained in Sec. II, namely,

$$\beta(\kappa)=\pi(-\lambda+b/\kappa)\ . \tag{4.21}$$

$D(\kappa)$. This is a "rescaling parameter" which multiplies and divides f^+ and f^-, respectively, at $\epsilon<0$, so that the resulting solutions $(Df^+,D^{-1}f^-)$ have a comparable amplitude near the outer classical turning point r_2. As Eq. (4.17) shows, D is analogous to A in that it depends on a tunneling integral from r_2 outward to infinity. The WKB expression for $D(\kappa)$ is again not identical to the exact expression in Table II for a Coulomb field, but the agreement becomes exact as the energy approaches threshold for all λ.

We conclude this WKB analysis with the identification of the function $g^0(r)$ which was introduced in Sec. II (d) as a second element of a base set $\{f^0,g^0\}$. No difficulty arises in selecting $g^0(r)$ "90° out of phase" with respect to f^0 *within the WKB approximation,* whose phase function $\varphi_{\rm II}(r)$ is common to all solutions in the classically allowed range. We set then simply

$$g^0(\epsilon,\lambda,r)=-A^{1/2}\left[\frac{2}{\pi k_{\rm II}}\right]^{1/2}\times\cos\left[\int_{r_1}^{r}k_{\rm II}(r')dr'+\frac{\pi}{4}\right]. \tag{4.22}$$

Continuation of g^0 into region I or III is provided by connection formulas. An energy-normalized base pair is now obtained from $\{f^0,g^0\}$ as

$$f=A^{1/2}f^0\ ,\quad g=A^{-1/2}g^0\ ,\quad \epsilon<0\ , \tag{4.23}$$

where A is replaced by B at $\epsilon>0$. The parameter $\mathcal{G}$ of Eq. (2.5) vanishes in the WKB approximation when b is nonzero in Eq. (3.1). For potentials having a single minimum, the parameter $\mathcal{G}$ of Eq. (2.5) vanishes in the WKB approximation. However, $\mathcal{G}$ is generally nonzero when there are more than two classical turning points. Physically this results from the fact that two solutions which differ in phase by 90° in one potential well (e.g., at small r) will no longer differ in phase by exactly 90° after tunneling[6] [see Eq. (4.41)].

B. Full treatment by a phase-amplitude method

The WKB approach has illustrated the interplay between the amplitude and phase of a wave function and its contribution to the determination of QDT parameters. This approach is developed into an *exact,* if generally numerical, treatment by phase-amplitude procedures which yield the phase and amplitude of a wave function as solutions of a pair of coupled differential equations. These methods preserve the features that have afforded a direct and interpretable construction of QDT parameters.

The common element of these methods is to represent the solution of a wave equation in the form

$$y(r)=\alpha(r)\sin\phi(r)\ . \tag{4.24}$$

A subsidiary relation, necessary to determine the amplitude $\alpha(r)$ and the phase $\phi(r)$, can take different forms, characteristic of alternative methods. The most familiar of these methods, described by Babikov[11] and by Calogero,[12] originates from classical mathematical physics.[13] It reduces the wave equation to a pair of first-order nonlinear equations which are readily interpreted. For our purposes it presents two shortcomings, namely, that it does not conserve the phase difference of a base pair of solutions $\{f^0,g^0\}$ as r varies and that its phase function $\phi(r)$ becomes imaginary in the classically inaccessible regions. These difficulties are avoided by an alternative method, which we shall follow even though it is more complicated numerically.

Its key element is a transformation of the Sturm-Liouville differential equation first discovered by

Milne,[14] but interpreted and greatly clarified in later papers by Young[15] and Wheeler.[16] The present treatment relies heavily on the recent study by Korsch and Laurent.[17] In 1930 Milne showed that *all* independent solutions $y(r)$ to an equation

$$y''(r)+k^2(r)y(r)=0 \tag{4.25}$$

can be expressed in terms of *any* particular solution of the nonlinear equation

$$\alpha''(r)+k^2(r)\alpha(r)=\frac{1}{\alpha^3(r)} . \tag{4.26}$$

The general solution $y(r)$ is given by

$$y(r)=a\alpha(r)\sin\left[\int^r \alpha^{-2}(r')dr'+b\right] , \tag{4.27}$$

where a and b are arbitrary constants. Observe that the phase and amplitude of (4.27) are related precisely as they are in the WKB approximation by

$$\phi(r)=\int^r \alpha^{-2}(r')dr' . \tag{4.28}$$

The fact that $\alpha^{-2}(r)$ is the rate of phase accumulation prompted the labeling of

$$K(r)\equiv\alpha^{-2}(r) \tag{4.29}$$

as the "quantum momentum".[17] Once α and ϕ have been calculated from Eqs. (4.26) and (4.28), it is quite clear how to find two solutions 90° out of phase. That is, we can simply choose them to be

$$\begin{aligned} f&=(2/\pi)^{1/2}\alpha \sin\phi , \\ g&=-(2/\pi)^{1/2}\alpha \cos\phi . \end{aligned} \tag{4.30}$$

This formulation lends itself to a geometrical interpretation of pairs of independent solutions to any such Sturm-Liouville differential equation. The independent solutions (f,g) can be thought of as the real and imaginary parts of a complex number

$$-i(2/\pi)^{1/2}\alpha(r)\exp[i\phi(r)] .$$

The modulus of this function is just the amplitude

$$(2/\pi)^{1/2}\alpha(r)=(f^2+g^2)^{1/2} , \tag{4.31}$$

while its phase is expressed as

$$\phi(r)=-\tan^{-1}(f/g) . \tag{4.32}$$

Equation (4.28) implies that the complex number rotates at a speed proportional to $\alpha^{-2}(r)$. In this representation two solutions 90° out of phase are mapped onto points at right angles which rotate at equal speed.

There is, of course, no unique way to require two solutions to differ in phase by 90°, although this phase difference implies orthogonality of solutions that extend to large distances $r\to\infty$ at $\epsilon>0$. This ambiguity appears in the Milne approach through the fact that $\alpha(r)$ in Eq. (4.27) can be *any* solution to the nonlinear Eq. (4.26). Accordingly the definition of (f,g) is not complete until the boundary conditions used to find $\alpha(r)$ are spelled out. One sensible way to do this, though far from the only one, is to let our classical ideas guide us by choosing $\alpha(r_c)\sim k^{-1/2}(r_c)$ at some critical radius r_c. This boundary condition was used by Korsch and Laurent[17] in their treatment of potentials with a single minimum at $r=r_c$, and they showed that the resulting amplitude and phase functions vary quite smoothly with r as desired. Caution must be exercised if a much different boundary condition is adopted to define α, as it may result in a highly oscillatory amplitude which is usually *not* the desired behavior.

In any event, once $\alpha(r)$ has been chosen and $\phi(r)$ determined also, then each of the three QDT basis pairs of independent radial functions can be calculated at once. In the following we stipulate that the phase function ϕ vanishes at $r=0$, which yields the explicit form

$$\phi(r)=\int_0^r \alpha^{-2}(r')dr' \tag{4.33}$$

to be entered in (4.27). The condition for a bound level in the purely long-range potential then resembles the Bohr-Sommerfeld quantization condition

$$\begin{aligned} \phi(\infty)&=(n+1)\pi \\ &=\int_0^\infty \alpha^{-2}(r')dr' , \quad n=0,1,\ldots . \end{aligned} \tag{4.34}$$

Moreover, comparison with the definition of β in Eq. (4.12) shows that β is simply

$$\beta=\phi(\infty) , \quad \epsilon<0 . \tag{4.35}$$

Now we have the tools to construct (possibly numerically) the alternative base pairs of independent solutions (f^0,g^0), (f,g), and (f^+,f^-), and the coefficients B, η, A, β, D, and $\mathcal{G}$ which interrelate them. Section II emphasized that $f^0(\epsilon,\lambda,r)$ can be determined at all energies by taking the small-r form $f^0\approx r^{\lambda+1}$ and integrating outward to larger radii. When this is done at positive energies, the amplitude $B^{-1/2}$ and phase η of the asymptotic oscillations can be extracted. The energy-normalized regular solution f is then simply

$$f=B^{1/2}f^0 \underset{r\to\infty}{\longrightarrow} (2/\pi k)^{1/2}\sin[kr+(b/k)\ln r+\eta] . \tag{4.36a}$$

Next the energy-normalized irregular solution, g is constructed for $\epsilon>0$ by integrating inward from $r=\infty$, starting from the larger-r form of g:

$$g(\epsilon,\lambda,r)\underset{r\to\infty}{\to}-(2/\pi k)^{1/2}\cos[kr+(b/k)\ln r+\eta]\ . \tag{4.36b}$$

The construction of $f^{\pm}$ is then trivial. Each of these functions is thus determined in a straightforward manner without any reference to the phase-amplitude method.

At negative energies a phase-amplitude procedure serves to define all the QDT functions and parameters unambiguously. Once Eq. (4.26) for the amplitude function $\alpha(r)$ is solved with suitable boundary conditions at small radii (as discussed above), $\phi(r)$ is given by quadrature in Eq. (4.33). The energy-normalized solutions (f,g) are then *defined* by Eq. (4.30) for $\epsilon<0$ only. The coefficient A is thus defined as the ratio

$$A=(2/\pi)\alpha^2(r)\sin^2\phi(r)/[f^0(\epsilon,\lambda,r)]^2\ , \tag{4.37}$$

which can be evaluated at any convenient radius.

The second function of the base pair $\{f^0,g^0\}$ is now taken to be proportional to $\alpha\cos\phi$ at all energies, with the coefficient required to yield the Wronskian

$$W(f^0,g^0)=2/\pi\ . \tag{4.38}$$

We have

$$g^0(\epsilon,\lambda,r)=-(2/\pi)^{1/2}A^{1/2}\alpha(r)\cos\phi(r)\ . \tag{4.39}$$

This amounts to setting $\mathcal{G}=0$ below threshold, as in the WKB approach. At positive energies g^0 and g need not be proportional and hence $\mathcal{G}\neq0$ in general. Since the Eqs. (4.36) and (4.38) imply that

$$g^0(\epsilon,\lambda,r)\underset{r\to\infty}{\to}-B^{1/2}(2/\pi\kappa)^{1/2}\sec(\eta^0-\eta)\times\cos[kr+(b/k)\ln r+\eta^0]\ , \tag{4.40}$$

the parameter $\mathcal{G}$ is expressed in terms of the asymptotic phase difference between g and g^0:

$$\mathcal{G}=B\tan(\eta-\eta^0)\ ,\quad \epsilon>0\ . \tag{4.41}$$

Finally, we determine the exponential solutions $f^{\pm}$ and the rescaling parameter D for $\epsilon<0$. The decaying solution f^+ is identified as the result of numerical integration of the radial Schrödinger equation inward from $r=\infty$, starting from the boundary condition (3.7). On the other hand, we know the rescaled solutions are expressed in terms of (f,g) by

$$\begin{bmatrix} Df^+ \\ D^{-1}f^- \end{bmatrix}=-(\pi\kappa)^{1/2}\begin{bmatrix} \cos\beta & \sin\beta \\ -\sin\beta & \cos\beta \end{bmatrix}\begin{bmatrix} f \\ g \end{bmatrix}, \tag{4.42}$$

with the value of β given by Eq. (4.35). The constant parameter D and the radial solution $f^-(\epsilon,\lambda,r)$ are thus obtained from (4.42) in terms of the known solutions f, g,, and f^+, and of β.

This completes the phase-amplitude specification of the radial solutions and of their QDT parameters. To illustrate these ideas further, we show in Appendix C how they are applied to a specific problem.

V. QDT PARAMETERS AT THE CORE BOUNDARY

Sections III and IV have dealt with the construction of the QDT parameters that relate the amplitude and phase of radial wave functions at short and at very large radial distances. This section deals, instead, with the characterization and construction of the parameters that serve to match a radial wave function at the core boundary $r=r_0$ to the wave function of the whole system, inclusive of the core. As noted in Sec. II (g), the system of interest has generally several alternative channels of excitation. Its wave function at $r\sim r_0$ takes, accordingly, the general form

$$\Psi=\Sigma_i\{M_i(r)\Phi_i(\omega)\}\ , \tag{5.1}$$

where $M_i(r)$ is the radial function of the particle (or fragment) that separates from the core residue in the ith channel and $\Phi_i(\omega)$ represents the ith state of the core residue as well as the spin of the particle and the angular motion of particle plus residue about their center of mass. The braces in (5.1) represent the appropriate antisymmetrization. Equation (5.1) coincides with (1.4) of paper I, whose description includes additional details.

Parameters that connect the system's wave function $\overline{\Psi}$ at $r\le r_0$ with the wave function (5.1) appropriate to $r\ge r_0$, can be determined in two steps: One should first calculate $\overline{\Psi}$ by a procedure that requires it to be regular at the origin and to take the form (5.1) at $r\sim r_0$ and then cast each $M_i(r)$ in the form (2.4),

$$M_i(r_0)=C_i[f^0(r_i)\cos\delta_i^0-g^0(r_0)\sin\delta_i^0]\ ,$$

with an amplitude C_i and a phase shift δ_i^0 to be determined by matching the value and derivative of M_i. The values of C_i and δ_i^0 depend here, of course, upon the definition of the irregular function $g^0(r)$, which is not unique, as noted in Sec. II (e) and elsewhere in this paper. However, any arbitrariness in the definition of $g^0(r)$ is compensated in the determination of the corresponding $\mathcal{G}(\epsilon)$ function and has *no effect upon the physically significant relationship* between $\overline{\Psi}$ and the observable QDT parameters at $r\to\infty$, namely, β_i or η_i. That is, the QDT parameters pertaining to the pairs $\{f^0,g^0\}$ serve only as standard stepping stones for connecting parameters of $\overline{\Psi}$ at the core boundary, $r=r_0$, to observable parameters at $r\to\infty$.

The R-matrix variational methods[18,19] of calculating energy eigenfunctions within a limited volume of space represent $\overline{\Psi}$ as a superposition of Slater determinants akin to those of multiconfiguration (analytical) Hartree-Fock calculations. These wavefunctions are readily cast in the form (5.1) near the volume boundary $r=r_0$. In particular, the approach of Ref. 19 seeks eigenchannel solutions $\overline{\Psi}_\alpha$ characterized by having identical values of the normal derivative

$$\left.\frac{\partial \ln M_i^{(\alpha)}(r)}{\partial r}\right|_{r=r_0}, \quad i=1,2,\ldots, \tag{5.2}$$

in all channels. The value of the boundary parameter (5.2) can be determined as the eigenvalue of the variational procedure for any given total energy, instead of the familiar procedure that seeks an energy eigenvalue for a fixed value of the boundary parameter. Channel amplitudes

$$\{M_1^{(\alpha)}(r_0),M_2^{(\alpha)}(r_0),\cdots\} \tag{5.3}$$

are provided as eigenvectors of the variational procedure. At any given energy, the number of mutually orthogonal eigenchannels α equals the number of channels i. As outlined above, QDT represents each of the amplitudes (5.3) in the form

$$M_i^{(\alpha)}(r_0)=C_i^{(\alpha)}[f_i^0(\epsilon_0,r_0)\cos\delta_{\alpha i}^0 - g_i^0(\epsilon_i,r_0)\sin\delta_{\alpha i}^0]\,. \tag{5.4}$$

The parameters $\delta_{\alpha i}^0$ and $C_i^{(\alpha)}$ can then be determined to fit simultaneously the values of $M_i^{(\alpha)}(r_0)$ and of its derivative (5.2).

A variant of this procedure, convenient for QDT applications, has been utilized by Lee.[20] The variant selects eigenchannels α by requiring the phase shifts $\delta_{\alpha i}^0$ to be equal in all channels i, instead of requiring the derivatives (5.2) to be equal. The representation (5.4) of the eigenchannel radial functions takes then the form

$$\begin{aligned} M_i^{(\alpha)}(r_0) &= U_{i\alpha}^0[f_i^0(\epsilon_i,r_0)\cos\delta_\alpha^0 - g_i^0(\epsilon_i,r_0)\sin\delta_\alpha^0]\,, \\ &= U_{i\alpha}^0[f_i^0(\epsilon_i,r_0) - g_i^0(\epsilon_i,r_0)\tan\delta_\alpha^0]\cos\delta_\alpha^0\,, \end{aligned} \tag{5.5}$$

where the sets of coefficients $U_{i\alpha}^0$, with fixed α and different i, represent an eigenvector of a modified R matrix, $K^{(s0)}$. In the last expression (5.5) the factor $\cos\delta_\alpha^0$ has been separated out as a normalization coefficient, while $\tan\delta_\alpha^0$ is an eigenvalue of the reaction matrix that was indicated by $-\pi K^{(s0)}$ in Sec. IV C of paper I. [Reference 20 did not in fact proceed as indicated here, but represented $M_i^{(\alpha)}(r_0)$ in terms of radial wave functions $\{f,g\}$ normalized per unit energy at $r\to\infty$ instead of $\{f^0,g^0\}$. The corresponding eigenphases are then indicated by δ_α, eigenvalues of $\tan^{-1}(-\pi K^{(s)})$, and the eigenvectors by $U_{i\alpha}$; the parameters $\mu_\alpha=\delta_\alpha/\pi$ are the usual eigenquantum defects of QDT.]

The eigenfunctions Ψ_α, with the structure (5.1), serve in QDT as a base set whose elements are superposed with coefficients A_α to generate eigenfunctions

$$\Sigma_{i\alpha}\{\Phi_i(\omega)[f_i^0(\epsilon_i,r)\cos\delta_\alpha^0 - g_i^0(\epsilon_i,r)\sin\delta_\alpha^0]\}\,U_{i\alpha}^0 A_\alpha\,, \quad r\geq r_0 \tag{5.6}$$

that satisfy boundary conditions at $r\to\infty$ appropriate to any specific problem. To this end an initial step replaces the base pairs $\{f_i^0,g_i^0\}$ by $\{f_i,g_i\}$ or $\{f_i^+,f_i^-\}$ normalized at $r\to\infty$, using the transformations of Secs. III and IV. This casts (5.6) into an expression involving superpositions of $f_i^\pm$ with coefficients which are combinations of $U_{i\alpha}^0,\delta_\alpha^0$ and of the QDT parameters of the external region. Asymptotic boundary conditions appropriate to a given physical situation can then be applied.[21] As a first step, one sets to zero the coefficients of the radial functions $f_i^-(\epsilon_i,r)$ in all channels with $\epsilon_i<0$, the so-called "closed" channels. For the "open" channels with $\epsilon_i>0$, outgoing- or ingoing-wave boundary conditions on the coefficients of their $f^\pm$ will determine, respectively, scattering or photoabsorption cross sections in terms of the combined set of QDT parameters.

TABLE I. QDT parameters for a general angular momentum $\lambda\equiv\lambda_R+i\alpha$ and strength b of the Coulomb field in Eq. (3.1). The parameters (X,x), (Y,y), and (Z,z) represent the modulus and phase of ratios of gamma functions defined in Appendix A and simplify in special cases as tabulated in Table II.

$\delta(\epsilon,\lambda)\equiv$	$\eta(k,\lambda)=-\frac{1}{2}\pi\lambda_R+(b/k)\ln(2k)+\phi(k,\lambda)$, $\phi=\frac{1}{2}(z_\alpha+z_{-\alpha})+\tan^{-1}\left[\frac{Z_\alpha-Z_{-\alpha}}{Z_\alpha+Z_{-\alpha}}\tan[\frac{1}{2}z_\alpha-\frac{1}{2}z_{-\alpha}-\alpha\ln(2k)]\right]$, for $\epsilon>0$ $\beta(\kappa,\lambda)=\pi(b/\kappa-\lambda_R)+\tilde{\phi}(\kappa,\alpha)$, $\tilde{\phi}=\tan^{-1}\{\tanh(\pi\alpha)\tan[y-\alpha\ln(2\kappa)]\}$, for $\epsilon<0$
$D^2(\kappa,\lambda)\equiv$	$\frac{1}{2}X(2\kappa)^{2b/\kappa}(\cos x\cosh(\pi\alpha)\sin[\pi(b/\kappa-\lambda_R)]+\sin x\sinh(\pi\alpha)\cos[\pi(b/\kappa-\lambda_R)]$ $+\{\sinh^2(\pi\alpha)+\sin^2[\pi(b/\kappa-\lambda_R)]\}^{1/2})$, for $\epsilon<0$ only
$\mathscr{B}(\epsilon,\lambda)\equiv$	$B(k,\lambda)=(4/\pi)(2k)^{2\lambda_R+1}e^{b\pi/k}\{Z_\alpha^2+Z_{-\alpha}^2+2Z_\alpha Z_{-\alpha}\cos[z_\alpha-z_{-\alpha}-2\alpha\ln(2k)]\}^{-1}$, for $\epsilon>0$ $A(\kappa,\lambda)=(4/Y^2)(2\kappa)^{2\lambda_R+1}\{\cosh(2\pi\alpha)+\cos[2y-2\alpha\ln(2\kappa)]\}^{-1}$, for $\epsilon<0$.

ACKNOWLEDGMENTS

We thank S. Watanabe for discussions and especially for access to his unpublished results on WKB interpretations. This work has been supported by the U. S. Department of Energy, Office of Basic Energy Sciences (Chicago) and the National Science Foundation (L.S.U.)

APPENDIX A: QDT PARAMETERS FOR COULOMB AND DIPOLE POTENTIALS

This appendix works out the details of the analytical reduction of the Jost functions $J^{\pm}(\epsilon,\lambda,b)$ to the QDT parameters B, η, A, β, and D for the potential in (3.1). It deals, therefore, with the subject addressed in Sec. II of paper I, but whereas I considered Coulomb, dipole, and zero fields separately, we deal with them here as a combined entity. Also, in amalgamating them into a single whole, we have found it necessary to depart somewhat from the definitions of these parameters in I, particularly for the case of dipole and zero fields.[22]

As a generalization of (3.3), we set

$$\lambda=\lambda_R+i\alpha\ , \tag{A1}$$

so that appropriate choices for (λ_R,α) reduce to dipole $(\lambda_R=-\frac{1}{2},\alpha\neq0)$ and to Coulomb and zero fields with real angular momentum $(\alpha=0)$, respectively. We begin with $\epsilon>0$ and from (3.9) and (3.10a), and have

$$J^{\pm}=\frac{1}{2}\left[\frac{1}{2k}\right]^{\lambda_R+1\mp\nu}\exp\{\mp i\tfrac{1}{2}\pi[\nu\mp(\lambda_R+1)]\}\times\left[\frac{\Gamma(2\lambda+2)}{\Gamma(\lambda+1\pm\nu)}e^{(1/2)\pi\alpha}e^{-i\alpha\ln(2k)}+(\alpha\leftrightarrow-\alpha)\right], \tag{A2}$$

where the last symbol in the large parentheses means that the previous term is repeated with α replaced everywhere by $-\alpha$. It is convenient to separate the ratio of gamma functions into its modulus and argument

$$\frac{\Gamma(2\lambda+2)}{\Gamma(\lambda+1+\nu)}=Z_\alpha e^{-(1/2)\pi\alpha}e^{iz_\alpha}\ , \tag{A3}$$

where Z_α and z_α are real functions of λ_R and ν. From (2.1) and (3.8), it follows that the expression in (A2) coincides with

$$(2\pi k)^{-1/2}B^{-1/2}\exp[\pm i(\eta-\tfrac{1}{2}\pi)]\ .$$

Separating out the modulus and argument of (A2) leads, therefore, to the identification of B and η. Table I records the general form and Table II the specific expressions in the limiting cases of the three kinds of long-range fields.

TABLE II. QDT parameters for zero, Coulomb, and dipole fields. This table is similar to the one in paper I except for some redefinition. The parameter ϕ defined here equals $-\frac{1}{2}\pi$ plus the ϕ of paper I.

	Parameter	Zero	Coulomb	Dipole
	λ_R	λ	λ	$-\frac{1}{2}$
	α	0	0	α
	b	0	b	0
$\epsilon<0$	X	1	$\Gamma(\lambda+1-b/\kappa)/\Gamma(\lambda+1+b/\kappa)$	1
	x	0	0	0
	D^2	$\sin(-\pi\lambda)$	$\pi(2\kappa)^{2b/\kappa}/[\Gamma(\lambda+1+b/\kappa)\Gamma(-\lambda+b/\kappa)]$	$\cosh(\pi\alpha)$
	Y	$\Gamma(\lambda+\frac{3}{2})2^{2\lambda+1}/\sin^{1/2}(-\pi\lambda)$	$\Gamma(2\lambda+2)[\Gamma(-\lambda+b/\kappa)/\Gamma(\lambda+1+b/\kappa)]^{1/2}$	$[2\pi\alpha/\sinh(2\pi\alpha)]^{1/2}$
	y	0	0	$2\alpha\ln2-\chi_\alpha,\ \chi_\alpha=\arg\Gamma(1-i\alpha)$
	β	$-\pi\lambda$	$\pi(b/\kappa-\lambda)$	$\frac{1}{2}\pi+\tilde{\phi}$ $\tan\tilde{\phi}=\tanh(\pi\alpha)\tan\left[\alpha\ln\frac{2}{\kappa}-\chi_\alpha\right]$
	A	$\frac{2}{[\Gamma(\lambda+\frac{3}{2})]^2}(\frac{1}{2}\kappa)^{2\lambda+1}\sin(-\pi\lambda)$	$\frac{2\Gamma(\lambda+1+b/\kappa)}{[\Gamma(2\lambda+2)]^2\Gamma(-\lambda+b/\kappa)}(2\kappa)^{2\lambda+1}$	$\frac{2\sinh(2\pi\alpha)}{\pi\alpha\{\cosh(2\pi\alpha)+\cos[2\alpha\ln(2/\kappa)-2\chi_\alpha]\}}$
$\epsilon>0$	Z_α	$\pi^{-1/2}2^{2\lambda+1}\Gamma(\lambda+\frac{3}{2})$	$\Gamma(2\lambda+2)/\lvert\Gamma(\lambda+1+ib/k)\rvert$	$[\alpha e^{\pi\alpha}/\sinh(\pi\alpha)]^{1/2}$
	z_α	0	$\arg\Gamma(\lambda+1-ib/k)\equiv\sigma_\lambda$	$2\alpha\ln2-\chi_\alpha$
	ϕ	0	σ_λ	$\tan\phi=\tanh(\frac{1}{2}\pi\alpha)\tan\left[\alpha\ln\frac{2}{k}-\chi_\alpha\right]$
	B	$[\Gamma(\lambda+\frac{3}{2})]^{-2}(\frac{1}{2}k)^{2\lambda+1}$	$\frac{\lvert\Gamma(\lambda+1+ib/k)\rvert^2}{[\Gamma(2\lambda+2)]^2}\frac{e^{\pi b/k}}{\pi}(2k)^{2\lambda+1}$	$\frac{2\sinh(\pi\alpha)}{\pi\alpha\{\cosh(\pi\alpha)+\cos[2\alpha\ln(2/k)-2\chi_\alpha]\}}$

For $\epsilon<0$, the unraveling of the two parameters $J^{\pm}$ in (3.10b) into the three parameters A, D, and β in (2.2) is made unique by specifying that β represents the accumulation of phase over the whole range of r. This criterion coincides with the conventional choice of $\beta=\pi(\nu-\lambda)$ for the Coulomb field as in paper I. It differs from I for the other fields. The apparent asymmetry of the last term in (3.10b) between J^+ and J^- is removed through the identity

$$\Gamma(z)\Gamma(1-z)=\pi/\sin\pi z\ .$$

Thus (3.10b) can be written

$$J^{\pm}=\mp\pi^{-1/2}\Gamma(2\lambda+2)\left[\frac{\Gamma(\nu-\lambda)}{\Gamma(\lambda+1+\nu)}\right]^{1/2}\times\left[\frac{\Gamma(\lambda+1-\nu)}{\Gamma(\lambda+1+\nu)}\sin\pi(\nu-\lambda)]\right]^{\pm1/2}\times\left[\frac{1}{2\kappa}\right]^{\lambda+1\mp\nu}\begin{Bmatrix}\cos[\pi(\nu-\lambda)]\\ \sin[\pi(\nu-\lambda)]\ .\end{Bmatrix} \tag{A4}$$

This form permits a ready comparison with (3.11b) to identify D and β. In particular, the rescaling parameter D follows immediately from the terms involving $\pm$ in (A4), that is,

$$D^2=\mathrm{Re}\left[(2\kappa)^{2\nu}\frac{\Gamma(\lambda+1-\nu)}{\Gamma(\lambda+1+\nu)}\sin[\pi(\nu-\lambda)]\right]\ . \tag{A5}$$

In forming the real part when λ is complex, it is convenient to define

$$\Gamma(\lambda+1-\nu)/\Gamma(\lambda+1+\nu)\equiv Xe^{ix}\ , \tag{A6}$$

which reduces to unity in the absence of a Coulomb field, that is, when $b=0$. The explicitly real form of D^2 is given in Table I with special forms for Coulomb, zero, and dipole fields in Table II.

Once D has been so defined, the "rescaled Jost functions," $J^{\pm}/D^{\pm1}$ are equal to

$$\mp(\pi\kappa A)^{-1/2}\begin{Bmatrix}\cos\beta\\ \sin\beta\end{Bmatrix}$$

from (2.2). Upon forming the real part of the rescaled Jost functions, we have

$$A^{-1/2}\begin{Bmatrix}\sin\beta\\ \cos\beta\end{Bmatrix}=2^{-(\lambda_R+2)}\kappa^{-(\lambda_R+1/2)}Y\,\mathrm{Re}\left[e^{i[y-\alpha\ln(2\kappa)]}\begin{Bmatrix}\sin[\pi(\nu-\lambda_R-i\alpha)]\\ \cos[\pi(\nu-\lambda_R-i\alpha)]\end{Bmatrix}\right]\ , \tag{A7}$$

with

$$Ye^{iy}\equiv\Gamma(2\lambda+2)[\Gamma(\nu-\lambda)/\Gamma(\lambda+1+\nu)]^{1/2}\ . \tag{A8}$$

Table I gives the general expressions and Table II the specific values of A and β for the three limiting cases.

The entries in Table II differ from the results in I in the following respects. The phase parameters ζ and η are unchanged and $B(k,\lambda)$ differs only trivially from paper I because of our removal of the normalization factor $a_0(\lambda)$ in (3.2). Likewise, for the Coulomb field, $A(\kappa,\lambda)$ differs only in this factor, and D and β are exactly as before. All three parameters are, however, substantially different for the dipole and zero fields.

A point to be stressed is that the parameters A, D, and $\sin\beta$ vanish for the case of zero field when λ is an integer; their reciprocals appear in physically significant combinations, e.g., in (2.2), but these expressions remain finite. Similarly imaginary values of parameters may emerge in the tables, but their significant combinations are nevertheless real. The definitions of QDT parameters adopted in paper I may prove to be preferable for actual calculations when only the long-range dipole field or zero field are present.

APPENDIX B: DEFINITION OF $g^0(r)$ AND $\mathcal{G}(\epsilon)$

The notation introduced at the end of Sec. III, which sets $f^0(\epsilon,\lambda,b,r)\equiv f^0_\lambda$, etc., will be followed here. The reciprocal linear relations between the base pair $\{f^0_\lambda,g^0_\lambda\}$ normalized at $r\to0$ and the energy-normalized pair $\{f_\lambda,g_\lambda\}$, implied by (2.1), (2.5), etc., are then

$$\begin{bmatrix} f_\lambda \\ g_\lambda \end{bmatrix} = \begin{bmatrix} \mathcal{B}_\lambda^{1/2} & 0 \\ \mathcal{B}_\lambda^{-1/2}\mathcal{G}_\lambda & \mathcal{B}_\lambda^{-1/2} \end{bmatrix} \begin{bmatrix} f_\lambda^0 \\ g_\lambda^0 \end{bmatrix},$$

$$\begin{bmatrix} f_\lambda^0 \\ g_\lambda^0 \end{bmatrix} = \begin{bmatrix} \mathcal{B}_\lambda^{-1/2} & 0 \\ -\mathcal{B}_\lambda^{-1/2}\mathcal{G}_\lambda & \mathcal{B}_\lambda^{1/2} \end{bmatrix} \begin{bmatrix} f_\lambda \\ g_\lambda \end{bmatrix}. \tag{B1}$$

In accordance with Sec. III, and following the original approach of Seaton[1] adhered to in paper I, we start here from the base pair $\{f_\lambda^0, f_{-\lambda-1}^0\}$ of independent solutions of Eq. (3.1). Recall that this pair's independence fails for a set of values of $\lambda=\lambda_c$, as noted at the end of Sec. III, and that the λ_c include (for $b\neq 0$) the integer orbital quantum numbers $\lambda_c=l$ of greatest physical significance. We also recall, from Sec. II (e), that the definition of $g^0(r)$ involves rather arbitrary conventions which serve to articulate the theory but do not affect its final results. Since no single convention appears definitely preferable, we treat here separately the two cases $\lambda\neq\lambda_c$, $\lambda=\lambda_c$, without attempting to blend the separate results into a unified convention.

Properties of the base pair $\{f_\lambda^0, f_{-\lambda-1}^0\}$ emerge by calculating and comparing their Wronskian evaluated from their limiting forms at $r\to 0$ and $r\to\infty$,

$$W(f_\lambda^0, f_{-\lambda-1}^0) = -(2\lambda+1), \tag{B2}$$

from $f^0(r)\to r^{\lambda+1}$,

$$W(f_\lambda^0, f_{-\lambda-1}^0) = \mathcal{B}_\lambda^{-1/2}\mathcal{B}_{-\lambda-1}^{-1/2}(2/\pi)\sin(\delta_\lambda-\delta_{-\lambda-1}), \tag{B3}$$

from (2.1). The expression (B3) vanishes at $\lambda=\lambda_c$, where the phase difference is a multiple of π. This phase difference has a simple expression for potentials represented by (3.1) with *real* λ, namely,

$$\delta_\lambda-\delta_{-\lambda-1} = \begin{cases} \eta_\lambda-\eta_{-\lambda-1} = -\pi\lambda-\tan^{-1}\{\tan[\pi(\lambda+1)]\tanh(\pi b/k)\}, & \epsilon>0 \\ \beta_\lambda-\beta_{-\lambda-1} = -\pi(2\lambda+1), & \epsilon<0. \end{cases} \tag{B4}$$

[It should be pointed out that (B3) is valid at $\epsilon<0$ only if $D_\lambda=D_{-\lambda-1}$, which is satisfied in Table II.] The nonindependence of $\{f_\lambda^0, f_{-\lambda-1}^0\}$ at $\lambda=\lambda_c$ will manifest itself through the occurrence of the expression (B3) in the denominators of our results. Indeed we shall start from the expression of the solution $g_\lambda(r)$ in terms of $\{f_\lambda^0, f_{-\lambda-1}^0\}$,

$$g_\lambda(r) = \frac{f_{-\lambda-1}(r)-\cos(\delta_\lambda-\delta_{-\lambda-1})f_\lambda(r)}{\sin(\delta_\lambda-\delta_{-\lambda-1})}, \tag{B5}$$

an expression that lags 90° in phase behind $f_\lambda(r)$ with the Wronskian $W(f_\lambda, g_\lambda)=2/\pi$. Formulas applying at $\lambda=\lambda_c$ are then obtained by de l' Hospital's limiting procedure for $\lambda\to\lambda_c$, which is familiar, e.g., from the definition of the Neuman function

$$N_n(z) = \lim_{\nu\to n} N_\nu(z) = \lim_{\nu\to n} \frac{J_{-\nu}(z)-\cos(\pi\nu)J_\nu(z)}{-\sin(\pi\nu)}. \tag{B6}$$

(a) $\lambda\neq\lambda_c$. The function g_λ^0 can now be defined by entering in Eq. (B1) the functions $f_\lambda=\mathcal{B}_\lambda^{1/2}f_\lambda^0$ and g_λ from (B5),

$$g_\lambda^0 = -\mathcal{G}_\lambda f_\lambda^0(r) + \mathcal{B}_\lambda^{1/2}\frac{\mathcal{B}_{-\lambda-1}^{1/2}f_{-\lambda-1}^0-\cos(\delta_\lambda-\delta_{-\lambda-1})\mathcal{B}_\lambda^{1/2}f_\lambda^0}{\sin(\delta_\lambda-\delta_{-\lambda-1})}. \tag{B7}$$

The function f_λ^0 is seen to drop out of this expression upon setting

$$\mathcal{G}_\lambda = -\mathcal{B}_\lambda\cot(\delta_\lambda-\delta_{-\lambda-1}), \tag{B8}$$

the formula anticipated in (3.17). Comparison of (B2) and (B3) permits us now also to eliminate $\mathcal{B}_\lambda^{1/2}\mathcal{B}_{-\lambda-1}^{1/2}$ from (B7), thus reducing that equation to

$$g_\lambda^0(r) = [-\pi(\lambda+\tfrac{1}{2})]^{-1}f_{-\lambda-1}^0(r). \tag{B9}$$

Recalling Eq. (B2), we see that this definition leads to the appropriate Wronskian

$$W(f_\lambda^0, g_\lambda^0) = 2/\pi. \tag{B10}$$

When $\lambda=\lambda_R+i\alpha$ is complex in (3.1), the analog of f_λ^0 is its real part and the analog of $f_{-\lambda-1}^0$ is

$$\mathrm{Im} f_\lambda^0 = \frac{1}{2i}(f_\lambda^0 - f_{\lambda^*}^0) \ . \tag{B11}$$

This function serves then as g_λ^0, with the additional normalization factor required to fit (B10). Equation (B8) remains a valid definition of $\mathscr{G}_\lambda$ with the values of $\mathscr{B}_\lambda$ and δ_λ drawn from Appendix A and appropriate to the treatment of complex λ. No complex values of λ belong to the special set λ_c, but numerical difficulties may arise for $\lambda=-\frac{1}{2}+i\alpha$ and $\alpha \ll 1$.

(b) $\lambda=\lambda_c$. The limiting form of (B5) for $\lambda \to \lambda_c$ is

$$g_{\lambda_c}(r) = \frac{(-1)^m[\partial f_{-\lambda-1}(r)/\partial\lambda]_{\lambda_c} - [\partial f_\lambda(r)/\partial\lambda]_{\lambda_c}}{[\partial(\delta_\lambda - \delta_{-\lambda-1})/\partial\lambda]_{\lambda_c}} \tag{B12}$$

with

$$m = (\delta_{\lambda_c} - \delta_{-\lambda_c-1})/\pi \ . \tag{B13}$$

[The denominator of (B12) would reduce to -2π for $\epsilon<0$, according to (B4), but its general expression is preserved in the following.] The definition of $g_{\lambda_c}^0$ from the g_{λ_c} in (B12) must identify and separate out the parts of (B12) that depend on ϵ sensitively at threshold. At $\epsilon>0$ such parts include $\delta_\lambda - \delta_{-\lambda-1}$. Our procedure will be simplified by defining $g_{\lambda_c}^0$ at $\epsilon<0$, with a smooth dependence on ϵ, extrapolating the result to $\epsilon>0$, and obtaining separate expressions of $\mathscr{G}_\lambda$ in the two energy ranges.

To this end we return to the initial expression (B5) of g_λ and transcribe it, using Eqs. (B2) and (B3) at $\epsilon<0$, into

$$g_\lambda(r) = \frac{A_\lambda^{-1/2}\bar{f}^0_{-\lambda-1}(r) - A_\lambda^{1/2}\cos(\beta_\lambda - \beta_{-\lambda-1})f_\lambda^0(r)}{\sin(\beta_\lambda - \beta_{-\lambda-1})} \ , \quad \epsilon<0 \tag{B14}$$

where

$$\bar{f}^0_{-\lambda-1}(r) = -\frac{\sin(\beta_\lambda - \beta_{-\lambda-1})}{\pi(\lambda+\frac{1}{2})} f^0_{-\lambda-1} \underset{\lambda\to\lambda_c}{\longrightarrow} A_{\lambda_c}\cos(\beta_{\lambda_c} - \beta_{-\lambda_c-1})f_{\lambda_c}^0 \tag{B15}$$

remains finite as $\lambda\to\lambda_c$ unlike $f^0_{-\lambda-1}$ and g_λ^0 itself in (B9). The limiting form of (B14) at λ_c, which will replace (B12), is

$$g_{\lambda_c}(r) = \frac{(-1)^m A_{\lambda_c}^{-1/2}(\partial \bar{f}^0_{-\lambda-1}/\partial\lambda)_{\lambda_c} - A_{\lambda_c}^{1/2}(\partial f_\lambda^0/\partial\lambda)_{\lambda_c} - A_{\lambda_c}^{-1/2}(\partial A_\lambda/\partial\lambda)_{\lambda_c} f_{\lambda_c}^0}{[\partial(\beta_\lambda - \beta_{-\lambda-1})/\partial\lambda]_{\lambda_c}} \ , \quad \epsilon<0 \ . \tag{B16}$$

The factor $(\partial A/\partial\lambda)_{\lambda_c}$ in the numerator of (B16) is now identified as the only one that depends sensitively on ϵ near threshold and hence is to be incorporated in $\mathscr{G}_{\lambda_c}$. Indeed $\partial f_\lambda^0/\partial\lambda$ and $\partial\bar{f}^0_{-\lambda-1}/\partial\lambda$ are smooth functions of energy by definition. The same hold for A_{λ_c} itself which equals $(-1)^m\bar{f}^0_{-\lambda_c-1}/f_{\lambda_c}^0$ according to (B15) and for $\beta_\lambda - \beta_{-\lambda-1}$ which is energy independent. Hence we define

$$\mathscr{G}_{\lambda_c} = -\frac{(\partial A_\lambda/\partial\lambda)_{\lambda_c}}{[\partial(\beta_\lambda - \beta_{-\lambda-1})/\partial\lambda]_{\lambda_c}} \ , \quad \epsilon<0 \ . \tag{B17}$$

The irregular energy-smooth solution $g_{\lambda_c}^0$ is accordingly defined by

$$g_{\lambda_c}^0(r) = \frac{(-1)^m(\partial\bar{f}^0_{-\lambda-1}/\partial\lambda)_{\lambda_c} - A_{\lambda_c}(\partial f_\lambda^0/\partial\lambda)_{\lambda_c}}{[\partial(\beta_\lambda - \beta_{-\lambda-1})/\partial\lambda]_{\lambda_c}} \ , \tag{B18}$$

initially at $\epsilon<0$ and by extrapolation at all ϵ. This extrapolation does not extend to $\mathscr{G}_{\lambda_c}$, however, which is nonanalytic at $\epsilon=0$. The combination of Eq. (B1) with (B5) and (B18) yields instead

$$\mathscr{G}_{\lambda_c} = -\left[\frac{\partial[B_\lambda\sin(\beta_\lambda - \beta_{-\lambda-1})/\sin(\eta_\lambda - \eta_{-\lambda-1})]/\partial\lambda}{\partial(\beta_\lambda - \beta_{-\lambda-1})/\partial\lambda}\right]_{\lambda_c} \ , \quad \epsilon>0 \ . \tag{B19}$$

APPENDIX C: QDT PARAMETERS FOR FREE-PARTICLE WAVE FUNCTION

To illustrate the interplay of different elements of this paper, we identify here the QDT parameters for a trivially simple example. The wave equation $\frac{1}{2}y''+\epsilon y=0$ has the regular solution

$$f^0(r)=\begin{cases} k^{-1}\sin(kr)\ , & \epsilon>0 \\ \kappa^{-1}\sinh(\kappa r)\ , & \epsilon<0 \end{cases} \tag{C1}$$

normalized to $f^0\to r$ for $r\to 0$.

The parameters B and η are given by (3.12a) and (3.11a) for $\epsilon>0$ in terms of the Jost functions $J^{\pm}$, which are in turn read off the representation (2.5) of (C1). Thus one finds in our example,

$$B=2k/\pi\ ,\quad \eta=0\ . \tag{C2}$$

For $\epsilon<0$, we avoid here the possible ambiguity in separating D from β by applying procedures from Sec. IV B. In general $\alpha(r)$ must be calculated numerically from (4.26), but for the present example a real solution with the energy-independent small-r normalization $\alpha(r=0)=1$ is given in closed form by

$$\alpha(r)=\kappa^{-1}[\sinh^2(\kappa r)+\kappa^2\cosh^2(\kappa r)]^{1/2}\ . \tag{C3}$$

The corresponding phase function (4.34) is then

$$\phi(r)=\tan^{-1}[\kappa^{-1}\tanh(\kappa r)]\ , \tag{C4}$$

from which follows, by (4.35),

$$\beta=\tan^{-1}(1/\kappa)\ ,\quad \sin\beta=(1+\kappa^2)^{-1/2}\ . \tag{C5}$$

The energy-normalized base pair for $\epsilon<0$ is now given by (4.30) in terms of α and ϕ,

$$\begin{aligned} f(r)&=\left[\frac{2}{\pi\kappa^2}\right]^{1/2}[\sinh^2(\kappa r)+\kappa^2\cosh^2(\kappa r)]^{1/2} \\ &\quad\times\sin\left[\tan^{-1}\left[\frac{1}{\kappa}\tanh(\kappa r)\right]\right] \\ &=\left[\frac{2}{\pi\kappa^2}\right]^{1/2}\sinh(\kappa r)\ , \\ g(r)&=-\left[\frac{2}{\pi}\right]^{1/2}\cosh(\kappa r)\ . \end{aligned} \tag{C6}$$

The final parameter D is generally obtained from (4.42), in terms of f, g, and β (previously found) and of f^+. In our case f^+ coincides with $e^{-\kappa r}$ at all r, removing any need of integrating it inward from $r=\infty$; Eq. (4.42) then gives

$$D=\left[\frac{2\kappa}{1+\kappa^2}\right]^{1/2}\ . \tag{C7}$$

The same Eq. (4.42) gives also the singular function

$$f^-(r)=e^{\kappa r}-\frac{1-\kappa^2}{1+\kappa^2}e^{-\kappa r}\ , \tag{C8}$$

which does include a decaying term of specified amplitude.

In accordance with Sec. IV B we set here $\mathcal{G}=0$ for $\epsilon<0$, whence

$$g^0(r)=A^{1/2}g(r)=-(2/\pi)\cosh(\kappa r)\ ,\quad \epsilon<0 \tag{C9}$$

where $A=2/\pi$ is obtained using Eq. (4.37). This function continues smoothly across the threshold into

$$g^0(r)=-(2/\pi)\cos(kr)\ ,\quad \epsilon>0\ . \tag{C10}$$

Section IV B regards (f,g) and (f^0,g^0) at $\epsilon>0$ as connected nontrivially by (2.5), with $\mathcal{G}$ given by (4.41); however, $\mathcal{G}$ vanishes in our example because $\eta^0=0$, too.

Some of the QDT parameters obtained by this procedure differ from those that result from Sec. III and Appendix A for the same example. The differences stem in part from alternative specifications of the singular function f^-.

[1]M. J. Seaton, Proc. Phys. Soc. London 88, 801 (1966); J. Phys. B 11, 4067 (1978).
[2]U. Fano, J. Opt. Soc. Am. 65, 979 (1975).
[3]C. H. Greene, U. Fano, and G. Strinati, Phys. Rev. A 19, 1485 (1979), referred to as paper I.
[4]S. Watanabe and C. H. Greene, Phys. Rev. A 22, 158

(1980).

[5]R. Colle, J. Chem. Phys. 74, 2910 (1981); A. Giusti, J. Phys. B 13, 3867 (1980).

[6]D. A. Harmin, Phys. Rev. A 24, 2491 (1981); 26, 2656 (1982); Phys. Rev. Lett. 49, 128 (1982).

[7]R. Jost, Helv. Phys. Acta 20, 256 (1947); R. G. Newton, *Scattering Theory of Waves and Particles* (McGraw-Hill, New York, 1966), Sec. 12.1.2; L. I. Schiff, *Quantum Mechanics,* 3rd ed. (McGraw-Hill, New York, 1968), Sec. 39. Our choice of convention in defining Jost functions is closest to Newton's in the standardization of both f^0 and $f^{\pm}$ in (3.2) and (3.7). We differ, however, in the $\pm$ labels on the Jost functions where our choice follows the original work of Jost, whereas Newton reverses the symbols. Also, we have not included an overall factor of $(2ik)^{-1}$ on the right-hand side of (3.8), in contrast to Newton whose Jost functions coincide with the Wronskian in (3.10). Dropping such a factor also necessitates a relative change in sign of the two terms in (3.8) in order to maintain the basic symmetry property (2.7). This extra sign, therefore, also occurs in our (3.11a) and (3.11b). Newton renormalizes his Jost functions by a factor $k^{-l}e^{\mp i(1/2)\pi l}$ so as to obtain the S matrix as the simple ratio of Jost functions. We do not follow this practice here because λ may be complex. Our S matrix thus includes phase factors explicitly.

[8]T. F. O'Malley, L. Spruch, and L. Rosenberg, J. Math. Phys. 2, 491 (1961).

[9]R. E. Langer, Phys. Rev. 51, 669 (1937).

[10]A. Messiach, *Quantum Mechanics* (North-Holland, Amsterdam, 1961), Vol. I, Chap. VI; L. I. Schiff, *Quantum Mechanics,* 3rd ed. (McGraw-Hill, New York, 1968), Sec. 34.

[11]V. V. Babikov, *Method Phazovi'kh Funktsii vi Kvantovoi Mekhanike* (Nauka, Moscow, 1968); Usp. Fiz. Nauk 92, 3 (1967) [Sov. Phys.—Usp. 10, 271 (1967)].

[12]F. Calogero, *Variable Phase Approach to Potential Scattering* (Academic, New York, 1967).

[13]R. Courant and D. Hilbert, *Methods of Mathematical Physics* (Interscience, New York, 1953), Vol. I, p. 332.

[14]W. E. Milne, Phys. Rev. 35, 863 (1930).

[15]E. Young, Phys. Rev. 38, 1612 (1931); 39, 445 (1932).

[16]J. A. Wheeler, Phys. Rev. 52, 1123 (1937).

[17]H. J. Korsch and H. Laurent, J. Phys. B 14, 4213 (1981).

[18]P. G. Burke, in *Invited Papers and Progress Reports of the Tenth ICPEAC, Paris, 1977,* edited by G. Watel (North-Holland, Amsterdam, 1978); P. G. Burke and D. Robb, Adv. At. Mol. Phys. 11, 144 (1975).

[19]U. Fano, in *Invited Papers and Progress Reports of the Tenth ICPEAC, Paris, 1977,* edited by G. Watel (North-Holland, Amsterdam, 1978); U. Fano and C. M. Lee, Phys. Rev. Lett. 31, 1573 (1973).

[20]C. M. Lee, Phys. Rev. A 10, 584 (1974).

[21]See, for instance, C. H. Greene, Phys. Rev. A 22, 149 (1980).

[22]Combined Coulomb and dipole fields were treated by J. Dubau, J. Phys. B 11, 4095 (1978).

1984 *J. Chem. Phys.* **80** 2514–25

A multichannel quantum defect analysis of diatomic predissociation and inelastic atomic scattering

Frederick H. Mies
Molecular Spectroscopy Division, National Bureau of Standards, Washington, D.C. 20234

(Received 15 August 1983; accepted 22 November 1983)

Given an $N_T \times N_T$ interaction matrix $\mathbf{W}^\infty(R)$ which describes the dissociation of a diatomic molecule into N_T asymptotic atomic channel states, we can generate exact numerical solutions to the close-coupled scattering equations. At total energies E above the highest dissociation threshold we obtain an $N_T \times N_T$ scattering matrix $\mathbf{S}(E)$ which defines the asymptotic structure of the N_T-fold degenerate multichannel scattering or continuum wave functions. This matrix varies rapidly with energy and is nonanalytic at thresholds. However, based on a multichannel quantum defect analysis (MCQDA) of the coupled equations we find that the numerical $\mathbf{S}(E)$ matrix can be made to yield a real, symmetric matrix $\mathbf{Y}(E)$ which is analytic in E. This matrix can then be analytically continued across threshold to provide rigorous analytic descriptions of the multichannel diatomic wave functions in the predissociating and bound-state regions of the energy spectrum. Since the extraction of $\mathbf{Y}(E)$ is predicated on assigning a reference potential $\mathscr{V}_\gamma(R)$ to each channel, the detailed energy variation of $\mathbf{Y}(E)$ is dependent on the choice of potentials. Fortunately, the physics contained in $\mathbf{W}^\infty(R)$ generally dictates an obvious set of reference potentials which usually make $\mathbf{Y}(E)$ slowly varying. As E is reduced below threshold we can use $\mathbf{Y}(E)$, often well represented as a constant over a wide range of energies, to provide a description of the predissociating molecule, including such observable properties as linewidths, level shifts, and branching ratios. At still lower energies, when all channels are closed, $\mathbf{Y}(E)$ offers a complete, nonperturbative description of the configuration interaction between bound electronic-rotational states of the molecule. Although many insights in this paper are provided by a WKB analysis of the reference solutions associated with each reference potential, the MCQDA yields a complete quantum mechanical description of the exact close-coupled wave functions. The quality of these wave functions is limited only by the accuracy of the molecular potentials and interaction matrix elements that are used to construct $\mathbf{W}^\infty(R)$, and by the number N_T and specific set of channels we choose to include in the close-coupled theory. The analysis is equally valid when applied to either adiabatic avoided crossings or diabatic curve crossings. More importantly, the formal structure of the close-coupled wave functions dictated by MCQDA yields a rigorous framework for the analysis of strongly interacting continuum, predissociating, and bound state channels without recourse to perturbation theory.

I. INTRODUCTION

A complete theory of diatomic predissociation can be developed from atomic scattering theory[1] using the channel state representation of the total wave function

$$\Psi_\gamma(E) = \sum_{\gamma'=1}^{N_T} \psi_{\gamma'}(\mathbf{r},\hat{R})F_{\gamma',\gamma}(E,R)/R. \quad (1)$$

The channel states $\{\psi_{\gamma'}\}$ are a select set of electronic-rotational states which conform to the Hund's case (e)[1–3] coupling scheme of diatomic spectroscopy. They are explicit functions of the molecular electronic coordinates $\mathbf{r}$ and interatomic polar coordinates $\hat{R}$, and are *defined* such that the interaction matrix $W^\infty_{\gamma',\gamma}(R)$ obtained from the total Hamiltonian H,

$$\langle\psi_{\gamma'}|H|\psi_\gamma\rangle = -\frac{\hbar^2}{2\mu R}\frac{\partial^2}{\partial R^2}R\delta_{\gamma',\gamma} + W^\infty_{\gamma',\gamma}(R) \quad (2)$$

obeys the asymptotic condition

$$W^\infty_{\gamma',\gamma}(R) \underset{R\to\infty}{\sim} \left[E^\infty_\gamma + \frac{\hbar^2 l(l+1)}{2\mu R^2}\right]\delta_{\gamma',\gamma} + \mathscr{O}(R^{-3}). \quad (3)$$

Each channel γ is associated with a particular pair of atomic states which defines the threshold energy $E^\infty_\gamma = E^\infty_A + E^\infty_B$ for the molecular dissociation. As $R\to\infty$ the total Hamiltonian becomes separable into the radial kinetic operator in Eq. (2), the isolated electronic Hamiltonians for atoms A and B, and an orbital angular momentum operator for the relative motion of the separating atoms with reduced mass $\mu = m_A m_B/(m_A + m_B)$. Thus the channel states ψ_γ must approach *exact* product eigenfunctions of the internal electronic states of the dissociating atoms with eigenvalue E^∞_γ, and they must also become eigenfunctions of the total nuclear angular momentum with eigenvalue $\hbar l$.

For a given total energy E the channel states are classified as open or closed depending on whether the channel wave number

$$k_\gamma = \sqrt{2\mu(E - E^\infty_\gamma)/\hbar^2} \quad (4)$$

is real and positive, with $E > E^\infty_\gamma$, or pure imaginary $k_\gamma = i|k_\gamma|$, with $E < E^\infty_\gamma$. The coupled equations generated by Eqs. (1) and (2), i.e.,

$$\left[-\frac{\hbar^2}{2\mu}\frac{\partial^2}{\partial R^2} - E\right]F_{\gamma,\gamma} + \sum_{\gamma'} W^\infty_{\gamma,\gamma'}F_{\gamma',\gamma} = 0, \quad (5)$$

yield N_T solution vectors which vanish as $R\to 0$ and form an $(N_T \times N_T)$ matrix of radial functions $\mathbf{F}(E,R)$ defining N_T independent solutions to the total Hamiltonian $(H - E)\Psi_\gamma(E) = 0$. However, when $N_T = N_o + N_c$ consists of N_o open channels, and N_c closed channels, the number of well-behaved, i.e., normalizable, solutions is reduced to N_o, i.e., with $\gamma,\gamma' = 1,N_o$,

$$\langle \Psi_\gamma(E) | H | \Psi_{\gamma'}(E') \rangle = E\delta(E - E')\delta_{\gamma,\gamma'}. \tag{6}$$

Thus the dimensionality of the physically meaningful set of radial functions is reduced to $(N_T \times N_o)$.

Similarly, the all important scattering matrix $S_{oo}(N_o \times N_o)$ which defines the *observable* properties of the diatomic system is reduced to $N_o \times N_o$. As we shall see, this is not to mean that the subset of closed channels do not influence the properties of the system. It merely implies that the *asymptotic* components $F_{\gamma',\gamma}(E,R)$ with $\gamma = 1$, N_o and $\gamma' = N_o + 1$, N_T must vanish.

Our objective is to obtain a symmetric matrix of real, energy-insensitive parameters $\mathbf{Y}(N_T \times N_T)$ which can describe the properties of the diatomic molecule as we pass through dissociation limits. If we block the matrix as follows:

$$\mathbf{Y}(N_T \times N_T) = \left(\begin{array}{c|c} \mathbf{Y}_{oo}(N_o \times N_o) & \mathbf{Y}_{oc}(N_o \times N_c) \\ \hline \mathbf{Y}_{co}(N_c \times N_o) & \mathbf{Y}_{cc}(N_c \times N_c) \end{array} \right), \tag{7}$$

our requirements are twofold. First, $\mathbf{Y}$ should remain continuous as we pass through thresholds and the number of open N_o and closed N_c channels is modified. Second, the variation of $\mathbf{Y}(E)$ with the total energy E should be sufficiently small that meaningful extrapolations of the matrix can be made across thresholds.

It should be apparent at this point that our intent is to develop a multichannel quantum defect[4–6] approach (MCQDA) to diatomic predissociation.[7–9] The philosophy of the approach is as follows. *Given exact solutions* of the coupled radial equations in Eq. (5) at one, or a few, energies above threshold when all channels are open $N_T = N_o$, we obtain $\mathbf{Y}(N_T \times N_T)$ from the scattering matrix $\mathbf{S}_{OO}(N_T \times N_T)$. We may then use the exact relationships we shall develop to obtain $\mathbf{S}_{oo}(N_o \times N_o)$ below threshold when the subset of N_c channels becomes closed. Actually the procedure is completely reversible. From an analysis of the resonant structure in $\mathbf{S}_{oo}(N_o \times N_o)$ below threshold we may also obtain the complete $\mathbf{Y}(N_T \times N_T)$ matrix and predict the inelastic atomic scattering cross sections above threshold.

Aside from the practical application of avoiding tedious and expensive calculations of the predissociating wave functions over a fine mesh of energies, the $\mathbf{Y}$ matrix gives direct insight into the predissociating widths, principal value shifts, and modifications in the background scattering.[10–13] Note that the MCQDA is rigorous for all coupling strengths and is not predicated on any application of perturbation theory. Also it can be extended to many-channel scattering phenomena without difficulty, initiating even more substantial savings in computational time. The particular strength of the MCQDA is greatest when we are confronted with dissociation limits involving degenerate, or near-degenerate multiplet state atoms—here the complex spectroscopy and dynamics of the system is concisely summarized in the irreducible $\mathbf{Y}$ matrix.

We have previously applied a generalized version of MCQDT[4–6] to diatomic continuum wave functions[23] using adiabatic electronic-rotational (AER) states.[1,2] In this paper we present a refined version of the theory which allows for analytic continuation of the continuum wave functions across dissociation thresholds. Further, we relax any commitment to the use of AER channel states. Thus we obtain a complete description of the diatomic wave functions and derive a unified theory of atom–atom scattering, diatomic predissociation, and the configuration interaction of molecular states in the vicinity of dissociation limits.

Ours is not the first attempt at developing a MCQDA for diatoms.[7,8] Further, it should be emphasized that a complete and *general* form of MCQDT has already been presented by Greene, Rau, and Fano (GRP)[9] which is very similar in spirit, and in detail, to the theory we have evolved. Our mutual exploitation of the WKB structure of the wave function in the classically accessible regions to help define unique, fully quantal reference functions is especially striking. The insight and the power of a WKB assisted analysis has already been demonstrated by Colle[7] who developed a completely semiclassical description of diatomic predissociation. The justification for the present paper is the very explicit application we have made of a completely *rigorous* form of MCQDA to describe dissociating molecules and their associated phenomena. To assist students of the GRF paper we include a dictionary of mutually equivalent parameters and functions in the Appendix.

The analytic properties of the various single channel solutions which are used as a basis to define $\mathbf{Y}(E)$ are presented in Sec. II. The $\mathbf{Y}(E)$ matrix and its relationship to the scattering matrix $\mathbf{S}_{oo}(E)$ and resonant phenomena is derived in Sec. III. A summary and conclusions are contained in Sec. IV. Results of exact close-coupled calculations are compared to the predictions of the MCQDA in paper II. Threshold and cusp behavior in the scattering cross sections are given special consideration in the Appendix of paper II.

II. SINGLE CHANNEL THEORY: A WKB ASSISTED ANALYSIS

A. Introduction

We shall associate a reference potential $\mathscr{V}_\gamma(R)$ with each channel. This potential is constrained to reproduce the *asymptotic* behavior of the interaction matrix in Eq. (3), i.e.,

$$\mathscr{V}_\gamma(R) \underset{R\to\infty}{\sim} E_\gamma^\infty + \frac{\hbar^2 l(l+1)}{2\mu R^2} + \mathscr{O}(R^{-3}), \tag{8}$$

but otherwise we are free to choose $\mathscr{V}_\gamma(R)$ to suit our conveniences. Obviously we would attempt to employ a potential which would most closely reproduce the expected elastic scattering, or bound state structure in the given channel, e.g., for a single channel expansion the only sensible choice would be $\mathscr{V}_\gamma(R) \equiv W_{\gamma,\gamma}(R)$. For multichannel coupling the various choices are discussed more fully in the next section. For the moment let us merely assume that $\mathscr{V}_\gamma(R)$ is given and analyze the case when the reference potential is attractive and can support a number, preferably a large number, of bound states.

We shall postulate that for a given total energy E there is a classically accessible region of interatomic distance for

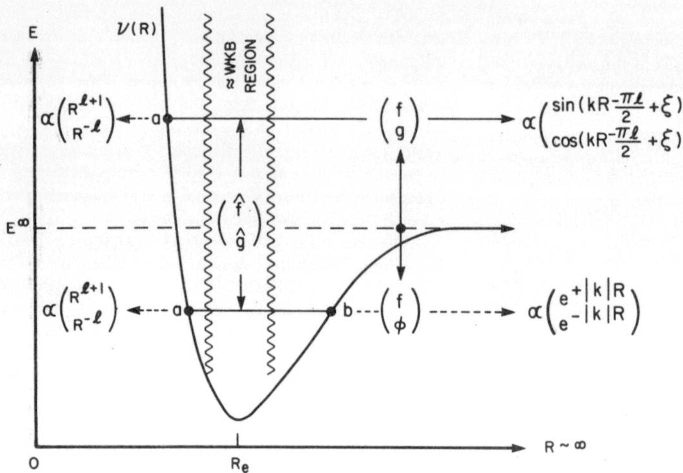

FIG. 1. Schematic figure of reference potential $\mathscr{V}(R)$ with indication of range, and limiting properties of the various reference functions. Our construction of the analytic functions $(\hat{f},\hat{g})$ is assisted by the visualization of a narrow, WKB-valid region in the vicinity of the equilibrium position R_e. This concept yields the approximate definitions in Table IV. However, the exact quantal functions are constructed according to the prescriptions in Table III.

each channel employed in the expansion of Eq. (1) such that the effective wave number $K_\gamma(E,R)$,

$$K_\gamma(E,R) = \sqrt{2\mu[E - \mathscr{V}_\gamma(R)]/\hbar^2} \tag{9}$$

is real and positive.[14] For open channels this region extends from some inner turning point $R = a(\gamma)$, where $K_\gamma(E,a) = 0$ to $R = \infty$.[15] For attractive potentials, with $E < E_\gamma^\infty$, there is a finite region extending from the inner $R = a(\gamma)$ to outer $R = b(\gamma)$ turning point defined by $K_\gamma(E,b) = 0$. These are shown schematically in Fig. 1. Beyond $R = b$ the effective wave number $K_\gamma = i|K_\gamma|$ is pure imaginary and approaches $i|k_\gamma|$ as $R \to \infty$. This identifies the channel as closed and introduces asymptotic divergences in the single channel solutions.

The function $f_\gamma(E,R)$ defined by the two boundary conditions listed in Table I is but one of an infinite variety of particular solutions which satisfy the following second-order equation

$$\left[\frac{\partial^2}{\partial R^2} + K_\gamma^2(E,R)\right] f_\gamma(E,R) = 0. \tag{10}$$

A total of five such functions are defined in Table I each of which perform a useful role in MCQDA. The pair of *independent* functions f_γ and g_γ, which are real, with an energy-independent Wronskian normalized to unity, i.e.,

$$W[g_\gamma, f_\gamma] \equiv g_\gamma \frac{\partial f_\gamma}{\partial R} - f_\gamma \frac{\partial g_\gamma}{\partial R} = 1 \tag{11}$$

are used *above* threshold to describe the scattering in the potential field $\mathscr{V}_\gamma(R)$. Three of the four boundary conditions needed to define this pair are imposed at $R = \infty$; the remaining condition

$$f_\gamma(E,R) \underset{R\to 0}{\propto} R^{l+1} \tag{12}$$

is introduced to insure that f_γ is well behaved as $R \to 0$. *Below* threshold both f_γ and g_γ become asymptotically divergent,

TABLE I. Definition of reference solutions.

Function (domain)[d]	Log. deriv.	Boundary condition[a] at $R = 0$	Boundary condition[a,b,d] at $R = R_e$	Boundary condition[a,c,d] at $R \sim \infty$
$f(E,R)$	$s = f'/f$	$\boxed{s \sim (l+1)/R}$	$f = C^{-1}\hat{f}$	$\boxed{f^2(k^2+s^2) \sim k}$
$(E>0)$		$\therefore f \propto R^{l+1}$		
$g(E,R)$	$u = g'/g \underset{E\to 0}{\sim} 0$	$g \propto R^{-l}$	$g = C\{\hat{g} + \hat{f}\tan\lambda\}$	$\boxed{g \sim sf/k}$
$(E>0)$		$\therefore$ irregular		$\boxed{u \sim -k^2/s}$
$\phi(E,R)$	$t = \phi'/\phi \underset{E\to 0}{\sim} 0$	$\phi \propto R^{-l}$, irregular	$\phi = \mathscr{N}\{\hat{f}\cos\nu - \hat{g}\sin\nu\}$	$\boxed{t = -\lvert k\rvert}$
$(E<0)$		unless $\nu(E) \equiv 0$		$\boxed{\phi \sim e^{-\lvert k\rvert R}/2\sqrt{\lvert k\rvert}}$
$\hat{f}(E,R)$	$\hat{s} = \hat{f}'/\hat{f}$	$\boxed{\hat{s} \sim (l+1)/R}$	$\boxed{\hat{f}^2(K^2+\hat{s}^2) \sim K}$	generally divergent,
(all E)	$\equiv s$	$\therefore \hat{f} \propto R^{l+1}$		$E<0$
$\hat{g}(E,R)$	$\hat{u} = \hat{g}'/\hat{g}$	$\hat{g} \propto R^{-l}$, irregular	$\boxed{\hat{g} \sim \hat{s}\hat{f}/K}$	generally divergent
(all E)		for all E	$\boxed{\hat{u} \sim -K^2/\hat{s}}$	$E<0$

[a] The two conditions which uniquely define each function are indicated by the boxed expressions.
[b] $K = K(E,R) = \sqrt{2\mu(E - \mathscr{V}(R))/\hbar^2}$. Parameters $\lambda = \lambda(E)$ and $\nu = \nu(E)$ are independent of R.
[c] $k \equiv K(E,\infty) = \sqrt{2\mu E/\hbar^2}$.
[d] Energy chosen such that $E = 0$ indicates dissociation threshold, with $\mathscr{V}(\infty) \equiv 0$.

and the asymptotically well-behaved function

$$\phi_\gamma(E,R) \underset{R\to\infty}{\sim} e^{-|k_\gamma|R}/2\sqrt{|k_\gamma|} \tag{13}$$

is introduced to facilitate the calculation of resonance and bound states in closed channels.

Although these three functions satisfy all the boundary conditions we would need to completely define the properties in any channel, they are not well suited to a MCQDA. Our intent is to analytically continue the exact close-coupled solutions in Eq. (5) across thresholds for selected channels: unfortunately the branch point associated with $k_\gamma = 0$ in Eq. (4) causes f_γ, g_γ, and ϕ_γ to be nonanalytic at $E = E_\gamma^\infty$. The problem originates from choosing boundary conditions at $R = \infty$. For this reason we introduce the independent pair of functions $\hat{f}_\gamma$ and $\hat{g}_\gamma$ which are constructed analogous to f_γ and g_γ, i.e.,

$$W[\hat{g}_\gamma,\hat{f}_\gamma] = 1, \tag{14a}$$

$$\hat{f}_\gamma(E,R) \underset{R\to 0}{\propto} R^{l+1}. \tag{14b}$$

However in this case we impose the three remaining boundary conditions at the position of the minimum $R = R_e$ for the potential $\mathscr{V}_\gamma(R)$, rather than at $R = \infty$. Using the conditions listed in Table I we obtain a *pair of solutions which remain analytic in E for all energies which exceed the minimum in the potential*, i.e., $E > \mathscr{V}_\gamma(R_e)$. This set of solutions will permit us to analytically continue the close-coupled solutions across thresholds into the classically accessible regions associated with closing channels.

Given two *independent* solutions such as $\hat{f}_\gamma(E,R)$ and $\hat{g}_\gamma(E,R)$ which suit our convenience we may uniquely specify any other particular solution of Eq. (10) at energy E as a linear combination of the chosen set. In this case we find,

$$f_\gamma(E,R) = C_\gamma^{-1}(E)\hat{f}_\gamma(E,R), \quad E \geqslant E_\gamma^\infty, \tag{15}$$

$$g_\gamma(E,R) = C_\gamma(E)\{\hat{g}_\gamma(E,R) + \tan\lambda_\gamma(E)\hat{f}_\gamma(E,R)\},$$
$$E \geqslant E_\gamma^\infty, \tag{16}$$

$$\phi_\gamma(E,R) = \mathscr{N}_\gamma(E)\{\cos\nu_\gamma(E)\hat{f}_\gamma(E,R)$$
$$- \sin\nu_\gamma(E)\hat{g}_\gamma(E,R)\}, \quad E_\gamma^\infty \geqslant E. \tag{17}$$

The coefficients which appear in the matrix of Wronskian relationships in Table II exercise a prominent role in the application of the MCQDA, and it is imperative that we devise numerically reliable algorithms to obtain these energy-dependent parameters. These are summarized in Table III.

The two parameters $C_\gamma(E)$ and $\tan\lambda_\gamma(E)$ which are defined above threshold by the scattering radial functions f_γ and g_γ allow us to predict the cusp and threshold behavior of the multichannel scattering cross sections[16] (see Appendix in paper II). It will be shown that, as $E \to E_\gamma^\infty$,

$$C_\gamma(E) \propto k_\gamma^{-l-1/2}, \tag{18}$$

$$\tan\lambda_\gamma(E) \underset{E\to E_\gamma^\infty}{\sim} -\cot\nu_\gamma(E_\gamma^\infty). \tag{19}$$

The properties of the scattering functions f_γ and g_γ are discussed in detail in Sec. II B. Note that the particular definitions chosen in Table I insure that at some energy increment Δ above threshold the solutions, (f_γ,g_γ) and $(\hat{f}_\gamma,\hat{g}_\gamma)$ will coincide, i.e.,

$$\left.\begin{array}{l} C_\gamma(E)\sim 1 \;, \\ \tan\lambda_\gamma(E)\sim 0 \;, \end{array}\right. E \geqslant E_\gamma^\infty + \Delta. \tag{20}$$

We will find that Δ is generally of the order of 10–20 cm^{-1} for the reference potentials used in this study and except for analysis of data very close to threshold we may use these two sets of independent functions interchangeably to represent *open* channel properties.

The evaluation of the energy-dependent phase $\nu_\gamma(E)$ in Eq. (17) is of paramount importance in the development of the MCQDT [we shall have no particular need for $\mathscr{N}_\gamma(E)$]. In general the asymptotically well-behaved function $\phi_\gamma(E,R)$ is divergent at $R = 0$ and approaches $\phi \propto R^{-l}$. However, if we evaluate the Wronskian between ϕ_γ and the function $\hat{f}_\gamma(E,R)$, which is well behaved at the origin, we obtain

$$W[\phi_\gamma,\hat{f}_\gamma] = -\mathscr{N}_\gamma(E)\sin\nu_\gamma(E) \tag{21}$$

and we find that the two functions are equivalent at specific energies $E \equiv E_n^\gamma < E_\gamma^\infty$ such that

$$\nu_\gamma(E) \underset{E\to E_n^\gamma}{\sim} n\pi. \tag{22}$$

This energy coincides with the vibrational eigenvalues defined by the attractive reference potential $\mathscr{V}_\gamma(R)$. The function $\hat{f}_\gamma(E_n^\gamma,R)$ (or alternately ϕ_γ) is well behaved at *both* $R = 0$ and $R = \infty$ for this particular energy and can be normalized[1] to yield an orthonormal set of vibrational wave functions,

$$|E_n^\gamma\rangle \equiv \left(\frac{4\mu}{\hbar^2\pi}\right)^{1/2}\left(\frac{\partial\pi E}{\partial\nu_\gamma(E)}\right)^{1/2}_{E_n^\gamma}\hat{f}_\gamma(E_n^\gamma,R). \tag{23}$$

The function $\phi_\gamma(R)$, and the parameter $\nu_\gamma(E)$ are discussed in Sec. II C.

The analytic properties of $\hat{f}_\gamma$ and $\hat{g}_\gamma$ and the relationships presented in Table III are discussed in Sec. II D. Although our results are completely quantum mechanical, the *interpretation* of the MCQDA analysis is greatly facilitated by recognizing the approximate WKB relationships sum-

TABLE II. Matrix of mixed Wronskians $W(a,b) = ab' - a'b$ between five reference solutions.

$W(a,b)$	$\hat{f}$	$\hat{g}$	f	g	ϕ
$\hat{f}$	0	-1	0	$-C(E)$	$\mathscr{N}(E)\sin\nu(E)$
$\hat{g}$	$+1$	0	$C^{-1}(E)$	$C(E)\tan\lambda(E)$	$\mathscr{N}(E)\cos\nu(E)$
f	0	$-C^{-1}(E)$	0	-1	undefined
g	$C(E)$	$-C(E)\tan\lambda(E)$	$+1$	0	undefined
ϕ	$-\mathscr{N}(E)\sin\nu(E)$	$-\mathscr{N}(E)\cos\nu(E)$	undefined	undefined	0

TABLE III. Definition of MCQDT parameters.

Function	Energy domain	
f	$E>0$	$f=C^{-1}\hat{f}$
g	$E>0$	$g=C\{\hat{g}+\tan\lambda\hat{f}\}$
ϕ	$E<0$	$\phi=\mathcal{N}\{\cos\nu\hat{f}-\sin\nu\hat{g}\}$

All parameters[a)] are to be evaluated in the limit as $R\to R_e$, i.e.,
$C(E)^{-2}=(f^2K+f'^2/K)$, $\mathcal{N}(E)^2=(\phi^2K+\phi'^2/K)$

$\cot\lambda(E)=\frac{K(s-u)}{(K^2+su)}\equiv\frac{K(f'g-g'f)}{(K^2fg+f'g')}$

$\tan\nu(E)=\frac{K(t-s)}{(K^2+st)}=\frac{K(f\phi'-\phi f')}{(K^2f\phi+f'\phi')}\equiv$meromorphic function E

[a] $\tan\lambda\tan\nu\underset{E\to 0}{\sim}-1$.

marized in Table IV. These are derived by assuming, as in Fig. 1, that a narrow region of internuclear distance which spans R_e is amenable to semiclassical analysis. This requires that

$$K_\gamma(E,R_e)=|K_\gamma(E,R_e)|\gg\sqrt{|\partial K_\gamma(E,R_e)/\partial R_e|}\,. \tag{24}$$

For example, in many instances we can employ an approximation for the phase $\nu_\gamma(E)$ that can be derived from the WKB connection formula,[17,18] i.e.,

$$\nu_\gamma(E)\approx\int_{a(\gamma)}^{b(\gamma)}K_\gamma(E,R)dR-\pi/2. \tag{25}$$

When used in Eq. (22) this yields the Bohr–Sommerfeld approximation[19] for bound states. This can be used as a first pass in deriving the exact quantum mechanical phase discussed in Sec. II C. Actually the MCQDA would be of limited value unless the semiclassical interpretation were valid. Breakdown of the WKB approximation at $R\cong R_e$ would only occur for very shallow potentials which supported very few bound states and exact close coupled wave functions could be easily and economically generated without recourse to the MCQDA.

TABLE IV. Relationships between exact quantal and approximate WKB solutions[a] in the vicinity of $R\cong R_e$.

Function[a,b,c]	Logarithmic derivative
$\hat{f}\cong\frac{\sin\beta}{K^{1/2}}$	$\hat{s}\cong K\cot\beta$
$\hat{g}\cong\frac{\cos\beta}{K^{1/2}}$	$\hat{u}\cong-K\tan\beta$
$f\cong\frac{C^{-1}\sin\beta}{K^{1/2}}$	$s\equiv\hat{s}$
$g\cong\frac{C}{\cos\lambda}\frac{\cos(\beta-\lambda)}{K^{1/2}}$	$u\cong-K\tan(\beta-\lambda)$
$\phi\cong\frac{\mathcal{N}\sin(\beta-\nu)}{K^{1/2}}$	$t\cong K\cot(\beta-\nu)$

[a] Derived assuming $K'(E,R)\cong 0$ in region $R\cong R_e$ such that $\beta(E,R)\cong\int_a^R K(E,R')\,dR'+\pi/4$, $\beta'(E,R)\cong K(E,R)$, where a = inner classical turning point defined by $K^2(E,a)=2\mu[E-\mathcal{V}(a)]/\hbar^2=0$.

[b] The analytic function $\hat{f}$ can be related to β as follows, proving that β is analytic in E, $\partial\beta(E,R)/\partial E\cong 2\mu/\hbar^2\int_0^R\hat{f}^2(E,R')\,dR'$.

[c] Below threshold there is a second classical turning point at $R=b>a$, such that $\nu(E)\approx\int_a^b K(E,R)\,dR-\pi/2$.

Readers who are content to accept the definitions in Tables I, II, and III and are not interested in the mathematical niceties may proceed to Sec. III.

B. Scattering solutions f_γ, g_γ above threshold

The pair of solutions f_γ and g_γ are used above threshold to describe the scattering behavior in open channels. They are characterized by an energy-dependent phase $\xi_\gamma(E)$ which defines the asymptotic properties of these independent functions. The asymptotic boundary condition in Table I is equivalent (to within an arbitrary phase factor) to the condition

$$\begin{aligned}f_\gamma(E,R)&\underset{R\to\infty}{\sim}\cos\xi_\gamma(E)J_\gamma(E,R)+\sin\xi_\gamma(E)N_\gamma(E,R)\\&\sim k_\gamma^{-1/2}\sin[k_\gamma R-\pi l/2+\xi_\gamma(E)].\end{aligned} \tag{26}$$

The corresponding limit for the independent function $g_\gamma(E,R)$ is

$$\begin{aligned}g_\gamma(E,R)&\underset{R\to\infty}{\sim}-\sin\xi_\gamma(E)J_\gamma(E,R)+\cos\xi_\gamma(E)N_\gamma(E,R)\\&\sim k_\gamma^{-1/2}\cos[k_\gamma R-\pi l/2+\xi_\gamma(E)].\end{aligned} \tag{27}$$

The spherical Bessel functions $J_\gamma(E,R)$ and $N_\gamma(E,R)$[1] in Eqs. (26) and (27) are the analogous solutions obtained for the pure centrifugal potential such that $W[N_\gamma,J_\gamma]=1$ and J_γ is well behaved as $R\to 0$. The parameter $\xi_\gamma(E)$ is the scattering phase shift[21] associated with the potential $\mathcal{V}_\gamma$. In the absence of any interaction, when $\mathcal{V}_\gamma$ is equivalent to Eq. (8) for all R, the phase shift $\xi_\gamma=0$ and $f_\gamma\equiv J_\gamma$. Otherwise, $f_\gamma(E,R)$ only asymptotically approaches Eq. (26) and ξ_γ yields the elastic scattering cross section[21]

$$\sigma_\gamma(E)=\frac{(2l+1)4\pi}{k_\gamma^2}\sin^2\xi_\gamma(E). \tag{28}$$

Note that $\xi_\gamma(E)$ is indeterminant by modular π. However, if we define $\xi_\gamma(\infty)=0$, then Levinson's theorem[1,20,21] predicts that $\xi_\gamma(E_\gamma^\infty)=(n^*+1)\pi$, where (n^*+1) is the number of bound states supported by the potential $\mathcal{V}_\gamma(R)$. The phase approaches this threshold value as

$$\xi_\gamma(E)\underset{E\to E_\gamma^\infty}{\sim}(n^*+1)\pi+(k_\gamma R_0)^{2l+1}. \tag{29}$$

The distance R_0 is a range parameter characteristic to each particular potential; for scattering in an $l=0$ potential R_0 is the scattering length.[9] We shall have need of Eq. (29) in the Appendix of paper II. The scattering phase shift in Fig. 4 of paper II conforms to Eq. (29) with $l=0$.

Because of the irregular behavior of $g_\gamma(E,R)\propto R^{-l}$ as $R\to 0$, this solution of Eq. (10) is not square integrable and would be rejected as "unphysical" if $\mathcal{V}_\gamma$ represented the exact elastic scattering potential and there were no further couplings to other channels. However, in the analysis of the close-coupled wave functions $g_\gamma(E,R)$ will play a role equally prominent with $f_\gamma(E,R)$. These functions must be recognized as mathematically useful reference functions which may be used to *replace* the Bessel functions $J_\gamma(E,R)$ and $N_\gamma(E,R)$ which are normally used to represent the asymptotic properties of the exact, physically meaningful total wave functions in Eq. (1). With well-chosen references potentials $\mathcal{V}_\gamma$ the validity of the expansion may hopefully be drawn into *finite* distances where analytic continuations across thresholds can be performed.

As noted in Eq. (20), the functions (f_γ, g_γ) become equivalent to the analytic functions $(\hat{f}_\gamma, \hat{g}_\gamma)$ at some energy Δ above threshold. This energy must be sufficient to insure that the WKB criterion in Eq. (24) remains valid from $R = R_e$ to $R = \infty$. In this case the logarithmic derivative $s_\gamma(E,R)$ approaches $K_\gamma \cot \beta_\gamma$ in Table IV, and $C_\gamma(E)$ approaches one in Eq. (20) as the following integral approaches zero:

$$\ln C_\gamma^2(E) \underset{E > E_\gamma^\infty}{\approx} \int_{R_e}^{\infty} \frac{K'_\gamma(E,R)}{K_\gamma(E,R)} \cos 2\beta_\gamma(E,R)\, dR. \tag{30}$$

As k_γ increases both $2K'_\gamma/K_\gamma \equiv -\mathscr{V}'_\gamma/(E - \mathscr{V}_\gamma)$ and the period of oscillation for $\cos 2\beta_\gamma$ decrease, causing a rapid decline in the magnitude of the integral as we leave the threshold region. As mentioned earlier, the deviation of Eq. (30) from zero becomes entirely negligible at energies of the order of 10–20 cm^{-1} above threshold and f_γ and $\hat{f}_\gamma$ are essentially equivalent. Similar arguments apply to the phase $\tan \lambda_\gamma(E)$ in Eq. (16), which approaches the following limit as we pass into this energy regime:

$$\tan \lambda_\gamma(E) \underset{E > E_\gamma^\infty}{\approx} \frac{1}{K_\gamma^2(E,R_e)} \int_{\infty}^{R_e} K_\gamma(E,R) K'_\gamma(E,R) \times \sin 2\beta_\gamma(E,R)\, dR. \tag{31}$$

Thus the irregular functions g_γ and $\hat{g}_\gamma$ also become equivalent, and can be used interchangeably except close to threshold. The threshold properties of $C_\gamma(E)$ and $\tan \lambda_\gamma(E)$ are discussed in Sec. II D.

C. The Asymptotically decaying reference solution $\phi_\gamma(E,R)$ below threshold: Evaluation of $\nu_\gamma(E)$

Below threshold both J_γ and N_γ in Eqs. (24) and (25) become asymptotically divergent,

$$J_\gamma(E,R) \sim \{e^{|k|R} - (-1)^l e^{-|k|R}\}/2\sqrt{|k|}, \tag{32a}$$

$$N_\gamma(E,R) \sim \{e^{|k|R} + (-1)^l e^{-|k|R}\}/(-1)^l 2\sqrt{|k|}, \tag{32b}$$

and it is numerically impossible to integrate Eq. (10) with sufficient accuracy to maintain account of the decaying components in Eqs. (32) and (26). Asymptotically $s(E,R) \sim |k|$ and both f_γ and $\hat{f}_\gamma$ are proportional to $e^{|k|R}$ with *undetermined* contributions from $e^{-|k|R}$ terms which make the evaluation of the phase $\xi_\gamma(E < E_\gamma^\infty)$ impossible. This is equivalent to our inability to normalize f_γ as $k^2 + s_\gamma^2 \sim 0$. [Note that the analytic function $\hat{f}_\gamma(E,R)$ is normalized at $R = R_e$ and this particular solution can be numerically evaluated for all E and all R.] However, without ξ_γ and $f_\gamma(E,\infty)$ we cannot uniquely define or numerically evaluate $g_\gamma(E,R)$ below threshold. Our definitions in Table I pinpoint the problem. With $s_\gamma(E,R)$ approaching a constant, $|k_\gamma|, u_\gamma(E,R)$ also equals $|k_\gamma|$ and g_γ becomes proportional to $f_\gamma(E,R)$ at large distances since we have lost track of the small components.

To complement $f_\gamma(E,R)$ below threshold we can introduce the particular solution $\phi_\gamma(E,R)$ is Eq. (21) which isolates the decaying components in Eq. (32), i.e.,

$$\phi_\gamma(E,R) \underset{R\to\infty}{\sim} [N_\gamma(E,R) - (-1)^l J_\gamma(E,R)]/2 \sim e^{-|k|R}/2\sqrt{|k|}. \tag{33}$$

In this case the logarithmic derivative in Table I is *chosen* to equal

$$t_\gamma(E,R) \equiv \phi'_\gamma(E,R)/\phi_\gamma(E,R) \underset{\substack{R\to\infty \\ E < E_\gamma^\infty}}{\sim} -|k| \tag{34}$$

and, together with the normalization prescribed in Eq. (33), we can always numerically evaluate this *unique* solution below threshold.

The definition of $\tan \nu_\gamma(E)$ in Table III, i.e.,

$$\tan \nu_\gamma(E) \underset{R\to R_e}{=} \frac{K_\gamma(\hat{f}_\gamma \phi'_\gamma - \phi_\gamma \hat{f}'_\gamma)}{K_\gamma^2 \hat{f}_\gamma \phi_\gamma + \hat{f}'_\gamma \phi'_\gamma} \equiv \frac{K_\gamma(t_\gamma - s_\gamma)}{K_\gamma^2 + t_\gamma s_\gamma}, \tag{35}$$

yields a meromorphic function of E which can be accurately evaluated at all energies below threshold using existing numerical codes.[16] The WKB solutions in Table IV yield the approximation in Eq. (25) which is often sufficiently accurate in itself, or can be used as a guide in evaluating the exact quantal expression (35). The solutions $\hat{f}_\gamma$ and ϕ_γ are independent, except when $\tan \nu_\gamma' = 0$, and Eq. (22) defines the vibrational eigenvalues for the potential $\mathscr{V}_\gamma(R)$.

In some of the numerical examples of MCQDA we shall employ a Morse potential,

$$\mathscr{V}_\gamma(R) = D_e(e^{-\alpha(R - R_e)} - 1)^2 - D_e \underset{R\to\infty}{\sim} 0, \tag{36a}$$

for which an analytic expression for $\nu_\gamma(E)$ can be derived,

$$\nu_\gamma(E) + \pi/2 = \frac{\pi}{2} \Sigma_\gamma \left\{ 1 - \sqrt{\frac{-E}{D_e}} \right\}. \tag{36b}$$

This is plotted in Fig. 3 of paper II for a specific value of the parameter

$$\Sigma_\gamma \equiv \sqrt{8\mu_{AB} D_e/\hbar^2 \alpha^2}.$$

The modular π values of ν_γ conform to the exact quantum mechanical eigenvalues

$$E_n^\gamma = -D_e \left\{ \left(\frac{2n+1}{\Sigma_\gamma} \right) - 1 \right\}^2, \quad n = 0,1, \ldots, n^*, \tag{36c}$$

where the number of bound states $(n^* + 1)$ supported by $\mathscr{V}_\gamma$ is prescribed by the condition $(n^* + 1) \leqslant (\Sigma_\gamma + 1)/2$. Of course the exact agreement between the quantal and WKB eigenvalues is a special feature of Morse (and harmonic) potentials, but in general, the analytic properties of $\nu_\gamma(E)$ can be derived from Eq. (25) and fine tuned with the exact values derivable from Eq. (35). Alternately we may simply use the $(n^* + 1)$ eigenvalues obtained for any given potential and use the functional form suggested by Eq. (36b) to interpolate $\nu_\gamma(E)$ or obtain a spline fit to extract intermediate energy values. The threshold value $\nu_\gamma(E_\gamma^\infty)$ to which we must extrapolate beyond the last eigenvalue $E_{n^*}^\gamma$ can be obtained from the condition $\tan \nu_\gamma(E_\gamma^\infty) \equiv -\cot \lambda_\gamma(E_\gamma^\infty)$ given in Eq. (19). This procedure was used to obtain the results in Fig. 4 of paper II.

D. The analytic solutions $\hat{f}_\gamma(E,R)$ and $\hat{g}_\gamma(E,R)$

Seaton[4–6] has thoroughly explored the analytic solutions associated with the Coulomb potential, and Greene *et al.*[12] have generalized the MCQD theory to asymptotic potentials proportional to R^{-p} with $p = 0,2$. As in GRF,[9] we shall present an analysis for arbitrary potentials. This has already been accomplished by Colle[7] for a two channel analysis of predissociation using a complete WKB interpretation of the reference radial functions for all internuclear distances, including the classical turning points $K_\gamma^2(E,R) = 0$, and the nonclassical regions $K_\gamma^2(E,R) < 0$, defined by Eq. (9). Although the structure of the boundary conditions *chosen* in Table I have been influenced by the WKB analysis of these functions in the vicinity of $R = R_e$, the present theory, as well as GRF[9] is rigorous and *not* committed to the approximation in Eq. (24).

Let us first consider the two functions $f_\gamma(E,R)$ and $\hat{f}_\gamma(E,R)$ which are well behaved at the origin. If we assign the following logarithmic derivative to $f_\gamma(E,R)$, i.e.,

$$s_\gamma(E,R) = f'_\gamma(E,R)/f_\gamma(E,R) \underset{R\to 0}{\sim} (l+1)/R, \tag{37}$$

we insure that Eq. (12) is satisfied. With this particular boundary condition we find $s_\gamma(E,R)$ is a meromorphic function of E and R with simple poles, which satisfies the Ricardi equation[22]

$$s'_\gamma(E,R) = -(s_\gamma^2 + K_\gamma^2). \tag{38}$$

In practice we are permitted to begin the numerical integration of Eq. (38) at any distance $R^* < a$, well within the nonclassical region such that $\mathscr{V}_\gamma(R^*) \gg E$. Assuming that

$$|K_\gamma(E,R)|_{R=R^*} \gg \sqrt{|\partial K_\gamma(E,R)/\partial R|}, \tag{39}$$

we may choose $s_\gamma(E,R^*) = +\sqrt{-K_\gamma^2(E,R)}$ and obtain the required solution for $R > R^*$. The resultant solution for $f_\gamma(E,R)$ obtained from Eq. (37) is

$$f_\gamma(E,R) = f_\gamma(E,R^*)\exp\left[\int_{R^*}^{R} s_\gamma(E,R')dR'\right]. \tag{40}$$

Note that any poles in the function $s_\gamma(E,R) \propto [R - R_n(E)]^{-1}$ are associated with zeros in the function $f_\gamma(E,R) \propto [R - R_n(E)]$ and the exponential factor in Eq. (40) is an entire function of R. At finite values of R the position of the poles $R_n(E)$ are analytic functions of E and, except for the factor $f_\gamma(E,R^*)$, $f_\gamma(E,R)$ is an entire function of E.

Obviously to uniquely specify $f_\gamma(E,R)$ we must apply a second boundary condition which can define the integration constant $f_\gamma(E,R^*)$. In scattering theory it is useful to apply the boundary conditions at $R \sim \infty$ for energies $E > E_\gamma^\infty$ when k_γ in Eq. (4) is real and positive, i.e.,

$$f_\gamma^2(k_\gamma^2 + s_\gamma^2) \underset{R\to\infty}{\sim} k_\gamma. \tag{41}$$

Unfortunately this causes $f_\gamma(E,R)$, and hence $f_\gamma(E,R^*)$ to be nonanalytic in E at the branch point $k_\gamma = 0$. However, any solution of Eq. (10), such as $\hat{f}_\gamma(E,R)$ in Table I, which shares the inner boundary condition imposed by Eq. (37) is well behaved at the origin and is in fact proportional to $f_\gamma(E,R)$ for all R and all E, and we obtain Eq. (15). We shall specify $\hat{f}_\gamma$ by its value at $R = R_e$, i.e.,

$$\hat{f}_\gamma^2(E,R)[K_\gamma^2(E,R) + s_\gamma^2(E,R)] \underset{R\to R_e}{\sim} K_\gamma(E,R). \tag{42}$$

This insures that $\hat{f}_\gamma(E,R^*)$ and, hence, $C_\gamma(E)f_\gamma(E,R^*)$ in Eq. (5) is analytic at threshold, and for all energies in excess of $\mathscr{V}_\gamma(R_e)$. The threshold value for C_γ^2 in Eq. (18) is obtained by comparing the behavior of the two functions $\hat{f}_\gamma$ and f_γ in the regions of interatomic distance where $\mathscr{V}_\gamma$ approaches Eq. (8), and $k_\gamma R$ remains less than unity,[21] i.e.,

$$\hat{f}_\gamma = \text{const}\times\left\{A\left(\frac{R}{R_0}\right)^{l+1} + B\left(\frac{R}{R_0}\right)^{-l}\frac{\tan\xi_\gamma(E)}{(k_\gamma R_0)^{2l+1}}\right\}, \tag{43}$$

$$f_\gamma \underset{\substack{R\to\infty\\ k_\gamma R<1}}{=} \cos\xi_\gamma(k_\gamma R_0)^{l+1/2}\times\{\qquad\}.$$

The scattering length R_0 is determined from the threshold behavior of the phase shift, i.e.,

$$\frac{\tan\xi_\gamma(E)}{(k_\gamma R_0)^{2l+1}} \underset{k_\gamma\to 0}{\sim} 1, \tag{44}$$

consistent with Eq. (29). The constraint

$$C_\gamma(E)\cos\xi_\gamma(k_\gamma R_0)^{l+1/2} \underset{E\to E_\gamma^\infty}{=} \text{const} \tag{45}$$

derived from Eqs. (15) and (43) yields the threshold condition in Eq. (18).

The irregular function $g_\gamma(E,R)$ may be obtained from a numerical integration of Eq. (10) using the asymptotic boundary conditions imposed in Table I. However since $s_\gamma(E,R)$ and $f_\gamma(E,R)$ are ill defined at $R = \infty$ for $E < E_\gamma^\infty$, the function is indeterminate below threshold. However, the function $\hat{g}_\gamma(E,R)$ is defined for all energies $E > \mathscr{V}_\gamma(R_e)$ and can be easily integrated in either direction using the initial conditions prescribed in Table I at $R = R_e$. Examining the high energy behavior of Eqs. (30) and (31) we see that the two functions g_γ and $\hat{g}_\gamma$ become equivalent. The threshold condition in Eq. (20) for the $\tan\lambda_\gamma(E)$ coefficient in Eq. (17) is obtained by recognizing that, as $E\to E_\gamma^\infty$, the logarithmic derivative functions for both $g_\gamma(E_\gamma^\infty,R)$ and $\phi_\gamma(E_\gamma^\infty,R)$ are identical for all R, i.e.,

$$\mu_\gamma(E_\gamma^\infty,R) = t_\gamma(E_\gamma^\infty,R). \tag{46}$$

Using this in the definitions of $\tan\nu_\gamma$ and $\tan\lambda_\gamma$ given in Table III yields the threshold constraint in Eq. (19).

III. FORMULATION OF THE MCQDA THEORY

The essential step in developing a MCQDA theory[4–6,9,23] is to select an appropriate reference potential $\mathscr{V}_\gamma(R)$ for each channel state $\psi_\gamma(\mathbf{r},\hat{R})$ in Eq. (1). Each channel radial component $F_{\gamma',\gamma}(E,R)$, whether open or closed, can then be uniquely specified in terms of the two independent radial functions $\hat{f}_\gamma(E,R)$ and $\hat{g}_\gamma(E,R)$ which describe the unperturbed solutions in the given reference potential. Our purpose is to obtain a rigorous matrix representation of the solution vectors of the form

$$\mathbf{F}(E,R) \equiv \mathscr{U}(R)[\hat{\mathbf{f}}^0(E,R) + \hat{\mathbf{g}}^0(E,R)\mathscr{Y}(E,R)]\hat{\mathbf{A}}(E,R), \tag{47}$$

where the superscript 0 is used to designate a diagonal matrix, e.g.,

$$\hat{\mathbf{f}}^0(E,R) \equiv \{\hat{f}_\gamma(E,R)\delta_{\gamma',\gamma}\}. \tag{48}$$

The R-dependent matrix $\mathscr{Y}(E,R)$, and its asymptotic value

$$\mathbf{Y}(E) \equiv \mathscr{Y}(E,\infty) \tag{49}$$

are analytic functions of E and, in particular, we can continue these matrices across dissociation limits to predict the structure of the solution vectors $\mathbf{F}(E,R)$ as various channels open or close. Our hope is to find a set of potentials $\mathscr{V}^0$ which minimize the energy dependence of $\mathbf{Y}(E)$ and allow extensive extrapolation of the resultant solutions.

The only restriction we place on $\mathscr{V}_\gamma(R)$ is that it must reproduce the exact asymptotic behavior of the interaction matrix $\mathbf{W}^\infty(R)$ in Eq. (3), as imposed in Eq. (8). This insures that the *orthogonal* matrix $\tilde{\mathscr{M}}(R)$ in Eq. (47) approaches the unit matrix $\mathbf{1}^0$ as $R \to \infty$,

$$\tilde{\mathscr{M}}(R) \underset{R\to\infty}{\sim} \mathbf{1}^0. \tag{50}$$

Since $\hat{\mathbf{g}}^0(E,R)$ is irregular at the origin, the *real symmetric* matrix $\mathscr{Y}(E,R)$ must vanish as $R \to 0$ to prevent the solutions $\mathbf{F}(E,R)$ from becoming ill behaved, i.e.,

$$\mathscr{Y}(E,R) = \mathscr{Y}^*(E,R) = \tilde{\mathscr{Y}}(E,R) \underset{R\to 0}{\sim} 0. \tag{51}$$

As $R \to \infty$ the complete scattering behavior of the system is contained in $\mathbf{Y}(E)$ which can be related to the reactance matrix $\mathbf{K}(E)$ and the scattering matrix $\mathbf{S}(E)$ which are obtained from the close-coupled solutions of Eq. (5). The matrix $\hat{\mathbf{A}}(E,R)$ effects the normalization of the radial functions and is necessary if we wish to evaluate expectation values or matrix elements involving the total wave function $\Psi_\gamma(E)$ and its interactions induced by perturbations that are *external* to H, such as radiative couplings. However to uniquely describe the dynamics of the atom–atom scattering embodied in the interaction matrix $\mathbf{W}^\infty(R)$ we merely require $\mathbf{Y}(E)$, and have no need to determine $\hat{\mathbf{A}}(E,R)$ or the explicit R dependence of $\mathscr{Y}(E,R)$. The role of the matrix $\mathscr{M}(R)$ which is independent of energy, will be clarified below.

The reference solutions $\hat{f}_\gamma(E,R)$ and $\hat{g}_\gamma(E,R)$ were constructed to be entire functions of E for all energies $E > \mathscr{V}_\gamma(R_e)$ in excess of the potential minimum which exists at $R = R_e$. This insures that $\mathscr{Y}(E,R)$ is also *analytic throughout the same energy domain*. This can be shown from the analysis of the coupled first-order nonlinear equations one can derive for $\mathscr{Y}(E,R)$, i.e.,

$$\mathscr{Y}'(E,R) = \mathbf{P}^{\hat{f}\hat{f}} + \mathbf{P}^{\hat{f}\hat{g}}\mathscr{Y} + \mathscr{Y}\mathbf{P}^{\hat{g}\hat{f}} + \mathscr{Y}\mathbf{P}^{\hat{g}\hat{g}}\mathscr{Y}, \tag{52}$$

where $\mathbf{P}^{ab} \equiv \tilde{\mathbf{P}}^{ba}$, i.e.,

$$\mathbf{P}^{ab}(E,R) = \{\mathbf{a}^0\mathscr{M}'\tilde{\mathscr{M}}\mathbf{b}^0 - \mathbf{a}^0\mathscr{M}'\tilde{\mathscr{M}}\mathbf{b}^0\} + \mathbf{a}^0\left\{\frac{2\mu}{\hbar^2}(\mathscr{V}^0(R) - \mathscr{M}\mathbf{W}^\infty(R)\tilde{\mathscr{M}})\right\}\mathbf{b}^0. \tag{53}$$

If $\mathbf{a}^0$ and $\mathbf{b}^0$ are analytic then $\mathbf{P}^{ab}(E,R)$ and hence the solutions $\mathscr{Y}(E,R)$ must also be analytic functions of E. We are not necessarily proposing to solve Eq. (52) directly, although the structure of this equation does lead to some useful approximations.[23] However it does assure us that $\mathbf{Y}(E) = \mathscr{Y}(E,\infty)$, which can be extracted from the direct numerical solutions of the coupled equations (5), can be analytically continued across thresholds.

If we were content to equate $\mathscr{V}_\gamma$ to the left-hand side of Eq. (8) then our reference functions would simply be solutions of Bessel's equation, and correspond to the usual spherical Bessel functions, i.e., $f_\gamma \equiv J_\gamma(E,R)$ and $g_\gamma \equiv N_\gamma(E,R)$, that are normally used to describe the asymptotic properties of a scattering wave function, as in Eq. (26). Thus, at $R = \infty$, the matrix $\mathbf{Y}(E) \equiv \mathbf{K}(E)$ would directly equal the reactance matrix one obtains from Eq. (5) when all channels are open. This choice yields the usual kinematic restraints imposed by formal scattering theory, such as conservation of flux, energy, momentum, etc., but completely ignores the dynamics imposed by the interaction matrix $\mathbf{W}^\infty(R)$ at *finite* R. Our chore is to find a more useful basis which takes cognizance of the physics contained in $\mathbf{W}^\infty(R)$ as the collision partners interact.

Obviously one choice would be to equate $\mathscr{V}_\gamma(R)$ to the diagonal elements $W^\infty_{\gamma,\gamma}(R)$. This is the practice used to develop the distorted wave approximation in scattering theory.[21] Generally, a set of Born–Oppenheimer electronic states are selected[1,2] and coupled by a variety of operators.[24] In this case $\mathscr{M}(R) = \mathbf{1}^0$ and only the second group of terms in Eq. (53) contribute to the solution of $\mathscr{Y}(E,R)$.

Alternately we may prefer to use the adiabatic approximation to generate a representation. For example, we can introduce an orthogonal transformation of the channel states to obtain a set of adiabatic electronic-rotational (AER) states which are implicit functions of R,

$$\chi_\gamma = \Sigma_{\gamma'}\psi_{\gamma'}\mathscr{M}_{\gamma',\gamma}(R), \tag{54}$$

such that $\mathscr{V}_\gamma(R)$ are the eigenvalues of the interaction matrix, i.e.,

$$\mathscr{M}(R)\mathbf{W}^\infty(R)\tilde{\mathscr{M}}(R) = \mathscr{V}^0(R), \tag{55}$$

and the orthogonal matrix $\mathscr{M}(R)\cdot\tilde{\mathscr{M}}(R) = \mathbf{1}^0$ satisfies the asymptotic condition in Eq. (50). Defining a transformed set of radial functions,

$$\mathbf{G}_{\gamma',\gamma}(E,R) = \sum_{\gamma''=1}^{N_T}\mathscr{M}_{\gamma',\gamma''}(R)F_{\gamma'',\gamma}(E,R), \tag{56}$$

we obtain the equivalence,

$$\begin{aligned}\Psi_\gamma(E) &= \sum_{\gamma'=1}^{N_T}\psi_{\gamma'}F_{\gamma',\gamma}(E,R)/R \\ &= \sum_{\gamma'=1}^{N_T}\chi_{\gamma'}G_{\gamma',\gamma}(E,R)/R.\end{aligned} \tag{57}$$

In view of Eq. (50) the asymptotic properties of $\mathbf{G}(E,R)$,

$$\mathbf{G}(E,R) = \{\hat{\mathbf{f}}^0 + \hat{\mathbf{g}}^0\mathscr{Y}(E,R)\}\hat{\mathbf{A}}(E,R) \tag{58}$$

are identical to $\mathbf{F}(E,R)$ in Eq. (47). In this representation only the first term in Eq. (53) contributes to $\mathbf{Y}(E)$.

Only the asymptotic properties of ψ_γ and $\mathbf{W}^\infty$, and hence $\mathscr{V}^0$ are uniquely specified. The success of the MCQDA depends on our cleverness, or luck in choosing $\mathscr{V}_\gamma$ in Eq. (8) which generates $\hat{f}_\gamma$ and $\hat{g}_\gamma$. Fortunately the choice is generally rather obvious, and, as we shall see, the quality of the results are quite insensitive to the representation. It is conceivable that a mixture of adiabatic and diabatic reference potentials would be required for different subsets of channels, and only a partial transformation such as in Eq. (54) might be employed which would not completely diagonalize $\mathbf{W}^\infty$ in Eq. (55).

Although the choice of $\mathscr{V}_\gamma$ is arbitrary, some care should be taken to insure that $\hat{f}^0$ yields a reasonable approximation to the exact solutions as $R \to 0$. Otherwise there will be a rapid rise in the contribution of $\mathbf{g}^0\mathscr{Y}(E,R)$ to Eq. (48) as we integrate Eq. (5) or (52) and a concomitant increase in the energy variation of $\mathbf{Y}(E)$ which is what we want to avoid. In most cases $\mathbf{W}_{\gamma,\gamma}(R)$ can be expected to be exponentially repulsive at short distances and this behavior should be reflected in $\mathscr{V}_\gamma(R)$. The explicit values of $\mathbf{Y}(E)$ are very sensitive functions of the reference functions, but the ability of the resultant extrapolations to reproduce the exact close-coupled results at neighboring energies above and below the dissociation limit is very forgiving of the choice of basis functions. One virtue of MCQDA is that we may *iterate* on the choice of $\mathscr{V}^0$ until a sufficiently energy-insensitive $\mathbf{Y}(E)$ matrix is obtained.

When *all* channels in Eq. (1) are open, i.e., $N_T = N_o$, we obtain N_T well-behaved solution vectors, such that $\mathbf{F} = \mathbf{F}(N_T \times N_T)$. We can express Eq. (47) in the equivalent form

$$\mathbf{F}(E,R) = \tilde{\mathscr{M}}(R)[\mathbf{f}^0(E,R) + \mathbf{g}^0(E,R)\mathbf{r}(E,R)]\mathbf{A}(E,R), \tag{59}$$

using the pair of independent references solutions f_γ and g_γ defined in Table I since these are well defined above threshold. The scattering matrix $\mathbf{S}(E)$ obtained from the solution of Eq. (5) is simply related to the asymptotic value of $\mathbf{r}(E,R)$ which is designated as

$$\mathscr{R}(E) \equiv \mathbf{r}(E,\infty). \tag{60}$$

Using the elastic scattering phase shift $\xi_\gamma(E)$ defined for each f_γ in Eq. (26), we construct a diagonal matrix

$$e^{i\xi^0} \equiv \{e^{i\xi_\gamma(E)}\delta_{\gamma',\gamma}\}. \tag{61}$$

The prescribed asymptotic form for the irregular solution g_γ in Eq. (27) gives us the relationship

$$\mathbf{S}(E) = e^{i\xi^0}[\mathbf{1}^0 + i\mathscr{R}(E)][\mathbf{1}^0 - i\mathscr{R}(E)]^{-1}e^{i\xi^0}. \tag{62}$$

We cannot continue $\mathscr{R}(E)$ across thresholds since it is derived using the nonanalytic functions f_γ and g_γ. However, using the relationships in Table III we can relate Eq. (59) to Eq. (47) and obtain the analytic matrix $\mathbf{Y}(E)$, i.e.,

$$\mathbf{Y}(E) = \{[\mathbf{C}^0\mathscr{R}(E)\mathbf{C}^0]^{-1} + \tan\boldsymbol{\lambda}^0\}^{-1}, \tag{63}$$

$$\mathscr{R}(E) = \mathbf{C}^{0-1}[\mathbf{Y}(E)^{-1} - \tan\boldsymbol{\lambda}^0]^{-1}\mathbf{C}^{0-1}. \tag{64}$$

Except near thresholds, as noted in Eqs. (18) and (20), the limiting values $\mathbf{C}^0 = \mathbf{1}$ and $\tan\boldsymbol{\lambda}^0 = 0$ can be employed to yield

$$\mathbf{Y}(E) \equiv \mathscr{R}(E), \quad E > E_\gamma^\infty + \Delta, \ \Delta > 0. \tag{65}$$

The analytic behavior of the scattering matrix in the vicinity of the threshold is discussed elsewhere[11] using the properties extracted from Eqs. (62) and (64).

Although the resultant matrix $\mathbf{S}(E)$ is rigorously defined by Eq. (62), it must be remembered that the individual $\xi^0(E)$, $\mathscr{R}(E)$, and hence $\mathbf{Y}(E)$ matrices which reproduce the scattering matrix are *not* unique, but depend on our choice of $\mathscr{V}^0(R)$.

If we envision the development of the N_T vectors in Eq. (47) as we generate solutions to the coupled equations (5) from $R = 0$, where $\mathscr{Y}(E,0) = 0$, we must eventually reach some internuclear distance at which $\mathscr{Y}(E,R)$ becomes constant. In general this distance will approach infinity. However, depending first on the intrinsic physics contained in $\mathbf{W}^\infty(R)$, and secondarily on our wise choice of $\mathscr{V}^0(R)$, the $\mathscr{Y}(E,R)$ matrix may become constant at a finite distance $R = R^*$.

The maximum utility of the MCQDA occurs when $\mathscr{Y}(E,R)$ approaches a constant at some finite R^* which lies *entirely within the classically accessible regions* $a < R^* < b$ in Fig. 1 *for all channels* $N_T = N_o + N_c$, whether open N_o or closed N_c, i.e.,

$$\mathbf{Y}(E) \equiv \mathscr{Y}(E,\infty) = \mathscr{Y}(E,R^*), \quad a < R^* < b. \tag{66}$$

If R^* is sufficiently removed from the closest turning point defined by any of the N_T channels, we may express the bracketed portion of Eq. (47) in terms of the approximate WKB forms for $\hat{\mathbf{f}}^0$ and $\hat{\mathbf{g}}^0$ in Table IV,

$$[\hat{\mathbf{f}}^0 + \hat{\mathbf{g}}^0\mathscr{Y}] \approx \frac{\sin\beta^0(E,R)}{\sqrt{|\mathbf{K}^0(E,R)|}} + \frac{\cos\beta^0(E,R)}{\sqrt{|\mathbf{K}^0(E,R)|}}\mathbf{Y}(E), \quad R^* < R < b. \tag{67}$$

We already know that $\hat{\mathbf{f}}^0$, $\hat{\mathbf{g}}^0$, and $\mathbf{Y}$ remain analytic in E even as selected channels become closed. This is apparent in Eq. (67) since in the integration from $R = 0$ to $R = R^*$ we take no cognizance of the fact that the N_c subset of closed channels will enter a nonclassical region when $R > b$. However if Eq. (67) is fulfilled we can often expect that, in addition to being continuous, $\mathbf{Y}(E)$ will be a very slowly varying function of E. If we have to penetrate deep into the nonclassical region beyond $R = b$ before $\mathscr{Y}(E,R)$ approaches a constant, then we can expect $\mathbf{Y}(E)$ to be a more vigorous function of E with a more restricted range of extrapolation.

The *particular* set of N_T-degenerate solutions we have defined in Eqs. (57), (47), and (58) can be used to construct an infinity variety of equally exact solutions to Eq. (5), all of which are well behaved at $R = 0$,

$$\Psi_\gamma^T(E) = \sum_{\gamma'=1}^{N_T} \Psi_{\gamma'}(E)C_{\gamma',\gamma}^T(E), \tag{68}$$

as long as $(C^T)^{-1}$ exists. Such transformations lead to our identification of the $\mathbf{S}(N_T \times N_T)$ scattering matrix in Eq. (62) and the relationships in Eqs. (63) and (64) by isolating appropriate asymptotic components of the reference functions when all channels are open, and Eqs. (26) and (27) apply. However as soon as a single channel, or a N_c set of channels, becomes closed we can not accept *any* of the N_T solution vectors offered by Eqs. (47) or (58) as physically meaningful. Using the notation in Eq. (7) for $\mathbf{Y}$, and all other matrices in Eq. (47), we see that the asymptotically divergent reference functions $\mathbf{f}_{cc}^0$ and/or $\mathbf{g}_{cc}^0$ contribute to every vector unless $\mathbf{Y}_{cc}$ vanishes. Thus we must find an appropriate transformation in Eq. (68) which will segregate these offending terms and yield N_0 asymptotically well-behaved solutions, which are entitled to be called wave functions for the system, leaving N_c ill-behaved solution vectors that we can reject as mathematically valid but nonphysical.

Given the phase $\nu_\gamma(E)$ which is defined for each reference potential $\mathscr{V}_\gamma(R)$ by Eq. (35), we can utilize the definitions in Table III to find a linear combination of $\hat{f}_\gamma$ and $\hat{g}_\gamma$ functions which decays asymptotically when $E < E_\gamma^\infty$, i.e.,

$$\{\hat{\mathbf{f}}^0_{cc} - \tan \nu^0_{cc} \hat{\mathbf{g}}^0_{cc}\} \propto \boldsymbol{\phi}^0_{cc} \underset{R\to\infty}{\sim} 0, \tag{69}$$

where we use the notation

$$\tan \nu^0_{cc} = \{\tan \nu_\gamma(E)\delta_{\gamma',\gamma}\}_{cc}. \tag{70}$$

This relationship leads to a transformation which yields N_0 wave functions with radial functions which approach

$$\mathbf{F}^T_{oo} \underset{R\to\infty}{\sim} [\hat{\mathbf{f}}^0_{oo} + \hat{\mathbf{g}}^0_{oo}\mathbf{Y}^T_{oo}], \tag{71}$$

$$\mathbf{F}^T_{co} \underset{R\to\infty}{\sim} \phi^0_{cc}\mathbf{Y}^T_{co} \sim 0, \tag{72}$$

where

$$\mathbf{Y}^T_{oo} \equiv \mathbf{Y}_{oo} - \mathbf{Y}_{oc}(\mathbf{Y}_{cc} + \tan \nu^0_{cc})^{-1}\mathbf{Y}_{co}. \tag{73}$$

Asymptotically the decaying components in Eq. (72) can be ignored, and we can obtain the resultant $(N_o \times N_o)$ *unitary* scattering matrix $S_{oo}(N_o \times N_o)$ completely from the $\mathbf{Y}^T_{oo}(N_o \times N_o)$ matrix which we construct from the *full* $\mathbf{Y}$ $(N_T \times N_T)$ matrix in Eq. (7), together with $\tan \nu^0_{cc}$. The same relationships used for the completely open scattering matrix in Eqs. (62) and (64) exist here, i.e.,

$$\mathscr{R}^T_{oo} = \mathbf{C}^{0-1}_{oo}[\mathbf{Y}^{T-1}_{oo} - \tan \lambda^0_{oo}]^{-1}\mathbf{C}^{0-1}_{oo}, \tag{74}$$

$$\mathbf{S}_{oo} = e^{i\xi^0_{oo}}[\mathbf{1}^0_{oo} + i\mathscr{R}^T_{oo}][\mathbf{1}^0_{oo} - i\mathscr{R}^T_{oo}]^{-1}e^{i\xi^0_{oo}}. \tag{75}$$

Equation (74) is unduly complicated by the generality of including possible threshold effects in the open channels. In general we can expect $\mathbf{C}^0_{oo} = \mathbf{1}^0_{oo}$ and $\tan \lambda^0_{oo} = 0$ in the vicinity of newly closed channels, and therefore,

$$\mathscr{R}^T_{oo}(E) \approx Y^T_{oo}(E). \tag{76}$$

Note that in the special case, when only one open ($\gamma \equiv 0$) and one closed ($\gamma \equiv 1$) channel are involved, the exact scattering phase shift $\eta_o(E)$ obtained from $S_{oo}(1 \times 1) = e^{i2\eta_o(E)}$ is given by the following expression:

$$\eta_o(E) = \xi_o(E) + \tan^{-1}\left[Y_{oo} - \frac{Y_{01}Y_{10}}{\tan \nu_1(E) + Y_{11}}\right]. \tag{77}$$

This gives an *exact* analytic representation of the predissociation level widths, shifts, and background interferences, which we associate with the approximate bound states in potential $\mathscr{V}_o(R)$ that are located by the condition $\nu_1(E) \sim n\pi$ in Eq. (22). The use of Eq. (77) is only limited by our willingness and ability to obtain the true energy dependence of $\mathbf{Y}(E)$ over extended ranges of E. Just below threshold it is often adequate to treat $\mathbf{Y}$ as a constant, or to employ linear extrapolations to penetrate deeper into the predissociating potential.

If all channels are closed, i.e., $N_T = N_C$, there is no general transformation matrix $\mathbf{C}^T(E)$ in Eq. (68) that can yield a well-behaved solution vector for arbitrary energies. However, at specific eigenvalues $E = E_n$ we can find a normalizable bound state wave function. These energies are determined by the transcendental equations derived from the determinant

$$|\tan \nu^0(E_n) + \mathbf{Y}(E_n)| = 0. \tag{78}$$

A constant $\mathbf{Y}$ and knowledge of $\tan \nu_\gamma(E)$ for each channel makes the analysis quite simple. A (2×2) matrix is equivalent to performing a *complete* CI between two interacting electronic-rotational states in conventional spectroscopic analysis, and implicitly contains the entire summation over bound *and* continuum vibrational states. If $\mathbf{Y} = 0$ we merely retrieve the eigenvalue spectrum in each unperturbed potential $\mathscr{V}_\gamma(R)$. The matrix $\mathbf{Y}$ obtained above the dissociation limit at a single arbitrary energy summarizes the multichannel configuration interactions among the bound states to *infinite* order without recourse to perturbation theory, or neglect of continuum state contributions. Obviously this analysis is most applicable near threshold when we expect $\mathbf{Y}(E)$ to be only modestly dependent on energy. However, this is just the region that is most difficult to treat by conventional spectroscopic analysis, and the MCQDA offers a complimentary approach which is critically needed when the atomic fragments are degenerate.

IV. SUMMARY AND CONCLUSIONS

A complete description of diatomic wave functions can be developed using a channel state expansion of the total wave function. This formulation is rigorous and accurate numerical methods are available for solving the close-coupled equations in Eq. (5) which yield the radial or vibrational portion of the total wave functions. The theory incorporates the correct asymptotic properties and readily describes the continuum states (open channels) of the diatom above the dissociation limits. The resonant structure which influences the atomic scattering below the threshold for a subset of closed channels is just a reflection of the bound-free coupling which leads to predissociation in the spectroscopic view of diatomic systems, and this can be described without recourse to perturbation theories or golden rules. If one pursues the close-coupled theory into the energy regime where all channels are closed then one obtains a nonperturbative analysis of bound-state interactions which form a natural analytic continuation of couplings which prevail in continuum regimes of the spectrum.

For a given total energy E there are only N_o independent solutions of the total Hamiltonian in Eq. (6) which are well behaved asymptotically. For convenience in scattering theory we often choose one particular set of energy-normalized solutions which exhibit boundary conditions that are easily relatable to observable cross sections. However, this set is not well designed to examine the variation of the wave functions as the total energy approaches, and passes through thresholds. The MCQDA attempts to find an equally exact set of particular solutions which best permits analytic continuation of the wave functions as subsets of channels become closed.

Using the analytic reference potential radial functions $\hat{f}_\gamma$ and $\hat{g}_\gamma$ in the classically accessible portions of the open channels, and suitably attractive portions of any closed channels, the form of Eqs. (47) and (67) suggest conditions when we may expect the critical matrix $\mathbf{Y}(E)$ to be reasonably independent of E. Given $\mathbf{Y}(E)$ we may then use Eq. (75) to extrapolate the scattering matrix across thresholds.

The usefulness of the MCQDA depends on two conditions. First, and foremost, the contribution of interchannel coupling to the scattering cross section must diminish to a negligible magnitude somewhere *within* the classically accessible regions for all coupled channels. This condition is outside the control of the analyst and is simply determined by the nature of the physics contained in the interaction matrix $\mathbf{W}^{\infty}(R)$ in Eq. (2).

Distant channels which are classically inaccessible at all interatomic distances and do not contribute any resonant structure to the scattering may be included as long as their contributions to the inelastic couplings among the dominant channels has expired. This is apparent in the explicit numerical calculations of $\mathscr{R}(E)$ using an adiabatic-electronic-rotational basis to evaluate the Hg_2, Cd_2, Zn_2 scattering.[11,23] In any particular application the region of dominant coupling must be carefully scrutinized although a qualitative examination of the diagonal interaction energies $W^{\infty}_{\gamma,\gamma}(R)$ is generally sufficient to label systems as viable candidates for MCQDA or not.

Given satisfactory molecular processes the second condition is to obtain an energy-insensitive $\mathbf{Y}(E)$ matrix which can permit extensive extrapolation over wide ranges of energy, especially in the vicinity of thresholds. This then permits quantitative determination of cross sections and resonance structure without need of repeated and expensive calculations, and also offers qualitative insight into the molecular observables and spectroscopy. Again the success of the MCQDA is ultimately limited by the innate physics of the interaction. However the explicit behavior of $\mathbf{Y}(E)$ is dependent on the *choice* of reference potentials one uses to derive this quantity from the exact scattering matrix $\mathbf{S}(E)$ and the usefulness of the results can be enhanced or restricted by the chosen $\mathscr{V}^{0}(R)$. Generally the choice is self-evident, and, as seen in paper II, the results are remarkably insensitive to the potentials. In any case, the analysis is rigorous and exact for any reasonable set of $\mathscr{V}^{0}$ which conform to Eq. (8) although undesirable variation of $\mathbf{Y}(E)$ with E may result.

It may be useful to conclude with a summary of the procedures that are required in the MCQDA.

(1) For a given interaction matrix $\mathbf{W}^{\infty}(R)$ calculate the exact multichannel close-coupled wave functions and scattering matrix $\mathbf{S}(E)$ at some total energy E—generally above threshold when all channels are open.

(2) Choose a reference potential $\mathscr{V}_{\gamma}(R)$ for each channel.

(3) Obtain the elastic scattering phase shifts $\xi_{\gamma}(E)$ for each channel. (These will be required not only at E, but over the entire range of energies involved in the extrapolations.)

(4) Use $\xi_{\gamma}(E)$ to extract a unique, irreducible, real-symmetric matrix $\mathbf{Y}(E)$ from the scattering matrix.

(5) $\mathbf{Y}(E)$ may be analytically continued across thresholds to describe the predissociation of channels which become closed, or the configuration interaction among the resultant bound states when *all* channels are closed. A well-chosen set of reference potentials will minimize the energy variation of $\mathbf{Y}(E)$, but the results are rigorous even for ill-chosen potentials.

(6) Calculate the energy dependent phase $\nu_{\gamma}(E)$ at energies below the dissociation limits of all attractive closed channels such that bound states exist in the reference potentials at $\mathscr{V}_{\gamma}(E^{\gamma}_{n}) = n\pi$. This can be done exactly from solutions of the single potential Schrödinger equation, or is generally well approximated using the WKB Bohr-Sommerfeld condition.

(7) The above ingredients may be used in Eq. (75) to obtain $\mathbf{S}_{oo}(E)$ at arbitrary energies, often using a constant $\mathbf{Y}$ matrix determined many hundreds of cm^{1} away. If necessary additional values of $\mathbf{Y}(E)$ can be derived at broadly spaced energies and extrapolated values of $\mathbf{Y}(E)$ can be used to obtain more precise evaluations of $S_{oo}(E)$. Equation (77) demonstrates the concise analytic form that can be used to fit the *exact* single-open channel phase shift when resonant interaction with a single-closed channel is present. More complicated multichannel interactions yield a similar economy of expression.

ACKNOWLEDGMENT

This work is supported in part by the Air Force Office of Scientific Research, Contract No. AFOSR-ISSA-83-00038.

APPENDIX

In view of the close similarity between the present theory for diatoms and the general MCQDT presented by Greene, Rau, and Fano (GRF)[9] it may be useful to delineate the approximate equivalences in our notations.

Present	GRF
(f,g)	$(f,-g)$
$(\hat{f},\hat{g})$	$(f^{0},-g^{0})$
C_{γ}^{-1}	$B^{-1/2}$ or $A^{-1/2}$
$\tan\lambda_{\gamma}$	G
ν_{γ}	β
η_{γ}	D^{-1}
ξ_{γ}	η
$\mathbf{Y}(E)$	$-\pi\mathbf{K}^{(so)}$

[1]F. H. Mies, Mol. Phys. **41**, 953 (1980).
[2]F. H. Mies, Mol. Phys. **41**, 973 (1980).
[3]G. Herzberg, *Spectra of Diatomic Molecules*, 2nd ed. (Van Nostrand Reinhold, Princeton, 1950).
[4]M. J. Seaton, Rep. Prog. Phys. **46**, 167 (1983).
[5]M. J. Seaton, J. Phys. B **11**, 4067 (1978).
[6]M. J. Seaton, Proc. Phys. Soc. London **88**, 801, 815 (1966).
[7]R. Colle, J. Chem. Phys. **74**, 2910 (1981).
[8]A. Giusti, J. Phys. B **13**, 3867 (1980).

[9]C. H. Greene, A. R. P. Rau, and U. Fano, Phys. Rev. A **26**, 2441 (1982).
[10]A. Guisti and U. Fano, J. Phys. B (submitted).
[11]F. H. Mies and P. S. Julienne, J. Chem. Phys. **80**, 2526 (1984).
[12]C. H. Greene, U. Fano, and G. Strinati, Phys. Rev. A **19**, 1485 (1978).
[13]U. Fano, Phys. Rev. **124**, 1866 (1961).
[14]Distant channels which are classically inaccessible at *all* interatomic distances and do not contribute resonance structure to the scattering may be included (Refs. 2 and 23) as long as their contributions to the *inelastic* couplings among the dominant channels has expired.
[15]If $K_\gamma(E,R) = 0$ is multivalued we may encounter more than one accessible region, as when tunneling through rotational barriers.
[16]T. Y. Wu and T. Ohmura, *Quantum Theory of Scattering* (Prentice-Hall, Englewood Cliffs, 1962).
[17]M. S. Child, J. Mol. Spectrosc. **53**, 280 (1974).
[18]L. I. Schiff, *Quantum Mechanics* (McGraw–Hill, New York, 1955).
[19]L. D. Landau and E. M. Lifshitz, *Quantum Mechanics, Non-Relativistic Theory* (Pergamon, London, 1958).
[20]F. H. Mies and P. S. Julienne, J. Chem. Phys. **77**, 6162 (1982).
[21]N. F. Mott and H. S. W. Massey, *The Theory of Atomic Collisions*, 3rd ed. (Oxford, London, 1965).
[22]H. T. Davis, *Introduction to Nonlinear Differential and Integral Equations* (U. S. Atomic Energy Commission, Washington, D.C., 1960).
[23]P. S. Julienne and F. H. Mies, J. Phys. B **14**, 4335 (1981).
[24]J. B. Delos, Rev. Mod. Phys. **53**, 287 (1981).

Part C
Molecular problems

1977 *J. Chem. Phys.* **66** 5584–609

Rovibronic interactions in the photoabsorption spectrum of molecular hydrogen and deuterium: An application of multichannel quantum defect methods

Ch. Jungen*

Laboratoire de Photophysique Moléculaire du CNRS, Université Paris-Sud, Orsay, France and Centre de Mécanique Ondulatoire Appliquée, Paris, France

O. Atabek†

Laboratoire de Photophysique Moléculaire du CNRS, Université Paris-Sud, Orsay, France

(Received 13 December 1976)

Multichannel quantum defect theory (MQDT) is applied to the treatment of electron motion in molecular Rydberg states and its coupling to the vibrational and rotational motions of the nuclei. The full rovibronic Hamiltonian is taken into account and the theory formulated in terms of parameters which relate to the standard Born–Oppenheimer (BO) theory. The development is an extension of Seaton's quantum defect theory and relies on Fano's frame transformation method. MQDT is not restricted to the BO *approximation*, but includes the so-called adiabatic corrections and nonadiabatic effects (i.e., vibronic coupling and rotational l uncoupling), however strong, without requiring the explicit evaluation of the interactions between individual levels. The general formalism can be simplified for the H_2 and D_2 molecules in the range below or near the ionization threshold and for vibrational excitation well below the dissociation energy. The only input data required are the potential energy curve of the ion core in its ground state and the ionization potential (both of which are accurately known), plus two BO potential functions of Σ and Π symmetry, respectively. The theory has been tested in detail using recent experimental and theoretical data on the first five Rydberg absorption transitions of H_2 and D_2. Theoretical rovibronic levels have been obtained for the B and C states over an extended range of vibrational and rotational quantum numbers on the basis of the best *ab initio* potential functions currently available. They are found to agree with the experimental levels to within a few cm^{-1} in all cases, and represent a considerable improvement over previous calculations carried out in the BO and adiabatic approximations with the *same* potential curves. Conversely, MQDT has been employed to extract BO curves of comparable accuracy for the B', D, and B'' states directly from the observed levels, without prior evaluation of molecular constants. The higher vibrational levels of the B'' and D states are subject to numerous perturbations by levels belonging to higher Rydberg states: the MQD calculation accounts for these as well, and predicts accurately all Rydberg levels with their manifold interactions, up to the ionization limit. The detailed discussion of these results is deferred, along with the extension to other molecules and to higher energies, into ranges where rotational, vibrational, or electronic preionization, predissociation, or dissociative ionization become important.

I. INTRODUCTION

The methods of multichannel quantum defect theory (MQDT, Seaton[1]) have been applied in the past to the spectra of atoms[1,2] (and to a limited extent of molecules also[3–5]), primarily in the discrete and continuous ranges *near ionization thresholds*. Fano[6] has recently reviewed this work. The theory deals with strongly perturbed Rydberg series in the discrete range and relates them to the broad autoionization profiles appearing in the continuum; the parameters used—quantum defects—are nearly independent of energy throughout a series and the adjoining continuum.

The present paper deals with the application of MQDT to the interpretation of the *lowest Rydberg states* of molecules. We have recently reported[7] a quantum defect calculation of some of the Rydberg levels of H_2 which was based on very accurate *ab initio* potential energy curve for the $2p\pi C$ state calculated by Kołos and Wolniewicz.[8] Somewhat surprisingly, this calculation gave level positions which were in essential agreement with experiment and were a considerable improvement over the values that Kołos and Wolniewicz themselves obtained[9] when they integrated numerically over the same potential energy curve using the Born–Oppenheimer approximation. From the point of view of the Born–Oppenheimer approach, the discrepancy between accurate *ab initio* Born–Oppenheimer levels and observed levels arises from the neglect of a number of terms in the molecular Hamiltonian. These terms are proportional to the inverse of the nuclear mass and represent the so-called adiabatic and nonadiabatic corrections. In the first excited states of molecular hydrogen, these corrections are seen in high-resolution optical spectra as a considerable breakdown of the familiar molecular isotope shift rules (Dabrowski and Herzberg[10]).

In this paper, it is shown how MQDT gets around the need to calculate the adiabatic and nonadiabatic corrections explicitly. Rather, these corrections arise as one aspect of vibration–electron coupling in a unified treatment which also accounts for the bonding effect of the Rydberg electron in its various Rydberg states and for "local" vibronic perturbations.

The basic concepts underlying the multichannel quantum defect method are as follows. The theory is an extension of the methods of collision theory into the range of negative electron energies where the "scattered" electron actually becomes a bound electron. When the principal quantum number n is large, the electron is likely to be found far away from the positively charged core, in a region (called B by Fano[6]) where its classical motion is slow compared to the nuclear motions in the core. The point of view of collision theory can then be exploited by utilizing a *space-fixed* coordinate frame for the Rydberg electron and expanding the eigenfunction of the system in terms of the eigenstates of the residual ion, characterized in a molecule by the vibrational and rotational quantum numbers v^+N^+ of the molecular core in its electronic state $|n^+\Lambda^+\rangle$. As n decreases along a Rydberg series, a growing proportion of the probability amplitude of the Rydberg electron is contained in an inner region (called A by Fano) closer to the core. Here the strong Coulomb attraction causes the electron to move much faster than the nuclei so that the Born–Oppenheimer approximation becomes increasingly valid. In this limiting situation the Rydberg electron is more appropriately described in terms of a *molecule-fixed* coordinate system. The electron takes part in the rotational motion in the sense that it now has a well-defined orbital angular momentum component, Λ, along the rotating molecular axis. It also assists in determining the potential energy curve $U(R)$ for nuclear vibrational motion which therefore differs somewhat from the curve $U_+(R)$ of the ion: the vibrational wavefunctions are now $|v\rangle$ rather than $|v^+\rangle$.

Fano has given the following picture of how the outer electron becomes coupled to the internal motions of the core.[3,6] As long as it is in Region B, the electron does not exchange energy or angular momentum with the core. Such exchanges do occur, however, because even with high n or with positive energy, the electron will spend some of its time in the inner region A where a different physical situation prevails. On the other hand, the low Rydberg levels reflect the physical characteristics of just this inner region and provide relevant information on electron–core coupling at short range. It is for this reason that based on the shapes $U(R)$ of potential curves in the first Rydberg states of H_2, one can predict the characteristics of vibrational autoionization in levels above threshold.[5,11] An important tool in making use of this point of view is the frame transformation which connects the coordinate frames appropriate to the regions A and B. Such transformations were first studied systematically in Fano's group, in numerous applications to neutral systems and negative ions.[6] The transformation matrix used in the present context gives the connection between states $|v^+N^+\rangle$ of the ionic core and states $|v\Lambda\rangle$ of the neutral molecule.

The most striking advantage of MQDT is that it entirely eliminates the usual laborious steps of evaluating the interactions between a large number of neighboring and distant electronic states in terms of off-diagonal rovibronic matrix elements derived from the Hamiltonian. Instead, the various terms of the Hamiltonian coupling the Rydberg electron to the core are *replaced by their net effect on its radial function outside the core.* For a bound electron this effect is embodied in the quantum defect μ; for a free electron it is reflected by the closely related asymptotic phase shift $\pi\mu$. By focusing attention on the radial function *outside* the core, MQDT replaces[6] the concept of individual interacting *molecular states* by that of interacting *channels*, each channel being defined by the orbital quantum number l of the emerging electron and by a particular rovibronic state of the residual core. (For example, a whole Rydberg series of levels together with the adjoining continuum corresponds to a single channel.) Accordingly, the boundary conditions at infinity for the radial function of the Rydberg electron are not specified in the channel interaction treatment, i.e., the total energy E is regarded as a continuous parameter; the general wavefunction is then represented as a superposition of channel wavefunctions at one energy E. Boundary conditions leading to quantization are imposed on the radial function only in a last step and yield an algebraic linear system whose size corresponds to the number of channels included and whose solutions predict the level positions. The power of the method derives largely from the fact that two alternative sets of quantum defects are introduced, representing the electron–core interaction in space-fixed and in molecule-fixed coordinate frames, respectively. These are, the defects $\bar{\mu}$ which reflect the effect of the *full rovibronic* Hamiltonian and relate directly to the observed levels through a Rydberg-like formula, and the defects μ which reflect the effect of the purely *electronic* Hamiltonian and relate to the hypothetical fixed-nuclei electronic energies $U(R)$. The main task of molecular MQDT is to establish the explicit relationship between the two sets of parameters by means of the frame transformation and thus to account for *all* aspects of electron–nuclear coupling in terms of the *same* quantities—fixed-nuclei potential curves—as underlie the much less accurate Born–Oppenheimer approximation.

The application of quantum defect methods to the fine structure of low Rydberg states requires the extension of the existing theory in many respects. Most importantly, it is necessary to develop the formalism from the full molecular Hamiltonian in order to understand in detail how the various terms are built into the quantum defect theory and in order to avoid overlooking some of the smaller terms. This approach differs from that used in previous work,[3,4,11] where the treatment concentrated on the motion of the excited electron while the molecular core was regarded essentially as a "black box." Adiabatic and nonadiabatic effects in the first excited states of molecular hydrogen are quite small on a quantum defect scale. For example, the specific isotope shift, which is part of the adiabatic correction and which is usually neglected, contributes no more than 10^{-4} to the quantum defect; with our approach we shall find direct evidence for it. Further, the analysis of the full Hamiltonian leads to a treatment of rovibronic motion as a whole, whereas previous work[3–5] had dealt with rotation–electron and vibration–electron coupling in separate steps and had not explicitly considered electron–electron coupling. It will be shown that the present formulation contains the essential elements required for further extension of the theory, into the autoionization region or into yet higher energy ranges where electronic core excitation plays a role. On the other hand, in this paper the application is limited to levels below threshold, so that we shall be able to use a simplified formalism which is set up in terms of very few parameters only. Finally, the present approach brings out features of the theory which point towards the possibility of incorporating molecular dissociation into the framework of MQDT. Indeed, it will be shown that small residual discrepancies, which appear when the present formalism is applied to high vibrational levels, may be regarded as the signs of a beginning transition towards dissociation.

A simplifying aspect of MQDT in practice is that the electronic quantum defects are generally smooth functions of the energy and of the vibrational coordinate. The number of parameters entering each quantum defect calculation can therefore be reduced greatly by the use of interpolation procedures, and it is possible to make effective use even of incomplete data. This feature will be important in the application in Sec. IV, where theoretical calculations based on *ab initio* potential curves are combined with data fitting procedures in a flexible manner. It will be shown how accurate electronic energies (potential curves) can be extracted directly from the experimental level positions, without passing through the intermediate steps involving determination of molecular constants (Dunham coefficients) and evaluation of adiabatic and nonadiabatic shifts. One thus attains a position similar to that of atomic spectroscopy where the electronic energy levels are directly accessible to experiment. The passage from the observed rovibronic levels to the underlying hypothetical fixed-nuclei energies is achieved without reliance on the validity of the Born–Oppenheimer approximation for any observed level. The Born–Oppenheimer model serves here for the description of *one* limiting situation, and nonadiabatic coupling is just the beginning departure towards the opposite limit, i.e., ionization.

II. QUANTUM DEFECT THEORY OF ELECTRON-NUCLEAR COUPLING

A. The rovibronic Hamiltonian

The nonrelativistic Hamiltonian of molecular hydrogen expressed in terms of molecule-fixed electron coordinates is (Kołos and Wolniewicz,[12] Bunker[13])

$$H=-\frac{\hbar^2}{2\mu}\nabla^2_{R\Theta\Phi}-\frac{\hbar^2}{2\mu_\alpha}\nabla_{R\Theta\Phi}\cdot(\nabla_1+\nabla_2)$$
$$-\frac{\hbar^2}{8\mu}\sum_{i,j=1}^{2}\nabla_i\cdot\nabla_j-\frac{\hbar^2}{2m}(\nabla_1^2+\nabla_2^2)+V(R,\mathbf{r}_1,\mathbf{r}_2)\,. \qquad (1)$$

Here m is the electron mass, $\mu=m_am_b/(m_a+m_b)$ is the nuclear reduced mass, and $\mu_\alpha^{-1}=(m_a-m_b)/m_am_b$ vanishes for symmetrical isotopes. The coordinates $\mathbf{r}=(x, y, z)$ of the electrons 1 and 2 refer to the midpoint between the nuclei a and b. The orientation of the Cartesian system has been defined in Ref. 13. The nuclear spatial arrangement is defined by the polar coordinates $\mathbf{R}=(R,\Theta,\Phi)$, where R is the internuclear distance. The explicit expressions for the Laplacian $\nabla^2_{R\Theta\Phi}$ in polar coordinates and the operator $\nabla_{R\Theta\Phi}(\nabla_1+\nabla_2)$ depend (Kronig[14]) on whether the electron coordinate system rotates with the molecule or is fixed in space, and can be found, for example, in Ref. 13 for the molecule-fixed frame implied by Eq. (1). The potential energy V is given by

$$V(R,\mathbf{r}_1,\mathbf{r}_2)=e^2\left[-\frac{1}{r_{1a}}-\frac{1}{r_{1b}}-\frac{1}{r_{2a}}-\frac{1}{r_{2b}}+\frac{1}{R}+\frac{1}{r_{12}}\right], \qquad (2)$$

where r_{1a} is the distance between electron 1 and nucleus a, etc. The first and fourth terms of the Hamiltonian (1) are the kinetic energy operators of the nuclei and electrons, respectively. The third term is the finite mass correction for combined motion of nuclei and electrons around the center of mass.[15] The second term, which vanishes for symmetrical isotopes, is an additional coupling of nuclei and electrons arising from the displacement of the nuclear center of mass from the molecular midpoint.

It is customary to partition the Hamiltonian equation (1) into two parts, the electronic Hamiltonian H^e which consists of the two last terms of Eq. (1) and does not depend on the nuclear masses, and a nuclear part consisting of the first three terms. Instead, we start out by breaking the Hamiltonian up into three parts, and shall refer to the Born–Oppenheimer partitioning only at a later stage. The first part is the complete rovibronic Hamiltonian of the ion H_2^+, the second part is the Hamiltonian for the motion of the excited electron 1 in a Coulomb field, whose solutions are known analytically.[1] The third part contains the remaining terms of the complete Hamiltonian. We convert to atomic units by multiplying by the electron mass and setting $\hbar=e=m=1$, except when m appears in mass ratios. We then have

$$H=H^{\text{ion}}+H'^{\text{Coulomb}}+H'^{\text{residual}}\,, \qquad (3a)$$

where

$$H^{\text{ion}}=\left[-\frac{1}{2}\nabla_2^2-\frac{1}{r_{2a}}-\frac{1}{r_{2b}}+\frac{1}{R}\right]-\frac{m}{2\mu}\nabla^{+2}_{R\Theta\Phi}-\frac{m}{8\mu}\nabla_2^2$$
$$-\frac{m}{2\mu_\alpha}\nabla^{+}_{R\Theta\Phi}\cdot\nabla_2\,, \qquad (3b)$$

$$H'^{\text{Coulomb}}=\left[-\frac{1}{2}\nabla_1^2-\frac{1}{r_1}\right]-\frac{m}{8\mu}\nabla_1^2\,, \qquad (3c)$$

$$H'^{\text{residual}}=\left[\frac{1}{r_1}+\frac{1}{r_{12}}-\frac{1}{r_{1a}}-\frac{1}{r_{1b}}\right]-\frac{m}{4\mu}\nabla_1\cdot\nabla_2$$
$$-\frac{m}{2\mu_\alpha}\nabla^{+}_{R\Theta\Phi}\cdot\nabla_1\,. \qquad (3d)$$

The primes on H in Eqs. (3c) and (3d) indicate that electron 1 now is described in a system (x_1', y_1', z_1') which does not rotate with the nuclear frame. As a consequence of this choice, the explicit expression for $\nabla^{+2}_{R\Theta\Phi}$ in Eq. (3b) differs from that of $\nabla^2_{R\Theta\Phi}$ in Eq. (1) by omission of the components l_{x_1} and l_{y_1} from all terms involving the electron orbital angular momentum components $L_x=l_{x_1}+l_{x_2}$ and $L_y=l_{y_1}+l_{y_2}$. (This is because the differentiations with respect to R, Θ, and Φ are now carried out with the molecule-fixed coordinates of electron 2 held fixed, and with the space-fixed coordinates of electron 1 held fixed.[14])

The coupling between the motion of the outer elec-

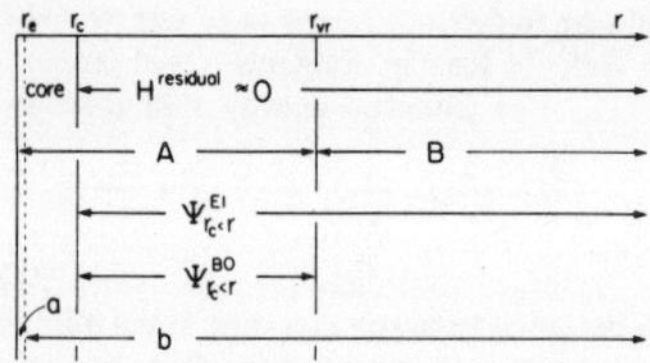

FIG. 1. Diagram showing the ranges in which the electron–ion and Born–Oppenheimer wavefunction expansions are defined. Note that Ψ^{EI} in the figure should read Ψ^{CC}.

tron and the individual core particles is contained in the residual potential H'^{residual}, through the dependence of the expression inside the bracket in Eq. (3d) on the positions of the core electron and the nuclei.[16] Note how outside the core, i.e., when $r_1 \sim r_{12} \sim r_{1a} \sim r_{1b}$, the coupling potential reduces to a small perturbation on the Coulomb potential r_1^{-1}. This is the circumstances which underlies the widely used Rydberg formula, and is also the basis for Seaton's more general MQDT.

In order to solve the Schrödinger equation with H, MQDT proceeds as follows. Electronic and nuclear motion of the free ion core are considered first (Sec. II.B). Molecular wavefunctions are then constructed as a close-coupling expansion of products of the ion core eigenfunctions with the eigenfunction of the Coulomb Hamiltonian H^{Coulomb} (Sec. II.C). In a next step, MQDT exploits the smallness of H^{residual} at large r_1, and actually assumes[1] that H'^{residual} vanishes for $r_1 > r_c$. The radius r_c defines the size of the "core" (see Fig. 1) and depends on the probability amplitude distribution of the core particles. (The problems arising from a polar core are not dealt with here.) Outside the core it is then possible to express the wavefunction *exactly* as a superposition of the regular and irregular solutions of the radial Schrödinger equation in a Coulomb field, $f(\nu, r_1)$ and $g(\nu, r_1)$.[1,3] The effective principal quantum number ν is defined in terms of the electron energy ϵ by $\nu = (-2\epsilon)^{-1/2}$; ν is real for bound ($\epsilon < 0$) and imaginary for continuum states ($\epsilon > 0$).

The admixture of the irregular solution takes account of the presence of a core. This representation of the radial wavefunction outside the core amounts to introducing a phase shift in the eigenfunctions of H^{Coulomb} when ϵ is positive. The relative amplitude of the irregular and regular components is directly related to the asymptotic phase shift, as follows from the asymptotic behavior of f and g as sine and cosine functions, respectively. For negative ϵ, application of the bound state boundary conditions restricts ν to the discrete set of values $\nu = n - \mu$, with the quantum defect μ appearing as the analogue of the continuum phase shift, divided by π.

MQDT does not explicitly consider the radial electronic wavefunction inside the core (for $r_1 < r_c$); all information about the function inside is summarized by the phase shift or the quantum defect outside. The fact that at small r_1 the energy-normalized[1,3] hydrogenic radial functions are nearly *energy independent* suggests that the phase shift $\pi\mu$ of the radial function as it emerges from the core must be a slowly varying function of energy. In the multichannel level pattern, on the other hand, the occurrence of "local" interchannel perturbations depends on the accidental coincidence of quantized Rydberg levels associated with different rotational, vibrational, or electronic series limits (see, for example, Fig. 3 below), and thus depends on boundary conditions at infinity, i.e., on details of electronic motion at long distances from the core. As a result of the perturbations, the actual level positions are characterized by *effective* quantum defects, for which we shall reserve the notation $\bar{\mu}$. The effective quantum defects in general vary substantially from one series member to the next, i.e., they are strongly *energy dependent*. A central feature of MQDT, shown in Secs. II.C and II.D, is to relate the quantities $\bar{\mu}$ to substantially *energy-independent* quantum defects, μ, which pertain to the electron–core interaction at short range, independently of asymptotic boundary conditions.

B. Electronic and nuclear motion in H_2^+

We use the standard adiabatic approximation for the description of nuclear motion in the H_2^+ core. The treatment is briefly outlined here as a contrast to the collision treatment carried out subsequently for the excited electron. The Schrödinger equation for electronic motion,

$$\left[-\frac{1}{2}\nabla_2^2 - \frac{1}{r_{2a}} - \frac{1}{r_{2b}} - E(R)\right]\psi_{n^+\Lambda^+}(\mathbf{r}_2, R) = 0 \qquad (4)$$

(or rather its analogue in elliptic coordinates) has been solved numerically by a number of authors,[17–19] and highly precise electron energies $E_{n^+\Lambda^+}(R)$ are nowadays available in the literature for a number of states $n^+\Lambda^+$ and over an extended range of R values (n^+ numbers the electronic states of given $|\Lambda^+|$). The Born–Oppenheimer potential energy curves for H_2^+ are

$$U_{n^+\Lambda^+}(R) = E_{n^+\Lambda^+}(R) + R^{-1} . \qquad (5)$$

The solutions of the rovibronic Schrödinger equation for H_2^+ are expanded in terms of the electronic functions $\psi_{n^+\Lambda^+}$, and a system of coupled differential equations for nuclear motion is obtained [cf Eq. (5) of Ref. 12, or Eq. (17) of Ref. 13]. In the adiabatic approximation, all coupling between electronic states is neglected and only the corrections within each state are retained. These corrections arise (i) from the noncommutation of the nuclear kinetic energy operator $\nabla^{+2}_{R\Theta\Phi}$ with the electronic Hamiltonian of Eq. (4), and (ii) from the last two terms in Eq. (3b). In the notation of Van Vleck,[20] the adiabatic corrections to the Born–Oppenheimer potential are

$$Q_{n^+\Lambda^+,n^+\Lambda^+}(R) = -\frac{m}{2\mu}\int\int\int \psi^*_{n^+\Lambda^+}(\mathbf{r}_2, R)\frac{\partial^2}{\partial R^2} \times \psi_{n^+\Lambda^+}(\mathbf{r}_2, R)\,d\mathbf{r}_2 \,, \qquad (6a)$$

$$P_{n^+\Lambda^+,n^+\Lambda^+}(R) = \frac{m}{2\mu}\frac{1}{R^2}\int\int\int \psi^*_{n^+\Lambda^+}(\mathbf{r}_2, R) \times (l_{x2}^2 + l_{y2}^2)\,\psi_{n^+\Lambda^+}(\mathbf{r}_2, R)\,d\mathbf{r}_2 \,, \qquad (6b)$$

$$S_{n^+\Lambda^+,n^+\Lambda^+}(R) = -\frac{m}{8\mu}\int\int\int \psi^*_{n^+\Lambda^+}(\mathbf{r}_2,R)\nabla_2^2\psi_{n^+\Lambda^+}(\mathbf{r}_2,R)\,d\mathbf{r}_2\ , \tag{6c}$$

and the adiabatic potential energy curve becomes

$$V_{n^+\Lambda^+}(R) = U_{n^+\Lambda^+}(R) + Q_{n^+\Lambda^+,n^+\Lambda^+}(R) + P_{n^+\Lambda^+,n^+\Lambda^+}(R) + S_{n^+\Lambda^+,n^+\Lambda^+}(R)\ . \tag{7}$$

The adiabatic correction terms have been evaluated for the electronic ground state from first principles with high precision and have been tabulated for a large number of R values.[18] The adiabatic rotation–vibration levels are obtained by solving

$$\left[-\frac{m}{2\mu}\nabla^{+2}_{R\Theta\Phi} + V_{n^+\Lambda^+}(R) - E\right]\phi^{n^+\Lambda^+}(R,\Theta,\Phi) = 0\ . \tag{8}$$

After separation of the rovibrational wavefunction into symmetric rotor functions[21] and vibrational factors

$$\phi^{n^+\Lambda^+}_{v^+N^+M^+} = R^{-1}\chi^{n^+\Lambda^+N^+}_{v^+}(R)\sqrt{(2N^++1)/4\pi}\ \mathfrak{D}^{(N^+)}_{M^+\Lambda^+}(\Phi,\Theta,0)^*\ , \tag{9}$$

this equations reduces to

$$\left[-\frac{m}{2\mu}\frac{d^2}{dR^2} + \frac{m}{2\mu}\frac{1}{R^2}[N^+(N^++1) - \Lambda^{+2}] + V_{n^+\Lambda^+}(R) - E\right]\chi^{n^+\Lambda^+N^+}_{v^+}(R) = 0\ , \tag{10}$$

where N^+ is the rotational quantum number of the core and Λ^+ is the orbital angular momentum component. Equation (10), solved numerically with the adiabatic potential $V_{n^+\Lambda^+}(R)$ from the *ab initio* data, yields[18,22] rotation–vibration levels $E = E(n^+\Lambda^+v^+N^+)$ for the electronic ground state ($n^+\Lambda^+ = X\,^2\Sigma_g^+$) which agree with the experimentally determined levels[5,23] ($v^+ = 0$–5) for H_2^+ and D_2^+ to within the experimental uncertainty of ± 0.5 cm^{-1}. It has therefore been concluded that nonadiabatic effects neglected in the adiabatic approximation must be small, and indeed theoretical calculations predict[24] that they reduce the vibrational intervals in the ground state of H_2^+ by no more than about 0.2 cm^{-1}. This is small compared to the vibronic interactions affecting the Rydberg electron, and we will in the following assume that Eq. (10) represents the ionization limits of H_2 accurately.

C. The rovibronic wavefunction

A basis set of rovibronic molecular wavefunctions of H_2 is obtained by forming antisymmetrized products of the adiabatic functions for the ion, with hydrogen atom radial functions multiplied by spherical harmonics for the excited electron. The electron–ion functions have the form

$$\Psi^{CC}_{n^+\Lambda^+v^+N^+M^+,\nu lm} = 2^{-1/2}\{\psi_{n^+\Lambda^+}(\mathbf{r}_2,R)\,\phi^{n^+\Lambda^+}_{v^+N^+M^+}(\mathbf{R})f_l(\nu_{n^+\Lambda^+v^+N^+},r_1)Y_{lm}(\vartheta_1',\varphi_1') \pm \psi_{n^+\Lambda^+}(\mathbf{r}_1,R)\phi^{n^+\Lambda^+}_{v^+N^+M^+}(\mathbf{R})f_l(\nu_{n^+\Lambda^+v^+N^+},r_2)\,Y_{lm}(\vartheta_2',\varphi_2')\}\ , \tag{11}$$

where the + sign holds for singlet states and the − sign holds for triplet states. The spin function is not explicitly indicated. The label CC anticipates the use of Eq. (11) in close-coupling expansions. The subscripts $n^+\Lambda^+v^+N^+$ on ν indicate that, for a given energy E, the electron energy ϵ and hence ν depend on the energy of the core. l and m are the orbital quantum numbers of the excited electron referred to space-fixed angles ϑ', φ'. Note particularly that the quantum defect approach uses radial functions f which are energy normalized and contain ν as a continuous parameter for all energies. The values of ν are restricted to a discrete spectrum for negative electron energies by boundary conditions at infinity; this restriction will be ignored until Sec. III. A.

Owing to the coupling between core and electron motion through H'^{residual}, the wavefunction is generally a superposition of wavefunctions of different energies E and quantum numbers,

$$\Psi^{CC} = \sum_{n^+\Lambda^+v^+N^+M^+,\nu lm} C_{n^+\Lambda^+v^+N^+M^+,\nu lm}\,\Psi^{CC}_{n^+\Lambda^+v^+N^+M^+,\nu lm}\ . \tag{12}$$

The summation sign implies integration over the energy E, which is represented by the values of ν, n^+, v^+, and N^+; note that n^+ and v^+ extend over a continuous range. For isolated molecules in field-free space the total angular momentum J and its projection $M = M^+ + m$ are conserved, and we can introduce eigenstates of J and M by specifying that the coefficients C in Eq. (12) depend on M^+ and m through a Clebsch–Gordan coefficient,

$$\Psi^{CC,JM} = \sum_{n^+\Lambda^+v^+N^+,\nu l} C'_{n^+\Lambda^+v^+N^+,\nu l} \times \sum_m (lm, N^+M-m\,|\,lN^+JM)\ \Psi^{CC}_{n^+\Lambda^+v^+N^+M^+,\nu lm}\ . \tag{13}$$

It is at this point that a central feature of collision theory is exploited. The result of the summation over energy in Eqs. (12) and (13) is represented by a new close-coupling expansion into products of core wavefunctions and of radial functions with ν corresponding to a *fixed* energy E. This expansion is *appropriate only for radial distances of the outer electron larger than the effective core radius* r_c, beyond which H'^{residual} is disregarded. In this range the radial function is then represented as a general Coulomb function which is a superposition of a regular component f and an irregular component g. (The approximation involved in disregarding H'^{residual} is related to the value of r_c. If one wishes to restrict r_c to a particular finite value, one may have to allow[25] for a variation of the relative amplitude of f and g outside the core as a function of r; this is not done in this work.) Further, there is no contribution to exchange interaction arising from electron positions outside the core, and we introduce no error if we omit antisymmetrization in this range. The coordinates of the excited electron 1 will from here on be denoted as $\mathbf{r} = (r, \vartheta', \varphi')$. Equation (13) leads thus to

$$\Psi^{CC,JM}_{r_c<r} = \sum_{lv^+N^+}{}' \bar{B}^J_{lv^+N^+}\,\Psi^{CC,JM}_{lv^+N^+,r_c<r}\ , \tag{14a}$$

with

$$\Psi^{CC,JM}_{lv^+N^+,r_c<r} = |n^+(R)\Lambda^+v^+N^+\rangle^{lJM}[f_l(\nu_{n^+\Lambda^+v^+N^+},r) \times\cos\pi\bar{\mu}_{lv^+N^+} - g_l(\nu_{n^+\Lambda^+v^+N^+},r)\sin\pi\bar{\mu}_{lv^+N^+}]\ , \tag{14b}$$

where the ket notation

$$|n^+(R)\Lambda^+v^+N^+)^{lJM} = \psi_{n^+\Lambda^+}(\mathbf{r}_2, R)\, R^{-1}\chi_{v^+}^{n^+\Lambda^+N^+}(R)$$
$$\times \sum_m [Y_{lm}(\vartheta', \varphi')\sqrt{(2N^+ + 1)/4\pi}\, \mathfrak{D}^{(N^+)}_{M-m,\Lambda^+}(\Phi, \Theta, 0)^*$$
$$\times (lm, N^+M - m | lN^+JM)] \qquad (14c)$$

emphasizes the quantum numbers of the ion core by which each individual term of the close-coupling expansion is characterized. The incides lJM indicate the dependence of the function on the l value of the electron coupled to the core and on the resultant angular momentum JM formed; i denotes each set of quantum numbers $\{n^+\Lambda^+l\}$ over which the summation is carried out. The coefficients $\bar{B}$ and $\bar{\mu}$ represent the effect of the interaction H'^{residual} and differ accordingly for singlet and triplet states.

These coefficients will be evaluated in Sec. IV through a mixture of *ab initio* and data-fitting procedures. As noted before, the energy ϵ and the quantum number ν which represents it are still regarded as continuous parameters at this stage. They will be related to the coefficients $\bar{\mu}$ in Sec. III by application of boundary conditions which determine the energy $-\epsilon = 1/2(n - \bar{\mu}_{iv^+N^+})^2$ by which a given level lies lower than the ionization limit $E(n^+\Lambda^+v^+N^+)$. In this Rydberg formula the quantities $\bar{\mu}$ play the role of effective channel quantum defects. The coefficients $\bar{B}$ will ultimately carry the detailed information concerning the mixing of different channels iv^+N^+ induced by H'^{residual} at a given energy. Both, $\bar{B}$'s and $\bar{\mu}$'s will in general be substantially energy dependent.

As noted in Sec. II. A, it will be important to replace the effective quantum defects $\bar{\mu}$ by short range quantum defects which are much more nearly energy independent. The choice of these parameters depends essentially on how far the energy range of application is to extend, or, in other words, how many channels will be included explicitly in the treatment. Specifically, in our application we seek quantum defects that remain constant over the range of a few rotational and vibrational quanta. In fact they will be expressed independently of the rotational and vibrational quantum numbers and will be related to the molecular Born–Oppenheimer (fixed-nuclei) potential energy curves $U_{n\Lambda}(R)$.

To do this, we introduce *another critical radius*, r_{vr}, within which $(0<r<r_{vr})$ the potential and kinetic energies of the Rydberg electron are far larger than the rovibrational level spacings of the core. The critical radius r_{vr} separates the Region A from Region B, as discussed in the introduction.[26] The magnitude of the first H_2^+ vibrational intervals ($\lesssim 10^{-2}$ a.u.) implies that in Region A $r \ll 1/(2\times10^{-2})$, i.e., this inner region extends to r_{vr} ~ 5 a.u. This radius is substantially larger than the equilibrium internuclear distance $R_e = 2$ a.u. (which, very qualitatively, may be taken as a measure of the radius of the vibrationless core), and for this reason r_{vr} has been placed *outside* the core in the schematic Fig. 1. Discrete Rydberg levels will approximately conform to the Born–Oppenheimer approximation if the bulk of their probability amplitude lies within Region A. In order to determine the range of ν values for which this is the case we consider the effective principal quantum numbers $\nu_{n^+\Lambda^+v^+N^+}$ in Eq. (14). These are interrelated since for each channel the total energy E is the sum of the core energy $E(n^+\Lambda^+v^+N^+)$ and the Coulomb energy $-(2\nu^2_{n^+\Lambda^+v^+N^+})^{-1}$, i.e.,

$$E(n^+\Lambda^+v^+N^+) - \frac{1}{2\nu^2_{n^+\Lambda^+v^+N^+}} = E(n^{+\prime}\Lambda^{+\prime}v^{+\prime}N^{+\prime}) - \frac{1}{2\nu^2_{n^{+\prime}\Lambda^{+\prime}v^{+\prime}N^{+\prime}}} \qquad (15)$$

for any pair of core states $n^+\Lambda^+v^+N^+$ and $n^{+\prime}\Lambda^{+\prime}v^{+\prime}N^{+\prime}$. The Born–Oppenheimer approximation holds (except for two additional mass-dependent corrections, see Sec. III. B) if the spacings between core rovibrational levels $E(n^+\Lambda^+v^+N^+)$ in a given electronic state are negligible compared to the binding energy $-\epsilon$ of the Rydberg electron. It is seen from Eq. (15) that under these circumstances the dependence of ν on v^+ and N^+ disappears,[27] i.e., for $2\nu^2(E-E') \ll 1$ we have

$$\nu_{n^+\Lambda^+v^+N^+} \sim \nu_{n^+\Lambda^+v^{+\prime}N^{+\prime}} \sim \nu_{n^+\Lambda^+}\,. \qquad (16)$$

Specifically in molecular hydrogen Eq. (16) holds when $\nu < 3$.

We shall at this point return to the Hamiltonian of Eq. (1) and, using Eq. (16), derive a rovibronic wavefunction $\Psi^{\text{BO},JM}$ as an expansion in terms of Born–Oppenheimer products. This function is valid in the limited range $r_c < r < r_{vr}$ and will define the corresponding short range quantum defects. This procedure of introducing short range (or eigen-) quantum defects by focusing attention on a suitably chosen region near the core is the essence of Fano's[3] original paper.

D. The Born–Oppenheimer expansion and the frame transformations

In the Born–Oppenheimer approach a molecule-fixed coordinate frame is adopted for both electrons, and in a first step one solves the Schrödinger equation for purely electronic motion with the nuclei held fixed at each R value. In our quantum defect approach, the electronic wavefunction is obtained in the range $r_c < r < r_{vr}$ by a procedure analogous to that used in the derivation of the rovibronic wavefunction equation (14). The difference is that now we eliminate at the outset all mass dependent terms in Eq. (3) and describe electron 1 in a system which rotates with the molecule [i.e., we omit the primes in Eqs. (3c) and (3d) *and* on the coordinates of the excited electron]. Equation (11) is then obtained without the rovibrational factor $\phi(\mathbf{R})$, and Eq. (14) accordingly takes the form

$$\psi_\Lambda(\mathbf{r}, \mathbf{r}_2, R)_{r_c<r<r_{vr}}$$
$$= \sum_l \{\bar{b}_{l\Lambda}(R)\psi_{n^+\Lambda^+}(\mathbf{r}_2, R) Y_{l,\Lambda-\Lambda^+}(\vartheta, \varphi)\, e^{i\Lambda^+\varphi}[f_l(\nu_{n^+\Lambda^+}, r)$$
$$\times \cos\pi\bar{\mu}_{l\Lambda}(R) - g_l(\nu_{n^+\Lambda^+}, r)\sin\pi\bar{\mu}_{l\Lambda}(R)]\}\,. \qquad (17)$$

The quantum numbers J, M of Eq. (14) have been replaced here by the projection Λ of the orbital angular momentum on the molecular axis. The angle ϑ is taken with respect to the molecular axis, and the difference in azimuth of the two electrons, φ, appears both in the orbital factor Y and in the subsidiary factor $e^{i\Lambda^+\varphi}$.[14] Restriction of l to even or odd values for a given $n^+\Lambda^+$, such as to yield the transformation property g or u, is implied.

Born–Oppenheimer rovibronic products are formed by multiplication of the electronic wavefunction (17) with rovibrational factors ϕ^{Λ}_{vJM}, analogous to the functions $\phi^{n^+\Lambda^+}_{v^+N^+M^+}$ defined in Sec. II. B for the free ion. Each product is characterized by the quantum number J, M, v, Λ and has the form

$$\Psi^{\mathrm{BO},JM}_{v\Lambda,r_c<r<r_{vr}} = \psi_\Lambda(\mathbf{r},\mathbf{r}_2,R)_{r_c<r<r_{vr}}\,\phi^{\Lambda}_{vJM}(R,\Theta,\Phi)$$

$$= \sum_i \{|n^+(R)\Lambda^+ v\Lambda\rangle^{lJM}\,\bar{b}_{i\Lambda}(R)\,[f_l(\nu_{n^+\Lambda^+},r)\cos\pi\bar{\mu}_{i\Lambda}(R) - g_l(\nu_{n^+\Lambda^+},r)\sin\pi\bar{\mu}_{i\Lambda}(R)]\}\,, \tag{18a}$$

where

$$\phi^{\Lambda}_{vJM}(R,\Theta,\Phi) = R^{-1}\chi^{\Lambda J}_v(R)\sqrt{(2J+1)/4\pi}\,\mathcal{D}^{(J)}_{M\Lambda}(\Phi,\Theta,0)^* \tag{18b}$$

and

$$|n^+(R)\Lambda^+ v\Lambda\rangle^{lJM} = \psi_{n^+\Lambda^+}(\mathbf{r}_2,R)\,Y_{l,\Lambda-\Lambda^+}(\vartheta,\varphi)\,e^{i\Lambda^+\varphi}\,\phi^{\Lambda}_{vJM}(R,\Theta,\Phi)\,. \tag{18c}$$

The vibrational wavefunctions $\chi^{\Lambda J}_v(R)$ are not necessarily those of the free ion core, as will be specified later. A general rovibronic wavefunction valid in the outer part of Region A is now written as an expansion in terms of Born-Oppenheimer products

$$\Psi^{\mathrm{BO},JM}_{r_c<r<r_{vr}} = \sum_{v\Lambda} A^J_{v\Lambda}\,\Psi^{\mathrm{BO},JM}_{v\Lambda,r_c<r<r_{vr}}\,. \tag{19}$$

The two expansions equations (19) and (14) must be equivalent in the range $r_c<r<r_{vr}$, and the connection between the two can be established by re-expanding Ψ^{BO} in terms of the close-coupling expansion (14). Recalling Eq. (16), we find

$$\bar{B}^J_{iv^+N^+}\cos\pi\bar{\mu}_{iv^+N^+} = \sum_{v\Lambda} A^J_{v\Lambda}\,\langle n^+(R)\Lambda^+v^+N^+|\,\bar{b}_{i\Lambda}(R)\cos\pi\bar{\mu}_{i\Lambda}(R)\,|n^+(R)\Lambda^+v\Lambda\rangle^{lJM}$$

$$= \sum_{v\Lambda} A^J_{v\Lambda}\,\langle N^+|\Lambda\rangle^{l\Lambda^+J}\,\langle v^+|\,\bar{b}_{i\Lambda}(R)\cos\pi\bar{\mu}_{i\Lambda}(R)\,|v\rangle^{n^+\Lambda^+N^+,\Lambda J} \tag{20}$$

for each iv^+N^+, together with an analogous equation with sine functions. These two equations determine the two parameters $\bar{B}_{iv^+N^+}$ and $\mu_{iv^+N^+}$. Here $\langle N^+|\Lambda\rangle$ refer to integration over ϑ, φ (or ϑ', φ') Θ, and Φ, and $\langle v^+|v\rangle$ refer to integration over R. Equation (20) establishes the connection between states $|v\Lambda\rangle$ and $|v^+N^+\rangle$ and thus constitutes the frame transformation.

Equation (20) and its analog express the coefficients $\bar{B}_{iv^+N^+}$ and effective quantum defects $\bar{\mu}_{iv^+N^+}$ of the general rovibronic wavefunction Eq. (14) in terms of the alternative coefficients $A_{v\Lambda}$ and electronic quantum defects $\bar{\mu}_{i\Lambda}(R)$. With these new coefficients, Eq. (14) is still valid everywhere outside the core, for positive as well as negative electron energies. Equation (20) ensures that the general function (14) will reduce to the more appropriate form Eq. (19) at short range. At the same time Eq. (14) has been expressed in terms of the quantum defects $\bar{\mu}_{i\Lambda}(R)$, i. e., electronic quantities which can be expected to be energy independent at least over the range of a few quanta of vibrational energy.

Each of the electronic wavefunctions ψ_Λ on which the Born–Oppenheimer expansion has been based is itself represented by Eq. (17) as the result of a purely electronic channel interaction. Accordingly, the quantum defects $\bar{\mu}_{i\Lambda}(R)$ are regarded in this and the following equations as effective quantities and are denoted by bars although, as noted, their variation with excitation energy is expected to be much smoother than that of the rovibronic quantum defects of Eq. (14). In the particular application of Secs. III and IV below, we shall focus on the relationship between the functions $\bar{\mu}_{i\Lambda}(R)$ and the Born–Oppenheimer potential energy curves of discrete electronic states.

The boundary conditions at infinity applied to the wavefunction (17) for negative energy will then restrict $\nu_{n^+\Lambda^+}$ to the set of values $n-\bar{\mu}_{i\Lambda}(R)$ and yield the (Λ, R)-dependent Rydberg equation (in a. u.)

$$U_{n\Lambda}(R) = U_{n^+\Lambda^+}(R) - [2(n-\bar{\mu}_{i\Lambda}(R))^2]^{-1}\,, \tag{21}$$

in which the $U_{n\Lambda}(R)$ are the fixed-nuclei potential energy curves of the neutral molecule, and the $U_{n^+\Lambda^+}(R)$ refer to the ion as in Eq. (5). The importance of Eq. (21) is that it provides the interface between the multichannel quantum defect treatment on the one hand, and *ab initio* multiconfiguration calculations on the other hand, in which the electronic wavefunction is evaluated for discrete states throughout the full range, $0<r<\infty$.

Having introduced the principal quantum number n in Eq. (21) we can now specify that the vibrational eigenfunctions $\chi^{\Lambda J}_v(R)$ introduced in Eq. (18) should be calculated for each value of n with the potential $U_{n\Lambda}(R)$. Accordingly, we use in the following the more appropriate labeling $\chi^{n\Lambda J}_v(R)$, $\phi^{n\Lambda}_{vJM}$, and $|v^{(n)}\rangle$ for Eqs. (18)–(20).

E. The electronic wavefunction at large R

The following remarks bear on the evolution of the electronic wavefunction along the vibrational coordinate towards separated atom form, and on how this evolution is built into our Eq. (17). The coupling of the Rydberg electron to the fixed core due to H^{residual} is reflected in the sum over i in Eq. (17) and in the parameters $\bar{b}_{i\Lambda}$ and $\bar{\mu}_{i\Lambda}$, i.e., in the coefficients of the regular and irregular Coulomb components. In turn, the coupling potential H^{residual} in the Hamiltonian contains the electron–electron repulsion term responsible for configuration mixing and exchange interaction. Similarly, H^{residual} also accounts for the anisotropy of the core field. These two interactions, which we take up in turn, cause a complete restructuring of the electronic wavefunction, as R increases to infinity.

It is known that the molecular orbital (MO) method breaks down in many instances when the internuclear separations become large (see, for example, Coulson and Fischer[28]). At large R the core electron is no longer distributed evenly around both nuclei but tends to go with one of them, forcing the outer electron to adjust accordingly. The wavefunction then becomes of the LCAS (linear combination of atomic substate) type. Mulliken has referred to this as "major MO configuration mixing" (CM),[29] and he showed that the Rydberg states of molecular hydrogen (or of other even-electron molecules) must in the separated atom limit be represented properly as 1 : 1 mixtures of singly and doubly excited MO configurations. According to him, major MO–CM is appreciable in molecular hydrogen already near $R = 3.5$ a.u. The Rydberg orbitals involved have at this stage still essentially united-atom character.[30]

From the point of view of quantum defect theory, major MO–CM can be discussed with reference to two regions, a and b, defined analogously to Regions A and B introduced earlier. The electron energy ϵ is now compared with the spacings of *electronic* core states, and the radius r_e which separates Regions a and b derives from the requirement that in Region a, $\nu_{n^+\Lambda^+} \sim \nu_{n^{+\prime}\Lambda^{+\prime}}$ [cf. Eq. (16)], where $n^{+\prime}\Lambda^{+\prime}$ includes at least the first excited state of H_2^+. A relevant point here is that the size of Region a changes substantially with R. At $R = 0$ it is contained entirely within the core ($r_e < r_c$), as follows from the position of the first excited state of the (He^+) core, 1.5 a.u. above the ground state. The radius r_e is indicated in Fig. 1 according to this situation. As the nuclei separate, the $2p\sigma$ component of the $n = 2$ He^+ state comes down to lower energies and eventually merges with the ground state into the common asymptote $H^+ + H(1s)$. As a consequence of the decreasing level separation in the core, the boundary $r = r_e$ between Regions a and b moves to increasingly larger r values until, near $R = 10$ a.u., Region a begins to extend beyond the core. For all smaller R the motion of the Rydberg electron remains independent of the core electron everywhere outside the core, so that the wavefunction of the core electron is just that of the alternative free ion states.

Anisotropic interaction vanishes in the united atom limit. At small R, its primary effect is to displace the atomic nl complexes and to split them into molecular $nl\lambda$ components, but to an increasing extent it also prevents conservation of the orbital angular momentum l of the Rydberg electron. As the molecule dissociates, l mixing is strong, and anisotropic interaction is responsible for the *reformation* of conserved orbital angular momentum, centered on the individual nuclei rather than the molecular midpoint.

III. APPLICATION TO p RYDBERG SERIES OF MOLECULAR HYDROGEN

A. The discrete state boundary conditions

In his monograph[31] on the spectrum of molecular hydrogen published in 1934, Richardson assigned the levels then known to be excited by photoabsorption as being the first members of a p Rydberg series. The series converges towards the electronic ground state of the ion and is split into molecular $p\sigma$ and $p\pi$ components. Richardson's assumption has been confirmed by the more recent work in which the series have been observed up to high n values.[5,23] We shall now describe a simplified quantum defect treatment which in fact corresponds to this picture.

Two approximations are introduced. First, configuration interaction is excluded from the multichannel treatment, on the grounds that for all $r > r_c$ the kinetic and potential energies of the Rydberg electron remain far smaller than required for electronic excitation of the core. This is so, provided both the total energy E and the internuclear distance R do not exceed certain limits. [The same physical situation was described in Sec. II.E in terms of the smallness of Region a, which unlike Region A is included in the core near $R = 0$ ($r_e \ll r_c$).] The close-coupling expansion equation (14a) can therefore be restricted to the electronic ground state of the ion. Second, l mixing is excluded from the multichannel treatment since we shall assume that outside the core the angular dependence of the wavefunction is the same as it would be in the united atom with $l = 1$. This approximation will have to be tested by comparison with experiment.[32] As a result of this restriction of the number of ionization channels one finds a corresponding reduction in the number of quantum defects $\bar{\mu}_{i\Lambda}(R)$. The set $i = \{n^+\Lambda^+ l\}$ is now replaced by $\{X^2\Sigma_g^+,\ l = 1\}$, and the index i may therefore be omitted. $\mu_\Lambda(R)$ [with $\Lambda = \Sigma$ or Π (i.e., 0 or 1)] should henceforth be taken to designate the electronic quantum defects, since we no longer regard them as effective parameters. Similarly, the potential curve designated as $U_+(R)$ or $V_+(R)$ will denote that of the ground state of H_2^+ in Eqs. (5), (8), (10), and (21).

In spite of these seemingly crude approximations, it must be noted that the electronic quantum defects $\mu_\Lambda(R)$ still embody contributions from *all* interactions inside the core. They still represent the exact effective functions, related to the Born–Oppenheimer potential energy curves by Eq. (21). These curves, if available, for example, from accurate *ab initio* calculations, account for configuration and anisotropic interaction insofar as these contribute to the electronic *energy*. Therefore we introduce no approximation here insofar as the electronic energy alone determines the rovibronic levels (i.e., the Born–Oppenheimer approximation holds). On the other hand, the nonadiabatic contributions to the energy levels will be evaluated only approximately by the simplified treatment, because taking a single term i in Eq. (14) prevents us from representing the evolution of the H_2 electronic *wavefunction* as R increases. As noted, this evolution is represented by the coefficients $\bar{b}_{i\Lambda}(R)$ in Eq. (20), which are now set to unity. By their R dependence these coefficients contribute to vibration–electron coupling. For $R>0$ they include the mixing of various l waves, and therefore cause also a modification of rotation–electron coupling. With $\bar{B}_{iv^+N^+}=\bar{B}_{v^+N^+}\delta_{ii'}$, and after omission of the indices i and n^+ everywhere, the two equations (20) for $\bar{B}$ and $\bar{\mu}$ now simplify to

$$\bar{B}^J_{v^+N^+}\cos\pi\bar{\mu}_{v^+N^+}=\sum_{v\Lambda}A^J_{v\Lambda}\langle N^+|\Lambda\rangle^{l=1,\Lambda^+=0,J}\times\int\chi^{\Lambda^+=0,N^+}_{v^+}(R)\cos\pi\mu_\Lambda(R)\chi^{n\Lambda J}_v(R)\,dR \tag{22}$$

for each v^+N^+, and an analogous expression with cos replaced by sin. Equation (14) can be substituted accordingly.

For $l=1$ and a given value of $J>0$, the core rotational angular momentum N^+ can take the three values $J-1$, J, $J+1$. Correspondingly, at short range, excitation of a p electron gives nondegenerate Σ^+_u and doubly degenerate $\Pi^\pm_u$ [i.e., $(2)^{-1/2}(|\Lambda=+1\rangle\pm|\Lambda=-1\rangle)$] excited states, so that the elements $\langle N^+|\Lambda\rangle$ in Eq. (22) form a 3×3 matrix. As explained, for example, in Refs. 3 and 5, the level parity for $N^+=J$ is $(-1)^{J+1}$, while it is $(-1)^J$ for the other two levels, such that the 3×3 transformation matrix separates into a 2×2 matrix and a single element. This single element is unity and in the two representations denotes $N^+=J$ or Π^-_u states, respectively. The elements of the 2×2 matrix transforming Hund's case b states into case d states are[3,5]

$$\begin{array}{c|cc} & |\Lambda=0\rangle & 2^{-1/2}[|\Lambda=+1\rangle+|\Lambda=-1\rangle] \\ \langle N^+=J-1| & \sqrt{\dfrac{J}{2J+1}} & \sqrt{\dfrac{J+1}{2J+1}} \\ \langle N^+=J+1| & -\sqrt{\dfrac{J+1}{2J+1}} & \sqrt{\dfrac{J}{2J+1}} \end{array}\;,\quad(l=1,\ \Lambda^+=0)\;. \tag{23}$$

For $J=0$ only one level occurs, which has pure Σ^+ character, i.e., $\Lambda=0$, $N^+=1$. The vibrational integral involves the vibrational wavefunctions of the core in its electronic ground state and of the molecule in a particular Rydberg state; they are obtained by numerical integration of the differential equation (10) or of its molecular analogue for H_2. At this stage, therefore, only the quantum defect curves $\mu_\Lambda(R)$, $\Lambda=0$, and 1 remain as unknown parameters in the problem, in addition to the $A_{v\Lambda}$ coefficients which are determined by application of boundary conditions at infinity.

We now apply the boundary conditions at infinity for discrete levels. The Coulomb functions f and g for negative electron energies can be represented asymptotically as superpositions of exponentially rising and falling components (see, for example, Ref. 3), and for bound levels the coefficient of the rising component must be set to zero in each channel. With Eq. (14) this leads to the requirement that for each channel v^+N^+

$$[\sin\pi\nu_{v^+N^+}\cos\pi\bar{\mu}_{v^+N^+}+\cos\pi\nu_{v^+N^+}\sin\pi\bar{\mu}_{v^+N^+}]\bar{B}^J_{v^+N^+}=0\;, \tag{24}$$

whence the condition $\nu_{v^+N^+}=n-\bar{\mu}_{v^+N^+}$ follows immediately. As noted before, the quantities $\bar{\mu}$ play the role of effective channel quantum defects. The Rydberg expression (21) for potential curves is obtained from the electronic wavefunction (17) in a similar manner. In order to relate the rovibronic level energies to the electronic quantum defects $\mu_\Lambda(R)$ we make use of the frame transformation relation (22); we do this here in two different ways, appropriate for the description of low or very high Rydberg levels, respectively. After insertion of the two equations (22) into Eq. (14), the boundary conditions lead now to the requirement that for each channel v^+N^+

$$\sum_{v\Lambda}[\sin\pi\nu_{v^+N^+}\mathfrak{C}^J_{v^+N^+,v\Lambda}+\cos\pi\nu_{v^+N^+}\mathfrak{S}^J_{v^+N^+,v\Lambda}]A^J_{v\Lambda}=0\;, \tag{25}$$

where

$$\mathfrak{C}^J_{v^+N^+,v\Lambda}=\langle N^+|\Lambda\rangle^{1,0,J}\int\chi^{0N^+}_{v^+}(R)\cos\pi\mu_\Lambda(R)\chi^{n\Lambda J}_v(R)\,dR\;, \tag{26}$$

and $\mathfrak{S}_{v^+N^+,v\Lambda}$ is given by a similar equation with cos replaced by sin. Alternatively, we can introduce the following expansions:

$$A^J_{v\Lambda}=\sum_{N^{+\prime}}\langle\Lambda|N^{+\prime}\rangle^{1,0,J}\beta^J_{vN^{+\prime}}\;, \tag{27a}$$

$$\chi^{n\Lambda J}_v(R)=\sum_{v^{+\prime}}\langle v|v^{+\prime}\rangle^{n\Lambda J,0N^{+\prime}}\chi^{0N^{+\prime}}_{v^{+\prime}}(R)\;,$$

which yield

$$\sum_v A^J_{v\Lambda}\chi^{n\Lambda J}_v(R)=\sum_{v^{+\prime}N^{+\prime}}B^J_{v^{+\prime}N^{+\prime}}\langle\Lambda|N^{+\prime}\rangle^{1,0,J}\chi^{0N^{+\prime}}_{v^{+\prime}}(R)\;, \tag{27b}$$

with

$$B^J_{v^{+\prime}N^{+\prime}}=\sum_v\beta^J_{vN^{+\prime}}\langle v|v^{+\prime}\rangle^{n\Lambda J,0N^{+\prime}}\;. \tag{27c}$$

The linear system then takes the form

$$\sum_{v^{+\prime}N^{+\prime}}[\sin\pi\nu_{v^+N^+}\mathfrak{C}^J_{v^+N^+,v^{+\prime}N^{+\prime}}+\cos\pi\nu_{v^+N^+}\mathfrak{S}^J_{v^+N^+,v^{+\prime}N^{+\prime}}]B^J_{v^{+\prime}N^{+\prime}}=0 \tag{28}$$

for each channel v^+N^+, where

$$\mathfrak{C}^J_{v^+N^+,v^{+\prime}N^{+\prime}}=\sum_\Lambda\left\{\langle N^+|\Lambda\rangle^{1,0,J}\int\chi^{0N^+}_{v^+}(R)\times\cos\pi\mu_\Lambda(R)\chi^{0N^{+\prime}}_{v^{+\prime}}(R)\,dR\,\langle\Lambda|N^{+\prime}\rangle^{1,0,J}\right\}\;, \tag{29}$$

and similarly for $\mathfrak{S}$. The two linear systems equations (25) and (28) each determine the effective principal quantum numbers ν as functions of the quantum defects $\mu_\Lambda(R)$. The two systems are equivalent; they differ in that each emphasizes one particular aspect of rotation/vibration–electron coupling. Equation (25) lends itself to the determination of levels whose principal quantum

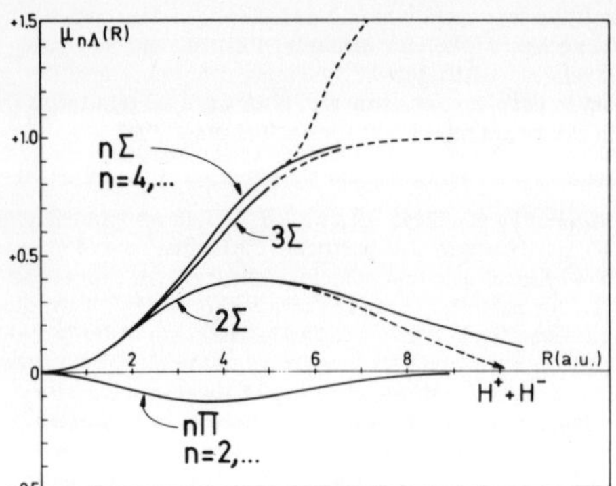

FIG. 2. Quantum defect curves $\mu_{n\Lambda}(R)$ of the lowest ungerade singlet Rydberg states of molecular hydrogen (see Sec. IV for details).

number is low, $2 \lesssim n \lesssim 4$. For these low Rydberg levels the Born–Oppenheimer classification according to $v\Lambda$ is approximately valid, so that usually one coefficient $A_{v\Lambda}$ will be predominant, and near unity. Furthermore, as is implied by Eq. (21), the Born–Oppenheimer curves $U_{n\Lambda}(R)$ at low n may have shapes rather different from the ion curve $U_+(R)$ when $\mu_\Lambda(R)$ varies strongly with R, so that the vibrational functions $\chi_v^{n\Lambda J}$ and $\chi_{v^+}^{0N^+}$ also are different. The infinite linear system is then more nearly diagonal when both sets of vibrational functions are introduced explicitly into the calculation. Equation (28) is more appropriate for the determination of high Rydberg levels with $n > 10$, and it should be used when the ionization continuum is involved. High Rydberg levels do not conform to the Born–Oppenheimer approximation. Their overall structure, apart from the perturbations due to electron–nuclear coupling, is that of Rydberg series characterized by v^+N^+ and nearly constant quantum defects $\bar{\mu}_{v^+N^+} = (\pi)^{-1}\arctan(\mathfrak{S}_{v^+N^+,v^+N^+}/\mathfrak{C}_{v^+N^+,v^+N^+})$. Correspondingly, the elements $\mathfrak{C}$ and $\mathfrak{S}$ in Eq. (29) are set up as nondiagonal matrices labeled by the channels v^+N^+ and $v^{+\prime}N^{+\prime}$ which they connect. The electronic quantum defects $\mu_\Lambda(R)$ occur here only implicitly as the basic quantities from which the matrices $\{\mathfrak{C}\}$ and $\{\mathfrak{S}\}$ are constructed by the successive transformations $\{\langle N^+|\Lambda\rangle\}$ and $\{\langle v^+|R\rangle\}$.

Expressions similar to Eqs. (25) and (28) have been given previously in different forms. Suppression in Eq. (25) of all parameters pertaining to vibration leads to the 2×2 homogeneous system for l uncoupling of a p electron obtained by Fano a few years ago [Eqs. (28) and (30a) in Ref. 3)]. An extension of Fano's equations to describe combined rotational/vibrational–electronic motion was given by Chang, Dill, and Fano,[4] who introduced the effective short-range quantum defect on a more phenomenological basis. Their equation is based on the description $v^+\Lambda$ intermediate between our equations (25) and (28), and is expressed with tan functions [which diverge when the continuous variable $\pi\mu_\Lambda(R)$ passes through $\pi/2$]. A less complete expression for pure vibration–electron coupling between two channels was used in Ref. 5 and more recently in Ref. 33.

B. Dependence of the quantum defect on internuclear distance, energy, and mass

With the definition equation (21) of the electronic quantum defects $\mu_\Lambda(R)$, we can derive their overall behavior directly from the dissociation limits of the fixed-nuclei curves $U_{n\Lambda}(R)$. In the first approximation we can neglect configuration interaction and relate the curves $\mu_\Lambda(R)$ to the potential curves of the prototype system H_2^+ which has one single electron moving in the field of two fixed positive charges. Weizel[34] has given a qualitative diagram for this case, correlating united atom with separated atom states. According to his correlation diagram, all $np\sigma$ states of the united atom dissociate with the principal quantum numbers of the electron lowered by one unit, into $n-1$. In Weizel's terminology, the united atom $np\sigma$ electron is said to be singly promoted. The $np\pi$ states, on the other hand, dissociate without change of the principal quantum number, i.e., the $np\pi$ electron is not promoted. If we now attribute the change of the effective principal quantum number to a corresponding variation of the quantum defect, we see that $\mu_\Sigma(R)$ must *increase* by one unit as R goes from zero to infinity, and therefore will have a substantial R dependence. Conversely, it is possible that $\mu_\Pi(R)$ has only a weak R dependence because it has the same value (zero) both at $R=0$ and ∞.

Figure 2 displays the curves $\mu_\Lambda(R)$ in the two-electron system H_2 for several values of n, based on the results of Sec. IV below. The figure shows that the overall behavior of $\mu_\Lambda(R)$ is indeed as expected: a strong R dependence for Σ and a weak R dependence for Π curves. Deviations from the general pattern occur also and show up as marked splittings of the μ_Σ curves for different n at large R. The variation of μ with n, in other words its *energy dependence*, is designated by $\mu_{n\Lambda}(R)$. This dependence largely reflects the presence of a second electron in the core and is a manifestation of the character of $\mu_{n\Lambda}(R)$ as an *effective* quantity, as pointed out in Sec. II.D.

This point of view enables us, in a very qualitative way, to foresee where the approximations made in this section may break down. We have assumed, in essence, that the electronic ionization energy, $U_+(R) - U_{n\Lambda}(R)$ in Eq. (21), is directly related to the R dependence of the electronic wavefunctions for $r > r_c$. This is true to the extent that configuration mixing is weak or does not vary with R, and that anisotropic interaction is also weak. With the pronounced energy dependence of $\mu_\Lambda(R)$ being indicative of configuration mixing, we must expect that Eqs. (25) and (28) fail to account for the details of electron–nuclear coupling in the corresponding ranges of R values.

The key expressions of this paper, Eqs. (20), (25), and (28), describe the coupling between core particles and Rydberg electron through H'^{residual} in terms of the functions $\mu_\Lambda(R)$. The quantum defect functions have been introduced [cf. Eq. (21)] as purely electronic quantities *independent of nuclear mass*. In the derivation it was assumed that the Born–Oppenheimer approximation becomes *exactly* valid if only Eq. (16) holds. This is not quite true, however, since the terms $-(m/8\mu)\nabla_1^2$ and $-(m/4\mu)\nabla_1\nabla_2$ in Eqs. (3c) and (3d) are independent of vibration and rotation. These must therefore be taken into account separately. The first of the terms, $-(m/8\mu)\nabla_1^2$ in $H^{\rm Coulomb}$, gives rise to the analogue of the so-called "normal" correction for finite mass familiar from one-electron atoms.[15] The second, $-(m/4\mu)\nabla_1\nabla_2$ in $H^{\rm residual}$, is an electronic cross term and corresponds to the "specific" mass effect known from the spectra of light atoms with at least two electrons.[15] In the standard molecular approach both operators contribute[20] to the adiabatic correction $S_{n\Lambda,n\Lambda}(R)$, analogous to the correction Eq. (6c) for H_2^+, and their matrix elements have to be evaluated numerically from accurate electronic wavefunctions.

The advantage of the present partitioning of the Hamiltonian according to Eq. (3) is that the normal mass effect is included in the *R-independent* Coulomb Hamiltonian $H^{\rm Coulomb}$. We can therefore account for it as in an atom,[35] by setting the Rydberg constant in the Rydberg formula (in wave number units) $\epsilon = -Ry/\nu^2$ to the values $Ry_{H_2} = 109707.42$ and $Ry_{D_2} = 109722.36$ cm^{-1} rather than using $Ry_\infty = 109737.318$ cm^{-1}. Note that this applies to the relation between ϵ and ν in the "electron–ion" function equation (14) and to the equations (25) and (28) derived from it, but not to Eqs. (17), (18), and (21). These latter are derived from the Hamiltonian (1) and serve merely to re-express the coefficients $\bar{B}$ in terms of the A's.

The specific mass effect represents an electron–core interaction since the two-electron operator contributes to matrix elements only at points where both electrons have nonvanishing probability amplitude, i.e., inside the core. It gives rise to a *mass-dependent* component of the quantum defect. We must therefore write in the previous expressions (20), (22), (26), and (29), strictly,

$$\mu_\Lambda^{\rm total}(R) = \mu_\Lambda(R) + (m/\mu)\,\mu_\Lambda^{\rm specific}(R)\,, \tag{30}$$

where (m/μ) is the mass factor defined in Eq. (3). We shall find that only in the lowest two ungerade singlet states of molecular hydrogen is there experimental evidence for the specific mass effect. The small mass contribution disappears upon dissociation into H(1s) + H(nl), when the two-electron probability distributions no longer overlap appreciably.

The neglect of anisotropic interaction in Sec. III.A introduces an additional "apparent" mass dependence into $\mu_\Lambda(R)$, which arises when R becomes large and which persists to $R\to\infty$. This can be understood as follows. The standard adiabatic correction for H_2, analogous to the corrections equation (6) for H_2^+, includes in addition to $S(R)$ also the terms $P(R)$ and $Q(R)$ which arise from the noncommutation of $\nabla^2_{R\Theta\Phi}$ with the electronic Hamiltonian H^e. It has been known since Van Vleck's early work[20] that in the limit of dissociating atoms the corrections $P(R)+Q(R)$ become equal to the electronic correction $S(R)$ in a given state. With regard to the excited electron in our problem this means that in addition to the normal mass correction [due to the last term in Eq. (3c)], another contribution of equal magnitude must be present. The physical reason for this is that the electron is now moving around a nucleus with *one-half* of the mass of the united atom so that the normal mass correction is *twice* as large. In the Born–Oppenheimer theoretical framework this contribution arises[20] because in the limit of separated atoms a displacement of one nucleus, e.g., by $+\frac{1}{2}\Delta R$ along the internuclear axis, is equivalent as far as the atom is concerned to a displacement by $-\Delta z_1$ of the excited electron centered on the nucleus. The operators $\partial^2/\partial R^2$ and $\partial^2/\partial z_1^2$ therefore have the same effect on the electronic wavefunction,[36] and the corresponding contributions to Q and S become equal. Similar arguments apply with respect to the other coordinates. Quantum defect theory would account for this contribution through the persisting R dependence of the coefficients $\bar{b}_{i\Lambda}(R)$ describing re-expansion of the spherical harmonics $Y_{l,\Lambda-\Lambda^+}$ on the centers of the separating atoms (cf. Sec. II.E). As noted before, our equations (25) and (28) do not represent this evolution of the electron wavefunction at large R. Rather, the effective quantum defects become asymptotically R independent just like the potential curves $U_{n\Lambda}(R)$ to which they are related. Given an accurate Born–Oppenheimer potential energy curve, therefore, Eqs. (25) and (28) will yield slightly too low rovibronic levels near a dissociation limit. From the relation $P(\infty)+Q(\infty)=S(\infty)$ it follows that the discrepancy is $(Ry_\infty - Ry_{M_2})/n^2$ cm^{-1} ($M_2 = H_2$ or D_2), where n is the separated atom value corresponding to dissociation into H(1s) + H(nl). For $n=2$ the contribution amounts to 7.5 cm^{-1} in H_2 and to 3.8 cm^{-1} in D_2. Conversely, in fitting potential energy curves to experimental line positions it would be found necessary to include in addition to $\mu^{\rm specific}$ a small mass-dependent quantum defect at large R, of magnitude -2.7×10^{-4} for H$^+$ + H ($n=2$).

We conclude that the simplified quantum defect theory of Sec. III.A accounts only in part for those effects which in the standard approach are represented by the correction terms $P(R)$ and $Q(R)$. The theory should, however, fully include these effects at smaller R values (small to intermediate vibrational quantum numbers) where configuration- and l mixing are not important. In this range the treatment should also account for the so-called nonadiabatic or vibronic coupling. In the Born–Oppenheimer approach this appears in the familiar operator form

$$-\frac{m}{\mu}\left[\frac{\partial}{\partial R}\,\psi_{n\Lambda}(\mathbf{r}_1, \mathbf{r}_2, R)\right]\frac{\partial}{\partial R} \tag{31}$$

and couples the differential equations for nuclear motion in different electronic states. The MQDT proceeds here in some analogy with the so-called *diabatic* approach often used in problems of nuclear dynamics involving potential curve crossings. Namely, it starts out from an electronic Hamiltonian ($H^{\rm Coulomb}$) which does not completely describe the excited electron,[36] but on the other hand has the advantage of being R independent so that the terms of the type of Eqs. (31) and (6a) do not arise initially. Subsequent introduction of the electrostatic perturbation $H'^{\rm residual}$ enables the R dependence of the electronic wavefunction to be expressed explicitly,[37] through the parameters of Eq. (17). Contrary to customary treatments, where one is often forced to restrict the problem to just *two* interacting states or curves, Eq. (17) represents the R-dependent mixture of an *infinite* number of electronic states. In other words, for every $n^+\Lambda^+ l$ one would have to carry out an infinite summation and integration over electronic wavefunctions valid inside as well as outside the core in order to construct a wavefunction which matches with Eq. (17) at $r = r_c$. This electronic wavefunction enables the perturbation H^{residual} to be related to the Born–Oppenheimer potential energy curves [Eq. (21)]: the vibronic matrix elements $\mathfrak{C}$ and $\mathfrak{S}$ in Eqs. (26) and (29) can therefore be calculated by averaging the electronic quantum defect over the vibrational motion. The quantum defect then plays a role similar to that of the electronic perturbation function used in the so-called diabatic approach.[37] Quite similar arguments apply to the discussion of rotation–electron coupling (l-uncoupling interaction) which is related to the angle dependence of the electronic wavefunction; no details are given here (cf. Refs. 3 and 5).

We finally mention the electronic isotope shift which was early foreseen as a manifestation of the partial breakdown of the Born–Oppenheimer separation.[20] This quantity is determined from electronic spectra as the shift of the electronic "origin" between two isotopes, obtained by extrapolation in each isotope of the observed level positions to the hypothetical rotationless and vibrationless state. In theory one evaluates the difference between the total adiabatic corrections $P(R_e)+Q(R_e)+S(R_e)$ at equilibrium in the excited and ground states, respectively. The difference of this quantity between two isotopes then represents essentially (although not exactly[20]) the electronic isotope shift. Dabrowski and Herzberg[10] have experimentally determined the electronic isotope shift for the B, B', and C states of molecular hydrogen. This shift does not appear explicitly in the present treatment, which makes no reference to the molecular equilibrium position R_e.

C. Numerical details

Each quantum defect calculation has been carried out in two successive steps. The first involved evaluation of the matrices $\{\mathfrak{C}\}$ and $\{\mathfrak{S}\}$ on the basis of the potential energy curve of the ion and a given quantum defect curve $\mu_\Lambda(R)$ or $\mu_\Lambda^{\text{total}}(R)$. The potential $U_+(R)$ was that of Wind,[17] complemented by the adiabatic corrections of Bishop and Wetmore[18b] to give $V_+(R)$, and extended by a five-point Lagrange interpolation to a mesh of 2000 points ranging from 0.1 to 10.0 a.u. The quantum defect curves were defined by a set of distinct points, typically about 25, which also were interpolated by Lagrange's formula. The integration of the differential equation (10) with $V_+(R)$ and, when necessary, with $U_{n\Lambda}(R)$ was carried out using the Numerov–Cooley[38] method with an integration step of 0.005 a.u. The vibrational functions obtained for a given potential were checked to be mutually orthogonal to within 10^{-4} or better. The eigenenergies provided the ionization limits $E(v^+N^+)$ needed in the second stage of the calculation. Their values were found to agree to within at least 0.2 cm^{-1} in all cases with the values published in the literature.[18,22] The elements $\mathfrak{C}$ and $\mathfrak{S}$ were then obtained by integration using Simpson's rule. Typically, in a calculation based on Eq. (25), 15 vibrational channels were included, with two rotational channels for each, so that eight 15×15 matrices (with $\Lambda=0, 1$; $N^+=J-1$, $J+1$; sine and cosine) with, in all, 1800 independent elements had to be evaluated from 45 vibrational wavefunctions. A similar calculation based on Eq. (28) necessitated evaluation of eight symmetric matrices with 960 independent elements from 30 vibrational functions.

These numbers illustrate the decisive numerical advantage brought about by the frame transformation method. The older version[1] of MQDT gives an expression for discrete level positions which is formally equivalent to our equation (28), with the difference that the elements $\mathfrak{C}$ and $\mathfrak{S}$ (called I and J in Ref. 1) are treated as independent quantities which would have to be fitted to experiment *individually*. It is obvious that with this approach the rovibronic multichannel problem would become grossly underdetermined and consequently untreatable. The equation (29) of the present paper is of key importance since in practice it reduces the number of unknown parameters in the problem by as much as 2 orders of magnitude. To be sure, the curves $\mu_\Lambda(R)$ consist each of an infinite number of points (just as there is an infinite number of vibrational channels), but since these must form a smooth curve, we can pick just a few of them and obtain the rest by interpolation.

The second step of the quantum defect calculations involved the solution of the linear system, Eq. (25) or (28). For a given total energy E, the variables $\nu_{v^+N^+}$ were expressed by the Rydberg formula (in cm^{-1})

$$\nu_{v^+N^+}(E) = \left[\frac{Ry_{M_2}}{E(v^+N^+)-E}\right]^{1/2} \tag{32}$$

in all rows except the first, and in this way the fulfillment of all relations given by Eq. (15) was ensured. [M_2 in Eq. (32) denotes H_2 or D_2]. It is convenient for the following to renumber the rows and columns of the linear system by running indices i and k such that, for example, in Eq. (25) $\mathfrak{C}_{i=1,k=1} = \mathfrak{C}_{v^+=0,N^+=J-1;v=0,\Lambda=0}$, and so forth. The systems (25) and (28) then take identical form, and nontrivial solutions are obtained when

$$\det\left|\tan\pi\nu_i\mathfrak{C}_{ik}+\mathfrak{S}_{ik}\right| = 0. \tag{33}$$

Equation (33) can be solved for any ν_i in terms of the other variables ν. We express all ν_i $(i\neq 1)$ according to Eq. (32) and solve Eq. (33) for the effective quantum number ν_1. The result is

$$\nu_1(E) = n - \bar{\mu}_1(E) = -\frac{1}{\pi}\arctan\left(\frac{\sum_k \mathfrak{S}_{1k}C_{1k}(E)}{\sum_k \mathfrak{C}_{1k}C_{1k}(E)}\right), \tag{34}$$

where $C_{ik}(E)$ is the cofactor of the element of the ith row and kth column of the determinant (33). The function $-\nu_1(E)$ increases monotonically with E.[3] Taken modulo 1 it represents the effective rovibronic quantum defect $\bar{\mu}_1(E)$ introduced in Eq. (14) [cf. Ref. 11 for

TABLE I. Summary of calculation procedures.

State	Initial data	MQD calculation	Results[a]	Comparison
$2\Sigma(B)$	U_{theor}, $V_{\mathrm{theor}}^{\mathrm{specific}}$	$\rightarrow$ μ_{theor}, $\mu_{\mathrm{theor}}^{\mathrm{specific}}$	$\rightarrow$ $T_{\mathrm{theor}}^{J=0}$	T_{exptl}
		$]\rightarrow$	$\rightarrow$ $T_{\mathrm{theor}}^{J>0}$	T_{exptl}
$2\Pi(C)$	U_{theor}	$\rightarrow$ μ_{theor}	$\rightarrow$ $T_{\mathrm{theor}}^{J=1,\Pi^-}$	$T_{\mathrm{exptl}}(\rightarrow \mu_{\mathrm{exptl}}^{\mathrm{specific}})$
$3\Sigma(B')$	$T_{\mathrm{exptl}}^{J=0}$	$[\mu_{\mathrm{exptl}} \leftrightarrows T_{\mathrm{exptl}}^{J=0}$ (8 iterations)$]$	$\rightarrow$ U_{exptl}	U_{theor}
		$]\rightarrow$	$\rightarrow$ $T_{\mathrm{calc}}^{J>0}$	T_{exptl}
$3\Pi(D)$	U_{theor}	$\rightarrow \mu_{\mathrm{theor}} \rightarrow T_{\mathrm{theor}}^{J=1,\Pi^-} \rightarrow [\mu_{\mathrm{exptl}} \leftrightarrows T_{\mathrm{exptl}}^{J=1,\Pi^-}$ (3 iterations)$]$	$\rightarrow$ U_{exptl}	U_{theor}
		$]\rightarrow$ (higher Rydberg levels)[b]	$T_{\mathrm{calc}}^{J>0}$ [b]	T_{exptl} [b]
$4\Sigma(B'')$	$T_{\mathrm{exptl}}^{J=0}$	$[\mu_{\mathrm{exptl}} \leftrightarrows T_{\mathrm{exptl}}^{J=0}$ (17 iterations)$]$	$\rightarrow$ U_{exptl}	U_{theor}
		$]\rightarrow$ (higher Rydberg levels)[b]	$T_{\mathrm{calc}}^{J=0}$ [b]	T_{exptl} [b]

[a] T: spectroscopic level energies; U: Born–Oppenheimer potential energy curve; μ: quantum defect $\mu_\Lambda(R)$; $V^{\mathrm{specific}} = \langle -(m/4\mu)\nabla_1\nabla_2\rangle_{n\Lambda,n\Lambda}$.
[b] Not discussed in detail in this paper.

graphical representations of $\bar{\mu}_1(E)$]. Simultaneously with Eq. (34) the Rydberg equation relating to the first ionization limit E_1,

$$\nu_1(E) = n - \bar{\mu}_1(E) = \left[\frac{Ry_{M_2}}{E_1 - E}\right]^{1/2} , \qquad (35)$$

must also hold, giving the effective rovibronic quantum defect as a monotonically decreasing function of E. The eigenvalues are determined as in previous work[1,3,39] by the crossing points of the two curves with opposite slope. Crossings occur (i) once each time the principal quantum number n in Eq. (35) increases by one unit, or (ii) whenever the argument of the arctan function in Eq. (34) passes through a resonance. The solutions of the first type yield the "regular" Rydberg structure converging to the first ionization limit E_1; those of the second type correspond to the perturbing "interlopers" associated with higher rovibrational channels. In the calculations the energy E was varied in the region of interest by steps corresponding to the desired level precision, and $\bar{\mu}_1(E)$ was calculated according to Eqs. (34) and (35) at each E, until a change of sign of their difference indicated that crossing had occurred.

The resulting energy E was finally converted into a spectroscopic term value by referring it to the ground level of H_2 (or D_2), $X^1\Sigma_g^+$, $v=0$, $J=0$ according to

$$T = E - E(v^+=0,\ N^+=0) + \text{I. P.} \qquad (36)$$

The theoretical ionization potentials I. P.$_{\mathrm{theor}}(H_2)$ $=124417.3$ cm^{-1} and I. P.$_{\mathrm{theor}}(D_2) = 124745.2$ cm^{-1} [40] (which were used for reasons of consistency) agree with the experimental ones[5,23,41] to within less than the experimental error limits of ± 0.4 cm^{-1} (H_2) and ± 0.6 cm^{-1} (D_2). The theoretical I. P.'s contain small contributions accounting for radiative and relativistic effects in the ground states of H_2 and H_2^+. These effects, which had been neglected thus far in this paper, are thus included at this point.

Particular precautions were taken in order to avoid truncation errors arising from the limited number of vibrational channels included in the calculations. The number of vibrational channels was varied systematically, and usually it was found that the result converged quite rapidly with increasing number of channels, but some difficulties were encountered in the highest discrete vibrational levels of some states, where vibronic interaction with the vibrational continuum contributes appreciably. Vibrational continuum states have not been included in this work but should be taken into account in order to attain higher accuracy.

IV. RESULTS

The theoretical treatment of Sec. III has been applied to the five lowest ungerade singlet states of H_2 and D_2. For each state a somewhat different approach was chosen, depending on the amount of theoretical and experimental data available. Table I summarizes the procedures used. Figure 3 presents an overview of the energy levels belonging to the lower Rydberg states of H_2.

A. The $B\,^1\Sigma_u^+$ state

One might expect the first excited (singlet) state of molecular hydrogen to be the least suited for a test of the quantum defect method. Although it is the first ($n=2$) member of the $np\sigma$ series, the B state has properties that substantially deviate from those commonly found in molecular Rydberg states. For example, the vibrational frequency of 1357 cm^{-1} is not much more than *half* that of the ion H_2^+ (2322 cm^{-1}) and the rotational constant $B_e = 20.0$ cm^{-1} is also substantially smaller than in the isolated core (30.0 cm^{-1}). Mulliken[29] has classified the B state as "semi-Rydberg" state and he discussed how this state, which in the united atom limit corresponds to $2p\sigma$, changes its nature twice on its course to dissociation: between about 3 and 7 a. u. it acquires predominant ionic character and tends towards the $H^+ + H^-$ dissociation limit, but eventually it dissociates into H(1s) + $H(2l)$ because of avoided crossings with higher $^1\Sigma_u^+$ states.[42,43]

This behavior is reflected in the quantum defect curve $\mu_{2\Sigma}(R)$ shown in Fig. 2. In the fully separable

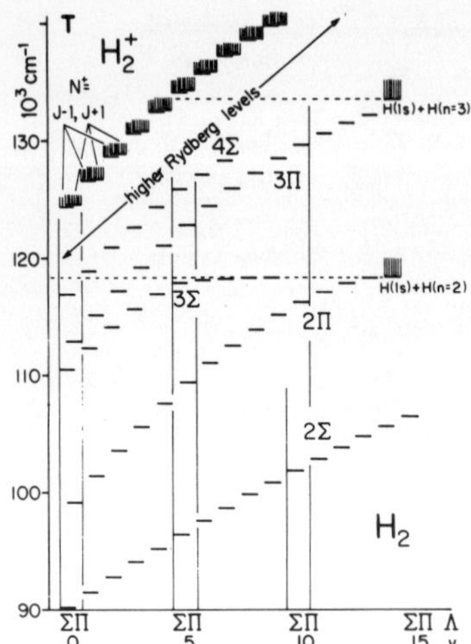

FIG. 3. Diagram of (ungerade, singlet) Rydberg levels of H_2. The figure is drawn to scale and corresponds to $J=1$ and level parity $(-1)^J$. For opposite parity only the levels labeled Π would be present (Π_u^-), and for $J=0$ only those labeled Σ. The quantum numbers $v\Lambda$ have no strict meaning as a consequence of the interactions which take place between all levels shown. The continua joining on to the individual Rydberg series and vibrational progressions are indicated by vertical hatching.

two-center problem the $2p\sigma$ state of the united atom dissociates into $H^+ + H(1s)$ (cf. Sec. III. B), and the quantum defect increases from zero at $R=0$ to unity at $R=\infty$. The curve $\mu_{2\Sigma}$ in Fig. 2 at first follows this route, but since a $^1\Sigma_u^+$ state cannot dissociate into two ground state H atoms [only a $^3\Sigma_u^+$ state arises from $H(1s)+H(1s)$], it is forced downward again and follows the ionic curve (dotted line) for some distance. This behavior results from the presence of the second electron in the core, and the ionic wavefunction formed is a particular LCAS function of the type mentioned in Sec. II. E. The interaction between the two electrons thus replaces, as it were, the forbidden dissociation limit $H(1s)+H(1s)$ by the additional one, $H^+ + H^-$. The quantum defect in this range does not refer to a H_2^+ core in any single electronic state, and ionization involves a considerable rearrangement of the core electronic structure.[29] The quantum defect curve has another inflection point at about 7 a.u. and swings back towards the zero line, indicating that the B state dissociates at the lowest "allowed" limit, $H(1s)+H(n=2)$.

Kołos and Wolniewicz have recently calculated a very accurate Born–Oppenheimer potential energy curve based on a 88-term wavefunction,[43] and they also have evaluated the corrections $P(R)$, $Q(R)$, and $S(R)$,[44] enabling them to determine theoretical band origins in the adiabatic approximation.[43] The present calculation is based on the $\mu_{2\Sigma}$ curve of Fig. 2 deduced from their Born–Oppenheimer potential $U_{2\Sigma}$ by use of Eq. (21). The theoretical adiabatic corrections did not need to be taken over into the quantum defect calculation except for the specific isotope effect, part of $S(R)$, listed by Kołos and Wolniewicz for five points. Their values were interpolated graphically to yield the same initial number of points as used in $\mu(R)$ (27), and then converted to a quantum defect scale to give $\mu_{2\Sigma}^{\text{specific}}(R)$.

Table II lists the $J=0$ levels obtained for two isotopes by the numerical procedures described in Sec. III. C. Figure 4 is an observed–calculated plot vs vibrational quantum number, in which both our MQDT results and the results of Kołos and Wolniewicz obtained by direct numerical integration are compared with experiment. Note that their values have been adjusted by a small overall shift to make them agree exactly with experiment for $v=0$ in H_2 and D_2, respectively. The experimental $J=0$ levels have been taken from Wilkinson[45] and Herzberg and Dabrowski.[10] The figure demonstrates that the quantum defect method gives as good agreement as the adiabatic approximation in this case, even though it does not necessitate the separate theoretical evaluation of the bulk of the adiabatic corrections. Indeed, the quantum defect results are seen to be somewhat superior for $0<v<6$ in H_2 and for $0<v<8$ in D_2. Quantum defect theory precludes the separate assessment of the various adiabatic and nonadiabatic corrections to a Born–Oppenheimer level—evaluating them all at once—but we may guess that the present improvement is due to inclusion of nonadiabatic (i.e., vibronic) mixing with

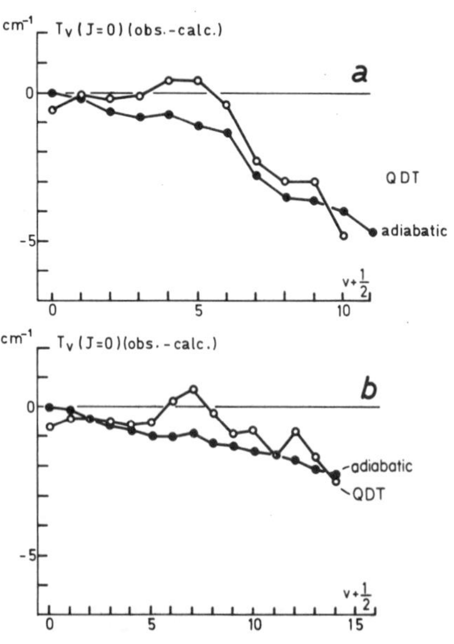

FIG. 4. Deviations of the observed levels T_v ($J=0$) in the $B\,^1\Sigma_u^+$ state from those derived *ab initio* in the adiabatic approximation (dots) and with quantum defect theory (circles): (a) for H_2, (b) for D_2. The $v+\frac{1}{2}$ scale for D_2 is smaller by the factor ρ than that for H_2.

TABLE II. Levels of the $B\,^1\Sigma_u^+$ state (cm^{-1}).

v[a]	MQDT	Δ	Adiabatic[b]	Δ	Exptl.[c,d]
			A. H_2, $J=0$		
0	90 204.1	−0.6	(90 203.5)	···	90 203.51
1	91 521.9	−0.1	91 522.0	−0.2	91 521.84
2	92 803.5	−0.2	92 803.9	−0.6	92 803.31
3	94 050.1	−0.1	94 050.8	−0.8	94 049.98[e]
4	95 262.8	+0.4	95 263.9	−0.7	95 263.18[e]
5	96 442.7	+0.4	96 444.2	−1.1	96 443.12
6	97 591.2	−0.4	97 592.1	−1.3	97 590.8[e]
7	98 707.6	−2.3	98 708.1	−2.8	98 705.34
8	99 792.1	−3.0	99 792.6	−3.5	99 789.14
9	100 845.4	−3.0	100 846.0	−3.6	100 842.38
10	101 869.5	−4.8	101 868.7	−4.0	101 864.65
			B. D_2, $J=0$		
0	90 634.5	−0.7	(90 633.8)	···	90 633.79[e]
1	91 576.3	−0.4	91 576.0	−0.1	91 575.94
2	92 499.2	−0.4	92 499.2	−0.4	92 498.78
3	93 403.8	−0.5	93 403.9	−0.6	93 403.26
4	94 290.9	−0.6	94 291.1	−0.8	94 290.30
5	95 160.6	−0.5	95 161.1	−1.0	95 160.05
6	96 013.2	+0.2	96 014.4	−1.0	96 013.38
7	96 849.6	+0.6	96 851.1	−0.9	96 850.2[e]
8	97 670.5	−0.2	97 671.5	−1.2	97 670.29
9	98 475.3	−0.9	98 475.7	−1.3	98 474.41
10	99 263.3	−0.8	99 264.0	−1.5	99 262.48
11	100 036.5	−1.7	100 036.4	−1.6	100 034.76
12	100 792.1	−0.8	100 793.1	−1.8	100 791.27
13	101 534.0	−1.7	101 534.4	−2.1	101 532.33
14	102 260.5	−2.5	102 260.3	−2.3	102 258.02

v	MQDT	Δ	Exptl.[d,f]
	C. H_2, $J=1$		
0	90 243.0	−0.7	90 242.32
1	91 558.8	0.0	91 558.82
2	92 838.8	−0.2	92 838.61
3	94 083.9	−0.2	94 083.72
4	95 295.3	+0.5	95 295.81
5	96 474.0	+0.3	96 474.27
6	97 621.3	+0.3	97 621.55
7	98 736.5	−2.6	98 733.92
8	99 821.9	−4.2	99 817.74
9	100 873.1	−3.7	100 869.39
	D. H_2, $J=3$		
0	90 435.4	−0.7	90 434.73
1	91 741.8	−0.1	91 741.67
2	93 013.7	−0.3	93 013.43
3	94 251.7	−0.2	94 251.52
4	95 456.5	+0.4	95 456.86
5	96 629.3	+0.2	96 629.47
6	97 770.8	−1.1	97 769.74
7	98 880.4	−1.6	98 878.75[g]
8	99 959.9	−2.8	99 957.11
9	101 006.2	−3.0	101 003.19
10	102 026.0	−3.7	102 022.27
	E. H_2, $J=5$		
0	90 774.5	−1.0	90 773.48
1	92 065.1	−0.8	92 064.29
2	93 323.2	−0.7	93 322.29
3	94 548.9	−0.5	94 548.36
4	95 742.4	−0.1	95 742.26
5	96 904.8	−0.6	96 904.18

TABLE II (*Continued*)

v	MQDT	Δ	Exptl.[d,f]
	E. H_2, $J=5$		
6	98 036.7	−2.1	98 034.61
7	99 136.6	−2.3	99 134.28
8	100 208.3	−2.3	100 206.02
9	101 245.0	−3.5	101 241.52

[a]The labels $v\Lambda$ used in this and the subsequent energy level tables refer to the component of the rovibronic wavefunction with the largest coefficient $A^J_{v\Lambda}$.
[b]After Kołos and Wolniewicz, Ref. 43.
[c]Obtained from the $P(1)$ rotational lines by adding the ground state rotational quantum $F''(1) = 118.50$ cm^{-1} (H_2) or 59.78 cm^{-1} (D_2).
[d]Values for $v = 0$–4 after Wilkinson (Ref. 45); for $v > 4$ after Dabrowski and Herzberg (Ref. 10).
[e]Band origin.
[f]Obtained from the $R(J-1)$ or $P(J+1)$ absorption lines by adding the appropriate ground state rotational energy, $F''(J-1)$ or $F''(J+1)$.
[g]From blended line.

higher states which pushes the levels to lower energies. Using a simple model Kołos and Wolniewicz estimated that nonadiabatic coupling shifts the origin of the $v=0$ band of H_2 down by about 0.4 cm^{-1} and that of the $v=5$ band by about 3 cm^{-1}, and their figures indeed agree qualitatively with the difference between their results and ours.

For $v>6$ in H_2 and $v>8$ in D_2, the agreement between experiment and quantum defect theory deteriorates and is no longer distinctly better than in the adiabatic calculation although, of course, on an absolute scale it is still exceedingly good. The $v=5$ level of H_2 and the $v=7$ level of D_2 both have their classical outer turning points near about 4.5 a.u., that is (as the curve $\mu_{2\Sigma}$ in Fig. 2 shows), in the region where configuration mixing has become important. The simplified quantum defect treatment of Sec. III obviously fails here partially since it does not include this additional source of vibronic interaction but on the contrary assumes vibronic mixing to diminish again, simply because the *derivative* of the quantum defect curve happens to pass through zero near 4.5 a.u.

Extension of the calculations to higher v values therefore did not seem warranted; we have instead calculated levels with higher values of angular momentum J. The results are included in Table II and are graphically represented in Fig. 5, which is analogous to Fig. 4. The similarity of the patterns obtained for different J values is striking. This is all the more so when one remembers that for $J>0$ the levels are subject not only to vibration–electron but also to rotation–electron coupling, so that there are now twice as many interacting channels (cf. Sec. III.A). It is again not possible to infer the magnitude of the effect of l uncoupling directly from the quantum defect calculations. We know from perturbation theory that in a first approximation the rotational nonadiabatic level shifts increase in proportion to $J(J+1)$. Julienne[46] has carried out calculations of the shifts in the B state using a two-state ($2p\sigma \sim 2p\pi$) "pure preces-

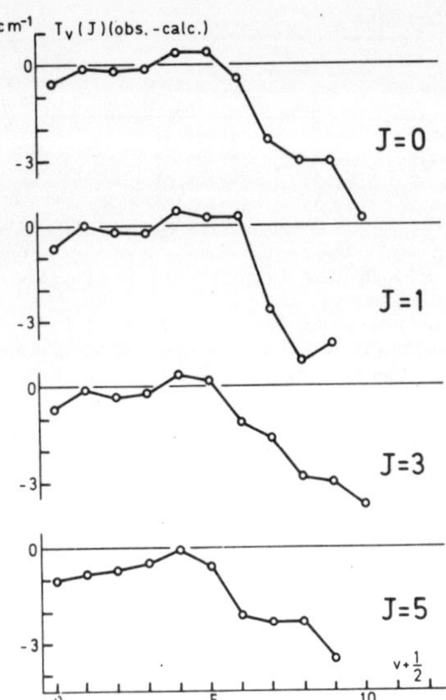

FIG. 5. Deviations of the observed levels $T_v(J)$ in the $B\,{}^1\Sigma_u^+$ state of H_2 from those derived *ab initio* using quantum defect theory.

sion" approximation, and he found that for $J=1$ they amount to about -0.5 cm^{-1} for the first vibrational levels (cf. his Fig. 4). For $J=5$ the effects of early l uncoupling must therefore be about 15 times as large, i.e., -7 cm^{-1}. Shifts of this magnitude, had they not been accounted for by the present treatment, would entirely spoil the agreement displayed in Fig. 5.

A further remark concerns the specific isotope effect which according to Eq. (30) gives a mass-dependent contribution to the quantum defect. If no theoretical value had been available, the effect could have been determined from the plot of Fig. 4 by the mass-dependent discrepancy caused by its neglect. Such a calculation was actually carried out and yielded the result $\mu_{2\Sigma}^{\text{specific}} = -0.24 \pm 0.02$ corresponding to a value $+10 \pm 1$ cm^{-1} for the specific effect in H_2 near equilibrium, in good agreement with the theoretical value $+9.825$ cm^{-1} at $R=2.43$ a.u.[44]

B. The $C\,{}^1\Pi_u$ state

The C state is the $n=2$ member of the $np\pi$ series. The quantum defect curve $\mu_{2\Pi}(R)$ shown in Fig. 2 recapitulates the behavior in the two-center system with one electron, where the united atom $2p\pi$ electron is unpromoted. Like the $\mu_{2\Sigma}$ curve the $\mu_{2\Pi}$ curve moves away from the zero line at intermediate values of R, but this time only slightly and to the negative side. This behavior is qualitatively understandable[5,47] because the nodal plane of the electronic wavefunction is now parallel rather than perpendicular to the molecular axis, so that pulling the nuclei apart influences the charge cloud less and raises the energy rather than lowering it. The slope of the quantum defect function near the equilibrium position $R=2.0$ a.u. of the core is also of opposite sign, making the π electron slightly bonding rather than antibonding.[5,47] The resulting vibrational frequency of the H_2^+-electron complex is larger by 117 cm^{-1} than in the free core. The rotational constant is larger by 1.1 cm^{-1} in $H_2({}^1\Pi_u)$, indicating a somewhat smaller equilibrium distance.

A very accurate Born–Oppenheimer potential energy curve for the C state has just recently been obtained by Kołos and Rychlewski[48] based on a 80-term wavefunction. The adiabatic corrections are unknown for this and for all higher states so that we depend entirely on MQDT for their evaluation. Calculations have been carried out in a similar manner as for the B state, including first the $J=N^+=1$ (Π_u^-) levels which are not subject to rotation–electron coupling, from $v=0$ up to the dissociation limit. The results for H_2 and D_2 are presented in Table III and illustrated on the observed–calculated plot of Fig. 6, where following Herzberg and Dabrowski[10] the abscissa scale for D_2 has been reduced by the factor $\rho = \sqrt{m_H/m_D} = 0.707379$.

The present improvement over the Born–Oppenheimer calculation of Ref. 48 is substantial. The adiabatic corrections are positive, so that the Born–Oppenheimer points referring to excited state *level energies* would appear in the positive range of Fig. 6. These deviations are quite large, e.g., $+97.8$ cm^{-1} for $v=0$, so that following Dabrowski and Herzberg we have preferred to display the deviations from experiment of the Born–Oppenheimer *transition energies*, by subtracting for each level the amount $+114.6$ of the adiabatic correction in the $v=0$, $J=0$ ground state of H_2.[10] The deviations from Born–Oppenheimer theory exhibit a systematic trend as functions of vibrational excitation and tend towards the value $(Ry_\infty - Ry_H)(1+\frac{1}{4}) - 114.6 = -39.9$ cm^{-1} (-20.0 cm^{-1} for D_2) representing the mass correction in the H(1s) + H(2l) separated atom limit. The present quantum defect calculation removes this trend together with the bulk of the discrepancy over an extended range of vibrational energy. The highest v values are the only exceptions; the residual deviations tend here towards the limiting values $+7.5$ cm^{-1} for H_2 and $+3.8$ cm^{-1} for D_2, in agreement with the prediction made in Sec. III. B.[49]

According to what has been said earlier, the remaining mass-dependent discrepancies at small vibrational energy must represent the specific isotope effect. The dotted straight line shown in Fig. 6 has been obtained by first drawing two straight lines parallel to the abscissa through the H_2 and D_2 points, respectively, and then extrapolating from H_2 [$\mu^{-1} \sim 2\,(\text{amu})^{-1}$] beyond D_2[$\mu^{-1} \sim 1(\text{amu})^{-1}$] to infinite mass ($\mu^{-1}=0$). The distance between the line and the H_2 data gives us directly the specific isotope effect in the C state of H_2 as $+4.4 \pm 0.3$ cm^{-1} near equilibrium, which corresponds to $\mu_{2\Pi}^{\text{specific}} = -0.16 \pm 0.01$. The position of the dotted line at a small negative value indicates that there is also a mass-independent contribution to the residual deviations. The cause may be sought in an incomplete convergence of

TABLE III. Levels of the $C\,^1\Pi_u^-$ state (cm^{-1}).

v	MQDT	Δ	A. H_2, $J=1$ BO[a]	Δ-114.6[b]	Exptl.[c,d]
0	99 147.8	+2.9	99 054.1	−18.0	99 150.72
1	101 453.4	+2.9	101 362.2	−20.5	101 456.33
2	103 624.9	+2.8	103 536.1	−23.0	103 627.74
3	105 664.9	···	105 578.4	···	···
4	107 575.3	+3.8	107 490.9	−26.4	107 579.06
5	109 356.8	+3.0	109 274.3	−29.1	109 359.77[f]
6	111 008.2	+2.7	110 927.6	−31.3	111 010.91
7	112 527.3	+3.0	112 448.4	−32.7	112 530.27
8	113 909.3	+4.3	113 832.2	−33.2	113 913.59[f]
9	115 147.5	+5.5	115 072.1	−33.7	115 152.99[f]
10	116 231.2	+4.5	116 157.4	−36.3	116 235.71
11	117 144.6	+5.5	117 072.3	−36.8	117 150.11[f]
12	117 861.7	+6.6	117 792.2	−38.5	117 868.27[f]
13					118 350.32[f]

v	MQDT	Δ	B. D_2, $J=1$ BO[a]	Δ-57.3[b]	Exptl.[c,e]
0	99 423.8	+0.8	99 377.3	−10.0	99 424.56
1	101 084.1	+0.8	101 038.6	−11.0	101 084.87
2	102 676.7	+0.8	102 632.1	−11.9	102 677.47
3	104 202.4	+1.2	104 158.9	−12.6	104 203.56[f]
4	105 662.5	+0.5	105 619.8	−14.1	105 662.96[f]
5	107 057.5	+2.0	107 015.5	−13.4	107 059.48
6	108 387.8	+1.2	108 346.6	−15.0	108 388.96
7	109 653.8	+1.1	109 613.3	−15.7	109 654.88
8	110 854.7	+1.2	110 814.8	−16.3	110 855.86
9	111 990.1	+1.4	111 950.9	−16.7	111 991.49
10	113 059.1	+1.3	113 020.5	−17.5	113 060.38
11	114 059.5	+1.3	114 021.6	−18.1	114 060.79
12	114 989.6	+2.0	114 952.3	−18.0	114 991.64
13	115 846.7	+1.0	115 809.9	−19.4	115 847.73
14	116 625.6	+1.6	116 589.4	−19.5	116 627.21
15	117 322.6	+1.0	117 286.6	−20.3	117 323.62
16	117 929.0	+1.3	117 893.4	−20.4	117 930.26
17	118 436.0	+1.3	118 401.1	−21.2	118 437.27
18					118 830.77
19					119 085.80

C. H_2, $J=3$

v	MQDT	Δ	Exptl.[c,d]
0	99 450.4	+3.0	99 453.36
1	101 740.3	+3.2	101 743.47
2	103 896.5	+3.2	103 899.66
3	105 921.5	+2.8	105 924.27
4	107 817.1	+2.8	107 819.93
5	109 583.6	+3.7	109 587.30
6	111 219.9	+2.8	111 222.74
7	112 723.4	+2.8	112 726.19
8	114 089.2	+3.6	114 092.84
9	115 310.3	+3.7	115 313.99
10	116 375.2	+4.1	116 379.34
11	117 267.0	+4.8	117 271.79[f]
12			117 964.88
13			118 413.11[f]

D. H_2, $J=5$

v	MQDT	Δ	Exptl.[c,d]
0	99 985.2	+2.6	99 987.84
1	102 247.3	+3.1	102 250.42
2	104 376.2	+3.3	104 379.49
3	106 374.4	+3.1	106 377.46
4	108 243.3	+4.1	108 247.37
5	109 983.3		
6	111 592.5	+1.7	111 594.19
7	113 068.2	+3.0	113 071.18
8	114 405.4	+2.5	114 407.87
9	115 594.0	+4.8	115 598.84

[a]Born–Oppenheimer approximation. Level positions obtained by solving the molecular analogue of Eq. (10), with $U_{2\Pi}$ from Ref. 48, and $J=\Lambda=1$.
[b]See text.
[c]Obtained from the $Q(J)$ absorption lines by adding the ground state rotational energy $F''(J)$.
[d]Values for $v=0$–4 after Dabrowski and Herzberg (Ref. 10); for $v>4$ after Namioka (Ref. 51).
[e]After Dabrowski and Herzberg (Ref. 10).
[f]From blended line.

the variational calculation of Kołos and Rychlewski, giving an energy which is too high by about 1.5 cm^{-1}. This is indeed the convergence error which they themselves have estimated. One should nevertheless be careful in drawing overly definite conclusions because, for example, the limited accuracy of the theoretical ionization potentials used in the present calculations might slightly affect the results. The smaller value of $\mu^{specific}$ in the C state as compared to the B state presumably reflects a diminished overlap of the two-electron distributions, due to the less favorable orientation of the π orbital with respect to the molecular axis mentioned already at the beginning of this paragraph.

The calculations have also been extended to levels with higher J values. The results are included in Table IV.A and illustrated by Fig. 7. The levels with $N^+=J$ (Π_u^-, circles in Fig. 7), which are rotationally unperturbed, give identical patterns as in the $N^+=J=1$ case

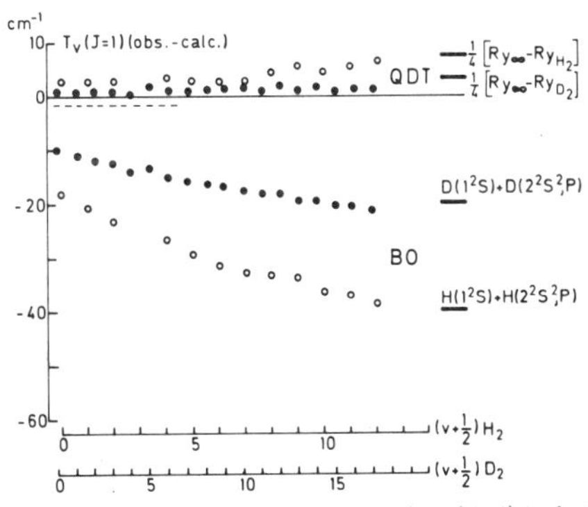

FIG. 6. Deviations of the observed levels $T_v(J=1)$ in the $C\,^1\Pi_u^-$ state from those derived *ab initio* for H_2 (circles) and D_2 (dots in the Born–Oppenheimer approximation (BO) and with quantum defect theory (QDT). The difference between the two results represents the adiabatic and nonadiabatic corrections. The remaining deviations in QDT correspond to the specific isotope effect (see text).

TABLE IV. Levels of the $C\,^1\Pi_u^+$ state of H_2 (cm^{-1}).

A. Levels

v	J	MQDT	Δ	Exptl. [a,b]
0	1	99 148.9	+3.0	99 151.92
0	3	99 456.3	+3.1	99 459.43
0	5	99 996.8	+4.1	100 000.88
1	1	101 454.5	+3.3	101 457.75
1	3	101 745.2	+3.8	101 749.02
1	5	102 230.5	−6.6	102 223.93
2	1	103 626.7	+2.2	103 628.92
2	3			103 896.21
2	5	104 394.6	+9.2	104 403.78

B. Λ doubling [c]

v	J	MQDT	Δ	Ford [d]	Δ	Exptl.
0	1	+1.1	+0.1	+1.2	0.0	+1.20
0	3	+5.9	+0.2	+6.6	−0.5	+6.07
0	5	+11.6	+1.4	+13.2	−0.2	+13.04
1	1	+1.1	+0.3	+1.2	+0.2	+1.42
1	3	+4.9	+0.7	+6.0	−0.4	+5.55
1	5	−16.8	−9.7	(+44.0) [e]	(−70.5) [e]	−26.49
2	1	+1.8	−0.6	+0.8	+0.4	+1.18
2	3			−3.0	−0.5	−3.45
2	5	+18.4	+5.9	+24.2	+0.1	+24.29

[a]Obtained from the $R(J-1)$ or $P(J+1)$ absorption lines by adding the ground state rotational energy $F''(J-1)$ or $F''(J+1)$.
[b]From Dabrowski and Herzberg (Ref. 10).
[c]Difference of Π_u^+ and Π_u^- components for a given J and v.
[d]Reference 50.
[e]The C–X (1–0) and B–X (10–0) bands exhibit an avoided crossing near $J=4$. The level calculated by Ford for $J=5$ belongs to the band which at $J=0$ is B–X (10–0). The corresponding absorption lines have not been observed (cf. Ref. 10).

and need no further comment. The remaining levels are related to the alternative ionization channels $N^+ = J\pm 1$ and have predominant Π_u^+ character (dots in Fig. 7); they have been obtained for $v=0$, 1, and 2 in the same calculation as the $J>0$ levels of the B state with which they interact rotationally. The agreement with experiment is similar to that for the $N^+=J$ levels for $J=1$ and 3 but is worse for $J=5$, somewhat surprisingly in view of the good agreement obtained for the B state levels.

The difference $\Pi_u^+ - \Pi_u^-$ of the levels for a given J is the Λ doubling, listed in Table IV.B. The Λ doubling in the C state depends not only on J but also on v as a consequence of the substantially different shapes of the potentials $U_{2\Sigma}$ and $U_{2\Pi}$ (which in turn are related to the different R dependences of $\mu_{2\Sigma}$ and $\mu_{2\Pi}$). An additional complication is that for $v>7$ the vibrational level structure of the B state is superimposed on that of the C state, giving rise to local rovibronic perturbations ("accidental l uncoupling") whose presence manifests itself by the sudden changes of sign of the Λ doubling for certain v and J.

It is interesting to compare the present results with the previous calculations of the C state Λ doubling by Julienne[46] and Ford.[50] Both authors treated l uncoupling as a pairwise interaction between the $2p\sigma$ and $2p\pi$ members of the np Rydberg series, and they evaluated the matrix elements of the operator cross term $-(m/\mu)R^{-2}(J_xL_x+J_yL_y)$ [i.e., the analogue of Eq. (31) for rotational motion] between zero-order vibronic wavefunctions. Julienne assumed unity overlap between the radial electronic wavefunctions ("pure precession) and thus reduced the problem to the evaluation of Franck–Condon factors weighted by R^{-2} between B and C state levels. His results gave too large values for the Λ doubling. Ford improved on this by calculating the electronic radial overlap with *ab initio* configuration interaction wavefunctions, and he was thus able to reduce the calculated Λ doubling by the correct amount. Both authors neglected the fact that when rotational interaction with the B state is reduced because of a reduced electronic overlap, the missing amount must turn up again as an additional interaction with *higher* states. Mulliken[47] had foreseen this effect years ago and called it "inhibited" l uncoupling, showing that the interaction with higher states reduces the Λ doubling further and may even cancel it when the $np\pi$ Rydberg member lies about halfway between the $np\sigma$ and $(n+1)p\sigma$ members (i.e., $\mu_\Sigma - \mu_\Pi \sim 0.5$). Quantum defect theory allows for just this behavior because it treats not the interaction between single Rydberg members, but between whole Rydberg series and ionization continua, gauging the difference of radial functions by the difference of the defects μ_Σ and μ_Π. In the present example $\mu_\Sigma - \mu_\Pi$ is about 0.25 near equilibrium, i.e., the situation is well on the way to the special case of inhibited l uncoupling, and this must be the reason for the small Λ doubling which we obtain. Our treatment possibly overestimates the degree of inhibition because it assumes $l=1$ and

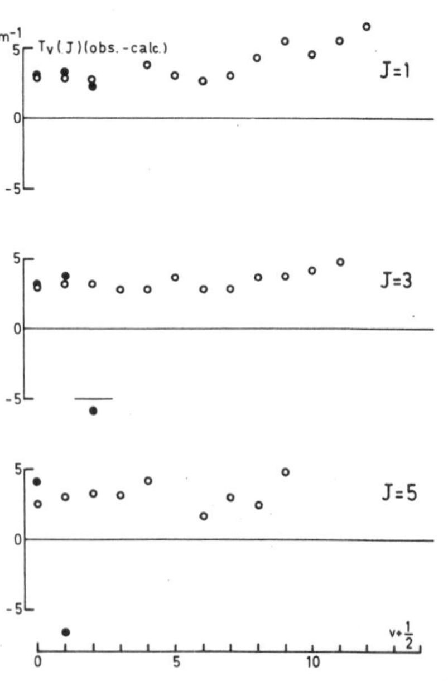

FIG. 7. Deviations of the observed $T_v(J)$ levels in the $C\,^1\Pi_u$ state of H_2 from those derived *ab initio* using quantum defect theory: Π_u^- component (circles), Π_u^+ component (dots).

TABLE V. Levels of the $D\,^1\Pi_u^-$ state (cm^{-1}).

			A. H_2, $J=1$		
			MQDT		
v	*ab initio*[b]	Δ	Present fit	Δ	Exptl.[a]
0	112 936.6	−4.8	112 931.0	+0.8	112 931.8
1	115 159.5	−5.1	115 154.0	+0.4	115 154.4[d]
2	117 252.9	−6.2	117 247.5	−0.8	117 246.7
3	119 220.8	−4.8	119 215.4	+0.6	119 216.0
4	121 066.1	−5.0	121 060.9	+0.2	121 061.1
5	122 791.1	−4.9	122 786.1	+0.1	122 786.2
6	124 397.1	−4.7	124 392.2	+0.2	124 392.4
7	125 883.0	−5.5	125 878.6	−1.1	125 877.5
8	127 252.1	−3.9	127 247.8	+0.4	127 248.2
9	128 499.9	−3.9	128 495.7	+0.3	128 496.0
10	129 624.1	−3.0	129 620.3	+0.8	129 621.1
11	130 620.5	−2.3	130 617.0	+1.2	130 618.2
12	131 485.6	−1.2	131 482.5	+1.9	131 484.4
13	132 212.3	−0.1	132 210.5	+1.7	132 212.2
			B. D_2, $J=1$		
0	113 228.1	−5.6	113 222.5	0.0	113 222.5
1	114 829.2	−5.5	114 823.7	0.0	114 823.7
2	116 364.5	−5.7	116 359.0	−0.2	116 358.8
3	117 835.4	−5.0	117 830.0	+0.4	117 830.4
4	119 243.6	−5.0	119 238.3	+0.3	119 238.6
5	120 590.1	−4.7	120 584.8	+0.6	120 585.4
6	121 875.7	−4.2	121 870.5	+1.0	121 871.5[d]
7	123 101.1	−4.8	123 096.0	+0.3	123 096.3
8	124 266.6	−4.7	124 262.5	−0.6	124 261.9
9	125 373.1	−5.1	125 368.4	−0.4	125 368.0
10	126 420.0	−4.5	126 415.4	+0.1	126 415.5
11	127 407.5	−4.4	127 403.0	+0.1	127 403.1
12	128 334.9	−4.3	128 330.6	0.0	128 330.6
13	129 201.8	−3.8	129 197.5	+0.5	129 198.0

	C. H_2, $J=3$		
v	MQDT[c]	Exptl.[a]	Δ
0	113 224.0	113 224.7	+0.7
1	115 431.6	115 432.0[d]	+0.4
2	117 510.2	117 510.3	+0.1
3	119 463.6	119 463.9	+0.3
	D. H_2, $J=5$		
0	113 741.6	113 741.4	−0.2
1	115 921.9	115 922.0	+0.1
2	117 974.0	117 977.2[d]	+3.2
3			

[a]From the $Q(1)$ absorption lines of Takezawa [Refs. 23(a) and 52].
[b]Using $U_{3\Pi}(R)$ of Kołos and Rychlewski (Ref. 48).
[c]Present fit.
[d]From blended line.

neglects higher partial waves as well as configuration mixing, thus concentrating the interaction of the C state with higher $^1\Sigma_u^+$ states in the nearby $3p\sigma B'$ state. Another shortcoming is that the quantum defects have been taken to be energy independent, that is, it was implicitly assumed that the $\mu_{2\Sigma}$ curve retains its validity for the higher $^1\Sigma_u^+$ Rydberg members, whereas this is not true, as Fig. 2 shows.

C. The $D\,^1\Pi_u$ state

The quantum defect curve of the $n=3\ ^1\Pi_u$ Rydberg state coincides with $\mu_{2\Pi}(R)$ so nearly that the two cannot be distinguished on the scale of Fig. 2. (The potential curves $U_{3\Pi}$ and $U_{2\Pi}$ on the other hand are not identical; in the D state the H_2 vibrational frequency differs by mere +34.4 cm^{-1} from that of the ion, as compared to +117 cm^{-1} in the C state[5]). It is therefore possible to calculate the levels of the D state in a good approximation by use of $\mu_{2\Pi}(R)$, without having to rely on a theoretical potential for the D state. Such a calculation was carried out and was found to predict the levels consistently about 5 cm^{-1} too low, with a slight trend in function of v.

Kołos and Rychlewski[48] have obtained a theoretical potential energy curve also for the D state. Using the quantum defect function $\mu_{3\Pi}(R)$ derived from their curve we have calculated the vibronic levels presented in Table V. The improvement over the Born–Oppenheimer calculation of Ref. 48 is comparable to that obtained for the C state. The observed–calculated plot of Fig. 8 shows that the theoretical levels obtained by MQDT are consistently about 5 cm^{-1} *higher* than the experimental levels of Namioka[51] and Takezawa,[23a, 52] with no systematic difference perceptible between the H_2 and D_2 values. We interpret the discrepancy as due to incomplete convergence of the theoretical variational calculation, and we have made use of the data shown in Fig. 8 in an attempt to derive an improved fixed-nuclei potential energy curve. The method essentially consisted of correcting the theoretical quantum defect curve in a trial and error procedure such that the resulting vibronic levels should come to lie on the abscissa in Fig. 8. The figure shows to what extent this has been possible. The underlying corrected curve $\mu_{3\Pi}(R)$ is listed in Table VI for 25 points together with the Born–Oppenheimer potential $U_{3\Pi}(R)$ derived from it by use of Eq. (21).

The $U_{3\Pi}$ curve thus obtained should deviate from the

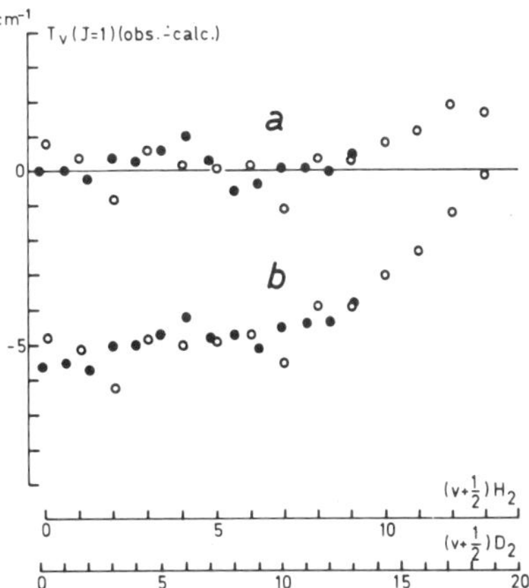

FIG. 8. Deviations of the observed $T_v(J=1)$ levels of H_2 and D_2 in the $D\,^1\Pi_u^-$ state from those obtained with quantum defect theory: (a) by fitting the potential function to the experimental levels, (b) by using the *ab initio* curve of Kołos and Rychlewski, Ref. 48.

TABLE VI. Potential energy curve for the $D\,^1\Pi_u$ state (a.u.).

	From experiment[a]		*ab initio*[b]	Δ
R	$\mu_{3\Pi}$	$U_{3\Pi}$	$U_{3\Pi}$	(cm^{-1})
1.00	−0.03990	−0.5058930	−0.50586485	−6.2
1.20	−0.04716	−0.5828237	−0.58279582	−6.1
1.40	−0.05512	−0.6235525	−0.62352515	−6.0
1.60	−0.06352	−0.6442128	−0.64418636	−5.8
1.80	−0.07205	−0.6532338	−0.65320792	−5.7
2.00	−0.08046	−0.6553255	−0.65529982	−5.6
2.20	−0.08858	−0.6532542	−0.65322903	−5.5
2.40	−0.09619	−0.6487109	−0.64868640	−5.4
2.60	−0.10316	−0.6427564	−0.64273290	−5.2
2.80	−0.10934	−0.6360731	−0.63604977	−5.1
3.00	−0.11464	−0.6291040	−0.62908115	−5.0
3.25	−0.11989	−0.6204249	−0.62040253	−4.9
3.50	−0.12349	−0.6121050	−0.61208510	−4.4
3.75	−0.12522	−0.6043545	−0.60433591	−4.1
4.00	−0.12504	−0.5972836	−0.59726659	−3.7
4.50	−0.11898	−0.5853379	−0.58532328	−3.2
5.00	−0.10627	−0.5762396	−0.57622687	−2.8
5.50	−0.08931	−0.5696214	−0.56961145	−2.2
6.00	−0.07098	−0.5649861	−0.56497868	−1.6
6.50	−0.05386	−0.5618286	−0.56182397	−1.0
7.00	−0.03949	−0.5597153	−0.55971440	−0.2
7.50	−0.028194[b]	−0.55831885[b]	−0.55831885	···
8.00	−0.019757[b]	−0.55740139[b]	−0.55740139	···
9.00	−0.009236[b]	−0.55641054[b]	−0.55641054	···
10.00	−0.003841[b]	−0.55599226[b]	−0.55599226	···

[a]See text. [b]After Kołos and Rychlewski (Ref. 48).

true Born–Oppenheimer binding energy by no more than ± 1 cm^{-1} for R values near equilibrium. The high quality of the experimental data is confirmed by the present calculations, which on the basis of *one single* potential reproduce with equal accuracy the absolute positions of vibronic levels in two isotopes whose masses differ by a factor of 2. It appears that the specific isotope effect has entirely collapsed at $n=3$. Extrapolation from $n=2$ to $n=3$, based on the shift in the C state and assuming a constant μ^{specific}, predicts a shift of 0.7 cm^{-1} between the H_2 and D_2 data points in Fig. 8. We find that with the present calculations we are unable to separate out so small an effect.

The quantum defect curve $\mu_{3\Pi}$ has also been used for the calculation of the first vibronic levels of H_2 with $J>1$ in the $N^+=J\,(\Pi_u^-)$ and $N^+\neq J$ (approximate Π_u^+) components. The $N^+\neq J$ levels were evaluated along with those of the $n=3$ $B'\,^1\Sigma_u^+$ state (see next section) in a calculation based on the two curves $\mu_{3\Sigma}$ and $\mu_{3\Pi}$. The results are included in Tables V and VII; Table VII also gives the values obtained for the Λ doubling. The agreement with experiment is good in all cases. One notes that the Λ doubling is much larger than for $n=2$ owing to more advanced l uncoupling at this higher energy, and yet is much better reproduced, although the predictions of MQDT are still slightly too small with few exceptions. The rotational perturbations which are responsible for the irregularity of the doubling involve here levels of the B' state. An example are the $D\,^1\Pi_u^+$, $v=1$, $J=5$ and $B'\,^1\Sigma_u^+$, $v=3$, $J=5$ levels which interact strongly.

Figure 3 shows that starting with $v=3$ the levels of the H_2 D state lie above the H(1s) + H(2l) dissociation limit, and for $v>7$ they lie higher than the ionization potential. From a theoretical point of view, one should therefore think of them not as bound states, but as resonances, broadened and shifted by predissociation and preionization. These processes are not in the scope of the present paper and will be studied in a forthcoming publication. In the calculations we have artificially eliminated discrete–continuum interactions by truncating the $\mathfrak{C}$ and $\mathfrak{S}$ matrices from both ends, i.e., by omitting the vibrational continuum and throwing away each ionization channel as soon as it opens, starting with $v^+=0$. Preliminary calculations of the absorption profiles in the ionization continuum indicate that the Π_u^- absorption peaks are shifted only by small fractions of a wavenumber unit as a consequence of the interaction with the continuum, so that hardly any error is introduced in the present potential curve determination.

D. The B' and $B''\,^1\Sigma_u^+$ states

When this work was begun no highly accurate theoretical potential energy curves existed for the $n=3$ B' and $n=4$ B'' members of the $^1\Sigma_u^+$ series.[53] Both curves have therefore been determined directly from experimental data. Comparison with *ab initio* theory became possible most recently when Kołos[54] obtained theoretical curves

TABLE VII. Levels of the $D\,^1\Pi_u^+$ state of H_2 (cm^{-1}).

A. Levels

v	J	MQDT[a]	Exptl.[b]	Δ
0	1	112 934.6	112 935.5	+0.9
0	3	113 242.7	113 244.3	+1.6
0	5	113 780.5	113 784.0[e]	+3.5
1	1	115 155.3	115 155.8	+0.5
1	3	115 436.9	115 438.2	+1.3
1	5	115 921.4	115 922.7	+1.3
2	1	117 251.3	117 251.6	+0.3
2	3	117 526.0	117 526.9	+0.9
2	5	117 992.4	117 989.9	−2.5
3	1	119 218.6	119 217.6[d]	−1.0
3	3		119 472.0[d]	
3	5		···	

B. Λ doubling[c]

v	J	MQDT	Exptl.	Δ
0	1	+3.6	+3.7	+0.1
0	3	+18.7	+19.6	+0.9
0	5	+38.9	+42.6	+3.7
1	1	+1.3	+1.4	+0.1
1	3	+5.3	+6.2	+0.9
1	5	−0.5	+0.7	+1.2
2	1	+3.8	+4.9	+1.1
2	3	+15.8	+16.6	+0.8
2	5	+18.4	+12.7	−5.7
3	1	+3.2	+1.6	−1.6
3	3		+8.1	

[a]See text.
[b]From the $R(J-1)$ or $P(J+1)$ lines of Takezawa [Ref. 23(a)].
[c]Difference of Π_u^+ and Π_u^- components for a given J and v.
[d]From diffuse line.
[e]From blended line.

TABLE VIII. Potential energy curve for the $B'\,^1\Sigma_u^+$ state (a.u.).

	From experiment[a]		*ab initio*[b]	Δ
R	$\mu_{3\Sigma}$	$U_{3\Sigma}$	$U_{3\Sigma}$	(cm^{-1})
1.20	0.0570	−0.5867029	−0.5866224	−17.7
1.40	0.0819	−0.6287013	−0.6286629	−8.4
1.60	0.1125	−0.6509061	−0.6508320	−16.3
1.80	0.1448	−0.6615870	−0.6615332	−11.8
2.00	0.1795	−0.6654860	−0.6654411	−9.9
2.20	0.2154	−0.6653225	−0.6652890	−7.4
2.40	0.2519	−0.6627608	−0.6627343	−5.8
2.60	0.2888	−0.6588548	−0.6588215	−7.3
2.80	0.3259	−0.6542782	−0.6542389	−8.6
3.00	0.3634	−0.6494881	−0.6494544	−7.4
3.50	0.4645	−0.6386310	−0.6386045	−5.8
4.00	0.5785	−0.6313558	−0.6313096	−10.1
4.50	0.6925	−0.6278446	−0.6278200	−5.4
5.00	0.7883	−0.6266360	−0.6266236	−2.7

[a]See text. [b]After Kołos (Ref. 54).

for the two states. The quantum defect curves $\mu_{3\Sigma}(R)$ and $\mu_{4\Sigma}(R)$ both have the same qualitative behavior as in the two-center system with one electron where the $np\sigma$ electron is singly promoted. While the $n=3$ and $n=4$ Σ curves are still distinct on the plot of Fig. 2, a clear tendency towards convergence of the combined (R, E) dependence is also seen. Indeed, the characteristic strong R dependence of the Σ curves is the source of the strong Σ–Σ vibronic perturbations occurring at higher n and is largely responsible for the vibrational preionization observed above the ionization potential.[5,33] An anomaly is seen in the $n=4$ curve near 5.5 a.u., where an avoided crossing occurs involving the curve (not shown in Fig. 2) which originates from the *doubly* promoted $4f\sigma$ united atom electron.[10,54] As a consequence, the $\mu_{4\Sigma}$ curve deviates and tends upward towards $\mu=2$ (dotted line, drawn from Kołos' data). The corresponding $U_{4\Sigma}$ curve has a repulsive part followed by a second minimum which is related to further interactions.[54] Molecular singlet levels arising from the $4f$ electron have escaped spectroscopic observation to this day. For the present purposes we have therefore neglected the avoided $p\sigma$~$f\sigma$ crossing, that is, we have implied that the curves *cross* as they would in the one-electron system (full line).

The B' state curve has been fitted to the rotationally unperturbed $J=0$ experimental levels of Namioka[51] (H_2) and Dabrowski and Herzberg[10](D_2). In the absorption spectrum these levels are represented by the $P(1)$ lines. A helpful circumstance in the fitting procedure was the *a priori* knowledge of the qualitative behavior of $\mu_{3\Sigma}(R)$. While in each calculation a full curve ranging from 0 to 10 a.u. was entered, the actual fine adjustment was started near equilibrium using the $v=0$ level of each isotope. We then worked both ways, to smaller and larger R values, and fitted successively higher levels up to $v=5$ for H_2 and $v=7$ for D_2. At $R=0$ the curves $\mu_{2\Sigma}$ and $\mu_{3\Sigma}$ must coincide, and we expect that at small $R>0$ they must remain very close: these additional constraints enabled the higher v levels to be fitted essentially by adjustment of just the outer limb of the curve.

The resulting curves $\mu_{3\Sigma}$ and $U_{3\Sigma}$ are presented in Table VIII for the range of R values which are classically accessible in the levels fitted. Table IX contains the calculated vibronic levels and compares them with the levels obtained from the theoretical potential and with the experimental levels. Figure 9 illustrates the the agreement with experiment obtained. Rather than discussing the magnitude of adiabatic and nonadiabatic effects once more, we make the comparison with *ab initio* theory now directly in terms of potential energy curves. Kołos evaluated the energy necessary to dissociate the $v=0$, $J=0$ level in the Born–Oppenheimer approximation, and he found disagreement with experi-

TABLE IX. Levels of the $B'\,^1\Sigma_u^+$ state (cm^{-1}).

A. H_2, $J=0$

v	MQDT Present fit	Δ	*ab initio*[b]	Δ	Exptl.[a]
0	110478.2	0.0	110486.7	−8.5	110478.18
1	112358.7	−0.8	112367.6	−9.7	112357.87
2	114075.8	−0.8	114084.7	−9.7	114074.98
3	115607.0	−1.4	115615.4	−9.8	115605.60[d]
4	116905.7	+1.5	116918.3	−11.1	116907.16
5	117860.7	+0.7	117865.5	−4.1	117861.41

B. D_2, $J=0$

v	MQDT Present fit	Δ	*ab initio*	Δ	Exptl.[c]
0	110815.8	−0.2	110824.2	−8.6	110815.62
1	112181.2	−0.2	112189.8	−8.8	112180.98
2	113467.2	−0.1	113476.4	−9.3	113467.09
3	114669.8	−1.1	114678.8	−10.1	114668.65[d]
4	115779.8	−0.3	115787.8	−8.3	115779.51
5	116784.4	+0.1	116792.6	−8.1	116784.48
6	117658.2	+1.4	117666.8	−7.2	117659.55
7	118356.4	+0.9	118363.7	−6.4	118357.28

C. H_2, $J=1$

v	MQDT, present work	Exptl.[a]	Δ
0	110529.3	110528.93	−0.4
1	112404.7	112403.75	−0.9
2	114119.0	114118.25	−0.7
3	115647.7	115646.78	−0.9

D. H_2, $J=3$

v	MQDT, present work	Exptl.[a]	Δ
0	110782.6	110781.39	−1.2
1	112634.5	112633.07	−1.4
2	114334.1	114332.62	−1.5
3	115850.3	115848.83	−1.5

E. H_2, $J=5$

v	MQDT, present work	Exptl.[a]	Δ
0	111231.4	111228.96	−2.4
1	113045.4	113042.52	−2.9
2	114716.4	114713.17	−3.2
3	116216.9	116215.40	−1.5

[a]Obtained from the $P(1)$ absorption lines of Namioka (Ref. 51).
[b]Using $U_{3\Sigma}(R)$ after Kołos (Ref. 54).
[c]Obtained from the $P(1)$ absorption lines of Dabrowski and Herzberg (Ref. 10).
[d]From blended line.

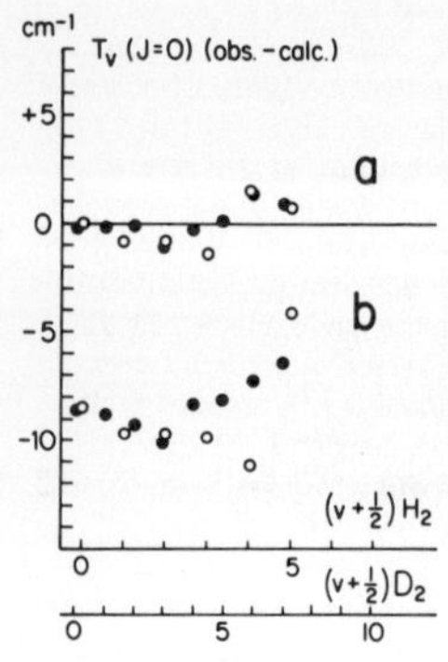

FIG. 9. Deviations of the observed $T_v(J=0)$ levels of H_2 and D_2 in the $B'\,{}^1\Sigma_u^+$ state from those obtained with quantum defect theory: (a) by fitting the potential function to the experimental levels, (b) by using the *ab initio* curve of Kołos, Ref. 54.

ment of the order of 40 cm^{-1}. The direct comparison of his theoretical with the present "experimental" potential, on the other hand, reveals a discrepancy of only 8 cm^{-1} near equilibrium, with his curve lying higher than ours, as expected. Kołos himself had estimated from the discrepancy found for $v=0$ that the convergence error of his calculation should amount to about 8 cm^{-1}. It is probably safe to conclude that the present experimental curve is correct to within 1 cm^{-1} near equilibrium, and to within a few cm^{-1} at the other R values.

Table IX includes a number of H_2 levels calculated for $J>0$, which again are in good agreement with experiment. A slight discrepancy increasing with J turns up nevertheless, and possibly is the counterpart of the too small Λ doublings calculated in the preceding section for the D state.

The present calculations do not yet include the last three H_2 and six D_2 vibronic levels. For these levels

TABLE X. Potential energy curve for the $B''\,{}^1\Sigma_u^+$ state (a.u.).

	From experiment [a]		*ab initio* [b]	Δ
R	$\mu_{4\Sigma}$	$U_{4\Sigma}$	$U_{4\Sigma}$	(cm^{-1})
1.20	0.0550	−0.56110	−0.56104043	−14
1.40	0.0775	−0.60248	−0.59858262	−855
1.60	0.1090	−0.62396	−0.62395625	−1
1.80	0.1475	−0.63394	−0.63385258	−20
2.00	0.1850	−0.63699	−0.63686585	−27
2.20	0.2217	−0.63586	−0.63575960	−23
2.40	0.2596	−0.63229	−0.63219466	−21
2.60	0.2988	−0.62733	−0.62722840	−23
2.80	0.3392	−0.62167	−0.62155022	−25
3.00	0.3810	−0.61574	−0.61558786	−33
3.25	0.4390	−0.60849	−0.60831826	−37
3.50	0.5000	−0.60167	−0.60149691	−38
3.75	0.5670	−0.59559	−0.59540666	−40
4.00	0.6320	−0.59016	−0.58964519	−114
4.25	0.6936	−0.58541	−0.58493813	−104
4.50	0.7500	−0.58128	−0.58121184	−14
4.75	0.7986	−0.57765	−0.57746183	−42
5.00	0.8400	−0.57449	−0.57407558	−91

[a]See text.
[b]After Kołos (Ref. 54).

TABLE XI. Levels of the $B''\,{}^1\Sigma_u^+$ state of $H_2(J=0)$(cm^{-1}).

v	Present fit [a]	Δ	*ab initio* [b]	Δ−114.6 [c]	Exptl. [d]
0	116885.2	+0.8	116785.2	−13.8	116886.0
1	118945.4	−0.4	118862.6	−32.2	118945.0
2	120869.9	−0.5	120806.4	−51.6	120869.4
3	122654.6	+1.8	122609.9	−68.1	122656.4
4	124307.7	0.0	124266.3	−73.2	124307.7 [e]
5	125855.7	−5.5	125774.6	−39.0	125850.2 [e]
6	127096.7	−2.2	127136.7	−156.8	127094.5
7	128311.3	−4.5			128306.8

[a]MQDT, see text.
[b]Born–Oppenheimer approximation. These values were obtained by subtracting from the theoretical $n=3$ dissociation limit in the Born–Oppenheimer approximation (=133543.8 cm^{-1} above the theoretical $v=0$, $J=0$ level of the ground state, corrected for adiabatic and nonadiabatic effects) the values given by Kołos (Ref. 54) in his Table III and IV, second column.
[c]Adiabatic correction at R_e in the ground state.
[d]Values from $v=0$–3 from the $P(1)$ lines observed by Takezawa [Ref. 23(a)]; for $v>3$ from the $P(1)$ lines listed in Ref. 5.
[e]From reassigned line (Ref. 56).

an abrupt decrease of the vibrational and rotational spacings is observed.[10,51] Correspondingly, the potential $U_{3\Sigma}$ which for $R<5$ a.u. follows a steep course towards the $n=2$ dissociation limit, flattens out suddenly near 5 a.u. and exhibits a long shallow "tail".[10,51] It is interesting to note that the quantum defect curve $\mu_{3\Sigma}(R)$ does not exhibit such a tail (Fig. 2; the dotted part of the curve is drawn from the data of Kołos). The curve $\mu_{3\Sigma}(R)$ embodies the effect of promotion of the σ electron as the dominant feature of electron–core interaction and singles it out in the Rydberg term of Eq. (21). It happens, for $n=3$ and $R>5$ a.u., that beginning promotion tends to raise the electronic energy above the asymptotic value by nearly as much as the attractive forces in the core lower it. The two contributions combined yield the peculiar shape of the potential energy curve.

The B'' state curve has been fitted on the basis of the experimental $J=0$ levels of Ref. 5. The quantum defect curve has been obtained in this case by a procedure of refinement of the algebraic four-parameter curve derived previously in Ref. 11. The resulting curves $\mu_{4\Sigma}$ and $U_{4\Sigma}$ are presented in Table X; Table XI lists the vibronic levels of the B'' state for $0<v<7$. Part of the theotheoretical curve calculated by Kołos[54] using a 76-term expansion of the electronic wavefunction is also included in Table X. The comparison shows that the theoretical curve lies again higher than the experimental one, but this time by 27 cm^{-1} near equilibrium. It appears that the bulk of this increased discrepancy derives from increased difficulties in the variational calculation. According to Kołos' own estimation, the convergence error must amount to about 24 cm^{-1} near equilibrium. The present experimental potential curve is therefore probably correct to within about 5 cm^{-1} near $R=2$ a.u.[55] A discrepancy orders of magnitude larger occurs at $R=1.4$ a.u., where the theoretical curve lies so high as to imply a negative quantum defect. This single point therefore appears to be out of line and may be the source of the surprisingly large discrepancies between theoretical Born–Oppenheimer levels and experiment (Table XI). The difference between the theoretical and experimental potential curves exceeds 100 cm^{-1} at a few more points near 4 a.u. and here possibly arises in part from shortcomings of the present fitting procedure. However, these deviations are probably not related to the use of a single-minimum instead of a double-minimum potential, as is shown by the test calculations of Kołos.[54]

The quantum defect treatment has not yet been applied to the data for the heavier isotope D_2. It turns out that for $n=4$ the extraction of the quantum defect curve from the experimental data is much more difficult than for $n=2$ or 3. One enters here the complicated spectral region where the Born–Oppenheimer approximation no longer holds even approximately for most of the observed levels. Rydberg states with higher n move closer and closer together, and the higher vibrational levels of the B'' state are tied to the dense vibrational-electronic Rydberg pattern by strong vibronic interaction. They can be separated from the surrounding Rydberg structure only by the quantitative quantum defect treatment itself, and the labels v in Table XI refer to the largest but not necessarily dominating component $A^{J=0}_{v\Sigma}$ of the vibronic wavefunction. The present potential energy curve determination necessitated a partial revision of the previous somewhat more empirical analysis of the absorption spectrum given in Ref. 5. The details of this study will be described in a later publication, which analyzes not only discrete level positions but simultaneously also intensities and preionization line profiles, all on the basis of MQDT.[56] The reader is referred to Ref. 11 for a preliminary account of the numerous perturbations affecting the B'' state levels and for a discussion of the specific role played by the preionizing $v>4$ levels of the B'' state in the photoionization process.

An important simplifying aspect in the analysis of these higher Rydberg states is that the variation of the quantum defect curve from one n value to the next becomes negligible so that in effect *two curves*, $\mu_{4\Sigma}$ and $\mu_{3\Pi}$, serve to interpret the *full* Rydberg structure from $n=3$ or 4 up to the rovibrational ionization limits and even beyond.[56] The potential curve determination for individual Rydberg states therefore comes to an end here.

V. DISCUSSION

The preceding section has demonstrated how quantum defect theory permits one to compare *ab initio* theory with experiment in great detail. Only a few years ago, the accurate experimental tests of quantum mechanical calculations were still limited to few molecular parameters, such as the dissociation energy or the ionization potential, and were only recently applied to the band origins of just one or two electronic states. They have been extended here to the full spectrum of bound rovibronic levels in many electronic states, in an approach which treats all states on one common basis and thereby reveals a high degree of consistency between numerous existing experimental as well as theoretical data.

It has already been mentioned that the quantum defect curves displayed in Fig. 2 will serve as a basis for the study of the higher Rydberg states of molecular hydrogen. Most of these highly excited levels have a short lifetime, since besides deexcitation by fluorescence they can also preionize and predissociate. Dehmer and Chupka[33] have recently observed the photoionization and photoabsorption cross sections under high resolution down to 716 Å (corresponding to the $v^+=6$ ionization limit). Their spectra yield information concerning the competition between the three processes at various excitation wavelengths, and they have thus been able to present for the first time an overall picture of the preferred deexcitation paths in this energy region. It turns out that these depend characteristically on the quantum numbers and parity of the quasistationary levels excited by photoabsorption. This is confirmed by the complementary data of Borrell, Glass, and Guyon,[57] who observed the fluorescence of H atoms produced by photoexcitation of H_2. The nuclear–electronic coupling which enables the molecule to pass from a discrete level into the electronic and/or vibrational continuum is the same which at lower energy governs the details of the bound level pattern. Preliminary calculations[56] show that the simplified approach of Sec. III indeed predicts correctly not only the positions but also the strongly varying profiles of various absorption lines, thus giving detailed insight into the mechanisms of vibrational and rotational preionization. The evaluation is based on the $\mu_{4\Sigma}$ and $\mu_{3\Pi}$ curves derived in Sec. IV (more precisely, on the matrices $\mathfrak{C}$ and $\mathfrak{S}$ derived from them), and its success confirms that the quantum defects $\mu_\Lambda(R)$ run smoothly through the ionization threshold and are virtually independent of energy over the range of a few H_2^+ vibrational intervals.

A substantial increase of the incident particle energy (photon or electron), from about 16 eV up to 30 eV or more, leads into the region where Crowe and McConkey[58] have observed a structured H^+ energy spectrum obtained from dissociative ionization of H_2. Excitation to these very high energies must involve formation of Rydberg states with an electronically excited ($1\sigma_u$) core. Dissociation may take place as a dissociative preionization process involving a two-electron jump, whereby the core is left in the vibrational continuum of its $1\sigma_g$ ground state (Crowe and McConkey,[58] Hazi,[59] Bottcher[60]). The intervening doubly excited Rydberg configurations are in fact the same which also at low energy are superposed on the singly excited configurations by "major" configuration mixing, except that at low energy they become important only at larger R where the $1\sigma_g$ and $1\sigma_u$ ion states approach each other (cf. the discussion in Sec. II. E). In other words, the new processes at high energy involve no interaction which has not already been included in the general electron–core coupling treatment outlined in Sec. II.

Extension of the present formalism to include a larger range of excitation energy therefore appears straightforward. First of all, the restriction to the electronic ground state made in Sec. III must be abandoned since now specific reference to alternative *electronic* ionization channels is necessary. The configurational structure of the electronic wavefunction has been formally described in Sec. II by the coefficients $\bar{b}_{i\Lambda}(R)$ first introduced in Eq. (17). The $\bar{b}$'s are now expanded as

$$\bar{b}_{i\Lambda}(R)\begin{Bmatrix}\cos\\ \sin\end{Bmatrix}(\pi\bar{\mu}_{i\Lambda}(R))$$

$$=\sum_\alpha a_{\alpha\Lambda}(R)\langle i|\alpha\rangle^{\Lambda,R}\begin{Bmatrix}\cos\\ \sin\end{Bmatrix}(\pi\mu_{\alpha\Lambda}(R)), \qquad (37)$$

in terms of an alternative set of coefficients a_α, a frame transformation matrix $\langle i|\alpha\rangle$ (where at present $i=n^+\Lambda^+$), and eigenphase shifts $\pi\mu_\alpha$. Note the analogy between this expansion and that of Eq. (20) which expresses the coefficients $B_{v^+N^+}$ of the rovibronic wavefunction in terms of the $A_{v\Lambda}$'s. The expansions used in atomic work are also of the same type.[39,6] The previous treatment of electron–nuclear coupling differs in that it utilizes the fact that the *good quantum numbers* v and Λ exist in the Born–Oppenheimer limit A, so that the elements $\langle N^+|\Lambda\rangle$ and $\langle v^+|v\rangle$ can be evaluated *a priori*. For higher excitation, instead, the elements $\langle i|\alpha\rangle$ in Eq. (37) are not known at the outset except for $R\to\infty$, and the details of their evolution must be determined along with the quantum defects $\mu_{\alpha\Lambda}(R)$, either theoretically or from experiment. Instead of the single (effective) quantum defect function $\mu_\Lambda(R)$ used for each Λ in the present work, there will be two or more, corresponding to the number of electronic ionization channels considered.

Whether the application of this concept to the problem of dissociative preionization just mentioned will yield useful results is to be shown by future work. It appears, in any case, that Eqs. (20) and (37) form the nucleus of a treatment which accounts for combined electron–electron and electron–nuclear coupling in molecular Rydberg states. It is known that in diatomic molecules composed of first row atoms the discrete part of the absorption spectrum is often dominated by the effects of configuration interaction. The best known examples are the spectra of NO and N_2.[61,62] The vibrational structure of several of the Rydberg series converging to the ground state of the ion is heavily perturbed in these molecules by vibronic interaction with non-Rydberg (valence) states of the same Λ value, whose configurations differ by two orbitals. The perturbations are so strong that the Rydberg series are not recognizable as such in the spectra, but can be deduced only by a procedure of deperturbation. The vibronic interaction has been interpreted[63,37] as the result of avoided crossings which occur between the Rydberg and valence state potential energy curves. The same idea can be transferred to higher energies where the Rydberg–valence interactions presumably are followed up by Rydberg–Rydberg interactions between series (or continua) with ground state and excited state cores, respectively. One then has to envision, for example, a pair of quantum defect curves $\mu_{\alpha\Lambda}(R)$ together with a 2×2 transformation matrix $\langle i|\alpha\rangle^{\Lambda,R}$. While these functions should generally vary smoothly with R, the eigenchannel coefficients $a_{\alpha\Lambda}(R)$ of Eq. (37) [and hence the $\bar{b}_{i\Lambda}(R)$] will exhibit a strong variation near the crossing point, indicative of the configurational change occurring in the wavefunction at this R value. This variation accounts by means of Eq. (20) for a coupling between vibrational channels or states.

A further point concerns anisotropic interaction. This interaction can be regarded as responsible for avoided crossings of curves (potential energy or quantum defect) which (at least in a first approximation) correspond to the same core state, with different l values of the Rydberg electron. An example is the $p\sigma\sim f\sigma$ avoided crossing mentioned in Sec. IV. D and partially seen in Fig. 2. Detailed spectroscopic evidence for molecular l mixing has just recently been obtained in the NO molecule by Miescher,[64] who observed near the first ionization limit the most complete known system of molecular Rydberg series including $ns\sigma$, $np\sigma$, π; $nd\sigma$, π, δ; $nf\sigma$, π, δ, ϕ components. By analyzing the rotational structure of these series near $n=7$ to 12, Miescher has been able to detect the symptoms of very weak $s\sim f$ and much stronger $s\sim d$ mixing. This is the region where the Born–Oppenheimer approximation for rotational motion begins to break down seriously. As a result, the rotational levels become ordered progressively according to N^+ rather than Λ, and the rotational branches in the absorption spectrum begin to move towards the positions expected for $N^+=J-l,\ldots,J+l$. Since at the same time the different l components for a given n come progressively closer, crossings occur, at first between certain rotational branches of different N^+ and l, and equal n. For example, the $N^+=J$ $(l=0)$ and $N^{+\prime}=J-3$ $(l=3)$ branches must cross over at some n value since the electronic f complex at low n lies higher than the nearby s component. In NO the crossings are found to be locally avoided in the $s\sim f$ case and strongly avoided in the $s\sim d$ case.

The connection between these observations and the present work is that the quantum defect method not only provides a suitable basis for the interpretation for such behavior, but also furnishes the link with the ionization continuum where the angular distribution of photoelectrons should be particularly sensitive to l mixing (Dill[65]). Both the general rovibronic wavefunction (14) and the expansion (20) of the $\bar{B}$ coefficients allow explicitly for a summation over partial waves l; this extension is therefore also built into the expansion (37) of the electronic coefficients $\bar{b}$. Neglecting vibrational motion and putting $i=l$ in Eq. (37) we find the following generalized expression for the elements $\mathfrak{S}$ in Eq. (29),

$$\mathfrak{S}^J_{lN^+,l'N^{+\prime}}=\sum_{\Lambda\alpha}\langle N^+|\Lambda\rangle^{l\Lambda^+J}\langle l|\alpha\rangle^\Lambda$$

$$\times\sin\pi\mu_{\alpha\Lambda}\langle\alpha|l'\rangle^\Lambda\langle\Lambda|N^{+\prime}\rangle^{l'\Lambda^+J}, \qquad (38)$$

and an analogous expression for the elements $\mathfrak{C}$. These elements are used for the description of the coupling between electronic and rotational motion in instances where l is not a constant of motion, i.e., where the assumption made originally by Fano[3,32] and also in Sec. III of this work is not valid. The elements $\langle l|\alpha\rangle$ denote the overlap between spherical harmonics $Y_{l,\Lambda-\Lambda^+}$ and the exact angular amplitude of the (fixed-nuclei) molecular electronic wavefunction. The μ_α's are the effective electronic quantum defects as they result from the anisotropic interaction near the core.

We finally mention molecular dissociation. Vibrational continuum states have been excluded throughout this work by the restriction of the $\mathfrak{S}$ and $\mathfrak{C}$ matrices to vibrationally bound channels. On the other hand, the evaluation by MQDT of bound levels in the present work has brought to light a variety of small symptoms pointing to the beginning breakdown of the simplified formalism introduced in Sec. III. We stress here that these symptoms can be regarded as the signs of a tendency of the molecular system towards *dissociation*, just as the nonadiabatic deviations from Born–Oppenheimer behavior mark the tendency towards *ionization*. Indeed, various discrepancies were attributed in Sec. IV to the neglect of "major" configuration mixing and/or l mixing in the Rydberg wavefunction expressions, that is, to the neglect of precisely those components of the wavefunction which bring about its transformation into a separated-atom form. From this standpoint, in a treatment of molecular dissociation the R-dependent matrices $\{\langle i\,|\,\alpha\rangle^R\}$ in Eq. (37) may be expected to play a role analogous to that of the frame transformation matrices $\{\langle N^+|\,\Lambda\rangle\}$ and $\{\langle v^+|R\rangle\}$ in this work. However, it is not clear at this time how the concept of a molecular core with finite radius r_c can be adapted to a situation where the nuclei move increasingly far apart.

ACKNOWLEDGMENTS

Foremost, thanks are due to Dr. Dan Dill (Boston) for his enthusiastic support given during all stages of the work. Without his help, in particular during the preparation of the manuscript, the writing of this article would hardly have been possible. From the early stages Professor U. Fano (Chicago) showed his strong interest and helped greatly by a number of discussions. We are deeply grateful to him for his detailed criticism of the manuscript and for numerous suggestions which have helped to improve the manuscript decisively.

Dr. S. Takezawa (Cambridge, MA USA) and Professor W. Kołos (Warsaw) made available their results before publication. Dr. A. J. Merer extended his hospitality to one of us during the final stages of the work at the University of British Columbia. He, as well as Mme. H. Lefebvre-Brion (Orsay), Dr. G. Herzberg (Ottawa), and Dr. R. Badger (Vancouver), gave us the benefit of their comments on the manuscript. To all of them we wish to express our sincere thanks.

We would like to thank the North Atlantic Treaty Organization for travel funds which made the frequent discussions with Dr. Dill possible.

*On leave (1976–1977) at the Department of Chemistry, University of British Columbia, Vancouver, Canada.
†Participated in the early stages of the work.

[1]M. J. Seaton, Proc. Phys. Soc. **88**, 801, 815 (1966).
[2]C. M. Lee and K. T. Lu, Phys. Rev. A **8**, 1241 (1973).
[3]U. Fano, Phys. Rev. A **2**, 353 (1970).
[4]E. S. Chang, D. Dill, and U. Fano, in *Abstracts of Papers, Proceedings of the Eighth International Conference on the Physics of Electronic and Atomic Collisions*, edited by B. C. Cobic and M. V. Kurepa (Institute of Physics, Belgrade, 1973), p. 536.
[5]G. Herzberg and Ch. Jungen, J. Mol. Spectrosc. **41**, 425 (1972).
[6]U. Fano, J. Opt. Soc. Am. **65**, 979 (1975).
[7]O. Atabek, D. Dill, and Ch. Jungen, Phys. Rev. Lett. **33**, 123 (1974).
[8]W. Kołos and L. Wolniewicz, J. Chem. Phys. **43**, 2429 (1965).
[9]W. Kołos and L. Wolniewicz, J. Chem. Phys. **48**, 3672 (1968).
[10]I. Dabrowski and G. Herzberg, Can. J. Phys. **52**, 1110 (1974).
[11](a) O. Atabek and Ch. Jungen, in *Electron and Photon Interactions with Atoms*, edited by H. Kleinpoppen and M. R. C. McDowell (Plenum, New York, 1976), p. 613; (b) O. Atabek, thesis, Université de Paris VI, 1976.
[12]W. Kołos and L. Wolniewicz, Rev. Mod. Phys. **35**, 473 (1963).
[13]P. R. Bunker, J. Mol. Spectrosc. **28**, 422 (1968).
[14]R. de L. Kronig, *Band Spectra and Molecular Structure* (Cambridge U. P., Cambridge, 1930).
[15]See, for example, (a) H. G. Kuhn, *Atomic Spectra* (Academic, London, 1969), 2nd ed., (b) I. I. Sobel'man, *An Introduction to the Theory of Atomic Spectra* (Pergamon, Oxford, 1972).
[16]The second, mass-independent term should not be forgotten and will be considered in Sec. III. B.
[17]H. Wind, J. Chem. Phys. **42**, 2371 (1965).
[18](a) W. Kołos, Acta Phys. Acad. Scient. Hung. **27**, 241 (1969); (b) D. M. Bishop and R. W. Wetmore, Mol. Phys. **26**, 145 (1973); **27**, 279 (1974); (c) G. Hunter, A. W. Yau, and H. O. Pritchard, At. Nucl. Data Tables **14**, 11 (1974).
[19](a) C. L. Beckel, M. Shafi, and J. M. Peek, J. Chem. Phys. **59**, 5288 (1973); (b) M. Shafi and C. L. Beckel, J. Chem. Phys. **59**, 5294 (1973).
[20]J. H. Van Vleck, J. Chem. Phys. **4**, 327 (1936).
[21]This choice of symmetric rotor functions follows the recent recommendation by J. M. Brown and B. J. Howard, Mol. Phys. **31**, 1517 (1976).
[22](a) H. Wind, J. Chem. Phys. **43**, 2956 (1965); (b) G. Hunter and H. O. Pritchard, J. Chem. Phys. **46**, 2146 (1967); (c) C. L. Beckel, B. D. Hansen, and J. M. Peek, J. Chem. Phys. **53**, 3681 (1970).
[23](a) S. Takezawa, J. Chem. Phys. **52**, 2575, 5793 (1970); (b) S. Takezawa and Y. Tanaka, J. Mol. Spectrosc. **54**, 379 (1975).
[24](a) P. R. Bunker, J. Mol. Spectrosc. **46**, 504 (1973); (b) D. M. Bishop, Mol. Phys. **28**, 1397 (1974).
[25]E. S. Chang and U. Fano, Phys. Rev. A **6**, 173 (1972).
[26]Fano[3] originally considered only rotation–electron coupling and defined Region A by the sphere of radius $r_r \sim 50$ a.u. obtained when the condition (16) is restricted to rotational states. An alternative region, a, was introduced by Chang and Fano (Ref. 25) and is analogous to the region called A here (cf. their Fig. 1).
[27]Since the energy spectrum of vibration is unlimited towards positive energies, there exist for every total energy E pairs of levels v^+N^+ and $v^{+\prime}N^{+\prime}$ such that Condition (16) is not fulfilled. In other words, the Born–Oppenheimer approximation may break down even when ν is small, provided the vibrational energy is large enough. The effects of such vibronic coupling are kept small by action of the Franck–Condon principle, which restricts the occurrence of appreciable nonzero vibrational matrix elements in Eq. (20) below, to a limited range of vibrational quantum numbers. They appear in the spectrum as interactions between levels of low ν/high v^+ and high ν/low v^+, respectively, which are close in energy. If the high v^+ component belongs to the vibrational continuum, predissociation occurs; if it belongs to the discrete spec-

trum, a local perturbation results of a type which plays a role in the photoionization process [cf. Ref. 11(a)].

[28]C. A. Coulson and I. Fischer, Philos. Mag. **40**, 386 (1949).

[29]R. S. Mulliken, J. Am. Chem. Soc. **88**, 1849 (1966).

[30]Mulliken points out that the rate at which the transformation of the wavefunction into LCAS (separated atom) form proceeds is governed by the ratio of the internuclear distance to the radius of maximum density of the electron orbital in the *core*. This means that for all but the lowest Rydberg states the transformation is completed *before* the core attains a size comparable to the radius of maximum density of the Rydberg orbital. Note that, for example, the $n=4$, $l=1$ hydrogenic Rydberg orbit has a diameter $2\bar{r}=46$ a.u., as compared with the classical outer turning point near $R\sim20$ a.u. of the last stable H_2^+ vibrational level.

[31]O. W. Richardson, *Molecular Hydrogen and Its Spectrum* (Yale U.P., New Haven, CT, 1934).

[32]Fano [Comments At. Mol. Phys. **1**, 140 (1969)] stressed the possible role of the centrifugal barrier in preserving l in homonuclear molecules. The barrier prevents electrons with $l\geq2$ from penetrating the anisotropic core region, and thereby preserves orbital angular momentum for all l outside the core. On the other hand, Jungen [J. Chem. Phys. **53**, 4168 (1970)] found evidence that in NO the "s" and "d" Rydberg states are nearly 1:1 mixtures of pure l orbital states. The degree of mixing appears to depend on the number and the nature of molecular core orbitals (precursors) present, cf. also Sec. V of this work.

[33]P. M. Dehmer and W. A. Chupka, J. Chem. Phys. **65**, 2243 (1976).

[34]W. Weizel, in *Handbuch der Experimentalphysik*, edited by W. Wien and F. Harms (Akademische, Leipzig, Supplement, Vol. 1, 1931), p. 42.

[35]The virial theorem can be applied to the motion of the Rydberg electron separately, because with $H^{\text{residual}}=0$ the motions of the core particles and of the outer electron are entirely decoupled, cf. Ref. 20, p. 335.

[36]The discussion is conducted here as if the effects due to the operators in Eqs. (31) and (6a) and the molecular analogue of Eq. (6b) could be separated into core and Rydberg parts. This is not strictly true since the operators $\partial^2/\partial R^2$ and $(L_x^2+L_y^2)$ can give rise to a small coupling between levels belonging to configurations which differ in both the core and the Rydberg orbital. Such interactions have been disregarded throughout this section.

[37]J. K. Lewis and J. T. Hougen, J. Chem. Phys. **48**, 5329 (1968).

[38]J. W. Cooley, Math. Comput. **15**, 363 (1961).

[39]K. T. Lu, Phys. Rev. A **4**, 579 (1971).

[40]B. Jeziorski and W. Kołos, Chem. Phys. Lett. **3**, 677 (1969); see also Refs. 5, 22(b), and 41.

[41]G. Herzberg, *Fundamental and Applied Laser Physics, Proceedings of the Esfahan Symposium*, edited by M. S. Feld, A. Javan, and N. A. Kurnit (Wiley, New York, 1973), p. 491.

[42]G. Herzberg and L. L. Howe, Can. J. Phys. **37**, 636 (1959).

[43]W. Kołos and L. Wolniewicz, Can. J. Phys. **53**, 2189 (1975).

[44]W. Kołos and L. Wolniewicz, J. Chem. Phys. **45**, 509 (1966).

[45]P. G. Wilkinson, Can. J. Phys. **46**, 1225 (1968).

[46]P. S. Julienne, J. Mol. Spectrosc. **48**, 508 (1973).

[47]R. S. Mulliken, J. Am. Chem. Soc. **86**, 3183 (1964); **91**, 4615 (1969).

[48]W. Kołos and J. Rychlewski, J. Mol. Spectrosc. **62**, 109 (1976).

[49]The pattern in Fig. 6 provides a direct measure of the vibrational energy at which the core attains a size comparable to the Rydberg orbit. The trend in the figure becomes marked for the H_2 points near $v=10$: this level has a classical outer turning point near $R\sim5$ a.u., i.e., at a distance just equal to the radius $\bar{r}=5$ a.u. of the hydrogenic $n=2$ orbit.

[50]A. L. Ford, J. Mol. Spectrosc. **53**, 364 (1974).

[51]T. Namioka, J. Chem. Phys. **40**, 3154 (1964); **41**, 2141 (1964).

[52]S. Takezawa (private communication).

[53]The B' curve calculated by L. Wolniewicz [Chem. Phys. Lett. **31**, 248 (1975)] with a 52-term wavefunction was used in a first stage, but was found to yield vibronic levels about 50 cm^{-1} higher than experimentally observed.

[54]W. Kołos, J. Mol. Spectrosc. **62**, 429 (1976).

[55]Namioka[51] points out that the B'' $v=0$, $J=0$ level is perturbed in H_2 by the nearby B' $v=4$, $J=0$ level, and as a consequence shifted to lower energy by 1.6 cm^{-1}. This type of vibronic interaction between successive Rydberg members is of course built into MQDT, but the present treatment nevertheless does not account for this particular accidental perturbation. The reason is that at each energy (or n value) the calculations were carried out with energy-independent quantum defect curves: the slight difference between $\mu_{4\Sigma}$ and $\mu_{3\Sigma}$ displayed in Fig. 2 was consequently neglected, and as a result it happens that the levels which in the spectrum are in exact resonance, are not quite in resonance in the calculations and therefore remain unperturbed. The B'' state potential determined here must consequently be about 2 cm^{-1} *too low* near equilibrium. Shortcomings of this type may become more widespread when vibrational motion covers the range where the energy dependence of μ_Σ is most pronounced, i.e., at large R and for low n. The R dependence of μ_Σ derived in the framework of the energy-independent formalism may then in part reflect the neglected E dependence, but insofar as the Born–Oppenheimer approximation is correct the treatment will also remain correct.

[56]Ch. Jungen, J. Chem. Phys. (to be published).

[57]P. Borrell, M. Glass-Maujean, and P. M. Guyon, J. Chem. Phys. (to be published).

[58]A. Crowe and J. W. McConkey, Phys. Rev. Lett. **31**, 192 (1973).

[59]U. Hazi, Chem. Phys. Lett. **25**, 259 (1974).

[60]C. Bottcher, J. Phys. B **7**, L352 (1974).

[61]E. Miescher and K. P. Huber, in *International Review of Science, Physical Chemistry* (Butterworths, London, 1976), Series 2, Vol. 3.

[62]K. Dressler, Can. J. Phys. **47**, 547 (1969).

[63]H. Lefebvre-Brion, Can. J. Phys. **47**, 541 (1969).

[64]E. Miescher, Can. J. Phys. **54**, 2074 (1976).

[65]D. Dill, Phys. Rev. A **6**, 160 (1972).

1980 *J. Chem. Phys.* **73** 3338–45

Calculation of rotational–vibrational preionization in H_2 by multichannel quantum defect theory

Ch. Jungen and Dan Dill[a)]

Laboratoire de Photophysique Moléculaire du CNRS, Université de Paris-Sud, 91405, Orsay, France

(Received 27 August 1979; accepted 11 October 1979)

Multichannel quantum defect theory is adapted to treat simultaneous rotational and vibrational preionization in H_2. The strongly preionized spectrum between the $N^+=0$ and $N^+=2$ rotational thresholds of photoionization of $H_2X^1\Sigma_g^+(J''=0, v''=0)$ to produce $H_2^+X^2\Sigma_g^+(N^+, v^+=0)$ is computed as example and good agreement is obtained with the photoionization data of Dehmer and Chupka.

I. INTRODUCTION

Jungen and Atabek[1] (hereafter referred to as JA) have used the multichannel quantum defect theory (MQDT), adapted by Fano[2,3] to molecular problems from techniques developed for atoms by Seaton,[4] to compute the positions of the low-lying $^1\Sigma_u^+$ and $^1\Pi_u^+$ Rydberg levels of H_2, extending and refining initial calculations of Atabek, Dill, and Jungen.[5] The energy range treated by JA was such that rotational or vibrational preionization channels were inaccessible. Further, accessible predissociation channels were not treated. Here we extend the MQDT to treat simultaneous rotational and vibrational preionization. In a companion paper[6] we make the analogous extension to incorporate low-lying predissociated states into the theory. A preliminary report of the work presented here has been given elsewhere.[7]

Fano treated rotational preionization in his original paper on H_2.[2] There, by neglecting vibration, matters were simplified by considering only the single $p(l=1)$ electronic open channel of the H_2^+ core in rotational state $N^+=0$ interacting with the single closed channel with the core in rotational state $N^+=2$. More generally, there are multiple interacting open and closed channels, corresponding to alternative rovibrational states of the H_2^+ core. The necessary extensions of the MQDT to treat this general situation were developed by Lu[8] and by Lee and Lu[9] in the context of atomic preionization. In Sec. II we adapt these extensions to treat molecular rovibrational continua.

Then in Sec. III we use the theory to calculate the photoionization spectrum[10,11] of $H_2\,X\,^1\Sigma_g^+(J''=0, v''=0)$ between the $N^+=0$, 2 rotational thresholds of $H_2^+\,X\,^2\Sigma_g^+(N^+, v^+=0)$. This calculation is a particularly significant example of the MQDT procedure in that vibrational interactions with the "levels" $5p\pi$, $v=2$ and $7p\pi$, $v=1$ of the photoexcited H_2 profoundly distort the rotationally preionizing levels of the Rydberg series $np2$ converging to the $N^+=2$ rotational threshold of H_2^+. Good agreement is obtained with the corresponding high-resolution photoionization spectrum measured by Dehmer and Chupka.[11]

The procedure developed here has been implemented such that any region of the H_2 photoionization spectrum can be calculated. In particular, we report elsewhere[12] results just above the $v^+=0-3$ vibrational thresholds of H_2^+, predicting rovibrational branching ratios of the well-resolved strongly preionizing peaks seen in these spectral regions. A special case of such calculations is the so-called open continuum, the spectral range in which closed channels can be neglected. Photoionization of H_2 above the H_2^+ dissociation limit can be so represented and corresponding analysis of vibrational branching ratios at 584 Å has been made.[13]

II. MQDT FOR OPEN CHANNELS

The essence of the MQDT[3,7] is to represent the molecular wave function in terms of the quantum defects μ_α which characterize the interaction of the photoelectron with the other electrons and the nuclei of the molecule at short range (region A) and the transformation coefficients $\langle i|\alpha\rangle$ which connect the short-range eigenchannels $|\alpha\rangle$ to the asymptotic (region B) decay channels $|i\rangle$. Because the μ_α and $|\alpha\rangle$ pertain to interactions at short range, where electronic kinetic energies are much greater than rotational and vibrational quanta, they change negligibly from one rovibrational threshold $|i\rangle$ to another. Rather, the enormous variation in spectral composition characteristic of preionization, in terms either of i-channel mixing coefficients B_i or of α-channel mixing coefficients A_α, emerges upon application of boundary conditions at long range appropriate to each spectral region. JA have written the general wave function for H_2 and obtained the linear system for discrete, low-lying levels (neglecting predissociation). Here we start from the JA wave function, impose boundary conditions appropriate when open (preionization) channels are present, and thereby obtain the linear system for the mixing coefficients B_i whose solution describes the rovibrational preionization spectrum of H_2.

The two kinds of ionization channels are defined in terms of the quantum numbers associated with all coordinates *except* for the radial distance r of the excited electron. The motion along r is the main object of the MQDT and is treated explicitly below. The decay channels are[1]

$$|i\rangle = |N^+v^+\rangle, \tag{1}$$

characterized by the state of the residual H_2^+ ion. In Eq. (1) only the rotational and vibrational state, with quantum numbers N^+ and v^+, respectively, are indicated explicitly since they are relevant to the present work. The short-range eigenchannels are

$$|\alpha\rangle = |\Lambda R\rangle, \tag{2}$$

[a)]On leave of absence from Department of Chemistry, Boston University, 685 Commonwealth Avenue, Boston, MA, 02215.

0021-9606/80/193338-08$01.00

characterized by the quantum number Λ of the projection of the total orbital momentum (Rydberg electron plus core) along the internuclear axis and the internuclear distance R, both well defined when the electron is close to the nuclei (region A, Hund's coupling case b). The (real) transformation coefficients are

$$\langle \alpha | i \rangle = \langle \Lambda R | N^+ v^+ \rangle = \langle \Lambda | N^+ \rangle^{J'} \chi_{v^+}^{N^+}(R) , \tag{3}$$

where the implied integration is carried out over all coordinates except r^1. The $\langle \Lambda | N^+ \rangle^{J'}$, Eq. (20) of Ref. 20, and given explicitly by JA Eq. (23), are coefficients of the transformation between Hund's case b and case d coupling of an $l=1$ electron and the $H_2^+ \, {}^2\Sigma_g^+$ core with total angular momentum J' and spin 0, and $\chi_{v^+}^{N^+}(R)$ are vibrational wave functions of the H_2^+ ion in its ground electronic state ${}^2\Sigma_g^+$ and rotational level N^+. The quantum defects are

$$\mu_\alpha = \mu_\Lambda (R) , \tag{4}$$

i.e., a Σ and Π quantum defect as a function of R, corresponding to the ${}^1\Sigma_u^+$ and ${}^1\Pi_u^\pm$ Rydberg series.

With these definitions the JA wave function can be written:

$$\Psi = \sum_i |i\rangle \sum_{i'} [f_{\nu_i}(r) \langle i | c | i' \rangle^{J'} - g_{\nu_i}(r) \langle i | s | i' \rangle^{J'}] B_{i'} , \tag{5}$$

where $|i\rangle$ is the combined core rovibronic and spin wave function of Eq. (1). This expression follows from JA Eqs. (14), (22), and (27b), and related discussion. The mixture of the regular (f) and irregular (g) Coulomb functions[2] is determined by [see JA Eq. (29)].

$$\langle i | c | i' \rangle^{J'} = \sum_\Lambda \langle N^+ | \Lambda \rangle^{J'} \langle N^+ v^+ | c_\Lambda | N^{+\prime} v^{+\prime} \rangle \langle \Lambda | N^{+\prime} \rangle^{J'} , \tag{6a}$$

$$\langle N^+ v^+ | c_\Lambda | N^{+\prime} v^{+\prime} \rangle = \int dR \, \chi_{v^+}^{N^+}(R) \cos \pi \mu_\Lambda (R) \chi_{v^{+\prime}}^{N^{+\prime}}(R) , \tag{6b}$$

with a similar expression for $\langle i | s | i' \rangle^{J'}$ with cos replaced by sin. Boundary conditions are applied to the wave function (5) differently, depending on whether a particular ionization channel is open or closed.

For closed channels $|i\rangle$ the electronic energy is negative, $\epsilon_i = -\frac{1}{2}\nu_i^2$, and the functions f and g each consist at large electron distances r of rising and falling exponentials.[14] The corresponding boundary condition is that the coefficients of the rising exponentials vanish. One thereby obtains for all such closed channels ($\forall i \in Q$)

$$\forall i \in Q \; \sum_{i'} [\sin \pi \nu_i \langle i | c | i' \rangle^{J'} + \cos \pi \nu_i \langle i | s | i' \rangle^{J'}] B_{i'} = 0 , \tag{7}$$

JA Eq. (28).

For open channels the electronic energy is positive, ν is imaginary, and the functions f and g respectively are at large r sin and cos oscillatory functions.[15] The corresponding boundary condition involves specification of phase shifts in all open channels to yield so-called incoming-wave normalization appropriate to photoelectron wave functions.[16,17] First the decay channel $i = \bar{i}$ that is being observed, e.g., by energy analysis of the photoelectrons, is specified. Then the sin and cos functions are rewritten in terms of incoming and outgoing spherical waves. Finally, the coefficients of the outgoing waves in all channels $i \neq \bar{i}$ are set equal to zero. One thereby ensures for large r an outgoing plane wave only in the channel $\bar{i}$ and incoming spherical waves in all channels. This boundary condition follows from the time-dependent representation of the photoionization process for which, if ionization takes place at $t=0$, the electron will be observed at $t=\infty$ only in the channel $\bar{i}$. [This is analogous to outgoing-wave normalization familiar in scattering theory, where an incident plane wave is specified in the channel $\bar{i}$ and outgoing spherical (scattered) waves are observed in all channels. Then, in a time-dependent picture, if the scattering takes place at $t=0$, the electron was incident on the target at $t=-\infty$ only in the channel $\bar{i}$.]

Relative to unit amplitude of the outgoing spherical wave in channel $\bar{i}$, the amplitude of the incoming spherical wave in channel i is by definition the $\bar{i}i$ element of the scattering matrix S. Thus, we require for a selected member $\bar{i}$ of the set P of all open channels ($\forall i \in P$)[18a,b]

$$\forall i \in P, \; \Psi^{\bar{i}} \underset{r\to\infty}{\propto} \sum_i (e^{i k_i r} \delta_{i\bar{i}} - S_{\bar{i}i}^* e^{-i k_i r}) | i \rangle / r . \tag{8}$$

Note that the angular function in $\hat{r}$ is included in $|i\rangle$ and that the phase $-l\pi/2 - k_i^{-1} \ln(2k_i r) + \arg\Gamma(l+1-i/k_i)$ has been suppressed because only the single partial wave $l=1$ contributes in this spectral range. More generally, a range of angular momenta will contribute, which in turn must be superposed to yield the required incoming-wave-normalized Coulomb plane wave.[18c] To relate the expression (8) to the standing-wave functions f and g appearing in Eq. (5), we introduce the eigenchannels $|\rho\rangle$ of the open-channel interaction, with respect to which the matrix S is diagonal, with eigenvalues $e^{i2\pi\tau_\rho}$. The quantities τ_ρ are eigenphases, in units of π radians, of the open-channel interaction. The unitary transformation

$$\forall \in P \; |\rho\rangle = \sum_i |i\rangle \langle i | \rho \rangle \tag{9}$$

connecting the decay channels and the eigenchannels diagonalizes the matrix S,

$$\forall i, i' \in P \; S_{\rho\rho'} = \delta_{\rho\rho'} e^{i 2\pi\tau_\rho} = \sum_{ii'} \langle \rho | i \rangle S_{ii'} \langle i' | \rho' \rangle . \tag{10}$$

Thus the boundary condition (8) can be written equivalently as

$$\forall \bar{i} \in P, \; \Psi^{\bar{i}} \underset{r\to\infty}{\longrightarrow} \sum_\rho e^{-i\pi\tau_\rho} \langle \rho | \bar{i} \rangle \Psi_\rho , \tag{11}$$

where the Ψ_ρ must be standing-wave eigenfunctions

$$\Psi_\rho \underset{r\to\infty}{\longrightarrow} \sum_i |i\rangle \langle i | \rho \rangle [f_{\nu_i}(r) \cos \pi\tau_\rho - g_{\nu_i}(r) \sin \pi\tau_\rho] \tag{12}$$

of the open-channel interaction. That is, the boundary condition (8) can be implemented by requiring that Eq. (5) take the asymptotic form Eq. (12), with a common asymptotic phase shift $\pi\tau_\rho$ in each open channel. With the Ψ_ρ so obtained (their number is equal to the number of open channels) we can form the required $\Psi^{\bar{i}}$ according to Eq. (11).

Note that if all channels were open, then the channels $|\alpha\rangle$ and $|\rho\rangle$ would be identical, i.e.,

$$\lim_{\text{all } i \text{ open}} \begin{cases} \langle i|\rho\rangle \to \langle i|\alpha\rangle & (13a) \\ \pi\tau_\rho \to \pi\mu_\alpha \,. & (13b) \end{cases}$$

as can be shown by using Eq. (6) together with Eq. (27) of JA and Eq. (15) below.

This limit applies to spectral regions where the effects of closed channels are sufficiently weak as to be ignored. With the relations (13) the wave function (8) is completely specified, and the spectral composition coefficients B_i can be determined directly by comparison to the wave function (5) without having to solve auxillary equations.[19] They are the basis of the MQDT treatment given elsewhere[13] of so-called open-continuum photoionization in H_2. In general, however, and especially in the case of preionization, interactions with closed channels will distort the open-channel eigenvectors and eigenphases from the open-continuum values (13). Thus, the deviations of $\langle i|\alpha\rangle - \langle i|\rho\rangle$ and $\mu_\alpha - \tau_\rho$ from zero are direct measures of the strength of the interaction with closed channels.

Application of the boundary condition (12) to all open-channel components of the wave function (5) yields the pair of equations

$$\forall i \in P \quad \sum_{i'} \langle i|c|i'\rangle B^\rho_{i'} = \langle i|\rho\rangle \cos\pi\tau_\rho \,, \tag{14a}$$

$$\forall i \in P \quad \sum_{i'} \langle i|s|i'\rangle B^\rho_{i'} = \langle i|\rho\rangle \sin\pi\tau_\rho \,, \tag{14b}$$

which can be combined to yield

$$\forall i \in P \quad \sum_{i'} (\sin\pi\tau_\rho\langle i|c|i'\rangle - \cos\pi\tau_\rho\langle i|s|i'\rangle)B^\rho_{i'} = 0 \,, \tag{15a}$$

$$\forall i \in P \quad \sum_{i'} (\cos\pi\tau_\rho\langle i|c|i'\rangle + \sin\pi\tau_\rho\langle i|s|i'\rangle)B^\rho_{i'} = \langle i|\rho\rangle \,. \tag{15b}$$

Equations (7) and (15a) form a homogeneous linear system which determines (to within an overall normalization) the coefficients B^ρ_i of the decay-channel spectral composition. The secular equation of this system establishes the eigenphases τ_ρ, there being as many τ_ρ and sets of coefficients B^ρ_i as there are open decay channels i. Then the elements $\langle i|\rho\rangle$ of the unitary transformation (9) are determined using Eq. (15b). The normalization requirement

$$\sum_i \langle\rho|i\rangle\langle i|\rho\rangle = 1 \tag{15c}$$

which ensures the unitarity of the transformation (9) fixes the magnitudes of the B^ρ_i. This wave function (5) appropriate to preionization, Eqs. (11) and (12), is thereby completely specified.

To compute the photoionization spectrum, matrix elements of the electric-dipole interaction must be taken between the wave function (11) and the initial state. The photoionization oscillator strength is given in atomic units by

$$\left.\frac{df}{dE}\right|_{\bar{i}} = 2h\nu(2J''+1)^{-1}\sum_{M''} |D^{J''M''}_{\bar{i}}|^2 \,, \tag{16}$$

where the dipole-length matrix element is defined by

$$D^{J''M''}_{\bar{i}} = \langle\Psi^{\bar{i}}|\mathbf{r}|0, J''M''\rangle \tag{17}$$

and $|0, J''M''\rangle$ denotes the initial state. For ground-state photoexcitation the initial state is localized within region A and hence the matrix element (17) can be expressed as a sum of rotationally invariant matrix elements $d_\Lambda(R)$ of excitation to alternative region-A components $|\Psi_\alpha\rangle$ of the final state $|\Psi^{\bar{i}}\rangle$. First, we use the complex conjugate of Eq. (11), i.e., $\Psi^{\bar{i}*}$, in Eq. (17) to obtain

$$D^{J''M''}_{\bar{i}} = \sum_\rho e^{i\pi\tau_\rho}\langle\bar{i}|\rho\rangle\langle\Psi_\rho|\mathbf{r}|0, J''M''\rangle \,, \tag{18a}$$

where the unitarity relation

$$\langle\bar{i}|\rho\rangle = \langle\rho|\bar{i}\rangle^* \tag{19}$$

has been used. The function Ψ_ρ has so far been considered only in its asymptotic form Eq. (12) and must now be written for all r outside the core,

$$\Psi_\rho = \sum_{i\in P} |i\rangle\langle i|\rho\rangle[f_{\nu_i}(r)\cos\pi\tau_\rho - g_{\nu_i}(r)\sin\pi\tau_\rho] + \sum_{i\in Q} |i\rangle\sum_{i'}[f_{\nu_i}(r)\langle i|c|i'\rangle - g_{\nu_i}(r)\langle i|s|i'\rangle]B^\rho_{i'} \tag{20}$$

which, by means of Eq. (14), is seen to have the required form Eq. (5), Equation (7) ensures that the additional closed-channel components in Eq. (20) fall off exponentially at large r so that the asymptotic form of Eq. (12) is preserved. Next, we consider the form of Ψ_ρ in region A where the overlap with the ground state function will be taken. We use Eq. (6) together with the fact that the alternative f_{ν_i} (and g_{ν_i}) differ negligibly from one another in the inner region and so may be factored out of the sum in Eq. (20), to rewrite Ψ_ρ as a sum over Born–Oppenheimer products Ψ_α.

$$\Psi_\rho \simeq \sum_{i'} B^\rho_{i'} \sum_\alpha \langle\alpha|i'\rangle\Psi_\alpha \tag{21a}$$

$$\Psi_\alpha = |\alpha\rangle[f_\nu(r)\cos\pi\mu_\alpha - g_\nu(r)\sin\pi\mu_\alpha] \,, \tag{21b}$$

where Ψ_α is equivalent to Eq. (18) of JA. We can now rewrite Eq. (18a) as

$$\langle\Psi_\rho|\mathbf{r}|0, J'', M''\rangle = \sum_{i'} B^\rho_{i'} \sum_\alpha \langle i'|\alpha\rangle\langle\Psi_\alpha|\mathbf{r}|0, J''M''\rangle = \sum_{i'} B^\rho_{i'} \sum_\Lambda \int dR\langle N^+|\Lambda\rangle^{J'} \chi^{N^+}_{v^+}(R)\langle\Psi_\Lambda(R)|\mathbf{r}|0, J''M''\rangle \,. \tag{18b}$$

Finally, following the procedure given by Dill[20] we obtain the explicit dependence on the matrix elements $d_\Lambda(R)$ as:

$$\langle\Psi_\Lambda(R)|\mathbf{r}|0, J''M''\rangle = d_\Lambda(R)\chi^{J''}_{v''}(R) \times\langle\Lambda|J''\rangle^{J'}\langle J'M''|J''M'', 10\rangle \,. \tag{18c}$$

In Eq. (18c) the product of the Clebsch–Gordan coefficient $\langle J'M''|J''M'', 10\rangle$ and the transformation coefficient $\langle\Lambda|J''\rangle^{J'}$ arises from the integrations over the nuclear angular coordinates.[20] The coefficient $\langle\Lambda|J''\rangle^{J'}$ is given by the same expression as $\langle\Lambda|N^+\rangle^{J'}$, an accident deriving from the facts that on absorption the dipole interaction contributes unit angular momentum to a Σ state, and on l uncoupling the electron departs with unit angular momentum leaving the H_2^+ core in a Σ state. The region-A dipole amplitude (18c) differs from the corresponding amplitude (18) of Ref. 20 by the dependence of

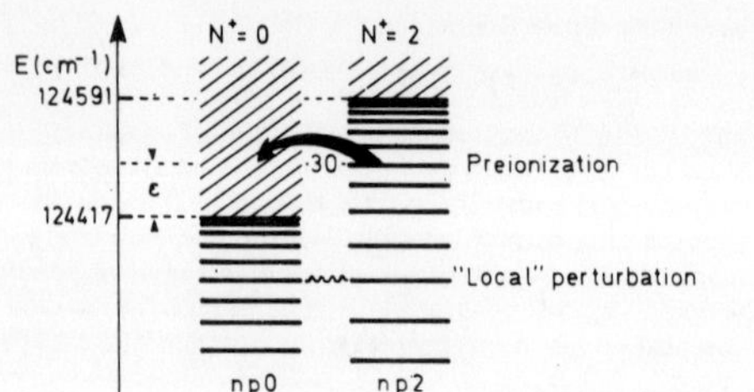

FIG. 1. Schematic illustration of rotational preionization in H_2.

d_Λ on R and by the appearance of the initial-state vibrational function $\chi_{v''}^{J''}(R)$. Averaging over the initial-state magnetic sublevels yields

$$(2J''+1)^{-1}\sum_{M''}|\langle J'M''|J''M'',10\rangle|^2=\frac{1}{3}\frac{(2J'+1)}{(2J''+1)}, \qquad (22)$$

so that we finally obtain

$$\left.\frac{df}{dE}\right|_{\bar{i}}=\frac{2h\nu}{3}\frac{(2J'+1)}{(2J''+1)}\left|\sum_\rho e^{i\pi\tau_\rho}\langle\bar{i}\,|\rho\rangle\sum_i B_i^\rho\sum_\Lambda\int dR\langle N^+|\Lambda\rangle^{J'}\right.$$
$$\left.\times\chi_{v^+}^{N^+}(R)d_\Lambda(R)\chi_{v''}^{J''}(R)\langle\Lambda\,|J''\rangle^{J'}\right|^2 \qquad (23)$$

for the oscillator strength for photoionization in the presence of preionization.

Equation (23) completely specifies the photoionization spectrum. The phases τ_ρ, transformation coefficients $\langle\bar{i}\,|\rho\rangle$ (and $\langle i\,|\rho\rangle$), and the spectral-composition coefficients B_i^ρ are given by Eqs. (7) and (15a)–(15c). The vibrational wave functions can be obtained by integration in the initial- and ionic-state potential curves, and the coefficients $\langle N^+|\Lambda\rangle^{J'}$ and $\langle\Lambda|J''\rangle^{J'}$ are known analytically. Only the $d_\Lambda(R)$ remain to be determined. They contain the fundamental information about the photon-absorption process. They may be calculated from first principles or, alternatively, can be viewed as parameters to be determined by fit to experiment. In the example treated below we take the intermediate approach of estimating them from absolute oscillator-strength measurements, and leave detailed fitting to future work.

Backx *et al.*[21] have measured the total oscillator strength for threshold photoionization of ground-state H_2 to be 0.144 eV^{-1}. Summing over ion rotational levels and ignoring vibration, the oscillator strength is given by Eq. (49d) of Ref. 20, i.e.,

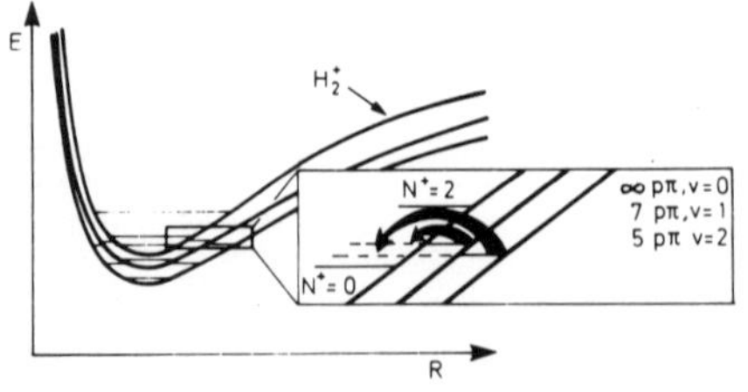

FIG. 2. Schematic illustration of vibrational preionization in H_2.

TABLE I. Quantum defect function $\mu_{5\Pi}(R)$ obtained from the $Q(1)$ lines[25] of the state $5p\,^1\Pi_u^-$.

R	$\mu_{5\Pi}$
1.00	−0.039 77
1.20	−0.046 96
1.40	−0.054 84
1.60	−0.063 14
1.80	−0.071 56
2.00	−0.079 76
2.20	−0.087 73
2.40	−0.095 09
2.60	−0.101 86
2.80	−0.107 64
3.00	−0.112 54
3.25	−0.117 19
3.50	−0.119 99
3.75	−0.120 72
4.00	−0.120 04
4.50	−0.113 98
5.00	−0.101 27
5.50	−0.085 31
6.00	−0.067 98
6.50	−0.050 86

$$\frac{df}{dE}=2h\nu(d_\Sigma^2+sd_\Pi^2)/3\ . \qquad (24)$$

We know that in the united atom limit ($R=0$) $d_\Sigma=d_\Pi$ and Herzberg and Jungen[22] have accounted for discrete absorption intensities by taking $d_\Sigma/d_\Pi=1$ independently of R. Then, taking $h\nu=124417.2\ \text{cm}^{-1}$, the H_2 ionization potential,[22] we obtain

$$d_\Sigma=d_\Pi=1.86\ \text{a.u.} \qquad (25)$$

from the measurement of Backx *et al.* and Eq. (24). More generally, the d_Λ depend both on R and the local kinetic energy $\epsilon(R)$ of the photoelectron. As discussed elsewhere,[13] the reasonableness of Eq. (25) derives in part from the fact that these two dependences approximately offset one another. The agreement obtained below with the experimental spectrum further confirms Eq. (25) as a good starting point.

III. ROTATIONAL/VIBRATIONAL PREIONIZATION IN H_2

Photoabsorption[22] or photoionization[10,11] of para-H_2 in its rotational and vibrational ground state yields H_2^+ in various vibrational levels and rotational states $N^+=0$ and $N^+=2$. For $v^+=0$ in the 174.3 cm^{-1} spectral range between the $N^+=0$ and $N^+=2$ rotational thresholds there is seen[23,24] a rich structure due to rotational and vibrational preionization. At the simplest level we expect a Rydberg series $np2$ of the p photoelectron converging to the $N^+=2$ rotational threshold. But also the electron can collide with the H_2^+ core, take up the core rotational energy, and escape into the continuum of the $N^+=0$ Rydberg series $np0$. This is illustrated schematically in Fig. 1. Such rotational preionization accounts for the broadening of the $np2$ lines above the $H_2^+\,X\,^2\Sigma_g^+$ ($N^+=0$, $v^+=0$) limit.[24] The Beutler–Fano profiles characteristic of preionization are clearly seen in the photoion-

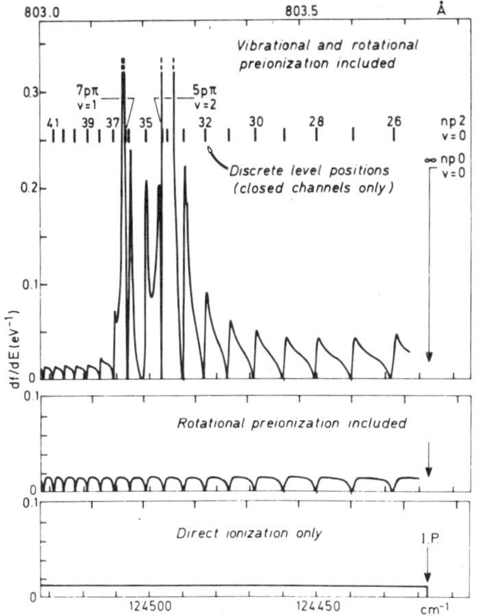

FIG. 3. MQDT calculation of photoabsorption of $H_2\ X\ ^1\Sigma_g^+\ (J''=0,\ v''=0)$ near the ionization threshold.

ization data.[23] In a similar way, collisions can interchange vibrational and electronic energy, as sketched in Fig. 2. Such vibrational preionization accounts for the two very intense lines between the rotational thresholds.[23,24] These correspond to the levels $5p\pi, v=2$ and $7p\pi, v=1$. Finally, the apparent lack of structure between these lines and the $N^+=2$ limit is due to the interplay of the rotational and vibrational preionization processes, the vibrational decay redistributing spectral intensity from the higher to the lower members of the rotational series. It is this spectral range that we now treat by MQDT.

The calculation was performed using the Σ quantum defect function given in Table X of JA. The Π function is given here in Table I. It was determined as described in JA from the $Q(1)$ lines of the $5p\pi$ state measured by Takezawa.[25] This function differs only in the third figure from the function $\mu_{3\Pi}(R)$ given in Table VI of JA. The linear system, Eqs. (7) and (15a), was constructed by taking 22 decay channels, $N^+=0, 2$ for each $v^+=0, 1, \ldots, 10$. The results are unchanged by increasing this basis further. They are given in Fig. 3.

In the bottom panel of Fig. 3 we show the oscillator strength obtained from Eqs. (24) and (25) by multiplying Eq. (24) with the Franck–Condon factor, i.e., neglecting preionization altogether. On the other hand, in the top panel we show the level positions one would obtain, by the procedure of JA, if all decay channels were closed, i.e., neglecting the $N^+=0$, $v^+=0$ ionization continuum. We see there the members of the $np2$ Rydberg series and the two interlopers $5p\pi, v=2$ and $7p\pi, v=1$. In Fano's original work[2] the interaction between the $np2$ levels and the $N^+=0$ continuum was treated; interactions with vibration and hence the two interlopers were not considered. The resulting pure rotational preionization spectrum of Beutler–Fano profiles is shown in the middle panel. The profiles have the appearance of

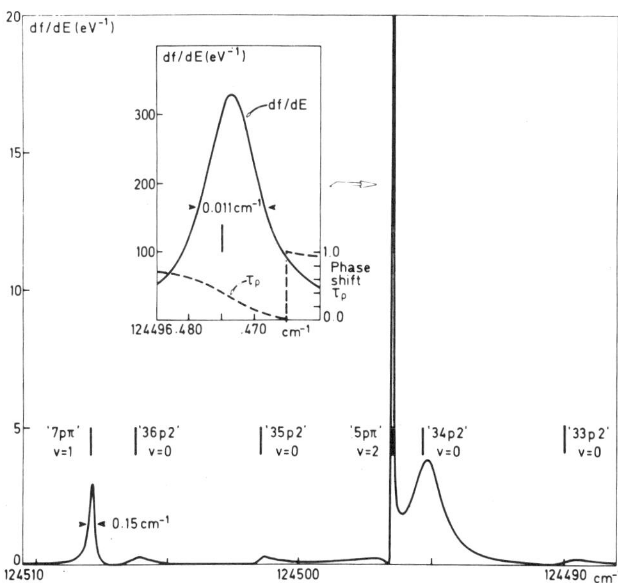

FIG. 4. MQDT results of Fig. 3 on an expanded energy scale and a compressed oscillator strength scale, showing the width of the $5p\pi$, $v=2$ and $7p\pi$, $v=1$ interlopers and the spectral variation of the eigenphase shift τ_ρ for the former.

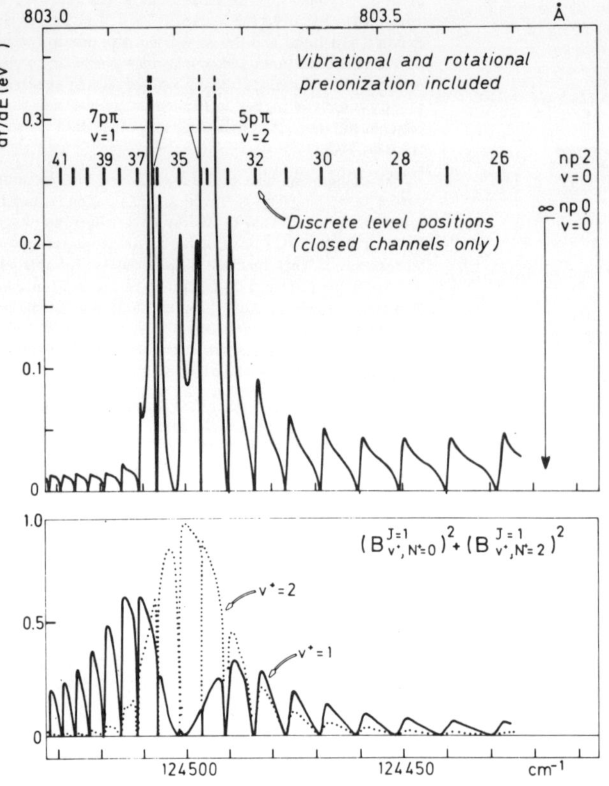

FIG. 5. Spectral composition with respect to ionic vibrational level v^+ of the MQDT results of Fig. 3. Rotational components have been summed for each level v^+. Levels $v^+ > 2$ contribute negligibly. The $v^+ = 0$ contribution is the defect from unity of the sum of the $v^+ = 1$, 2 contributions.

"emission windows," in other words, the positions of the preionized $np2$ levels (top panel) coincide approximately with the intensity minima.[22] Finally, when vibrationally preionizing channels are also introduced, via Eqs. (7) and (15), then the two interlopers preionize and distort the rotationally preionizing levels as well. The resulting full rotational/vibrational preionization spectrum is shown in the top panel of Fig. 3.

It is seen that the effect of vibrational preionization on the ionization cross section is profound and that it affects the whole range shown, corresponding to about 100 cm^{-1}. Indeed, if the fine variations of the cross section are neglected, the whole spectrum can be viewed as one "giant" resonance of about 50 cm^{-1} width which causes a global transfer of intensity from the high energy to the low energy side of the vibrational peaks. This transfer leads to further modifications of the fine structure. For example, for $n = 26$ and 27 the intensity *minima* still correspond nearly to the discrete $np2$ levels, but for higher n the profiles become progressively distorted until, near $n = 32$ to 35, it is the intensity *maxima* which coincide with the discrete level positions. In other words, there exists no longer a simple relationship between the extrema of the ionization curve and the positions of the preionizing levels.

In Fig. 4 the spectrum in the region of the vibrational interlopers is shown on a greatly expanded energy scale. Thereby the widths are seen to be surprisingly narrow, namely, 0.011 cm^{-1} for the $5p\pi$, $v = 2$ level and 0.15 cm^{-1} for the $7p\pi$, $v = 1$ level. The difference between $v = 1$ and $v = 2$ is the expected variation with Δv for an approximately linear quantum defect function. The widths reflect directly the rise by unity of the eigenphase τ_ρ,[26] as illustrated for $5p\pi$, $v = 2$ in the inset to Fig. 4.

We conclude that despite their narrowness the vibrationally preionizing lines affect the neighboring spectrum over a much larger spectral range. This can be seen again in Fig. 4, where the intensity of the $34p2$ Rydberg member is enhanced by an order of magnitude, relative to other $np2$ lines, by the close-lying $5p\pi$, $v = 2$. The global modification of the continuum by the vibrationally preionizing levels is related to the spectral composition of the full rovibronic wave function. This is illustrated in Fig. 5, where we reproduce in the upper panel the closed-channel level positions and the calculated spectrum from Fig. 3. In the lower panel we show the spectral composition in terms of the $v^+ = 0, 1, 2$ channels (summed over their rotational components for clarity of presentation). We have plotted the squared

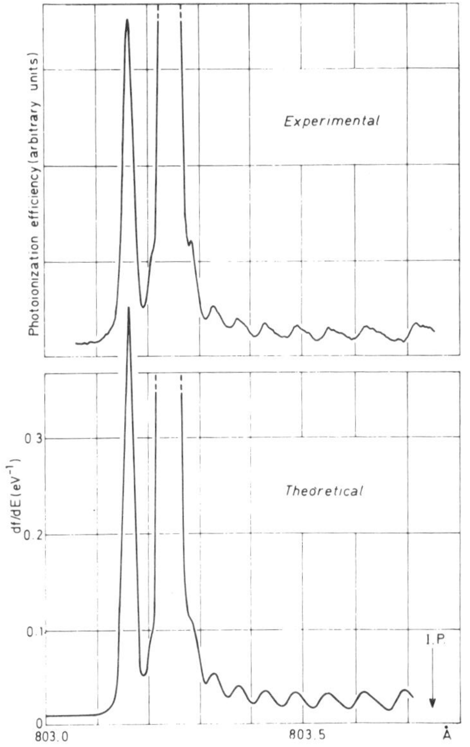

FIG. 6. MQDT results of Fig. 3 broadened to a resolution of 0.022 Å to correspond to the measurements of Dehmer and Chupka, Fig. 2 of Ref. 11.

modulus of the coefficients B renormalized such that the $v^+=0$ contribution is the defect from unity of the sum of the $v^+=1$ and $v^+=2$ contributions (ionic levels with $v^+>2$ are negligibly populated in this spectral range). We see that significant population of vibrationally excited ionic states occurs over many wave number units and thereby gives rise to the giant resonance profile mentioned above. In view of these complexities it is clear that vibrational and rotational preionization cannot be meaningfully treated as separate processes in this spectral region. A key feature of the MQDT is that it is based on no such assumed separability, i.e., it is applicable independently of coupling strengths between alternative decay mechanisms.

Finally, in Fig. 6 the calculated spectrum is compared with the Dehmer and Chupka photoionization spectrum.[23] The calculated spectrum was convoluted with a triangular apparatus function of width 0.022 Å. This width was determined by comparison of the $7p\pi$, $v=1$ peak to the experimental peak. It is somewhat larger than the width 0.016 Å given by Dehmer and Chupka.[11] The difference derives from random slippage in the stepping of the grating motor, which was operated at its limit of one step per measurement to obtain the best possible definition of the rich H_2 spectrum.[27] The width 0.022 Å can be interpreted as an effective width which takes this into account, at least in an average way. Note, however, that where no slippage occurs in the grating motor, the effective resolution is better than this effective width. This can be seen, for example, near 803.28 Å, where the peak associated with $33p2$, $v^+=0$ (cf. Figs. 3 and 4) is just resolved by the experiment but appears merely as a shoulder in our calculation. Since Dehmer and Chupka give only *relative* intensities, we have adjusted their curve by a global factor in order to match it with our calculations. We matched the two spectra at the peak at 803.48 Å, which is broad and therefore not sensitive to the choice of the convolution width. Figure 6 shows that the overall agreement between the calculated and observed spectra is very gratifying. In particular, we see that even the experimental background ionization is correctly reproduced.

IV. CONCLUSION

Extension of such calculations to treat H_2 rovibrational preionization in other spectral ranges is straightforward and, as we have said, results for the relatively broad peaks seen[10,11] just above the $v^+=0, 4$ thresholds will be presented elsewhere.[12] Another aspect which has been investigated is the angular distribution of the photoelectrons. While there is as yet no measurement of H_2 photoelectron angular distributions in preionization spectral regions, the theory has been set down[20] and calculations have shown the angular distribution to be even more sensitive to resonance processes than the corresponding oscillator strength owing to interference between alternative dipole amplitudes.[3,28,29] The spectral region treated here is a special case for which the photoelectrons have a $\cos^2\theta$ (asymmetry parameter $\beta=2$) distribution along the electric vector of the light independently of preionization, because ionization is from the J'' state of H_2 to the $N^+=0$ state of H_2^+.[20] This restriction does not apply when $J''\neq 0$, and elsewhere we report angular distribution calculations in the analogous spectral region between the $N^+=1$ and $N^+=3$ thresholds of H_2 ($J''=1, v''=0$) photoionization.[30]

In conclusion, we emphasize that the calculations reported here have been purposely limited to the spectral region just above the lowest photoionization threshold of H_2. This spectral region provides a severe test of the MQDT due to the complicated interaction between rotational and vibrational degrees of freedom, especially against the excellent data available.[23] A surprising result of our calculations is that preionizing lines, while influencing the spectrum over a considerable range, can remain very narrow themselves. Indeed, even the high resolution spectra of Refs. 11 and 22 do not yet fully resolve the portion of the spectrum shown in Fig. 4. In earlier photoionization spectra of H_2 only the giant resonance is seen near 803 Å, whose apparent width is about 3 orders of magnitude larger than the width of $5p\pi$, $v=2$ and $7p\pi$, $v=1$. This is the case in the early spectrum of Comes and Wellern,[31] while in the spectrum of Chupka and Berkowitz[10] one is just beginning to see some of the fine structure in the wing of the giant resonance. From a general point of view our example shows that caution should be used even in high-resolution photoionization spectra, when one attempts to interpret observed spectral features in terms of decay widths.[32]

Finally, we stress that the quantum defect functions $\mu_{\Sigma}(R)$ and $\mu_{\Pi}(R)$ contain the essence of the electron–rovibrational coupling responsible for preionization, and the connection between this information and the observed spectrum is provided by the MQDT. The results shown in Figs. 3–6 demonstrate the power of the MQDT in the description of the interaction between electronic and rovibrational degrees of freedom. This point of view is in no way special to H_2 and should considerably aid the unravelling of the complexities of molecular preionization spectra.

We thank M. Raoult (Orsay) for helpful discussions. We also thank the North Atlantic Treaty Organization for travel funds.

[1]Ch. Jungen and O. Atabek, J. Chem. Phys. **66**, 5584 (1977).
[2]U. Fano, Phys. Rev. A **2**, 353 (1970).
[3]U. Fano, J. Opt. Soc. Am. **65**, 979 (1975).
[4]M. J. Seaton, Proc. Phys. Soc. London **88**, 801 (1966).
[5]O. Atabek, D. Dill, and Ch. Jungen, Phys. Rev. Lett. **33**, 123 (1974).
[6]D. Dill and Ch. Jungen (to be published).
[7]D. Dill and Ch. Jungen, "Quantum Defect Functions. Interconverters of Electronic and Nuclear Motion," invited talk presented at a symposium in celebration of Professor Robert S. Mulliken's 50th anniversary on the University of Chicago Faculty, University of Chicago, October 27 and 28, 1978, J. Phys. Chem., to be published.
[8]K. T. Lu, Phys. Rev. A **4**, 579 (1971).
[9]C. M. Lee and K. T. Lu, Phys. Rev. A **8**, 1241 (1973).
[10]W. A. Chupka and J. W. Berkowitz, J. Chem. Phys. **51**, 4244 (1969).
[11]P. M. Dehmer and W. A. Chupka, J. Chem. Phys. **65**, 2243 (1976).
[12]M. Raoult and Ch. Jungen, J. Chem. Phys. (to be published).
[13]M. Raoult and Ch. Jungen (to be published).
[14]Equations (10)–(12) of Ref. 2.
[15]Equations (9) of Ref. 2.
[16]G. Breit and H. A. Bethe, Phys. Rev. **93**, 888 (1954).
[17]See, e.g., M. L. Goldberger and K. M. Watson, *Collision Theory* (Wiley, New York, 1964), pp. 188–190.
[18](a) N. F. Mott and H. S. W. Massey, *The Theory of Atomic Collisions* (Clarendon, Oxford, 1965), 3rd ed., pp. 388–391; (b) R. J. W. Henry and L. Lipsky, Phys. Rev. **153**, 51 (1967), Eq. (3); (c) Matching to Coulomb plane waves is discussed in detail in, for example, D. Dill and J. L. Dehmer, J. Chem. Phys. **61**, 692 (1974), especially Eqs. (32) and (39)–(41).
[19]See Eqs. (21)–(23) of Ref. 2.
[20]D. Dill, Phys. Rev. A **6**, 160 (1972).
[21]C. Backx, G. P. Wight, and M. J. Van der Wiel, J. Phys. B **9**, 315 (1976). We have taken the oscillator strength at 0.73 eV below threshold, corresponding to vertical excitation with $h\nu$ = I. P. at R = 1.4 a.u.
[22]G. Herzberg and Ch. Jungen, J. Mol. Spectrosc. **41**, 425 (1972).
[23]Figure 2 of Ref. 11.
[24]Figure 1 of Ref. 22.
[25]S. Takezawa, J. Chem. Phys. **52**, 2575 (1970).
[26]There is only one open decay channel and hence only one eigenchannel $|\rho\rangle$ in the spectral range treated here. More generally a particular spectral feature will correspond to an increase by unity of the sum of all eigenphases $|\rho\rangle$, though the bulk of this variation can be due to a single eigenchannel.
[27]P. M. Dehmer (private communication).
[28]D. Dill, Phys. Rev. A **6**, 1976 (1973).
[29]For a general discussion, see D. Dill, in *Photoionization and Other Probes of Many-Electron Interactions*, edited by F. J. Wuilleumier (Plenum, New York, 1976), p. 387.
[30]M. Raoult, Ch. Jungen, and D. Dill (to be published).
[31]F. J. Comes and H. O. Wellern, Z. Naturforsch. Teil A **23**, 881 (1968).
[32]This is similar to the problem of the content of the boa constrictor in Antoine de Saint Exupéry's story *The Little Prince* (Harcourt, Brace Jovanovich, New York, 1971).

1984 *Phys. Rev. Lett.* **52** 1681–4
Reprinted with permission from the American Physical Society

Stroboscopic Effect between Electronic and Nuclear Motion in Highly Excited Molecular Rydberg States

P. Labastie, M. C. Bordas, B. Tribollet, and M. Broyer
Laboratorie de Spectrométrie Ionique et Moléculaire, Université Lyon I, F-69622 Villeurbanne Cédex, France

(Received 14 February 1984)

The Na_2 Rydberg states have been observed for $15 \leqslant n \leqslant 90$ and $4 \leqslant J \leqslant 45$. The evolution of the spectra is studied as a function of the relative values of ω_{el}, the electronic rotational frequency, and ω_N, the nuclear rotational frequency. The molecular eigenfunction is found to be pure Hund's case *a* when $k\omega_{el} = 2\omega_N$ (k being an integer). This may be explained in terms of stroboscopic effects arising from the movement of the Rydberg electron relative to the ionic core.

PACS numbers: 33.80.Eh, 33.10.−n

In highly excited molecular Rydberg states, the rotational frequency ω_{el} of the Rydberg electron in its orbital motion can become smaller than the rotational frequency ω_N of the nuclei. Under these conditions, the Born-Oppenheimer approximation is no longer valid and the electron angular momentum $\vec{l}$ tends to be decoupled from the internuclear axis.

The interpretation of these effects was rather involved,[1] until the experimental work of Herzberg and Jungen[2] on H_2 whose interpretation by quantum-defect theory was due to Fano.[3] However, in their experiment they observed only Rydberg levels with $n \leqslant 40$ and $J = 0,1,2$.

Recently optical-optical double resonance[4-8] has proved to be a powerful technique to study molecular Rydberg states. Using this method we have observed Na_2 molecular Rydberg states for $15 \leqslant n \leqslant 90$ and $4 \leqslant J \leqslant 45$. These new results enable us to study in detail the molecular spectrum when $\omega_{el} \simeq \omega_N$ or $\omega_{el} << \omega_N$.

In the intermediate region $\omega_{el} \simeq \omega_N$, where the spectra should be complicated, some simplification occurs when $k\omega_{el} = 2\omega_N$ (k an integer), because of stroboscopic effects between the two frequencies.

The experimental setup consists of a supersonic sodium beam crossed at right angles by two superimposed tunable dye lasers pumped by the same pulsed yttrium aluminum garnet laser.[9] The first laser selects a well-defined rovibrational level (v',J') of the Na_2 $A\,^1\Sigma_u^+$ state. The second laser excites the (v,J) levels in the *nd* complex. Beyond the ionization limit, these states with $v \geqslant 1$ are autoionizing by vibrational coupling. The electrons are then collected by a pulsed electric field which lags behind the laser pulses by ~ 30 ns, and they are then detected by means of an electron multiplier.

At low n values $\omega_{el} >> \omega_N$, which is the usual condition for validity of the Born-Oppenheimer approximation. The projection Λ of the electronic angular momentum $\vec{l}$ onto the internuclear axis is a good quantum number. The lines have the usual molecular structure, *P, Q, R*. However, Λ doubling occurs and the *P, Q, R* lines are not equidistant. We observe $nd\,^1\Sigma_g$, $nd\,^1\Pi_g$, $nd\,^1\Delta_g$ series. The occurrence of $\Sigma \rightarrow \Delta$ transitions is explained by the mixing of wave functions, due to Λ doubling.[9]

As n increases, rotational perturbations become important, and the regular structure of the spectrum is lost. For the interpretation of these spectra we have used the multichannel quantum-defect theory introduced by Seaton,[10] and applied to molecular Rydberg states by Fano.[3] The theoretical aspects are described elsewhere.[11] Let us summarize here the relevant equations of the problem.

At given J and $l = 2$, we have five ionization channels: $N^+ = J-2, \ldots, J+2$ (where N^+ is the angular momentum of the ion). These five channels split into two for levels of minus symmetry ($N^+ = J-1, J+1$) and three for levels of plus symmetry ($N^+ = J-2, J, J+2$). The eigenchannels are those of Hund's case *a*, i.e., $nd\,^1\Pi_g^-$, $nd\Delta_g^-$ for the minus series and $nd\,^1\Sigma_g^+$, $nd\,^1\Pi_g^+$, $nd\,^1\Delta_g^+$ for the plus series. The eigen quantum defects have been found in Ref. 9 to be: $\mu_{d\Sigma} = 0.21$, $\mu_{d\Pi} = -0.04$, and $\mu_{d\Delta} = 0.43$. The transformation matrix from ionization to eigenchannels is the Hund's case-*d* to Hund's case-*a* transformation matrix $U^{Jl}_{\Lambda N^+}$ which is known by standard Racah algebra. We define the effective quantum numbers ν_{N^+} as

$$E = T_\infty(v) + BN^+(N^+ + 1) - R/(\nu_{N^+})^2, \quad (1)$$

where E is the energy of the state of interest, B the rotational constant of the molecular ion, and R the

Rydberg constant. This equation gives $i-1$ relations between the ν_{N^+}'s, where i is the number of channnels.

If we write the wave function in the case-a basis

$$|\psi\rangle = \sum_\Lambda A_\Lambda |l\Lambda J\rangle, \tag{2}$$

the multichannel quantum-defect theory gives the following set of equations[12]:

$$\sum_\Lambda A_\Lambda U^{Jl}_{\Lambda N^+} \sin\pi(\nu_{N^+} + \mu_\Lambda) = 0 \tag{3}$$

for each N^+. This system has nonzero solutions when

$$\det[U^{Jl}_{\Lambda N^+} \sin\pi(\nu_{N^+} + \mu_\Lambda)] = 0. \tag{4}$$

Equation (4) together with relation (1) determines the discrete energy levels. The A_Λ's are then found by resolution of Eq. (3). The results of multichannel quantum-defect theory analysis permit a precise reproduction of experimental spectra to be calculated,[11] as illustrated by Figs. 1, 2, and 3. We emphasize here some peculiar features of these spectra.

As pointed out by Lu,[13] when all the ν_{N^+}'s are equal modulo 1, Eqs. (4) and (3) have a simple solution

$$\begin{aligned} &\nu_{N_1^+} = n - \mu_{\Lambda_0} \equiv \nu_{N_2^+} \equiv \ldots (\mathrm{mod}\ 1), \\ &A_\Lambda = \delta_{\Lambda\Lambda_0}. \end{aligned} \tag{5}$$

The eigenfunctions are then pure Hund's case a functions.

Now, let us consider the case of series of minus symmetry (the same analysis may be pursued for plus symmetry). Equation (1) gives

$$\nu_{J-1} = \nu_{J+1}[1 - (4B/R)(J+\tfrac{1}{2})\nu^2_{J+1}]^{-1/2}. \tag{6}$$

While

$$(4B/R)(J+\tfrac{1}{2})\nu^2_{J+1} \ll 1, \tag{7}$$

the derivative $d\nu_{J-1}/d\nu_{J+1} \simeq 1$, so that the difference $\nu_{J-1} - \nu_{J+1}$ is a constant for variations of ν_{J+1} over a few integer values. However, even when (7) is fulfilled, this difference may be an integer value, since

$$\nu_{J-1} - \nu_{J+1} \simeq (4B/R)(J+\tfrac{1}{2})\nu^3_{J+1}. \tag{8}$$

This fact is important: The difference $\nu_{J-1} - \nu_{J+1}$ may be an integer and keep this value during a few levels, so that a whole region of the spectrum satisfies Eq. (5). In this region, the spectra are strongly simplified, because $^1\Delta$ lines are forbidden, the $^1\Pi$ lines are much more intense than the $^1\Sigma$ ones, and the P, Q, R lines lie at the same energy. This is illustrated in Fig. 1. Indeed, a "periodic return" of the system to Hund's case a occurs when $\nu_{J-1} - \nu_{J+1}$ is an integer.

The classical interpretation of this effect is illuminating. If ω_{el} and ω_N are the classical frequencies of the electronic and nuclear motion, respectively, we have

$$\hbar\omega_{el} \simeq 2R/3, \quad \hbar\omega_N \simeq 2B(J+\tfrac{1}{2}). \tag{9}$$

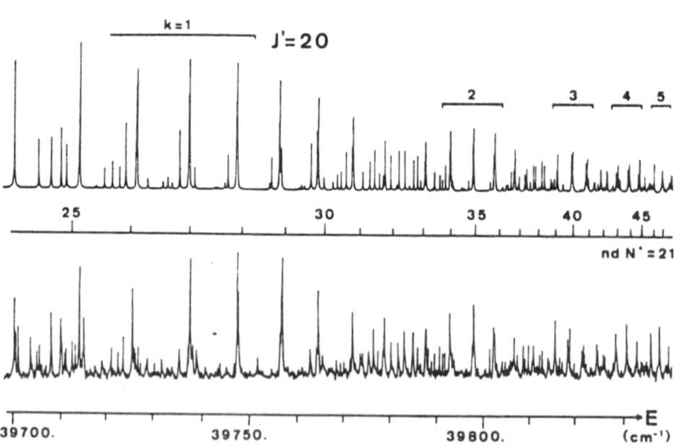

FIG. 1. Theoretical (upper) and experimental (lower) spectra are shown for the $A\,^1\Sigma_u^+$, $v'=3$, $J'=20$ intermediate level. Simplifications corresponding to $k=1,2,3,4,5$ are clearly observed. The levels of the series $nd\ N^+ = J+1$ are also given.

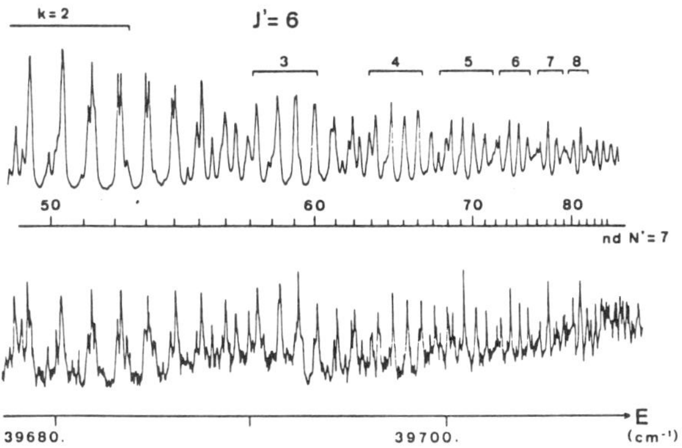

FIG. 2. Theoretical (upper) and experimental (lower) spectra are shown for the $A\ ^1\Sigma_u^+$, $v'=2$, $J'=6$ intermediate level. In the region $50 \leq n \leq 90$, stroboscopic effects always occur corresponding to $k=2,3,4,5,6,7,8$.

In fact the electron "sees" the detailed structure of the core (in particular the internuclear axis) only when it passes through it. Hence, it "lights" the core like a stroboscope at frequency ω_{el}. The internuclear axis seems at rest when it makes an integer number of half turns during an electron period, i.e., when $2\omega_N = k\omega_{el}$. This is precisely the condition $\nu_{J-1} - \nu_{J+1} = k$. Since the axis seems at rest, the projection Λ of $\vec{l}$ onto this axis is a good quantum number, and the coupling is Hund's case *a*.

The observation of these stroboscopic effects is strongly dependent on the value of $\hbar\omega_N$, i.e., of $B(J+\frac{1}{2})$: It must be possible to satisfy simultaneously (7) and $2\omega_N/\omega_{el} = k$, which imposes

$$[R/B(2J)k]^{1/3} >> 1. \tag{10}$$

Condition (10) is valid at low *BJ* values, and was not fulfilled in H_2,[2] even at $J=1$, so that these stroboscopic effects have never previously been observed.

Stroboscopic effects are observed when $k\omega_{el} = 2\omega_N$ but k is not allowed to be too high, since the condition (10) would no longer hold. In fact, a new aspect of the spectra appears in the region of few $h\omega_N$ below the ionization limit. The main effects

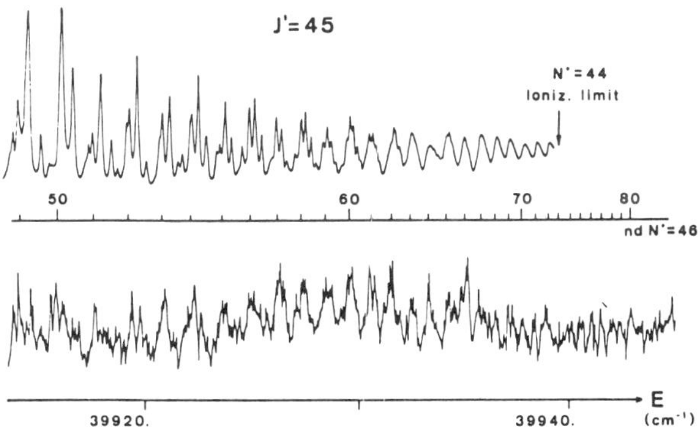

FIG. 3. Theoretical (upper) and experimental (lower) spectra are shown for the $A\ ^1\Sigma_u^+$, $v'=2$, and $J'=45$ intermediate state. In the region $50 < n < 90$ quasi-Fano profiles are observed, but no stroboscopic effects. The signal-to-noise ratio is too low to observe experimentally the lines of the $nd\ N^+ = J-1$ series inside the quasi-Fano profiles.

in this region can be explained by the interaction of two Rydberg series: a very dense series of levels converging towards the *nd* $N^+ = J' - 1$ limit which forms a quasicontinuum, and the levels of the other *nd* $N^+ = J' + 1$ series. Indeed these two series are the only ones allowed by the selection rule $\Delta N^+ = 0$, i.e., $N^+ = J' \pm 1$. Hence, each line of the second series is a quasi-Fano profile filled with the lines of the other one.

This Fano profile region is not obtained on our experimental spectra if J' is too low. Figure 2 shows a $J' = 6$ spectrum in the range $50 < n < 90$. Many stroboscopic oscillations are seen, but no Fano profiles. On the other hand, Fig. 3 shows a $J' = 45$ spectrum in the same range. Here quasi-Fano profiles appear as broad lines, since the individual lines of the *nd* $N^+ = J' - 1$ series are not resolved. Note that the experimental spectrum extends also beyond the $N^+ = J' - 1$ ionization limit, where quasi-Fano profiles become true ones.

Laboratoire de Spectrométrie Ionique et Moléculaire is a laboratoire associé au Centre National de la Recherche Scientifique.

[1]Y. N. Chiu, J. Chem. Phys. **41**, 3235 (1964); R. S. Mulliken, J. Am. Chem. Soc. **86**, 3183 (1964), and **88**, 1849 (1966), and **91**, 4615 (1969).

[2]G. Herzberg and Ch. Jungen, J. Mol. Spectrosc. **41**, 425 (1972).

[3]U. Fano, Phys. Rev. A **2**, 353 (1970).

[4]N. W. Carlson, A. J. Taylor, and A. L. Schawlow, Phys. Rev. Lett. **45**, 18 (1980).

[5]S. Leutwyler, A. Herrmann, L. Wöste, and E. Schumacher, Chem. Phys. **48**, 253 (1980).

[6]S. Martin, J. Chevaleyre, S. Valignat, J. P. Perrot, M. Broyer, B. Cabaud, and A. Hoareau, Chem. Phys. Lett. **87**, 235 (1982).

[7]M. Broyer, J. Chevaleyre, G. Delacretaz, S. Martin, and L. Wöste, Chem. Phys. Lett. **99**, 206 (1983).

[8]M. Seaver, W. A. Chupka, S. D. Colson, and D. Gauyacq, J. Phys. Chem. **87**, 2226 (1983).

[9]S. Martin, J. Chevaleyre, M. Chr. Bordas, S. Valignat, and M. Broyer, J. Chem. Phys. **79**, 4132 (1983).

[10]M. J. Seaton, Proc. Phys. Soc., London **88**, 801 (1966).

[11]M. C. Bordas, M. Broyer, P. Labastie, and B. Tribollet, to be published.

[12]C. Greene, U. Fano, and G. Strinati, Phys. Rev. A **19**, 1485 (1979).

[13]K. T. Lu, Phys. Rev. A **4**, 579 (1971).

1980 *J. Phys. B: At. Mol. Phys.* **13** 3867–94

A multichannel quantum defect approach to dissociative recombination

A Giusti

Department d'Astrophysique Fondamentale, Observatoire de Meudon, 92190 Meudon, France

Received 19 March 1980, in final form 29 May 1980

Abstract. The molecular multichannel quantum defect theory, initially developed to describe the escape of an electron from the molecular core region, is extended to include dissociation of the molecule into two atoms. The mixing between the two types of channels results from electronic interaction between a Rydberg series and dissociative valence states of the neutral molecule. This interaction is introduced as an additional phaseshift into the continuum parts of the wavefunctions. The resulting unified treatment of (pre)ionisation and (pre)dissociation is applied to the process of dissociative recombination, following the approach of C M Lee. The 'indirect' process proceeding via electron capture in vibrationally excited Rydberg states is taken into account fully by the inclusion of closed electron–ion channels in the calculation. Model calculations are presented and the influence of various physical parameters on the dissociative recombination profile is tested.

1. Introduction

The dissociative recombination (DR) of a diatomic molecular ion can be viewed as a collision process which proceeds through the following two reaction mechanisms:

$$\begin{array}{ccc} e+AB^{+} & \longrightarrow & AB^{**} \rightarrow A+B. \\ & \searrow \quad \nearrow & \\ & AB^{*} & \end{array}$$

Both mechanisms involve formation of a doubly excited dissociative state AB^{**} of the neutral molecule (often referred to as the 'resonant' state) which after dissociation yields the atomic products A and B. In the first pathway of the above reaction the electron + ion continuum is directly coupled to the repulsive state AB^{**}. The second pathway proceeds through an additional intermediate step, corresponding to electron capture in a Rydberg level AB^{*} associated with a vibrationally excited state of the initial ion AB^{+}; the intermediate Rydberg level is 'resonant' in the sense that it yields resonances in the otherwise smooth DR cross section. These two reactions mechanisms, illustrated in figures 1(*a*) and (*b*), are often called 'direct' and 'indirect', respectively.

Bardsley (1968) previously formulated the theory of DR in the configuration mixing framework. His theory takes account of electron departure (autoionisation) which may occur after the capture in the repulsive state AB^{**} and reduce the dissociation rate; it

0022-3700/80/193867+28$01.50

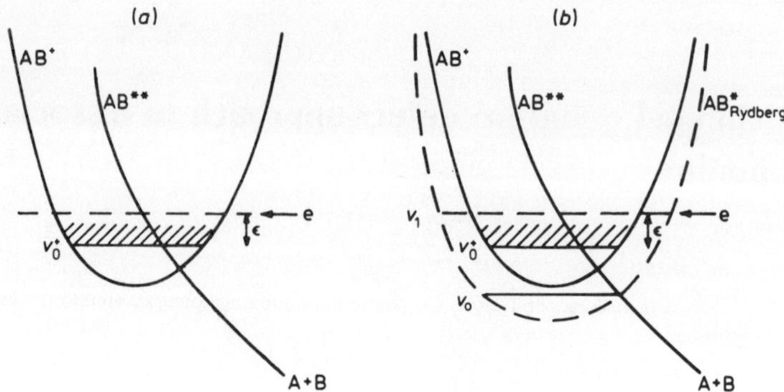

Figure 1. Dissociative recombination of a diatomic molecular ion; (*a*) direct process (electronic capture in the dissociative state AB^{**}), (*b*) indirect process via vibrational capture in a Rydberg state AB^* predissociated by AB^{**}.

avoids the 'local approximation' (Bardsley *et al* 1964, O'Malley 1966, Bottcher 1976) which fails at low energy or when the dissociative potential curve does not intersect the curve of the molecular ion initial state. However, the indirect process is treated separately by the usual resonance theory which leads to a two-continua Breit and Wigner formula, and the interference with the direct process is neglected. Lee (1977) has recently proposed an approach in which both processes are treated simultaneously on the basis of multichannel quantum defect theory (MQDT) so that their interference is taken into account. The present work retains the MQDT approach but overcomes some inconsistencies of the previous treatment. In order to simplify the theory we neglect in this paper the rotation of the molecule, i.e. we treat only the indirect process in which the incident electron gives energy to the vibrational motion of the nuclei: capture by rotational excitation is much less efficient, even for H_2 which has a large rotational constant (see e.g. Bottcher 1976). In any case, extension of the MQDT formalism for dissociative recombination to include rotational capture is straightforward, following the treatment of rovibronic perturbations in the H_2 Rydberg series by Jungen and Atabek (1977)† or, more specifically, the treatment of rovibronic preionisation (the inverse process of resonant capture) by Jungen and Dill (1980)‡. Let us also point out that our treatment rests on a *quasidiabatic* representation (O'Malley 1967, Sidis and Lefebvre-Brion 1971) of the electronic molecular states, in which the electronic interaction between Rydberg and dissociative valence states is not diagonalised: their potential curves exhibit crossings instead of having the numerous avoided crossings of the adiabatic Born–Oppenheimer curves (see figure 1 of O'Malley 1967). The remaining interaction, responsible for the predissociation of the discrete Rydberg and electron–ion continuum states, constitutes the kernel of the following treatment.

Section 2 recalls the physical and theoretical basis of the quantum defect theory in its generalised form. The MQDT equations for dissociative recombination are written in § 3. Results of model calculations are presented in § 4, where the influence of Rydberg states on the DR process is tested as a function of parameters characterising the Rydberg series, the dissociative state, or their mutual interaction.

† To be referred to in the following as JA.
‡ To be referred to in the following as JD.

2. The main characteristics of the generalised QDT

2.1. Physical background

Quantum defect theory unifies the theoretical description of several physical problems which customarily are considered separately. Thus it treats bound and continuous energy ranges simultaneously. The original QDT (Seaton 1958) and MQDT (Seaton 1966) proceeded by extrapolating the properties of atomic Rydberg series to electron–atomic ion collisions, the *quantum defect* becoming at positive energies the *phaseshift* (divided by π) of the scattered electron wavefunction. In this approach a 'channel' is formed by a set of states consisting of a common ionic core state and an electron of arbitrary energy, positive or negative. Later this treatment was extended to molecular Rydberg series (Fano 1970) and has recently been generalised by Fano (1978) and by Greene *et al* (1979) to long-range potentials other than the Coulomb potential, that is, to 'collision' partners other than electron–ionic core. Their generalised quantum defect theory allows a unified treatment of different continua arising in molecular processes, as in DR, for example, where both electron–molecule and atom–atom continua are involved. Such processes may then be considered as *rearrangement collisions* whatever the nature of the dissociation products is and the MQDT treatment yields a generalised S matrix including all the open channels.

The basic concept allowing this unified approach to molecular processes is the distinction between short- and long-range interactions, which results in the subdivision of the configuration space into several regions. The innermost region, called the 'reaction zone' by Lee (1975, 1977), generalises the concept of the core region of the electron–ion case: it contains all short-range 'strong' interactions, in a sense which must be redefined in each particular case. Actually, QDT does not deal with the wavefunction inside the reaction zone, but only with the net effect short-range interactions have on the wavefunctions in the external zone: their effect is cast in the form of a *phaseshift* for each long-range wavefunction, and of *mixing coefficients* between them. These quantities serve as input parameters (equivalent to initial conditions on the boundary of the reaction zone) for the MQDT calculations in the external zone. They vary slowly with the 'collision' energy on both sides of the threshold because they result from strong interactions involving the exchange of large amounts of energy. Hence they may be obtained alternatively from scattering data or from molecular structure data. In both cases, these data come from an *ab initio* calculation or from fitting to experimental results (semi-empirical approach).

The external zone wavefunction is expressed as a linear superposition of long-range potential eigenfunctions, e.g. Coulomb functions in the case of electron–ion dissociation. When several channels are still coupled in that region due to extra terms in the Hamiltonian (typically rovibronic or spin–orbit interactions which become important only at large distances), MQDT subdivides the external zone into different regions corresponding to different coupling schemes between the collision partners. Instead of introducing explicitly the operators responsible for transitions between channels, one performs 'frame transformations' (Fano 1970, Chang and Fano 1972) between eigenbasis sets adapted to each region. Boundary conditions applied at infinity (i.e. on the eigenchannels of the most external region) allow the mixing coefficients to be determined analytically. We stress that it is only at this stage that bound and continuum states are distinguished which correspond to closed or open channels respectively.

2.2. Theoretical implementation of MQDT

The simplicity of MQDT equations results from a specific representation of continuum wavefunctions in the external zone. This will be recalled very briefly with emphasis on the connection with the alternative approach of configuration mixing (CM).

(*a*) *One-channel case.* Let H_0 be a one-particle radial Hamiltonian with a continuous spectrum, and V an additional potential, not necessarily small nor local but of limited range:

$$V(r)=0 \qquad r>r_0.$$

Outside the range r_0 of this potential, the wavefunction may be written exactly as a linear superposition of two independent eigenfunctions of H_0 for the same energy:

$$\varphi_E(r) \underset{r>r_0}{=} f_E(r)\cos\eta(E)-g_E(r)\sin\eta(E). \tag{1}$$

The f function used by QDT is regular at the origin and behaves asymptotically (for positive energies) as a sine function; g lags in phase by $\frac{1}{2}\pi$ with respect to f such that the asymptotic behaviour of the total wavefunction is

$$\varphi_E(r) \underset{r\to\infty(E>0)}{\sim} A(E)\sin(kr+\delta(E)+\eta(E))$$

where $A(E)$ is a normalisation constant chosen so that the f function is energy independent in the limit $r\to 0$; k is the collision wavenumber and $\delta(E)$ the phaseshift due to the potential terms other than $V(r)$.

The ratio $\tan\eta(E)$ of the coefficients in the standing wave (1) may be interpreted as the element of the reaction matrix K for the collision energy E:

$$\tan\eta(E)=-\pi K(E). \tag{2}$$

This is the basic connection formula between QDT and CM treatments; the latter expands the total wavefunction in the form:

$$\varphi_E(r)\propto f_E(r)+\mathscr{P}\int \mathrm{d}E'\frac{1}{E-E'}K(E',E)f_{E'}(r) \tag{3}$$

where the matrix element $K(E',E)$ verifies the integral equation

$$K(E',E)=\langle f_{E'}|V|f_E\rangle+\mathscr{P}\int \mathrm{d}E''\frac{\langle f_{E'}|V|f_{E''}\rangle}{E'-E''}K(E'',E) \tag{4}$$

which is the real form of the Lippmann–Schwinger equation (see e.g. Levine 1969).

The continuum expansion (3), valid for all distances r, has been shown by Fano (1978) and Greene *et al* (1979) to coincide for $r>r_0$ with the QDT wavefunction (1), proportional to the unnormalised form

$$\varphi_E(r)\propto f_E(r)+\pi K(E)g_E(r)$$

which follows from equation (2) and where $K(E)$ is the 'on the energy shell' element $K(E,E)$. QDT uses the fact that for $r>r_0$ the effect of the potential V amounts to adding to the regular eigenfunction f_E of H_0 the irregular solution g_E *with the same energy* leading to a much more tractable expression than expansion (3).

(*b*) *Many-channel case.* In the case of a multichannel system, we shall denote the total wavefunction in a given channel by φ_{iE}, consisting of a 'core' wavefunction multiplied by a one-particle continuum wavefunction: the theory is presently restricted to the case of a single escaping particle or equivalently to dissociation of the system into two fragments only, thus involving a single continuum in each channel. Nevertheless, the continua may be of different nature in alternative channels, for example electronic or nuclear in the case of DR (see § 3). MQDT takes account of an additional interchannel interaction V by adjoining to each channel wavefunction φ_{iE} a superposition of 'conjugated' wavefunctions $\bar{\varphi}_{jE}$ with the same *total* energy E deduced from the φ_j's by dephasing each continuum factor by $-\frac{1}{2}\pi$. Starting from N orthonormal channel wavefunctions one obtains a new orthogonal set not yet normalised:

$$\psi_{iE} = \varphi_{iE} + \sum_j \pi K_{ji}(E)\bar{\varphi}_{jE}. \tag{5}$$

The matrix elements $K_{ji}(E)$ constitute the $N \times N$ 'on the energy shell' reaction matrix associated with the interaction V; they satisfy the integral coupled equations which are a generalisation of equation (4):

$$K_{ji}(E) = \langle \varphi_{jE}|V|\varphi_{iE}\rangle + \sum_k \mathscr{P}\int \mathrm{d}E' \frac{\langle \varphi_{jE}|V|\varphi_{kE'}\rangle}{E-E'} K_{ki}(E', E). \tag{6}$$

In order to cast the result of the short-range interaction in terms of phaseshifts as in the one-channel case, one introduces the (real) eigenvalues of the K matrix which may be written $-\pi^{-1}\tan\eta_\alpha$ ($\alpha = 1, 2, \ldots, N$). They are obtained by solving the algebraic system

$$\sum_j \pi K_{ij} U_{j\alpha} = -\tan\eta_\alpha U_{i\alpha} \qquad \alpha = 1, 2, \ldots, N \tag{7}$$

where U is the unitary eigenvector matrix of K; for simplicity the energy index has been omitted, i.e. $K_{ij}(E) = K_{ij}$.

The set of functions ψ_i of equation (5) is thus replaced by the set of the N eigenchannel functions $\psi_\alpha = \Sigma_i U_{i\alpha}\psi_i B_\alpha$. Using equation (7) together with the normalisation relations $B_\alpha = \cos\eta_\alpha$ they may be expressed in the form

$$\psi_\alpha = \sum_i U_{i\alpha}(\varphi_i \cos\eta_\alpha - \bar{\varphi}_i \sin\eta_\alpha). \tag{8}$$

These eigenchannels, defined on the boundary of the reaction zone, are used by MQDT in the external zone where they are mixed by change of coupling schemes. Each ψ_α consists of a superposition 'on the energy shell' of standing waves with a *common* phaseshift η_α due to the short-range interaction V and added to the phaseshifts resulting from other short- or long-range interactions.

We close this brief presentation of MQDT with a few remarks which are relevant to the treatment of DR presented in the next section.

(i) The QDT approach is very suitable for a stepwise treatment of different parts of the interaction. Taking for simplicity the one-channel case, let K' be the K matrix associated with a second short-range potential V', and $\eta'(E) = -\tan^{-1}\pi K'(E)$ the corresponding phaseshift. The final wavefunction may be expressed by adjoining to φ_E (equation (1)) the 'conjugated' function $\bar{\varphi}_E$:

$$\begin{aligned}\varphi'_E(r) &= \varphi_E(r)\cos\eta'(E) - \bar{\varphi}_E(r)\sin\eta'(E)\\ &= f_E(r)\cos(\eta(E)+\eta'(E)) - g_E(r)\sin(\eta(E)+\eta'(E))\\ &\underset{r\to\infty}{\sim} A(E)\sin(kr+\eta(E)+\eta'(E)).\end{aligned}$$

(ii) This stepwise procedure permits a practical use of the connection formula between QDT and CM treatments: part of the phaseshift needed for a QDT treatment may be calculated by using the CM equations (2) and (4) (with $E' = E$) for the one-channel case, or (6) and (7) for the many-channel case. This method is particularly suitable for a weak interaction allowing a perturbation expansion (Born series) of equation (4) or (6), for example for equation (4):

$$K(E) = \langle f_E|V|f_E\rangle + \mathscr{P}\int dE' \frac{\langle f_E|V|f_{E'}\rangle}{E-E'}\langle f_{E'}|V|f_E\rangle + \dots \tag{9}$$

in which the principal part integrations of second and higher order terms may be performed by a discretisation technique (see e.g. Atabek and Lefebvre 1972). The 'exact' resolution of the integral equations (4) or (6) is also possible by discretisation methods leading to algebraic systems (Altick and Moore 1966).

(iii) The success of QDT for the electron–ion case (f and g then being Coulomb functions) results from the strength of the Coulomb attraction: it screens rapidly with increasing distance all the other interactions, such that the reaction zone is small. Moreover, it supports many bound levels (infinite Rydberg series) allowing use of spectroscopic data. In the case of interaction between an electron and a neutral the long-range interactions are no longer screened and the reaction zone may become too large for the practical application of the method. Two ways for overcoming this difficulty are now described.

Firstly, one may use simple Bessel functions as basis functions (f, g) and consider the phaseshifts η (equation (2)) or η_α (equation (7)) as unknown functions of the radial distance r: this corresponds to the *variable phase method* (Calogero 1967) which requires the solution of only first-order differential equations and is successfully used in the external zone in connection with R-matrix calculations (Vo Ky Lan and Le Dourneuf 1980).

Secondly, one may use more sophisticated basis functions, including in H_0 part of the interaction potential: this is the generalised quantum defect approach recently developed by Greene *et al* (1979). We will proceed in that way by including dissociative channels in MQDT: the (F_{d_i}, G_{d_i}) nuclear wavefunctions introduced in equations (12) and (15) below correspond to a Hamiltonian H_0 which includes all the electronic interactions except the Rydberg valence mixing.

(iv) Finally we point out that the exact dimensions of the reaction zone or of the different regions of the external zone do not matter for strict MQDT calculations: they involve only analytical equations obtained by identifying *constant* expansion coefficients of the wavefunction in alternative basis sets (see § 3.2), such that the numerical values of the boundary radii do not intervene. This is of course not the case when one is to perform *ab initio* calculations in order to obtain the reaction zone parameters (the value of r_0 must not be underestimated) or when the variable phase method is to be used in the external zone.

3. MQDT treatment of dissociative recombination

3.1. *The reaction zone and the eigenchannels at its boundary*

We consider a DR process involving N_{el} ionisation channels and N_d dissociative channels. The first type of channels corresponds to an external electron with energy ε

ranging in a limited domain on both sides of the threshold, added to a molecular ion in a given stable electronic state (generally the ground state) and in different vibrational levels. Neglecting molecular rotation (see § 1) we may use the molecular body frame throughout the calculations, with the orbital angular momentum l of the external electron coupled to the molecule to yield a state with well defined total electronic angular momentum component Λ with respect to the internuclear axis. Besides neglecting l uncoupling, we do not introduce the l mixing due to molecular anisotropic interactions (Jungen 1970, Raseev *et al* 1978) for the sake of simplicity, although this effect could be adequately treated in the multichannel theory. As a result of these approximations we deal with a single electronic partial wave l and a single molecular symmetry Λ at a time. In other words, the angular part of the wavefunctions can be disregarded from here on. Similarly, the radial factors r^{-1} and R^{-1} will be understood to be included in the radial wavefunctions of the external electron and of the nuclei. The dissociative channels correspond to highly excited states† of the molecule, whose potential curves intersect or approach closely that of the ion electronic state and dissociate below the ion initial vibrational level. The electronic Hamiltonian is supposed to be separately diagonalised within both sets of channels but has non-vanishing matrix elements between the electronic continuum configuration and the valence one:

$$V_{i,l}^{\mathrm{el}}(\epsilon, R) = \langle \mathscr{A}(\Phi_{\mathrm{core}}(q^+, R)\varphi_\epsilon^l(q, R))|H_{\mathrm{el}}|\Phi_{\mathrm{d}_i}(q, R)\rangle \tag{10}$$

where Φ_{core} and Φ_{d_i} are the electronic wavefunctions of the ion initial state and of the neutral molecule dissociative state, respectively, and φ_ϵ is the radial wavefunction of the external electron. The integration is performed over the electronic coordinates (denoted collectively by q for the neutral molecule, q^+ for the molecular ion) and extends only to r_0 because the valence orbitals are confined at short range. Thus depending weakly on the collision energy, these couplings will be the building blocks for the reaction zone parameters (see § 2.1). They may be obtained either by direct scattering calculations (Bottcher 1976, Raseev *et al* 1978) or from predissociation studies of the Rydberg states using the scaling law

$$n^{*3/2}V_{i,n}(R) = V_i^{\mathrm{el}}(\epsilon_n, R) \qquad \text{(in atomic units)}$$

where $V_{i,n}$ is the analogy of (10) for the bound Rydberg state with effective principal quantum number n^*, and $\epsilon_n = -1/2n^{*2}$.

In its electronic dimension (r coordinate) the reaction zone will be defined, as in the usual QDT, by the range r_0 of electron–core interactions other than the Coulombic attraction. They result for $r > r_0$ in a phaseshift $\pi\mu(R)$ which depends parametrically on the internuclear distance, according to the Born–Oppenheimer (BO) approximation. The region of validity of the BO approximation is determined by the relative velocity of the external electron and of the nuclei: typically it extends beyond the boundary r_0 of the reaction zone and defines the so-called 'region A', the inner part of the external zone. The ionisation channel wavefunctions are there (see e.g. IA):

$$\psi_v^{\mathrm{A}} \underset{r>r_0}{=} \chi_v(R)\Phi_{\mathrm{core}}(q^+, R)(f_l(\nu, r)\cos\pi\mu(R) - g_l(\nu, r)\sin\pi\mu(R)). \tag{11}$$

The χ_v are bound vibrational wavefunctions whose choice will be specified in § 3.3 and (f_l, g_l) are Coulomb functions with $v = (-2\epsilon)^{-1/2}$. The kinetic energy ϵ of the external

† They will be called 'valence states' in the following for brevity, but they may also be Rydberg states of a series corresponding to an excited dissociative ion state.

electron may be regarded in region A as independent of the channel owing to the fact that the energy gaps between the thresholds corresponding to alternative channels are small in comparison with the potential energy in region A.

For the dissociative channels, we use the BO wavefunctions

$$\psi_{d_i} = \Phi_{d_i}(q, R) F_{d_i}(\kappa_i, R) \tag{12}$$

where $\kappa_i = \sqrt{2M(E - E_{d_i})}$ is the wavenumber for the relative motion of the two constituent atoms of reduced mass M in the dissociative state i with the dissociation energy E_{d_i}. F_{d_i} is the energy normalised solution of the nuclear Schrödinger equation in the repulsive potential U_{d_i}:

$$\left(-\frac{1}{2M}\frac{d^2}{dR^2} + U_{d_i}(R) - E\right) F_{d_i}(\kappa_i, R) = 0 \tag{13}$$

with the asymptotic behaviour

$$F_{d_i}(\kappa_i, R) \underset{R\to\infty}{\sim} \sqrt{2M/\kappa_i\pi}\,\sin(\kappa_i R + \delta_i(E)).$$

In its nuclear dimension (R coordinate), the reaction zone will be defined as the region ($R < R_0$) where the Rydberg–valence total interactions

$$V_{v,d_i} = \int_0^\infty \chi_v(R) V_i^{el}(\epsilon, R) F_{d_i}(\kappa_i, R)\, dR \simeq \int_0^{R_0} (\ldots)\, dR \tag{14}$$

are built. (For $R > R_0$ one has either $V_i^{el}(R) = 0$ or $\chi_v(R) = 0$, or both.)

Starting from the wavefunctions (11) and (12), we have to take account of the interactions (14) by constructing the associated K matrix. In doing this, we make use of the stepwise procedure allowed by the QDT formulation, as stated above. The K matrix is calculated from its CM expression (6) or rather, if possible, by the perturbation expansion (9). The V_{v,d_i} (equation (14)) are the only non-zero elements of the interaction matrix (i.e. $V_{vv'} = V_{d_i d_j} = 0$); nevertheless, extensions of the formalism for which these conditions are relaxed will be mentioned below in § (3.3.2). Note however that two electronic or two dissociative channels may still become indirectly coupled when the K matrix is calculated beyond first order ($K_{vv'} \neq 0$, $K_{d_i d_j} \neq 0$). Finally one defines the set of $N = N_{el} + N_d$ eigenchannels at the boundary of the reaction zone by diagonalising the K matrix following the general relation (8) with the same definitions for the notations $U_{i\alpha}$ and η_α:

$$\begin{aligned}\psi_\alpha &= \sum_v U_{v\alpha}(\psi_v^A \cos\eta_\alpha - \bar{\psi}_v^A \sin\eta_\alpha) + \sum_i U_{d_i\alpha}(\psi_{d_i}\cos\eta_\alpha - \bar{\psi}_{d_i}\sin\eta_\alpha)\\ &= \sum_v \chi_v(R)\Phi_{core}(q^+, R) U_{v\alpha}[f_l(\nu, r)\cos(\pi\mu(R) + \eta_\alpha)\\ &\quad - g_l(\nu, r)\sin(\pi\mu(R) + \eta_\alpha)]\\ &\quad + \sum_i \Phi_{d_i}(q, R) U_{d_i\alpha}(F_{d_i}(\kappa_i, R)\cos\eta_\alpha - G_{d_i}(\kappa_i, R)\sin\eta_\alpha).\end{aligned} \tag{15}$$

According to the general definition for 'conjugate' wavefunctions $\bar{\psi}$ (see equation (5)), $\bar{\psi}_v$ and $\bar{\psi}_{d_i}$ are deduced from ψ_v^A (equation (11)) and ψ_{d_i} (equation (12)) by the substitutions $f \to g$, $g \to -f$ and $F_d \to G_d$, $G_d \to -F_d$, respectively. G_{d_i} is the solution of the nuclear Schrödinger equation (13) with the asymptotic behaviour

$$G_{d_i}(\kappa_i, R) \underset{R\to\infty}{\sim} -\sqrt{2M/\kappa_i\pi}\cos(\kappa_i R + \delta_i(E)).$$

The wavefunctions (15) differ from the eigenchannel functions of JA and JD in two respects, first by the addition of the dissociative channels and second by the additional phaseshifts η_α and transformation elements $U_{v\alpha}$ which appear in the ionisation channels as a consequence of the interactions V_{vd_i} not considered in the earlier works. These new parameters introduce in the MQDT treatment the non-adiabatic interactions responsible for the (pre)dissociation of the molecule in Rydberg or electronic continuum states.

The eigenchannel functions ψ_α are used to expand the energy eigenfunction ψ outside the reaction zone as

$$\psi \underset{r>r_0, R>R_0}{=} \sum_\alpha A_\alpha \psi_\alpha. \tag{16}$$

The MQDT equations start from this expansion with the aim of determining the unknown coefficients A_α by means of the boundary conditions at infinity.

3.2. The MQDT equations in the external zone

In the asymptotic region ($r, R \to \infty$) one has to deal with the *decay channels* corresponding to the dissociation of the molecular complex into either a molecular ion plus a free electron or into two separate atoms. When r increases, the external electron is no longer able to exchange energy with the core. The *ionisation* decay channel functions are then expressed as products of an ion core wavefunction multiplied by a radial factor for the outer electron:

$$\psi^{B}_{v^+} = \chi_{v^+}(R)\Phi_{\text{core}}(q^+, R)(c_{v^+} f_l(\nu_{v^+}, r) - d_{v^+} g_l(\nu_{v^+}, r)) \tag{17}$$

with

$$\nu_{v^+} = (-2\epsilon_{v^+})^{-1/2}.$$

The χ_{v^+} are vibrational wavefunctions of the molecular ion and the coefficients c_{v^+}, d_{v^+} depend on the vibrational quantum number v^+ rather than on the internuclear distance R (cf equation (11)): this means that in the outer part of the external zone 'region B' the BO approximation does not hold for the system as a whole although it is conserved for the core as well as for the valence states. Note that the kinetic energy $\epsilon_{v^+} = E - E_{v^+}$ (positive or negative) is now specified for each ionisation channel v^+ with threshold energy E_{v^+}.

The *dissociative* decay channels on the other hand are still represented by the wavefunctions (12) at infinity since here we do not take account of any coupling in the external zone. Note that rotational or spin–orbit coupling could be treated by frame transformations between basis sets adapted to different coupling schemes, which amounts to subdividing the external zone into several regions in its R dimension, exactly as for the electronic dimension (this approach is closely related to the use of Hund's coupling schemes in atom–atom collisions, see Nikitin 1974).

The next step of the treatment is analogous to the usual MQD treatment: the energy eigenfunction (16) is re-expanded in a form adapted to region B as a superposition of the various decay channel functions (equations (12) and (17)). Thereby one uses the fact that the wavefunctions f_l(resp. g_l) in alternative channels v^+ are nearly identical at short distances. The result is

$$\psi=\sum_{v^+}\chi_{v^+}\Phi_{\rm core}\left(f_l(\nu_{v^+},r)\sum_\alpha A_\alpha\mathscr{C}_{v^+\alpha}-g_l(\nu_{v^+},r)\sum_\alpha A_\alpha\mathscr{S}_{v^+\alpha}\right)$$
$$+\sum_i\Phi_{{\rm d}_i}(F_{{\rm d}_i}(\kappa_i,R)\sum_\alpha A_\alpha\mathscr{C}_{{\rm d}_i\alpha}-G_{{\rm d}_i}(\kappa_i,R)\sum_\alpha A_\alpha\mathscr{S}_{{\rm d}_i\alpha}) \tag{18}$$

with

$$\mathscr{C}_{v^+\alpha}=\sum_v U_{v\alpha}\langle\chi_{v^+}|\cos(\pi\mu(R)+\eta_\alpha)|\chi_v\rangle \qquad \mathscr{C}_{{\rm d}_i\alpha}=U_{{\rm d}_i\alpha}\cos\eta_\alpha$$
$$\mathscr{S}_{v^+\alpha}=\sum_v U_{v\alpha}\langle\chi_{v^+}|\sin(\pi\mu(R)+\eta_\alpha)|\chi_v\rangle \qquad \mathscr{S}_{{\rm d}_i\alpha}=U_{{\rm d}_i\alpha}\sin\eta_\alpha. \tag{19}$$

The expansion (18) allows us to specify the boundary conditions in each decay channel, leading to MQDT equations. Let P and Q be the set of the $N_{\rm P}$ open and $N_{\rm Q}$ closed ionisation channels, respectively, with $N_{\rm el}=N_{\rm P}+N_{\rm Q}$, and let $N_{\rm op}=N_{\rm P}+N_{\rm d}$ be the total number of open decay channels. Our aim now is to calculate the matrix elements of the generalised S matrix (of dimension $N_{\rm op}$) between the open ionisation channels and the dissociative ones from which the DR cross sections will be obtained. We must then construct a wavefunction having the following asymptotic behaviour:

$$\psi(v_0^+\underset{r\to\infty}{\sim}\sum_{v^+\in{\rm P}}\chi_{v^+}(R)\Phi_{\rm core}\{\exp[-{\rm i}(k_{v^+}r+\sigma_l)]\delta_{v^+v_0^+}+\exp[{\rm i}(k_{v^+}r+\sigma_l)]S_{v^+v_0^+}\} \tag{20a}$$

$$\underset{R\to\infty}{\sim}\sum_i\Phi_{{\rm d}_i}(q,R)S_{{\rm d}_iv_0^+}\exp[{\rm i}(\kappa_iR+\delta_i)] \tag{20b}$$

where $k_{v^+}={\rm i}/\nu_{v^+}=\sqrt{2(E-E_{v^+})}$. σ_l is the l Coulomb phaseshift, δ_i the phaseshift due to the molecular potential $U_{{\rm d}_i}$, and v_0^+ denotes the initial vibrational level of the molecular ion.

These asymptotic conditions imply, of course, that the closed channel components of the wavefunction must die away at large distances and they are not indicated in the asymptotic expressions (20). The open channel portion of the total wavefunction is expressed here in terms of *running* waves instead of the *standing* waves used for the eigenchannel functions of equation (15). As a consequence direct identification of equations (20) with the expansion (16) would introduce a large number of complex coefficients A without clear physical meaning. It is more suitable to reduce the problem to the usual scattering situation by first eliminating explicit reference to the closed channels. This can be done as in the multichannel preionisation studies of Lee (1971), Lee and Lu (1973) or Jungen and Dill (1980) (JD) since both open and closed decay channels are also involved in this case. We construct first $N_{\rm op}$ 'collision eigenchannels' whose open channel portions are the eigenstates of the open channel S matrix introduced in equations (20). The eigenchannel functions ψ_ρ are characterised asymptotically by a standing wave behaviour with a *common* phaseshift τ_ρ for all the open channel components (corresponding to decay into atom–atom as well as electron–ion products) and by exponential decrease for all the closed channel components.

Starting from the expansions

$$\psi_\rho=\sum_\alpha A_\alpha^\rho\psi_\alpha=\sum_{v^+}\langle v^+|\rho\rangle\psi_{v^+}^{\rm B}+\sum_i\langle{\rm d}_i|\rho\rangle\psi_{{\rm d}_i}$$

the unknown coefficients A_α^ρ are shown to satisfy the homogeneous system (see e.g. JD, equations (7) and (15a)):

$$\sum_\alpha A_\alpha^\rho(\mathscr{C}_{v^+\alpha} \sin \pi\nu_{v^+} + \mathscr{S}_{v^+\alpha} \cos \pi\nu_{v^+}) = 0 \qquad (v^+ \in \mathrm{Q})$$

$$\sum_\alpha A_\alpha^\rho(\mathscr{C}_{v^+\alpha} \sin \tau_\rho - \mathscr{S}_{v^+\alpha} \cos \tau_\rho) = 0 \qquad (v^+ \in \mathrm{P}) \tag{21}$$

$$\sum_\alpha A_\alpha^\rho(\mathscr{C}_{\mathrm{d}_i\alpha} \sin \tau_\rho - \mathscr{S}_{\mathrm{d}_i\alpha} \cos \tau_\rho) = 0 \qquad (i = 1, 2, \ldots, N_\mathrm{d}).$$

Its compatibility condition

$$\det \begin{vmatrix} \mathscr{C}_{v^+\alpha} \sin \pi\nu_{v^+} + \mathscr{S}_{v^+\alpha} \cos \pi\nu_{v^+} \\ \vdots \\ \mathscr{C}_{v^+\alpha} \sin \tau_\rho - \mathscr{S}_{v^+\alpha} \cos \tau_\rho \\ \vdots \\ \mathscr{C}_{\mathrm{d}_i\alpha} \sin \tau_\rho - \mathscr{S}_{\mathrm{d}_i\alpha} \cos \tau_\rho \end{vmatrix} = 0 \tag{22}$$

determines the N_{op} eigenphases τ_ρ and the coefficients A_α^ρ are obtained from (21) to within a normalisation factor (without importance for the S-matrix elements). The components $\langle v^+|\rho\rangle$ and $\langle \mathrm{d}_i|\rho\rangle$ are then calculated using the relations (analogous to equation (15b) of JD):

$$\langle v^+|\rho\rangle = \sum_\alpha A_\alpha^\rho(\mathscr{C}_{v^+\alpha} \cos \tau_\rho + \mathscr{S}_{v^+\alpha} \sin \tau_\rho) \qquad v^+ \in \mathrm{P}$$

$$\langle \mathrm{d}_i|\rho\rangle = \sum_\alpha A_\alpha^\rho(\mathscr{C}_{\mathrm{d}_i\alpha} \cos \tau_\rho + \mathscr{S}_{\mathrm{d}_i\alpha} \sin \tau_\rho)$$

and the S-matrix elements of equations (20) are obtained from the eigenvalues $\exp(2\mathrm{i}\tau_\rho)$ of S by a simple change of basis:

$$S_{v^+v_0^+} = \sum_\rho \langle v^+|\rho\rangle \exp(2\mathrm{i}\tau_\rho)\langle\rho|v_0^+\rangle \tag{23a}$$

$$S_{\mathrm{d}_iv_0^+} = \sum_\rho \langle \mathrm{d}_i|\rho\rangle \exp(2\mathrm{i}\tau_\rho)\langle\rho|v_0^+\rangle. \tag{23b}$$

Equations (23) constitute the main result of this paper. We stress that they involve explicitly only the N_{op} open decay channels, but the effect of the closed channels is embodied in the eigenphases τ_ρ. The $S_{v^+v_0^+}$ elements can be used to calculate vibrational excitation cross sections, both direct and resonant. $S_{\mathrm{d}_iv_0^+}$ yields the partial cross section for dissociation in the channel d_i out of the initial vibrational level v_0^+ of the ion and relative to a single electronic partial wave l:

$$\sigma_{\mathrm{d}_iv_0^+}^l = \tfrac{1}{2}(\pi/k^2)r_i|S_{\mathrm{d}_iv_0^+}^l|^2 \tag{24}$$

where r_i stands for the multiplicity ratio between the final and initial electronic states (of the neutral molecule and the molecular ion, respectively) and the factor two at the denominator takes account of the incident electron spin multiplicity.

The total cross section for scattering of an electron on the ion in its vibrational level v_0^+ is

$$\sigma_{v_0^+} = (\pi/2k^2)\sum_{l,i} r_i|S_{\mathrm{d}_iv_0^+}^l|^2.$$

If several vibrational levels v_0^+ are initially populated with relative populations $g(v_0^+)$, the experimental cross section (without vibrational resolution) is to be compared with the quantity:

$$\sigma = \sum_{v_0^+} g(v_0^+)\sigma_{v_0^+}.$$

3.3. Complements and remarks

3.3.1. The vibrational basis set in region A. The choice of the vibrational wavefunctions χ_v in equation (11) has not been specified so far. Jungen and Atabek (1977) (JA) have proposed two alternative choices, namely the set χ_v^n of vibrational wavefunctions of the neutral molecule in a given Rydberg state, or the set χ_{v^+} of the molecular ion vibrational functions as in equation (17) for the region B. The basis χ_v^n is more appropriate within the BO framework, but the χ_{v^+} may be used even in region A in as much as one takes account of the R dependence of the quantum defect μ. The first vibrational basis is calculated with the Rydberg potential curve

$$U_n(R) = U^+(R) - 1/2(n - \mu(R))^2 \qquad \text{(in au)}$$

which involves both the ion potential curve $U^+(R)$ and the quantum defect function $\mu(R)$. The second basis is obtained from the potential U^+ alone, but the quantum defect function is also involved through the matrix elements $\mathscr{C}_{v^+\alpha}$ and $\zeta_{v^+\alpha}$ of equation (19).

The best choice is indicated by numerical considerations and depends on the energy range. In the following section we are concerned with highly excited preionised states and we have used the basis χ_{v^+}. When studying predissociation effects in the low part of the spectrum where vibronic perturbations rather than preionisation arise, a basis χ_v^n is certainly more convenient, especially if μ varies strongly with R such that the curves U_n and U^+ differ notably for low n. Two remarks are in order here.

(i) The above discussion may be related to the choice of vibrational wavefunctions for the R-matrix formulation of electron–molecule vibrational excitation (Schneider *et al* 1979): when the potential curve of the compound state AB^- differs notably from that of the initial electronic state of the neutral molecule AB, it is very useful to introduce the vibrational wavefunctions of the AB^- state, corresponding to our χ_v^n wavefunctions.

(ii) The frame transformation matrix elements $\mathscr{C}_{v^+\alpha}$ and $\mathscr{S}_{v^+\alpha}$ of equation (19) may be recast in the form

$$\mathscr{C}_{v^+\alpha} = \sum_v U_{v\alpha}(c_{v^+v}\cos\eta_\alpha - s_{v^+v}\sin\eta_\alpha) \qquad \mathscr{S}_{v^+\alpha} = \sum_v U_{v\alpha}(c_{v^+v}\sin\eta_\alpha + s_{v^+v}\cos\eta_\alpha)$$

with (25)

$$c_{v^+v} = \int \chi_{v^+}(R)\cos\pi\mu(R)\chi_v(R)\,\mathrm{d}R \qquad s_{v^+v} = \int \chi_{v^+}(R)\sin\pi\mu(R)\chi_v(R)\,\mathrm{d}R$$

where the c_{v^+v} and s_{v^+v} are the matrix elements involved in vibronic perturbation or preionisation studies (see JA and JD). The expressions (25) show clearly that vibronic coupling arises due to the R dependence of the quantum defect, whose variations allow the external electron to 'feel' the nuclear motion and thus to exchange energy with it (a constant quantum defect would result in identity between the sets χ_v^n and χ_{v^+} and therefore in the selection rule $\delta_{vv^+} = 0$).

3.3.2. Some extensions of the formalism. We have stressed in the introduction that our treatment rests on a quasidiabatic representation of the molecular states in which some weak and localised electronic interactions have not been diagonalised such that some potential curves of identical symmetries may cross. This representation may be used not only for the Rydberg valence interaction responsible for the DR process, but also for an electronic coupling between two crossing dissociative states of the same symmetry. The branching ratio between the two dissociation limits E_{d_i} is different when the curves are followed diabatically (neglecting the electronic coupling) or adiabatically (neglecting the radial coupling at the avoided crossing) as was found, for example, by Lee (1977) for the two dissociative states A′ $^2\Sigma$ and I $^2\Sigma$ of the NO molecule. A more reliable branching ratio would be obtained within a small extension of the present MQDT treatment. Using the quasidiabatic crossing curves, we may account for their residual electronic coupling $V^{el}_{12}(R)$ by including in the reaction zone interaction matrix (defined analogously to equation (14)) the element

$$V_{d_1d_2} = \langle F_{d_1} | V^{el}_{12}(R) | F_{d_2} \rangle \tag{26}$$

which was assumed to be zero above. Note that the convergence of the integrals (14) giving the matrix elements V_{v,d_i} is ensured by the bound character of the vibrational wavefunctions χ_v, even when the electronic couplings $V^{el}_i(R)$ do not vanish rapidly as R increases. Here on the opposite side neither of the two nuclear continuum wavefunctions F_{d_1} and F_{d_2} vanish at large R so that the integral converges only if the electronic coupling vanishes for $R > R_0$, in accordance with the quasidiabatic representation. Moreover the electronic couplings conserved in this approach can always be chosen to be weak (in the other case the diagonalisation of the electronic Hamiltonian would lead to 'smooth' curves and the adiabatic picture would be better), such that the K matrix will be generally calculated using perturbation formula (9) in first order.

Other possible extensions of the formalism consist in modifying the boundary conditions in the different channels so as to study processes other than dissociative recombination. For example, *associative ionisation*, which is the inverse process of DR

$$A + B^* \rightarrow AB^{**} \rightarrow AB^+ + e \tag{27}$$

has been recently studied experimentally by Ringer and Gentry (1979) for NO^+ formation, by Neynaber and Tang (1979) for rare-gas association, or by Poulaert *et al* (1978, 1979) for H_2^+ formation in H^+–H^- collisions. The present formalism can be adapted to this process essentially by the corresponding modification of the asymptotic wavefunction (20). Note that associative ionisation may also proceed both by direct and indirect (via vibrationally preionised Rydberg states) processes such that narrow resonances would be seen in high resolution experimental cross sections.

The present theory can also be applied to molecular photoabsorption in the spectral range where both rovibronic preionisation (or perturbations) and electronic predissociation occur, corresponding to the processes

$$AB + h\nu \begin{cases} \nearrow AB^+ + e \\ \searrow A + B. \end{cases} \tag{28}$$

Our treatment in this case adds dissociation as a new channel to the MQDT treatments of JA and JD. Experimental evidence of this competition between preionisation and predissociation has been found, for example, in several molecules (O_2, N_2, CO, CO_2) by Lee *et al* (1974, 1975) who used synchrotron radiation, and most recently in NO by Miescher *et al* (1978).

Finally, *electronic* perturbations or preionisation in a Rydberg series may be treated by a similar approach, starting from a quasidiabatic representation of the molecular states; this extension will be discussed in a separate paper by Giusti and Lefebvre-Brion (1980).

We conclude by recalling a limitation of our formalism which has already been noted in § 3.3.2: at present we can treat only processes in which a *single* continuum is involved in each decay channel. For example, the present theory may not be directly applied to Penning ionisation which is closely related to associative ionisation:

$$\mathrm{A}+\mathrm{B}\rightarrow\mathrm{AB}^{**}\rightarrow\mathrm{A}+\mathrm{B}^{+}+\mathrm{e} \tag{29}$$

but where the molecular system dissociates into three fragments. A similar limitation is already present in *atomic* QDT, which is restricted to a single escaping electron. Steps to extend the theory in this direction have been put forward (see Fano 1977 and references therein, Clark and Greene 1979) but further work is necessary, particularly in the case where the fragments have masses of very different magnitudes such as in the above Penning ionisation reaction (29).

3.3.3. Comparison with the MQDT *treatment of Lee (1977).* We now compare our theory with the earlier treatment of DR by Lee (1977) who was the first author to undertake a MQDT approach to this problem. The main difference with the treatment presented here lies in the choice of eigenchannels for the region A of the external zone. We consider here the case $N_{\mathrm{d}}=1$ to which Lee's treatment is limited. Instead of the $N_{\mathrm{el}}+1$ eigenstates ψ_α(equation (15)) of the reaction zone K matrix, Lee defines N_{el} pairs of wavefunctions ψ_v^1 and ψ_v^2, mixing each vibrational channel *separately* with the dissociative channel. He then constructs a set of $2N_{\mathrm{P}}$ collision eigenchannels ψ_{ρ_1} and ψ_{ρ_2} associated with $2N_{\mathrm{P}}$ eigenphases τ_{ρ_1} and τ_{ρ_2}, whereas we have only $N_{\mathrm{P}}+1$ eigenchannels and eigenphases (the ψ_ρ and τ_ρ defined by equations (21)–(23)) which is the exact number physically required. Our introduction of the K-matrix eigenstates thus avoids the indetermination arising in the equations of Lee who must determine $2N_{\mathrm{P}}$ coefficients with only $N_{\mathrm{P}}+1$ boundary conditions. Moreover, our formulation lends itself much more easily to calculation beyond the first order of perturbation or to simultaneous inclusion of several dissociative channels. Incidentally, we can note that in the case where several dissociative channels of the same symmetry are present, even though they are not directly coupled, one will obtain different results depending on whether a separate calculation is performed for each of them or whether the global treatment proposed above is used (see equations (A.10) and (A.11) of the appendix).

Numerical tests have shown that even in the first order of perturbation the two treatments yield different values for the cross sections. In particular, we find that the resonance shapes depicted in § 4 are different when we use Lee's equations. A difference arises also in the analytical formula established in the appendix for the simple case where only the direct process contributes in the presence of several open ionisation channels: our expression (A.5)

$$\sigma_{\mathrm{d}v_0^+}=\frac{\pi}{k^2}\frac{r}{2}\frac{4(\xi_{v_0^+})^2}{[1+\Sigma_v\,(\xi_v)^2]^2}\qquad\qquad \xi_v=\pi\langle\chi_v|V^{\mathrm{el}}(R)|F_{\mathrm{d}}\rangle \tag{30}$$

is replaced by

$$\sigma_{dv_0^+} = \frac{\pi}{k^2} \frac{r}{2} \frac{4(\xi_{v_0^+})^2}{[1+(\xi_{v_0^+})^2]^2} \tag{31}$$

when using equation (25) of Lee's paper at the same level of approximation. In these expressions the deviation of the denominators from unity is a measure of the autoionisation of the dissociative state, as will be shown in the next paragraph. It appears that equation (31) does not account completely for it since it neglects ionisation processes in which the ion is left in a vibrational level v^+ different from the initial level v_0^+.

3.3.4. Comparison with the CM *treatment of Bardsley* (*1968*). The usual theory of dissociative recombination in the configuration mixing framework treats the direct and indirect processes separately. In the evaluation of the direct process one uses the CM approach effectively only for the electronic part of the molecular wavefunction, i.e. for studying the mixing between discrete and continuum *electronic* states. The mixing between discrete and continuum *vibrational* states is taken into account in a separate step by modification of the homogeneous Schrödinger equation (13) for the dissociative vibrational wavefunction (as above, a single dissociative state is included): a second member is added, which consists in a 'feeding' term taking account of the coupling with the entrance channel v_0^+ and in a complex autoionisation term involving the dissociative wavefunction in an integral. The Schrödinger equation then becomes an integro-differential equation, for which Bardsley has given an elegant algebraic method of solution. Nevertheless, the 'local approximation' is used most of the time: in the autoionisation term, the vibrational quantisation of the ion final state is omitted such that the integral character of the nuclear Schrödinger equation disappears (see the appendix of Bardsley's paper). This approximation is certainly valid at high energy where numerous vibrational channels are open and where the energy of the ejected electron depends weakly on the vibrational level of the residual ion (see Parlant and Fiquet-Fayard 1974), but it becomes poor for the low energies generally involved in DR experiments. The MQDT treatment does not use this approximation since each important vibrational channel is explicitly included of the calculation.

Another approximation frequently used in DR calculations by the CM method consists of the complete neglect of the autoionisation term. One thus gets the 'capture cross section',

$$\sigma_{\text{capt}} = \frac{4\pi^3}{k^2} \frac{r}{2} |\langle \chi_{v_0^+} | V_{\text{el}}(R) | F_d \rangle|^2. \tag{32}$$

The ratio between the DR cross section and the capture cross section is called the survival factor $S (S \leqslant 1)$. Comparing (32) and (A.5) we obtain the MQDT expression within the approximations of the appendix (see comments herein):

$$S = \left[1 + \sum_{v^+ \in P} (\xi_{v^+})^2\right]^{-2}. \tag{33}$$

More exact calculations do not allow explicit factorisation of the DR cross section in the form of a product $\sigma_{\text{capt}} S$, but we stress that MQDT accounts automatically for autoionisation, including via the preionised Rydberg states.

The treatment of the indirect dissociative recombination process in the traditional method (Bardsley 1968, Bottcher 1976) differs notably from the treatment of the direct process presented above: the configuration mixing approach is used not only for the electronic parts of the wavefunctions but also for their nuclear parts. The intermediate Rydberg state which corresponds to a well defined excited vibrational level is regarded as a discrete resonant state (both for electronic and vibrational motions) mixed with two continua: the electronic continuum of the initial scattering state ($e+AB^+$) and the vibrational one of the dissociative state. Neglecting the interaction between both continua which is responsible for the direct process, one obtains the usual Breit and Wigner formula with two decay modes:

$$\sigma_{\mathrm{res}} \propto \frac{\pi}{k^2} \frac{\Gamma_{\mathrm{a}}\Gamma_{\mathrm{d}}}{(E-E_{\mathrm{res}})^2+\frac{1}{4}(\Gamma_{\mathrm{a}}+\Gamma_{\mathrm{d}})^2} \tag{34}$$

where Γ_{a} is the vibrational preionisation (or capture) width of the Rydberg level and Γ_{d} its predissociation width. Two failures of this treatment are overcome by the MQDT approach.

(i) The vibrational preionisation width

$$\Gamma_{\mathrm{a}} = 2\pi \left| \left\langle \psi_i \left| \frac{1}{R^2} \frac{\mathrm{d}}{\mathrm{d}R} \left(R^2 \frac{\mathrm{d}}{\mathrm{d}R} \right) \right| \psi_f \right\rangle \right|^2 \tag{35}$$

where ψ_i and ψ_f are the total wavefunctions corresponding to the initial (discrete) and final (continuum) states, involves the derivative with respect to R of electronic quantities. These tedious calculations are avoided in the MQDT formalism which does not introduce the concept of an autoionisation width. As in the calculation of molecular photoionisation by MQDT (see e.g. JD, to be compared with the traditional treatment of vibrational preionisation—Berry (1966), Berry and Nielsen (1970), Duzy and Berry (1976)), the present work enables one to calculate directly cross section energy profiles including resonant and non-resonant contributions.

(ii) The CM treatment does not account for the interference between direct and indirect processes since it neglects here the continuum–continuum interaction (similarly, the vibronic coupling was neglected when treating the direct process). On the contrary, the MQDT approach starts from the interaction between electronic and vibrational continua, represented by the matrix elements $V_{\upsilon d_i}$ (equation (14)), and treats simultaneously discrete and continuous Rydberg configurations: the resonances due to temporary capture in discrete Rydberg states arise automatically when the boundary conditions are specified. We will see in § 4 that the interferences considerably alter the Lorentzian profile given by the Breit and Wigner formula (34).

4. Model calculations and discussion

In this section we present numerical tests in order to get some idea of the energy dependence to be expected for the DR cross section. We are mainly interested here in the resonances due to the indirect process where the electron capture involves vibrationally excited Rydberg states, rather than in the overall behaviour of the direct process cross sections: apart from the resonances, our predicted cross sections decrease essentially proportional to E^{-1}. Small deviations from this law arise when the relative positions of the dissociative state and ion initial state potential curves are varied and the energy dependence of the Franck–Condon factor is hence changed. The same behaviour has been found in all previous DR calculations and is in agreement with the experimental results (see for example McGowan *et al* 1979).

4.1. Description of the test calculations

Our model includes the minimal set of channels, namely a single dissociative channel and two electron–ion channels which are associated with different vibrational levels of the molecular ion: the $v^+=0$ initial level which corresponds to an open channel with a continuum electron and the $v^+=1$ level defining the closed channel with a bound Rydberg electron. The electronic couplings given by equation (10) and the quantum defects have been assumed to be energy independent. The 3×3 K matrix associated with the configuration interaction responsible for the dissociation has been calculated using the Born expansion (9) in first order.

In all calculations we have kept fixed the potential curve of the ion initial state (figure 2) and the R dependence of the electronic couplings reported in table 1. We used numerical values which pertain to the CH^+ radical: the ion curve is that of the x $^1\Sigma^+$ ground state calculated by Green *et al* (1972) and the electronic couplings were obtained by Raseev *et al* (1978) when studying the electron capture in a $^2\Pi$ dissociative

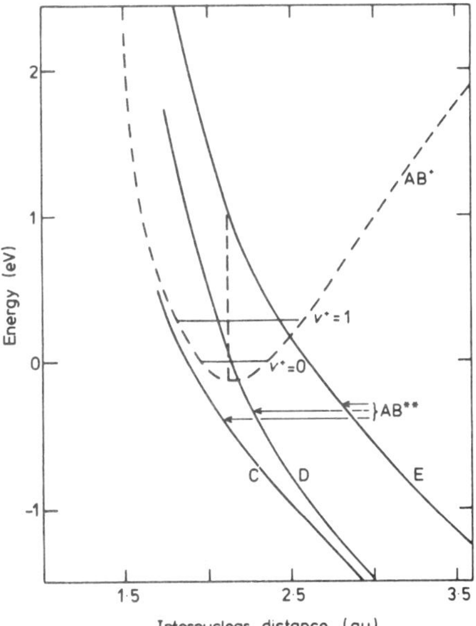

Figure 2. The potential curves used in the test calculations. The dissociation limits are at −4·92 eV for the dissociative curves C and D and −3·83 eV for curve E.

Table 1. Values of the electronic couplings used for the test calculations.

R (au)	$V^{el}_{l=1}(R)$ (au)	$V^{el}_{l=2}(R)$ (au)	$\Gamma_{el}(R)=2\pi\,\Sigma_l\,\lvert V^{el}_l(R)\rvert^2$ (eV)
1·5	0·0167	0·0056	0·053
2·0	0·0165	0·0065	0·054
2·5	0·0210	0·0044	0·078
3·0	0·0250	0·0015	0·107

state of CH. These parameters have previously served in a theoretical study of CH^+ dissociative recombination (Giusti 1979) and are taken here as a point of departure. Two partial waves (pπ and dπ) are taken into account but in order to simplify the resonance structure only the p wave was assumed to contribute to the resonant process. The d-wave contribution to the direct process (two open channel calculation) is included to yield a realistic profile with a background due to the non-resonant partial wave.

The quantum defect for the pπ series involved in the indirect process has been represented by the linear function

$$\mu(R)=\mu(R_e)+\left(\frac{d\mu}{dR}\right)_{R_e}(R-R_e) \qquad (36)$$

where the distance $R_e=2$ au was kept fixed and the parameters $\mu(R_e)$ and $(d\mu/dR)R_e$ have been varied to test the influence of magnitude and slope of the quantum defect on the resonant part of the cross section. The influence of the relative disposition of the dissociative curve and the ion curve has also been studied by using successively the different dissociative curves shown in figure (2). Finally, the strength of the electronic coupling has been varied by multiplying the numerical values of table 1 by R-independent factors.

The MQDT proper calculations (see § 3.2) have been performed as described in § III-C of JA. The inclusion of the dissociative channels adds only the step of diagonalising the K matrix in order to calculate the eigenchannel parameters $U_{i\alpha}$ and μ_α (equation (15)). The results of the test calculations are shown on a logarithmic scale in figures (3) and (6) and will be discussed in turn. A general point to be noted is that the

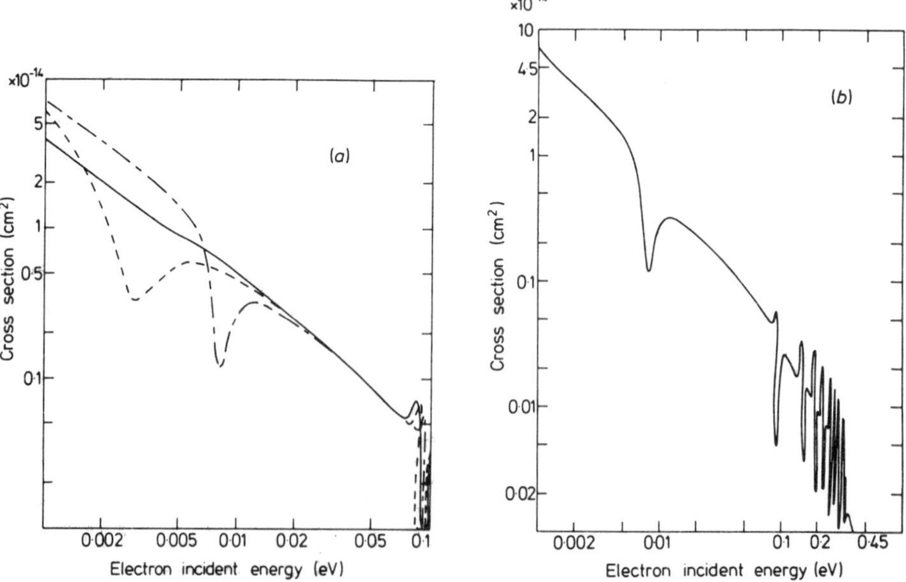

Figure 3. The DR cross section for different values of the quantum defect $\mu(R_e)$ (equation (36)). (*a*) — — —, $\mu(R_e)=0{\cdot}59$; - - - -, $\mu(R_e)=0{\cdot}64$; ———, $\mu(R_e)=0{\cdot}69$. This set of calculations corresponds to $(d\mu/dR)_{R_e}=-0{\cdot}02$, the dissociative curve C of figure 2 and the couplings V^{el} of table 1. (*b*) $\mu(R_e)=0{\cdot}59$ at higher energy.

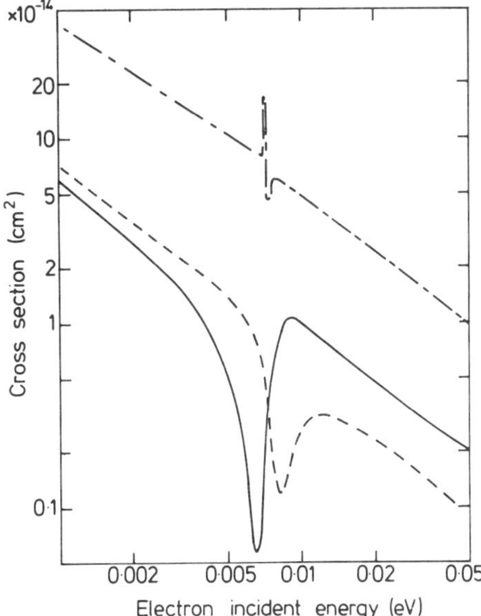

Figure 4. The DR cross section for different dissociative curves. - - - -, curve C of figure 2; — - — -, curve D of figure 2; ———, curve E of figure 2. This set of calculations corresponds to $\mu(R_e)=0{\cdot}59$, $(d\mu/dR)_{R_e}=-0{\cdot}02$ and the couplings V^{el} of table 1.

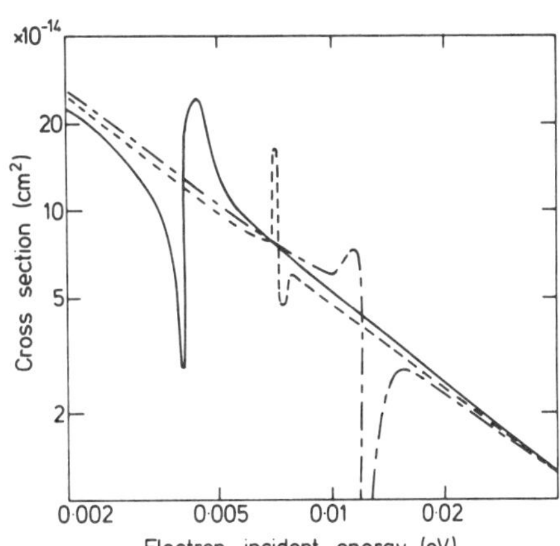

Figure 5. The DR cross section for different values of the slope $(d\mu/dR)_{R_e}$. — - — -, $(d\mu/dR)_{R_e}=-0{\cdot}2$; - - - -, $(d\mu/dR)_{R_e}=-0{\cdot}02$; ———, $(d\mu/dR)_{R_e}=+0{\cdot}1$. This set of calculations corresponds to $\mu(R_e)=0{\cdot}59$, the dissociative curve D of figure 2 and the couplings V^{el} of table 1.

magnitude of the cross section may change considerably from one case to the other, depending on the nuclear overlap $\langle\chi_{v_0^+}|F_d\rangle$ and on the strength of the electronic coupling (recall the expression (32) of the capture cross section).

4.2. Results and discussion

The calculated cross sections display a converging Rydberg series of narrow resonances corresponding to vibrational capture of the electron in successive Rydberg states having $v=1$ (cf figure 1(*b*)). Their positions are approximately given by the formula

$$\epsilon_n=\hbar\omega\Delta v-2(n-\bar{\mu})^{-2} \qquad \text{(in au)} \tag{37}$$

where Δv is the change of vibrational quantum number ($\Delta v=1$ in the tests) and $\bar{\mu}$ represents the quantum defect value averaged on the vibrational motion:

$$\bar{\mu}=\mu(R_e)+\left(\frac{d\mu}{dR}\right)_{R_e}\langle\chi_1|R-R_e|\chi_1\rangle. \tag{38}$$

The first resonance, on which the following tests will be centred, corresponds generally to the Rydberg state $n=7$, as expected for a vibrational frequency of about 2700 cm^{-1}, but for large values of the quantum defect (e.g. $\mu(R_e)=0{\cdot}69$ in figure 3(*a*)) this resonance slips below the threshold and electron capture involves first the $n=8$ Rydberg state. The series terminates when the electron incident energy equals the

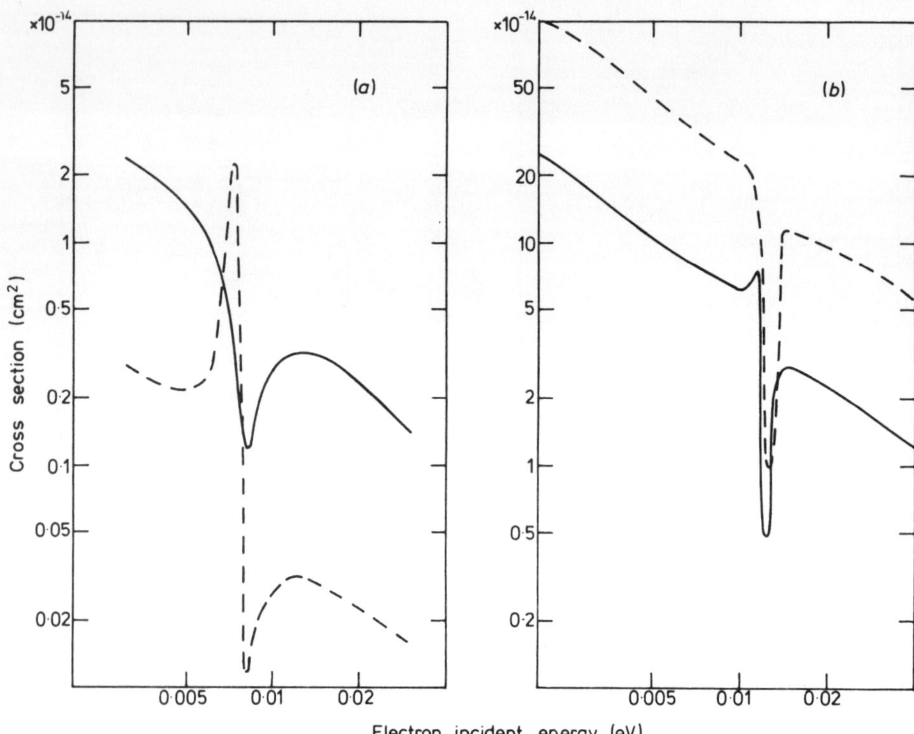

Figure 6. The DR cross section for different values of the electronic couplings. (*a*) ———, V^{el} of table 1; - - - -, the same divided by $\sqrt{10}$. This set of calculations corresponds to $\mu(R_e) = 0{\cdot}59$, $(d\mu/dR)_{R_e} = -0{\cdot}02$ and the dissociative curve C of figure 2. (*b*) ———, V^{el} of table 1; - - - -, the same multiplied by $\sqrt{10}$. This set of calculations corresponds to $\mu(R_e) = 0{\cdot}59$, $(d\mu/dR)_{R_e} = -0{\cdot}2$ and the dissociative curve D of figure 2.

vibrational spacing $\hbar\omega = 0{\cdot}34$ eV after which the $v^+ = 1$ level of the ion state corresponds to an open channel. We remark that the resonance positions depend not only on the magnitude of the quantum defect (figure 3(*a*)) but also on its variations with R (figure 5); this results from the anharmonicity of the potential curve of the ion, which leads to a non-vanishing second term in equation (38).

The resonance widths, and especially their shapes vary considerably depending on the slope of $\mu(R)$, the position of the dissociative curve and the strength of the electronic coupling. This behaviour may be understood qualitatively when the resonances are analysed in terms of the parameters Γ and q introduced by Fano (1961) for describing the resonance profiles in photoionisation (or more generally in any process involving the excitation of an isolated discrete state embedded in a continuum). The resonance width is $\Gamma = 2\pi|V|^2$ where V is the coupling strength between the discrete and continuum configurations. The profile index q may be written approximately as

$$q \simeq \langle\phi|T|i\rangle/\pi V\langle\psi_E|T|i\rangle \tag{39}$$

where ϕ and ψ_E are the discrete and continuum wavefunctions respectively and T is the transition operator (e.g. the dipolar momentum for the case of photoabsorption) responsible for the excitation process out of the initial state $|i\rangle$ (see figure 7(*a*)).

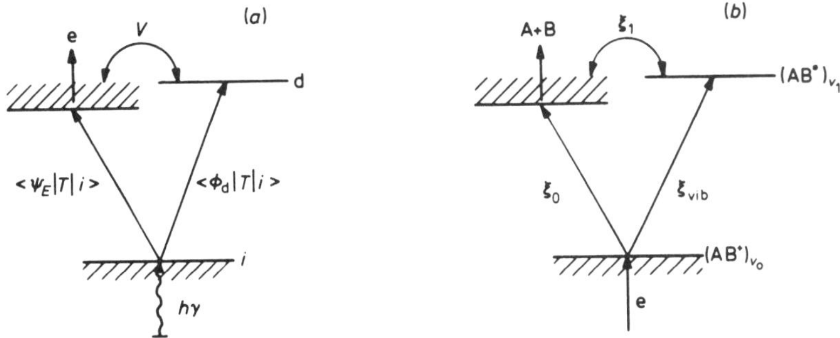

Figure 7. Schematic illustration of (*a*) the photoionisation process, (*b*) the dissociative recombination process.

In the context of dissociative recombination, we may consider the resonant Rydberg state as being embedded in the *dissociative* continuum, both being excited through electron capture (rather than photon capture) by the ion initial state (figure 7(*b*)). In this picture, the resonance width is the predissociation width of the resonant Rydberg level

$$\Gamma = \frac{1}{n^{*3}} \frac{2}{\pi} |\xi_1|^2 \qquad \text{in au} \tag{40}$$

where ξ_1 is defined by equation (30) with $v = 1$ and n^* is the effective quantum number of the Rydberg state. The profile index becomes

$$q \simeq \xi_{vib}/\xi_1\xi_0 \tag{41}$$

where ξ_0 (given by equation (30) with $v = 0$) is the electronic coupling responsible for direct electron capture in the dissociative state and ξ_{vib} stands for the vibronic coupling which allows capture into the Rydberg state.

We recall that the vibronic coupling is not explicitly introduced in the MQDT treatment but is accounted for by a frame transformation, and results from the variations of the quantum defect with R (see § 3.3.1). Herzberg and Jungen (1972) (see also Raoult and Jungen 1980) have shown that in the case of a two-channel vibrational interaction one can nevertheless define a parameter ξ_{vib} as

$$\xi_{vib} = -\pi(\mathrm{d}\mu/\mathrm{d}R)_{R_e}\langle\chi_0|R - R_e|\chi_1\rangle. \tag{42}$$

Here we have chosen the same normalisation as for ξ_1 (in the ratio $\pi n^{*3/2}$ with the interaction parameter of Herzberg and Jungen) since the index q given by equation (41) involves only the ratio ξ_{vib}/ξ_1.

Several features of the tests may be explained by means of the two parameters q and Γ.

(*a*) Firstly, weak variations of the quantum defect ($(\mathrm{d}\mu/\mathrm{d}R)_{R_e} = -0{\cdot}02$ au^{-1}) lead to a *window* resonance ($q \simeq 0$, see figure 3(*a*) and 4) except for the dissociative curve E of figure 2 for which the nuclear overlap $\langle\chi_1|F_d\rangle$ and thus the coupling term ξ_1 are very small. The dip in the cross section appears because at the resonance energy the excited state wavefunction is almost purely that of the discrete level and the vibrational capture is very weak: both direct and indirect processes are thus almost suppressed. One obtains a *dispersion profile* (whose orientation changes with the sign of $\mathrm{d}\mu/\mathrm{d}R$) when the two mechanisms may interfere, i.e. when

(i) the slope $(d\mu/dR)_{R_e}$ is increased (figure 5)
(ii) the Franck–Condon overlap $\langle\xi_1|F_d\rangle$ is decreased (figure 4)
(iii) the electronic coupling is decreased (figure 6(*a*)).

(*b*) Secondly, the evolution of the resonance shape as a function of the principal quantum number deserves special attention: one might expect a nearly constant profile along a series as for atomic photoionisation (Fano and Cooper 1965) because the index q of equation (41) is independent of the effective quantum number n^*. However, in our case the profile is modified by the energy dependence of the *nuclear overlaps* in the coupling terms ξ_0 and ξ_1. For example, the first resonance of figure 3(*b*) ($n = 7$) induces a dip in the cross section, whereas all higher Rydberg resonances appear with a dispersion profile due to the decrease of the nuclear factors in the product $\chi_1 V^{el} F_d$ in the case of the dissociative curve C of figure 2.

(*c*) Thirdly, the resonance width depends as the predissociation width both on the position of the dissociative curve (figure 4) and on the strength of the electronic coupling (figure 6). Nevertheless, figure 5 shows clearly that for the dissociative curve (D) the resonance width varies with the value of $(d\mu/dR)_{R_e}$, i.e. with the strength of the vibrational preionisation. In this case, corresponding to a very small predissociation width (of the order of 10^{-4} eV), the resonance width is certainly not given by equation (40) but rather by the preionisation width Γ_a (equation (35)) represented in the framework of the present discussion by the quantity $2/(n^{*3}\pi)|\xi_{vib}|^2$ (cf equation (42)). This indicates that the resonance width should be given in the general case by the sum of the preionisation and predissociation widths as in the Breit and Wigner formula (34) which in this respect is more correct that the above interpretation in terms of simple Beutler–Fano profiles. The shortcomings of the analogy with the photoionisation process are easily understood when one realises that the usual photoionisation treatment takes account of the excitation process in first order only, neglecting the radiative decay of the discrete and continuum excited state due to higher order couplings by the transition operator *T*. This approximation is generally quite justified in the case of photoionisation because photon emission (fluorescence) is usually several orders of magnitude smaller than electron emission (note however that attempts to avoid this approximation have been made by Nitzan (1974) and by Armstrong *et al* (1978)). In the case of DR, the fluorescence is replaced by electron emission and ionisation by molecular dissociation (figure 7). Both processes may occur with comparable decay times: for the continuum state a survival factor (see § 3.3.4) smaller than one results while the discrete state may have a width which is larger than expected for predissociation alone, as in the case of the dissociative curve D. The electronic continuum, unlike the photon continuum, must be treated on the same footing as the dissociation continuum: this principle underlies our MQDT treatment which takes full account of autoionisation, not only for the discrete level but also for the dissociative state. Finally, we insist once more on the basic difference between the MQDT treatment and the traditional CM treatment which underlies the (Γ, q) parametrisation: instead of treating each resonant state individually and assuming that all such states are well separated from each other, MQDT treats the Rydberg series responsible for the resonance features as a whole, together with the adjoining continuum.

4.3. The experimental situation

Although DR has been studied in a great number of experiments, very little evidence for the presence of resonance structure has emerged so far: to our knowledge, the only resonances reported are those obtained in the merged beam experiments on H_2^+ and H_3^+ (McGowan *et al* 1976, Auerbach *et al* 1977). However, the author has been informed by J W McGowan (1979, private communication) that difficulties have been encountered in reproducing the resonances in the H_2^+ DR cross section; even in this case further work is therefore needed to clarify the situation.

Several reasons may explain the apparent lack of resonance structure in previous experiments which are now discussed.

(i) First of all, in order to detect such structures one should have an experimental resolution of the order of a few meV at one's disposal, which is rarely the case at present. Among the various techniques, the merged beam method certainly offers an advantage, especially at low energy, over the trapped ion technique (see e.g. Waals and Dunn 1974) or the inclined beam method (Peart and Dolder 1971) (we do not mention plasma afterglow experiments because they do not measure DR cross sections but only reaction rate coefficients).

(ii) The molecules which have been studied so far in most detail are those for which the DR process is strong. However, they may not be the most favourable cases for detecting resonance structure: a large Franck–Condon overlap of the dissociative wavefunction with the initial level v_0 of the ion corresponds to a much smaller overlap with the level v_0+1 at low energy (recall the tests relative to the curve D of figure 2). The predissociation width of the intermediate Rydberg level will then be very small and the resonance may escape detection. On the other hand, if the repulsive curve does not intersect the v_0 vibrational level within its Franck–Condon region (as for example the curves C and E of figure 2) a large overlap with the v_0+1 level may still be possible and yield broad resonances. These are likely to appear as *dips* in the cross section especially when the energy is low (see figures 3 and 4), except when the quantum defect varies strongly with R.

(iii) A further experimental problem lies in the selection of one or at most a few vibrational levels for the molecular ions: if many vibrational levels are populated, the resonance structures corresponding to alternative initial levels v_0 will appear superimposed onto each other and yield an averaged smooth cross section profile unless the resolution is very high. The trapped ion method is in this respect more suitable than the merged beam technique because the ions may spend a period in the trap before they recombine so that vibrational decay can occur (Waals and Dunn 1974, Mathur *et al* 1979). The most efficient experimental way for studying the influence of the Rydberg states on DR would be perhaps to determine the cross sections for the inverse process, the associative ionisation (equation (27)), by using merged atomic beams and measuring the energy of the escaping electron: one might thus reconcile a very high resolution (up to $0{\cdot}6\times10^{-3}$ eV in the experiments of Poulaert *et al* (1979)) with the possibility of selecting a single vibrational level of the residual ion. At the present time, spectroscopic studies which display the competition between autoionisation and predissociation seem to be the best field of application for the present theory because the energy resolution currently achieved in spectroscopy is very high.

More generally, the growing experimental ability to excite molecules at any chosen energy (e.g. by using synchrotron radiation) and subsequently measuring its decay in several channels simultaneously (fluorescence, ionisation and dissociation—see for example Guyon (1979)) calls for a unified theoretical treatment of the alternative decay modes for a given molecular complex. Such a global treatment of several types of continua is likely to rest on the same physical and theoretical basis as the MQDT approach used in this work. Semi-empirical methods may be combined with *ab initio* calculations restricted to the reaction zone (e.g. R-matrix calculations). The present work is intended as a first step in this direction.

Acknowledgments

Foremost, thanks are due to Dr Ch Jungen who introduced me to multichannel quantum defect theory. I can only inadequately express my deep gratitude for his constant support during all stages of this work and especially during the preparation of the manuscript. In addition I am indebted to him and to Maurice Raoult for having supplied their MQDT subroutine.

I thank also very sincerely Dr H Lefebvre-Brion and Dr D Dill for the strong interest they have shown in this work and for the many discussions which were very decisive for its completion.

Appendix. First-order analytical expression for direct dissociative recombination

When only open vibrational channels are included in the calculation, we have to treat a real scattering problem with $N = N_{\mathrm{el}} + N_{\mathrm{d}}$ channels. Expansion (18) may now be directly used for determining the running waves with asymptotic behaviour (20). The conditions on the incoming waves yield the linear system

$$\sum_\alpha A_\alpha(\mathscr{C}_{v^+\alpha} - \mathrm{i}\mathscr{S}_{v^+\alpha}) = \delta_{v^+v_0^+} \qquad v^+ = 0, 1, \ldots, N_{\mathrm{el}}$$

$$\sum_\alpha A_\alpha(\mathscr{C}_{v^+\alpha} - \mathrm{i}\mathscr{S}_{\mathrm{d}_i\alpha}) = 0 \qquad i = 1, 2, \ldots, N_{\mathrm{d}}$$

which may be written in the form (using equations (19))†

$$\sum_\alpha A_\alpha\Big(\sum_v \langle v^+|\exp[-\mathrm{i}(\pi\mu(R) + \eta_\alpha)]|v\rangle U_{v\alpha}\Big) = \delta_{v^+v_0^+}$$

$$\sum_\alpha A_\alpha U_{\mathrm{d}_i\alpha}\exp(-\mathrm{i}\eta_\alpha) = 0.$$

The solutions are

$$A_\alpha = \exp(\mathrm{i}\eta_\alpha)(\tilde{U}^{-1})_{\alpha v_0^+} \qquad \alpha = 1, 2, \ldots, N$$

where $\tilde{U}$ is the unitary matrix defined by the relations

$$\tilde{U}_{v^+\alpha} = \sum_v \langle v^+|\exp(-\mathrm{i}\pi\mu(R))|v\rangle\, U_{v\alpha}$$

$$\tilde{\mathrm{U}}_{\mathrm{d}_i\alpha} = U_{\mathrm{d}_i\alpha}.$$

The coefficient of the outgoing wave in the dissociative channel d_i (equation (23*b*)) is thus

$$S_{\mathrm{d}_iv_0^+} = \sum_\alpha U_{\mathrm{d}_i\alpha}\exp(2\mathrm{i}\eta_\alpha)\tilde{U}^{-1}_{\alpha v_0^+} = \sum_{\alpha,v} U_{\mathrm{d}_i\alpha}\exp(2\mathrm{i}\eta_\alpha)U^*_{\alpha v}\langle v|\exp(\mathrm{i}\pi\mu(R))|v_0^+\rangle. \qquad \text{(A.1)}$$

† Although the sets of wavefunctions χ_v and χ_{v^+} are identical in the calculations (see § 3.3.1), we keep distinct notations for vibrational wavefunctions in regions A and B in order to clarify the equations.

One also gets the coefficient of the outgoing wave in each ionisation channel v^+ (equation (23a)) from which the vibrational excitation cross section would be calculated

$$S_{v^+v_0^+} = \sum_{\alpha,v,v'} \langle v^+| \exp(\mathrm{i}\pi\mu(R))|v'\rangle U_{v'\alpha} \exp(2\mathrm{i}\eta_\alpha) U^*_{\alpha v} \langle v| \exp(\mathrm{i}\pi\mu(R))|v_0^+\rangle. \tag{A.2}$$

In the expressions (A.1) and (A.2), the respective contributions of electronic and vibronic interactions appear clearly. They can be reduced to explicit expressions in cases where the matrix U and the eigenphases η_α may be expressed analytically in first order of perturbation. We consider the following two simple cases.

(a) *Case of a single dissociative channel* ($N_\mathrm{d} = 1$). The first-order K matrix (including the factor π) is

$$\pi K = \begin{bmatrix} 0 & \xi_0 & \xi_1 & \cdots & \xi_{N_\mathrm{el}-1} \\ \xi_0 & 0 & & 0 & \\ \xi_1 & & \ddots & & \\ \vdots & 0 & & & \\ \xi_{N_\mathrm{el}-1} & & & & 0 \end{bmatrix} \tag{A.3}$$

with $\xi_v = \pi \int \chi_v(R) V_\mathrm{el}(R) F_\mathrm{d}(R)\,\mathrm{d}R^*$ (cf equation (14)). Its eigenvalues are $\lambda = 0$ (multiplicity $N_\mathrm{el} - 1$) and $\lambda = \pm\xi$, with $\xi^2 = \Sigma_v\, \xi_v^2$. The corresponding eigenvectors form the unitary matrix

$$U = \frac{1}{\sqrt{2}\,\xi} \begin{bmatrix} 0 & 0 & \cdots & 0 & \xi & -\xi \\ & & & & \xi_0 & \xi_0 \\ & A & & & \vdots & \vdots \\ & & & & \xi_{N_\mathrm{el}-1} & \xi_{N_\mathrm{el}-1} \end{bmatrix}.$$

The matrix elements $U_{v\alpha}$ denoted collectively by A constitute $N_\mathrm{el} - 1$ orthogonal columns satisfying the relations

$$\sum_v U_{v\alpha}\xi_v = 0 \qquad \alpha = 1, 2, \ldots, N_\mathrm{el} - 1.$$

The S-matrix elements (A.1) become here

$$S_{\mathrm{d}v_0^+} = (\mathrm{i}/\xi) \sin 2\eta \sum_v \xi_v \langle v| \exp(\mathrm{i}\pi\mu(R))|v_0^+\rangle \qquad \eta = -\tan^{-1}\xi$$

and the DR cross section (equation (24)) is

$$\sigma_{\mathrm{d}v_0^+} = \frac{\pi}{k^2}\frac{r}{2} 4 \frac{(\Sigma_v\, \xi_v \langle v| \cos \pi\mu(R)|v_0^+\rangle)^2 + (\Sigma_v\, \xi_v \langle v| \sin \pi\mu(R)|v_0^+\rangle)^2}{[1 + \Sigma_v\, (\xi_v)^2]^2}. \tag{A.4}$$

Neglecting the R dependence of the quantum defect, we obtain the simple expression

$$\sigma_{\mathrm{d}v_0^+} = \frac{\pi}{k^2}\frac{r}{2}\frac{4(\xi_{v_0^+})^2}{[1 + \Sigma_v\, (\xi_v)^2]^2} \tag{A.5}$$

which is compared with the treatment by Lee in § 3.3.3 (see equation (31)) and also permits evaluation of the survival factor (see equation (33)). If many vibrational channels are open we may make use of the closure relation and write the survival factor as

$$S = \left(1 + \pi^2 \int F_\mathrm{d}(R) V_\mathrm{el}^2(R) F_\mathrm{d}(R)\,\mathrm{d}R\right)^{-2} \simeq \left(1 + \pi^2 \bar{V}_\mathrm{el}^2 \int_{R_\mathrm{T}}^{R_\mathrm{c}} F_\mathrm{d}^2(R)\,\mathrm{d}R\right)^{-2} \tag{A.6}$$

where the electronic interaction has been approximated by the step function

$$V_{el}(R) = \bar{V}_{el} \qquad R_T \leqslant R \leqslant R_c$$
$$= 0 \qquad R_c < R.$$

The cut off radius R_c coincides with the high-energy reaction zone dimension R_0 defined by equation (14) for large values of the vibrational quantum number v. R_T is the turning point for the radial motion in the repulsive potential U_d, defined such that the local momentum (in atomic units)

$$\kappa(E, R) = \sqrt{2M(E - U_d(R))} \tag{A.7}$$

vanishes for $R = R_R$ (in (A.6) we neglect the centrifugal energy term according to our approximation of neglecting the molecular rotation). A physical interpretation of the survival factor (A.6) may be given in the quasiclassical framework, using the WKB wavefunction

$$F_d(\kappa, R) = \left(\frac{2M}{\pi\kappa(R)}\right)^{1/2} \sin\left(\int_{R_T}^{R} \kappa(R')\,dR' + \frac{\pi}{4}\right).$$

The integral at the denominator of (A.6) becomes

$$I = \int_{R_T}^{R_c} \frac{2M}{\pi\kappa(R)} \sin^2\left(\int_{R_T}^{R} \kappa(R')\,dR' + \frac{\pi}{4}\right) dR = \int_{R_T}^{R_c} \frac{M}{\pi} \frac{1 + \sin(2\int_{R_T}^{R} \kappa(R')\,dR')}{\kappa(R)}\,dR.$$

The rapidly oscillating sinusoidal term in the integrand may be neglected so that one gets

$$I \simeq \pi^{-1} \int_{R_T}^{R_c} (M/\kappa(R))\,dR \simeq \pi^{-1} \int_{R_T}^{R_c} (dR/dt)\,dt \simeq \pi^{-1}\tau_s$$

where τ_s is the stabilisation time beyond which autoionisation in $AB^+ + e$ becomes impossible.

Let us denote by $\Gamma_a = 2\pi|\bar{V}^{el}|^2$ the approximate electronic width for autoionisation and by $\tau_a = \Gamma_a^{-1}$ the mean autoionisation time. Expression (A.6) may then be written as

$$S = (1 + \tfrac{1}{2}\Gamma_a\tau_s)^{-2} = (1 + \tfrac{1}{2}\tau_s/\tau_a)^{-2}. \tag{A.7}$$

For weak autoionisation ($\tau_a \gg \tau_s$) we obtain the approximate expression

$$S \simeq (1 + \Gamma_a\tau_s)^{-1} = \tau_a/(\tau_a + \tau_s) \tag{A.8}$$

which was suggested by Bates (1950). Another approximate form in the same limit ($\tau_a \gg \tau_s$) is

$$S \simeq 1 - \Gamma_a\tau_s$$

which coincides with the first-order expansion of the survival factor

$$S = \exp\left(-\int \Gamma_a(R)\,dt\right) \simeq \exp(-\Gamma_a\tau_s) \tag{A.9}$$

obtained by Bardsley *et al* (1964) for the dissociative attachment process at high energy. Expressions (A.7) and (A.9) differ notably when Γ_a increases, but it must be noted that both of them require Γ_a to be small in order to be valid: (A.7) has been obtained in a first-order perturbation treatment of the capture (or autoionisation) interaction and

(A.9) rests on a first-order development of the quantity $(E-U_d(R)-i\Gamma_a(R))^{1/2}$ with respect to $\Gamma_a/(E-U_d(R))$ of the order of $\Gamma_a\tau_s$.

(*b*) *Case of a single vibrational ionisation channel.* The first-order K matrix is now

$$\pi K=\begin{bmatrix} 0 & 0 & \dots & 0 & \xi_{d_1} \\ 0 & & & & \xi_{d_2} \\ \vdots & & & & \vdots \\ & & & & \xi_{N_d} \\ \xi_{d_1} & \xi_{d_2} & \dots & \xi_{N_d} & 0 \end{bmatrix} \qquad \xi_{d_i}=\pi\langle\chi_v|V_{el}(R)|F_{d_i}\rangle.$$

Its eigenvalues are the same as for the matrix (A.3), with the new definition of $\xi=(\Sigma_i\,\xi_{d_i}^2)^{1/2}$. Proceeding as above, we arrive easily at the partial cross section for dissociation in the channel d_i:

$$\sigma_{d_i}=\frac{\pi}{k^2}\frac{r_i}{2}\frac{4\xi_{d_i}^2}{(1+\xi^2)^2}. \tag{A.10}$$

The total dissociative recombination cross section is then

$$\sigma=\frac{2\pi}{k^2}\frac{\Sigma_i\,r_i(\xi_{d_i})^2}{(1+\xi^2)^2} \tag{A.11}$$

whereas a separate treatment for each dissociation channel would lead to the expression

$$\sigma=\frac{2\pi}{k^2}\sum_i\frac{r_i(\xi_{d_i})^2}{[1+(\xi_{d_i})^2]^2}. \tag{A.12}$$

Note added in proof. It has been pointed out to the author by Dr T F O'Malley that *broad* resonances due to high vibrational levels of very low Rydberg states (e.g. $n=3$) may appear in the DR cross sections. These will be in window form similar to those in the case of very low vibronic capture ($\Delta v^+\gg 1$, see § 4.2). We have verified that they appear in our MQDT model calculations as soon as the corresponding vibrational channels are included.

References

Altick P L and Moore E N 1966 *Phys. Rev.* **147** A59

Armstrong L, Theodosiou C E and Wall M J 1978 *Phys. Rev.* A **18** 2538

Atabek O and Lefebvre 1972 *J. Chem. Phys. Lett.* **17** 167

Auerbach D, Cacak R, Caudano R, Gaily T D, Keyser J, McGowan J W, Mitchell J B A and Wilk S J 1977 *J. Phys. B: Atom. Molec. Phys.* **10** 3797

Bardsley J N 1968 *J. Phys. B: Atom. Molec. Phys.* **1** 349, 365

Bardsley J N, Herzenberg A and Mandl F 1964 *Proc. 3rd Int. Conf. on Physics of Electronic and Atomic Collisions* (Amsterdam: North-Holland) p 415

Bates D R 1950 *Phys. Rev.* **78** 492

Berry R S 1966 *J. Chem. Phys.* **45** 1228

Berry R S and Nielsen S E 1970 *Phys. Rev.* A **1** 383

Bottcher C 1976 *J. Phys. B: Atom. Molec. Phys.* **9** 2899

Calogero F 1967 *Variable Phase Approach to Potential Scattering* (New York: Academic)

Chang E S and Fano U 1972 *Phys. Rev.* A **1** 173

Clark C W and Greene C H 1979 *Proc. 11th Conf. on Physics of Electronic and Atomic Collisions* (Kuoto: Society for Atomic Collision Research) Abstracts p 244

Duzy C and Berry R S 1976 *J. Chem. Phys.* **64** 2421, 2431

Fano U 1961 *Phys. Rev.* **124** 1866
—— 1970 *Phys. Rev.* A **2** 353
—— 1975 *J. Opt. Soc. Am.* **65** 979
—— 1977 *Proc. 10th Int. Conf. on Physics of Electronic and Atomic Collisions* Invited papers and progress reports, ed G. Watel (Amsterdam: North-Holland)
—— 1978 *Phys. Rev.* A **17** 93
Fano U and Cooper J W 1965 *Phys. Rev.* **137** 1364
Giusti A 1979 *Thèse de Doctorat d'Etat* Université Paris-Nord
Giusti A and Lefebvre-Brion H 1980 *Chem. Phys. Lett.* to be published
Green S, Bagus P S, Liu B, McLean A D and Yoshimine M 1972 *Phys. Rev.* A **5** 1614
Greene C H, Fano U and Strinati G 1979 *Phys. Rev.* A 1485
Guyon P M 1979 *Proc. 4th EPS General Conference* (York) p 489
Herzberg G and Jungen C 1972 *J. Molec. Spectrosc.* **41** 425
Jungen C 1970 *J. Chem. Phys.* **53** 4168
Jungen C and Atabek O 1977 *J. Chem. Phys.* **66** 5584
Jungen C and Dill D 1980 *J. Chem. Phys.* in press
Lee C M 1975 *Phys. Rev.* A **11** 1692
—— 1977 *Phys. Rev.* A **16** 109
Lee C M and Lu K T 1973 *Phys. Rev.* A **8** 1241
Lee L C, Carlson R W, Judge D L and Ogawa M 1974 *J. Chem. Phys.* **61** 3261
—— 1975 *J. Chem. Phys.* **63** 3987
Levine R D 1969 *Quantum Mechanics of Molecular Rate Processes* (Oxford: Clarendon)
Mathur D, Hasted J B and Khan S U 1979 *J. Phys. B: Atom. Molec. Phys.* **12** 2043
McGowan J W, Caudano R and Keyser J 1976 *Phys. Rev. Lett.* **36** 447
McGowan J W, Mul P M, D'Angelo V S, Mitchell J B A, Defrance P and Froelich H R 1979 *Phys. Rev. Letters* **42** 373
Miescher E, Lee Y T and Gürtler P 1978 *J. Chem. Phys.* **68** 2753
Neynaber R H and Tang S Y 1979 *Proc. 11th Int. Conf. on Physics of Electronic and Atomic Collisions* (Kyoto: Society for Atomic Collision Research) Abstracts p 874
Nikitin E E 1974 *Theory of Elementary Atomic and Molecular Processes in Gases* (Oxford: Clarendon)
Nitzan A 1974 *Molec. Phys.* **27** 65
O'Malley T F 1967 *Phys. Rev.* **162** 98
Parlant G and Fiquet-Fayard F 1974 *Proc. 9th Int. Conf. of Electronic and Atomic Collisions* (Seattle: University of Washington) Abstracts p 279
Peart B and Dolder K T 1971 *J. Phys. B: Atom. Molec. Phys.* **4** 1496
Poulaert G, Brouillard F, Claeys W, Defrance P and McGowan J W 1979 *Proc. 11th Int. Conf. on Physics of Electronic and Atomic Collisions* (Kyoto: Society for Atomic Collision Research) abstracts p 876
Poulaert G, Brouillard F, Claeys W, McGowan J W and Van Vassenhove G 1978 *J. Phys. B: Atom. Molec. Phys.* **11** L671
Raoult M and Jungen C 1980 *J. Chem. Phys.* in press
Raseev G, Giusti A and Lefebvre-Brion H 1978 *J. Phys. B: Atom. Molec. Phys.* **11** 2735
Ringer G and Gentry W R 1979 *J. Chem. Phys.* **71** 1902
Schneider B I, Le Dourneuf M and Burke P G 1979 *J. Phys. B: Atom. Molec. Phys.* **12** L365
Seaton M J 1958 *Mon. Not. R. Astron. Soc.* **118** 504
—— 1966 *Proc. Phys. Soc.* **88** 801, 815
Sidis V and Lefebvre-Brion H 1971 *J. Phys. B: Atom. Molec. Phys.* **4** 1040
Vo Ky Lan and Le Dourneuf M 1980 *Proc. 11th Int. Conf. on Physics of Electronic and Atomic Collisions* Invited papers and progress reports, ed N Odo and N Takayanagi (Amsterdam: North Holland) p 751
Waals F L and Dunn G H 1974 *Phys. Today* **27** 30

1992 *J. Phys. B: At. Mol. Opt. Phys.* **25** 5257

Corrigenda

The role of Rydberg states in H_2^+ dissociative recombination with slow electrons

I F Schneider, O Dulieu and A Giusti-Suzor 1991 *J. Phys. B: At. Mol. Opt. Phys.* **24** L289

Two errors in the figure labelling have been noted.

In figure 2(a), the range of internuclear distance (abscissa R) for the quantum defects should read between 0 and 10 au, instead of 0 and 4 au which holds only for the electronic interactions in figure 2(b). All the calculations have been performed with the correct values of the quantum defects.

In figure 3(a), the resonance labels 8s and 8d should be inverted, as they are in figure 3(b) below.

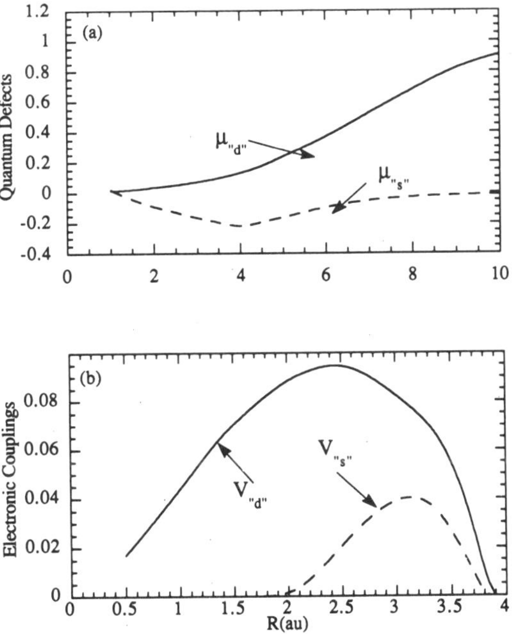

Figure 2. (corrected)

Non-coplanar (e, 2e) reactions

S Jetzke and F H M Faisal 1992 *J. Phys. B: At. Mol. Opt. Phys.* **25** L269

During the production process, figures 1(a) and 2(a) were inadvertently interchanged. Figure labelled 1(a) should therefore be 2(a), and the figure labelled 2(a) should be 1(a). The figure captions are correct.

0953-4075/92/235257+01$07.50

1991 *J. Phys. B: At. Mol. Opt. Phys.* **24** L289–97

LETTER TO THE EDITOR

The role of Rydberg states in H_2^+ dissociative recombination with slow electrons

I F Schneider†§, O Dulieu† and A Giusti-Suzor‖

† Laboratoire des Collisions Atomiques et Moléculaires (URA 281 du CNRS), Bâtiment 351, Université Paris-Sud, 91405 Orsay, France
‡ Laboratoire de Chimie-Physique (URA 176 du CNRS), Université Paris 6, 11 rue Pierre et Marie Curie, 75235 Paris Cedex 5, France

Received 16 April 1991, in final form 9 May 1991

Abstract. The cross section for H_2^+ dissociative recombination with slow electrons is calculated using the multichannel quantum defect theory, with the set of molecular data used for the recent study of H + H* associative ionization by Urbain *et al.* Excited vibrational levels of $s\sigma_g$ and $d\sigma_g$ Rydberg states show up as broad resonances, mostly dips, in satisfying agreement with the vibrationally selected measurements of Mitchell and his collaborators. We analyse the mechanisms for resonance formation and discuss the adequacy of the theoretical treatment.

The role of Rydberg states in dissociative recombination (DR) has been questioned since the first low energy measurements of the process

$$H_2^+ + e \rightarrow H + H^* \tag{1}$$

by McGowan *et al* (1976), using the merged beam technique. The numerous structures found in the cross section were attributed to the temporary formation of bound Rydberg states which can decay by autoionization and predissociation:

$$H_2^+ + e \rightarrow H_2^*(\text{Rydberg}) \begin{matrix} \nearrow H_2^+ + e \\ \searrow H + H^*. \end{matrix} \tag{2}$$

This 'indirect' process was first predicted by Bardsley (1968), who evaluated its contribution with a Breit–Wigner formula leading to a Lorentzian resonance shape. What this first experiment revealed is that the indirect process strongly interferes with the direct one (reaction (1)) and leads to more complex structures including 'window' resonances.

This experimental result prompted a unified treatment of direct and indirect processes. The multichannel quantum defect theory (MQDT) (Seaton 1983, Greene and Jungen 1985) seemed the most suitable approach to account for the numerous series of resonances due to excited vibrational levels of Rydberg molecular states. Inclusion of dissociative channels in this theory was achieved by Giusti (1980) and first applied to H_2^+ dissociative recombination (Giusti-Suzor *et al* 1983). Both interference between direct and indirect recombination processes and overlapping effects between closely

§ Permanent address: Institute of Atomic Physics Bucarest, Laser Department, 76900 Màgurele, Romania.
‖ Also at: Laboratoire de Photophysique Moléculaire, Bâtiment 213, Université Paris-Sud, 91405 Orsay, France.

0953-4075/91/120289+09$03.50

lying resonances were taken into account, leading to a complex resonance structure in the cross section, mostly in the form of dips.

Two other calculations were performed in the following years. Nakashima *et al* (1987) used the MQDT approach with an improved set of molecular data. Hickman (1987) used the 'configuration mixing' formalism of Bardsley (1968). Autoionizing Rydberg levels were included but treated within an 'independent resonance approximation' that neglects the overlapping effects. It was not clear at that time if the large disagreement (one order of magnitude) between his results and the MQDT ones, for the same set of molecular data, were due to this approximation or to the first-order nature, with respect to the electronic interaction between the dissociative and Rydberg states, of the MQDT calculations. Comparison with experiment was not decisive, mostly because several vibrational levels were populated in the H_2^+ beam, with unknown proportions.

Important improvements have been achieved recently, on three fronts.

(i) Hus *et al* (1988) performed a new merged beam experiment using a radio-frequency trap which confines the ions long enough before they are extracted for almost all the excited vibrational levels to decay. The resolution and energy calibration have also been upgraded recently (Mitchell 1990) and the last results now seem adequate for comparison with theoretical calculations.

(ii) A new set of data (quantum defects, electronic coupling, dissociative potential curve) has been extracted (Urbain 1990) from recent molecular calculations. They have been used for the calculation of the cross section for the associative ionization (AI) process:

$$H + H^*(n = 2) \rightarrow H_2^+ + e \tag{3}$$

the inverse of the DR process, leading to a good agreement with experimental values (Urbain 1990, Urbain *et al* 1991).

(iii) Finally, the MQDT treatment can now include second-order effects of the Rydberg–valence electronic interaction (Hickman 1989, Guberman and Giusti-Suzor 1991).

The purpose of the present calculations is not to obtain a semi-empirical determination of the molecular parameters. Rather, we start from the set of data used for the inverse process (3), very slightly adjusted to get the right number of resonances in the energy range of the latest experiment (20–100 meV). From the comparison between the theoretical and measured cross sections, we are first able to assign a dominant character to each of the observed resonances. In particular, we demonstrate the importance of the 's' series of Rydberg states, overlooked in the previous DR calculations. Then, we can analyse the different mechanisms for electron capture, direct and indirect, electronic and vibronic, and their influence on the DR cross section. Finally, we discuss the implications for related processes.

The dissociative recombination of H_2^+ with slow electrons is dominated by the first doubly excited state $(2p\sigma_u)^2$ of the neutral molecule (figure 1). We neglect rotational coupling such that only ${}^1\Sigma_g^+$ states will be considered. In a quasidiabatic representation (Sidis and Lefebvre-Brion 1971) this repulsive state defines the *dissociation channel* in which the incoming electron is captured via the dielectronic coupling. Below threshold, the doubly excited state interacts also with the monoexcited Rydberg states of the same symmetry, which thus undergo electronic predissociation. In addition, these Rydberg states are coupled together, and also to the electronic continuum, by vibronic interactions.

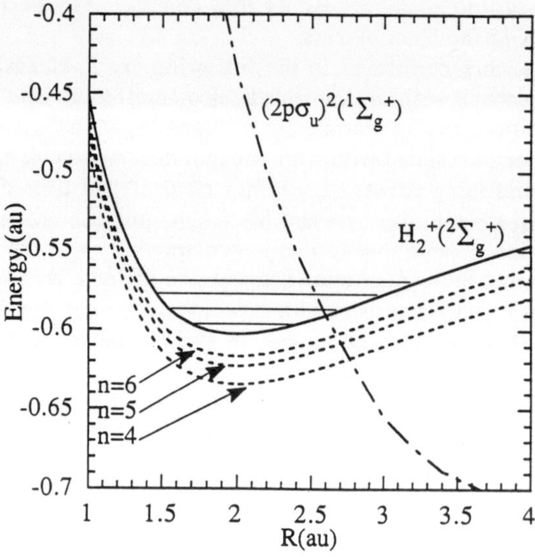

Figure 1. Diabatic potential energy curves involved in the low-energy dissociative recombination of H_2^+. The broken curves symbolize a few $(1s\sigma_g nl\sigma_g)\ ^1\Sigma_g^+$ Rydberg states, with $l=0$ or 2.

Molecular MQDT organizes the monoexcited states in sets of *ionization channels.* Each of them is characterized by the angular quantum numbers $l\lambda$ of the external electron, either in a Rydberg or a continuum orbital. Only $l=0$ and 2 ('s' and 'd') orbitals with σ_g symmetry (which will be understood in most of the following formulae) have been shown to contribute. These channels are further defined by their ionization thresholds, which are the successive vibrational levels v of the ion ground state. Finally, each ionization channel corresponds to a quantum defect $\mu_l(R)$ (or phaseshift $\pi\mu_l$ in the continuum) and scaled electronic coupling $V_l(R)$ with the doubly excited state (Giusti 1980). Resulting from short-range interactions, these quantities vary slowly with the external electron energy and will be considered as energy independent.

Although little is known from spectroscopy about the $^1\Sigma_g^+$ molecular states in H_2, they have been the subject of numerous calculations. First, *ab initio* calculations of the low-lying states (Wolniewicz and Dressler 1977, 1979) provide high accuracy adiabatic potential curves. Ross and Jungen (1987) have extracted diabatic parameters using a global diabatization method based on MQDT. Second, several electron–H_2^+ scattering calculations (Hazi *et al* 1983, Schneider and Collins 1983, Takagi and Nakamura 1983, Shimamura *et al* 1990) yield resonant phaseshifts from which the energy and electronic width of the doubly excited state may be deduced, together with diabatic quantum defects. The set of molecular data used for the study of H+H* associative ionization has been determined from the comparison between these various calculations (Urbain 1990). The scattering results for the doubly excited state are restricted to $R \leqslant 2.6\ a_0$, the crossing point between the $(2p\sigma_u)^2$ potential curve and that of the ion ground state (figure 1). In contrast, the MQDT diabatization of *ab initio* calculations is most precise in the region of the avoided crossings in the double-well states EG, GK and H$\bar{\text{H}}$, that is for $R > 2.5\ a_0$. The two sets of data join smoothly in the intermediate region, yielding the quantum defects and electronic couplings shown on figure 2, together with the $(2p\sigma_u)^2$ dissociative curve of figure 1. The restricted range of the 's' interaction corresponds to the observation that the s-wave practically does not interact with the doubly excited configuration in the continuum ($R < 2.6\ a_0$), while the adiabatic $(1s\sigma_g ns\sigma_g)$ Rydberg states show marked avoided crossings at larger distance.

With these molecular data, we have performed a MQDT calculation of the cross section for reaction (1), with indirect process (2) automatically included. The successive steps of the treatment have been recently reviewed (Guberman and Giusti-Suzor 1991). Briefly, the vibronic interaction between the ionization channels is included through the channel mixing coefficients:

$$C_{lv,l'v'} = \delta_{ll'} \int \chi_v(R) \cos \pi\mu_l(R) \chi_{v'}(R)\, \mathrm{d}R \tag{4a}$$

$$S_{lv,l'v'} = \delta_{ll'} \int \chi_v(R) \sin \pi\mu_l(R) \chi_{v'}(R)\, \mathrm{d}R \tag{4b}$$

where $\chi_v(R)$ is the radial wavefunction for the nuclear motion in the ion vibrational level v. The electronic K-matrix is then built, just equal at first order to the off-diagonal interaction matrix with dimensionless elements:

$$V_{lv,\mathrm{d}} = \int \chi_v(R) V_l(R) F_\mathrm{d}(E, R)\, \mathrm{d}R \tag{5}$$

resulting from nuclear integration of the electronic couplings $V_l(R)$. The continuum nuclear wavefunction F_d in the dissociation potential of the $(2p\sigma_u)^2$ state is normalized to energy. Second-order terms in the Born expansion of the Lippman–Schwinger equation for the K matrix introduce indirect electronic couplings between the ionization

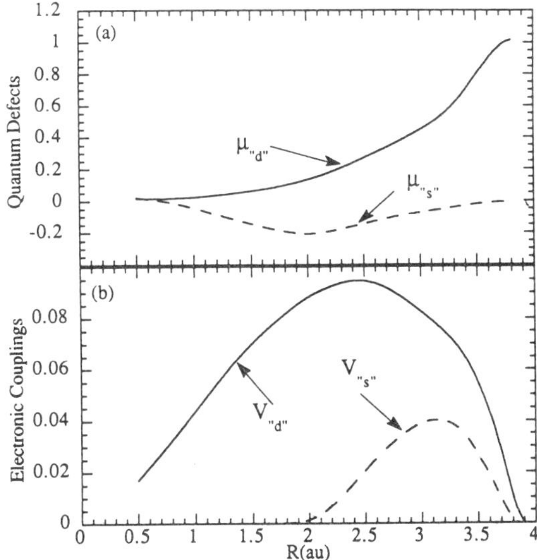

Figure 2. Quantum defects (*a*) and electronic couplings (*b*) used in the calculations; 's' and 'd' stand for the $(1s\sigma_g ns\sigma_g)\ ^1\Sigma_g^+$ and $(1s\sigma_g nd\sigma_g)\ ^1\Sigma_g^+$ Rydberg series.

channels, measured by the matrix elements

$$K_{lv,l'v'}=\frac{1}{W}\iint \chi_v(R)V_l(R)F_d(E,R_<)G_d(E,R_>)V_l(R')\chi_v(R')\,dR\,dR' \tag{6}$$

where G_d is an irregular solution of the nuclear Schrödinger equation in the dissociative potential, lagging in phase by $\pi/2$ with respect to F_d and W is the Wronskian of the pair (F_d, G_d). These second-order terms were neglected in the previous MQDT calculations, and their possible importance was first pointed out by Hickman (1987, 1989). Note that l-mixing between s and d series or partial waves, neglected in the molecular data, is introduced here through this indirect channel coupling.

Vibronic and electronic interactions are combined in a generalized scattering matrix that encompasses ionization and dissociation channels, open and closed, and varies smoothly with energy. Finally, this matrix is contracted into the usual scattering matrix S reduced to open channels (Seaton 1983, Nakashima *et al* 1987). This step introduces rapid energy variations in the vicinity of the bound Rydberg levels, resulting in the resonant behaviour observed for the DR cross section:

$$\sigma=\frac{\pi}{2\varepsilon}\frac{1}{4}\sum_l |S_{lv_0,d}|^2 \tag{7}$$

with v_0 the ion initial vibrational level (assumed to be in the lowest rotational level). The factor $\frac{1}{4}$ is the multiplicity ratio between the final ${}^1\Sigma_g^+$ state and the $({}^2\Sigma_g^+ + e)$ composite initial state, and ε is the kinetic energy of the colliding electron (in atomic units).

Our results are shown in figure 3 for $v_0=0$ and 1, together with the cross sections for the direct process only. The calculations have been performed in the range (0.2–120 meV) on a 0.2 meV energy mesh. A total of 31 channels were included, corresponding to ionization channels with $v=0$ to 14 and s and d characters, besides the dissociation channel. In this low energy range only one (for $v_0=0$) or two (for $v_0=1$) ionization thresholds are open (see figure 1) such that the calculations involve only 3 (for $v_0=0$) or 5 (for $v_0=1$) open channels. Convolution with a 2.5 meV triangular apparatus function leads to the results on figure 4, where the experimental curve (Mitchell 1990) is reported with a small shift in the energy scale, to be justified below.

The first conclusion to be drawn from figure 4 is that practically all the excited vibrational (and probably also rotational) levels of the recombining ions have been removed, proving the efficiency of the new RF ion trap (Hus *et al* 1988): the resonance pattern of the experimental curve could not be reproduced with a mixture of ions in $v_0=0$ and $v_0=1$ levels. With this in mind, the agreement in magnitude between the experimental and theoretical cross sections proves the quality of the electronic couplings and dissociative potential curve used here, since the efficiency of the direct DR process is dominated by the interaction matrix elements (5), with $v=0$.

Analysis of our results leads to the resonance assignment given on figure 3. Most of the resonances in that energy range are due to $s\sigma_g$ Rydberg levels, in spite of the weak interaction of the corresponding continuum electron partial wave with the dissociative state (figure 2(b)). According to the $(n-\mu)^{-3}$ scaling for the width and shift along a Rydberg series, the low n resonances are broad and very sensitive to the molecular data. A slight change in the quantum defect values ($\Delta\mu_{\cdot s\cdot}=-0.0056$, $\Delta\mu_{\cdot d\cdot}=0.015$) with respect to the values used for the AI calculations brings the low n resonances into better agreement with the observed ones, without altering the AI cross section for which the nuclear partial wave summation washes out the detail of the resonance

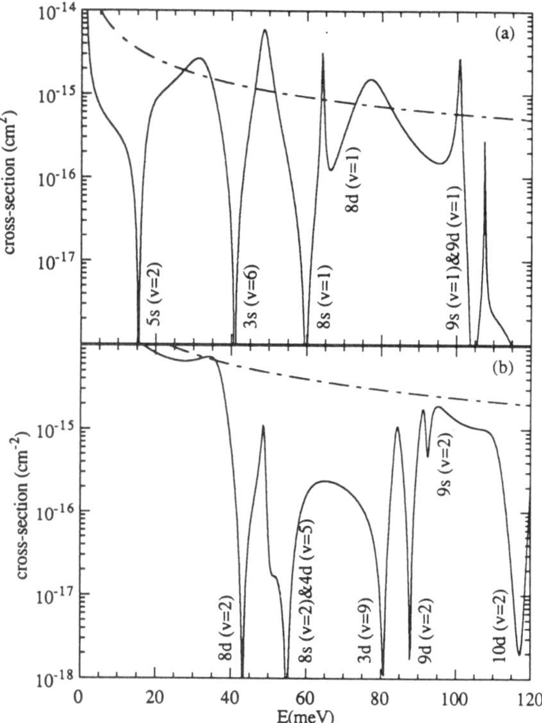

Figure 3. Cross sections for dissociative recombination of H_2^+ ground state, in vibrational levels $v_0 = 0$ (a) and $1(b)$. The broken curves correspond to the cross section for the *direct* process. Each resonance is labelled by the vibrational (v), principal (n) and dominant orbital (l) quantum numbers of the Rydberg state, with the orbital symmetry σ_g understood.

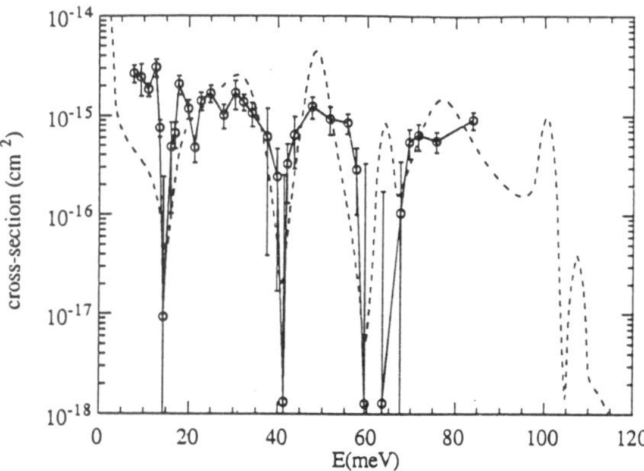

Figure 4. Comparison between the merged beam measurements of Mitchell (1990) and the present calculations for H_2^+ dissociative recombination. The broken curve is the theoretical cross section for the initial ion level $v_0 = 0$ (cf figure 3(a)) convoluted with a 2.5 meV triangular apparatus function.

structure. On the contrary, the high n resonances depend weakly on the quantum defect values such that the s and d ($n=8, v=1$) resonances overlap and lead to the broad dip observed at about 60 meV. The systematic shift between the experimental resonance energies (originally 5 meV higher than reported on figure 4) and the calculated ones is probably due to a small inaccuracy in the energy calibration of the experiment, since any further adjustment of the quantum defect values would not lead to a uniform shift for all the resonances.

A detailed analysis of the resonance shape in DR has been given by Guberman and Giusti-Suzor (1991) in terms of the profile index q of Fano (1961), for isolated resonances. Here we stay in the real overlapping situation and discuss the role of the different types of interactions.

The dotted curve in figure 5 shows the cross section obtained with the quantum defect functions $\mu_{\cdot s^{\cdot}}$ and $\mu_{\cdot d^{\cdot}}$ assumed to be constant (equal to their values at the internuclear equilibrium distance), and the electronic K matrix calculated at first order. Therefore, both the vibronic capture into bound Rydberg levels (measured by the mixing coefficients (4)) and the indirect electronic capture (measured by the second-order matrix elements (6)) are suppressed. The resonances show up as pure 'windows' (profile index $q=0$) and their width is just due to the predissociation of the Rydberg levels by the repulsive $(2p\sigma_u)^2$ state. The local depression in the DR cross section associated with each Rydberg level results from destructive interference between direct and indirect processes rather than from competition between autoionization and predissociation, since this model suppresses autoionization as well as electron capture for the bound states.

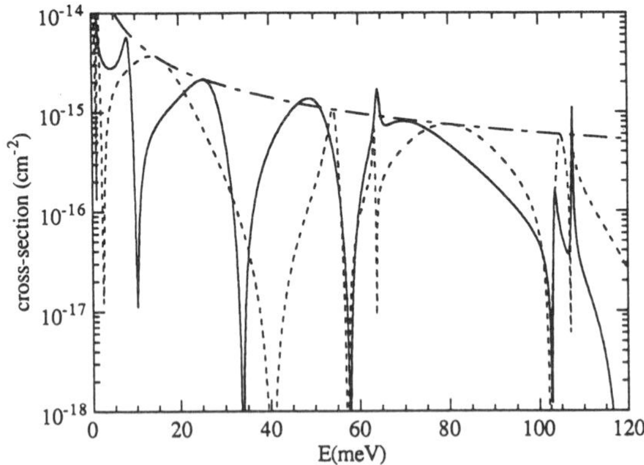

Figure 5. Analysis of the mechanisms for electron capture, with initial ion level $v_0=0$: ·—·—, direct process only (closed channels not included); all the other calculations include three open and 28 closed channels; ----, no capture in Rydberg levels (constant quantum defects and first order electronic K matrix, see text); ——, *vibronic* capture included, via the R-dependence of the quantum defects.

When vibronic interactions are introduced through the R dependence of the quantum defects (full curve in figure 5), the resonances are notably shifted, attesting for a substantial change of the 'mean' quantum defect value. A slight increase of the cross section is observed due to vibronic capture, mostly for the lowest v resonances ($v=1$) for which the mixing coefficients with $v'=0$ in (4) are largest. Nevertheless, destructive interference still dominates the resonance pattern.

Inclusion of the second-order terms (6) of the K matrix leads to the full curve in figure 3(a). Definite spikes appear in the cross section due to indirect electronic capture in the Rydberg states, but broad dips are still present. This step of the calculation is responsible for the two features in the final cross sections which depart clearly from experiment (figure 4): the sharp peak around 50 meV associated with the ($3s\sigma$, $v = 6$) level and the broad depression near threshold, partly due to second-order interaction with the ($3d\sigma$, $v = 7$) level that we locate about 15 meV below threshold.

It is not surprising that our perturbative treatment of electronic interaction is less satisfying for the lowest Rydberg states, subjected to stronger couplings. This has also been noted in AI calculations for sodium atoms (Dulieu *et al* 1990, 1991). We intend to combine the present MQDT approach, very efficient for high Rydberg levels, with a coupled equation treatment for the lowest states. A first step in this direction has been achieved for AI by Urbain *et al* (1991). The low energy DR process, very sensitive to the molecular data and other aspects of the theory, provides a good test case since well defined high resolution experiments can now be performed. The present comparison between experiment and theory already confirms that the indirect DR process tends to reduce the recombination rate rather than to increase it, contrary to the predictions of Hickman (1987) within the 'independent resonance approximation'. This behaviour extends to other diatomic molecules with a low-lying repulsive state leading to efficient direct dissociation, and may have important implications for ionospheric recombination (Guberman and Giusti-Suzor 1991).

We are very grateful to J B Mitchell for providing us with experimental data prior to their publication, and to X Urbain for making available the set of molecular data used in this work. We thank F Masnou-Seeuws and J Weiner for helpful discussions. IFS acknowledges the French 'Ministère de la Recherche et de la Technologie' for a research grant and the 'Comité d'Echanges Scientifiques Orsay-Roumanie' for support during his visit at the Université Paris-Sud. AGS thanks NATO for a collaborative research grant.

References

Bardsley J N 1968 *J. Phys. B: At. Mol. Phys.* **1** 349, 365

Dulieu O, Giusti-Suzor A and Masnou-Seeuws F 1990 *8th Eur. Conf. Dynamics of Molecular Collisions* Abstracts p 139

—— 1991 *J. Phys. B: At. Mol. Opt. Phys.* submitted

Fano U 1961 *Phys. Rev.* **124** 1866

Giusti A 1980 *J. Phys. B: At. Mol. Phys.* **13** 3867

Giusti-Suzor A, Bardsley J N and Derkits C 1983 *Phys. Rev.* A **28** 682

Greene C H and Jungen Ch 1985 *Adv. At. Mol. Phys.* **21** 51

Guberman S L and Giusti-Suzor A 1991 *J. Chem. Phys.* in press

Hazi A, Derkits C and Bardsley J N 1983 *Phys. Rev.* A **27** 1751

Hickman A P 1987 *J. Phys. B: At. Mol. Phys.* **20** 2091

—— 1989 *Dissociative Recombination: Theory, Experiment and Applications* ed J B A Mitchell and S L Guberman (Singapore: World Scientific) p 35

Hus H, Yousif F, Noren C, Sen A and Mitchell J B A 1988 *Phys. Rev. Lett.* **60** 1006

McGowan J W, Caudano R and Keiser C J 1976 *Phys. Rev. Lett.* **36** 1447

Mitchell J B A 1990 Private communication

Nakashima K, Takagi H and Nakamura H 1987 *J. Chem. Phys.* **86** 726

Ross S and Jungen Ch 1987 *Phys. Rev. Lett.* **59** 1297

Schneider B I and Collins L A 1983 *Phys. Rev.* A **28** 166

Seaton M J 1983 *Prog. Phys.* **46** 167

Shimamura I, Noble C J and Burke P G 1990 *Phys. Rev.* A **41** 3545
Sidis V and Lefebvre-Brion H 1971 *J. Phys. B: At. Mol. Phys.* **4** 1040
Takagi H and Nakamura H 1983 *Phys. Rev.* A **27** 691
Urbain X 1990 *Thèse* Université de Louvain-la-Neuve (unpublished)
Urbain X, Cornet A, Brouillard F and Giusti-Suzor A 1991 *Phys. Rev. Lett.* **66** 1685
Wilniewicz L and Dressler K 1977 *J. Mol. Spectrosc.* **67** 416
—— 1979 *J. Mol. Spectrosc.* **77** 286

1984 *Phys. Rev. Lett.* **53** 2394–7
Reprinted with permission from the American Physical Society

Unified Treatment of Dissociation and Ionization Processes in Molecular Hydrogen

Ch. Jungen
Laboratoire de Photophysique Moléculaire, Centre National de la Recherche Scientifique, Université de Paris-Sud, F-91405 Orsay, France
(Received 20 August 1984)

A theoretical procedure is introduced which yields eigenvalues and eigenvectors of the reactance matrix pertaining to competing dissociation and ionization processes in molecular Rydberg states. A calculation is set up in terms of adiabatic (nuclear-coordinate dependent) quantum defects $\mu(R)$ which contain the essence of the physics of the excited molecular complex at short range. The application to preionized and predissociated resonances in the H_2 spectrum yields good agreement with experiment.

PACS numbers: 33.10.Cs, 33.80.Eh, 33.80.Gj, 34.10.+x

Four years ago Dill and I[1] used multichannel quantum defect theory to calculate the photoionization spectrum of molecular hydrogen near threshold. The calculation included effects of vibrational-rotational preionization and reproduced very closely the complicated resonance structure observed by Dehmer and Chupka[2] in a high-resolution experiment. The calculation was straightforward because of a greatly simplifying physical circumstance whose importance had been stressed by Fano some years earlier. Fano[3] had suggested that even a highly excited bound or continuum electron conforms to the Born-Oppenheimer approximation so long as it is moving inside or near the molecular core. The Born-Oppenheimer separation of rovibrational and electronic motion breaks down only when the outer electron roams far from the core where its motion is comparable to or slower than the nuclear motion. This picture implies that nuclear-coordinate–dependent (body-frame) electron phase shifts $\pi\mu(R)$ [or the equivalent quantum defects $\mu(R)$ in the discrete range] provide all that is needed to calculate preionization, which involves a transfer of rovibrational energy from the core to the excited electron. The nondiagonal reactance or scattering matrices which account for this process can be constructed from the adiabatic phase shifts by means of an analytic transformation to the laboratory frame, which is applied to the wave function at some point outside the core, in a zone which one may term the "recoupling" zone.

In this Letter I show how this concept can be extended further to include *molecular dissociation* and *predissociation* occurring within a molecular Rydberg channel. Indeed, predissociation may be regarded[4] as the counterpart of vibrational-rotational preionization since here energy is transferred from the Rydberg electron to the nuclei, enabling the system to dissociate into atomic fragments. An analogous problem is encountered in dissociative attachment or associative detachment collisions involving hydrogen-halides.[5] The present extension aims at calculating the *competition* between preionization and predissociation, a problem which hitherto has been solved only for cases where weakly avoided crossings occur between Rydberg states and a dissociating valence state, i.e., where there is electronic rearrangement in the core.[6] The present work is still based on adiabatic quantum defects as the only dynamical parameters. It demonstrates that quantum defects, familiar from atomic physics, have a relevance far beyond their original context: Here they serve for the description of the *chemical transformation of matter* in a half-collision.

Figure 1 shows a prototype set of molecular Rydberg potential energy curves, namely, the lowest states of molecular hydrogen which can be excited by photoabsorption. The curves can be represented

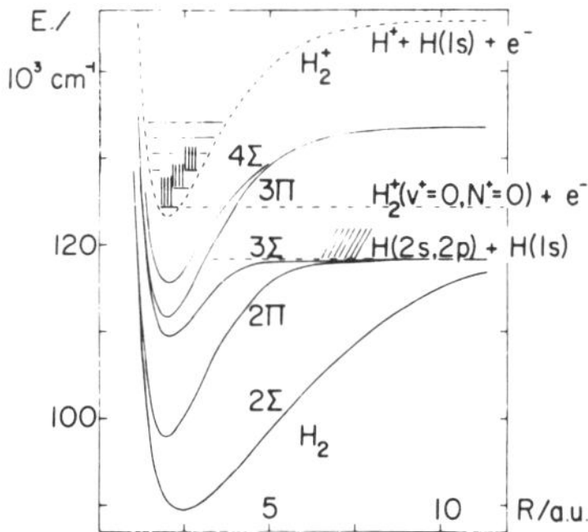

FIG. 1. Potential energy curves of singlet *ungerade* symmetry for H_2 (solid lines). The curve for H_2^+ ($X^2\Sigma_g^+$) is also shown (dashed line). Continua are indicated by hatching.

by the Rydberg equation

$$U_{n\Lambda}(R) = U^+(R) - \{2[n-\mu_\Lambda(R)]^2\}^{-1} \text{ a.u.} \tag{1}$$

with a *single* quantum defect function $\mu_\Lambda(R)$ for each symmetry Λ. (An exception is the large-R portion of the lowest curve whose associated quantum defect function differs significantly from that representing the other Σ states.) $U^+(R)$ is the potential function of the ion and n is the principal quantum number. The set of curves represents the lower discrete portion of a molecular *p*-wave *ionization channel* associated with the electronic ground state of the ion. (s and d electrons yield *gerade* states which do not appear in the dipole absorption spectrum.) Upon ionization the core remains in a particular vibration-rotation level v^+N^+. Dissociation and predissociation may take place *within* the Rydberg series because for energies higher than the first dissociation limit each of the lowest states $n\Lambda$ acts simultaneously as a *dissociation channel.*

In order to analyze the alternative fragmentation processes consider the two-dimensional configuration space spanned by the radial coordinate r of the Rydberg electron and the internuclear distance R (Fig. 2). Several zones are defined by the following critical values of r and R: r_0 corresponds to the boundary of the ion core, r_1 is the distance where the Born-Oppenheimer approximation is beginning to break down, and r_2 is the distance where all relevant bound Rydberg components have fallen exponentially to a negligibly small value. R_0 is the nuclear separation beyond which all relevant bound vibrational wave functions have become negligibly small. Typical values of r_0, r_1, r_2, and R_0 are, respectively, 4, 6, 30, and 7 a.u. Consider now a close-coupling expansion of the total wave function Ψ outside the reaction zone with components of general form $|i\rangle[f_i(r)\cos\delta - g_i(r)\sin\delta]$. The specific form of the expansion terms appropriate to each zone is indicated in Fig. 2. The kets $|i\rangle$ are the wave functions for all degrees of freedom of the system except the radial coordinate corresponding to fragmentation. They include the angular and spin wave functions of the escaping particle (electron or atom), coupled so as to yield a given total angular momentum J, space-fixed component M, and parity. The f_i and g_i are regular and irregular radial functions for the relative motion of the fragments, i.e., Coulomb functions (f,g) or vibrational continuum functions (F,G). The scattering phase shift δ represents the net effect of the short-range interactions in each channel. Note that in the inner part of region I each expansion term is a Born-

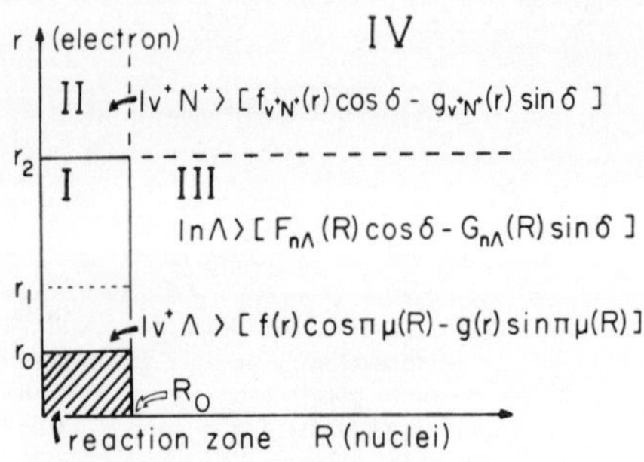

FIG. 2. Close-coupling expansion terms appropriate in different regions of configuration space.

Oppenheimer product with a core function $|v^+\Lambda\rangle$ characterized by a well-defined electronic angular momentum component Λ in the body frame. Accordingly δ is set equal to the adiabatic electron phase shift $\pi\mu_\Lambda(R)$. In region IV Ψ is set equal to zero and thereby the method is at present limited to the range below the threshold for dissociative ionization.

The three forms of the close-coupling expansion indicated in Fig. 2 are equivalent to boundary conditions imposed on Ψ along the boundary of region I. Thereby attention is focused on the recoupling zone in which the dynamics of the system changes drastically from adiabatic to entirely nonadiabatic motion.

In this finite volume a multichannel rovibronic wave function Ψ_E is set up for a given total energy E, similar to the expansion used in Ref. 1, and which satisfies the specific boundary conditions just outlined. The difference from the earlier work is, firstly, that the truncated set of vibrational wave functions, $\{\chi_{v^+}(R)\}$, included in the calculation is extended to contain numerous functions reaching out to the boundary R_0. Secondly, a *common* phase shift $\pi\tau_\rho$ is imposed on both the various electronic radial components at $r=r_2$ and the vibrational radial components at $R=R_0$. Thereby an eigenchannel R-matrix problem of the type introduced by Fano and Lee[7] is defined, in which a multiparticle Schrödinger equation is to be solved within a finite volume. More specifically, the rovibronic multichannel quantum defect treatment in the recoupling zone leads to an eigenvalue problem of the type described first by Seaton,[8] namely,

$$\sum_{i'}[\sin\beta_i(E)\mathscr{C}_{ii'}(E) + \cos\beta_i(E)\mathscr{S}_{ii'}(E)]B_{i'} = 0. \tag{2}$$

Here i is an index for asymptotic channels which now may be closed or open. The β_i express the condition imposed on each asymptotic component at large distance, while the matrices $\mathscr{C}$ and $\mathscr{S}$ reflect the conditions imposed on the wave function at the boundary of the reaction zone.

Consider now for simplicity a situation where a single dissociation channel, $d = n\Lambda$, is open and molecular rotation can be disregarded. In the first stage of the calculation the dissociation channel does not appear explicitly. Instead, three classes of ionization channels $|v^+\rangle$ are distinguished, P, Q_1, and Q_2, respectively, which correspond to low, medium, and high vibrational excitation of the ion core. P contains the n_p open ionization channels. Q_1 contains n_{q_1} "weakly" closed ionization channels, namely, those which yield Rydberg resonances in the range of interest near E. Q_2 contains n_{q_2} "strongly" closed ionization channels corresponding to vibrational wave functions with a nonnegligible amplitude at R_0. This third set represents the dissociation continuum. The linear system (2) is then set up by putting $i = v^+$, $i' = v^{+\prime}$. At this stage open-channel boundary conditions are applied to P and Q_1 and closed-channel conditions to Q_2, respectively. The choice of the unphysical condition for Q_1 at this point takes advantage of the flexibility of quantum defect theory, in which the physical conditions need be applied only in the final stage. The present choice eliminates the resonances arising from Q_1 and is tantamount to reducing the radius r_2. As a result, a reactance matrix is obtained which varies smoothly with energy. (I am indebted to Ch. H. Greene for suggesting this procedure.) One puts

$$\beta_{v^+}(E) = -\pi\tau_\rho, \quad v^+ \in P, Q_1, \tag{3a}$$

$$\beta_{v^+}(E) = \pi\nu_{v^+}(\tau_\rho), \quad v^+ \in Q_2. \tag{3b}$$

The $\pi\tau_\rho$ are the eigenphases on the boundary of zone I to be determined. The $\nu_{v^+}(\tau_\rho)$ are the effective principal quantum numbers relating to the ionization limits v^+, viz., $\nu_{v^+}(\tau_\rho) = \{-2[E - E(v^+, \tau_\rho)]\}^{-1/2}$. The key elements here are the following: (i) the energy spectrum $E(v^+)$ of the vibrational wave functions calculated in the finite range $0 < R < R_0$ is *discrete*, because of imposition of a definite value of the logarithmic derivative on the functions $\chi_{v^+}(R)$ at $R = R_0$; (ii) this spectrum, and hence the effective principal quantum numbers ν in Eq. (3b), *depends on the phase* $\pi\tau_\rho$ since the corresponding logarithmic derivative $-b(\tau_\rho)$ is chosen. The coupling matrices $\mathscr{C}$ and $\mathscr{S}$ also depend on the phase since their elements are vibrational integrals of the type

$$\mathscr{C}_{v^+v^{+\prime}}(\tau_\rho) = \int_0^{R_0} dR\, \chi_{v^+}(R, \tau_\rho) \cos\pi\mu(R) \chi_{v^{+\prime}}(R, \tau_\rho) \tag{3c}$$

(and an analogous expression for $\mathscr{S}$ with $\sin\pi\mu$) familiar from the theory of vibrational preionization.[1]

In the R-matrix procedure of Fano and Lee τ_ρ is varied until an eigenvalue of the Hamiltonian in the finite volume coincides with the desired total energy E. The condition (3a) together with the linear system (2) generates $n_p + n_{q_1}$ such solutions. An additional eigenvalue arises through Eq. (3b), namely, when a low-n Rydberg level associated with the set Q_2 passes through E because of the variation of the ionization energies $E(v^+, \tau_\rho)$ as functions of τ_ρ. This level represents the dissociation continuum, and its mixing with vibrationally bound components through the linear system (2) provides the mechanism for the *conversion of energy*.

Each of the $n_p + n_{q_1} + 1$ eigenfunctions $\Psi^{(\rho)}(r_0 < r < r_2, R < R_0)$ thus found finally has to be projected at the surface of the recoupling zone onto the energy-normalized asymptotic channel functions outside. Following Refs. 1 and 7 one finds

$$\langle\rho|v^+\rangle = \sum_{v^{+\prime}} [\mathscr{C}_{v^+v^{+\prime}}(\tau_\rho)\cos\pi\tau_\rho + \mathscr{S}_{v^+v^{+\prime}}(\tau_\rho)\sin\pi\tau_\rho] B^{(\rho)}_{v^{+\prime}}, \tag{4a}$$

$$\langle\rho|d\rangle = \frac{\sum_{v^{+\prime}} \chi_{v^{+\prime}}(R_0, \tau_\rho) B^{(\rho)}_{v^{+\prime}}}{[n - \mu(R_0)]^{-3/2}[F_{n\Lambda}(R_0)\cos\pi\tau_\rho - G_{n\Lambda}(R_0)\sin\pi\tau_\rho]}. \tag{4b}$$

Equation (4) yields the eigenvectors of the desired reactance matrix. Its nondiagonal form is

$$\mathscr{R}_{ii'} = \sum_\rho \langle i|\rho\rangle \tan\pi\tau_\rho \langle\rho|i'\rangle, \quad i = v^+ \in P, Q_1 \quad \text{or} \quad i = d. \tag{5}$$

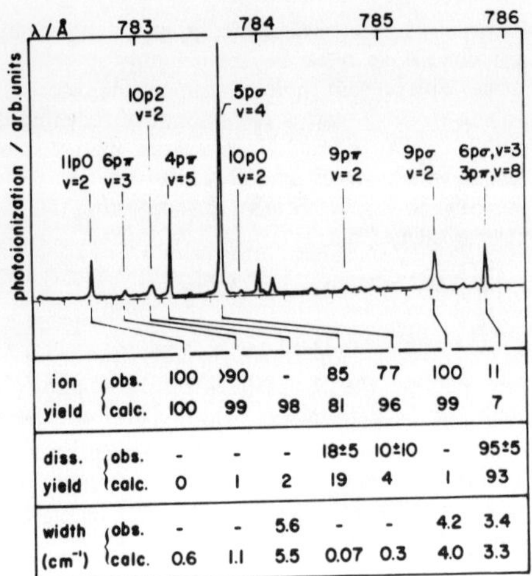

ion yield	obs.	100	>90	-	85	77	100	11
	calc.	100	99	98	81	96	99	7
diss. yield	obs.	-	-	-	18±5	10±10	-	95±5
	calc.	0	1	2	19	4	1	93
width (cm^{-1})	obs.	-	-	5.6	-	-	4.2	3.4
	calc.	0.6	1.1	5.5	0.07	0.3	4.0	3.3

FIG. 3. Top: Photoionization efficiency curve $[J'=1$ (negative parity) $\leftarrow J''=0]$ of parahydrogen near 784 Å (Ref. 2). The unassigned peaks correspond to $J''=1$ transitions. Bottom: Observed (Refs. 2 and 9) and calculated ionization and dissociation yields and resonance widths for the individual resonances.

This matrix now has dimension $n_p + n_{q_1} + 1$ and includes the dissociation channel *explicitly*, condensed into a *single* row and column. In a final stage, then, the photodissociation and photoionization cross sections including resonances can be calculated from this reactance-type matrix by applying the physical boundary conditions at infinity, with use of standard quantum defect methods.[1,8]

Figure 3 illustrates the application of the method to photoabsorption by H_2. At the top of the figure is shown a section of the high-resolution photoionization efficiency curve obtained by Dehmer and Chupka from cold parahydrogen. Complementary photodissociation excitation spectra have been recorded by Guyon, Breton, and Glass-Maujean[9] but are not shown here. The spectrum is quite perturbed because of multiple interactions between the various Rydberg levels. This is revealed by the irregular intensity pattern and is confirmed by the calculations. In this range the $H_2^+(v^+=0$ and 1, $N^+=0$ and 2) $+e$ and $H(1s)+H(2s,2p)$ channels are open and accessible by dipole absorption from $H_2(J''=0)$. The multichannel quantum defect calculation yielded directly all seven partial-cross-section spectra evaluated point by point on a convenient energy mesh. In the lower part of the figure are given the total dissociation and ionization yields obtained by integration over each Rydberg resonance, and also the resonance width, along with the available experimental data from Refs. 2 and 9. Experiment and theory agree quite well. The observed yields for $5p\sigma, v=4$ are inconsistent since their sum is less than 100% depsite the fact that the molecular fluorescence decay channel is inactive.[10] The calculation agrees with Ref. 9 in this case. A more detailed account of this work will be given elsewhere.

I am greatly indebted to Dr. Dan Dill for participating in the early stages of this work, and to Dr. Ch. H. Greene and Dr. A. Giusti-Suzor for many discussions and for their comments on the manuscript.

[1]Ch. Jungen and Dan Dill, J. Chem. Phys. **73**, 3338 (1980).

[2]P. M. Dehmer and W. A. Chupka, J. Chem. Phys. **65**, 2243 (1976), and private communication.

[3]U. Fano, Phys. Rev. A **2**, 353 (1970).

[4]E. S. Chang, Dan Dill, and U. Fano, in *Abstracts of Papers, Proceedings of the Eighth International Conference on the Physics of Electronic and Atomic Collisions,* edited by B. C. Cobic and M. V. Kurepa (Institute of Physics, Belgrade, 1973), p. 536.

[5]J. P. Gauyacq, J. Phys. B **15**, 2721 (1982).

[6]A. Giusti-Suzor and Ch. Jungen, J. Chem. Phys. **80**, 986 (1984).

[7]U. Fano and C. M. Lee, Phys. Rev. Lett. **31**, 1573 (1973).

[8]M. J. Seaton, Proc. Phys. Soc., London **88**, 801, 815 (1966), and Rep. Prog. Phys. **46**, 167 (1983).

[9]P. M. Guyon, J. Breton, and M. Glass-Maujean, Chem. Phys. Lett. **68**, 314 (1979), and **63**, 591 (1979), and private communication.

[10]J. Breton, P. M. Guyon, and M. Glass-Maujean, Phys. Rev. A **21**, 1909 (1980).

1994 *Phys. Rev.* A **49** 4364–77
Reprinted with permission from the American Physical Society

Multichannel quantum-defect theory of double-minimum $^1\Sigma_g^+$ states in H_2. II. Vibronic-energy levels

S. C. Ross
Department of Physics, University of New Brunswick, P.O. Box 4400, Fredericton, New Brunswick, Canada E3B 5A3

Ch. Jungen
Laboratoire Aimé Cotton du CNRS, Université de Paris–Sud, 91405 Orsay, France

(Received 11 November 1993)

In the preceding paper [Phys. Rev. A **49**, 4353 (1994)] we obtained the quantum-defect matrix of the strongly interacting double-minimum states of H_2 by fitting to the *ab initio* clamped-nuclei electronic energies of Wolniewicz and Dressler [J. Chem. Phys. **82**, 3292 (1985), and private communication]. Yu, Dressler, and Wolniewicz have calculated the vibronic energies of the corresponding states using an approach involving the state-by-state evaluation of vibronic coupling, and the solution of a set of coupled equations. Here we calculate the vibronic energies using our quantum-defect matrix in a version of scattering theory known as multichannel quantum-defect theory (MQDT). This less traditional treatment involves both singly and doubly excited channels and reproduces the vibronic energies to almost the same precision as the coupled-equations approach. In addition, several refinements have been made to MQDT.

PACS number(s): 33.10.Cs, 33.10.Lb, 34.10.+x, 34.80.Kw

I. INTRODUCTION

In the previous paper in this series [1] (which we refer to as "RJ-I") we showed that the highly accurate *ab initio* clamped-nuclei potential-energy curves of Wolniewicz and Dressler [2,3] can be represented by a set of smooth quantum-defect functions. In this paper, we further show that these quantum-defect functions contain all of the information needed to calculate the energies of the vibrational levels in the corresponding electronic states.

The study of the $^1\Sigma_g^+$ electronic states of H_2 using more traditional molecular-dynamics techniques has been pursued in greatest detail by Dressler and co-workers (Ref. [4] and references therein, and, more recently, Refs. [5] and [6]). Their investigations involve, as a first step, performing highly accurate clamped-nuclei *ab initio* calculations of those electronic states chosen for inclusion in the calculations. Adiabatic corrections are then determined for each state and a set of nonadiabatic coupling functions is determined for each pair of states. In the $^1\Sigma_g^+$ states considered here the adiabatic corrections are as large as 860 cm^{-1} [2], and the vibrational levels are shifted by nonadiabatic effects by up to 190 cm^{-1} [4]. These strong interactions, and the resulting mixing in character of the states involved, means that it is inappropriate to refer to these states as being vibrational levels of particular electronic states. Instead we must refer to them as "vibronic" states. The final step in the traditional treatment is to include these interactions in a single global coupled-equations calculation.

These *ab initio* calculations have been very successful in reproducing the known energy levels of the *EF*, *GK*, and *H* states of H_2 and thus serve as a benchmark against which any other study of these levels must be compared. An important feature of these calculations is that they proceed on a state-by-state basis, explicitly accounting for the interactions between each pair of states. For Rydberg series, however, such an approach cannot be extended indefinitely because the number of mutually interacting states is infinite, with the density of states growing explosively as the total energy approaches the ionization limit.

In multichannel quantum-defect theory (MQDT), however, the notion of distinct states is replaced by the concept of a channel which incorporates an entire series of electronic Rydberg states in one entity. The quantum-mechanical functions describing the channels and their evolution with internuclear spacing is specified by quantum-defect functions. The quantum-defect functions thus provide a single uniform description of all the states in the channel, thereby allowing the theory to account for effects arising from the entire Rydberg series and the continuum lying above. The fact that the quantum defects describe the dependence of the electronic channel functions on the vibrational coordinate leads to their accounting for the strong adiabatic effects, and for the nonadiabatic interactions that occur between the vibrational levels of *different* electronic states, with no further information required. The vibronic energies that we obtain for the 2, 3, and 4 $^1\Sigma_g^+$ electronic states using the MQDT technique are of almost the same quality as those Dressler and co-workers obtained using the full molecular-dynamics apparatus. The MQDT results have an rms error of 6.0 cm^{-1} which compares very well to the value of 3.0 cm^{-1} obtained from the traditional approach [3,6].

II. THEORY

In 1977 Jungen and Atabek [7] (hereinafter referred to as "JA") presented the MQDT theory for rovibronic states of diatomic molecules, including provision for a direct treatment of electronic interaction. Ten years were

1050-2947/94/49(6)/4364(14)/$06.00

to pass until in 1987 we performed a calculation accounting for this interaction [8]. As described in RJ-I our 1987 calculation served as a "test of principle." In the following we show in detail how to use the quantum-defect matrix determined in RJ-I to calculate the energies of the first 39 vibronic levels of the excited $^1\Sigma_g^+$ states H_2, adapting the treatment to account for the energy dependence of the quantum-defect matrix. Many parts of the following development have previously been given, but are scattered throughout the literature. Our aim here is to draw the various pieces together and present a consistent overall picture of the method.

In MQDT the energy of the Rydberg electron is initially a free parameter so that rather than introducing basis functions corresponding to particular eigenstates of a molecule at specific bound-state energies, MQDT instead introduces basis functions in which all properties of the system are specified *except* the energy of the Rydberg electron. The resulting energy-dependent basis functions each correspond to what we call a "channel." These channel functions describe the asymptotic behavior of the molecule when the Rydberg electron is far from the ion core; they form a natural path to ionization and are called the "long-range" or "asymptotic" channels. It is only at the end of the theoretical development that boundary conditions are imposed on the channel functions and the energy eigenstates are obtained.

For a given total angular momentum, J, and Z-axis projection, M, the asymptotic channels are labeled by the electronic state of the combined ion+electron system i, and the vibrational-rotational state $v_i^+N_i^+$ of the ion core. Thus each molecular channel function contains a factor $|iv_i^+N_i^+\rangle^{JM}$ describing the degrees of freedom of the molecule other than the radial coordinator r, of the Rydberg electron. The vibrational motion in the ion core is assumed to be adiabatic, with nonadiabatic effects arising from interaction with the colliding Rydberg electron. It is also assumed that once the Rydberg electron is further than some finite distance from the core its motion occurs on a simple Coulomb potential-energy surface centered on the ion core and can, therefore, be described by a linear combination of a base pair of Coulomb functions. As in RJ-I we use the Coulomb base pair $f(r)$ and $h(r)$ (Ref. [9]) with which unphysical states having $n<l+1$, such as $1p$, $1d$, $2d$, etc., can usually be avoided in the MQDT. The Coulomb functions depend on the energy ϵ of the Rydberg electron, as well as on its l value. For a given total energy E of the molecule, the energy ϵ of the Rydberg electron is the difference between E and the energy of the ion core, i.e., when the molecule is in the channel $|iv_i^+N_i^+\rangle$ we may write

$$\epsilon_{iv_i^+N_i^+}=E-E^+_{iv_i^+N_i^+} \ . \tag{1}$$

Because the l value of the electron is part of the electronic state i of the molecule we write it as l_i, and thus the Coulomb functions we use are $f_{l_i}(r,\epsilon_{iv_i^+N_i^+})$ and $h_{l_i}(r,\epsilon_{iv_i^+N_i^+})$.

The MQDT channel functions are now written as

$$\Psi^{JM}_{iv_i^+N_i^+}(E)=f_{l_i}(r,\epsilon_{iv_i^+N_i^+})|iv_i^+N_i^+\rangle^{JM} + \sum_{jv_j^+N_j^+} K_{iv_i^+N_i^+,jv_j^+N_j^+} \times h_{l_j}(r,\epsilon_{jv_j^+N_j^+})|jv_j^+N_j^+\rangle^{JM} \ . \tag{2}$$

They describe a molecule initially in the channel $|iv_i^+N_i^+\rangle^{JM}$ which, upon the collision of the Rydberg electron with the core, enters a state involving a mixture of all channels $|jv_j^+N_j^+\rangle^{JM}$. (Note that J and M remain good quantum numbers and are preserved in the collision so that the problem for each JM state is independent.) The collision induced admixture of channels in the MQDT basis functions is given by the real and symmetric full rovibronic reaction matrix K. In atomic MQDT the K matrix is often determined by fitting to experimental energy levels. For molecular problems the additional degree of freedom introduced by the vibrational motion means that such a fitting is not feasible due to the number of independent elements in the matrix. In the calculations described in the present work, we use a 200×200 K matrix to calculate the first 39 vibronic levels of the excited electronic states of the $^1\Sigma_g^+$ manifold. It is evident that 39 experimental energies are not sufficient to determine the 20 000 independent elements of the full K matrix.

JA, following earlier work in Refs. [10–12], showed that the solution to this problem lies in an examination of the physics of the actual collision. As the Rydberg electron falls towards the ion core it accelerates under the influence of the attractive Coulomb potential-energy function of the positive H_2^+ ion core. The resulting electron-ion collision is consequently very rapid with the result that during the collision the Rydberg electron interacts with a core that appears "frozen" at some particular configuration of the nuclei. The region wherein this is true we call the Born-Oppenheimer region because once the Rydberg electron is this close to the core the motion of *all* the electrons (rather than just the ion-core electrons) is separable from the vibrational-rotational motion. This means that the vibrational quantum number v_i^+ appropriate for the description of the asymptotic channels is replaced in the Born-Oppenheimer region by the internuclear spacing R seen by the Rydberg electron at the moment of collision. Similarly the total orbital angular momentum Λ around the molecular axis becomes a good quantum number, replacing the rotational angular momentum of the ion core, N_i^+, appropriate in the asymptotic region. The appropriate functions for describing the molecule in the Born-Oppenheimer region may, therefore, be written as $|iR\Lambda\rangle^{JM}$, instead of $|iv_i^+N_i^+\rangle^{JM}$, in Eq. (2). These are referred to as "short-range" or "Born-Oppenheimer" functions. The fact that it is not necessary to introduce a strong energy (i.e., state) dependence in the quantum-defect matrix validates this picture and also validates the derivation in RJ-I of the quantum-defect matrix from clamped-nuclei (or Born-Oppenheimer) *ab initio* calculations.

By equating, in the Born-Oppenheimer region, the total wave function for the molecule expanded in asymptotic channel functions with its expansion in Born-Oppenheimer channel functions, JA were able to determine the full K matrix of Eq. (2) in terms of the R-dependent electronic $K^{\Lambda}(R)$ matrix. This procedure amounts to carrying out a transformation from the laboratory to the molecular frame. A detailed discussion of the concept of the vibrational frame transformation and its relation to the energy dependence of the quantum defects has been presented by Greene [13]. The electronic quantum-defect matrix $\eta^{\Lambda}(R)$, determined in RJ-I for $\Lambda=0$, is connected to $K^{\Lambda}(R)$ via the relation between their elements

$$K_{ij}^{\Lambda}(R)=\tan\pi\eta_{ij}^{\Lambda}(R)\ . \tag{3}$$

The relationship between the electronic K^{Λ} matrices of Eq. (3) and the full K rovibronic matrix of Eq. (2) is implicit in Eq. (29) of JA, and is also given in Eq. (71) of Greene and Jungen [14]. Here we write it as

$$K_{iv_i^+N_i^+,jv_j^+N_j^+}=\int dR\ {}^{JM}\langle iv_i^+N_i^+|\left\{\sum_{\Lambda}|iR\Lambda\rangle^{JM}K_{ij}^{\Lambda}(R)\,{}^{JM}\langle jR\Lambda|\right\}|jv_j^+N_j^+\rangle^{JM}\ . \tag{4}$$

(The superscript Λ on the electronic K matrix distinguishes it from the full rovibronic matrix.)

Equation (4) directly reflects the physical picture underlying the frame transformation. The left-hand factor of the integrand corresponds to the system initially being in the asymptotic channel $|iv_i^+N_i^+\rangle^{JM}$. When the electron is close enough to the core the collision state satisfies the Born-Oppenheimer approximation and we enter the part of Eq. (4) in parenthesis. In this region the collision channels $|iR\Lambda\rangle^{JM}$ are appropriate for the description of the molecule. The incoming state is projected over the set of different collision channels states, as indicated by the summation over Λ and the integration over R. Once in this collision state the energy of the colliding Rydberg electron can lead to electronic rearrangement in the core via the off-diagonal elements of the electronic reaction matrix $K_{ij}^{\Lambda}(R)$, which connect different electronic states, i and j. Because the details of this rearrangement depend on the situation in the core at the moment of the collision the K_{ij} elements are functions of R as well as of Λ. When the Rydberg electron leaves the collision region the system is projected on to the asymptotic channel $|jv_j^+N_j^+\rangle^{JM}$, on the right of the integrand in Eq. (4). The full rovibronic reaction matrix of Eq. (4) thus leads to and accounts for a host of interactions between different rotational, vibrational, and electronic channels. At energies below threshold these interactions become apparent as perturbations of the rovibronic levels. A source of the power of MQDT resides in the fact that the limits corresponding to a coupled and uncoupled Rydberg electron are both directly and explicitly incorporated in the theory. The transformation between these two limits is embedded in the bra-kets ${}^{JM}\langle iv_i^+N_i^+|iR\Lambda\rangle^{JM}$ of Eq. (4) which will be specified below.

Once Eq. (4) has been used to perform the transformation from the electronic reaction matrix to the full rovibronic reaction matrix, the total MQDT wave function may then be written as a linear combination of the MQDT basis functions of Eq. (2)

$$\Psi^{JM}(E)=\sum_{iv_i^+N_i^+}Z_{iv_i^+N_i^+}\Psi^{JM}_{iv_i^+N_i^+}(E)\ . \tag{5}$$

An important aspect of this expansion for the total wave function is that it only involves channel functions evaluated "on the energy shell," i.e., at the energy E. This is possible because the channel functions of Eq. (2) are defined at all energies and already account for interchannel mixing *via* the incorporation of h Coulomb functions in their definitions through the rovibronic K matrix. Imposition of bound-state boundary conditions on the total wave function $\Psi^{JM}(E)$ of Eq. (5) results in the following well-known [9] determinantal condition for the bound-state energies

$$\left|K+\frac{\tan(\pi\nu)}{A(\nu)}\right|=0\ , \tag{6}$$

where ν is the effective quantum number

$$\nu_{iv_i^+N_i^+}=\left[-\frac{1}{2\epsilon_{iv_i^+N_i^+}}\right]^{1/2} \tag{7}$$

(see Sec. III C 2 for the slight modification of this relation actually used in our calculations), and $A(\nu)$ arises from our use of the h Coulomb function [9]

$$A(\nu_{iv_i^+N_i^+})=\prod_{j=0}^{l_i}\left[1-\frac{j^2}{\nu^2_{iv_i^+N_i^+}}\right]\ . \tag{8}$$

$\tan(\pi\nu)$ and $A(\nu)$ are the diagonal matrices

$$\begin{aligned}[\tan(\pi\nu)]_{iv_i^+N_i^+,jv_j^+N_j^+}&=\delta_{ij}\delta_{v_i^+v_j^+}\delta_{N_i^+N_j^+}\tan(\pi\nu_{iv_i^+N_i^+})\ ,\\ [A(\nu)]_{iv_i^+N_i^+,jv_j^+N_j^+}&=\delta_{ij}\delta_{v_i^+v_j^+}\delta_{N_i^+N_j^+}A_{l_i}(\nu_{iv_i^+N_i^+})\ .\end{aligned} \tag{9}$$

Equation (6) is just the rovibronic generalization of the equivalent electronic Eq. (9) of RJ-I and it specifies the rovibronic bound-state energies of the molecule. The procedure is to first use Eq. (3) to calculate the electronic K matrix from the electronic η quantum defect we obtained in RJ-I. (For states involving values of J other than 0 the appropriate η matrices for Λ other than 0 have to be determined. This poses no problem.) The electronic K matrix is then substituted into Eq. (4) and the full rovibronic K matrix is calculated as described in Sec. III. With the full rovibronic K matrix the determinant in Eq.

(6) may be calculated for any given total energy E by use of Eqs. (7) to (9), with the energy ϵ of the Rydberg electron as seen in each channel given by Eq. (1). If the determinant is zero then the trial energy E corresponds to the energy of a rovibronic bound state of the molecule. If it differs from zero then the energy does not correspond to a rovibronic bound state and another energy may be tried. This search procedure is automated in the computer program.

The energy search procedure was guided by experimental results, using a grid spacing of several wavenumber units in the region of each experimental energy. Because it is possible to miss levels care must be taken. Two independent formulations of the problem were used, one in terms of the *tangent* function, the other in terms of *sine* and *cosine* functions as in JA. These have different behaviors around the eigenenergies and facilitate the search for problematic levels. Occasionally "false" levels corresponding to high lying vibrational states of the unphysical $2d$ state appear. This is presumably due in part to the finite basis set used in the calculations. The false levels, however, are easily identified by examination of the eigenvectors, and are then discarded.

III. CALCULATION OF THE ROVIBRONIC-ENERGY LEVELS

In this section we describe the details of the procedure used to calculate the full rovibronic K matrix of Eq. (4) and subsequently the rovibronic bound-state energies using Eq. (6)

A. Frame transformation

We begin by specifying the channel factors that appear in Eq. (2). These are given by

$$|iv_i^+N_i^+\rangle^{JM}=|n_i^+\Gamma_i^+\Lambda_i^+\rangle|v_i^+;iN_i^+\rangle\sum_m(N_i^+M-m,l_im|JM)|N_i^+\Lambda_i^+M-m\rangle|l_im\rangle' . \tag{10}$$

Here $|n_i^+\Gamma_i^+\Lambda_i^+\rangle$ is the electronic core state, with n_i^+ numbering core states of angular-momentum projection Λ_i^+ and of symmetry Γ_i^+. Thus, n_i^+, Γ_i^+, and Λ_i^+, along with l_i, are contained in the compound index i. In the present example, the core states involved are the $1\sigma_g$ and $1\sigma_u$ states of H_2^+. $|v_i^+;iN_i^+\rangle$ are the adiabatic vibrational core wave functions which depend on both the electronic state implicit in the index i and on the core rotational quantum number N_i^+. Finally, the angular momenta of the ion core and of the Rydberg electron are coupled by standard Clebsch-Gordon coefficients to yield a given J and M. The ion-core rotational functions $|N_i^+\Lambda_i^+M_i^+\rangle$ correspond to the isomorphic linear molecule Hamiltonian [15], whereas the Rydberg electron is described by standard spherical harmonics $|l_im\rangle'$. The prime denotes quantization with respect to the space-fixed Z axis.

The short-range functions that appear in Eq. (4) for the rovibronic K matrix are given by

$$|iR\Lambda\rangle^{JM}=|n_i^+\Gamma_i^+\Lambda_i^+\rangle|R\rangle|l_i\Lambda-\Lambda_i^+\rangle|J\Lambda M\rangle . \tag{11}$$

Here R is the internuclear distance as before and $|J\Lambda M\rangle$ is a rotational wave function which now describes the rotational of the whole molecule, including the Rydberg electron. Correspondingly the spherical harmonic $|l_i\Lambda-\Lambda_i^+\rangle$ refers to quantization in the molecular frame, and Λ is the total angular-momentum component along the molecular axis. The core electronic wave functions in Eq. (11) are the same as in the asymptotic expression Eq. (10).

The two expressions, Eqs. (10) and (11), must now each be combined to yield states of definite total parity. Letting $\rlap{/}p$ represent the parity ($\pm$) of the total wave function and $\rlap{/}p^+$ the electronic parity of the core (-1 for Σ^- states, $+1$ otherwise), the symmetric combinations for the asymptotic region are found to be *symmetrized asymptotic basis functions,*

$$|iv_i^+N_i^+;\rlap{/}p\rangle^{JM}=\frac{1}{\sqrt{2(1+\delta_{\Lambda^+0})}}\sum_m(N_i^+M-m,l_im|JM)$$
$$\times\{|n_i^+\Gamma_i^+\Lambda_i^+\rangle|N_i^+\Lambda_i^+M-m\rangle+\rlap{/}p\rlap{/}p^+(-1)^{N_i^++l_i}|n_i^+\Gamma_i^+\overline{\Lambda_i^+}\rangle|N_i^+\overline{\Lambda_i^+}M-m\rangle\}|l_im\rangle'|v_i^+;iN_i^+\rangle , \tag{12}$$

where all quantum numbers other than m and M are nonnegative, and $\overline{\Lambda_i^+}=-\Lambda_i^+$.

For the Born-Oppenheimer region there are two distinct types of symmetrized basis functions as follows: *symmetrized Born-Oppenheimer basis functions,*

$$|iR\Lambda;\oplus\rlap{/}p\rangle^{JM}=\frac{1}{\sqrt{2(1+\delta_{\Lambda^+0}\delta_{\Lambda 0})}}\{|n_i^+\Gamma_i^+\Lambda_i^+\rangle|l_i\Lambda-\Lambda_i^+\rangle|J\Lambda M\rangle+\rlap{/}p\rlap{/}p^+(-1)^J|n_i^+\Gamma_i^+\overline{\Lambda_i^+}\rangle|l_i\overline{\Lambda-\Lambda_i^+}\rangle|J\overline{\Lambda}M\rangle\}|R\rangle , \tag{13}$$

$$|iR\Lambda;\ominus\not{p}\rangle^{JM}=\frac{1}{\sqrt{2(1+\delta_{\Lambda^+0}\delta_{\Lambda 0})}}\{|n_i^+\Gamma_i^+\overline{\Lambda_i^+}\rangle|l_i\Lambda+\Lambda_i^+\rangle|J\Lambda M\rangle+\not{p}\not{p}^+(-1)^J|n_i^+\Gamma_i^+\Lambda_i^+\rangle|l_i\overline{\Lambda+\Lambda_i^+}\rangle|J\bar{\Lambda}M\rangle\}|R\rangle .$$

Here, the $\oplus$ functions correspond to parallel angular-momentum projections of the ion core and the Rydberg electron, while the $\ominus$ functions correspond to antiparallel projections. The $\ominus$ functions only occur for non-Σ cores combined with non-σ electrons and, thus, do not play a role in the current work.

Using these expressions the evaluation of the rotational frame transformation required for the full K matrix of Eq. (4), is performed following the procedure of Chang and Fano [16]. The resulting elements of the frame transformation are

$$\begin{aligned}&{}^{JM}\langle iv_i^+N_i^+;\not{p}|iR\Lambda;\oplus\not{p}\rangle^{JM}\\&\quad=\frac{1}{\sqrt{(1+\delta_{\Lambda_i^+0})(1+\delta_{\Lambda_i^+0}\delta_{\Lambda 0})}}\left[\frac{2N_i^++1}{2J+1}\right]^{1/2}[1+\delta_{\Lambda_i^+0}\not{p}\not{p}^+(-1)^{N_i^++l_i}](N_i^+\Lambda_i^+,l_i\Lambda-\Lambda_i^+|J\Lambda)\langle v_i^+;iN_i^+|R\rangle ,\\&{}^{JM}\langle iv_i^+N_i^+;\not{p}|iR\Lambda;\ominus\not{p}\rangle^{JM}\\&\quad=\frac{1}{\sqrt{(1+\delta_{\Lambda_i^+0})(1+\delta_{\Lambda_i^+0}\delta_{\Lambda 0})}}\left[\frac{2N_i^++1}{2J+1}\right]^{1/2}[\delta_{\Lambda_i^+0}+\not{p}\not{p}^+(-1)^{N_i^++l_i}](N_i^+\overline{\Lambda_i^+},l_i\Lambda-\overline{\Lambda_i^+}|J\Lambda)\langle v_i^+;iN_i^+|R\rangle .\end{aligned}\tag{14}$$

This frame transformation is diagonal in i, J, M, and $\not{p}$, and corresponds to that given by Child and Jungen [17]. Note that these authors corrected an error in the original derivation by Chang and Fano.

B. Evaluation of full rovibronic reaction matrix

In order to evaluate the full rovibronic K matrix of Eq. (4) one must substitute the frame transformation of Eqs. (14) into (4). Factoring out the terms that do not depend on the internuclear distance R and using Eq. (3) we find that the integral to be performed is

$$\begin{aligned}&\int\langle v_i^+;iN_i^+|R\rangle K_{ij}^\Lambda(R)\langle R|v_j^+;jN_j^+\rangle dR\\&\quad=\langle v_i^+;iN_i^+|\tan\pi\eta_{ij}^\Lambda(R)|v_j^+;jN_j^+\rangle .\end{aligned}\tag{15}$$

A difficulty arises in the evaluation of these elements because of the appearance of the tangent function. Since the η defects can, and in the present case do, cross half integer values the integrand is singular. In the case of a single electronic channel, Du and Greene [18] use a technique of evaluation based on the spectral decomposition of the $\eta(R)$ operator which results in the evaluation of integrals involving $\eta(R)$, instead of $\tan\pi\eta(R)$. Because of the smoothness with R of the quantum defects [see, for example, Fig. 2 of JA or Fig. 3(a) of RJ-I] these integrals pose no numerical difficulty. In this work, we generalize Du and Greene's technique to additionally account for electronic channel mixing. Derived in Appendix A, this technique involves the following three steps.

(1) For each Λ, the electronic matrix $K_{ij}^\Lambda(R)=\tan\pi\eta_{ij}^\Lambda(R)$ is diagonalized on a relatively coarse grid of R values by the matrix $V^\Lambda(R)$. The arctangent of the resulting diagonal matrix is calculated (equivalent, for a diagonal matrix, to taking the arctangent of each diagonal element), thereby removing the poles that occur in the electronic reaction matrix $K^\Lambda(R)$, and the result is divided by π to yield the diagonal matrix of "eigenquantum defects" η^Λ. Care is taken in evaluating the arctangent to ensure continuity with R of the diagonal elements η_α^Λ. The inverse transformation is applied to the matrix of eigenquantum defects to yield the smooth electronic matrix

$$\begin{aligned}M^\Lambda(R)&=V^\Lambda(R)[\eta^\Lambda(R)]V^{\Lambda^\dagger}(R)\\&=V^\Lambda(R)\left[\frac{1}{\pi}\arctan(V^{\Lambda^\dagger}(R)K^\Lambda(R)V^\Lambda(R))\right]\\&\quad\times V^{\Lambda^\dagger}(R) ,\end{aligned}\tag{16}$$

which is splined onto a fine grid of R values in preparation for numerical integration in step 2.

(2) The smoothness with R of the $M^\Lambda(R)$ matrix allows the straightforward Simpson's rule evaluation of the matrix elements of $M^\Lambda(R)$ in terms of the adiabatic ionic vibrational functions $|v_i^+;iN_i^+\rangle$. The matrix elements are multiplied by the appropriate rotational parts of the frame transformation of Eq. (14) and summed over Λ to yield the complete rovibronic quantum-defect matrix M (where for clarity we neglect symmetry)

$$\begin{aligned}M_{iv_i^+N_i^+,jv_j^+N_j^+}=\sum_\Lambda&{}^{JMi}\langle N_i^+|\Lambda\rangle^{JMi}\\&\times\langle v_i^+;iN_i^+|M_{ij}^\Lambda(R)|v_j^+;jN_j^+\rangle\\&\times{}^{JMj}\langle\Lambda|N_j^+\rangle^{JMj} .\end{aligned}\tag{17}$$

(3) The full rovibronic K matrix of Eq. (4) is then recovered by diagonalizing M with the unitary matrix U, taking tangentπ of the diagonal result, and then applying the inverse transformation

$$K=U\tan(\pi U^\dagger MU)U^\dagger .\tag{18}$$

The appearance of the tangentπ in this step is a formal

reversal of the $1/\pi$arctangent of step 1, and is therefore what returns us to the K matrix.

Thus, the integration in Eq. (15) of the possibly ill-behaved $\tan\pi\eta(R)$ functions is replaced by Eq. (17) by integrals involving well behaved R dependent functions. The price paid for this simplification is the diagonalization in Eq. (18) of the large (200×200 here) rovibronic quantum-defect matrix of Eq. (17), a step not needed in a direct evaluation of the integrals involving the tangent function.

The quantum-defect matrix determined in RJ-I, and which we use here, involves a linear dependence on energy of the η_{ss} defect. A particular advantage of the evaluation technique just described is that if the energy dependence of the quantum defects is smooth then this smoothness carries through to the M matrix evaluated in step 2 [Eq. (17)]. The energy dependence can, therefore, be accounted for by performing a preparatory calculation of M on a coarse grid of Rydberg electron energies, ϵ. Because this matrix is a smooth function of energy, interpolation to determine its elements at other energies is straightforward and accurate. Thus, in the energy search calculation, each element of M is determined by interpolating between its precalculated values. Following Gao and Greene [19], we use the energy-modified vibrational frame transformation, i.e., we use the arithmetic mean, $\bar{\epsilon}=(\epsilon_i+\epsilon_j)/2$, of the channel energies in evaluating $M_{ij}^{\Lambda}(R)$ when performing the integrals of Eq. (17). In this way steps 1 and 2, which include the numerical integrations, only need to be performed on the coarse energy grid. In the current work of a grid of three points over the energy range $\epsilon=-0.25$–0.00 a.u. combined with quadratic interpolation resulted in essentially identical results as obtained by a cubic spline interpolation over five points. Note, however, that in the case of an energy dependence the disadvantage of having to diagonalize the large M matrix in Eq. (18) is repeated for each energy.

Our results show that this method of accounting for the energy dependence of the quantum-defect matrix represents a significant improvement over that used in our proof of principle calculation [8], wherein the η_{ss} quantum defect was taken as a step function of energy, having a single step between the 2s and 3s diabatic states. The linear energy dependence of the quantum defect used here is also more appropriate than the step energy dependence used there. This better treatment, combined with the improvement in the fitted quantum-defect functions obtained in RJ-I, is the origin of the significant improvement in the vibronic results obtained below.

C. Remaining effects

In their paper on the use of MQDT to account for rovibronic interaction in H_2, JA begin their development of the theory by partitioning the molecular Hamiltonian in a manner different from that of the traditional quantum-mechanical treatment [see Eq. (3) of JA]. Their partitioning involved identifying the parts of the Hamiltonian relating to the H_2^+ ion core, the Rydberg electron, and to the interaction between the ion core and the Rydberg electron. MQDT, applied without correction, accounts for the most important of the terms in each of these parts of the Hamiltonian. The remaining terms can generally be accounted for by slight modifications of the MQDT procedure. Using the notation of JA, we may identify the terms neglecting in the "pure" MQDT for homonuclear H_2 as the following (in a.u.):

$$H_{\text{missing}}^{\text{Ion}}=\frac{1}{2\mu}\nabla_{R\Theta\Phi}^{+2}-\frac{1}{8\mu}\nabla_2^2\,,$$

$$H_{\text{missing}}^{\text{Rydberg}}=\frac{1}{8\mu}\nabla_1^2\,, \tag{19}$$

$$H_{\text{missing}}^{\text{Interaction}}=\frac{1}{4\mu}\nabla_1\cdot\nabla_2\,,$$

where the subscripts "1" and "2" represent the Rydberg electron and the core electron, respectively, and μ is the reduced mass of the H_2 molecule. Because these "missing" terms are all relatively small we consider them in a diagonal (or adiabatic) approximation by accounting for their expectation values. In Secs. III C 1–III C 3 we discuss each of these terms in turn, and how they are accounted for in the MQDT treatment. Finally, we discuss the consequences of the neglect of l mixing at large R and of relativistic effects in the H_2^+ ion in Secs. III C 4 and III C 5, respectively.

1. $H_{\text{missing}}^{\text{Ion}}$

As described in JA, the elements missing from the ionic part of the MQDT procedure, and given above as $H_{\text{missing}}^{\text{Ion}}$, can be accounted for by a standard adiabatic approximation in which the expectation value $\langle H_{\text{missing}}^{\text{Ion}}(R)\rangle$ is determined for each electronic state of the ion as a function of R, neglecting coupling with other states. These expectation values are the R-dependent adiabatic corrections which are added to the ionic potential curves for the determination of the adiabatically corrected vibrational wave functions and energies of the ion. It is these adiabatic vibrational wave functions which were used in JA and which we use to calculate the full rovibronic K matrix of Eq. (4), using the procedure of Sec. III B. The adiabatic vibrational energies are used in Eq. (1).

The present problem involves two electronic states of the H_2^+ ionic core, the $1\sigma_g$ and the $1\sigma_u$ states. Although the adiabatic corrections we use for the $1\sigma_g$ state are well known [20], those of the $1\sigma_u$ state of the H_2^+ ion core are apparently not available in tabular form in the literature and we calculated them ourselves as described in Appendix B.

This term is responsible for quite uniform increases in the energies of the calculated ionic levels, of the order of roughly 60 cm^{-1} around equilibrium for the $1\sigma_g$ state, and around 120 cm^{-1} at the same $R=2$ a.u. value for the $1\sigma_u$ state. These increases carry through to the calculated molecular levels.

2. $H_{\text{missing}}^{\text{Rydberg}}$

This term, the "normal mass effect," is clearly proportional to the kinetic-energy operator for the Rydberg electron,

$$H_{\text{missing}}^{\text{Rydberg}}=\frac{1}{8\mu}\nabla_1^2=\frac{1}{4\mu}\left[\frac{1}{2(m_e=1)}\nabla_1^2\right]=\frac{1}{4\mu}T_1 . \quad (20)$$

We account for this term in essentially the same way as we did for $H_{\text{missing}}^{\text{Ion}}$, by including its expectation value in the MQDT treatment. In the present work, we do this in a more complete way than has previously been done. In JA, for example, the expectation value $\langle H_{\text{missing}}^{\text{Rydberg}}\rangle$ was determined using the virial theorem for the electron alone, as is done for atoms [21]. Here the fact that our MQDT determination of the quantum defects in RJ-I relied on the Born-Oppenheimer approximation, in which the nuclei are held "clamped," is also taken into account in the calculation of the electronic expectation value of $\langle H_{\text{missing}}^{\text{Rydberg}}\rangle$.

The virial theorem and its application to the clamped-nuclei approximation is well known from standard texts [22]. Using the virial theorem the expectation value of the kinetic-energy operator is given by

$$\langle T\rangle=-E(R)-R\frac{dE(R)}{dR} , \quad (21)$$

where $E(R)$ is the electronic energy of the state. We may also apply this result to the H_2^+ ion core, denoting the related core quantities with a superscript "+,"

$$\langle T^+\rangle=-E^+(R)-R\frac{dE^+(R)}{dR} . \quad (22)$$

Subtracting this equation from the previous one, remembering that Rydberg electron has total energy ϵ and is indicated by the index "1," we obtain,

$$\begin{aligned}\langle T_1\rangle&=\langle T\rangle-\langle T^+\rangle\\&=-[E(R)-E^+(R)]-R\frac{d}{dR}[E(R)-E^+(R)]\\&=-\left[\epsilon(R)+R\frac{d\epsilon(R)}{dR}\right] , \end{aligned}\quad (23)$$

and hence, in atomic units ($m_e=1$),

$$\langle H_{\text{missing}}^{\text{Rydberg}}\rangle=-\frac{1}{4\mu}\left[\epsilon(R)+R\frac{d\epsilon(R)}{dR}\right] . \quad (24)$$

This quantity represents the amount by which the calculated energies must be increased. Due to the different forms of the two terms appearing in Eq. (24) their incorporation into our MQDT treatment is handled in different ways and so discussed separately.

The first term, $-\epsilon(R)/4\mu$, corresponds to a proportional change in the energy of the Rydberg electron

$$\epsilon^{\text{Corrected}}(R)=\left[1-\frac{1}{4\mu}\right]\epsilon(R) . \quad (25)$$

It is accounted for in the usual manner by decreasing the Rydberg energy by a factor of $(1-1/4\mu)$ throughout the rovibronic treatment [7]. This amounts to shifting the Rydberg constant in Eq. (7) from the value $\frac{1}{2}$ a.u., appropriate for infinite mass, to the value $\frac{1}{2}(1-1/4\mu)$ a.u., a difference of just under 30 cm^{-1} for H_2.

The second term of Eq. (24) is not simply proportional to ϵ and can therefore not be absorbed into the Rydberg constant. Instead, we adjust the energies obtained in the MQDT by changing the quantum defects. In Appendix C we show that this small electronic correction can be treated in a diagonal approximation by correcting the diagonal quantum defects $\eta(R)$ according to

$$\eta^{\text{Corrected}}=\eta+\Delta\eta=\eta-\frac{1}{4\mu}R\frac{\partial\eta}{\partial R} . \quad (26)$$

In this work, the derivative correction of Eq. (26) is applied. Its effect on the calculated rovibronic energies was determined by comparing the results of including or not including it. Such a comparison shows that this correction improves the results for the lowest levels of the *F* state by up to 9 cm^{-1}. The overall rms error of the calculated levels is only slightly improved by the inclusion of this correction, but the average error decreases from 3.9 to 0.3 cm^{-1}.

3. $H_{\text{missing}}^{\text{Interaction}}$

This term is called the "specific mass effect" and is discussed in some detail in JA. The integrand in $H_{\text{missing}}^{\text{Interaction}}$ involves a coupling between the wave function of the core electron and that of the Rydberg electron. For Rydberg states, these wave functions only overlap slightly and, thus, this term is expected to be small. Wolniewicz and Dressler [2] refer to this term as H_3' and have calculated it for the *EF*, *GK*, and $H\bar{H}$ electronic states and even for these low Rydberg states it is quite small (absolute values <2.3 cm^{-1} for the *EF* state, <7.2 cm^{-1} for the *GK* state, and <1.9 cm^{-1} for the $H\bar{H}$ state). This term is, therefore, of approximately the same size as the precision of the fitting to the Born-Oppenheimer potentials in RJ-I. At higher energy the overlap between the core and Rydberg electron wave functions becomes even smaller and, therefore, this term will rapidly become even less important. We therefore neglect it.

4. *Neglect of l mixing at large R*

Another shortcoming of the present treatment is the restriction of the partial wave expansion to $l\leq 2$ which clearly becomes inadequate as *R* becomes large. JA have shown that because the quantum defects still reproduce the accurate Born-Oppenheimer potential-energy functions at large *R*, the consequences of this restriction are limited to the adiabatic corrections. However, the physical content of the quantum-defect functions as representing the partial-wave expansion is lost to some extent. Another feature at large *R* is that the physically appropriate description of the Rydberg electron is that it orbits one or the other of the dissociating atoms, rather than an H_2^+ ion core. Thus, in this region, the mass appearing in Eq. (25) should now be that of the H nucleus, rather than that of the united atom (He). In the limit of large *R* this corresponds to a missing correction $\frac{1}{4}(\mathcal{R}_\infty-\mathcal{R}_{H_2})$ to the energy, where $\mathcal{R}_\infty$ is the Rydberg constant for infinite mass and $\mathcal{R}_{H_2}$ that corrected for the finite mass of H_2. JA showed that for $n=2$ this causes levels near dissociation, and thus sampling large R geometries, to be calculated 7.5 cm^{-1} too low. They indeed found this to be the case in the $2p\pi\, C\,^1\Pi_u$ state (cf. Fig. 6 of JA). This same behavior occurs here for high lying vibrational levels of the $F\,^1\Sigma_g^+$ state, as will be discussed in the Discussion of Results.

5. *Relativistic effects in* H_2^+

Relativistic effects in the H_2^+ ion core can be accounted for by simply using vibrational energies incorporating these effects when evaluating Eq. (1) for use in Eq. (7). Such energies, however, are only available for the physical vibrational levels, which form only a small subset of the 200 vibrational levels used in our calculation. Furthermore, these relativistic corrections are small [23] and their inclusion or noninclusion has only a small effect on the energies obtained in our calculations. The results quoted in Table I do not include the relativistic effects of the ion.

D. Computational details

The determination of the full rovibronic reaction matrix of Eq. (4), by the procedure of Sec. III B, requires vibrational wave functions for the $1\sigma_g$ and $1\sigma_u$ states of the H_2^+ ion. These we determine using Numerov-Cooley numerical integration [24] over the range $R=0.1-12.0$ a.u., on a grid of 999 points, incorporating the adiabatic corrections corresponding to $\langle H^{\mathrm{Ion}}_{\mathrm{missing}}\rangle$ in the ionic potential-energy curves. For the $1\sigma_g$ state, this procedure is straightforward and in our calculations we have included the levels $v=0-44$ for both the s and d channels built on this ionic state. The p channel, however, is built on the repulsive $1\sigma_u$ state which does not support bound vibrational levels in the region in which we require vibrational basis functions. We handle this situation as before [8], by fixing the logarithmic derivative of the vibrational functions to some constant value at the outer R value. This results in an orthonormal vibrational basis set that samples the vibrational continuum over a finite range of R and at discrete energies. We chose to include those vibrational functions that are zero at the R value forming the outer bound for our calculation and have included the levels $v=0-109$. These vibrational basis sets ensure a complete coverage of the R range important for the states considered in this work.

In RJ-I we determined the η^{Λ} matrix for $\Lambda=0$. In the present work, we restrict our attention to $J=0$ levels with the consequence that the summation over Λ in Eq. (4) collapses to this single value. In addition, we have restricted the treatment of $l\leq 2$. Higher l states are nonpenetrating and, thus, are not normally strongly involved in configuration mixing for moderate values of R. Under these circumstances the symmetrized Born-Oppenheimer and asymptotic basis functions of Eqs. (12) and (13) are

$$\begin{array}{ll} \textit{Born-Oppenheimer} & \textit{Asymptotic} \\ |1\sigma_g\rangle|J=0\ \Lambda=0\ M\rangle|l=0\ \sigma\rangle: & |1\sigma_g\rangle\sum_m (N^+=0\ M-m,\ l=0\ m)|N^+=0\ 0\ M-m\rangle|l=0\ m\rangle' , \\ |1\sigma_g\rangle|J=0\ \Lambda=0\ M\rangle|l=2\ \sigma\rangle: & |1\sigma_g\rangle\sum_m (N^+=2\ M-m,\ l=2\ m)|N^+=2\ 0\ M-m\rangle|l=2\ m\rangle' , \\ |1\sigma_u\rangle|J=0\ \Lambda=0\ M\rangle|l=1\ \sigma\rangle: & |1\sigma_u\rangle\sum_m (N^+=1\ M-m,\ l=1\ m)|N^+=1\ 0\ M-m\rangle|l=1\ m\rangle' . \end{array} \tag{27}$$

Because the rotational frame transformation is diagonal in l it reduces for $J=0$ to the value unity and there is, therefore, a one-to-one correspondence between the Born-Oppenheimer and asymptotic basis functions as indicated in Eq. (27). Thus, the incoming asymptotic channels each project onto a single Born-Oppenheimer channel in the present case. In the core region, the nondiagonal quantum-defect matrix leads to configuration mixing between different Born-Oppenheimer states with $\Lambda=0$ *via* the full rovibronic reaction matrix of Eq. (4). For $J\neq 0$ the one-to-one correspondence between asymptotic and Born-Oppenheimer rotational channels is lost and further mixing due to the rotational part of the frame transformation becomes possible as will be discussed with regard to the $3d$ complex in the next paper in this series.

Physical constants used are from Ref. [25]. The quantum-defect calculations yield vibronic energies relative to the $v^+=0$, $N^+=0$, $1\sigma_g$ level of H_2^+. We convert these to energies above the $v=0$, $J=0$, level of neutral H_2 by adding the theoretical ionization potential of H_2 of Kołos, Szalewicz, and Monkhorst, viz., 124 417.512 cm^{-1} [26]. The $1\sigma_g$ and $1\sigma_u$ potential-energy curves of H_2^+ are from the work of Wind [27] and Bates, Ledsham, and Stewart [28], respectively.

IV. DISCUSSION OF RESULTS

The comparison with experiment of the $J=0$ vibronic energies that we have calculated is presented in Table I. For comparison this table also includes results obtained by Yu, Dressler, and Wolniewicz [3,6] using the coupled equations technique. Both calculations are seen to be in extremely good agreement with experiment. The comparison is illustrated by Fig. 1. In each part of this figure we show the observed minus calculated energies in wave numbers of the first 39 vibrational levels of the excited electronic states of ${}^1\Sigma_g^+$ symmetry as functions of the total energy of the vibrational state relative to the $J=0$, $v=0$ level of the $\tilde{X}\,{}^1\Sigma_g^+$ ground state. This is done for different theoretical approaches and approximations. Only the first 39 states are considered because the 40th state is not observed experimentally, and higher states involve large values of R and were not appropriate to consider here.

Figure 1(a) compares the results obtained in the Born-Oppenheimer calculation and those obtained from the present vibronic MQDT calculations. Remember that both of these calculations are based on the same *ab initio* Born-Oppenheimer (or "clamped-nuclei") potential-energy curves. In the Born-Oppenheimer calculation the

TABLE I. Comparison of observed and calculated vibronic energies of H_2 (cm^{-1}).

Vibronic state			Vibronic energy[a]	(Observed) − (Calculated) MQDT	(Observed) − (Calculated) Coupled equations[a]
$E0$			99 164.782	4.4	−0.5
		$F0$	99 363.92	1.3	−0.4
		$F1$	100 558.92	0.1	−0.4
$E1$			101 494.749	5.0	−1.0
		$F2$	101 698.93	0.3	−0.5
		$F3$	102 778.28	−1.2	−0.4
$E2$			103 559.59	−2.3	−1.4
		$F4$	103 838.54	0.0	−0.8
		$F5$	104 730.61	−3.1	−0.9
$EF9$			105 384.90	−0.1	−2.1
$EF10$			105 966.16	3.0	−1.8
$EF11$			106 713.07	2.6	−1.7
$EF12$			107 425.87	4.8	−2.0
$EF13$			108 098.56	6.1	−2.1
$EF14$			108 793.55	5.7	−2.4
$EF15$			109 493.90	6.3	−2.3
$EF16$			110 163.38	7.3	−1.5
$EF17$			110 794.19	6.2	−1.1
$EF18$			111 370.69	−4.6	−4.7
	$GK0$		111 628.81	−7.5	−3.6
	$GK1$		111 812.665	1.1	−2.9
$EF19$			112 106.09	7.2	−1.6
$EF20$			112 711.80	5.2	−3.6
		$H0$	112 957.57	−10.3	−1.0
	$GK2$		113 258.24	2.7	−7.0
$EF21$			113 393.50	2.5	−2.0
$EF22$			113 861.40	4.7	−1.7
	$GK3$		114 044.66	−1.3	−10.2
$EF23$			114 510.539	4.6	−5.5
$EF24$			115 024.834	1.3	−1.3
	$GK4$		115 099.84	−16.0	−9.2
		$H1$	115 251.52	5.1	−6.2
$EF25$			115 563.70	−1.4	−5.8
$EF26$			116 041.59	−1.4	−6.5
	$GK5$		116 164.81	1.7	−7.8
$EF27$			116 508.24	−5.2	−4.5
$EF28$			116 915.36	−16.8	−9.8
	$GK6$		117 081.43	11.4	−10.4
		$H2$	117 297.02	6.7	−8.0

[a]Values from Ref. 3.

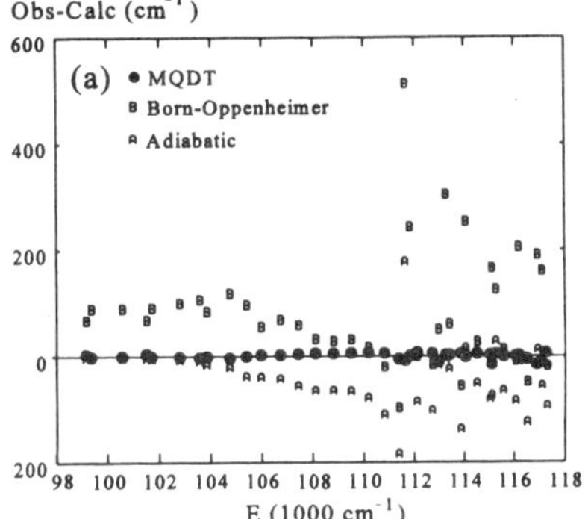

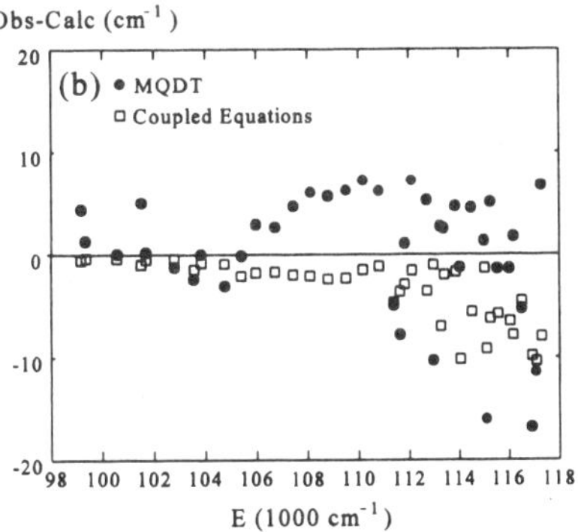

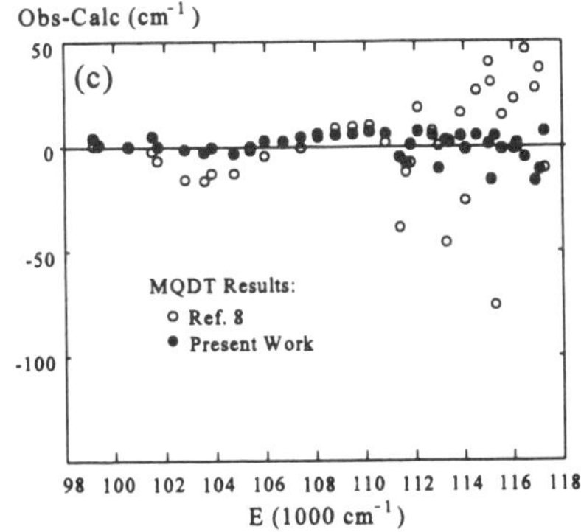

FIG. 1. Comparison of observed minus calculated vibronic energies as a function of the total vibronic energy relative to the $J=0$, $v=0$ level of the $\tilde{X}\,{}^1\Sigma_g^+$ ground state. (a) Comparison of the Born-Oppenheimer ("*B*"), Adiabatic ("*A*"), and MQDT (●) results. (b) Comparison of the coupled equations (Ref. [3]) (□) and MQDT (●) results. (c) Comparison of our "proof of principle" results (Ref. [8]) (○) and the present results (●).

wave functions and energies are obtained by numerical integration of the vibrational Schrödinger equation. The size of the non-Born-Oppenheimer effects in these states of H_2 is readily apparent in this figure. The root-mean-square error of the Born-Oppenheimer calculated energies for these states is about 110 cm^{-1}, compared to 6.0 cm^{-1} for the MQDT calculations. Indeed, one of the Born-Oppenheimer levels is wrong by over 500 cm^{-1}. Quadrelli, Dressler, and Wolniewicz [3,4] calculated adiabatic corrections to the potential-energy curves for these states. These corrections were as large as 800 cm^{-1} and result in a significant improvement in the calculated vibronic energies as is also seen in Fig. 1(a). However, even taking these adiabatic corrections into account resulted in an rms error of 58 cm^{-1}, with the energies of some levels calculated more than 180 cm^{-1} away from the experimental values. It is clear from this figure that the MQDT results are very much superior to those obtained in either the simple Born-Oppenheimer approximation, or in an adiabatic approximation.

Figure 1(b) compares results of nonadiabatic coupled-equations calculations performed by Yu, Dressler, and Wolniewicz [3,6] with the current MQDT results. Note the twentyfold increase in scale in Fig. 1(b) compared to 1(a). The coupled-equations results are in better agreement with experiment than are the MQDT results, however not overwhelmingly so. The rms errors are 6.0 cm^{-1} for the MQDT results *versus* 3.0 cm^{-1} for the coupled-equations results. The ranges of observed minus calculated values are -16.8–$+7.3$ cm^{-1} for the MQDT results and -10.4–-0.4 cm^{-1} for the coupled-equations results. In comparing these calculations, it must be remembered that the MQDT results are based on the quantum-defect matrix obtained in RJ-I, which reproduces the Born-Oppenheimer potential-energy curves to within several wave-number units. We do not expect our vibronic results to be any better than this.

Figure 1(b) also shows that the MQDT results scatter more or less uniformly above and below the observed energies. This is not true of the coupled-equations results which *must* lie to higher energy than the observed levels [and so below the axis in Fig. 1(b)]. Indeed the average of the observed minus calculated energies is only $+0.3$ cm^{-1} for the MQDT results, as opposed to -3.5 cm^{-1} for the coupled-equations calculations. Examination of the figure shows that the quality of the coupled-equations results degrades more dramatically at higher energy than do the MQDT results. Dividing the results into two regions at 111 000 cm^{-1} the rms error of the MQDT degrades by a factor of only 2.2 in the higher-energy range, whereas the coupled-equations results degrade by a factor of 4.3. Indeed based on the same division of the energy range the average error improves from 2.6–-1.6 cm^{-1} for the MQDT results, but for the coupled-equations approach degrades by a factor of 4, from -1.3 to -5.4 cm^{-1}. This is possibly due to the fact that the MQDT results take into account the effect of the entire Rydberg series lying to higher energy, whereas the coupled-equations results only take into account the first five excited $^1\Sigma_g^+$ electronic states.

Figure 1(b) and Table I further show that the levels between 106 000 and 111 000 cm^{-1}, which sample large R values corresponding to the outer limb of the F state are systematically calculated about 7 cm^{-1} too low. This clearly reflects the missing correction $\frac{1}{4}(\mathcal{R}_\infty - \mathcal{R}_{H_2})$ discussed in Sec. III C 4.

In the preceding paper, we discussed how our η matrix results in a good prediction for the 4*s* diabatic state, but a much poorer one for the 4*d* state. This is also reflected in our vibronic calculations for the $v=0$ levels of the O ($\approx 4s$) and P ($\approx 4d$) $^1\Sigma_g^+(J=0)$ states wherein our prediction for the O state is in much better agreement with the recent unpublished assignments of Dressler [3] than is that for the P state. As described in RJ-I this is attributable to our present neglect of the energy dependence of η_{dd}.

V. CONCLUSION

The results displayed in the three parts of Fig. 1 demonstrate how two entirely different methods of treating strong vibronic coupling, both starting out from the same Born-Oppenheimer clamped-nuclei potential-energy curves, reproduce an extended set of experimental level energies with near-spectroscopic accuracy. The recent coupled-equations calculations by Yu, Dressler, and Wolniewicz [6] represent a significant improvement over those performed by Quadrelli, Dressler, and Wolniewicz [4]. The improvement in the Born-Oppenheimer potential curves is responsible for some of the improvement in both methods. However, for the coupled-equations method the improvement is most striking for the higher lying vibronic levels and is due to the fact that the $5\,^1\Sigma_g^+$ and $6\,^1\Sigma_g^+$ electronic states are now included, which was not possible in the earlier work. The previous MQDT calculations of Ross and Jungen [8] already implicitly contained *all* of the $^1\Sigma_g^+$ levels, including the electronic continuum lying above. The improvement obtained in the MQDT calculations, evident in Fig. 1(c), is due instead to the more careful treatment of the energy dependence of the η_{ss} defect, and possibly also due to the inclusion of the η_{sd} defect. The possibility for further improvement of the current MQDT results lies in using the new *ab initio* calculations of Wolniewicz and Dressler [5] to determine the energy dependence of the η_{dd} defect and possibly also that of the η_{pp} defect, as discussed in RJ-I.

For a more fundamental point of view, the present MQDT calculations are a striking demonstration of the power of the frame-transformation method, which has allowed us to produce the tens of thousands of elements of the rovibrational reaction matrix from the clamped-nuclei electronic quantum-defect matrix which is composed of only six smooth functions $\eta_{ij}(R)$. This success indicates that the physical picture underlying our treatment is basically correct:

(i) The doubly excited repulsive state which is usually considered as arising from the $(1\sigma_u)^2$ electronic configuration can be viewed realistically as the lowest member of the $(1\sigma_u)(np\sigma_u)$ series.

(ii) The collision between the Rydberg electron and the core is *fast*, that is, there is no trapping of the electron inside the core region. This behavior contrasts with that found in electron-molecule collisions, such as in $e^- + N_2$, which are often dominated by shape resonances whereby the incident electron is temporarily trapped within a potential barrier.

Our immediate goal is to use the quantum-defect matrix of RJ-I in calculations for states having J other than 0, and to continue our calculations to higher energy where the coupled-equations approach becomes more difficult to implement. The calculations for $J \neq 0$ will form the subject of the next paper in this series.

ACKNOWLEDGMENTS

S.C.R. thanks the Natural Sciences and Engineering Research Council of Canada for its support of this work. He also thanks both the ETH, Zürich, Switzerland and the late Professor Fraser Birss for support during the crucial early stages of this work. In addition, he thanks the Laboratoire Aimé Cotton du CNRS, Orsay, France, for its hospitality and assistance during many working visits. Professor Kurt Dressler is again thanked for his helpful comments and continuing provision of useful data in advance of publication, Dr. Annick Suzor-Weiner is thanked for her critical examination of this manuscript, and the assistance of the late Professor Fraser Birss with regard to the numerical evaluation of the determinants is gratefully remembered.

APPENDIX A

In this appendix we derive the three-step technique of Sec. III B for the calculation of the full rovibronic K matrix. This technique is based on the standard spectral representation of the function of an operator A, with eigenvalues a and eigenfunctions $|a\rangle$,

$$f(A) = \sum_a f(a)|a\rangle\langle a| \ . \tag{A1}$$

From the spectral representation it is straightforward to show that

$$\begin{aligned} &P = \sum_i |i\rangle\langle i|, \quad f(0)=0 \ , \\ &P \text{ and } A \text{ operate in different spaces} \\ &\qquad \Longrightarrow f(PA) = Pf(A) \ , \end{aligned} \tag{A2}$$

where the $|i\rangle$ are orthonormal. For particular application to electronic channel mixing the following result, also based on the spectral representation, is used to handle the summation over Λ in Eq. (4)

$$\begin{aligned} &f(0)=0, \quad P_\Lambda P_{\Lambda'}=0 \quad \text{for } \Lambda \neq \Lambda' \\ &\qquad \Longrightarrow f\left[\sum_\Lambda P_\Lambda\right] = \sum_\Lambda f(P_\Lambda) \ . \end{aligned} \tag{A3}$$

Letting $[M]$ denote the matrix representation of the operator M then the matrix representation $[f(M)]$, of a function of M can be expressed by

$$U \text{ unitary} \Longrightarrow [f(M)] = Uf(U^\dagger[M]U)U^\dagger \ . \tag{A4}$$

Equation (A4) gives rise to the technique of Sec. III B, allowing the evaluation of the full rovibronic K matrix, while bypassing the problematic integrals involving the tangent function in Eq. (15). Thus, instead of evaluating the matrix elements of $f(M)$ on the left-hand side of the equality in Eq. (A4), the matrix elements of the "bare" operator M appearing on the right-hand side of the equality are evaluated. The resulting matrix $[M]$ is then diagonalized by the unitary matrix U and the function f is applied to the diagonal matrix $U^\dagger[M]U$ (achieved by simply evaluating the function of the diagonal elements). The final step is to use U to transform back.

In situations involving molecular vibration the MQDT reaction matrix is nothing other than the matrix representation in the appropriate basis of the function, $f(M) = \tan\pi M$, of an operator of the form $M = \sum_\Lambda P_\Lambda A_\Lambda$. That is $K = [f(M)]$, where the operator M depends on the particular problem. Once M has been identified application of Eq. (A4) allows the evaluation of the matrix elements of $f(M)$ and thus of K. In the remainder of this section we consider three particular cases, in order of increasing complexity, so as to more clearly illustrate the technique. We give M for each case, and verify that it satisfies $K = [f(M)]$.

Case (a). Pure vibronic interaction (no rotation, no electronic interaction). In this case $M = \eta^\Lambda(R)$, so that $P_\Lambda = \mathbb{I}$ and $A_\Lambda = \eta^\Lambda(R)$, where Λ only takes on one value. The basis set here consists of the vibrational wave functions $|v_i^+\rangle$ of the ion core, so that the matrix representation of tangent πM is

$$\begin{aligned} &[f(M)] = [f(\eta^\Lambda(R))] = [\tan\pi\eta^\Lambda(R)] \ , \\ &\text{which has elements} \\ &\qquad \langle v_i^+|\tan\pi\eta^\Lambda(R)|v_j^+\rangle = K_{v_i^+ v_j^+} \ . \end{aligned} \tag{A5}$$

These are indeed the elements of the pure vibronic K matrix, verifying that M has been correctly chosen.

Case (b). Rovibronic interaction (no electronic interaction). This is the case for which Du and Greene [18] introduced this technique. Here $M = \sum_\Lambda |\Lambda\rangle\eta^\Lambda(R)\langle\Lambda|$, and thus $P_\Lambda = |\Lambda\rangle\langle\Lambda|$ and $A_\Lambda = \eta^\Lambda(R)$. The basis functions now include a rotational part and are $|v_i^+ N_i^+\rangle$. Because $P_\Lambda P_{\Lambda'} = 0$ if $\Lambda \neq \Lambda'$, Eq. (A3) may be applied to M. Then, using Eq. (A2), the matrix representation of tangent πM is

$$\begin{aligned} [f(M)] &= \left[\!\left[f\left[\sum_\Lambda P_\Lambda A_\Lambda\right]\right]\!\right] \\ &= \left[\!\left[\sum_\Lambda f(P_\Lambda A_\Lambda)\right]\!\right] = \left[\!\left[\sum_\Lambda P_\Lambda f(A_\Lambda)\right]\!\right] \ , \end{aligned} \tag{A6}$$

which has elements

$$\langle v_i^+ N_i^+ | \left\{ \sum_\Lambda |\Lambda\rangle \tan\pi\eta^\Lambda(R) \langle\Lambda| \right\} | v_j^+ N_j^+ \rangle$$

$$= \sum_\Lambda \langle N_i^+|\Lambda\rangle \langle v_i^+;N_i^+|\tan\pi\eta^\Lambda(R)|v_j^+;N_j^+\rangle \langle\Lambda|N_j^+\rangle$$

$$= K_{v_i^+ N_i^+, v_j^+ N_j^+} ,$$

which are indeed those of the rovibronic K matrix of Eq. (5) of Du and Greene [18] [equivalent to Eq. (29) of JA], verifying the choice of M.

Case (c). Rovibronic interaction with electronic interaction (present work). In this case $M=\sum_\Lambda |\Lambda\rangle\{\sum_\alpha |\alpha\rangle^\Lambda \eta_\alpha^\Lambda(R)^\Lambda\langle\alpha|\}\langle\Lambda|$, so that $P_\Lambda=|\Lambda\rangle\langle\Lambda|$ and $A_\Lambda=\sum_\alpha |\alpha\rangle^\Lambda \eta_\alpha^\Lambda(R)^\Lambda\langle\alpha|$, where $|\alpha\rangle^\Lambda$ are the eigenfunctions and η_α^Λ the related eigenquantum defects of the electronic K^Λ matrix obtained in Eq. (16). The basis functions now additionally include an electronic part and are $|iv_i^+N_i^+\rangle$. Again, because $P_\Lambda P_{\Lambda'}=0$ when $\Lambda\neq\Lambda'$, Eqs. (A3) and (A2) may be applied to M, resulting in an expression containing $f(A_\Lambda)$. The special feature here is that A_Λ itself has the form $A_\Lambda=\sum_\alpha P_\alpha^\Lambda \eta_\alpha^\Lambda(R)$, where $P_\alpha^\Lambda=|\alpha\rangle^{\Lambda\Lambda}\langle\alpha|$, and thus Eqs. (A3) and (A2) can be applied once more, this time to $f(A_\Lambda)$. Doing this the matrix representation of tangent πM is

$$[f(M)]=\left[\!\left[f\left(\sum_\Lambda P_\Lambda A_\Lambda\right)\right]\!\right]=\left[\!\left[\sum_\Lambda f(P_\Lambda A_\Lambda)\right]\!\right]=\left[\!\left[\sum_\Lambda P_\Lambda f(A_\Lambda)\right]\!\right]$$

$$=\left[\!\left[\sum_\Lambda P_\Lambda f\left(\sum_\alpha P_\alpha^\Lambda \eta_\alpha^\Lambda\right)\right]\!\right]=\left[\!\left[\sum_\Lambda P_\Lambda \sum_\alpha P_\alpha^\Lambda f(\eta_\alpha^\Lambda)\right]\!\right] ,$$

which has elements

$$\langle iv_i^+N_i^+| \sum_\Lambda |\Lambda\rangle \left\{\sum_\alpha |\alpha\rangle^\Lambda \tan\pi\eta^\Lambda(R)\,^\Lambda\langle\alpha| \right\} \langle\Lambda||jv_j^+N_j^+\rangle$$

$$= \sum_\Lambda \langle iv_i^+N_i^+||\Lambda\rangle|i\rangle \left\{\sum_\alpha \langle i|\alpha\rangle^\Lambda \tan\pi\eta^\Lambda(R)\,^\Lambda\langle\alpha|j\rangle\right\} \langle j|\langle\Lambda||jv_j^+N_j^+\rangle$$

$$= \sum_\Lambda \langle iv_i^+N_i^+||\Lambda\rangle|i\rangle K_{ij}^\Lambda(R) \langle j|\langle\Lambda||jv_j^+N_j^+\rangle$$

$$= \int dR\, \langle iv_i^+N_i^+| \left\{\sum_\Lambda |iR\Lambda\rangle K_{ij}^\Lambda(R)\langle jR\Lambda|\right\} |jv_j^+N_j^+\rangle$$

$$= K_{iv_i^+N_i^+, jv_j^+N_j^+} , \tag{A7}$$

which are indeed those of the full rovibronic K matrix of Eq. (4) for arbitrary JM values, thereby verifying that M has been correctly chosen. (The introduction of $|i\rangle\langle i|$ and $|j\rangle\langle j|$ in the second step is valid due to the presence of $\langle i|$ and $|j\rangle$ in $\langle iv_i^+N_i^+|$ and $|jv_j^+N_j^+\rangle$.) $\langle i|\alpha\rangle^\Lambda$ is the $i\alpha$ element of the diagonalizing transformation V^Λ of Eq. (16).

Having determined M for each of these cases the evaluation of the rovibronic K matrix proceeds using Eq. (A4),

$$K=[f(M)]=Uf(U^\dagger[M]U)U^\dagger ,$$

where $[M]$ has elements

$$\text{case (a)}\quad \langle v_i^+|\eta^\Lambda(R)|v_j^+\rangle ,$$

$$\text{case (b)}\quad \sum_\Lambda \langle N_i^+|\Lambda\rangle\langle v_i^+;N_i^+|\eta^\Lambda(R)|v_j^+;N_j^+\rangle\langle\Lambda|N_j^+\rangle , \tag{A8}$$

$$\text{case (c)}\quad \sum_\Lambda \langle N_i^+|\Lambda\rangle\langle v_i^+;iN_i^+| \left\{\sum_\alpha \langle i|\alpha\rangle^\Lambda \eta_\alpha^\Lambda(R)\,^\Lambda\langle\alpha|j\rangle\right\} |v_j^+;jN_j^+\rangle\langle\Lambda|N_j^+\rangle ,$$

and U is chosen in each case so that $U^\dagger[M]U$ is diagonal. The right-hand side of the equality in Eq. (A8) is used to evaluate the vibronic K matrix. This is done by first evaluating the elements of the matrix representation of M. The result is the matrix $[M]$ in Eq. (A8), which, for case (c) is also given by Eq. (17). This is step 2 of the three-step process of Sec. III B. Step 3 is then to diagonalize $[M]$ by the unitary matrix U, take tangentπ of each element of the resulting diagonal matrix $U^\dagger[M]U$, and finally to transform back with the U matrix again, as shown at the top of Eq. (A8) and in Eq. (18). In this way the K matrix may be evaluated without having to integrate over the tangent function.

APPENDIX B

In this appendix we briefly describe our *ab initio* calculation of the adiabatic corrections for the $1\sigma_u$ state of H_2^+. Because of the modest nature of our requirements *vis a vis* precision (the order of a wave-number unit is satisfactory) we did not implement the procedure of Ponomarev, Puzynina, and Truskova [29], and instead proceeded by Simpson's rule integration, using the H_2^+ electronic wave functions of Bates, Ledsham, and Stewart [28] in Eqs. (4) and (6) of Bishop *et al.* [30] [with the correction that the sum on j in their expression for $M(\mu)$ should be from $\Lambda+l$ and not from Λ. This correction is needed in order to obtain equal adiabatic corrections for the $1\sigma_g$ and $1\sigma_u$ states at $R=\infty$, expected as these two steps converge to the same $H(1s)+H(2s)$ limit]. In this integration the $\partial/\partial\mu$ derivatives were evaluated numerically and the $\partial/\partial\lambda$ derivatives analytically. The calculation was done on a coarse grid of R values from 0.8–8.0 a.u., with other values obtained by splining, as was done for the $1\sigma_g$ state. The number of points and the range of integration over λ were adjusted to achieve reasonable numerical convergence. As a check we used the same program to calculate the adiabatic corrections for the $1\sigma_g$ state to compare with the values obtained by Bishop and Wetmore [20]. In the range $R=0.6$–6.0 a.u. our calculations agree with theirs to within better than 1 μa.u. (0.2 cm^{-1}), thus providing some confidence in our calculations for the $1\sigma_u$ state. The $R\to\infty$ value of the adiabatic correction for both states, $-m_eE(\infty)/M_p$ (M_P denotes mass of proton), was used to fix the adiabatic corrections at large R. For the purposes of splining the asymptotic value was used to fix the corrections at $R=100$ a.u.

The expectation values of the two terms of $H^{\text{Ion}}_{\text{missing}}$ in the limit $R\to\infty$ provides another check on the precision of our adiabatic corrections. JA point out that since the work of Van Vleck [21] it is known that in this limit the expectation values of these two terms become equal, and thus each becomes equal to half of the total asymptotic value. For the $1\sigma_g$ and $1\sigma_u$ states, the asymptotic limit for each of these terms is thus 136 μa.u. At $R=8$ a.u. Bishop *et al.* [30] obtain 135 and 133 μa.u., whereas our more approximate calculations give 137 and 134 μa.u. This comparison serves to indicate both the precision of our calculations at large R and also roughly how close to equality the two terms should be. Thus, our results of 131 and 136 μa.u. for these two terms for the $1\sigma_u$ state at $R=8$ a.u. are seen to be quite reasonable.

APPENDIX C

The derivative correction to the quantum defect, forming part of $\langle H^{\text{Rydberg}}_{\text{missing}}\rangle$, is obtained in an electronically diagonal approximation. Additionally putting aside any energy dependence of the quantum defects, the single electronic channel Eq. (4) of RJ-I

$$0=\tan\pi\nu+A(\nu)\tan\pi\eta \;, \tag{C1}$$

establishes the relationship between the energies ϵ [through ν in Eq. (3) of RJ-I] and the quantum defects η. To effect a change of $\Delta\epsilon$ on an energy satisfying Eq. (C1) would require a change of η of approximately

$$\Delta\eta\approx\frac{\partial\eta}{\partial\epsilon}\Delta\epsilon \;, \tag{C2}$$

where $\partial\eta/\partial\epsilon$ does not refer to any energy dependence of the quantum defects (which we neglect here), but instead to the relation between η and ϵ implicit in Eq. (C1). The change desired in ϵ is that given by the second term of Eq. (24),

$$\Delta\epsilon=-\frac{1}{4\mu}R\frac{\partial\epsilon}{\partial R} \;, \tag{C3}$$

and, thus, the correction to the diagonal quantum defects is given by

$$\Delta\eta=-\frac{1}{4\mu}R\frac{\partial\eta}{\partial\epsilon}\frac{\partial\epsilon}{\partial R}=-\frac{1}{4\mu}R\frac{\partial\eta}{\partial R} \;. \tag{C4}$$

[1] S. C. Ross and Ch. Jungen, preceding paper, Phys. Rev. A **49**, 4353 (1994).

[2] L. Wolniewicz and K. Dressler, J. Chem. Phys. **82**, 3292 (1985).

[3] K. Dressler (private communication).

[4] P. Quadrelli, K. Dressler, and L. Wolniewicz, J. Chem. Phys. **92**, 7461 (1990).

[5] L. Wolniewicz and K. Dressler, J. Chem. Phys. **100**, 444 (1994).

[6] S. Yu, K. Dressler, and L. Wolniewicz (unpublished).

[7] Ch. Jungen and O. Atabek, J. Chem. Phys. **66**, 5584 (1977).

[8] S. Ross and Ch. Jungen, Phys. Rev. Lett. **59**, 1297 (1987).

[9] M. J. Seaton, Rep. Prog. Phys. **46**, 167 (1983).

[10] U. Fano, Phys. Rev. A **2**, 353 (1970).

[11] G. Herzberg and Ch. Jungen, J. Mol. Spectrosc. **41**, 425 (1972).

[12] E. S. Chang, D. Dill, and U. Fano, in *Abstracts of Papers, Proceedings of the Eighth International Conference on the Physics of Electronic and Atomic Collisions*, edited by B. C. Cobic and M. V. Kurepa (Institute of Physics, Belgrade, 1973), p. 536.

[13] C. H. Greene, Comments At. Mol. Phys. **23**, 209 (1989).

[14] C. H. Greene and Ch. Jungen, Adv. At. Mol. Phys. **21**, 51 (1985).

[15] P. R. Bunker, *Molecular Symmetry and Spectroscopy* (Academic, New York, 1979).

[16] E. S. Chang and U. Fano, Phys. Rev. A **6**, 173 (1972).

[17] M. S. Child and Ch. Jungen, J. Chem. Phys. **93**, 7756 (1990).

[18] N. Y. Du and C. H. Greene, J. Chem. Phys. **85**, 5430 (1986).

[19] H. Gao and C. H. Greene, J. Chem. Phys. **91**, 3988 (1989).
[20] D. M. Bishop and R. W. Wetmore, Mol. Phys. **26**, 145 (1973).
[21] J. H. Van Vleck, J. Chem. Phys. **4**, 327 (1936).
[22] John C. Slater, *Quantum Theory of Molecules and Solids*, Electronic Structure of Molecules Vol. I (McGraw-Hill, New York, 1963).
[23] L. Wolniewicz and T. Orlikowski, Mol. Phys. **74**, 104 (1991).
[24] J. W. Cooley, Math. Comp. **15**, 363 (1961).
[25] E. R. Cohen and B. N. Taylor, *The 1986 Adjustment of the Fundamental Physical Constants*, report of CODATA Task Group on Fundamental Constants, CODATA Bulletin 63 (Pergamon, Elmsford, New York, 1986).
[26] W. Kołos, K. Szalewicz, and H. J. Monkhorst, J. Chem. Phys. **84**, 3278 (1986).
[27] H. Wind, J. Chem. Phys. **42**, 2371 (1965).
[28] D. R. Bates, K. Ledsham, and A. L. Stewart, Phil. Trans. R. Soc. London, Ser. A **246**, 215 (1953).
[29] L. I. Ponomarev, T. P. Puzynina, and N. F. Truskova, J. Phys. B **22**, 3861 (1978).
[30] D. M. Bishop, Shing-Kuo Shih, C. L. Beckel, Fun-Min Wu, and J. M. Peek, J. Chem. Phys. **63**, 4836 (1975).

1988 *Phys. Rev. Lett.* **61** 2538–41
Reprinted with permission from the American Physical Society

Application of Generalized Quantum-Defect Theory to Photodissociation Processes: Predissociation of Ar-H_2

M. Raoult
Laboratoire de Photophysique Moleculaire du Centre National de la Recherche Scientifique, Bâtiment 213, Université de Paris-Sud, 91405 Orsay CEDEX, France

G. G. Balint-Kurti
School of Chemistry, University of Bristol, Bristol BS8 1TS, United Kingdom
(Received 18 July 1988)

The application of generalized quantum-defect theory to photodissociation processes is described. The theory is of particular value when sharp resonance behavior is present in the photodissociation cross section. We show how the theory is capable of correctly predicting very sharp resonance features without the necessity of repeatedly solving systems of coupled differential equations at a very closely spaced sequence of energies. All the essential quantities occurring in the theory are smoothly varying functions of the energy despite the sharp resonances they predict.

PACS numbers: 33.80.Gj, 31.10.+z

Multichannel quantum-defect theory has proved itself as a valuable tool in the analysis of electronic Rydberg spectra of both atoms[1] and molecules.[2,3] It has recently been generalized to treat systems other than those involving Coulomb forces.[4] In this generalized form it constitutes an alternative way of handling the exact theory of scattering or continuum processes. Its principle advantages are that the central quantities computed in the theory are smoothly varying with energy. They may therefore be easily interpolated over energy intervals which are large compared with the widths of resonance features occurring in the scattering cross sections or spectra being studied. Our aim in the present work has been to apply this generalized quantum-defect theory to predissociation processes in van der Waals molecules. These processes have been extensively studied both experimentally[5] and theoretically.[6,7] The more exact theoretical treatments of the van der Waals predissociation process involve the solution of large systems of coupled differential equations. As the line shapes we discuss below typically have a width of the order of 0.02 to 0.002 cm^{-1} (10^{-6} to 10^{-7} eV), conventional methods require these difficult calculations to be carried out very many times, at very closely spaced energies, both to locate and to map out the shapes of the resonances.

Generalized multichannel quantum-defect theory (MQDT) overcomes this problem by distinguishing two different regions of importance. First, there is the inner or core region, in which the coupling between the different channels in the set of coupled equations is nonnegligible. The coupled differential equations are solved exactly by numerical methods in this inner region. The diagonal potential matrix elements are, in general, much more long ranged than the matrix elements which couple the different channels, so substantial elastic scattering may still take place outside the inner strong-coupling region. This residual elastic scattering is taken into account by expansion of the wave function at the boundary of the inner region ($R=R_0$) in terms of the *exact* or numerically determined solutions of the uncoupled channel equations (i.e., the set of coupled differential equations in which the coupling between the different channels has been ignored). The essential aspect in which the MQDT theory differs from regular scattering theory is that when the wave function is expanded in terms of the channel wave functions at the inner-region boundary, both asymptotically open and *weakly* closed channels are treated on an equal footing. In particular, at this stage, the condition that a channel is closed at $R=\infty$ is not imposed for those channels which are only weakly closed (i.e., channels which are nearly open). As we will see below, this analysis of the wave function yields a **K** matrix whose elements are slowly varying functions of the energy. The sharp energy dependence characteristic of resonance behavior comes about only through the imposition of the asymptotic boundary conditions or, stated in an alternative manner, through the elimination of the closed channels. Note, however, that the quantities which are combined to yield this resonance are all individually smoothly varying functions of the energy.

The integral photodissociation cross section, which gives the probability of radiation being absorbed by the Ar-H_2 complex and of the complex subsequently fragmenting to yield Ar+$H_2(v,j)$, may be written as[6,7]

$$\sigma(Evj\,|\,E_iJ_i)=\frac{8\pi^3\nu}{3c(2J_i+1)}\sum_{J,Jl}|\langle\Phi^-(EJvjl)\,|\,\mu\,|\,\Phi_0(E_iJ_i)\rangle|^2, \tag{1}$$

where $R^{-1}\Phi_0(E_iJ_i)$ is the radial part of the initial bound-state wave function and $R^{-1}\Phi^-(EJvjl)$ is that of the continuum wave function. The radial Schrödinger equation which must be solved for the continuum wave function in the space-fixed coordinate system is[6,8]

$$\left[-\frac{\hbar^2}{2\mu}\frac{d^2}{dR^2}+V^J_{vjl,vjl}(R)+\frac{l(l+1)\hbar^2}{2\mu R^2}+(\epsilon_{vj}-E)\right]\Phi^{Jvjl}_{vjl}(R)=-\sum_{v'j'l'}V^J_{vjl,v'j'l'}(R)\Phi^{Jvjl}_{v'j'l'}(R). \quad (2)$$

The set of coupled differential equations represented by Eq. (2) is solved exactly (numerically) by the renormalized Numerov method.[9] We include in the solution all channels considered to be important, in particular, some strongly closed channels, whose asymptotic energies lie well above the energy under investigation.

The next essential step in MQDT treatment of the scattering process is the computation of the exact channel wave functions. These are the solutions of Eq. (2) with the interchannel coupling on the right-hand side of the equation put to zero. We denote the regular and irregular solutions of the uncoupled channel equations by $f_{vjl}(R)$ and $g_{vjl}(R)$, respectively. These functions may conveniently be found by our numerically solving the uncoupled differential equations using the Milne[10] method. In his approach the regular and irregular channel wave functions of the open and weakly closed channels are represented in an amplitude-phase form:

$$f_{vjl}(R)=\alpha_{vjl}(R)\sin[\phi_{vjl}(R)],$$
$$g_{vjl}(R)=-\alpha_{vjl}(R)\cos[\phi_{vjl}(R)]. \quad (3)$$

When this *Ansatz* [Eq. (3)] for the wave function is substituted into the Schrödinger equation, an equation for $\alpha_{vjl}(R)$ is obtained and also a relationship for $\phi_{vjl}(R)$ in terms of $\alpha_{vjl}(R)$. These equations are solved to obtain the channel wave functions. The particular advantage of using this formulation is that the required, energy normalized, regular and irregular solutions may be generated without difficulty at any required energy E. A key quantity needed in generalized quantum-defect theory is also derived at this stage from the channel amplitudes for the weakly closed channels, namely,

$$\beta_{vjl}=\int_0^\infty\frac{1}{\alpha_{vjl}(R)^2}dR. \quad (4)$$

This quantity, β_{vjl}, is the generalization of the effective quantum number ν used in standard quantum-defect theory.[4]

At the boundary of the inner region (i.e., when $R=R_0$), the wave functions, which have been computed by a numerical solution of the coupled equations, are expanded in terms of the f and g channel wave functions:

$$\Phi^{Jvjl}_{v'j'l'}(R)=f_{vjl}(R)\delta_{v'j'l',vjl}+\sum_{v'j'l'}g_{v'j'l'}(R)\mathcal{K}_{v'j'l',vjl}. \quad (5)$$

The expansion of the wave function in this form yields the $\mathcal{K}$ matrix. This matrix is of dimension N_T and includes all the channels which enter the problem (i.e., the asymptotically open and both the asymptotically weakly and strongly closed channels). Having once obtained the full $\mathcal{K}$ matrix at the boundary of the inner region, we retain only its most important section, corresponding to the asymptotically open and weakly closed channels. The new, reduced-dimensionality **K** matrix consists of the upper left $N_R\times N_R$ block of the full $\mathcal{K}$ matrix.

Except for the asymptotic analysis of the wave function which is carried out using the well established techniques of multichannel quantum defect theory, we have now completely solved the scattering problem. The regular solutions of the Schrödinger equation, with specified outgoing boundary conditions, are given by Eq. (5). We now propagate these solutions inwards using the renormalized Numerov method to obtain the required regular solutions at all R values. These are needed to compute the dipole moment integrals occurring in the equation for the photodissociation cross section [Eq. (1)]. The ground-state wave function itself is computed by diagonalization of the Hamiltonian matrix in a suitably chosen basis. The most important point to note at this stage is that the elements of the **K** matrix all vary slowly with energy. The **K** matrix is now partitioned into two blocks corresponding to the asymptotically open and (weakly) closed channels

$$\mathbf{K}=\begin{pmatrix}K^{oo} & K^{oc}\\ K^{co} & K^{cc}\end{pmatrix}. \quad (6)$$

The asymptotic analysis of the wave function which involves the elimination of the weakly closed channels is now performed, yielding the expression[2]

$$K(E)=K^{oo}-K^{oc}[K^{cc}+\tan\beta(E)]^{-1}K^{co}, \quad (7)$$

where $\beta(E)$ is a diagonal matrix with elements β_{vjl} [see Eq. (4)]. This is the standard **K** or reactance matrix[2] and gives the asymptotic form of the true wave function in terms of the regular and irregular channel wave functions of the open channels only. It displays the expected resonance features near the quasibound states of the closed channels, and gives rise to the predissociative resonance line shapes.

The theory outlined above has been applied to the problem of photodissociation of the Ar-H_2 van der Waals molecule. In this Letter we study, in particular, the resonances in the photodissociation cross section associated with rotational predissociation processes. These and other processes have been studied previously with other methods.[6] The same potentials and dipole moment matrix elements are used here as in earlier work. The ground-state ($J_i=0$) wave function is calculated by our first computing the lowest bound-state channel wave functions using the Milne method for the lowest three channels. Matrix elements of the full Hamiltonian are then evaluated in this basis and the eigenfunction corresponding to the lowest eigenvalue of the matrix is taken to be the ground-state wave function of the molecule.

The results of the present calculations have been compared with those of Ref. 6 for several resonances, but space restrictions do not allow us to include a full comparison here. For the lowest rotational predissociation resonance (see Table IV of Ref. 6), the photon energy at line center agrees to within 0.002 cm^{-1} [332.823 cm^{-1} (Ref. 6) and 332.821 cm^{-1} (present work)]; the half widths differ by 6% (0.0200 and 0.0188 cm^{-1}), and the total photodissociation cross sections at line center agree to within 0.4% (5.34×10^{-6} and 5.36×10^{-6} Å^2). Figure 1 gives a graphical presentation

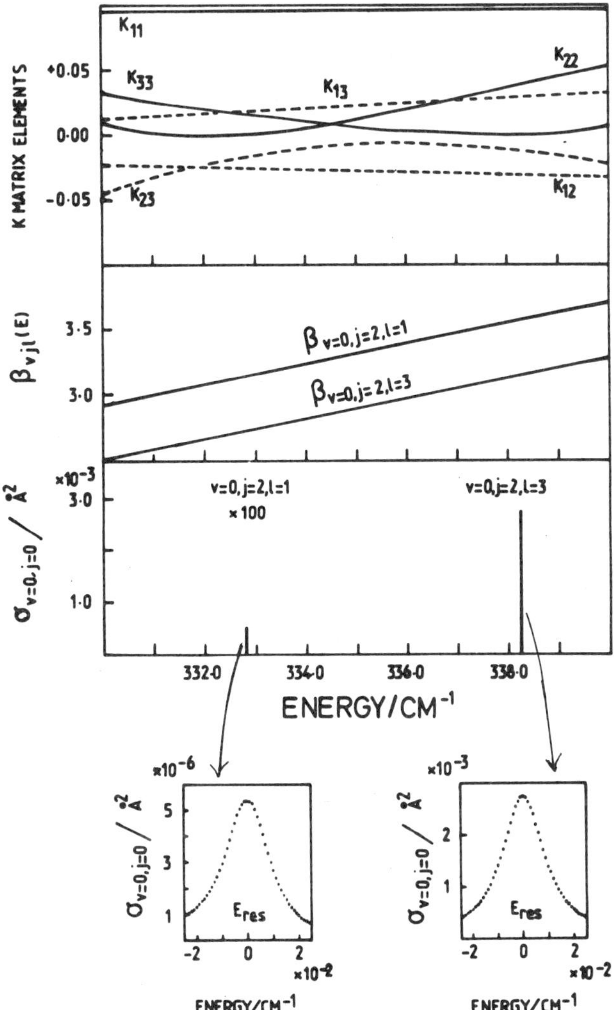

FIG. 1. Photodissociation cross sections for the two lowest-energy rotational predissociation processes in Ar-H_2. The top two panels show the (slow) variation of the **K** matrix elements [Eq. (6)] and of β_{vjl} [Eqs. (4) and (7)], while the lower panel and the two windows show the form of the predissociation line shape as a function of energy.

of our results for the first two rotational predissociation resonances. The energy range of the diagram spans about 10 cm^{-1}. In contrast, the resonances themselves, shown in the form of photodissociation cross sections in the windows at the bottom of the figure, have an energy width of the order of only 2.0×10^{-2} cm^{-1}. The topmost panel of the figure shows the smooth variation of the elements of the **K** matrix [Eq. (6)] over the whole range of the figure. The second panel similarly shows how smoothly the β_{vjl} quantities vary over the same large energy range. The resonance behavior, seen in the extremely narrow line shapes of the cross sections, comes about from the combination of these smoothly varying quantities to eliminate the asymptotically closed channels by the recipes of multichannel quantum-defect theory.

In this work multichannel quantum-defect theory has been applied for the first time to a heavy-particle breakup process in a triatomic system. The power of the MQDT method lies in the fact that it correctly reproduces resonance features in continuum processes, in this case the rotational predissociation line shapes of van der Waals molecules, but nevertheless involves the calculation of basic quantities which vary slowly with energy. This feature of the theory is exemplified in Fig. 1, where the slow variation of the computed **K** matrix elements are contrasted with the extremely sharp resonance line shapes calculated from them.

We thank C. H. Greene, Ch. Jungen, and M. S. Child for helpful discussions. G.G.B.-K. thanks the Science and Engineering Research Council for the provision of computer time on the Rutherford and Cray computers.

---

[1]M. J. Seaton, Rep. Prog. Phys. **46**, 167 (1983).

[2]C. H. Greene and Ch. Jungen, Adv. At. Mol. Phys. **21**, 51 (1985).

[3]M. Raoult, J. Chem. Phys. **87**, 4736 (1987); A. Giusti-Suzor and Ch. Jungen, J. Chem. Phys. **80**, 986 (1984).

[4]C. H. Greene, A. R. P. Rau, and U. Fano, Phys. Rev. A **26**, 2441 (1982); F. H. Mies, J. Chem. Phys. **80**, 2514 (1984); F. H. Mies and P. S. Julienne, J. Chem. Phys. **80**, 2526 (1984).

[5]A. R. W. McKellar, Faraday Discuss. Chem. Soc. **73**, 89 (1982); K. C. Janda, Adv. Chem. Phys. **60**, 201 (1985); J. C. Drobits and M. I. Lester, J. Chem. Phys. **88**, 120 (1988).

[6]I. F. Kidd and G. G. Balint-Kurti, J. Chem. Phys. **82**, 93 (1985).

[7]G. G. Balint-Kurti and M. Shapiro, Adv. Chem. Phys. **60**, 403 (1985); J. M. Hutson and R. J. LeRoy, J. Chem. Phys. **83**, 1197 (1985).

[8]G. G. Balint-Kurti, in *International Review of Science*, edited by A. D. Buckingham and C. A. Coulson (Butterworths, London, 1975), Ser. 2, Vol. 1.

[9]B. R. Johnson, J. Chem. Phys. **69**, 4678 (1978); P. S. Dardi, S. Shi, and W. H. Miller, J. Chem. Phys. **83**, 575 (1985).

[10]W. E. Milne, Phys. Rev. **35**, 863 (1930); H. J. Korsch and H. Laurent, J. Phys. B **14**, 4213 (1977).

MQDT Bibliography 1982–95

Note

For the years preceding 1982 the following list complements the references given in M J Seaton's review reprinted in this volume

(a) Books and review articles

Aymar M 1984 Rydberg series of alkaline-earth atoms Ca through Ba. The interplay of laser spectroscopy and multichannel quantum defect analysis *Phys. Rep.* **110** 163–200

Aymar M 1994 R-matrix calculation of quantum defects *Topics in Atomic and Nuclear Collisions* ed B Remaud, Calboreanu and V Zoran (New York: Plenum) 271–90

Aymar M, Greene C H and Luc-Koenig E 1996 Multichannel Rydberg spectroscopy of complex atoms *Rev. Mod. Phys.* (in press)

Dulieu O 1994 Application of quantum defect theory to the associative ionisation. Reaction between two laser-excited sodium atoms *Topics in Atomic and Nuclear Collisions* ed B Remaud, Calboreanu and V Zoran (New York: Plenum) 291–301

Fano U 1975 Unified treatment of perturbed series, continuous spectra and collisions *J. Opt. Soc. Am.* **65** 979–87

Fano U 1988 General features of collisions and spectra *Fundamental Processes of Atomic Dynamics* ed J S Briggs, H Kleinpoppen and H O Lutz (New York: Plenum) 41–9

Fano U and Rau A R P 1986 *Atomic Collisions and Spectra* (Orlando: Academic) 409 pp

Friedrich H 1990 *Theoretische Atomphysik* (Berlin: Springer) 316 pp

Friedrich H 1991 *Theoretical Atomic Physics* (Berlin: Springer) 316 pp

Giusti-Suzor A and Schneider I F 1994 Applications of multichannel quantum defect theory to collision processes *Topics in Atomic and Nuclear Collisions* ed B Remaud, Calboreanu and V Zoran (New York: Plenum) 303–12

Greene C H and Jungen Ch 1985 Molecular applications of quantum defect theory *Adv. At. Mol. Phys.* **21** 51–121

Jungen Ch 1988 Quantum defect theory for molecules *Fundamental Processes of Atomic Dynamics* ed J S Briggs, H Kleinpoppen and H O Lutz (New York: Plenum) 79–103

Lefebvre-Brion H 1995 Rotationally resolved autoionization in molecular Rydberg states *High Resolution Laser Photoionization and Photoelectron Studies* ed I Powis, T Baer and C Y Ng (New York: Wiley) 171–93

Nakamura H 1991 What are the basic mechanisms of electronic transitions in molecular dynamic processes? *Int. Rev. Phys. Chem.* **10** 123–88

O'Mahony P F 1988 Atoms in external fields *Fundamental Processes of Atomic Dynamics* ed J S Briggs, H Kleinpoppen and H O Lutz (New York: Plenum) 197–216

Rau A R P 1988 A unified view of collisions and spectra *Fundamental Processes of Atomic Dynamics* ed J S Briggs, H Kleinpoppen and H O Lutz (New York: Plenum) 51–77

Ross S 1991 An MQDT primer *Half Collision Phenomena in Molecules* ed M Garcia-Sucre, G Raseev and S C Ross (New York: American Institute of Physics) 73–110

(b) Research papers

Abramov D I and Gusev V V 1992 The Coulomb two-centre problem near the boundary of the continuous spectrum *J. Phys. B: At. Mol. Opt. Phys.* **25** 2445–57

Abutaleb M, de Graaff R J, Ubachs W and Hogervorst W 1991 $6p_{1/2,3/2}nf$ $J = 1$–5 autoionizing series of barium *Phys. Rev.* A **44** 4187–202

Abutaleb M, de Graaff R J, Ubachs W, Hogervorst W and Aymar M 1991 The 5dnp and 5dnf $J = 1$ autoionizing series of Ba in one-photon laser excitation from the $6s^2$ $^1S_0$ ground state *J. Phys. B: At. Mol. Opt. Phys.* **24** 3565–73

Alber G and Zoller P 1984 Structure of autoionizing Rydberg series in strong laser fields: a multichannel-quantum-defect-theory approach *Phys. Rev.* A **29** 2290–3

Alber G and Zoller P 1988 Near-threshold excitation of Rydberg series by strong laser fields *Phys. Rev.* A **37** 377–89

Alber G, Cooper J and Rau A R P 1984 Unified treatment of radiative and dielectronic recombination *Phys. Rev.* A **30** 2845–8

Alijah A, Hinze J and Broad J T 1990 Photoionisation of hydrogen in a strong magnetic field *J. Phys. B: At. Mol. Opt. Phys.* **23** 45–60

Armstrong D J and Greene C H 1994 Evolution of barium autoionizing Stark spectra with energy and field *Phys. Rev.* A **50** 4956–74

Armstrong D J and Robicheaux F 1993 Photoionization of the scandium atom. III. Experimental and theoretical spectra from an excited state *Phys. Rev.* A **48** 4450–60

Armstrong D J, Wood R P and Greene C H 1993 Photoionization of the 5d6p $^3D_1$ state of barium *Phys. Rev.* A **47** 1981–8

Assimopoulos S, Bolovinos A, Jimoyiannis A, Tsekeris P, Luc-Koenig E and Aymar M 1994 3dnl $J = 0^e$–2^e autoionizing levels of calcium: observation by laser optogalvanic spectroscopy and theoretical analysis *J. Phys. B: At. Mol. Opt. Phys.* **27** 2471–96

Atabek O and Jungen Ch 1976 Quantum defect theory of excited $^1\Sigma_u^+$ levels of H_2 *Electron and Photon Interactions with Atoms* ed H Kleinpoppen and M R C McDowell (New York: Plenum) 613–20

Atabek O, Dill D and Jungen Ch 1974 Quantum-defect theory of excited $^1\Pi_u^-$ levels of H_2 *Phys. Rev. Lett.* **33** 123–6

Aymar M 1985 Stabilisation of 5dnd autoionising states of barium by destructive interference effects between ionising channels *J. Phys. B: At. Mol. Phys.* **18** L763–9

Aymar M 1987 Eigenchannel R-matrix calculations of the photoabsorption spectrum of strontium *J. Phys. B: At. Mol. Phys.* **20** 6507–29

Aymar M 1990 Eigenchannel R-matrix calculation of the $J = 1$ odd-parity spectrum of barium *J. Phys. B: At. Mol. Opt. Phys.* **23** 2697–716

Aymar M and Camus P 1983 Multichannel quantum-defect analysis of the bound even-parity spectrum of neutral barium *Phys. Rev.* A **28** 850–7

Aymar M and Lecomte J M 1989 R-matrix calculation of photoionisation from Rydberg states in strontium *J. Phys. B: At. Mol. Opt. Phys.* **22** 223–45

Aymar M and Telmini M 1991 Eigenchannel R-matrix study of the $J = 0$ and $J = 2$ even parity spectra of calcium below the Ca^+ $3d_{3/2}$ threshold *J. Phys. B: At. Mol. Opt. Phys.* **24** 4935–56

Aymar M, Camus P and El Himdy A 1982 The $6p^2$ $^1S_0$ autoionised level of neutral barium *J. Phys. B: At. Mol. Phys.* **15** L759–63

Aymar M, Grafström P, Levison C, Lundberg H and Svanberg S 1982 *J. Phys. B: At. Mol. Phys.* **15** 877–82

Aymar M, Luc-Koenig E and Lecomte J M 1994 R-matrix study of the 6dng, 6dni and 7sni $J = 6$ autoionizing Rydberg series of barium *J. Phys. B: At. Mol. Opt. Phys.* **27** 2425–45

Aymar M, Luc-Koenig E and Watanabe S 1987 R-matrix calculations of eigenchannel multichannel quantum defect parameters for strontium *J. Phys. B: At. Mol. Phys.* **20** 4325–45

Aymar M, Robaux O and Wane S 1984 Central-field calculations of photoionisation cross sections of excited states of Rb and Sr^+ and analysis of photoionisation cross sections of excited alkali atoms using quantum defect theory *J. Phys. B: At. Mol. Phys.* **17** 993–1007

Baig M A 1983 High Rydberg transitions in the principal and intercombination series of mercury *J. Phys. B: At. Mol. Phys.* **16** 1511–23

Baig M A and Connerade J P 1984 A high-resolution study of the principal series of Sr I *J. Phys. B: At. Mol. Phys.* **17** L271–4

Baig M A and Connerade J P 1984 Centrifugal barrier effects in the high Rydberg states and autoionising resonances of neon *J. Phys. B: At. Mol. Phys.* **17** 1785–96

Baig M A and Connerade J P 1986 Autoionising Rydberg series in the spectrum of N_2 *J. Phys. B: At. Mol. Phys.* **19** L605–12

Baig M A, Ahmad S, Akram M, Connerade J P and Hormes J 1992 The absorption spectrum of Tl I in the vacuum ultraviolet: single and double excitation of 5d, 6s and 6p electrons *J. Phys. B: At. Mol. Opt. Phys.* **25** 1719–34

Baig M A, Ahmad S, Griesmann U, Connerade J P, Bhatti S A and Ahmad N 1992 Inner shell and double excitation spectrum of ytterbium involving the 4f and 6s subshells *J. Phys. B: At. Mol. Opt. Phys.* **25** 321–41

Baig M A, Akram M, Bhatti S A, Sommer K and Hormes J 1994 Autoionizing resonances in the 4d subshell excitation spectrum of cadmium *J. Phys. B: At. Mol. Opt. Phys.* **27** 1693–708

Baig M A, Connerade J P and Hormes J 1982 Autoionisation resonances in the $4p\pi$ spectrum of methyl bromide *J. Phys. B: At. Mol. Phys.* **15** L5–10

Baig M A, Connerade J P and Hormes J 1986 Autoionisation resonances in the photoabsorption spectrum of C_2H_5I *J. Phys. B: At. Mol. Phys.* **19** L343–7

Baig M A, Connerade J P, Mayhew C, Noeldeke G and Seaton M J 1984 On the principal series of helium *J. Phys. B: At. Mol. Phys.* **17** L383–7

Baig M A, Rashid A, Ahmad I, Rafi M, Connerade J P and Hormes J 1990 High-resolution photoabsorption study of the 3d spectrum of chromium *J. Phys. B: At. Mol. Opt. Phys.* **23** 3489–509

Baig M A, Rashid A, Iqbal Z and Hormes J 1991 High resolution absorption spectrum of palladium in the 4d subshell excitation region *J. Phys. B: At. Mol. Opt. Phys.* **24** 2295–304

Balashov E M, Golubkov G V and Ivanov G K 1984 Radiative transitions between Rydberg states of molecules *Sov. Phys.–JETP* **59** 1188–94

Bartschat K and Greene C H 1993 Short-range correlation and relaxation effects on the $(6p^2)$ $^1S_0$ autoionizing state of atomic barium *J. Phys. B: At. Mol. Opt. Phys.* **26** L109–16

Bartschat K, McLaughlin B M and Hoversten R A 1991 Photoionization and excitation of atomic barium from the (6s6p) $^1P_1^o$ state *J. Phys. B: At. Mol. Opt. Phys.* **24** 3359–71

Baruch M C, Cai L, Jones R R and Gallagher T F 1992 Barium $5d_j nf_{j'}$ $J = 3$ autoionizing series *Phys. Rev.* A **45** 6395–403

Bente E A J M and Hogervorst W 1989 The 5dnf $J = 4$ and 5 autoionising Rydberg series in barium: experiment and MQDT analysis *J. Phys. B: At. Mol. Opt. Phys.* **22** 2679–704

Berrington K A and Seaton M J 1985 Use of the R-matrix method for bound-state calculations: II. Results for energy levels of C^+ *J. Phys. B: At. Mol. Phys.* **18** 2587–99

Berrington K A, Burke P G, Butler K, Seaton M J, Storey P J, Taylor K T and Yan Y 1987 Atomic data for opacity calculations: II. Computational methods *J. Phys. B: At. Mol. Phys.* **20** 6379–97

Biernacki D T, Colson S D and Eyler E E 1988 Rotationally resolved double resonance spectra of NO Rydberg states near the first ionization limit *J. Chem. Phys.* **88** 2099–107

Bi-ru W, You-feng Z, Yun-fei X, Li-gang P, Ji L and Jian-wei Z 1991 The 6snp $^3P_{0,2}$ Rydberg series of neutral ytterbium *J. Phys. B: At. Mol. Opt. Phys.* **24** 49–55

Blum R D and Pradhan A K 1991 Fine structure and resonance transitions in C^+ *Phys. Rev.* A **44** 6123–6

Bohn J L and Fano U 1994 Multichannel quantum mechanics as a Hamiltonian phase flow *Phys. Rev.* A **50** 2893–8

Bordas C and Helm H 1991 Predissociation of H_3 near its ionization threshold induced by very weak electric fields *Phys. Rev.* A **43** 3645–52

Bordas C, Brevet P F, Broyer M, Chevaleyre J, Labastie P and Perrot J P 1988 Electric-field-hindered vibrational autoionization in molecular Rydberg states *Phys. Rev. Lett.* **60** 917–20

Bordas C, Labastie P, Chevaleyre J and Broyer M 1989 MQDT analysis of rovibrational interactions and autoionization in Na_2 Rydberg states *Chem. Phys.* **129** 21–39

Bordas M Ch, Broyer M, Chevaleyre J, Labastie P and Martin S 1985 Multichannel quantum defect theory analysis of the Rydberg spectrum of Na_2 *J. Physique* **46** 27–38

Bordas M C, Lembo L J and Helm H 1991 Spectroscopy and multichannel quantum-defect theory analysis of the np Rydberg series of H_3 *Phys. Rev.* A **44** 1817–27

Borodin V M, Fabrikant I I and Kazansky A K 1993 Collisional broadening of Rydberg atoms perturbed by ground-state sodium atoms *Z. Phys.* D **27** 45–7

Bounakhla M, Lemoigne J P, Grandin J P, Husson X, Kucal H and Aymar M 1993 Laser spectroscopy of even parity Rydberg series of krypton *J. Phys. B: At. Mol. Opt. Phys.* **26** 345–61

Branchett S E and Tennyson J 1992 Transition moments for excitation to Rydberg states

of molecules using the R-matrix method: H_2 with $n \leqslant 5$ *J. Phys. B: At. Mol. Opt. Phys.* **25** 2017–26
Brevet J P, Bordas Ch, Broyer M, Jalbert G and Labastie P 1991 MQDT operatorial formalism analysis of molecular Rydberg states in weak electric fields: application to Na_2 *J. Phys. II France* **1** 875–97
Brown C M, Tilford S G and Ginter M L 1975 Extended identifications of odd energy levels of Si I: Lu–Fano graphical analysis *J. Opt. Soc. Am.* **65** 385–8
Büchner M, Raseev G and Cherepkov N A 1992 Cross section and spin polarization for photoionization of the HI molecule: rotationally resolved theoretical results in the spin-orbit autoionization region *J. Chem. Phys.* **96** 2691–702
Bühler B, Cremer C and Gerber G 1985 Investigation of the $6p^2$ ($^3P_0$) np Rydberg series of bismuth by multiphoton excitation *Z. Phys.* A **320** 71–9
Camus P, Pillet P and Boulmer J 1984 Spectroscopic studies of 9dn′d doubly excited autoionising states of neutral barium *J. Phys. B: At. Mol. Phys.* **18** L481–7
Carré B, d'Oliveira P, Fournier P R and Gounand F 1990 $6p_{1/2,3/2}np$; $J = 0, 1, 2$ autoionizing series in atomic barium *Phys. Rev.* A **42** 6545–59
Carre B, Fournier P R, Porterat D, Lagadec H, Gounand F and Aymar M 1994 Photoionization of the $6p^2$(LS)J state in barium: measurement and calculation of the partial cross sections *J. Phys. B: At. Mol. Opt. Phys.* **27** 1027–49
Cavagnero M J 1995 Floquet analysis of inelastic collisions of ions with Rydberg atoms *Phys. Rev.* A **52** 2865–75
Chang E S 1984 Theory of rotational-vibronic excitation in linear molecules by slow electrons *J. Phys. B: At. Mol. Phys.* **17** 3341–51
Chang E S and Fano U 1972 Theory of electron–molecule collisions by frame transformations *Phys. Rev.* A **6** 173–85
Chang J J 1993 General form of the Dirac quantum-defect theory *Phys. Rev.* A **48** 1769–79
Chen C T and Robicheaux F 1994 Photoabsorption cross section of O, S, Se and Te *Phys. Rev.* A **50** 3968–79
Child M S and Jungen Ch 1990 Quantum defect theory for asymmetric tops: application to the Rydberg spectrum of H_2O *J. Chem. Phys.* **93** 7756–66
Christensen R B and Norcross D W 1985 Electron-impact excitation of Li II: A model study of wave-function and collisional approximations and of resonance effects *Phys. Rev.* A **31** 142–51
Christensen R B, Norcross D W and Pradhan A K 1985 Electron-impact excitation of ions in the magnesium sequence: Fe XV *Phys. Rev.* A **32** 93–104
Clark C W 1977 Electron scattering from diatomic polar molecules. I. The limitation of the Born approximation *Phys. Rev.* A **16** 1419–22
Clark C W 1979 Electron scattering from diatomic polar molecules. II. Treatment by frame transformations *Phys. Rev.* A **20** 1875–89
Clark C W 1983 Low-energy electron-atom scattering in a magnetic field *Phys. Rev.* A **28** 83–90
Clark C W and Taylor K T 1982 Eigenchannel analysis of neon-negative-ion resonances *J. Phys. B: At. Mol. Phys.* **15** L213–9
Colle R 1981 MQDT analysis of predissociation. I. A two channel problem with semiclassical scattering parameters *J. Chem. Phys.* **74** 2910–9

Connerade J P 1982 Quantum defect theory in a double-valley potential *J. Phys. B: At. Mol. Phys.* **15** L881–6

Connerade J P 1983 On Rydberg series of autoionising resonances *J. Phys. B: At. Mol. Phys.* **16** L329–35

Connerade J P 1983 On the "giant resonance" in the 4d spectra of Ba, Ba^+ and Ba^{2+} *J. Phys. B: At. Mol. Phys.* **16** L257–62

Connerade J P 1985 Autoionising line shapes *J. Phys. B: At. Mol. Phys.* **18** L367–71

Connerade J P 1991 On the perturbation of an autoionizing series by an antiresonance *J. Phys. B: At. Mol. Opt. Phys.* **24** L513–20

Connerade J P 1993 K-matrix theory and quantum defect plots for autoionizing series *J. Phys. B: At. Mol. Opt. Phys.* **26** 4041–56

Connerade J P and Lane A M 1985 Vanishing points of particle widths of resonances: an atomic parallel of a well known nuclear phenomenon *J. Phys. B: At. Mol. Phys.* **18** L605–10

Connerade J P and Lane A M 1987 The interaction between a Rydberg series and a shape or giant resonance *J. Phys. B: At. Mol. Phys.* **20** L181–6

Connerade J P, Baig M A and Sweeney M 1987 Centrifugal barrier effects in the 3p spectrum of calcium *J. Phys. B: At. Mol. Phys.* **20** L771–5

Connerade J P, Lane A M and Baig M A 1985 Rydberg structure within a broad resonance *J. Phys. B: At. Mol. Phys.* **18** 3507–27

Cooke W E and Cromer C L 1985 Multichannel quantum-defect theory and an equivalent N-level system *Phys. Rev.* A **32** 2725–38

Cooke W E and Cromer C L 1986 Closed-form expression for two-photon core excitation of Rydberg states *Phys. Rev.* A **33** 3529–30

Cornaggia C, Giusti-Suzor A and Jungen Ch 1987 Photoionization of the EF excited state of H_2: calculation of vibrational branching ratios and photoelectron angular distributions *J. Chem. Phys.* **87** 3934–41

Dagata J A, Findley G L, McGlynn S P, Connerade J P and Baig M A 1981 Molecular Rydberg transitions. Multichannel approaches to electronic states: CH_3I *Phys. Rev.* A **24** 2485–90

Dagata J A, Scott M A and McGlynn S P 1988 A multichannel quantum defect treatment of the Rydberg states of CH_3I *J. Chem. Phys.* **88** 9–14

Dai C J 1995 Spectroscopic properties of Mg 3pns autoionizing states *Phys. Rev.* A **51** 2951–6

Dai C J 1995 Perturbed 5snd $^{1,3}D_2$ Rydberg series of Sr *Phys. Rev.* A **52** 4416–24

Dai C J, Schinn G W and Gallagher T F 1990 Mg 3pnd ($J = 3$) autoionization spectra using isolated-core excitation *Phys. Rev.* A **42** 223–35

de Graaf R J, Ubachs W, Hogervorst W and Abutaleb M 1990 Narrow $6p_{1/2,3/2}np$ $J = 1$ autoionizing Rydberg series in barium and their interaction with $6p_{3/2}nf$ states *Phys. Rev.* A **42** 5473–80

de Graaff R J, Ubachs W and Hogervorst W 1992 4fnf doubly excited autoionizing series in barium *Phys. Rev.* A **45** 166–78

Dill D and Jungen Ch 1980 Quantum-defect functions. Interconverters of electronic and nuclear motion *J. Phys. Chem.* **84** 2116–22

Domcke W 1983 Analytic theory of resonances and bound states near Coulomb thresholds *J. Phys. B: At. Mol. Phys.* **16** 359–80

Domcke W, Sobolewski A L and Lin S H 1988 Resonances in molecular photoionization. IV. Theory of one-color and two-color near-threshold photoionization of molecules *J. Chem. Phys.* **89** 6209–19

Drescher M, Brockhinke A, Böwering N, Heinzmann U and Lefebvre-Brion H 1993 Rotationally resolved single-photon ionization of HCl and DCl *J. Chem. Phys.* **99** 2300–6

Drew J E and Storey P J 1982 Photoionization of C^{2+} and C^{3+} *J. Phys. B: At. Mol. Phys.* **15** 2357–66

Du N Y and Greene C H 1986 Quantum defect analysis of HD photoionization *J. Chem. Phys.* **85** 5430–6

Du N Y and Greene C H 1989 Multichannel Rydberg spectra of the rare gas dimers *J. Chem. Phys.* **90** 6347–60

Dubau J 1976 Quantum defect theory: photoionization *Electron and Photon Interactions with Atoms* ed H Kleinpoppen and M R C McDowell (New York: Plenum) 99–107

Dubau J and Seaton M J 1984 Quantum defect theory XIII. Radiative transitions *J. Phys. B: At. Mol. Phys.* **17** 381–403

Dubs R L and Julienne P S 1991 A multichannel quantum defect half-collision analysis of K_2 photodissociation through the $B^1\Pi_u$ state *J. Chem. Phys.* **95** 4177–87

Dulieu O, Giusti-Suzor A and Masnou-Seeuws F 1991 Theoretical treatment of the associative ionization reaction between two laser-excited sodium atoms. Direct and indirect processes *J. Phys. B: At. Mol. Opt. Phys.* **24** 4391–408

Dulieu O, Magnier S and Masnou-Seeuws F 1994 Doubly-excited states for the Na_2 molecule: application to the dynamics of the associative ionization reaction *Z. Phys.* D **32** 229–40

Fabrikant I I 1983 Generalised quantum defect theory for electron scattering by polar molecules *J. Phys. B: At. Mol. Phys.* **16** 1253–67

Fabrikant I I 1983 Frame transformation effective-range theory: application to interaction of polar molecules with electrons *J. Phys. B: At. Mol. Phys.* **16** 1269–82

Fabrikant I I 1986 R-matrix theory of inelastic processes in low-energy electron collisions with HCl molecule *Z. Phys.* D **3** 401–10

Fano U 1970 Effect of the centrifugal force on electron–molecule interactions *Comments At. Mol. Phys.* **1** 140–3

Fano U 1988 Sharp and diffuse interlopers in excitation spectra *Phys. Rev.* A **37** 4037–9

Fano U and Lee C M 1973 Variational calculation of R matrices. Application to Ar photoabsorption *Phys. Rev. Lett.* **31** 1573–6

Fano U and Lu K T 1984 Herzberg's impact on the physics of Rydberg states *Can. J. Phys.* **62** 1264–73

Fano U and Sidky E Y 1992 Semianalytic theory of Rydberg electron diamagnetism *Phys. Rev.* A **45** 4776–91

Fano U, Theodosiou C E and Dehmer J L 1976 Electron-optical properties of atomic fields *Rev. Mod. Phys.* **48** 49–68

Farooqi S M, Connerade J P, Greene C H, Marangos J, Hutchinson M H R and Shen N 1991 Laser-based VUV spectroscopy of doubly excited Ca I *J. Phys. B: At. Mol. Opt. Phys.* **24** L179–84

Farooqi S M, Connerade J P and Aymar M 1992 New observations of the doubly-excited spectrum of Sr I *J. Phys. B: At. Mol. Opt. Phys.* **25** L219–23

Feldman U, Brown C M, Doschek G A, Moore C E and Rosenberg F D 1976 XUV spectrum of C I observed from Skylab during a solar flare *J. Opt. Soc. Am.* **66** 853–9

Fielding H H 1994 Rydberg electron wavepacket dynamics in molecular hydrogen *J. Phys. B: At. Mol. Opt. Phys.* **27** 5883–91

Fielding H H and Softley T P 1992 Observation of the Stark effect in autoionizing Rydberg states of Ar with a multichannel quantum-defect analysis *J. Phys. B: At. Mol. Opt. Phys.* **25** 4125–39

Fielding H H and Softley T P 1994 Multichannel-quantum-defect-theory analysis of the Stark effect in autoionizing Rydberg states of H_2 *Phys. Rev.* A **49** 969–81

Fink M G J and Zoller P 1989 Quantum-defect parametrization of perturbative two-photon ionization cross sections *Phys. Rev.* A **39** 2933–47

Fredin S, Gauyacq D, Horani M, Jungen Ch, Lefevre G and Masnou-Seeuws F 1987 s and d Rydberg series of NO probed by double resonance multiphoton ionization. Multichannel quantum defect analysis *Mol. Phys.* **60** 825–66

Frey P, Lawen M, Breyer F, Klar H and Hotop H 1982 Partial cross sections for Rb^- photodetachment in the region of the $Rb(5p^2P_{1/2,3/2})$ thresholds and their analysis by multichannel quantum defect theory *Z. Phys.* A **304** 155–65

Frey P, Lawen M, Breyer F, Klar H and Hotop H 1982 Partial cross sections for Rb^- photodetachment in the region of the $Rb(5p^2P_{1/2,3/2})$ thresholds and their analysis by multichannel quantum defect theory (Erratum) *Z. Phys.* A **306** 185

Friedrich H and Chu M 1983 Autoionizing states of the hydrogen atom in strong magnetic fields *Phys. Rev.* A **28** 1423–8

Friedrich H and Wintgen D 1985 Interfering resonances and bound states in the continuum *Phys. Rev.* A **32** 3231–42

Gangopadhyay P, Tang X, Lambropoulos P and Shakeshaft R 1986 Theory of autoionization of Xe under two- and three-photon excitation *Phys. Rev.* A **34** 2998–3002

Ganz J, Raab M, Hotop H and Geiger J 1984 Changing the Beutler–Fano profile of the Ne(ns′) autoionizing resonances *Phys. Rev. Lett.* **53** 1547–50

Gao H 1992 Vibrational excitation of H_2 by electron impact: an energy-dependent vibrational-frame-transformation approach *Phys. Rev.* A **45** 6895–8

Gao H and Greene C H 1989 Energy-dependent vibrational frame transformation for electron–molecule scattering with simplified models *J. Chem. Phys.* **91** 3988–98

Gao H and Greene C H 1990 Alternative vibrational frame transformation for electron–molecule scattering *Phys. Rev.* A **42** 6946–9

Gao H and Greene C H 1990 Quantum defect description of non-Rydberg molecules *J. Chem. Phys.* **93** 1791–8

Gao H, Jungen Ch and Greene C H 1993 Predissociation of H_2 in the $3p\pi D^1\Pi_u^+$ state *Phys. Rev.* A **47** 4877–84

Gauyacq D, Roche A L, Seaver M, Colson D and Chupka W A 1990 s and d Rydberg complexes of NO probed by double resonance multiphoton ionisation in the region $n^* = 5$ to $n^* = 25$; multichannel quantum defect analysis. Part II *Mol. Phys.* **71** 1311–31

Ginter D S and Ginter M L 1980 The spectrum and structure of the He_2 molecule. Multichannel quantum defect analyses of the triplet levels associated with $(1\sigma_g)^2(1\sigma_u)$ npλ *J. Mol. Spectrosc.* **82** 152–75

Ginter D S and Ginter M L 1983 Spectrum and structure of the He_2 molecule: characterization of the triplet states associated with $(1\sigma_g)^2(1\sigma_u)ns\sigma$, $nd\sigma$, $nd\pi$, and $nd\delta$ $(n = 5\text{–}12)$ *J. Mol. Spectrosc.* **101** 139–60

Ginter D S and Ginter M L 1986 Transitions to and three-limit, four-channel representations for the msmp^3, mpnd $J = 3^o$ levels in Si I and Sn I *J. Chem. Phys.* **85** 6536–43

Ginter D S and Ginter M L 1987 Energy dependence in coupled channel models for electronic structures in atoms *J. Chem. Phys.* **86** 1437–44

Ginter D S and Ginter M L 1988 Multichannel interactions in the $(1\sigma_g)^2(1\sigma_u)$nsσ, ndλ $(^3\Sigma_u^+, ^3\Sigma_u^+, ^3\Pi_u, ^3\Delta_u)$ Rydberg structures of He_2 *J. Chem. Phys.* **88** 3761–74

Ginter D S, Ginter M L and Brown C M 1984 Multichannel interactions in the $(1\sigma_g)^2(1\sigma_u)$npλ $(^3\Pi_g, ^3\Sigma_g^+)$ Rydberg structures of He_2 *J. Chem. Phys.* **81** 6013–25

Giusti-Suzor A and Fano U 1984 Alternative parameters of channel interactions: II. A Hamiltonian model *J. Phys. B: At. Mol. Phys.* **17** 4267–75

Giusti-Suzor A and Fano U 1984 Alternative parameters of channel interactions: III. Note on a narrow band in the Ba $J = 2$ spectrum *J. Phys. B: At. Mol. Phys.* **17** 4277–83

Giusti-Suzor A and Jungen Ch 1984 Theoretical study of competing photoionization and photodissociation processes in the NO molecule *J. Chem. Phys.* **80** 986–1000

Giusti-Suzor A and Lefebvre-Brion H 1980 A multichannel quantum defect approach to molecular autoionization *Chem. Phys. Lett.* **76** 132–5

Giusti-Suzor A and Lefebvre-Brion H 1984 Theoretical study of complex resonances near ionization thresholds: application to the N_2 photoionization spectrum *Phys. Rev.* A **30** 3057–65

Giusti-Suzor A and Zoller P 1987 Rydberg electrons in laser fields: a finite-range-interaction problem *Phys. Rev.* A **36** 5178–88

Giusti-Suzor A, Bardsley J N and Derkits C 1983 Dissociative recombination in low-energy e-H_2^+ collisions *Phys. Rev.* A **28** 682–91

Glab W L and Qin K 1993 Stark mapping of H_2 Rydberg states in the strong-field regime with dynamical resolution *Phys. Rev.* A **48** 4492–9

Glab W L, Qin K and Bistransin M 1995 Rotational and vibrational interactions of singlet gerade Rydberg states of H_2 near the ionization limit *J. Chem. Phys.* **102** 2338–50

Goforth T L and Watson D K 1992 Multichannel quantum defects calculated using a smooth reaction matrix *Phys. Rev.* A **46** 1239–47

Goforth T L, Snitchler G L and Watson D K 1987 Determination of quantum defects from a negative-energy reaction matrix *Phys. Rev.* A **35** 904–7

Golubkov G V and Ivanov G K 1981 Vibrational and rotational excitation of molecular ions by slow electrons *Chem. Phys. Lett.* **81** 110–2

Golubkov G V and Ivanov G K 1984 Near-threshold photoionisation theory for diatomic molecules *J. Phys. B: At. Mol. Phys.* **17** 747–61

Golubkov G V and Ivanov G K 1984 Transitions between Rydberg states of diatomic molecules in slow collisions with atoms *Z. Phys.* A **319** 17–23

Golubkov G V and Ivanov G K 1988 Near-threshold associative ionisation reactions of atoms *J. Phys. B: At. Mol. Opt. Phys.* **21** 2049–63

Golubkov G V and Ivanov G K 1993 Dissociative recombination of electrons and molecular ions in monochromatic IR light *Sov. Phys.–JETP* **77** 574–86

Gounand F, Carré B, Fournier P R, d'Oliveira P and Aymar M 1991 The ($5d_{3/2,5/2}$; np and nf) $J = 1$ series in atomic barium *J. Phys. B: At. Mol. Opt. Phys.* **24** 1309–19

Gounand F, Gallagher T F, Sandner W, Safinya K A and Kachru R 1983 Interaction between two Rydberg series of autoionizing levels in barium *Phys. Rev.* A **27** 1925–38

Goutis S, Aymar M, Kompitsas and Camus P 1992 The perturbed even-parity $J = 1^e$, 2^e autoionizing spectra of strontium below the $4d_{5/2}$ threshold: observation and theoretical analysis *J. Phys. B: At. Mol. Opt. Phys.* **25** 3433–61

Grafström P, Levinson C, Lundberg H, Svanberg S, Grundevik P, Nilsson L and Aymar M 1982 Zeeman effect in the perturbed $6snd\,^{1,3}D_2$ sequences of Ba-I: test of MQDT wavefunctions *Z. Phys.* A **308** 95–101

Greene C H 1980 Dependence of photoabsorption spectra on long-range fields *Phys. Rev.* A **22** 149–57

Greene C H 1980 Interpretation of Feshbach resonances in H^- photodetachment *J. Phys. B: At. Mol. Phys.* **13** L39–44

Greene C H 1981 Doubly excited states of the akaline-earth atoms *Phys. Rev.* A **23** 661–78

Greene C H 1985 Channel-interaction theory in a finite volume *Phys. Rev.* A **32** 1880–2

Greene C H 1987 Negative-ion photodetachment in a weak magnetic field *Phys. Rev.* A **36** 4236–44

Greene C H 1989 Interaction between electronic and vibrational motions *Comments At. Mol. Phys.* **23** 209–28

Greene C H 1990 Photoabsorption spectra of the heavy alkali-metal negative ions *Phys. Rev.* A **42** 1405–15

Greene C H and Jungen Ch 1985 Vibrational frame transformation for electron–molecule scattering *Phys. Rev. Lett.* **55** 1066–9

Greene C H and Kim L 1988 Streamlined eigenchannel treatment of open-shell spectra *Phys. Rev.* A **38** 5953–6

Greene C H and Rau A R P 1985 Dipole threshold laws for single and double detachment from negative ions *Phys. Rev.* A **32** 1352–6

Greene C H and Theodosiou C E 1990 Photoionization of the Ba 6s6p $^1P_1$ state *Phys. Rev.* A **42** 5773–5

Greene C H, Fano U and Strinati G 1979 General form of the quantum-defect theory *Phys. Rev.* A **19** 1485–509

Guberman S L 1994 Dissociative recombination without a curve crossing *Phys. Rev.* A **49** R4277–80

Guberman S L and Giusti-Suzor A 1991 The generation of $O(^1S)$ from the dissociative recombination of O_2^+ *J. Chem. Phys.* **95** 2602–13

Harmin D A 1981 Hydrogenic Stark effect: properties of the wavefunctions *Phys. Rev.* A **24** 2491–512

Harmin D A 1982 Theory of the Stark effect *Phys. Rev.* A **26** 2656–81

Harmin D A 1984 Analytical study of quasidiscrete Stark levels in Rydberg atoms *Phys. Rev.* A **30** 2413–28

Harris N A and Jungen Ch 1993 Rydberg states of calcium fluoride *Phys. Rev. Lett.* **70** 2549–52

Harth K, Ganz J, Raab M, Lu K T, Geiger J and Hotop H 1985 On the s–d interaction in neon *J. Phys. B: At. Mol. Phys.* **18** L825–32

Hasegawa S and Suzuki A 1996 Resonance ionization spectroscopy and analysis on even-parity states of Pb I *Phys. Rev.* A **53** 3014–22

Heckenkamp Ch, Schäfers F, Schönhense G and Heinzmann U 1986 Experimental characterization of the Xe 5p photoionization by angle- and spin-resolved photoelectron spectroscopy *Z. Phys.* D **2** 257–74

Henle W A and Zoller P 1987 Multichannel quantum defect parametrisation of resonant multiphoton ionisation *J. Phys. B: At. Mol. Phys.* **20** 4007–25

Henriet A and Masnou-Seeuws F 1988 Two-electron calculations for the intermediate Rydberg states of Na_2: molecular quantum defects *J. Phys. B: At. Mol. Opt. Phys.* **21** L339–46

Herzberg G and Jungen Ch 1972 Rydberg series and ionization potential of the H_2 molecule *J. Mol. Spectrosc.* **41** 425–86

Herzberg G and Jungen Ch 1986 The 4f states of He_2: a new spectrum of He_2 in the near infrared *J. Chem. Phys.* **84** 1181–92

Hieronymus H, Neukammer J and Rinneberg H 1992 Line narrowing and reversal in profile symmetry caused by intrachannel mixing of $[5dnd]_{J=0}$ autoionizing Ba Rydberg series *J. Phys. B: At. Mol. Opt. Phys.* **25** 3463–74

Hill III W T, Sugar J, Lucatorto T B and Cheng K T 1987 Analysis of the $5p^6$–$5p^5$nl ($J = 1$) Rydberg series in Ba^{2+} *Phys. Rev.* A **36** 1200–6

Huang W, Zou Y, Tong X M and Li J M 1995 Atomic energy levels and Landé g factors: a theoretical study *Phys. Rev.* A **52** 2770–7

Huber K P and Jungen Ch 1990 High-resolution jet absorption study of nitrogen near 800 Å *J. Chem. Phys.* **92** 850–61

Huber K P, Jungen Ch, Yoshino K, Ito K and Stark G 1994 The f Rydberg series in the absorption spectrum of N_2 *J. Chem. Phys.* **100** 7957–72

Ito K, Masuda H, Morioka Y and Ueda K 1993 Stark effect for the Rydberg states of the krypton atom near the ionization threshold *Phys. Rev.* A **47** 1187–96

Ivanov G K and Golubkov G V 1984 Coupling of the processes of dissociative recombination and scattering of slow electrons by molecular ions *Chem. Phys. Lett.* **107** 261–4

Ivanov G K and Golubkov G V 1985 A simple version of the multichannel quantum defect analysis of inelastic atomic processes involving molecular Rydberg states *J. Phys. B: At. Mol. Phys.* **18** L383–7

Ivanov G K and Golubkov G V 1986 The photodissociation of molecules near the ionisation threshold *Z. Phys.* D **1** 199–206

Jaffe S M, Kachru R, van Linden van den Heuvell H B and Gallagher T F 1985 Ba 6png, $J = 3$ and $J = 5$ autoionizing states *Phys. Rev.* A **32** 1480–8

Jalbert G, Labastie P, Brevet P F, Bordas C and Broyer M 1989 Density-of-states matrix for Rydberg systems: application to the Stark effect *Phys. Rev.* A **40** 784–94

Jimoyiannis A, Bolovinos A, Tsekeris P and Camus P 1993 Two-step laser optogalvanic spectroscopy of high J momentum 4dnd and 4dng autoionizing states of strontium *Z. Phys.* D **25** 135–44

Jones R R, Dai C J and Gallagher T F 1990 Ba $6p_j nf_{j'}$, autoionizing series *Phys. Rev.* A **41** 316–26

Jones R R, Fu P and Gallagher T F 1991 Channel interactions in the Ba $6d_jnl_{j'}$, autoionizing series for $l = 2$, 3 and 4 *Phys. Rev.* A **44** 4265–79

Julienne P S and Mies F H 1981 A multichannel distorted-wave approximation *J. Phys. B: At. Mol. Phys.* **14** 4335–47

Jungen Ch 1980 Molecular preionization and predissociation *J. Chim. Phys. (Paris)* **77** 27–33

Jungen Ch 1982 The impact of Fano's ideas on molecular spectroscopy *Physics of Electronic and Atomic Collisions* ed S Datz (Amsterdam: North Holland) 455–66

Jungen Ch and Raoult M 1981 Spectroscopy in the ionisation continuum: vibrational preionisation in H_2 calculated by multichannel quantum-defect theory *Farad. Discuss. Chem. Soc.* **71** 253–71

Jungen Ch, Dabrowski I, Herzberg G and Kendall D J W 1989 High orbital angular momentum states in H_2 and D_2 II. The 6h–5g and 6g–5f transitions *J. Chem. Phys.* **91** 3926–33

Jungen Ch, Dabrowski I, Herzberg G and Vervloet M 1990 High orbital angular momentum states in H_2 and D_2. III. Singlet–triplet splittings, energy levels, and ionization potentials *J. Chem. Phys.* **93** 2289–98

Jungen Ch, Pratt S T and Ross S C 1995 Multichannel quantum defect theory and double-resonance spectroscopy of autoionizing levels of molecular hydrogen *J. Phys. Chem.* **99** 1700–10

Kachru R, Tran N H, Pillet P and Gallagher T F 1985 Angular distributions and branching ratios of the electrons ejected from the Ba $[6p_{1/2(3/2)}ns_{1/2}]_{J=1}$ autoionizing states *Phys. Rev.* A **31** 218–34

Kachru R, van Linden van den Heuvell H B and Gallagher T F 1985 Resolution of the $Ba(6p_jnd_j)_1$ and $(6p_jng_j)_3$ autoionizing states and their mixing with the $(6p_jns_{1/2})_1$ and $(6p_jng_j)_3$ states *Phys. Rev.* A **31** 700–8

Kaufmann K, Baumeister W and Jungen M 1987 The use of the Lanczos procedure for calculating continuum orbitals with an L^2 method *J. Phys. B: At. Mol. Phys.* **20** 4299–308

Kim L and Greene C H 1987 Two-electron excitations in atomic calcium. II. Fine-structure effects *Phys. Rev.* A **36** 4272–9

Kim L and Greene C H 1989 Stable negative ions of the heavy alkaline-earth atoms *J. Phys. B: At. Mol. Opt. Phys.* **22** L175–82

Koeckhoven S M, Buma W J and de Lange C A 1994 Three-photon excitation of autoionizing states of Ar, Kr and Xe between the $^2P_{3/2}$ and $^2P_{1/2}$ ionic limits *Phys. Rev.* A **49** 3322–32

Komninos Y and Nicolaides C A 1987 Multi-channel reaction matrix theory and configuration-interaction in the discrete and in the continuous spectrum. Inclusion of closed channels and derivation of quantum defect theory *Z. Phys.* D **4** 301–12

Kompitsas M, Cohen S, Nicolaides C A, Robaux O, Aymar M and Camus P 1990 Observation and theoretical analysis of the odd $J = 3$ autoionising spectrum of Sr up to the 4d threshold *J. Phys. B: At. Mol. Opt. Phys.* **23** 2247–67

Kompitsas M, Goutis S, Aymar M and Camus P 1991 The even-parity $J = 0$ autoionizing spectrum of strontium below the $4d_{5/2}$ threshold: observation and theoretical analysis *J. Phys. B: At. Mol. Opt. Phys.* **24** 1557–74

König A, Neukammer J, Hieronymus H and Rinneberg H 1991 Field-ionization Stark spectra of Ba Rydberg atoms at high density of states *Phys. Rev.* A **43** 2402–15

Lane A M 1984 Photoionisation through isolated Rydberg states interacting with a broad state *J. Phys. B: At. Mol. Phys.* **17** 2213–25

Lane A M 1986 Observational aspects of threshold continuity theorems *J. Phys. B: At. Mol. Phys.* **19** L601–4

Lane A M 1986 The application of Wigner's R-matrix theory to atomic physics 1986 *J. Phys. B: At. Mol. Phys.* **19** 253–7

Lange V, Aymar M, Eichmann U and Sandner W 1991 Barium 6pns ($J = 1$) autoionizing Rydberg states: comparison between experiment and R-matrix calculations *J. Phys. B: At. Mol. Opt. Phys.* **24** 91–109

Lecomte J M, Telmini M, Aymar M and Luc-Koenig E 1994 R-matrix analysis of the $6d^2$ and $7p^2$ autoionizing resonances of barium *J. Phys. B: At. Mol. Opt. Phys.* **27** 667–97

Lee C M 1974 Spectroscopy and collision theory. III. Atomic eigenchannel calculation by a Hartree–Fock–Roothaan method *Phys. Rev.* A **10** 584–600

Lee C M 1974 Spin polarization and angular distribution of photoelectrons in the Jacob–Wick helicity formalism. Application to autoionization resonances *Phys. Rev.* A **10** 1598–604

Lee C M 1975 Multichannel photodetachment theory *Phys. Rev.* A **11** 1692–9

Lee C M and Pratt R H 1975 Radiative capture of high-energy electrons *Phys. Rev.* A **12** 1825–9

Lefebvre-Brion H 1988 Influence of the autoionization in the calculation of the fluorescence polarization for the $A^2\Pi_u$–$X^2\Pi_g$ transition in O_2^+ *J. Chem. Phys.* **89** 2691–6

Lefebvre-Brion H and Keller F 1989 Competition between autoionization and predissociation in the HCl and DCl molecules *J. Chem. Phys.* **90** 7176–83

Lefebvre-Brion H, Dehmer P M and Chupka W A 1986 Vibrational effects in the spin-orbit autoionization of HBr *J. Chem. Phys.* **85** 45–50

Lefebvre-Brion H, Dehmer P M and Chupka W A 1988 Spin-orbit and electronic autoionization in HCl *J. Chem. Phys.* **88** 811–7

Lefebvre-Brion H, Giusti-Suzor A and Raseev G 1985 Theoretical study of the spin-orbit autoionization in molecules. Application to the HI photoionization spectrum *J. Chem. Phys.* **83** 1557–66

Lefebvre-Brion H, Salzmann M, Klausing H W, Müller M, Böwering N and Heinzmann U 1989 Influence of autoionisation and predissociation on the photoelectron parameters in HBr *J. Phys. B: At. Mol. Opt. Phys.* **22** 3891–900

Le Rouzo H and Raseev G 1984 Finite-volume variational method: first application to direct molecular photoionization *Phys. Rev.* A **29** 1214–23

Leuchs G and Smith S J 1985 Excited electron correlations in resonant multiphoton ionization via barium Rydberg states *Phys. Rev.* A **31** 2283–90

Leyh B and Raseev G 1986 Theoretical study of electronic autoionization in CO: vibrationally resolved results between 17 and 18.3 eV *Phys. Rev.* A **34** 2920–35

L'Huillier A, Tang X and Lambropoulos P 1989 Multiphoton ionization of rare gases using multichannel-quantum-defect theory *Phys. Rev.* A **39** 1112–22

Lindsay M D, Cai L T, Schinn G W, Dai C J and Gallagher T F 1992 Angular distribution of ejected electrons from autoionizing 3pns states of magnesium *Phys. Rev.* A **45**

231–41

Lindsay M D, Dai C J, Cai L T, Gallagher T F, Robicheaux F and Greene C H 1992 Angular distributions of ejected electrons from autoionizing 3pnd states of magnesium *Phys. Rev.* A **46** 3789–806

Liu L and Li J M 1991 The electronic structure of molecular Rydberg states of diatomic molecules *J. Phys. B: At. Mol. Opt. Phys.* **24** 1893–8

Liu X W and Wang Z W 1989 Multichannel quantum-defect theory of lifetimes for highly excited states of atoms: calculation of Yb lifetimes in the perturbed 6snd $^{1,3}D_2$ sequences *Phys. Rev.* A **40** 1838–42

Lombardi M, Labastie P, Bordas M C and Broyer M 1988 Molecular Rydberg states: classical chaos and its correspondence in quantum mechanics *J. Chem. Phys.* **89** 3479–90

Lu K T 1971 Spectroscopy and collision theory. The Xe absorption spectrum *Phys. Rev.* A **4** 579–96

Lu K T 1977 High-lying levels in uranium atomic vapor near the ionization limit *Phys. Rev.* A **16** 2184–6

Lu K T and Mansfield M W D 1976 The perturbed series, $2p^5(^2P_{3/2,1/2})$ns, of ions along the Ne I isoelectronic sequence *Electron and Photon Interactions with Atoms* ed H Kleinpoppen and M R C McDowell (New York: Plenum) 627–31

Lu K T and Rau A R P 1983 Multichannel quantum-defect theory of perturbed Rydberg atoms in external fields *Phys. Rev.* A **28** 2623–33

Lu K T, Sun J Q and Beigang R 1988 Autoionization structures induced by hyperfine interaction: a generalized Fermi–Segrè–Goudsmit formula *Phys. Rev.* A **37** 2220–3

Luc-Koenig E and Aymar M 1991 Eigenchannel R-matrix calculation of the even-parity 6pnp and 6pnf $J = 0, 1, 2$ autoionizing levels of barium *J. Phys. B: At. Mol. Opt. Phys.* **24** 4323–40

Luc-Koenig E, Aymar M and Lecomte J M 1994 R-matrix predictions on $4f^2$ levels and 4fnf Rydberg series of barium with $J = 4$–6 *J. Phys. B: At. Mol. Opt. Phys.* **27** 2447–69

Luc-Koenig E, Aymar M, Van Leeuwen R, Ubachs W and Hogervorst W 1995 Polarization effects in autoionization processes: the 5d5g states in barium *Phys. Rev.* A **52** 208–15

Luc-Koenig E, Bolovinos A, Aymar M, Assimopoulos S, Jimoyiannis A and Tsekeris P 1994 3dnl $J = 3^e$ autoionizing levels of calcium: observation by laser optogalvanic spectroscopy and theoretical analysis *Z. Phys.* D **32** 49–59

Luc-Koenig E, Lecomte J M and Aymar M 1994 R-matrix analysis of positions and widths of perturbed 7sng $J = 4$ series of barium *J. Phys. B: At. Mol. Opt. Phys.* **27** 699–713

Luo D and Pradhan A K 1990 R-matrix calculations for electron-impact excitation of C^+, N^{2+}, and O^{3+} including fine structure *Phys. Rev.* A **41** 165–73

Lyras A 1990 Rydberg series of autoionizing states in strong laser fields *Z. Phys.* D **15** 3–12

Maeda K, Ueda K and Ito K 1993 High-resolution measurement for photoabsorption cross sections in the autoionization regions of Ar, Kr and Xe *J. Phys. B: At. Mol. Opt. Phys.* **26** 1541–55

Maleki N 1982 Schwinger's variational principle for bound states *Phys. Rev.* A **26** 644–6

Mank A, Drescher M, Huth-Fehre T, Böwering N, Heinzmann U and Lefebvre-Brion H 1991 Photoionization of jet-cooled HI with coherent vacuum ultraviolet radiation: evidence for Hund's case (e) *J. Chem. Phys.* **95** 1676–87

Martin I and Karwowski J 1991 Quantum defect orbitals and the Dirac second-order equation *J. Phys. B: At. Mol. Opt. Phys.* **24** 1539–42

Martins M and Zimmermann P 1993 Analysis of autoionizing Rydberg series Cu I $3d^9$ 4snl with the multichannel quantum defect theory *Z. Phys.* D **27** 115–21

Mathews C W, Ginter M L, Ginter D S and Brown C M 1989 Absorption spectrum of Bi I in the 2022- to 1307-Å region *J. Opt. Soc. Am.* B **6** 1627–43

Mellinger A, Vidal C R and Jungen Ch 1996 Laser-reduced fluorescence study of the carbon monoxide nd triplet Rydberg series: experimental results and multichannel quantum defect analysis *J. Chem. Phys.* **104** 8913–21

Miecznik G and Greene C H 1996 Calculation of near-threshold O^- photodetachment, including fine-structure effects *Phys. Rev.* A **53** 3247–52

Miecznik G, Greene C H and Robicheaux F 1995 Spin-orbit effects in aluminum photoionization *Phys. Rev.* A **51** 513–27

Mies F H 1980 A scattering theory of diatomic molecules. Expansion in adiabatic electronic-rotational states *Mol. Phys.* **41** 973–86

Mies F H 1980 A scattering theory of diatomic molecules. General formalism using the channel state representation *Mol. Phys.* **41** 953–72

Mies F H and Julienne P S 1984 A multichannel quantum defect analysis of two-states couplings in diatomic molecules *J. Chem. Phys.* **80** 2526–36

Monteiro T S and Taylor K T 1989 Rydberg states of the H_2 molecule in a magnetic field: the linear Zeeman term *J. Phys. B: At. Mol. Opt. Phys.* **22** L191–200

Monteiro T S and Taylor K T 1990 H_2 molecule in a magnetic field *J. Phys. B: At. Mol. Opt. Phys.* **23** 427–39

Moores D L 1989 The analysis of resonance structures in electron impact ionisation cross sections by many-channel quantum defect theory *J. Phys. B: At. Mol. Opt. Phys.* **22** 1395–409

Mullins O C, Zhu Y, Xu E Y and Gallagher T F 1985 Channel interaction of the three 6pnd $J = 3$ autoionizing series in barium *Phys. Rev.* A **32** 2234–41

Murphy J E, Friedman-Hill E and Field R W 1995 A multichannel quantum defect fit to the $n^* = 6$–8 core-penetrating s≈p≈d supercomplexes of CaF *J. Chem. Phys.* **103** 6459–66

Nahar S N and Pradhan A K 1991 Photoionization and electron–ion recombination: the carbon sequence *Phys. Rev.* A **44** 2935–48

Nahar S N and Pradhan A K 1994 Unified treatment of electron–ion recombination in the close-coupling approximation *Phys. Rev.* A **49** 1816–35

Nakamura H 1983 Semiclassical theory of predissociation induced by rotational (Coriolis) coupling *Chem. Phys.* **78** 235–41

Nakashima K, Nakamura H, Achiba Y and Kimura K 1989 Autoionization mechanism of the NO molecule: calculation of quantum defect and theoretical analysis of multiphoton ionization experiment *J. Chem. Phys.* **91** 1603–10

Neukammer J and Rinneberg H 1982 Configuration interaction and singlet–triplet mixing probed by hyperfine structure of 6s14d $^1D_2$ and $^3D_{1,2,3}$ barium *J. Phys. B: At. Mol. Phys.* **15** 3787–99

Nikitin S I 1984 Asymptotic calculation of the Rydberg series quantum defects for the doubly excited states of two-electron atoms *J. Phys. B: At. Mol. Phys.* **17** 4459–76

O'Mahony P F and Greene C H 1985 Doubly excited states of beryllium and magnesium *Phys. Rev.* A **31** 250–9

O'Mahony P F and Watanabe S 1985 Channel coupling in the $^1D^e$ spectrum of beryllium *J. Phys. B: At. Mol. Phys.* **18** L239–44

Osanai Y, Noro T and Sasaki F 1989 R-matrix study on the perturbed Rydberg series of atoms: I. The 2sns ^1S and the 2snd ^1D Rydberg series of beryllium *J. Phys. B: At. Mol. Opt. Phys.* **22** 3615–30

Osanai Y, Noro T and Sasaki F 1989 R-matrix study on the perturbed Rydberg series of atoms: II. The 3sns ^1S and the 3snd ^1D Rydberg series of magnesium *J. Phys. B: At. Mol. Opt. Phys.* **22** 3631–45

Osanai Y, Noro T and Sasaki F 1991 R-matrix study on perturbed Rydberg series of atoms: III. The 4sns ^1S and the 4snd ^1D Rydberg series of calcium *J. Phys. B: At. Mol. Opt. Phys.* **24** 2641–52

Pan S and Lu K T 1988 Triatomic molecular H_3 frame transformation theory of the l-uncoupling effect *Phys. Rev.* A **37** 299–302

Pan S and Mies F H 1988 Rydberg-like properties of rotational-vibrational levels and dissociation continuum associated with alkali-halide charge-transfer states *J. Chem. Phys.* **89** 3096–103

Pan X 1991 Electron–ion collisions by the R-matrix method with multichannel quantum-defect theory *Phys. Rev.* A **44** 7269–82

Pellarin M, Vialle J L, Carré M, Lermé J and Aymar M 1988 Even parity series of argon Rydberg states studied by fast-beam collinear laser spectroscopy *J. Phys. B: At. Mol. Opt. Phys.* **21** 3833–49

Post B H, Vassen W, Hogervorst W, Aymar M and Robaux O 1985 Odd-parity Rydberg levels in neutral barium: term values and multichannel quantum defect theory analysis *J. Phys. B: At. Mol. Phys.* **18** 187–206

Pradhan A K 1983 Resonance and intermediate-coupling effects in electron scattering with highly charged ions. II. Autoionization and dielectronic recombination *Phys. Rev.* A **28** 2128–36

Pratt S T, Jungen Ch and Miescher E 1989 Two-photon spectroscopy of Rydberg states of NO *J. Chem. Phys.* **90** 5971–81

Qin K, Bistransin M and Glab W L 1993 Stark effect and rotational-series interactions on high Rydberg states of molecular hydrogen *Phys. Rev.* A **47** 4154–9

Radojevic V and Talman J D 1990 Calculation of Beutler–Fano autoionising resonances in neon: $1s^2 2s^2 2p^5 (^2P_{1/2})$ ns and $nd_{3/2}$ series *J. Phys. B: At. Mol. Opt. Phys.* **23** 2241–6

Raoult M 1987 A unified treatment of $^2\Pi$–$^2\Pi$ Rydberg-valence state interactions in NO *J. Chem. Phys.* **87** 4736–61

Raoult M and Balint-Kurti G G 1990 Frame transformation theory for heavy particle scattering: application to the rotational predissociation of Ar-H_2 *J. Chem. Phys.* **93** 6508–19

Raoult M and Jungen Ch 1981 Calculation of vibrational preionization in H_2 by multichannel quantum defect theory: total and partial cross sections and photoelectron angular distributions *J. Chem. Phys.* **74** 3388–99

Raoult M, Guizard S and Gauyacq D 1991 Rydberg states of NO in a magnetic field: multichannel quantum defect approach of the linear Zeeman effect *J. Chem. Phys.* **95** 8853–65

Raoult M, Jungen Ch and Dill D 1980 Photoelectron angular distributions in H_2: calculation of rotational-vibrational preionization by multichannel quantum defect theory *J. Chim. Phys. (Paris)* **77** 599–604

Raoult M, Le Rouzo H, Raseev G and Lefebvre-Brion H 1983 *Ab-initio* approach to the multichannel quantum defect calculation of the electronic autoionisation in the Hopfield series of N_2 *J. Phys. B: At. Mol. Phys.* **16** 4601–17

Raseev G, Leyh B and Lefebvre-Brion H 1986 Autoionization in diatomic molecules: an example of electrostatic autoionization in CO *Z. Phys.* D **2** 319–26

Rau A R P and Fano U 1971 Theory of photodetachment near fine-structure thresholds *Phys. Rev.* A **4** 1751–9

Robicheaux F 1991 Driving nuclei with resonant electrons: *ab initio* study of $(e{+}H_2)^2\Sigma_u^+$ *Phys. Rev.* A **43** 5946–55

Robicheaux F and Gao B 1993 Multichannel quantum-defect approach for two-photon processes *Phys. Rev.* A **47** 2904–12

Robicheaux F and Greene C H 1992 Regularities in calculated photoionization cross sections for the halogens *Phys. Rev.* A **46** 3821–33

Robicheaux F and Greene C H 1993 Partial and differential photoionization cross sections of Cl and Br *Phys. Rev.* A **47** 1066–74

Robicheaux F and Greene C H 1993 Photoionization of the scandium atom. I. General features *Phys. Rev.* A **48** 4429–40

Robicheaux F and Greene C H 1993 Photoionization of the scandium atom. II. Classifications *Phys. Rev.* A **48** 4441–9

Robicheaux F and Greene C H 1993 Valence-shell photoabsorption spectra of C, Si, Ge, and Sn *Phys. Rev.* A **47** 4908–19

Robicheaux F, Fano U, Cavagnero M and Harmin D A 1987 Generalized WKB and Milne solutions to one-dimensional wave equations *Phys. Rev.* A **35** 3619–30

Robicheaux F, Gorczyca W, Pindzola M S and Badnell N R 1995 Inclusion of radiation damping in the close-coupling equations for electron–atom scattering *Phys. Rev.* A **52** 1319–33

Roche A L and Jungen Ch 1993 Multichannel quantum defect analysis of preionizing Rydberg states of Li_2 including rovibronic interactions *J. Chem. Phys.* **98** 3637–41

Rosenberg L 1996 Bound-state methods for low-energy electron–ion scattering *Phys. Rev.* A **53** 791–7

Ross S and Jungen Ch 1987 Quantum-defect theory of double-minimum states in H_2 *Phys. Rev. Lett.* **59** 1297–300

Ross S C and Jungen Ch 1994 Multichannel quantum-defect theory of double-minimum $^1\Sigma_g^+$ states in H_2 I. Potential-energy curves *Phys. Rev.* A **49** 4353–63

Ross S C and Jungen Ch 1994 Multichannel quantum-defect theory of $n = 2$ and 3 gerades states in H_2: rovibronic energy levels *Phys. Rev.* A **50** 4618–28

Rottke H and Welge K H 1992 Singlet gerade Rydberg states of molecular hydrogen *J. Chem. Phys.* **97** 908–26

Sadeghpour H R, Greene C H and Cavagnero M 1992 Extensive eigenchannel R-matrix study of the H^- photodetachment spectrum *Phys. Rev.* A **45** 1587–95

Sandner W, Eichmann U, Lange V and Völkel M 1986 Barium autoionising Rydberg series in the vicinity of a shape resonance *J. Phys. B: At. Mol. Phys.* **19** 51–71

Sarpal B K and Tennyson J 1992 Electronic transitions in the HeH molecule: a MQDT approach *J. Phys. B: At. Mol. Opt. Phys.* **25** L49–55

Sarpal B K, Branchett S E, Tennyson J and Morgan L A 1991 Bound states using the R-matrix method: Rydberg states of HeH *J. Phys. B: At. Mol. Opt. Phys.* **24** 3685–99

Sarpal B K, Tennyson J and Morgan L A 1991 Vibrationally resolved electron HeH^+ collisions using the non-adiabatic R-matrix method *J. Phys. B: At. Mol. Opt. Phys.* **24** 1851–66

Schäfers F, Schönhense G and Heinzmann U 1983 Analysis of Ar $3p^6$ and Kr $4p^6$ photoionization from photoelectron-spin-polarization data *Phys. Rev.* A **28** 802–14

Schinn G W, Dai C J and Gallagher T F 1991 Mg 3pns and 3pnd ($J = 1$) autoionizing series *Phys. Rev.* A **43** 2316–27

Schwarz M, Duchowicz R, Demtröder W and Jungen Ch 1988 Autoionizing Rydberg states of Li_2: analysis of electronic–rotational interactions *J. Chem. Phys.* **89** 5460–72

Seaton M J 1982 Bound channels and quantum defect theory *J. Phys. B: At. Mol. Phys.* **15** 3899–913

Seaton M J 1984 Discussion of a complex-potential model for di-electronic recombination *J. Phys. B: At. Mol. Phys.* **17** L531–3

Seaton M J 1985 Use of the R-matrix method for bound-state calculations: I. General theory *J. Phys. B: At. Mol. Phys.* **18** 2111–31

Seipp I and Taylor K T 1994 The combined R-matrix and complex coordinate method applied to the Stark and Stark diamagnetic Rydberg spectra of Na *J. Phys. B: At. Mol. Opt. Phys.* **27** 2785–99

Sidky E Y 1993 Semianalytic theory of Rydberg electron diamagnetism. II *Phys. Rev.* A **47** 2812–8

Sil N C, Crees M A and Seaton M J 1984 Integrals involving products of Coulomb functions and inverse powers of the radial coordinate *J. Phys. B: At. Mol. Phys.* **17** 1–19

Sobolewski A L 1987 Resonances in molecular photoionization. II. Autoionization effects in the NO molecule *J. Chem. Phys.* **87** 331–8

Sobolewski A L and Domcke W 1987 Resonances in molecular photoionization. I. Model calculations and analysis of general phenomena *J. Chem. Phys.* **86** 176–87

Sobolewski A L and Domcke W 1988 Resonances in molecular photoionization. III. Multichannel extension and application to polyatomic molecules *J. Chem. Phys.* **88** 5571–9

Softley T P and Hudson A J 1994 Multichannel quantum defect theory simulation of the zero-kinetic-energy photoelectron spectrum of H_2 *J. Chem. Phys.* **101** 923–8

Sohl J E, Zhu Y and Knight R D 1990 Two-color laser photoionization spectroscopy of Ti I: multichannel quantum defect theory analysis and a new ionization potential *J. Opt. Soc. Am.* B **7** 9–14

Solov'ev E A 1986 Modification of molecular Rydberg states by superpromotion of the diabatic term to the continuous spectrum *Sov. Phys.–JETP* **63** 678–81

Sorensen S L, Åberg T, Tulkki J, Rachlew-Källne E, Sundström G and Kirm M 1994 Argon 3s autoionization resonances *Phys. Rev.* A **50** 1218–30

Staib A and Domcke W 1990 Analysis of the Jahn-Teller effect in the np^2 E′ Rydberg series of H_3 and D_3 *Z. Phys.* D **16** 275–82

Staib A and Domcke W 1991 Jahn-Teller coupling in Rydberg series of benzene *J. Chem. Phys.* **94** 5402–13

Staib A and Domcke W 1993 Vibronic coupling in the pE″ Rydberg series of NH_3 *Chem. Phys. Lett.* **204** 505–10

Staib A, Domcke W and Sobolewski A L 1989 MQDT analysis of radiationless decay rates of autoionizing Rydberg states of polyatomic molecules *Chem. Phys. Lett.* **162** 336–41

Staib A, Domcke W and Sobolewski A L 1990 Jahn-Teller effect in Rydberg series: a multi-state vibronic coupling problem *Z. Phys.* D **16** 49–60

Steenman-Clark L and Faucher P 1984 The effect of resonances on the forbidden line of He-like ions *J. Phys. B: At. Mol. Phys.* **17** 73–81

Stephens J A and Greene C H 1994 Quantum defect description of H_3 Rydberg state dynamics *Phys. Rev. Lett.* **72** 1624–7

Stephens J A and Greene C H 1994 Vibrational autoionization in H_2 above the $v^+ = 3$ ionization limit *J. Chem. Phys.* **100** 7135–43

Stephens J A and Greene C H 1995 Rydberg state dynamics of rotating, vibrating H_3 and the Jahn-Teller effect *J. Chem. Phys.* **102** 1579–91

Stephens J A and Greene C H 1995 Competing predissociation and preionization in the photoabsorption of H_2 above the ionization threshold *J. Chem. Phys.* **103** 5470–5

Stephens J A and McKoy V 1992 Studies of molecular Rydberg states by Schwinger variational-quantum defect methods: application to molecular hydrogen *J. Chem. Phys.* **97** 8060–72

Story J G and Cooke W E 1989 Shake-off measurements of electron–ion-scattering phase shifts *Phys. Rev.* A **39** 4610–4

Strömholm C, Schneider I F, Sundström G, Carata L, Danared H, Datz S, Dulieu O, Källberg A, af Ugglas M, Urbain X, Zengin V, Suzor-Weiner A and Larsson M 1995 Absolute cross sections for dissociative recombination of HD^+: comparison of experiment and theory *Phys. Rev.* A **52** R4320–3

Sun H and Nakamura H 1990 Theoretical study of the dissociative recombination of NO^+ with slow electrons *J. Chem. Phys.* **93** 6491–501

Sun J Q 1989 Multichannel quantum defect theory of the hyperfine structure of high Rydberg states *Phys. Rev.* A **40** 7355–8

Sun J Q and Lin C D 1992 Single-channel quantum-defect theory for the description of doubly excited states of helium *Phys. Rev.* A **46** 5489–96

Sun J Q and Lu K T 1988 Hyperfine structure of extremely high Rydberg msns $^1S_0$ and msns $^3S_1$ series in odd alkaline-earth isotopes *J. Phys. B: At. Mol. Opt. Phys.* **21** 1957–67

Sun J Q and Lu K T 1989 Crossing in odd alkaline-earth isotopes induced by hyperfine interaction *J. Phys. B: At. Mol. Opt. Phys.* **22** 2963–7

Takagi H 1993 Theoretical problems in the dissociative recombination of H_2^+ + e *Dissociative Recombination* ed B R Rowe *et al* (New York: Plenum) 75–85

Takagi H and Nakamura H 1981 Autoionization of highly excited Rydberg states of diatomic molecules *J. Chem. Phys.* **74** 5808–13

Takagi H and Nakamura H 1986 Dissociative recombination of H_2^+ by collisions with slow electrons *J. Chem. Phys.* **84** 2431–2

Takagi H and Nakamura H 1988 Theoretical study of associative ionization of hydrogen atoms *J. Chem. Phys.* **88** 4552–3

Tang J Z, Watanabe S and Matsuzawa M 1993 Very narrow doubly excited $2(1,0)^-n$ and $2(-1,0)^0n$ $^1P_0$ states of helium *Phys. Rev.* A **48** 841–4

Taylor K T and Norcross D W 1986 Alkali-metal negative ions. IV. Multichannel calculations of K^- photodetachment *Phys. Rev.* A **34** 3878–91

Taylor K T, Zeippen C J and Le Dourneuf M 1984 The photoionisation of the $Si^+(^2P^o)$ ground state: a combined application of the R-matrix and quantum defect theories *J. Phys. B: At. Mol. Phys.* **17** L157–63

Telmini M, Aymar M and Lecomte J M 1993 Eigenchannel R-matrix calculation of the even-parity 6pnf $J = 2, 3, 4$ autoionizing levels of barium *J. Phys. B: At. Mol. Opt. Phys.* **26** 233–54

Tennyson J 1987 Fully vibrationally resolved photoionisation of H_2 and D_2 *J. Phys. B: At. Mol. Phys.* **20** L375–8

Tran N H, Kachru R and Gallagher T F 1982 Multistep laser excitation of barium: apparent interferences due to overlap-integral variations *Phys. Rev.* A **26** 3016–9

Tran N H, Pillet P, Kachru R and Gallagher T F 1984 Multistep laser excitation of autoionizing Rydberg states *Phys. Rev.* A **29** 2640–50

Ueda K 1987 Spectral line shapes of autoionizing Rydberg series *Phys. Rev.* A **35** 2484–92

Urbain X, Cornet A, Brouillard F and Giusti-Suzor A 1991 Dynamics of an elementary bond-forming process: associative ionization in H(1s) + H(2s) collisions *Phys. Rev. Lett.* **66** 1685–8

van Leeuwen K A H and Hogervorst W 1983 High-resolution studies of the Stark effect in even-parity Rydberg states of Ca I *J. Phys. B: At. Mol. Phys.* **16** 3873–93

van Leeuwen K A H, Hogervorst W and Post B H 1983 Stark effect in barium 6snd $^1D_2$ Rydberg states; evidence of strong perturbations in the $^1F_3$ series *Phys. Rev.* A **28** 1901–7

van Leeuwen R, Ubachs W and Hogervorst W 1995 Interference effects in the autoionization of 4f7h and 6dni states of barium *Phys. Rev.* A **52** 4567–77

Vassen W and Hogervorst W 1988 Hyperfine structure and multichannel quantum defect theory analysis of the 6sng Rydberg series of barium *Z. Phys.* D **8** 149–61

Vrakking M J J, Lee Y T, Gilbert R D and Child M S 1993 Resonance-enhanced one- and two-photon ionization of the water molecule: preliminary analysis by multichannel quantum defect theory *J. Chem. Phys.* **98** 1902–15

Wane S and Aymar M 1987 Excited-state photoionisation and radiative recombination for ions of the potassium isoelectronic sequence *J. Phys. B: At. Mol. Phys.* **20** 2657–75

Wang Q and Greene C H 1989 Quantal description of diamagnetic quasi-Landau-resonances *Phys. Rev.* A **40** 742–51

Wang Q and Greene C H 1991 Parameter-dependent multichannel Rydberg spectra *Phys. Rev.* A **44** 1874–83

Wang Q and Greene C H 1991 R-matrix calculation of atomic hydrogen photoionization in a strong magnetic field *Phys. Rev.* A **44** 7448–58

Wang Y and Chupka W A 1992 Prescription for preparation of molecular ions in selected rotational states *Chem. Phys. Lett.* **200** 192–8

Wang Z and Lu K T 1985 Eigenchannel quantum-defect theory of open-shell atoms. I. Autoionization resonances and eigenphase shifts of chlorine *Phys. Rev.* A **31** 1515–20

Wang Z and Lu K T 1985 Eigenchannel quatum-defect theory of open-shell atoms. II. Calculation of $3p^4(^3P)ns$ Rydberg spectra of the chlorine atom *Phys. Rev.* A **31** 1521–8

Wang Z W and Lu K T 1988 Eigenchannel quantum-defect theory of open-shell atoms. III. Calculation of $3p^4(^3P)ns$ Rydberg spectra of the chlorine isoelectronic ions from Ar II to Zn XIV *Phys. Rev.* A **37** 2941–6

Watanabe S 1986 Multichannel photodetachment theory revisited from the hyperspherical viewpoint *J. Phys. B: At. Mol. Phys.* **19** 1577–94

Watanabe S and Greene C H 1980 Atomic polarizability in negative-ion photodetachment *Phys. Rev.* A **22** 158–69

Watanabe S, Fano U and Greene C H 1984 Spin correlations in photodetachment *Phys. Rev.* A **29** 177–82

Watson D K 1983 *Ab initio* determination of quatum defects by calculation of the poles of the Schwinger T matrix *Phys. Rev.* A **28** 40–4

Werij H G C, Greene C H, Theodosiou C E and Gallagher A 1992 Oscillator strengths and radiative branching ratios in atomic Sr *Phys. Rev.* A **46** 1248–60

Wintgen D and Friedrich H 1987 Perturbed Rydberg series of autoionizing resonances *Phys. Rev.* A **35** 1628–33

Wong H Y, Rau A R P and Greene C H 1988 Negative-ion photodetachment in an electric field *Phys. Rev.* A **37** 2393–403

Wood R P, Greene C H and Armstrong D 1993 Photoionization of the barium 6s6p $^1P_1^o$ state: comparison of theory and experiment including hyperfine-depolarization effects *Phys. Rev.* A **47** 229–35

Xingye L, Wanfa L and Zhankui J 1994 Test of the multichannel quantum-defect wave function by a Landé-factor (g_J) investigation in the perturbed 6snp $^{1,3}P_1$ sequences of Yb I *Phys. Rev.* A **49** 4443–7

Xu E Y, Helm H and Kachru R 1989 Autoionizing np Rydberg states of H_2 *Phys. Rev.* A **39** 3979–91

Xu E Y, Zhu Y, Mullins O C and Gallagher T F 1986 Sr $5p_{1/2}ns_{1/2}$ and $5p_{3/2}ns_{1/2}$ $J = 1$ autoionizing states *Phys. Rev.* A **33** 2401–9

Xu E Y, Zhu Y, Mullins O C and Gallagher T F 1987 Sr 5pnd $J = 3$ autoionizing series and interference of excitation due to bound-state perturbation *Phys. Rev.* A **35** 1138–48

Yan Y and Seaton M J 1985 R-matrix theory of the hydrogen atom *J. Phys. B: At. Mol. Phys.* **18** 2577–85

Yencha A J, Hopkirk A, Grover J R, Cheng B M, Lefebvre-Brion H and Keller F 1995 Ion-pair formation in the photodissociation of HF and DF *J. Chem. Phys.* **103** 2882–7

Yencha A J, Kaur D, Donovan R J, Kvaran A, Hopkirk A, Lefebvre-Brion H and Keller F 1993 Ion-pair formation in the photodissociation of HCl and DCl *J. Chem. Phys.* **99** 4986–92

Yoo B and Greene C H 1986 Implementation of the quantum-defect theory for arbitrary long-range potentials *Phys. Rev.* A **34** 1635–41

Zaki Ewiss M A, Hogervorst W, Vassen W and Post B H 1985 The Stark effect in the 6snf Rydberg series of barium I. $^3F_2$ series *Z. Phys.* A **322** 211–9

Zhankui J, Xingye L and Min S 1993 Multichannel-quantum-defect-theory analysis of natural radiative lifetimes in the perturbed Rydberg sequence $4f^{14}6snp$ $^3P_2$ of Yb I *Phys. Rev.* A **48** 2451–2

Zhu Y F, Grant E R and Lefebvre-Brion H 1993 Spin-orbit and rotational autoionization in HCl and DCl *J. Chem. Phys.* **99** 2287–99

Zhu Y, Xu E Y and Gallagher T F 1987 Energy and angular distributions of autoionization electrons and channel interaction effects for the Sr 5pns $J = 1$ autoionizing series *Phys. Rev.* A **36** 3751–67

(c) Database

'Atomic engineering system, based on MQDT' *Electronic database management software system* Atomic Engineering Corp., PO Box 3342, Gaithersburg, MD 20885–3342, USA